INDUSTRIAL AUTOMATED SYSTEMS

INSTRUMENTATION AND MOTION CONTROL

Terry Bartelt

Fox Valley Technical College

Cengage

Australia • Brazil • Canada • Mexico • Singapore • United Kingdom • United States

Cengage

Industrial Automated Systems: Instrumentation and Motion Control
Terry Bartelt

Vice President, Career and Professional Editorial: Dave Garza

Director of Learning Solutions: Sandy Clark

Acquisitions Editor: Stacy Masucci

Managing Editor: Larry Main

Senior Product Manager: John Fisher

Senior Editorial Assistant: Dawn Daugherty

Vice President, Career and Professional Marketing: Jennifer Baker

Marketing Director: Deborah Yarnell

Marketing Coordinator: Mark Pierro

Associate Marketing Manager: Scott A. Chrysler

Production Director: Wendy Troeger

Manufacturing Buyer: Beverly Bresliin

Production Manager: Mark Bernard

Production Project Manager: Dewanshu Ranjan

Production Service/Compositor: Pre-Press PMG

Senior Art Director: David Arsenault

Permission SME Image/Media: Deanna Ettinger

Permission SME Text: Mardell Glinski schultz

Library of Congress Control Number: 2009941551

ISBN-13: 978-1-4354-8888-5
ISBN-10: 1-4354-8888-1

Book Only:
ISBN-13: 978-1-111-32190-1
ISBN-10: 1-111-32190-6

Cengage
200 Pier 4 Boulevard
Boston, MA 02210
USA

Notice to the Reader

Publisher does not warrant or guarantee any of the products described herein or perform any independent analysis in connection with any of the product information contained herein. Publisher does not assume, and expressly disclaims, any obligation to obtain and include information other than that provided to it by the manufacturer. The reader is expressly warned to consider and adopt all safety precautions that might be indicated by the activities described herein and to avoid all potential hazards. By following the instructions contained herein, the reader willingly assumes all risks in connection with such instructions. The publisher makes no representations or warranties of any kind, including but not limited to, the warranties of fitness for particular purpose or merchantability, nor are any such representations implied with respect to the material set forth herein, and the publisher takes no responsibility with respect to such material. The publisher shall not be liable for any special, consequential, or exemplary damages resulting, in whole or part, from the readers' use of, or reliance upon, this material.

Printed in Mexico
Print Number: 03 Print Year: 2022

Contents

Section 3 Electric Motors

Section 8 Motion Control

Section 9 Industrial Networking

Lab.*Source* Contents

The Lab Manual to Accompany *Industrial Automated Systems: Instrumentation and Motion Control,* is available for free in the back-of-the-book CD.

Preface

Intended Audience

This book is intended for use in electronics-based industrial courses. It is designed for either two- or four-year technology programs, such as Electronics Technology, Electronics Engineering Technology, Instrumentation Technology, Industrial Maintenance, Mechatronics, Electromechanical Technology, and Automated Manufacturing Systems, as well as for university technology programs.

The book provides comprehensive coverage of components, circuits, instruments, and control techniques used in industrial automated systems. The focus is on operation, rather than mathematical design concepts. It is designed to be used for several different courses, such as electrical motors, programmable logic controllers, sensors, industrial control, variable-speed drives, servomechanisms, industrial solid-state electronics, and various instrumentation and process control classes. To understand the material, recommended prerequisites include DC/AC fundamentals, solid-state devices, and digital electronics.

Textbook Organization

The book is organized in a logical sequence that first examines a block diagram of an industrial control system and then expands on the function of measurement devices, instruments, and control techniques.

SECTION 1: Industrial Control Overview

Chapter 1, *Introduction to Industrial Control Systems,* introduces the basic concepts of open- and closed-loop systems common to motion and process control.

Chapter 2, *Interfacing Devices,* describes the operation of various IC devices such as operational amplifiers, D/A and A/D converters, 555 timers, and optocouplers used in circuits throughout the text.

Chapter 3, *Thyristors,* describes the operation of solid-state components and circuits used to provide power to actuator devices.

SECTION 2: The Controller

Chapter 4, *The Controller Operation,* describes On-Off, proportional-integral-derivative (PID), and time-proportioning operations performed by the control block of a closed-loop system.

SECTION 3: Electric Motors

Chapter 5, *DC Motors,* introduces the fundamentals of DC motor theory and explains the operation of series, shunt, and compound motors.

Chapter 6, *AC Motors,* explains the fundamentals of AC motor theory and describes the operation of various single-phase and three-phase motors.

Chapter 7, *Servo Motors,* describes the operation of several motors used in motion-control operations.

SECTION 4: Variable-Speed Drives

Chapter 8, *DC Drives,* explains the internal electronic circuitry and control parameters of a DC variable-speed drive.

Chapter 9, *AC Drives,* describes the converter, intermediate, and inverter circuitry of an AC drive, including the control parameters and adjustments.

SECTION 5: Process Control and Instrumentation

Chapter 10, *Pressure Systems,* describes the scientific fundamentals of pressure and the operation of instruments used to make measurements.
Chapter 11, *Temperature Control,* describes the scientific principles of temperature and the operation of instruments used to make measurements.
Chapter 12, *Flow Control,* describes the scientific fundamentals of flow and the operation of instruments used to make measurements.
Chapter 13, *Level-Control Systems,* describes the scientific principles of level and the operation of instruments used to make measurements.
Chapter 14, *Analytical Instrumentation,* describes the chemical properties of solutions, such as pH, conductivity, combustion, humidity, and instruments used to make measurements.
Chapter 15, *Industrial Process Techniques and Instrumentation,* provides information on production processes and the instruments used to monitor, control, and manipulate process variables, such as pressure, temperature, level, and flow.
Chapter 16, *Instrumentation Symbology,* provides information about drawings used in process control called P&IDs (Piping and Instrumentation Diagrams), such as symbols, Tag Numbers, Functional Identifiers, Line Symbols, and Title Blocks.
Chapter 17, *Process-Control Methods,* describes various control techniques, such as On-Off, PID, feed-forward, ratio, cascade, and adaptive.
Chapter 18, *Instrument Calibration and Controller Tuning,* describes how to perform calibration procedures on instruments, and how to properly tune controllers using the Ziegler-Nichols method.

SECTION 6: Detection Sensors

Chapter 19, *Industrial Detection Sensors and Interfacing,* describes the operation of proximity and optical sensors that detect the presence of an object in both motion- and process-control applications.
Chapter 20, *Industrial Wireless Technologies,* addresses wireless devices that are used to sense and transmit data about the status of variables in the field. Information about architectural schemes, wireless technologies, network topologies, wireless standards, security, and power management are presented.

SECTION 7: Programmable Controllers

Chapter 21, *Introduction to Programmable Controllers,* provides an introduction to control components and wiring configurations used in ladder logic circuits and includes the hardware and addressing of programmable controllers.
Chapter 22, *Fundamental PLC Programming,* explains the fundamental instructions used for basic PLC applications, such as Examine-On and Examine-Off, timers, counters, latching, manipulation, and arithmetic functions.
Chapter 23, *Advanced Programming, PLC Interfacing, and Troubleshooting,* describes advanced programming techniques, such as sequencing, jumps, and subroutines. It explains how to connect field devices to I/O modules and provides a procedure for troubleshooting PLCs and the devices to which they are connected.

SECTION 8: Motion Control

Chapter 24, *Elements of Motion Control,* provides detailed information on various instruments and equipment used in motion-control applications.
Chapter 25, *Motion-Control Feedback Devices,* explains the operation of sensors used for velocity and position in motion-control applications.
Chapter 26, *Fundamentals of Servomechanisms,* describes the operation of various servo circuits and control parameters that need to be controlled.

SECTION 9: Industrial Networking

Chapter 27, *Industrial Networking,* provides an overview of networking that is incorporated into the industrial sector.

Features

- Systems approach to understanding industrial measurement devices, instruments, and control techniques. Organization begins with a block diagram of an industrial control system and then expands on the function of each block in detail. Elements of a closed-loop network and its individual functions are discussed.
- Broad and comprehensive coverage of instrumentation, process control, and servomechanisms helps the reader gain a better understanding of the entire spectrum of industrial control.
- Thorough coverage of proximity and optical sensors, sensor interfacing, sensor troubleshooting procedures, and industrial wireless technologies.
- PLC programming and applications are taught using a generic approach to programmable logic controllers, while using the Allen-Bradley SLC 500 as a sample.
- Follows a nonmathematical approach to industrial control techniques, such as On-Off, PID, ratio, cascade, feed-forward, adaptive, and time proportioning.
- Application examples are integrated throughout the book to gain practical insights. Examples of practical applications of functional industrial systems are on the back-of-the-book CD.

Supplements

Lab Manual

- **Introducing new free lab exercises with Lab.*Source!* A complementary CD includes the Lab Manual for free in the back of the book.** Save your students from the added expense of purchasing a separate, printed lab manual. Lab.*Source* CD provides learners with the valuable hands-on experience they need to reinforce concepts covered in the text.

  In today's economy, where every dollar counts, Lab.*Source* is a sensible alternative to expensive printed lab manuals. Lab.*Source* CD in the back of the book includes 46 experiments written by the author. All answers to the lab manual experiments are provided for instructors on the *Instructor Resource* CD.
- The back-of-the-book CD also includes 196 animated multimedia presentations on topics covered throughout the book, and a chapter on practical applications.

Instructor Resource

This electronic Instructor's Management System is an educational resource that creates a truly electronic classroom. The CD contains tools and instructional resources that will enrich your classroom and make your preparation time shorter. The elements of the *Instructor Resource* link directly to the text and tie together to provide a unified instructional system. (ISBN: 1435488873)

Features contained in *Instructor Resource* include:

- **Instructor's Guide:** This comprehensive instructor's guide contains solutions to all end-of-chapter problems. It also includes recommendations on videos that can be used and a list of vendors who sell training equipment to perform laboratory experiments.
- **Lab Manual Solutions:** Solutions to the Lab Manual's Procedure and Experiment Questions are provided.
- **Lab.Builder:** This template gives instructors the advantage of being able to create their own labs electronically.

- **PowerPoint Presentation:** These slides provide the basis for a lecture outline that helps you to present concepts and material. Key points and concepts can be graphically highlighted for student retention.
- **Computerized Testbank:** This computerized testbank includes approximately 1,400 questions such as true/false, multiple-choice, completion, and short-answer, provided in multiple formats to assess student comprehension.
- **Image Library:** Images from the textbook allow you to customize PowerPoint presentations, or to use them as transparency masters. Image Library enables the user to browse and search images by using key words. This is a quick and easy tool for enhancing teaching and research projects.
- **Electronics Technology Web site:** Additional online resources are available at ***http://www.electronictech.com***.

Additional Resources

The author has participated in a National Science Foundation project to create lessons on a computer called *learning objects*. Learning objects are brief lessons that use Flash software to provide animation to describe a concept. In 2005, Terry Bartelt was awarded an NSF grant to continue the project for three more years to create more learning objects, many of which will cover concepts in this textbook. These learning objects are accessible on the back-of-the-book CD.

Acknowledgments

The author would like to express his appreciation to John Casey, Craig Hemken, Terry Fleischman, Mark Miller, Glen Schneider, Lawrence Ortner, and students who provided valuable information and suggestions for the manuscript. The author and Delmar Cengage Learning gratefully acknowledge the contributions of the following reviewers who reviewed much of the content in this text:

Russell Bowker
Northeast Community College
Norfolk, NE

Frank Claude
Dunwoody College of Technology
Minneapolis, MN

Leei Mao
Greenville Technical College
Greenville, SC

Daniel Saine
DeVry University
Phoenix, AZ

SECTION

1

Industrial Control Overview

OUTLINE

Section 1 introduces key concepts in industrial control. Chapter 1 introduces the student to the ways in which industrial control systems are classified. It then provides an introductory overview of the elements that make up an industrial control loop.

Chapter 2 describes the operation of discrete components and integrated circuits that are used throughout the book.

The remaining sections describe each element of a control loop in detail so that the entire spectrum of industrial control is addressed.

CHAPTER

1

Introduction to Industrial Control Systems

OBJECTIVES

At the conclusion of this chapter, you should be able to:

- List the classifications of industrial control systems.
- Describe the differences among industrial control systems and provide examples of each type.
- Define the following terms associated with industrial control systems:

Servos	Batch	Instrumentation
Servomechanisms	Continuous	

- Describe the differences between open- and closed-loop systems.
- Define the following terms associated with open- and closed-loop systems:

Negative Feedback	Error Detector	Disturbance
Controlled Variable	Error Signal	Measured Variable
Measurement Device	Controller	Manipulated Variable
Feedback Signal	Actuator	Controller Output Signal
Setpoint	Manufacturing Process	

- List the factors that affect the dynamic response of a closed-loop system.
- Describe the operation of feed-forward control.
- List three factors that cause the controlled variable to differ from the setpoint.

INTRODUCTION

The industrial revolution began in England during the mid-1700s, when it was discovered that productivity of spinning wheels and weaving machines could be dramatically increased by fitting them with steam-powered engines. Further inventions and new ideas in plant layouts during the 1850s enabled the United States to surpass England as the manufacturing leader of the world. Around the turn of the twentieth century, the electric motor replaced steam and water wheels as a power source. Factories became larger; machines were improved to allow closer tolerances; and the assembly line method of mass production was created.

Between World Wars I and II, the feedback control system was developed, enabling manually operated machines to be replaced by automated equipment. The feedback control system is a key element in today's manufacturing operations. The term **industrial controls** is used to define this type of system, which automatically monitors manufacturing processes being executed and takes appropriate corrective action if the operation is not performing properly.

During World War II, significant advances in feedback technology occurred due to the sophisticated control systems required by military weapons. After the war, the techniques used in military equipment were applied to industrial controls to further improve the quality of products and to increase productivity.

Because many modern factory machines are automated, the technicians who install, troubleshoot, and repair them need to be highly trained. To perform effectively, these individuals must understand the elements, operational theory, and terminology associated with industrial control systems.

Industrial control theory encompasses many fields, but uses the same basic principles, whether controlling the position of an object, the speed of a motor, or the temperature and pressure of a manufacturing process.

In this chapter, the various types of industrial control systems, their characteristics, and important terminology will be studied.

1-1 Industrial Control Classifications

IAU306
Industrial Control
Classifications

Motion and Process Controls

Industrial control systems are often classified by *what* they control: either motion or process.

Motion Control

A **motion control** system is an automatic control system that controls the physical motion or position of an object. One example is the industrial robot arm that performs welding operations and assembly procedures.

There are three characteristics that are common to all motion control systems. First, motion control devices control the position, speed, acceleration, or deceleration of a mechanical object. Second, the motion or position of the object being controlled is measured. Third, motion devices typically respond to input commands within fractions of a second, rather than seconds or minutes, as in process control. Hence, motion control systems are faster than process control systems.

Motion control systems are also referred to as *servos,* or *servomechanisms.* Other examples of motion control applications are computer numeric controlled (CNC) machine tool equipment, printing presses, office copiers, packaging equipment, and electronics parts insertion machines that place components onto a printed circuit board.

Process Control

The other type of industrial control system is **process control**. In process control, one or more variables are regulated during the manufacturing of a product. These variables may include temperature, pressure, flow rate, liquid and solid level, pH, or humidity. This regulated process must compensate for any outside disturbance that changes the variable. The response time of a process control system is typically slow, and can vary from a few seconds to several minutes. Process control is the type of industrial control system most often used in manufacturing. Process control systems are divided into two categories, *batch* and *continuous.*

Batch Process **Batch processing** is a sequence of timed operations executed on the product being manufactured. An example is an industrial machine that produces various types of cookies, as shown in Figure 1-1. Suppose that chocolate-chip cookies are made in the first production run. First, the oven is turned on to the desired temperature. Next, the required ingredients in proper quantities are dispensed into the sealed mixing chamber. A large blender then begins to mix the contents.

After a few minutes, vanilla is added, and the mixing process continues. After a prescribed period of time, the batter is the proper consistency, the blender stops turning, and the compressor turns on to force air into the mixing chamber. When the air pressure reaches a

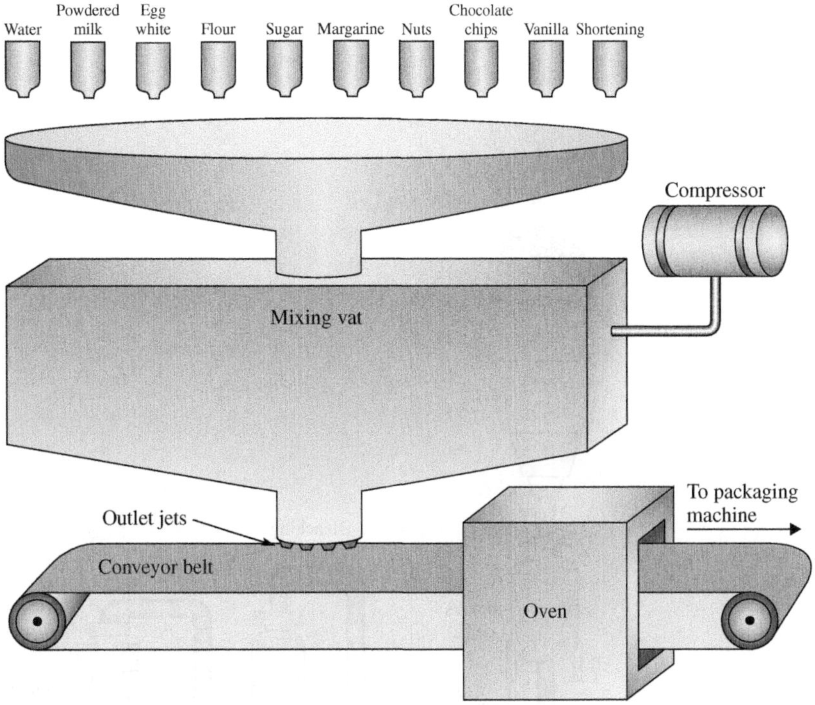

FIGURE 1-1 Batch-processing cookie machine

certain point, the conveyor belt turns on. The pressurized air forces the dough through outlet jets onto the belt. The dough balls become fully baked as they pass through the oven. The cookies cool as the belt carries them to the packaging machine.

After the packaging step is completed, the mixing vat, blender, and conveyor belt are washed before a batch of raisin-oatmeal cookies is made. Products from foods to petroleum to soap to medicines are made from a mixture of ingredients that undergo a similar batch process operation.

Batch process is also known as *sequence* (or *sequential*) *process.*

Continuous Process In the **continuous process** category, one or more operations are being performed as the product is being passed through a process. Raw materials are continuously entering and leaving each process step. Producing paper, as shown in Figure 1-2, is an example of continuous process. Water, temperature, and speed are constantly monitored and regulated as the pulp is placed on screens, fed through rollers, and gradually transformed into a finished paper product. The continuous process can last for hours, days, or even weeks without interruption. Everything from wire to textiles to plastic bags is manufactured using a continuous manufacturing process similar to the paper machine's.

Other examples of continuous process control applications are wastewater treatment, nuclear power production, oil refining, and natural gas distribution through pipe lines.

Another term commonly used instead of process control is *instrumentation.*

The primary difference between process and motion control is the control method that is required. In process control, the emphasis is placed on sustaining a constant condition of a parameter, such as level, pressure, or flow rate of a liquid. In motion control, the input command is constantly changing. The emphasis of the system is to follow the changes in the desired input signal as closely as possible. Variations of the input signal are typically very rapid.

Open- and Closed-Loop Systems

The purpose of any industrial system is to maintain one or more variables in a production process at a desired value. These variables include pressures, temperatures, fluid levels, flow rates, composition of materials, motor speeds, and positions of a robotic arm.

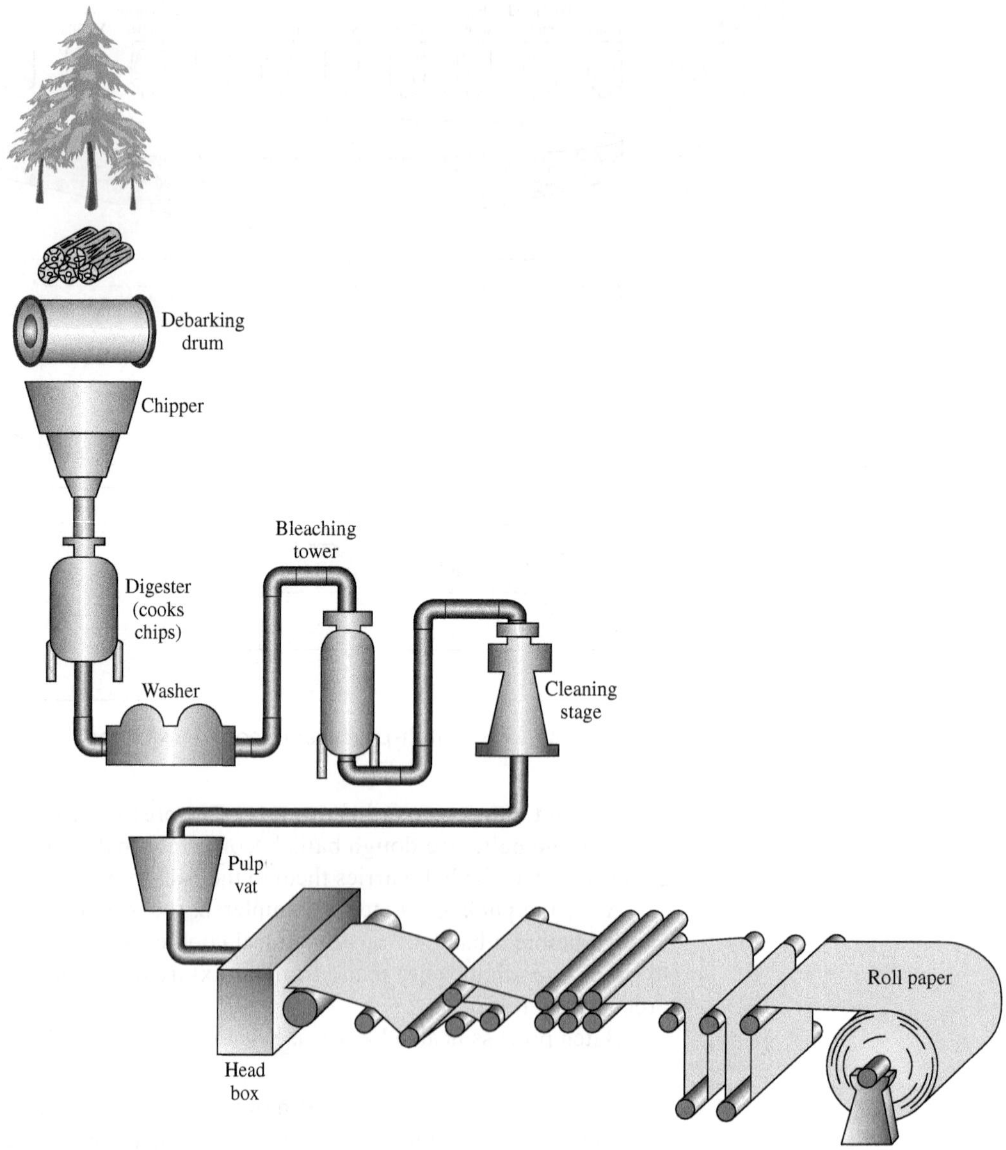

FIGURE 1-2 A pulp and paper operation is a process control application

Industrial control systems are also classified by how they control variables, either manually in an **open-loop** system or automatically in a **closed-loop** system.

Open-Loop Systems

An open-loop system is the simplest way to control a system. A tank that supplies water for an irrigation system can be used to illustrate an open-loop (or manual control) system. The diagram in Figure 1-3 shows a system composed of a storage tank, an inlet pipe with a manual control valve, and an outlet pipe. A continuous flow of water from a natural spring enters the tank at the inlet, and water flows from the outlet pipe to the irrigation system. The process variable that is maintained in the tank is the water level. Ideally, the manual flow control valve setting and the size of the outlet pipe are exactly the same. When this occurs, the water level in the tank remains the same. Therefore, the process reaches a steady-state condition, or is said to be *balanced*. The problem with this design is that any change or disturbance will upset the balance. For example, a substantial rainfall may occur, causing additional water to enter the storage tank from the top. Since there is more water entering the tank than exiting, the level will rise. If this situation is not corrected, the tank will eventually

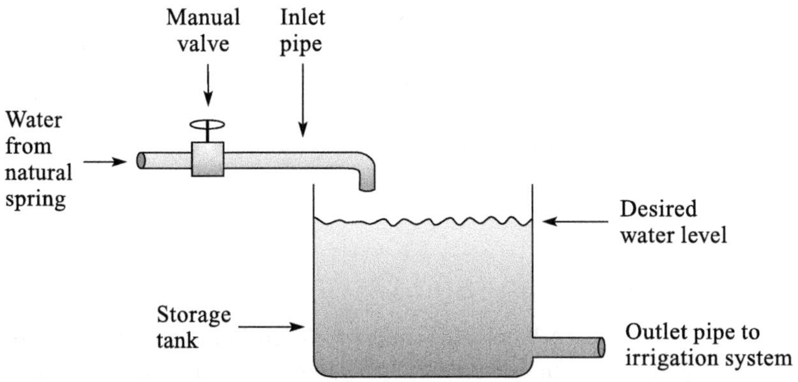

FIGURE 1-3 An open-loop reservoir system that stores water for an irrigation system

overflow. Excessive evaporation will also upset the balance. If it occurs over a prolonged period of time, the water level in the tank may become unacceptably low.

A human operator who periodically inspects the tank can change the control valve setting to compensate for these disturbances.

An example of a manually operated open-loop system is the speed of a car being controlled by the driver. The driver adjusts the throttle to maintain a highway speed when going uphill, downhill, or on level terrain.

Closed-Loop Systems

There are many situations in industry where the open-loop system is adequate. However, some manufacturing applications require continuous monitoring and self-correcting action of the operation for long periods of time without interruption. The automatic closed-loop configuration performs the self-correcting function. This automatic system employs a feedback loop to keep track of how closely the system is doing the job it was commanded to do.

The reservoir system can also be used to illustrate a closed-loop operation. To perform automatic control, the system is modified by replacing the manually controlled valve with an adjustable valve connected to a float, as shown in Figure 1-4. The valve, the float, and the linkage mechanism provide the feedback loop.

If the level of the water in the tank goes up, the float is pushed upward; if the level goes down, the float moves downward. The float is connected to the inlet valve by a mechanical linkage. As the water level rises, the float moves upward, pushing on the lever and closing the valve, thus reducing the water flow into the tank. If the water level lowers, the float moves

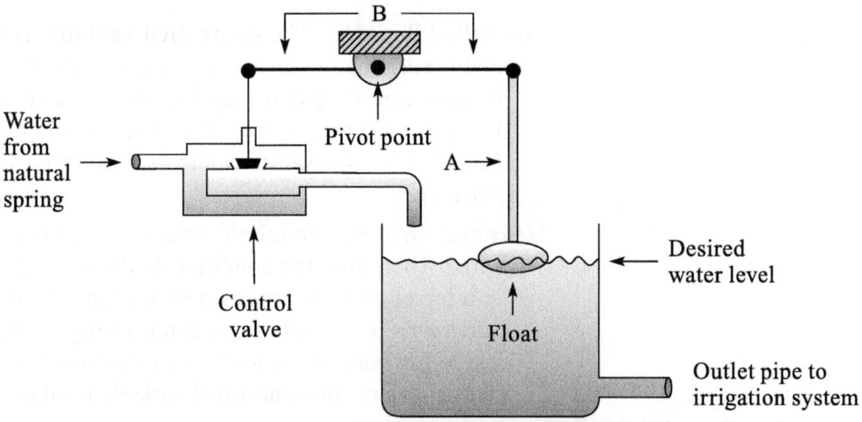

FIGURE 1-4 A closed-loop system that uses a linkage mechanism as a feedback device to provide self-correcting capabilities

downward, pulling on the lever and opening the valve, thus allowing more water into the tank. To adjust for a desired level of water in the tank, the float is moved up or down on the float rod A.

Most automated manufacturing processes use closed-loop control. These systems that have a self-regulation capability are designed to produce a continuous balance.

1-2 Elements of Open- and Closed-Loop Systems

IAU3306
Elements of a
Closed-Loop System

A block diagram of a closed-loop control system is shown in Figure 1-5. Each block shows an element of the system that performs a significant function in the operation. The lines between the blocks show the input and output signals of each element, and the arrowheads indicate the direction in which they flow.

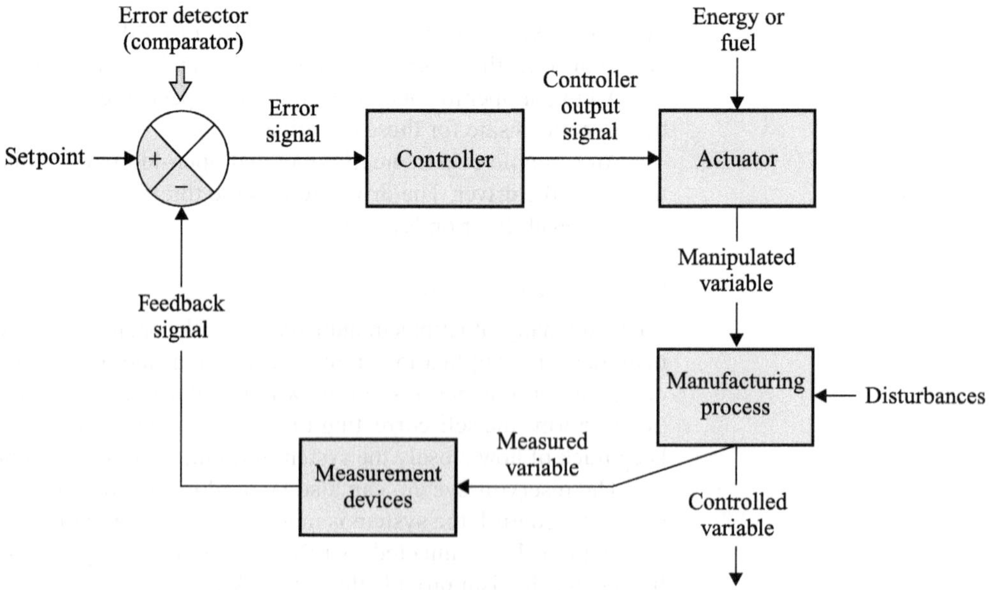

FIGURE 1-5 Closed-loop block diagram that shows elements, input/output signals, and signal direction

This section describes the functions of the blocks, their signals, and common terminology used in a typical closed-loop network:

Controlled Variable. The **controlled variable** is the actual variable being monitored and maintained at a desired value in the manufacturing process. Examples in a process control system may include temperature, pressure, and flow rate. Examples in a motion control system may be position or velocity. In the water reservoir system (Figure 1-4), the water level is the controlled variable. Another term used is *process variable.*

Measured Variable. To monitor the status of the controlled variable, it must be measured. Therefore, the condition of the controlled variable at a specific point in time is referred to as the **measured variable.** Various methods are used to make measurements. One method of determining a controlled variable such as the level of water, for example, is to measure the pressure at the bottom of a tank. The pressure that represents the controlled variable is taken at the instant of measurement.

Measurement Device. The **measurement device** is the "eye" of the system. It senses the measured variable and produces an output signal that represents the status of the controlled variable. Examples in a process control system may include a thermocouple to measure temperature or a humidity detector to measure moisture. Examples in a motion control system may be an optical device to measure position or a tachometer to measure rotational speed. In the water reservoir system, the float is the measurement device. Other terms used are *detector, transducer,* and *sensor.*

Feedback Signal. The **feedback signal** is the output of the measurement device. In the water reservoir system, the feedback signal is the vertical position of member A in the linkage mechanism (see Figure 1-4). Other terms used are *measured value, measurement signal,* or *position feedback* if in a position loop, or *velocity feedback* if in a velocity loop.

Setpoint. The **setpoint** is the prescribed input value applied to the loop that indicates the desired condition of the controlled variable. The setpoint may be manually set by a human operator, automatically set by an electronic device, or programmed into a computer. In the water reservoir system, the setpoint is determined by the position at which the float is placed along rod A. Other terms used are *command* and *reference.*

Error Detector. The **error detector** compares the setpoint to the feedback signal. It then produces an output signal that is proportional to the difference between them. In the water reservoir system, the error detector is the entire linkage mechanism. Other terms used are *comparator* or *comparer* and *summing junction.*

Error Signal. The **error signal** is the output of the error detector. If the setpoint and the feedback signal are not equal, an error signal proportional to their difference develops. When the feedback and setpoint signals are equal, the error signal goes to zero. In the reservoir system (Figure 1-4), the error signal is the angular position of member B of the linkage mechanism. Other terms used are *difference signal* and *deviation.*

Controller. The **controller** is the "brain" of the system. It receives the error signal (for closed-loop control) as its input, and develops an output signal that causes the controlled variable to become the value specified by the setpoint. Most controllers are operated electronically, although some of the older process control systems use air pressure in pneumatic devices. The operation of an electronic controller is performed by hardwired circuitry or computer software. The controller produces a small electrical signal that usually needs to be conditioned or modified before it is sent to the next element. For example, it must be amplified if it is applied to an electrical motor, or connected to a proportional air pressure if it is applied to a pneumatic positioner or a control valve. The control function is also performed by programmable logic controllers (PLCs) and panel-mounted microprocessor controllers.

Actuator. The **actuator** is the "muscle" of the system. It is a device that alters some type of energy or fuel supply, causing the controlled variable to match the desired setpoint. Examples of energy or fuel are the flow of steam, water, air, gas, or electrical current. A practical application is a commercial bakery where the objective is to keep the temperature in an oven at 375 degrees. The temperature is the controlled variable. The temperature is determined by how much gas is fed to the oven burner. A valve in the gas line controls the flow by the amount it opens or closes. The valve is the actuator in the system. In the reservoir system, the actuator is the flow control valve, connected to the inlet pipe. Other terms used are the *final control element* and *final correcting device.* Common types of actuators are louvers, hydraulic cylinders, pumps, and motors.

Manipulated Variable. The amount of fuel or energy that is altered by the actuator is referred to as the **manipulated variable.** The amount by which the manipulated variable is changed by the actuator affects the condition of the controlled variable. In the commercial oven example, the gas flow rate is the manipulated variable, and the temperature is the controlled variable. In the reservoir system, the flow is the manipulated variable. The flow rate is altered by the control valve (actuator), which affects the condition of the controlled variable (level).

Manufacturing Process. The **manufacturing process** is the operation performed by the actuator to control a physical variable, such as the motion of a machine or the processing of a liquid.

Disturbance. A **disturbance** is a factor that upsets the manufacturing process being performed, causing a change in the controlled variable. In the reservoir system, the disturbances are the rainfall and evaporation that alter the water level.

A block diagram of an open-loop system is shown in Figure 1-6. The controller, actuator, and manufacturing process blocks perform the same operations as the closed-loop system shown in Figure 1-5. However, instead of the error signal being applied to the controller, the setpoint provides its input. Also, there is no feedback loop, and a comparator is not used by the open-loop system.

It is possible for an open-loop system to perform automated operations. For example, the washing machine that launders clothes in your home uses a timer to control the wash cycles. An industrial laundry machine also uses timing devices to perform the same functions but on a larger scale. However, there is no feedback loop that monitors and takes corrective action if the timer becomes inaccurate, the temperature of the water changes, or a major problem arises that requires the machine to shut down.

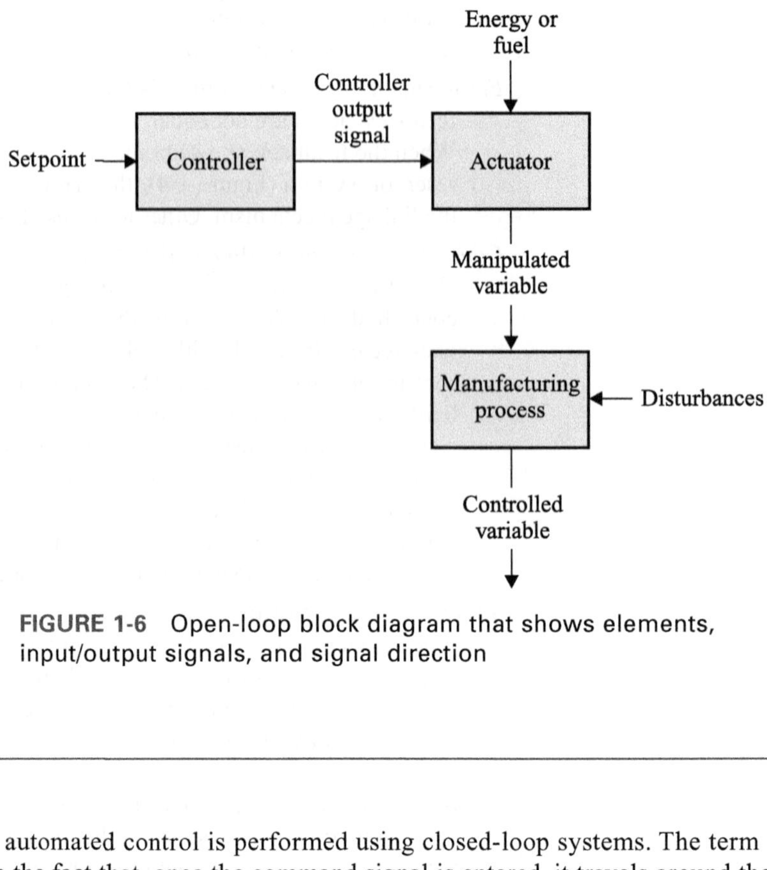

FIGURE 1-6 Open-loop block diagram that shows elements, input/output signals, and signal direction

1-3 Feedback Control

Industrial automated control is performed using closed-loop systems. The term *loop* is derived from the fact that, once the command signal is entered, it travels around the loop until equilibrium is restored.

To summarize the operation of a closed-loop system, the objective is to keep the controlled variable equal to the desired setpoint. A measurement device monitors the controlled variable and sends a measurement signal to the error detector that represents its condition along the feedback loop. An error detector compares the feedback signal to the setpoint and produces an error signal that is proportional to the difference between them. The error signal is fed to a controller, which determines which kind of action should occur to make the controlled variable equal to the setpoint. The output of the controller causes the actuator to adjust the manipulated variable. Altering the manipulated variable causes the condition of the controlled variable to change to the desired value.

The basic concept of feedback control is that an error must exist before some corrective action can be made. An error can develop in one of three ways:

1. The setpoint is changed.
2. A disturbance appears.
3. The load demand varies.

In the reservoir system of Figure 1-4, the setpoint is changed by adjusting the position of the float along linkage A. A disturbance is caused when rain supplies additional water to the tank or evaporation lowers the level. The water flowing out of the tank to the irrigation system is referred to as the *load*. If the level of the water in the irrigation system suddenly lowers, the back pressure on the outlet pipe will decrease and cause the fluid to drain more rapidly. This downstream condition is referred to as a *load change*. The setpoint and load demand are changes that normally occur in a system. The disturbance is an unwanted condition.

Feedback signals may be either positive or negative. If the feedback signal's polarity aids a command input signal, it is said to be positive or regenerative feedback. Positive feedback is used in radios. If the radio signal is weak, an automatic gain control (AGC) circuit is activated. Its output is a feedback signal that boosts the radio signal's overall strength.

However, when positive feedback is used in industrial closed-loop systems, the input usually loses control over the output. If the feedback signal opposes the input signal, the system is said to use negative or degenerative feedback. By combining negative feedback values from the command signal, a closed-loop system works properly.

An example of closed-loop control that uses negative feedback is the central heating system in a house. The thermostat in Figure 1-7 monitors the temperature in the house and compares it to the desired reference setting. Suppose the room temperature drops to 66 degrees from the reference setting of 72 degrees. The measured feedback value is subtracted from the setpoint command and causes a 6-degree discrepancy. The thermostat contacts will close and cause the furnace to turn on. The furnace supplies heat until the temperature is back to the reference setting. When the negative feedback is sufficient to cancel the command, the error no longer exists. The thermostat then opens and switches the furnace off until the house cools down below the reference. As this cycle repeats, the temperature in the house is automatically maintained without human intervention.

The speed of an automobile can also be controlled automatically by a closed-loop system called a cruise control. The desired speed is set by an electronic mechanism usually

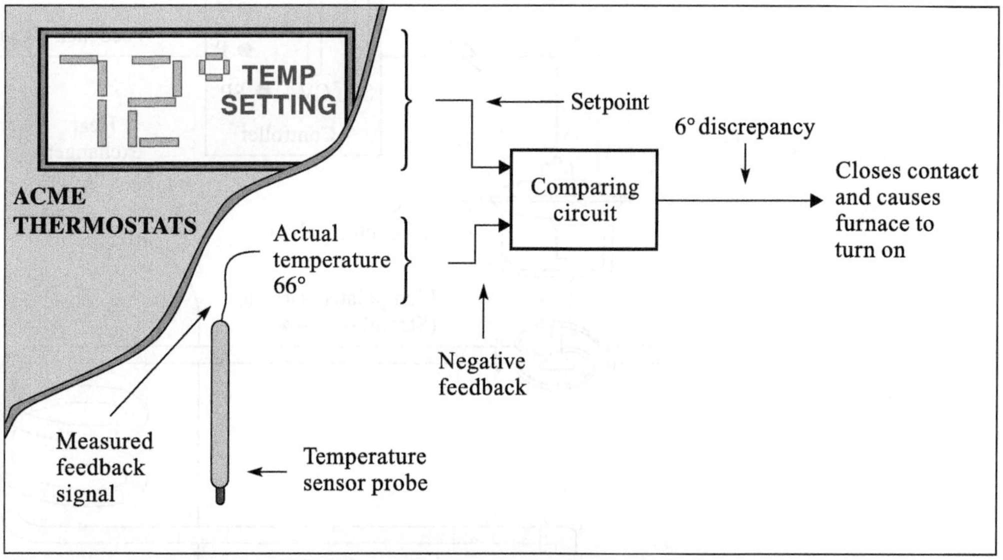

FIGURE 1-7 A thermostat uses a negative feedback signal to control the temperature of a house

placed on the steering wheel assembly. A Hall-effect speed sensor connected to the front axle generates a signal proportional to the actual speed. An electronic error detector compares the actual speed to the desired speed, and then sends a signal representing the difference between them to a controller. The controller sends a signal that causes the fuel flow to the engine to vary. If a car that is traveling on a level road suddenly encounters an uphill grade, it begins to slow down. Because the actual speed is lower than the desired speed, the error detector sends a signal to the controller that causes more fuel to flow to the engine. The additional fuel causes the car to accelerate until it reaches the desired speed.

1-4 Practical Feedback Application

IAU206
Feedback System Application

An actual practical application of a feedback system used in a manufacturing process is shown in Figure 1-8. The diagram shows a heat exchanger. Its function is to supply water at a precise elevated temperature to a mixing vat that produces a chemical reaction. Cold water enters the bottom of the tank. The water is heated as it passes through steam-filled coils and leaves the tank through a port located at the top.

This example illustrates how the elements of a closed-loop feedback system provide automatic control. The elements consist of a thermal sensor, controller, and actuator. Together, they keep the temperature of the water that leaves the tank as close as possible to the setpoint when process conditions change.

There are three factors that can cause the condition of the controlled variable to become different from the setpoint. Two of the three factors are intentional. One intentional factor is changing the setpoint to a new desired temperature level. Another intentional factor is a *load change.* An example of a load change in the heat exchanger is an increase in the pump's flow rate so that the water leaves the top port of the tank much more rapidly than usual. This condition would cause the water to flow through the tank more quickly. As a result, the water will not be heated as much as it flows through the coils, causing the outgoing temperature to be lower. An unintentional factor is a *disturbance.* One example of a disturbance in the heat exchanger is a decrease in the temperature of the water entering the tank. When this condition exists, the temperature of the water in the tank will drop below setpoint. This situation occurs because the water entering the tank is colder. Since the temperature of the heating (steam) coils remains unchanged, the temperature of the water leaving the tank will be lower.

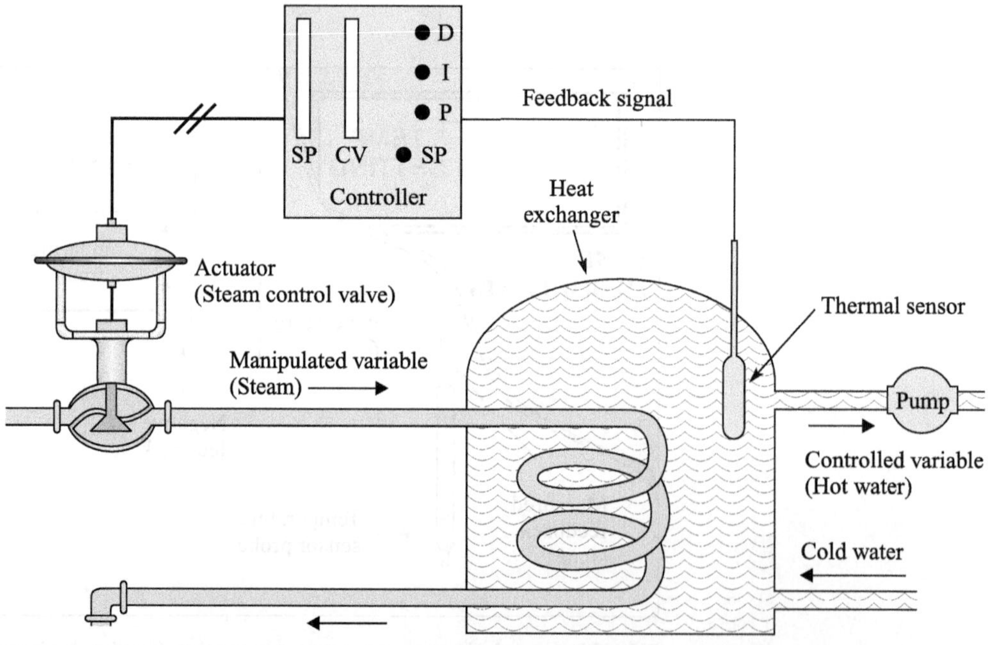

FIGURE 1-8 Closed-loop temperature control system

Whenever there is a difference between the setpoint and the condition of the controlled variable, the control system with feedback compensates for any error. For example, suppose that the temperature of the water leaving the heat exchanger falls below the setpoint. Thermal energy, which is the measured variable, is detected by the sensor. The sensor produces an electrical signal, which is the feedback signal to the controller. The controller compares the measured value to the setpoint. The amount of the deviation determines the value of the controller output signal. This output signal goes to the final control element, which is a steam control valve. To return the water temperature back to the setpoint, the valve is opened further by the actuator, allowing more steam, which is the manipulated variable, to enter the coils. As the coils become hotter, the temperature of the water, which passes through them, also rises.

As the water temperature returns to the setpoint, the deviation becomes smaller. The controller responds by changing its output signal to the valve. The new output signal causes the valve to reduce the flow of steam through the coils and causes the water to be heated at the proper rate.

1-5 Dynamic Response of a Closed-Loop System

IAU12408
The Dynamic Response of a Closed-Loop System

The objective of a closed-loop system is to return the controlled variable back to the condition specified by the command signal when a setpoint change, a disturbance, or a load change occurs. However, there is not an immediate response. Instead, it takes a certain amount of time delay for the system to correct itself and re-establish a balanced condition. A measure of the loop's corrective action, as a function of time, is referred to as its **dynamic response.** There are several factors that contribute to the response delay:

- The **response time** of the instruments in the control loop. The instruments include the sensor, controller, and final control element. All instruments have a *time lag.* This is the time beginning when a change is received at its input and ending at the time it produces an output.
- The **time duration** as a signal passes from one instrument in the loop to the next.
- The **static inertia** of the controlled variable. When energy is applied, the variable opposes being changed and creates a delay. Eventually, the energy overcomes the resistance and causes the variable to reach its desired state. This delayed action is referred to as **pure lag.** The amount of lag is determined by the capacity (physical size) of the material; the lag is proportional to the amount of its mass. The type of material of which a controlled variable consists also affects the lag. For example, the temperature of a gas will change more quickly than that of a liquid when exposed to thermal energy. The chemical properties of the controlled variable can also affect the amount of delay.
- The elapsed time between the instant a deviation of the controlled variable occurs and the corrective action begins. This factor is referred to as **dead time.** A pipeline that passes fluid can be used to illustrate an example of dead time. The control function of the closed-loop system is to regulate the temperature of the fluid flowing through the pipe. If the temperature of the fluid entering the pipe suddenly drops, there is a brief time period that passes before the fluid reaches a sensor downstream. The time from when the fluid enters the pipe until the sensor begins to initiate the closed-loop response is the dead time.

1-6 Feed-Forward Control

IAU3406
The Feed-Forward Control System

Two conditions can minimize the effectiveness of feedback control. The first is the occurrence of large magnitude disturbances. The second is long delays in the dynamic response of the control loop. To compensate for these limitations of feedback control, **feed-forward** control can be used.

The operation of feed-forward control is very different from feedback control. Feedback control takes corrective action after an error develops. The objective of feed-forward control is

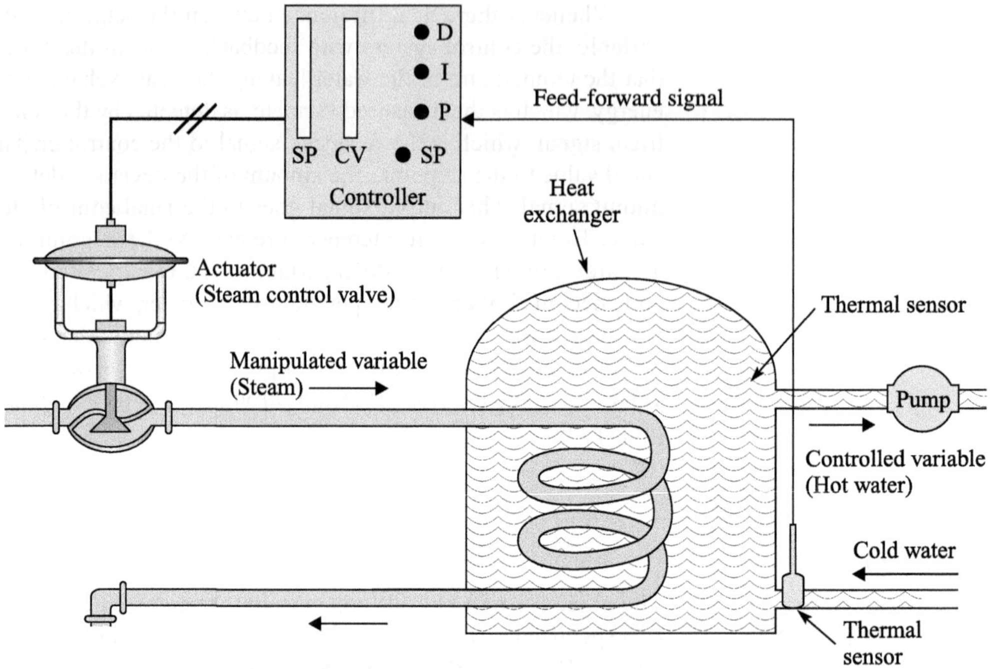

FIGURE 1-9 Feed-forward control of a temperature control system

to prevent errors from occurring. Typically, feed-forward cannot prevent errors. Instead, it minimizes them.

The heat exchanger system described in Section 1-4 can be modified for feed-forward control, as shown in Figure 1-9. Instead of placing the thermal sensor inside the tank to detect a temperature deviation of the heated water, a thermal sensor is placed in the inlet pipe. As soon as there is a change in the temperature of the incoming cold water, it is detected before entering the tank. The controller responds by adjusting the position of the steam valve. By varying the steam through the coil at this time, corrective action occurs before the controlled variable leaving the outlet pipe can deviate from the setpoint temperature.

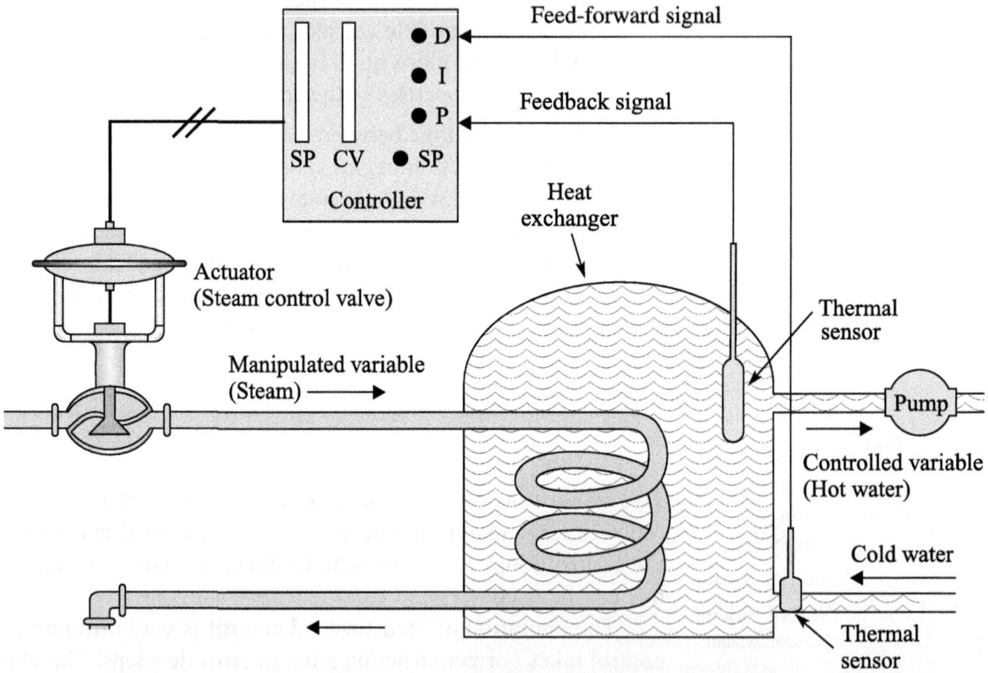

FIGURE 1-10 Feed-forward control loop with a feedback control loop

The feed-forward control system does not operate perfectly. There are always unmeasurable disturbances that cannot be detected, such as a worn flow valve, a sensor out of tolerance, or inexact mathematical calculations processed by the controller. Over a period of time, these unmeasurable disturbances affect the operation and eventually the water temperature in the tank, finally causing the water to reach an unacceptable temperature level. Due to the inaccuracy of feed-forward control, it is seldom used by itself. By adding feedback control to the system, corrections by the controller can be made if the controlled variable deviates from the setpoint due to unmeasurable disturbances.

Figure 1-10 shows a heat exchanger system that uses both feed-forward control and feedback control. The controller receives input signals from two sensors. The sensor in the inlet line provides the feed-forward signal, and the sensor near the outlet provides the feedback signal.

In summary, feed-forward control adjusts the operation of the actuator to prevent changes in the controlled variable. Feed-forward controllers must make very sophisticated calculations to compute the changes of the actuator needed to compensate for variations in disturbances. Since they require highly skilled engineers, they typically are used only in critical applications within the plant.

Problems

1. The two classifications of industrial control systems are ________ control and ________ control.
2. List another name for each of the following terms.
   Motion Control
   Process Control
   Batch Process
3. A closed-loop industrial system typically uses ________ (negative, positive) feedback.
4. List two examples of controlled variables for motion control applications and two examples for process control applications.
   Motion Control
   Process Control
5. List one example of a measurement device for a motion control application and one example for a process control application.
   Motion Control
   Process Control
6. The control method used in ____ control applications is to sustain a constant condition of the controlled variable.
   a. servo b. process
7. An open-loop system does not have a ____.
   a. controller c. feedback loop
   b. final control element d. none of the above
8. T/F The measured variable represents the condition of the controlled variable.
9. The output of the measurement device is called the ________ ________.
10. Define *setpoint*.
11. The difference between the setpoint and feedback signal is referred to as the ________ signal, and is produced by the ________ detector.
12. T/F The controller can be considered the brain of a closed-loop system.
13. Altering the ________ variable causes the condition of the ________ variable to change.
    a. controlled b. manipulated
14. The device that provides the muscle to perform work in the closed-loop system is referred to as the ________.
15. The ____ is sent to the final control element.
    a. measured variable c. error signal
    b. feedback signal d. control signal
16. Which of the following influences causes a controlled variable to change? ____
    a. A disturbance occurs. c. The setpoint is adjusted.
    b. A load demand varies. d. all of the above
17. Which of the following factors contributes to the dynamic response of a single control loop? ____
    a. the instrument in a control loop
    b. the inertia of the controlled variable
    c. dead time
    d. all of the above
18. T/F The manipulated variable and controlled variable are synonymous terms in a closed-loop system.
19. T/F The basic concept of feedback control is that an error must exist before some corrective action can be made.
20. A pressurized tank must maintain a gas at 325 psi. A pressure sensor is used to measure the condition of the controlled variable. As the gas cools, the pressure in the tank decreases. When it drops to 300 psi, a valve is opened, which allows steam to flow to a heat exchanger inside the tank. The additional steam heats the gas and causes pressure to rise.
    ____ What is the controlled variable in this process?
    ____ What is the manipulated variable in this process?
    ____ What is the setpoint?
    ____ What is the measured variable?
    a. gas pressure d. 300 psi
    b. steam flow e. pressure
    c. 325 psi f. heat
21. T/F Feed-forward control is seldom used except in combination with feedback control.

22. Which of the following conditions is compensated for by using feed-forward control? ____
    a. excessive lag time
    b. large disturbances
    c. an error signal
    d. feedback signal
23. The objective of ____ control is to prevent the controlled variable from deviating from the setpoint.
    a. feedback
    b. feed-forward
24. When feedback and feed-forward control are performed together, the primary function of feed-forward is to make corrections for ____ disturbances, and feedback control to make corrections for ____ disturbances.
    a. measurable
    b. unmeasurable

CHAPTER

2

Interfacing Devices

OBJECTIVES

At the conclusion of this chapter, you should be able to:

- Identify the schematic diagrams, describe the operations, and calculate the outputs of the comparator, inverting, summing, noninverting, and difference operational amplifiers (op amps).
- Identify the schematic diagrams of the integrator and differentiator op amps and draw the output waveforms they produce when various input signals are applied.
- Given applied input signals, indicate the resulting output of the digital comparator device.
- Describe the wave-shaping capability and operating characteristics of a Schmitt trigger.
- Determine how optoelectronic devices are switched and explain the isolation function they perform.
- Explain the operation of analog-to-digital and digital-to-analog converters, determine their resolution, and make the proper wiring connections to their integrated circuit packages.
- Assemble monostable and astable multivibrators using a 555 monolithic integrated circuit and use calculations to determine their output.

INTRODUCTION

In Chapter 1, a block diagram was used to describe the operation of the elements of a closed-loop system. Each element plays a significant role in the operation of the system. One of the requirements of any system is the successful interfacing or connecting together of the various blocks. In a block diagram, an interface is represented by the lines between two blocks, indicating that some type of signal passes from one to the other.

There are many components and circuits that perform the functions of each element. Sometimes, the signals processed in one element are incompatible with those that can be used in the next element. To make the elements compatible, various types of conversion components are used to interface them together.

To help the reader understand the material covered in later chapters, this chapter describes the basic operation of discrete components and integrated circuits that are used within and between the elements. The components covered here have been selected based on many of the circuits covered in the remainder of the book.

2-1 Fundamental Operational Amplifiers

A very versatile amplifier device is the **operational amplifier** (op amp). One of the most popular op amps is the uA741, which is fabricated inside an 8-pin integrated circuit package. There are three important characteristics of op amps that make them ideal amplifiers:

1. High input impedance

2. High voltage gain
3. Low output impedance

Figure 2-1 shows the standard schematic symbol of the uA741 op amp. Represented by a triangle, the op amp has two input terminals located at the base on the left and a single output located at the apex of the triangle. There are also two separate power-supply lines. The one located at the top base is connected to a positive potential, and the other located at the bottom base is connected to a negative potential. These two power supplies allow the output voltage to swing to either a positive or a negative voltage with respect to ground.

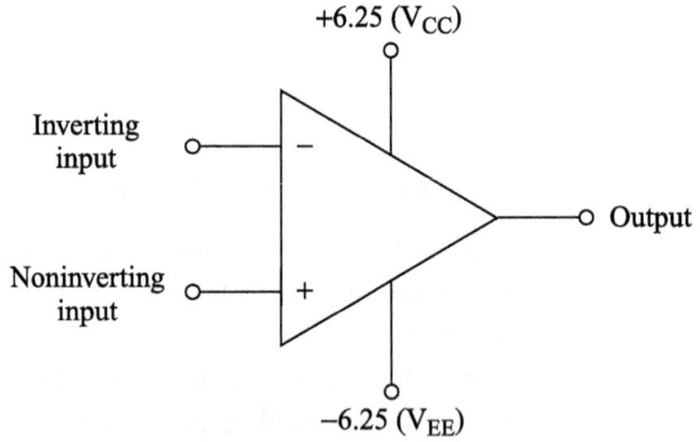

FIGURE 2-1 Standard symbol of an operational amplifier

One of the inputs has a minus sign. This is called the *inverting input,* because any DC or AC signal applied to its input produces an output voltage of the opposite polarity. The other input has a plus sign and is called the *noninverting input.* Any DC or AC signal applied at this input produces an output of the same polarity.

When external components are connected to the input and output leads, the op amp is capable of performing several functions. How the components are connected determines which function the op amp performs.

Operational Amplifier Comparator

SSE4603
Op Amp Comparator

Figure 2-2 shows an op amp configuration that operates as a voltage comparator. This device compares the voltage applied to one input to the voltage applied at the other input. Any difference between the voltages drives the op amp output into either a positive- or a negative-volt saturation condition. Saturation is about 80 percent of the supply voltage. Therefore, 5 volts is produced if the power supply is 6.25 volts. The polarity of the output is determined by the polarity of the voltages applied at the inputs. When the voltage applied to the inverting

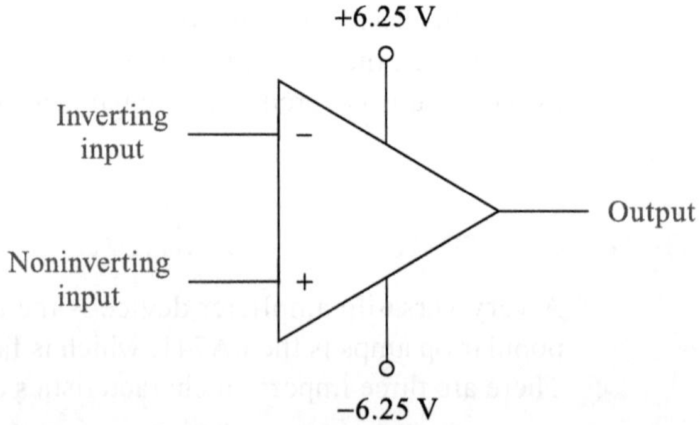

FIGURE 2-2 Op amp comparator

input is more positive than the voltage at the noninverting terminals, the output swings to a –5-volt saturation potential. Likewise, when the voltage applied to the inverting input is more negative than the voltage at the noninverting input, the output swings to the +5-volt saturation potential. However, when the input voltages are the same amplitude, the output is zero. The following equations provide a summary of the operation for the voltage comparator:

Inverting input voltage < noninverting input voltage = positive output voltage

Inverting input voltage > noninverting input voltage = negative output voltage

Inverting input voltage = noninverting input voltage = zero output voltage

Table 2-1 provides examples of how the op amp, operating as a comparator, responds to several input voltages.

TABLE 2-1 Operation of an Op Amp Comparator

Inverting Input Terminal (Volts)	*Noninverting Input Terminal (Volts)*	*Output Saturation Voltage (Volts)*
+1	–1	–5
+1	+2	+5
+2	+1	–5
0	0	0
–1	+1	+5
0	–1	–5
0	+1	+5
+3	+3	0

Inverting Operational Amplifier

SSE5403
Inverting Op Amp

A typical op amp can have a voltage gain of approximately 200,000. However, the output voltage level cannot exceed approximately 80 percent of the supply voltage. For example, the maximum output voltages of the op amp in Figure 2-1 are +5 and –5 volts because the power-supply potentials are +6.25 and 6.25 volts. Therefore, it only takes a 25-uV input to result in a positive or negative 5-volt output voltage, depending on the input-signal polarity and the terminal to which it is applied.

However, the op amp is used for many applications that require a voltage gain less than 200,000. A technique called *feedback* is used to control the gain of this device, and it is accomplished by connecting a resistor from the output terminal to an input lead. A negative-feedback circuit is shown in Figure 2-3. Its operation is as follows:

- Both input terminals have high impedances; therefore, they do not allow current to flow into or out of them.
- The potential at the inverting input lead is called *0-volt virtual ground* (that is, it acts like a 0-volt ground). The positive input lead is connected to an actual 0-volt ground potential.

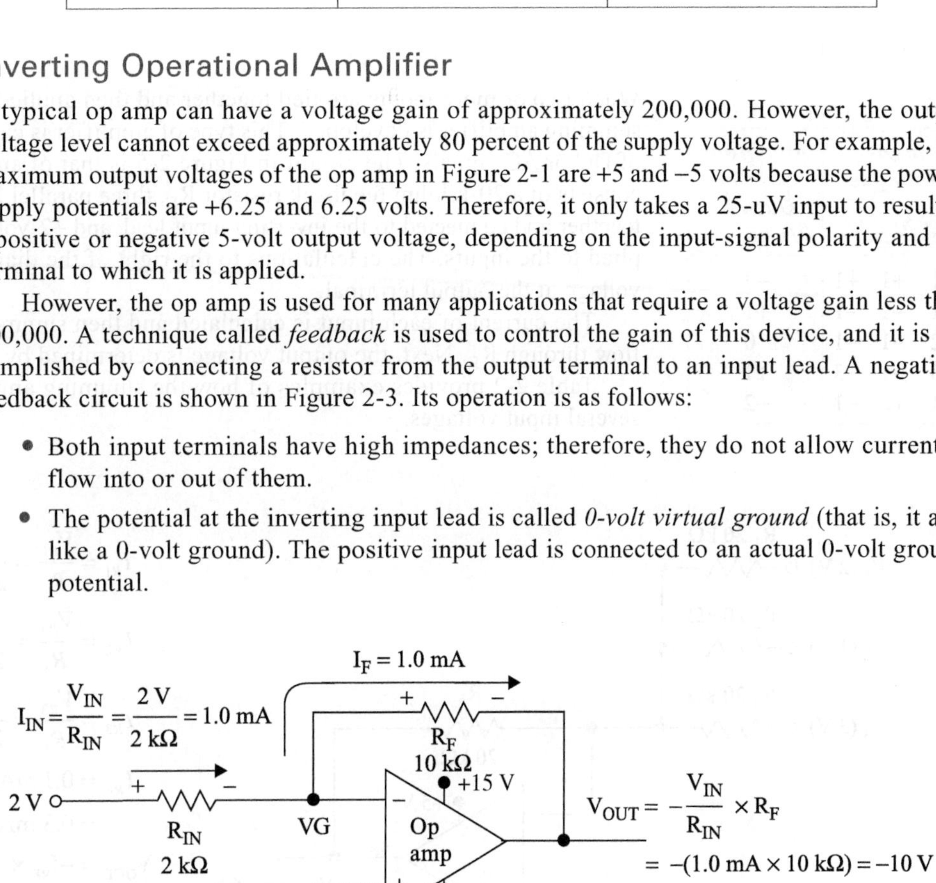

FIGURE 2-3 Inverting op amp

- Because point VG is 0 volts, there is a voltage drop of 2 volts across the 2-kilohm resistor, R_{IN}. 1 mA flows through it.
- The 1 mA cannot flow into the op amp. Therefore, it flows up through the 10-kilohm feedback resistor R_F, developing a 10-volt drop across it.
- Because V_{OUT} is measured with respect to the virtual ground, its voltage is −10 volts.

The voltage *gain* of the op amp is determined by:

$$V_{GAIN} = \frac{V_{OUT}}{V_{IN}}$$

FIGURE 2-4 Input and output voltages of an inverting op amp with a gain of 10

V_{IN}	V_{OUT} *(Volts)*
+0.2	−2
−0.4	+4
0	0
+0.32	−3.2

The gain of the inverting op amp in Figure 2-3 is 5 because a 2-volt signal is applied to the input and an inverted −10-volt signal is at the output. A negative input voltage applied to this amplifier produces a positive output. The gain is influenced by the resistance ratio of R_F compared to R_{IN}. The larger R_F becomes compared to R_{IN}, the larger the gain.

The output voltage can also be determined by:

$$V_{OUT} = \left[-\frac{R_F}{R_{IN}}\right](V_{IN})$$

SSE7306
Summing Op Amp

Figure 2-4 provides examples of how the inverting amplifier with a gain of 10 responds to several input voltages.

TABLE 2-2 Operation of an Inverting Summing Amplifier

Input Voltages			Algebraic Sum of Output Voltages
V_1	V_2	V_3	
+1	+1	+1	−3
+1	−1	−1	+1
+2	−1	−1	0
−3	−1	+3	+1
+1	+2	−1	−2

Summing Amplifier

When two or more inputs are tied together and then applied to an input lead of an op amp, a summing amplifier is developed. This type of amplifier is capable of adding the algebraic sum of DC or AC signals. The circuit in Figure 2-5 is that of an inverting summing amplifier. It consists of a 20-kilohm feedback resistor R_F, three parallel 20-kilohm summing resistors tied together and connected to the inverting input lead, and +2-volt, +1-volt, and +3-volt signals applied to the inputs. The calculations to the right of the diagram show how to determine the voltage at the output terminal.

The current of each input is calculated and then summed to obtain the resulting current flow through R_F. Next, the output voltage is determined by multiplying $-I_{RF}$ times R_F.

Table 2-2 provides examples of how the summing amplifier in Figure 2-5 responds to several input voltages.

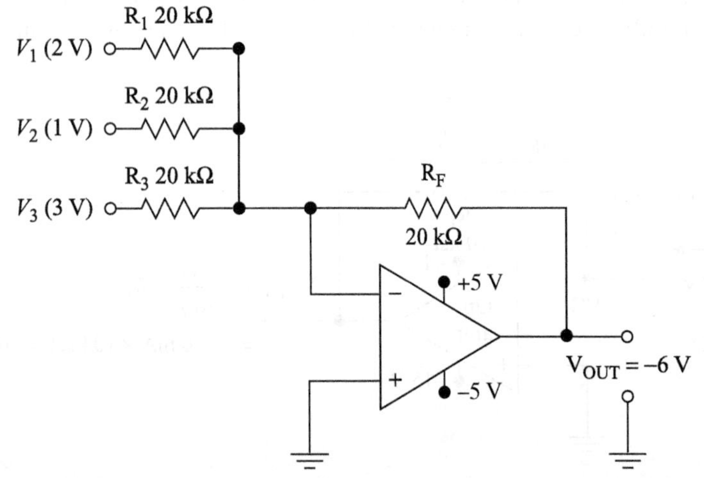

$$I_{R1} = \frac{V_{R1}}{R_1} = \frac{2\text{ V}}{20\text{ k}\Omega} = 0.1\text{ mA}$$

$$I_{R2} = \frac{V_{R2}}{R_2} = \frac{1\text{ V}}{20\text{ k}\Omega} = 0.05\text{ mA}$$

$$I_{R3} = \frac{V_{R3}}{R_3} = \frac{3\text{ V}}{20\text{ k}\Omega} = 0.15\text{ mA}$$

$$I_{RF} = 0.1\text{ mA} + 0.05\text{ mA} + 0.15\text{ mA}$$

$$= 0.3\text{ mA} \quad \text{or} \quad -0.3\text{ mA (inverted)}$$

$$V_{OUT} = -I_{RF} \times R_F$$

$$= -0.3\text{ mA} \times 20\text{ k}\Omega = -6\text{ V}$$

FIGURE 2-5 Inverting summing amplifier

Noninverting Amplifier

SSE7906
Noninverting Op Amp

Some applications require that an amplified output signal be in phase with the input. Using an operational amplifier, this is accomplished by applying the input signal to the noninverting input, while the feedback to control gain is still provided by connecting the output terminal to the inverting input through a resistor (R_F). One lead of resistor R_{IN} is also connected to the inverting input. The other lead of R_{IN} is connected to a 0-volt ground potential.

Figure 2-6 shows the schematic diagram of a noninverting amplifier. The gain of the circuit is influenced by resistors R_{IN} and R_F. The equation used to determine the gain of the noninverting amplifier is derived by adding 1 to the resistance ratio of R_F and R_{IN}. Thus:

$$\text{Gain} = 1 + \frac{R_F}{R_{IN}}$$

The output voltage is determined by:

$$V_{OUT} = \left[1 + \frac{R_F}{R_{IN}}\right](V_{IN})$$

The gain will always be greater than 1.

Table 2-3 provides examples of how the noninverting amplifier shown in Figure 2-6 responds to several input voltages.

TABLE 2-3 Operation of a Noninverting Op Amp

V_{IN}	V_{OUT}
+0.3	+3.3
−1.0	−11
+.75	+8.25
−.52	−5.72

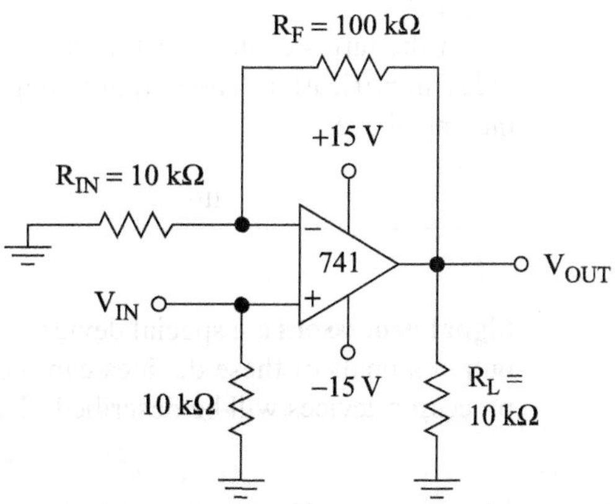

FIGURE 2-6 Noninverting operational amplifier

Difference Operational Amplifier

SSE8006
Difference Op Amp

The **difference operational amplifier** (shown in Figure 2-7) finds the algebraic difference between two input voltages. Neither the inverting input nor the noninverting input is grounded. Instead, signals are applied to both inputs at the same time, and the difference between them is amplified. If the signals are the same, the output voltage is zero.

Note that the circuit uses the closed-loop feedback configuration, which results in a controlled amplified output voltage. If all the external resistors are equal, no amplification takes place. Instead, the voltage difference op amp performs the arithmetic operation of subtraction. For example, suppose that 3 volts are applied to the inverting input V_1, and 6 volts to the noninverting input V_2. The voltage difference between these inputs is 3 volts, which is developed at the op amp output.

The nonamplified output can be calculated using the following formula if all of the resistors are of the same value (10 kΩ):

$$\begin{aligned} V_{OUT} &= V_2 - V_1 \\ &= 6V - 3V \\ &= +3V \end{aligned}$$

If the voltage at the inverting input (V_1) is more negative than the voltage at the noninverting input (V_2), the polarity of the output will be positive, and vice versa. Table 2-4 provides

TABLE 2-4 Operation of a Difference Operational Amplifier

Input Voltage		*Output Voltage Algebraic Difference (Inverted)*
V_1	V_2	
+2	+4	+2
+4	+2	−2
+4	−2	−6
−2	+4	+6
−4	−2	+2
−2	−4	−2

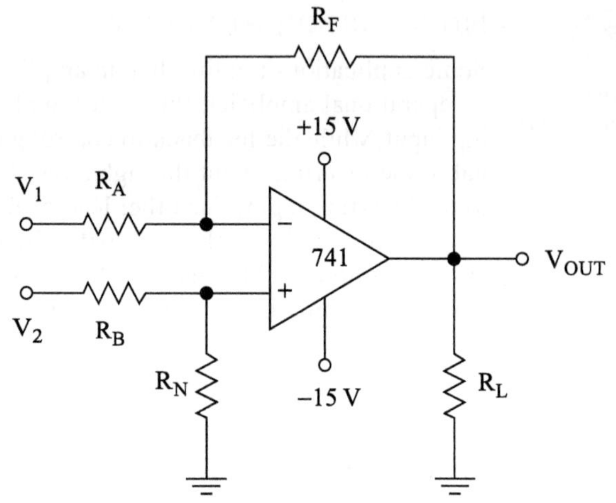

All resistors are 10 kΩ.

FIGURE 2-7 Difference op amp

examples of various input conditions and the resulting output voltages for the circuit in Figure 2-7.

If the ratios of the resistor values in the circuit are changed, the difference op amp provides amplification. The output voltages can be determined by using a different formula than the one above.

2-2 Signal Processors

Signal processors are special devices that change or modify signals applied to their inputs. The output signals of these devices can then be used to perform specific functions. Three signal processor devices will be described: the *integrator,* the *differentiator,* and the *Schmitt trigger.*

Integrator Operational Amplifier

SSE5303
The Integrator
Op Amp

An **integrator** is an amplifier circuit that continuously increases its gain over a period of time. The magnitude of the output is proportional to the period of time that a constant DC input signal is present. Figure 2-8(a) shows the schematic diagram of the op amp integrator. The circuit resembles that of an inverting op amp. The difference is that a capacitor replaces the resistor as the feedback element. The waveform diagrams in Figure 2-8(b) illustrate the operation of the circuit when different DC voltages are applied to the input.

When the input voltage changes from 0 to +5 volts, at T_1 of the waveform, the capacitor initially has a low impedance because it is discharged. The gain of the op amp is zero because the ratio of the feedback resistance to the input resistance is zero. This action is expressed by the formula for the inverting op amp: $V_{OUT} = R_{FB}/R_{IN}$.

As the capacitor begins to charge, the impedance path to current flow increases. Because the feedback resistance rises, the R_{CFB} ratio increases. The result is that the output of the op amp increases in a linear fashion. Since the inverting input is used, the output will be a negative-going waveform. Eventually, the waveform levels off because the op amp reaches saturation, as shown at T_2 of the diagram.

At T_3, the input voltage changes from +5 to 0 volts. The capacitor discharges and causes the output to return to 0 volts. If a negative voltage is applied to the input, a positive-going signal develops at the output. If a square wave is applied to the input, a sawtooth waveform will develop at the output, as shown in Figure 2-8(c). The rate at which the output changes is determined by the capacitor and resistor values.

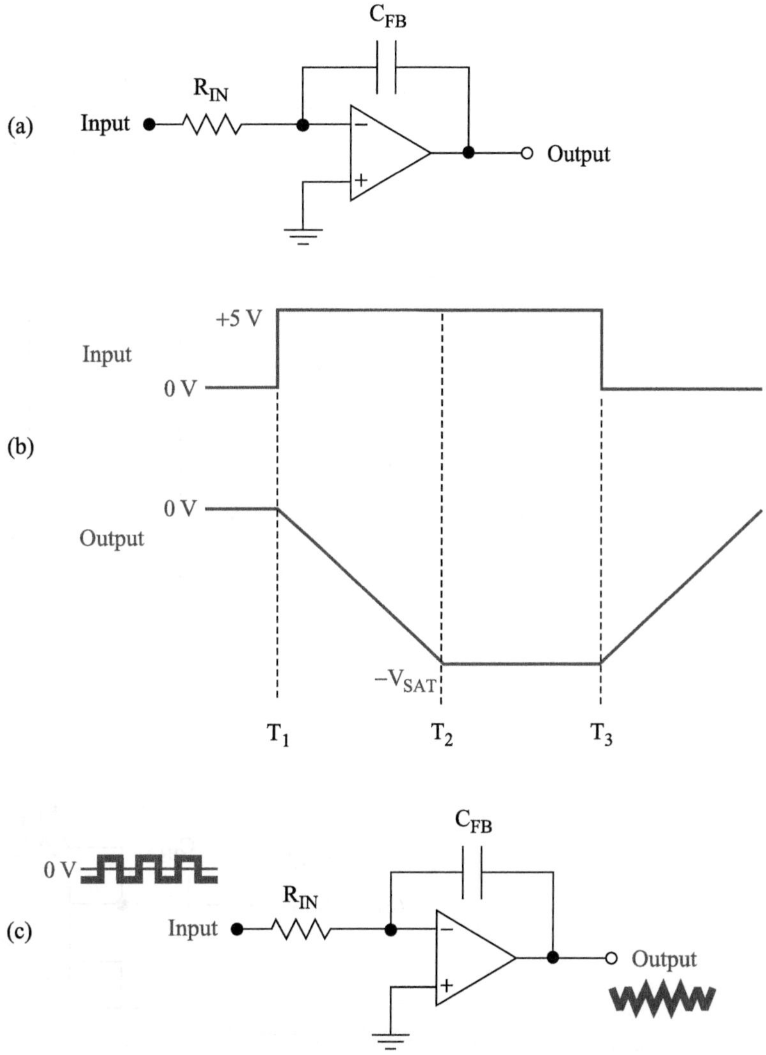

FIGURE 2-8 Integrator operational amplifier

SSE5203
The Differentiator
Op Amp

Differentiator Operational Amplifier

A **differentiator** is an amplifier circuit that produces an output proportional to the rate of change of the input signal. Figure 2-9(a) shows the schematic diagram of the op amp differentiator. Its configuration is opposite to that of the integrator because the capacitor replaces the input resistor instead of the feedback resistor. The waveform diagrams in Figure 2-9(b) illustrate how the differentiator responds to different input signals. Since the inverting lead is used, the output signal that develops will be in the opposite direction as the rate of change of the signal applied to the input.

When the input voltage is DC and remains constant, as shown from T_1 to T_2 on the waveform, the output of the differentiator is 0 volts. If the voltage changes at a slow, steady rate, the output will be a small constant DC voltage, as shown from T_2 to T_3. If the voltage changes at a fast, steady rate, as shown from T_3 to T_4, the output will be a high constant DC voltage. When a sawtooth is applied to the input, a square-wave signal is produced, as shown in Figure 2-9(c). As the sawtooth goes in the positive direction, the square-wave alternation is negative. A negative-going sawtooth produces a positive alternation of the square wave. Figure 2-9(d) shows that when a square wave is applied to the input, a series of spikes is produced at the output. The polarity of each spike is determined by the positive- or negative-going transition of the square wave.

Integrators and differentiators are used to control output actuators in closed-loop automated systems.

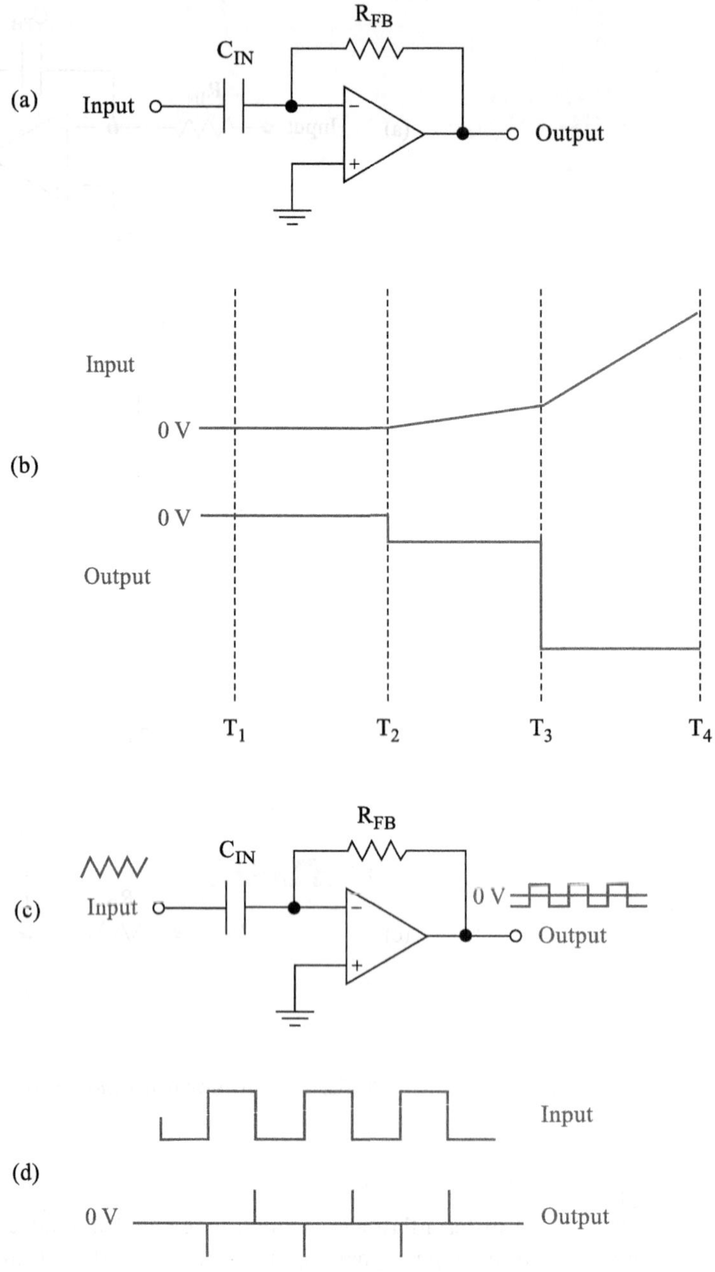

FIGURE 2-9 Differentiator operational amplifier

SSE8507
Schmitt Trigger

Wave-Shaping Schmitt Trigger

The **Schmitt trigger** is a device that produces rectangular wave signals. It is often used to convert sine waves or arbitrary waveforms into crisp, square-shaped signals. It is also used to restore square waves, which sometimes become distorted due to electromagnetic interference (called *noise*) during transmission, back to their required square-shaped waveforms. The Schmitt trigger uses positive feedback internally to speed up level transitions. It also utilizes an effect called *hysteresis,* which means that the switching threshold on a positive-going input signal is at a higher voltage level than the switching threshold on a negative-going signal. Schmitt triggers can also be used to transform the following waveforms into rectangular-shaped signals:

- A low-voltage AC wave
- Signals with slow rise times, such as those produced from charging and discharging capacitors, and temperature-sensing transducers

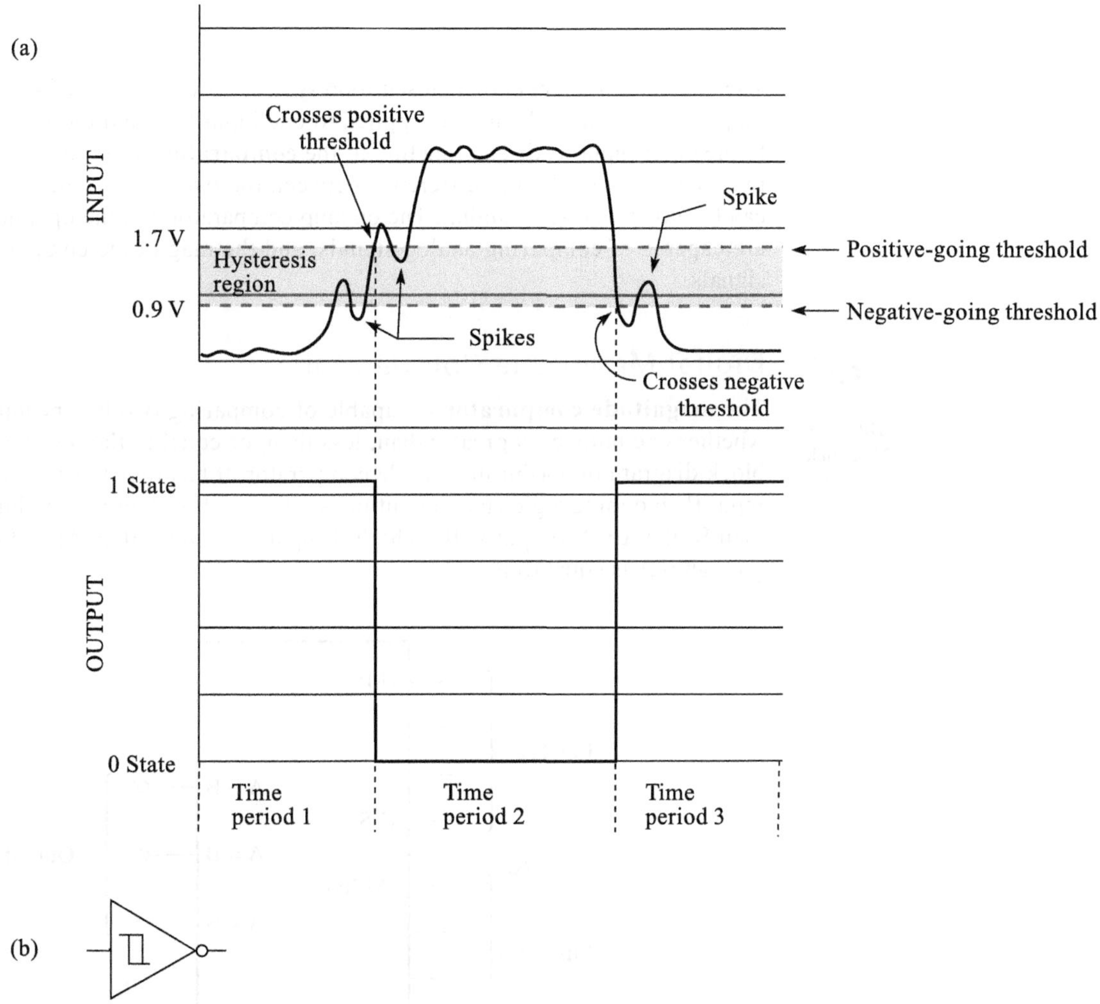

FIGURE 2-10 Schmitt trigger

Figure 2-10(a) illustrates the switching action of a Schmitt trigger inverter and also shows how the hysteresis characteristics reconstruct a distorted square wave.

Operation

Time Period 1. A logic 0 is recognized at the input, and a 1 state is generated at the inverting output.

Time Period 2. A logic 1 at the input is recognized if the input voltage exceeds the 1.7-volt positive-going threshold level that causes the output to snap to a logic 0 value. Note that the ragged spike on the input signal caused from noise drops below 1.7 volts into the hysteresis region during period 2. The output does not change unless the input drops below the 0.9-volt negative-going threshold level.

Time Period 3. A logic 0 at the input is recognized if the voltage drops below the 0.9-volt negative-going threshold level, which causes the output to snap to a logic 1 value. Note that a spike on the input rises above 0.9 volts into the hysteresis region during time period 3. The output does not change unless the input reaches the 1.7-volt positive-going threshold level.

The logic symbol for a Schmitt trigger inverter is shown in Figure 2-10(b). It includes a miniature hysteresis waveform inside the symbol to indicate that it is a Schmitt trigger instead of a regular inverter.

2-3 Comparator Devices

The comparator element of a closed-loop system shown in Figure 1-5 has two inputs and one output. The command signal is applied to one input lead and the feedback signal is applied to the other input lead. The function of the **comparator** is to produce an output error signal that is determined by the difference between the two inputs. The input and output signals can be either analog or digital. The op amp comparator and the op amp difference amplifier are capable of comparing analog signals, and the magnitude comparator compares digital signals.

DIG3403
Magnitude
Comparators

Digital Magnitude Comparator

The **magnitude comparator** is capable of comparing two binary numbers and indicating whether one number is greater than, less than, or equal to the other. Figure 2-11 shows the block diagram of a 4-bit magnitude comparator. It has four lines for input A, four lines for input B, and three logic state output lines. The A > B output will go high if input A is larger than B; the A < B output will go high if input B is larger than A; and the A = B output will go high if A is equal to B.

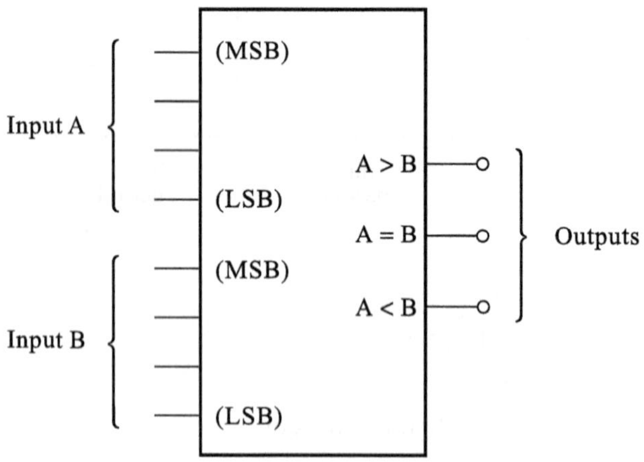

FIGURE 2-11 Block diagram of a magnitude comparator

A TTL 7485 magnitude comparator integrated circuit (IC) is shown in Figure 2-12(a). Its operation is identical to the block diagram circuit in Figure 2-11. It also has three inputs called the expansion (also cascade) input lines, labeled $I_A < B$, $I_A = B$, and $I_A > B$. If only 4-bit words are being compared, the $I_A = B$ input should be wired to a high and the $I_A < B$ and $I_A > B$ inputs should be wired to a low. Figures 2-12(b) and 2-12(c) show the pin diagram and the truth table of the 7485 IC.

Several 7485 ICs can be connected together to compare binary numbers larger than 4 bits. The block diagram of Figure 2-13 on page 28 shows how two 7485 ICs are cascaded to make an 8-bit comparator. The four least significant bits of each 8-bit word are connected to inputs A_0–A_3 and B_0–B_3 of the comparator on the left. The four most significant bits of each 8-bit word are connected to inputs A_0–A_3 and B_0–B_3 of the comparator on the right. The A > B, A = B, and A < B outputs of the least significant comparator are connected to the expansion inputs of the most significant comparator. The expansion lines of the least significant comparator should be wired as if it were comparing only two 4-bit words. The comparison results of the two 8-bit words are generated at the three output lines of the most significant comparator.

DIG3703
Cascading Magnitude
Comparators

The cascaded inputs resulting from the comparison of the low-order numbers are always overridden by the high-order numbers. The only time the cascaded input affects the output is when the two high-order numbers are equal.

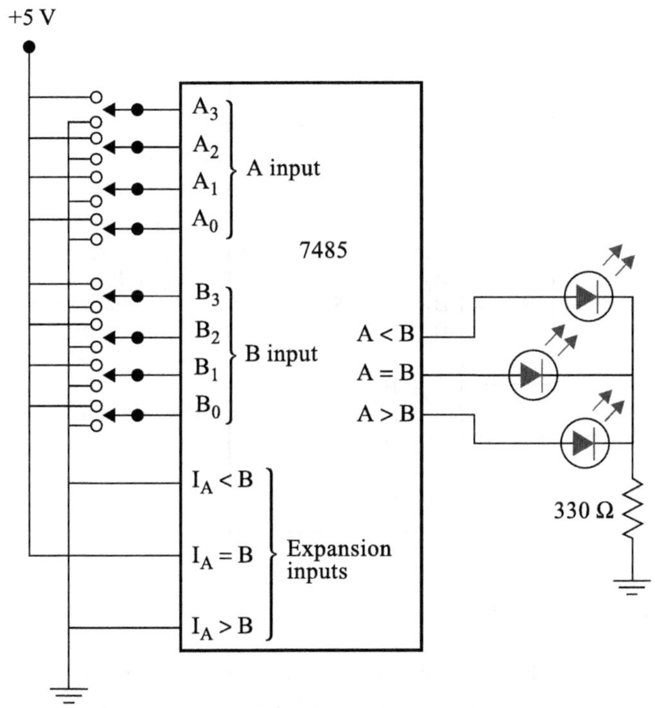

(a)

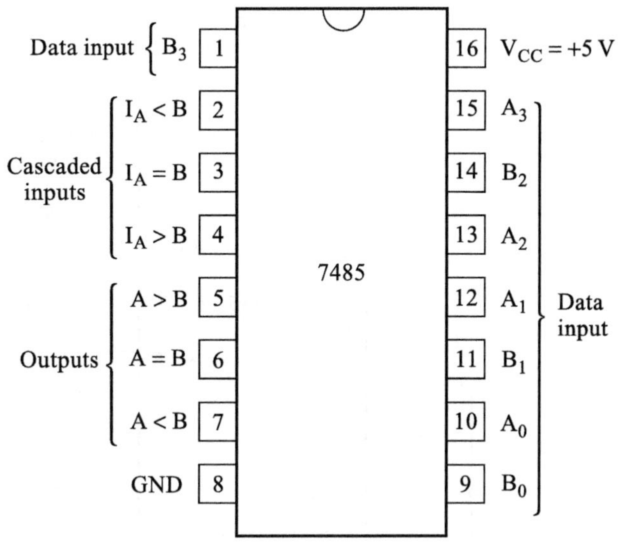

(b)

FUNCTION TABLE (X = Don't care)

Comparing Inputs				*Cascading Inputs*			*Outputs*		
A3, B3	**A2, B2**	**A1, B1**	**A0, B0**	$I_A > B$	$I_A < B$	$I_A = B$	**A > B**	**A < B**	**A = B**
A3 > B3	X	X	X	X	X	X	H	L	L
A3 < B3	X	X	X	X	X	X	L	H	L
A3 = B3	A2 > B2	X	X	X	X	X	H	L	L
A3 = B3	A2 < B2	X	X	X	X	X	L	H	L
A3 = B3	A2 = B2	A1 > B1	X	X	X	X	H	L	L
A3 = B3	A2 = B2	A1 < B1	X	X	X	X	L	H	L
A3 = B3	A2 = B2	A1 = B1	A0 > B0	X	X	X	H	L	L
A3 = B3	A2 = B2	A1 = B1	A0 < B0	X	X	X	L	H	L
A3 = B3	A2 = B2	A1 = B1	A0 = B0	H	L	L	H	L	L
A3 = B3	A2 = B2	A1 = B1	A0 = B0	L	H	L	L	H	L
A3 = B3	A2 = B2	A1 = B1	A0 = B0	X	X	H	L	L	H
A3 = B3	A2 = B2	A1 = B1	A0 = B0	H	H	L	L	L	L
A3 = B3	A2 = B2	A1 = B1	A0 = B0	L	L	L	H	H	L

X = Does not apply

(c)

FIGURE 2-12 The 7485 magnitude comparator IC

2-4 Optoelectronic Interface Devices

SSE5003
The Optocoupler

The voltage used by one element of a closed-loop system may need to be isolated from the voltage used by another element. Therefore, they cannot be directly connected to each other. Optoelectronic devices are used to make the output of one section compatible with the input of another section. Optoelectronic devices pass electrical signals from one element to another by means of light energy and semiconductors. An optoelectronic device consists of a light source and a photo detector, as shown in Figure 2-14(a). The light source converts electrical energy to light. The detector converts light energy to electrical energy.

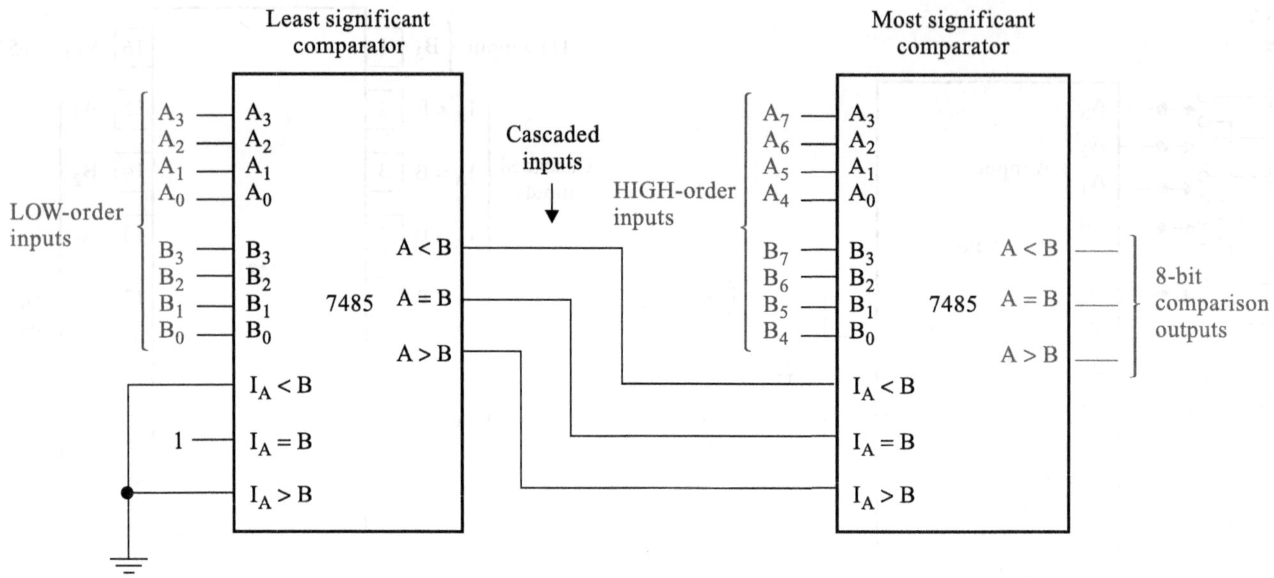

FIGURE 2-13 Magnitude comparison of two 8-bit strings (or binary words)

The light source is usually a semiconductor light emitting diode (LED). In the forward-biased state, light emission occurs when electrons combine with holes around the PN junction, as shown in Figure 2-14(b). During this process, the electrons fall to a lower energy level and energy in the form of photons is released. Photons are light particles that travel in a waveform pattern.

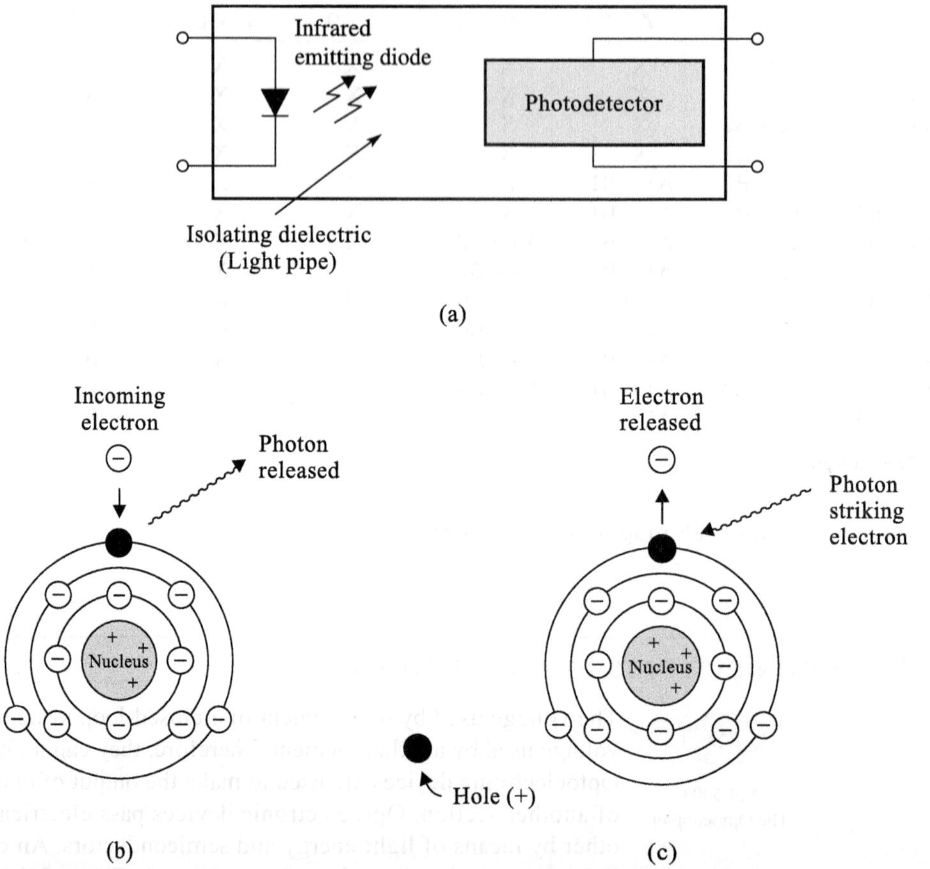

FIGURE 2-14 (a) Block diagram of an optoisolator; (b and c) Atomic structure of photoelectronic devices

Light detection of photons is accomplished by semiconductor devices. As the photons strike the semiconductor material at the PN junction, valence electrons are released, as shown in Figure 2-14(c). The valence electrons available in the semiconductor then enable current to pass through the PN junction.

The detection of light and conversion into current is performed by light-activated devices such as photodiodes, phototransistors, photo SCRs (silicon-controlled rectifiers), and photo triacs.

Photodiodes

The **photodiode,** shown in Figure 2-15, is a PN-junction device that operates in the reverse-bias mode. When a PN junction is reverse-biased, heat causes the freeing of minority carriers in the depletion layer, which contributes to a small leakage current. High-energy photons strike the PN junction of the diode when it is exposed to light from the LED. The impact of photons causes electrons to be dislodged from their orbit and leave holes behind. This action of electron-hole pairs due to light exposure causes minority carriers to increase. The resulting current flow through the diode increases proportionately with light intensity. Photodiodes are used in applications that require quick response time for fast switching detection. Their primary limitation is that they allow current only in the microampere range between 50 μA and 500 μA to flow.

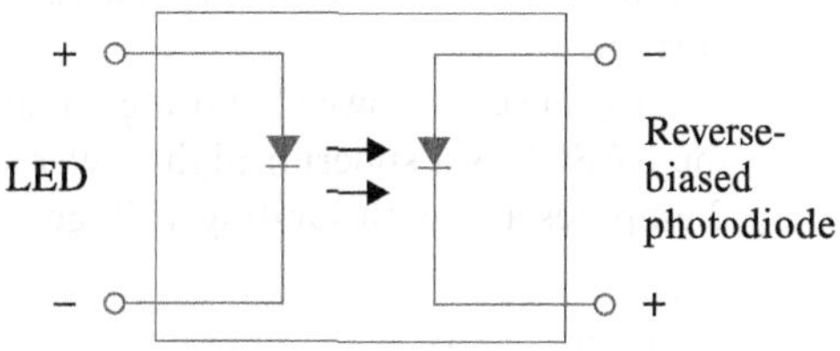

FIGURE 2-15 Photodiode in optocoupler

Phototransistors

The **phototransistor** depends on a light source for its operation. Typically, the phototransistor has no external base lead, as shown in Figure 2-16. Therefore, there is no bias source for external control. Instead, a light source operates the transistor in the same manner as a bias source. When photons from the LED strike the transistor's collector–base junction, the flow of minority carriers increases. This action causes the emitter–collector current to rise. If light intensity increases, more emitter–collector current will flow. Because the transistor amplifies, the amount of the output current it produces is much higher than that of the photodiode under the same illuminating conditions. However, its response time is slower than that of the photodiode.

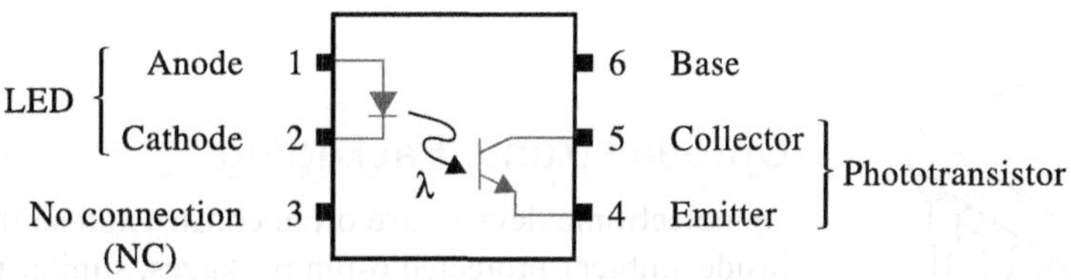

FIGURE 2-16 Phototransistor in optocoupler

Photo SCR

The **photo SCR** is also referred to as a light-activated SCR, or LASCR. The operation of the LASCR is similar to the conventional SCR except that it is usually activated by light instead of by a gate voltage that draws gate current. The LASCR symbol is shown in Figure 2-17.

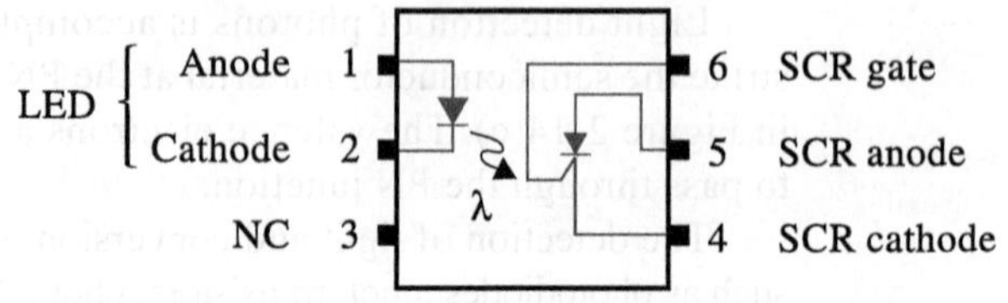

FIGURE 2-17 Photo SCR (LASCR) in optocoupler

The SCR is normally in the off condition. Its three leads enable the SCR to be triggered in one of three ways:

1. By light shining on the PN junction;
2. By a positive voltage drawing gate current applied to the gate; and
3. By a combination of the gate voltage and light intensity.

The output power an SCR controls is much higher than the amount required to trigger it. The level of light intensity used to turn on the LASCR can be controlled by adjusting the gate-cathode bias resistance. For example, a larger value resistance prevents the LASCR from turning on until a large amount of light intensity is reached. The LASCR remains on even after the light or the gate voltage is removed. When the current flowing through it is reduced below its holding current value, the SCR turns off and effectively blocks any current.

Because its power handling capacity is far beyond that of other optoelectronic devices, the LASCR is a superior high-power switch. Photo SCRs are capable of switching current of 2 amperes and withstanding voltages as high as 200 volts.

Photo Triac

The **photo triac** is a bidirectional device designed to switch AC signals and pass current in both directions. Its symbol is shown in Figure 2-18. The photo triac is normally off if its PN junction is not exposed to light radiation of a certain intensity. During each alternation, it turns on when triggered by a specified light intensity, and turns off when the conducting current falls below a certain level. The current capacity of the photo triac is not as high as the LASCR.

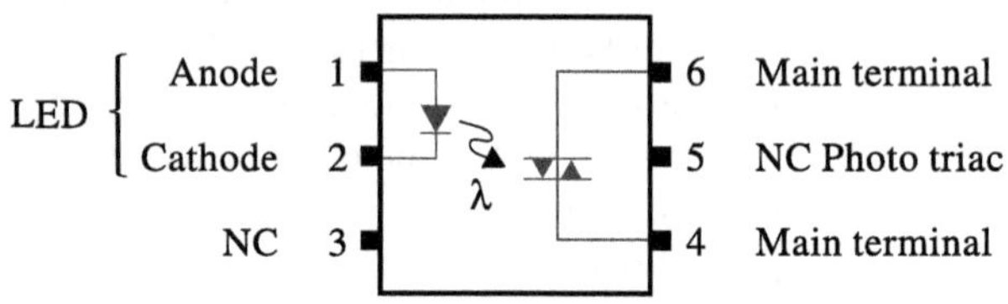

FIGURE 2-18 Photo triac in optocoupler

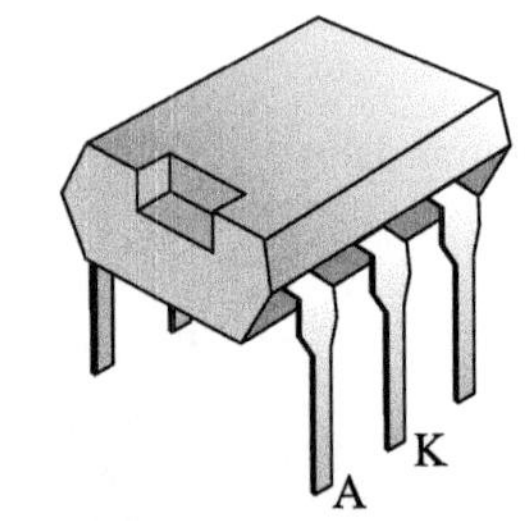

FIGURE 2-19 Photoelectronic (optoisolator) package

Optoelectronic Packaging

Optoelectronic devices are often constructed so that the light emitter and detector are sealed inside ambient-protected 6-pin packages, similar to the one in Figure 2-19 used for ICs. This package, often referred to as an *optocoupler*, does not allow any external light to enter. The input, usually a +5 volts, is applied to two pins of the IC package. These two pins are connected to the terminals of the internal LED. A different voltage source, for example, +12 volts, +100 volts, or 120 VAC, is connected to two detector output leads of the IC. If the LED is turned on, its light illuminates the photodetector, which initiates an output current. The insulation resistance between the emitter and the detector is great enough to withstand an output voltage 5000 times greater than the input voltage. Some devices are capable of operating as high as 100 kHz. Since

CIS5908
Analog and
Digital Converters

there is no electrical connection between the emitter and the detector, the package is often called an *optoisolator.* IC packages are often used as an interface between a low-voltage microprocessor and a high-voltage AC motor that the microprocessor controls. They also protect against unwanted signals being induced into control circuitry due to power line noise that can improperly turn on a machine.

2-5 Digital-to-Analog Converters

CIS6208
Digital-to-Analog
Converters

Digital-to-analog converters (**DAC**s or D/A converters) are used to convert digital signals representing binary numbers into proportional analog voltages. Although these devices are now available in IC packages, they are analyzed here in a discrete form to better describe how they function.

A 4-bit input DAC is shown in Figure 2-20. It consists of a summing amplifier with its feedback resistor (R_F), four summing resistors, and four switches that are used to provide a 4-bit binary input. A switch in the open position represents the 0 state. In the closed position, it represents the 1 state. The placement of each switch corresponds to the same 8-4-2-1 weighted values of a 4-bit binary number. Resistors R_1 through R_4 are also selected with a weight proportional to the next. The 12.5-kilohm resistor R_4 is connected at the MSB (most significant bit) input line. The values of the remaining resistors are selected by making each progressive resistor twice the size of the preceding one. The analog voltage is always at the op amp output. The circuit is designed to operate so that a 4-bit binary number represented by the four switches is converted into voltages. Because 16 different combinations of switch positions are possible, (0–15), 16 different analog voltage levels proportional to the digital number applied are produced. The circuit in Figure 2-20 is designed so that it develops an analog output voltage equivalent to the binary number applied. For example, when all switches are in the open position to represent a binary input of 0000, the output is 0 volts. If SW1 is moved to the closed position (binary 0001), the op amp output will be –1 volts. If SW1 and SW3 are in the closed position (binary 0101), the op amp output will be –5 volts. If all four switches are in the closed position (binary 1111), the analog output voltage will be –15 volts. The analog output voltage for each combination of switch setting can be determined by the same formula used for the summing operational amplifier.

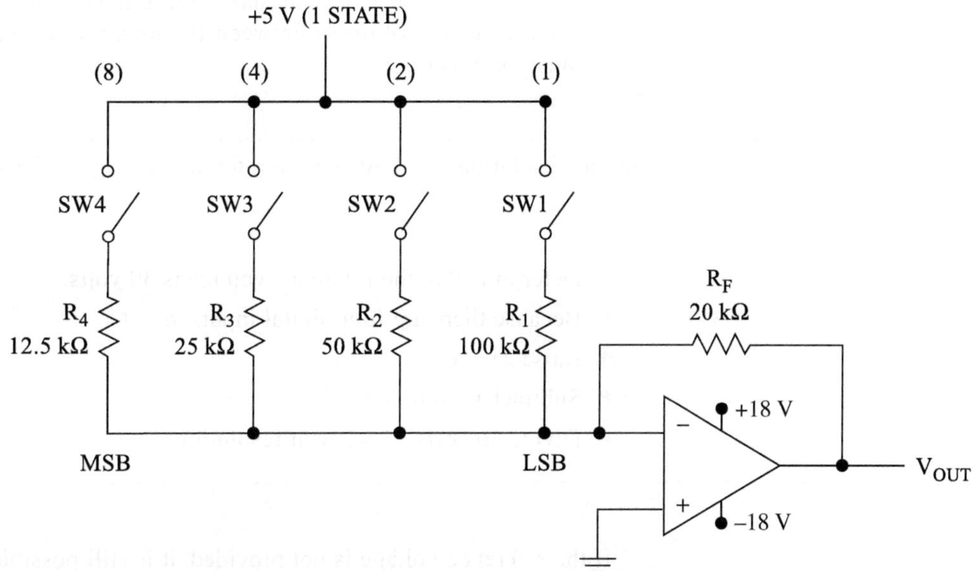

FIGURE 2-20 Binary-weighted D/A converter

EXAMPLE 2-1

What is the analog output voltage of the DAC in Figure 2-20 when a binary 1001 is applied?

Solution

$$I_{R_1} = \frac{V_{R_1}}{R_1} = \frac{5\text{ V}}{100\text{ kohms}} = .05\text{ mA}$$

$$I_{R_4} = \frac{V_{R_4}}{R_4} = \frac{5\text{ V}}{12.5\text{ kohms}} = .4\text{ mA}$$

$$I_{R_F} = .05\text{ mA} + .4\text{ mA} = .45\text{ mA}$$

$$V_{OUT} = -I_{R_F} \times R_F$$

$$= -.45\text{ mA} \times 20\text{ kohms} = -9\text{V}$$

Figure 2-21(a) provides all possible digital inputs and the corresponding output voltages for the circuit in Figure 2-20. Figure 2-21(b) provides the same information in a graphic format. The 4-bit DAC divides the reference analog output into 15 equal divisions.

DACs in IC form are available with 8, 12, and 16 binary inputs. As the number of inputs increases, the reference analog voltage is divided into smaller divisions. For example, 8-bit DACs divide the analog output voltage into 255 equal parts, 12-bit converters into 4095 equal parts, and 16-bit converters into 65,535 equal divisions.

The number of equal divisions into which a DAC divides the reference voltage is called the *resolution*. The resolution of a DAC can be determined by the following formula:

$$\frac{V_{REF}}{2^n - 1}$$

- The 2 in the formula represents the binary number system.
- The *n* is the exponent that specifies to what power 2 is raised. It is determined by the number of binary inputs used at the input of the DAC. By taking 2 to the *n*th power, the maximum binary (equivalent decimal) number is determined.
- A 1 is subtracted from the maximum binary number to determine the number of equal steps (resolution) between the maximum binary number and the minimum binary number.

EXAMPLE 2-2

Find the resolution of a DAC with a reference voltage of 30 volts and 4 inputs.

Solution

- Determine that the reference voltage is 30 volts.
- Because there are four digital inputs, $n = 4$.
- Raise $2^4 = 16$.
- Subtract 1 from 16 = 15.
- Divide 30 volts/15 = 2-volt resolution.

If the reference voltage is not provided, it is still possible to determine the resolution if the analog output voltage and the decimal input values are both provided.

(a)

SW4 8	SW3 4	SW2 2	SW1 1	V_{OUT} (−V)
0	0	0	0	0
0	0	0	1	1
0	0	1	0	2
0	0	1	1	3
0	1	0	0	4
0	1	0	1	5
0	1	1	0	6
0	1	1	1	7
1	0	0	0	8
1	0	0	1	9
1	0	1	0	10
1	0	1	1	11
1	1	0	0	12
1	1	0	1	13
1	1	1	0	14
1	1	1	1	15

(b)

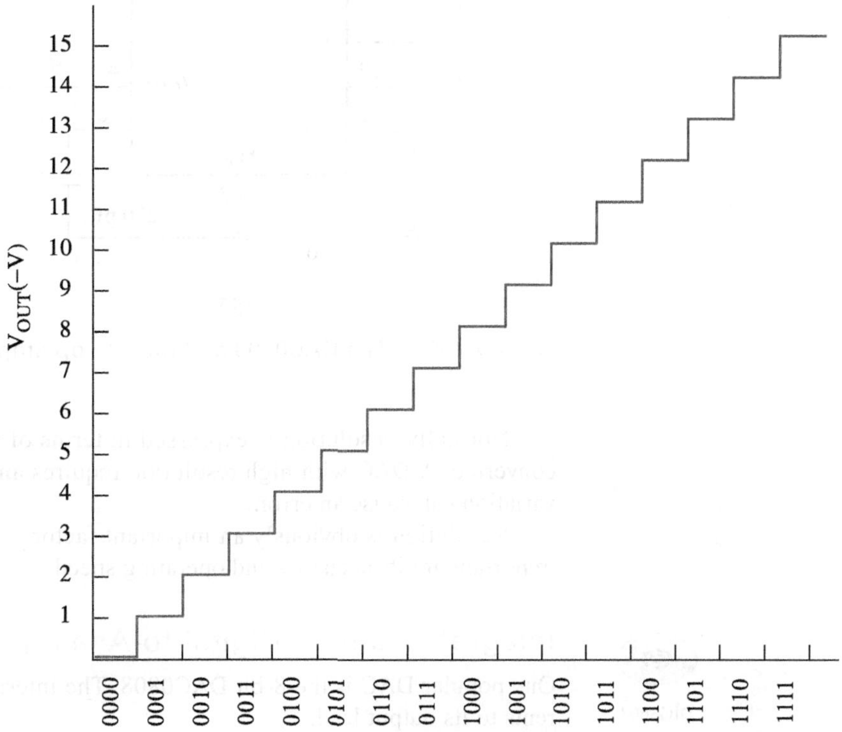

FIGURE 2-21 Analog output vs. digital input for the circuit shown in Figure 2-20

EXAMPLE 2-3

A 6-bit DAC produces an output voltage of 2.1 volts when the binary number applied to its input is 101010_2. What is the output voltage if the binary number changes to 111000_2?

Solution

Step 1: Convert the original binary number to the equivalent decimal value.

$$101010_2 = 42_{10}$$

Step 2: Determine the resolution by dividing the original analog output voltage by 42.

$$2.1/42 = .05 \text{ V (resolution)}$$

Step 3: Convert the new binary input number to its equivalent decimal value.

$$111000_2 = 56_{10}$$

Step 4: Multiply the resolution times 56.

$$.05 \times 56 = 2.8\text{ V}$$

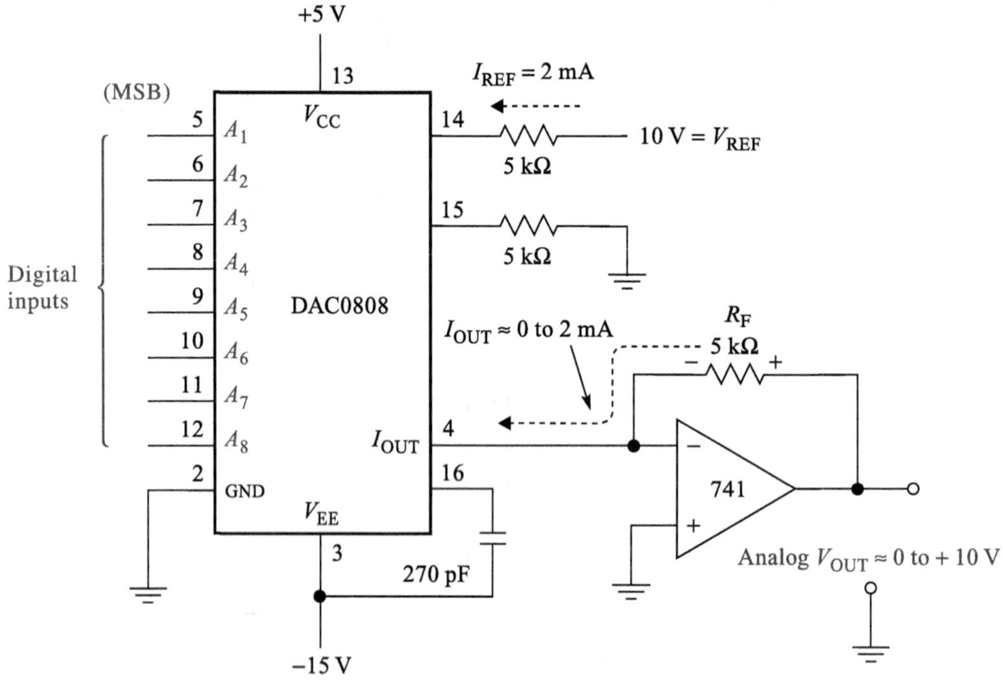

FIGURE 2-22 The DAC0808 and the 741 op amp connected to form a DAC

Normally, resolution is expressed in terms of the number of binary input bits that are converted. A DAC with high resolution requires an accurate reference voltage because any variation can cause an error.

Resolution is obviously an important factor to consider when purchasing a DAC. Also important are its accuracy and operating speed.

Integrated-Circuit Digital-to-Analog Converter

DIG3603 Troubleshooting a Digital-to-Analog Converter

One popular DAC is the 8-bit DAC0808. The internal components supply proportional currents to its output lead.

Figure 2-22 shows the DAC0808 connected to an external 741 op amp. The current range is dictated by the 10-volt 5-kilohm combination connected to pin 14. The 2 mA flowing through resistor R_f is the maximum amount of current that can flow through output pin 4 (I_{OUT}). When the digital input is $0000\ 0000_2$, the minimum current of 0 mA flows through pin 4. When the digital input is $1111\ 1111_2$, the maximum current of 2 mA flows through pin 4. By using a 5-kilohm feedback resistor (R_f), the analog output voltage at the op amp output ranges from 0 to 10 volts. The 10 volts is produced when I_{OUT} is 2 mA. If a different analog output voltage range is desired, the gain of the op amp is adjusted by changing resistor R_f to a different value.

2-6 Analog-to-Digital Converters

The analog-to-digital converter (**ADC** or A/D converter) is capable of converting analog input voltages into proportional digital numbers. Analog-to-digital converters that operate at high speeds employ a circuit called a *successive-approximation register* (SAR).

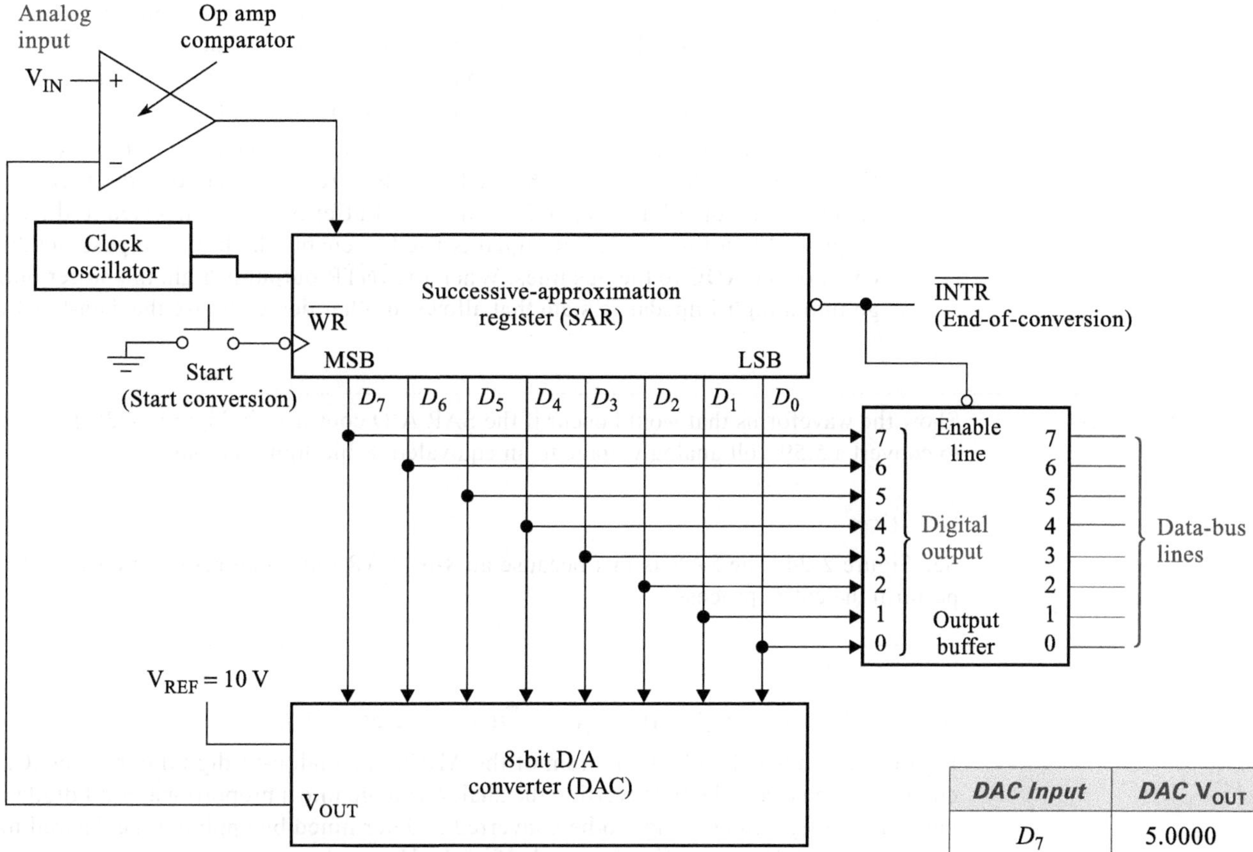

(a)

(b)

DAC Input	DAC V_{OUT}
D_7	5.0000
D_6	2.5000
D_5	1.2500
D_4	0.6250
D_3	0.3125
D_2	0.15625
D_1	0.078125
D_0	0.0390625

FIGURE 2-23 Simplified SAR A/D converter

Figure 2-23(a) shows a simplified block diagram of an ADC that uses an SAR. Its eight output lines D_0–D_7 cause the D/A converter to produce different voltages, as shown in Figure 2-23(b). These voltages will result if 10 volts is supplied to the V_{REF} input line. Its operation is as follows:

1. When the START button is pressed, the SAR is reset on the negative edge of the pulse applied to the $\overline{WR}$ input.
2. The conversion is begun on the leading edge of the conversion pulse after the START button is released.
3. When the positive transition of the first clock pulse occurs, the SAR produces a high at its MSB output, D_7. This causes the D/A converter to produce an analog voltage that is one-half its maximum value.
4. If the D/A converter output is higher than the unknown analog voltage (analog V_{IN}), the SAR output returns low. If the D/A converter output is lower than the analog input voltage, the SAR leaves bit 7 high.
5. The second clock pulse causes the next lower bit, D_6, to produce a high. If it causes the D/A converter output to be higher than the analog input, it returns to a low. If not, the SAR leaves D_6 high.
6. This process continues with the remaining six bits, D_5 to D_0.

7. At the end of the process, the SAR contains an 8-bit binary output that causes the D/A converter to produce an analog output equal to the unknown analog input. This occurs at the end of the eighth clock pulse. The 8-bit binary number contained by the SAR represents the analog input present at the eight output lines.
8. At the moment the eight-step conversion process is complete, the end-of-conversion $\overline{\text{INTR}}$ line goes low. Because the ADC outputs are often shared with other devices on a common data-bus line, an 8-bit tri-state buffer is often connected to the digital outputs. When low, the $\overline{\text{INTR}}$ signal is used to enable the buffer to pass the digital count of the ADC to the bus lines. When the $\overline{\text{INTR}}$ output is high, the buffer outputs go into a high-impedance state that allows another device to use the data-bus lines.

EXAMPLE 2-4

Show the waveforms that would occur if the SAR A/D converter in Figure 2-23(a) were used to convert a 5.59-volt analog voltage to an equivalent 8-bit digital output.

Solution

See Figure 2-24. The SAR is fast because an 8-bit SAR only requires eight clock pulses to perform the entire process.

Integrated-Circuit Analog-to-Digital Converter

Figure 2-25 shows the block diagram of the ADC0804 analog-to-digital converter IC. The circuit shown is capable of converting the analog voltage into a proportional 8-bit digital output. The analog voltage range to be converted is determined by applying the desired maximum voltage to V_{DC}. For fine tuning, half of the V_{DC} voltage is applied to input $V_{REF/2}$. If necessary, a slight voltage change at $V_{REF/2}$ will then bring the ADC into calibration. By applying 5.12 volts to V_{DC} and 2.56 volts to $V_{REF/2}$, the circuit is capable of converting an analog voltage connected across $V_{IN}(+)$ and $V_{IN}(-)$ ranging from 0 to 5.12 volts. With 8 output leads, there are 256 different analog voltage levels that are converted into digital outputs. Therefore, the resolution of this device is 0.39 percent (1/255 = .0039 = 0.39%). With 5.12 volts as the maximum input voltage, each 0.02-volt (5.12 × .0039) increase causes the binary count to increase by 1.

EXAMPLE 2-5

If the binary count produced by the ADC is 01011101_2, what is the analog voltage applied to the input?

Solution

Step 1: Convert the binary number to an equivalent decimal value.

$$01011101_2 = 93_{10}$$

Step 2: Multiply 93 times the resolution.

$$93 \times 0.02 = 1.86\text{V}$$

It is possible to determine the binary output when the resolution and the analog input voltage are known.

EXAMPLE 2-6

If the analog voltage applied to the input is 3.04 volts, what is the binary count at the digital output?

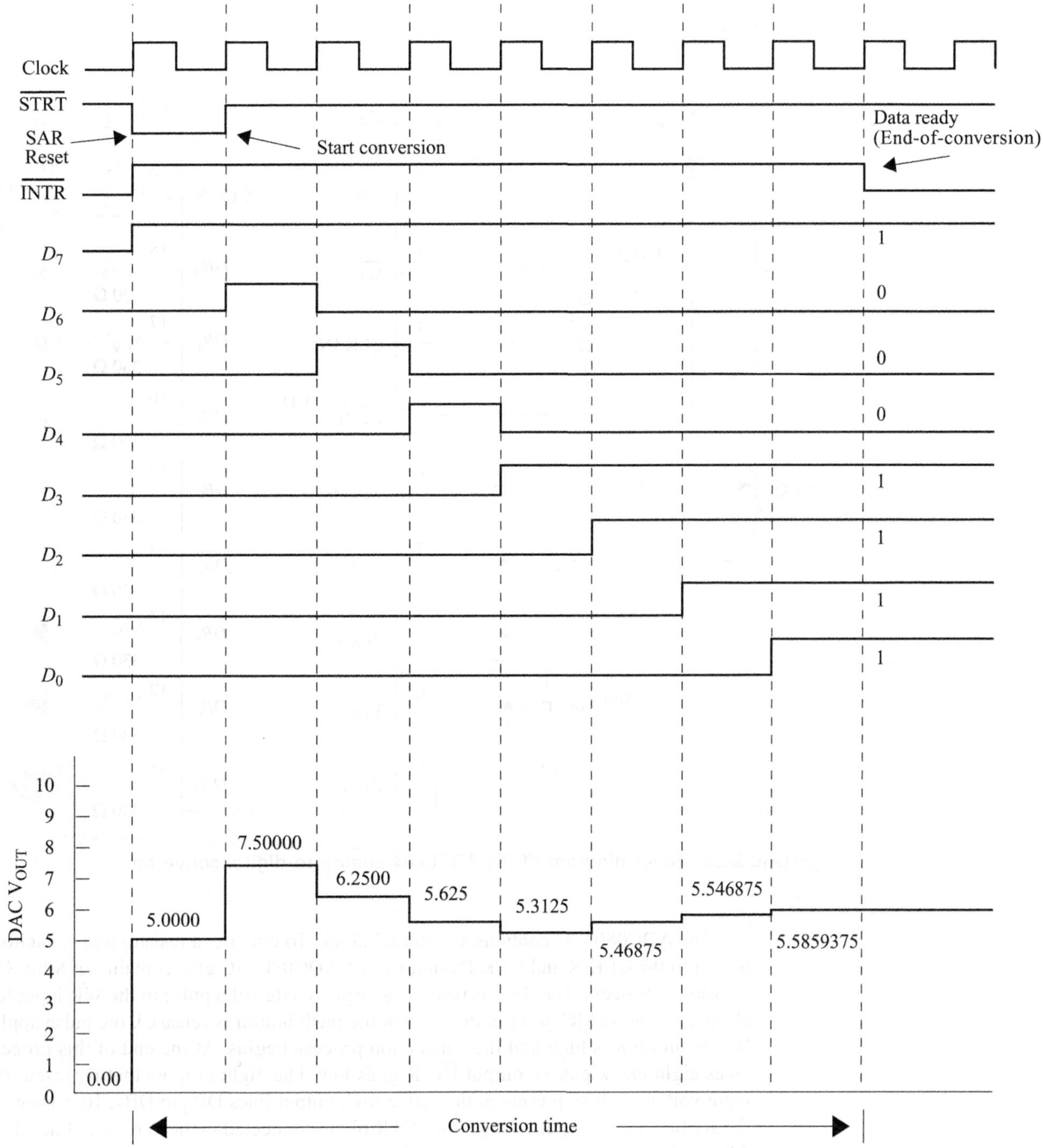

FIGURE 2-24 Timing diagram for an SAR ADC

Solution

Step 1: Divide the input voltage by the resolution.

$$3.04/.02 = 152$$

Step 2: Convert the decimal value into the equivalent binary number.

$$152_{10} = 10011000_2$$

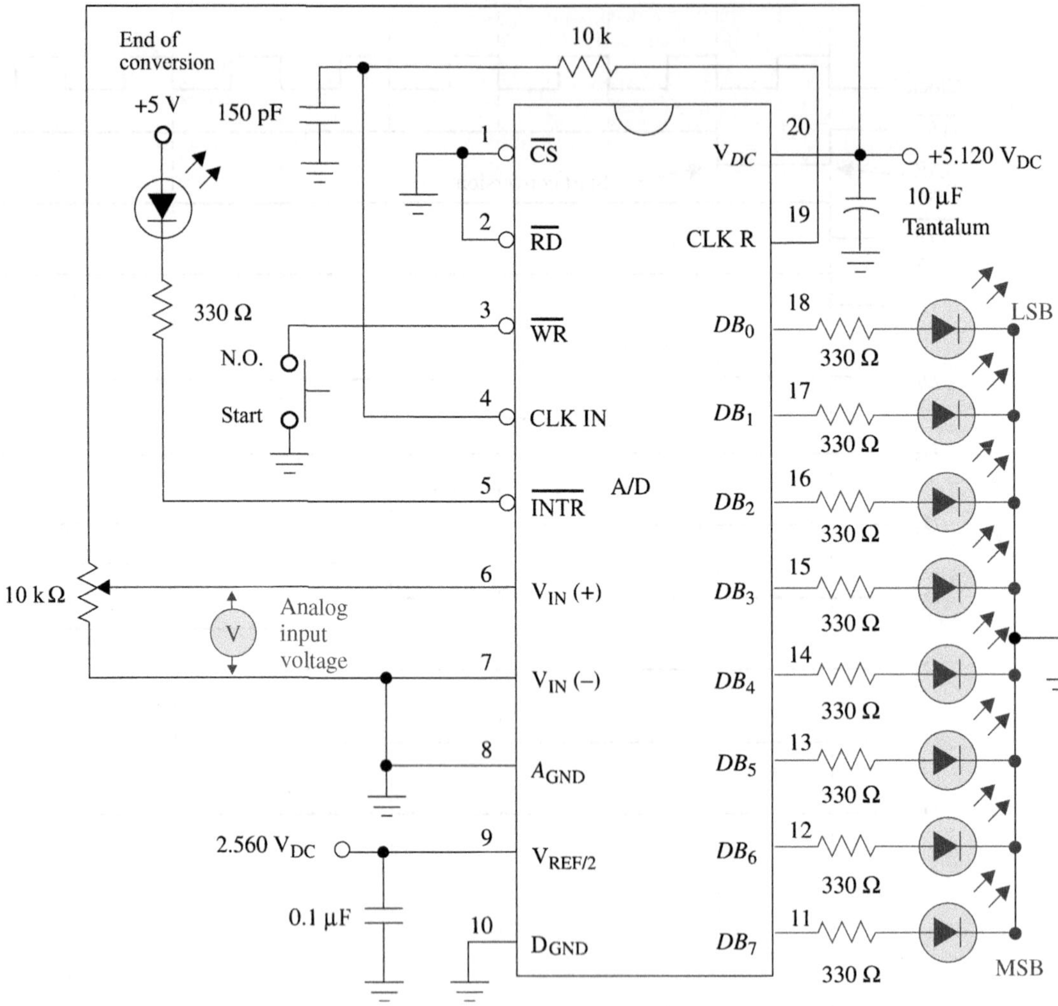

FIGURE 2-25 Block diagram of the ADC0804 analog-to-digital converter

The ADC0804 IC contains an internal clock. To operate, a resistor and capacitor are connected to the CLK R and CLK IN inputs. The ADC0804 IC also contains an 8-bit SAR for its conversion process. The SAR is reset on a negative edge of a pulse to the $\overline{WR}$ input lead by the closure of the START push button. When the push button is released, the pulse applied to the $\overline{WR}$ input returns high and the conversion process begins. At the end of this process, which takes eight clock pulses, output $\overline{INTR}$ goes low. The eight outputs that represent the analog input voltage will be present at the active-high output lines DB_0 to DB_7. To continue updating the applied analog input voltage, the $\overline{INTR}$ pin is connected to the WR input line. By doing so, 5000 to 10,000 conversions can be made per second.

The ADC0804 IC is a CMOS (Complementary metal-oxide semiconductor) device that is designed to interface directly with some types of microprocessors. Therefore, some of its pins, such as $\overline{RD}$, $\overline{WR}$, $\overline{CS}$, and $\overline{INTR}$, correspond to leads of the similarly labeled microprocessors.

2-7 Timing Devices

Timing devices are used to produce rectangular signals referred to as *square-wave signals.* Timing devices may generate either a single pulse or a continuous string of pulses. Single pulses are used to preset data into memory devices or to clear data. These signals are produced by **monostable multivibrators.** Continuous pulses are used as clock signals that are the heartbeat in computer devices. As they are fed through computer-based equipment, all events throughout the computing systems are properly timed and synchronized. These signals are produced by **astable multivibrators.**

A linear IC specifically designed for timing applications is the 555 monolithic IC chip. A pin diagram of this chip is shown in Figure 2-26.

When a minimal number of external resistors and capacitors are connected to various pins of the 555 IC, it operates as an astable or monostable multivibrator. Figure 2-27 shows a schematic diagram of the 555 IC. It consists of the following sections:

SSE7806
Internal Elements of a 555 Timer

Voltage Divider Network. Resistors R_1, R_2, and R_3 are all 5 kilohms. They form a voltage divider which biases the inverting (–) input of comparator A at 2/3 the power

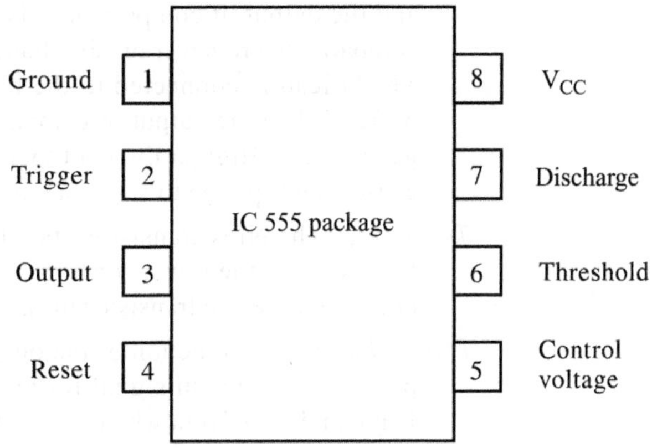

FIGURE 2-26 555 IC package

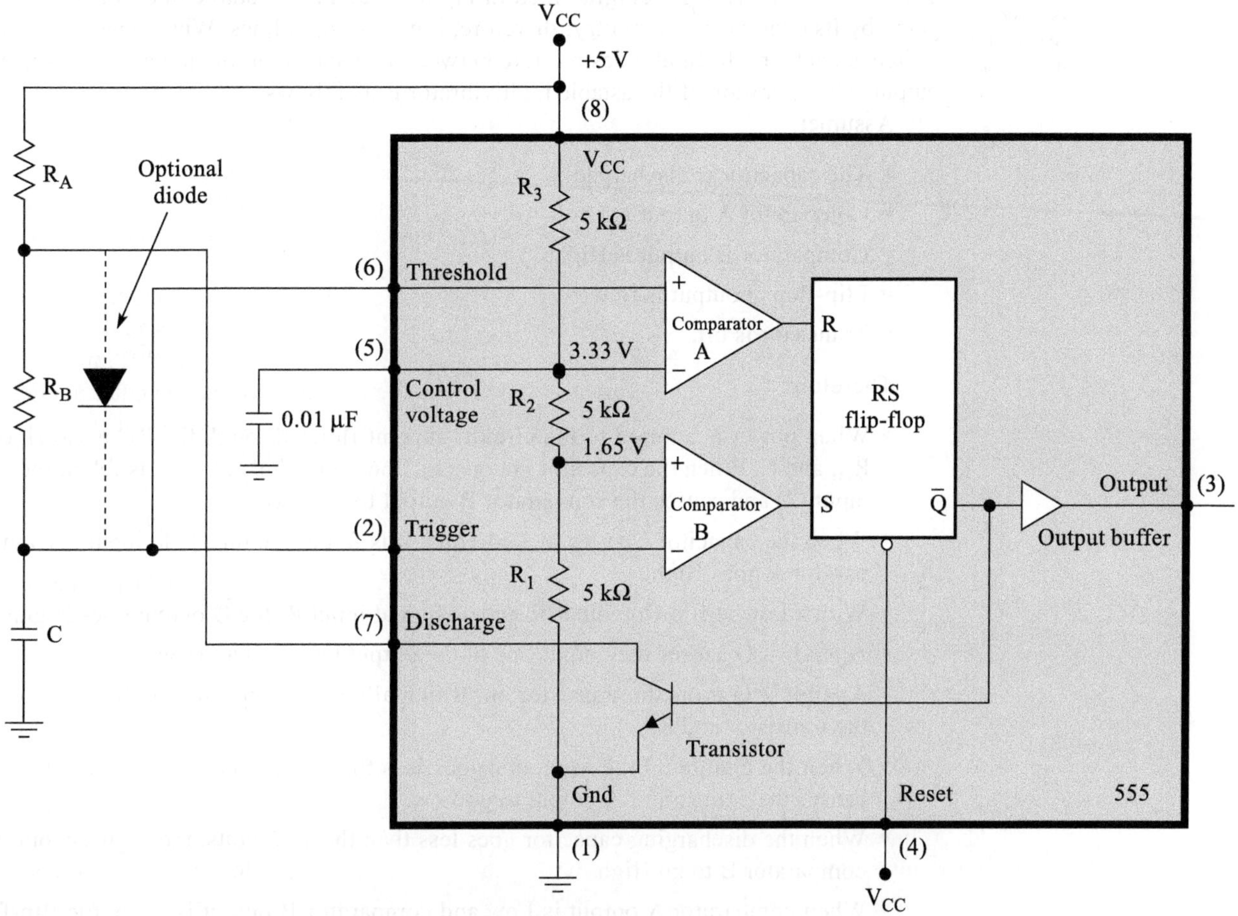

FIGURE 2-27 Schematic diagram of the 555 IC astable timer

supply voltage (3.33 V), and the noninverting (+) input of comparator B at 1/3 the power supply voltage (1.65 V).

Voltage Comparators. Each comparator has one of its inputs connected to an external pin. The noninverting input of comparator A is connected to external pin 6, called the *threshold terminal.* The inverting input of comparator B is connected to external pin 2, called the *trigger terminal.* The output of comparator A is Low if the voltage at the threshold terminal is lower than 3.33 volts. The output of comparator B is Low if the voltage at the trigger terminal is greater than 1.65 volts. The logic levels at the comparator outputs control the flip-flop.

RS Flip-Flop. The output of comparator A is connected to the R input of the flip-flop, and the output of comparator B is connected to the S input. The outputs of the two comparators are never on simultaneously. Only output $\overline{Q}$ of the RS flip-flop is used. The $\overline{Q}$ lead is connected to the base of the transistor and the input of the output buffer. When the output of comparator A goes High, it causes the flip-flop to reset, generating a High at the $\overline{Q}$ output. When the output of comparator B goes High, it causes the flip-flop to set, generating a Low at output $\overline{Q}$.

Transistor. The NPN transistor operates like a switch. When the $\overline{Q}$ output of the flip-flop is High, the transistor turns on and operates like a closed switch. When the $\overline{Q}$ output is Low, the transistor turns off.

Output Buffer. The function of the output buffer is to produce a high-current voltage to provide a sufficient signal for external circuitry. The buffer goes Low when $\overline{Q}$ is High, and goes High when $\overline{Q}$ is Low because it is an inverting amplifier.

555 Astable Multivibrator

SSE8106
The 555 Astable Multivibrator

The astable multivibrator diagrammed in Figure 2-27 has no stable output state. It is triggered by its own internal circuitry; therefore, it has no input lines. When power is applied, it switches back and forth at a desired rate between two states, producing a square wave at its output. The operation of the astable multivibrator is as follows:

Assume:

- The capacitor is discharged.
- Comparator A output is Low.
- Comparator B output is High.
- Flip-flop $\overline{Q}$ output is Low.
- Transistor is off.

Therefore:

- When power is applied to the circuit, current flows through the RC network of R_A, R_B, and C. When the capacitor charges to 1.66 volts, this potential is felt at the trigger input (2) and causes the comparator B output to go Low.
- When the capacitor charges to 3.34 volts, it is felt at the threshold input (6) and comparator A goes High.
- With a Low at flip-flop input S, and a High at input R, the $\overline{Q}$ output goes High.
- A High at $\overline{Q}$ causes the output line of the output buffer to go Low.
- A High at $\overline{Q}$ turns the transistor on, which allows the capacitor to discharge through the transistor and R_B.
- When the charge on the capacitor goes less than 3.33 volts, the threshold potential causes the comparator A output to go Low.
- When the discharging capacitor goes less than the 1.65 volts, the trigger input causes comparator B to go High.
- When comparator A output is Low and comparator B output is High, the flip-flop $\overline{Q}$ output goes Low.
- A $\overline{Q}$ Low output causes the output line of the output buffer to go High.

- A Low turns the transistor off, which opens the discharge path of the capacitor and starts the charging phase of the next cycle.

The rate at which the IC's internal components turn on and off is determined by the values of the external components connected to the IC.

The frequency of the output can be determined by the following formula:

$$f = \frac{1.44}{(R_A + 2R_B)C}$$

EXAMPLE 2-7

What is the frequency of the astable multivibrator with the following values of external components? $R_A = 4.7\ \text{k}\Omega$, $R_B = 270\ \Omega$, and $C = 0.47\ \mu\text{fd}$.

Solution

$$f = \frac{1.44}{(R_A + 2R_B)C}$$

$$= \frac{1.44}{(4.7\ \text{k}\Omega + 540\ \Omega)0.47\ \mu\text{fd}}$$

$$= \frac{1.44}{0.0024628}$$

$$= 585\ \text{Hz}$$

SSE8206
The Duty Cycle of a Multivibrator

Initially, the external capacitor charges through R_A and R_B and then discharges through R_B. These charging and discharging times affect what is called a *duty cycle.* The duty cycle is the ratio of time the output terminal is High to the total time of one cycle. The duty cycle is set precisely by the ratio of these two resistors. The charging time (output buffer is High) is T_1. The discharging time (output buffer is Low) is T_2. The total period of time for one cycle is T. These values are calculated as follows:

$$T_1 = 0.693(R_A + R_B)(C)$$

$$T_2 = 0.693(R_B)(C)$$

$$T = T_1 + T_2 = 0.693(R_A + 2R_B)(C)$$

The duty cycle is:

$$\text{DC} = \frac{T_1}{T} \quad \text{or} \quad \text{DC} = \frac{R_A + R_B}{R_A + 2R_B}$$

EXAMPLE 2-8

What is the duty cycle of the astable multivibrator with the following values of external components? $R_A = 10\ \text{k}\Omega$, $R_B = 4.7\ \text{k}\Omega$.

Solution

$$\text{DC} = \frac{R_A + R_B}{R_A + 2R_B}$$

$$= \frac{10\ \text{k}\Omega + 4.7\ \text{k}\Omega}{10\ \text{k}\Omega + 9.4\ \text{k}\Omega}$$

$$= \frac{14.7\ \text{k}\Omega}{19.4\ \text{k}\Omega}$$

$$= .76, \text{ or } 76\%$$

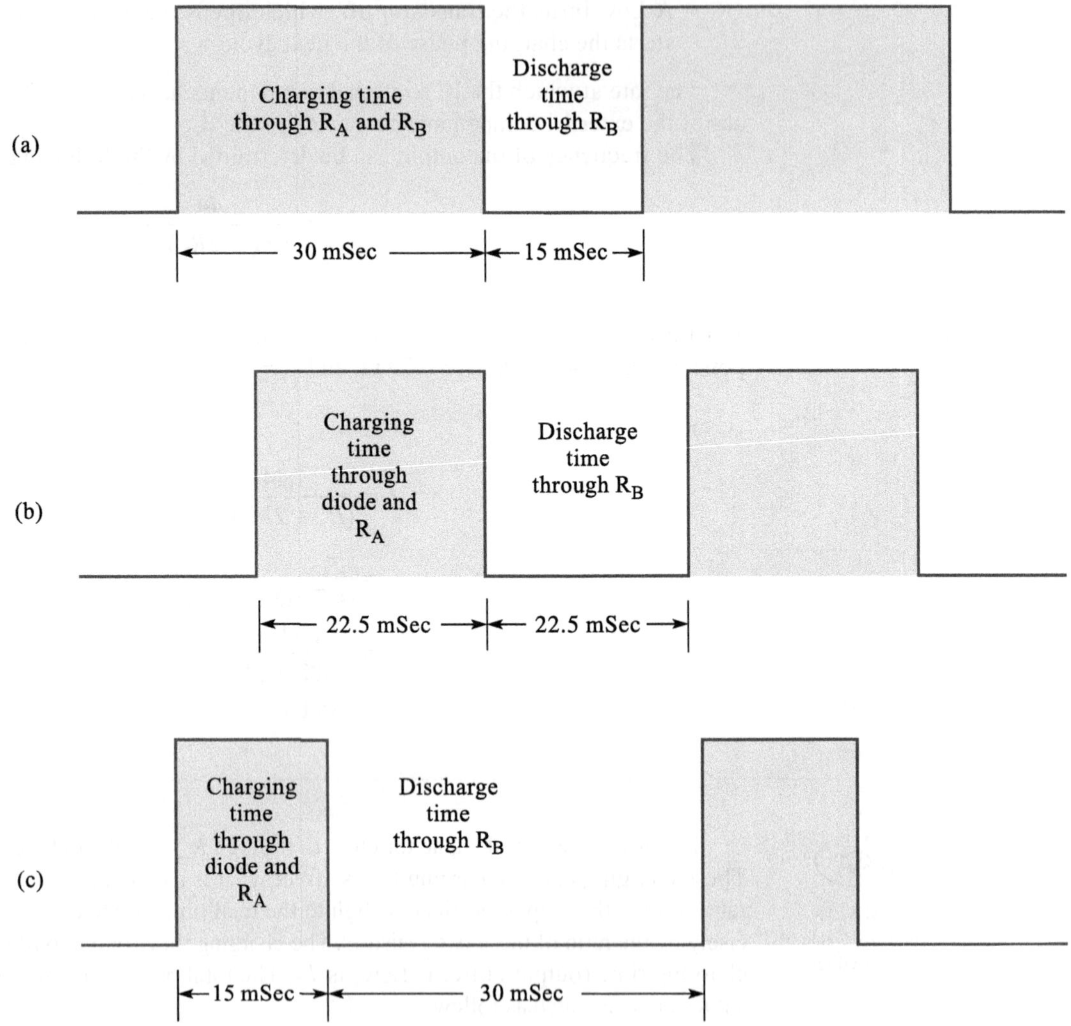

FIGURE 2-28 Waveforms showing duty cycles

Because the capacitor charges up through R_A and R_B and then discharges only through R_B, the duty cycle is always greater than 50 percent, as shown in the top waveform of Figure 2-28(a). However, it may be desirable to have a symmetrical square wave, which means that the time duration of the positive alternation equals that of the negative alternation, as shown in Figure 2-28(b). This would result if the duty cycle is 50 percent. This situation is possible only if the charging and discharging time durations of the capacitor are the same. By making R_A and R_B the same, and placing a diode across R_B with the anode connected to pin 7, and the cathode to pin 6, a symmetrical square wave is possible. The placement of the diode bypasses R_B and allows the capacitor to charge only through R_A. When the capacitor discharges, its current path is blocked by the reverse-biased diode, and only flows through R_B. Therefore, the charge and discharge paths are through resistances of the same value. Depending on the resistance ratios of R_A and R_B, this configuration allows the duty cycle to vary over a range of 5 to 95 percent, as shown in Figure 2-28(c).

555 Monostable Multivibrator

SSE8306
The 555 Monostable Multivibrator

The **monostable multivibrator,** also known as a *one-shot,* is characterized as having only one stable state. Its output is normally 0. When a triggering signal is applied to its input, the output changes from its normal stable state to a logic 1 (unstable state) for a specified length of time before automatically returning to its stable state. The triggering signal comes from either a mechanical switch or another circuit. The period of time the monostable multivibrator remains

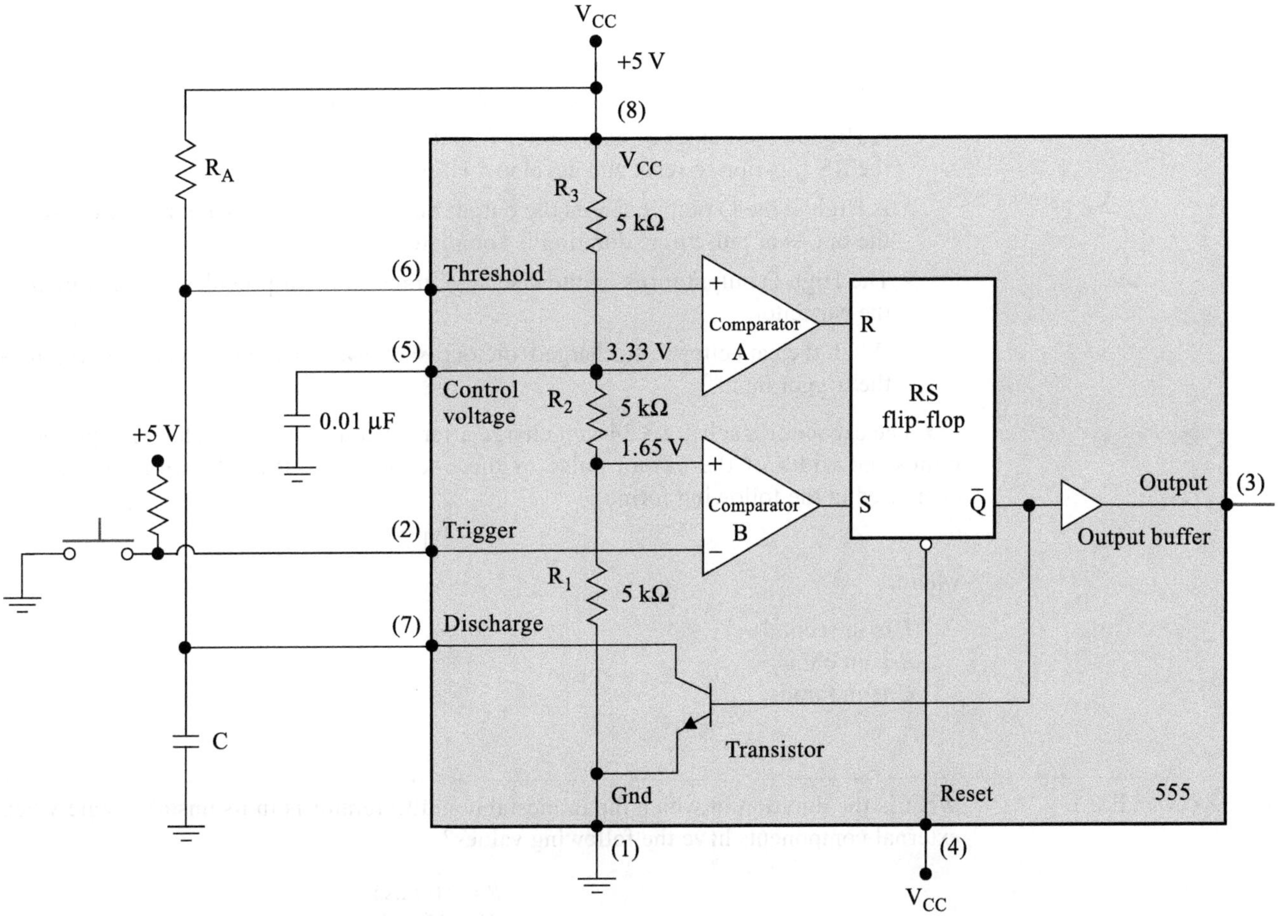

FIGURE 2-29 Schematic diagram of a 555 timer with the external timing components to form a monostable multivibrator

in its unstable state is determined by an external RC timing circuit. The output pulse generated can be either longer or shorter than the input pulse.

Figure 2-29 shows the required connections for a 555 IC to operate as a one-shot. Its operation is as follows:

Assume:

- The capacitor is discharged.
- Comparator A output is Low.
- Comparator B output is Low.
- Flip-flop $\overline{Q}$ output is High.
- The transistor is on.
- The output buffer is Low.
- A +5-volt High is applied to the trigger input.

Therefore:

- While the trigger signal is brought from a High to a temporary 0-volt potential by a push button closure, the comparator B output goes High. The comparator B output returns to a Low when the push button is released.
- A Low applied to the flip-flop's R input from comparator A and a temporary High applied to the flip-flop's S input from comparator B cause the flip-flop's $\overline{Q}$ output to go Low.
- A Low at the $\overline{Q}$ output of the flip-flop causes the buffer output to go High.

- A Low at the $\overline{Q}$ output turns off the discharge transistor that enables the capacitor to begin charging up toward $+V_{CC}$.
- When the capacitor charges to 3.34 volts, the comparator A output goes High.
- A High at the output of comparator A and a Low at the output of comparator B cause the RS flip-flop to reset and develop a High at its $\overline{Q}$ lead.
- A High at the $\overline{Q}$ output causes the output buffer to go back to a normal Low state, and the one-shot pulse time duration is complete.
- The High $\overline{Q}$ output turns on the discharge transistor that provides a discharge path for the capacitor.
- When the capacitor is discharged, the one-shot awaits another negative-going pulse at the trigger input.

The capacitor reaches a 3.34-volt charge after 1.1 time constants. This time period determines the width of the output pulse of the one-shot. The time duration of the pulse is expressed in the following formula:

$$T = 1.1\ RC$$

where,

T is in seconds,
R is in ohms,
C is in farads.

EXAMPLE 2-9

What is the duration in which the monostable multivibrator is in its unstable state when the external components have the following values?

$$R = 100\ \text{k}\Omega$$
$$C = 47\ \mu\text{fd}$$

Solution

$$\begin{aligned} T &= 1.1\ RC \\ &= 1.1 \times 100\ \text{k}\Omega \times 47\ \mu\text{fd} \\ &= 5.17\ \text{seconds} \end{aligned}$$

The one-shot pulse duration can range from microseconds to several minutes.

Problems

1. An inverting op amp circuit has R_F = 5 kilohms and R_{IN} = 1 kilohm. What is the gain of this circuit?
2. What is the output voltage of the circuit in Figure 2-30?
3. A noninverting op amp with an input resistor of 10 kilohms and a feedback resistor of 50 kilohms has 0.4 volts applied to its input. What is the output voltage?
4. At the moment an input signal applied to an integrator changes from 0 to +3 volts, it has a ________ (minimum, maximum) gain.
5. When a sawtooth-shaped signal is applied to the input of a differentiator, a ________-shaped signal is produced at the output.
6. The process where the switching threshold on a positive-going input signal applied to a Schmitt trigger is higher than the negative-going signal is referred to as ________.
7. Fill in the parentheses of each of the following equations with <, >, or = symbols to describe how an op amp comparator operates:
   a. Inverting input voltage (____) noninverting input voltage = positive output voltage.
   b. Inverting input voltage (____) noninverting input voltage = zero output voltage.
   c. Inverting input voltage (____) noninverting input voltage = negative output voltage.

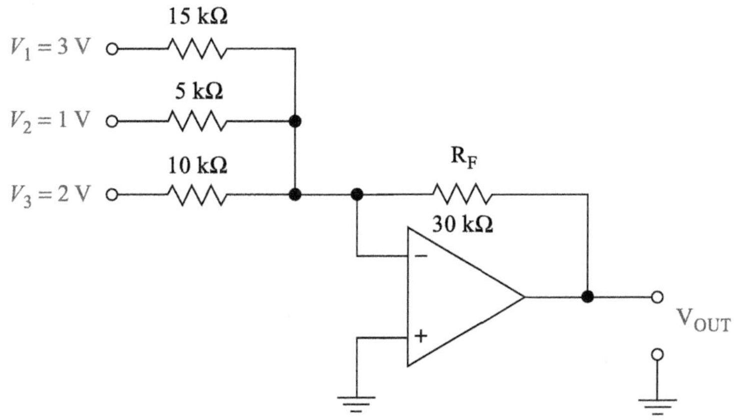

FIGURE 2-30 Circuit for Problem 2

8. Assuming the values of all resistors connected to a difference op amp are the same, what is the output voltage when +3 volts is applied to the inverting input and +1 volt is applied to the noninverting input?
9. What three decisions does a magnitude comparator perform when comparing two different binary numbers?
10. Describe the meaning of the term *optoisolator*.
11. When light strikes the base of a ________ (forward, reverse)-biased photoelectric detector, it turns on.
12. DAC resolution is determined by the number of digital ________ lines available.
13. What is the resolution of a DAC with a maximum voltage of 15 volts and 5 input lines?
14. A 5-bit DAC produces an output of 9.2 volts when the binary number applied to its input is 10111_2. What is the output voltage if the binary number changes to 10101_2?
15. What is the resolution of a 5-bit ADC that has a maximum analog input voltage of +10 volts?
16. An 8-bit "successive-approximation-register" ADC requires how many clock pulses for each conversion?
17. In Figure 2-25, if the output reads 10000000, the analog voltage applied to the input is ________ volts.
18. What is the duty cycle of the astable multivibrator with the following values of external components? $R_A = 3.3$ kΩ, $R_B = 6.8$ kΩ.
19. The duty cycle of a square wave is the ratio of time the output terminal is ________ (low, high) to the total time period of one cycle.
20. How can a 555 astable multivibrator be constructed to produce a square wave that ranges from a 5 to 95 percent duty cycle?
21. Referring to Figure 2-27, what is the output frequency if $R_A = 100$ kΩ, $R_B = 10$ kΩ, and $C = 10$ μfd?
22. Referring to Figure 2-29, what is the amount of time the one-shot is in the unstable state if $R_A = 2$ kΩ and $C = 1$ μfd?

CHAPTER

3

Thyristors

OBJECTIVES

At the conclusion of this chapter, you should be able to:

- Describe the characteristics of thyristors as compared to other semiconductor devices.
- Draw the schematic symbol of the SCR, UJT, Diac, Triac, and IGBT.
- Explain the characteristics and the operation of SCRs, UJTs, Diacs, Triacs, and IGBTs.
- Draw the current–voltage characteristic curve for an SCR, a UJT, a Diac, and a Triac.
- Describe ways in which voltages at the leads of SCRs, UJTs, Diacs, Triacs, and IGBTs turn them on and off.
- Draw the waveforms produced at the output of SCRs, UJTs, Diacs, and Triacs when various voltages are applied to their inputs.
- Describe the operation of various sample circuits that have SCRs, UJTs, Diacs, and Triacs.

INTRODUCTION

Thyristors are switching devices that have the characteristics of either being fully turned on or being fully turned off. They are not capable of gradually changing the current flow through them. Under normal conditions, a thyristor remains off and acts like an open component until it is turned on when a breakover voltage is reached. When turned on, it becomes a low-resistance component until the current flowing through it decreases to below the rated holding-current value of the device. The name *thyristor* is derived from the thyratron gas tube, predecessor of this semiconductor component with the same characteristics.

Thyristors are extensively used in high-current circuits that provide electricity to devices such as lamp dimmers, motors, and other devices that require high power and variable control. The main advantage thyristors have over other semiconductor devices that are capable of switching on and off, such as a transistor, is that they can control very large amounts of load power with very little control power needed to trigger them. Also, unlike transistors that require a high and continuous base control current to maintain saturation, thyristors need only a relatively low and brief current to turn them into a saturation state. Thyristors have a greater power handling capacity than transistors. Also, in comparison to transistors, thyristors power both DC and AC loads.

Thyristor devices and other semiconductor components that are used with them, including silicon-controlled rectifiers (SCRs), unijunction transistors (UJTs), triacs, diacs, and insulated gate bipolar junction transistors (IGBTs), will be covered in this chapter. Examples of various circuits that are powered by these devices will also be provided.

3-1 Silicon-Controlled Rectifiers

SSE4803
The Silicon-Controlled Rectifier

One of the most common types of thyristors used is the **silicon-controlled rectifier (SCR).**

SCR Construction

An SCR is constructed from four alternately doped silicon layers of P- and N-type semiconductor materials, as shown in Figure 3-1(a). They are sandwiched together to form three junctions, and have three terminals that are attached to three of the four layers. The terminal leads are called the anode (A), the cathode (K), and the gate (G), as shown in Figure 3-1(b).

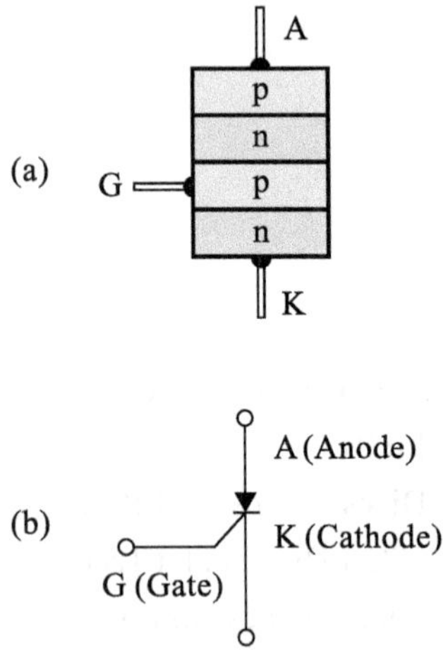

FIGURE 3-1 Silicon-controlled rectifier

SCR Operation

The SCR operates similarly to a standard diode, where current flows only in the forward-biased condition and is blocked when reverse biased. Therefore, the SCR is able to perform rectification. However, the conduction of the SCR does not begin until two conditions exist simultaneously.

- It is forward biased.
- A triggering signal is applied to the gate terminal.

Once the SCR is turned on, it conducts current like a regular diode and acts like a latched switch. During conduction, the signal applied to the gate no longer has an effect on the SCR's operation, and it remains on as long as the current level does not fall below a predetermined value, which is called the *holding current*. When the current drops below the holding-current level, the SCR will turn off. It cannot turn on again by applying a positive voltage to the gate unless there is sufficient forward-bias voltage across its anode and cathode leads.

SCR Equivalent Circuit

An equivalent circuit of the SCR is shown in Figure 3-2(a). It consists of both an npn and a PNP transistor, which are interconnected. The upper three layers act as a PNP transistor, Q_1, and the lower three layers act as an npn transistor, Q_2. Two DC power supplies are added to the equivalent circuit in Figure 3-2(b) for biasing the transistor.

With the switch connected to the base of Q_2 open, the npn transistor will not conduct because its emitter–base junction is not forward biased. The nonconducting Q_2 will, in turn, prevent Q_1 from conducting because there is no base current that can pass through the npn transistor to the base of the PNP transistor.

If switch 1 is momentarily closed, the gate of Q_2 is made positive in respect to its emitter; the junction between them is forward biased and turns the npn transistor on. This conduction causes the negative potential of power supply B to be applied to the base of Q_1. With the positive potential of power supply B at the emitter of Q_1, its emitter–base junction is forward biased, causing it to turn on also. As the PNP transistor is conducting, its collector current draws a continuous base current from the npn transistor. The two transistors hold each other in the on, or conducting state, even if the voltage applied to the gate of Q_2 was momentary and the switch is now open. This condition represents how the SCR is latched on as current flows from the cathode to its anode lead.

To switch the equivalent SCR circuit back off, it is necessary to reduce the voltage of power supply B to almost zero. When this situation occurs, both transistors will turn off. This condition represents how reducing the cathode–anode voltage of an activated SCR, and consequently dropping the holding current to near zero, will cause it to turn off. Both transistors, or the SCR they represent, will remain off until both the anode–cathode voltage increases and a positive gate voltage is reapplied.

Figure 3-3 shows the equivalent circuit replaced by an SCR that is properly biased by two power supplies. The switch, S_1, is used to apply or remove the gate-triggering voltage obtained from the power supply V_{GG}. A second switch, S_2, is used to apply or remove the anode–cathode voltage supplied by V_{AA}. The function of resistor R_G is to limit the gate current; resistor R_L limits the anode current (I_A). When S_2 is closed, the SCR's anode–cathode junction is forward biased. However, it will not turn on until S_1 is also closed. At the

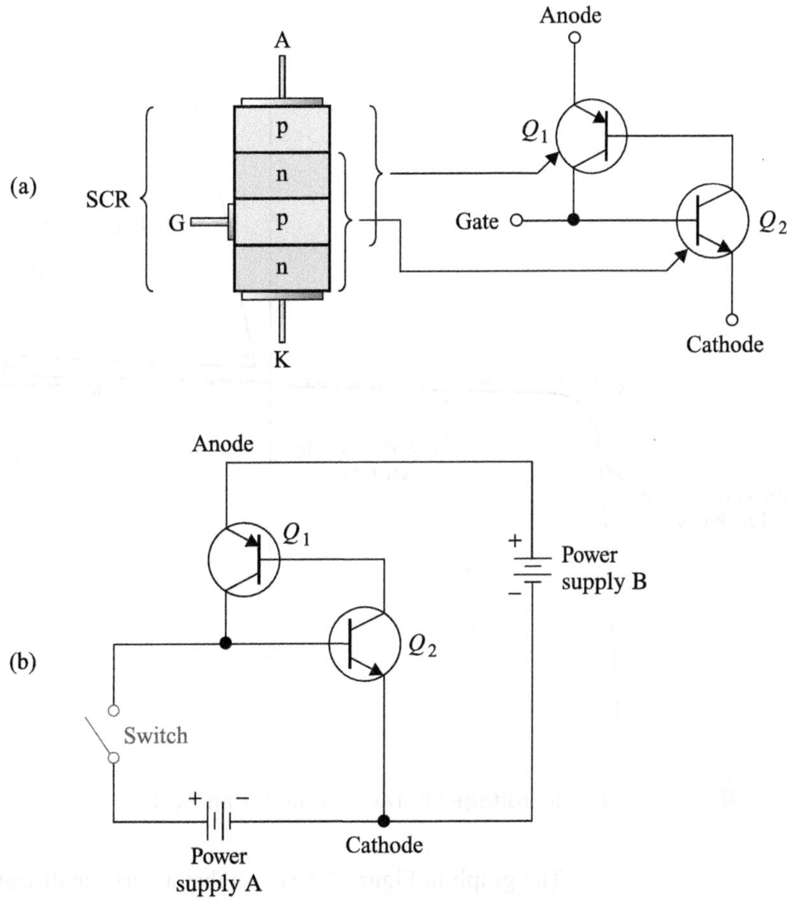

FIGURE 3-2 (a) SCR equivalent circuit; (b) Switching operation of an SCR

moment S_1 applies a voltage to the gate, the SCR turns on. The SCR remains on even after S_1 is reopened. The only condition under which it will turn off is when the current through the SCR drops below its minimum holding-current (I_H) value. This condition occurs when S_2 is opened.

Current–Voltage Characteristics of an SCR

The operation of an SCR is similar to that of standard diodes when they are reverse biased. They have a very high internal impedance and block current flow unless a voltage level is reached that causes conduction to occur. This level is called the *reverse breakdown voltage*.

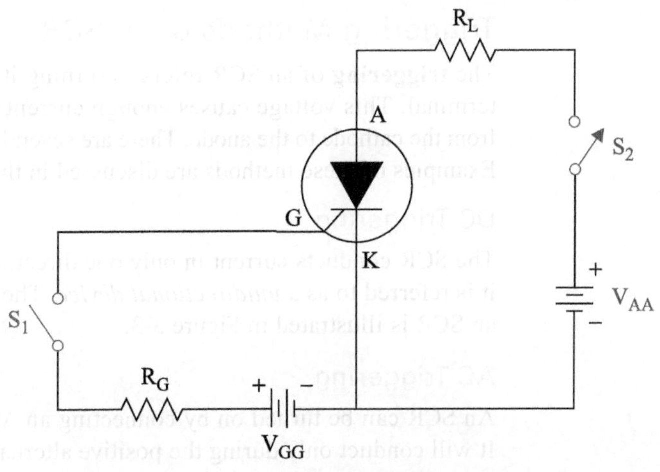

FIGURE 3-3 An SCR properly biased

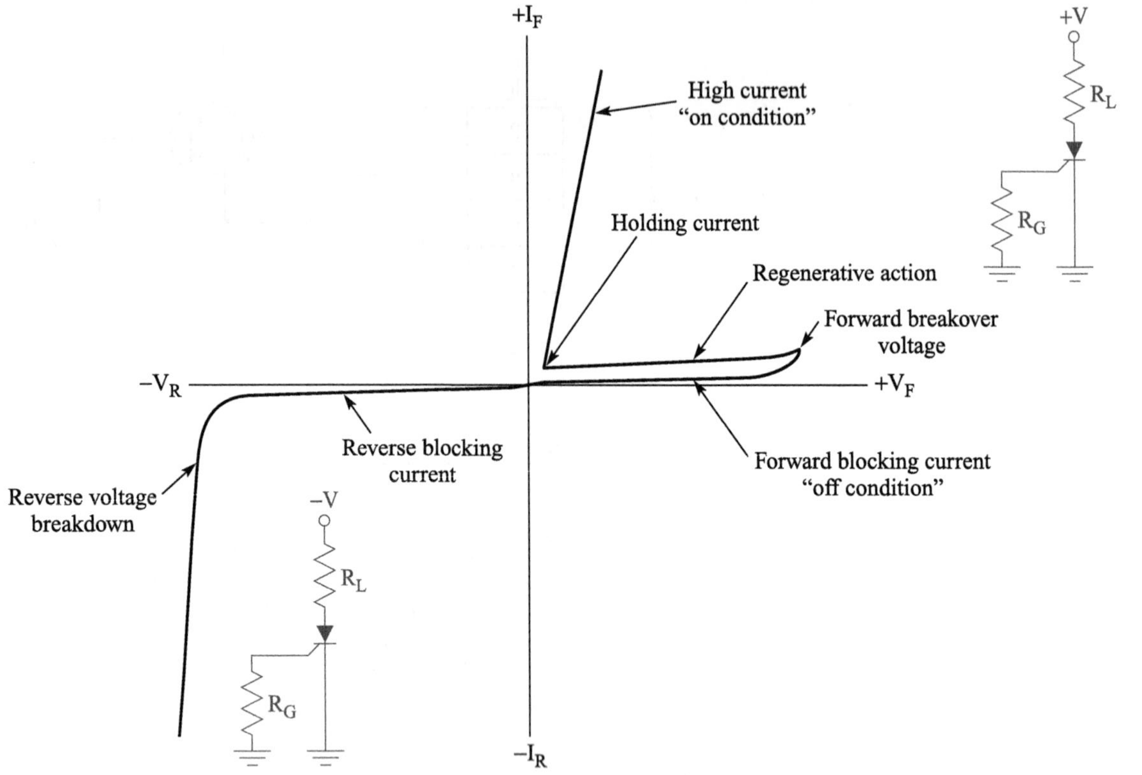

FIGURE 3-4 Current–voltage characteristics for an SCR

The graph in Figure 3-4 shows that a very small leakage current flows in the reverse-biased condition until the reverse breakdown voltage is reached. When exceeded, the reverse current rapidly increases to a large value and may destroy the SCR.

When the SCR is forward biased and the gate is connected to ground, the internal impedance of the SCR is very high and a small current, called *forward-blocking current,* flows. If the voltage is increased beyond a level called *forward breakover voltage,* a regenerative action occurs with the PN junction, and the internal impedance of the SCR decreases. The result is that the voltage across the SCR drops, and the current from the cathode to the anode increases rapidly, as shown in Figure 3-4. The SCR will remain on as long as this current does not fall below the holding-current rating.

Under normal operation, the voltage applied across the anode and cathode will be less than the forward breakover point. It can be reduced by applying a positive voltage at the gate to cause gate current to be increased. The greater the amount of gate current that flows, the lower the point at which forward breakover occurs.

Triggering Methods of an SCR

The **triggering** of an SCR refers to turning it on by applying a positive voltage to the gate terminal. This voltage causes enough current to flow through the gate to begin conduction from the cathode to the anode. There are several ways in which to provide the positive potential. Examples of these methods are discussed in the following paragraphs.

DC Triggering

The SCR conducts current in only one direction—from the cathode to the anode. Therefore, it is referred to as a *unidirectional device.* The circuit conditions that cause DC triggering of an SCR is illustrated in Figure 3-3.

AC Triggering

An SCR can be turned on by connecting an AC voltage across its anode and cathode leads. It will conduct only during the positive alternation, when the SCR is forward biased. During

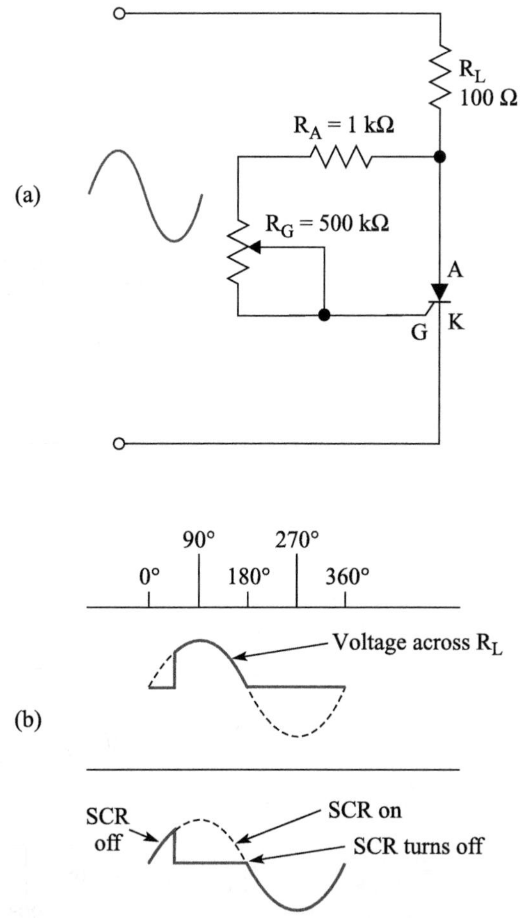

FIGURE 3-5 (a) AC triggering circuit that turns the SCR on from 0° to 90°; (b) Waveforms

the negative alternation, it will be reverse biased and, like a regular diode, will block current flow.

Figure 3-5(a) shows fixed resistor (R_A) and rheostat (R_G) connected between the anode and the gate leads of the SCR. The rheostat is sufficiently large enough to limit and control the gate current that turns on the SCR. The function of R_A is to prevent the gate from being shorted to the anode when the wiper arm of R_G is moved to the top.

As the positive alternation of the AC voltage increases, the positive voltage across the SCR also rises, and the gate begins to draw current. A point is reached when enough gate current causes the SCR to fire. The SCR stays on during the remainder of the alternation until the voltage decreases to a level where the holding current it produces can no longer sustain conduction. Figure 3-5(b) shows the waveform across the SCR and the load resistor when it is triggering at 45 degrees.

The drawback of this circuit is that the SCR can only be triggered between 0 and 90 degrees of the alternation. If it is not turned on by the time it reaches 90 degrees, the positive alternation begins to descend and the voltage at the gate decreases. To extend the time at which the trigger pulse will occur up to 180 degrees, a capacitor is added, as shown in Figure 3-6(a). The capacitor and resistors, R_A and R_G, become an RC circuit. The rate at which the capacitor charges is controlled by the resistance of the rheostat. As the wiper arm of R_G is lowered and its resistance increases, it takes longer for a positive voltage at the capacitor's top plate to become great enough to fire the SCR. If the SCR is not triggered by the time the positive alternation of the AC supply reaches 90 degrees, the SCR can still be turned on because the capacitor will continue to charge during the descending portion of the positive alternation. Figure 3-6(b) shows the waveform across the SCR and the load resistor when it is triggered at 135 degrees.

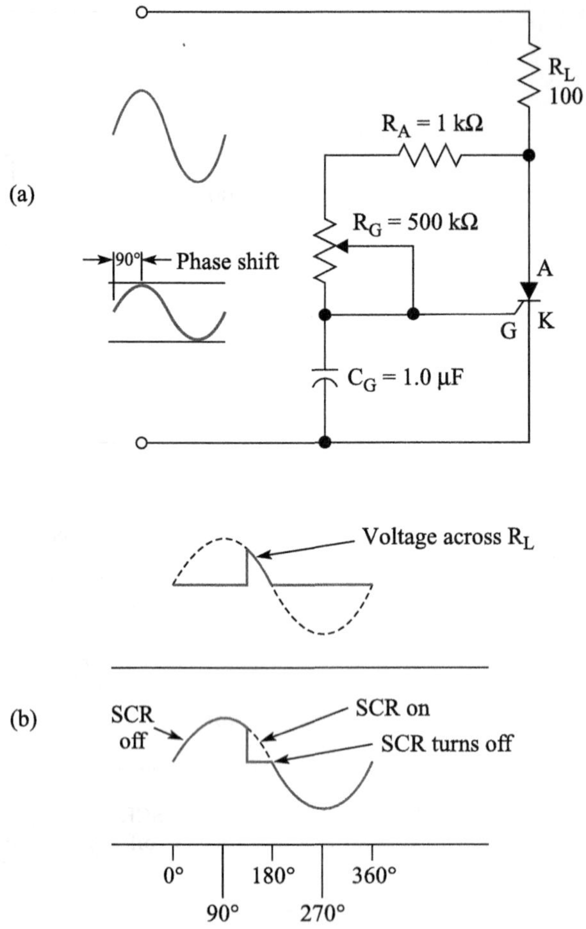

FIGURE 3-6 (a) AC triggering circuit that turns the SCR on from 0° to 180°; (b) Waveforms

SSE4903
Waveforms of an SCR Circuit

Phase Control SCR Circuit

When an AC voltage is applied across the anode–cathode leads, the SCR rectifies the alternating current just like a conventional diode. It blocks one alternation of the cycle when it is reverse biased, and it is capable of turning on and conducting current during any portion of the other half of the cycle while it is forward biased. The time at which it begins to conduct depends on when a trigger voltage is applied during this alternation. If the SCR is turned on early in the half-cycle, it will conduct more average power than if it is triggered later in the alternation.

Figure 3-7 shows how an SCR controls the power delivered to a universal motor. A technique called *phase control* is used to vary the SCR's conduction angle, thereby causing an SCR to switch in such a way that the ratio of on (conducting) time to off (nonconducting) time may be varied, allowing average power to the load to be changed. The term *phase control* refers to the time relationship between two events. In this case, it is the time relationship between the occurrence of the trigger pulse and the point at which the conduction alternation begins.

When the negative alternation occurs, the SCR is reverse biased and will not conduct. Diode D_2 is forward biased, allowing current to bypass R_1 and charge the capacitor. Due to the high opposition of the charging capacitor, the current through the motor is small enough to be ineffective.

When the positive alternation occurs, the alternation begins to rise, the SCR becomes forward biased, and the capacitor begins to discharge the polarities it received during the negative alternation. Once discharged, it recharges at the opposite polarity. The discharging and charging rates are determined by the component values of the RC network. When the positive charge at the top plate of the capacitor is high enough, D_1 forward-biases and allows current to flow through the gate and fire the SCR. The result is that the supply current flows through

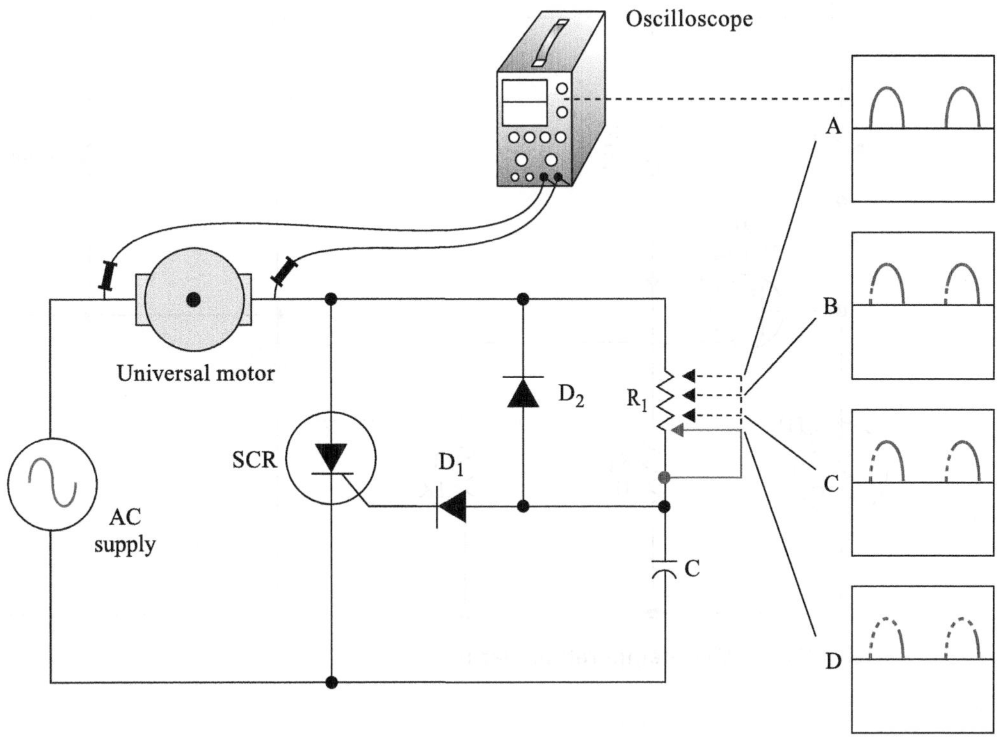

FIGURE 3-7 The phase control of an SCR varies the speed of a motor

the motor during the remaining portion of the alternation. During the next negative alternation, the SCR is reverse biased and is off, and the voltage drop across the motor is reduced. The SCR turns on again at the point when the next positive alternation occurs. If the variable resistor is in position A, there is zero resistance and the capacitor discharges and charges almost immediately. The result is that current will flow through the motor for nearly 100 percent of the positive alternation. As the variable resistor arm is moved downward, resistance increases, which causes the capacitor to discharge and charge more slowly. This results in a firing time delay, which decreases the amount of current delivered to the motor. Diagrams A through D in Figure 3-7 show the waveform patterns displayed by the oscilloscope as the RC time constants are changed by varying the wiper arm position of R_1.

In practical applications, control devices such as computers, programmable controllers, and sensors produce a low-voltage DC signal to turn on the SCR at a precise moment.

SCRs are used for such applications as driving DC motors, lighting, welding machines, heating operations, and other equipment that requires the high variable-voltage control capability that this semiconductor device can provide. SCRs can control current as high as several hundred amperes at voltages greater than 1000 volts. For this reason, they are commonly used in industrial control electronic equipment that requires their specific control capabilities.

Temperature-Operated Alarms

Temperature-operated alarms can be used as automatic overheat alarms, or underheat alarms. Two practical temperature-operated alarm circuits are described in Figures 3-8 and 3-9. These circuits use inexpensive *negative temperature coefficient thermistors* as temperature-sensing elements. These devices act as temperature-sensitive resistors that present a high resistance at low temperatures and a low resistance at high temperatures.

Over-Temperature Alarm

Figure 3-8 shows the practical circuit of an over-temperature alarm. R_1, R_2, R_3 and the thermistor, TH_1, are wired in the form of a Wheatstone bridge; Q_1 is used as a bridge balance detector and SCR driver. R_1 is adjusted so that the bridge is balanced at a desired temperature.

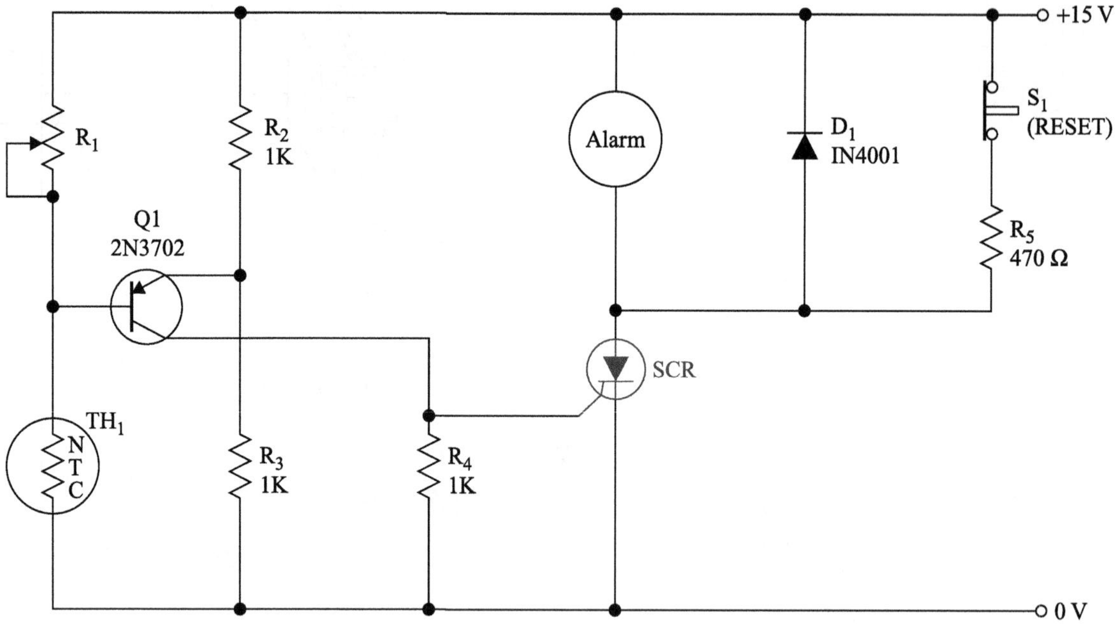

FIGURE 3-8 Over-temperature alarm

The base of Q_1 and its emitter are at equal potentials under this condition, so Q_1 and the alarm are off. As the temperature to which the thermistor is exposed rises, the resistance of TH_1 decreases; therefore, the voltage at the base of Q_1 becomes less positive (more negative) than the emitter, causing the transistor to turn on. The result is that current flows from R_4 through the collector, out of the emitter, and through R_2. As the current through R_4 increases, the voltage drop across it also increases to a level that causes enough of a positive potential to turn on the gate of the SCR. As the SCR fires, current through the alarm flows, and it turns on. The alarm will remain on even after the over-temperature condition no longer exists. To reset the alarm, the reset push button (S_1) must be pressed. This opens one of the current paths that flow through the SCR. By eliminating one of two parallel paths, the resistance to the current flow through the SCR increases. If the alarm coil resistance is 470 Ω, the resistance to current flow doubles when S_1 is opened and the SCR's current will be reduced in half. When current flows only through the alarm coil, the SCR current drops below its minimum holding-current level and it stops conducting.

Diode D_1 is a suppressor for the alarm coil. When the SCR turns off, current stops flowing through the alarm coil. The coil is an inductor, which tries to avoid a change in current flow. To maintain current flow, the magnetic field developed around the coil collapses and induces a voltage into the coil in an effort to continue current flow through it. The collapsing field of the alarm coil produces a counter-EMF voltage, which is negative at the top of the coil compared to the more positive potential at the bottom of the alarm coil. As a result, the diode, D_1, is forward biased. The diode provides a path for the collapsing field current. Without D_1, the momentary counter-EMF voltage could be as high as several hundred volts. Besides creating electrical noise for other circuits, the large voltage spike could damage the SCR.

Under-Temperature Alarm

The action of the over-temperature alarm can be reversed so that the alarm turns on when the temperature falls below a preset level by simply transposing the positions of R_1 and TH_1, as shown in Figure 3-9. This circuit can be used as a frost or under-temperature alarm. As the temperature to which the thermistor is exposed decreases, its resistance increases. The rise in its resistance creates a larger voltage drop across it and a lower voltage drop across R_1. The result is that the voltage at the base of Q_1 becomes less positive (or more negative) than the emitter, causing the transistor to become more forward biased and to turn on with more intensity. As current flow through the transistor from R_4 is increased, the voltage at the gate of the SCR becomes more positive. At a given level, the positive voltage becomes great

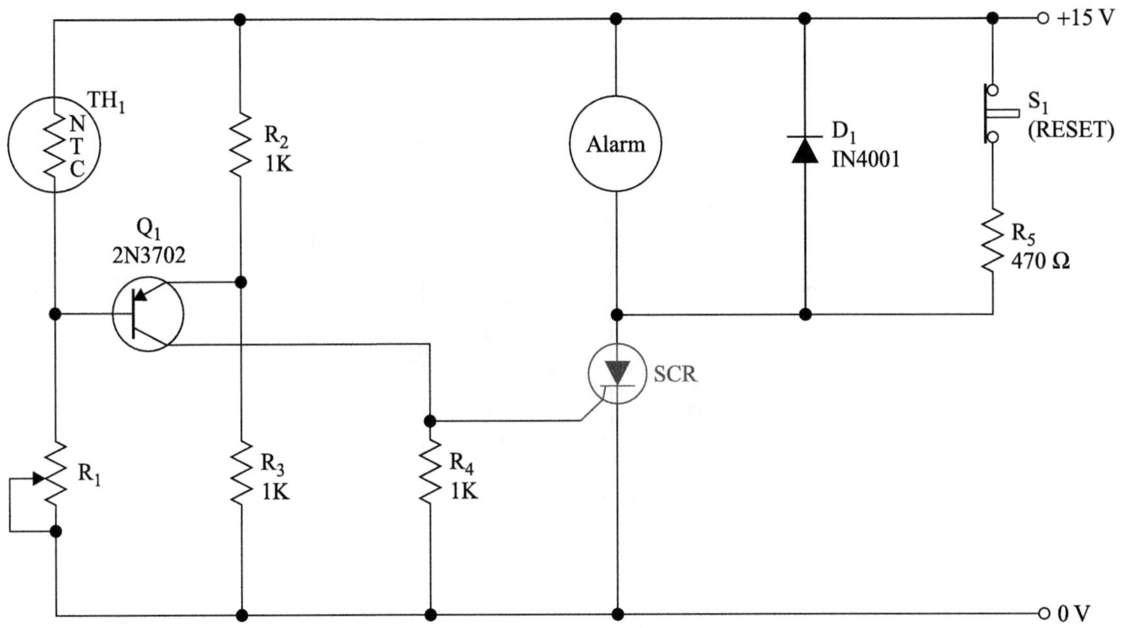

FIGURE 3-9 Under-temperature alarm

enough to fire the SCR. The alarm will remain on after the temperature returns to normal until its holding current is reduced by pressing the reset switch, S_1.

3-2 Unijunction Transistors

A **unijunction transistor (UJT)** is a thyristor component that is used exclusively for switching and not for amplification. Its switching capabilities make it an ideal device for turning on an SCR.

The structure of a UJT is shown in Figure 3-10(a). It is made of an N-type silicon bar called a *base*, to which a small amount of P-type material is diffused. The base terminals connected to each end are labeled B_1 (base 1) and B_2 (base 2). The P material is the emitter and is labeled E, as shown in Figure 3-10(b). A power supply is connected across the base with the negative lead at B_1 and the positive lead at B_2. The internal resistance of the base causes a voltage gradient to form, as shown in Figure 3-10(c). Since the bar is a semiconductor, it allows only a small current to flow. Together, the emitter and B_1 form the single PN junction of the UJT. An equivalent circuit in Figure 3-10(d) presents the internal characteristic of the UJT. The resistance between B_1 and point X is r_{B_1}, and the resistance between B_2 and point X is r_{B_2}. The sum of these two resistances is R_{BB}. The diode represents the PN junction between the emitter and the N material.

A bias voltage is connected across the junction with the emitter more positive than B_1. In its normal state, there is no current flow from B_1 to the emitter. When the bias voltage reaches 7.7 V ($V_{r_{B1}}$ + diode breakover), the UJT turns on. An increased current flows from B_1 to B_2, and from B_1 to the emitter.

When the UJT fires, the r_{B_1} resistance does not stay constant. Instead, it decreases to a minimum value, and when it does, V_E decreases to a minimum value, even though the emitter current (I_E) increases. This condition is referred to as the *negative-resistance region*. Current continues to flow until the emitter voltage drops below the point where the pn junction is no longer forward biased. The UJT turns off until the bias voltage again becomes 7.7 volts.

Current–Voltage Characteristics of a UJT

The graph in Figure 3-11(b) shows the current and voltage characteristics of the UJT. The area on the graph before the UJT fires is called the *cutoff region*. When the voltage, V_E,

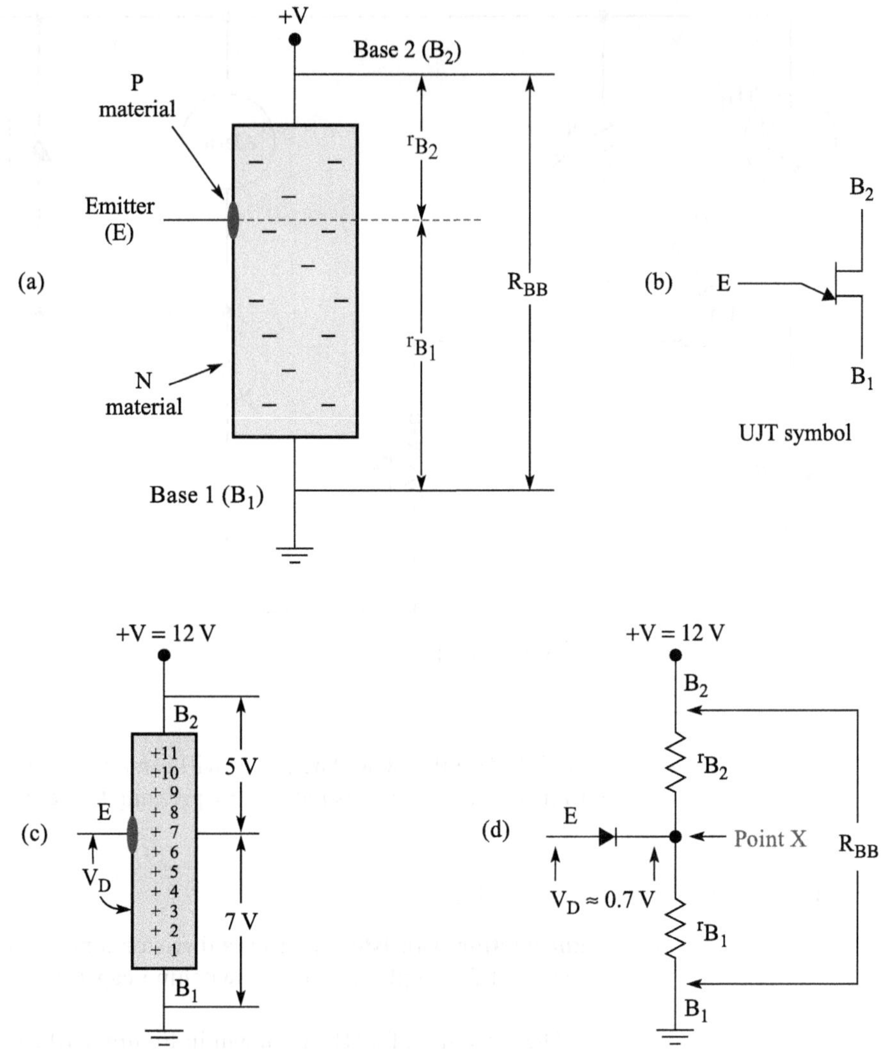

FIGURE 3-10 Unijunction transistor

reaches the time at which the UJT fires, I_E increases, I_{B_1} increases, and I_{B_2} increases. Because of the negative-resistance condition that develops, V_E decreases. The point on the curve that represents the emitter voltage just before the UJT fires is referred to as the *peak voltage* (V_P). The minimum voltage that forms on the emitter after the UJT fires is called the *valley voltage* (V_V). At this point, the PN junction is no longer forward biased, and the UJT turns off. The area on the graph between V_P and V_V is called the *negative-resistance region*. When more current flows through the emitter after the UJT fires, it is in the saturation region. The firing voltage of 7.7 V at the emitter described in Figure 3-10 may vary. The actual point at which the UJT reaches its peak voltage (V_P) and fires is determined by the *intrinsic stand-off ratio*. The abbreviation R_{BB} represents the specific amount of intrinsic resistance of the N-type material between the B_1 and the B_2 terminals. The intrinsic resistance from B_1 to E is called r_{B_1}, and the intrinsic resistance from E to B_2 is labeled r_{B_2}. Therefore, $R_{BB} = r_{B_1} + r_{B_2}$. The point at which the UJT reaches its peak voltage and fires is determined by the ratio of r_{B_1} to R_{BB}. Intrinsic standoff ratio, represented by the Greek letter η (eta), is used to determine the UJT's firing point. It can be expressed mathematically as:

$$\eta = \frac{r_{B_1}}{R_{BB}}$$

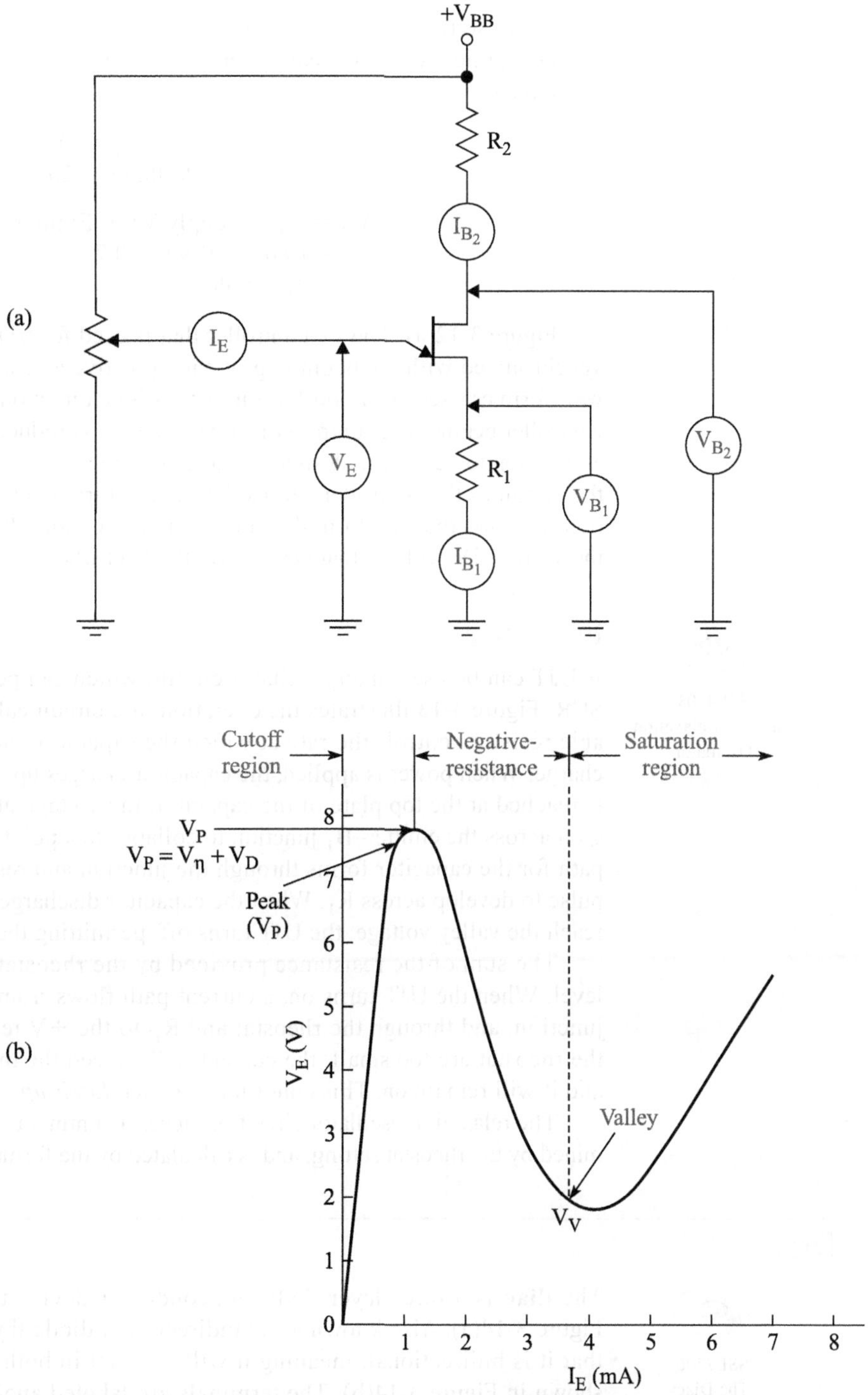

FIGURE 3-11 UJT current–voltage characteristics: (a) I–V nomenclature; (b) Emitter characteristic curve

Typical values of η will range from 0.5 to 0.8. Since the intrinsic resistances that affect η are determined by the physical construction of the device, their values cannot be measured, nor can they be controlled by external components. Therefore, the intrinsic standoff ratio of each UJT can only be provided by the manufacturer.

Suppose the η value of a UJT is 0.6 and the DC voltage supply is 20 volts. The peak voltage applied to V_E that causes the UJT to fire is calculated as follows.

Given:

$$\eta = 0.6$$
$$\text{Supply Voltage} = 20 \text{ V}$$

$$\begin{aligned} V_P &= (\eta \times \text{Supply V}) + \text{Emitter Junction Breakover} \\ &= (0.6 \times 20 \text{ V}) + 0.7 \\ &= 12.7 \text{ volts} \end{aligned}$$

Figure 3-12(a) shows a controller that is used for triggering an SCR. The control signal is synchronized with the beginning of each positive alternation applied to the SCR. As the AC waveform crosses the 0-volt level in the positive direction, a timer in the microprocessor-based controller begins. The controller is programmed to produce the pulse at a specific moment in the alternation to generate a desired average voltage. Figure 3-12(b) provides waveform samples of the circuit as the controller causes the SCR to turn on at various points of the positive alternation. It shows the waveform that drops across the motor before and after the SCR turns on, and the average DC voltage that develops at the load when a pulse is applied to the gate.

Relaxation Oscillator

SSE4703
The UJT Relaxation Oscillator

A UJT can be used in an oscillator circuit, which can perform the triggering function of the SCR. Figure 3-13 illustrates the operation of a circuit called a *relaxation oscillator*. The variable resistor controls the rate at which the capacitor charges. Initially, the capacitor has no charge. When power is applied, the capacitor charges up exponentially. When the peak voltage is reached at the top plate of the capacitor, the voltage at the emitter causes the depletion region across the emitter–B_1 junction to collapse to a point where the UJT turns on. A discharge path for the capacitor forms through the junction and resistor R_1. The current surge causes a pulse to develop across R_1. When the capacitor discharges to a point that causes the emitter to reach the valley voltage, the UJT turns off, permitting the capacitor to begin charging again.

The sum of the resistance provided by the rheostat and R_E need to be above a certain level. When the UJT turns on, a current path flows from ground through R_1, the B_1–emitter junction, and through the rheostat and R_E to the +V terminal. If the resistances of R_E and the rheostat are too small, the current will exceed the minimum holding current of the UJT, and it will remain on. This condition is called *latch-up*.

The relaxation oscillator circuit produces a continuous train of pulses. Its frequency is determined by the rheostat setting, and is calculated by the formula located to the left of Figure 3-13.

3-3 Diac

SSE7706
The Diac

The **diac** is a three-layer PNP semiconductor device that has two junctions, as shown in Figure 3-14(a). Also known as a "bidirectional diode thyristor," this official name indicates that it is bidirectional, meaning it will conduct in both directions. Its schematic symbol is shown in Figure 3-14(b). The terminals are labeled anode 1 and anode 2. The diac operates similar to two zener diodes placed back-to-back in series, as shown in Figure 3-14(c).

Diac Operation

When a small positive voltage is applied across the diac terminals, as shown in Figure 3-15(a), junction J_1 is forward biased and junction J_2 is reverse biased. Since current is blocked by J_2, there is no conduction through J_1. As the voltage increases, the reverse-voltage breakover point is reached at J_2 as a result of avalanche breakdown, allowing the conduction of current through the diac, as shown in Figure 3-15(b). The action at J_2 is similar to what happens with a zener diode when it begins to conduct in the reverse-biased condition.

When the opposite polarity is applied across the diac, as shown in Figure 3-15(c), junction J_1 is reverse biased and junction J_2 is forward biased. As the applied voltage is increased more

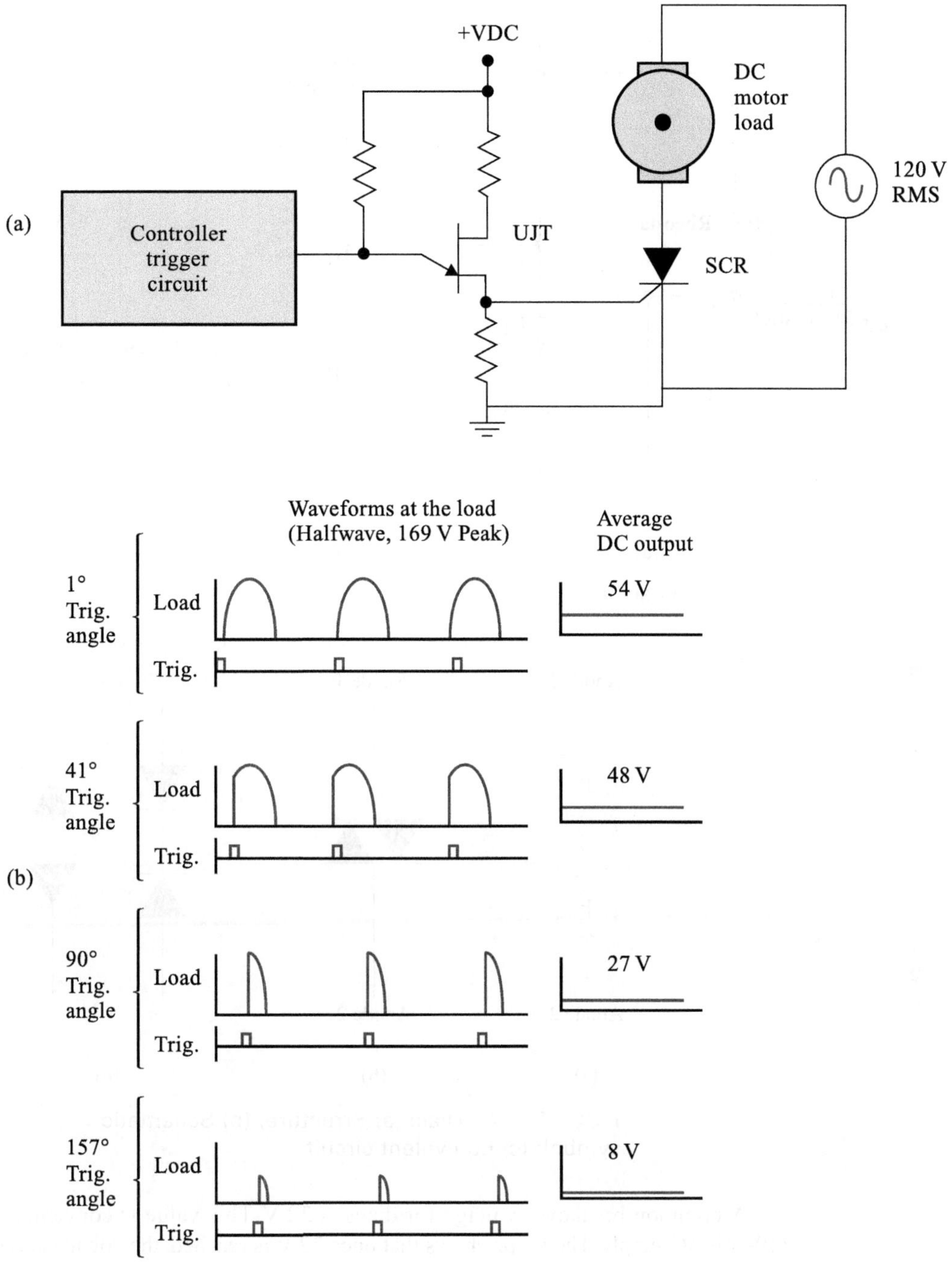

FIGURE 3-12 UJT controller

negatively, the reverse-voltage breakover point is reached at J_1 as a result of avalanche breakdown, which causes conduction to begin in the opposite direction, as shown in Figure 3-15(d).

Current–Voltage Characteristics of a Diac

The graph in Figure 3-16 shows the current–voltage characteristics of a diac. As a positive voltage across the diac is increased, a very small blocking current flows even though the device is in the "off condition." When the breakover point is reached, the diac turns on and a high current flows through it. This characteristic is the result of negative resistance or the regeneration of the device. To protect the diac from excessive current, a resistor is placed in series with it.

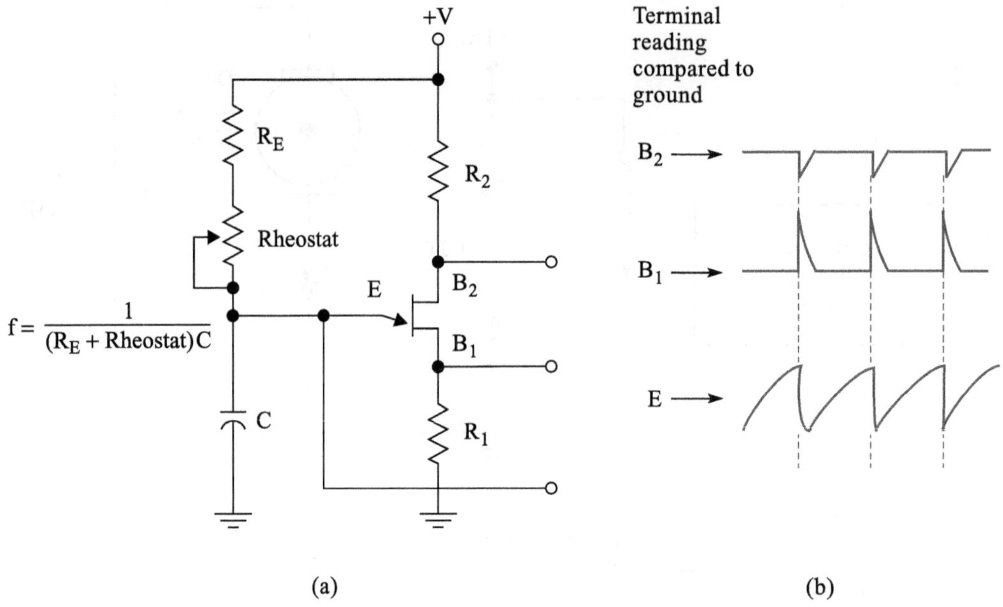

FIGURE 3-13 Relaxation oscillator

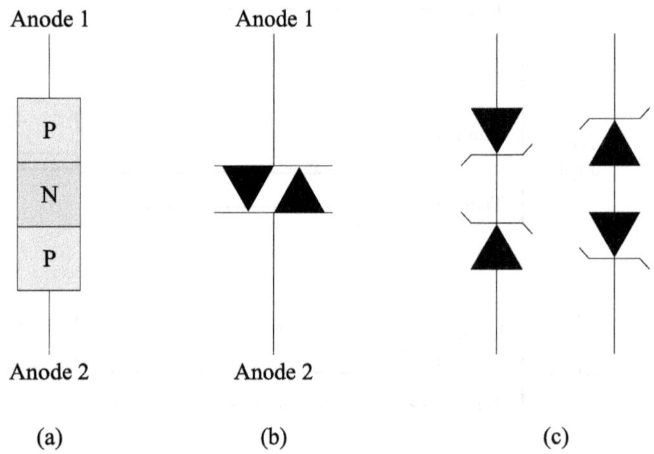

FIGURE 3-14 Diac: (a) Structure; (b) Schematic symbol; (c) Equivalent circuit

A common breakover voltage for diacs is 32 V. This value is convenient for use with a 110-volt AC supply. The graph shows that once 32 V is reached, the voltage across the terminals decreases to a value lower than the voltage breakover point. The surge of current that begins to flow accounts for a sharp trigger pulse that is used to turn on other types of thyristors.

The operation of the diac is the same as when a voltage of the opposite polarity is applied across it, as shown on the left portion of the graph in Figure 3-16. Since both junctions are equally doped, there is very little difference in magnitude between the breakover voltages when the negative or positive voltages are applied across the diac. The difference is typically less than 1 volt. Therefore, the diac is able to provide a pulse at nearly the same point for both half-cycles of the AC supply.

3-4 Triac

SSE7406
The Triac

The fact that the SCR will conduct current in only one direction makes it a unidirectional device. Some types of loads use AC power, which requires a variable supply current that flows in both directions.

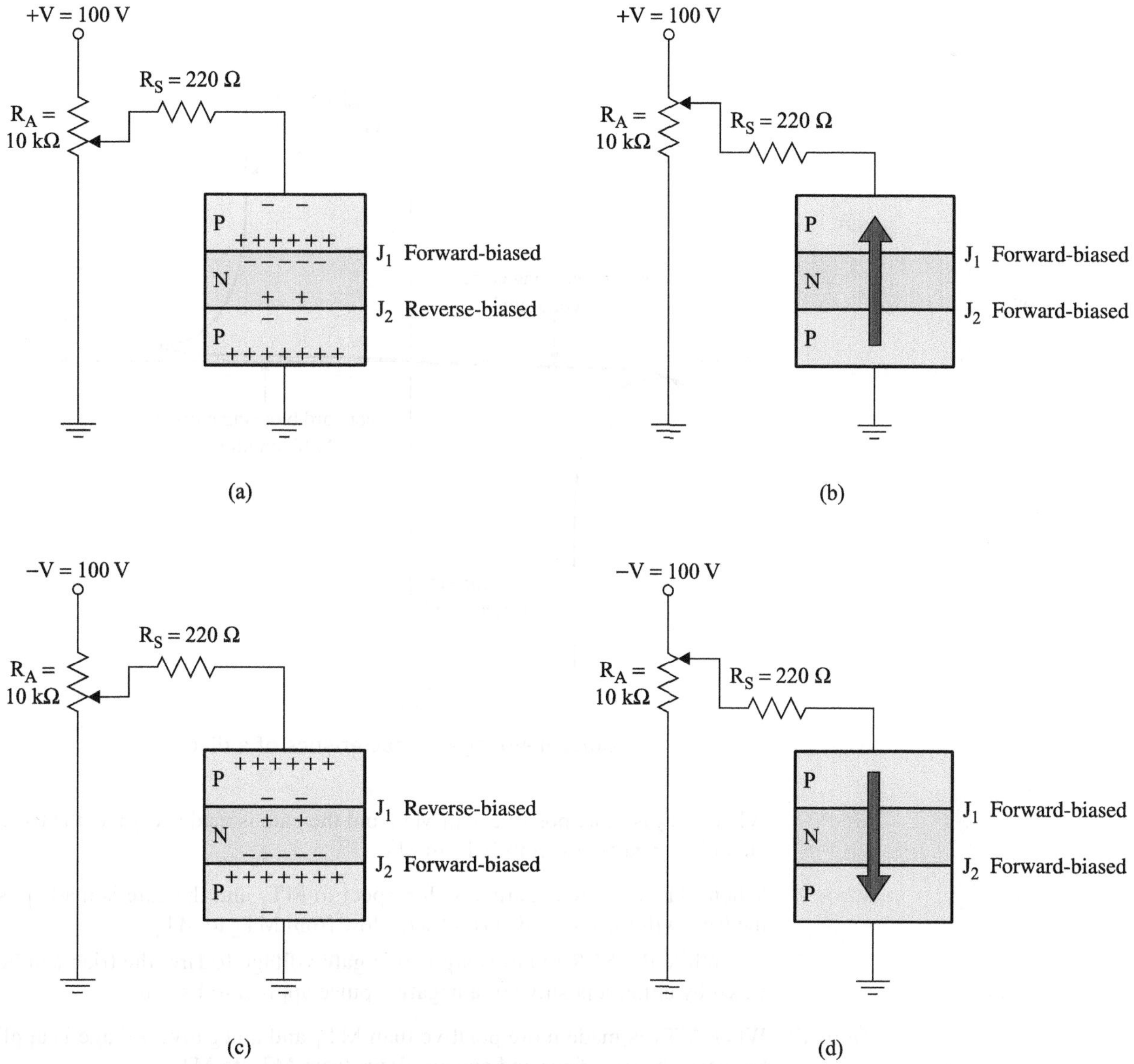

FIGURE 3-15 Diac operation: (a) Before forward breakover; (b) Current flow after forward breakover; (c) Before reverse breakover; (d) Current flow after reverse breakover

In applications where full variable control of an AC signal is required, a bidirectional thyristor device called a **triac** is used. A triac is basically a two-way SCR with one gate that passes current in both directions. A triac symbol is shown in Figure 3-17(a). Its terminals are labeled MT_1 (main terminal 1), MT_2 (main terminal 2), and G (gate). When the triac is on, the primary current passes between MT_1 and MT_2. The triac symbol is sometimes placed within a circle, and the leads may or may not be labeled.

The triac is a three-layer PN semiconductor device with added N regions, as shown in Figure 3-17(b). These added regions enable each terminal to be connected to both P- and N-type materials. This structure allows current to pass between MT_1 and MT_2 through an NPNP series of layers or a PNPN series of layers. In effect, these NPNP and PNPN devices are similar to the two parallel SCRs that are connected in opposite directions shown in Figure 3-17(c).

The triac begins conduction when the voltage across its main terminals exceeds the breakover point. The breakover voltage is essentially the same at each polarity. The triac can be turned on at a lower voltage if it is triggered by a voltage pulse applied to the gate. Once it is triggered on, the gate loses control, and the triac will continue to conduct until the current drops below the holding-current level of the device.

There are four modes of triggering for a triac, as shown in Figure 3-18.

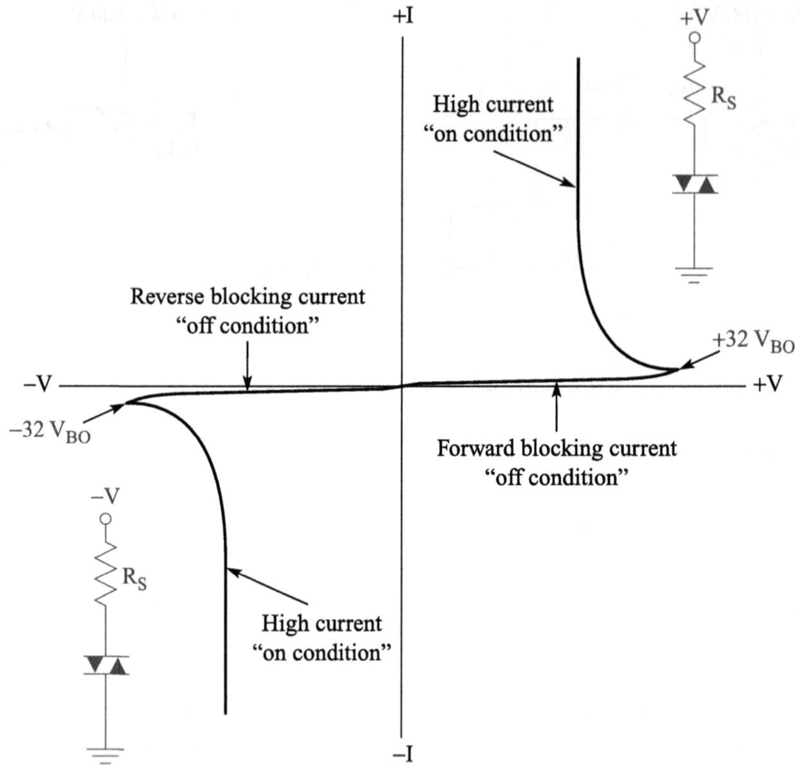

FIGURE 3-16 Current–voltage characteristics of a diac

Mode 1: When MT_2 is more positive than MT_1 and the gate is made positive, the triac turns on and current flows from MT_1 to MT_2.

Mode 2: When MT_2 is more negative with respect to MT_1 and the gate is made positive, the triac will turn on and current will flow from MT_2 to MT_1.

While the SCR requires a positive gate voltage to fire, the triac can be triggered by either a positive or a negative pulse applied to its gate.

Mode 3: When MT_2 is made more positive than MT_1 and a negative voltage is applied to the gate, the triac fires and current flows from MT_1 to MT_2.

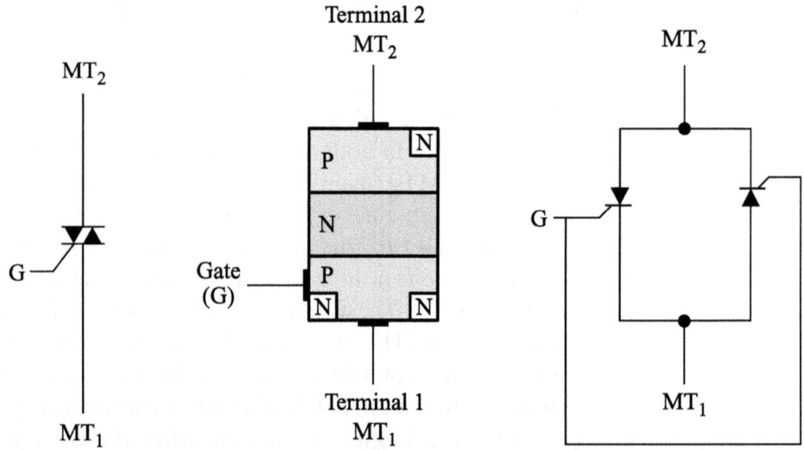

FIGURE 3-17 Triac: (a) Schematic symbol; (b) Structure; (c) Equivalent circuit

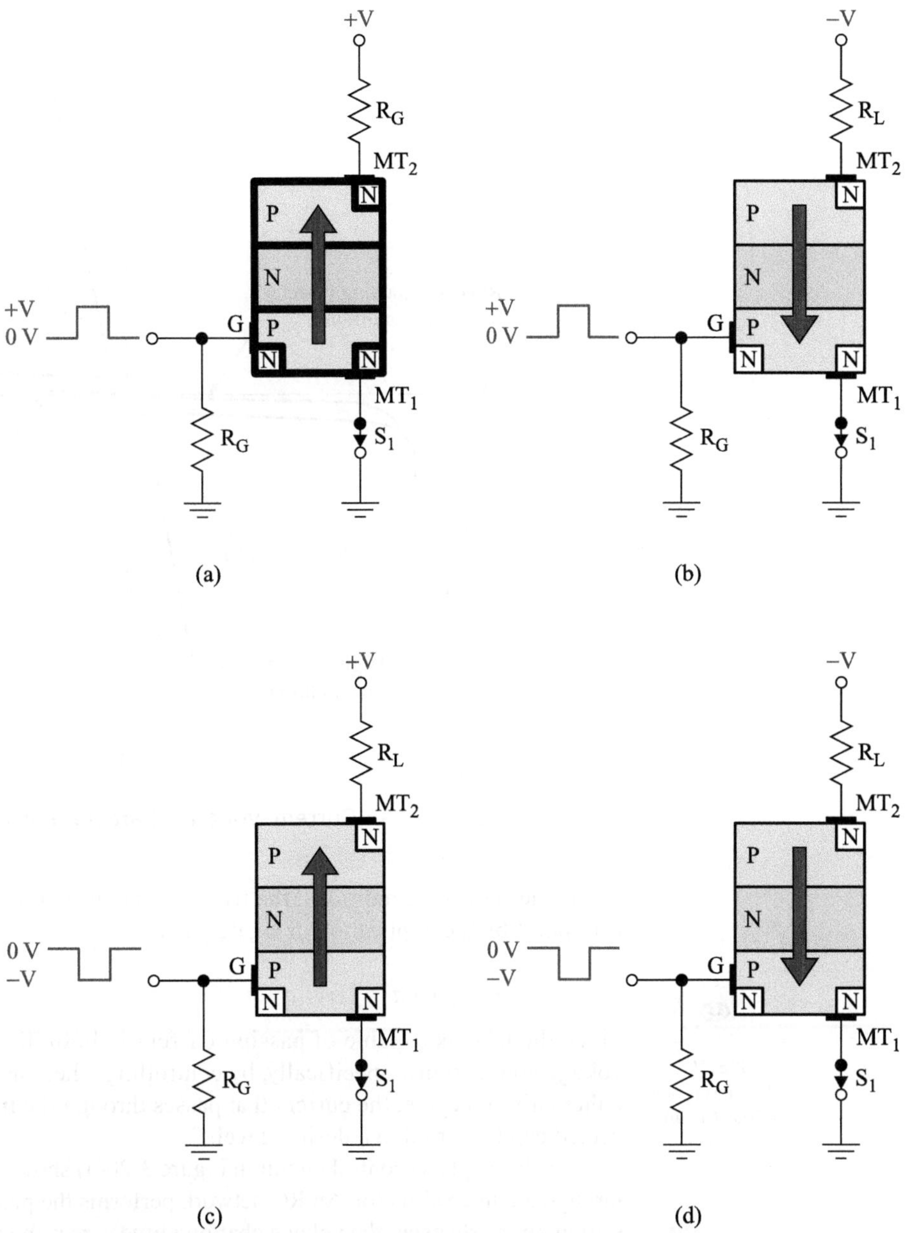

FIGURE 3-18 Four modes of triggering for a triac: (a) Mode 1; (b) Mode 2; (c) Mode 3; (d) Mode 4

Mode 4: When MT_2 is made negative in respect to MT_1 and a negative gate voltage is applied, the triac turns on and current flows from MT_2 to MT_1.

Current–Voltage Characteristics of a Triac

The graph in Figure 3-19 shows the current–voltage characteristics of a triac. They are similar to an SCR, except that conduction occurs in both directions instead of only one. As a voltage at either polarity increases, a very small blocking current flows even though the device is in the off condition. When the breakover point is exceeded, the triac turns on and its internal impedance decreases. The result is that the voltage across the main terminals drops and the current flows through it rapidly increase. The triac will remain on as long as this current does not drop below the holding-current value.

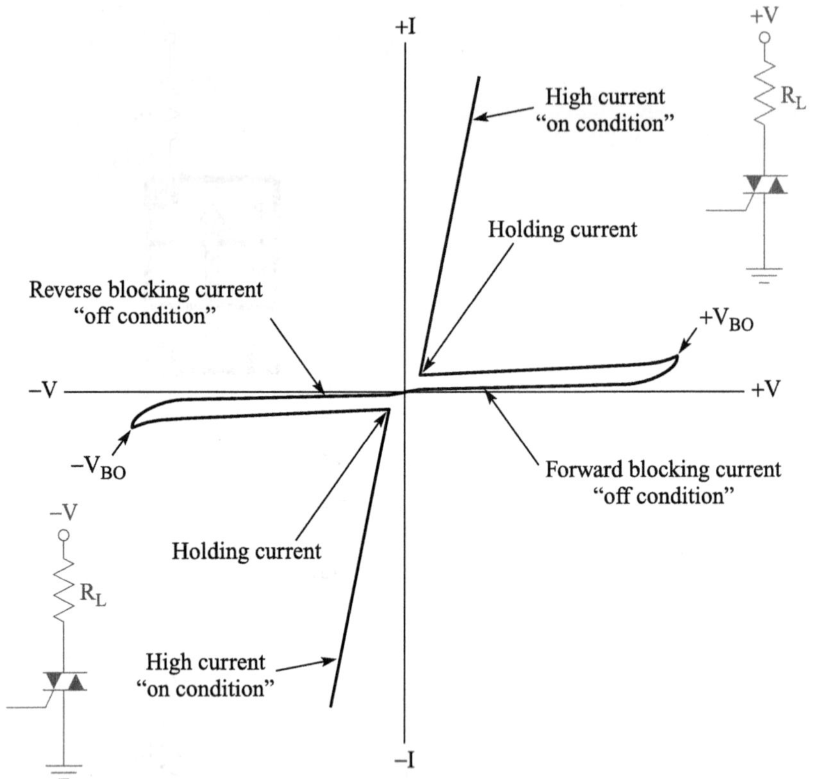

FIGURE 3-19 Current–voltage characteristics of a triac

Under normal conditions, the triac will be turned on at a voltage lower than the breakover point by the application of a gate pulse.

Gate Control of a Triac

SSE7506 Waveforms of a Triac Circuit

Since the triac is capable of passing current in both directions, it is primarily used in AC voltage applications. Specifically, by controlling when the pulse is applied to the gate during either half of a cycle, the current that passes through the triac and the load to which it is connected can be varied to a desired level.

The basic phase control circuit in Figure 3-20(a) shows how a diac and triac control current through a universal motor. An RC network performs the phase control function. By varying the setting on the rheostat, the voltage charging time across the capacitor changes. When the charge on the capacitor reaches the breakover voltage of the diac, it fires. When the diac turns on, it triggers the triac into conduction. This action occurs during both the negative and the positive alternations. The approximate circuit waveforms at strategic points in the circuit are shown in Figure 3-20(b).

The waveforms that are shown represent when the triac is turned on at 90° during the positive alternation, and at 270° in the cycle, which is during the negative alternation. At these firing times, the motor will run at half speed. If the resistance of the rheostat is increased, it will take longer for the capacitor to charge, causing both the diac and triac to fire later during each half-cycle, thereby allowing less current to flow through the motor. The motor will run at full speed when the rheostat setting is 0 ohms, causing the capacitor to charge quickly. Maximum current will flow through the motor because the diac and triac fire early during each half-cycle.

Circuits similar to this are used in light dimmer-control circuits. An incandescent light replaces the motor, and the dimmer knob is connected to the shaft of the rheostat that adjusts the wiper arm. Since the diac is commonly used to trigger a triac, manufacturers make a combined diac/triac device that is available in a single package. Figure 3-21 on page 66 shows the schematic symbol for this unit.

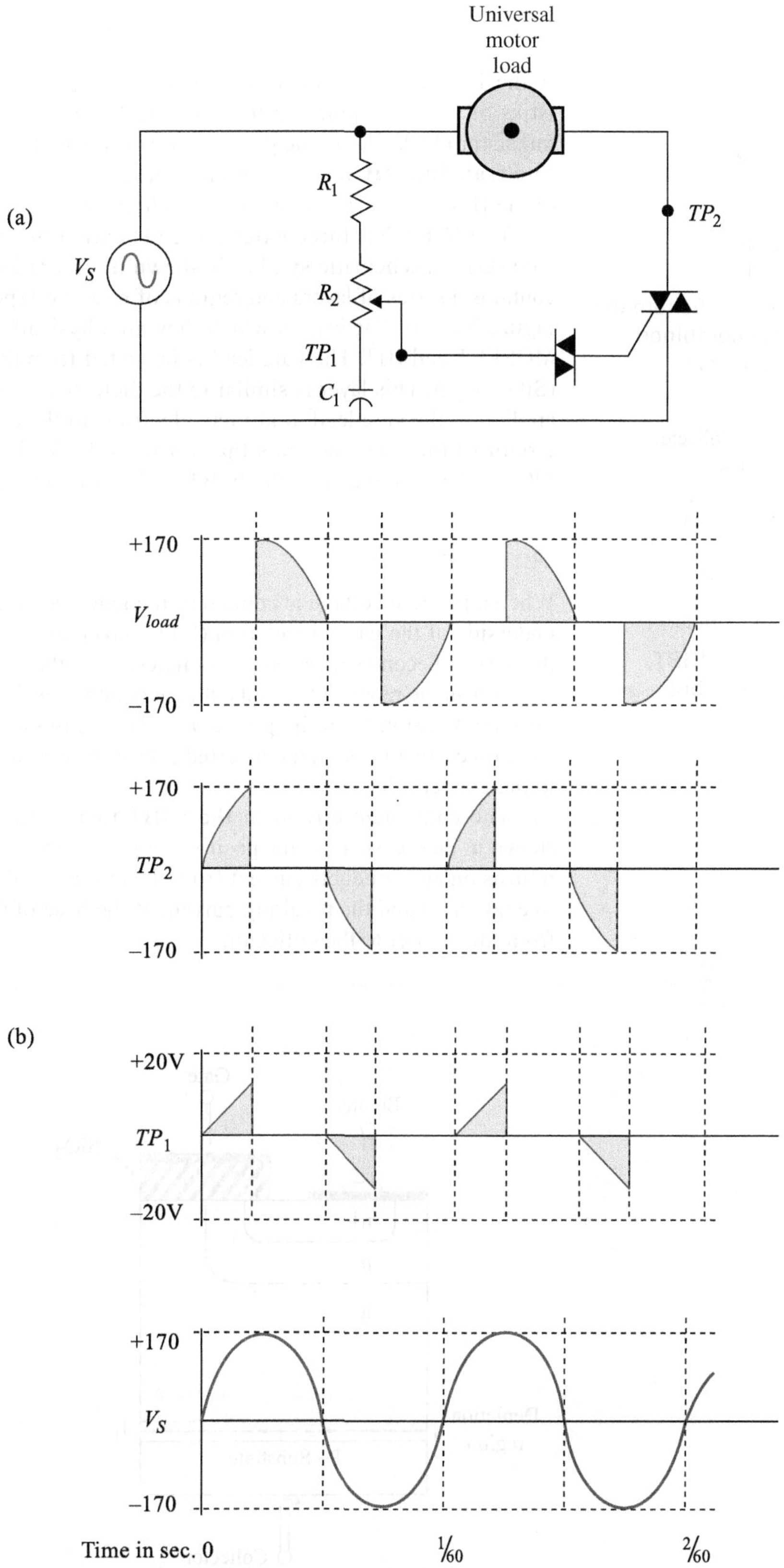

FIGURE 3-20 Diac–triac motor control by phase control

3-5 IGBTs

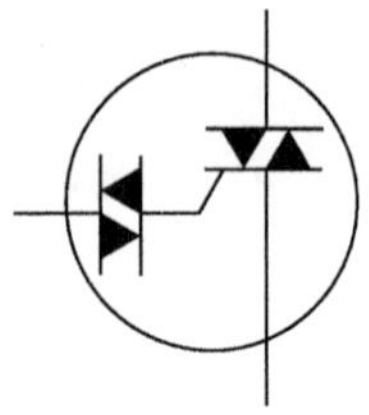

FIGURE 3-21 Schematic symbol for combined diac–triac device

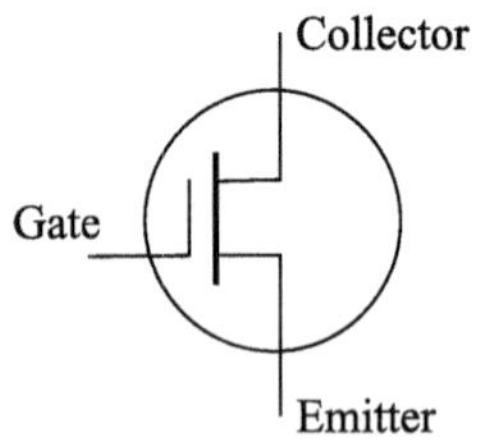

FIGURE 3-22 IGBT schematic symbol

In the 1980s, semiconductor designers developed a component that combines the characteristics of a bipolar junction transistor (BJT) and a metal-oxide semiconductor field-effect transistor (MOSFET). The device that was created is called an **insulated gate bipolar junction transistor (IGBT).** Combining these two devices provides the high-current capabilities of the BJT and the excellent fast-switching capabilities of the MOSFET.

The IGBT has three external leads called the emitter (E), the collector (C), and the gate (G). Its schematic symbol is shown in Figure 3-22. The internal structure of the IGBT contains a series of layers and regions of n- and p-type semiconductor materials, as shown in Figure 3-23(a). The way in which they are physically arranged creates a combination of the MOSFET and BJT. The gate lead is insulated from the p- and n-regions by a silicon dioxide (SiO_2) layer. This layer is similar to the dielectric of a capacitor. When a positive potential is applied to the gate lead, it attracts electrons to the opposite side of the layer. An equivalent circuit of these components that form the IGBT is shown in Figure 3-24. An n-channel MOSFET is at the gate of the IGBT, and its output is applied to the base of a PNP transistor.

Turn-On

When a positive voltage is applied to the gate lead, electrons flow from the p-substrate to the underside of the gate's insulation (SiO_2) layer, as shown in Figure 3-23(b). The result is that the p-layer becomes inverted (an n-material). If the voltage caused by the inverted p-substrate region is in the range of 0.7 volts, then junction J_1 is forward biased and some holes are injected into the n-region from the p+-region. The combined increase of hole current at J_1 and the electron current through the inverted p-layer causes an increase in conduction from the emitter to the collector leads.

The equivalent circuit of the IGBT that contains BJT and MOSFET components is shown in Figure 3-24. As the positive voltage is applied to the gate of the n-channel MOSFET, it turns on and conducts current from the emitter to the base of the PNP transistor. The negative potential and the resulting current at the base of Q_1 causes it to turn on and pass current from the emitter to the collector.

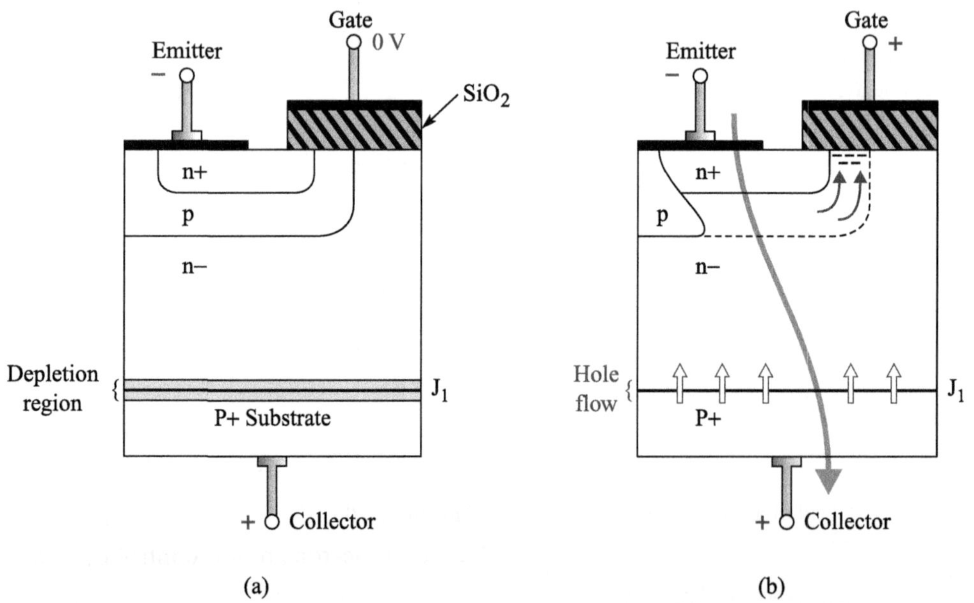

FIGURE 3-23 P and N semiconductor regions of an IGBT: (a) In the off state; (b) In the on state

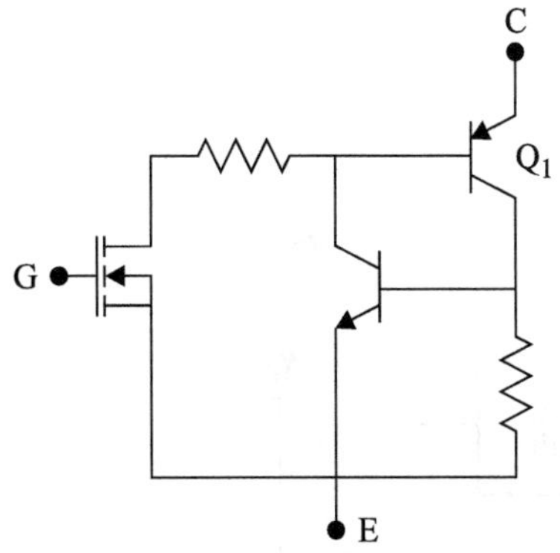

FIGURE 3-24 Equivalent circuit of the IGBT

Turn-Off

Applying a negative-bias voltage at the gate does not cause the inversion of the p-layer to form. The supply of holes into the n-region is blocked, at which time the turn-off process begins. However, the turn-off condition cannot be quickly completed due to the high concentration of minority carriers that were injected into the n-region during conduction. Eventually, the density of electrons and holes decays due to recombination. This slow switch action is called "tail off," and in some applications is considered a drawback of the IGBT.

The negative voltage at the gate affects the equivalent circuit by turning off the MOSFET. This condition creates an open at the base of Q_1 and turns the transistor off. The result is a high impedance across the emitter–collector leads. The IGBT is capable of operating at switching speeds up to 20 kHz. It is used in circuit applications up to 1000 volts.

Problems

1. Thyristors are primarily used as ____ devices.
   a. amplification c. switching
   b. power supply rectifiers d. all of the above
2. When the holding current drops below its minimum value, the thyristor will ____.
   a. be destroyed c. turn off
   b. not turn off
3. The term "trigger" refers to turning ____ a thyristor device.
   a. on b. off
4. The SCR has ____ semiconductor layers and ____ junctions.
5. The trigger time for an SCR with a capacitor at its gate ranges from ____.
   a. 0° to 45° c. 0° to 180°
   b. 0° to 90° d. 180° to 360°
6. The SCR is triggered on by a(n) ____ voltage applied to its gate.
   a. negative c. either a or b
   b. positive
7. The gate lead of an SCR is used to turn it ____.
   a. on c. both a and b
   b. off
8. A ____ voltage can be applied across an SCR.
   a. DC c. both a and b
   b. AC
9. Once the SCR is triggered on, it can be turned off in a DC circuit when a ____.
   a. positive voltage is removed from the gate
   b. negative voltage is removed from the gate
   c. holding current falls below the minimum point
   d. all of the above
10. The ____ is a bidirectional device.
    a. SCR c. diac
    b. UJT d. IGBT
11. Given the sine wave applied to an SCR circuit in Figure 3-25, draw the 360° waveform across its anode and cathode leads if it is triggered at 90°.
12. In a particular UJT, $r_{B_1} = 6\ k\Omega$ and $r_{B_2} = 3\ k\Omega$. Therefore, its intrinsic standoff ratio is about ____.
    a. 0.5 c. 2.0
    b. 0.67 d. 1.2
13. Indicate if the following values increase or decrease when negative resistance occurs in a UJT:
    V_E : B_1:
    I_E: B_2:
14. The voltage on the emitter of a UJT just before it fires is called the ____
    a. cutoff region c. valley voltage
    b. saturation region d. peak voltage

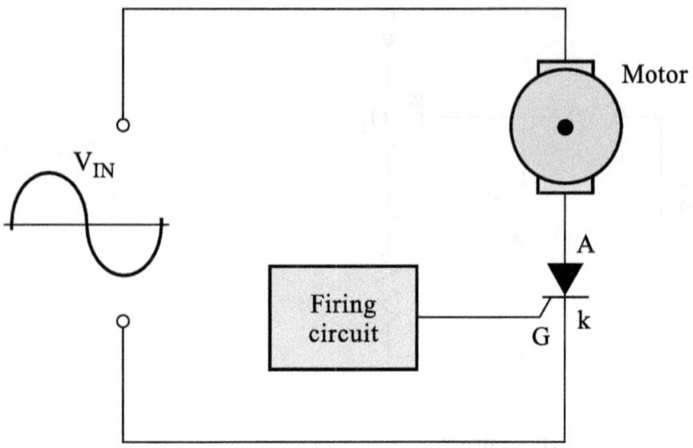

FIGURE 3-25 Problem 11

15. The frequency of a relaxation oscillator when R_E = 100 kΩ and C_E = 0.1 μfd is ____ Hz.
16. A latch-on condition will develop with a UJT relaxation oscillator if the value of resistor R_E is too ____ .
    a. small b. large
17. The ____ is primarily used to turn on a triac.
    a. diac c. SCR
    b. UJT
18. The breakover voltage of a diac is ____ in one direction as compared to in the opposite direction when an AC voltage is applied to its circuit.
    a. essentially the same b. higher
19. The correct output voltage waveform for Figure 3-26 is ____.

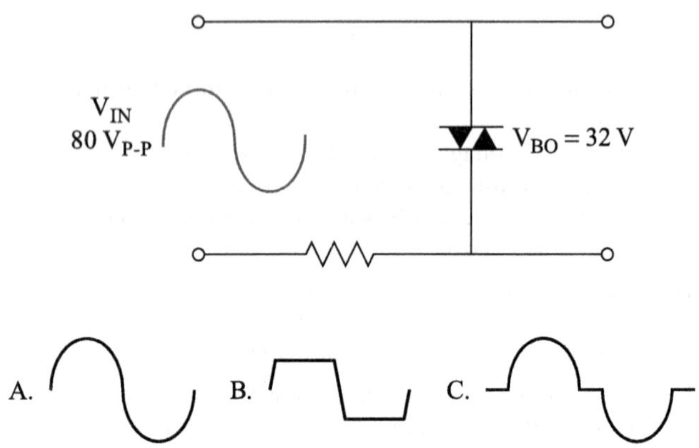

FIGURE 3-26 Problem 19

20. The voltage across the ____ drops to almost 0 volts after it fires.
    a. diac b. triac
21. A triac can be triggered into conduction when the voltage on its T_2 terminal is ____ .
    a. + and the gate is + d. – and the gate is –
    b. + and the gate is – e. all of the above
    c. – and the gate is +
22. The trigger time for a triac without a capacitor ranges from ____ .
    a. 0° to 45° and from 180° to 225°
    b. 0° to 90° and from 180° to 270°
    c. 0° to 90° and from 180° to 270°
    d. 0° to 180° and from 180° to 360°
23. The waveform in Figure 3-27 is dropped across the leads of ____ .
    a. an SCR c. a triac
    b. a diac d. an IGBT

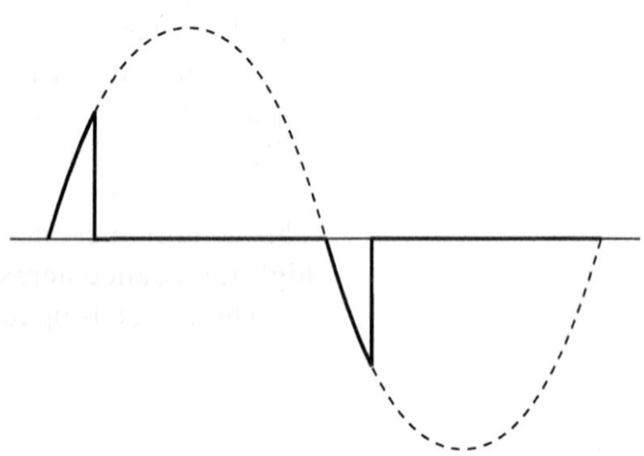

FIGURE 3-27 Problem 23

24. The ____ is capable of handling more current than the ____.
    a. triac b. SCR
25. The IGBT is turned on when its ____ voltage is positive.
    a. gate b. base
26. The ____ switching transition of an IGBT is faster than when it turns ____ .
    a. on-to-off b. off-to-on

SECTION

2

The Controller

OUTLINE

Section 1 introduced key concepts in industrial control, namely, control system classifications, interfacing, and thyristor operations. It provided an overview of the elements of an industrial control loop and described the operation of discrete components and various integrated circuits.

Section 2 consists of a single chapter, which describes the operation of the "brain" of the industrial control loop: the controller element. It addresses the operational techniques performed by controllers that use discrete components or computer software to perform their functions. The chapter in this section describes the operation of various control modes such as On-Off, PID (proportional-integral-derivative), and time proportioning.

CHAPTER

4

The Controller Operation

OBJECTIVES

At the conclusion of this chapter, you should be able to:

- List the four control modes used by the controller section.

On-Off Control	Proportional-Integral Control
Proportional Control	Proportional-Integral-Derivative Control

- Define the following terms associated with control modes of a closed-loop system:

Hysteresis	Proportional Band	Proportional Gain
Differential Gap	Stable/Unstable	Offset
PID	Steady-State Error	Deadband

- Describe the operation of each type of mode control function.
- Explain the operation of the operational amplifier circuitry that performs each of the three PID mode functions.
- List a practical application of a PID and a time-proportioning control system.
- Describe the operation of time-proportioning control.

INTRODUCTION

The controller is an element of the closed-loop system that processes information needed to perform the decision-making function. The controller can be considered the brain that enables automated systems to operate without human intervention.

The input applied to the controller is the error signal, which is proportional to the difference between the desired setpoint and the feedback signal. The controller calculates changes needed in the controlled variable to compensate for disturbances that upset the process, or changes in the setpoint. The controller responds to these changes by producing an output signal that drives the actuator to alter the controlled variable until the error signal is reduced toward zero.

The controller may be as simple as a spring-balanced mechanical lever or as complicated as a computer. It may control one process or several simultaneously. It may be analog, digital, or a combination of both.

4-1 Control Modes

Figure 4-1 illustrates the operation of the controller. The input applied to the controller is called the *error signal.* It represents the difference between the setpoint signal and the feedback signal. The input error signal is expressed by the following formula:

$$\text{Error Signal } e(t) = \text{Setpoint} - \text{Feedback Signal}$$

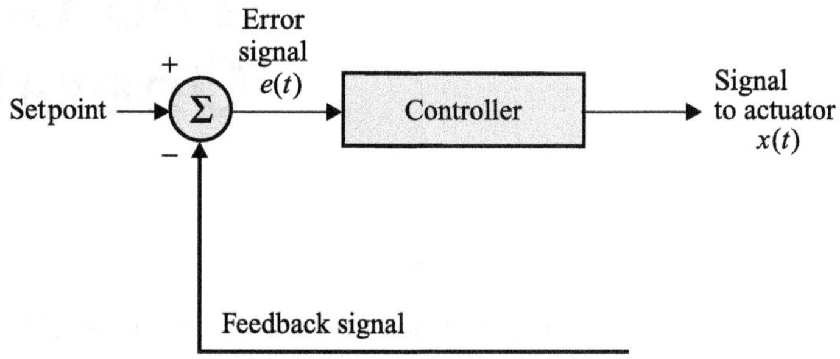

FIGURE 4-1 Controller representation

The error signal is not constant; it changes through time. Therefore, (t) is used with symbol e in the equation.

The controller output signal is expressed as $x(t)$. Because the controller output also changes with time, (t) is used with the output symbol x. The time required for the controller to respond to the error signal depends on the control mode used.

There are four control modes of operation that are commonly performed by the controller section of a closed-loop system:

1. On-Off
2. Proportional
3. Proportional-Integral
4. Proportional-Integral-Derivative

All four modes of control respond to error signals. They differ in the speed and accuracy with which they eliminate the error between the setpoint and the controlled variable.

4-2 On-Off Control

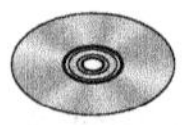

IAU2308
On-Off Control of a Feedback System

The **On-Off control** mode is the most basic type of control system. Its output has only two states, usually fully on and fully off. One state is used when the controlled variable (e.g., temperature, fluid level, voltage) is above the desired value (setpoint). The other state is used when the controlled variable is below the setpoint. The On-Off controller is also referred to as the *two-position*, or *bang-bang*, control.

The home heating system shown in Figure 4-2 illustrates this mode of control. The thermostat is the measurement device. When the room temperature (controlled variable) falls below the setting (setpoint), the thermostat closes a switch that is connected to a fuel valve in the furnace, as shown in Figure 4-2(a). With the switch closed, the valve is fully opened. The furnace turns on and begins to generate heat. When the room temperature rises above the setting, the thermostat opens the switch connected to the furnace fuel valve. An open switch closes the fuel valve to extinguish the flame. With the furnace off, the temperature in the room begins to fall. When the temperature has gone low enough, the furnace turns back on.

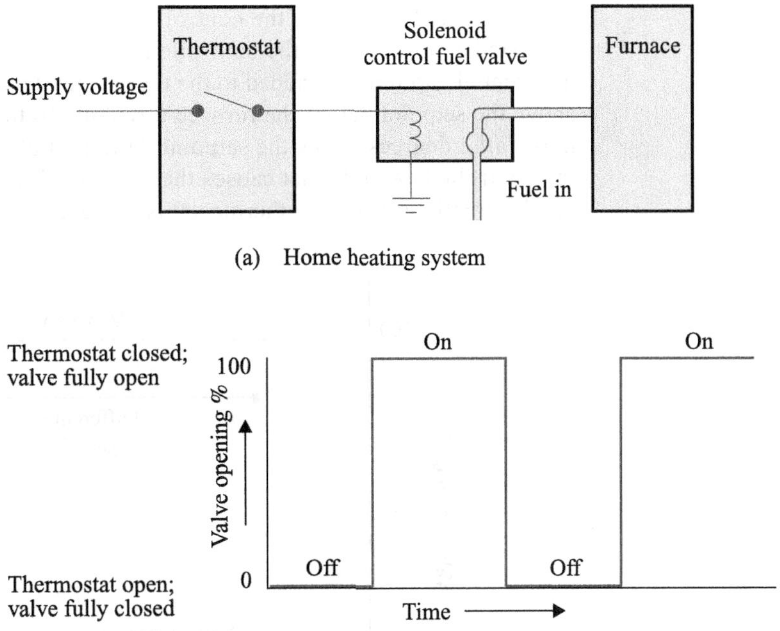

(a) Home heating system

(b) Graph illustrating operation of thermostat and furnace fuel valve

FIGURE 4-2 On-Off controller

Since the controlled variable must deviate from the setpoint to cause control action, the process response will continually cycle. The cycling occurs because of two factors:

1. Process disturbances cause the output to deviate from the setpoint.
2. The corrective action of the On-Off controller cannot adjust the output to exactly match the process demand. Instead, by being either fully on or fully off, the actuator's response is too large to return the process to the setpoint. The temperature is said to oscillate as it continually rises above and below the setpoint, as graphically shown in Figure 4-3.

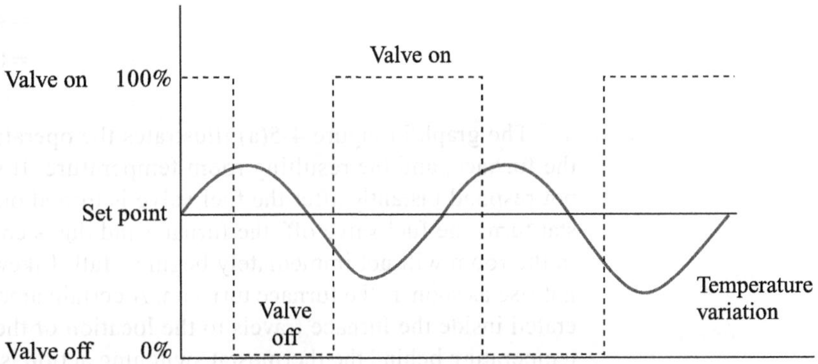

FIGURE 4-3 Graph illustrating temperature oscillation above and below setpoint as fuel valve is opened and closed

The inherent cycle condition is detrimental to most final correcting devices, such as the fuel valve, pumps, relays, etc. By turning the output on and off so frequently, the rapid oscillation wears equipment and shortens its life. This condition is often referred to as **short cycling.**

To prevent rapid cycling, the time between the oscillations can be lengthened by adding an On-Off **differential gap** to the controller function. Also referred to as the **deadband,** the differential gap forces the controlled variable to move above or below the setpoint by a specified amount before the controlled action will change again. Figure 4-4 illustrates the differential gap function added to the thermostat device. The temperature must rise 2 degrees above the setpoint before the furnace turns off. To turn the furnace back on, the temperature must fall 2 degrees below the setpoint. Differential gap is defined as the smallest change in the controlled variable that causes the value to shift from on to off, or off to on. Therefore, the differential gap for the thermostat is 4 degrees.

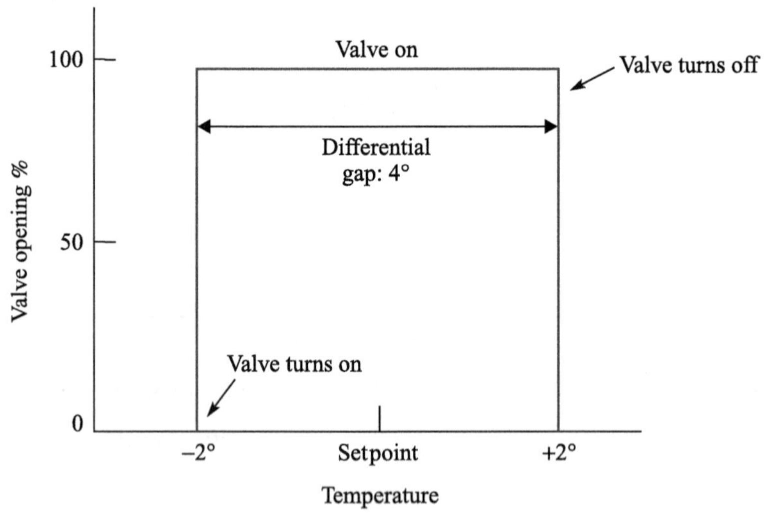

FIGURE 4-4 Differential gap of a thermostat

The differential gap is also expressed as a percentage of the full range of the controlling device. For example, the temperature range on a typical home thermostat is between 40 and 90 degrees. Therefore, the full range is 50 degrees (90 – 40 = 50). A temperature variance of 4 degrees represents 8 percent of full control range, because

$$
\begin{aligned}
\text{\% Differential Gap (Deadband)} &= \frac{\text{Differential Gap}}{\text{Total Control Range}} \\
&= 4/50 \\
&= 0.08 \text{ or } 8\%
\end{aligned}
$$

The graph in Figure 4-5(a) illustrates the operation of the thermostat, the fuel valve in the furnace, and the resulting room temperature. It shows that the room temperature does not respond instantly after the fuel valve is turned on or off. For instance, after the thermostat turns the fuel valve off, the furnace and ducts contain enough heat that the temperature in the room will not immediately begin to fall. Likewise, the temperature in the room does not rise as soon as the furnace turns on. A certain amount of time passes before the heat generated inside the furnace travels to the location of the thermostat. This lagging effect of the temperature behind the thermostat switching action is called **process lag time.** Figure 4-5(b) shows the effects of narrowing the differential gap. The narrow gap causes rapid cycling with a small deviation from setpoint. The wider differential shown in Figure 4-5(a) causes less frequent cycling, but at the expense of greater deviation from the setpoint. A compromise is made between frequency of cycling and amplitude.

Because the On-Off control mode is simple, inexpensive, and inherently reliable, it is the most common type of feedback system. On-Off controllers are widely used in applications that can tolerate the cycling and deviation from setpoint. For example, they control thermostatic furnaces, refrigerators, and solenoids to open or close flow valves that pass liquids to a tank. They would never be used to control precision devices, such as a robot.

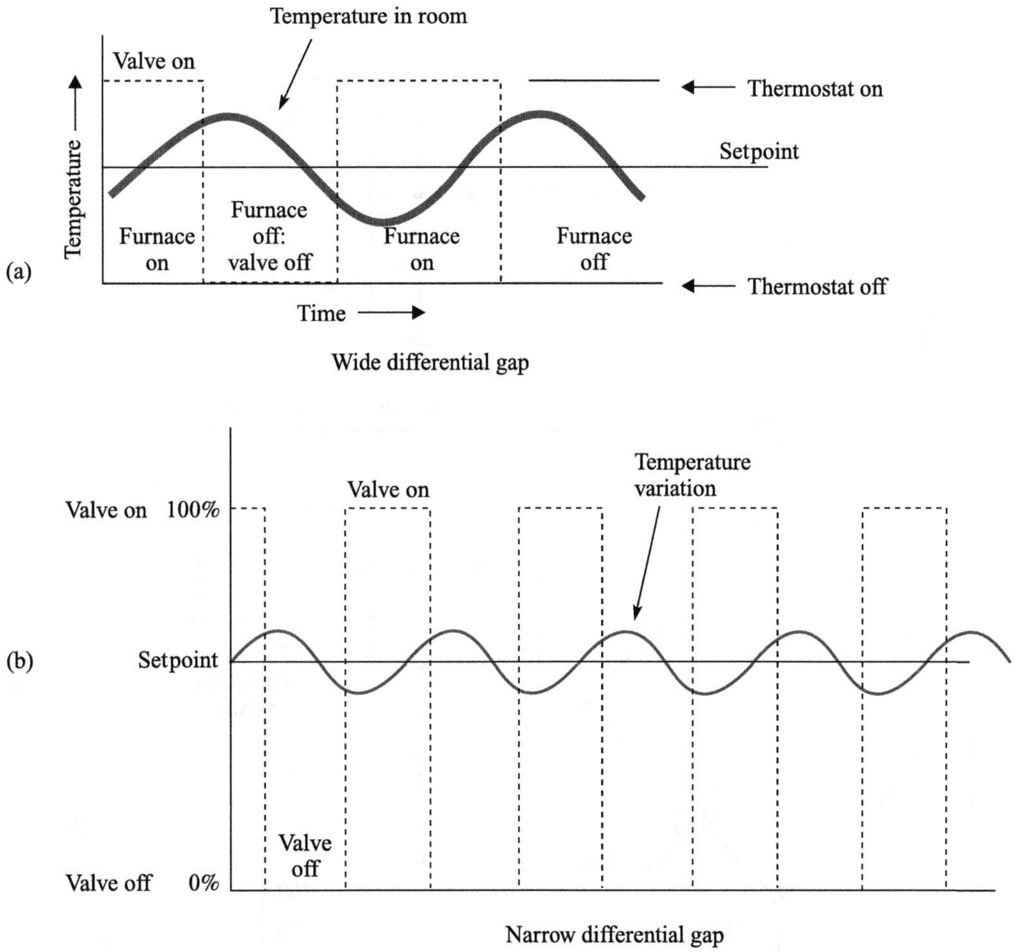

FIGURE 4-5 Comparison of systems with wide and narrow differential gaps

4-3 Proportional Control

IAU12008
Proportional Control
Amplifier

Some situations require tighter control of the process variable than On-Off control can provide. **Proportional control** provides better control because its output operates linearly anywhere between fully on and fully off. As its name implies, its output changes proportionally to the input error signal. The greater the error, the more the output responds. This action returns the controlled variable to the desired setpoint value without the rapid cycling of On-Off control.

To illustrate the operation of proportional control, the operation of a furnace is again used. To obtain this type of control, two modifications must be made to an On-Off system. First, the On-Off switch in the thermostat is replaced by a thermistor in a bridge network. The output of the bridge produces a variable voltage in response to temperature changes. Second, the solenoid-type fuel valve in the furnace must be replaced by a proportional valve. The proportional valve opens proportionally to the input voltage from the bridge. The larger the voltage, the more fuel it supplies so that a higher temperature is produced.

Figure 4-6(a) shows the proportional control furnace system. The graph in Figure 4-6(b) plots the percentage of the proportional valve opening versus room temperature. The temperature of 70 degrees is the setpoint for the system. At a 70-degree room temperature, the proportional valve is 50 percent open. At this point, the bridge is balanced and a 0-volt feedback signal is produced. The positive voltage error signal produced by the summing op amp causes the proportional valve to be half-open when the setpoint of −5 volts is not offset by the feedback signal. If the temperature drops below 70 degrees, the resistance of the thermistor increases. As a result, the voltage at the inverting input of the difference op amp becomes greater than the voltage at the noninverting input. The feedback signal produced by the difference op amp

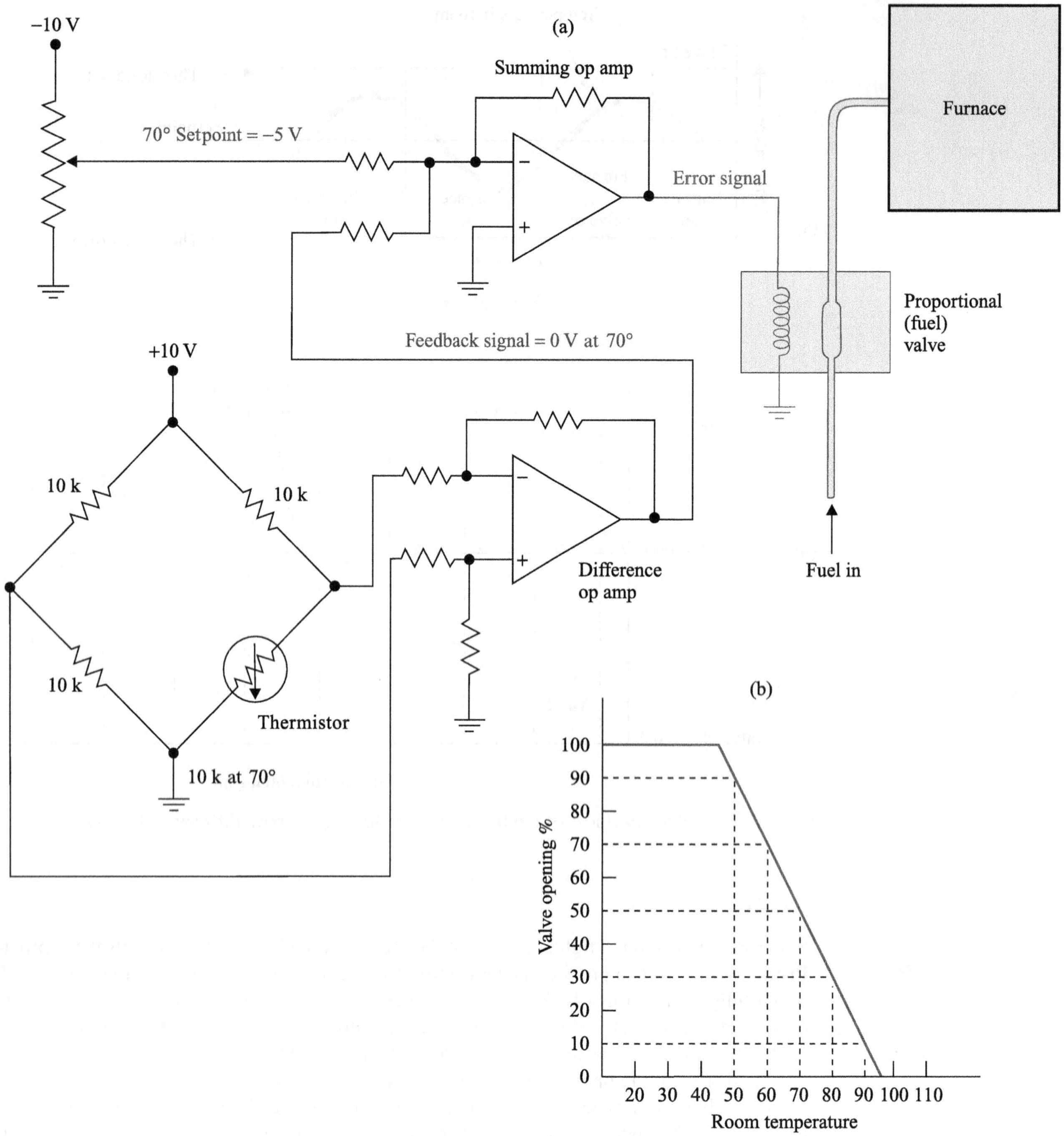

FIGURE 4-6 Proportional control heating system

becomes a negative voltage that is added to the negative setpoint voltage at the summing op amp. Therefore, as the error signal produced by the summing op amp goes more positive, the proportional valve opens more than 50 percent. For example, if the room temperature suddenly drops to 60 degrees, the error signal goes higher and causes the valve to open from 50 percent to 70 percent. The room temperature rises back to its setpoint. If the temperature rises above 70 degrees, the thermistor resistance decreases, causing the voltage at the inverting input to become less than the noninverting input at the difference op amp. Therefore, a positive voltage feedback signal is produced that cancels the negative setpoint voltage at the summing op amp. The canceling effect produces a less positive error signal, causing the proportional valve to be less than 50 percent open. As the valve closes to less than 50 percent, the room temperature decreases back toward the setpoint.

Controller Amplification

The controller has the capability of amplifying the amount at which its output changes in proportion to the change applied to its input. Controllers typically have a way in which the magnitude of the amplification can be adjusted. Amplification by a proportional controller is referred to as either **proportional gain** or **proportional band (PB).** The example of the furnace can be used to describe the difference between them.

The thermistor is the sensor, the summing op amp functions as both the comparator and the controller, and the fuel valve is the actuator. The graph in Figure 4-6(b) shows that the full operating temperature range over which the valve can be controlled is 50 degrees (95 – 45 = 50). The desired setpoint temperature of 70 degrees is exactly in the middle of the span. If the temperature drops to 45 degrees, the proportional valve fully opens. When the temperature rises to 95 degrees, the valve becomes fully closed. Within the 50-degree temperature span, the valve response is proportional to the temperature change.

Proportional Gain

Gain is the ratio of change in output to change in input, as described mathematically by the following formula:

$$\text{Gain} = \frac{\text{Percentage of Output Change}}{\text{Percentage of Input Change}}$$

In the furnace example, the input to the controller is a feedback signal that represents the temperature. The output of the controller is applied to the fuel valve, which controls the flow of gas to the furnace. Whenever the temperature changes by 1 degree, or 2 percent of the span, the valve opening varies by 2 percent of its span. According to the formula, the gain is 1.

$$\text{Gain} = \frac{2\%\ \text{Output Change}}{2\%\ \text{Input Change}} = 1$$

By increasing the gain of the controller to 2, a temperature change of 1 degree (2 percent of the span) will cause the valve to vary by 4 percent of its span. The result is that the controlled variable is restored to a desired value more quickly.

Proportional Band

Amplification is also expressed as *proportional band (PB).* PB is defined as the percentage change in the controlled variable that causes the final control element to go through 100 percent of its range. The PB can be determined mathematically by using the following formula:

$$\text{PB} = \frac{\text{Controlled Variable \% Change}}{\text{Final Control Element \% Change}} \times 100$$

The width of the PB setting on a controller determines how much controlled variable change is required to cause a given amount of movement by the final control element. For example, to cause a final control element to move through 100 percent of its range, a controller with a PB setting of 100 requires that the controlled variable change twice as much as it does in one having a PB setting of 50.

In the furnace example, assume that the PB setting, which causes the operation shown on the graph in Figure 4-6(b), is 100. Whenever the temperature (controlled variable) changes by 1 degree, or 2 percent of the span, the valve opening varies the final control element by 2 percent of its span. By reducing the PB setting to 50, a temperature change of 1 degree (2 percent of the span) will cause the valve to vary by 4 percent of its span.

In a system that has a narrow PB, the response to a disturbance is rapid. The temperature is adjusted to the setpoint quickly. The response to a system with a wider PB will take a longer time.

It would appear that the system with the narrow PB is better because the setpoint temperature would be restored more quickly. However, the characteristic of a narrow PB is that a system has a tendency to oscillate. When the system responds quickly, it tends to overshoot the setpoint. The system tries to correct itself by shifting the valve in the opposite direction. However, the system overshoots again in the opposite direction. The oscillations normally die out, at which time the system becomes **stable.** If the PB is too small, oscillations will not stop. A PB of 0 percent will cause the system to operate almost the same as the On-Off control. When the system continues to oscillate, it is **unstable.**

The size of the PB is simply the inverse of the proportional gain. The following formulas show how to convert between gain and PB values:

Note: PB is in percent.

$$PB = \frac{1}{Gain} \times 100 \quad \text{and} \quad Gain = \frac{1}{PB} \times 100$$

EXAMPLE 4-1

Calculate the gain of the process if the PB setting is at 25 percent.

Solution

$$\begin{aligned} Gain &= \frac{1}{PB} \times 100 \\ &= \frac{1}{25} \times 100 \\ &= 4 \end{aligned}$$

EXAMPLE 4-2

Calculate the PB setting if the gain of the process is 8.

Solution

$$\begin{aligned} PB &= \frac{1}{Gain} \times 100 \\ &= \frac{1}{8} \times 100 \\ &= .125 \times 100 \\ &= 12.5\% \end{aligned}$$

Steady-State Error

The proportional controller is tuned so that the setpoint causes the proportional valve to open 50 percent with a given load. The 50-percent figure is desirable because the controller has equal amounts of corrective action from the setpoint to the maximum and minimum temperature settings. When the temperature produced by the furnace is at the 70-degree setpoint, the voltage supplied to the proportional valve is 50 percent.

If the load changes, the 50-percent valve position can no longer maintain the same temperature. Figure 4-7 illustrates what action then occurs. Suppose a disturbance causes the temperature to drop. A more positive error signal voltage is produced. The condition causes

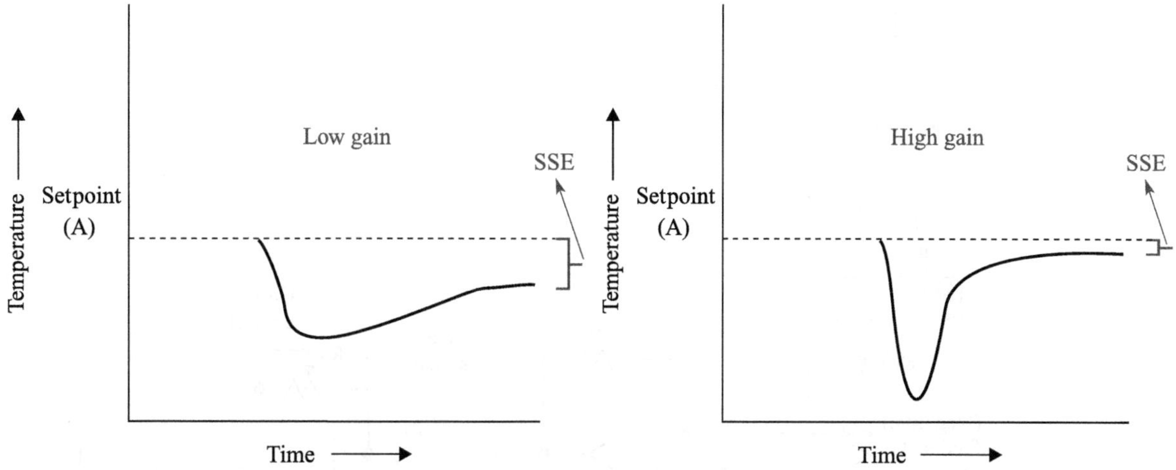

FIGURE 4-7 Relationship between gain and steady-state error

the proportional valve to open more than 50 percent. As the temperature rises, the thermistor resistance decreases and the proportional valve starts to close. If the disturbance continues for a long time, the proportional valve cannot return to the 50-percent open position. Instead, it must remain open more than 50 percent to offset the disturbance. For the valve to remain open above 50 percent, the error signal must be slightly more positive. Thus, the actual measured temperature could never climb to the 70-degree setpoint. Instead, it may stop at approximately 69.9 degrees in order to maintain the error voltage necessary to keep the valve open slightly more than 50 percent. The difference between the setpoint and the measured value is called **steady-state error,** or **offset.**

The example of steady-state error in the heating system is similar to that found in many process control applications. Some process control systems allow some degree of steady-state errors. In other systems—especially motion control applications—proportional control does not provide the necessary level of control. In a position type of motion control application, steady-state error cannot exist because precision is required.

Figure 4-8 shows the schematic diagram for a proportional position control robotic system. The output of the system is connected to the arm of a robot. The arm is attached mechanically to the wiper of a potentiometer. As the robot's arm moves, the output voltage of the potentiometer at the wiper varies. The potentiometer is the feedback device that supplies the negative feedback signal in the system. The voltage produced indicates the position of the arm and is applied to the inverting input of an op amp. The command setpoint signal is supplied by a computer. Since a computer's output signal is digital, a D/A converter (DAC) is needed to change the value to an equivalent analog voltage. The output of the D/A converter is connected to the noninverting input of the op amp. The op amp used is a difference type that functions as the comparator. Its function is to compare the command (setpoint) signal with the feedback signal and produce an appropriate error signal.

The output of the difference op amp is connected to an inverting op amp which amplifies the error signal. This is called a *proportional op amp* because its gain is proportional to the ratio of the resistor values for R_{IN} and R_f. Since the output power of a standard op amp is seldom high enough to drive a motor, the error signal is further amplified by a power amp. The output of the inverting power amp is connected to the motor that drives the robot's arm.

When the computer digital output is zero, the voltage of the potentiometer will be zero volts, and the robot arm will be in the lowest position. Suppose the computer supplies a new position command to move the arm upward. The computer data and the resulting voltage of the D/A converter are shown in Figure 4-9. The computer outputs a series of numbers that increment until a value is reached that represents the desired position. The analog output voltage of the D/A converter ramps upward in small steps in a positive direction. The voltage change stops when the computer stops incrementing. This signal is compared to the feedback signal from the potentiometer by the difference op amp. Since the feedback signal lags behind the setpoint signal (because it does not respond immediately), a positive error signal voltage is produced by the

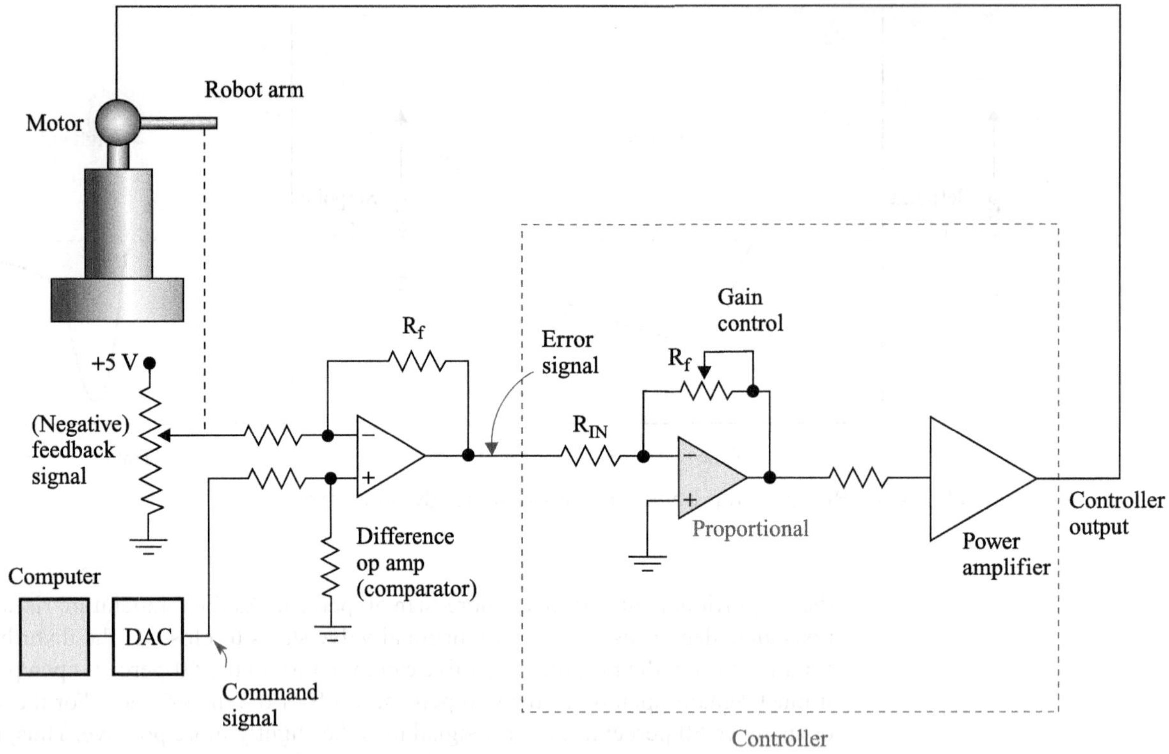

FIGURE 4-8 Proportional mode control system

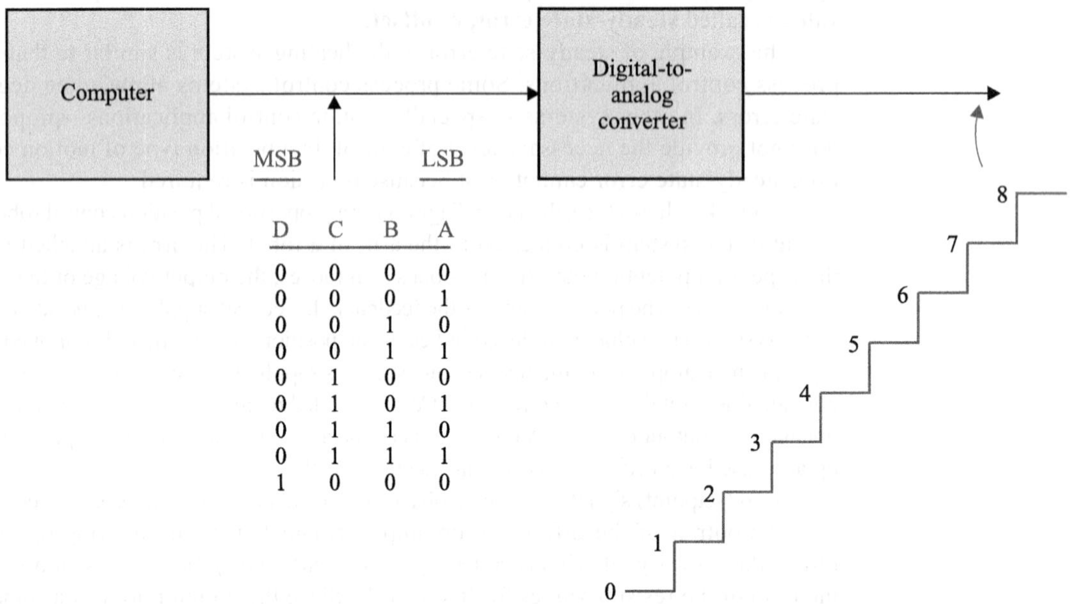

FIGURE 4-9 Computer command output signal converted to a proportional analog voltage

difference op amp. The error signal is amplified by the proportional op amp and is also inverted to a negative voltage. The output of the proportional op amp is further amplified by the power amp and is also inverted to a positive voltage. With a positive voltage applied to the motor, the arm moves upward. The arm stops when it reaches the command position. At that position, the voltage of the potentiometer will equal the voltage of the D/A converter. This condition causes the comparator output to go to 0 volts, which stops the motor from turning the arm.

A closed-loop proportional motion control system is unlikely to have precise accuracy. The arm might never reach the desired position because the closer it approaches the location,

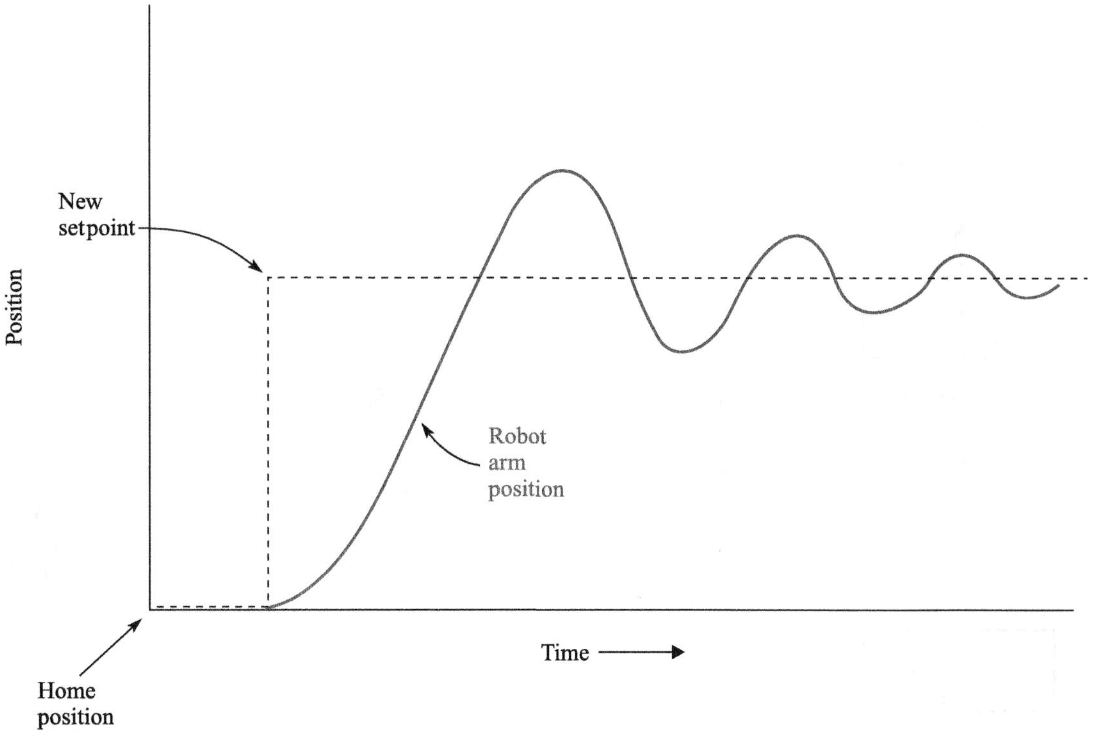

FIGURE 4-10 Instability of a proportional mode control system when gain is too high

the closer the voltage on the wiper comes to the command signal voltage. Therefore, the output voltage applied to the difference op amp becomes very small and causes the motor to slow down. Eventually, the mechanical friction of the robot arm cannot be overcome by the small amount of current that flows through the motor from the power amp. It falls short of its desired position, and a residual condition of steady-state error exists.

To reduce offset, the PB can be made smaller by increasing the gain of the proportional op amp. This adjustment will also speed response to the command signal. However, the PB can be narrowed only so far before *instability* occurs. Instability exists when the device being positioned oscillates because of overshooting. The system will try to correct the overshoot error by reversing the direction of the arm. The arm oscillates above and below the position before it dampens out and stops, as illustrated by the graph in Figure 4-10.

The friction of the load is not the only cause of steady-state error. Offset depends on three factors:

1. Load or demand on the process
2. The low gain or wide PB of the controller
3. The setpoint at which the controller is set

Changes in any of these three factors can result in some offset. To overcome offset, the control mode known as **integral control** is used.

4-4 Proportional-Integral Control

IAU10708
Integral Reset Setting

The **integral** (or *reset*) mode of control is designed to eliminate the offset inherent in proportional mode control. It develops a control signal that depends on the absolute value of the offset. The integral mode does not function by itself. It is used along with the proportional control mode in the controller section of a closed-loop system.

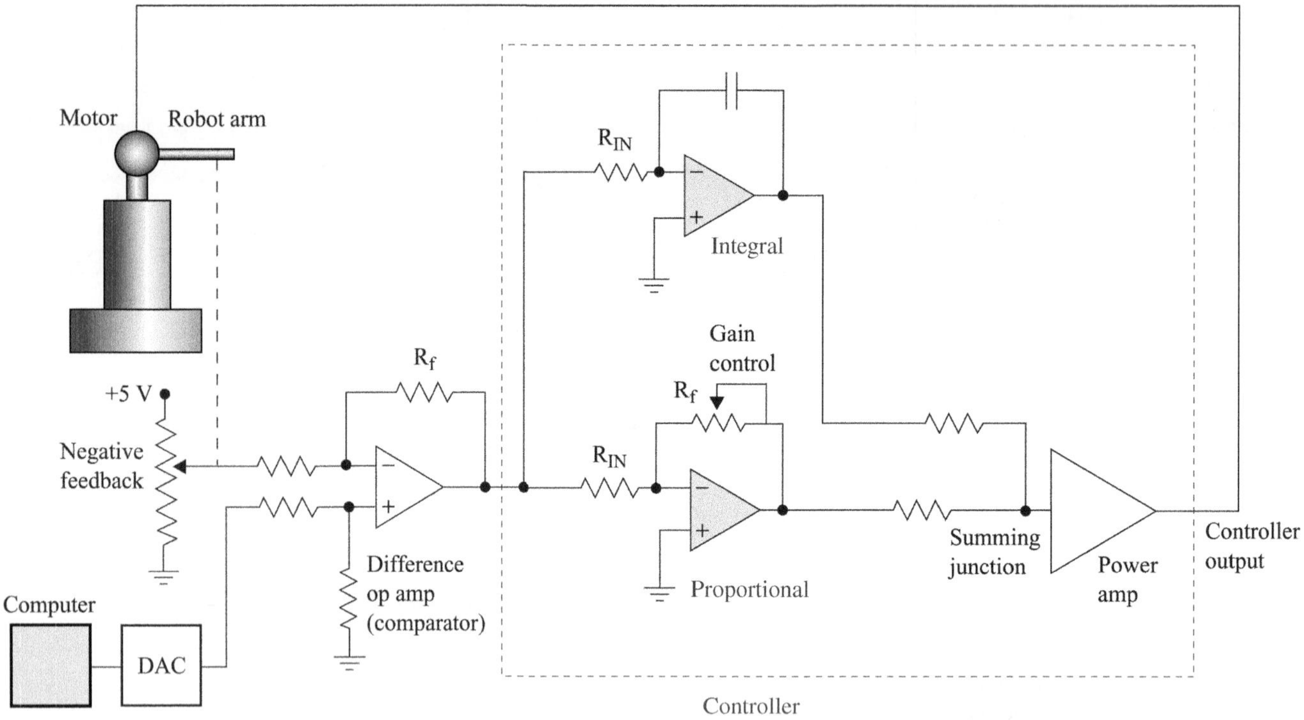

FIGURE 4-11 Proportional-integral mode control system

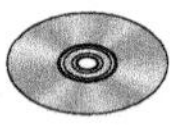

IAU12108
Integral Control Mode Application

When an error signal first appears, the controller is tuned so that the proportional control signal returns the process to the desired control point. This proportional control signal is immediate and fast-acting. If a deviation between the setpoint and the controlled variable is present after the operation of the proportional control mode is completed, an additional corrective signal is required, which is supplied by the integral control mode function. A small corrective action is developed slowly to reduce the deviation to zero only after it is certain that there is a definite steady-state error.

The operation of the integral mode of control is illustrated in the robotic closed-loop system shown in Figure 4-11. It shows the same circuitry as the proportional-only controller, with an additional amplifier that performs the integral action. This second op amp is called an *integrator.*

The integrator resembles an inverting op amp. The difference is that the feedback resistor is replaced by a capacitor. At the first instant a DC voltage is applied to its input, the capacitor operates like a short circuit. Recall that the gain of an inverting op amp is dependent on the ratio of the feedback resistance (R_f) and the input resistance (R_{IN}): Gain = R_f/R_{IN}. Therefore, since the capacitor initially provides low impedance in the feedback loop, the gain of the integrator is very low. The output voltage is also low. However, as the capacitor begins to charge, the current charging the capacitor decreases. Its impedance increases until it is fully charged, at which time it acts like an open switch. The result is that the R_F/R_{IN} ratio increases, the gain of the op amp increases, and the output voltage reaches saturation. The magnitude of the integrator output is proportional to the input voltage and the length of time the voltage is applied.

The operation of an integrator is further illustrated in Figure 4-12. This shows a graph that compares the input voltage with the output voltage. At T_1, a positive DC voltage is applied to the inverting input of the integrator. Its inverted output voltage increases in a negative direction until saturation is reached at T_2.

Suppose the computer sends out a command signal for the robot to move. The proportional function of the controller immediately responds to the setpoint change and drives the motor. The robot arm moves in the direction commanded by the computer, but stops just short of the desired position. The proportional mode has completed its response to the command setpoint change. Since the arm is out of position and does not achieve the desired

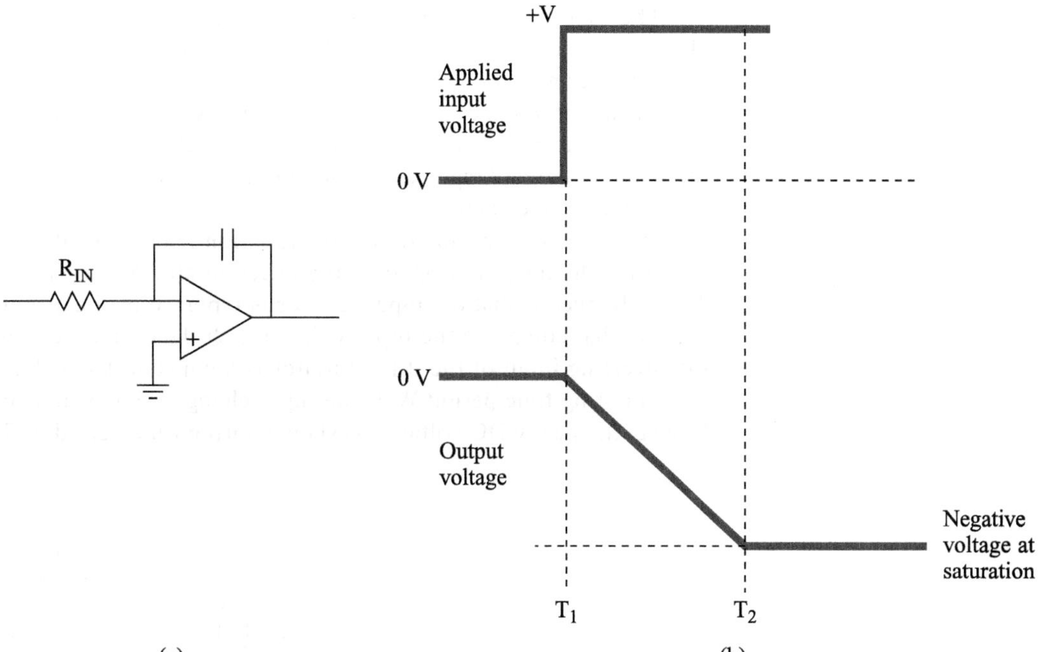

FIGURE 4-12 Operation of an integrator op amp: (a) Schematic diagram; (b) Waveform diagram

location, the setpoint voltage and the feedback voltage are not the same. The result is that a steady-state error is present at the difference op amp output. This voltage is not sufficient to drive the robot arm the remaining distance. At this point, the integral control mode takes over.

The steady-state error is also fed to the integrator op amp. The longer the error exists, the greater the output voltage of the integrator op amp becomes. The increasing integrator error signal is summed with the small amplified offset voltage of the proportional amp output. In time, the power amp receives enough input energy to turn the motor shaft until the robot arm attains the desired position. The steady-state error is then eliminated. Since the integral action continues to reset the amplifier gain until the process variable equals the setpoint value, it is also referred to as the *reset* mode of control. Although the two names are synonymous, *reset* is the older term.

The proportional-integral control mode is used in applications where load disturbances occur frequently and setpoint changes are infrequent. It is also used when load changes are slow, to allow enough time to elapse before it is necessary for the integral function to aid the proportional operation.

4-5 Proportional-Integral-Derivative Control

IAU9808
Derivative control Mode Analogy

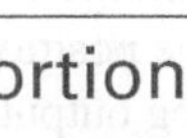

ELE5308
PID Control

It is usually desirable to move the robotic arm quickly from one position to another. However, rapid movements are not possible using the proportional mode without excessive gain. If the gain setting of a proportional amp is too high, an instability condition develops where overshoot and subsequent oscillations occur. To reduce the overshoot and bring the controlled variable to the setpoint rapidly, a control mode called **derivative** or *rate* control is used.

The term *derivative* refers to the rate of change. A derivative controller produces an output that is proportional to the rate that the error signal changes. If the error signal is changing very rapidly, the derivative output is large. When the error signal is changing slowly, the derivative output is small. If the error signal is stable, the derivative output is zero.

The function of the derivative controller is to provide a proportional correction to a changing error signal. For example, if the error signal gap increases, the derivative mode control gives a boost to the system to stop the error from increasing any further. The more rapidly the error signal increases, the larger the boost. When the error signal gap decreases, the derivative mode control provides braking action. The more rapidly the error gap closes, the stronger the braking action. The braking action reduces overshoot and dampens out any oscillations of the controlled variable.

In Figure 4-13(a), the derivative function is performed by a differentiator op amp circuit. Like the integrator op amp, the differentiator op amp resembles an inverting op amp. The difference is that the input resistor is replaced by a capacitor. Figure 4-13(b) provides a graph that compares the input voltage with the output voltage. If the voltage applied to the inverting input of the differentiator is a constant DC voltage, the output is 0 volts, as shown during time period W. If the input changes slowly at a constant rate, the output will be a small, steady DC voltage, as shown during time period X. Time period Y shows that if

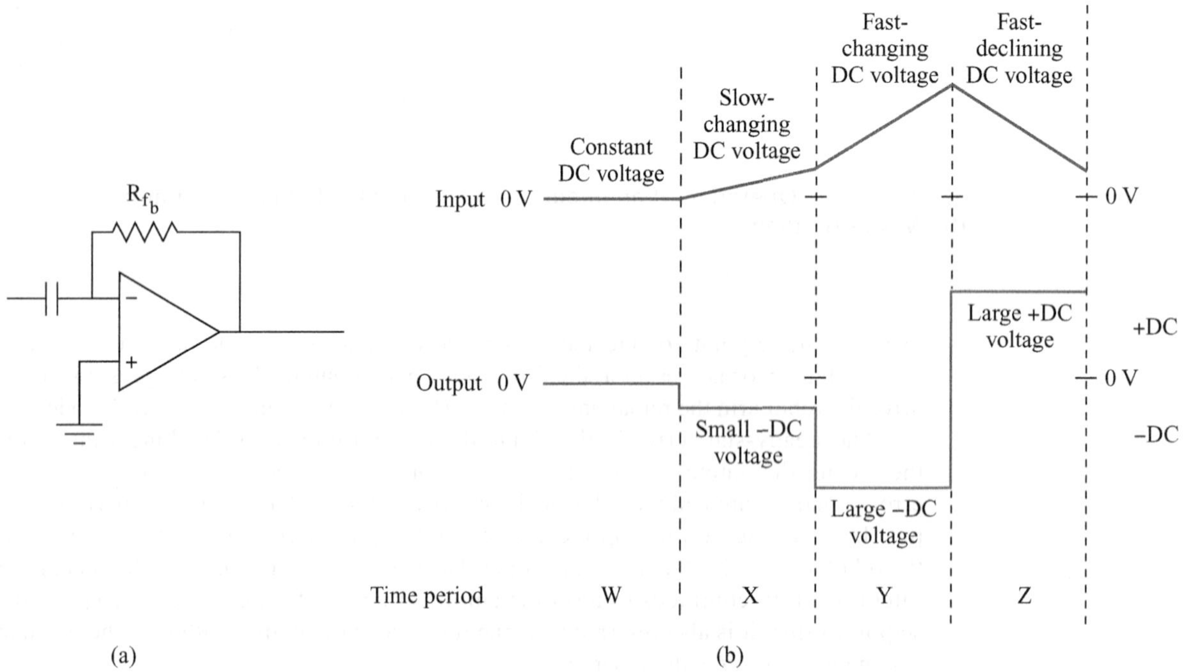

FIGURE 4-13 Operation of a differentiator op amp

the input voltage rises at a rate that is constant, but more rapidly, the differentiator output amplitude will be higher. Time period Z shows that if the input voltage decreases at the same rate, the differentiator will produce an equal output voltage of opposite polarity.

Figure 4-14 shows a robotic control system that contains proportional, integral, and derivative (PID) circuits. Figure 4-15 graphically shows the operation of the PID mode control system. Suppose an application requires that the robot arm move quickly from one position to another. The computer outputs a rapidly incrementing series of numbers. The analog output waveform of the D/A converter shown between points A and D rises quickly at a steady rate.

The analog signal is fed to the noninverting input of the difference op amp. Since the arm of the robot initially does not move, the output of the difference amp starts to rise and develop an error signal. As the error is fed to the input of the proportional amp, it is further amplified by the summing power amp. This action causes the motor to drive the robot arm toward the desired position. As it does, a voltage from the potentiometer, which is the feedback signal, begins to rise, as shown soon after time period A begins. However, the amplitude of the error signal continues to grow in the positive direction, as shown between points A and B. This happens because the stationary inertia of the robot arm has to be overcome, causing it to move slowly at the start.

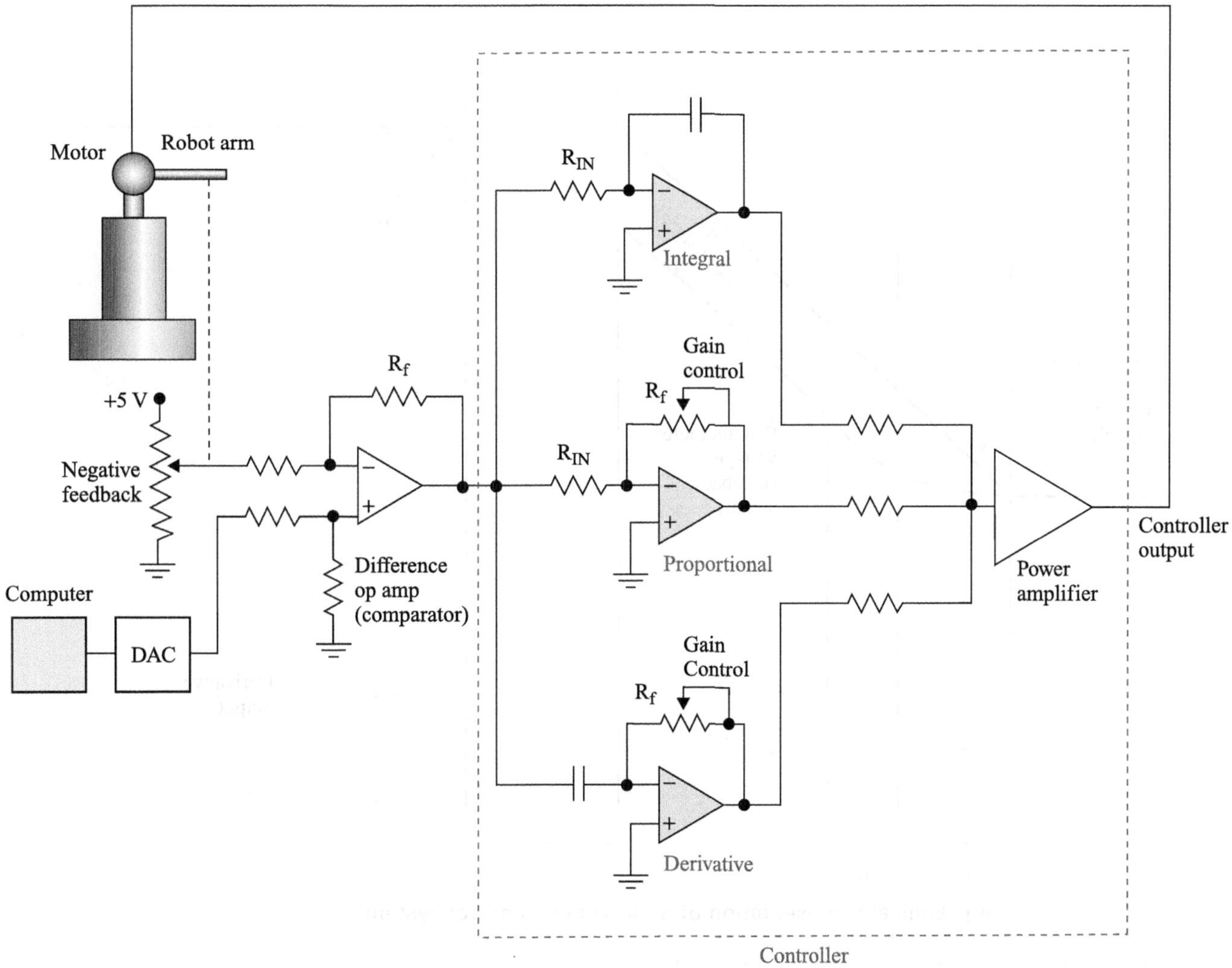

FIGURE 4-14 PID mode control system

Therefore, the measured variable from the feedback potentiometer does not change as quickly as the command signal from the computer.

The output of the difference op amp is also feeding the derivative amp. As the error signal voltage increases its amplitude, as shown between points A and B, a negative voltage is created by the derivative network. The derivative voltage is added to the proportional voltage by the summing power amp. The combined voltages cause the power amp output to increase, which makes the robot arm move faster. Eventually, it moves rapidly enough that the measured variable is changing as quickly as the command setpoint signal, as shown at time period C. This boost by the derivative function prevents the error signal from increasing any further.

Between points C and D, the error signal does not change. The output of the derivative amp goes to zero and the proportional function operates alone.

When the command signal from the computer reaches the value that represents the desired position, it stops changing. The output voltage of the D/A converter also stops increasing, as shown at time period D. Since the arm has not yet reached the desired position, the setpoint and measured variable are unequal. Therefore, the difference amp continues to produce a voltage, causing the arm to continue moving. Because the error signal decreases in amplitude, as shown between time periods D and E, a positive voltage is produced by the differentiator op amp. This voltage is subtracted from the proportional output by the summing power amp. Since the combined voltages cancel, the power amp output decreases. The result is that the motor causes the arm to slow down enough that it does not overshoot.

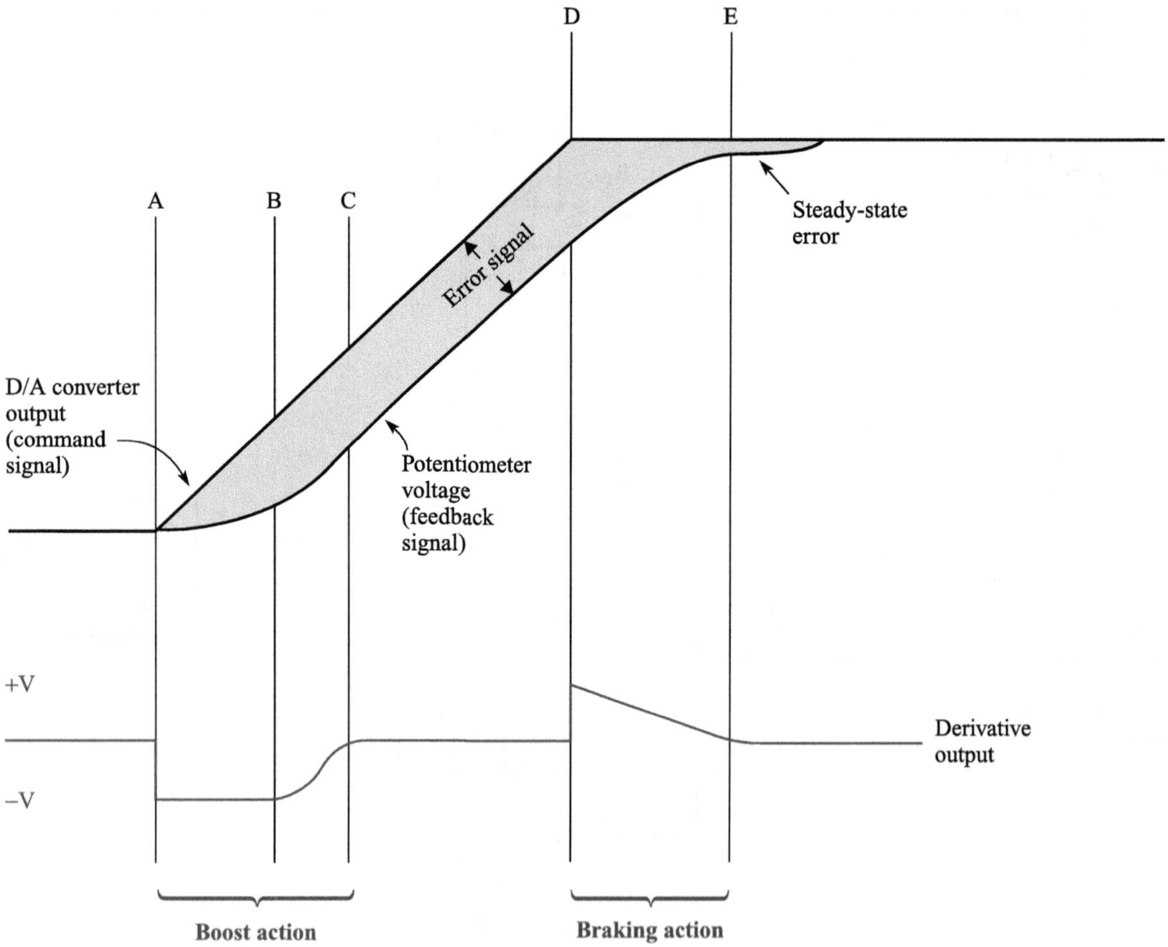

FIGURE 4-15 A graphical representation of a PID robotic control system

As the arm nears the desired position, the error signal stops changing and approaches zero. The result is that the proportional and derivative outputs go to zero. Since the setpoint and measured variable are not exactly equal, as shown in part E, a steady-state error exists and the integral op amp takes over to cause the arm to move the remaining distance.

Whenever there is a large setpoint change, the controlled variable will usually lag behind and cause a rapid change of the error signal. Because the derivative controller detects this trend, it responds by compensating for large system changes before they fully develop. Therefore, derivative control is sometimes referred to as *anticipatory* or *predictive* control. Because derivative control tends to reduce system oscillation, the proportional gain can be set at higher values to further increase the speed of response of the controller to system disturbances.

The derivative mode is used only when the controlled variable lags behind a setpoint change and an error signal develops. In the robot example, which is a motion control application, the operation is relatively fast. However, there is a lagging condition that exists because the arm position cannot respond as quickly as the command signal. Therefore, the derivative mode is used by the controller to minimize the error signal that develops from this condition. The derivative mode is often used in a slow-acting process control application, such as regulating the temperature of a liquid in a large tank. If a new setpoint is set, the static inertia of the liquid does not allow the temperature to immediately change. Therefore, a lagging condition develops. An example where derivative control would not be used is in an airflow control application. Whenever a new setpoint is set, a flow control valve changes and immediately alters the flow rate of the air. Since the response of the system is very fast, a lagging condition does not develop and the derivative control action is not required.

Derivative mode control is never used alone. It is usually combined with the proportional and integral modes for systems that cannot tolerate offset error and require a high degree of stability. This type of system—which combines the advantages of proportional, integral, and derivative action—is known as a three-mode **PID** controller. PID controllers are best suited for systems that need to react quickly to large disturbances. They are also recommended in systems where load changes frequently occur.

Derivative control is rarely combined with proportional control. When it is, proportional-derivative action is used in applications where lag times vary and offset error is tolerated.

To obtain the best possible PID control for a particular application, the gain settings for each mode must initially be made. These settings are different for each system. While the system is actually running, *tuning* adjustments are often made to the gain settings to attain optimal performance. Gain adjustments can be performed by trial and error, or automatically by autotune controllers.

In a conventional PID system, the process being controlled is seldom performed by op amp circuits. Instead, these three modes are performed by computer software packages that calculate a set of differential equations. When a setpoint change or disturbance occurs, feedback devices send signals that cause the numerical values of the mathematical equations to change. Calculations are made to produce a solution that tells the PID controller how to adjust the system parameters to meet the new requirements.

PID control is an industrial standard well understood by many control engineers. It is a popular control technique that has been proven through many years of use. PID control can presently be performed with a personal computer or a programmable controller. PID controllers can be purchased as prebuilt, preprogrammed assemblies.

4-6 Time-Proportioning Control

A common method of controlling a DC voltage actuator device, such as a heating element or a DC motor, is to vary the DC voltage of an amplifier that drives it. This method is called the *proportional* mode because the magnitude at which the actuator is driven is proportional to the amplitude of the applied voltage. For example, by doubling the applied voltage produced by an amplifier, the temperature of the heating element is doubled if the gain is 1.

Another way of controlling an actuator is to use an operation called **time proportioning.** Also called the pulse width modulation (PWM) technique, time proportioning is a method in which the amplifier output is switched alternately to fully on and fully off. Changing the ratio of signal-on to signal-off varies the average voltage produced. Figure 4-16 illustrates various time-proportioning output signals that are in the form of a square wave. The square wave is at +10 volts when the amplifier is on, and 0 volts when it is off. The ratio of the time the square wave is on to the total time period of one cycle is called the *duty cycle.*

If the duty cycle is 25 percent, as shown in Figure 4-16(a), the on-time occurs for 25 percent of the time, and the off-time occurs for 75 percent of the time.

The average DC voltage produced, which affects how much power is applied to the actuator, can be determined by multiplying the duty cycle times the on-state DC amplitude of the square wave.

EXAMPLE 4-3

What is the average voltage at a 25-percent duty cycle when the on-state DC amplitude of the square wave is +10 volts?

Solution

$$\text{Average Voltage} = \text{Duty Cycle} \times \text{On-State DC Voltage}$$
$$= .25 \text{ x} + 10 \text{ V}$$
$$= 2.5_{\text{AVE}} \text{ DC volts}$$

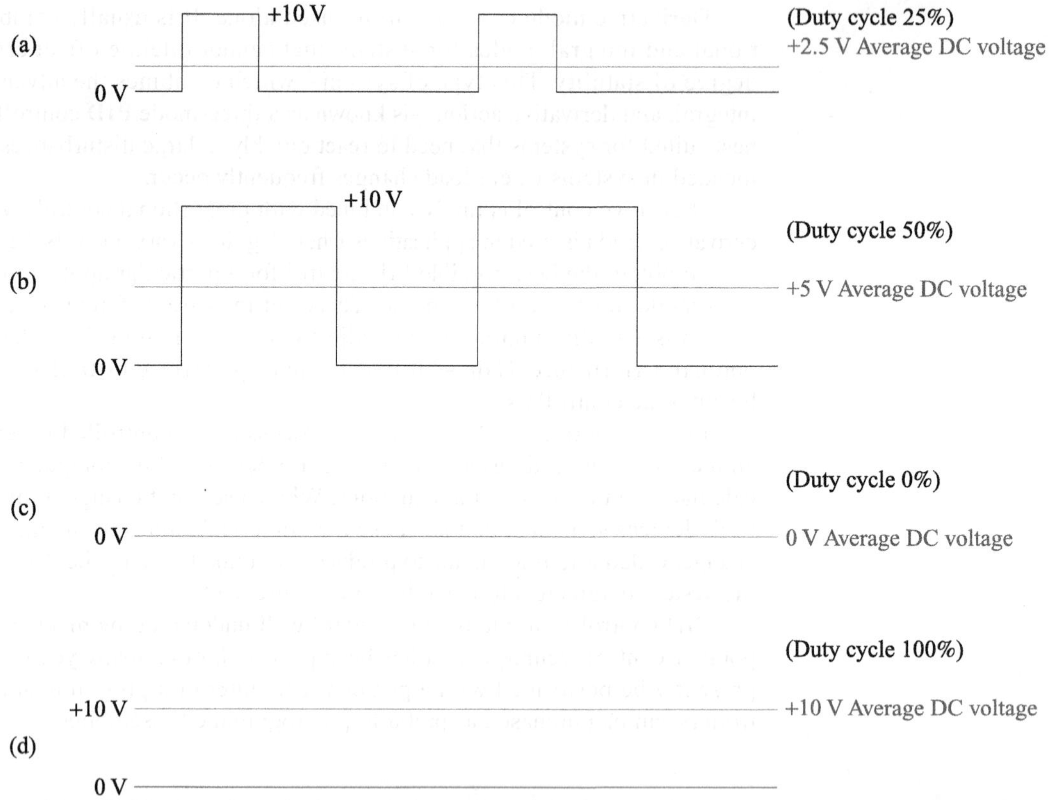

FIGURE 4-16 Waveforms of average DC voltages at different duty cycles: (a) 25-percent duty cycle; (b) 50-percent duty cycle; (c) 0-percent duty cycle; (d) 100-percent duty cycle

When the duty cycle is 50 percent, as shown in Figure 4-16(b), the on-time/off-time ratio is 50 percent, so the average voltage is 5 volts. If the duty cycle is 0 percent, the amplifier is not turned on, so the average voltage is 0 volts, as shown in Figure 4-16(c). At a duty cycle of 100 percent, the amplifier is always on, and the average DC voltage is +10 volts, as shown in Figure 4-16(d).

4-7 Time-Proportioning Circuit

SSE4503
The Time-Proportioning Op Amp

A voltage-level detector op amp, as shown in Figure 4-17, can be used as a time-proportioning circuit. The output of the op amp is 0 volts at any moment when the noninverting input is less than the inverting input. When the voltage at the noninverting input is greater than the voltage at the inverting input, the op amp goes into saturation and produces +10 volts.

A 0- to +10-volt peak-to-peak sawtooth signal is applied to the inverting (–) input of the op amp. A 0- to +10-volt signal is applied to the noninverting (+) input of the op amp by varying the wiper arm of a potentiometer. When the wiper arm is at the 0-volt ground position, the sawtooth at the (–) input is always equal to or greater than the voltage at the (+) input. Therefore, the output of the op amp is at a constant 0-volt potential, as shown in Figure 4-17(b).

When the wiper arm is in the middle position, a +5-volt potential is applied to the (+) input. The sawtooth potential applied to the (–) input does not become greater than the (+) input until it goes above +5 volts, halfway up the ascending portion of the waveform, and remains that way until it drops below +5 volts, halfway down the descending portion. As a result, the op amp produces a square wave with a 50-percent duty cycle, as shown in Figure 4-17(c). The resulting average DC voltage is +5 volts ($0.5 \times 10\text{ V} = 5\text{ V}$). When the wiper arm is positioned at the top, 10 volts is applied to the (+) input. The voltage applied to the (–) input is never greater than the (+) input, so the op amp output is always +10 volts, as shown in Figure 4-17(d).

SSE4403
Time-Proportioning Application

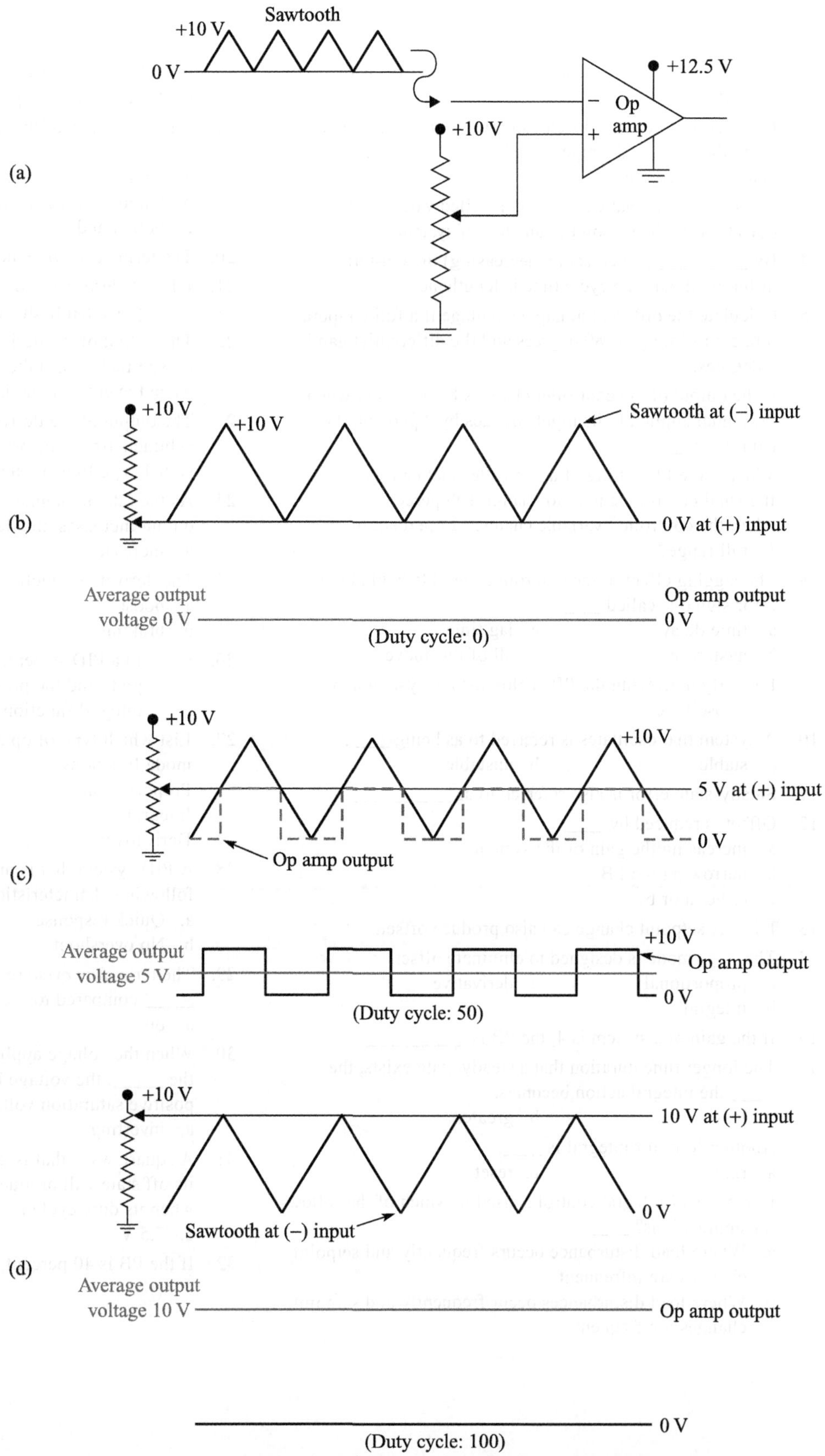

FIGURE 4-17 (a) Time-proportioning circuit; (b) 0-percent duty cycle; (c) 50-percent duty cycle; (d) 100-percent duty cycle

Problems

1. Which section of the closed-loop system performs the four control modes?
2. In an On-Off heating system, an error signal is produced when the measured temperature is __________ (above, below) the setpoint.
3. List two factors that cause the controlled variable to deviate from the setpoint in an On-Off system.
4. By __________ (increasing, decreasing) the On-Off differential gap, the cycle time is lengthened.
5. Calculate the differential gap percentage if a full temperature control range is 80 degrees and the differential gap is 8 degrees.
6. If the output of the controller changes by 6 percent when the signal applied to its input changes by 2 percent, the gain is __________.
7. What is the PB setting of a controller that causes the final control element to change 100 percent when the controlled variable changes 25 percent of its full range?
8. The lagging effect of the controlled variable behind the error signal is called ____.
   a. time delay
   b. hysteresis
   c. lag time
   d. all of the above
9. T/F By narrowing the PB, a closed-loop system may oscillate.
10. A system that oscillates is referred to as being ____.
    a. stable
    b. unstable
11. Steady-state error is also referred to as __________.
12. Offset is reduced by ____.
    a. increasing the gain of the system
    b. narrowing the PB
    c. either a or b
13. T/F A setpoint change can also produce offset.
14. The ____ mode is designed to eliminate offset.
    a. proportional
    b. integral
    c. derivative
15. If the gain of a system is 4, the PB is __________.
16. The longer time duration that a steady-state exists, the ____ the integral action becomes.
    a. less
    b. greater
17. Another term for integral is ____.
    a. rate
    b. reset
18. Proportional-integral control is used in which of the following applications? ____
    a. Where load disturbance occurs frequently and setpoint changes are infrequent.
    b. Where load disturbances occur frequently and setpoint changes are frequent.
    c. Where load changes are slow.
    d. Where load changes are fast.
19. The magnitude of the integral output is proportional to the ____.
    a. applied input voltage
    b. length of time the error signal exists
    c. both a and b
20. The term *derivative* means __________ of change.
21. T/F A derivative controller produces an output that is proportional to the amplitude of the error signal.
22. The output of the derivative function __________ (adds to, subtracts from) the output of the proportional output when the error signal is getting larger.
23. The output of the derivative function __________ (adds to, subtracts from) the proportional output when the error signal is getting smaller.
24. As the rate of change of an error signal at the controller's output increases, the derivative signal ____.
    a. increases
    b. decreases
25. The derivative function gives the actuator a ____ action.
    a. boost
    b. braking
    c. both a and b
26. T/F In a PID system, after the derivative function is complete and the proportional signal is ineffective, the integral function is performed.
27. List which type of op amp performs the following PID mode functions:
    Proportional:
    Integral:
    Derivative:
28. A PID system that is properly tuned provides which of the following characteristics? ____
    a. Quick response
    b. No overshoot
    c. No offset
    d. All of the above
29. The term *duty cycle* refers to the amount of time a signal is _____ compared to the period of one complete cycle.
    a. on
    b. off
30. When the voltage applied to the _____ input is greater than the _____, the voltage level detector op amp produces a positive saturation voltage at its output.
    a. inverting
    b. noninverting
31. A square wave that is 20 volts at its on state and 0 volts at its off state will produce an average DC voltage of _____ when its duty cycle is 75.
    a. 7.5 V
    b. 10 V
    c. 15 V
32. If the PB is 40 percent, what is the gain?

SECTION 3

Electric Motors

OUTLINE

The electric motor is the most common device used to perform the actuator function in an industrial control loop. It converts electrical energy into mechanical power. The electric motor is the workhorse in both commercial and industrial applications. In the home, the furnace, refrigerator, washer, and dryer are all powered by electric motors. They also drive manufacturing industry. It is estimated that over 60 percent of all electrical power generated is used to supply industrial electric motors. Because they are used so extensively, electric motors are an important area of study in the field of industrial electronics.

CHAPTER

5

DC Motors

OBJECTIVES

At the conclusion of this chapter, you should be able to:

- Describe the operating principle of a DC motor.
- List the major components of a DC motor.
- Define the following terms:

Motor Action	Holding Torque	Full Load
Main Field	Speed Regulation	Overload
Commutation	Armature Reaction	Partial Load
Rotary Motion	Neutral Plane	No Load
Torque Force	CEMF	

- Make the following calculations for a DC motor:

Speed Regulation	Work	Horsepower
Torque	Power	Efficiency

- Describe the operation of the following DC motors and identify their characteristics:

Shunt	Series	Compound

- Reverse the direction of a DC motor.
- Choose the types of DC motors needed for specific applications.

INTRODUCTION

A direct current (DC) motor converts DC electrical energy into mechanical energy. As direct current is used by the motor, it produces a mechanical rotary action at the motor shaft. The shaft is physically coupled to a machine or other mechanical device to perform some type of work.

DC motors are highly versatile mechanisms. They are well suited for many industrial applications. For example, they are used where accurate control of speed or position of the load is required. They can be accelerated or decelerated quickly and smoothly, and their direction easily reversed. This makes them very useful in machine tool operations and in robotics. They provide higher starting torque than other motor types. Because the DC battery is the best portable power supply, DC motors are used for electric tools, carts, tow motors, and other forms of mobile equipment. In this chapter, the operation and characteristics of the DC motor are described.

5-1 Principles of Operation

The conversion of electrical energy to mechanical energy is accomplished by a principle called **motor action.** There are two requirements for motor action to exist. The first requirement is that there is a current flow through a conductor. As it does, a circular magnetic

IAU2708
Motor Action

IAU13208
Fundamentals of a
DC Motor

field develops around the wire. These magnetic flux lines go in a direction described by the *left-hand rule* shown in Figure 5-1. The thumb points in the direction of electron current flow. The fingers point in the direction of the circular magnetic flux lines around the wire.

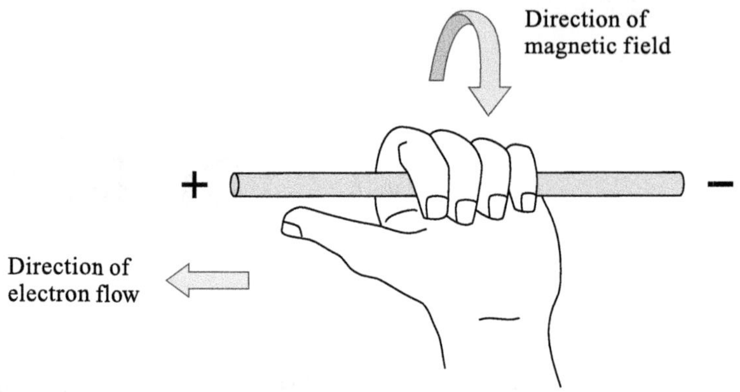

FIGURE 5-1 The "left-hand rule," showing the direction of electron flow and the magnetic field around a conductor

The second requirement is that a force on the conductor develops. The force is produced when the conducting wire is placed inside the magnetic field formed between two magnetic poles, as shown in Figure 5-2. This magnetic field is referred to as the **main field.** The direction of the force depends upon the direction of current through the wire and the direction of the flux lines between the poles.

Figure 5-2 illustrates this concept. The main field develops between two poles of either a permanent magnet or an electromagnet. Normally, these flux lines are straight and go in a north-to-south direction. However, when the conductor is placed between the poles, the lines become distorted. On one side of the wire, the flux lines of the conductor combine with the main field and become very concentrated. On the other side, the flux lines of the conductor and the main field go in the opposite direction. The effect is that they cancel each other, making a weak force. The side with the concentrated flux lines is elastic like rubber bands. Since they are stretched, they tend to straighten out. Straightening exerts a force on the conductor and pushes in the direction of the weak side until it moves out of the field. Figure 5-3 illustrates the *right-hand rule* for motors. It shows the direction in which a conductor carrying current will be moved in a magnetic field. The index finger points in the direction of the magnetic field lines (north-to-south). The middle finger points in the direction of the current in the wire. The thumb points in the direction of the wire movement. This is the fundamental principle of motor action.

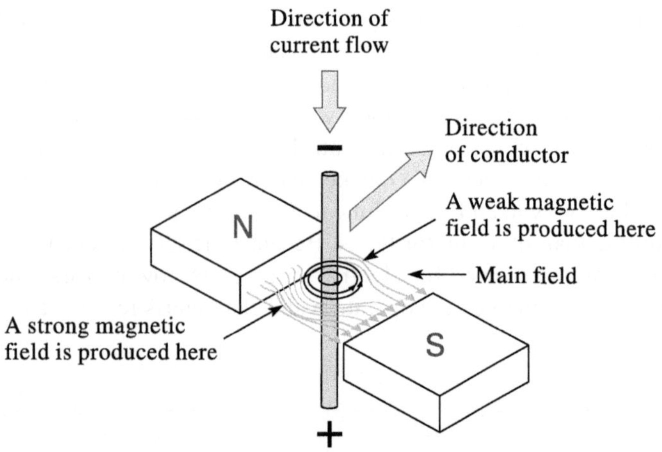

FIGURE 5-2 Interaction of a conductor inside a magnetic field causing movement of the wire

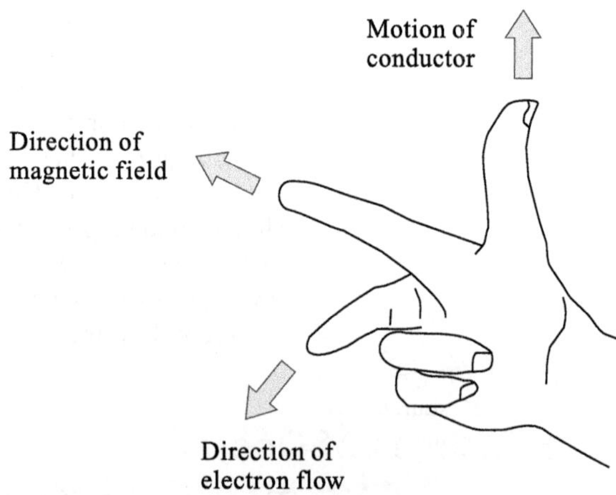

FIGURE 5-3 The "right-hand rule"

5-2 Rotary Motion

A current-carrying conductor in a magnetic field tends to move at right angles to the field. Once it moves out of the field, the force is reduced to zero and no further action takes place. Since a motor produces a continuous rotary motion, it is necessary to make the current-carrying conductor into a single loop of wire. Figure 5-4 shows the loop placed between two magnetic poles. When the loop is connected to a DC supply, current flows from point A to point B on one side of the loop, and from point C to point D on the other side of the loop; that is, the current flows in opposite directions through the sides of the loop across from each other. Therefore, one side is pushed upward and the other side is pushed downward. Because the loop is designed to pivot on its axis, the combined force results in a twisting action called **torque.**

This action is illustrated by the cross-sectional view in Figure 5-5. The ⊗ indicates the point at which current flows into the page. The dot indicates the point at which current flows out of the page. The large arrows show the direction of each wire segment. The loop rotation is counterclockwise (CCW). When it reaches a position perpendicular to the field, there is no interaction of the magnetic fields. This is called the **neutral plane.** Due to inertia, the loop continues to spin CCW. However, since the direction of current flow through the loop does not change, the interaction between the conductor segments and the flux lines develops a force in the opposite direction. Instead of continuing in the CCW direction, the loop stops and then changes direction. An oscillating motion is produced until the armature settles at the neutral plane.

A continuous rotation is achieved in Figure 5-6 by reversing the direction of current through the loop the instant it passes through the neutral plane. The current change is accomplished by a switching device called a **commutator.** Sometimes referred to as a *mechanical rectifier,* the commutator is in the shape of a ring that is split into two segments. Each segment is connected to an end of the loop. The commutator and loop rotate together and are referred to as the **armature.** A pair of carbon brushes supply current to the armature windings. The brushes are sliding connectors that make contact with the commutator segments as the armature rotates. Each brush is connected to a terminal of the DC supply.

The drawings in Figure 5-7 illustrate how the switching action of the brushes and commutator causes the armature to rotate one revolution. The direction of the flux lines in

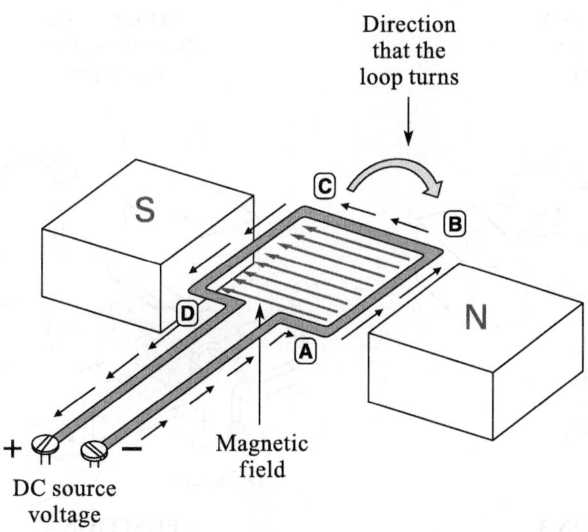

FIGURE 5-4 Direction of torque developed by a loop of wire

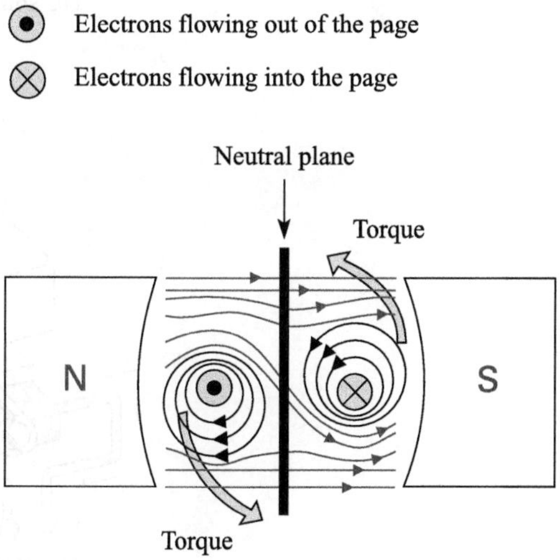

FIGURE 5-5 Cross-sectional view of the loop of wire inside the main field. The distortion of this field creates the force that causes the rotation

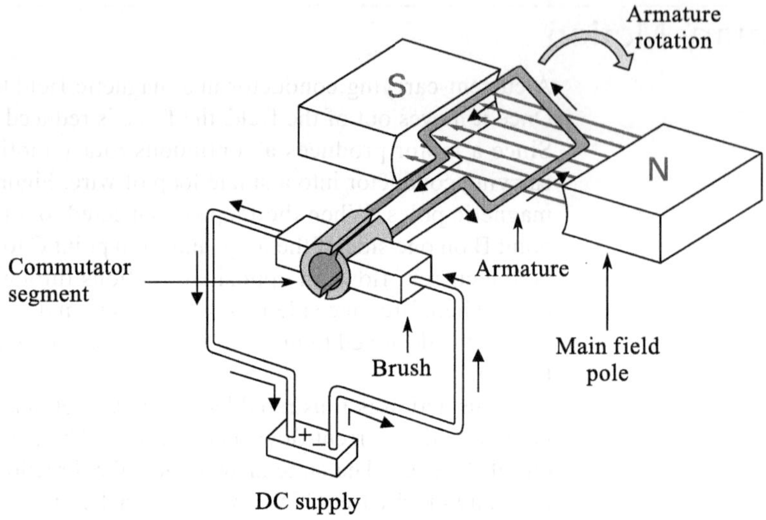

FIGURE 5-6 Commutator and brush arrangement for a simple DC motor

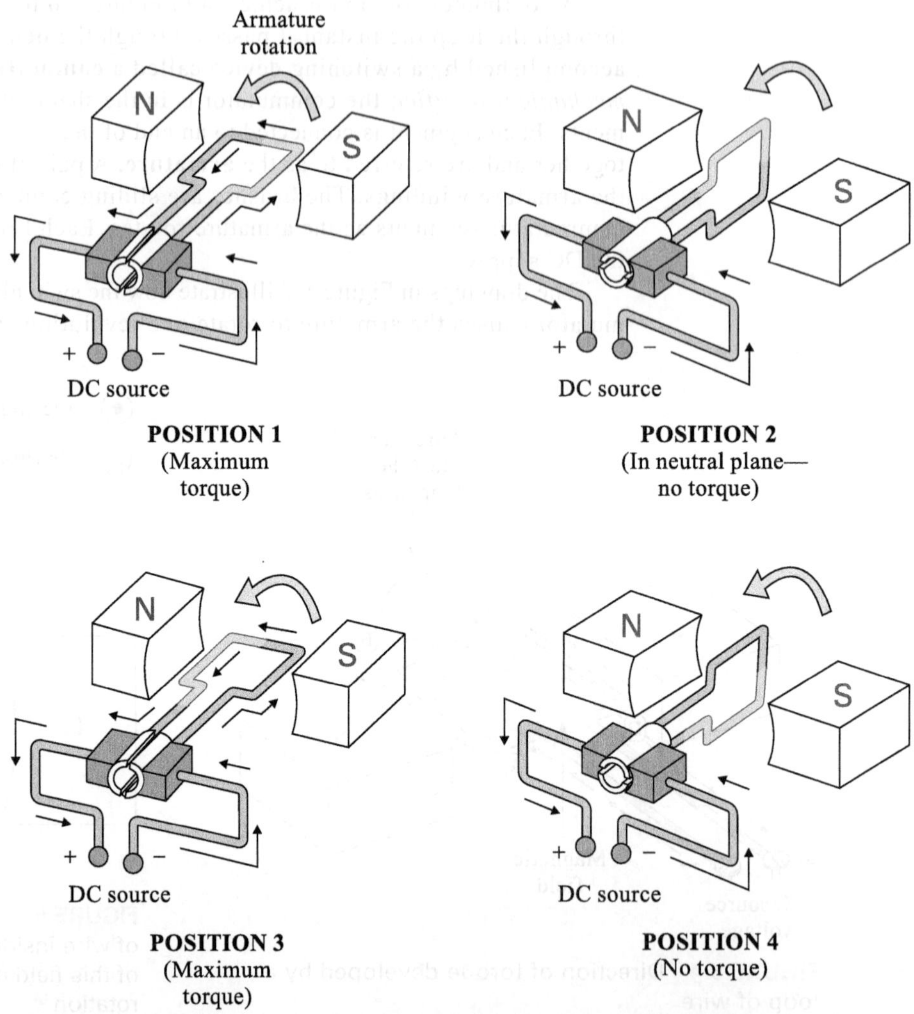

FIGURE 5-7 The operation of a simple DC motor as it rotates 360 degrees

the main field is from the north pole to the south pole. The switching action is called **commutation.**

1. In position 1, the current enters the loop through the negative brush and exits through the positive brush. The torque developed causes the armature to rotate in a CCW direction.
2. When the armature is in position 2, the brushes make contact with both commutator segments. The armature loop shorts out and current flows from one brush to the other through the commutator segments. The result is that no torque is produced. However, inertia causes the armature to continue rotating past this position.
3. When the armature rotates past the neutral position in position 3, the sides of the loop are in the opposite position from where they were in position 1. The switching action of the commutator reverses the direction of current flow through the armature loop. This causes current to flow into the armature segment closest to the south pole, as it did in position 1. The torque developed causes the armature to continue rotating in the CCW direction.
4. In position 4, the armature is again in the neutral position. Since inertia carries the armature toward the position shown in position 1, the cycle is repeated.

IAU13008
The Multi-Loop Armature of a DC Motor

The rotation of the armature continues in one direction because the commutation keeps reversing the current direction through the loop. This way, the armature always interacts the same way with the main field to maintain a continuous torque in one direction.

There are two disadvantages of using a motor with one armature loop. One problem is starting the motor when the armature is in the neutral position. Since the armature loop is shorted, no torque is developed to cause movement. To start, the armature must be physically moved out of the neutral position. The other disadvantage is that when the motor runs, its speed is erratic because its torque is irregular. Maximum torque is produced when the armature loop is parallel to the main field, and minimum when it is located in the neutral plane.

Both problems are corrected by using a two-loop armature with four commutator segments, as shown in Figure 5-8. The ends of the loop are connected to opposite segments of the commutator, and the loops are electrically connected in parallel. When one loop is in the neutral position, the other is in the position of maximum torque. As the armature turns, the commutator switches current to the loop that approaches the neutral plane. The disadvantage of this configuration is that during brief moments of the revolution only one loop is connected, while the other rotates as dead weight. This situation occurs at the moment when commutation takes place. The loop that is horizontal in the diagram creates the torque. The other loop is in the neutral plane and is dead weight because there is no interaction of magnetic fields.

By connecting loops of the armature to adjacent commutator segments, this problem is corrected. One commutator segment per loop is used instead of two segments per loop. Electrically,

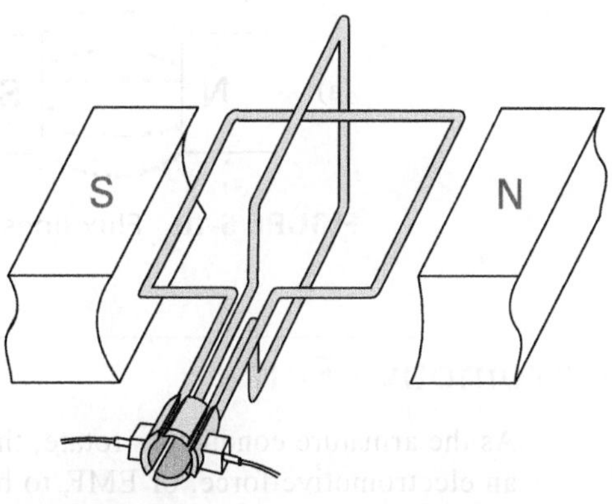

FIGURE 5-8 A two-loop armature provides self-starting and steadier and stronger torque

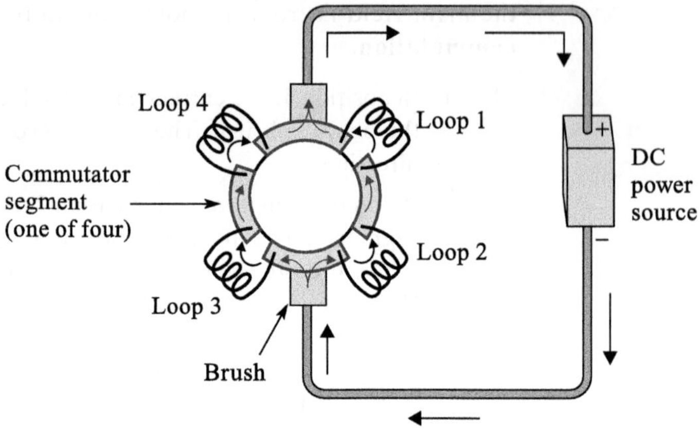

FIGURE 5-9 A four-loop armature that uses four commutator segments

the armature is two series circuits connected in parallel, as shown in Figure 5-9. When current flows through the brushes, all four loops carry the current and contribute to the torque.

5-3 Practical DC Motors

In a practical motor, more than four armature loops and commutator segments are used. Since each conducting wire develops torque, a larger number of loops and commutator segments produce more turning force. Additional torque is also developed by adding more turns to each armature loop. A further improvement is the use of more than one set of field poles. Adding more poles makes it possible for an armature conductor to develop maximum torque several times during a revolution.

5-4 Control of Field Flux

Magnetic flux lines have a tendency to repel each other, even if they run parallel. Figure 5-10(a) shows how flux lines between two poles of a magnet bow away from each other. To eliminate bowing, magnets curved at the ends of the poles are used, as shown in Figure 5-10(b). The outer magnetic lines have a greater intensity. Therefore, the stronger flux lines force the other lines inward so that they run straight between the poles.

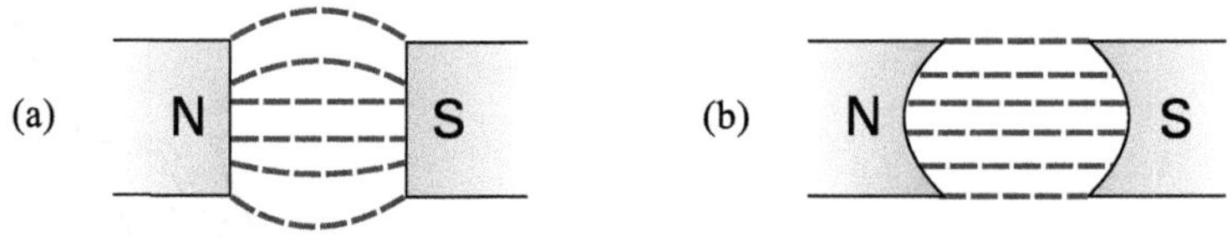

FIGURE 5-10 Flux lines between poles

5-5 Counterelectromotive Force

IAU9908
Counterelectromotive
Force

As the armature conductors rotate, they cut through the main field. These conditions cause an electromotive force, or EMF, to be induced into the armature coils in the same way in which voltage is produced by a generator. The more rapidly the armature turns, the more EMF it generates. The induced EMF opposes the EMF applied to the armature by the DC power source. For this reason, it is called a **counterelectromotive force** (CEMF), or *back EMF.* To the power source, the CEMF appears as another power source connected series-opposing. The CEMF does not supply an opposing current; however, it reduces the current that flows through the armature. The CEMF cancels out a portion of the applied voltage, and the difference between the two forms a net voltage that affects how much current flows through the armature. The voltage applied by the DC source is always greater than the CEMF.

The amount of CEMF produced is not constant. It varies according to three factors:

1. *The physical properties of the armature.* These include the number of turns in the coil, its diameter, and its length. The induced EMF increases as the size gets larger.
2. *The strength of the magnetic field supplied by the field poles.* The induced EMF increases as the flux becomes stronger.
3. *The rotational speed of the armature.* A rapidly moving conductor will induce more CEMF than a slowly moving wire.

5-6 Armature Reaction

IAU13108
Armature Reaction

At the moment the switching action of the commutator takes place, the armature loop is at a right angle to the field flux lines and midway between the pole pieces. This axis is called the *geometric neutral plane.* Because the loop is not cutting flux lines, it will not generate a CEMF.

In practice, the actual neutral plane of the motor shifts from the geometric neutral plane, as shown in Figure 5-11. The shift takes place because there are two magnetic fields between the poles. One is the main field, and the other is the flux lines built up around the armature conductors. Their interaction distorts the main field. The perpendicular neutral plane becomes shifted in the direction opposite the armature rotation. This shifting of the neutral plane is known as **armature reaction.** Armature reaction varies depending on the armature current and speed of the motor. As more current is applied, the more rapidly the motor runs and the larger the armature reaction becomes.

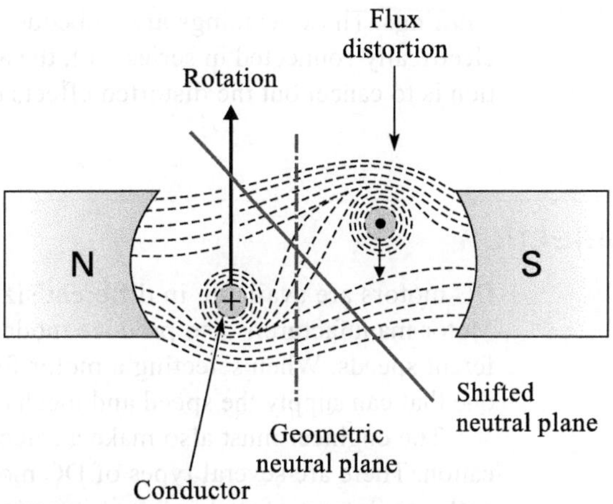

FIGURE 5-11 Shifted neutral plane due to armature reaction

With the neutral plane shifted, commutation is disrupted because it no longer takes place when the corresponding armature loop is perpendicular to the main field. Instead, the armature cuts through the tilted flux lines the moment the brushes make and break contact with the commutator segments. As a result, EMF is induced into the loop, which causes arcing to occur at the commutator segments that move under the brush. Sparking causes the brushes and the commutator to pit, increasing the wear on both.

The arcing due to armature reaction adversely affects the motor in three ways:

1. It reduces torque.
2. It makes the motor less efficient.
3. The continuous sparking shortens the life of the brushes and damages the commutator.

Interpoles

The effect of armature reaction is corrected using special windings called **interpoles,** sometimes called *commutating poles.* Shown in Figure 5-12(a), they are smaller poles placed between the main poles. Interpole windings are connected in series with the armature windings. The magnetic fields formed around the interpoles oppose the magnetic field around the armature coils and push back the distorted flux lines so that they are in a straight line between the poles. Therefore, the neutral plane is shifted back to the original position, as shown in Figure 5-12(b). Interpole windings are self-regulating, because they are in series with the armature. If the armature current increases, so does the canceling effect of the interpoles.

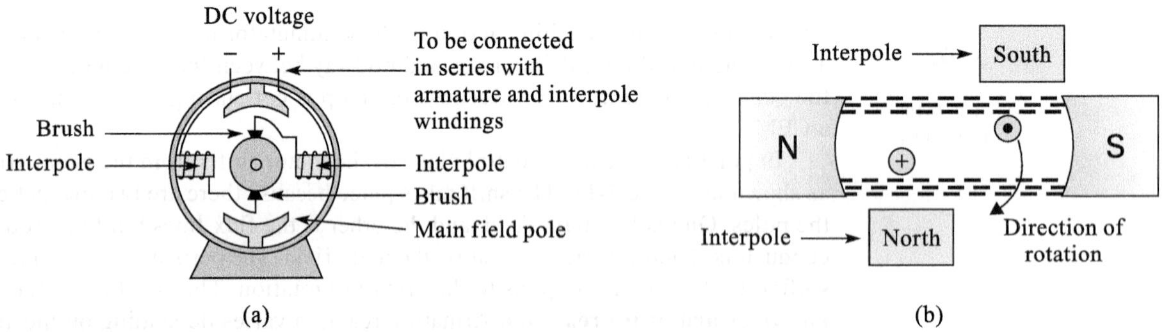

FIGURE 5-12 Interpoles

Another method sometimes used to correct armature reaction is by using **compensating windings.** These windings are embedded into the metal core of the main field poles and are electrically connected in series with the armature conductors. Like the interpoles, their function is to cancel out the distorted effects of the armature field.

5-7 Motor Selection

DC motors are available in different sizes. The larger they are, the more power they have. Motor manufacturers also produce models of DC motors that are designed to operate at different speeds. When selecting a motor for a particular application, the engineer should find one that can supply the speed and mechanical power required by the load being driven.

The engineer must also make a selection based on the motor type that best fits an application. There are several types of DC motors to choose from. Because they are wired differently, each type of motor has its own operating characteristic. Operational requirements determine what type of motor to use.

Two characteristics used in this selection process are:

Speed Regulation. How much the motor speed will vary with a change in the mechanical load.

Torque. How much torque is available when starting a motor, or how much it will vary with a sudden change in load.

Each type of motor will operate differently when subjected to the various load conditions it encounters.

Speed Regulation

A motor is designed to operate at **full load.** Full load is the maximum power it can provide to run its rated mechanical load all of the time. It is possible for the motor to run above full load, but not for a sustained period of time: It will overheat and likely become damaged. This situation is called an **overload** condition. The overload condition becomes excessive if the motor stalls because it is unable to move the load. If it stops, the current drawn from the power source is maximum and a circuit-protection device will deactivate the motor. When the physical load is reduced from the full-load condition, this situation is called a **partial-load** condition. The motor operates at **no load** when the physical load is disconnected from the motor shaft.

When the mechanical load connected to the motor is reduced, the motor speed will increase. The amount it increases depends on the type of motor employed. The ability of a motor to maintain its speed when the load is changed is called **speed regulation**. The speed regulation of a motor is calculated by comparing its no-load speed to its full-load speed. It is usually expressed as a percentage of its full-load speed by using the following formula:

$$\text{Speed Regulation} = \frac{\text{No-Load Speed} - \text{Full-Load Speed}}{\text{Full-Load Speed}} \times 100$$

EXAMPLE 5-1

The no-load speed of a motor is 1800 RPM. When the rated load is connected to the shaft, the speed drops to 1720 RPM. What is the speed regulation in percent?

Solution

$$\frac{1800 - 1720}{1720} \times 100 = 4.65\%$$

An example of speed regulation is the operation of a hand drill that uses permanent magnets to develop the main field. When the drill is turned on, it runs at no load. As the drill bit cuts through the material, it slows down to the full-load condition. The amount it slows down at full load compared to the no-load speed is its speed regulation value.

If the speed of the motor is relatively constant over its normal operating range, the motor has good speed regulation. It will perform well as a constant-speed motor. A motor whose speed varies greatly from no load to full load has poor speed regulation.

Torque

Force is a push or pull that can cause motion. When the forces on an object do not act through a common point, there is a tendency to rotate. This twisting action that causes an object to rotate is called *torque*. Torque causes a motor shaft to turn.

The load a motor is driving may rotate like a fan or a pump. It may also be a mechanism that moves in a straight line, like a conveyor belt. Even though these mechanical loads move differently, they are all powered by the turning action of a motor.

When a load is pushed or pulled in a straight line, the force that moves it is measured in pounds. Torque, which is a combination of force and leverage, is measured in pound-feet (lb-ft). The amount of torque a motor produces is calculated by multiplying the force it will exert by the distance between the center of the shaft and the point where the force is being applied, as determined by the following formula:

$$T = F \times r, \text{ where}$$

F is the tangential magnetic force acting on the conducting armature, measured in pounds.

r is the radius in feet, measured from the axis of rotation to the point where the force is applied.

T is the rotary action exerted by the motor shaft, measured in pound-feet (*lb-ft*).

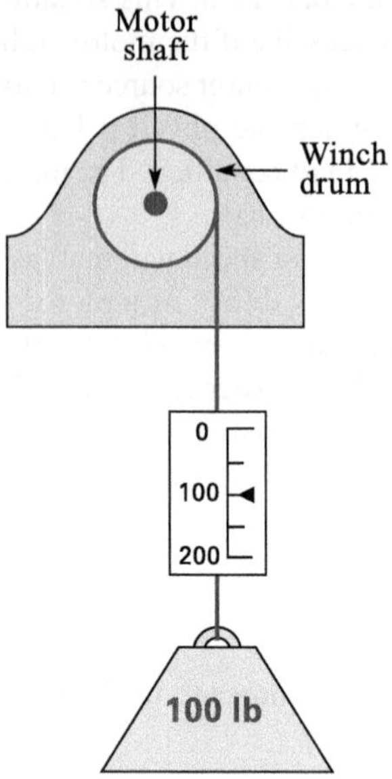

FIGURE 5-13 Winch lifting a 100-pound weight

When a load is connected to the shaft of the motor, it exerts a resistance, or opposing torque, in the opposite direction from the one in which the motor turns. If the torque produced by the motor is greater than the counter torque of the load, the motor shaft will turn. If the counter torque is greater than the torque produced by the motor, the shaft will not turn, and may rotate backward if it is large enough. Even though the load does not move, torque is still produced.

The magnitude of the torque produced by the motor is determined by the following factors:

- Strength of the main field, ϕ.
- The strength of the armature field. This value is expressed by the value of the armature current, I_a.
- The physical construction of the motor, K. These include:
  1. The active length of the conductors
  2. The number of active conductors
  3. The radius of the armature.

The physical properties of the motor are a fixed constant because they are left unchanged. The torque of the motor can therefore be controlled by changing the magnetic strengths of the main field and the armature field.

Figure 5-13 illustrates the concept of torque. It shows a motor turning a winch to lift a weight of 100 pounds. To determine how much torque is required to lift the weight, the radius of the winch drum is multiplied by the force exerted by the weight being lifted. If the diameter of the winch drum is 3 feet, 150 pound-feet of torque is needed.

$$\begin{aligned} \text{Torque} &= \text{Force} \times \text{Radius} \\ &= 100 \text{ lb} \times 1.5 \text{ ft} \\ &= 150 \text{ lb-ft} \end{aligned}$$

EXAMPLE 5-2

Find the torque required to produce a tangential force of 240 lb at the surface of a pulley 6 inches in diameter.

Solution

$$\begin{aligned} T &= F \times r \\ &= 240 \times 3/12 \\ &= 60 \text{ lb-ft} \end{aligned}$$

Suppose the load in Figure 5-13 is doubled. The motor will respond to the change by producing just enough torque to satisfy the demands of the new load. In this situation, the motor will have to exert 300 pound-feet of torque. As long as the torque requirements of the load are within the capabilities of the motor, it will always move the load.

The torque developed by the motor when driving its rated mechanical load is called the *rated load torque.* This is a constant torque that drives the load at a steady speed.

When starting the motor from a dead stop, it takes more effort to get it started than to keep it running. The same concept applies to starting a car. First gear is used when starting to provide the extra torque needed to overcome the inertia of starting. Less torque is required to keep the motor or car moving.

Electric motors are designed to supply the extra torque needed to start the load. The starting torque of DC motors ranges from 150 to 500 percent of the rated load torque. Speed-torque curves for different types of DC motors will be provided throughout the remainder of the chapter.

Work

The primary function of a motor is to perform work. The motor does mechanical work when it supplies a force to move a physical object across a distance. The force acting on the object must overcome some resisting force. For example, work is done when the weight in Figure 5-14 is pulled. Work is calculated by multiplying distance times force, as shown by the following formula:

$W = D \times F$
where, D = Distance in feet
F = Force in pounds
W = Work in foot-pounds

Work is not done unless the load is moved a distance.

The concept of work is illustrated in Figure 5-14. A motor exerts a torque to lift a 250-pound weight a distance of 10 feet. Therefore, the motor performs 2500 foot-pounds of work.

$$\begin{aligned} W &= D \times F \\ &= 10 \text{ ft} \times 250 \text{ lb} \\ &= 2500 \text{ ft-lb} \end{aligned}$$

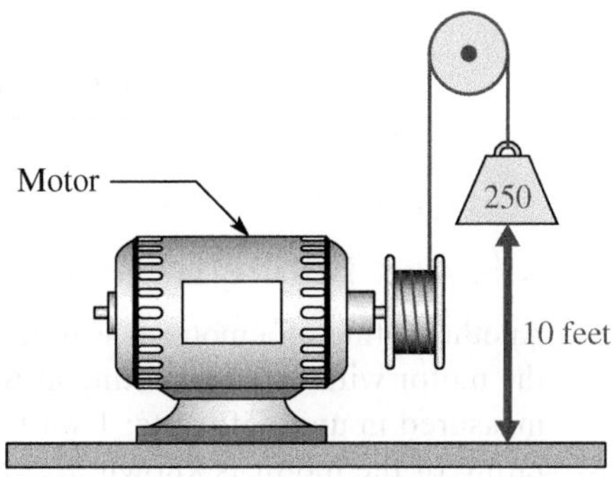

FIGURE 5-14 A motor lifting a 250-pound weight a distance of 10 feet

If a small motor operates long enough, it will perform a lot of work. A powerful motor will do the work quickly.

Power

Power is defined as the rate of doing work. It describes how rapidly a particular amount of work is accomplished and is calculated by the following formula:

$$Power = \frac{Work}{Time}$$

where, *Work* is in foot-pounds
Time is in minutes
Power is in foot-pounds per minute

Suppose 5000 pounds of cargo is lifted by a winch to a height of 50 feet. The work required is 5000 × 50, or 250,000 foot-pounds. If the time it takes to raise the cargo is 2 minutes, the power required is:

$$\text{Power} = \frac{250{,}000 \text{ ft-lb}}{2 \text{ min}} = 125{,}000 \text{ ft-lb/min}$$

Horsepower

Placed on the housing of a motor is a nameplate that provides relevant information for the engineer or technician. It does not contain information about the torque the motor exerts or the amount of work it will perform. Instead, it lists a power rating in units of horsepower that the motor delivers to the load.

This information is useful in determining if the motor is large enough to drive the load. When 33,000 pounds are moved 1 foot in 1 minute (or an equivalent combination), 1 horsepower (hp) of work is performed. In the example of the cargo winch, the horsepower required to raise the load is:

$$\frac{125{,}000 \text{ ft-lb/min}}{33{,}000} = 3.79 \text{ hp}$$

The combination of the speed at which the motor runs and the output torque it exerts determines the horsepower it is capable of producing. In the example of the cargo winch, twice the horsepower is required to move a load of twice the weight in 2 minutes, or the same weight at twice the speed.

Suppose a pulley is connected to the end of a motor shaft that produces an output torque of 10 lb-ft at a rate of 1000 RPM. The formula for determining the horsepower of a rotary output is:

$$\text{hp} = \frac{\text{Speed (RPM)} \times 2\pi \times \text{Torque (lb-ft)}}{33{,}000}$$

$$\text{hp} = \frac{1000 \text{ RPM} \times 2\pi \times 10 \text{ lb-ft}}{33{,}000} = \frac{62{,}800}{33{,}000} = 1.9 \text{ hp}$$

Kilowatt Rating of a Motor

Another rating of a motor is *wattage.* This value identifies the amount of power consumed by the motor without overheating at its rated voltage and speed as it performs work. Power is measured in units of watts; 1 watt equals 0.737 foot-pounds per second. If the horsepower rating of the motor is known, its value can be converted to watts by multiplying it by 746, since there are 746 watts to a horsepower. Until recently, motor power ratings were given exclusively in hp. Now, more manufacturers are rating motors by kilowatts (kW).

The units of horsepower can be converted to kilowatts by the following formula:

$$P_{\text{kW}} = \text{hp} \times 0.746$$

EXAMPLE 5-3

Determine the kW rating of a 11.42-hp motor.

Solution

$$P_{\text{kW}} = 11.42 \times 0.746$$
$$= 8.52 \text{ kW}$$

Motor Efficiency

The mechanical output power of a motor used to drive a load is always less than the power supplied to its input. A part of the energy supplied to the motor is dissipated into heat and is therefore wasted.

The heat losses of motors consist of copper losses and mechanical losses. Examples of both types of losses are as follows:

1. Copper losses
   a. Armature I^2R losses
   b. Field losses
      (1) Shunt field I^2R losses

(2) Series field I^2R losses
(3) Interpole field I^2R losses

2. Mechanical losses
   a. Iron losses
      (1) Eddy-current
      (2) Hysteresis
   b. Friction losses
      (1) Bearing friction
      (2) Brush friction
      (3) Windage (air friction)

These unavoidable losses are expressed as **efficiency.** The efficiency rating of a motor is simply the ratio of the power produced by the output shaft to the power supplied by the source. It is expressed in percentage by the formula:

$$\text{Percent Efficiency} = \frac{\text{Power Out}}{\text{Power In}} \times 100$$

EXAMPLE 5-4

Suppose the wattage consumed by the motor is 3.75 kW and it produces an output of 4.75 hp. What is the efficiency of the motor?

Solution

1. Calculate the wattage at the output by multiplying

$$4.75 \text{ hp} \times 746 = 3543.5 \text{ W}$$

2. Use the efficiency formula:

$$\frac{3543.5 \text{ W}}{3750 \text{ W}} \times 100 = 94.5\%$$

5-8 Interrelationships

The ultimate function of the motor is to drive a mechanical load. The energy required to run the motor is drawn from the power source at almost the same rate at which mechanical power is being used. Therefore, the rate of electrical power consumption is directly proportional to the mechanical requirements of the load plus heat losses.

A change in the mechanical load has an effect on armature current, torque, speed, and CEMF, all of which are related to one another. The armature current produces a magnetic field around the armature. The interaction with the main field causes the armature to turn. The rotating armature produces a CEMF. The CEMF regulates the armature current. At any normal operating speed, the exact amount of CEMF produced will limit the armature current to a value just sufficient to produce the torque required to drive the load.

The motor is also a self-regulating device. If the load varies, the speed changes, which affects the CEMF. The new CEMF adjusts the armature current until the torque matches the load's new requirements. With all of the factors balanced, the motor is in a state of equilibrium.

5-9 Basic Motor Construction

IAU9508
Basic DC Electrical
Motor Construction

Mechanically, all motors have two main parts or assemblies: the armature and the field poles. The horsepower developed by a motor results from the reaction between the magnetic fields created by these two parts.

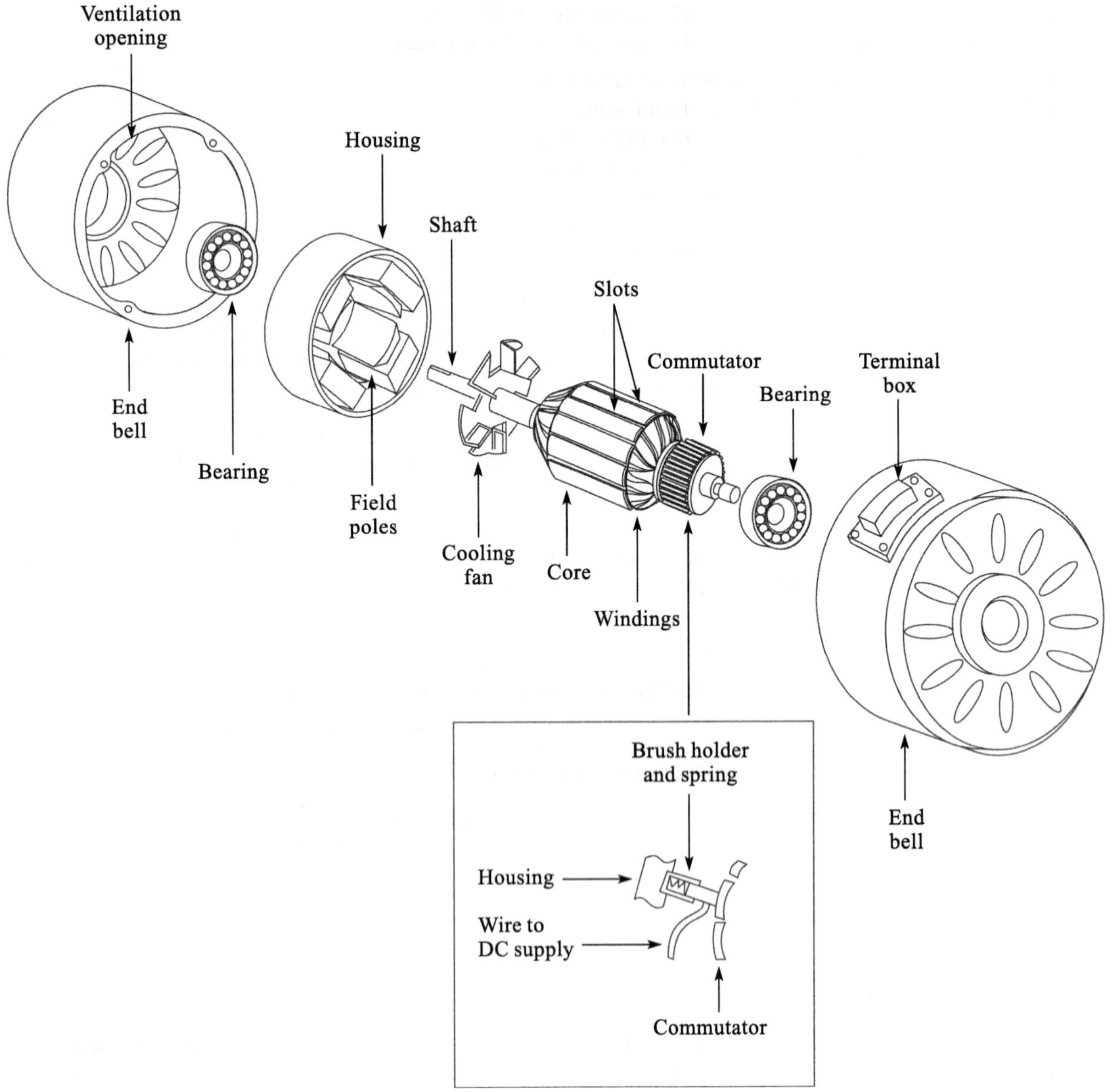

FIGURE 5-15 Parts of a DC motor

The main parts of a DC motor are shown in Figure 5-15. The field poles are core pieces mounted inside a nonmoving, hollow, drum-shaped housing. These field pole pieces are either permanent magnets or electromagnets. If interpoles are used, they are placed between the main pole pieces. End covers (also called *bells*) that support bearings are placed at each end of the housing. Together, all of these parts make up the field pole assembly. The housing is made of steel, which conducts magnetic flux better than air and allows stronger magnetic fields to be established. Its strength also physically supports the stresses that develop inside the motor as it drives the load. The field pole assembly is also referred to as the *stator*.

The moving portion of the motor is the armature, which rotates inside the housing. It consists of a cylindrical core made of sheet-steel laminations that are attached to the shaft. The outer surface of the core has slots where the armature loops are placed. The armature windings are soldered to the commutator, which is also mounted on the shaft. A fan attached to the end of the shaft keeps the internal parts of the motor cool as the armature rotates. The bearings mounted on the stator's end plates support the shaft at both ends. The brushes are pressed against the commutator by specially designed tension springs. The brushes and springs are placed inside holders that mount to the stator housing. The armature assembly is also called the *rotor*.

The flux produced by the field windings passes through the motor housing, field poles, armature core, and any air gaps, all of which is known as the *magnetic circuit* of a motor. Electric circuits of a DC motor are made up of the armature winding, commutator, brushes, and field winding (if it is an electromagnet).

5-10 Motor Classifications

IAU11508
The DC Motor

The most common way to classify DC motors is by describing how the flux lines of the main field are supplied. For example, the motor described in Figure 5-7 uses a permanent magnet.

In the other types of DC motors, the field is supplied by an electromagnet. The field assembly consists of coils wrapped around laminated pole pieces that are mounted on the inside of the drum housing. Therefore, these types of motors are often referred to as *wound-field* motors. The power source that supplies the electromagnetic current for the armature is also used for the field coils. The flux lines supplied by electromagnets are much stronger than those of permanent magnets. Also, the field strength can be varied to achieve desired results. There are three principal types of wound-field DC motors: shunt, series, and compound. They are classified by how their field windings are connected to the DC supply in relation to the armature.

The Shunt Motor

IAU13708
The DC Shunt Motor

The **shunt motor,** shown in Figure 5-16, gets its name from the fact that the field winding is connected in parallel—or shunt—with the armature windings. This configuration provides an independent path for current flow through each coil. Because the two windings are in parallel, the applied voltage connected to each of them is the same. The shunt field coil is wound with many turns of fine wire. Therefore, the shunt field has a higher resistance than the armature circuit and draws less current. Because the current is low, the field coil requires a large number of turns to produce a magnetic field of sufficient strength.

The interaction between the magnetic fields of the shunt field coil and the armature produces the torque that causes the motor shaft to rotate. The strength of the shunt field with respect to the armature field will determine both the motor's torque and the speed at which it rotates. The stronger the magnetic fields, the greater the torque.

Since the shunt coil is connected across the fixed-line voltage terminals, its magnetic field strength is constant. Even though the CEMF in the armature varies as the speed

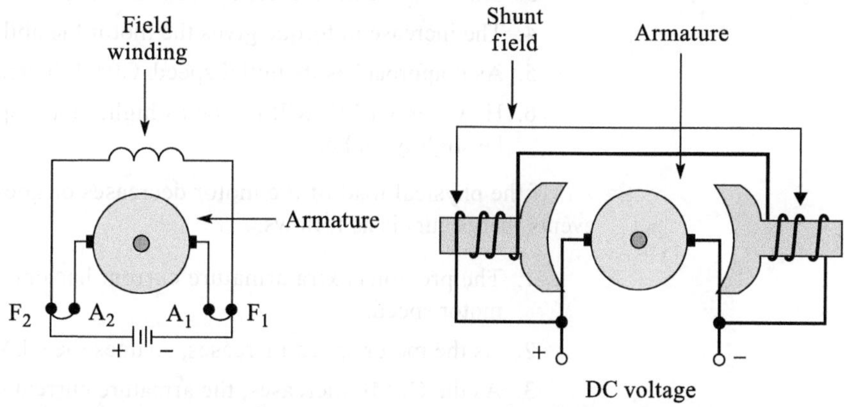

SCHEMATIC DIAGRAM **WIRING DIAGRAM**

FIGURE 5-16 DC shunt motor with the field connected in parallel with the armature

changes, it has no effect on the field strength of the shunt coil. Therefore, the flux field does not change significantly as the physical loading conditions vary. The speed of the motor is mostly proportional to the applied voltage, as described in the formula below:

$$\text{Speed} = V_A - (I_A \times R_A)/K_E \times \phi$$

where, V_A = Applied Armature Voltage
I_A = Armature Current
R_A = Armature Resistance
K_E = Motor Design Constants
ϕ = Field Flux

If we look closely at the formula and consider the motor, the copper resistance of the armature is very small. So, relatively speaking, any amount of armature current, I_A, multiplied by a very small armature resistance, R_A, is going to be a small number. K_T represents the motor design constants, which include (among other things) the number of turns of wire for the field and armature windings, the size of the wires, and the magnetic air gap between the armature and the field. Since these variables are different from motor to motor, but not within an individual motor, they will not affect the speed of an individual motor. Most shunt-field motor systems maintain a constant voltage to the field, so the shunt field flux, ϕ, remains constant. The result is that the speed of a shunt-field motor is mostly dependent on the applied voltage and represented in the following approximation formula:

$$\text{Speed} \approx \frac{V_A}{K_E \times \phi}$$

Another important concept to understand is that CEMF produced by the armature windings rotating through the magnetic field is also proportional to the speed. If we consider that CEMF is equal to $V_A - (I_A \times R_A)$ under normal conditions then CEMF is a good measure and is proportional to the speed of the motor. This principle is based on the assumption that CEMF is equal to $V_A - (I_A \times R_A)$ under normal operating conditions. Many DC motor speed-control systems depend on monitoring CEMF of an armature as feedback to control the speed of the motor.

DC shunt-field motors have some ability to self-regulate their speed. Suppose that the motor is operating in a no-load condition and changes to a loaded condition. When the physical load increases:

1. The motor begins to slow down.
2. The reduction in speed proportionally reduces the CEMF.
3. Since V_A – CEMF increases, I_A increases, which increases torque.
4. The increase in torque gives the motor the ability to approach its initial speed.
5. As it approaches its initial speed, CEMF increases and I_A decreases.
6. However, CEMF will not be as high; motor speed will be slightly less; and I_A will be slightly higher.

If the physical load of the motor decreases or goes to a no-load condition, the chain of events that occurs is as follows:

1. The previous extra armature current has enough torque to start an increase in the motor speed.
2. As the motor speed increases, so does the CEMF.
3. As the CEMF increases, the armature current decreases.
4. The no-load torque-friction equilibrium is quickly reached at only a slightly greater speed.

The no-load speed is slightly higher than the rated speed.

Since the speed regulation from no load to full load of the shunt motor does not exceed 12 percent, it is considered a constant-speed motor. Enough armature current is left over to bring the motor speed almost back to where it was before the extra load was applied. Because of these constant speed characteristics, DC shunt motors are used for applications requiring exact control, such as numerical control machines.

Torque Characteristics of Shunt Motors

The amount of starting torque the motor produces determines how rapidly it accelerates. It speeds up as long as the developed torque is more than the load's resistance.

When the motor is turned on, the shaft is not rotating. Since the RPM is zero, there is no CEMF. The net voltage equals the applied voltage, so the current flow through the armature is as high as possible. At starting, all types of DC motors produce their maximum torque because the interaction between the rotor and the stator magnetic fields is at the highest level.

Consider a shunt motor with an armature resistance of 5 ohms and 115 volts applied. The armature current at zero speed (no CEMF) is equal to the applied voltage divided by armature resistance.

$$I_A = \frac{V_A}{R_A} = \frac{115\text{ V}}{5\,\Omega} = 23\text{ A}$$

The magnitude of torque for a shunt motor is illustrated by the torque formula:

$$T = K_T \times I_A \times \phi$$

$$\text{Motor Constant } K_T = .5$$

$$\text{Armature Current } I_A = 23\text{ A}$$

$$\text{Field Flux } \phi = 1$$

$$T = .5 \times 23\text{ A} \times 1$$

$$= 11.5\text{ lb-ft}$$

The starting armature current is the highest of all DC motors because the opposition to the supply current is primarily the armature resistance. Yet, the shunt type has the lowest torque of all wound-field DC motors.

The reason for the low torque lies in the construction of the field coil. Its resistance is very high because it has many turns of fine wire. Therefore, the field current and field strength are very low. Since torque is proportional to armature current I_A and field strength ϕ, the resultant torque produced is relatively small. The starting torque of a shunt motor is approximately 150 percent of its full-load torque rating.

Motor Speed Control in Shunt Motors

In many applications, the speed of the motor must be varied. The intentional control of shunt motor speed is accomplished by three methods: field flux control, terminal voltage control, and armature voltage control.

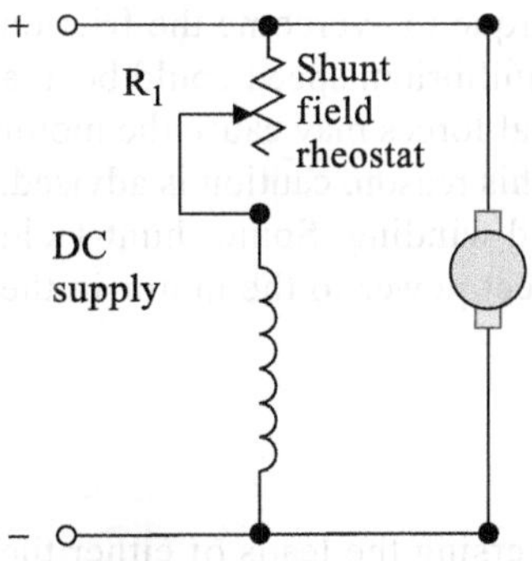

FIGURE 5-17 Shunt field rheostat added for speed control

Field Flux Control The RPM of a DC shunt motor can be controlled beyond its rated base speed by changing the strength of the main field flux. The field is varied by placing a rheostat in series with the shunt field, as shown in Figure 5-17. When resistance is increased, the speed goes up. Conversely, as resistance is reduced, the speed goes down. Therefore, the shunt field acts as a magnetic brake on the armature.

Although it may seem more logical that a reduction in field flux will also reduce speed, the opposite occurs. The speed of the motor actually increases because the reduced field flux causes the CEMF in the armature circuit to decrease. We can recall from the previously described formula

$$\text{Speed} = \frac{V_A - (I_A \times R_A)}{K_E \times \phi}$$

that the shunt field flux density (ϕ) inversely affects the motor speed. As the voltage to the shunt field decreases, the current going through the field windings decreases, causing a reduction in the magnetic field flux density. That decrease in flux density results in a higher speed for the motor and less torque. Essentially, what is happening is:

1. The armature is spinning through few magnetic lines of force.
2. Therefore, less CEMF is developed.
3. With less CEMF, armature current begins to increase.
4. This increases the torque initially.
5. The motor is forced to spin more rapidly.
6. As the motor speed increases, so does CEMF, and reduces the armature current again.

The end result is that the magnetic field is weaker, and the available torque is lower. Therefore, the motor speed continues to rise until the torque and mechanical load balance.

This method of speed control is used only for applications that require a constant horsepower in a partial-load condition. Other limitations of field flux control include relatively low starting torque and poor speed regulation.

Terminal Voltage Control The RPM of a DC shunt motor can be controlled below normal speed by varying the terminal voltage. This method is seldom used because a reduction in speed is accompanied by a substantial loss of torque.

Armature Voltage Control When the field is connected to the same power supply as the armature, it is called a *self-excited* DC shunt motor. It is also possible to connect separate power supplies to the field coil and armature circuit, called a *separately excited* DC shunt motor.

The preferred method of controlling the speed of a separately excited DC shunt motor is by adjusting the armature voltage while maintaining a constant field voltage. This technique is used to decrease the motor speed below its rated base speed. Speed regulation and starting torque are generally not affected, except at the very lowest speeds.

Open Field Condition

IAU13808
The Shunt Motor
Open Field

If the field coil branch opens, the flux strength produced around it decreases to a level supplied only by the residual magnetism of the iron core. This condition causes the CEMF in the armature to drop drastically, and causes its current to rise to a very high level. The net voltage of the applied voltage minus the CEMF rises to a high level and with it, the armature current.

The armature will continue to rotate more rapidly and the CEMF will increase as the applied armature current decreases toward that equilibrium point where the speed reaches maximum. The armature current is then producing enough torque to overcome the friction and other losses of the motor. However, this torque-friction equilibrium speed could be at a very high speed of rotation, a speed high enough that centrifugal forces may cause the motor to fly apart. This condition is called a *runaway condition*. For this reason, caution is advised. Do not operate a shunt-field motor without power to the field winding. Some shunt-field motor systems have a field-loss circuit for safety to disconnect power to the motor in the event that the shunt field is lost.

Direction of Rotation

The direction that a shunt motor turns can be changed by reversing the leads of either the field coil or the armature branch, but not both. However, if compensating windings or interpoles are used to counteract armature reaction, they are placed in series with the armature. Therefore, the standard practice is to reverse the polarity of the armature leads.

The Series Motor

IAU11608
Series Motors

The **series motor,** shown in Figure 5-18, gets its name from the fact that the field winding is connected in series with its armature. The field coil develops little resistance because it is wound with few turns. The small resistance allows a high current to flow through the windings. The field coil is wound with a large gauge (size) wire to handle the current that passes through the armature. Even though the coil has a small number of turns, the magnetic field that forms around the windings develops an adequate torque because its flux lines are concentrated by the pole pieces, and because the current is high.

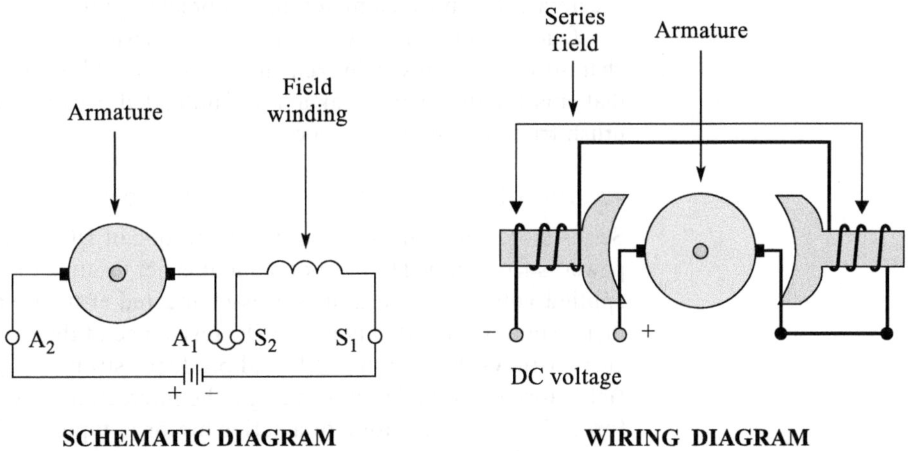

FIGURE 5-18 DC series motor with the field connected in series with the armature

Since the field coil and armature are connected in series, the same current flows through both coils. As the physical loading conditions change the speed, the CEMF causes the armature current to vary, affecting the magnetic field around each coil. Therefore, torque is proportional to the square of the current, and speed is inversely proportional to current. These conditions prevent the motor from maintaining a constant speed under changing load conditions. Therefore, a DC series motor is classified as a poor speed-regulation machine.

Suppose the load the series motor is driving increases.

1. The motor will slow down.
2. Because the armature turns more slowly, it does not cut the field flux lines as rapidly, and less CEMF will be induced.
3. A lower CEMF causes more current to flow through both the armature and field windings, thus strengthening the magnetic flux.
4. The stronger field around the armature and field coils provides the torque necessary to turn the increased load.
5. Equilibrium is reached when the given amount of CEMF generated limits the armature current to produce the right amount of torque for the load.

IAU12508
Series Motor
Runaway

If the load is decreased, the same conditions occur in reverse order.

If the load coupled to a series motor is disconnected, it goes into a no-load condition called *runaway.* In this situation, the motor will accelerate until it physically breaks apart. For example, suppose a normally loaded motor is running. The current flow through the armature and series coil develops a flux that produces just enough torque to turn the load.

1. At the moment the load is removed, the current flow is larger than that required by the load. Therefore, the motor speed increases.
2. As the motor speed increases, the CEMF gets larger.
3. A greater CEMF causes the current through the armature and field to diminish. If the resultant field strength reduction were directly proportional to the armature current, it

would decrease at the same rate at which the speed increased. Therefore, the CEMF would stop increasing, the current would become constant, and the speed would stabilize.

4. However, because the series field coil has few turns of heavy wire, its flux strength decreases more rapidly than the armature current decreases. This condition keeps the CEMF from building as quickly as the speed increases.
5. The CEMF is unable to reduce armature current rapidly enough to stop the motor from increasing its speed.

Even though the armature current continues to decrease, the torque it produces is enough to accelerate the unloaded motor until it breaks apart.

Due to their runaway characteristics, series motors are not recommended for belt- or chain-driven systems. A broken chain or belt could result in a no-load condition. It is unlikely that small motors will break apart if unloaded because there is usually enough bearing and brush friction to limit their speed.

Torque Characteristics of Series Motors

Series motors have the highest starting torque of DC motors. The reason for this is that when power is first applied to the motor and it is not turning, there is no CEMF produced. The entire applied voltage is across the series-connected armature and field windings. The current at that instant is limited only by the DC resistance of the two windings. The result is that a high current flows through the coils and produces strong magnetic fields. With strong magnetic fields formed around both windings, the interaction between them creates a large amount of force. The starting torque of a series motor is typically 350 to 500 percent of its full-load torque rating.

The starting armature current is lower in the series motor than in the shunt motor because it is opposed by two series coils. The opposition to the supply current is the armature coil and the field coil. Consider a series motor with an armature resistance of 5 ohms, a field coil resistance of 10 ohms, and 115 volts applied. The armature current at zero speed (no CEMF) is equal to the applied voltage divided by the total resistance of the armature coil and field coil:

$$I_A = \frac{V_A}{R_A + R_F} = \frac{115\text{ V}}{5+10} = 7.67\text{ A}$$

The magnitude of torque for a series motor is illustrated by the torque formula:

$$T = K_T \times I_A \times \phi$$

Since the field strength also depends on armature current, the torque equation is rewritten:

$$\begin{aligned} T &= C_T \times I_A^2 \\ \text{Motor Constant } C_T &= .5 \\ \text{Armature Current} &= 7.67\text{ A} \\ T &= .5 \times 7.67^2 \\ &= 29.4\text{ lb-ft} \end{aligned}$$

Note: C_T is a new constant that combines K_T with the ratio of field strength to armature current.

Compare this result to the torque calculation for shunt motors. The torque of the series motor is greater than that of a shunt motor even though its starting armature current is less. Therefore, one characteristic of a series DC motor is that it can provide a very high torque when starting, or when a sudden heavy load is encountered, as with cranes and railway installations.

The Compound Motor

IAU13908
DC Compound
Motors

A **compound motor,** shown in Figure 5-19, has both a series field and a shunt field. Both the series and shunt coils contribute to the field flux and are wound around the same pole pieces. The series field coil is connected in series with the armature circuit. The shunt field coil is

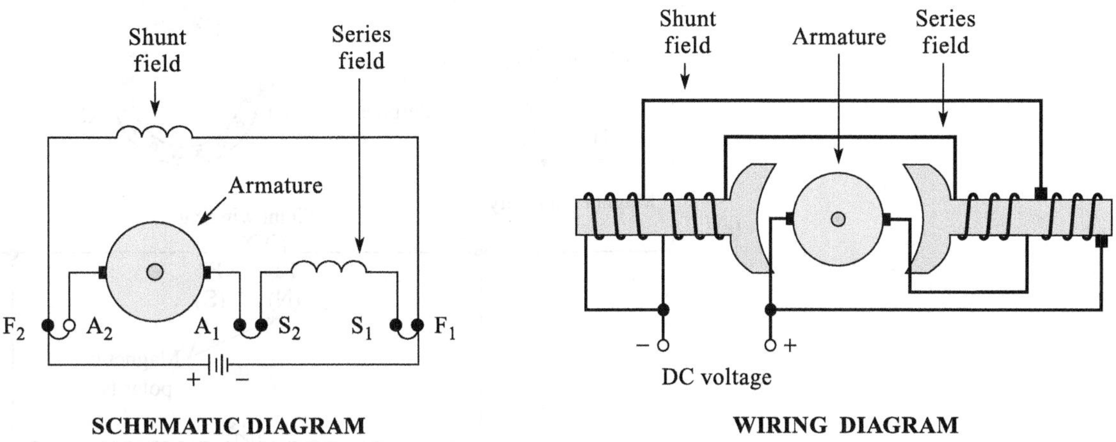

FIGURE 5-19 DC compound motor with the field connected in both series and parallel with the armature

connected in parallel with the armature circuit. There are two types of compound motors, *cumulative,* and *differential.* Each one has different characteristics.

Cumulative Compound Motor

The series and shunt windings of the cumulative compound motor are connected so that their magnetic fields have the same polarity, as shown in Figure 5-20(a). In this configuration, the magnetic fields of both windings are additive (cumulative).

The series field has the most impact on the operation when the motor is first turned on. Since the armature is not turning and producing a CEMF, a large amount of current flows through the series coil and a strong field is established immediately. This provides a high torque when starting, or when the load demand suddenly increases.

Suppose that the cumulative compound motor is operating in an under-load condition when the physical load is increased. The motor slows down, the CEMF decreases, and the armature and series field currents increase. However, since the armature is cutting the shunt field that has a constant strength, the amount that the CEMF weakens is limited. When enough current flows through the armature to create the necessary magnetic interaction (between the armature, series, and shunt coils) to match the torque demand of the increased load, the motor speed stops changing. The change in speed is more than that of a shunt motor, but less than that of a series motor. The speed regulation of a cumulative compound motor is about 25 percent.

If the cumulative compound motor encounters a no-load condition, the armature will speed up. However, it does not have the runaway characteristics of the series motor because a large enough CEMF is developed as the armature cuts through the series and shunt coil fields. When the armature current decreases to a certain level, the torque decreases so that it can no longer accelerate the motor, and the speed stabilizes.

Differential Compound Motor

The series and shunt windings of the differential compound motor are connected so that their magnetic fields have opposite polarities, as shown in Figure 5-20(b). In this configuration, the magnetic fields of both windings are opposite each other.

This configuration causes the series field to oppose the shunt field when a load is applied. The resulting decrease of the field flux will make the CEMF decrease and the armature current increase, causing the speed to stay relatively constant. Therefore, excellent speed regulation can be provided by a differential compound motor. However, if overloaded, the series field may become strong enough to override the shunt field, which causes the motor to stop and sometimes to reverse itself. Therefore, differential compound motors should not be used if it is likely that an overload condition will be encountered.

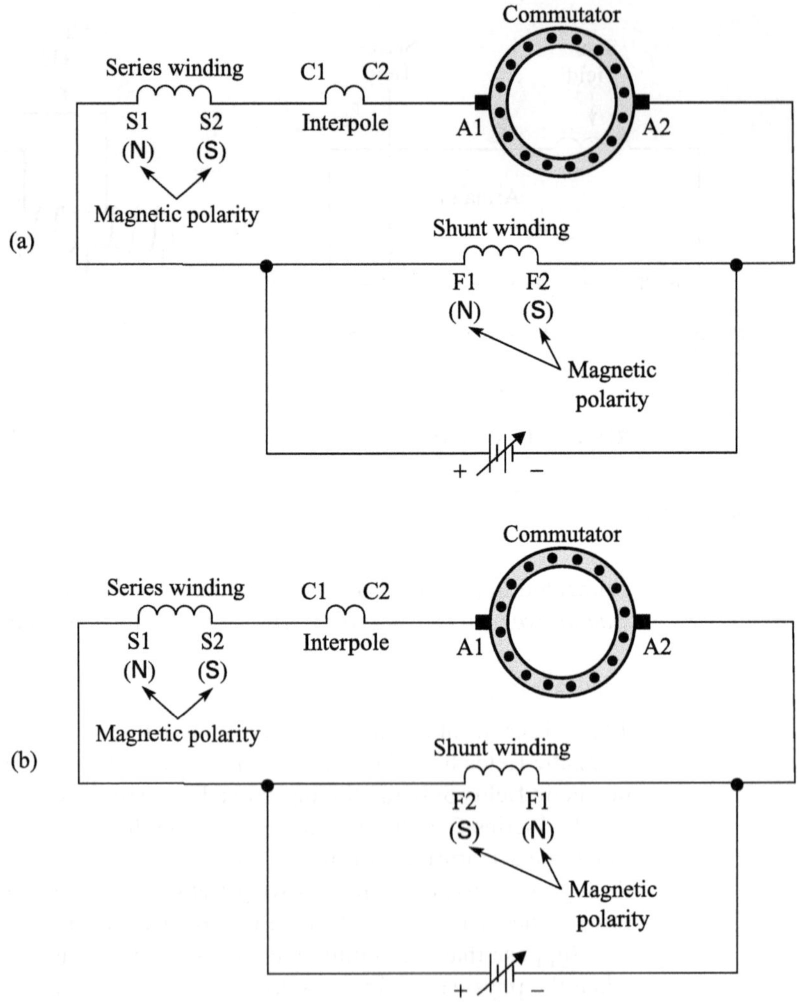

FIGURE 5-20 (a) A cumulative DC compound motor; (b) A differential DC compound motor

Torque Characteristics of Compound Motors

The starting torque of the cumulative compound motor is approximately 300 to 400 percent of its full-load rating, which is greater than that of the shunt motor but less than that of the series motor. It is not as strong as the series motor because of the influence of both the series field and the shunt field. The differential compound motor has less torque than either the series motor or the shunt motor. Because the series and shunt magnetic fluxes cancel, its overall field flux is weak.

If the physical load of a running cumulative compound motor is increased, its speed will reduce slightly more than with a shunt motor. The speed stabilizes when the increase in armature current causes an increase in torque to handle the added load. The speed of the differential compound motor, however, may rise as the load is increased before it stabilizes. Figure 5-21 shows the torque and speed characteristics of the three types of motors.

Compound motors are used in various industrial applications, such as freight elevators, stamping presses, rolling mills, and metal shears.

IAU9708 Reversing the Rotation of DC Motors

Reversing DC Motors

Reversing the direction of rotation of wound-field DC motors is achieved by changing the direction of electron flow through the field (or fields) relative to the electron flow through the armature. Therefore, the direction of rotation cannot be changed by simply reversing the

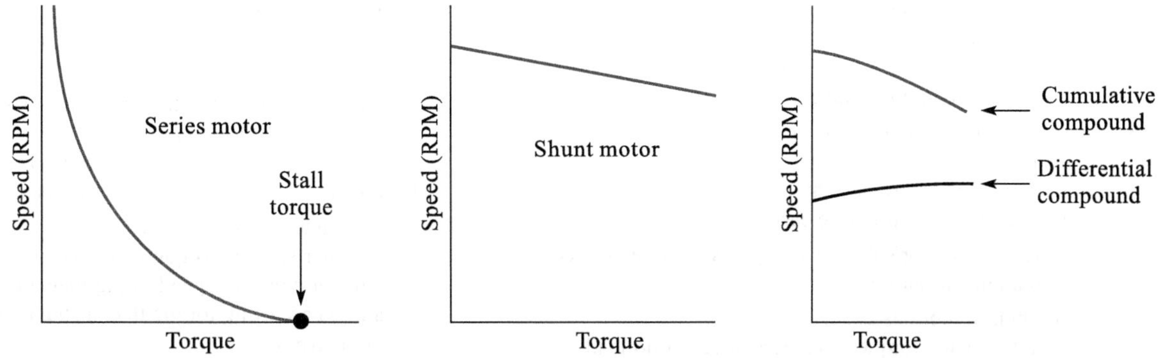

FIGURE 5-21 Graph of torque produced versus armature speed with line voltage held constant

negative and positive leads of the DC power source that feeds the motor. Instead, either the field windings or armature windings can be reversed, but not both. In a compound motor, both the shunt and series field coils must be changed, or else the motor will be switched from a cumulative to a differential configuration (or vice versa).

In industrial practice, to reverse the direction of motor rotation, it is standard to change the armature connections. If the motor has compensating windings, they are considered a part of the armature circuit. Therefore, current flow through them will also be reversed by changing the armature connections.

5-11 Coil Terminal Identification

IAU9608
Labeling DC Motor Connections

The electrical parts of a DC motor consist of different types of windings that are marked for identification. The shunt field winding, which consists of many turns of fine wire with a resistance between 100 and 500 ohms, is marked as F_1 and F_2. The series field winding, which consists of a few turns of a heavier gauge wire with a resistance of 1 to 5 ohms, is marked S_1 and S_2. The armature winding, which has a very low resistance, is marked A_1 and A_2. If the motor includes a commutating winding or interpole winding as part of the armature circuit, it is marked C_1 and C_2.

Problems

1. The twisting effect of a DC motor called ________ is produced primarily by the interaction of magnetic fields.
2. The main magnetic field in a motor comes from ____.
   a. the field coil
   b. the armature coil
   c. the commutator and brushes
3. Maximum field interaction occurs at the moment the armature conductors are moving ____ the main field.
   a. in the same direction as
   b. at a right angle to
4. Which of the following are functions of the brushes and commutator? ____
   a. To provide a path for armature current flow
   b. To connect and disconnect armature coils in sequence
   c. To provide a path for field current flow
5. T/F In a DC motor, there is a large inrush of current at first, which then drops off as the armature begins to rotate, generating CEMF.
6. T/F The interpoles are always connected in parallel with the armature.
7. When the motor speed increases, the CEMF in the armature ________ (increases, decreases).
8. Armature reaction is corrected by ________, which is/are used to shift the neutral plane back to the proper position.
9. Torque is produced by a motor when the load it is driving ____.
   a. moves
   b. does not move
   c. either a or b
10. Work is performed by a motor when the load it is driving ____ moved a distance.
    a. is
    b. is not
    c. either a or b
11. A motor rated at ¾ hp can also be rated at ____ watts of output power.
    a. 384
    b. 559.5
    c. 2.5
    d. 1.253

12. In an operating DC motor, the armature current depends on the applied voltage _____.
    a. minus the CEMF c. both a and b
    b. and the armature resistance
13. In a DC motor, an increasing mechanical load _____.
    a. increases armature current
    b. decreases armature current
    c. has no effect on armature current
14. T/F A shunt motor's field winding has more resistance than the armature.
15. A DC shunt motor has _____.
    a. a high starting torque c. zero speed at no load
    b. a constant speed rating d. all of the above
16. To change direction of a DC shunt motor with compensating windings, you must interchange leads _____ or _____.
    a. A_1 and F_1 d. F_1 and F_2
    b. S_1 and S_2 e. F_1 and C_1
    c. C_1 and C_2 f. A_1 and A_2
17. When the load on a DC shunt motor is increased, its speed will _____ and the amount of torque developed will _____.
    a. increase, increase c. increase, decrease
    b. decrease, decrease d. decrease, increase
18. A DC shunt motor operating at 240 V draws 4.5 A. It has an output of 1 hp. Its efficiency is _____.
    a. 69 c. 58
    b. 73 d. 43
19. T/F A series DC motor should never be connected to a load by a belt or chain drive.
20. The series field winding of a DC series motor has __________ (low, high) resistance.
21. A DC series motor has _____.
    a. low starting torque c. high starting torque
    b. low no-load speed d. zero speed at no-load
22. In a DC series motor, if the armature current is reduced to one-half of its full-load rating, the torque is _____.
    a. constant c. reduced
    b. doubled d. increased
23. A compound motor has _____.
    a. a higher starting torque than a DC motor
    b. a better constant speed rating than a shunt motor
    c. a higher starting torque than a shunt motor
    d. no interpoles
24. When connecting a compound motor for operation, which leads are wired in parallel with the power supply? _____
    a. A_1–A_2 c. S_1–S_2
    b. C_1–C_2 d. F_1–F_2
25. A differential compound motor has _____ than a cumulative compound motor.
    a. a higher starting torque c. more constant speed
    b. a higher RPM at no load
26. A cumulative compound motor has _____ than a differential compound motor.
    a. a higher starting torque b. a better speed regulation
27. T/F Neither a series nor a compound motor can be reversed simply by changing the input power leads.
28. The magnetic fields of the series and shunt motors of a cumulative compound motor _____.
    a. are additive b. cancel

CHAPTER

6

AC Motors

OBJECTIVES

At the conclusion of this chapter, you should be able to:

- Describe the principles of the alternating and rotating magnetic fields.
- List the different types of rotors and stators in AC motors.
- List the factors that determine the speed of an AC motor.
- Calculate the following for an AC motor:
  Synchronous Speed Slip
- List the different types of AC motors and describe their operation.
- Identify the characteristics of each type of AC motor.
- Reverse the direction of an AC motor.
- Choose the type of AC motor needed for specific applications.
- List and describe the types of information provided on a typical motor nameplate.

INTRODUCTION

An alternating current (AC) motor converts AC electrical energy into mechanical energy, producing a mechanical rotary action that performs some type of work. Because alternating current is the standard power generated and distributed, AC motors are the most common type of motors used in commercial and industrial applications.

Generators at power plants develop **three-phase power,** which is delivered to industrial plants. Huge motors use the three-phase electricity to provide the mechanical power for many types of production machinery, for example, pumps, cranes, and paper machines. Single-phase power is also delivered from the three-phase distribution to industry, residential, and small business customers. AC motors that use single-phase electricity typically produce less horsepower than three-phase motors. They drive such things as furnaces, air conditioners, washing machines, ovens, clocks, and fans.

There are many types of AC motors. Each one has different operating characteristics that provide the speed and torque capabilities for specific applications. Their durability enables them to operate 24 hours a day for many years without maintenance.

6-1 Fundamental Operation

Figure 6-1 shows a simplified diagram of an AC motor. It has two pole pieces with a permanent magnet placed between them. The coil of wire that wraps around the pole pieces forms electromagnets. The electromagnets are stationary and are called the **field poles** or the **stator.** The permanent magnet is free to turn and is called the **rotor.**

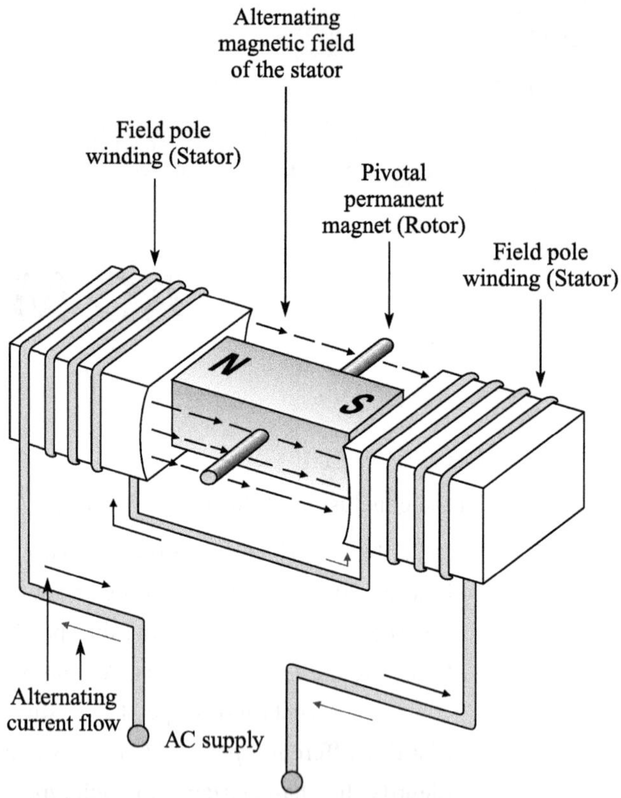

FIGURE 6-1 Fundamental AC motor

Alternating Field

The stator windings in Figure 6-1 are excited by AC power. The resultant field generated between the poles alternates with the applied alternating power. As the rotor magnet interacts with the poles of the stator, it pivots on its axis. The rotor will make one complete revolution for each complete AC cycle applied to the stator, as shown in Figure 6-2(a)–(e).

(a) At time T_0, no field is developed between the stator poles because there is no current.

(b) During time period T_1, the positive alternation of AC voltage occurs. As the field builds up around each stator piece, the polarities of the rotor ends closest to them are alike. The rotor begins to turn because the like poles are repelled. After the rotor goes past a quarter turn, it is attracted to the opposite poles of the stator. It continues to rotate until the N and S poles of the rotor are aligned with the opposite poles of the stator.

(c) At time T_2, the applied current is zero, and there is no field between the poles. Due to inertia, the rotor continues to turn past 180 degrees.

(d) During time period T_3, the AC current changes direction through the field coils. The polarity of the stator magnetic poles is reversed and the rotor is again repelled.

(e) After the rotor goes past three quarters of a turn, the rotor ends are attracted to the unlike stator poles. Also, the AC current and resultant field strength drops until it reaches zero. The inertia carries the rotor past 360 degrees as it begins another rotation and the next AC cycle is repeated.

Rotary Field

There are two disadvantages to the AC motor described in Figure 6-2. First, if the rotor was exactly parallel to the stator's flux lines, the magnetic repulsion would be equal and it probably would not rotate. It would start to turn only if the rotor was slightly offset.

Second, the rotor might not run in the desired direction. The direction it was offset from the stator's flux lines would determine the direction it turned.

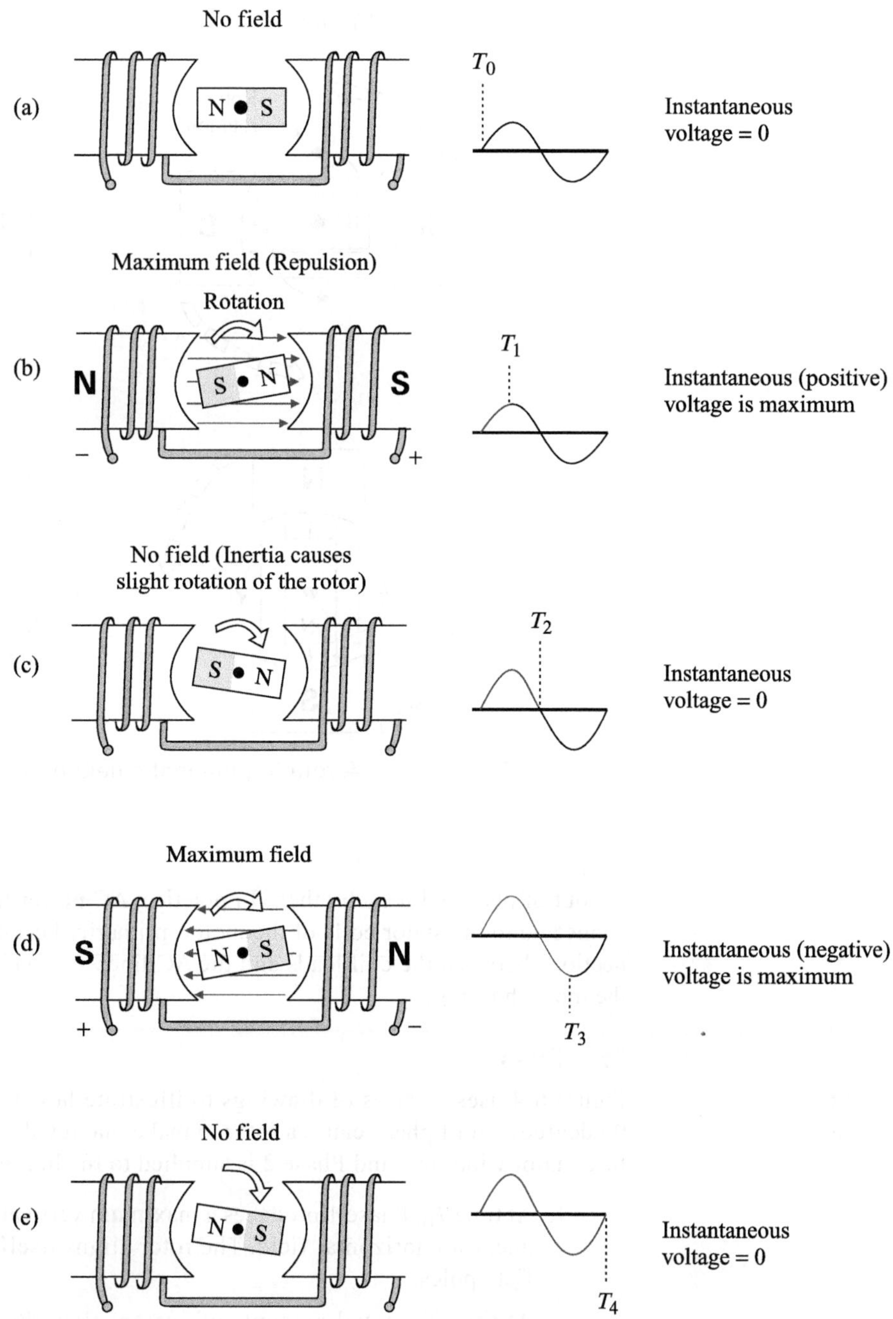

FIGURE 6-2 One revolution of an AC motor with one cycle of AC power applied

Both of these disadvantages are corrected by making the stator's magnetic field rotate instead of alternate, as shown in Figure 6-3. As the field poles revolve in a clockwise direction, they attract the opposite poles of the rotor. The result is that the rotor turns by following the rotating field.

6-2 Stator Construction and Operation

Stator Construction

It is impractical to physically rotate the stator field poles, as illustrated in Figure 6-3. However, it is possible to rotate the fields electronically by applying two or three sine waves that

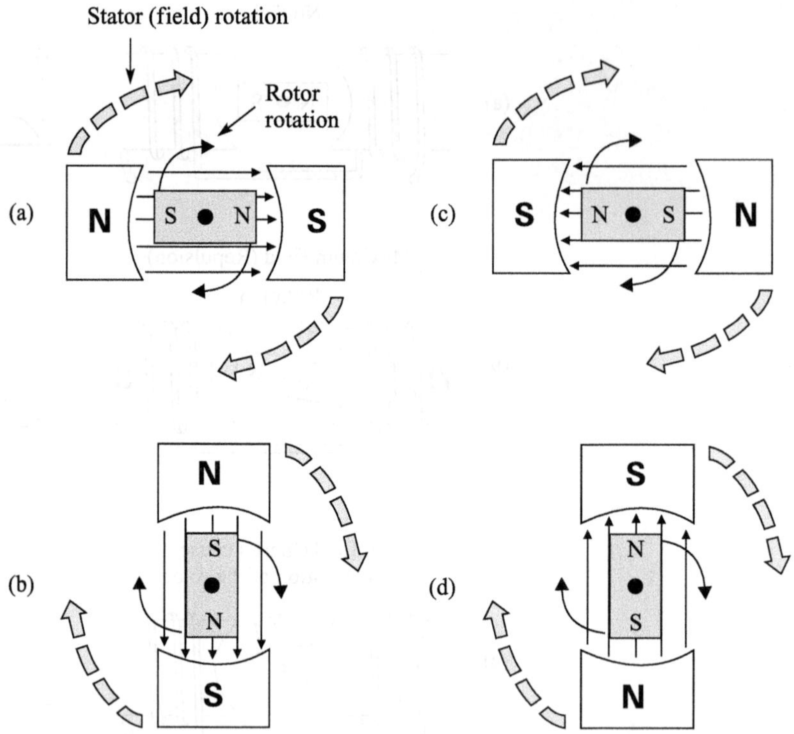

FIGURE 6-3 A rotating magnetic field of a stator

are out of phase with each other. In practice, AC power lines that supply the sine waves are connected to the stator coils of the motor, primarily because they are stationary. These connections between the coil leads and the AC lines are made inside a terminal box located on the motor housing.

Two-Phase

Figure 6-4 uses a series of drawings to illustrate how two AC current sine waves that are 90 degrees out of phase cause a rotor to make one revolution. Phase 1 is supplied to the vertical stator windings, and Phase 2 is supplied to the horizontal stator windings.

1. At time T_0, Phase 1 produces a maximum vertical magnetic field, while Phase 2 produces no horizontal field. The rotor aligns itself vertically with the two energized field poles.
2. At time T_1, equal amounts of current flow through both vertical and horizontal windings. A resultant flux develops between adjacent poles, which causes the rotor to turn 45 degrees counterclockwise (CCW).
3. At time T_2, no current flows through the vertical windings, while maximum current flows through the horizontal coils. The rotor turns another 45 degrees CCW and aligns itself between the horizontal poles.
4. At time T_3, current flow decreases through the horizontal windings. Meanwhile, the current flow through the vertical coils reverses direction. The resultant flux causes the rotor to turn another 45 degrees CCW.

Between time periods T_4 and T_8, the process continues and the rotor turns as it follows the rotating stator field. After the 360-degree rotation is completed, the next revolution will begin in the same direction. The rate at which the magnetic field in the stator rotates is called the **synchronous speed.**

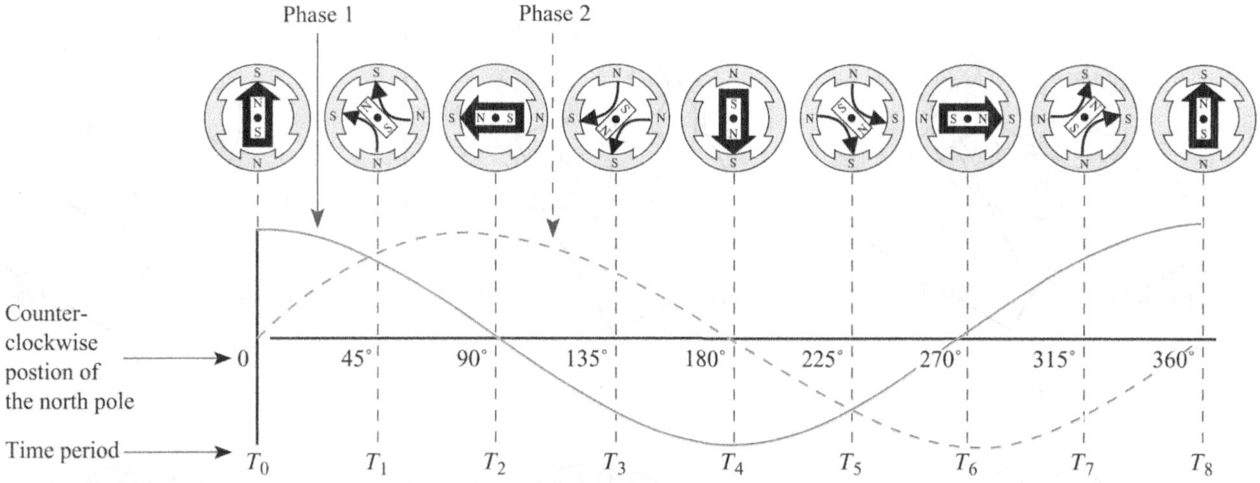

FIGURE 6-4 Two sine waves used to create a rotating magnetic field

Three-Phase

Industrial factories commonly use three-phase power, in addition to single-phase voltages. Three-phase AC power consists of three alternating currents of equal frequency and amplitude, but each differing in phase from the others by one-third of a period. This characteristic makes three-phase AC power ideal for developing rotating stator fields for motors.

The creation of a rotating stator field using three-phase power is illustrated in Figure 6-5. The three phases of alternating current can be thought of as three different single-phase power supplies. These three-phase currents reach maximum values at different times. Each phase supplies one of three separate pairs of coils wound around stator poles. The phases are designated as A, B, and C. Phase A supplies poles A_1 and A_2, Phase B supplies poles B_1 and B_2, and Phase C serves poles C_1 and C_2. Each set of windings is equidistant from the others. Because the three-phase currents are displaced in time by 120 electrical degrees, and the three-phase windings are equally spaced 60 mechanical degrees apart, the resulting magnetic field will rotate in space as though the poles are rotating mechanically.

Figure 6-5(a)–(f) illustrates the sequence of events that occurs during one 360-degree rotation of the stator field with three-phase power supplied.

- T_1 Figure 6-5(a) shows the resultant magnetic field from all three currents during time period T_1. Because Phase A has the greatest amplitude, the greatest concentration of magnetic flux lines is between stator poles A_1 and A_2.
- T_2 During time period T_2, Phase C has the greatest magnitude, causing the field to shift from poles A_1 and A_2 to poles C_1 and C_2.
- T_3 During time period T_3, Phase B has the largest amplitude, and the field shifts another 60 degrees between poles B_1 and B_2.
- T_4 During time period T_4, Phase A has the greatest amplitude, but current flow is in the opposite direction than it was during time period T_1. The field develops between poles A_1 and A_2, but at the opposite polarity.
- T_5 During time period T_5, the field develops between poles C_1 and C_2, but in the opposite direction than it was during time period T_2.
- T_6 During time period T_6, Phase B has the greatest amplitude and causes the stator field to rotate another 60 degrees between poles B_1 and B_2, but in the opposite direction than it was during time period T_3.

The changes in amplitude and direction of the current flow always occur in the same order, and at the same time interval, to create the rotating field. The direction of field rotation can be changed by reversing any two of the three-phase lines connected to the coils.

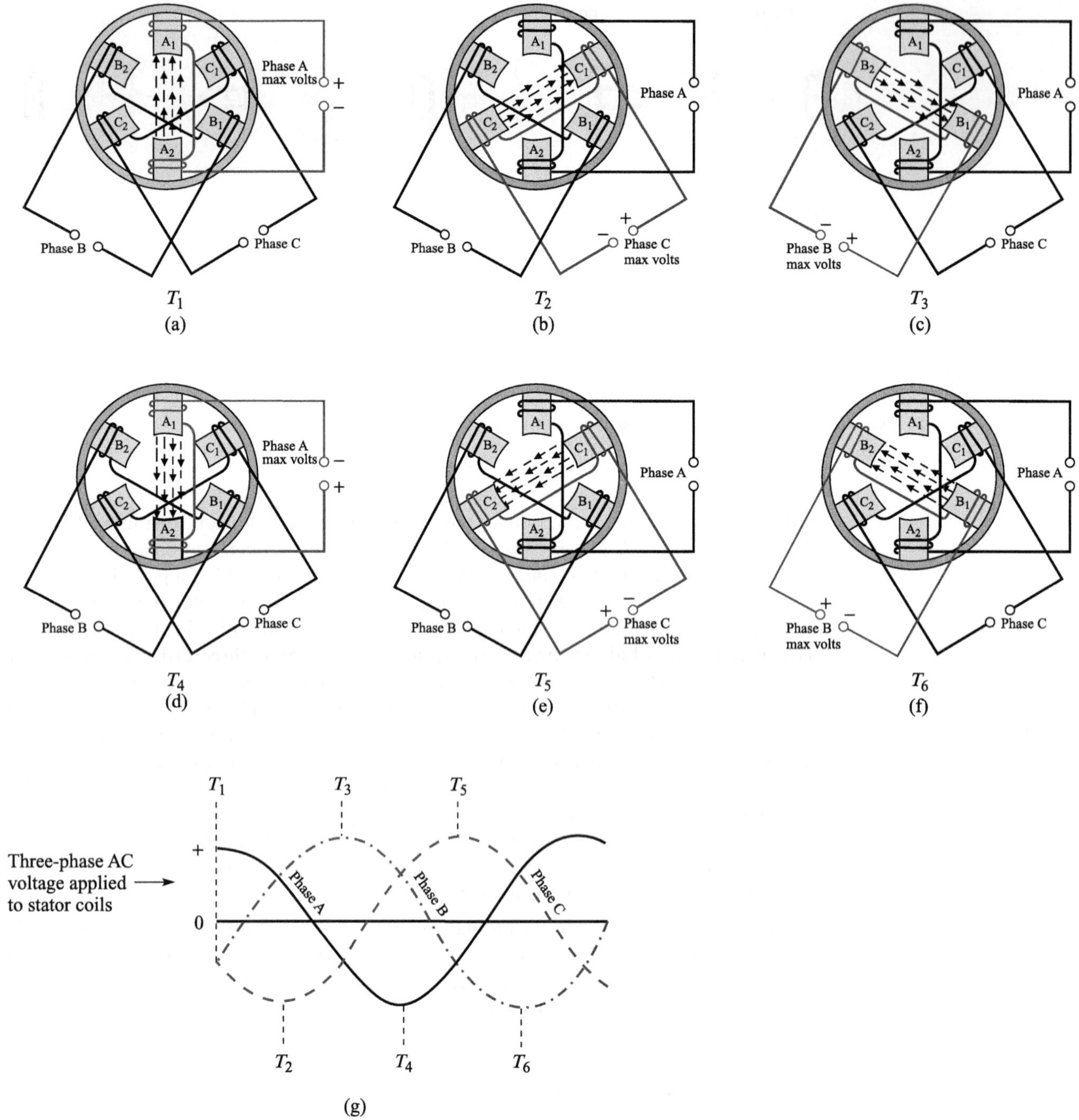

FIGURE 6-5 Three-phase AC power: magnetic fields and waveform

IAU13608
Synchronous Speed of an AC Motor

IAU11208
RPM of AC Motors

Synchronous Speed

The instant power is applied to the motor, current flows through the stator coils. The stator's magnetic field begins to revolve at synchronous speed. Three factors determine the speed at which the magnetic field rotates:

1. The frequency of the applied voltage
2. The number of stator poles per phase
3. Changing the inductance of the stator coils

The higher the frequency, the more rapidly the motor runs. The more poles a motor has, the more slowly it runs. The smallest number of poles possible in an AC motor is two.

One cycle of the applied voltage is required for each pair of poles to cause the rotor to turn 360 degrees. For example, in a 2-pole motor (1 pair), the stator field makes 1 revolution per cycle of 60 Hz power, or 3600 RPM. The formula for determining the synchronous speed of the stator field is:

$$N = \frac{f \times 60}{P}$$

where,

N = RPM
P = Number of pole pairs (per phase)
f = Applied frequency
60 = Formula constant based on seconds per minute

EXAMPLE 6-1

Find the synchronous speed of a 4-pole motor (2 pole pairs) with 60 Hz applied.

Solution

$$N = \frac{60\text{ Hz} \times 60\text{ seconds/minute}}{2\text{ Pole Pairs}} = 1800\text{ RPM}$$

AC motors are wound for synchronous speeds, as shown in Table 6-1.

TABLE 6-1 AC Motor Synchronous Speeds

Number of Poles	*60 Hz Synchronous Speed*
2	3600
4	1800
6	1200
8	900
10	720
12	600

The reason the synchronous speed of a four-pole motor is half that of a two-pole motor is described in Figure 6-6(a). The current from the AC source flows through all four coils

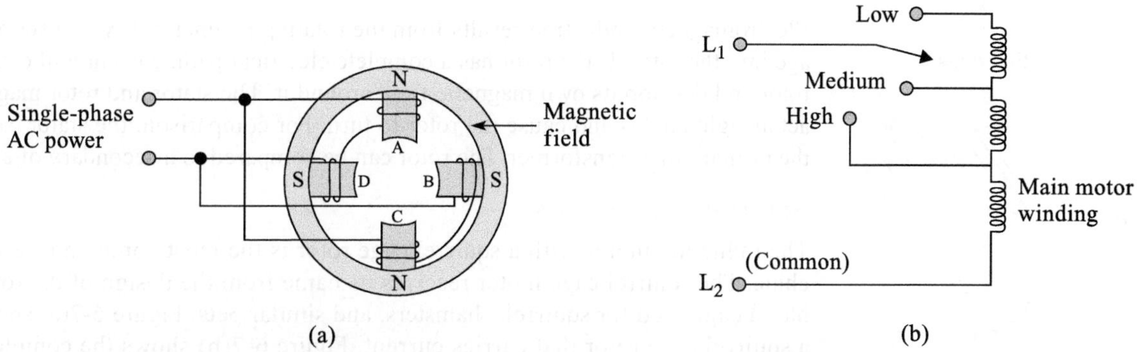

FIGURE 6-6 A four-pole stator: (a) With complementary poles placed 90 degrees apart; (b) A multispeed control

simultaneously. Because the coils are wound around adjacent poles in the opposite direction, complementary north and south poles are formed 90 degrees apart.

Suppose that during a positive alternation, windings A and C develop north poles and windings B and D form south poles, as shown in Figure 6-6(a). As a reference, consider the north pole at winding A. When the negative alternation occurs, the current through the coils reverses and the polarity at each pole changes. Windings A and C develop south poles and windings B and D form north poles. The effect is that the north pole at coil A rotates 90 degrees clockwise to coil B. During the next alternation, currents reverse and the north pole rotates clockwise to winding C. Every 180-degree alternation causes the field to rotate 90 degrees. Therefore, it requires two cycles of AC power to rotate the stator field one 360-degree revolution. The rotor follows the field in an attempt to lock in on it. An AC motor that can run at more than one fixed speed makes use of this principle. To change the speed, a switch is moved to connect the stator windings to a different number of poles.

In addition to changing motor speed, the reason for having more than two stator poles in a motor is to make the field stronger, causing the motor to run more smoothly.

The third method of varying the speed of a single-phase AC motor is by changing the inductance of the motor windings. The schematic diagram in Figure 6-6(b) shows a three-speed motor. The permanent main motor winding is located between the terminals marked *Common* and *High.* When the rotary switch is changed from the *High* position to the *Medium* position, additional coils are inserted. The increased inductive reactance reduces the current flow through the winding. The result is that the magnetic field is reduced and the motor produces less torque, causing the rotor to turn more slowly than the field, so its speed decreases. The torque and speed decrease further by adding more turns when the rotary switch is changed to the low-speed position. This type of speed control is generally used only to operate low-torque loads such as fans and blowers.

6-3 Types of AC Motors

IAU13408
Armature Action of AC Motors

There are two basic types of single-phase and three-phase AC motors, **induction motors** and **synchronous motors.** They differ in the way the magnetic fields at their rotor poles are created.

Rotor

If a permanent magnet were used as the rotor, it would turn as the magnetic field is rotated around the stator. AC motors do not use permanent magnets for their rotors. Instead, they use electromagnets. There are two methods of energizing the rotor so that it creates its own magnetic field. The first is to connect an electrical current to the rotor windings. This type of rotor is used for AC synchronous motors. In the second method, the rotor is not connected to any electrical source. Instead, it becomes an electromagnet through electromagnetic induction. This type of rotor is used for AC induction motors.

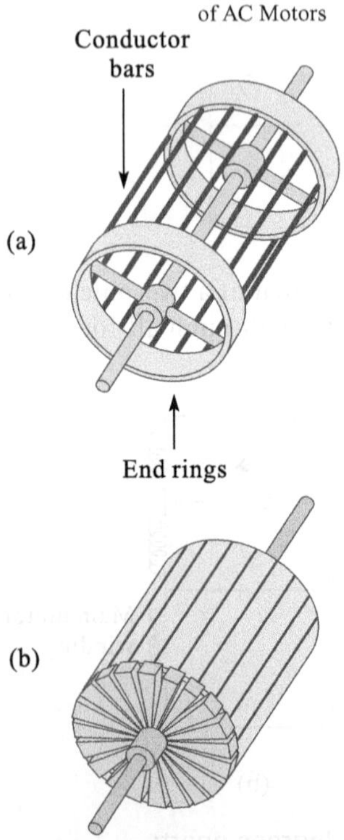

FIGURE 6-7 A squirrel cage induction motor rotor

Rotor Construction

Induction Motor Rotors

Electromagnetic induction results from the rotating magnetic flux of a stator inducing a voltage into the rotor. If the rotor has a complete electrical path, current will circulate through the rotor and develop its own magnetic field around it. The stator and rotor magnetic fields interact at right angles and cause the rotor to turn. For comparison, the stator can be described as the primary of a transformer. The rotor can be compared to a secondary of a transformer.

Squirrel Cage Rotors

The induction motor with a squirrel cage rotor is the most common rotating electrical machine. The squirrel cage motor receives its name from the design of the rotor, which resembles a cage used for squirrels, hamsters, and similar pets. Figure 6-7(a) shows the portion of a squirrel cage rotor that carries current. Figure 6-7(b) shows the complete rotor with the iron core in place. The "cage" portion is made of aluminum or brass bars, embedded just below the surface of the core. They are joined to conducting *end rings* that are placed at each end of the core. The end rings short-circuit the bars and provide a complete circuit path for current to flow through, regardless of the rotor's position. Note that a bar always forms a pair with another bar directly opposite it in the rotor. Along with the end rings, these pairs resemble the one-loop rotor that was described in Chapter 5.

Wound Rotors

Another type of rotor used by AC motors is the **wound rotor.** As its name implies, the rotor is constructed using wound coils of wire in place of the conducting bars of the squirrel cage motor. Current flows through the wound coils and creates a surrounding magnetic field. The rotor turns as its magnetic field interacts with the stator field. The current that flows through the rotor is either induced by the rotating stator field or is provided by an external DC power source. The number of rotor poles must be the same as the number of stator poles. Each winding terminates at slip rings that are mounted on the shaft of the motor. The currents are carried by brushes that ride on the slip rings to an external connection. The brushes connect either to a DC power source or to an external resistor bank (if the currents are induced). This type of rotor is commonly used in three-phase motors.

IAU10108
The Induction Motor slip

Principles of Operation When the stator winding is energized by a two- or three-phase supply, a rotating magnetic field develops at synchronous speed. As the field sweeps across the rotor, an electromotive force (EMF) is induced in the conducting bars by transformer action. The resultant current flows through the complete circuit loops consisting of the bars and end rings. Because these loops are short circuits with very low resistance, the current flow is high, producing a strong magnetic field. As the rotor and stator fields interact, motor action is created and the rotor turns, as shown in Figure 6-8. Unlike a DC motor, which has a stationary main field, the AC motor has a rotating field. Instead of the rotor turning 45 degrees until it is out of the main field, as in the DC motor, it follows the main rotating field. However, though it will chase the main rotating field, it will never catch it.

IAU13508
The Torque of an Induction Motor

Induction Motor Rotor Slip When the motor is turned on, the rotor is stationary and the stator field rotates. At this time, the relative positions of the two are as different as possible; maximum current is induced into the rotor, a strong magnetic field forms, and a large amount of starting torque develops. As the rotor approaches the synchronous speed of the stator field, less current is induced in the motor, and the rotor exerts less torque.

The rotor of an induction motor cannot run at synchronous speed. If it were possible for the rotor to attain the same speed as the rotating field, the flux lines of the stator could not be cut by the rotor. There would be no EMF induced into the rotor and no rotor current. Because its flux would be lost, there would be no torque developed to turn the rotor. However, this condition is not possible because there will be friction and windage losses. To induce an EMF, the rotor speed must be less than synchronous speed. This difference between rotor speed and synchronous speed is called **slip.**

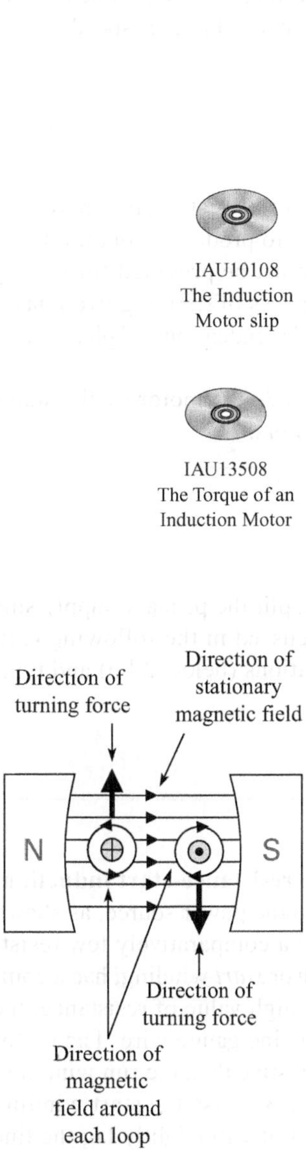

FIGURE 6-8 The interaction of the rotor and stator magnetic fields creates motor action, which causes the rotor to turn

If no weighted load is connected to the rotor shaft, the rotor and the stator rotating magnetic fields will spin at nearly the same rate. A minimal induced voltage will produce a very small amount of torque. In practical motors, the no-load slip is 2 to 10 percent. If a load is added to the motor shaft or the load is increased, the rotor will slow down and the slippage will increase. A larger amount of induced voltage will be developed and torque will increase. As the load increases, the percentage of slip increases. The amount of slip is also affected by the type of rotor bars used in the construction of the rotor.

The slip of an induction motor is expressed as the percentage of synchronous speed. The percentage of slip is determined by subtracting the speed of the rotor from the synchronous speed, and dividing the difference by the synchronous speed. Take, for example, a two-phase motor that has a synchronous speed of 3600 RPM and a rotor speed of 3450. The percent slip can be determined by using the following formula:

$$\text{Percent Slip} = \frac{\text{Synchronous Speed} - \text{Rotor Speed}}{\text{Synchronous Speed}} \times 100$$

$$= \frac{150 \text{ RPM}}{3600 \text{ RPM}} \times 100$$

$$= 4.16\%$$

At a certain point when the load is great enough to cause the slip to become excessive, an induction motor will reach a breakdown condition. This situation will develop when the slip is somewhere between 10 and 30 percent, depending on the motor. When breakdown occurs, the motor will suddenly stall because the rotor cannot slow down any more to produce enough torque. When a motor is stalled, it still produces torque. However, if it is stalled or runs well below its rated speed for too long, it will draw excessive current, overheat, and destroy the motor's insulation. Most motors have overload protection, but the protection cannot be totally relied upon.

Synchronous Motor Rotors

In a synchronous motor, the rotor poles are not produced by inductance. Instead, they are formed by an electromagnetic coil that receives current from a source of direct current. The current often is supplied by a controller located outside the motor. A rectifier inside the controller takes AC line voltage and converts it to DC. Brushes and slip rings are used to make the electrical connection between the controller's output and the rotating rotor coils. When it is at full speed, the rotor follows the stator's magnetic field, so it runs at synchronous speed.

6-4 Single-Phase Induction Motors

IAU11008
Single-Phase Motors

IAU10908
Phase-Splitting of Single-Phase Motors

Typically, single-phase commercial power is supplied to residential customers. Therefore, two- or three-phase AC voltages cannot be used by motor stators to produce a rotating field. To start a single-phase motor used in the home, some means must be provided for getting two phases from the standard single-phase AC source. One process of deriving two phases from a single-phase source is *phase-splitting*. The two sets of independent out-of-phase magnetic fluxes develop the rotating magnetic field.

The process of phase-splitting is performed electrically inside the motor by the stator circuitry. There are several popular types of single-phase AC motors:

1. Resistance-start induction-run motor
2. Capacitor-start induction-run motor
3. Shaded-pole motor

Their names are derived from the types of components used to split the primary supply sine wave into a simulated secondary phase. These motors are discussed in the following text. Single-phase motors are most often used for low-power applications (below 2 hp) and typically where three-phase power is not available.

6-5 Resistance-Start Induction-Run Motor

One of the most widely used types of single-phase motors is the **resistance-start induction-run motor.** It has two separate windings connected in parallel to the power source, as shown in Figure 6-9(a). One coil, called the *main* or *run* winding, has a comparatively low resistance and a high inductance. The second coil, called the *auxiliary* or *start* winding, has a comparatively high resistance and lower inductance. To achieve the high value of resistance, the start winding is made of fewer turns than the run winding with fine gauge wire. The motor receives its name from the fact that the start winding is more resistive than the run winding.

When power is first applied, both windings are energized. Because the start winding has low inductance and high resistance, the current flow through it will slightly lag the line voltage. Since the run winding is more inductive, the current flow through it will appreciably lag the applied voltage, as in any inductive circuit. The resultant two out-of-phase currents resemble a two-phase power source. Ideally, the phase difference should be 90 degrees, because maximum starting torque is developed in this situation. In practical motors, however, the phase difference is much less. In the resistance-start motor, the phase difference is 35 to

40 degrees (Figure 6-9(b)). Nevertheless, the phase difference of the currents is enough to create two magnetic fields that are out of phase to form an overall rotating magnetic field in the stator. This condition applies torque to the rotor, thereby starting the motor.

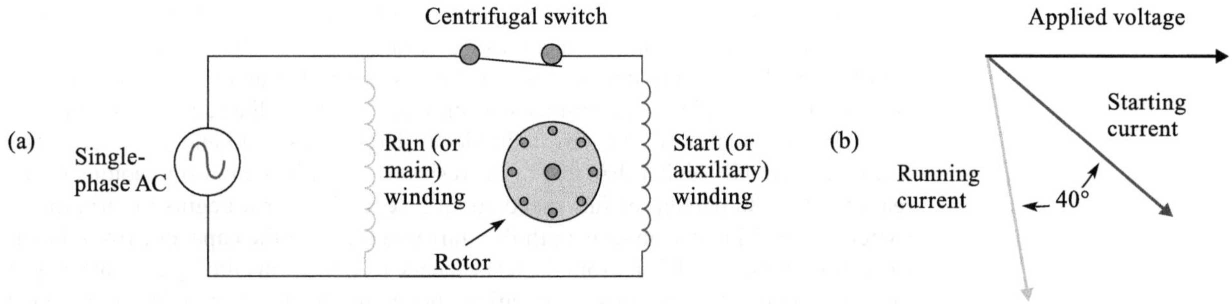

FIGURE 6-9 The resistance-start split-phase motor

The rotating field in the stator is necessary only to start the rotor turning. Once the rotor accelerates to about 80 percent of its normal speed, it is able to follow the alternating magnetic field created by the run winding. Since the field of the start winding is no longer required, it is removed from the circuit by a mechanical device called a *centrifugal switch,* which is connected in series with the start winding. The centrifugal switch contains a set of spring-loaded weights, which push a fiber washer against a movable switch contact. At start-up, the contacts are closed, which electrically connects the start coil to the power source. As the shaft accelerates, the centrifugal force causes the weights to overcome the force of the springs. The washer retracts and the contacts open, which disconnects the start winding from the circuit. It is necessary for the start winding to be disconnected because, if it is not, its high resistance will generate enough heat to burn out the coil.

Operating Characteristics

The primary advantages of the resistance-start motor are that it is inexpensive, requires very little maintenance, and has constant speed characteristics. The no-load current is usually 60 to 80 percent of the current drawn by the motor at full load. Most of the no-load current consumed by the motor is used to produce the magnetic fields around the motor's coils. Only a small portion is used to overcome the mechanical friction and the copper and iron losses.

One disadvantage of a resistance-start induction-run motor is its low starting torque. Two conditions cause this characteristic. The first is that the start windings are made up of thin wire that has high resistance, which limits current and causes a relatively small magnetic field to form. The second condition is that the main winding current lags behind the auxiliary winding by a small amount, resulting in a weak rotating field. These conditions limit the starting torque to only 150 to 200 percent of the motor's rated running torque at full load.

Since the high starting current decreases almost instantly, this is not a major problem. However, resistance-start motors larger than 1/3 hp are usually not approved by power companies for applications that require frequent starting and stopping.

Another disadvantage of this type of motor is its noise. Because of the varying magnitude of the magnetic fields that cut the rotor, the torque developed under load is pulsating and causes a 120-cycle vibration. This vibration can be reduced by using resilient rubber mounting supports.

Resistance-start motors are most commonly manufactured in sizes from 1/30 hp to 1/2 hp. They are widely used to drive loads that are fairly easy to start, do not require reversing, and do not need to be started and stopped frequently. For example, they are well suited for small machines such as drill presses, oil burners, sump pumps, some washing machines, and a number of other household appliances. These motors run on both 115 and 230 VAC.

6-6 Capacitor-Start Induction-Run Motor

Another type of split-phase motor is the **capacitor-start motor.** Like the resistance-start motor, the capacitor-start motor has two windings: a start winding and a run winding. They are both connected across the line, and both are in parallel. However, the capacitor-start motor has a low reactance electrolytic capacitor in series with the start winding. The capacitor value ranges from 150 to 180 μfd. To keep it at a reasonable size, an electrolytic type made for intermediate duty is used. It is usually mounted in a metal casing located on top of the motor. A centrifugal switch is connected in series with the capacitor and start windings, as shown in Figure 6-10(a).

The purpose of the capacitor is to produce a larger phase shift and a resultant starting torque that is substantially higher than that of the resistance-start induction-run motor. When the motor reaches 70 to 80 percent of full speed about 3 seconds after it begins to turn, the centrifugal switch opens. This disconnects both the start winding and the capacitor from the circuit. The capacitor-start motor differs from the resistance-start motor only during the starting period. After the machine reaches its normal operating speed and the auxiliary winding is removed, their performance becomes almost identical. Therefore, the combination of a high starting torque with the constant RPM capabilities of the resistance-start motor gives the capacitor-start motor the ability to maintain excellent speed regulation under a wide range of load conditions.

There are two ways in which the capacitor improves starting torque:

1. The capacitor causes the start-winding current to lead the applied voltage. Because the run-winding current lags the applied voltage the same way it does in the resistance-start motor, the phase shift between the currents of the two windings is nearly 90 degrees, as shown in Figure 6-10(b). Under this condition, the motor approaches two-phase operation.
2. In a resistance-start motor, the number of turns in the auxiliary winding must be kept low so that the current in the start winding is nearly in phase with the applied voltage. The result is that during starting, the surge current is high and gives the motor about 150 percent more than its rated base speed running torque.

   During start-up with a capacitor-start motor, the reactance of the capacitor cancels the inductive reactance of the start winding. This effect allows a high surge current for a brief period of time, causing a large magnetic field to form around the starting coil. The auxiliary coil of a capacitor-start motor has a greater number of turns than the auxiliary coil of a resistance-start motor. Therefore, its coil has a greater number of ampere-turns, which produces a larger rotating flux and a stronger starting torque. Despite having a larger inductive reactance due to more coil turns, the phase shift is kept close to 0 degrees because X_C and X_L cancel each other.

It is very important that the centrifugal switch operates properly. If the capacitor is kept in the circuit too long, it will be damaged or its life shortened appreciably. Also, it should not be used to start a motor more than eight times per hour; frequent starting can cause it to overheat. It is important to use a replacement capacitor with a proper microfarad rating. If the capacitor is too small, the starting current will be less than 90 degrees out of phase with the run current. If the capacitor is too large, the starting current will be more than 90 degrees out of phase with the run current. In both cases the torque will be reduced.

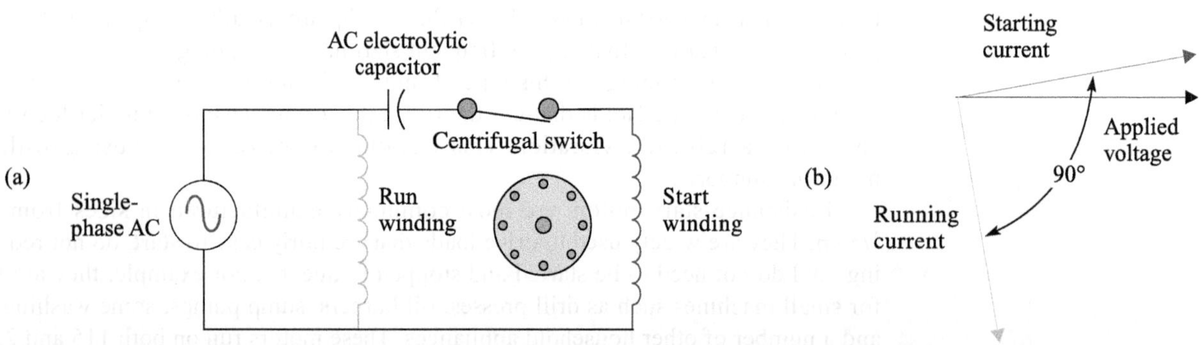

FIGURE 6-10 The capacitor-start induction-run motor

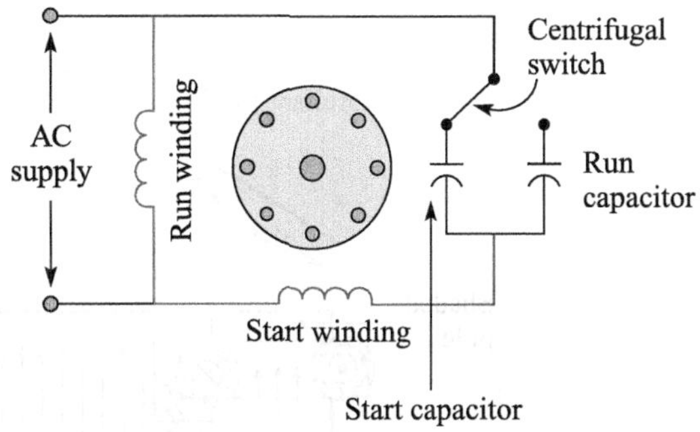

FIGURE 6-11 Capacitor-start, capacitor-run single-phase AC motor

A capacitor-start motor has a starting torque that ranges from 225 to 400 percent of its rated full-load running torque, or roughly 2.5 times greater than a resistance-start motor. Capacitor-start motors are manufactured in both fractional and integral sizes, up to 7.5 hp. They are well suited for applications that require relatively frequent starting and hard-to-start loads, such as pumps, conveyors, and compressors used by refrigeration systems and air conditioners. They also drive machine tools that require single-phase power.

The direction of rotation of a capacitor-start motor is changed by reversing the connections of either the main winding or the auxiliary winding, but not both. In practice, the start winding circuit leads are interchanged.

Some larger capacitor-start motors use two capacitors in the start winding, as shown in Figure 6-11. With this type of motor, called a *capacitor-start capacitor-run motor*, its start winding is not disconnected from the line. One capacitor is larger than the other. The larger one is used when the motor starts. When the motor reaches about 75 percent of its operating speed, a centrifugal switch disconnects the start capacitor and connects the run capacitor. The purpose of keeping the start winding connected is to maintain split-phase power. This enables the motor to have excellent starting and running torque, good speed regulation, and a power factor of nearly 100 percent at rated load, and causes it to run quietly and efficiently.

The capacitor-start capacitor-run motor is normally manufactured in sizes from 5 to 20 hp. Practical applications are oil burners, fans, and metal and woodworking machines that run on single-phase power.

6-7 Shaded-Pole Motor

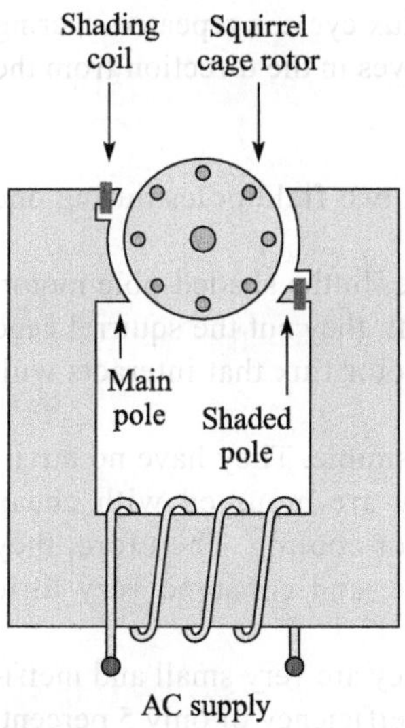

FIGURE 6-12 Shaded-pole motor

The **shaded-pole motor** is a type of induction motor that uses a squirrel cage rotor and a main stator winding. However, it differs from other types of induction motors in the manner in which it develops the required rotating field. Figure 6-12 shows that the stator poles are divided into two parts. The smaller segment is called the shaded pole and is surrounded by a metal ring called a *shading coil.* The larger segment is called the *unshaded pole* or *main pole.* The shading coil forms a complete circuit and operates in the same manner as a transformer with a shorted secondary winding. Its function is to delay the flux lines from passing through the shaded pole until they are about 90 degrees behind the applied voltage.

The movement of flux around the stator poles is described by the following explanations and shown in Figure 6-13:

1. Figure 6-13(a): When the current of the AC waveform increases from zero toward a positive peak, the flux builds up throughout the stator pole. As the flux lines cut through the shaded pole, an EMF is induced in the short-circuited ring, which causes current to flow. The current develops flux lines around the conducting ring in the direction shown by the curved arrows. Note that the ring coil flux and main pole flux lines are in opposite directions within the inner area of the shaded pole. The effect is that they cancel each other, which weakens the main pole flux lines that pass through the shaded pole. Some of the shaded pole flux is diverted, which causes the bulk of the magnetism to pass through the unshaded side.

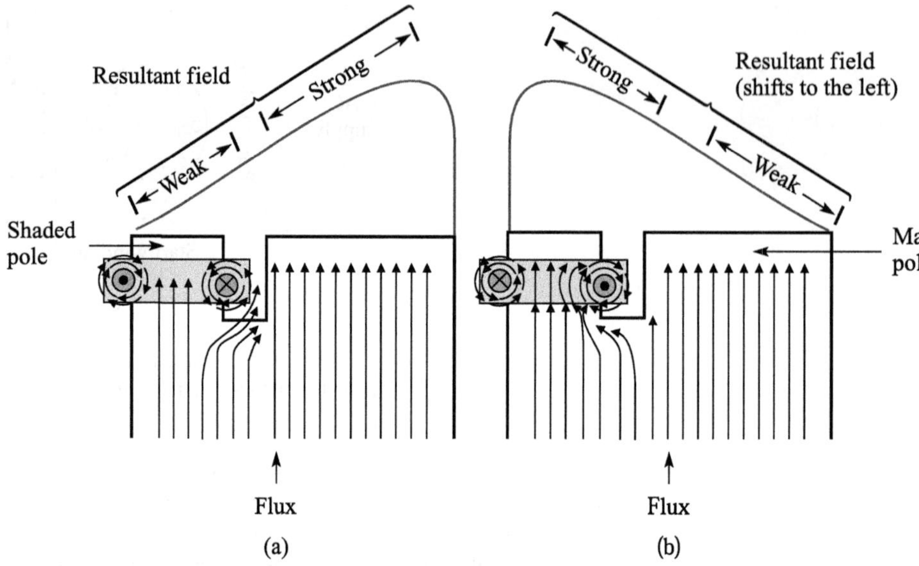

FIGURE 6-13 Shaded-pole motor flux

2. When the AC current reaches its peak value, the flux does not change. With a constant flux strength, the shading coils will not be cut by moving flux lines. There will be no flux lines developed around the ring to affect the main field because no EMF is induced. At this time, the main pole flux lines will be distributed more uniformly over the entire stator pole piece.
3. Figure 6-13(b): When the stator voltage decreases, the flux also decreases through the stator pole. The flux lines cut through the ring coil in the opposite direction. The induced EMF and resulting current reverse polarity and the flux lines around the ring coil change direction, as shown by the curved arrows. The ring coil and main pole flux lines are aligned in the same direction within the inner area of the shaded pole. The effect is that they reinforce each other and tend to oppose the decrease in flux. This delays the collapse of the stator pole flux lines that pass through the shaded poles. Some of the main pole flux is diverted and concentrated through the shaded pole.
4. When the AC current passes through zero and increases toward the negative peak, the flux lines in the stator pole change direction as the flux cycle is repeated. During either alternation of the AC cycle, the resultant field moves in the direction from the unshaded pole toward the shaded pole.

Figure 6-14 shows the effect of the shifting flux lines between two field poles during one alternation of an AC cycle.

In the other AC motors described, the magnetic fields rotate. In the shaded-pole motor, the field merely shifts across the pole face. As the flux lines shift, they cut the squirrel cage rotor bars and induce an EMF. The resulting current creates a rotor flux that interacts with the stator flux to develop the torque needed to turn the rotor.

The construction of shaded-pole induction motors is very simple. They have no auxiliary winding, no capacitor, and no centrifugal switch. They are mounted with cheap sleeve bearings and are designed to have air pass over them for cooling. Therefore, they are very inexpensive, rugged, require very little maintenance, and consume very little electricity.

However, shaded-pole motors have several disadvantages. They are very small and inefficient. The smallest size, which produces $^1/_{120}$ hp, operates at an efficiency of only 5 percent. A motor of $^1/_{20}$ hp is 35 percent efficient, and the starting torque is only 40 to 50 percent of

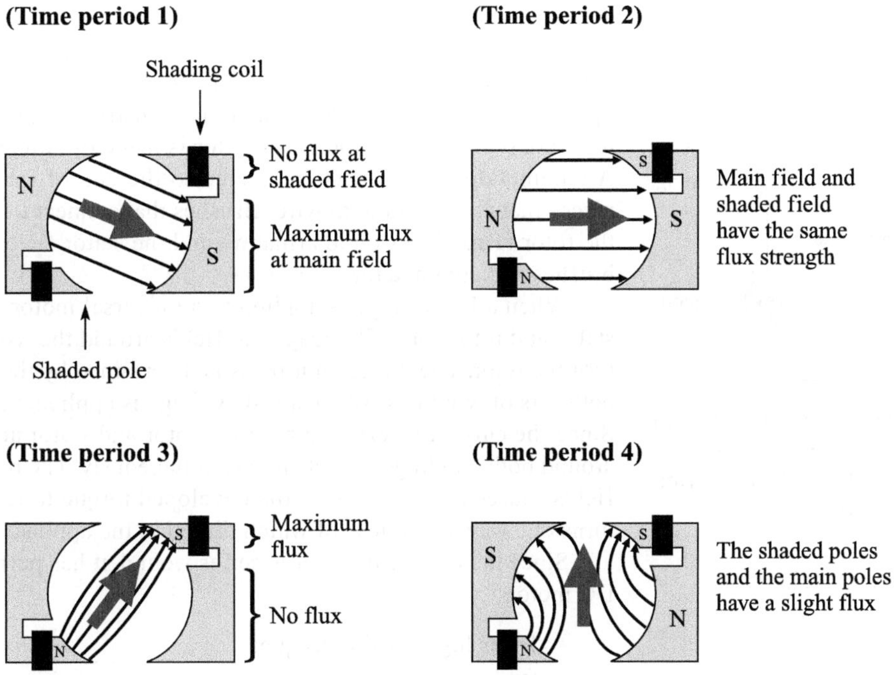

FIGURE 6-14 The movement of the overall magnetic field between two field poles during one alternation of an AC cycle in a shaded-pole motor

full-load torque, though they handle overloading very well. Rotation is usually restricted to one direction.

These types of motors are used for light load applications that require small output horsepower, such as clocks, fans, blowers, pumps, toys, and other items that are inexpensive to make and operate. The most frequent cause of failure is dry bearings.

6-8 Troubleshooting Split-Phase AC Motors

The most frequent cause of malfunction in capacitor-start motors is a defective capacitor. If it opens, the start winding is disconnected, the motor begins to hum, and it will not start. If a short exists, the current will become high and will either blow a circuit breaker or burn out the start winding.

If a resistance-start motor does not turn when power is applied, give the rotor shaft a spin by hand. If it runs, either the centrifugal switch or the start winding is open. An ohmmeter can be used to check an open winding or a defective switch.

Thermal Protection

Some split-phase motors have built-in thermal overload protection. When a predetermined temperature is reached, a thermal switch made of a bimetal strip opens. It protects the run and start windings from overheating by removing them from the power source. The overheating can be caused by a lack of proper ventilation or an ambient temperature that is too high. It is also caused by excessive current that results from either a motor load that is too high or one that prevents rotation. Some thermal protection devices automatically reconnect the motor to the line after the motor cools off. Other devices are reactivated by a reset button mounted on the frame.

6-9 Universal Motors

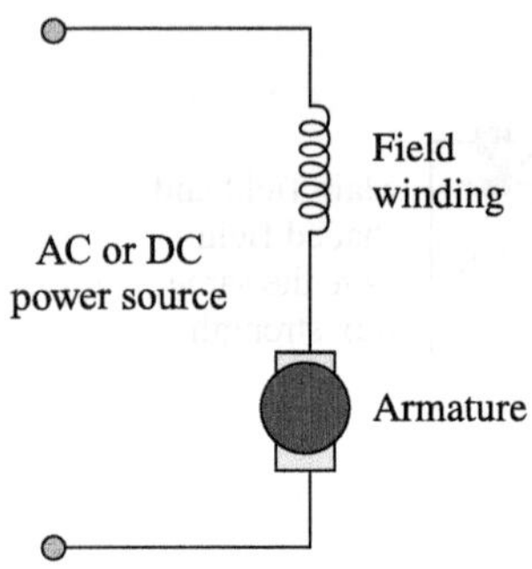

FIGURE 6-15 The circuit for a universal motor, similar to that of a series DC motor

A **universal motor** is usually categorized as an AC motor. It gets its name from its ability to operate on either AC or DC voltage. Its construction, shown in Figure 6-15, is very similar to a series-wound DC motor. The rotor is made of laminated iron wound with loops of wire. A commutator segment is connected to the end of each loop. The stator is made of pole pieces that are wound with wire. Brushes that connect to the stator ride on the commutator as the rotor turns. The wound armature and the stator field are connected in series through the brushes and commutator.

When a DC voltage is applied to a universal motor, the same current flows through the stator and rotor coils. The magnetic fields around the winding interact and develop torque to turn the rotor. The direction it turns is determined by the direction the current flows through both sets of windings. When an AC voltage is applied, the direction of current will alternate. Since the current reverses in both the rotor and stator at the same time, the magnetic fields around both windings also change simultaneously. The result is that the interaction of the two fields causes the direction of the developed torque to remain the same. Therefore, the rotor turns one way, regardless of which direction the applied current flows.

Since the universal motor is series wound, it has performance characteristics similar to a DC motor.

- It has high starting torque.
- The no-load speed is very high, but not high enough to break apart.
- It has poor speed regulation when the load to which it is connected changes. To minimize the effect of the load, most universal motors are designed to overcome this by operating at a very high speed of 3500 RPM or greater. In a router application, for example, the universal motor runs at 18,000 RPM.

There are three significant differences between the DC series motor and the universal motor:

1. In a DC motor, the iron cores are made of solid iron. The universal motor uses laminated iron to reduce energy loss from excessive heating due to eddy currents that are created from magnetic fields constantly reversing direction.
2. In a universal motor, the magnitude of the fields will fluctuate at twice the line frequency (120 times a second). This condition creates a reduction in output torque compared to a DC motor, which has constant field strength. This reduction of output power is partially compensated for by using more armature loops.
3. There is an excessive voltage drop across the series field windings of a DC series motor due to the high inductive reactance that develops when an AC voltage is applied. To minimize the voltage drop across the series coil of a universal motor, it is wound with a small number of turns on a low reluctance core.

In a universal motor, a third set of coils called *compensating windings* is connected in series with the rotor and stator. These windings perform two functions: They correct the neutral plane position that is distorted by armature reaction, and they compensate for undesirable reactive voltages attributed to the inductance of the armature.

The speed of a universal motor is determined by the following factors:

- A change in the applied voltage. As the voltage increases, the current rises to create a stronger magnetic field, which causes the speed to increase.
- A change in the load. As the physical load the motor is driving becomes harder to turn, the speed decreases.
- A change in frequency of the power supply. As the frequency increases, inductive reactance of the series coil rises, which causes current flow through both the field and armature windings to decrease. The result is that the magnetic fields around them become weaker, creating a weaker torque, and the motor slows down.

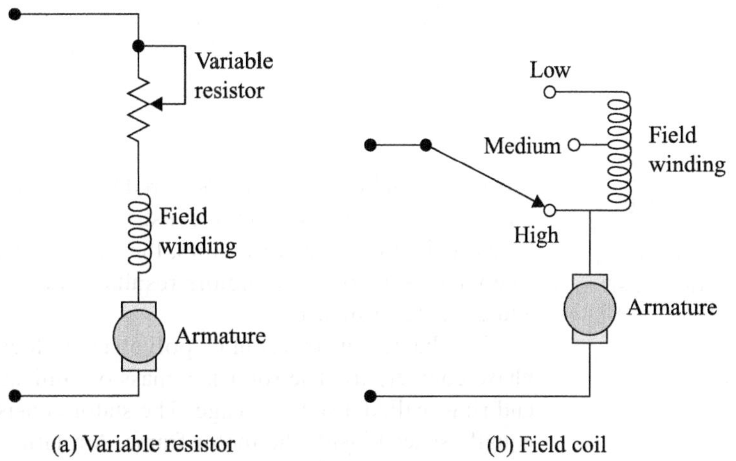

FIGURE 6-16 Methods used to control speed of a universal motor: (a) Variable resistor (b) Tapping off various points of a field coil

Two methods commonly used for speed control of the universal motor are as follows:

1. By inserting a variable resistor in series with the field coil and armature, as shown in Figure 6-16(a), the voltage across them can be changed. An application example is a sewing machine. Its speed is varied by a foot pedal, which contains a variable resistor. Another example is a variable-speed hand drill. As the trigger is pulled, a resistance in series with the motor is changed to vary the RPM.
2. By tapping a field coil at various points, as shown in Figure 6-16(b), different inductive reactance values are developed. When the switch is set at the low-speed position, the entire winding is used and the maximum inductive reactance forms and causes minimum current to flow. At the high-speed position setting, the entire coil is bypassed and the inductive reactance is reduced, causing a higher current to flow. An application example of this principle is a blender. A different field pole connection is made for each speed position setting.

Universal motors are used in many portable applications that require high horsepower for the size. A few examples are vacuum cleaners, polishers, hedge trimmers, circular saws, and mixers. One of the most frequent malfunctions encountered with universal motors is worn-out brushes. They typically require replacement after 300 to 1000 hours of use. Another fault that develops is shorted armature windings, which occurs when excessive currents break down the insulation of the armature wire when the motor is overloaded.

The direction of rotation can be reversed by changing the connections to either the field coil or the armature, but not both. In actual practice, only the armature connections are interchanged.

6-10 Three-Phase Motors

Most of the motors used in industry operate directly on three-phase power. Also called *polyphase* motors, they have several advantages over single-phase motors. They are simpler in construction, more efficient, and less likely to become defective. Also, by using three phases, a more powerful machine can be built into a smaller frame.

There are three types of three-phase motors:

1. Induction motor
2. Wound-rotor motor
3. Synchronous motor

All three motors use the same basic design of the stator winding, but differ in the type of rotor used. They are discussed in Sections 6-11, 6-12, and 6-13.

6-11 Induction Motor

IAU11808
The Three-Phase Motor Stator Field

Most industrial machines are powered by three-phase squirrel cage **induction motors.** These motors are efficient, simple in construction, and require very little maintenance. When compared with other types of motors, their physical size is small for a given horsepower rating. The wide usage of these motors results from their relatively low cost, rugged construction, and good performance.

The basic construction of polyphase induction motor is much like that of the single-phase counterpart. The rotor is a mass of laminated iron with embedded conductor bars and end rings called a squirrel cage. The stator consists of a group of coils wound on pole cores equally spaced inside the motor frame. A rotating field moves around the stator.

The concept of the rotating magnetic field is illustrated in Figure 6-17. There are six pole cores 60 degrees apart. There are three sets of paired coils labeled A, B, and C. Each set is connected to one phase of the three-phase power source. The coils in each pair are wound around cores that are located opposite each other. Since current flows through one coil of a pair from the outer to the inner connection, while current flows through the other coil from the inner to the outer connection, the magnetic fields developed at the poles facing each other are of different polarities. The diagram illustrates how a single resultant magnetic field is created by combined currents that flow through the three pairs of stator coils. Seven different time intervals are shown during one three-phase cycle.

T_0 At this time, Phase B is at the peak of the positive alternation and the current through the B coils is at a maximum value of 20 amps. Phases A and C are negative and at an amplitude of –10 amps. Most of the strength of this field is produced by the current flowing through the pair of B coils, which is at maximum strength. This field is aided by the adjacent A and C fields. The resultant field is in the direction shown by the inside arrow. It forms inside the pole pieces where the rotor is located.

T_1 At time period T_1, the current in Phase B decreases to +10 amps, Phase C reverses direction to +10 amps, and Phase A increases to –20 amps. Most of the field is

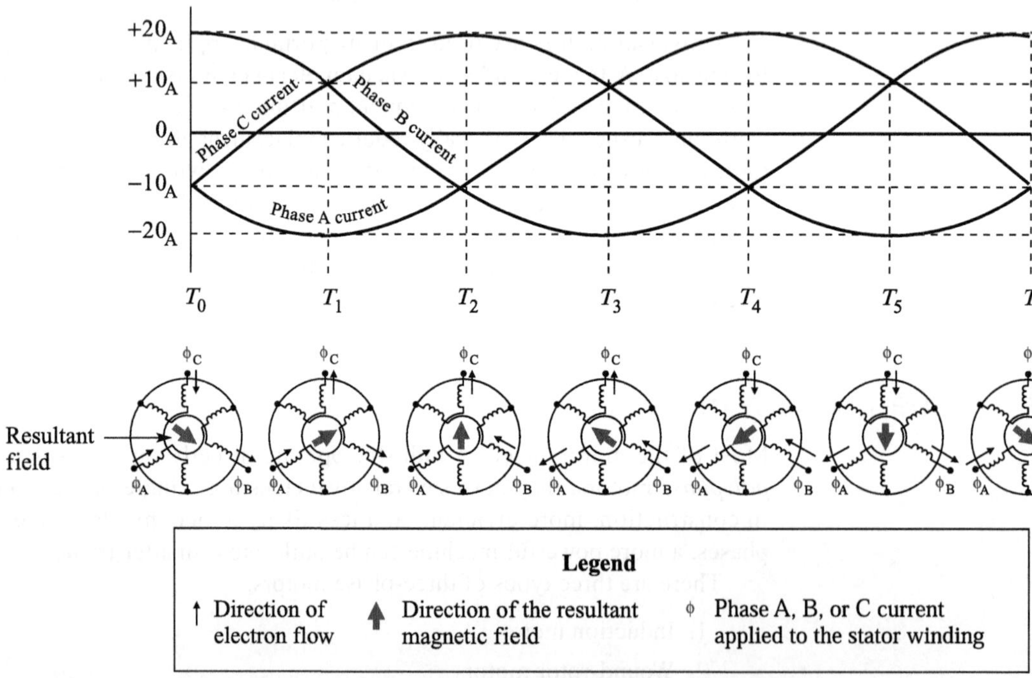

FIGURE 6-17 The rotating field of a three-phase induction motor

produced by current flowing through the pair of A coils, which is at maximum strength. The resultant field shown by the arrow rotates CCW.

T_2 At time period T_2, the current in Phase B reverses direction to −10 amps, Phase C increases to a maximum value of +20 amps, and Phase A changes to −10 amps. The resultant magnetic field rotates another 60 degrees CCW.

The remaining time periods illustrate that the resultant field continues to rotate CCW.

Note that for every 60-degree change in the AC waveform, the resultant field also rotates 60 degrees. If the frequency of the line voltage is a typical 60 Hz, the resultant field rotates 60 times a second, or 3600 revolutions per minute. This is the synchronous speed. By adding another pair of windings to each phase of the AC applied voltage, four poles per phase are used. Since a four-pole field will rotate at half the speed of a two-pole field, the stator field will rotate at 1800 RPM. The same calculations are used as with single-phase induction motors to determine speed, slip, and speed regulation.

The revolving field produced by the stator currents cuts the squirrel cage conducting bars of the rotor. The induced voltages that cause currents to flow develop magnetic fields around the bars. The interaction between the rotor and the stator fields produces a torque that causes the rotor to turn in the same direction as the stator field movement. At starting, the three-phase induction motor produces 100 to 275 percent of its rated full-load torque.

One favorable characteristic of a three-phase induction motor is that it is a constant-speed machine with excellent speed regulation. Its behavior under load is similar to its single-phase counterparts. When the load is increased, speed slightly decreases, which causes the slip to increase. This action causes a greater EMF to be induced into the rotor. The higher current increases the torque to accommodate the new load requirement and prevent the speed from lowering further. If the load on the motor is increased beyond the full-load rating of the machine, it becomes overloaded. At a certain point, the larger rotor current and torque cannot prevent the motor from stalling. This condition is called *pull-out torque*. The three-phase induction motor may be severely overloaded for short periods of time. However, a prolonged overload will increase the temperature and damage the motor. The pull-out torque of a three-phase motor is much greater than starting torque.

Sometimes, one of the motor's stator leads connected to the three-phase power line becomes open. This situation may result from a blown fuse or broken connection. A motor in this condition is called *single-phased.* A polyphase motor will not start when it is single phased. Instead, it will hum because the remaining stator currents set up an alternating magnetic field instead of a rotating field. If the motor is running when the single-phase condition develops, it will continue to rotate as a single-phase motor, but its performance will not be normal.

Three-phase induction motors are available in sizes that range from fractional to 50,000 hp, or 37,000 kilowatts. Applications of these motors include large conveyor systems, large pumps, and machine tools, all of which require good speed regulation.

6-12 Wound-Rotor Motor

IAU13308
Wound-Rotor Motors

The **wound-rotor induction motor** (WRIM) is another type of polyphase motor. Its rotor consists of a set of three coils in place of the conducting bars of the squirrel cage rotor. The coils are preformed and are placed in the slots of a laminated iron core. It has many of the same characteristics as the squirrel cage motor. For example, the stator of a wound-rotor motor is the same as the stator of a squirrel cage motor. As the stator field rotates, it induces an alternating voltage into the rotor windings just as it would in the squirrel cage's shorting bars.

A squirrel cage motor's rotor resistance is fixed. Therefore, its speed–torque characteristics are fixed at full-load operation. Some motor applications require that the speed and torque of an induction motor be varied. This is possible with a wound-rotor induction motor. To perform these operations, certain design requirements must be met:

1. The rotor construction must be similar to that of the stator. It is wound for three-phase power in a wye-connected configuration and must have the same number of poles as the stator. The windings must be highly insulated. Since windings are used

instead of copper or aluminum bars, there will be a stronger interaction between the stator and the rotor magnetic fields. The result is a greater amount of torque.

2. There must be a resistive network connected to the rotor windings made up of three rheostat resistors, one for each wye coil leg. Since the rotor will be spinning and the resistor network will be fixed, they must be connected by slip rings, as shown in Figure 6-18. By varying the amount of resistance, the induced rotor currents and the magnetic flux lines they create can be regulated. This feature allows for variable speed and torque control.

Start-Up

During the start-up phase of the wound-rotor induction motor, the external resistance connected to the rotor is set to the maximum value. This resistance limits the amount of rotor current. It also causes the rotor to become more resistive and less inductive. Therefore, the stator flux and rotor flux are more closely in phase with each other. This causes the flux strengths of both fields to be at their maximum values during the peak of the alternation, producing a large amount of torque. This starting torque can be made equal to the maximum torque if a high starting torque is desired.

As the motor speed increases, the induced voltage will decrease because of less cutting action between the rotor coils and the rotating stator field. The decrease in induced voltage produces less current flow and a smaller torque. To provide maximum torque throughout the acceleration range, the resistance is gradually reduced as the motor speeds up, either manually or automatically. Once the motor reaches operating speed, the external resistance is reduced to zero and each rotor coil is short circuited. The rotor windings are then electrically equivalent to a squirrel cage rotor. At full speed, the two types of motors have similar characteristics. Since the maximum torque can be maintained throughout the acceleration period, the wound-rotor motor is desirable when starting high-inertia loads.

Many power companies require that motors draw only a small amount of current when they are started, to avoid voltage fluctuations or the flickering of lights. For this reason, wound-rotor motors are often selected because they develop a starting torque of 150 percent

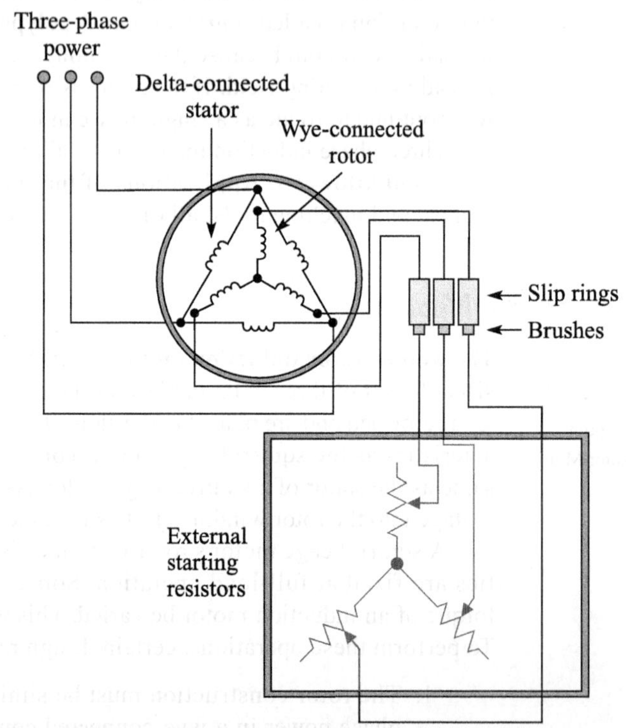

FIGURE 6-18 Wound-rotor motor system

of full-load torque with a starting current of 150 percent of full-load current. By comparison, a squirrel cage motor draws 600 percent of full-load current to develop a starting torque of 150 percent of full-load torque.

Speed Control

Even though the squirrel cage motor has excellent speed regulation, its speed cannot be varied without using complex electronic devices called *AC drives*. Wound-rotor motors also have excellent speed regulation, along with superior speed control capabilities. The speed cannot be made to run faster than synchronous speed, but can be slowed down by as much as 50 to 75 percent. The greater the resistance inserted into the rotor circuit, the more slowly it will turn.

Speed control is made possible by the external resistor controller. If the resistance is increased, the rotor current decreases. Since the stator current is proportional to rotor current because of transformer action, it also decreases. The magnetic field strength of both coils decreases, which causes the torque to decrease. As the rotor slows down, the slip increases, and the induced voltage increases. The rotor current will increase until it is sufficient to develop the torque necessary to turn the load at the slower speed.

The motor rotates at a slower speed, but with the same current and torque that it had before the resistance was increased. Because this slip is greater, the induced slip frequency is higher, which causes the rotor reactance to increase. The reactance is proportional to resistance, which causes the power factor to remain constant.

To change the direction of a wound-rotor motor, interchange any two of the three stator terminals. There will be no change in direction if the rotor terminals are interchanged because the rotor is connected only to external resistors.

The wound-rotor induction motor has been largely replaced by solid-state AC drives for varying speed control of AC motors. However, there are some applications where the AC drives cannot be used. For example, WRIMs are used in applications where they are exposed to sudden loads that bog down during the starting period, such as in a rock crusher. If the motor is locked by being jammed against rock fragments, the slip shoots up, which causes a surge of high rotor current. However, the heat does not build up inside the motor and destroy the windings. Instead, most of the heat is dissipated by the external resistors.

Other applications of this motor are pulp chippers in paper mills, automobile crushers in junkyards, hammer mills, and printing presses, where frequent and smooth starting, stopping, and reversing of high-inertia loads and speed control are required.

The disadvantages of WRIMs compared to squirrel cage motors are as follows:

1. Decreased efficiency due to losses in the external resistance
2. Poorer speed regulation at low RPMs
3. Increased maintenance due to brushes and slip rings
4. Higher manufacturing cost because of insulated rotor windings, slip rings, and brushes

6-13 Synchronous Motor

IAU10208
Synchronous Motors

The **synchronous motor** is the third type of polyphase motor. It gets its name from the term *synchronous speed,* which describes the rotating speed of the stator's magnetic field. Unlike induction motors, which have to run at less than their synchronous speed, synchronous motor rotors turn at the same RPM as the stator's magnetic field. The synchronous motor performs two primary functions: It converts AC electrical energy into mechanical power at accurate speeds, and it performs power factor correction.

The stator windings of synchronous motors are excited with a three-phase voltage to establish a rotating magnetic field. The most common type of synchronous motor has two different rotor circuits. One circuit is a set of squirrel cage bars with shorting rings that resemble those used by induction motors. They are called *damper* or *amortisseur* windings. Instead of being embedded in the rotor core, the damper windings are locked on the outer periphery of the pole core called the *pole face.* The other circuit contains coils that are wound

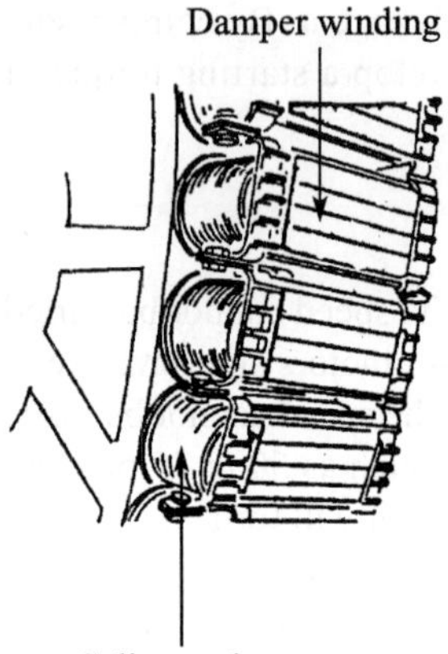

FIGURE 6-19 The rotor of a synchronous motor

on laminated core bodies called *salient poles.* (The word *salient* means projecting out.) Therefore, as the name implies, the pole pieces project outward from the shaft. Figure 6-19 shows the rotor of a synchronous motor with damper windings and salient poles. When the motor approaches operating speed, direct current is fed to the coils through brushes and slip rings mounted on the motor shaft. Each coil makes the salient pole become electromagnetic. The number of rotor poles equals the number of stator coils.

Starting Synchronous Motors

To start a three-phase synchronous motor, a three-phase voltage is applied to the stator, while no DC power is applied to the salient-pole windings. A rotating magnetic field revolves around the stator and cuts across the rotor coils. Because the rotating magnetic field exerts opposing forces on each of the salient poles, they cancel and there is no torque created by them.

The starting torque is provided by the amortisseur windings. Therefore, during the start-up period, the synchronous motor operates as a squirrel cage induction motor. At the beginning of the start-up period, the resultant rotor field lags behind the stator field by almost 180 degrees. As the motor speeds up, the field shifts until it is a little more than 90 degrees behind the stator field.

Reaching Synchronous Speed

When the rotor has accelerated to a speed close to the synchronous speed of the stator field, it is ready for synchronizing. Direct current is applied to the rotor and makes each salient pole an electromagnet. Since each coil is wound in the opposite direction from the adjacent coils, the adjacent poles around the rotor have different polarities. Each rotor pole of fixed polarity is attracted to the rotating magnetic poles of the stator. At a certain point, there is enough force to lock the rotor poles to the magnetic poles as they shift from one stator pole to the next. The amount of force needed to pull the rotor into synchronization is called *pull-in torque.* Because the rotor turns as quickly as the field rotates, there is no cutting action between the stator field and amortisseur coils. This causes the current flow in the damper winding to cease so that it no longer operates as a squirrel cage motor.

It is critical to have the DC power supply applied to the rotor at the precise moment at which the rotor and stator fields are synchronized. If it is done too soon, the pull-in torque is too low. The rotor will slip back and be attracted to the previous stator field. If it continues to slip back, the motor jerks, which may shake it enough to cause damage. Programmable industrial controllers are used to apply excitation current to the rotor coils at the proper moment. Before programmable controllers existed, one person, usually a foreman, was trained to perform this function.

Sometimes a Pony motor is used to help get the rotor up to synchronous speed, especially when the rotor is heavily loaded. The Pony motor is a small DC motor mounted on the same shaft as the synchronous motor, as shown in Figure 6-20. After its armature reaches synchronous speed, the Pony motor becomes a DC generator. The DC output voltage supplies the excitation current for the salient rotor poles through the brushes. Figure 6-21 shows the diagram of the motor/generator circuit. The DC motor/generator is sometimes called an *exciter.* Excitation current can also be supplied by a rectifier or a direct connection to a DC bus line.

When the motor runs at no load, the center of the rotor field is aligned with the center of the stator field, as shown in Figure 6-22(a). This is known as *torque angle.* As the motor is loaded, the angle increases, producing more torque. At full load, the torque angle is around 30 degrees, as shown in Figure 6-22(b). The motor continues to turn at the same speed as before. At about 150 to 200 percent of full load, the motor becomes overloaded. The torque angle becomes too great and the motor pulls out of synchronization, which is called *pull-out torque.* In this condition, the rotor slows down and lags behind the rotating stator field. A point is reached where the flux link between the rotor and the stator is broken. This situation occurs when the rotor pole is halfway between the stator pole to which it is attached and the one behind it. As the rotor speed slips behind the rotating stator field, two conditions develop. First, it becomes an induction motor again. However, the cage is intended for starting only. The motor will draw high current and overheat if it does not pull into synchronous speed fairly quickly. The second condition that

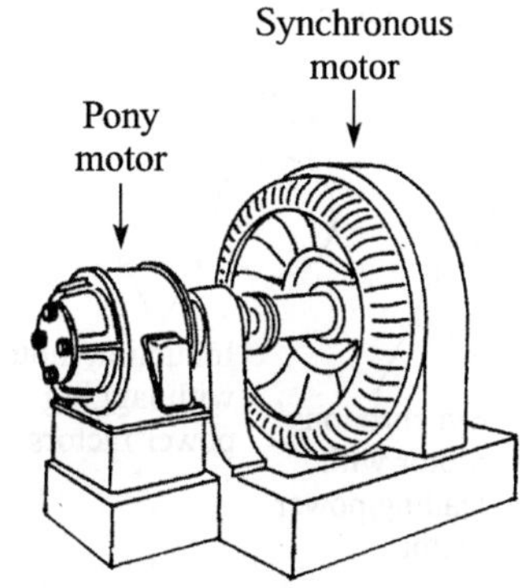

FIGURE 6-20 The Pony motor connected to the shaft of the synchronous motor

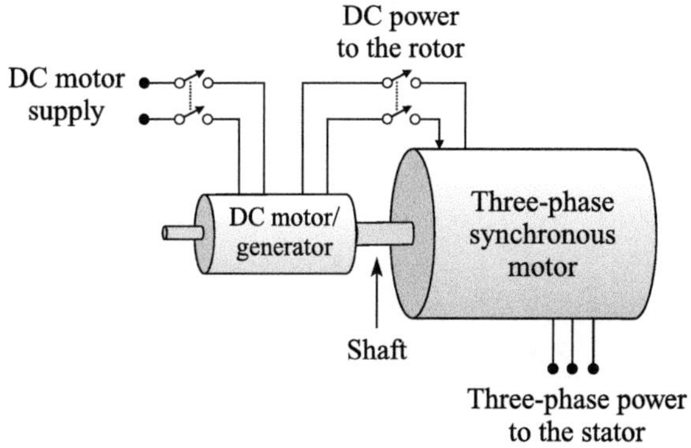

FIGURE 6-21 The schematic diagram of a motor/generator circuit

develops is the rotor attempts to lock onto each stator field that passes. The motor may vibrate enough to become damaged. The motor must be shut down immediately and the load reduced before being restarted. Another factor that may cause these symptoms is defective components, such as stator windings, rotor windings, brushes, or the rotor power supply.

Synchronous motors are durable, dependable, efficient, and insensitive to line voltage variations. They are used in applications that require constant speed, such as pumps, or in applications that require a rating that exceeds 1 hp per RPM. Because they do not have good starting torque, they should not be used to run equipment that frequently starts and stops, such as conveyors. Another important function of synchronous motors is power factor correction.

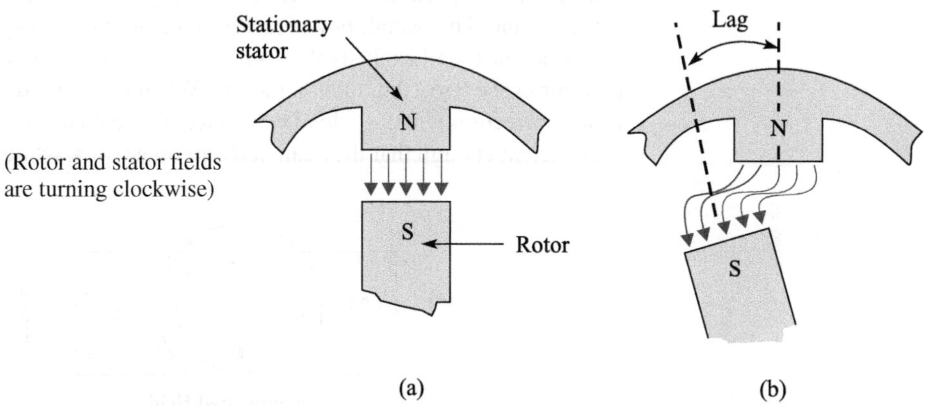

FIGURE 6-22 Rotor-stator position

Power Factor Correction

ACE2903
Power Factor

ACE3303
Power Factor
Correction

Most motors used in industrial plants are induction motors. The total power supplied to induction motors consists of true power and reactive power. The actual work performed is produced by the true power supplied to the resistive components of the motor. The reactive power produces the magnetic fields in the stator, rotor, and air gap. Therefore, induction motors run with a lagging power factor. If the motors are lightly loaded, the power factor becomes low, which implies that the reactive component is large. The power factor of the entire power distribution system within the plant becomes low if several of these induction motors are connected to the same line.

There are several reasons why low power factors are undesirable. The voltage regulation of generators, transformers, and supply lines becomes low. The current-carrying capabilities of supply lines are reduced to a lower level than their rated values. For example, a system can only supply 60 percent of rated power at 0.6 power factor. Power companies may assess a penalty charge for industrial sites that run at a power factor below a certain value.

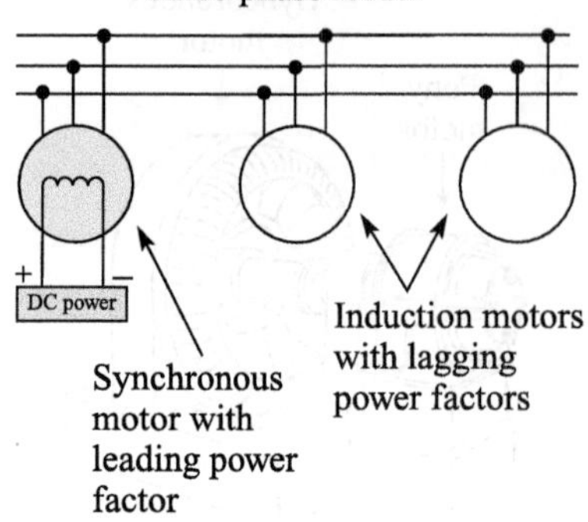

FIGURE 6-23 Synchronous motor used to correct the power factor of other motors

The synchronous motor has the ability to improve the power factor when connected to the same distribution lines as induction motors, as shown in Figure 6-23. The amount of power factor correction is controlled by the amount of DC current applied to the rotor. An adjustment of the rotor current causes the stator current to lead, lag, or be equal in phase with the applied voltage. If the current supplied to the rotor is low, the motor is *underexcited,* and it will have a lagging power factor like an induction motor, as shown in Figure 6-24(a). As current to the rotor is increased, a counter electromotive force (CEMF) is induced into the stator. The CEMF causes the applied voltage and incoming alternating current to be in phase. At a certain point, the rotor is *normally excited,* and the power factor of the motor is at unity, as shown in Figure 6-24(b). The current supplied to the motor is at its lowest level. If the rotor current is further increased, the rotor becomes *overexcited.* The CEMF induced into the stator is so high that it causes the stator current to lead the applied voltage, just like a capacitor, as shown in Figure 6-24(c). In this condition, the synchronous motor can supply reactive power to induction motors connected to the same line. The overall power factor in the distribution system then approaches unity and true power is consumed exclusively. To provide power factor correction, one synchronous motor is used for every 6 to 10 induction motors. When the synchronous motor is used strictly for power factor correction without a load connected, it is called a *synchronous condenser.* Newer designs are efficient enough that they can perform other work while correcting power factor.

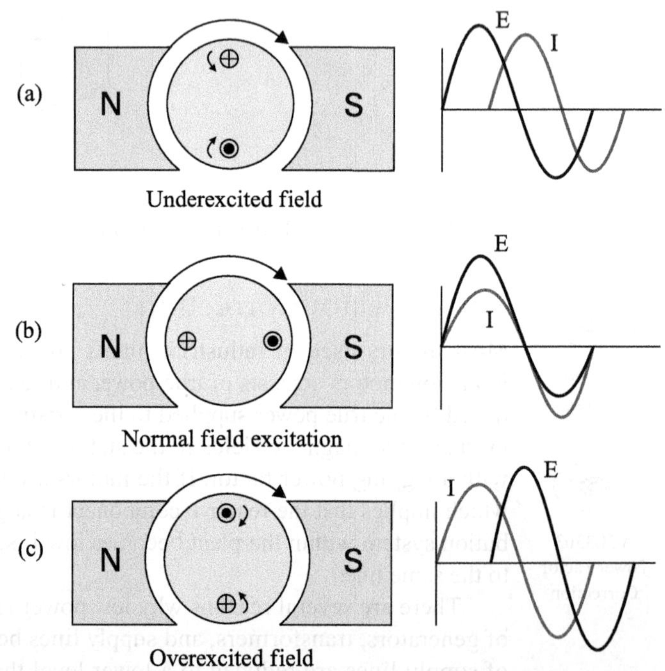

FIGURE 6-24 Field excitation in a synchronous motor

Power factor is also corrected by using capacitor banks connected to the supply lines and by loading the induction motors as fully as possible.

6-14 Motor Nameplate

NEMA Standards

There are many manufacturers that make the different types of AC motors described in this chapter. When producing a motor that will be used in the United States, manufacturers must comply with certain standards. The regulation agency that establishes and enforces the required specifications is the National Electrical Manufacturing Association, also known as NEMA. These standards include:

- Electrical voltage and current ratings
- Dimensions of the mounting bolt holes
- Diameter of the motor shaft
- Distance between the center of the shaft and the mounting plate

Compliance to these standards ensures interchangeability between motors built by different companies.

The Nameplate

Electric motors have a metal plate mounted on the housing. Its purpose is to provide pertinent information about the motor, especially if it becomes defective and needs to be replaced. The data printed on the motor must comply with NEMA standards.

The sample in Figure 6-25 will be used as a reference to illustrate the types of information on a typical nameplate.

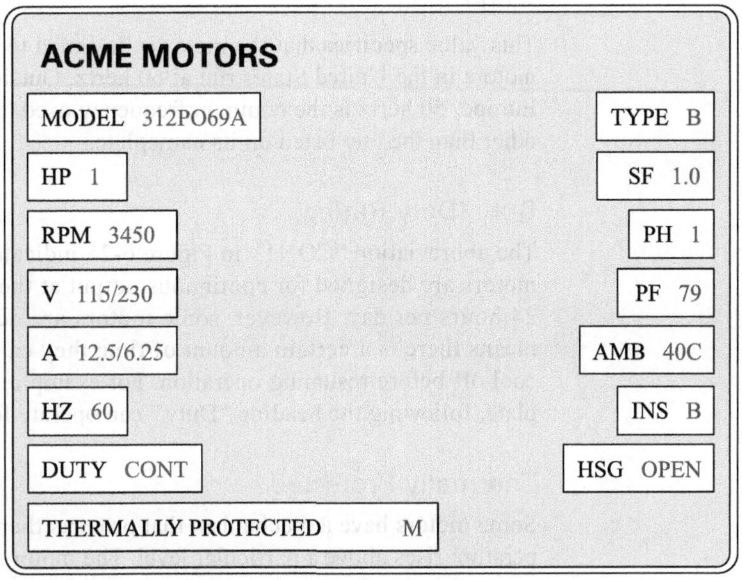

FIGURE 6-25 Motor nameplate

Model (Number)

The model number identifies a motor that has specific operating characteristics. It is also a reference tool when using a manufacturer's catalog that has all the information available about the motor. This number is particularly useful when returning a motor under warranty or for finding an exact replacement.

HP (Horsepower)

This is the horsepower that the motor is designed to produce. If it is a fractional motor, such as a half-horsepower motor, 0.5 will be printed on the plate. A replacement motor should have the exact horsepower rating. In an emergency, a motor with a higher horsepower can be used as a substitute, but because it would not run at full load, it would be inefficient.

RPM (Speed)

The speed of a motor is the RPM at which the shaft rotates. The number on the nameplate is usually the nominal full-load speed. In this example, it is 3450 RPM. If it becomes overloaded, the motor will slow down to below the rated speed. When unloaded, it may run above 3450 RPM, but never higher than the synchronous speed of 3600 RPM. Common rated speeds of AC motors are 1750, 1175, and 890 RPM.

V (Volts)

This is the voltage at which the motor is designed to operate. Common voltages for single-phase motors are 115 and 230 volts. Common voltages for three-phase motors are 230 and 460 volts. Less common voltage ratings for motors are 208, 550, 660, and 2300 volts. Motors are designed to operate within 10 percent of their rated voltages to compensate for line losses. The sample in Figure 6-25 shows two voltages listed, which means this motor can operate at 115 or 230 volts.

A (amps)

This value refers to how many amps the motor will draw from each phase when producing its rated output power. The sample shows two amperage ratings. The higher value is the current draw when connected to the lower voltage listed on the nameplate (115 volts). This information is useful for determining the required size of the wires, brushes, and contactors through which current from the supply lines flows.

Hz (Frequency)

This value specifies that the motor is designed to operate at a particular AC frequency. Most motors in the United States run at 60 hertz. Outside the United States, especially in Western Europe, 50 hertz is the common frequency used. Some motors cannot operate at a frequency other than the one listed on its nameplate.

Duty (Duty Rating)

The abbreviation "CONT" in Figure 6-25 indicates that it is a continuous duty motor. Most motors are designed for continuous output at their rated power, which means they can run 24 hours per day. However, some motors are designed for intermittent operation, which means there is a certain amount of time they can run before they need to be shut down to cool off before resuming operation. For example, a motor with the number 20 on the nameplate, following the heading "Duty," can operate for 20 minutes before it must be turned off.

Thermally Protected

Some motors have a mechanism that protects them by creating an electrical open if the temperature rises above a particular level. The motor shuts off when this condition occurs. Some motors have an automatic reset feature, which turns the motor back on after it cools off. The letter M on the sample nameplate indicates that the reset mechanism must be reset manually to resume its operation.

Type

Motors have design ratings designated by letters. The most common are the letters A, B, C, and D. Each one is determined by the way the motor is wound, which affects the start and run characteristics.

Type A	Special
Type B	Normal Starting Torque
Type C	High Starting Torque
Type D	High Starting Torque and High Slip (used for punch presses)

Type B will fulfill most industrial applications, so it is the most common.

SF (Service Factor)

This value indicates if a motor can operate above its rated horsepower. If it has an SF number of 1.0, it cannot operate above the horsepower rating listed on the nameplate. If it has an SF number of 1.25, for example, it can produce 1.25 times its rated horsepower without damage. The drawback of running in this condition is that the efficiency and power factor of the motor will be lower than when it runs at its rated horsepower.

PH (Phase)

This value indicates if the motor is a single-phase or a three-phase motor.

PF (Power Factor)

This value indicates the percentage of apparent power used by the motor. All motors have a power factor, which is affected by their inductive load. A high PF number is desirable.

AMB (Ambient Temperature)

This value is the maximum temperature of the surrounding air within which a motor can safely operate. If it is exposed to a higher temperature environment, it will overheat and become damaged. The value on the sample nameplate is 40°C, or 104°F. A replacement motor should not have an ambient temperature rating lower than the one it is replacing.

INS (Insulation Class)

Motors are constructed with wires that have insulation materials that can withstand different temperatures. There are four common classes with different temperature ratings:

Class A	105°C
Class B	130°C
Class C	155°C
Class D	180°C

A motor can be replaced with either one that is in the same class or one with a higher temperature rating.

HSG (Motor Enclosure)

This information indicates the ventilation requirements of the motor. Examples are as follows:

Open to Air. This motor has an internal fan that pulls air from one end and pushes it through the other end. The term "OPEN" on the sample nameplate indicates this type of motor housing enclosure.

Open Drip-Proof. A vent is placed at the bottom to prevent drops of liquid or solids from falling on the motor at an angle of not greater than 15° vertical.

Totally Enclosed. This motor is cooled by convection as an external fan blows air over its housing, and it is used in dusty, dirty, and corrosive atmospheres.

EP—Explosion-Proof. This motor is used in hazardous environments that contain gas vapors, coal dust, or alcohol. Its housing is made to be extra strong to prevent an internal explosion from igniting a gas or vapor in the surrounding atmosphere.

Problems

1. Which components of a DC motor are not found in an AC motor? ____
   a. armature
   b. brushes and commutator
   c. field windings
   d. fan assembly
2. Which of the two types of motor requires less maintenance? ____
   a. DC
   b. AC
3. The stationary portion of the AC motor is called the __________ and the rotating part is called the __________.
4. The speed of an AC motor depends on the __________ of the power supply and the __________.
5. The RPM that the rotating field moves around the stator is called __________ __________.
6. What is the RPM speed of a single-phase 12-pole AC motor operating from a 60-Hz power source? ____
7. Calculate the percent slip of an induction motor that has a synchronous speed of 3600 RPM and a rotor speed of 3400 RPM.
8. T/F The rotor must turn at a slower RPM than the speed of the rotating synchronous field in the induction motor.
9. A(n) ____ must be established in a split-phase motor to produce the starting torque.
   a. rotating field
   b. alternating field
10. To change the rotation of a split-phase AC motor, ____.
    a. reverse the power leads
    b. reverse the main and auxiliary windings with respect to each other
    c. both a and b
11. Once the split-phase motor reaches ____ percent of its synchronous speed, the centrifugal switch ____ to ____ the ____ windings.
    a. 45–50
    b. 75–80
    c. 95–100
    d. opens
    e. closes
    f. connect
    g. disconnect
    h. main
    i. auxiliary
12. In a resistance-start motor, current in the auxiliary winding always __________ (leads, lags) the current in the main winding.
13. The capacitor-start AC motor has a capacitor in ____ with the ____ winding.
    a. parallel/auxiliary
    b. series/auxiliary
    c. parallel/main
    d. series/main
14. The ____-start motor has the larger starting torque.
    a. resistance
    b. capacitor
15. The capacitor-start capacitor-run motor has how many capacitors? ____
16. The ____-start motor is well suited for applications that require frequent starting loads, such as a refrigerator.
    a. resistance
    b. capacitor
17. The direction of the torque in a shaded-pole motor is toward the ____.
    a. main pole
    b. shaded pole
18. The shaded-pole motor is commonly ____-directional.
    a. uni
    b. bi
19. Motors that operate from either AC or DC power are called __________ motors.
20. The speed of a universal motor can be controlled by ____.
    a. using a variable resistor
    b. tapping a field coil at various points
    c. either a or b
21. In actual practice, the ____ leads are interchanged to reverse the direction of a universal motor.
    a. field coil
    b. armature
22. A three-phase induction motor uses a __________ (smaller, larger) frame than a single-phase induction motor of equal horsepower.
23. List two ways that a three-phase motor becomes single phased.
24. The direction of the wound-rotor induction motor can be reversed by interchanging ____ of the three stator leads.
    a. one
    b. two
    c. three
25. T/F During operation, if the resistance of a wound-rotor motor is set to zero, it performs very much differently than a squirrel cage motor.
26. A(n) __________ (increase/decrease) in the external resistance of a wound-rotor motor causes an increase in the rotor resistance and therefore a reduction in the rotor current and speed.
27. T/F The amortisseur windings of a synchronous motor perform a similar function to the bars and shorting rings of a squirrel cage induction motor.
28. A three-phase induction motor goes into a single-phase condition when power is disconnected from ____ of its stator leads.
    a. one
    b. two
29. AC motors that turn at the same speed as the rotating magnetic field are called __________ motors.
30. In a synchronous motor, the DC excitation is applied to the __________ (rotor's, stator's) magnetic field.
31. In a synchronous motor, stator current will lead the applied voltage if the DC motor field is __________ (overexcited, underexcited).
32. List two favorable features of synchronous motors.
33. T/F A Duty Rating of 20 on the nameplate of a motor indicates that it is designed to run continuously for 24 hours a day.
34. T/F A common voltage used to power a three-phase motor is 115 V.
35. The A (amps) value on the nameplate for a three-phase motor indicates how much current ______ will draw.
    a. the entire motor
    b. each phase

CHAPTER

7

Servo Motors

OBJECTIVES

At the conclusion of this chapter, you should be able to:

- Describe the operation of the following servo motors:
  - Wound Armature PM Motor
  - Moving Coil Motor
  - Brushless DC Motor
  - PM Stepper Motor
  - VR Stepper Motor
  - AC Servo Motor
- Define the following terms:
  - Servo Motor
  - Holding Torque
  - Stepping Rate
  - Step Angle
- List a practical application of the different types of servo motors.

INTRODUCTION

The traditional motors described in the two previous chapters are used in applications requiring moderate to high power. Typically, they turn in one direction, and their speeds can be varied only within a very limited range. When traditional motors are stopped, they usually coast until they no longer turn.

For motion control applications that require special performance characteristics such as precise speed or accurate positioning in both directions, specialty motors are often more suitable than traditional motors. These nontraditional motors are often used as **servo motors.** This term refers to any motor that uses a closed-loop feedback signal to monitor its velocity and position, or that uses open-loop digital equipment to provide precise input command signals.

The servo motors described in this chapter are bidirectional position devices that typically operate between low and moderate power.

7-1 DC Servo Motors

DC servo motors are controlled by direct current command signals that are applied to coils, which become electromagnets. The magnetic fields that form around the coils interact with permanent magnets (PMs) and cause the rotating member of the motor to turn. These DC servo motors are referred to as *PM motors* and are classified into two categories, depending on how the permanent magnet is used. One type of PM motor uses a wound armature and brushes like a conventional DC motor. The pole pieces, however, are permanent magnets instead of electromagnets. This category includes the wound armature motor and the moving coil motor. The other category of PM motors uses wound field coils and a permanent magnet rotor. This category includes brushless DC motors.

In the mid-1970s, the popularity of PM motors increased with the introduction of rare earth magnets. Rare earth magnets have greater flux strength than the ferrite magnets they replaced, which results in greater torque produced by the motor. PM motors are smaller and lighter than wound field DC motors that produce the same amount of torque. They are often used in applications that require portability and low maintenance requirements.

7-2 Wound Armature PM Motor

The **wound armature PM motor** is shown in Figure 7-1(a). It has permanent magnets in the outside of the motor that function as the stator, and current-carrying conductors in the rotor. It is similar in construction to a wound rotor motor. The armature contains wound coils that are placed in the slots of an iron core. Brushes and commutator segments, located at the end of the motor, switch the current as the armature turns so that the required magnetic field forms around the rotor coils. The interaction between the magnetic flux lines of the rotor and stator fields produces a continuous torque when a DC current is applied. The permanent magnets that form the stator typically are in either a two-pole or a four-pole structure, although six or more poles have been used.

Since permanent magnets are used for the pole pieces, the field flux remains constant. There is no electromotive force (EMF) induced into the field poles to cause the flux strength to vary. This gives the motor linear speed-torque characteristics similar to a conventional DC shunt motor, as shown in the chart in Figure 7-1(b). The speed is varied by changing the armature voltage. The direction is easily changed by reversing the current applied to the armature leads.

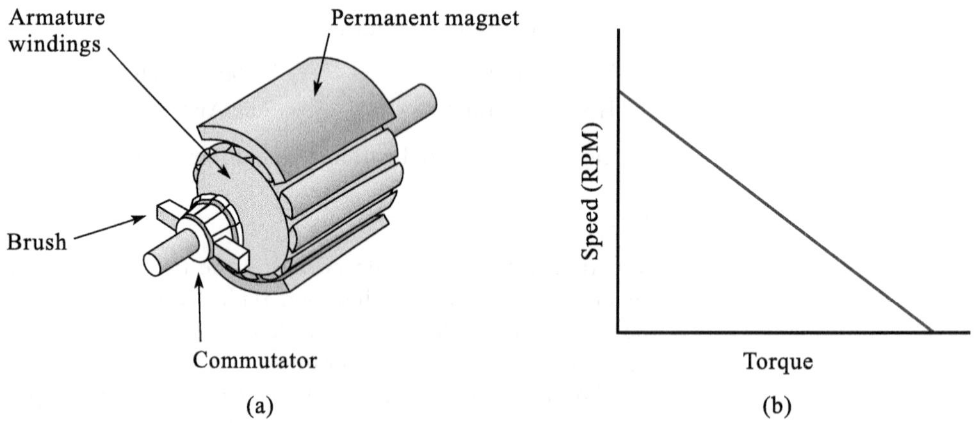

FIGURE 7-1 Wound armature PM motor

The wound armature PM motor does have some limitations. Because the current-carrying rotor coils, which are heat-producing, are located at the inside of the housing, this motor is thermally inefficient because the heat cannot readily escape. The result is that the motor cannot produce high torque for prolonged periods without overheating and becoming damaged. Also, the brushes periodically need replacement because of wear as they ride along the commutator segments. PM motors are commonly used in office machines, printers, and disk drives. Larger PM motors are used for manufacturing positioning equipment, such as an industrial robot.

To effectively operate in positioning applications, this motor must be controlled by a closed-loop servo system that consists of a controller, an amplifier, and a position sensor that operates as a feedback device.

7-3 Moving Coil Motor

The **moving coil motor (MCM)** is designed very differently from other types of motors. The stator field is provided by eight pairs of permanent magnets that are on each side of the disk and parallel to the motor shaft. These magnets are placed around the perimeter of the motor housing. They are arranged so that they provide alternating magnetic fields as shown in Figure 7-2. By fitting as many magnets as possible around the circumference of the motor, the maximum number of stator flux lines is provided to produce the highest possible torque.

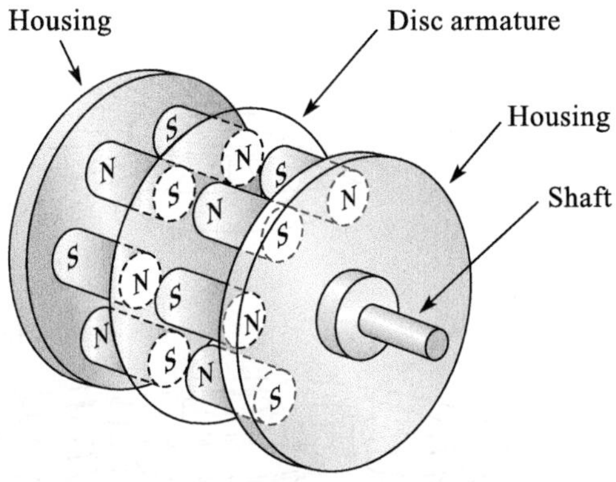

FIGURE 7-2 Permanent magnets are arranged to produce alternate magnetic polarities

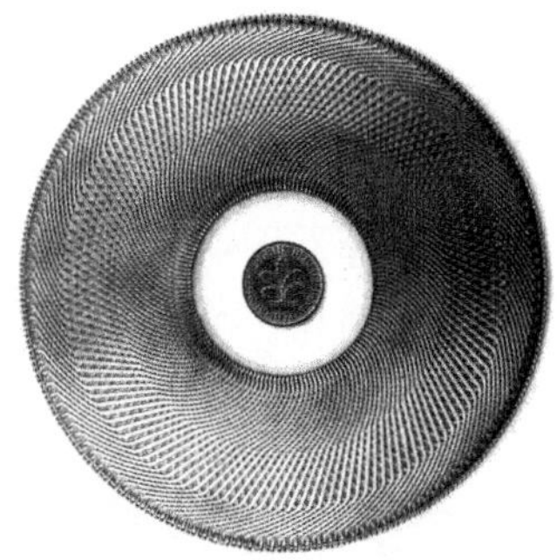

FIGURE 7-3 A thin disk armature made of fiberglass with copper conductors placed on each side

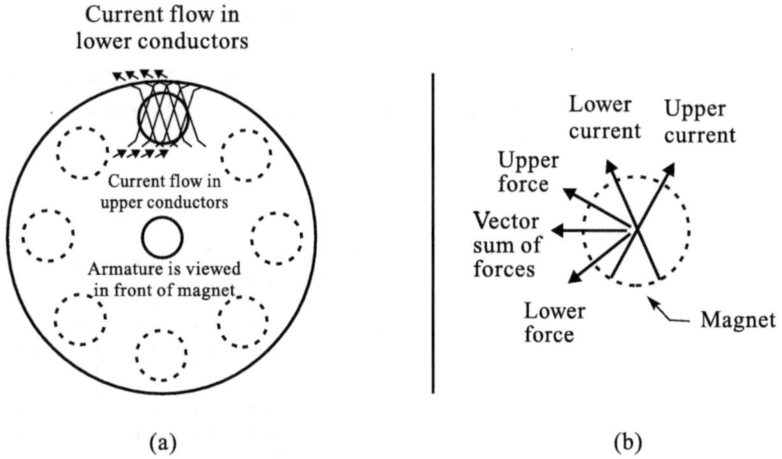

FIGURE 7-4 The interaction of magnetic fields creates the forces that turn the motor

The armature, shown in Figure 7-3, is a thin disk made of fiberglass. Two layers of copper conductors are formed on each side of the fiberglass in much the same way as in a printed circuit. One layer, called the *upper conductor,* is placed on top of the other layer, called the *lower conductor*. The conductor paths of each layer are arranged at 30-degree angles to each other, as shown in Figure 7-4(a). The armature interacts with the permanent magnet field to produce a force tangent to each magnetic pole, as shown in Figure 7-4(b). Enough torque is provided to turn the armature.

The ends of the conductors are located at the center and at one side of the disk in the shape of commutator segments. As the disk turns, brushes ride on the commutator to provide direct current to the conductors, as shown in Figure 7-5. As current flows through the upper and lower conductors, a resultant magnetic field is produced. Because the armature is in the shape of a disk, it does not use iron. This provides two advantages. First, the disk is light, so it has low inertia. This enables the armature to accelerate rapidly (from 0 to 3000 RPM in $^1/_6$ of a revolution), stop quickly, and reverse direction easily. Second, the brush life is extended because its armature's low inductance does not cause arcing. Also, the large number of conductors enables the MCM to run smoothly at speeds as low as 1 RPM, unlike conventional DC motors, which tend to cog at low speeds.

The speed of the motor is varied by changing the amount of voltage supplied to the armature. The voltage is in the form of DC pulses at a frequency of about 20 kHz. The average voltage varies by changing the width of the pulses. The ratio of time the pulses are on to the time they are off determines the amount of average voltage. For example, Figure 7-6 shows that when the pulses are turned on longer, the average voltage will be higher.

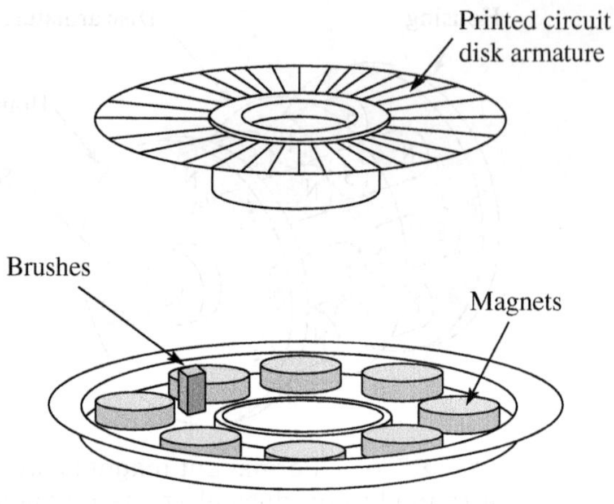

FIGURE 7-5 Brushes ride on the commutator located in the center of the disk to provide direct current to the conductors

The highest voltage possible is 24 volts, which occurs when there are no pulses and the voltage is a constant 24 volts DC. The rotor turns at its top speed, which is over 4000 RPM for MCMs.

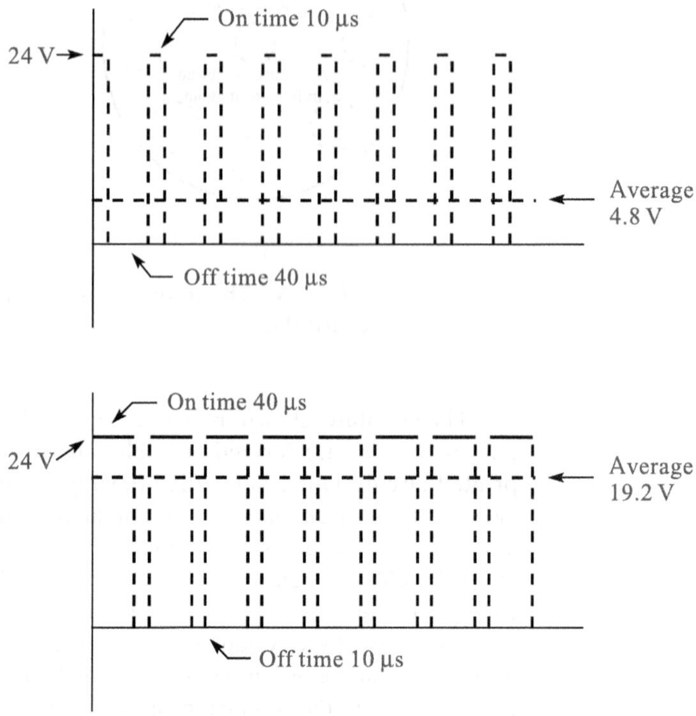

FIGURE 7-6 Square-wave pulses supplied to the armature

MCMs, also known as *pancake motors*, are used in applications that require high torque, fast acceleration, and small size, such as tape transport systems and computer peripheral devices.

7-4 Brushless DC Motors

Brushless DC motors (BDCM) contain a powerful permanent magnet rotor and fixed stator windings. The stationary stator windings are usually three-phase, which means that three separate voltages are supplied to three different sets of windings.

The BDCM also contains a converter and a rotor position sensor. The converter is an electronic commutator that changes direct current into pulsating DC voltages. The pulses are applied to the stator windings to create a rotating magnetic field. This field attracts the permanent magnet rotor. As it follows the rotating field, the rotor turns. The rotor position sensor provides feedback signals to the converter so that it switches current pulses through the stator coils in the proper sequence and at the proper time.

The operation of a BDCM is illustrated in Figure 7-7. Three transistor switching devices are connected to a DC power source. When a transistor is turned on, it supplies a phase current to a stator field coil. When current flows through a coil, a south pole is created at the end of the pole face. While SW2 is in the down position, no more than one transistor can be turned on at any given moment. The north pole of the permanent magnet rotor aligns itself to a stator pole that is energized.

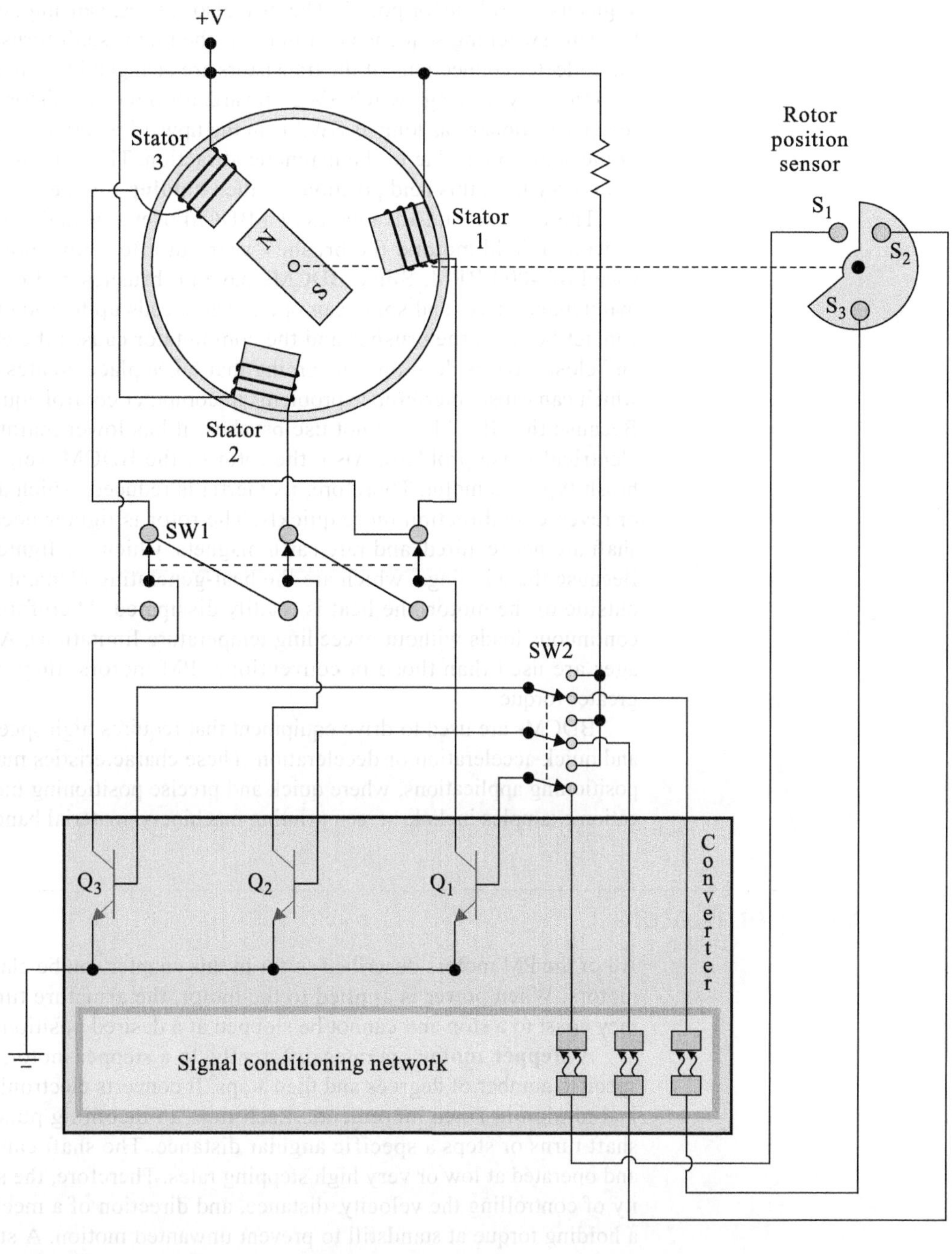

FIGURE 7-7 BDCM with converter and sensor

A rotor position sensor that consists of a round disk with 120 degrees cut away is mounted on the shaft. Three proximity position sensors (such as Hall-effect detectors) are mounted 120 degrees apart around the shaft, within sensing distance of the disk. Therefore, the disk is always being sensed by two of the detectors. Each detector output is connected to a switching transistor through a signal conditioning network. A detector that does not sense the disk causes the emitter of the optocoupler to which it is connected to turn on. The receiver of the optocoupler sends a positive voltage to the base of the transistor to which it is connected and turns it on. Figure 7-7 shows the north pole of the rotor aligned with stator pole 3. With the disk in the position shown, sensor S_1 turns on transistor Q_1, causing current to flow through stator coil 1. The south pole created at the pole face will attract the rotor and cause it to turn 120 degrees clockwise. When the rotor aligns itself to stator pole 1, the disk will be turned so that it is sensed by detectors S_1 and S_3. Detector S_2 will then turn on transistor Q_2 and cause current to flow through stator coil 2. The rotor turns another 120 degrees clockwise as its north pole aligns itself with stator pole 2. The disk also rotates, causing detector S_3 to turn on transistor Q_3. This switching sequence continues as the motor shaft turns in the clockwise direction. If the collector connections of the transistors are changed by switch SW1, the motor will reverse direction. By moving switch SW2 upward, all three transistors turn on. The rotor stops and remains stationary as long as SW2 is in the upward position. The rotor remains aligned to one of the stator poles due to the magnetic attraction. The amount of force required to move the rotor away from this held position is called **holding torque.**

There are several advantages of a BDCM over a wound-field DC motor that has brushes. In wound-field motors, the brushes wear out after only about 2000 hours and have a top speed of 4000 RPM. Since BDCMs have no brushes and commutator, they are virtually maintenance-free, and some can operate at speeds up to 100,000 RPM. Also, as the surface contact between the brushes and the commutator causes the electrical connections to open and close, sparks develop. The arcing that takes place creates magnetic fields called *noise,* which can cause interference problems in computer control equipment placed near the motor. Because the BDCM does not use brushes, it has lower maintenance requirements and no electrical noise problem. Also, the rotor of the BDCM weighs less and is smaller than a brush-type DC motor. Therefore, its inertia is reduced, which allows the motor to accelerate or reverse its direction more quickly. The rotor is lighter because brushes mounted to the shaft are not required, and rare earth magnets, which are lighter than wound coils, are used. Because the windings, which are the heat-generating element of the motor, are close to the outside of the motor, the heat is readily dissipated. Therefore, it is able to handle heavier continuous loads without exceeding temperature limitations. Also, since higher supply voltages are used than those in conventional PM motors, they operate at higher speeds and greater torque.

BDCMs are used to drive equipment that requires high speeds, high peak torque capacity, and quick acceleration or deceleration. These characteristics make them well suited for servo positioning applications, where quick and precise positioning movements are required. Application examples include screen printing machinery, material handling equipment, and grinders.

7-5 Stepper Motors

All of the PM motors described so far in this chapter can be classified as continuous rotation motors. When power is applied to the motor, the armature turns. When power is removed, they coast to a stop and cannot be stopped at a desired position.

A **stepper motor** operates differently. In a stepper motor, the armature turns through a specific number of degrees and then stops. It converts electronic digital signals into mechanical motion in fixed increments. Each time an incoming pulse is applied to the motor, its shaft turns or steps a specific angular distance. The shaft can be driven in either direction and operated at low or very high stepping rates. Therefore, the stepper motor has the capability of controlling the velocity, distance, and direction of a mechanical load. It also produces a holding torque at standstill to prevent unwanted motion. A stepper motor is typically used as an actuator in motion control applications that require accurate positioning.

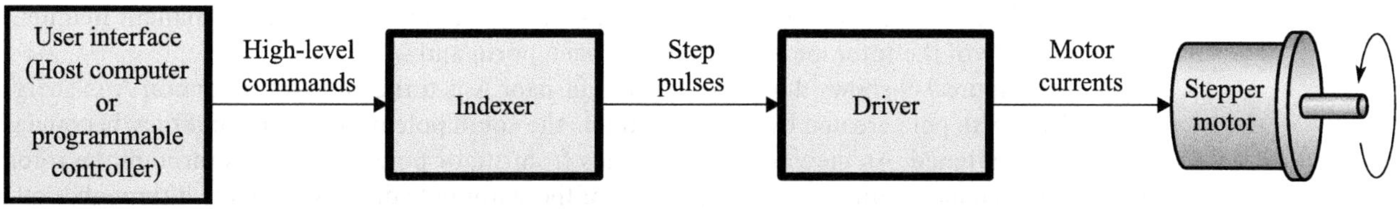

FIGURE 7-8 Elements of a typical stepper motor system

One attractive feature of the stepper motor is that it responds to digital signals. Therefore, it can be controlled by computers or a computer peripheral device. The elements that make up a typical stepper motor system are shown in Figure 7-8. The function of each element is as follows:

User Interface

The device used by the operator to communicate to the stepper motor system is called the *user interface*. Data entered by this device are high-level commands, which include speed, acceleration, distance, direction, and ramping routines. These devices include computers, programmable logic controllers, switch panels, thumbwheel switches, and handheld terminals.

Indexer

The primary function of the indexer is to convert the desired motion information entered into the user interface device into move signals that are applied to a driver that powers the motor. Indexers can include features such as data communication, I/O lines, memory for storing motion programs, and encoder feedback for closed-loop positioning.

Driver

The driver converts the step input signals from the indexer into current pulses that power and drive the motor. Depending on the driver that is selected, they produce signals that range from low-power full- or half-stepping to high-performance microstepping.

Stepper Motor

The stepper motor converts the signals from the driver into fixed mechanical increments of motion. The movements they make are very accurate and can be used in precision motion applications without a closed loop to verify accuracy.

Encoder

For applications that require extreme accuracy, a measurement device that provides feedback information is used, such as an incremental encoder. It verifies that the motor is in the desired position specified by the indexer. If it is not, corrective action is taken. The feedback information is sent by the measurement device to the indexer. Another function of the feedback device is to detect whether there is a stall condition.

Many types of stepper motors exist, of which the permanent magnet and the variable reluctance (VR) motors are the most popular.

7-6 Permanent Magnet Stepper Motor

The operation of a stepper motor is based on the magnetic principle that like poles repel and unlike poles attract. Its operation is illustrated by the simplified view of a stepper motor in Figure 7-9. It is constructed of four electromagnets located 90 degrees apart in the stator. An external transistor is connected to each of the coils. The function of the transistors is to electronically perform the switching (commutating) action that creates a moving magnetic field in the stator. Current is passed through a particular coil when the transistor to which it is connected is turned on. When it is energized, the coiled wire is wrapped so that its pole becomes magnetic north. A cylindrical rotor with six magnetic teeth located 45 degrees apart is placed inside the four coils.

The rotor has no armature windings. Instead, each tooth is a permanent magnet. The polarity of the rotor teeth alternate between north and south.

Figure 7-9 shows the motor when transistor A is turned on and stator coil A is energized. The north pole created by the coil attracts the south pole of tooth (1), and they become physically aligned. As they do, the flux lines from stator pole face A pass through the rotor and continue their path by entering the face of the stator pole directly opposite it through tooth (4).

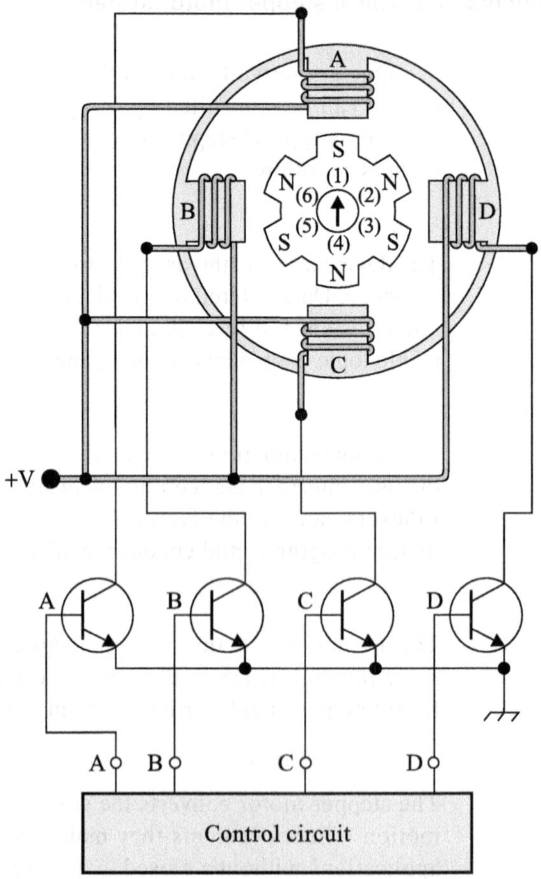

FIGURE 7-9 Permanent magnet stepper motor

The reason why the flux lines pass through tooth (4) is because it is the only one aligned with another stator pole face. Even though coil C is not energized by transistor C, the pole still becomes magnetized because of the flux lines that pass through it. The result is that its pole face becomes a south pole, which adds to the torque of the rotor as it attracts the north pole of tooth (4). The flux lines complete the magnetic circuit through the frame of the motor. Pole C is referred to as being a *passive* magnet because it is not magnetized by current flowing through its own coil.

As a reference, the position of the rotor shown in Figure 7-9 will be considered 0 degrees and is indicated by the arrow pointing straight upward. To show how the rotor turns, imagine that transistor A turns off and transistor B turns on simultaneously. Pole B of the stator energizes and becomes a north pole, and pole D becomes a passive south pole. The rotor turns clockwise until tooth (5) aligns with the north pole of stator B, and the north pole of tooth (2) becomes aligned with pole D. The rotor stops after turning clockwise by 30 degrees.

The next step of the motor occurs when transistor B in the control circuit turns off and transistor C simultaneously turns on. Pole C energizes and becomes a north pole, and pole A becomes a passive south pole. The south pole of tooth (3) is attracted to and aligns with the north stator pole C, and tooth (6) aligns with the passive south stator pole A. The rotor stops after turning clockwise by another 30 degrees.

TABLE 7-1 Transistor Switching Sequence:
(a) To produce 30-degree clockwise rotor steps;
(b) To produce 30-degree counterclockwise rotor steps

Shaft Position (Degrees)	***Transistor Turned On***
0	*A*
30	*B*
60	*C*
90	*D*
120	*A*
150	*B*
180	*C*
210	*D*
240	*A*
270	*B*
300	*C*
330	*D*
360	*A*

(a)

Shaft Position (Degrees)	***Transistor Turned On***
0	*A*
−30	*D*
−60	*C*
−90	*B*
−120	*A*
−150	*D*
−180	*C*
−210	*B*
−240	*A*
−270	*D*
−300	*C*
−330	*B*
−360	*A*

(b)

Table 7-1(a) shows the sequence in which the transistors turn on to cause the steps that take place during one complete clockwise revolution of the rotor. Reversing the switching sequence of the transistor causes the rotor to turn in the counterclockwise (CCW) direction. The sequence of steps that occur during one complete CCW rotation is shown in Table 7-1(b).

Half-Stepping

The step of the rotor's rotation can be reduced from 30 degrees to smaller movements of 15 degrees by altering the switching sequence of the control circuit. Cutting the movements

TABLE 7-2 Switching Sequence to Produce 15-Degree Half-Steps

Shaft Position (Degrees)	***Transistor Turned On***
0	*A*
15	*C* and *D*
30	*B*
45	*A* and *D*
60	*C*
75	*A* and *B*
90	*D*
105	*B* and *C*
120	*A*
135	*C* and *D*
150	*B*
165	*A* and *D*
180	*C*

in half is called *half-stepping*. Table 7-2 shows how this is accomplished. Starting on line one, transistor A energizes coil A and causes the rotor to be in the reference 0-degree position. During the next step of the sequence, transistor A is turned off, and transistors C and D turn on to energize coils C and D, thereby causing pole faces C and D to become active north poles. A combined net magnetic north pole field is created in the middle of the space between stator poles C and D. This causes the closest south pole rotor tooth (3), which is 15 degrees away, to move into alignment. The result is that the rotor makes a 15-degree clockwise step. During the next step in the sequence, transistors C and D turn off, and transistor B simultaneously turns on and energizes coil B. Its pole face becomes a north pole and attracts the south pole of rotor tooth (5) by turning another 15 degrees clockwise as it becomes aligned. The remainder of the lines in Table 7-2 show the remaining transistor switching sequence that causes the rotor to move until it turns 180 degrees. Notice that the sequence repeats at 120 degrees. The sequence will occur three times during each 360-degree revolution.

In practice, there are multiple teeth with alternating north and south poles on the rotor. There are also additional coil windings that alternate in the same sequence as the four-pole stator shown in Figure 7-9. This allows for better stepping resolution.

The physical size and weight of a cylindrical-toothed rotor is relatively large; therefore, it is more difficult to move, which causes the motor to react slowly. This limits the stepping rate of the motor. To overcome potential inertia problems, some PM stepper motors use a flat disk rotor instead, which weighs 60 percent less than the cylinder type.

The flat disk rotor is shown in Figure 7-10. It is supported on a nonmagnetic hub and placed inside two C-shaped electromagnetic cores. The outer edge of the disk is composed of tiny individual magnets. The magnets are evenly spaced and are polarized with alternating north and south poles. Although the C-shaped electromagnets appear to be placed across from each other, they actually are offset from each other by half a rotor pole. Each electromagnet is energized by a different phase.

As one electromagnet is energized, the rotor aligns itself to the magnetic field it produces. Next, the first phase is turned off and the second electromagnet is energized. The disk will turn one-half of a half rotor pole to align itself to the magnetic field produced by the second phase. The rotor continues to turn by simultaneously de-energizing one coil and energizing the other.

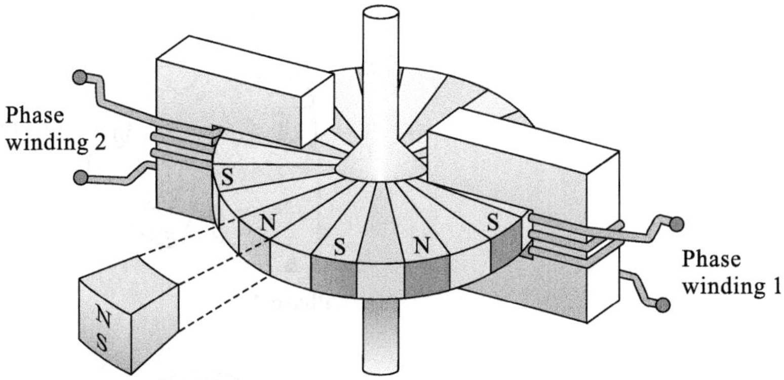

FIGURE 7-10 Disk-type permanent magnet stepper motor

A typical rotor has 100 magnets of equal size. Therefore, the angular distance between poles around the outside portion of the 360-degree disk is 3.6 degrees. Each time the disk rotates when the C-shaped electromagnet is energized, it moves one-half the distance of the magnetic pole segments, or 1.8 degrees.

7-7 Variable Reluctance Stepper Motor

IAU14208
The VR Stepper Motor

The **variable reluctance (VR) stepper motor** uses electromagnetic stator poles. Its rotor is in the shape of a disk with teeth and slots around the outer edge. The soft-iron rotor is unmagnetized.

The operation of a VR stepper motor is illustrated in Figure 7-11. It uses six sets of stator windings to form six pole pairs. The poles in each pair are located directly across from each other and are energized at the same time because their coils are connected in series. Because the coils are wound in the opposite direction, the poles in each pair will always have different polarity. There are 12 stator poles that are equally spaced apart by 30 degrees. The rotor has eight teeth that are equally spaced at 45-degree intervals. Therefore, the alignment of the stator and rotor is different by 15 degrees (45° − 30° = 15°). When the rotor rotates one step, it moves 15 degrees.

Figure 7-11(a) shows that the teeth on the rotor will align themselves with the flux lines created by the north and south poles of coils A and A′ when they are energized. Note that the teeth are not lined up under the unenergized stator poles. To step the rotor 15 degrees CCW, coils A and A′ are de-energized and coils B and B′ are energized. The teeth closest to the B coils will align themselves with the flux lines of both B poles. The rotor turns because it is made of soft iron, which has a very low reluctance rating. The magnetic flux passes through the iron much more readily than through air. The difference of reluctance between iron and air creates a force that causes the rotor to turn so that the flux lines can pass through the iron teeth rather than the air slots between them. The next 15-degree step is made by de-energizing the B and B′ coils and energizing the C and C′ coils. This action continues by repeatedly energizing the coil pairs in the sequence of A, B, and C. Therefore, it takes 24 steps to turn the rotor a full 360 degrees. The more rapid the sequence, the faster the rotor turns. The VR stepper motor can run as high as 100,000 RPM. The number of signal changes determines the distance the rotor travels. By reversing the sequence, the rotor will rotate in a clockwise direction. By using the pulse width modulation (PWM) method, the torque of the motor can be controlled by varying the width of the pulses.

In practice, there are multiple teeth on the stator. There are also eight coil windings that alternate in the same sequence as the four-pole stator, as shown in Figure 7-11(e). There are 50 teeth machined into the rotor. Each pulse moves the rotor a distance of 1/4 tooth. Since it takes four steps to advance the width of one tooth and the adjacent space between the teeth, it takes 200 steps to complete one revolution. The step angle (also resolution) can be computed by dividing 360 degrees by 200 steps: 360/200 = 1.8 degrees.

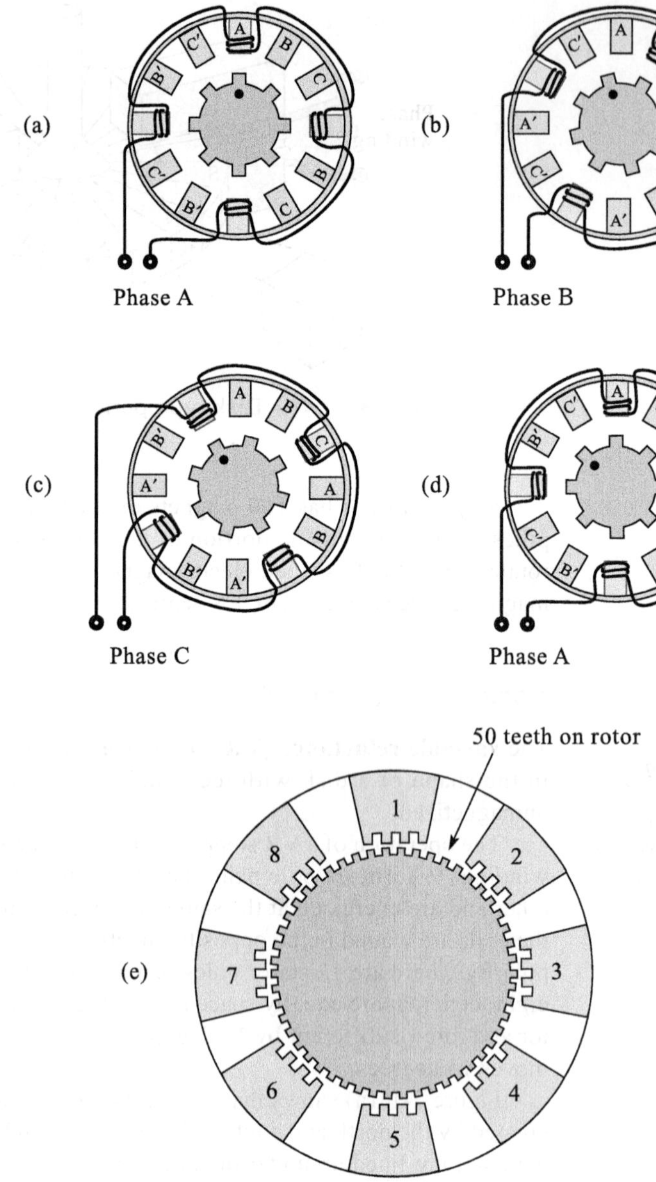

FIGURE 7-11 VR stepper motor: (a–d) Four step positions; (e) High-resolution VR stepper motor

Because the rotors of VR stepper motors do not have to be magnetized, they can be small and light. The rotor's small size gives it low inertia, so it can respond quickly to control signal changes. Applications include variable-speed fans, blowers, and hazardous environments where the conditions are very hot.

The name *variable reluctance (VR) stepper motor* is derived from the principle of its operation. When the rotor turns and a tooth aligns with a stator coil, it moves to a position that minimizes the magnetic reluctance of the overall flux.

Stepper Motor Terminology

Two important terms relating to stepper motors are *stepping rate* and *step angle.* **Stepping rate** is the maximum number of steps the motor can make in a second. The number of degrees per arc that the motor moves per step is called **step angle.** The step angle is determined by the number of rotor teeth and stator poles used.

The actual speed of the rotor depends on the step angle and stepping rate. It can be calculated by using the following formula:

$$n = \frac{Y \times S}{6}$$

Where, n = Speed in RPM
Y = Step angles in degrees
S = Steps per second
6 = A formula constant

Microstepping

IAU14508
Using the Microstepping Technique for Stepper Motors

There are some undesirable characteristics associated with stepper motors. At low speeds, the motor jerks as it steps, which can cause rough running in the mechanical mechanism to which it is attached. If a smooth operation is required at low speed, the motor speed is kept high and a gear reduction transmission is connected between the motor and the load. Another drawback of the stepper motor is limited resolution. Some applications may require accurate positioning under 1 degree.

A technique that overcomes low speed and resolution problems is microstepping. Instead of square waves energizing the stator coils that cause the rotor to start and stop as it turns, simulated sine waves created by small steps are used instead. Figure 7-12(a) shows conventional

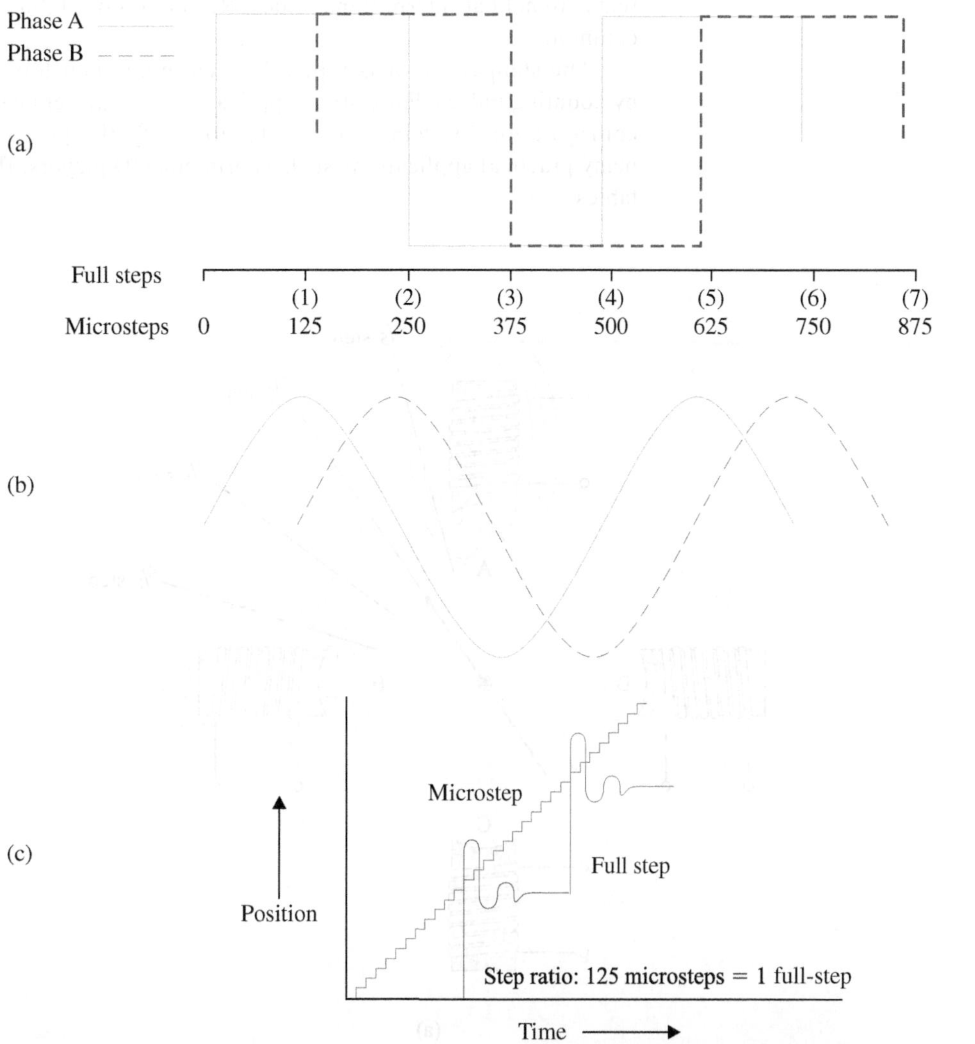

FIGURE 7-12 Microstepping vs. full-step systems

half-step signals that cause a stepper motor to move from one position to the next. Stator coils are either fully turned on or completely turned off to produce motion. Figure 7-12(b) shows the microstepping technique using two sine waves. The current is increased in one phase while the current in the other phase gradually decreases. The sine waves are created by a varying voltage that increments or decrements in small steps called *microsteps,* shown in Figure 7-12(c). Figure 7-13(a) is used to show the concept of the microstepping technique. Similar to half-steps described earlier, two adjacent stator field poles are energized at the same time. However, with microstepping, their coils are driven with different voltage levels that change gradually. The current through one coil increases while the current through the other one decreases.

The motor shown in Figure 7-13 is capable of making fine microstep movements between poles A and B instead of one full step. When pole A is energized by 5 volts and pole B is de-energized by 0 volts, the rotor is aligned with stator A. During the next microstep, pole A is reduced to 4 volts, and 1 volt is applied to pole B. A resultant south pole forms that causes the north pole of the rotor to align with it 1/5 of a step between the poles. When the voltage at coil A reduces to 3 volts and coil B increases to 2 volts, the rotor makes another movement 2/5 of a step toward pole B. Figure 7-13(b) shows the voltage for poles A and B to get 5 microsteps between each full step. In actual applications, there are as many as 125 microsteps to each full step in a conventional stepper motor. These signals are developed by a microprocessor located in the drive circuitry. These sinusoidal signals cause the rotor to move smoothly without jerking, and accurate positioning is attained if the currents are held at intermittent values. Resolution of 20,000 to 50,000 steps per revolution is common.

The stepper motor is typically used in an open-loop system. Position is determined by counting pulses. For critical applications, the system can be modified into a closed-loop configuration by using an encoder to verify the position. Stepper motors are used in many practical applications such as printers, CD players, floppy disks, and X-Y positioning tables.

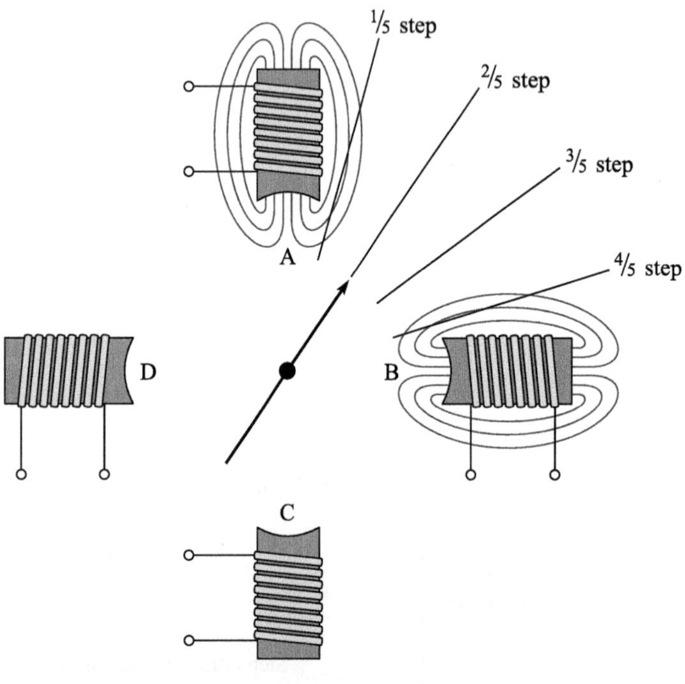

(a)

Pole A	Pole B	Position
5 V	0 V	Pole A (full step)
4 V	1 V	1/5 step
3 V	2 V	2/5 step
2 V	3 V	3/5 step
1 V	4 V	4/5 step
0 V	5 V	Pole B (full step)

(b)

FIGURE 7-13 Microstepping

7-8 AC Servo Motors

AC servo motors are controlled by AC command signals that are applied to coils, which become electromagnets. The magnetic fields that form around the coils interact with other electromagnets and cause the rotating member of the motor to turn.

AC Brushless Servo Motor

Recall that in a single-phase induction motor, two separate stator windings are energized by two AC voltages that are 90 degrees out of phase. One phase, called the *main phase,* is taken directly from the AC supply. The other phase, called the *auxiliary phase,* is tapped off the supply source and is shifted 90 degrees by a capacitor or inductor. These types of motors are sometimes referred to as *split-phase motors*. The purpose of the two phases is to create a resultant magnetic field that rotates around the rotor. As the field rotates, a voltage is induced into the rotor, causing it to turn due to the magnetic interaction between the rotor and the stator fields.

The AC brushless servo motor operates on the same principle as the single-phase induction motor. Both motor types have squirrel cage rotors and two sets of stator windings that are energized by two AC voltages that are 90 degrees out of phase. Instead of using a capacitor or inductor to develop the auxiliary phase, the AC servo motor uses an electronic circuit to perform this function. This circuit, shown in Figure 7-14, is referred to as an *AC servo drive amplifier*.

The AC line source supplies power to the main winding. It also provides power to the servo drive amplifier. A feedback signal from a transducer is used to indicate the actual position. Another input of the amplifier is a command signal that indicates the desired position. The difference between the measured feedback signal and the command signal creates an error signal. This error signal (V_e) is used to control the firing angles of phase-shift circuits, such as those that are powered by a triac. By firing the triac at different points during each alternation, the RMS value supplied to the amplifier output, labeled V_a, can be varied.

When the difference between the actual and the desired position is great, output V_a will be large. The strong magnetic field fed to the auxiliary winding will cause the servo motor to run at a high speed. As the difference between the actual position and the desired position is reduced, the V_e also decreases. Therefore, the magnetic field in the auxiliary winding weakens and makes the motor run more slowly. When the object is in the desired position, V_a goes to zero, there is no magnetic field from the auxiliary winding, and the motor stops. Even

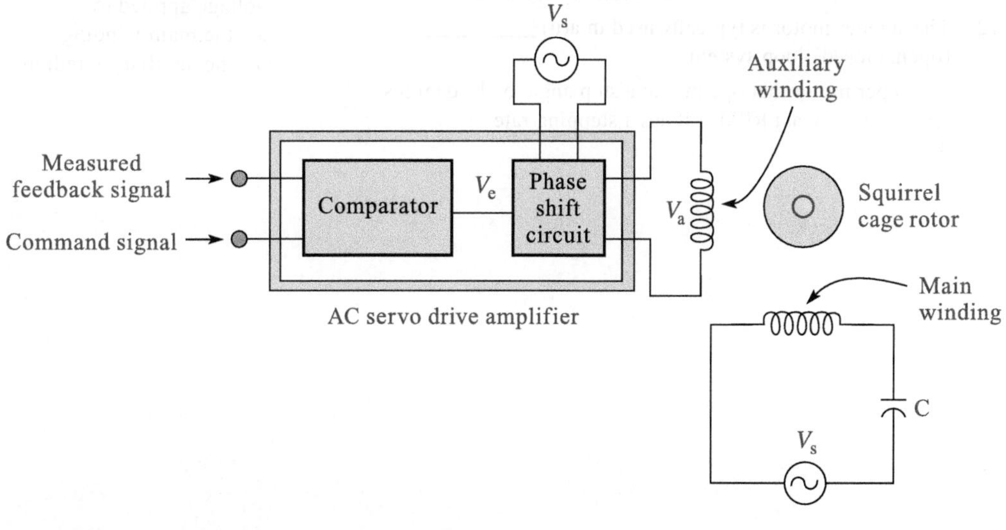

FIGURE 7-14 AC brushless servo motor

though the voltage supplied to the main winding does not change, the motor will not turn because there is no rotating magnetic field due to the single-phase condition.

Recall that it is possible for some split-phase motors to continue turning when single phased. To prevent rotation under this condition, the rotor of the servo motor has high-resistance conducting bars.

Problems

1. T/F The difference between the two classifications of PM motors is whether the rotor and field are made of permanent magnets or coils.
2. The speed of a wound armature PM motor is varied by changing the current applied to the ____.
   a. armature b. stator
3. List three applications of a wound armature PM motor.
4. The speed at which a moving coil motor runs is controlled by varying the ____.
   a. amplitude of the DC voltage applied to the armature
   b. width of the pulses applied to the armature
5. List two applications of MCM.
6. T/F BDCM usually use three-phase power.
7. The direction a BDCM turns can be reversed by switching two of the three connections to the ____ of the transistor.
   a. base b. collector
8. The amount of force required to move the rotor of the BDCM away from the held position is called __________ __________.
9. A __________ motor converts electronic digital signals into fixed increments of positional movement.
10. T/F The operation of a variable reluctance motor is based on the ease with which flux lines pass through soft iron.
11. If the stator's magnetic field of the VR stepper motor moves clockwise, the rotor turns ____.
    a. clockwise b. counterclockwise
12. The stepper motor is typically used in a(n) __________ (open, closed)-loop system.
13. A stepper motor that operates at a step angle of 12 degrees will rotate at what RPM if it has a stepping rate of 360? ____
14. In a variable reluctance motor, if the salient poles are 15 degrees apart and the rotor poles are 20 degrees apart, the step angle is ____ degrees.
    a. 5 c. 20
    b. 15 d. 35
15. The maximum number of steps a stepper motor can make in a second is called __________ __________.
16. The number of degrees a stepper motor turns per step is called __________ __________.
17. The speed at which a stepper motor turns is increased by ____ applied to the stator.
    a. increasing the stepper rate applied to the stator
    b. increasing the voltage
    c. both a and b
18. A stepper motor with twice the step angle of a second stepper motor will turn at ____ the RPM if the stepper rate applied to both is the same.
    a. half b. twice
19. Microstepping is achieved by ____ .
    a. gradually increasing the voltage applied to one coil while decreasing the voltage at a second coil
    b. simultaneously applying simulated sine waves that are out of phase to different coils
    c. both a and b
20. In an AC servo motor, the phase-shifted voltage is supplied to the ____ winding.
    a. main b. auxiliary
21. The speed of an AC servo motor is changed by varying the voltage applied to ____.
    a. the main winding c. none of the above
    b. the auxiliary winding

SECTION

4

Variable-Speed Drives

OUTLINE

Most of the machines used in industry are powered by motors. Some production process applications require that a motor maintains constant speed or torque regardless of the physical load placed on it. Other situations require the motor to change speeds either very quickly or very slowly. Mechanical mechanisms were the first devices used to control motors. For example, gear reducers were used to slow the speed of motors. When a speed alteration was required, the motor system was shut down while gears and belts were changed. When a gear mechanism was used for controlling position, backlash limited the accuracy.

Technological advancements in electronics have resulted in the development of the present electronic motor drive system to control electric motors. Commonly

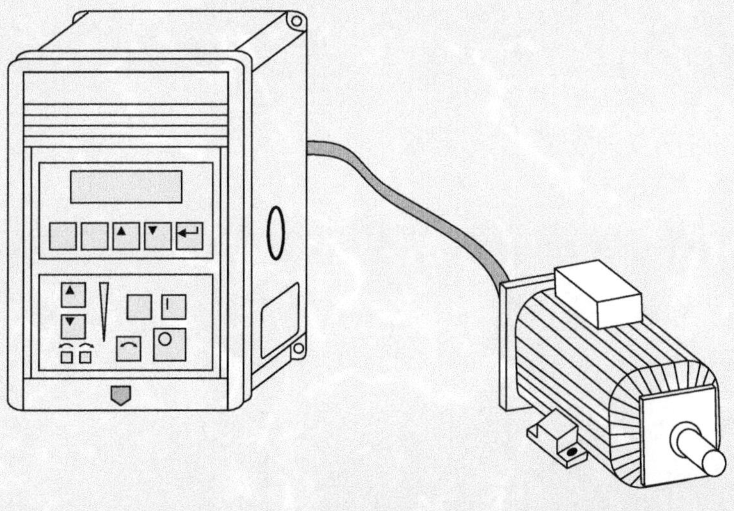

A variable-speed drive operates a motor

known as the *drive,* it uses integrated circuits and solid-state devices in the circuitry that supplies power to the motor. A drive connected to a motor is shown in the illustration on page 161. The drive is capable of controlling the speed, torque, horsepower, direction, and position of motors with more precision than mechanical devices. The benefits of accurately controlling a motor are energy savings, increased productivity, and better quality. Electronic devices are typically categorized into two groups, DC and AC, according to the type of motor they control.

The DC industrial drive was the first type of electronic drive system. It controlled the speed of a DC motor using variable resistors to regulate current flow through the armature. The resistors were replaced when high-current solid-state devices such as diodes, SCRs, and power transistors were developed. AC drives were not used until a method was devised to vary the frequency of the voltage applied to the AC motor. Not until the advent of the microprocessor integrated circuit was this possible. Today, AC drive systems are becoming increasingly sophisticated. They now perform many of the operations that previously could be performed only by DC systems. However, DC drive systems are not being totally replaced, because DC motors have some advantages over AC motors, such as higher starting torque and horsepower output.

The following two chapters discuss the most popular types of drives. Their basic parts are illustrated in block diagram form to describe the underlying theory of operation more clearly. Each block is further broken down into simplified discrete circuitry to show how the signals are developed, controlled, or processed. Typical applications of drives are given, as well as the drive's characteristics, strengths, and weaknesses.

The variable motor drive primarily performs the function of the controller element of an industrial control system. Some drives also use the measurement feedback device and comparer elements of an industrial control system to perform their operations.

CHAPTER

8

DC Drives

OBJECTIVES

At the conclusion of this chapter, you should be able to:

- List the reasons for using DC drives and describe their advantages over AC drives.
- Identify the types of control functions performed by DC drives.
- Describe the purposes of the following sections of a DC drive:

  Operator Control
  Drive Controller
  DC Motor
  Speed Regulator

- Explain the operation of the variable-voltage DC drive circuits that perform the following functions:

  Motor Speed Control
  Speed Regulation
  IR Compensation
  Current Limiting
  High/Low Speed Adjustment
  Field Current Speed Control

- List the load characteristic requirements of several types of equipment controlled by DC drives.
- Identify three classifications of motor braking techniques and describe the operation of each type.

INTRODUCTION

DC drives are used in both motion control and process control. They are ideal for material handling applications because they have high starting torque. They can start and stop DC motors quickly and efficiently. Additionally, they are capable of smoothly slowing down a motor to 0 RPM and then immediately accelerating it in the opposite direction. Soft starts and stops by controlled acceleration and deceleration save energy and reduce stress on expensive equipment. The newest generation of DC drives uses microprocessor technology to perform advanced operations. One function that it provides is monitoring the current drawn by the motor. If the current becomes excessive, the drive will protect the motor by reducing the voltage it supplies. A modern drive can also provide troubleshooting diagnostics for the technician when faults occur. In addition, drive controllers can exchange data information with other computerized devices, which is a requirement in some automation applications.

Contrary to the prediction that DC drive applications will eventually be replaced with AC drives and motors, there is evidence that this situation will not occur. Some applications still require the performance provided by DC motors. For example, the horsepower and constant-torque characteristics of a DC drive and motor are higher than those of an AC drive system. Therefore, they are often used in applications where the performance of an AC system is poor, such as starting a heavy load or maintaining a constant torque and speed when the physical load is suddenly changed. These types of conditions exist with conveyor systems and extruders.

8-1 DC Drive Fundamentals

DC drive systems are divided into three major sections, as shown in Figure 8-1: the operator controls, the drive controller, and the DC motor. A feedback configuration is also a part of the system.

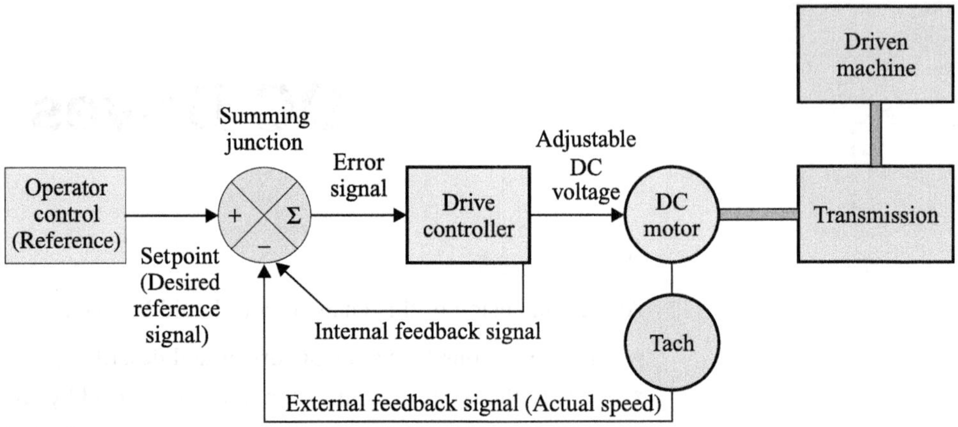

FIGURE 8-1 DC drive control system

Operator Control

The **operator control** section provides the operator a way to start, stop, and change the direction and speed of the motor. The controls used may be an integral part of the controller in the form of switches, push buttons, potentiometers, or a handheld keypad. They also may be remotely mounted in the form of programmable controllers or personal computers. The use of remote devices is increasing due to their greater flexibility in process or machine control. They also have diagnostic functions that provide information to identify a malfunction if it occurs.

Drive Controller

The **drive controller** converts an AC source voltage to an adjustable DC voltage, which is then applied to a DC motor armature. Regulation characteristics of the controller will run the motor at a desired speed, torque, and horsepower set by a reference input. Additional circuits can be used to protect the drive system from overloads and various circuit faults.

DC Motor

The *DC motor* converts the adjustable DC voltage from the drive controller to rotating mechanical energy. The RPM and direction of the motor shaft are determined by the magnitude and polarity of the adjustable voltage applied to the motor. The motor shaft is usually coupled to a transmission device, which is then connected to the driven machine.

The DC motor in a typical drive control system is usually a shunt wound or permanent magnet type. An adjustable voltage supply is connected to the armature of both types of motors to primarily control speed. A fixed or adjustable voltage is connected to the field coil of the shunt motor. An adjustable field control can be added to provide higher speed and better torque control of the motor.

Speed Regulator (Feedback Method)

The speed of the motor is regulated by a closed-loop circuit. The desired speed of the motor is determined by a variable voltage from the operator control input device. The voltage produced, called the *setpoint,* is applied to a summing junction. The actual speed of the motor is converted to a proportional voltage. This voltage, called the *feedback signal,* is also applied to the summing junction. The setpoint and feedback signals are continuously compared. Any difference causes an error signal to form at the summing junction output. The result is that a speed correction is made, the setpoint and feedback signals become matched, and the error signal goes to 0 volts.

Two methods are used to monitor the speed of the motor: internal and external. Each method produces a feedback voltage proportional to the motor velocity. The internal method uses armature current as the feedback signal. The external method uses a tachometer to produce a DC feedback voltage. Since the internal method monitors only the motor, the external method is more accurate because it monitors the entire closed-loop system. The external loop provides speed regulation as low as 1 percent.

The most popular DC drive is the **variable-voltage drive.** It controls motors by regulating the amount of voltage applied to the armature winding, shunt field winding, or both.

8-2 Variable-Voltage DC Drive

A simple speed control circuit is shown in Figure 8-2. It consists of a motor armature and an SCR. The SCR performs two functions. First, it rectifies the AC supply into a pulsating direct current, causing current to flow through the DC armature in one direction. Second, since it can be turned on at any time during the alternation, it controls the amount of current that flows through the DC motor armature. Therefore, the speed of the motor can be controlled, since it is directly proportional to the current that flows through the armature. The 120 VAC source is converted into 0 to 55 average DC volts.

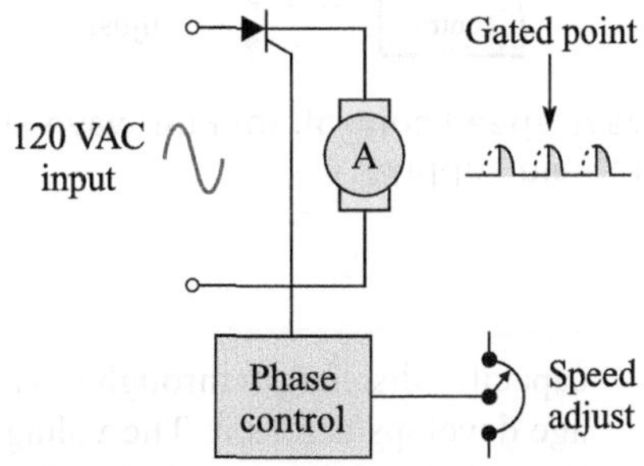

FIGURE 8-2 Half-wave speed control

The half-wave rectifier is rarely used because of motor noise and excessive heating. By supplying power to the motor armature with full-wave rectification, the overall performance of the motor is significantly improved. Figure 8-3(a) shows a full-wave rectifier that can supply current to the armature during each alternation. It consists of two rectifier diodes and two SCRs. During each alternation, current flows through one of the SCRs and the diode connected to it in series. Current that flows through the diode and armature is regulated by the SCR.

DC motor drives control either permanent magnet or shunt field wound motors. If the motor is shunt wound, a field supply is also required. Figure 8-3(b) shows two diodes, D_5 and D_6, and a field coil added to the circuit. Full-wave rectification is provided to the field coil by diodes D_5 and D_6, and D_3 and D_4 from the armature bridge rectifier. The field coil also functions as an inductor to provide filtering for the pulsating DC output of the rectifier.

A firing circuit, shown in Figure 8-4, is used to turn the SCRs on during the alternations. It consists of two parallel branches, one with an RC network and the other with a UJT pulse configuration. The capacitor and resistor in the RC branch remain constant, and the resistance of the transistor varies depending on how hard it is turned on. When the charge on the capacitor reaches a certain voltage, the depletion region of the UJT collapses and the

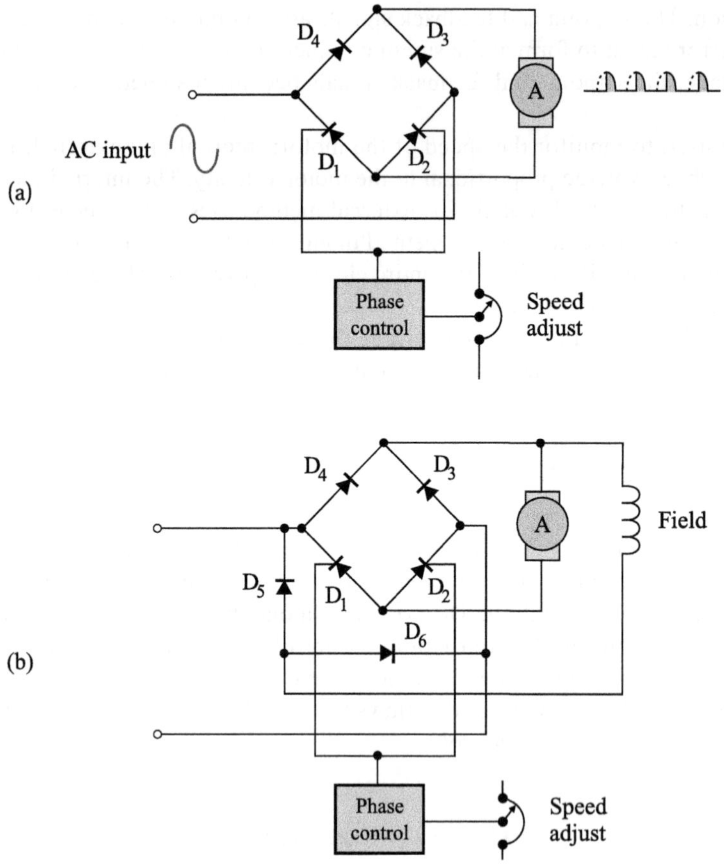

FIGURE 8-3 (a) Full-wave speed control; (b) Full-wave speed control with fixed shunt field supply

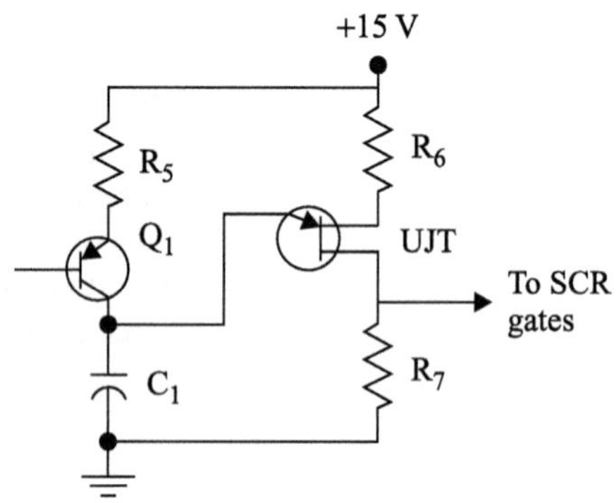

FIGURE 8-4 Firing circuit

capacitor discharges through it. As the pulse of current flows through R_7, a momentary voltage develops across it. The voltage formed at the top of R_7 is fed to the gate leads of SCRs D_1 and D_2 of Figure 8-3. The SCR that is anode/cathode forward biased will fire when the gate lead receives the pulse voltage. The SCR turns off when the voltage across the anode and cathode approaches 0 volts.

When the voltage at the transistor base of Q_1 in Figure 8-4 becomes more negative, it turns on harder and its resistance decreases. The time constant of the branch decreases, which causes the capacitor to charge more quickly. The result is that the firing delay angle decreases, which allows more current to flow through the armature.

Figure 8-5 shows the entire variable-voltage DC drive. Figures 8-6 through 8-8 focus on parts of the drive shown in Figure 8-5. Figures 8-9 and 8-10 are optional circuits that can be added to the drive for greater control capabilities.

Motor Speed Control

The speed of the motor is operator controlled by adjusting a knob connected to potentiometer R_1 in Figure 8-5.

The fixed terminals of the potentiometer are connected to the positive 15-volt supply. As the wiper arm is moved upward, the input lead connected to R_2 of the summing op amp goes more positive. The result is that the op amp output goes negative, turns Q_1 on harder, and causes the capacitor to charge more quickly. As it does, the UJT turns on sooner during the alternation and fires the SCR earlier. This action causes the current to increase through the armature. The rise in armature current increases the speed of the motor.

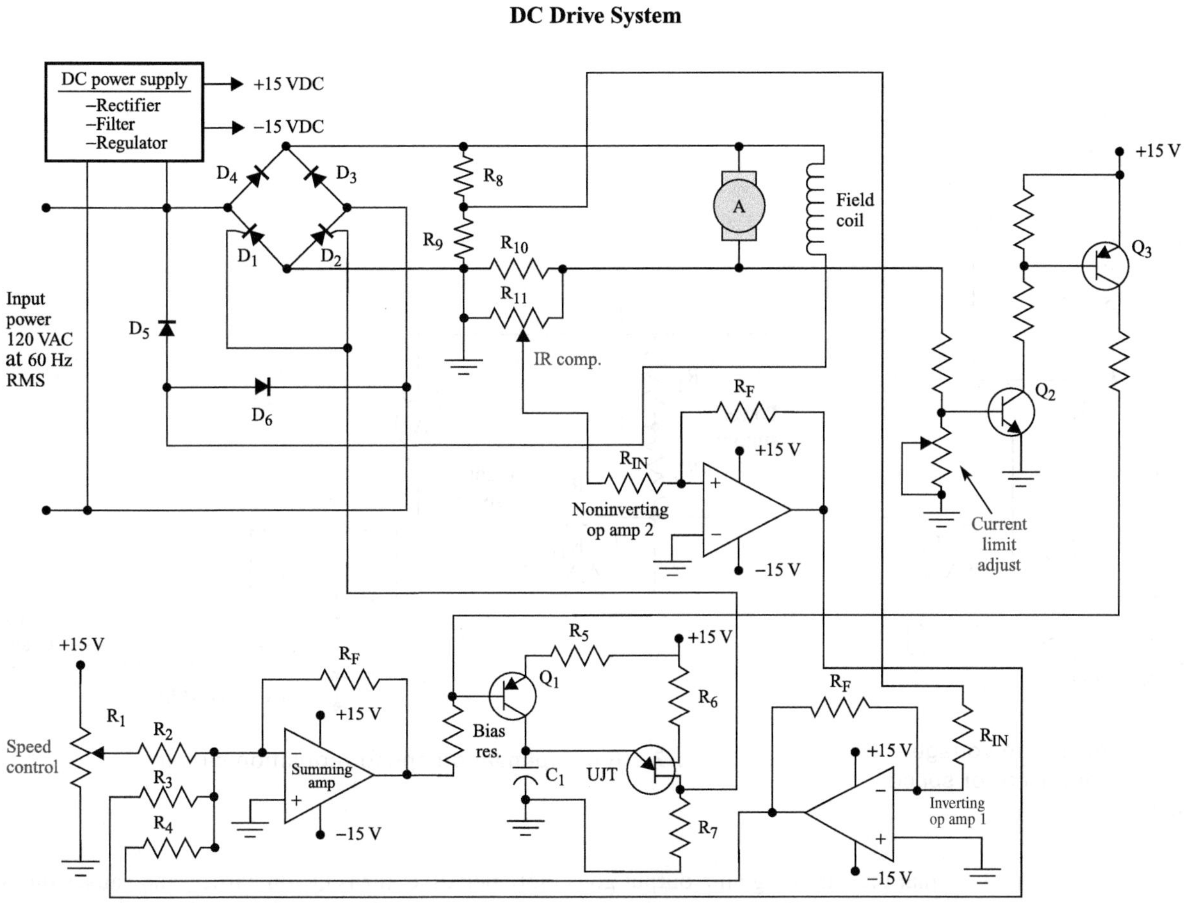

FIGURE 8-5 DC drive system

Speed Regulation

When the operator sets the speed control potentiometer, the motor rotates at the desired speed. However, there are two conditions that can cause the speed to go above or below the speed setting: a fluctuation in the applied voltage, and a change in the physical load the motor is driving. To compensate for these conditions, two feedback networks are added in the drive's circuitry. The function they perform is **speed regulation**. The speed regulation networks are called *voltage regulator* and *IR compensation*.

Recall from Section 5-10 that the speed of a DC shunt motor with a constant field is proportional to the CEMF, which is equal to $V_A - (I_A \times R_A)$. To provide speed regulation, the regulator network keeps V_A from fluctuating, and the IR compensation network keeps I_A constant.

Voltage Regulator

The voltage regulator feedback network, shown in Figure 8-6 on page 168, consists of two resistors, R_8 and R_9, that are connected across the output of the bridge rectifier. The feedback signal develops at the junction between the resistors and is amplified by inverting op amp 1 before it is applied to summing op amp input resistor R_4. Suppose that there is a significant voltage fluctuation in the AC line that supplies power to the drive. If this fluctuation is a reduction, the output of the bridge rectifier will also become smaller, causing the voltage at the armature (V_A) and the voltage at the junction between R_8 and R_9 to decrease. As the junction voltage decreases, it is inverted by op amp 1, which makes the voltage applied to summing amp resistor R_4 become less negative. This causes a reduction in the cancellation of the positive voltage applied to the other inputs of the summing amp. The result is

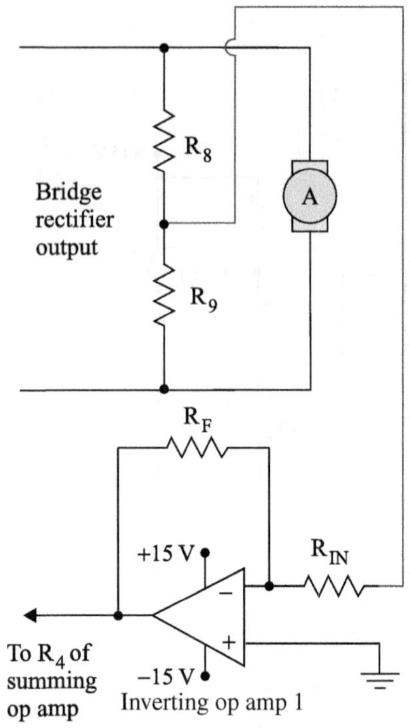

FIGURE 8-6 Armature voltage feedback network used for speed regulation

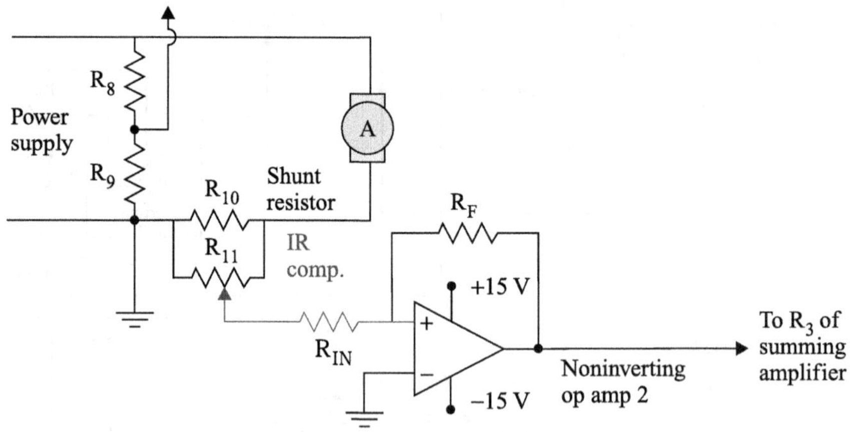

FIGURE 8-7 IR compensation speed regulation circuit

that the summing amp output goes more negative, turns Q_1 on harder, and causes the SCR to fire sooner. This opposes the change in armature voltage, which stabilizes the motor speed.

IR Compensation

An **IR compensation** circuit, shown in Figure 8-7, also improves the speed regulation of the DC drive. The IR circuit, which refers to *internal resistance* of the motor, is used to compensate for an internal voltage drop when the load on the motor increases. Paper being wound on a roll of a paper machine is an example of how the load on a motor increases. The more paper that is wound on the roll, the heavier it gets, slowing the motor down. The IR comp signal is obtained at R_{11}, which is the IR comp adjustment potentiometer. It divides the voltage developed across resistor R_{10}. The voltage across R_{10} is proportional to the motor current and the load. In the event that the motor slows down, the CEMF decreases, causing the current through the armature and R_{10} to increase. The IR voltage at R_{11} increases and is amplified by noninverting op amp 2 before being fed back to summing op amp input resistor R_3. The effect is that the timing circuit fires the SCRs sooner, which speeds the motor back to the desired RPM. Potentiometer R_{11} is used to adjust the feedback voltage at a level that properly compensates for any load variances. By moving the wiper arm of R_{11} to the right, increased compensation will be obtained.

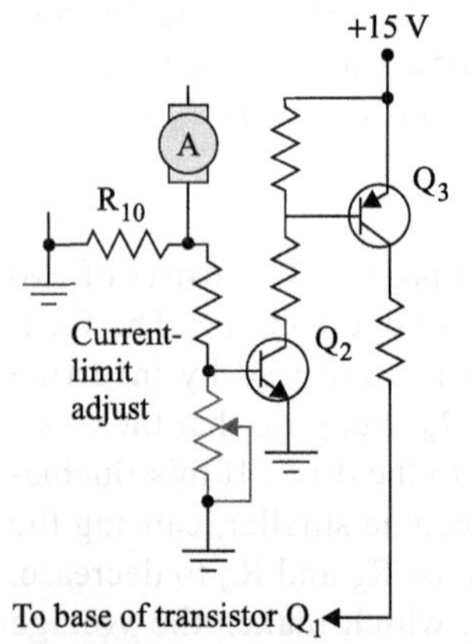

FIGURE 8-8 Current-limiting circuit

Current Limiting

If a malfunction develops that causes the motor to stall or slow down far below its rated speed, the current through the armature will become excessive. These large currents occur because the slow rotation of the motor reduces the CEMF so that the net voltage at the armature becomes very high. The current levels may rise above the maximum forward current rating of the semiconductors in the bridge assembly. Either the semiconductors or the armature windings are destroyed. To limit the current to a safe amount, a **current-limiting** network, shown in Figure 8-8, is added to the circuit. The current through the armature is monitored by connecting an input lead to the bottom terminal of the armature. If the current becomes too high, a greater voltage forms across R_{10}. The voltage at the top of the current-limit adjust rheostat goes more positive.

Transistor Q_2 turns on harder and causes the voltage at the base of transistor Q_3 to go in a negative direction. As PNP transistor Q_3 turns on harder, the potential at the base of Q_1 goes in the positive direction. The result is that as transistor Q_1 turns on less, it prevents the timing circuit from firing any sooner, and therefore limits the current flow through the armature. The current-limit adjust rheostat is used to adjust the network so that the current is limited to the proper level. Moving the wiper arm of the current-limit rheostat downward will decrease the amount of current permitted to flow through the armature.

High-Speed/Low-Speed Adjustment

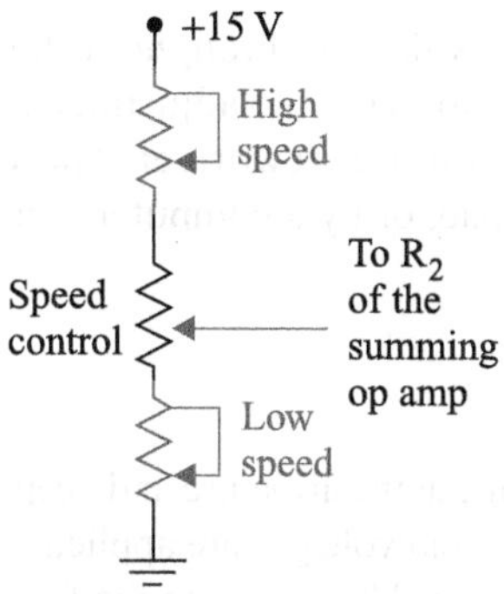

FIGURE 8-9 High- and low-speed adjustments

There are some applications that should avoid maximum motor speed when the speed control is adjusted to the maximum motor speed setting. An example of such an application is when the maximum rated speed of the motor is higher than the rated RPM of the load it is driving. Some applications should avoid 0 RPM when the speed control is set at minimum speed. Figure 8-9 shows **high-/low-speed adjustment.** A rheostat is connected between the speed control potentiometer and the +15-volt supply. Another rheostat is connected between the potentiometer and the ground. If the wiper arms of both rheostats are set all the way to the bottom, they are shorted out. The speed control setting varies from 0 to 15 volts, which enables the motor to rotate from 0 RPM to maximum speed. If the wiper arm at the top rheostat is moved slightly upward, the voltage divider changes. Therefore, when the speed control is set at maximum speed, the voltage at the wiper arm may be only +12 volts instead of +15 volts. Likewise, when the bottom rheostat wiper arm is moved slightly upward, the voltage at the potentiometer's wiper arm may be +3 volts instead of 0 the volts when the speed control setting is adjusted to minimum speed.

Acceleration/Deceleration Adjustment

Suppose that a paper machine is at 0 RPM when the speed control is suddenly adjusted from the minimum to the maximum speed setting. Maximum current will flow through the armature. If the adjustment is made too quickly, something may become damaged. For example, the surge of current may overheat the armature windings, since the CEMF is minimal because of the low motor RPM. Even if the armature windings withstand the large momentary current, the motor shaft coupled to the paper machine may sheer if the inertia of the paper roll is too great to overcome.

By placing a capacitor and rheostat between the speed adjust wiper arm and the ground, as shown in Figure 8-10, both problems can be avoided. If the speed control is suddenly changed from the minimum to maximum setting, the voltage of the wiper arm will not be felt immediately at the summing input. Instead, the voltage forms at the top plate of the capacitor at a rate determined by the setting of rheostat R_{12}. The charging time delays application of the voltage to the motor, causing it to accelerate at a slowed rate. This action is referred to as a *soft start.* If the speed control is suddenly adjusted to a lower voltage, the potential at the top plate of the capacitor decreases at a slower rate. The voltage on the capacitor lowers at a rate determined by the setting of rheostat R_{12}. As the capacitor discharges, the motor decelerates at a slowed rate.

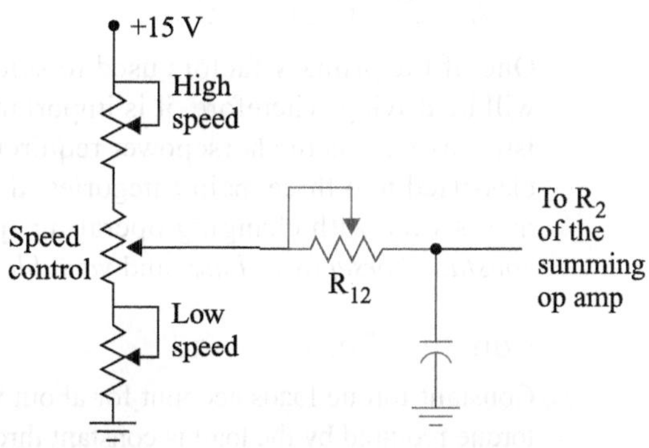

FIGURE 8-10 Acceleration/deceleration control

Methods Used to Adjust Parameters

The common adjustments that may be performed on typical DC drives are:

- Speed Control
- IR Compensation
- Current Limiting
- High Speed/Low Speed
- Acceleration/Deceleration

How these adjustments affect the operation of the drive was explained throughout this section of the chapter. Variable resistors were used in the diagram to make the adjustments. In an actual drive, resistance changes are made with a control knob or screwdriver. These same adjustments can be made on advanced drives by using a keypad, or by a computer connected with an interface cable.

Field Current Speed Control

The speed of a shunt motor is controlled primarily by varying the voltage at the armature and keeping the field voltage constant. When the maximum rated armature and field voltages are applied to the motor, it runs at *base speed.* It is possible to further control motor speed by changing the field current, called **field current speed control.** The following formula reinforces this statement:

$$N = \frac{V - IR}{K \cdot \phi}$$

where,

N = Speed
V = Applied armature voltage
IR = Internal motor resistance drop
K = Motor design constant
ϕ = Motor field flux

By reducing the field current to half its original value, the armature CEMF will decrease and allow more current flow through the armature windings. With the additional current, speed increases and CEMF builds back up to its original level. When equilibrium is reached, the motor RPM is doubled. It is possible to increase the motor speed up to five times the base speed. However, weakening the field by reducing the field voltage also reduces motor torque.

The field current may be varied manually or automatically. The control circuit uses a rheostat or an SCR bridge to regulate the amount of current flow. DC drives with this capability are frequently called *voltage drives* and *field drives.*

Load Characteristics

One of the primary factors used to select an adjustable drive is the type of load the motor will be driving. Therefore, it is important to understand both the speed and torque characteristics as well as the horsepower requirements of the load being considered. Motor loads are classified into three main categories, depending on how the torque and horsepower requirements vary with changing operating speed. These three groups are *constant-torque load, constant-horsepower load,* and *variable-torque load.*

Constant-Torque Load

Constant-torque loads account for about 90 percent of industrial applications. In this group, the torque required by the load is constant throughout the speed range. The amount of torque required at low speeds is the same as the amount required at high speeds. An example of this type of load is

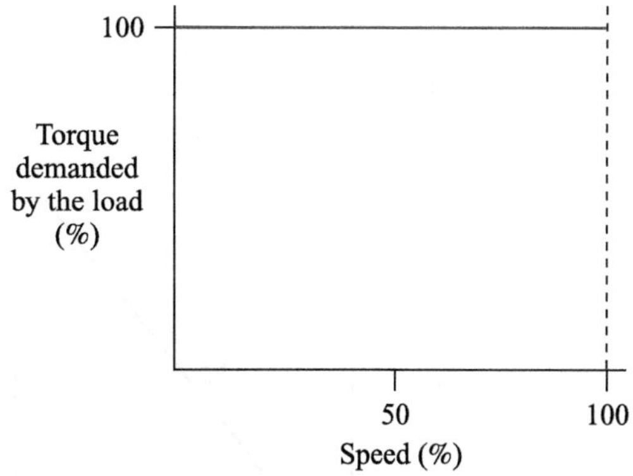

FIGURE 8-11 Constant-torque load graph

friction. Examples of friction loads are conveyors, extruders, and hoist. Motors that provide constant torque are also used when shock loads, overloads, or high inertia loads are encountered.

A graph showing the relatively constant torque demanded by the load over the rated speed range, 0 to base speed (100%), is illustrated in Figure 8-11. The speed is adjusted over the entire range by controlling the armature voltage from zero to the rated nameplate voltage and keeping the field strength at the full rated level.

A DC drive that controls a permanent magnet or shunt field motor is used for this type of load requirement. Speed is adjustable from 0 to 100 percent by controlling the armature voltage as the field strength is held constant.

Constant-Horsepower Load

In this category, the horsepower required by the load is constant. To hold the horsepower constant, it is necessary for the motor to run above the base speed. According to the following basic horsepower formula,

$$\text{hp} = \frac{\text{Torque} \times \text{Speed}}{5252}$$

where,

Torque is measured in lb-ft
Speed is measured in RPM
5252 is a proportional constant

to keep the horsepower constant the torque must decrease as the speed increases.

The RPM of the motor is controlled to stay at three or four times base speed by weakening the motor field. A DC motor torque and horsepower graph for constant-torque and constant-hp loads is shown in Figure 8-12 on page 172.

An application for a constant-horsepower load is a machine tool lathe. It requires slow speed and high torque for rough cuts, and high speed with less torque for fine cuts where little material is removed. DC drives that power and control shunt field motors are used for this type of load requirement. The control function that supplies a varying field current to enable the motor to run above base speed is not a standard feature on all DC drives. If the drive does not have this capability, a separate field supply module may be required.

Variable-Torque Load

In this category, the torque required by the load varies proportionally to the mathematical square of the speed. This characteristic is expressed by the following formula:

$$\text{Torque} = \text{Constant} \times \text{Speed}^2$$

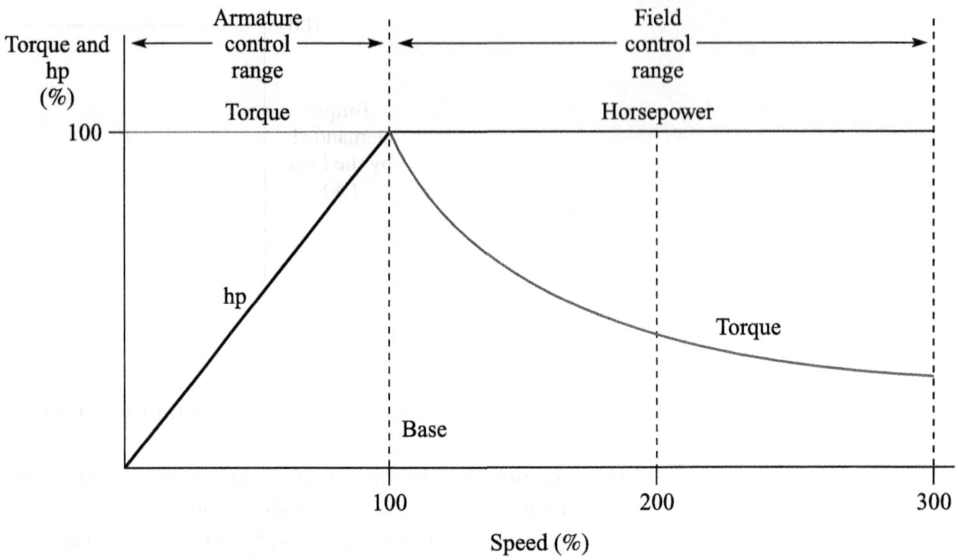

FIGURE 8-12 Constant-horsepower load graph

It is graphically shown in Figure 8-13. As the torque demanded by the load increases, the speed of the motor must be increased at a proportionally squared (speed^2) rate to supply the larger torque that is required. Examples of loads that demand variable torque from a motor are large centrifugal fans and pumps. Also included are punch presses, which are high-inertia machines that use flywheels to supply most of the operating energy.

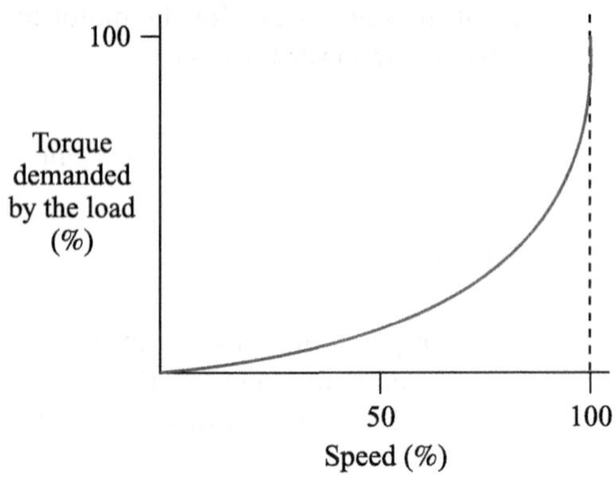

FIGURE 8-13 Variable-torque load graph

8-3 Motor Braking

Some industrial applications require that a rotary mechanism be stopped quickly. However, because of inertia, it may be difficult to stop a mechanism abruptly. Since DC electric motors are used as the prime mover for some types of rotating machines, certain types of motor braking techniques have been developed. Two popular methods are **dynamic braking** and **regenerative braking.**

Dynamic Braking

Dynamic braking uses a resistor to absorb the kinetic energy from the rotating armature and the load to which it is connected. Figure 8-14 shows the circuit configuration for dynamic braking. When the start button is closed, current flows through the control relay (CR) coil. All normally open (NO) contacts close and all normally closed (NC) contacts open. Current continues to energize the CR coil when the start button is released, as current flows through the contact that is parallel to it. Current flows through the motor armature from the negative to the positive terminal of the motor control circuitry. When the stop button is pressed, current no longer flows through the CR coil. The N.O. contacts open and the N.C. contacts close. The field coil remains energized, but current no longer flows through the armature from the motor control circuitry. The armature becomes connected in series with the load resistor. As it spins due to inertia, the armature coil cuts the shunt field flux lines and consequently becomes a generator. Current induced into the armature now flows in the opposite direction from the original current. The result is that a reverse torque is developed. At first, the current and torque are high, so the motor speed drops quickly. As the motor slows down, less current is induced into the armature coil. The reverse torque progressively gets smaller, which causes the motor to gradually slow to a halt. This braking action is the result of kinetic energy being extracted from the rotating machinery as dissipated heat by the resistor.

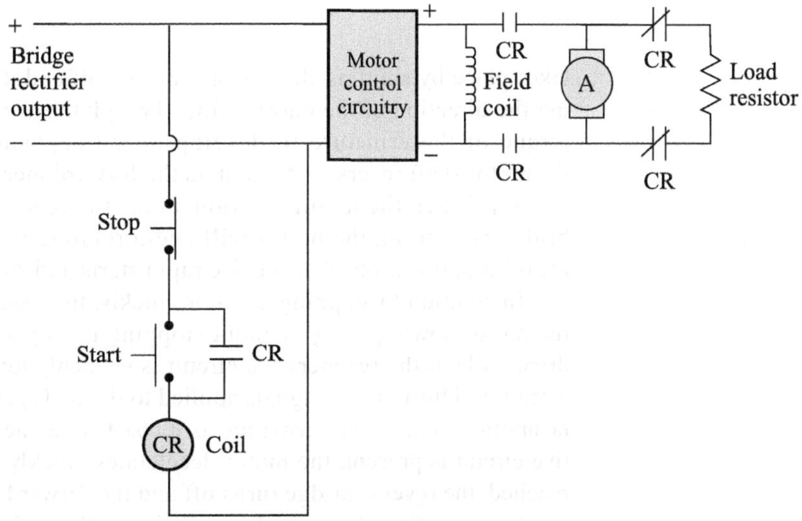

FIGURE 8-14 Dynamic braking circuit

The smaller the load resistance, the more current is drawn from the armature and the more rapidly it slows down. Dynamic braking resistors of small ratings are typically constructed of wire wound on ceramic. Resistors with greater ratings are usually made of fabricated steel stampings, a convoluted strip, cast iron, or liquid, all of which provide the large surface areas required to dissipate the heat produced during braking.

When selecting a load resistor, it is necessary to choose one with a rating based on the duty it is required to perform, taking account of load inertia and the number of stops per hour. In practice, a load resistor value is chosen so that the initial braking current is about twice the rated motor current. The initial braking torque is then twice the normal torque of the motor.

In newer equipment, it is more common to use solid-state devices and logic circuits instead of contactors to perform dynamic braking. The dynamic braking function used in positioning applications is called *indexing*. This function is needed when heavy loads are required to stop within a short amount of time. Dynamic braking action is only active during motor rotation, and therefore cannot be used as an electric holding brake.

Regenerative Braking

In *regenerative braking,* power from the kinetic energy is returned to the power supply. Figure 8-15(a) shows the circuit that causes the motor to stop quickly. One bridge supplies current to the motor armature in one direction; the other bridge supplies current in the opposite direction. Suppose the motor is running in the forward direction. The braking action

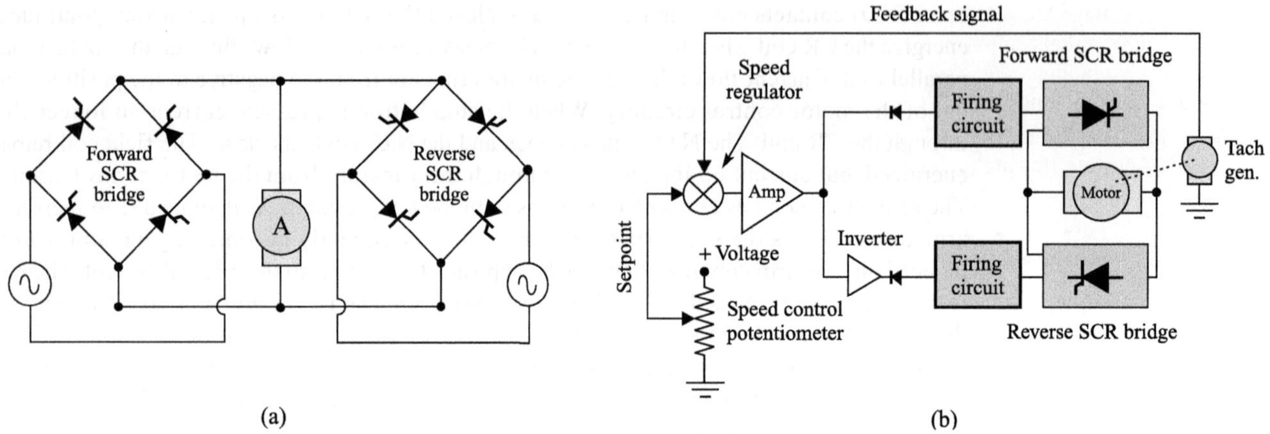

FIGURE 8-15 Regenerative braking

takes place by shutting the forward bridge off and then turning the reverse bridge on, causing the direction of the current flow through the armature to change. This action causes the torque of the armature to develop in the opposite direction. As the torque applied to the motor shaft reverses, it counters the forward inertia of the motor, causing it to decelerate quickly. When the motor rotation stops, the reverse SCR bridge turns off. If the reverse bridge is kept on, the motor will begin rotating in the opposite direction. This system is ideal for applications that require rapid starts and stops or frequent direction reversals.

In addition to stopping a motor quickly, the regenerative motor braking circuit can slow the motor down quickly without stopping it. Suppose that the speed adjust knob on a DC drive without the regenerative circuit is suddenly turned to a lower RPM setting. The firing circuit will turn off the signals applied to the SCR gates in the bridge circuit. No voltage will be applied to the motor armature as it coasts to its new speed setting. However, if a regenerative circuit is present, the motor decelerates quickly as it is braking. When the new speed is reached, the reverse bridge turns off and the forward bridge turns back on.

Regeneration also provides a precise method of speed regulation. Figure 8-15(b) shows a block diagram of the circuit that performs this function. A speed control potentiometer provides a setpoint voltage to a speed regulator to obtain the desired RPM. The output voltage of a DC generator/tachometer reliably indicates both the magnitude and direction of motor speed. This voltage is also applied to the speed regulator. The polarities of the set point signal and feedback signal applied to the speed regulator, are opposite. Assume that the forward SCR bridge is on and the motor is rotating as desired, when the inertia of the rotating load causes the motor to accelerate to a rate that is higher than the desired RPM. The feedback signal becomes greater than the setpoint. The output of the speed regulator changes polarity, turns the forward bridge off, and turns the reverse bridge on. The current change through the armature reverses the motor torque and the motor slows down until the setpoint is greater than the feedback voltage. The motor then runs at that speed.

Problems

1. List two advantages of DC drives over AC drives.
2. DC drives give the DC motor which of the following characteristics? ____
   a. controlled acceleration and deceleration
   b. high starting horsepower
   c. high running horsepower
   d. good speed regulation
   e. all of the above
3. List two advantages of soft starts and stops controlled by acceleration and deceleration.
4. Which of the following names are major sections of a DC drive system?
   a. Operating Controls
   b. Drive Controller
   c. DC Motor
   d. Feedback
   e. all of the above
5. List two applications DC drive systems are used for.

6. A variable-voltage DC drive controls motors by regulating the amount of voltage applied to the ____.
   a. armature windings c. both a and b
   b. shunt field windings
7. The current-limiting network of a DC drive keeps the amount of current flowing through the ____ below a required level.
   a. motor's armature b. field windings
8. Which DC drive adjustment provides a *soft start* to a DC motor? ____
   a. low speed adjustment c. speed control
   b. IR compensation d. acceleration
9. Which of the following motor functions can be manipulated by the operator control section? ____
   a. starting d. speed
   b. stopping e. all of the above
   c. change of direction
10. Which of the following types of circuits are used by the drive controller section? ____
    a. AC-to-DC converter c. protection
    b. regulation d. all of the above
11. T/F The armature is the primary part of the motor where an adjustable voltage is connected for speed control.
12. What are the two functions performed by the SCRs in a variable-voltage drive?
13. The function of Q_1 in Figure 8-4 is ____.
    a. to operate as a switching transistor
    b. to operate as a variable-resistance device
14. Referring to Figure 8-5, if the motor slows down, the voltage at the bottom input of the summing amplifier output __________ (increases, decreases), causing the summing amplifier to become more __________ (negative, positive), which turns Q_1 on __________ (less, harder), causing the SCR to fire __________ (sooner, later), allowing __________ (less, more) current to flow through the armature.
15. Referring to Figure 8-7, if the motor speed slows down, the IR voltage at R_{11} __________ (increases, decreases), causing the SCRs to fire __________ (sooner, later) in the alternation.
16. As the motor runs at base speed, the voltage applied to the armature and field is ____.
    a. low c. maximum
    b. moderate
17. By adjusting the DC drive so that the field current is reduced to half, the motor speed ____.
    a. is cut off c. stays the same
    b. doubles
18. For a DC variable-speed drive to cause the motor to provide a constant torque to the load, it varies the motor's ____, and it holds the ____ constant.
    a. armature voltage b. field strength
19. For a DC variable-speed drive to cause the motor to provide a constant horsepower to the load, it varies the motor's ____ to run it between 0 RPM and base speed, and it varies the ____ to run the motor above the base speed.
    a. armature voltage b. field strength
20. To provide variable torque to a load, the DC variable-speed drive causes the speed of the motor to ____ when an increase in torque is required.
    a. increase b. decrease
21. Match the application example with the appropriate motor load classifications.
    Constant-Torque Load ____
    Constant-Horsepower Load ____
    Variable-Torque Load ____
    a. machine tool lathe d. centrifugal fan
    b. pump e. extruder
    c. hoist
22. Match the proper term with the description that explains the method used to perform motor braking.
    ____ Uses a resistor to absorb kinetic energy from the rotating armature and load.
    ____ Returns power from kinetic energy to the power supply.
    a. dynamic b. regenerative

CHAPTER

9

AC Drives

OBJECTIVES

At the conclusion of this chapter, you should be able to:

- List the reasons for using AC drives to control the speed of AC motors.
- Explain the methods of controlling the speed of an AC motor and identify which one is most effective.
- Describe the purpose of the following sections of an AC drive:

  Operator Control | Drive Controller | AC Motor
- Explain the differences among volts/Hz, AC closed-loop vector, and sensorless vector control methods.
- Describe the operation of the following major sections of the VAC drive:

  Converter | Intermediate Circuit | Inverter
- Determine the V/H ratio when frequency changes and the voltage is constant.
- Describe the operation of the flux vector drive.
- List the types of parameters that are preset by a programmable drive unit.
- List the types of functions that are programmed on the control panel of an AC variable-speed drive.
- Explain the following drive functions of an AC variable-speed drive:

  Acceleration | Slip Compensation | Power Ride-Through
  Deceleration | Critical Frequency Rejection | Automatic Restart
  S-Curve | Self-Protection
- Identify the classifications of motor braking techniques used for stopping motors quickly, and describe their operation.
- Describe the four-quadrant operation of AC motors.
- List the factors to consider when selecting the proper type of AC drive for a particular application.
- Explain the difference between standard AC and inverter duty-induction motors.

INTRODUCTION

Much of the electricity produced by electric power utility companies is consumed by AC motors that run fans, blowers, and pumps. These devices are used in production applications for industry, for the temperature control and ventilation of large buildings, and for furnaces and appliances in the home. A recent U.S. government report stated that over 50 percent of the electrical energy produced is used by AC motors.

AC induction motors are sized for maximum loads and are operated at a constant full speed, because they are supplied with power from AC lines at a fixed sinusoidal voltage and fixed frequency. However, a high percentage of pumps, fans, and blowers have output flow requirements that fluctuate. To vary the flow of liquids or gases, companies often employ throttling and restrictive devices, such as valves, dampers, and orifices. Even though these mechanical restrictors provide effective control methods, up to 30 percent of the power consumed by the motors is not applied to the work they are meant to perform. The energy that is wasted is consumed by the flow restrictors as friction and heat diffusion.

Another drawback of using mechanical methods to reduce flow is that it develops stress on the flow system. Dampers, valves, fans, pumps, and motors wear out more quickly because they are always fighting one another. Back pressure that develops can cause weak locations in pipes or duct work to rupture, requiring high maintenance costs. An alternative to using mechanical restrictive devices to change flow rate is the AC variable-speed drive. Shown in Figure 9-1, an AC drive converts the fixed voltage and fixed frequency supplied by power lines to the variable frequency and voltage used by AC motors. Suppose a pumping application using a drive requires a 50 percent reduction in flow. The drive reduces the flow rate electronically by slowing the pump motor to half speed. As it does, the drive cuts its output frequency in half. It also puts out only the voltage and current required to drive the load, which is 50 percent of what it produces at full speed capacity. Therefore, the motor consumes only half the power it did when running at maximum speed.

By employing an AC drive that adjusts the speed of the motor to vary flow rate, throttling and restrictive devices can be eliminated. Since the speed of a motor is proportional to the electricity it consumes, using an AC drive offers the potential for significant energy savings. An example of calculated savings is shown in Table 9-1.

TABLE 9-1 Calculated Savings

A centrifugal pump supplies a flow of liquid through a piping system.

- The pump runs 24 hours per day, 7 days a week, 52 weeks per year (8736 total hours).
- The motor that drives the pump has a rating of 50 hp, and runs at full speed.
- To vary the flow of liquid through the pipes, a throttle valve is used.
- The electricity cost to run the motor is determined by the following calculation:

$$(50 \text{ hp} \times 0.746 = 37.3 \text{ kW}) \times (8736 \text{ hrs}) \times (0.07 \text{ cents/kW-hr})$$
$$= \$22{,}810 \text{ Annual Operating Cost}$$

Assuming that the average flow rate over a 1-year period is at 50 percent of full capacity, the motor will consume maximum energy because it runs at a constant full speed.

However, by replacing the throttle valve with a variable-speed drive to cut the flow rate in half, the annual operating cost will be reduced 50 percent. The savings will be:

$$0.5 \times \$22{,}810 = \$11{,}405$$

Because a motor may consume as much as 20 times its acquisition cost in electricity every year, it is financially and ecologically advantageous to improve the efficiency of the motor.

9-1 AC Drive Fundamentals

Unlike the DC motor, the speed of an AC motor is not controlled exclusively by varying the applied voltage. It is true that reducing the AC supply voltage will lower motor RPM. However, it will also reduce the torque, so that any changes in torque demand imposed by the mechanical load will cause the motor to run at erratic speeds.

The speed of an AC motor typically is determined by how many cycles of alternating current it receives each second. In the United States, the frequency applied is 60 hertz. Speed is also determined by the number of poles. Electric currents in the motor pole windings create electromagnets. Each magnet must have a north and a south pole. Therefore, magnetic poles always come in pairs. Motors are commonly wired to have two, four, six, or eight poles. The synchronous speed of a given AC motor is derived from the following equation:

$$N = \frac{60f}{P}$$

where,

N = Speed (RPM)

f = Frequency (Hz)

P = Number of pole pairs (per phase)

60 = To determine RPM, the frequency given in seconds is converted to the number of cycles in a minute by multiplying 60 times f.

Since the number of poles (P) of a motor remains constant, the speed (N) of the motor is more easily controlled by varying the applied frequency (f).

AC variable-speed drives take standard commercial AC power and convert it to a variable frequency and voltage, which is delivered primarily to three-phase motors, as shown in Figure 9-1. The output frequency of some drives ranges from 0 to 400 hertz, which smoothly controls the motor speed from a low RPM to the base speed and above.

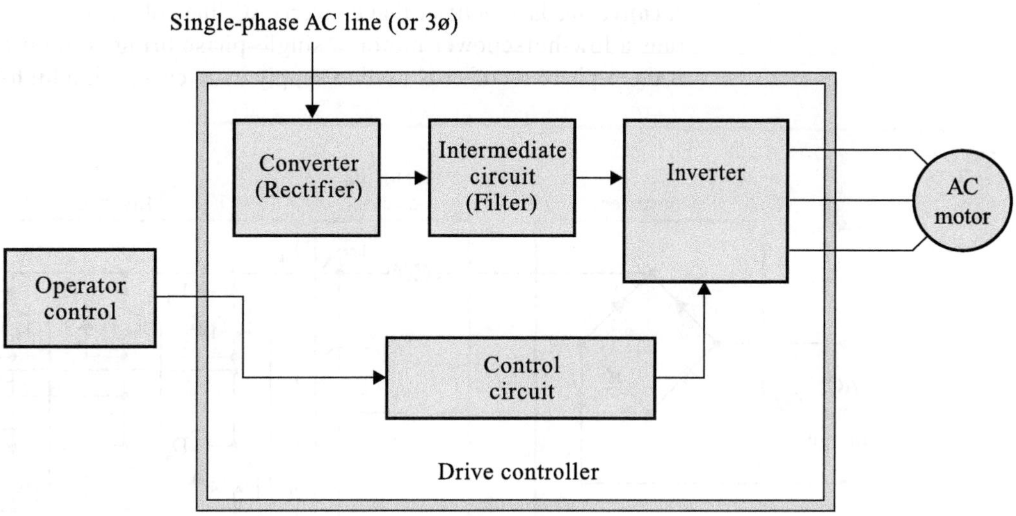

FIGURE 9-1 Block diagram of an AC drive system

9-2 AC Drive System

The block diagram of a typical AC adjustable drive system (Figure 9-1) has three basic sections: the *operator control,* the *drive controller,* and an *AC motor.*

Operator Control. The **operator control** provides the operator a way to control the AC motor. In addition to control functions, it also displays operating conditions of the drive and indicates fault signals. The fault signals displayed are followed by a drive-shutdown sequence.

Drive Controller. The **drive controller** converts the fixed voltage and frequency of an AC power line to an adjustable AC voltage and frequency.

AC Motor. The **AC motor** converts the adjustable AC frequency and voltage from the drive controller output to rotating mechanical energy. Three-phase squirrel cage induction motors are primarily used. AC drives can also control synchronous motors.

The part of the system that is actually the adjustable frequency drive is the *drive controller.* Also referred to as an *inverter,* a drive controller can operate motors ranging in size from a fraction of a horsepower to several thousand horsepower. Inverters can be sold separately because the operating control and the AC motor may already be in place.

The objective of most AC drive systems is to control an induction motor so that the system meets one or more of the following criteria:

1. Provides full-load torque from 0 RPM to low speed.
2. Prevents torque fluctuations at low speed.
3. Maintains a set speed when the load torque varies.
4. Provides a starting torque of at least 150 percent of the rated full-load torque.
5. Controls the rate of acceleration and deceleration.

9-3 Drive Controller Internal Circuitry

The internal circuitry of the drive controller consists of four main sections, the *converter,* the *intermediate circuit,* the *inverter,* and the *control circuit,* as shown in Figure 9-2.

1. Converter

A **converter** is a rectifier that converts AC line voltage to a pulsating DC voltage. If the drive runs a low-horsepower motor, a single-phase bridge circuit can be used for rectification. A three-phase rectifier is used to supply extra current if a high-horsepower motor is used.

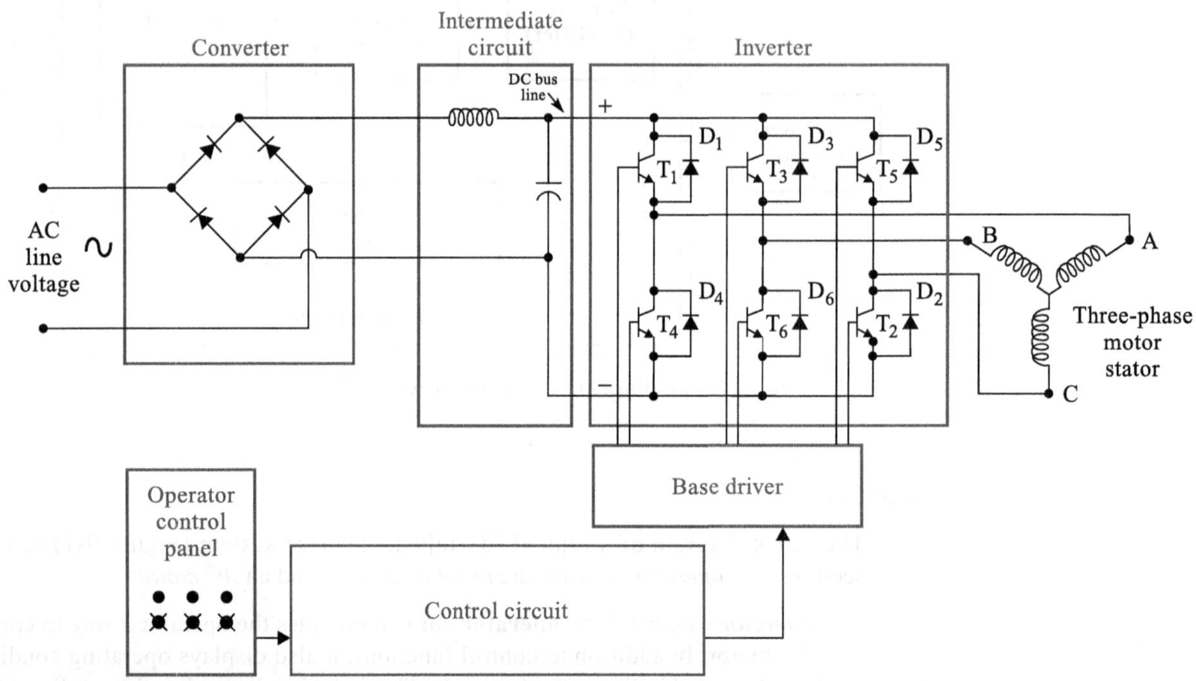

FIGURE 9-2 Internal circuitry of a drive controller

2. The Intermediate Circuit

The **intermediate circuit** transforms the pulsating DC signal from the converter to a smooth DC waveform. The output of the intermediate circuit is called the *DC bus line.*

3. Inverter

The function of the **inverter** is to "*invert*" the DC bus line voltage back to simulated AC power at the desired frequency. An inverter consists of six electronic switching devices used to develop an approximate three-phase AC signal for the motor. In Figure 9-2, switching transistors are used. The voltage across the transistor network is supplied from the DC bus line of the intermediate circuit. The transistors turn on and off in the proper sequence to create three sets of voltage waveforms between A–B, B–C, and C–A.

4. Control Circuit

The **control circuit** is the fourth main block of the AC variable-speed drive. It performs several functions:

- It receives the operating commands from the control panel or other programming device. These commands are used to determine controlled drive variables such as the output frequency, output voltage, and motor direction.
- It generates command pulses, which turn the semiconductors on and off in the inverter.
- It produces outputs for the control panel display to indicate the drive speed, operating status, and fault conditions.
- It performs an orderly shutdown procedure of the drive when an abnormal condition arises.

It also provides closed-loop control. It monitors various motor and load conditions by using feedback sensing devices and changes the firing signals applied to the switching transistors when adjustments are needed. The controller uses microcomputers and software to control operation in this way.

9-4 Circuit Operation of the AC Drive

The operation of each section of a generic AC drive's internal circuitry is as follows:

The Converter

Since most drives run high-horsepower motors, single-phase bridge rectifiers may not generate enough current. Most AC **drives** use a three-phase rectifier to supply the extra power required by large motors. A three-phase rectifier supplies more power than a single-phase rectifier because its pulsations are closer together and, therefore, the filtering required is minimized.

A full-wave three-phase rectifier circuit is shown in Figure 9-3. Its six diodes either block or pass current. A wye transformer is the source that supplies both positive and negative alternations of three-phase current to the rectifier.

Three facts about the operation of the rectifier must be known to analyze the circuit:

1. At any given time, two diodes are on and the remaining four are off.
2. One odd-numbered diode (D_1, D_3, or D_5) and one even-numbered diode (D_2, D_4, or D_6) are always on.
3. Electron current flows out of the A, B, or C phase transformer terminals from the most negative voltage, through an odd-numbered diode, through the load resistor, and through an even-numbered diode, to whichever phase terminal—A, B, or C—has the highest positive voltage.

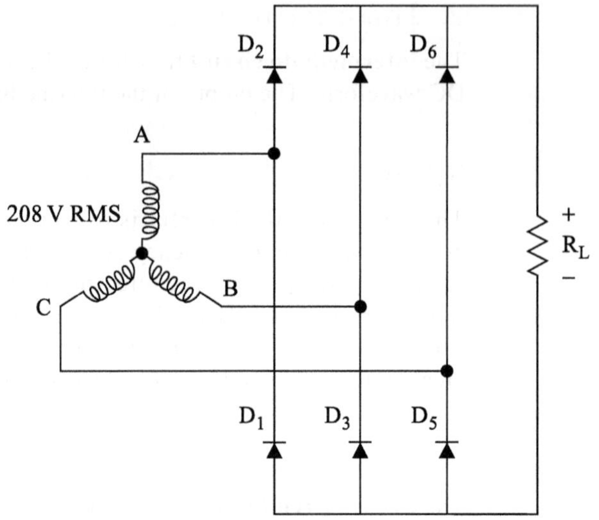

FIGURE 9-3 A wave source supplies power to a three-phase full-wave rectifier

Figure 9-4 graphically illustrates the rectification process in detail. Figure 9-4(a) shows the 208-volt RMS three-phase voltage supplied from the source terminals of the transformer to the rectifier. The timing chart in Figure 9-4(b) shows intervals when the corresponding diodes are on. Figure 9-4(c) uses closed switches to show which diodes are forward biased during 30- or 60-degree intervals.

Between 0 and 30 degrees, terminal C is most positive, which forward biases diode D_6, and terminal B is most negative, which forward biases D_3. The resulting 30-degree alternation is shown in Figure 9-5. From 30 to 90 degrees, terminal A is most positive, which turns D_2 on, and terminal B is most negative, which turns D_3 on. The resulting voltage pulses form across the load resistor during this, and every, 60-degree period. Since the magnitude of the pulses fluctuates approximately 59 volts (235 V to 294 V), the rectifier output requires only minimal filtering.

The Intermediate Circuit

The pulsating waveform from the converter is filtered by the intermediate section.

There are two typical circuit configurations that provide the filtering action required by this section. The configuration used depends on the type of converter and inverter to which it is interfaced.

The simplest type of intermediate circuit is shown in Figure 9-6(a). It consists of a single large coil that provides filtering action. This type of intermediate circuit is used to supply the bus line with a constant DC voltage. It is also capable of transferring energy back into the AC line when regeneration occurs due to a condition where the inertia of the load overcomes the torque of a motor (called *overhauling*), or when deceleration of a motor is too rapid.

The intermediate circuit can also consist of a coil and capacitor, as shown in Figure 9-6(b). This low-pass filter smooths the pulsating DC at any voltage providing filter, a constant DC voltage, at the bus line.

The Inverter Operation

By controlling the switching sequence of the inverter transistors so that they turn on and off very rapidly, simulated sine waves can be created. The type of variable-speed drive with this capability is called a *pulse width modulation* (PWM) drive, shown in Figure 9-7. Most drives currently use the PWM switching method.

The operation of the PWM inverter section can be illustrated by using the switch model in Figure 9-8(b). Three SPDT switches are connected in parallel across a DC supply. Depending

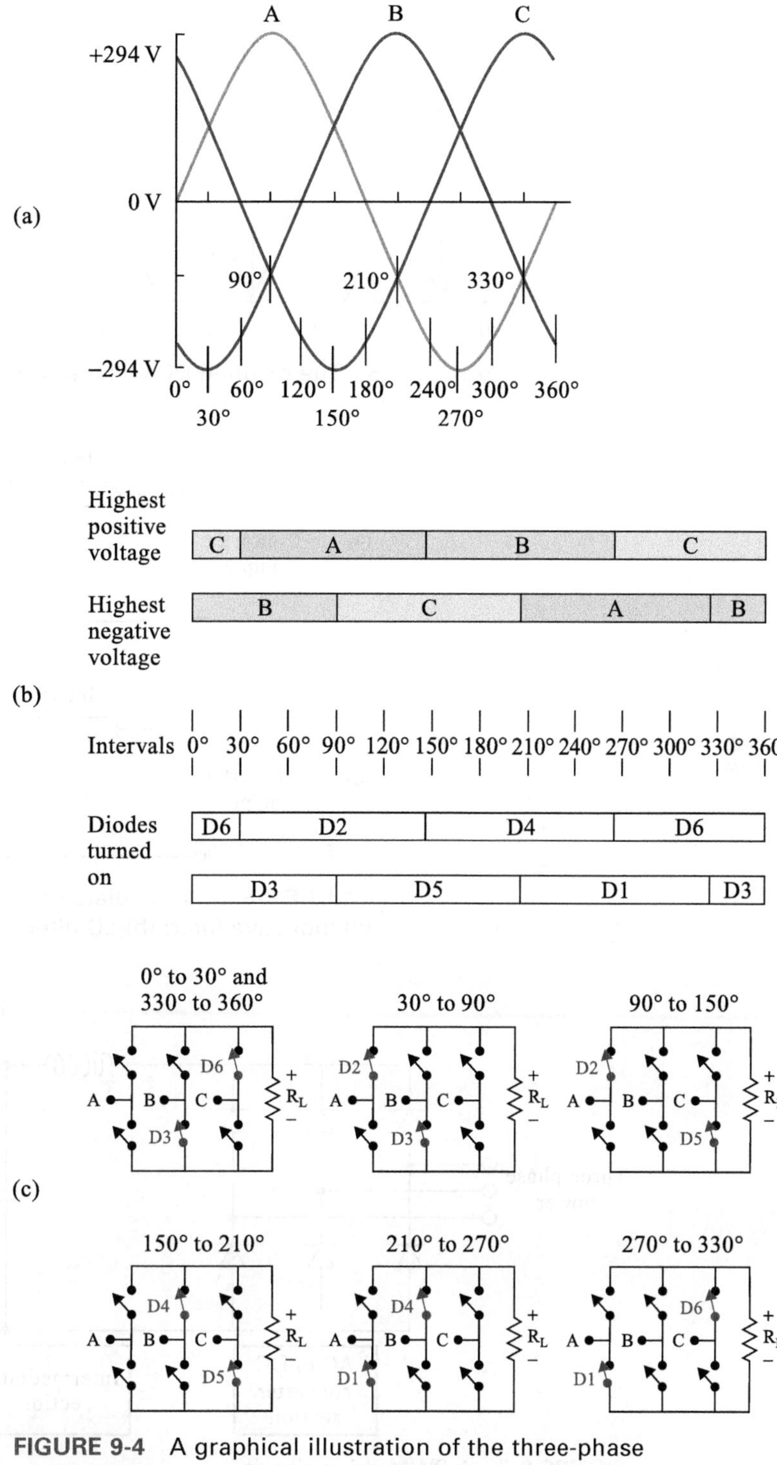

FIGURE 9-4 A graphical illustration of the three-phase rectification process

on the switch position, either a positive voltage or a zero-volt ground connects to the swivel terminal. The voltage is positive in the up position and zero in the down position. The wires connected to each swivel terminal are labeled A, B, and C, to identify each of the three-phase inverter output lines. Figure 9-8(a) shows how the settings of switches A and B create a resultant simulated AC single-phase line voltage. Consider output line B as the reference and compare output line A to it.

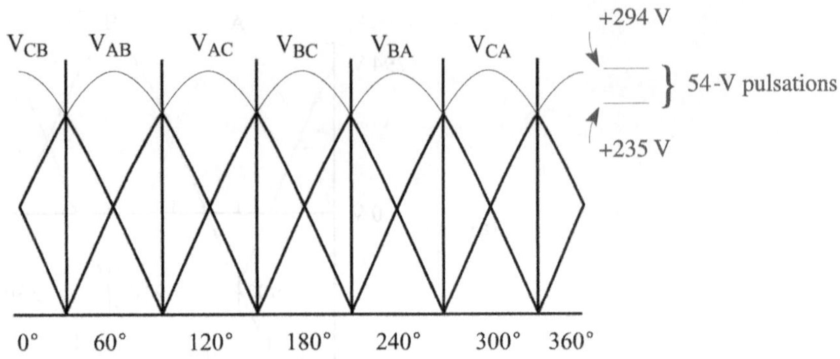

FIGURE 9-5 The rectified DC voltage waveform across R_L

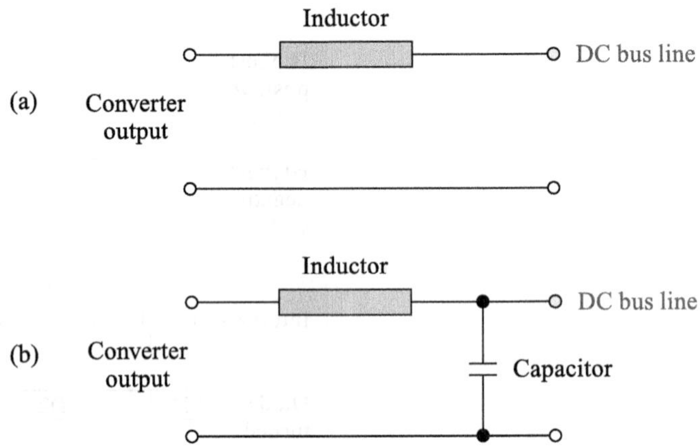

FIGURE 9-6 Intermediate section filter circuits: (a) Inductive filter; (b) LC filter

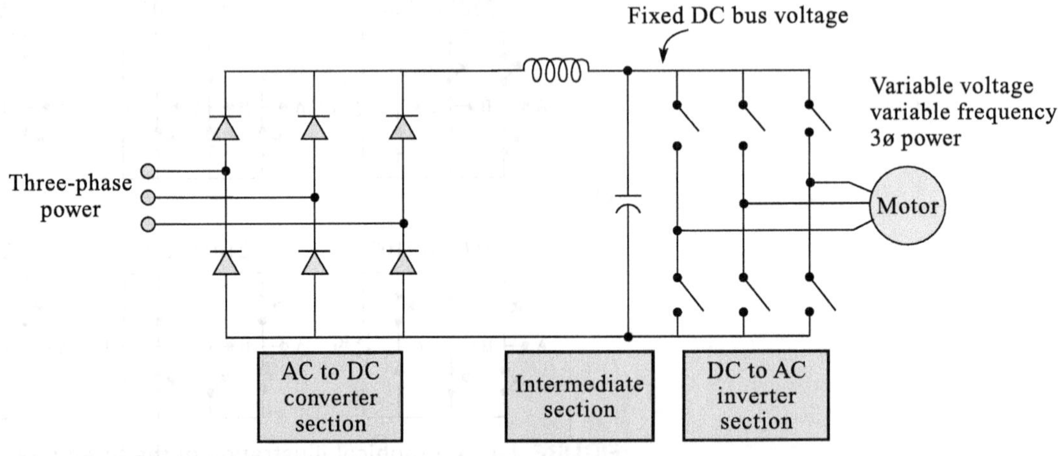

FIGURE 9-7 A PWM drive circuit

- During time periods a, g, i, m, o, and u, both switches are in the down position. Since both lines are the same ground potential, line voltage AB is zero.
- During time periods b, d, f, h, and j, switch A is up and switch B is down. Output A is positive in respect to B. The result is that line voltage AB becomes positive.
- During time periods c, e, k, q, and s, both switches are up. Because outputs A and B are the same positive potential, line voltage AB is zero.

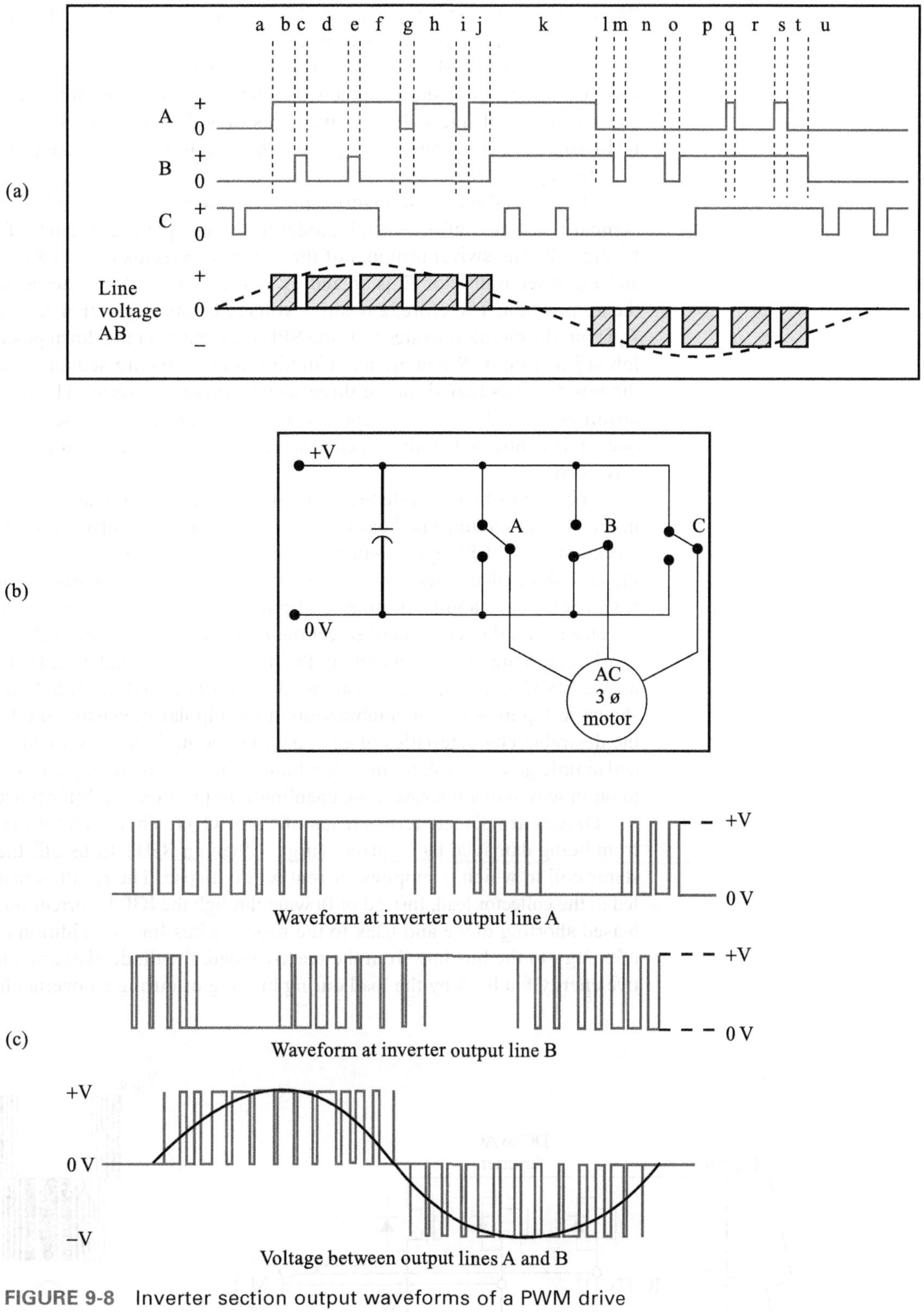

FIGURE 9-8 Inverter section output waveforms of a PWM drive

- During time periods l, n, p, r, and t, switch A is down and switch B is up. Output A is negative with respect to B. The result is that line voltage AB becomes negative.

The switching sequence forms a voltage pattern that consists of five pulses during each alternation. The dashed line is the RMS voltage created. The widths of the individual pulses are adjusted to produce a variable RMS voltage. The longer the pulse duration, the higher the

RMS voltage. The pulse groups are timed to provide variable frequency. The way in which the signals are produced is called *modulation.*

In a typical PWM inverter circuit, more than five pulses are developed during an alternation. Figure 9-8(c) shows an actual square-wave pattern on two inverter output terminals and the resultant line voltage waveforms they develop. The shapes of the pulses form an RMS voltage that becomes sinusoidal, which causes a current, approximate to sinusoidal, to flow through the motor.

The transistors in the inverter perform the function of SPDT switches. For example, compare switch A of the switch model to the left parallel branch of the inverter section in Figure 9-2. The swivel terminal of the switch represents the junction between transistors T_1 and T_4. When transistor T_1 is on and T_4 is off, the circuit operates as if the SPDT switch is in the up position. Therefore, a positive voltage exists at junction A. If transistor T_1 is off and T_4 is on, the circuit operates as if the SPDT switch is in the down position and zero voltage is felt at junction A. When all six switching transistors are sequenced properly, they develop the rapid pulses that simulate three-phase voltage patterns. The line voltages form across terminals AB, BC, and AC, and are 120 degrees out of phase with one another. The frequency is controlled by the speed at which the inverter transistors develop the simulated AC waveform.

The signals that activate the switching transistors originate from a modulator logic board in the control circuit. The board consists of low-power signal-generating circuits and microprocessor-based LSI logic circuits, which control the performance characteristics of the drive. Pulse width is decreased for lower RMS voltages and increased for higher voltages. Lower frequencies have a greater number of pulses in one alternation. As the frequency increases, the circuits on the logic board select the number of pulses per cycle.

Because the semiconductors in the inverter must switch at extremely high speeds to produce a PWM output, insulated gate bipolar transistors (IGBT) are frequently used. The IGBT, shown in Figure 9-9, is a combination of the bipolar transistor and the MOS-FET. It utilizes the desirable characteristics of each one, including high power capacity, good conductivity, and simple gate control, to provide a high switching frequency. An IGBT can switch from off to on in less than a microsecond, enabling it to produce a 1-MHz pulse.

Diodes are placed across leads of each IGBT. Their function is to protect each IGBT from being damaged by a current surge. When an IGBT turns off, the magnetic field of the stator coil to which it supplies current will collapse. The result is that an induced current is fed to the collector lead. Instead of flowing through the IGBT, current flows through the forward-biased shorting diode and back to the negative bus line. In addition to permitting the return of energy to the bus line from the reactive load, the diode also provides a path for regenerative energy fed back by the load during braking or during an overhauling condition.

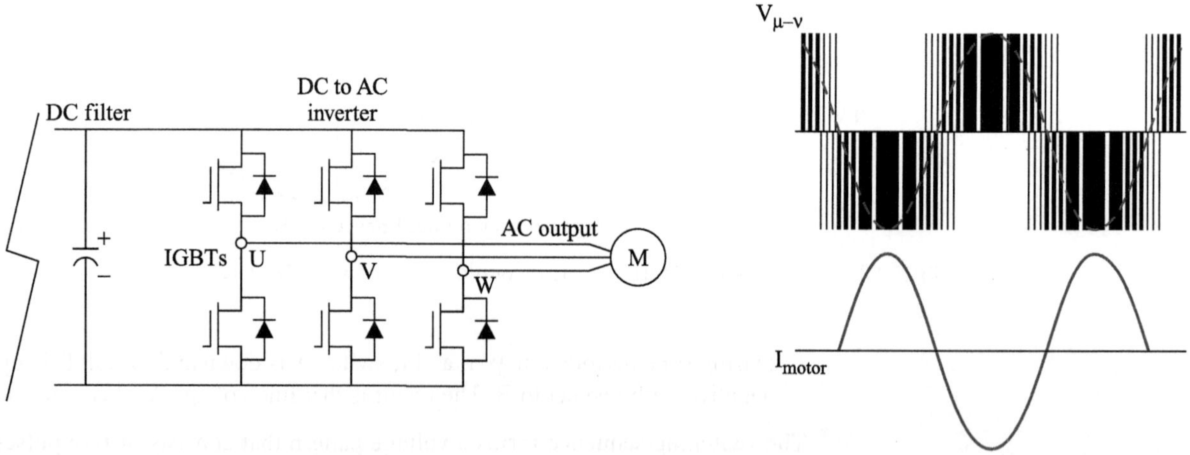

FIGURE 9-9 Inverter with IGBTs

9-5 Flux Vector Control

The operation of a PWM drive is adequate for steady-state conditions, or in applications where a slow-response speed change is adequate when load variations are encountered. However, there are some specialized applications where abrupt load, speed, and position changes are likely to occur. A PWM drive that uses **flux vector control** (also known as *field-oriented control*) is better suited to handle these conditions.

A flux vector control drive uses a high-speed digital processor in its control circuit. These processors, which have dropped to affordable prices, enable AC induction motors to perform in a way that could only be accomplished previously by DC or servo motors.

The term *vector* is derived from the method by which the drive controls the magnetic flux vector. A vector in mathematics indicates amplitude and phase angle. The vector drive controls the motor by varying the magnitude of the stator flux and adjusts the angle of the stator flux in reference to the rotor.

The primary reason for a DC motor's effective operating characteristics is its physical construction. The way in which the brushes make contact with the commutator causes the resultant flux of the rotor to be perpendicular to the stator field. Because the armature's magnetic field and the stator field are always quadrature (90 degrees to one another), maximum torque is always produced. The drawback of a DC motor is that it is a high-maintenance machine; the brushes require frequent replacement due to their physical contact with the commutator. The AC asynchronous motor is preferable to the DC motor for most industrial applications because of the low capital and maintenance cost, high efficiency, and long-term dependability.

The fixed 90-degree relationship between the rotor and the stator in a DC motor does not naturally occur in an AC induction motor. The induction motor rotor is basically two rings with bars placed between them like a wheel for a squirrel, hence the name *squirrel cage*. Unlike the DC motor, rotor flux is produced by having a rotating stator field induce current into the bars of the cage. As the stator and rotor fields interact, torque is produced, which causes the rotor to turn. However, the rotor lags behind, and rotates more slowly than, the stator field. This difference in rotational speed is referred to as *slip*. Slip is measured as an angular velocity expressed as a frequency. Under normal operating conditions, the (slip) frequency is about 3 hertz. When a greater load is placed on the rotor, slip increases, more current is induced into the armature, and torque increases. The maximum motor torque is produced when the rotor flux relative to the stator field is at 90 degrees. The drawback of the AC induction motor is that it responds slowly to load changes. It is acceptable for driving applications such as pumps, fans, and compressors. However, when rapid dynamic control of speed, torque, and response to a change in load is necessary, its operation is too crude.

The fundamental problem with controlling an AC induction motor is that the stator and rotor flux cannot be regulated separately. Instead, current induced into the rotor is only responsive to the variation of the frequency and current amplitude supplied to the stator. Recent developments in digital signal processing (DSP) used in flux vector control methods have overcome the difficulty of providing accurate control of the rotor by manipulating the stator flux. The control scheme of DSP is to constantly monitor and control the amplitude of the stator magnetomotive force (MMF) flux, as well as its position relative to the rotor. A high-performance microprocessor repeats 2000 or more computations per second by using a look-up table to make readjustments if they are required. This method is referred to as *flux vector control.*

Drives that use vector control consist of a power conversion section, which uses six input diodes that rectify the three-phase power to a fixed pulsating DC voltage. A capacitor filters the pulsating ripple from the rectifier to a smooth DC bus-line voltage. A set of six transistors in the inverter section convert the DC bus-line voltage into a synthesized three-phase PWM voltage to power the motor. The vector drive differs from the PWM drive in using two independent feedback control loops, as shown in Figure 9-10(a). The loop that controls amplitude is referred to as the *current feedback.* It uses a sensor such as a clamp-on ammeter or Hall-effect transducer to detect the magnitude of stator current at each of the three stator lines. When an increase in load occurs, the vector drive will not allow the slip to increase. The vector drive responds by increasing the amplitude of the current to the stator, as shown

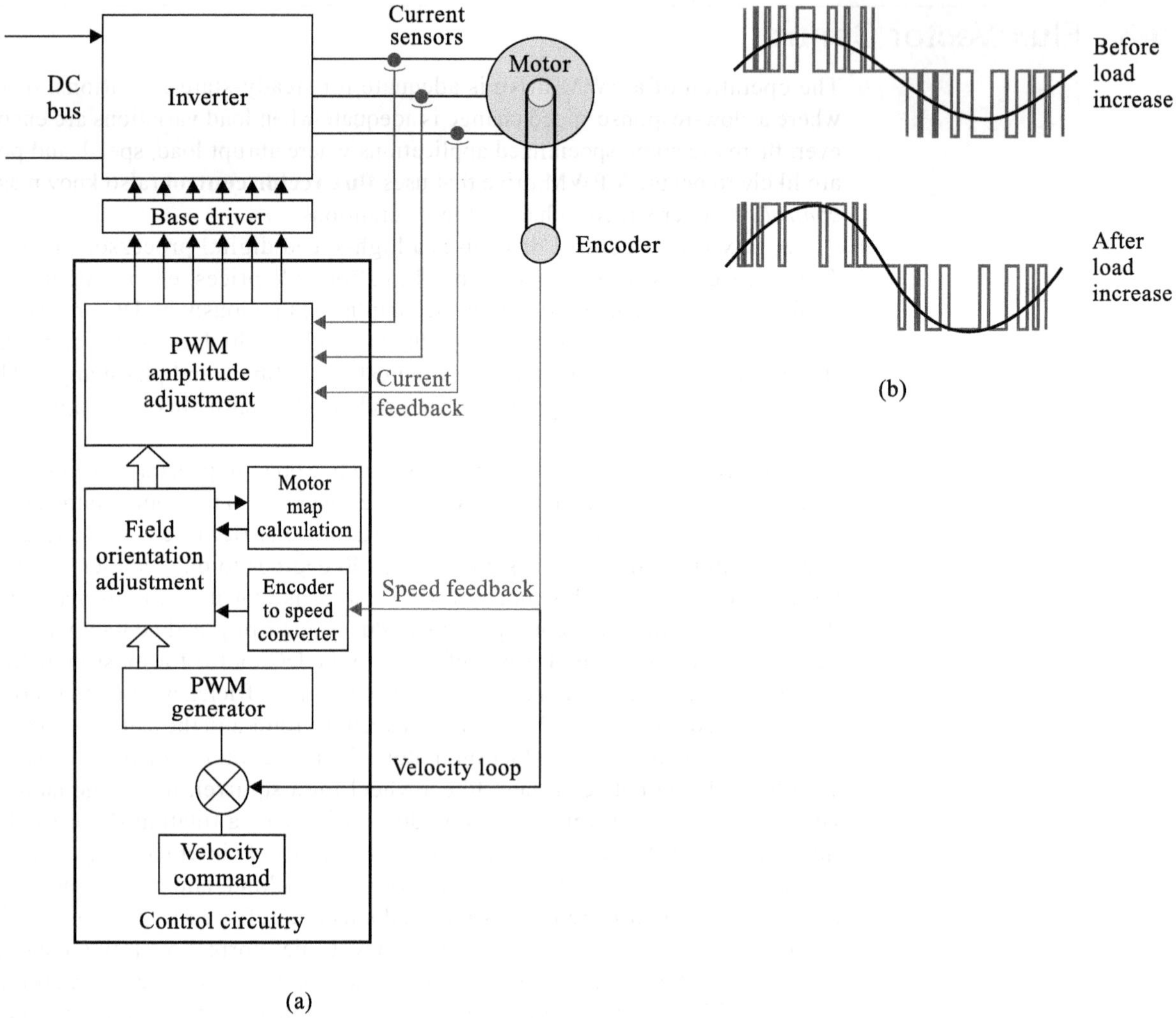

FIGURE 9-10 Block diagram of the vector drive control

in Figure 9-10(b). The result is that the torque increases to meet the demand of the load. The loop that controls relative positioning is referred to as the *speed feedback.* It uses a resolver, or encoder, to sense the velocity of the rotor. The vector drive's control technique of positioning causes a slip frequency to occur where the flux of the stator and rotor are always perpendicular, thus creating maximum torque. If the load changes, the rotor velocity is altered and the slip varies. When the new rotor RPM is sensed by the feedback detector, the vector drive changes the stator frequency until the slip causes the stator and rotor flux lines to return to the desired 90-degree angle. This method is referred to as *field orientation.*

Figure 9-11 illustrates how field orientation is achieved. Figure 9-11(a) shows half of one sinusoidal three-phase sine wave supplied to the stator coils. Figure 9-11(b) represents the 120-degree spacing of the stator windings and the direction of current flowing through them at successive 60-degree intervals. For example, at 0 degrees, currents from lines B and C flow to neutral and combine to flow through line A. The direction of these currents corresponds to the polarity of the instantaneous current. Figure 9-11(c) shows vectors positioned head-to-tail that represent current flow during each 60-degree interval. Vectors B and C at 0 degrees are shown with arrows pointing toward the neutral, while in vector A the arrow points away from the neutral. Also, the length of each vector varies to represent the magnitude of current. Figure 9-11(c) also shows the resulting magnetic field at the 0-degree interval. Because the stator coils are primarily inductive, the magnetic field lags the current by 90 degrees. As the diagram advances through each interval from 0 to 180 degrees, the resulting flux makes one-half revolution in a counterclockwise direction. Because

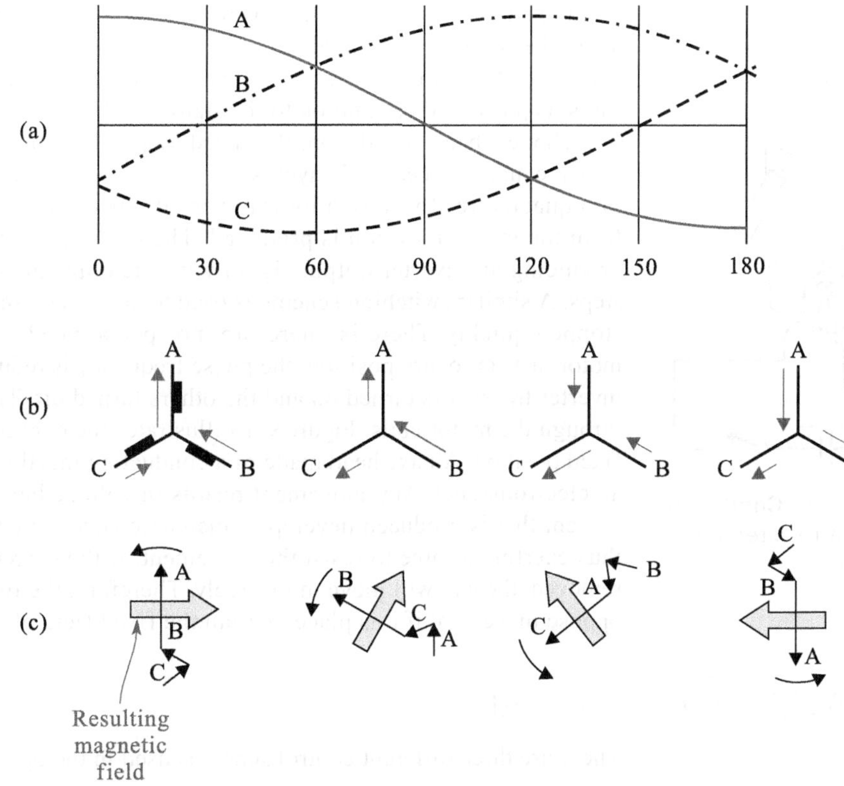

FIGURE 9-11 Development of a rotating stator field

the stator current shown in the diagram is supplied by a three-phase commercial line, the amplitude and frequency are constant. In a vector drive where the inverter stage artificially produces a synthesized waveform, the sine wave can be disrupted. For example, Figure 9-12(a) shows the stator current suddenly advanced by 30 degrees. The effect is that the stator flux also is instantaneously advanced through 30 degrees. This jump is referred to as a *step change.* Figure 9-12(b) shows the stator current held constant for 30 degrees. The result is

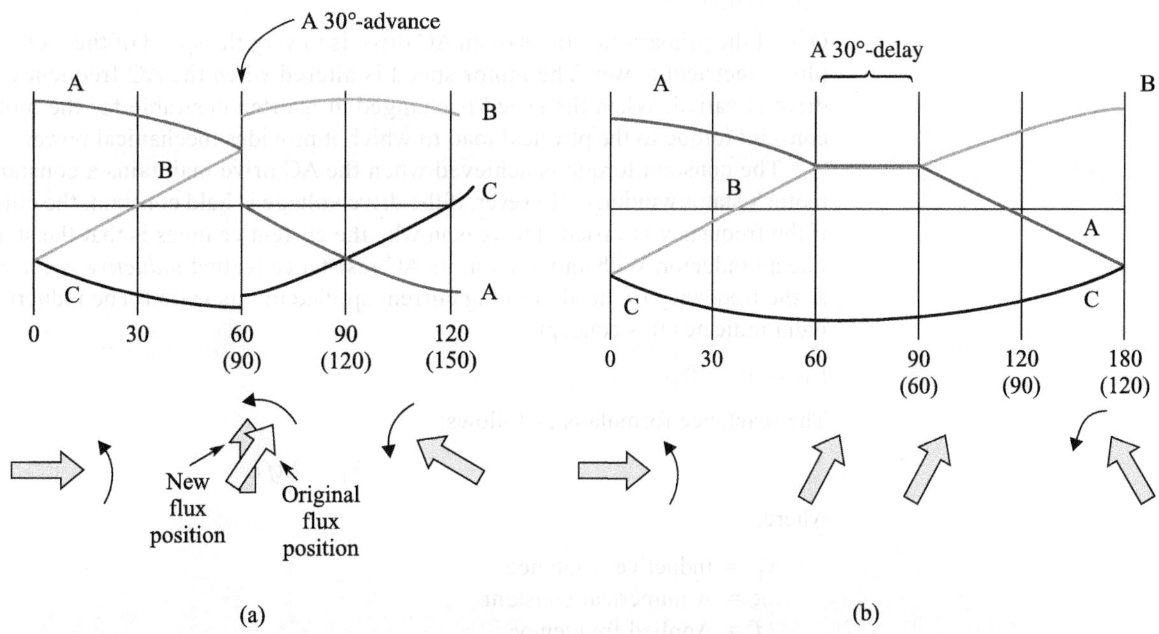

FIGURE 9-12 Effects of stator flux step changes: (a) Magnetic flux advanced 30 degrees; (b) Magnetic flux held stationary for 30 degrees

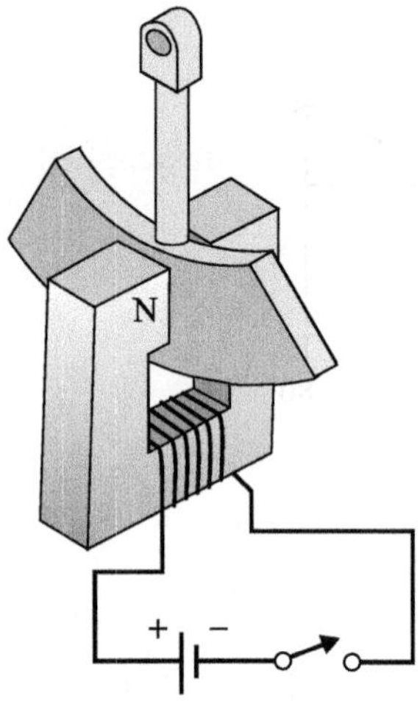

FIGURE 9-13 Conductor movement resisted by induction

that the stator flux is held stationary for one 30-degree interval. The ability to advance or delay the stator currents makes it possible to adjust the stator flux to maintain a 90-degree angle with the rotor, even when a change in the load causes the rotor speed to vary, and disrupts its positional orientation to the stator field.

The feedback signal from the speed loop is used by the DSP to cause the stator currents to make an advanced or delayed step change so that the stator and rotor fluxes return back into quadrature. To cause a rotor to vary its speed, a proportional output frequency change from the inverter section is produced. The direction of the motor's rotation is reversed by producing an inverter output signal with the opposite sequence of instantaneous voltage steps. A similar switching scheme is used to apply reverse torque when the motor needs to be stopped quickly. There is, therefore, no special need for independent braking. To hold a motor in a stationary position, the phase frequency is reduced to zero by keeping some of the inverter transistors turned on and the others turned off. The result is that a DC current flows through the motor coils. Figure 9-13 illustrates the concept of how the rotor can be held in a fixed position. An axe head made of a conducting metal is suspended within the flux lines of an electromagnet. Any movement results in voltage being induced into the conductor. The current that is produced develops a magnetic field, which is opposite to the electromagnet, thus exerting a force to resist the movement. If the current supplied to the electromagnet is removed, the axe will move more freely. Therefore, the rotor, which is a conductor, will also oppose movement if it is placed within the fixed field of the stator.

9-6 PWM Control Methods

There are three different control schemes used in the operation of PWM drives.

1. Volts/Hertz
2. AC Closed-Loop Vector
3. Sensorless Vector

Volts/Hertz Control

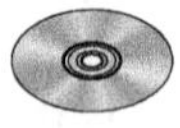

IAU14908
Volts per Hertz Ratio in AC Drives

The simplest control process used to operate a PWM drive is the **volts/hertz** method. The term V/Hz is the abbreviation for *volts per hertz* and refers to how the drive varies its voltage proportionally with frequency changes.

Volts/Hertz Ratio

One of the primary functions of an AC drive is to vary the speed of the motor to which it supplies electrical power. The motor speed is altered when the AC frequency supplied by the drive is varied. When the speed is changed, it is often desirable for the motor to produce a constant torque to the physical load to which it provides mechanical power.

The constant torque is achieved when the AC drive maintains a constant current to the motor's stator windings. However, if the drive voltage is held constant, the current will change if the frequency is varied. The reason why the current changes is that the stator windings act like an inductor. With an inductor, its AC resistance (called *inductive reactance*) will change as the frequency of the alternating current applied to it is varied. The inductive reactance formula indicates this concept.

Inductive Reactance

The reactance formula is as follows:

$$X_L = 2\pi f L$$

where,

X_L = Inductive reactance
2π = A numerical constant
f = Applied frequency
L = Inductance of the coil

The opposition to current flow by the coil is directly proportional to the frequency of the applied voltage. For example, if the frequency is decreased to half, the inductive reactance will also decrease by half.

If X_L is decreased to half and the drive's voltage is held constant, the current supplied to the motor will double, as shown by the following Ohm's law formula.

Ohm's Law

Ohm's law is as follows:

$$I = \frac{V}{X_L}$$

The increase in current will cause the motor's torque to rise by the same proportion. However, the motor may overheat because of the excessive current.

The current to the motor can be kept at a constant level if the voltage is changed at the same proportional rate as the frequency. For example, if the frequency is reduced by half, the voltage is decreased to 50 percent. Conversely, if the frequency is doubled, the voltage is also doubled. A variable-frequency drive maintains current to the motor's stator and keeps its torque at constant levels by varying both the applied voltage and frequency. This variation is called **voltage-to-frequency (V/Hz) ratio.** It is programmed with software to perform this function.

A V/Hz drive has no feedback device that monitors the load. Therefore, these drives have no self-correcting function. They are used for general purpose operations, such as running fans, pumps, mixers, and other devices where speed control is not critical. The advantages and disadvantages of a V/Hz drive are shown in Table 9-2.

TABLE 9-2 AC V/Hz Drives

Pros and Cons

Advantages

- Simple, "look-up table" control of voltage and frequency
- Good speed regulation (1%–3%)
- No motor speed feedback needed
- Multi-motor capability

Limitations

- Low dynamic performance on sudden load changes
- Limited starting torque
- Lacks torque reference capability
- Overload limited to 150%

Best for General Purpose and Variable Torque

- Centrifugal pumps and fans
- Conveyors
- Mixers and agitators
- Other light-duty nondynamic loads

Torque Boost Function

When a motor is being started, a standard drive might supply no voltage at 0 RPM, as shown in Figure 9-14. At a slow RPM, the voltage is low and provides minimal current. However, as

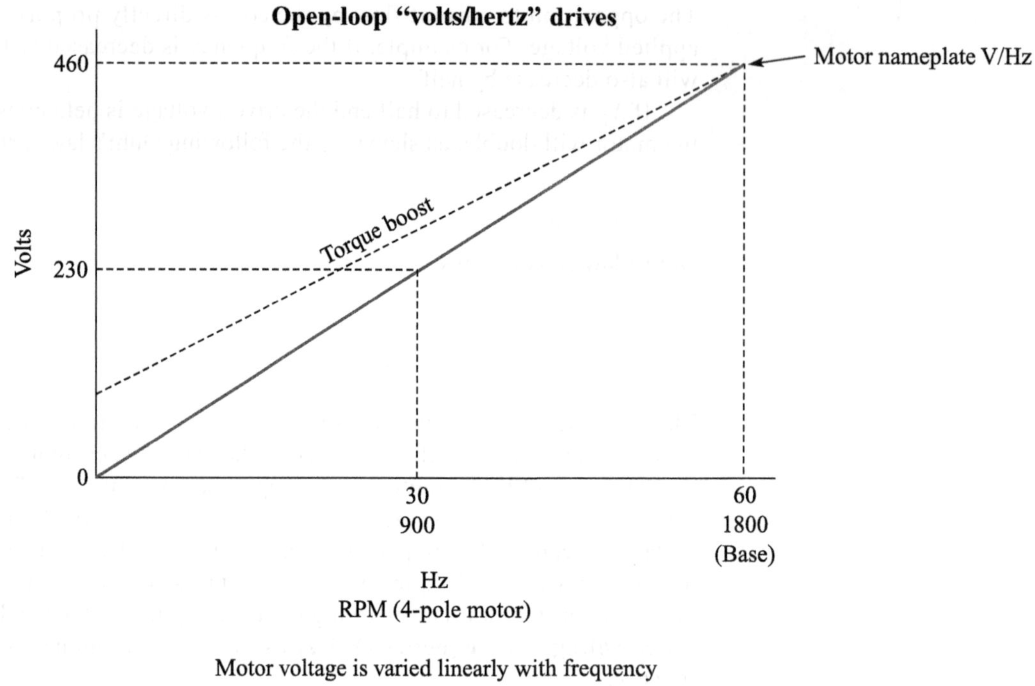

FIGURE 9-14 Open-loop "volts/hertz" drives

a motor begins to turn and accelerate, the drive must provide enough current to produce sufficient torque to overcome inertia of its load. Under this condition, some volts/hertz drives are capable of supplying the extra current by a function called *torque boost* (also known as *auto boost*). The graph shows how additional current is provided because a voltage greater than zero is produced at 0 RPM. The voltage continues at a higher-than-normal level and in a linear fashion as the motor increases to base speed.

AC Closed-Loop Vector

This type of drive uses both closed loops shown in Figure 9-10. The inner loop measures current drawn by the motor to monitor torque. The encoder in the outer loop is considered a sensor, which measures RPM and position, and produces digital signals to represent their values. Data from the inner and outer loop measurements are converted into numerical values and put into formulas for calculation by a mathematical model of the motor's algorithm. The model will use a lookup table to indicate what the motor should be doing at the given frequency, current, and voltage. The speed regulation is very accurate and is within 0.01 percent. It also can produce 100 percent torque down to 0 RPM. When the motor is not turning, a special cooling fan may be required. The advantages and disadvantages of an AC closed-loop vector drive are shown in Table 9-3.

Sensorless Vector Drives

The term *sensorless vector drive* is misleading because it uses sensing devices that monitor the operation, and provide feedback data in a closed-loop system. The sensor is located in the inner loop, as shown in Figure 9-10, to measure the current drawn from the motor for monitoring speed and torque. This feedback signal is converted into data and is used for mathematical calculations in the same manner as the AC closed-loop vector drive.

The advantages of sensorless vector drives over the V/Hz type are that they provide speed and torque control. The advantages and disadvantages of a sensorless vector drive are shown in Table 9-4.

TABLE 9-3 AC Closed-Loop Vector Drives

Pros and Cons

Advantages

- Ultra-high torque and speed loop performance and response
- Excellent speed regulation to .01%
- Full torque to zero speed
- Extra-wide speed range control

Limitations

- Requires encoder feedback
- Single motor operation only
- May require premium vector motor for full performance benefits
- 4-quadrant (regenerative) operation requires additional hardware

Best for High-Performance Applications

- Converting applications
- Spindles and lathes
- Extruders
- Other historically DC applications

TABLE 9-4 AC Sensorless Vector Drives

Pros and Cons

Advantages

- High starting torque capability (150% at 1 Hz)
- Improved speed regulation (<1%)
- No motor speed feedback needed
- Self-tuning to motor
- Separate speed and torque reference inputs

Limitations

- Speed regulation may fall short in certain high-performance applications
- Lacks zero-speed holding capability
- Multi-motor usage defaults to V/Hz operation
- Torque control in excess of 2 × base speed may be difficult

Suitable for all general-purpose, variable-torque, and moderate- to high-performance applications

- Extruders
- Winders and unwind stands
- Process lines

9-7 Control Panel Inputs and Drive Functions

AC drives have an operator station called a *control panel,* which is typically located on the front of the drive enclosure. It has adjustable inputs such as knobs and keypads that are used to set up, operate, and control the drive. It also has an alphanumeric display. The hardware and display can be used together for programming desired operating parameters during setup. The display also provides information about operating conditions while the drive runs the motor. This information includes showing the RPM of the motor that the drive is accelerating, or which direction it is turning. LEDs are also used to provide some of this information.

Technological advancements are constantly being made in the design of electronic motor drives. The most advanced drives use microprocessor and EEPROM memory chips. These "smart drives" are programmable. If proper preset adjustments are made, they are capable of optimizing their operation under various load conditions.

Figure 9-15 shows a control panel for an AC drive. The panel is referred to as the *human interface module (HIM)*. All parameters are entered by simply pressing its keys. The data are shown on the display and stored in its memory. Fault conditions are displayed if a malfunction develops. The drive also has terminals for making hardware connections to external devices, such as potentiometers to provide the standard 0 to ±10 VDC or 4 to 20 mA DC analog signals from remote locations. These terminals also provide interfacing connections to other equipment, such as programmable controllers or computer I/O cards that process digital signals.

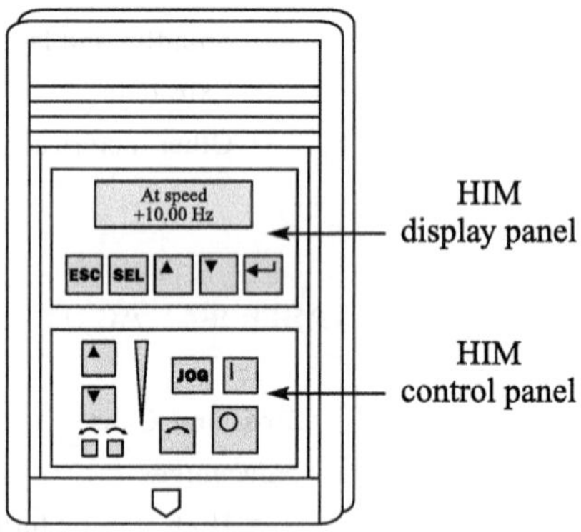

FIGURE 9-15 AC drive control panel

There are two sections of the HIM panel: the *display panel* and the *control panel*. The different types of input options on each section of the panel and the types of drive functions they can adjust for are listed as follows:

Input Options

Display Panel

Escape When pressed, the ESCape key will cause the programming system to go back one level in the menu structure.

Select Pressing the SELect key alternately causes the top or bottom line of the display to become active. A flashing first character indicates which line is active.

Up/Down Arrows These keys are used to increment and decrement a value or to scroll through different groups or parameters.

Enter When pressed, a group or parameter will be selected or a parameter value will be entered into memory. After a parameter has been entered into memory, the top line of the display will automatically become active, allowing another parameter (or group) to be chosen.

Examples of Information Provided by the HIM Display

Accelerating +10.00 Hz	Choose Mode Program
At Speed +60.00 Hz	Choose Mode DISPLAY
Decelerating +30.00 Hz	Overvolt Fault F 5
Stopped +0.00 Hz	Output Current 0.00 Amps

Control Panel

Start The start key will initiate the drive operation.

Stop When pressed, a stop sequence will be initiated causing the motor to "Coast," "Ramp," "Brake," or perform an "S-Curve" operation to stop. The type of stop operation depends on which parameter setting is programmed into the drive.

Jog When pressed, the drive will start and jog to the frequency set by the "Jog Frequency" parameter.

Reverse This key controls the direction of the motor rotation. When it is pressed while the motor is running in one direction, the drive will decelerate until it stops, change direction, and then accelerate until it reaches the set speed. This sequence is called *anti-plugging*. Without using a drive, this function is achieved by switching any two of the three stator leads of the induction motor, as shown in Figure 9-16. A drive does not switch two leads of the motor to change its direction. Instead, the control circuitry causes the inverter's solid-state switches to reverse the phase sequence.

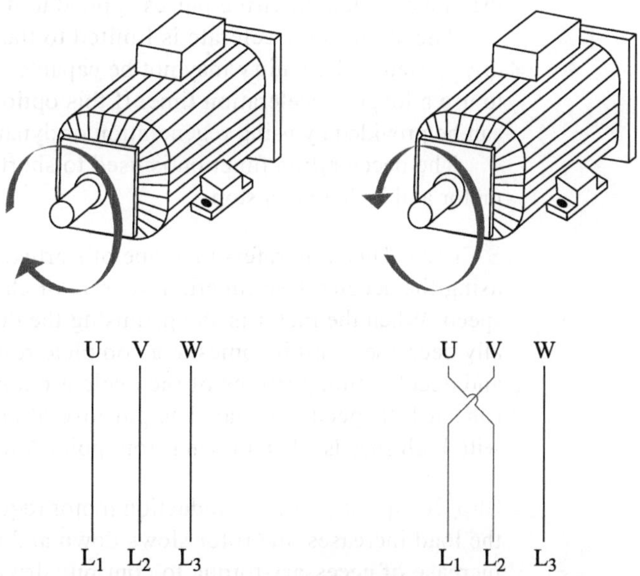

FIGURE 9-16 The motor reverses direction when the phase sequence is changed

Direction LEDs (Indicators) During normal operation, the appropriate LED will illuminate continuously to indicate commanded direction of motor rotation. When the reverse command is initiated, the flashing LED will indicate motor rotation while decelerating. The opposite LED will be illuminated continuously, indicating a change in commanded direction.

Up/Down Arrows Pressing either of these keys will increase or decrease the output frequency of the drive. A visual indication of the commanded frequency will be displayed on the Speed Indicator LEDs.

Speed Indicator LEDS This display uses LEDs to visually indicate, in steps, the commanded frequency.

Drive Functions

Acceleration This control is used for starting a load gently. This function is also known as *ramping* or *soft start*. The acceleration rate is determined by an adjustment setting that controls the rate at which the firing pulses applied to the inverter switching transistors increase. The control setting for this function adjusts for the time duration of the acceleration period. The minimum amount of acceleration time is less than 1 second. The maximum time is as high as several minutes. Minimum settings are usually for lighter loads and maximum settings are for heavy loads.

Without the ramping feature, a motor can demand an inrush of current that can reach 600 to 700 percent of the rated current value when starting heavy loads. By accelerating more slowly, an excess current surge is prevented, which protects the motor and inverter from being damaged. Variable-speed drives typically limit starting current and torque to approximately 150 percent of full load. Ramping is also desirable in some production applications, such as preventing bottles on a conveyor belt from falling when its drive motor starts.

Deceleration Normally, to stop a motor, the power is shut off and the motor coasts until it stops rotating. If the equipment connected to a motor has low friction and a lot of inertia, such as fans and centrifugals, it could coast for a long time. To stop the load more quickly, a controlled stop function called *deceleration* is provided. Deceleration is the process of reducing the synchronous speed of the drive output below the actual speed of the coasting motor and load. The deceleration rate is determined by an adjustment setting that controls the rate at which the firing pulses applied to the inverter switches decrease.

The ability to decelerate is limited to the size of the load. If the load is too large, the drive's standard circuitry may not be capable of bringing it to a fast stop. One alternative is to use a longer deceleration time. If this option is not acceptable, additional braking power can be provided by using regeneration or dynamic braking circuitry.

The deceleration function is used to shorten or lengthen the time required to bring the motor and its load to a stop.

S-Curve This term refers to a type of start/stop cycle of a motor. When the motor is started using the acceleration function, its rate of change gradually increases until it reaches full speed. When the motor is stopped using the deceleration function, its rate of change gradually decreases until it comes to a complete rest. The rates of change during the acceleration and deceleration portions of the cycle are nonlinear, and the portion of the cycle when it runs at full speed is linear. The purpose of an S-curve is to combine soft starts and stops with high speeds when moving from point A to B. An application is running an elevator.

Slip Compensation An induction motor requires slip to induce current into its rotor. When the load increases, the rotor slows down and the slip increases. The greater slip causes an increase of necessary torque to continue driving the load. However, the speed of the rotor is reduced as it continues to drive the load. Some applications cannot tolerate a reduction in speed.

Some AC drives have a slip-compensation function to prevent this situation. This type of drive uses a feedback signal to indicate when the rotor speed changes. When it senses a speed reduction, the drive responds by increasing the synchronous frequency of its output that is supplied to the stator windings. The result is that the rotor returns to its original speed while maintaining the needed slip increase. The correct slip-compensation setting is capable of maintaining a speed regulation of 0.5 percent over the entire operating speed range.

Critical Frequency Rejection In various industrial applications, especially those that use fans and pumps, a resonant frequency can develop. This condition can create severe vibrations in the mechanical drive train of the motor and cause damage. To avoid the resonant frequencies, the drive can be programmed to operate at frequencies above and below the resonant bandwidth. This feature is also referred to as a *skip frequency function*.

Power Loss Ride-Through If power is temporarily lost, 1 to 2 seconds for example, the AC drive can be programmed to keep running. It uses the inertia of the load and the stored energy from the inductive and capacitive components in the DC bus line to continue. The drive will go into a fault mode after 2 seconds when the bus can no longer supply enough voltage because the reserved energy has been consumed.

Automatic Restart Some AC drives are capable of restarting automatically. This feature may be desirable if the drive shuts down due to a temporary dip in the power supply level. The drive can be programmed to make several attempts at restarting. For example, four attempts can be made at 3-minute intervals. If the drive is unable to restart after the selected number of attempts, it will go into the fault condition mode until it is reprogrammed.

9-8 Inverter Self-Protection Function

Most drives have the capability to detect when a fault develops. A fault condition occurs when there is a defect in the drive circuitry, when there is a problem with the motor, when an extreme load condition arises, or when the power supply output deviates from its required parameters. To protect itself, the drive will stop running and allow the motor to coast to a stop. This feature is referred to as a *fault trip* function. The following types of fault trip functions protect the drive:

Overcurrent Protection

This function shuts off the semiconductors in the inverter if the rated motor current is momentarily exceeded. The following conditions cause an overcurrent situation:

- Inertia of the load is excessively large.
- A short circuit develops in the output leads or motor windings.
- A component inside the inverter section has short circuited.

Overload Protection

This function shuts off the inverter if currents in excess of 150 percent are drawn.

Overvoltage Protection

This function shuts off the inverter if the voltage at its output terminals elevates to a destructive level. This condition can develop if an overhauling load occurs or the load is decelerated too quickly. In either situation, regeneration occurs and energy is fed back into the drive. The DC bus-line voltage is monitored to determine if this condition develops.

Undervoltage Protection

When the incoming line voltage falls below a certain level—195 volts for a 220-volt three-phase inverter, as an example—the drive stops to prevent a misoperation condition.

Overheating Protection

This function stops the inverter operation when the temperature of the heatsink rises, and reduces the cooling effect to the switching transistors. The following conditions can cause overheating:

- The ambient temperature is too high.
- The heatsink is dirty.
- The drive's cooling fan has stopped.
- The ventilation slots are obstructed.

9-9 Motor Braking

One method of quickly stopping a motor and the load to which it is mechanically coupled is to use the deceleration function of the drive. However, if the inertia of the load is too large, it will overcome the magnetic field strength of the stator coils. The result is that it will rotate more rapidly than the synchronous speed of the rotating stator field. This action is referred to as *overhauling.*

When the synchronous speed drops below the actual speed of the coasting load, the motor becomes a three-phase generator and feeds the voltages back into the drive. These voltages forward-bias the diodes placed across the IGBTs of the inverter section. The current that results is fed back into the intermediate section of the drive. The intermediate circuit voltage can rise until the VFD trips out to protect its internal circuitry. As a result, the load will slowly coast to a stop. To provide a braking action that cannot be achieved by the deceleration function of the drive due to overhauling, two alternative methods can be used.

Regenerative Braking

Regenerative braking is a process by which energy is transferred from the motor that produces generator action back to the DC bus line. The current that charges the filter capacitor in the intermediate section is from the motor instead of the AC power line. Most of this energy is reused after the conversion of the regeneration action back to motoring action. From 10 to 15 percent of the energy is dissipated in losses by the capacitor, diodes, and motor windings. If the regeneration causes the DC bus voltage to become excessive, an overvoltage protection circuit activates.

Another type of drive that performs regeneration transfers energy from the motor back to the AC power line.

Common Bussing

An alternative method of returning energy to the AC power lines through regenerative braking is *common bussing.* This method has more than one motor connected across the same DC bus line in the intermediate circuit, as shown in Figure 9-17. When any of the motors encounters a condition in which the inertia of the load causes the motor's rotor to turn more rapidly than the synchronous speed of the stator's magnetic field, the motor generates a voltage current to be transferred by the bus line to the motors that are in the motoring operation. An example of when this condition exists is where an unwind motor and rewind motor are both connected to the same bus line. The primary advantage of this configuration is that the net power consumed is reduced due to the efficient use of returned energy.

Other conditions in which the braking operation is used are:

- Fast deceleration of high inertia loads is required
- Stopping on a timed ramp
- Cyclic loads
- Eccentric shaft loading

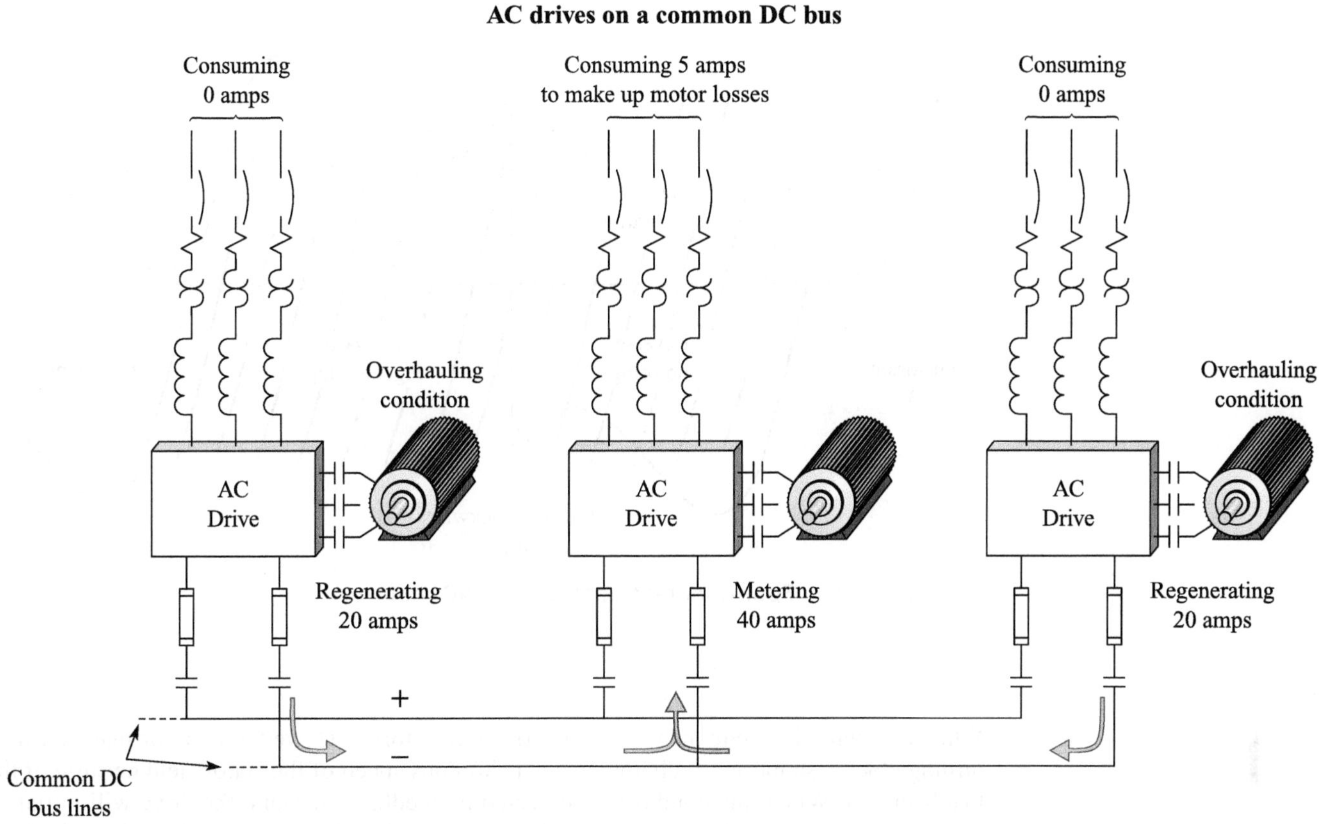

FIGURE 9-17 AC drives on a common DC bus

9-10 Four-Quadrant Operation of AC Motors

AC drives provide power to give a motor torque at constant or variable speeds under controlled conditions. However, when the inertia of the load is large enough, its momentum causes the motor rotor to run more rapidly than the synchronous speed of the stator when it is slowed, or it may continue to turn after power is shut off. This condition is called *overhauling*. When overhauling begins, the motoring action stops and the motor becomes a generator that produces voltage. An example of how an overhauling condition can develop is when too many people get on a descending escalator that is powered by a motor.

Modern AC drives are capable of providing full control of the motor in both clockwise and counterclockwise rotations, and preventing overhauling conditions in both directions. Overhauling is prevented by a braking action. A drive that provides these characteristics is referred to as having **four-quadrant control.** Figure 9-18 shows the four quadrants.

Quadrant 1

In quadrant 1, the switching sequence of the inverter transistors causes the rotor to turn clockwise. Both the torque produced by the rotor and its rotation are in the same direction.

Quadrant 3

In quadrant 3, the rotor turns counterclockwise because the sequence at which the inverter transistors are switched is opposite to that which occurs in the quadrant 1 operation.

A drive that operates in both quadrants 1 and 3 causes the motor to drive a load in both forward and reverse directions. Examples are the motor in an electric vehicle that runs forward *and* in reverse, or in a crane that must raise *and* lower a load.

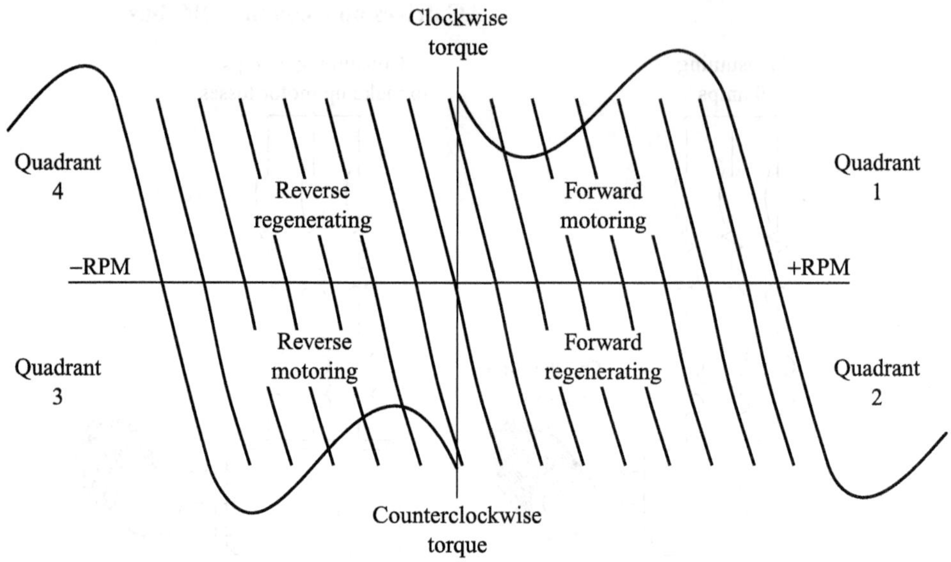

FIGURE 9-18 Four-quadrant operation of AC motors

Quadrant 2

When the switching sequence of the inverter transistors is slowed, the motor may continue turning clockwise more rapidly than the synchronous speed of the stator field because of the load's inertia. When this condition is detected by feedback sensors, the drive will initiate a braking action to slow the motor. Once the speed is reduced to the desired level, the braking action stops and the motoring operation is restored. In quadrant 2, the torque produced by the motor is opposite to its rotation.

Quadrant 4

If the switching sequence of the inverter transistors is slowed when the motor is turning counterclockwise, the inertia of the load may cause it to continue turning more rapidly than the stator field rotates. By detecting this condition with sensors, the drive will cause braking action to slow it down to the desired speed. In this quadrant, the torque produced by the motor is opposite to its rotation.

The quadrant 2 operation does not initiate braking action when the motor is operating in the quadrant 1 forward (clockwise) direction. Likewise, the quadrant 4 operation does not cause braking action when the motor is operating in the quadrant 3 reverse direction.

9-11 AC Drive Selection

The most important consideration when selecting the type of variable-speed drive, or drive configuration, is the application. The following factors should be used in the selection process:

Torque Requirements

There are generally two types of torque requirements, *variable* and *constant,* as shown on the speed vs. torque curves in Figure 9-19.

Figure 9-19 (a) shows a variable torque chart in which very little torque is required at starting or low speeds. The torque that is needed is not proportional to speed. Many types of fans, pumps, and other equipment with centrifugal loads require variable torque. Drives that are selected for these applications typically have low horsepower rating and are either the V/Hz or sensorless type. These types of applications provide the largest amount of energy savings.

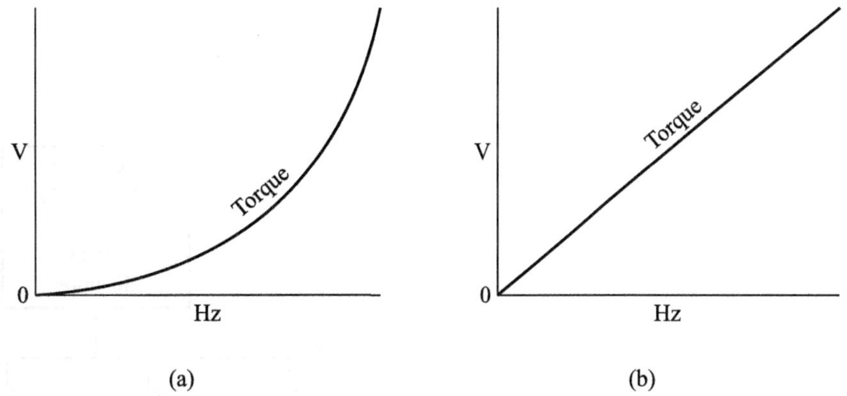

FIGURE 9-19 (a) Constant speed vs. torque curve; (b) Variable speed vs. torque curve

Figure 9-19 (b) shows a constant torque chart in which torque per unit speed remains constant from 0 to base speed. Applications include rewinders, laminators, and printing presses. Drives that are selected for these machines typically have high horsepower ratings because of the high current capabilities that are required to provide high torque. Sensorless vector and AC closed-loop flux vector drives with feedback capabilities that monitor the operation are used to provide constant torque. These applications usually do not provide high energy savings. Instead, their primary function is to provide soft starting or precision control.

Programming Capability Requirements

Each AC induction motor is designed to run at a particular RPM, called **base speed.** When power is applied, the motor accelerates from 0 RPM. As it does, the torque remains constant and the horsepower increases until the base speed is reached, as shown on the chart of Figure 9-20. Above base speed, the torque falls off while the horsepower remains constant.

It is possible to configure a drive to run above and below base speed for certain applications. One example is a paper winder. When it begins to wind, the empty core is light and does not require much torque to turn. Therefore, it can be turned above base speed at a reduced torque. As the roll is being wound and becomes heavier, the torque required to turn it becomes greater. Therefore, it is necessary for the drive to slow the speed and eventually run below the base speed so that torque is increased to handle the load. Eventually, the roll is heavy and requires the motor to provide maximum torque. The green line on the graph shows the speed and torque relationship for this application. By using this winding method, both

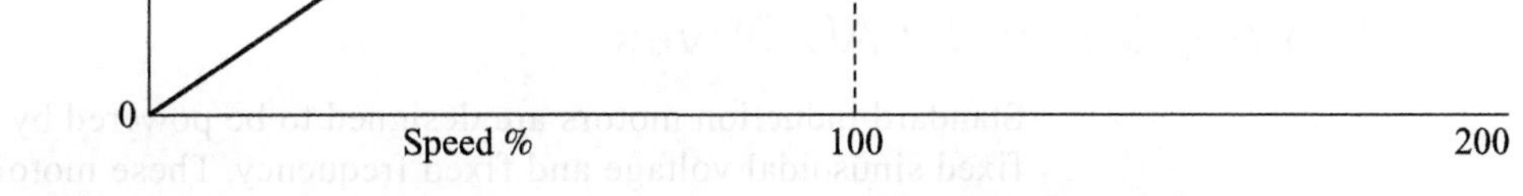

FIGURE 9-20 Torque vs. speed below, at, and above base speed

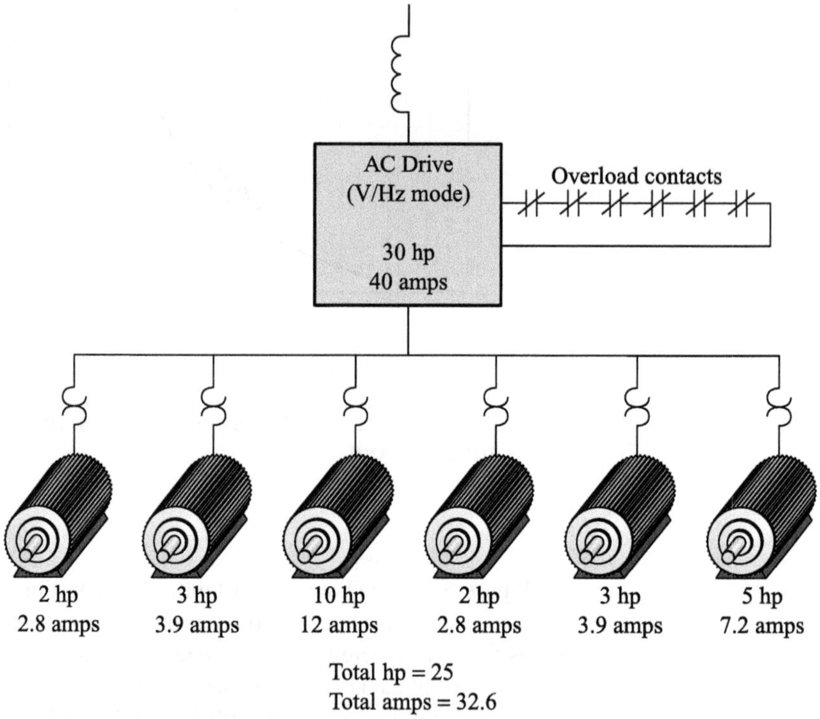

FIGURE 9-21 Multimotor applications

the motor and drive can be undersized to perform this operation, which results in cost savings because less power is consumed.

Multimotor Application

Whenever there are several motors that operate the same way, it may be possible to run them with one drive. Shown in Figure 9-21, motors that power the loads are connected to the drive output. An example of this type of operation is a knitting mill. Since there are a series of bobbins that move up and down together, their movements can be controlled simultaneously by one drive. The obvious advantage of this configuration is that one drive is less expensive than using individual drives for each load. However, there are several requirements that must be followed for this configuration.

- The amperes drawn by the collective motors must not exceed the capacity of the drive.
- Each motor must have its own overload.
- The motor speeds will be in slip speed range with respect to one another.
- Interlock output contactors to drive run logic, when used.

With more than one motor powered by the same drive, encoder feedback data from each motor cannot be used. For example, if only one motor required an adjustment, the drive would alter the operation of all the others, even when it was not required. Therefore, only V/Hz drives that do not have feedback capabilities are used in multimotor applications.

9-12 Motors Driven by AC Drives

Standard induction motors are designed to be powered by AC supply mains that provide a fixed sinusoidal voltage and fixed frequency. These motors are exposed to voltage spikes from a collapsing magnetic field only when the motor is turned off.

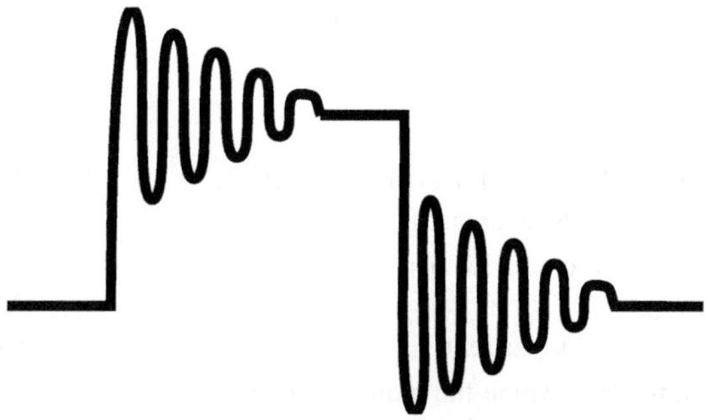

FIGURE 9-22 Spike voltage waveform

When an induction motor is powered by an AC drive, high-frequency pulses are applied to the stator windings to create simulated sine waves. Each time the rising or falling edge of a rectangular pulse occurs, the stator coil induces a voltage spike, as shown in Figure 9-22. Each spike can range from 900 to 1400 volts. These pulses stress the insulation in the first few turns of each winding. The coils beyond the first few turns are not affected because of the inductance they produce, the inductive reactance they produce, and the capacitance that develops from adjacent windings. Eventually, the voltage spikes can punch through the insulation.

The insulation of a standard induction motor can be inadequate when it is powered by a PWM drive. As an alternative, specially designed induction motors called *inverter duty* or *inverter rated motors* have been developed. A thicker layer of insulation is added between adjacent windings. This insulation, referred to as *phase paper, interphase insulation,* or *phase barrier,* is made of cambric, Nomex, Milar, or other insulating film. Slot paper is normally used to separate adjacent coils within the stator slots.

To suppress the amplitude of the spikes, inductors are often placed at the inverter output terminal of the drive. For severe spike conditions, LC filters can be added to the drive output.

Other drawbacks of standard AC motors are that the rotor bar and cooling impellers of the fan are designed for 60-hertz sinewave power and constant speed. Since inverter duty motors are required to run above base speed, their shafts are built stronger and the fan is designed to provide better cooling. Also, many include an end-mounting provision for encoder tachometers that supply feedback information for closed-loop operation.

Problems

1. T/F One common method of slowing the speed of an AC motor is to reduce the applied voltage.
2. Determine the RPM of an eight-pole AC motor when 60 Hz is applied.
3. AC drives cause the speed of an AC induction motor to change by ____.
   a. varying the bus voltage
   b. varying the armature current
   c. varying the applied frequency
   d. changing the number of poles
4. AC drives are also commonly referred to as __________.
5. Match the following names of the VFD sections with the appropriate circuit functions that they perform:
   a. converter c. inverter
   b. intermediate circuit
   ____ Rectifies AC into pulsating DC.
   ____ Filters pulsating DC signals into a pure DC voltage.
   ____ Converts a DC voltage into a simulating AC waveform.
6. To keep the torque of an AC motor constant when the speed increases, the current supplied by the drive must _________.
   a. increase c. stay the same
   b. decrease
7. A variable-frequency drive supplies current directly to the __________.
   a. rotor b. stator
8. T/F The auto boost circuit is used to help accelerate an AC motor when it is starting or running at a low speed.
9. The drive provides a constant torque for the motor by __________ the voltage as it decreases the frequency.
   a. increasing b. decreasing

10. The semiconductor switching device that is used in the inverter section of a PWM drive is the ____.
    a. bipolar transistor c. IGBT
    b. SCR
11. As the pulse width of a PWM signal decreases, the RMS voltage __________ (increases, decreases).
12. Which feedback loop in a vector drive performs the field orientation function? ____
    a. speed b. current
13. *Overhauling* is a condition in which the motor's armature rotates ____ than the rotating stator field.
    a. more slowly b. more rapidly
14. Name a method of producing the feedback signal for the following control loops in a vector drive. ____
    a. speed feedback b. current feedback
15. The ____ type drive has no self-correcting capability.
    a. volts/Hz c. AC closed-loop vector
    b. sensorless vector
16. List five parameters that are typically adjusted by programmable AC motor drives.
17. The angle at which the vector drive attempts to maintain relative positioning between the stator flux and the rotor position is ____ degrees.
    a. 0 c. 45
    b. 30 d. 90
18. If a standard induction motor designed to operate at 60 Hz is driven by an alternating current that is at a lower frequency, it may become damaged because ____.
    a. it will cog and vibrate
    b. excessive current will flow and ruin the insulation
19. The PWM drive used to power several motors simultaneously is the ____ type.
    a. volts/Hz c. sensorless vector
    b. AC closed-loop d. all of the above
20. The ____ loop of the AC closed-loop vector drive provides position information.
    a. inner b. outer
21. ____ induction motors are designed to be powered by AC drives.
    a. Standard b. Inverter duty
22. The braking action by which power from the motor is transferred to the AC power line is called ____.
    a. regenerative braking b. common bussing
23. The switching sequence of the inverter transistors, while the drive is operating in the quadrant ____ mode, is the same as when it is operating in the quadrant 1 mode.
    a. 2 b. 3 c. 4
24. The PWM drive that can produce a torque down to 0 RPM is the ____ type.
    a. volts/Hz c. sensorless vector
    b. AC closed-loop d. all of the above
25. T/F It is possible to transfer energy from a motor that is in an overhauling condition to another motor if they are connected to the same bus line.
26. During quadrant ____, the drive does not initiate braking action when the motor is operating in the quadrant 4 mode.
    a. 1 b. 2 c. 3
27. The ____ control function of an AC drive is used for starting a load gently.
    a. speed c. volts/Hz
    b. acceleration
28. The ____ control function of an AC drive is used for stopping the load quickly.
    a. speed c. deceleration
    b. braking
29. A/n ____ control function of an AC drive combines soft starts and stops.
    a. slip-compensation c. volts/Hz
    b. S-curve
30. T/F The critical frequency rejection control function of an AC drive avoids resonant frequencies that may damage a machine.
31. The ____ control function of an AC drive causes the stator frequency to increase when the rotor speed reduces to an undesirable RPM.
    a. S-curve
    b. critical frequency rejection
    c. acceleration
    d. slip-compensation
32. The ____ protection function of an AC drive causes the motor to coast to a stop and prevents it from being damaged.
    a. overcurrent d. undervoltage
    b. overload e. overheating
    c. overvoltage f. all of the above
33. A drive with____-torque capabilities provides the greatest amount of energy savings.
    a. variable b. constant
34. By using the programming capabilities of some variable speed drives, it is possible to undersize the ____ .
    a. motor b. drive c. both a and b

SECTION

5

Process Control and Instrumentation

OUTLINE

The branch of control engineering that involves the regulation of variables that affect products as they are being manufactured is known as **process control.** Specifically, these variables, called *process variables,* include pressure, temperature, flow, levels, and the product composition of gases, liquids, and solids. The term *process* refers to the manipulation of variables by production equipment to alter raw materials until they reach a desired condition. The number and types of variables being manipulated depend upon the product being manufactured. Products produced in process industries include chemicals, refined petroleum, treated foodstuffs, paper, plastics, and metals. Process control engineering is also involved in the public service fields that provide water purification, waste water treatment, and electrical power production.

During the early years of industrial manufacturing, the process variables were controlled manually. Today, the same operations are performed by automated systems that require only minimal human intervention. The hardware used in an automated system is called **instrumentation** equipment.

An instrument used in process control directly performs one or more of the following three functions:

Measurement During the manufacturing process, it is often necessary to monitor the existence or magnitude of process variables. Measuring instruments are the eyes of the automated system function. The measurements may be read and used in real time for control purposes, or stored as data for use at a later time.

Control Another function of an instrumentation device is to ensure that a process variable is maintained at a specific value or within specific limits. For example, if a disturbance causes the controlled variable to deviate from the setpoint, the controller must call for a corrective action. The control section is the brain of the automated system, which compares the setpoint to a feedback signal from the measuring device. In response, it provides an appropriate signal to the output actuator device, which makes any necessary changes.

Manipulation The final control element is the actuator, or muscle, of an automated system. Its function is to manipulate energy or flow of materials at a desired quantity or rate.

Process control is one of two branches of control engineering that use automation to produce a product. The other branch is referred to as *servo control,* or **servomechanisms.** A servomechanism is a machine that controls mechanical position or motion. Examples of servo machines are robotic welders, machine tool equipment, printing presses, packaging equipment, and electronic parts insertion machines, which place components onto a printed circuit board. The primary difference between these branches is the control method that is required. In process control, emphasis is placed on sustaining a constant controlled variable, such as temperature or pressure. The reference point (setpoint) is seldom changed. It may remain constant for several days. The objective of the system is to provide regulation when the controlled variable fluctuates due to a disturbance. Alterations of the controlled variable and the system reaction to correct a change are typically slow. In servo control, the setpoint is constantly changing. The emphasis of the system is to follow the changes in the desired input signal as closely as possible. Variations of the setpoint are typically very rapid. This does not mean that process control systems are never subjected to reference point changes or that servo systems never encounter disturbances. Both systems must be capable of responding to either disturbances or reference point changes.

An example of an automated process control system is a production machine that manufactures catsup. The primary objective of the process is to make premium quality catsup at \$2.50 a bottle. This goal is achieved when the temperature of the ingredients to be cooked is within 10 degrees above or below 160 degrees. If the temperature of the cooking vat rises above 170 degrees, as shown graphically in the illustration below, the catsup will have less viscosity and be watered down. The product no longer meets quality control standards for the premium line. It will either

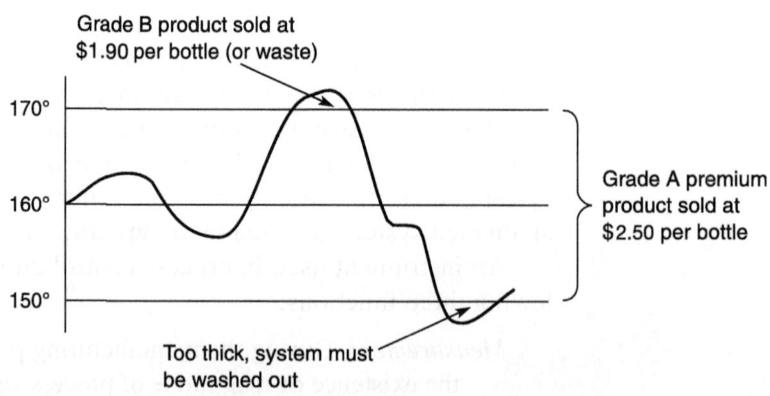

Catsup viscosity

be sold as a grade B product and the price will be reduced to $1.90 a bottle, or it will be discarded as waste. In either case, the profit will be reduced. If the cooking temperature of the vat decreases below 150 degrees, the viscosity of the catsup will increase and it will become thicker. In this situation, the valves in the outlet pipes will become clogged and the production line will be shut down while the system is washed out.

Through the continuous control of systems and processes, automated production provides the advantages of better product consistency, more precise tolerances, and reduced waste. The improved efficiency it offers cuts energy cost, which increases profit and minimizes conservation concerns.

This section of the book consists of nine chapters that provide information on the operation of process control and instrumentation equipment. Chapters 10 through 14 define the properties and characteristics of the process variables: pressure, temperature, flow, level, and product composition. They also describe various types of instrumentation equipment used for measuring and manipulating the variable to maintain a desired condition. Chapter 15 has two major parts: the first part describes various types of manufacturing methods used in process control; the second part identifies various types of instrumentation devices used throughout a closed loop. Chapter 16 provides coverage on how to use common process and instrumentation drawings (P&ID) diagrams. Chapter 17 describes various control techniques used in automated process control, including the characteristics of each one. Chapter 18 provides information about calibration and controller tuning procedures.

CHAPTER

10

Pressure Systems

OBJECTIVES

At the conclusion of this chapter, you should be able to:

- Define *pressure*.
- Define *fluid*.
- Given force and area, calculate the pressure exerted by a fluid.
- Identify five factors that affect the pressure exerted by a liquid.
- Calculate pressure by using specific gravity and depth values for a liquid in a container.
- Identify three factors that affect the pressure exerted by a gas.
- List the reference value for gage, absolute, and vacuum pressures.
- Convert psia to psig, and psig to psia.
- Calculate differential pressure.
- Identify the difference between direct and indirect measurements.
- Describe the operation of the following nonelectrical measuring devices:

Barometer	Bourdon Tube	Bellows
Manometer	Diaphragm	Capsular

- Describe the operation of the following electronic pressure sensors:

Semiconductor Strain Gauge	Transverse Voltage Strain Gauge	Variable Capacitor Pressure Detector

- Explain the operation of the following pressure systems:

Hydroelectric	Vacuum	Static Distribution
Pneumatic	Steam	

INTRODUCTION

Many products manufactured in industry result from a process that involves **pressure.** In fact, the number of pressure gauges and pressure controls in most plants is higher than the number of all the other instruments used for different types of processes combined.

Pressure is defined as force exerted by a **fluid** over a unit of surface area. Different forms of fluids include liquids, gases, steam, and air. Fluids can be the ingredients used to make a product, or the product itself. Fluids are also the medium used to transfer energy from one location to another within a confined network of pipes, tubing, and vessels. Various mechanical devices control the fluids that ultimately provide the power to perform some type of work. Pressures that fluids exert in industrial systems range from a near-vacuum level to 10,000 psig and above.

Various industrial applications involve increasing or decreasing the pressure exerted by a fluid. Reasons for increasing pressure in a process system are:

1. To perform work with a machine such as a hydraulic-powered press or a pneumatic air drill.
2. To move fluids through pipes or to a higher elevation by overcoming friction and gravity.
3. To transform a fluid into a desired physical state. For example, a nitrogen gas will become a liquid when the pressure is high enough.
4. To increase the boiling point of a product. The greater the amount of pressure exerted on a liquid, the greater the amount of heat that is required to bring it to its boiling point. By increasing the temperature and boiling point, a product can be cooked more quickly and the production rate is thus increased.

Reasons for decreasing pressure in a process system are:

1. To perform work, such as lifting and moving sheets of paper in a printing press operation.
2. To draw liquid into a vessel. Lowering the pressure inside a container helps to fill it with a liquid.
3. Changing the physical state of a fluid, such as causing a liquid to vaporize.

To perform these various operations, it is often necessary to control the pressure accurately. Closed-loop systems that perform this function can maintain the pressure at some specific value, or within a required range.

Several reasons to accurately measure and control pressure are:

1. To ensure that the pressure does not become too great and create an explosion, which could cause injury or damage to equipment. A controlled system will either set off an alarm or shut down the equipment responsible for the deviation.
2. To ensure quality of products that must be processed at specific pressures and temperatures.
3. To maximize efficiency by preventing pumps or compressors from wasting energy by raising or lowering pressures more than is needed to produce the desired result.

10-1 Pressure Laws

Pressure is measured as force per unit area. In the English system, force is measured in pounds, and unit area for pressure measurements is the square inch. Therefore, pressure is commonly expressed in terms of pounds per square inch, or *psi*.

Pressure is defined mathematically by the following formula:

$$P = \frac{F}{A}$$

where,

P = pressure (psi)
F = force (pounds)
A = area (square inches)

Figure 10-1 illustrates how both force and area affect pressure. The 1-square-inch object in Figure 10-1(a) weighing 1 pound exerts 1 psi of pressure on the surface on which it is resting. The pressure will double to 2 psi if twice the weight is resting on 1 square inch, as shown in Figure 10-1(b).

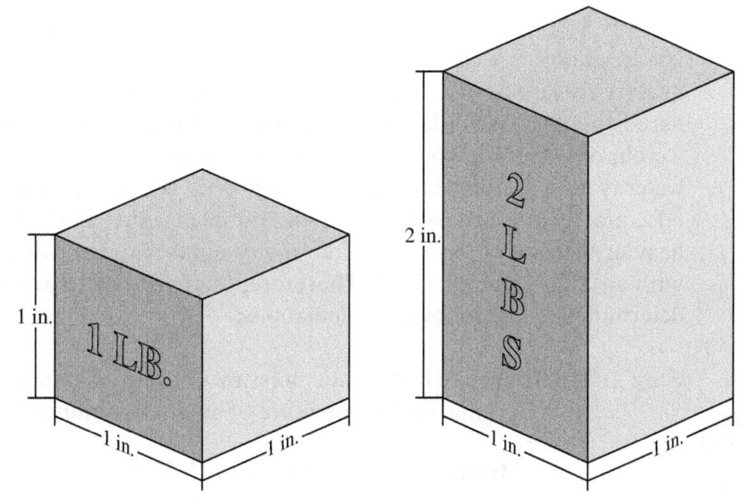

(a) (b)

FIGURE 10-1 Pressure

10-2 Properties of a Liquid

The molecules of a liquid are closely attracted to each other, which is why most substances in this state are incompressible. Its molecules are constantly moving, slipping, and sliding past one another. This enables liquids to flow and take the shape of a container. The ability of a liquid to accomplish this action is related to a property called **viscosity.**

The pressure exerted by products or ingredients in liquid form must be closely monitored and controlled. The amount of pressure is affected by several factors: height, weight, temperature, atmospheric pressure, and mechanical machines.

Height

The height of the liquid affects the pressure it exerts. The term *head* is commonly used to describe the height of a liquid above the measurement point. Head is given in inches, feet, or other units of distance.

Weight

Another factor that affects liquid pressure is its weight. Different liquids have different weights due to their **density.** Density is defined as the weight of a certain volume of liquid, expressed in pounds per unit volume.

Hydrostatic Pressure

ELE606
Hydrostatic Pressure

The pressure exerted by a material in a vessel is directly proportional to its height times its weight. This pressure is created from the weight of the material itself. The force that is produced by the material is called **hydrostatic pressure**, or *head pressure*. The formula for hydraulic pressure is:

$$\text{Pressure} = \text{Height} \times \text{Density}$$

For example, the head of water inside a tank can be determined by measuring the height and obtaining the weight density of water. Figure 10-2 shows a tank with a water level of 15 feet. The weight of 1 square inch of water 1 foot high is 0.433 lb. The hydrostatic pressure is computed by the following calculation:

$$\text{Hydrostatic Pressure (head)} = 15\text{ ft} \times 0.433\text{ psi/ft} = 6.495\text{ psi}$$

Hydrostatic Pressure

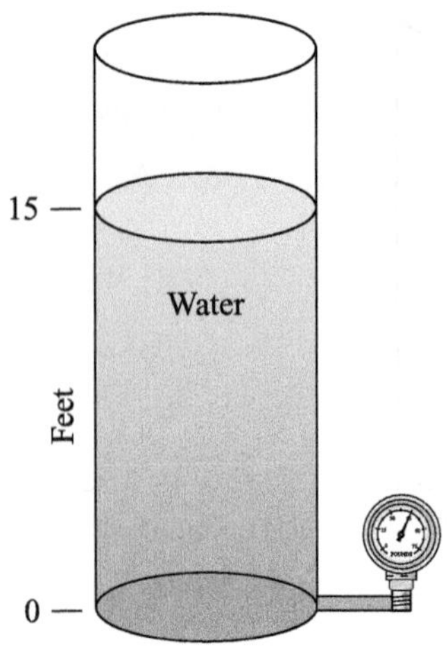

Specific Gravity

If a liquid other than water is in the container, calculations may be made for its **specific gravity** (SG). Specific gravity of a liquid indicates how much lighter or heavier it is compared to water at 60 degrees Fahrenheit. This relationship is calculated by dividing the weight of a specific volume of liquid by the weight of the same volume of water. Because water is used as a standard, its specific gravity is 1.0. If a liquid is lighter than water, its specific gravity will be less than 1.0. An example is ethyl alcohol, with an S.G. of 0.79. A liquid heavier than water has a specific gravity number greater than 1.0. An example is mercury with an S.G. value of 13.57. Therefore, the hydrostatic pressure of 10 feet of mercury is determined by the following calculations:

Step 1: Determine the hydrostatic pressure of a 1-foot column of mercury using its specific gravity value and the pressure of an equivalent height of water.

$$\text{Mercury Pressure (Density)} = 13.57\ (\text{S.G.}) \times 0.433\ \text{psi/ft (Water)} = 5.876\ \text{psi}$$

Step 2: Multiply the hydrostatic pressure of a 1-foot column of mercury times the level of the mercury.

$$\text{Pressure} = 5.876\ \text{psi/ft} \times 10\ \text{ft} = 58.76\ \text{psi}$$

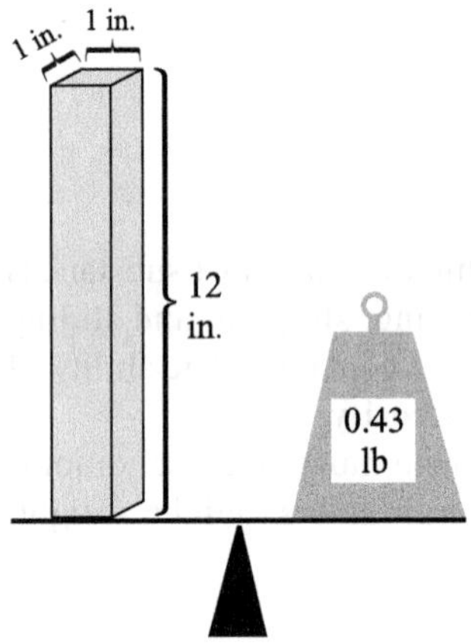

FIGURE 10-2 Hydrostatic pressure is developed by the weight of a column of liquid

Temperature

The temperature of a liquid affects the pressure it exerts. Increasing the temperature expands the liquid (i.e., its molecules move farther apart) and reduces its density. If the liquid is in an open vessel, the hydrostatic pressure remains the same, because the density reduces at the same rate at which the level rises; the decrease in density and the increase in volume cancel each other to keep the weight of the liquid constant. However, if the liquid is confined in a closed vessel, as shown in Figure 10-3, the pressure rises. As the temperature of the vapor inside the vessel increases, the vapor pressure also increases. The pressure due to the vapor is transmitted through the liquid and adds to the pressure due to the weight of the liquid itself.

Atmospheric Pressure

The earth is surrounded by its atmosphere, a layer of air approximately 100 miles thick that is pressed against the earth's surface by gravity. Under normal conditions, the weight of a 1-square-inch column of air from the top of the layer to sea level is 14.7 psi. At an elevation higher than sea level, such as Mexico City (altitude 7,800 feet), the atmospheric pressure is only 11.1 psi, because the column of air above the earth's surface at that location is shorter. Figure 10-4 illustrates the difference in hydrostatic pressure between liquids in Boston and Mexico City.

The pressure from the atmosphere also will exert a force on a liquid in an open vessel. Atmospheric weather conditions also affect the hydrostatic reading of the liquid in an open vessel. The reading will decrease if a low pressure front has moved into the region.

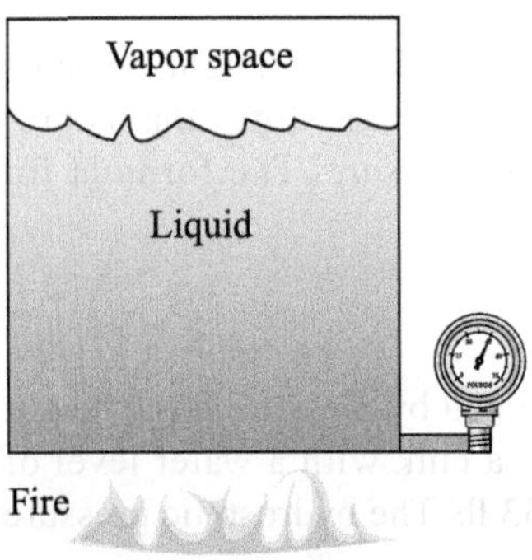

FIGURE 10-3 The heated vapor presses down on the liquid

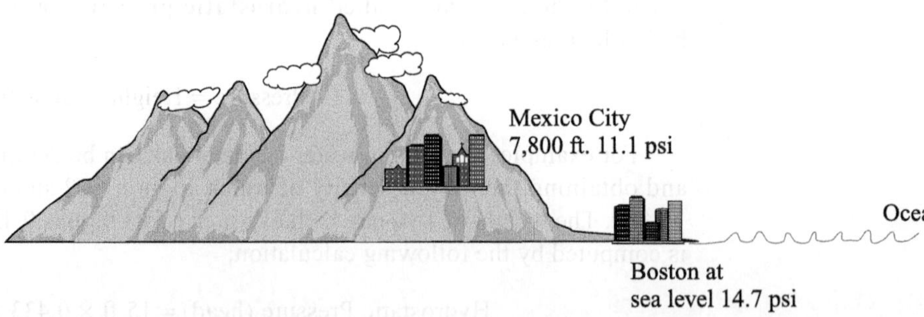

FIGURE 10-4 The higher elevations produce less atmospheric pressure

Mechanical Machines

A mechanical machine can change the pressure of a liquid. For example, pumps, such as the one shown in Figure 10-5, are used to move liquids through pipes, hoses, or into or out of containers, or to do work. When the liquid is pumped into a confined area, the pressure increases. The pressure of any confined liquid is called *hydraulic pressure*.

10-3 Properties of a Gas

Gases are another type of fluid used in industrial process applications. A gas can be air from the atmosphere, vapor, or steam.

Unlike molecules of solids and liquids, which remain attracted to one another, gas molecules remain separate. They are also constantly moving very rapidly, crashing into one another and into nearby surfaces. The molecular energy of the gas causes it to take the shape and fill the volume of its container. Unless the gas is confined in a container, it will disperse. Most of the volume occupied by a gas is space. If the space was eliminated, the gas molecules would be in contact with one another and become a solid or liquid.

Because of their high degree of molecular activity, gases take the shape of their container and exert equal pressure in all directions on its walls. The factors that affect this pressure are temperature of the gas, volume of the container, and air removal from the container.

Temperature of the Gas

In a gas, the molecules are constantly moving. In a confined vessel, billions of molecules collide each second. A pressure gauge placed in a gas container will interpret these collisions as a single pressure. As the temperature of the gas increases, its molecules move more rapidly and collide more frequently in a given span of time. The result is that the pressure increases proportionately with the rise in temperature.

When the temperature of a gas that is not confined in a sealed container is increased, the gas will expand proportionately and the pressure will remain constant.

Volume of the Gas Container

When the area of the enclosed container is decreased, the space between the gas molecules is reduced. This action is called **compression.** By decreasing the space between the molecules, a proportionately greater number of collisions will occur in a given span of time, resulting in a higher pressure. The temperature of the gas will also increase. Compressed gas is stored in a confined tank at a higher pressure than atmosphere. The pressure also increases as an additional amount of gas is pumped into the system. Compression can be accomplished by a mechanical compressor, as shown in Figure 10-5. This device is a cylindrical container with a close-fitting piston and

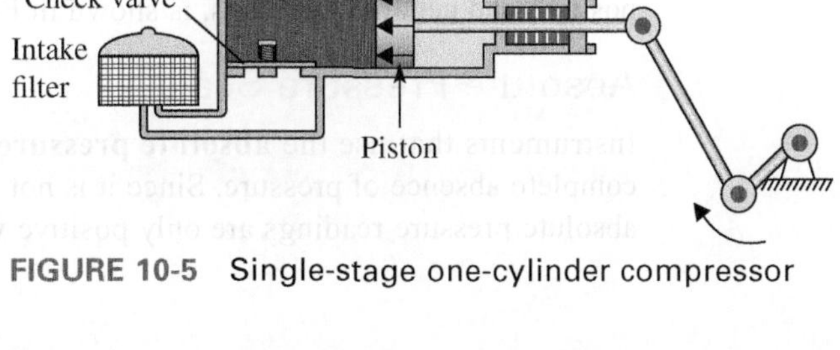

FIGURE 10-5 Single-stage one-cylinder compressor

a seal. This piston is driven in a reciprocating manner by a crankshaft connected to a motor. An inlet port allows air in from the atmosphere. The piston pushes the air out of the exhaust port into an enclosed system with a fixed volume.

Gas Removal from a Container

If any gas is removed from a sealed container, its pressure will become less than the atmospheric pressure surrounding the vessel. Any reduction of pressure compared to the atmospheric pressure is a partial **vacuum.** If the gas is completely removed, a full vacuum exists.

The vacuum is created by a pump, as shown in Figure 10-6. It operates on the same principle as the compressor. The difference is that the intake port is connected to the enclosed system, and the discharged air is pumped out of the exhausted port into the atmosphere. Most systems require only a partial vacuum to perform the desired operation.

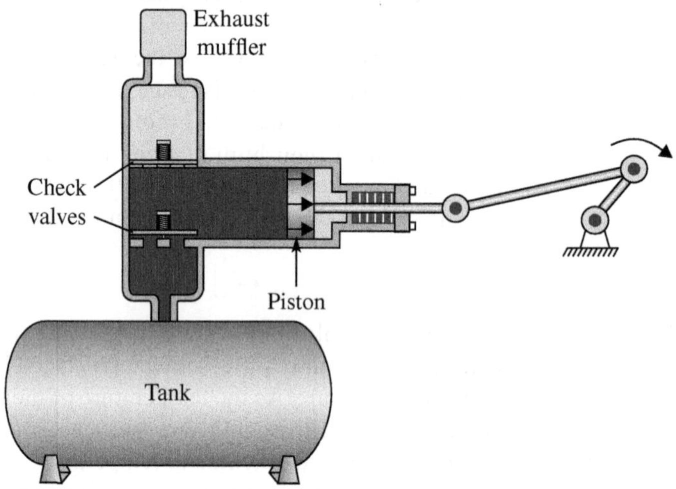

FIGURE 10-6 Single-stage one-cylinder vacuum pump

10-4 Pressure Measurement Scales

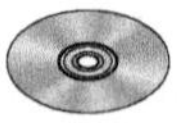

IAU3806
Pressure Measurement Scales

Pressure measurements always show the measured pressure as compared to a reference pressure. There are four basic scales, each distinguished by the reference pressure used. Instruments that measure pressure use one of these scales: *gage pressure, absolute pressure, differential pressure, vacuum pressure,* or *inches of water column.*

Gage Pressure Scale

Instruments with the **gage pressure** scale use atmospheric pressure as the reference point. If the sensing element is exposed to the atmosphere, the measurement scale records zero. The units of measurement are recorded in *psig* (pounds per square inch, gage).

Gage pressure is either positive or negative, depending upon its level above or below the atmospheric pressure reference. A gage pressure instrument will read +30 psi when measuring an inflated tire with 30 pounds per square inch of air pressure. This value indicates a positive pressure of 30 psi above atmospheric pressure. A negative gage pressure indicates a pressure in pounds per square inch below atmospheric pressure. A negative pressure of −14.7 psi indicates a full vacuum. A gage pressure scale that makes a positive reading only is shown in Figure 10-7(a). A compound gage pressure scale, which makes both positive and negative readings, is shown in Figure 10-7(b).

Absolute Pressure Scale

Instruments that use the **absolute pressure** scale are referenced to absolute zero, or the complete absence of pressure. Since it is not possible to have a pressure less than a vacuum, absolute pressure readings are only positive values. If the sensing element is exposed to the

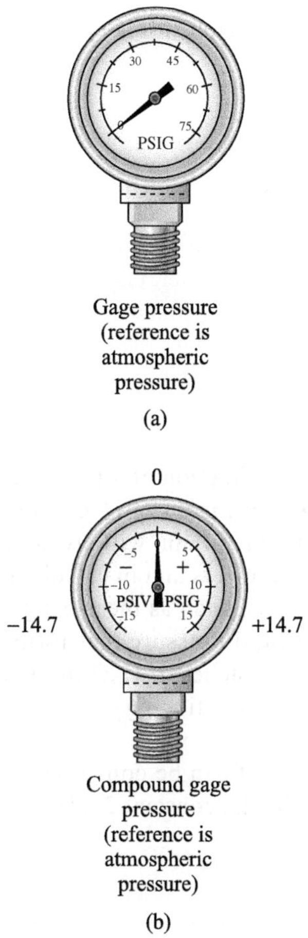

FIGURE 10-7 Gage pressure measurement scales

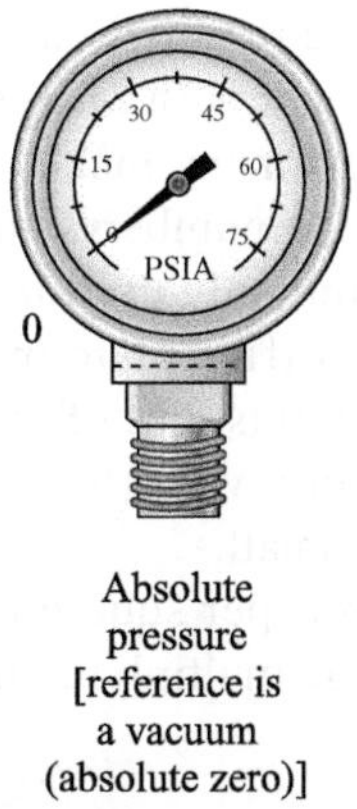

FIGURE 10-8 Absolute pressure measurement scale with a vacuum reading of zero as the reference

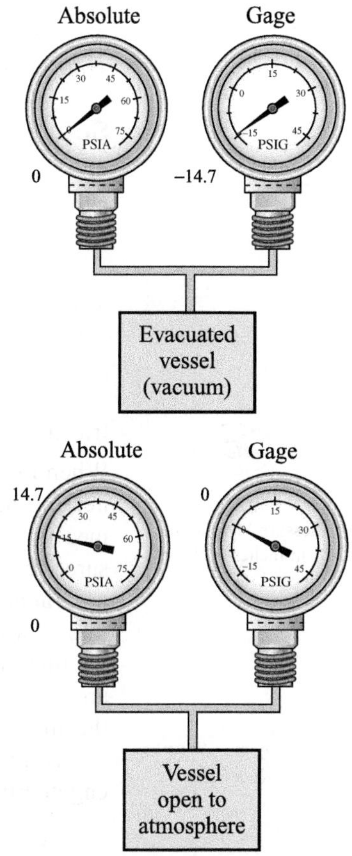

FIGURE 10-9 Comparison of readings by absolute and gage measurement scales at a vacuum and at atmospheric pressure

atmosphere at sea level, the measurement scale will read 14.7 pounds per square inch. The units of measurement are recorded in *psia* (pounds per square inch, absolute).

Absolute pressure readings are generally more accurate than gage readings. The reference of gage pressure instruments is not consistent: atmospheric pressure fluctuates with weather changes and altitude. With absolute pressure instruments, the reference point is consistent. A pure vacuum is the same at sea level as it is on top of Mt. Everest. An absolute pressure scale is shown in Figure 10-8.

To convert gage to absolute pressure, simply add atmospheric pressure to the psig pressure value. To convert absolute to gage, subtract atmospheric pressure from the psia measurement. The diagram in Figure 10-9 shows a comparison of both gauges when measuring pressures under various conditions.

EXAMPLE 10-1

A gage reading of 30 psig is taken. What will an absolute pressure instrument read under the same conditions?

Solution

$$\begin{aligned}\text{Absolute Pressure} &= \text{Gage Reading} + 14.7\\ &= 30 + 14.7\\ &= 44.7 \text{ psia}\end{aligned}$$

EXAMPLE 10-2

An absolute reading of 60 psia is taken. What will a gage pressure instrument read under the same conditions?

Solution

$$\text{Gage Pressure} = \text{Absolute Reading} - 14.7$$
$$= 60 - 14.7$$
$$= 45.3 \text{ psig}$$

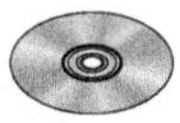

ELE506
Pressure Measurements in Inches of Water

Inches of Water Column

When making low pressure readings, such as those that are a fraction of a pound, the engineering unit of measurement is in *inches of water column,* or *inches of H_2O.* This unit of measurement refers to how many inches of water in a vertical column will create the pressure. The result is that larger numbers can be used in measurements and in calculations.

Inches of H_2O is commonly used in airflow applications, such as measuring the flow rate of a gas through an orifice plate in a pipe. A differential pressure that forms across the plate is very small. So this unit of measurement is used. Inches of H_2O measurements are also commonly used in HVAC (heating, ventilation, air-conditioning) systems because the air pressures are very small.

If a pressure in pounds per square inch (psi) is known, it can be converted into H_2O engineering units by simply multiplying the reading in psi by the constant 27.71.

EXAMPLE 10-3

Determine the inches in H_2O value if a measurement of 1.8 psi is made.

Solution

$$1.8 \times 27.71 = 50$$

Differential Pressure Scale

Pressure readings are also measured in units of **differential pressure,** given in *psid* (pounds per square inch, differential), or ΔP. Differential pressure is used to express the difference in pressure between two measured pressures. It can be determined by subtracting the lower reading from the higher reading. The calculation must be made by using values from the same type of measurement scale.

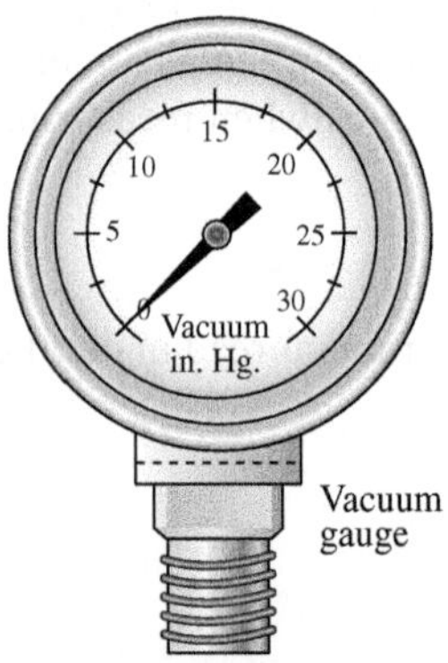

FIGURE 10-10 A vacuum pressure measurement scale

Vacuum Pressure Scale

Instruments that measure a vacuum use a scale that begins at atmospheric pressure, just as gage pressure, but works its way down to a vacuum. In the United States, the most common vacuum scale is listed in units of inches of mercury (*in Hg*), as shown in Figure 10-10. The gauge reads zero when measuring atmospheric pressure, and 29.92 in Hg when measuring a complete vacuum. This unit of measurement is based on a barometer tube that uses mercury to indicate atmospheric pressure. The operation of this device is described in the next part. Figure 10-11 shows a graphic comparison of the different pressure scales.

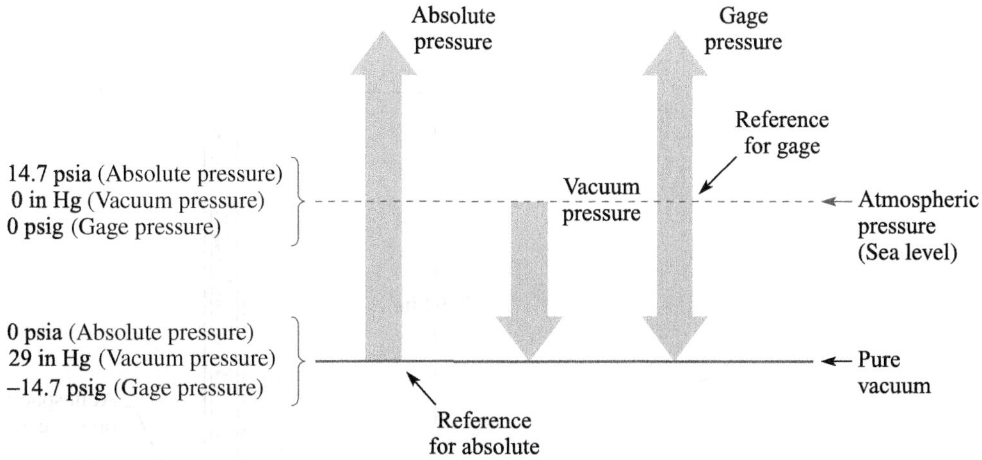

FIGURE 10-11 Comparison of different pressure scales

10-5 Pressure Measurement Instruments

Many manufacturing operations require that specific conditions exist. For example, to combine a powder and a liquid in preparation for a food-blending operation, a pressure condition within a few pounds may be required. Pressure above or below the required range can ruin the food. Excessive pressure in a system can rupture equipment or cause an explosion.

Measuring instruments are used to monitor pressure conditions so that corrective action can be taken if necessary. The forces created by the pressure produce a deflection, a distortion, or some change in volume or dimension of the instrument's sensing element. The physical alteration of the element provides the movement that is necessary to indicate the pressure reading.

Measuring instruments that detect pressure, temperature, level, and flow conditions are classified by whether they make the measurements *directly* or *indirectly.* For example, the level of liquid in a tank can be measured directly with a dipstick. When measurements are taken indirectly, some variable other than the liquid is read. For example, by making an indirect measurement of a tank's weight, the level of its contents can be determined. This method is also referred to as an **inferred measurement.** The weight is used to *infer* a level measurement.

Both nonelectrical measuring devices and electronic sensors used to measure pressure are described below.

10-6 Nonelectrical Pressure Sensors

Nonelectrical pressure sensors fall into two categories: liquid column gauges and mechanical gauges.

Liquid Column Gauges

As previously stated, it is possible to measure pressure by monitoring the height of liquid in a column. For example, a pressure of 0.433 psi will support a 12-inch column of water. This method, called *head pressure,* is very simple and accurate. For this reason, it is often used to calibrate other types of pressure gauges or instruments.

The operation of this measuring device is based on the same principle as the barometer that measures atmospheric pressure, shown in Figure 10-12. It consists of a glass tube filled with mercury that is sealed at one end and open at the other end. Mercury starts to drain out of the tube when it is positioned vertically with its open end inserted into a dish of mercury. When there is enough atmospheric pressure exerted on the mercury in the dish to support the mercury in the tube, it will stop draining. The gap that forms above the mercury is a vacuum. Changes in pressure cause the mercury in the tube to rise or fall. The larger the

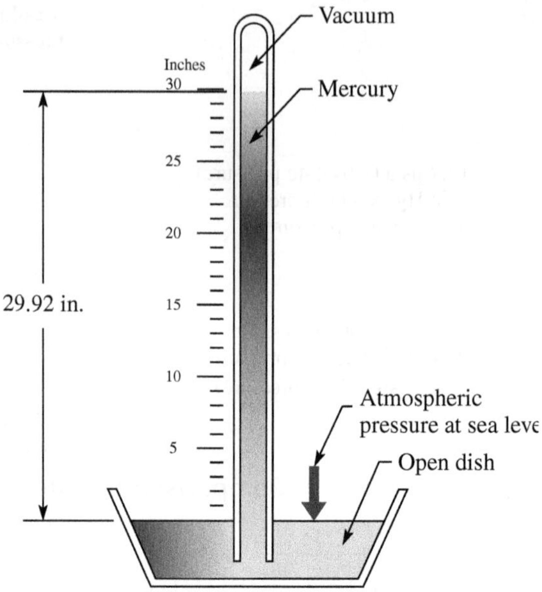

FIGURE 10-12 A barometer reading of 29.92 inches of mercury at sea level

pressure, the higher the mercury will rise. A scale on the side of the glass tube allows the height of the liquid to be read directly. At sea level, 14.7 psi of atmospheric pressure will support 29.92 inches of mercury.

In addition to being affected by pressure, the level of the liquid is also influenced by its density. Because it takes less pressure to support liquids lighter than mercury, water is often used in liquid column gauges to measure very low pressures, vapors, pressures below atmospheric pressure, or a vacuum. Mercury is typically used to detect and indicate **higher** pressures. Liquid column devices that use water as the medium should never be exposed to temperatures below freezing.

Manometer

A manometer is a liquid column device used to measure pressure. It consists of a glass tube bent in the shape of a U so that there are two columns. Each column is exposed to a different pressure source. To determine the amount of pressure, read the rise of liquid in one column and the drop in the other, and add them together. Manometers are primarily used to calibrate other pressure sensors because they are the most accurate and reliable devices to measure pressure. However, manometers are limited to low pressure readings at ambient temperatures. The units of linear measurements they provide are in inches or centimeters, and must then be converted to pressure units. There are several types of manometers used to measure various types of pressure.

Gage Pressure

A manometer that measures gage pressure uses one column as a reference. The end of the reference tube is open so that it is exposed to atmospheric pressure. The other column is connected to the process being measured. The pressure measurement is taken by reading the difference in height of the two columns on the scale. If the level in the reference column is higher than the liquid in the measurement column, a positive value above atmospheric pressure is read. The opposite situation indicates a negative value below atmospheric pressure. The maximum negative pressure reading indicates a vacuum. Figure 10-13 shows a U-shaped manometer and how a gage pressure manometer measures both positive and negative readings.

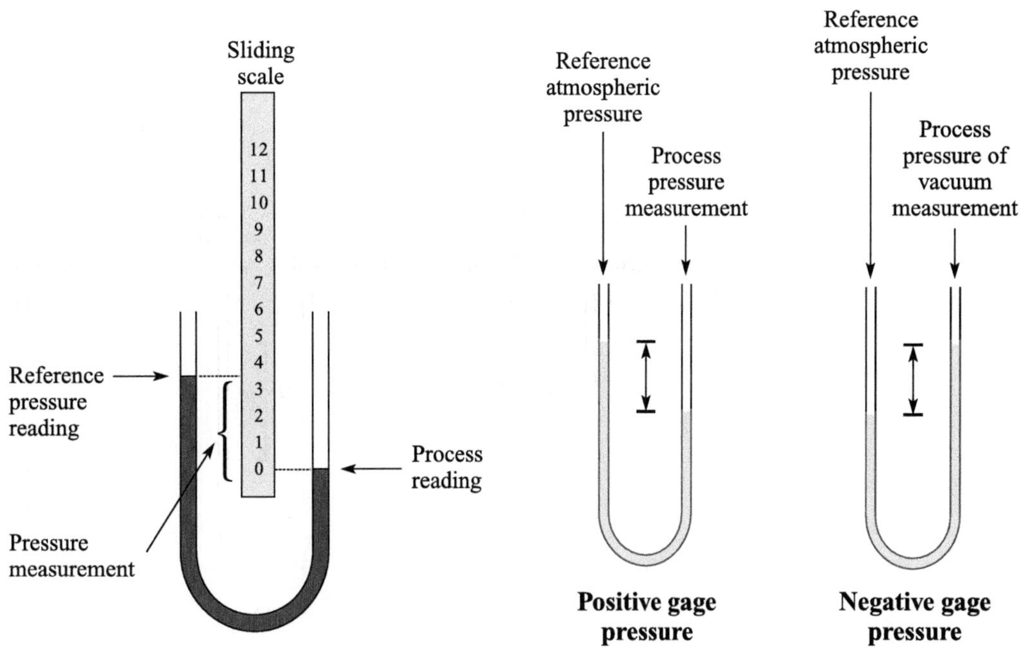

FIGURE 10-13 Gage pressure manometer

The formula for measuring gage pressure with a manometer is

$$\text{Gage Pressure (psig)} = \text{Total Deflection (inches)} \times \text{Liquid Constant (psi)}$$

EXAMPLE 10-4

The liquid in the reference column of a manometer rises 3 inches and the liquid in the pressure column lowers 3 inches. If mercury is used as the liquid, what is the gage pressure being measured? *Mercury has a constant of 0.491 psi per inch.*

Solution

$$\begin{aligned}\text{Gage Pressure} &= \text{Total Deflection} \times \text{Mercury Constant}\\ &= 6 \text{ inches} \times 0.491\\ &= 2.946 \text{ psig}\end{aligned}$$

Absolute Pressure A manometer that measures absolute pressure of a vacuum has a reference column with the end of the tube sealed. The closed end creates a vacuum and allows an evacuation space to form. The other column is connected to the measured pressure. If a high pressure exists, it will push the level in the measured column downward and the level in the reference column upward, thus creating a large level difference between both columns. If a vacuum is measured, the levels in both columns will be the same. Figure 10-14 shows how an absolute pressure manometer makes measurements of a vacuum (Figure 10-14(a)), standard atmospheric pressure (Figure 10-14(b)), and a positive pressure (Figure 10-14(c)).

Differential Pressure The manometer can also be used to measure differential pressure. Simply connect the open ends of each column to the two pressure sources being compared, as shown in Figure 10-15. The level differences of the liquid provide the pressure reading.

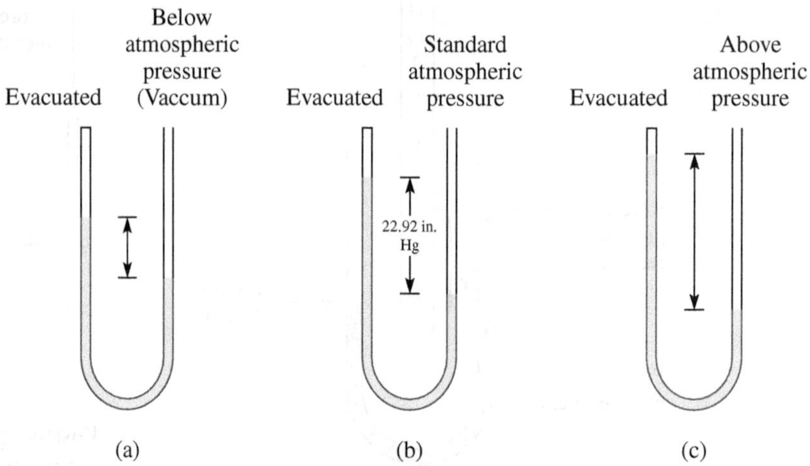

FIGURE 10-14 An absolute pressure manometer

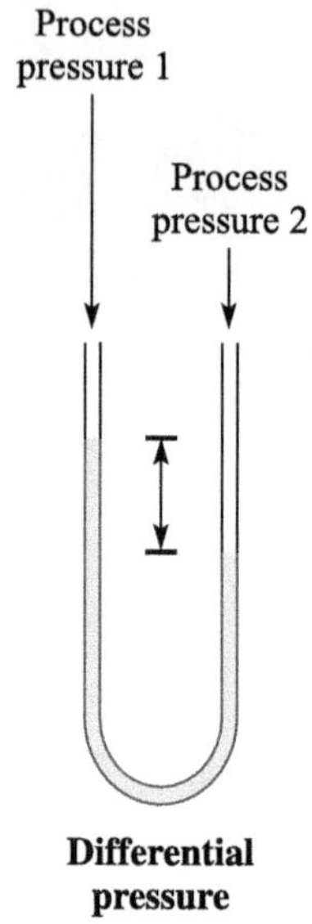

FIGURE 10-15 A differential pressure manometer

The formula used to calculate differential pressure by a manometer is:

$$\text{Differential Pressure (DP)} = (\text{Column 2 Movement} - \text{Column 1 Movement})\ \text{Liquid Constant}$$

EXAMPLE 10-5

The liquid in column 2 of a manometer rises 2.25 inches and the liquid in column 1 lowers 2.25 inches. If mercury is used as the liquid, what is the differential pressure being measured?

Solution

$$\begin{aligned}\text{DP} &= (\text{Pressure 2} - \text{Pressure 1})\ \text{Mercury Constant}\\ &= 4.5 \text{ inches} \times 0.491\\ &= 2.21 \text{ psid}\end{aligned}$$

Mechanical Gauges

Because of their durability, mechanical pressure gauges are often preferred over liquid-filled glass gauges. They are also inexpensive and reliably accurate. Mechanical gauges are constructed by using two major parts, a sensing element and an indicator. The element is elastic and changes form when exposed to pressure. The element is mechanically connected to an indicator device, such as a needle-shaped pointer. As the pressure changes, the element's shape changes and moves the pointer over a scale to indicate a reading.

There are several types of mechanical gauges. They differ primarily by the type of device each uses as the sensing element.

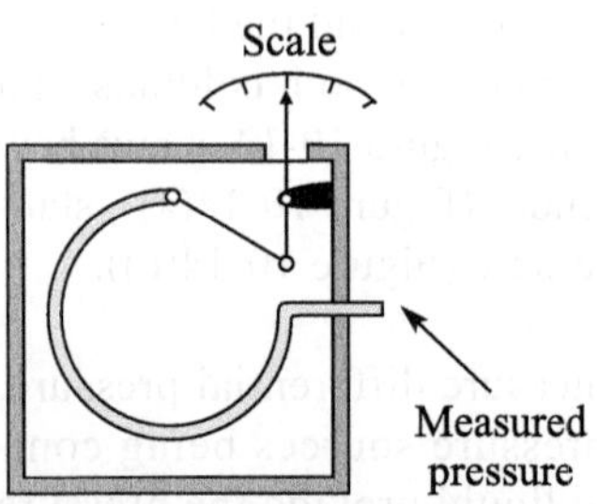

FIGURE 10-16 A Bourdon tube gauge

Bourdon Tube Gauge

The Bourdon tube gauge illustrated in Figure 10-16 shows a C-shaped metal tube as the element. The tube is hollow, sealed at one end, and open at the other end. The open end is exposed to the pressure being measured. An increase in process pressure creates a higher pressure inside the tube, while the pressure outside the tube remains the same. This increase in differential pressure causes the coil to unwind. As it becomes straighter, the needle coupled to the tube will move toward the right of the scale to indicate a higher pressure reading. A decrease in measured pressure reduces the differential pressure between the inside and the outside of the tube. The tube returns toward its original shape, and the needle moves to the left.

Bourdon tubes are available to measure pressure ranges from 0 to 15 psi and from 0 to 6000 psi. They are made from a variety of materials to provide compatibility with the measured fluids.

Diaphragm Gauge

The diaphragm gauge in Figure 10-17(a) measures absolute pressure. It uses a flat, flexible material that bends or flexes when exposed to pressure. One side of the diaphragm is connected to the process being measured and the other side is exposed to a vacuum, which is the reference pressure. The diaphragm element bends toward the side that has the lowest pressure and pushes against an opposing spring. Also referred to as a *load spring,* its strength determines the range and sensitivity of the instrument. The spring also helps to return the element to its original shape as the pressure is reduced. The diaphragm element is mechanically connected to a pointer that indicates the pressure reading.

Figure 10-17(b) shows a gage pressure measuring device. One side of the diaphragm is exposed to atmospheric pressure as a reference. The other side of the element senses the process pressure. The element is pushed toward the side with the least pressure. Figure 10-17(c) shows a differential pressure measuring device. The differential pressure gauge does not have a reference. It only measures the difference between two pressures. The construction of the differential gauge is identical to the gage device. Each side of the element is connected to one of the two pressures being measured. Instead of having a meter face with a gage scale, a differential scale is used.

For low pressure readings, a nonmetallic element such as Teflon is used, often without an opposing spring. Because of its resiliency, it is able to distort when lower pressures are applied. Metal elements are not as flexible, but bend enough when measuring high pressure. Because metals tend to withstand exposure to harmful elements, they are used in environments that are corrosive and where the temperatures are high. Diaphragm gauges sense pressure from 30 in Hg vacuum to 6000 psig.

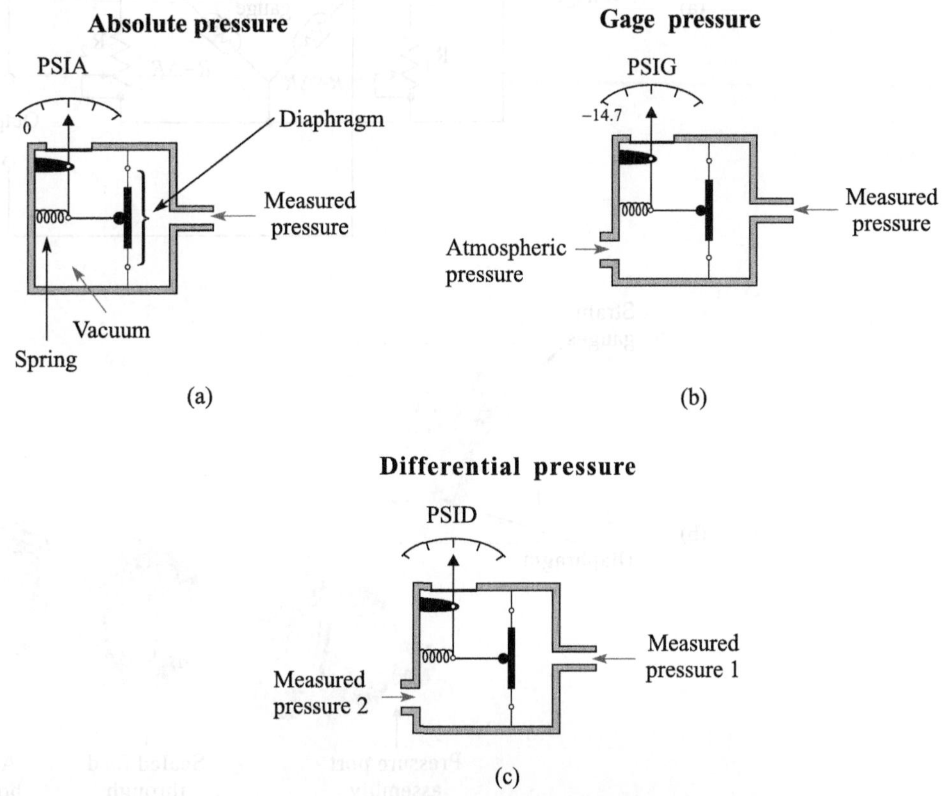

FIGURE 10-17 The diaphragm mechanical pressure gauge

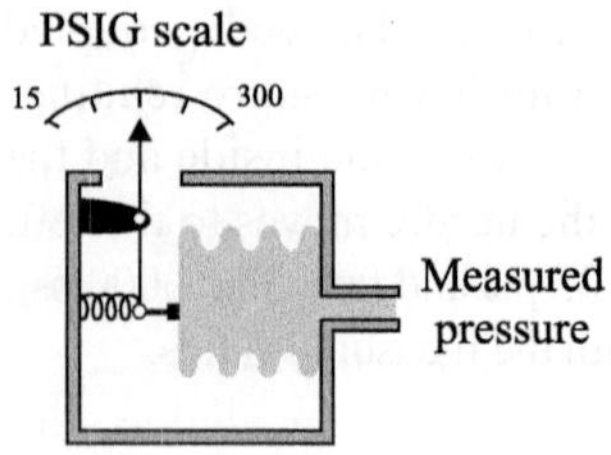

FIGURE 10-18 A bellows pressure gauge

Bellows Gauge

Flat diaphragm elements have a limited range of motion and produce nonlinear readings. By using sensing devices with pleated walls, the flexibility and movement increase. An elastic element that resembles an accordion bellows is shown in Figure 10-18. When applying the measurement pressure to the inside of the element, it expands and causes a larger needle deflection than the flat element. The bellows material may be brass, phosphor bronze, or stainless steel, depending on required range, sensitivity, corrosion resistance, and cost. An opposing spring is employed to control range and sensitivity. Bellows instruments measure pressures that range from 30 in Hg vacuum up to 500 psig.

10-7 Electronic Pressure Sensors

New advancements in electronic technology have resulted in the development of electronic sensors to measure pressure. These devices are more accurate, more reliable, and less expensive than many of the mechanical measuring instruments they are replacing.

Semiconductor Strain Gauges

The strain gauge is used to detect the strain on a body caused by force. This device is typically constructed with a Wheatstone bridge arrangement. It is shown in both schematic and exploded views in Figure 10-19. The four resistive elements that make up the bridge network are made of

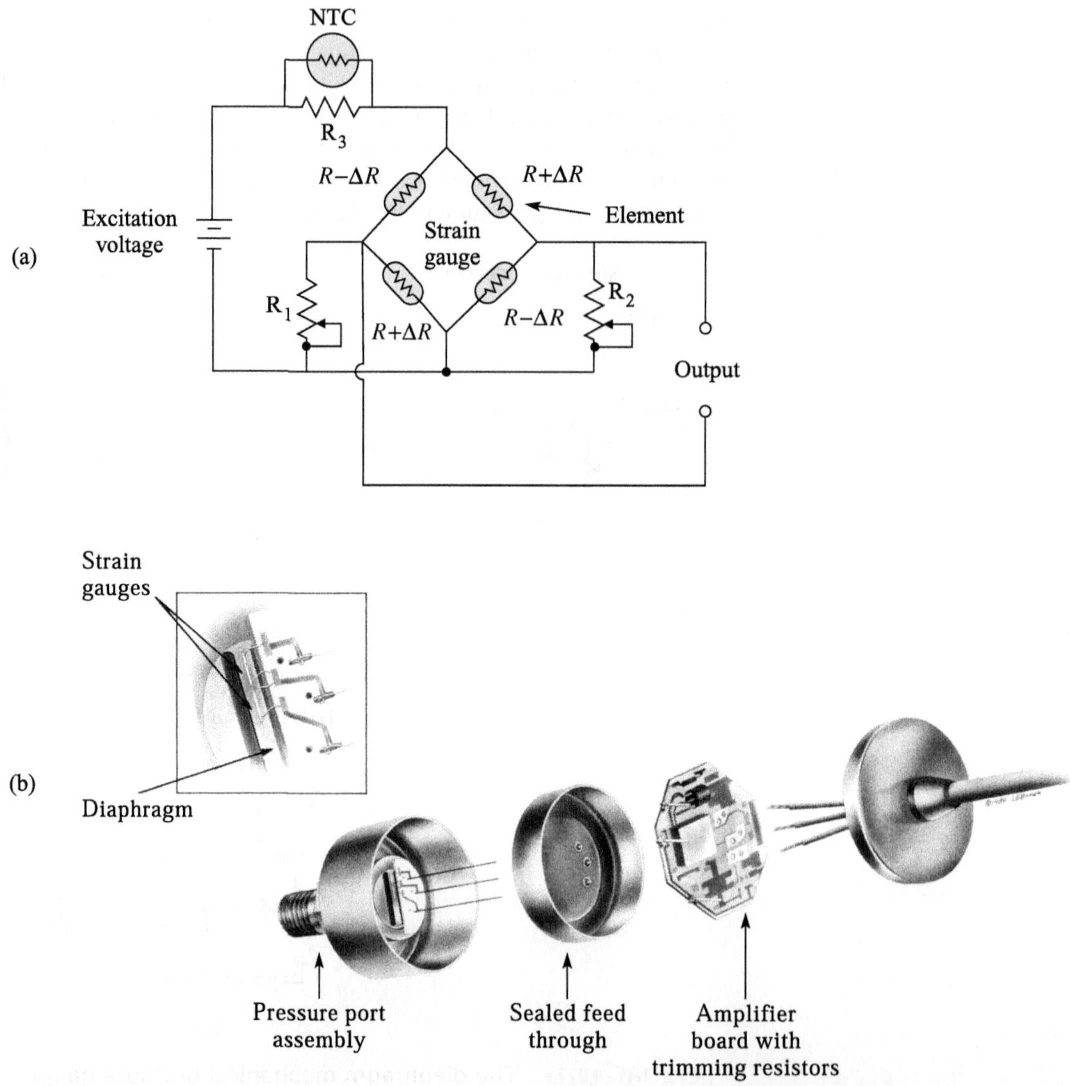

FIGURE 10-19 A semiconductor strain gauge configuration (Courtesy of Data Instruments)

piezoelectric semiconductors. The elements are bonded to a pressure-sensitive diaphragm. A pressure change causes the elements to expand. A compressive strain on the diaphragm will cause the element to contract. The distortion of the elements produces a differential resistance change, which is measured by applying a constant excitation voltage to the bridge. The diaphragm deflection is an analog output voltage up to 250 millivolts proportional to pressure. Without pressure, the bridge output is 0 volts because the four elements are balanced.

The formula $R + \Delta R$ and $R - \Delta R$ represents the elements' actual resistance value with pressure applied to the diaphragm. The R represents the resistor value without pressure applied to the diaphragm. The ΔR represents the change in resistance due to an applied pressure. When pressure is applied, all four resistors' elements change the same amount. The elements are mechanically connected to the diaphragm so that two of them compress while the other two expand. The result is that the resistance of two elements increases, and the resistance of the other two decreases.

Variable resistors R_1 and R_2 are trim pots. They are used to externally balance the resistance of any element changes due to component aging. The thermistor connected across R_3 is used to temperature-compensate the bridge network when the ambient temperature changes. The output voltage is applied to an amplifier and to the transducer's internal or external conditioning circuits. The sensitivity of semiconductor strain gauges is 100 times that of wire strain gauges. Silicon has very good elasticity, which makes it an ideal material for receiving an applied force. Since it is a perfect crystal, it does not become permanently stretched. Instead, it returns to its original shape after the force is removed.

For better temperature ranges and stability, semiconductor elements are epoxy bonded onto stainless steel diaphragms.

Transverse Voltage Strain Gauge

The transverse voltage strain gauge is a new configuration developed by Motorola. Figure 10-20 shows the top view of the sensor element. Current passes through a semiconductor piezoresistor from pins 1 to 3. Pressure that stresses the diaphragm is applied at a right angle to current flow. The deflection of the diaphragm causes the resistor element to bend. As it does, a transverse electric field is developed that is sensed as voltages at pins 2 and 4, which are located at the midpoint of the resistor.

The advantage of this type of strain gauge is that it uses only one element. A single element eliminates the need to closely match the four stress- and temperature-sensitive resistors of a Wheatstone bridge design. It also simplifies the additional circuitry necessary to accomplish calibration and temperature compensation.

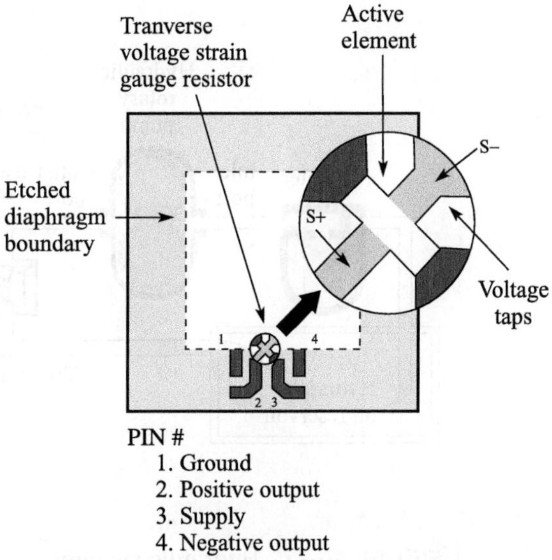

FIGURE 10-20 Transverse voltage strain gauge

Variable Capacitor Pressure Detector

The **variable capacitor pressure detector** uses two conductive plates oriented adjacent to each other and separated by air. Shown in Figure 10-21, one plate is fixed and the other plate is positioned by an elastic element, such as a bellows. It works on the principle that the capacitor's value is changed by varying the distance between the plates. As the measured pressure varies, it changes the shape of the element and causes the plate to move either closer to or farther away from the fixed plate. The change in capacitance is translated electronically into a control signal that represents the measured pressure.

The entire capacitor sensor is sealed in a capsule with the diaphragm exposed to the process pressure. These transducers are economical, small, rugged, vibration-resistant, and accurate to within 0.2 percent throughout their entire range. These features make them well suited for many applications, and they are one of the most widely used pressure transducers in the process industry.

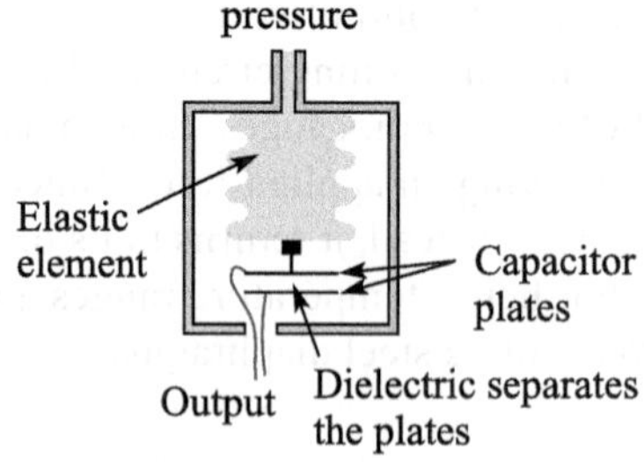

FIGURE 10-21 Variable capacitor pressure detector

10-8 Pressure Control Systems

The most common types of industrial process systems that employ pressure are hydraulic, pneumatic, vacuum, static, and steam pressure distribution systems.

Hydraulic Systems

Many types of machinery used in the manufacturing industry are powered by hydraulic pressure. Most types of hydraulic systems recirculate an oil-based fluid, as shown in Figure 10-22. The illustration shows a double-acting hydraulic system that controls a punch press.

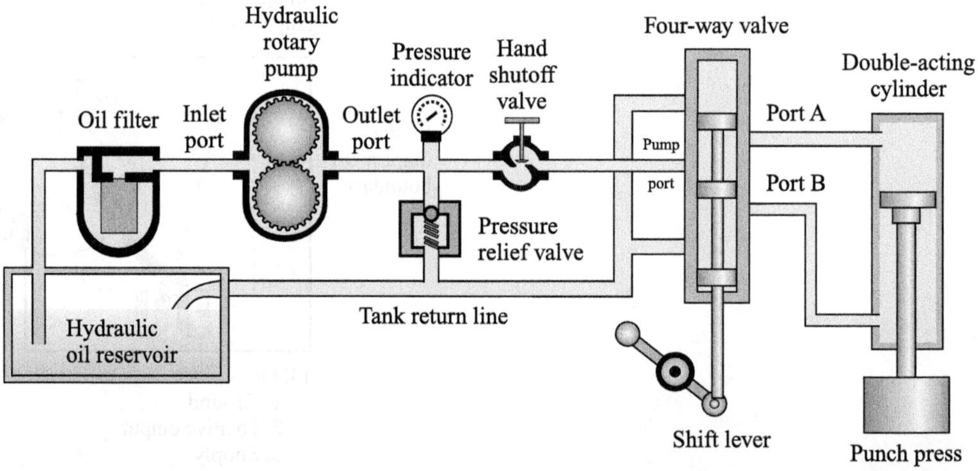

FIGURE 10-22 Hydraulic system

The figure shows a motor-driven rotary pump as the energy source of the system. It converts electrical energy into mechanical energy. With each revolution of the pump, a fixed amount of hydraulic fluid enters the inlet port from the reservoir. The liquid is set into motion by being forced through the outlet port into the system through transmission pipes or flexible tubing. This outlet port is called the *pressure line*. The hydraulic fluid is in the state of ready operation as it circulates throughout the system.

A filter in the feedline is placed between the reservoir and the pump. In this position, it removes any dirt particles or contaminants from the oil before it enters the system. Its purpose is to extend the life of the system.

As fluid flows through the system, it encounters resistance due to friction from surface areas of the transmission lines and from various components. The pressure is developed as fluid is forced against the surface areas. Indicator instruments such as pressure gauges and flowmeters are placed throughout the system to show different operating conditions, or to monitor the components in order to show they are functioning properly.

The mechanical control of the hydraulic system is achieved by a number of components, such as directional valves, flow control valves, and regulators. They provide either full or partial control of system fluid.

The directional control valve, or four-way valve, alters the directional flow path of the fluid. It consists of a valve body with four internal passages and a sliding spool that connects and disconnects the passages. When the spool is moved to the extreme bottom position, the pump port is connected to Port A, and the tank return line is connected to Port B. Pressurized fluid enters the cylinder at Port A and fluid is forced out of Port B into the return lines, causing the rod to extend. When the spool is in the extreme top position, the pump port is connected to Port B and the tank return line to Port A. Fluid enters Port B and exits Port A, causing the rod to retract.

The double-acting cylinder is used to control a punch press and serves as the load device of the system. It performs work by changing the mechanical energy of hydraulic fluid into linear motion that moves the ram of the press. Ports are located at each end of the cylinder body through which fluid can enter and exit, thus allowing the piston rod to move in two directions.

The hand shutoff is a flow control valve. By turning the knob, the amount of fluid flow in both directions can be adjusted between maximum flow and no flow. When the valve is fully open, maximum fluid will flow; when fully closed, no fluid will flow.

A pressure relief valve is connected between the pump output port and the reservoir. It consists of a valve body and a spool that is biased by a spring. When the pressure at the pump end of the spool opposite the spring is high enough, the spool is pushed open. A path is created for flow between the pump and the tank. The purpose of the relief valve is to prevent excessive pressure from building up in the system. This function is accomplished by providing a route for the fluid between the pump and the tank when the flow paths downstream are blocked. This situation would occur under the following conditions: the cylinder is fully extended or retracted, the flow control valve is fully closed, or the four-way directional control valve is in a position that blocks flow to and from the cylinder.

Pneumatic Systems

One of the major applications of pneumatic pressure systems is on mass production assembly lines. The compressed air of these systems provides the power for industrial processes that require high forces or high-impact blows to produce products.

Figure 10-23 illustrates a pneumatic system that operates a hand tool. An air compressor unit serves as the energy source of the system. It forces surrounding air into a tank under pressure. The tank where the compressed air is stored serves as a reservoir until it is eventually distributed into the system when needed. In an industrial setting, the compressor is driven by an electric motor. Portable pneumatic systems use internal combustion engines. The compressor forces the air into the tank, stopping when the pressure reaches a certain level. The pressure must be maintained during system operation. As the air is used, the pressure drops. If it falls below a predetermined level, the compressor turns on until enough air is replaced.

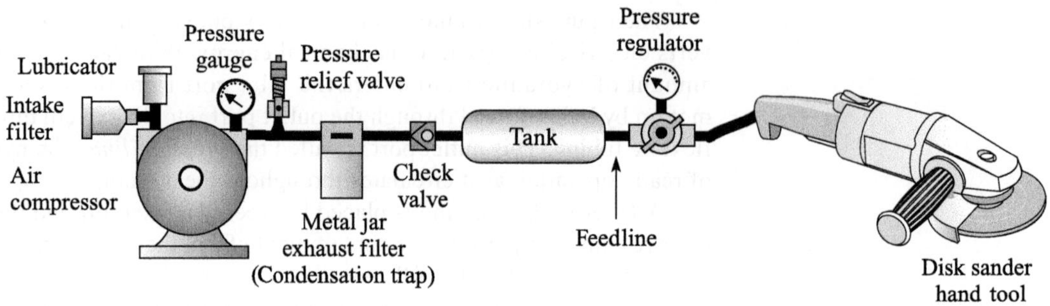

FIGURE 10-23 A pneumatic system that drives a rotary actuator

A feedline from the tank provides air for distribution throughout the plant. Solid pipes, tubing, and flexible hoses are used to transmit pressure in the system. The air is not recirculated and returned to the tank. Instead, it is released from the system back into the atmosphere.

Before the air enters the compressor, it must first be conditioned. Conditioning involves the removal of dirt by an air intake filter, the removal of moisture by a condensation trap, and the injection of a fine oil mist to provide lubrication for moving parts. Figure 10-23 shows the filter and lubricator at the inlet of the compressor and a metal jar exhaust that collects moisture from oil vapors produced by the lubricator. The pressure gauge monitors system pressure. The pressure relief valve vents air into the atmosphere if the pressure in the system becomes excessive. The check valve prevents high air pressure from returning to the compressor from the tank due to backflow. The tank is a reservoir that holds pressurized air for intermittent usage. The pressure regulator is a variable pressure valve that operates by restricting and blocking flow to the working portion of the circuit. The actuating speed of the load can be regulated to different speeds by adjusting the flow control setting of the valve.

The load of the system that performs work is a rotary actuator that is capable of variable speed control. This type of load device would be used in industrial applications such as buffing, drilling, grinding, and mixing. Pneumatic load devices are also designed to produce linear motion to perform work. For example, a double-acting cylinder is used in industrial applications such as hoisting, elevating, pile driving, and clamping.

The pneumatic system operation is monitored by pressure indicators placed at strategic locations. Their readings provide information on system operation and troubleshooting.

Vacuum Systems

Any enclosed space containing air or other gas at a pressure lower than atmospheric pressure is defined as a *vacuum.* Just as with compressed air, a vacuum condition can be used to perform various types of work applications.

Figure 10-24 shows a diagram of an On/Off cycling vacuum system. All of the components and pipes are enclosed and isolated from the outside atmosphere. A vacuum pump removes air from the system. A storage tank is used to accumulate vacuum for on-demand power needs. The vacuum pressure in the tank is monitored by a measuring sensor. If the vacuum pressure is not great enough, the sensor develops a signal that turns the pump on. The vacuum pressure lowers when actuator devices pull air into the system as they perform work, or from any leaks that may develop in the system. When the sensor detects that the vacuum pressure has reached a preset level, its output signal changes and turns off the pump. The usual range between the turn-on and turn-off points is about 5 to 15 in. Hg. The reason for using a reservoir tank is that it permits On/Off operation, which allows the vacuum pump time to cool down.

A vacuum relief valve provides protection from an excessive vacuum condition. The check valve installed between the vacuum pump and the tank allows airflow out of the tank. This one-directional component prevents the backflow of air into the tank. An intake filter is used to prevent foreign particles such as dust or sand from entering the pump mechanism. A bottle-like tank called a *liquid trap* uses gravity to prevent liquid materials from being sucked into the pump.

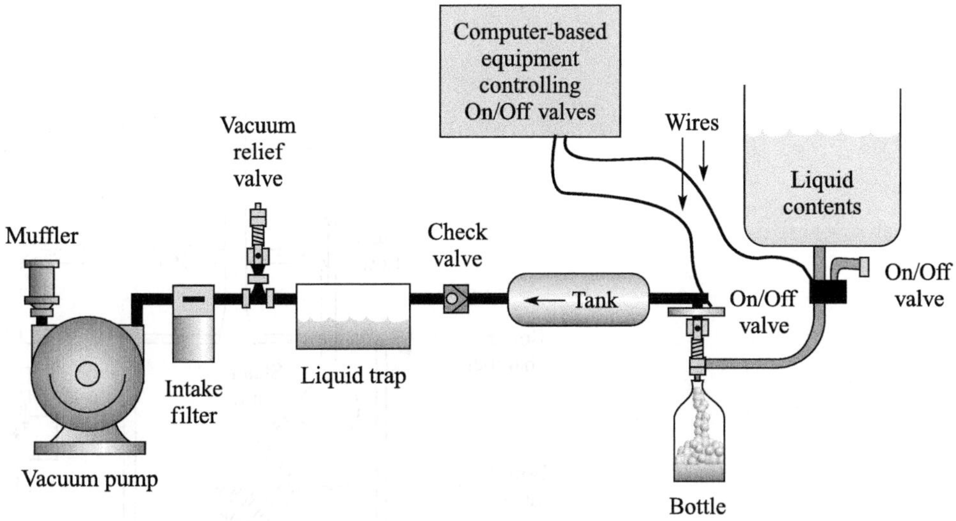

FIGURE 10-24 A vacuum system used to fill bottles with liquid

The vacuum system performs work by creating a pressure differential between the airtight surfaces of the equipment. As controlled actuators open vents, the outside air creates a mechanical force as it is sucked into the system. This force is able to produce different types of work operations.

The actuator in this system is an injector device that fills a bottle with liquid. A vacuum tube is placed in the center of a drain tap. At the moment liquid is dispensed into the container, the vacuum tube is activated to suck air out of the bottle. The filling operation is much faster because the liquid is literally forced into the bottle, rather than filling because of gravity.

Static Pressure Systems

Static pressure systems are also referred to as *hydrostatic.* They are used for industrial applications where fluids are distributed during the manufacturing process. These fluids can include liquids, gases, or solids (such as powders) that flow. Specifically, hydrostatic systems are used for batch processing applications such as mixing or blending operations that occur for a limited period of time.

The pressure developed is the result of the fluid source elevated above the working section of the system. The fluid is usually held in a storage tank where it is stored until it is needed. The depth and density of the fluid develop a force at the bottom of the tank called *static head.* When the fluid is released, this pressure is required for the distribution of fluids to locations throughout the system.

The variety of control components used in the static system is limited to flow control/shutoff valves. Since cylinders or other types of actuators are not used, there are no load components that perform work. Instead, the load is the resistance of the entire system and the only work function that exists is the result of heat generated due to friction. The operation of the system is monitored by strategically placed pressure and flow measurement instruments.

A batch process static pressure system that manufactures soft drinks of different flavors is illustrated in Figure 12-1 in the chapter on flow control. Ingredients are stored in elevated containers. As the ingredients are needed, valves open to allow drainage into a mixing tank.

Steam Pressure Systems

Steam pressure is used in industry for a variety of purposes. It is used as a heat source for food processing, chemical processing, refining, or simply for warming the plant facility. Steam pressure is also used to perform some types of work, such as driving a turbine to generate electricity. Steam pressure is developed by applying heat to water. Water is transformed into a vapor, which creates a pressure as it expands throughout the system.

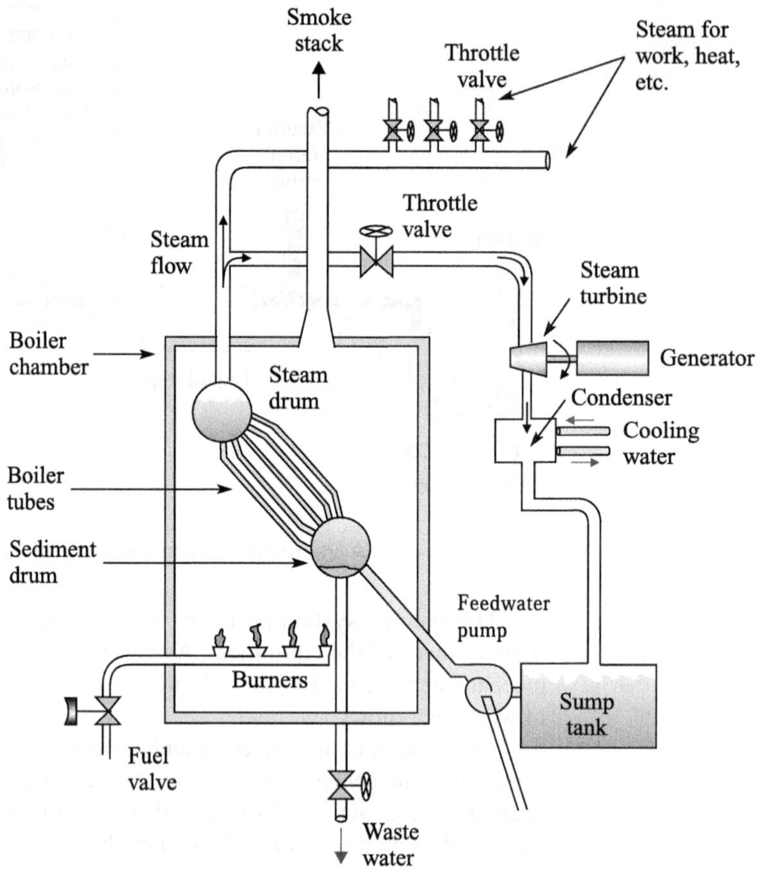

FIGURE 10-25 Steam boiler

The energy source of the steam pressure system is a boiler. The water tube boiler design illustrated in Figure 10-25 is a very popular method of developing steam pressure. Tubes that contain water are placed inside a sealed metal chamber that is heated by a furnace unit. The tubes are exposed to high temperatures by the surrounding air inside the chamber. Electricity or fossil fuel such as coal, oil, or natural gas is the energy source for the furnace.

As the water inside the tubes is heated, it changes into pressurized steam, which is then forced through pipes and tubes that serve as transmission lines. As the steam flows throughout the system during the production process, its pressure and flow are controlled by directional and flow control valves. Pressure gauges monitor the steam at strategic locations and may provide proportional electrical signals as input data to control equipment for automation purposes. The pressure in a steam system can be described as *head pressure.*

In some systems, the steam is recirculated. After passing through the actuator section, it is condensed back into water before returning to the boiler to be used again.

Problems

1. Which of the following is defined as fluid? ____.
   a. liquid
   b. gas
   c. steam
   d. air
   e. all of the above
2. To perform work such as that of a hydraulic press or pneumatic air drill, pressure must be __________ (increased, decreased).
3. To lift individual sheets from a stack of papers, pressure is ____.
   a. increased
   b. decreased
4. Liquids __________ (are, are not) compressible.
5. Water has __________ (more, less) density than an equal volume of mercury.
6. Pressure is measured as a force per unit of ____.
   a. area
   b. volume
   c. time
   d. mass
7. A decrease in pressure will change some types of ____ into a ____.
   a. gas(es)
   b. liquid(s)

8. The hydrostatic pressure of 1 square inch of water 100 feet deep is how many psi? If the liquid is ethyl alcohol instead of water, the hydrostatic pressure is how many psi?
9. A decrease in temperature causes the head pressure of a liquid in a confined container to ____.
   a. increase
   b. decrease
10. The head pressure of 1 liquid gallon in Mexico City is ____ the pressure of an identical open vessel in Miami.
    a. less than
    b. greater than
    c. the same as
11. Suppose a liquid at 60 degrees Fahrenheit has a specific gravity of 6.28. If the measured pressure at the bottom of the tank in which it is stored is 51.3 psi, what is the height of the liquid?
12. A liquid with a specific gravity less than 1 is ____ than water.
    a. lighter
    b. heavier
13. A temperature rise ____ the density of a gas.
    a. increases
    b. decreases
    c. has no effect on
14. The pressure exerted by a gas in an open container will ____ when its temperature rises.
    a. increase
    b. decrease
    c. stay the same
15. A condition where air is forced into a confined container is called ____.
    a. compression
    b. a vacuum
16. As air is removed from a confined container, the inside pressure __________ (increases, decreases).
17. List the reference points of the following pressure scales:
    __________ Gage
    __________ Absolute
18. Convert an absolute pressure of 64.7 psia to gage pressure.
19. At sea level, 14.7 psi of atmospheric pressure will support __________ inches of mercury in a glass tube.
20. Of the two columns in a U-shaped manometer, the one with the lowest level of liquid has the ____ pressure applied to it.
    a. lowest
    b. highest
21. What is the gage pressure measured by a manometer using mercury if the reference column rises 1.5 inches and the pressure column lowers 1.5 inches?
22. What are the inches in H_2O value if a measurement of 6.8 psi is made?
23. A (lower, higher) __________ pressure detected by a Bourdon tube causes the coil to unwind.
24. Describe the purpose of variable resistors R_1 and R_2 in Figure 10-19.
25. Identify the output terminals of the transverse voltage strain gauge in Figure 10-20.
26. What is the power source of a hydraulic system?
27. What is the power source of a pneumatic system?
28. Air is pumped __________ (out of, into) the storage tank of a vacuum system.
29. What is the power source of a static pressure system?
30. What is the energy source of a steam pressure system?

CHAPTER

11

Temperature Control

OBJECTIVES

At the conclusion of this chapter, you should be able to:

- Define *thermal energy*.
- Explain the law of thermodynamics.
- List the three types of heat transfer.
- Describe the operation of the following heat sources of thermal energy:

Blast Furnace	Arc
Electronic Heat Element	Resistance Induction

- Describe the operation of a cold thermal energy source (refrigeration system).
- Define *temperature*.
- Identify the Fahrenheit and Celsius scales and convert specific values from one scale to another.
- List several reasons for monitoring temperatures in process control applications.
- Define *BTUs*.
- Describe the principle of operation of each of the following temperature-indicator devices:

Crayons	Pellets	Liquid Crystals
Paints	Labels	

- Describe the principle of operation for each of the following mechanical temperature measurement instruments:

Bulb Thermometer	Bimetallic Thermometer

- Describe the principle of operation for each of the following electrical measurement instruments:

Thermocouple	Thermistor
Resistance Temperature Detectors (RTDs)	Radiation Thermometry

INTRODUCTION

Many products manufactured today are the result of a process that involves temperature control. High temperatures may be used to soften metals before they are formed into a desired shape or to melt plastic in an injection molding machine. Low temperatures are necessary to preserve dairy products in the food processing industry. Manufacturing processes that are affected by temperature are referred to as *thermal systems*.

11-1 Fundamentals of Temperature

Scientific theory states that molecules of matter are in continuous motion due to kinetic energy. Molecular movement creates heat known as **thermal energy.** Thermal energy is measured in temperature. Suppose one end of an object is exposed to the elevated temperature of a flame. The object's molecules in contact with the flame will move more rapidly and create heat. The heat transfers from the heated area to the cooler areas throughout the object. Thermal equilibrium is attained when the object's temperature reaches the elevated temperature. Thermal movement from hot to cold is called **thermodynamics.** Each type of matter has an ignition point at which a chemical reaction (called a *fire*) causes it to burn. When a cold theoretical temperature (called *absolute zero*) is reached, the molecular movement stops and no heat is generated.

11-2 Thermal Control Systems

Thermal systems supply thermal energy from a source, provide a path for its distribution, and convert the energy into some kind of work.

Temperature control is maintained at a desired level either manually by a human operator or automatically. In an automated system, the thermal energy is regulated by a controller. The control function is accomplished by altering the flow of thermal energy from the energy source to the load device that performs the work. There are two types of control methods: On/Off and proportional. The On/Off controller directs energy from the source when the load device's temperature falls below a certain level and stops the flow of energy to the load device when the desired temperature has been reached. An example is a home heating system. The proportional control method provides only the amount of energy from the source that the load needs. Both types of controllers respond to the temperature difference between the setpoint and the measured value.

11-3 Thermodynamic Transfer

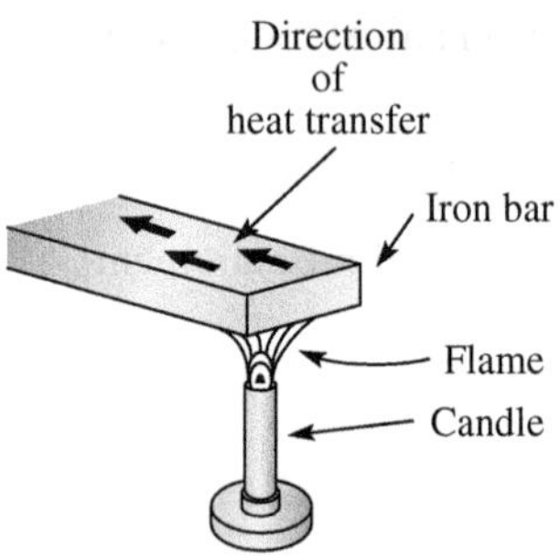

FIGURE 11-1 The heat conduction principle

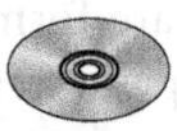

ELE706 Thermodynamic Transfer

The transmission path of thermodynamic transfer can be through materials that are solids, liquids, gases, or a vacuum. The process by which heat is transferred by a solid is called **conduction.** Figure 11-1 shows an example of conduction. One end of a metal bar is placed over an open-flame heat source. The molecules over the flame move more rapidly. The increased molecular velocity causes collisions with neighboring molecules, which in turn causes them to move more quickly. This action continues until the molecular movement throughout the bar has increased. The best solid-material thermal conductors are metals. Some types of non-metal solids are insulators.

Since most fluids are poor conductors of heat, very little thermodynamic transfer takes place through the process of conduction. The transfer of heat through fluids such as liquids and gases takes place through a process called **convection.** When a container of fluid is placed above a heat source, the bottom layer begins to expand. The warmer fluid becomes less dense than the fluid above it and therefore moves to the top of the container. The cooler fluid—which is heavier—goes to the bottom. In this manner, heat is transferred through the constant movement of circulating currents, as shown in Figure 11-2.

Thermal energy can also be transferred through a vacuum by a process called **radiation.** In theory, bundles of energy are radiated away from atoms in the heat source as wavelike patterns that travel at the speed of light. Radiated heat transfer also takes place through air. A prime example is the sun, which radiates heat through the vacuum of space and through the Earth's atmosphere, as shown in Figure 11-3.

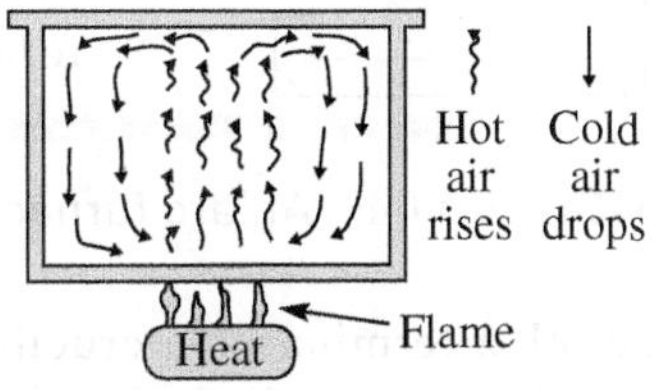

FIGURE 11-2 The heat convection principle

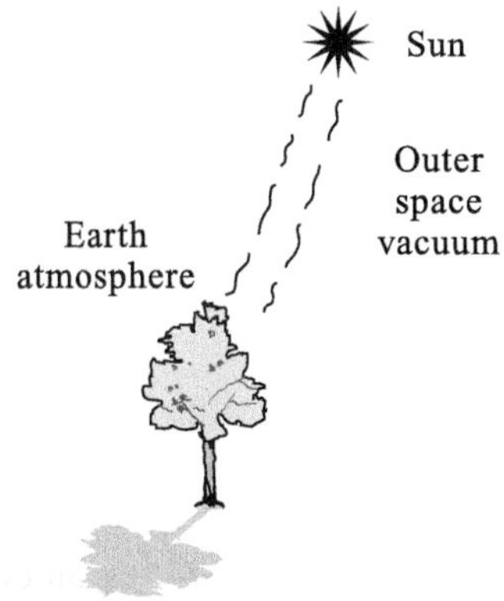

FIGURE 11-3 The heat radiation principle

11-4 Thermal Energy Source

Thermal energy is primarily produced by a change in the state of matter. For example, when a fuel such as coal is combined with oxygen at a high enough temperature, it ignites into a chemical reaction called a *fire*. As it burns, it changes into another material called *carbon dioxide*. The combining of carbon and oxygen causes heat to be released. Thermal energy is also released by matter changing form. For example, when a liquid changes to a gas, evaporation occurs and the gas becomes colder than when it was in liquid form. When a gas is converted to a liquid by being compressed, the liquid becomes warmer than when it was a gas.

This section describes several common types of thermal energy sources used in industrial applications. This study will include hot and cold temperature systems.

Industrial Furnaces

The most common heat source for many manufacturing operations is the furnace. It is usually built of metal and brick because these materials will withstand high temperatures. The shapes of industrial furnaces vary, but all of them confine the heat generated inside their chambers. Industrial furnaces provide high temperatures to harden metals by a process called *heat treating*. These furnaces are also designed to melt materials, such as iron in a foundry. Industrial furnaces provide heat to heat-treat materials, whether by cooking food or hardening metals. They are also designed to melt materials, such as butter to pour over popcorn or iron for foundry use.

Combustion Furnace

The most common heat source in industry is the combustion furnace. It usually combines fossil fuel and oxygen at a high temperature to produce heat. Smaller furnaces are ideal for heat-treating machined gears to make their metal harder. For a higher temperature, the blast furnace is used. The high temperature is achieved by forcing large amounts of air into the burning chamber. The blast furnace is used to manufacture products such as glass, steel, and cement.

Systems that use steam from boilers to develop pressure are also used to supply heat. Fossil fuels are often burned to convert the water into steam. Boiler systems provide very safe and accurate temperature control.

Electric Furnaces

Furnaces can also be heated by electricity. There are three types of electrical heating systems commonly used in industry: arc, resistance, and induction.

Arc Furnaces The arc furnace is used in the process of smelting steel in a foundry, as illustrated in Figure 11-4. The voltage is applied by connecting one power supply terminal to

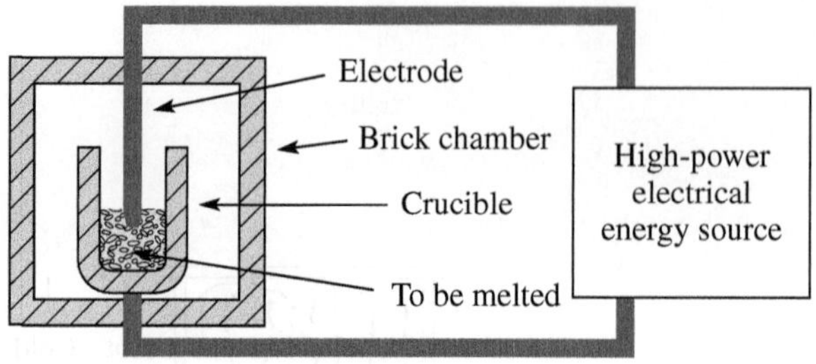

FIGURE 11-4 An arc furnace

an electrode and the other terminal to a crucible made of a conductive material such as graphite. When the voltage is applied, the electrode is in contact with the metal to be melted. Current begins to flow and an arc is formed at ignition. The electrode is placed a small distance from the metal. To keep the arc burning, the heat from the arc melts the metal to a liquid state.

Resistance Furnaces Resistance furnaces use heating elements similar to those in a kitchen oven. The element has large resistance, which produces heat when current flows through. The elements are placed inside an insulated chamber where all surfaces of the object or material being heated are uniformly exposed to heat. This type of furnace is used in batch processing for heat-treating purposes. It is also used in burn-in tests where integrated circuits are exposed to high temperatures for several hours to test their reliability.

Induction Furnaces Induction furnaces are also used to melt metals. Again, insulated chambers are used to hold the work to be heated. Coils of wire are wrapped around the chamber and AC current is applied. A magnetic field constantly expands and contracts around the coils. As the field sweeps across the iron, an eddy current is induced, which causes molecules to move around. As the molecules shift positions, an intense heat results in a very short period of time.

Cooling Systems

The most common type of refrigeration system is the household refrigerator. Refrigeration systems are used in the food processing industry to keep perishable products at low temperatures. They are also used in other manufacturing fields, such as chemical plants, to cool liquids to required levels before they are used in a blending operation.

The principle of refrigeration can best be explained by describing an old icebox used in the home 75 years ago. This unit was a wooden box insulated with cork or sawdust. Chunks of ice were placed in an upper compartment and the perishables underneath. The cooling process works on the principle that heat energy from warmer objects transfers to cooler objects nearby. In the icebox, heat from the food moves to the ice. The transfer of heat continues until everything inside reaches the same temperature.

The modern refrigerator operates on the principle that evaporation absorbs heat energy, and condensation gives off heat energy. By controlling these two reactions mechanically, refrigerators are able to perform the same functions as the iceboxes they replaced. Freon is evaporated and condensed in the refrigerator system. This refrigerant is used because it evaporates quickly when exposed to room temperature.

The refrigeration system is illustrated in Figure 11-5(a). Its components include a compressor, a condenser, a capillary tube, an evaporator, and two fans. The compressor is the device that pumps the refrigerant throughout the system. The freon in liquid form enters the capillary tube, shown in Figure 11-5(b), through a port with a small diameter. As the freon leaves the small capillary tube, it enters a larger tube called the *evaporator*. The sudden increase in tube size creates an abrupt drop in pressure which causes the liquid freon to evaporate into a gas and cool down. As the cold gas flows through the evaporation coils, it

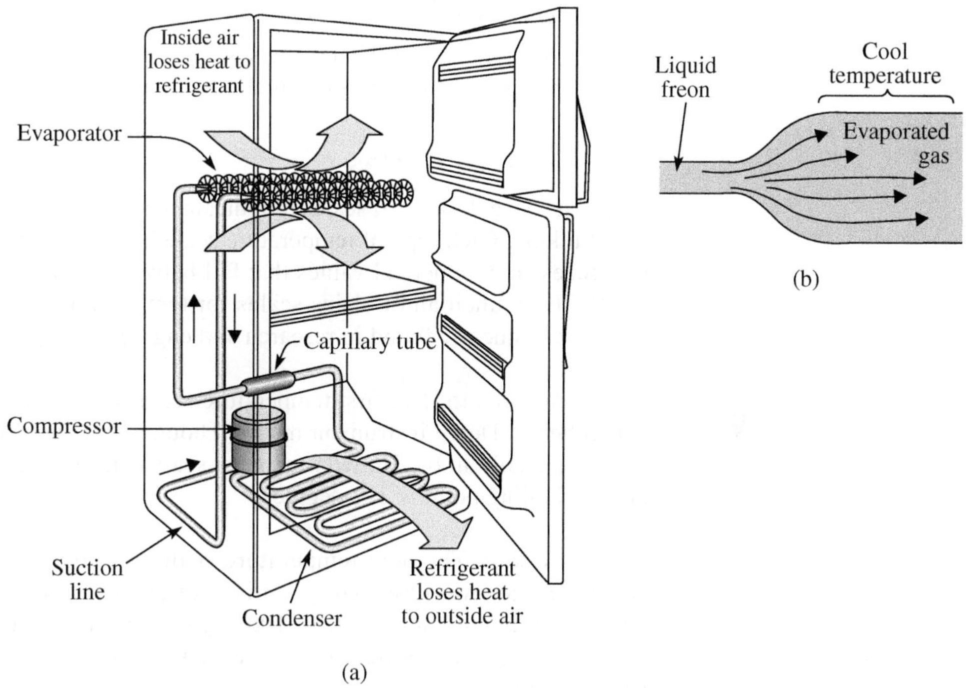

FIGURE 11-5 A refrigeration system

absorbs heat from the contents in the refrigerator compartment. The heat energy transfer is assisted by an evaporator fan. As the compressor runs, it creates a suction and draws the warmer gas into its inlet port. Inside the compressor, the gas is compressed into a liquid. The conversion from a gas to a liquid state also generates heat. As the liquid freon is pumped from the compressor outlet port, it flows through a series of folded tubes called a *condenser*. The condenser is a radiator that releases heat from the warm freon flowing through the tubes to the outside. The heat transfer is assisted by the fan that blows air over the condenser coils.

If the refrigerator compressor were to continue running, the food would freeze. To keep the temperature within a desired range, a thermostat is used to turn the compressor on and off. When the temperature rises to a certain level, the thermostat activates a switch to turn the compressor on. The thermostat turns the compressor off when the temperature reaches the desired lower level.

11-5 Temperature Measurements

There are many manufacturing applications that require precise measurements of temperature. This function is performed by instruments placed at the energy source, the controller, or the system load. These devices provide visual indication, or a mechanical or electrical feedback signal in a closed-loop system for automated control.

The components of nonelectrical instruments are physically altered as they respond to temperature changes. Electronic instruments are designed to produce electronic signals proportional to variations of temperature.

To provide good control of industrial processes, accurate measurements of temperature are essential. There are several reasons to monitor temperatures in process control applications:

1. Precise temperature conditions are required when combining two chemicals to form a compound.
2. Over-temperature conditions that could cause excessive pressure must be avoided in an enclosed system to prevent ruptures or explosions.

3. Temperatures must be kept below freezing to prevent stored food from spoilage.
4. By ensuring that the heating system is consuming energy efficiently, fuel costs can be minimized and environmental conservation concerns can be met.

Temperature Scales

Scientists have developed scales to indicate temperature, such as Fahrenheit, Celsius, Kelvin, and Rankine. Each type of temperature scale has fixed reference points at which water boils or freezes, and numerical values that fall between those points. Most industrial applications use the Fahrenheit and Celsius scales for temperature measurements. The other two scales are most frequently found in research and engineering applications.

ELE3108
Temperature Scale

Fahrenheit Scale The first temperature scale was developed in the early 1700s by Gabriel Fahrenheit, a Dutch instrument maker. Though modified from its original form, it is widely used, especially in the United States. At sea level, the freezing point of water is 32 degrees, and the boiling point is 212 degrees.

Celsius Scale The next temperature scale developed was designed by Anders Celsius. A similar scale was designed by Christin of Lyons, who named it the *Centigrade* scale. Both designs use a numerical value of 100 degrees for the boiling point of water and 0 degree for the freezing point of water at sea level. This scale is referred to as either the Centigrade or Celsius scale.

Each of these temperature scales can be converted to the other by the following equations:

$$°C = 5/9(°F - 32)$$
$$°F = (9/5 \times °C) + 32$$

EXAMPLE 11-1 Convert 77 degrees Fahrenheit into its equivalent temperature value in Celsius.

Solution

$$\begin{aligned} °C &= 5/9\ (°F - 32) \\ &= 5/9\ (77 - 32) \\ &= 5/9\ (45) \\ &= 25 \end{aligned}$$

EXAMPLE 11-2 Convert 10 degrees Celsius into its equivalent temperature value in Fahrenheit.

Solution

$$\begin{aligned} °F &= (9/5 \times °C) + 32 \\ &= (9/5 \times 10) + 32 \\ &= (18) + 32 \\ &= 50 \end{aligned}$$

Heat is thermal energy that has the ability to perform work. Thermal energy is rated in work units called *calories* and *BTUs* (British thermal units). A *calorie* is the heat required to raise the temperature of 1 gram of water 1 degree Celsius. A BTU is the amount of heat required to raise the temperature of 1 pound of water 1 degree Fahrenheit.

EXAMPLE 11-3

Calculate how many calories are required to raise the temperature of 5 grams of water 2 degrees Celsius.

Solution

$$\text{Calories} = (H_2O)\ \text{Weight} \times \text{Degrees (Celsius)}$$
$$= 5\ \text{grams} \times 2°\text{C}$$
$$= 10$$

EXAMPLE 11-4

Calculate how many BTUs are required to raise the temperature of 10 pounds of water 5 degrees Fahrenheit.

Solution

$$\text{BTUs} = \text{Weight}\ (H_2O) \times \text{Degrees (Fahrenheit)}$$
$$= 10\ \text{lb} \times 5°\text{F}$$
$$= 50$$

Differential Temperature

In some applications, the difference between two temperature measurements is used. Comparing one temperature to another is called *differential temperature.* An application of differential temperature measurements is the system in Figure 11-6, which monitors the efficiency of a heat exchanger used to heat water flowing through a pipe. Sensor T_1 measures the temperature of the water before it enters the exchanger, and sensor T_2 measures the hotter water that exits the exchanger. Signals from both sensors are sent to a differential temperature transmitter that measures the difference between T_1 and T_2 (ΔT). A flowmeter is placed in the piping system to measure the flow rate (Q) of the water.

Signals from the flowmeter (Q) and the differential transmitter (ΔT) are sent to a controller. The controller is programmed to multiply the gain in temperature ($T_2 - T_1$) by flow rate to indicate the relative efficiency of the system. The formula to determine relative efficiency, as a percentage, is

$$\text{Relative Efficiency} = \frac{Q\Delta T}{C} \times 100$$

An example of the heat exchange system in Figure 11-6 operating at 100 percent would be as follows:

> When the flow rate of the water is 20 cubic feet per minute, it rises from 180 to 200 degrees Fahrenheit as it passes through the heat exchanger. Suppose the flow rate doubles to 40 cubic feet per minute. Since the speed of the water doubles, it is inside the exchanger for half the time and therefore rises 10 instead of 20 degrees.

No heat exchanger system can operate at 100 percent efficiency. There is an acceptable limit as to how much the efficiency can deviate, usually established during its design phase. If this limit is exceeded, the system requires attention to what is causing it to stray.

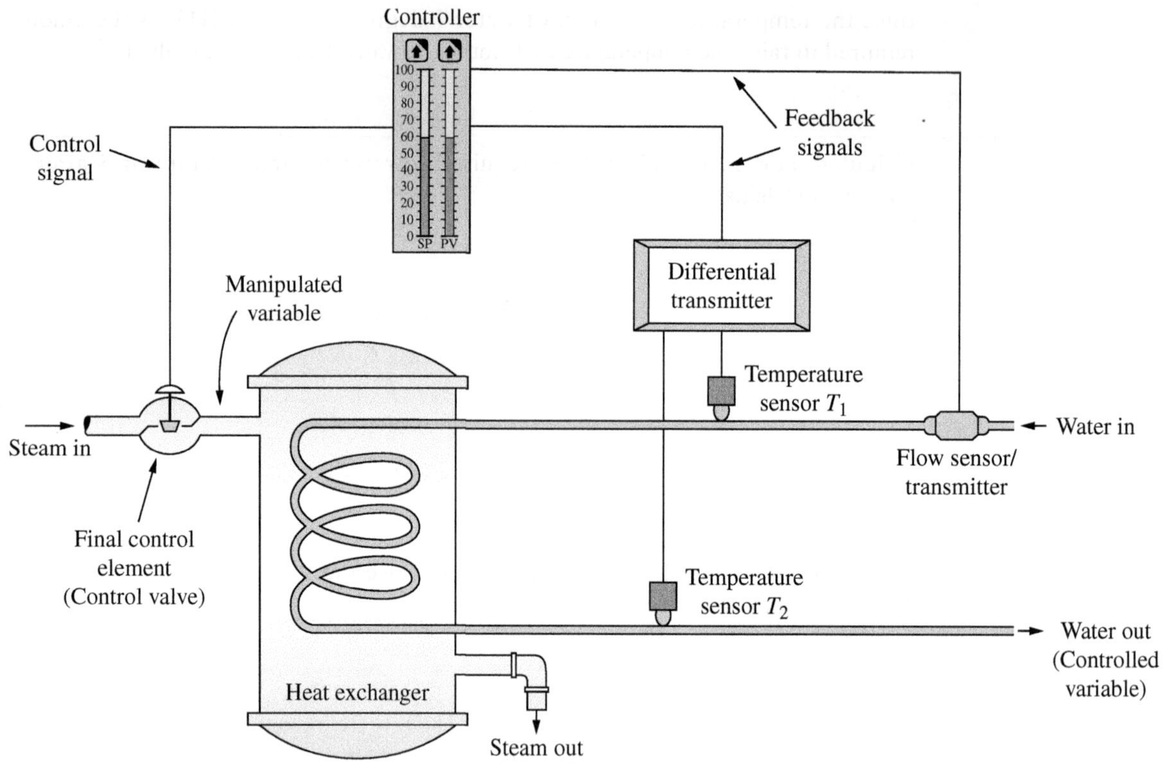

FIGURE 11-6 Differential temperature measurement

EXAMPLE 11-5

If the actual temperature of the water rises from 180 to 192 degrees Fahrenheit as 30 cubic feet per minute passes through the exchanger in the example given above, what is the relative efficiency?

Solution

Step 1:

$$C = (Q\Delta T)\text{ Ideal} = (20\text{ ft}^3/\text{min})(20°\text{F}) = (40\text{ ft}^3/\text{min})(10°\text{F}) = 400°\text{F} \cdot \text{ft}^3/\text{min}$$

Step 2:

$$(Q\text{ Actual})(\Delta T\text{ Actual}) = (30\text{ ft}^3/\text{min})(12°\text{F}) = 360°\text{F} \cdot \text{ft}^3/\text{min}$$

Step 3:

$$\text{Relative Efficiency} = \frac{(Q\text{ Actual})\ (\Delta T\text{ Actual})}{C} \times 100$$

$$= \frac{360}{400} \times 100 = 90\%$$

11-6 Temperature-Indicating Devices

A number of industrial situations require an indication that a predetermined temperature has or has not been reached. For such situations, several different types of heat-sensitive materials have been developed solely for indication and monitoring purposes. These temperature-sensing materials are made of crystalline solids.

The materials operate on the principle that when heating occurs, a temperature will be reached at which the solids change color or melt to a liquid. The change provides a visual indication that the necessary temperature has been reached. Temperature indicators are available in the form of crayons, paints, pellets, and labels. They are applied directly to or placed near the object being monitored.

These heat-sensitive indicators are accurate to within 1 percent and respond within a few tenths of a second. Because they are inexpensive, they are preferred in situations where they will burn, such as a heat zone of an industrial oven, ceramic kiln, or for products that travel on conveyors through a furnace.

Crayons

The crayon is available in stick form. They are manufactured in 100 different temperature ratings, ranging from 100 to 2500 degrees Fahrenheit. The workpiece is marked with the crayon. When the predetermined temperature is reached, the crayon liquefies, notifying the observer that the workpiece has reached that temperature.

Paints

A paint indicator is a lacquer that dries to a dull finish. When the predetermined temperature has been reached, its finish turns glossy and transparent. Paints are often used on very smooth surfaces to which crayons cannot stick.

Pellets

The pellet works on the same principle as crayons and paints. Pellets are used in applications where extended heating periods are involved or when oxidation of a workpiece might obscure a crayon or paint marking. Because pellets are bulkier than crayons or paint, they are used when visual indication must be observed at a distance.

Labels

The label shown in Figure 11-7(a) has one or more heat-sensitive indicators sealed under transparent, heat-resistant windows. Each indicator changes color at a specific temperature. Labels are available in nonreversible styles to show peak temperature and reversible styles to indicate changing temperatures.

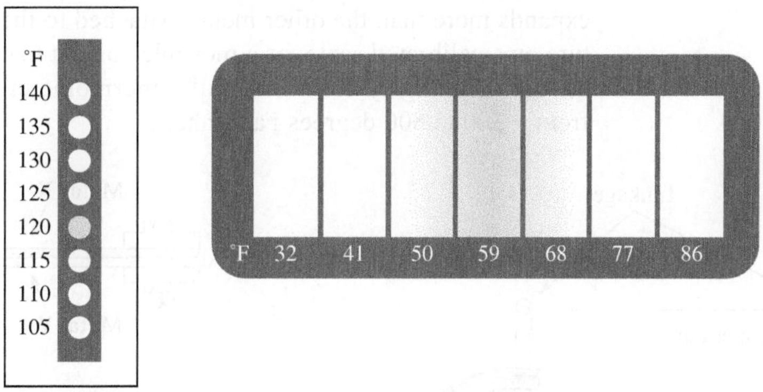

(a) (b)

FIGURE 11-7 Temperature indicators made of crystalline solids

Liquid Crystal Indicator

The liquid crystal indicator shown in Figure 11-7(b) uses crystal material sandwiched between an adhesive backing and transparent film. The crystals change to different colors indicating different temperatures. Temperature is read by observing which patch has changed to a specific color.

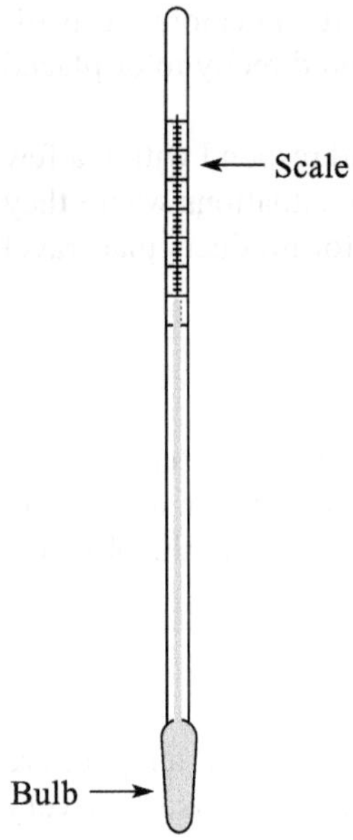

FIGURE 11-8 Liquid glass thermometer

Liquid-Filled Thermometers

Glass Thermometer

The thermometer shown in Figure 11-8 is a closed tube with a reservoir at the bottom and is partially filled with a liquid. It operates on the principle that materials expand when exposed to heat. For example, when the temperature increases, the liquid expands and rises in the glass tube. The temperature reading is taken by comparing the top level of liquid to the corresponding number on an adjacent temperature scale. Glass thermometers are very accurate and reliable. However, for industrial applications they are too fragile, are not capable of recording historical trends, and cannot provide feedback signals that are required for feedback control.

Filled-Bulb Thermometer

By modifying the thermometer principle, filled-bulb thermometers have been developed that are more durable than glass thermometers. They are also capable of providing feedback action for control purposes and recording temperature variations over a period of time. The filled-bulb thermometer is shown in Figure 11-9. The bulb is the primary sensor that detects changes in thermal energy. It also serves as the reservoir for the liquid or gas. For greater durability, the bulb is made of metal. The capillary is a tube that connects the bulb to the pressure–volume element. The element is a spiral tube that bends due to pressure changes in the filled-bulb thermometer system. The linkage is physically connected to the pressure–volume element. As the coil expands or contracts, it causes the linkage to move a needle over a temperature scale. The linkage may be attached to a pen that draws a line on a circulating chart with a temperature scale. The linkage can also be connected to an electrical component such as a potentiometer to provide feedback signals for a closed-loop system.

Bimetallic Thermometer

The bimetallic thermometer, as shown in Figure 11-10, is made of two dissimilar metal strips that are physically bonded together. Each metal has a different expansion ratio. As temperature changes, the strip will bend in the direction of the metal with the lower expansion rate. The deflection of the strip settles at a position that represents the temperature value. The strip can be attached to an indicator scale, recording chart, or linkage used to provide a feedback signal for a closed-loop system. To allow the bimetallic strip to be placed in a smaller space than a straight element requires, they are usually wound into a spiral, helix, or coil. For example, when the helix in Figure 11-11 is heated, it unwinds because one metal expands more than the other metal. Attached to the helix is a pointer that indicates temperature on a calibrated scale, or a movable contact that opens or closes a switch for a furnace or air-conditioning system. Bimetallic thermometers are capable of measuring temperatures from −300 to 800 degrees Fahrenheit.

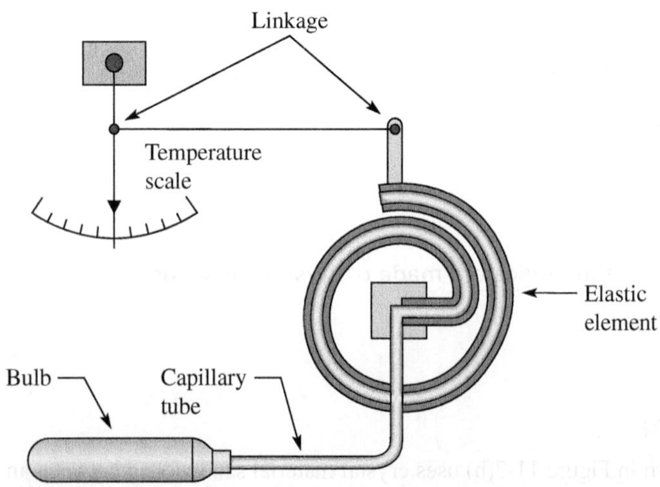

FIGURE 11-9 Filled-bulb thermometer

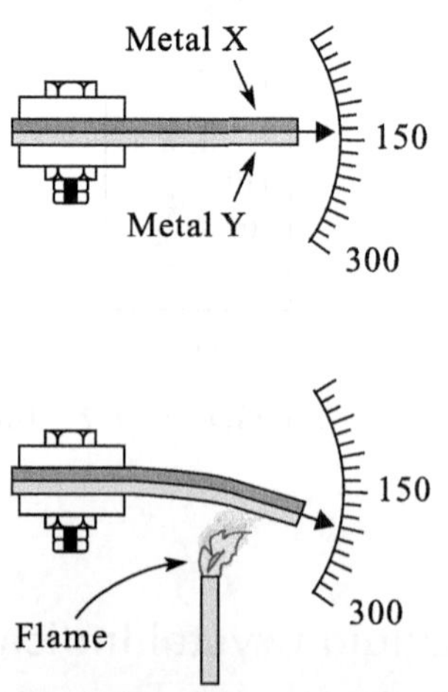

FIGURE 11-10 Bimetallic thermometer

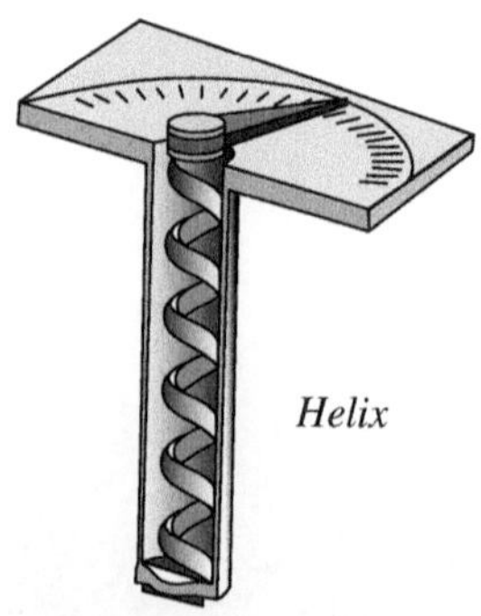

FIGURE 11-11 Helix-style temperature indicator

11-7 Electronic Sensors

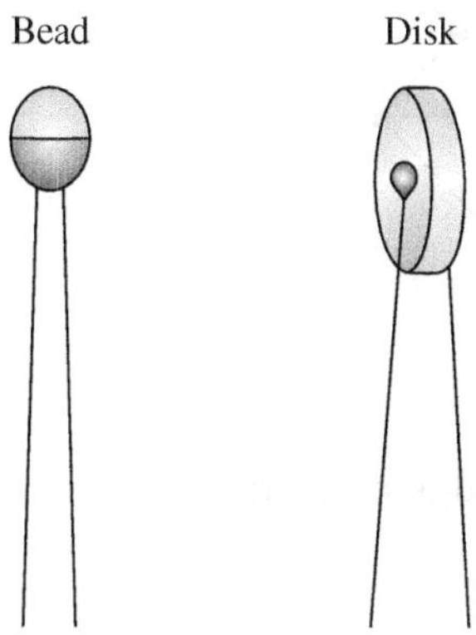

FIGURE 11-12 A thermistor body in the bead form

There are two general types of electrical sensors used to measure temperatures: thermoresistive and thermoelectric. Thermoresistive sensors—thermistors and resistance temperature detectors—change resistance as the ambient temperature varies. Thermoelectric sensors—thermocouples—produce a voltage proportional to the surrounding temperature. Table 11-1 (page 250) compares these temperature sensors.

Thermistor

One type of temperature sensor is the **thermistor.** Its name is a derivation of the term *thermal resistor.* The thermistor exhibits a large change in electrical resistance when subjected to a relatively small change in temperature. Temperature variations can be caused either by a change in the ambient temperature external to the thermistor or, internally, by a change in the current through the thermistor.

Thermistors are constructed by using a paste-like metal oxide mixture to form certain shapes, such as a bead (shown in Figure 11-12), disks, or rods. A set of two conducting wires is inserted into the paste. The mixture is hardened when exposed to heat by a process called *sintering.* The type of oxides, the proportions used, and the physical size of the thermistor determine the desired temperature and resistance ranges for the device.

Oxidized metals have characteristics similar to those of semiconductor materials. At lower temperatures, the valence electrons in the outer shell are strongly bound to each atom and function as good insulators. As the temperature rises, the thermal activity of the atom increases. Valence electrons gain sufficient energy to break away from the atoms. The electrons become free to take part in current that flows through the material. As temperature increases, more electrons become available and the resistance of the material decreases. This characteristic of the thermistor is called a **negative temperature coefficient.** The letters *NTC* are placed inside the thermistor symbol to indicate this characteristic.

How Thermistors Are Used

There are many circuits in which thermistors are used. A few of the more common applications are discussed below.

Temperature Measurement The primary function of the thermistor is to exhibit a change in resistance as a function of temperature. In measurement instruments, this resistance is often converted into a voltage reading by using a voltage divider as part of a voltage divider network. The diagram in Figure 11-13 shows that the output is taken across the fixed resistor.

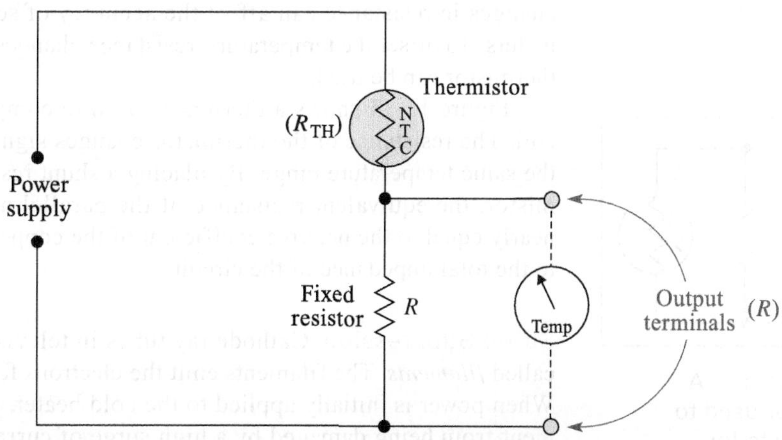

FIGURE 11-13 Temperature-measuring voltage divider

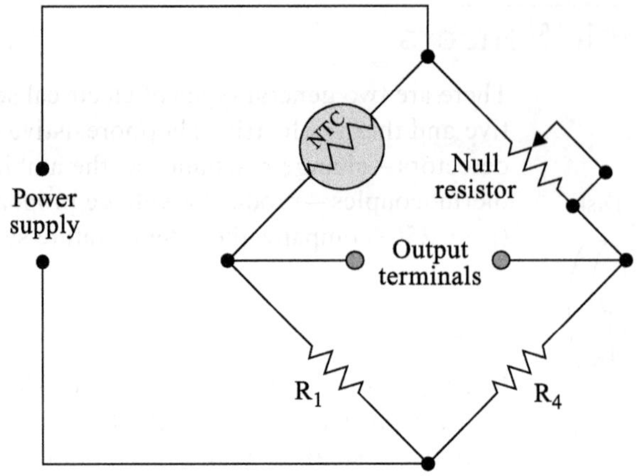

FIGURE 11-14 Temperature-measuring bridge network

As temperature increases, the thermistor resistance decreases. Therefore, the voltage across the resistor increases. If a meter with a scale that reads temperature is placed across the output terminals, its reading will increase as temperature increases. The output voltage as a function of temperature can be expressed by the following formula:

$$V_o = V_s \times \frac{R}{R + R_{TH}}$$

The resistance of variable R is the parallel equivalent resistance of the load connected to the output terminals and the fixed resistor.

When measuring ambient temperature, the circuit should be designed to keep the current through the thermistor low. If the current is too high, the thermistor will become much warmer than the surrounding temperature due to self-heating.

Another temperature measuring circuit is shown in Figure 11-14. R_1 and R_4 are precision resistors. Their values are selected to match the particular thermistor used. The variable null resistor is adjusted to balance the bridge so that the output is 0 volts. Based on bridge theory, a voltage at the bridge output will develop as the thermistor resistance changes.

Temperature Compensation The resistance of metals, such as copper, changes when subjected to temperature variations. These metals have a positive temperature coefficient. Such changes in resistance can affect the accuracy of sensitive measuring instruments, such as meters. To offset the temperature-resistance changes, the negative coefficient properties of a thermistor can be used.

Figure 11-15 shows a thermistor used to compensate for the resistance change of the coil. The resistance of the thermistor changes significantly more than that of the coil over the same temperature range. By placing a shunt resistor of the proper value across the thermistor, the equivalent resistance of the parallel network provides a negative coefficient nearly equal to the positive coefficient of the copper. The network adds less than 15 percent to the total impedance of the circuit.

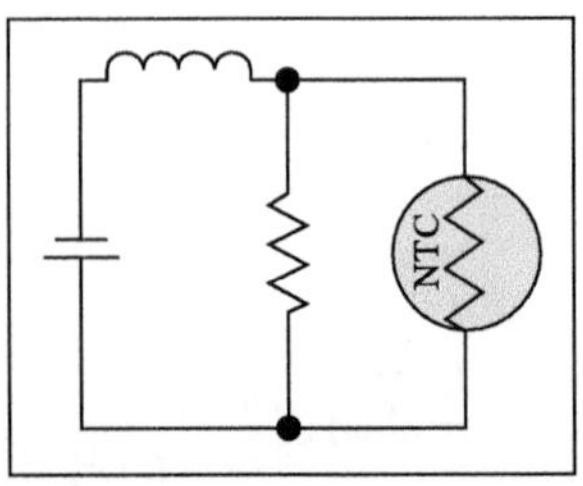

FIGURE 11-15 A thermistor used to compensate for temperature variations

Surge Suppression Cathode ray tubes in televisions and oscilloscopes use heater coils called *filaments*. The filaments emit the electrons for the beam that scans the display screen. When power is initially applied to the cold heater, its resistance is low. To prevent the filament from being damaged by a high surge of current, a thermistor is placed in series. The high starting resistance of the thermistor limits the current to a safe value. As current begins

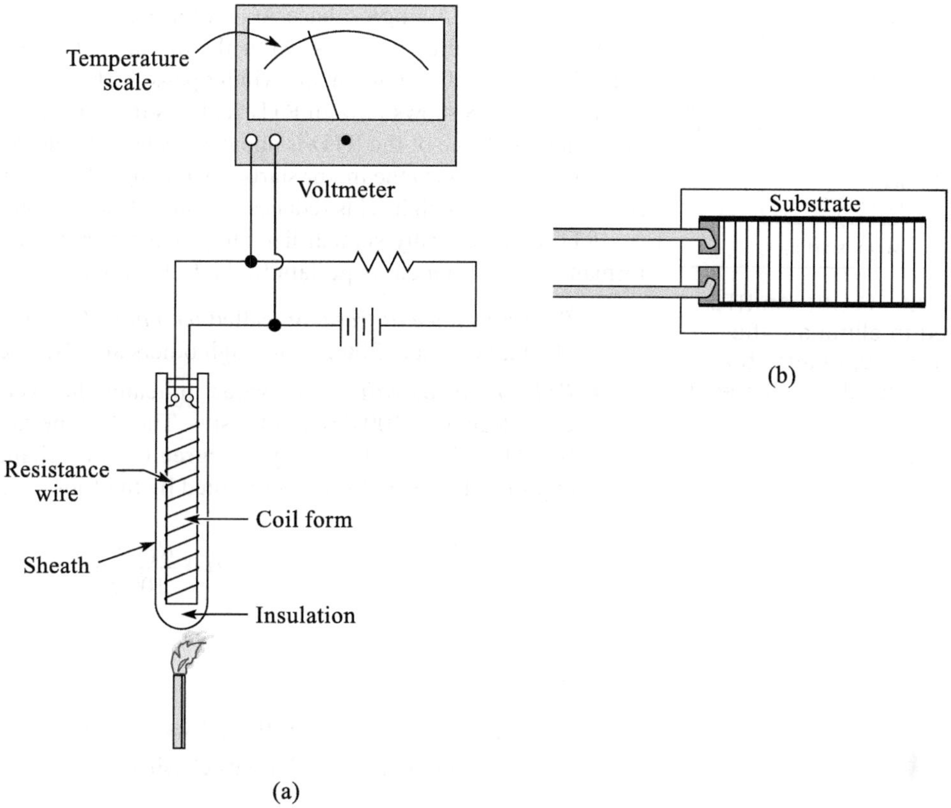

FIGURE 11-16 Types of RTDs

to flow, the thermistor self-heats and its resistance is reduced. At the same time, the temperature and resistance of the filament increase to normal operating values.

Resistance Temperature Detectors

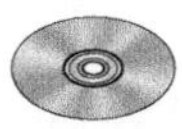

ELE2808
Resistance Temperature
Detector

The resistance of electrical conductivity metals varies directly with temperature. Therefore, metals have a **positive temperature coefficient (PTC).** This means that as their temperature increases, their resistance increases.

Some types of metals are used in a temperature-sensing device called a **resistance temperature detector (RTD).** Two metals commonly used in RTDs are nickel and platinum. Nickel is the most sensitive metal because it provides the greatest change of resistance for a given unit of temperature change. Platinum is used in applications that require a resistance change over wider temperature ranges.

RTDs are constructed by placing a coil of fine wire inside a housing, shown in Figure 11-16(a), to protect it from outside contamination. By connecting an RTD in series with a resistor, a constant voltage source, and a voltmeter, changes in temperature at the RTD can be determined by measuring a change in the voltage. The voltmeter uses a temperature scale instead of a voltage scale. RTDs are also constructed by placing a thin film on a ceramic substrate, as shown in Figure 11-16(b). A laser beam is used to burn away the film until its resistance is at a prescribed value. The complete assembly is then sealed in a protective enclosure.

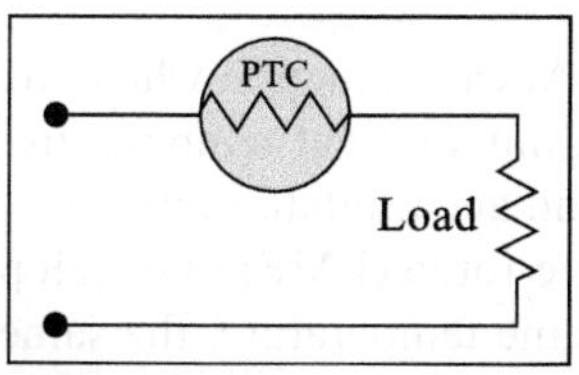

FIGURE 11-17 An RTD used as an overcurrent protection device

RTD Applications

Overcurrent Protection

The positive temperature coefficient characteristics of an RTD make it an ideal overcurrent protection device. Figure 11-17 shows an RTD connected in series with a load. During normal operating conditions, the RTD resistance is low. Therefore, its effect on current flow is minimal. When a short circuit or an overcurrent condition occurs, the RTD resistance goes high and limits the current to a low level.

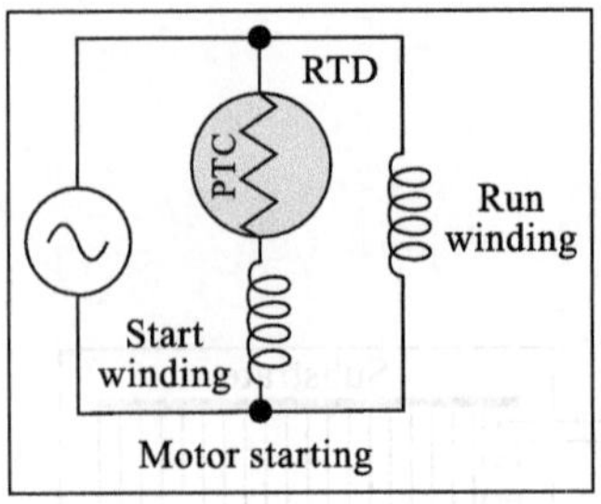

FIGURE 11-18 An RTD used to eliminate the start winding after the motor reaches full speed

Motor Starting A single-phase AC motor has a start winding and a run winding that are connected in parallel branches. After the motor is running, the start winding should not be used. At full speed, a centrifugal switch opens the branch with the start coil.

Figure 11-18 shows how an RTD replaces the centrifugal switch. At ambient temperature, the initial resistance of the RTD is about 100 ohms. It allows sufficient current to flow through the start winding when the motor starts. By the time the motor is at full speed, the RTD is heated and its resistance is high. This reduces the current flow through the start winding to near zero.

RTDs are generally standardized to provide easy interchangeability and predictable performance. These standards pertain to the following ratings.

- *RTD resistance at freezing,* called *ice point (0 degree).* Common resistance values are 100 Ω, 50 Ω, and 200 Ω, although others are also used.
- *RTD alpha (α) value.* This value indicates the average slope of the RTD resistance curve from 0 to 100 degrees Celsius. The slope pertains to the amount the resistance of the RTD changes for every 1 degree Celsius change in temperature to which it is exposed. The α factor is represented by the following formula:

$$\alpha = \frac{R_{100} - R_0}{100(R_0)}$$

where,

R_{100} = Resistance at 100 degrees Celsius

R_0 = Resistance at 0 degrees Celsius

EXAMPLE 11-6

What is the alpha value for an RTD that has 100 ohms of resistance at freezing, and 138.5 ohms at 100 degrees Celsius?

Solution

$$\alpha = \frac{R_{100} - R_0}{100(R_0)} = \frac{138.5 - 100}{100 \times 100} = 0.00385$$

The alpha of 0.00385 is actually a commonly accepted α value of RTDs used in process instrumentation.

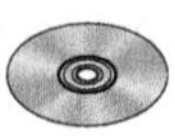

ELE2708 Thermocouples

Thermocouple

Figure 11-19(a) shows dissimilar metal wires joined at both ends. At each junction where the wires are in contact, they are exposed to heat from the surrounding ambient temperature. This causes a small number of electrons to drift from metal B and accumulate in metal A. The slight accumulation of electrons causes a small electromotive force (EMF) to develop between the metals. Because the junctions are subjected to the same temperature, the same amount of voltage develops across them.

Suppose that heat is applied to the junction on the left. A larger voltage potential will develop across it than across the one on the right. An equivalent circuit in Figure 11-19(b)

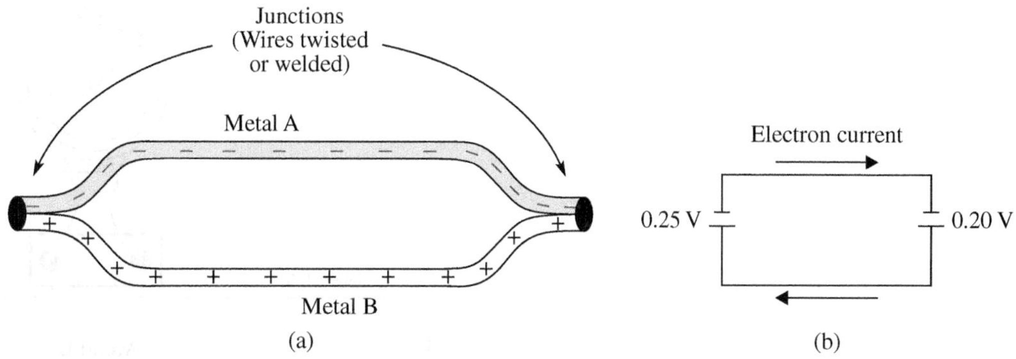

FIGURE 11-19 Thermocouple

is used to illustrate the result. Each battery represents the two junctions. The difference in voltage between the junctions forms a net voltage of 0.05 volts. This causes electron current to flow in the closed circuit formed by the wires. This phenomenon is called the *Seebeck effect*, named after its inventor, the German physicist Thomas Seebeck.

The device shown in Figure 11-19 is called a **thermocouple,** which is a transducer that converts heat into voltage. The amount of voltage developed by a thermocouple junction is affected by the amount of heat applied to it. The higher the temperature, the greater the voltage produced. The voltage–temperature characteristics of a thermocouple are nonlinear over its rated temperature range. Equal changes of temperature at the low end of its rated working temperature range produce different output voltages than do equal changes of temperature at its high end. An example of voltage changes at two temperature ranges for a J thermocouple, which is one type commonly used, is as follows:

	Temperature	*Output* (mV_{DC})	
Lower Temperature Range	−200°C	−7.890	mV = −0.002/°C
	−199°C	−7.868	
Higher Temperature Range	+1190°C	68.980	mV = +0.057/°C
	+1191°C	68.037	

Instrumentation using thermocouples must linearize this nonlinear output voltage in order to properly display the correct temperature. In modern digital instrumentation, this is achieved by a software lookup table stored in memory.

When using a very narrow temperature range within a thermocouple's working temperature specification, the voltage–temperature characteristics can approach a degree of linearity. The metals used determine the polarity and voltage range of the junction. The voltage–temperature characteristics are different for each type of thermocouple, and are shown by the graph in Figure 11-24. Each one of these thermocouple types uses a distinct pair of metals.

Figure 11-20 shows the loop opened at the top wire. The junction exposed to the temperature to be measured is called the *hot junction.* The other junction in the loop is called the *cold junction.* The output voltage that appears across the opening will be proportional to the temperature difference between the junctions. If the temperature of one of the junctions is known, the voltage across the opening can be used to calculate the temperature of the other junction. The cold junction is considered the reference junction in a thermocouple because the temperature applied to it is a known value. A temperature display instrument is usually connected across the open leads. The value it displays is determined by the voltage

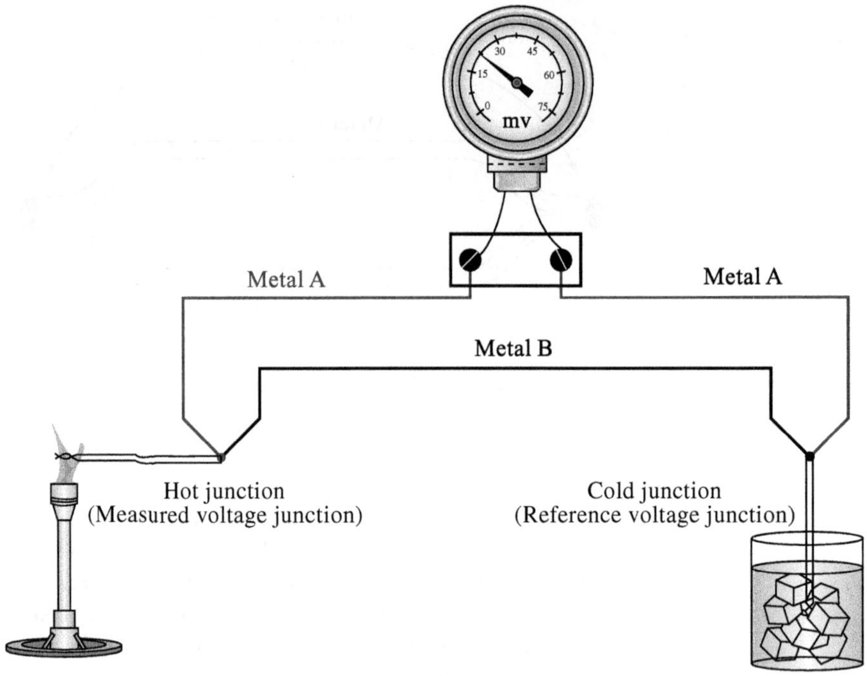

FIGURE 11-20 A thermocouple with hot and cold junctions

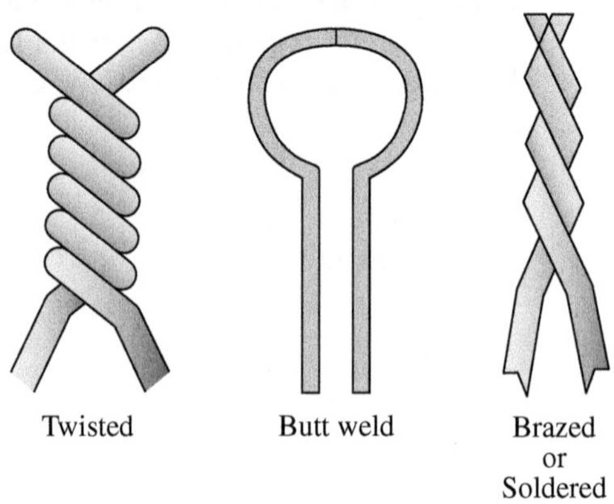

FIGURE 11-21 Types of thermocouple junctions

developed across the terminals. The reference junction is typically exposed to a temperature of 0 degrees Celsius (32 degrees Fahrenheit), because it provides a high degree of accuracy. Also, all thermocouple reference tables are designed with their cold junction at 0 degrees Celsius.

Several different methods, used to join the thermocouple wires at the junction, include twisting, soldering, brazing, and welding, as shown in Figure 11-21.

An automatic thermocouple network that does not have a 0 degrees Celsius cold junction is illustrated in Figure 11-22. It lists the voltages throughout the circuit. Both of the reference junctions and a resistor bridge network are integrated on a substrate. Therefore, they are subjected to the same ambient temperature. One leg of the bridge, R_3, is a thermistor with a negative coefficient. Resistors R_1, R_2, and R_4 are not temperature-sensitive. As the ambient temperature changes, R_3 varies proportionately and unbalances the bridge. The voltage at point A changes the same amount as the two cold junctions, but at the opposite polarity;

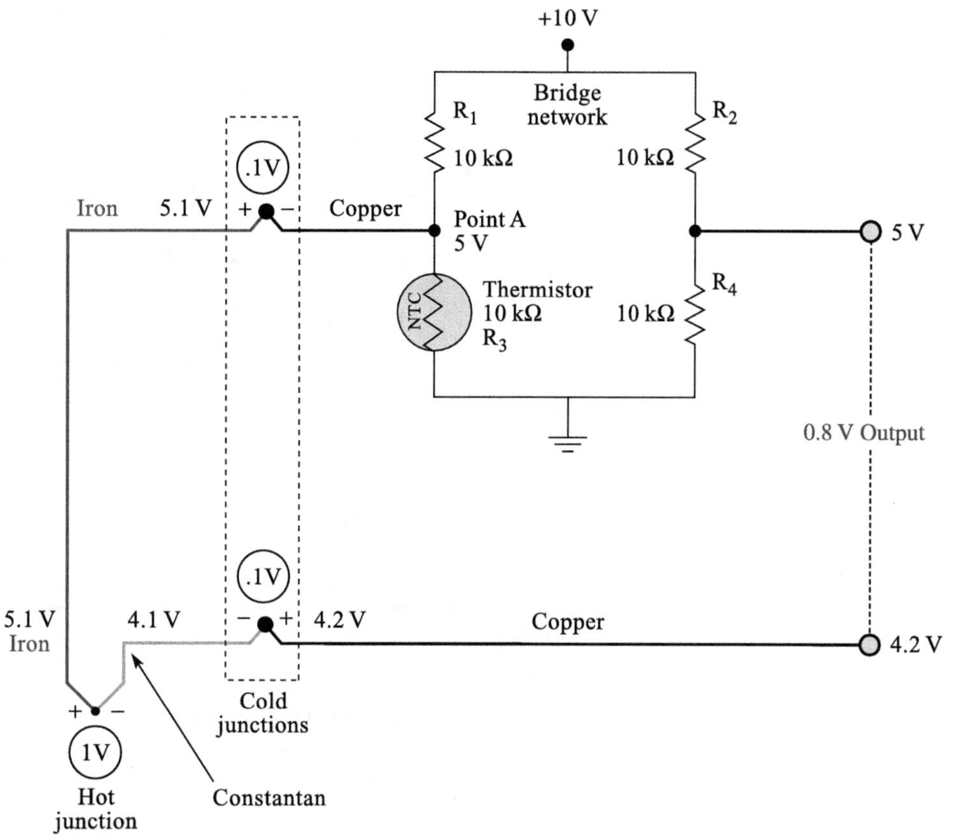

FIGURE 11-22 A thermocouple with a cold junction compensating network

therefore, both voltages cancel. The only voltage change at the output should be the result of the temperature variance detected at the hot junction.

EXAMPLE 11-7

Refer to Figure 11-22. Make five assumptions:

1. The voltage produced by the measured hot junction is 1 volt as the temperature remains constant.
2. Every 100 degrees Fahrenheit change of ambient temperature causes the thermistor to vary 750 ohms. At +100 degrees Fahrenheit, the thermistor resistance is 10 kΩ.
3. Every 100 degrees Fahrenheit change of ambient temperature causes each cold junction to change 0.1 volts.
4. The output of the thermocouple network is 0.8 volts.
5. At +100 degrees Fahrenheit, the thermistor resistance is 10 kΩ and each cold junction produces 0.1 volts.

Suppose that the circuit in Figure 11-22 is exposed to an ambient temperature of 100 degrees Fahrenheit. If the ambient temperature increases to 200 degrees Fahrenheit, determine whether the output voltage remains at 0.8 volts.

Solution

Figure 11-23 shows the voltage drop changes, which cause the output voltage drop to remain at 0.8 volts.

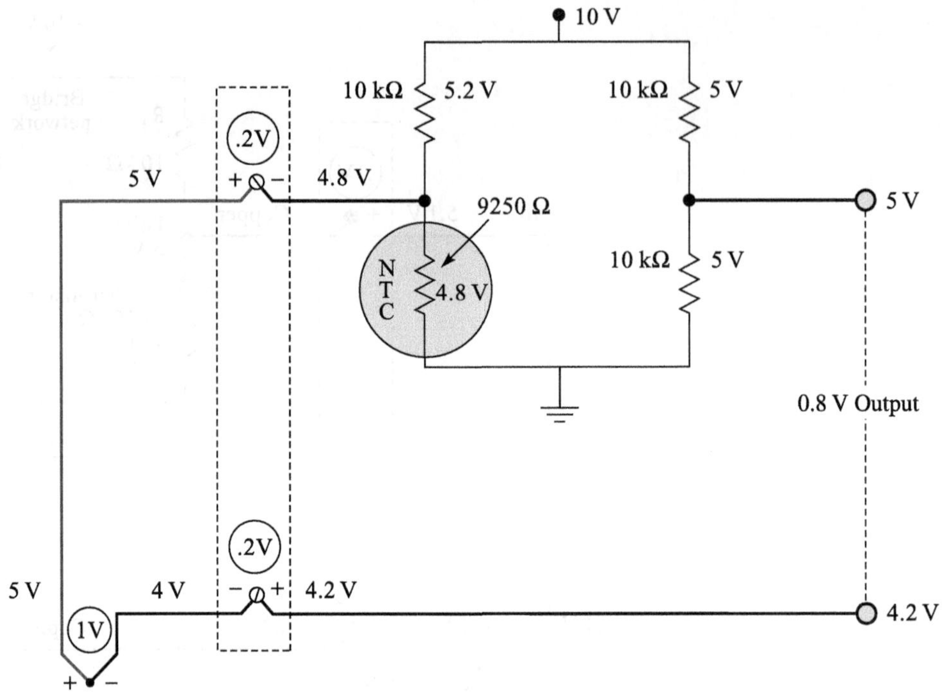

FIGURE 11-23 Example 11-7 solution

Devices that perform automatic compensation for thermocouples are commercially available. They are called *cold junction compensators*. In recent years, computers have also been used to compensate thermocouples.

Some thermocouple metal combinations are in common use throughout industry. They have become standardized by the American National Standards Institute (ANSI), and are identified by letter designations. The more common types include:

1. Iron–Constantan (Type J)
   The iron produces the positive voltage and the constantan produces the negative voltage. The materials are rugged, but the iron wire is susceptible to oxidation, especially at high temperatures. This type is recommended for applications in which reducing atmospheres exist. A reducing atmosphere refers to a reduced level of oxygen in the air. If the air has a high level of oxygen combined with a high temperature, certain metals, especially iron, will become oxidized. Therefore, a thermocouple with iron as one of its metals should only be exposed to a reducing atmosphere, otherwise it will deteriorate at an accelerated rate.
2. Copper–Constantan (Type T)
   The copper produces the positive voltage and the constantan produces the negative voltage. This type is used in low-temperature applications. It is capable of resisting moisture and is effective when exposed to reducing atmospheres and oxidizing conditions.
3. Chromel–Alumel (Type K)
   The chromel produces the positive voltage and the alumel produces the negative voltage. This type is recommended for high-temperature applications and can be used in high-oxidizing conditions, but not reducing atmospheres.
4. Chromel–Constantan (Type E)
   The chromel produces the positive voltage and the constantan produces the negative voltage. This type has the highest sensitivity because it produces the highest output and can be used over a wide range of temperatures. It should not be used in applications with reducing atmospheres.

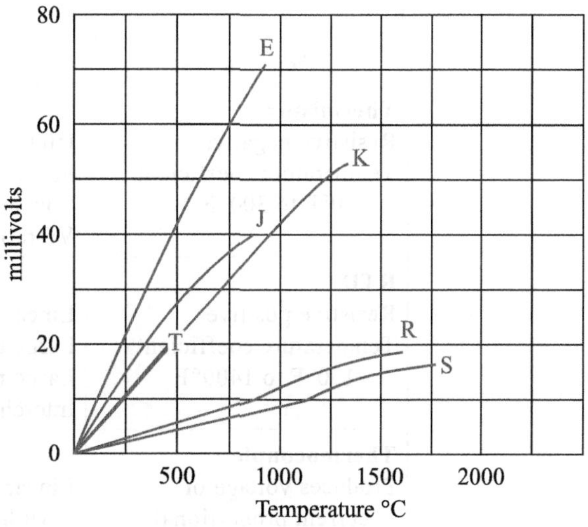

FIGURE 11-24 Millivolts produced by thermocouple types at various temperatures

5. Platinum–Rhodium, Platinum (Types R, S, and B. The difference depends on the ratios of the two metals.)
   The combined metals produce the positive voltage and the pure platinum produces the negative voltage. These types are recommended for very high-temperature measurements in which an oxidizing atmosphere exists and high accuracy is required.

The graph in Figure 11-24 shows the voltages produced at various temperatures by each type of thermocouple.

Each type of thermocouple has different characteristics, and its accuracy is affected by various conditions. The selection of a thermocouple is based on the following considerations:

1. Temperature range
2. Susceptibility to oxidation
3. Reducing atmosphere
4. Sensitivity
5. Accuracy
6. Cost

Thermocouples are used in industry to measure temperatures of ovens and furnaces, molten plastic vats, and nuclear reactor cores. Thermocouples respond quickly to temperature changes, are rugged, and have a wide temperature range.

Table 11-1 provides a comparison of the three types of temperature sensors.

Probe Assemblies

In most applications, temperature-sensing elements should be protected from the environment surrounding the point of measurement. For example, if the sensor is immersed in a liquid, the leads or body may become corroded. Conductive liquids may short the two leads connected to the sensor body, resulting in false readings. The measurement of air temperature can be misread if the humidity is high or if the probe is exposed to wind, which causes a cooling effect.

These false readings can be avoided by enclosing the sensor body within a housing. Protective housings such as the one shown in Figure 11-25 are made of glass, ceramic, epoxy, stainless steel, and other metals.

One type of protective device is the **thermowell.** An RTD or thermocouple is placed inside the device, which is a tube like that shown in Figure 11-26. The tube is usually threaded

TABLE 11-1 Temperature Sensor Comparisons

Type/Range	*Advantages*	*Disadvantages*
Thermistor Resistive negative Temperature coefficient −40°F to 300°F	High sensitivity Fast response Low cost Vibration-resistant	Narrow temperature span Nonlinear output
RTD Resistive positive Temperature coefficient −150°F to 1400°F	Linear output Large temperature span Large resistance range Interchangeability	Low sensitivity High cost Vibration
Thermocouple Produces voltage or current proportional to temperature −300°F to 4200°F	Linear output within a given temperature range	Least sensitive Requires reference

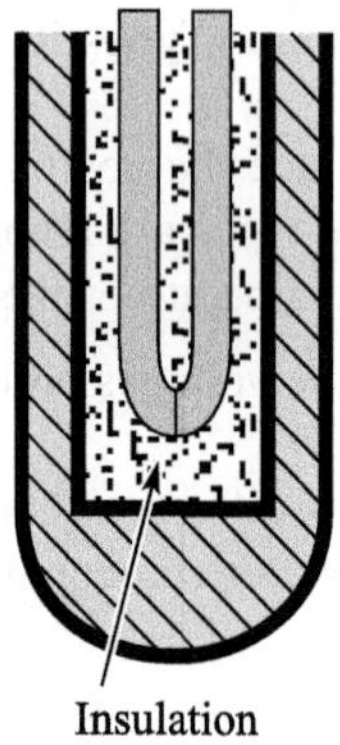

FIGURE 11-25 Protective housing for a thermocouple

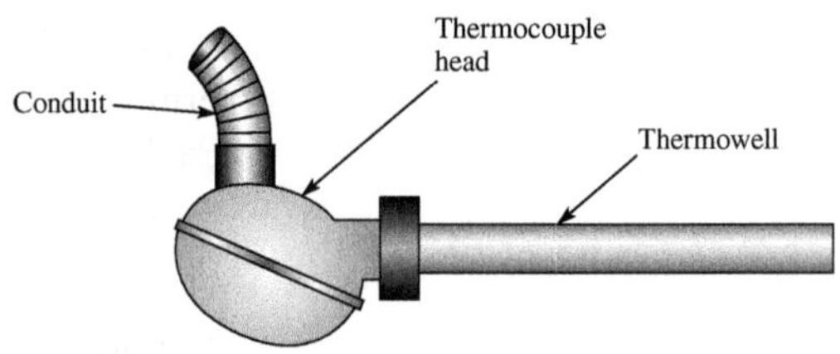

FIGURE 11-26 A thermowell

or welded into a vessel or pipeline. The disadvantage of using the thermowell is that the response time is longer because the thermoconduction through the tube must occur before reaching the sensor body. A spring inside the tube is often used to keep the sensor body firmly in contact with the thermowell tip to improve the conduction and response time.

Radiation Thermometry

ELE3308 Radiation Pyrometer Temperature Sensors

Most temperature instruments are invasive devices that make physical contact with the solids, liquids, or gases being measured. They make direct temperature readings as thermal energy is transferred by conduction to the sensing element.

It is also possible to take temperature readings without making physical contact by using a noninvasive device. A method called **radiation thermometry** infers temperature by measuring the thermal energy radiated from the surface of the measured body. The instrument used to make these readings is usually referred to as a *radiation pyrometer*. The term *pyrometer* is derived from this instrument's ability to measure high temperatures. Instruments of this type are ideally suited for applications where conventional sensors cannot be employed, such as:

1. When objects are moving, such as rolling mills in steel production, paper manufacturing, glass making, and conveyor belts.
2. Where temperatures are extremely hot, such as in furnace atmospheres.

3. Where noncontact measurements are required because of contamination, such as in food and pharmaceutical production.
4. Where corrosive and hazardous conditions exist, such as around high-voltage conductors.
5. Where measurements are taken from a distance.

ELE3208
Temperature Emissivity of Nonblackbodies

The principle of operation of radiation thermometry is based on the basic law of physics, which states that every object at a temperature above absolute zero radiates electromagnetic energy. The frequency range of the electromagnetic waves includes visible light and lower-frequency infrared light. As the temperature of the object changes, the frequency also changes. For example, if the temperature rises, the frequency increases and the wavelength becomes shorter. This principle is illustrated by observing metal being heated. As it gets hotter, the color changes from red, to yellow, to white. The color change is a result of the frequency increasing.

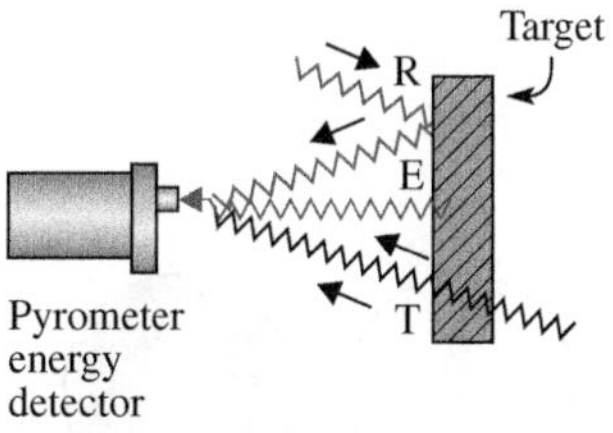

FIGURE 11-27 The target object radiating emitted (E), reflected (R), and transmitted (T) energy to the measurement detector

By focusing the electromagnetic energy waves emitted by the measured object (target) on a detector element, measurements are taken. The signal from the element is electronically processed and the frequency is converted into a proportional temperature readout for display. Radiation thermometry theory is based on the assumption that the total energy emitted by a body is the result of its temperature. An object with this capability is referred to as a *blackbody*. Most objects, however, do not radiate energy from temperature alone. Instead, they also reflect and transmit (as fiber optics do) radiant energy, as shown in Figure 11-27. Therefore, the total radiated energy is the sum of emitted energy (E), reflected energy (R), and transmitted energy (T).

$$E + R + T = \text{Radiated Energy}$$

For a perfect blackbody, $E = 1$, $R = 0$, and $T = 0$. For a non-blackbody object, the value for E is still 1.0, but the values for R and T are greater than 0. To account for the reflective and transmitted energy of an object, a term referred to as *emissivity* must be considered when making measurements. *Emissivity* is defined as the ratio of total energy radiated by an object to the emitted energy of a blackbody made of the same material at the same temperature. The emissivity of a blackbody is 1.0. The emissivity of non-blackbodies falls between 0.0 and 1.0. For example, the emissivity of non-blackbody carbon at 76 degrees Fahrenheit is approximately 0.8. This value means that the carbon is radiating more energy than a carbon blackbody at the same temperature. Therefore, a pyrometer detector measuring the temperature of this target will record a reading higher than the actual temperature because it detects three types of energy instead of one. Many radiation pyrometers have an emissivity adjustment knob that allows the operator to compensate for the emissivity ratio factor. By setting the knob to 0.8, for example, the instrument electronically calculates this value and the frequency it detects to indicate the correct temperature of carbon. The operator uses a specific emissivity reference table to determine the adjustment setting of the instrument required for various materials. This table is developed in a research lab where measurements of emitted energy from holes drilled into various materials and measurements of total radiated energy from non-blackbodies made of the same materials are taken at various different temperatures required for various materials. The holes become blackbodies because they are not capable of emitting reflective or transmitted energy. Some targets have very low emissivity. If the values are below 0.2, accurate measurements are not always possible. Examples of these objects are polished metallic surfaces that are reflective, and thin film plastic that transmits a high amount of energy.

Based on the different techniques used to measure radiant energy, there are three categories of instruments: broadband, optical detector, and ratio pyrometers.

Broadband Pyrometers

A broadband pyrometer, shown in Figure 11-28(a), uses a lens system or sight tube that directs the radiation onto a blackened reference surface inside the instrument. The filter is used to pass electromagnetic waves within a desired frequency range. The energy detector

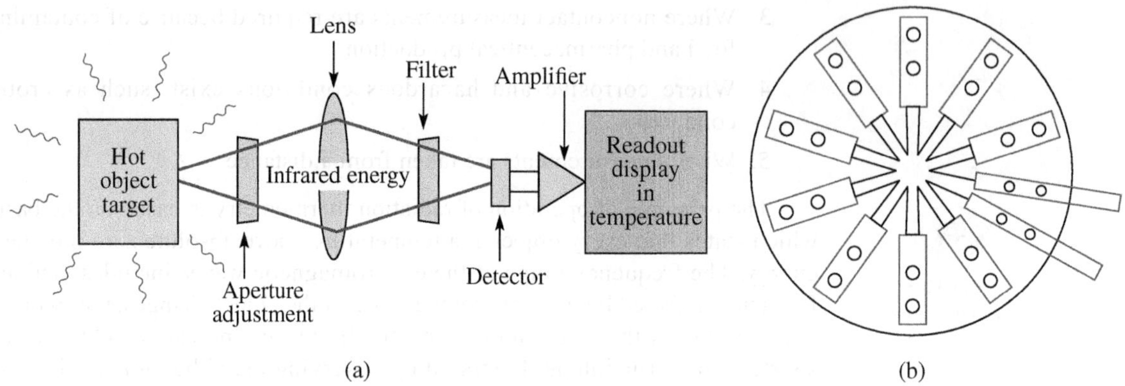

FIGURE 11-28 A simplified pyrometer temperature-measuring instrument and a thermopile

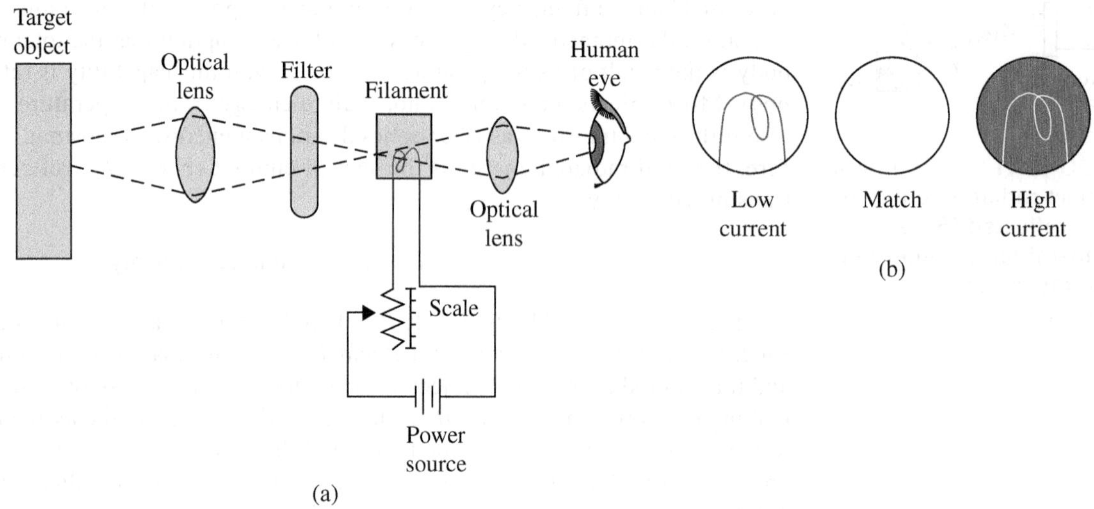

FIGURE 11-29 Optical pyrometer

employs a device known as a *thermopile* to measure the temperature of the reference area. A thermopile, shown in Figure 11-28(b), consists of several thermocouples connected in series to provide greater sensitivity to small changes in temperature. The composite output of this detector is a DC voltage that is proportional to the amount of energy at its surface.

Optical Pyrometers

ELE2908
Optical Pyrometer

The optical pyrometer is shown in Figure 11-29(a). A viewfinder is positioned to allow observance of the target and the filament of a lightbulb simultaneously, as shown in Figure 11-29(b). The object being measured is compared to the brightness of the filament. A current adjustment is made by the operator until they are both the same intensity, at which time the filament visually disappears into the background. When the brightness of both objects is equal, their temperatures are also the same. A scale on the current adjustment knob is calibrated to indicate the temperature of both the target and the reference filament.

Ratio Pyrometers

ELE3508
Ratio Pyrometers

At all temperatures, the target object radiates energy at different frequencies. Most pyrometers measure the dominant waves with the most energy.

The ratio pyrometer differs from other pyrometers by taking measurements of two different frequencies emitted by an object. First, the radiant energy of a blue wavelength is passed through a blue filter and its power strength is measured. Then the radiant energy of a red wavelength is passed through a red filter and measured. These measured values, along with the actual lengths of blue and red magnetic waves, are used in the following formula to determine a ratio quantity:

Red Wavelength = 1.0
Blue Wavelength = 0.8
RP = Power Reading of Red Wavelength
BP = Power Reading of Blue Wavelength

$$\text{Ratio} = \frac{\text{RP} - \text{BP}}{1.0 - 0.8}$$

The target temperature is inferred from the ratio value electronically calculated by the instrument. Figure 11-30 graphically illustrates the power readings taken at the blue and red wavelengths for two different temperatures.

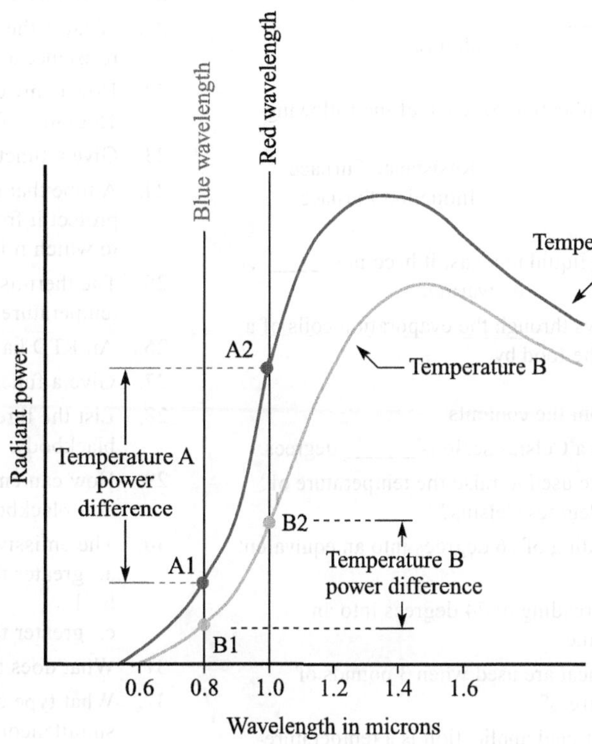

FIGURE 11-30 The power detection of two different wavelengths by a ratio pyrometer

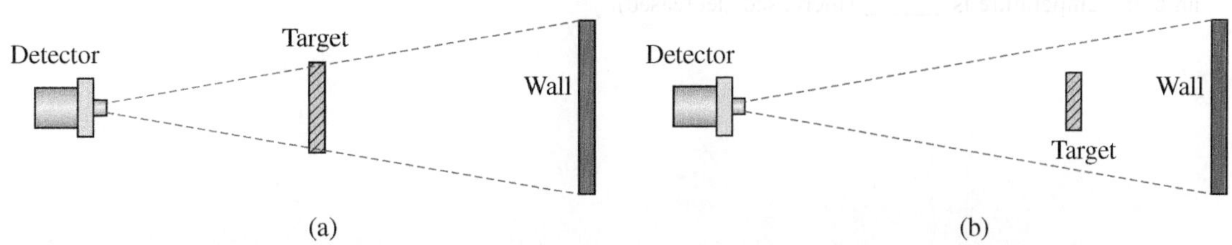

FIGURE 11-31 Field of view

The advantage of the ratio technique is that it is less susceptible than other measurement methods to dust, steam, and other factors that distort readings.

Field of View

When using a pyrometer, the target being measured should completely fill the view of the instrument. Figure 11-31(a) shows a proper technique for measuring the target. The reading in Figure 11-31(b) will not be accurate because the pyrometer will also detect the temperature of the wall surface in the background.

Problems

1. Molecular movement creates heat known as ______.
2. Thermal energy moves from ______ objects to ______ objects.
   a. warmer b. cooler
3. The process by which thermal energy transfers through a solid is called ______.
   a. conductance c. radiation
   b. convection
4. The type of thermal energy transfer that takes place through a vacuum is called ______.
   a. conductance c. radiation
   b. convection
5. Name a functional application for each of the following heating sources:
   Blast Furnace, Resistance Furnace
   Fossil Fuel Furnace, Induction Furnace
   Arc Furnace
6. As freon turns from a liquid to a gas, it becomes ______.
   a. cooler b. warmer
7. As the cold freon flows through the evaporation coils of a refrigerator, it cools the food by ______.
   a. emitting cold air
   b. absorbing heat from the contents
8. The freezing point on a Celsius scale is ______ degrees.
9. How many calories are used to raise the temperature of 5 grams of water 10 degrees Celsius?
10. Convert a Celsius reading of 16 degrees into an equivalent Fahrenheit value.
11. Convert a Fahrenheit reading of 74 degrees into an equivalent Celsius value.
12. How many BTUs of heat are used when 3 pounds of water are raised 6 degrees?
13. For what type of functional application is a temperature-sensing indicator made of crystalline solids used?
14. When heated, why does the liquid in a thermometer rise in the tube?
15. As a bimetallic thermometer element straightens, the ambient temperature is ______ (increased, decreased).
16. A thermocouple is a ______ device.
    a. thermoelectric b. thermoresistive
17. The ______ (hot, cold) junction of a thermocouple is the reference point.
18. Give a functional application of a thermocouple.
19. The resistance of an RTD is ______ (directly, inversely) proportional to temperature.
20. RTDs are considered ______ (linear, nonlinear) devices.
21. What is the alpha value for an RTD that has 50 ohms of resistance at freezing, and 69.25 ohms at 100 degrees Celsius?
22. How many calories are required to raise the temperature of 15 grams of water 3 degrees Celsius?
23. Give a functional application of an RTD.
24. A tube that encloses a temperature-sensing device to protect it from being damaged by environmental elements to which it is exposed is called a ______.
25. The thermistor has a ______ (negative, positive) temperature coefficient.
26. An RTD has a/an ______ (PTC, NTC).
27. Give a functional application of a thermistor.
28. List the three types of energy radiated from a non-blackbody.
29. How can emitted energy be radiated exclusively from a non-blackbody?
30. The emissivity value of a non-blackbody is ______.
    a. greater than 0 but less than 1
    b. 1
    c. greater than 1
31. What does a thermopile consist of?
32. What type of pyrometer views a lightbulb and target simultaneously in a viewfinder?
33. Which type of pyrometer is the least susceptible to dust and smoke?

CHAPTER

12

Flow Control

OBJECTIVES

At the conclusion of this chapter, you should be able to:

- Define *flow*.
- Describe the importance of measuring and controlling flow in industrial processes.
- List some types of materials measured for flow and how they are transferred.
- Explain the difference between volumetric flow rate and mass flow rate.
- List common measurement units of flow rate.
- Describe the method used for measuring the volumetric flow rate and mass flow rate of solid materials.
- List four factors that affect the flow rate of liquids.
- Calculate the Reynolds number for a liquid.
- Describe the operation of the following mechanical measurement instruments used to determine flow rate:

Differential Pressure	Rotary-Vane	Turbine Flowmeter
Rotameter	Lobed Impeller	

- Describe the operation of the following electronic sensors used to measure flow:

Coriolis Mass Flowmeter	Electromagnetic Flow Detector	Vortex Flowmeter
Rotor Flow Detector	Thermal Flowmeter	Ultrasonic Flowmeter
Time-of-Flight Flowmeter		Thermal Mass Meter

- State a rule that describes the placement of flow sensors in a pipe system.
- Select the most appropriate flow-measuring device for a particular application.

INTRODUCTION

Many types of industrial applications involve the flow of raw materials, feedstocks, products, or waste during the manufacturing process. **Flow** is the transfer of material from one location to another. The materials can be raw materials, products, or wastes in the form of solids, liquids, gases, suspensions, or slurries. Components such as pipes, hoses, channels, or conveyor belts are used to move materials. Flow can be continuous or sporadic, depending on the type of process being performed.

This chapter will first discuss the basic principles of flow, and then describe the operation of mechanical and electronic measuring instruments.

12-1 Systems Concepts

Automated systems that control flow first determine flow rates or volume by various measurement techniques, and then use the data to regulate the movement. These systems employ a source, a path, a control function, an actuator, and a measuring instrument to operate.

An automated flow control system is illustrated in Figure 12-1. It shows a batch process machine that makes soft drinks. A computer-based controller turns the valves on and off to direct the flow. Flow sensors that measure flow rate and volume are located on each pipe. They send feedback data to the controller.

At the beginning of a batch, valve V_1 opens and water drains into the batch tank. When the flow sensor S_1 registers that a certain volume of water has passed through the pipe where it is located, the controller closes V_1. V_2 then opens until sensor S_2 measures that the required amount of water enters the carbonation tank. As V_2 closes, V_3 opens to allow CO_2 into the carbonation tank. When sensor S_3 measures that a certain volume of gas has entered the tank, the controller closes V_3 and V_4 opens to fill the seltzer tank. Selected valves then open in a certain sequence to allow the necessary coloring and flavoring ingredients to drain into the batch tank. The volume of each coloring and flavoring fluid is monitored by sensor S_4.

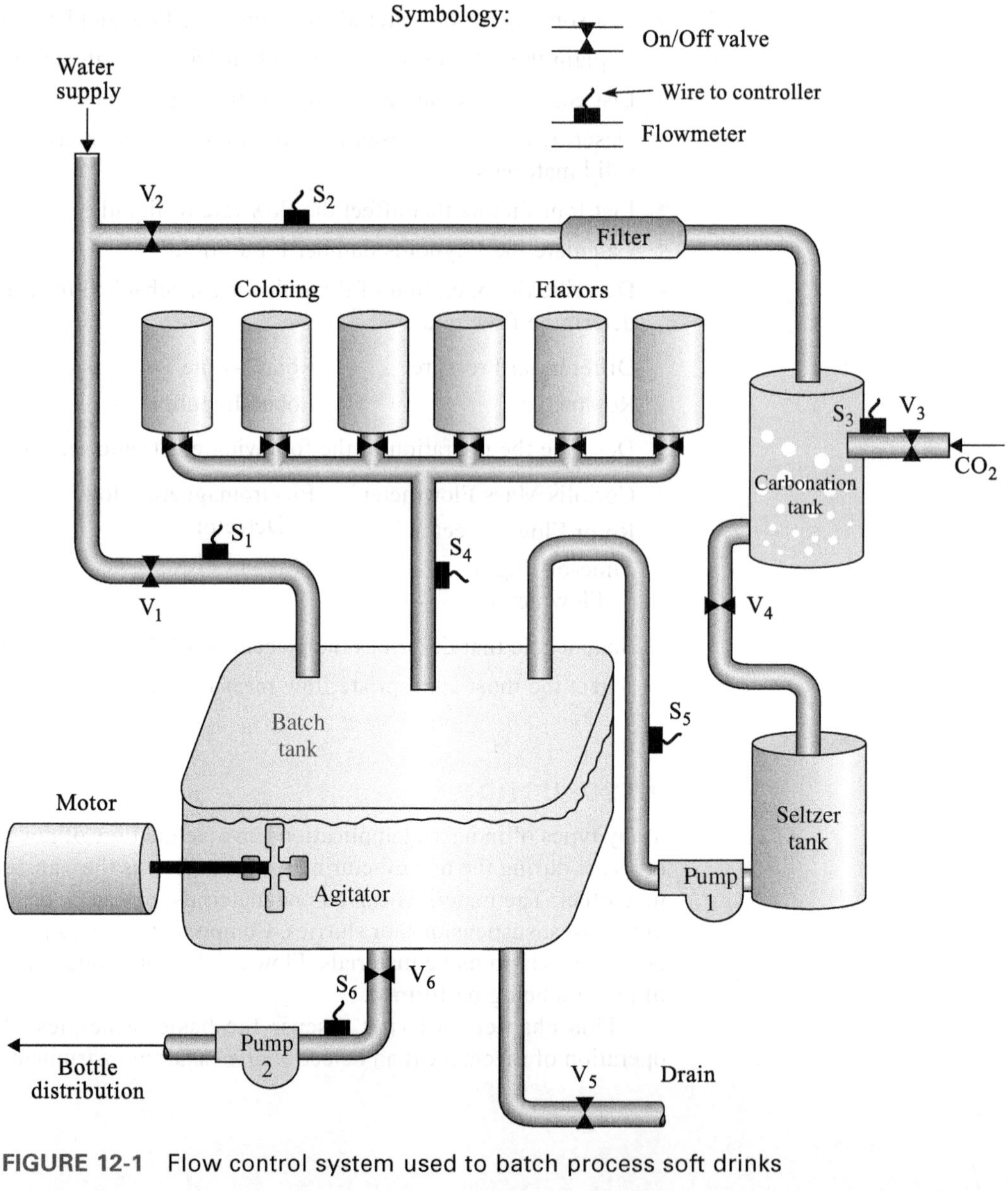

FIGURE 12-1 Flow control system used to batch process soft drinks

When the required amount passes S_4, the valve allowing an ingredient to flow closes, and the next valve in the sequence is opened. After the sequence is completed and the final valve closes, pump 1 turns on to transfer the seltzer ingredients at a desired flow rate into the batch tank. Sensor S_5 monitors the rate of flow as the carbonized fluid passes through its pipe. The controller will vary the speed of the pump so that the flow rate is at the required value. When the desired portions of the formula are present in the tank, pump 1 stops and the motor starts to run an agitator that mixes the contents for the required period of time. When the motor stops, pump 2 transfers the finished product to the bottle-filling stage.

After the batch run is completed, V_1 opens and fills the tank with a supply of fresh water to rinse the tank. Then the motor is activated to aid the rinsing process by agitating the water. When the agitation cycle is complete, V_5 opens to allow the batch tank to empty. After V_5 is closed, the tank is clean and ready for the next production run.

Reasons for Control

To provide good control in the process industries, accurate measurement of flow is essential. Three reasons to monitor the flow of materials are:

1. To ensure that the correct proportions of raw materials are combined during the manufacturing process.
2. To ensure that ingredients are supplied at the proper rate during the mixing and blending of the materials.
3. To prevent a high flow rate that might cause pressure or temperature to become dangerous, overspills to occur, or machines to overspeed.

Flow measurements are also used to determine how much of a product is passed from the supplier to the customer. This application is known as **custody transfer.** Measuring flow accurately is essential in keeping records for accounting purposes.

The flow of materials is a response to an applied force. The force may be produced by a motor that drives a pump or a conveyor belt. Force may be supplied by pressure in a hydraulic system or an air compressor. Force is also produced by static head pressure.

12-2 Flow Units of Measurement

Three common classifications that are used to determine flow measurements are **volumetric flow rate, flow velocity,** and **mass flow rate.**

ELE2307
Types of Flow Measurements

Volumetric Flow Rate

Volumetric flow rate instruments are used to determine the volume of material that flows during a specific period of time. The volume can be read as cubic feet, gallons, or liters. The time can be read per unit of time, such as seconds, minutes, or hours.

In many situations, volumetric flow rate is not measured directly. Instead, an *inferred* measurement is taken. **Inferred** means that some other variable is measured, and then translated into the reading that is required. For example, volumetric flow rate of a liquid can be determined by measuring the velocity at which it flows through a pipe. By using the velocity measurement and the area of the pipe in the following formula, the volumetric flow rate can be calculated:

$$Q = VA$$

where,

Q = Volumetric flow rate in units of volume per units of time
V = Velocity of the fluid
A = Cross-sectional area of the pipe

Most volumetric flow rate measurements are taken using an inferred method.

Flow Velocity

Flow velocity is the distance a material travels in a carrier per unit of time. Typical units of measurements are kg/h or lb/h. It is expressed by the formula,

$$V = \frac{Q}{A}$$

Mass Flow Rate

Mass flow rate is a measure of how much actual mass flows past a location within a specific time period. Since mass measurements are commonly made at a specific gravity, it has a definite weight. The weight can be read in pounds, tons, grams, or kilograms. The time can be read per unit of time, such as seconds, minutes, or hours. The mass of a material is determined by its density. Increasing density results in more mass being in the same volume of space. The temperature of a liquid will affect its mass (density). For example, a gallon of water at a temperature of 39.3 degrees Fahrenheit (temperature when water is most dense) weighs 8.345 pounds. At a higher temperature of 212 degrees Fahrenheit, a gallon of water weighs 7.998 pounds. Mass flow measurements are used instead of volumetric flow rate for certain applications, such as with some processes involving chemicals that react on the basis of mass relationships of ingredients. Mass flow rate is expressed by the formula,

$$M = pQ$$

where,

M = Mass
p = Mass density or weight density

EXAMPLE 12-1

Water is forced by a pump through a pipe with an inside diameter of 2 inches at a flow velocity of 5 feet per second. Find the volumetric flow rate and the mass flow rate. *The weight density of water is 62.4 lb/ft³.*

Solution

Step 1: Determine the pipe's inside area, where the diameter in feet equals,

$$D = \frac{2 \text{ in.}}{12 \text{ in./ft}} = 0.167 \text{ ft}$$

$$A = \frac{\pi d^2}{4}$$

$$= \frac{3.14\,(0.167)^2}{4} = 0.0218 \text{ ft}^2$$

Step 2: Calculate the volumetric flow rate.

$$Q = VA$$

$$= (5 \text{ ft/s})\,(0.0218 \text{ ft}^2)\,(60 \text{ s/1 min})$$

$$= 6.54 \text{ ft}^3\text{/min}$$

Step 3: Determine the mass flow rate.

$$M = pQ$$

$$= (62.4 \text{ lb} \cdot \text{ft}^3)\,(6.54 \text{ ft}^3\text{/min})$$

$$= 408 \text{ lb} \cdot \text{min}$$

12-3 Solid Flow Measurement

The solid materials that are measured for mass flow rate are typically in the form of small particles, such as powder, pellets, or crushed material. A conveyor belt is usually used to move these materials from one location to another.

Figure 12-2 shows one method of measuring the flow of powder as it is being transported by a conveyor system. The powder is released by the hopper and transferred to another location, such as a mixing vat, a storage tank, or the hold of a ship. A measurement is taken by a load cell device that determines the weight of a fixed length of the belt. Using inferred data, such as the weight measurement and speed of the belt, allows a calculation of the mass flow rate, as shown by the following formula:

$$F = \frac{WS}{L}$$

where,

F = Mass flow rate in lb/min
W = Weight of a material on a section of length
S = Conveyor speed in ft/min
L = Length of the weighing platform

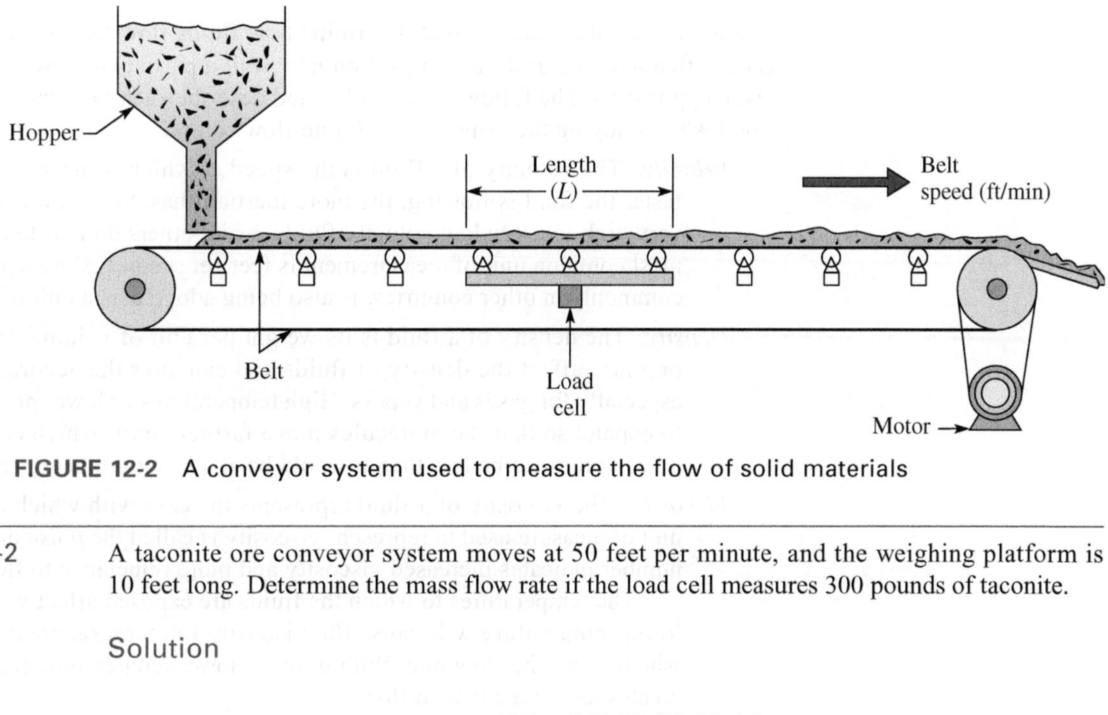

FIGURE 12-2 A conveyor system used to measure the flow of solid materials

EXAMPLE 12-2 A taconite ore conveyor system moves at 50 feet per minute, and the weighing platform is 10 feet long. Determine the mass flow rate if the load cell measures 300 pounds of taconite.

Solution

$$F = \frac{WS}{L} = \frac{(300 \text{ lb})(50 \text{ ft/min})}{10 \text{ ft}}$$
$$= 1500 \text{ lb/min}$$

The load cell that measures the weight is a strain gauge. Another instrument used to measure the ore weight is a linear voltage differential transformer (LVDT). The LVDT produces a variable voltage. Its amplitude is proportional to the amount at which its core moves within the coils of a transformer. Figure 12-3 shows a conveyor belt with a section that is allowed to droop due to the weight of the material it carries. The more the belt droops, the heavier the weight. The LVDT makes a weight measurement by reading the droop in the belt.

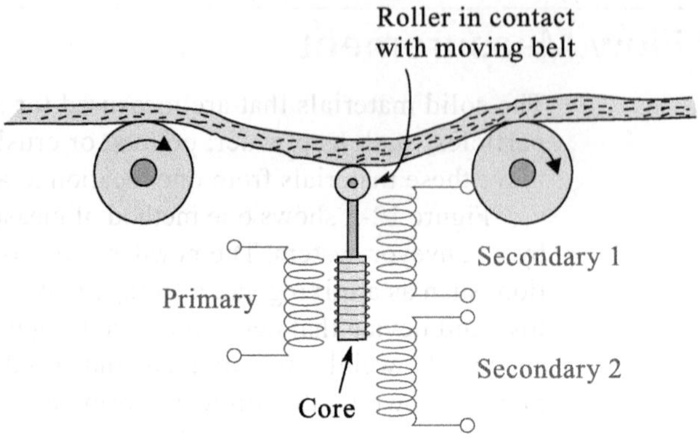

FIGURE 12-3 An LVDT used to measure the weight of materials flowing on the conveyor belt

12-4 Fluid Flow Measurement

Liquids, gases, and vapors are classified as fluids. The accurate measurement and control of fluid flow is essential in industrial processing plants that use water, steam, gases, petroleum, acids, base solutions, and other types of fluid materials.

Pipe Flow Principles

There are several influences that determine how fluids flow through a system. These influences often have a significant impact on how well a particular flowmeter will perform in a given application. The following terms list and describe each influence. They must be understood when studying the principles of liquid flow control:

Velocity. The velocity of a fluid is the speed at which it moves through the pipe. The faster the fluid is flowing, the more inertia it has. Some flowmeters work well with very high- or very low-velocity fluids, while others do not. In the United States, the most common unit of measurement is feet per second. Meters per second, used more commonly in other countries, is also being adopted as a unit of measurement.

Density. The density of a fluid is its weight per unit of volume. Both temperature and pressure affect the density of fluids and can alter the accuracy of measurements, especially for gases and vapors. High temperatures or lower pressures cause the fluid to expand so that the molecules move farther apart, which causes the weight of a given volume to be less than it would be at a lower temperature or higher pressure.

Viscosity. The viscosity of a fluid represents the ease with which it flows. A numerical unit of measure used to represent viscosity is called the *poise* or *centipoise*. A higher number indicates increased viscosity and more reluctance to flow.

The temperatures to which the fluids are exposed affect viscosity. With liquid, a lower temperature will cause the viscosity to increase, creating more reluctance, which slows the flow rate. With gases, a lower temperature decreases viscosity, and creates less reluctance to flow.

Pipe Size. The size of the pipe carrying a fluid affects the flow. The larger the diameter, the more easily the fluid will pass through.

Reynolds Number

ELE906
The Reynolds
Number

In 1883, Sir Osborne Reynolds, an English scientist, submitted a paper to the Royal Society that described the effects of the preceding four factors on fluid flow. He also presented a numerical scheme that assigned values to express the fluidity of a moving liquid based on the influence of these four factors. The numerical value, known as **Reynolds number** or **R number,** is determined by the following formula:

$$R = \frac{VDp}{u}$$

where,

V = Velocity
R = Reynolds number
D = Pipe inside diameter
p = Fluid density
u = Liquid viscosity

The R number represents the ratio of the liquid's inertial forces to its drag (viscous) forces. The velocity, pipe diameter, and fluid density are inertial forces, and viscosity is the drag force. The R number is used to identify the type of flow currents that are likely to occur. These currents are illustrated in Figure 12-4: *laminar* flow (Figure 12-4(a)), *transition* flow (Figure 12-4(b)), and *turbulent* flow (Figure 12-4(c)). The information supplied by the R number is useful when determining the proper flowmeter for a specific application. For example, some flowmeters are designed to read laminar flow, and would give erroneous readings if they were measuring turbulent flow. By knowing a fluid's R number, a suitable flowmeter with a rating of the same value can be selected.

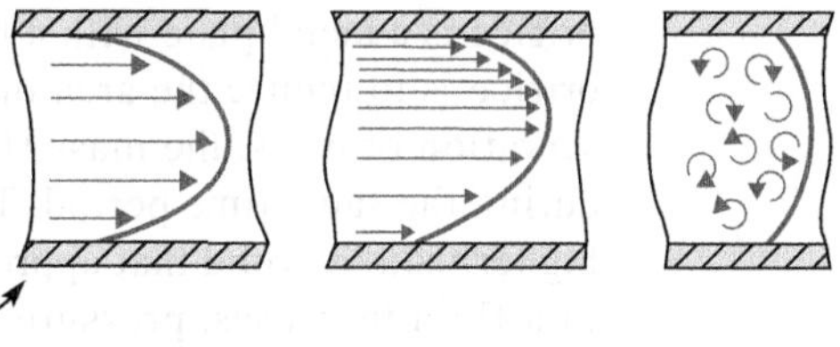

FIGURE 12-4 Flow currents of fluid

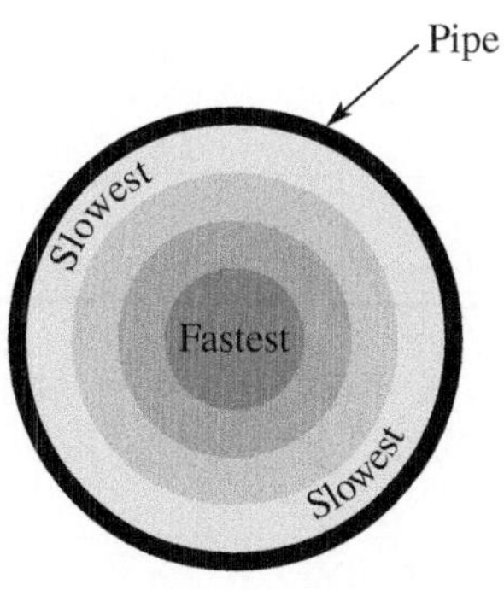

Figure 12-5 Fluid flow velocity layers

At very low velocities, R is low and laminar flow takes place. With this type of flow, liquid moves in layers. However, the fluids do not flow in uniform velocities across a given cross section of the pipe. The layers in contact with the pipe wall move at velocities close to zero because of friction. As drag forces decrease farther away from the pipe, the layers progressively travel at faster speeds as they near the center, as shown in Figure 12-5. The result is the parabolic shape of the velocity profile shown in Figure 12-4(a), which shows the laminar flow at an R value of 2000 or less. As the R number approaches 3000, the laminar flow becomes nonsymmetrical, as shown in Figure 12-4(b). At very high velocities or low viscosities, the R value is high. When the R number reaches the range of 7000 to 8000, the uniform layers break up and develop into turbulent eddies that travel in all directions in the fluid stream, as shown in Figure 12-4(c). This mixing action tends to produce a flow rate velocity that is constant across the full stream profile. The R number range of 3000 to about 7000 is the transition range between laminar and turbulent flow.

EXAMPLE 12-3

Determine the Reynolds number for the following values given for each factor used in the equation. Indicate if the flow is laminar, nonsymmetrical, or turbulent.

Solution

$$R = \frac{VDp}{u}$$

V = 2.3345 m/s
D = 0.00508 m (inside diameter)
p = 250 kg × m^3
u = 0.002 pa × s
R = 1482, therefore laminar

Fluid Flowmeter Classification

One method of classifying flowmeters is to divide them into the following four categories:

1. Differential Pressure
2. Positive Displacement
3. Velocity
4. Direct Reading Mass

Differential Pressure Flowmeter

ELE1407
Orifice Plate
Flowmeters

The **differential pressure flowmeter** is the most common type of instrument used to measure the flow of fluids through a pipe. These instruments account for well over 50 percent of all flow-measuring devices used in the process industry.

Differential Pressure Meter Figure 12-6 illustrates the operation of this device. A restriction on the flow, called an *orifice*, is installed in the pipe between two flanges. An orifice is a metal plate with a hole of a specified size bored through it. The purpose of the orifice is to reduce the area that the fluid can flow through. According to the law of conservation of mass, the mass (fluid) entering a pipe must equal the mass leaving the pipe during the same time period. Therefore, the velocity of the fluid that leaves the orifice is higher than the fluid that approaches it. According to Bernoulli's principle, as the velocity of a fluid increases, pressure decreases. The result is that there is more pressure on the

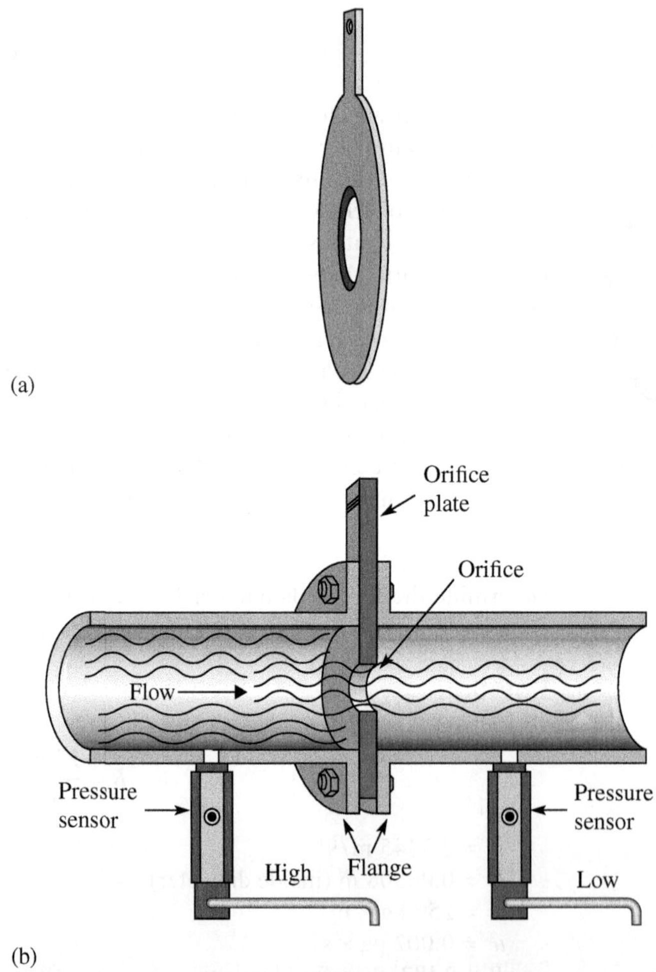

FIGURE 12-6 Differential pressure flowmeter: (a) Orifice plate; (b) Cutaway of pipe and orifice plate

incoming side of the orifice than on the outgoing side. As the fluid flow rate increases, back pressure on the incoming side increases as the orifice is restricting the flow. The pressure of the fluid on the outgoing side of the orifice also increases, but not as much as on the incoming side. That is why the differential pressure across the orifice plate increases. If the exact relationship between differential pressure and velocity is known, velocity can be calculated from an inferred differential pressure measurement and used to determine volumetric flow rate.

The flow rate of a liquid through an orifice plate increases in proportion to the square root of the pressure difference on each side. For example, if the flow rate doubles, the differential pressure is increased by 4. This relationship is shown mathematically by the formula,

$$Q = K\sqrt{\Delta P}$$

where,

Q = Flow rate
K = A constant determined by the orifice size and type of liquid
P = Differential pressure across the orifice plates

EXAMPLE 12-4

Suppose the differential pressure increases by 25. How much has the flow rate increased if $K = 1$?

Solution

Formula: $Q = K\sqrt{\Delta P}$

Original Flow Rate Value: $Q = 1\sqrt{1} = 1$

New Formula Value: $Q = 1\sqrt{25} = 5$

Therefore, the flow rate has increased by 5.

The signals from the sensors of each side of the orifice plate are sent to a transmitter. The output signal it produces varies in proportion to the square of the flow rate. Therefore, if the flow rate doubles, the signal increases by a factor of 4.

The signal can be made to vary in direct proportion to flow rate by connecting a device called a *square root extractor.* The device will produce a signal that will double if the flow rate doubles. The orifice that converts fluid flow into differential pressure is also referred to as a *primary element*. There are several types of primary elements used, as shown in Figure 12-7, but the orifice style shown in Figure 12-7(a) is the most popular. It is used to measure clean liquids and gases and produces its most accurate reading when measuring turbulent flow. A flowmeter also contains a secondary element. Its function is to convert the pressure difference into a measurement that is used to indicate fluid flow. Various types of detectors, such as piezoelectric sensors, are placed on either side of the plate to detect the pressure difference. The outputs of the detectors are compared, and their differences are converted electronically into the actual flow value.

The disadvantage of the primary element's design in Figure 12-7(a) is that the plate has sharp corners on which solid materials can catch. Therefore, it is not used to measure slurries, dirty fluids, or corrosive liquids. To measure these types of fluids, alternative restriction devices have been developed.

The *flow nozzle* type, shown in Figure 12-7(b), has a constriction with an elliptical contour shape. Since there are no sharp edges at the inlet side, there is less friction to the flow. Therefore, because of its slope, it is used to measure steam and high-capacity applications that deal with dirty or corrosive liquids.

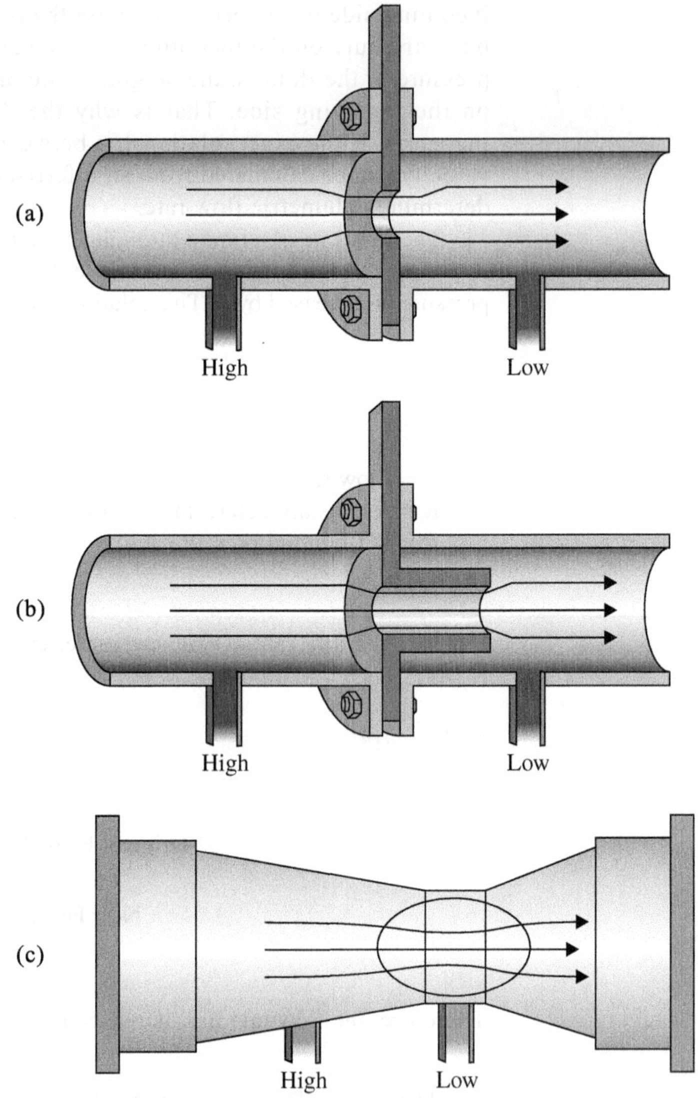

FIGURE 12-7 Three different types of flow restrictors used to convert pressure difference into flow measurements

The *Venturi tube* style in Figure 12-7(c) consists of a converging inlet section in which the cross section decreases in size. The high-pressure reading is taken at the incoming portion of the pipe just before it converges. As the diameter becomes smaller, the velocity of the fluid increases, resulting in a decrease of pressure. The low-pressure reading is taken at the location of the orifice with the smallest diameter. Since the Venturi tube has no sudden change in contour, solid particles tend to slide through its throat. Therefore, this style is recommended when measuring slurries and dirty fluids. The disadvantage of this style is that the resulting measurements are not as accurate as those based on plates with sharp-edged orifices.

The advantages of using differential pressure (DP) meters are that they are popular, well understood and inexpensive; they have no moving parts; and they are well suited for most gases and liquids. One limitation is that there is a nonlinear relationship between differential pressure and flow. Therefore, to linearize the signal, a transmitter or controller is used to extract the square root of the differential pressure measurement. Another limitation is that these meters present an obstruction to flow, which results in some unrecoverable pressure loss.

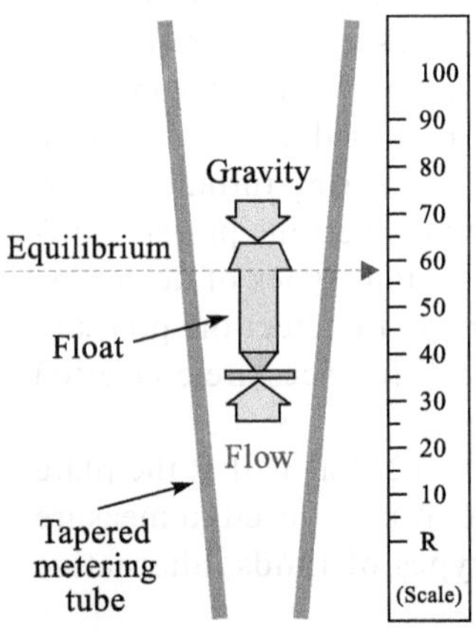

FIGURE 12-8 The rotameter used to measure the flow of liquids and gases

Rotameter The **rotameter** is a variation of the differential pressure flowmeter. It is also known as a *variable-area flowmeter*. The rotameter is shown in Figure 12-8. It consists of a tapered metering tube that is vertically mounted and a float that is free to move up and down

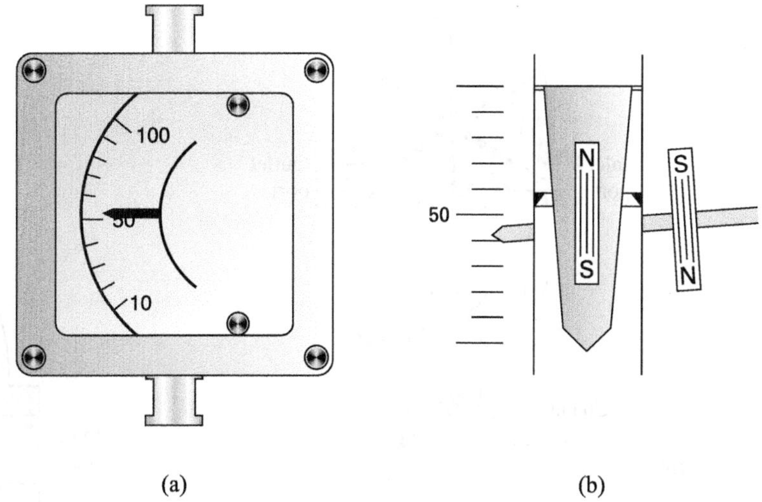

FIGURE 12-9 (a) Rotameter gauge; (b) Measurement principle

within the tube. The fluid to be measured enters the bottom of the tube and exits at the top. Its operation is based on the variable-area principle, where the flow raises the float to allow passage of the fluid.

When there is no fluid flowing through the meter, the float will settle at a location in the tube that has the same diameter as the float. As fluid begins to flow, it pushes the float upward, allowing passage between the tube and the float. The greater the flow, the higher the float is raised, and the area through which the fluid can pass is increased. The movement of the float is directly proportional to the flow rate. A marker on the float is used to identify a number on a measurement scale that indicates the flow rate.

Figure 12-9(a) shows a rotameter with an analog gauge that displays flow rate. A magnet is embedded into the float, as shown in Figure 12-9(b). As the flow rate increases and causes the float to rise, this increase causes the adjacent magnet connected to the pointer to rise and indicate a higher corresponding reading, By connecting the end of the pointer, opposite to the graph, to an electromechanical device, it will enable an electronic transmitter to produce a 4 to 20 mA signal. This signal can then be used to provide a feedback signal for closed-loop control.

When liquid flow is measured, the float is raised by a combination of the buoyancy of the liquid and the velocity force of the fluid. With gases, the float responds only to the velocity force of the gas. The float reaches a stable position when the upward force exerted by the fluid equals the downward gravitational force exerted by the weight of the float. Differential pressure meters are suitable for use with most types of liquids and gases. They are simple in construction, have no moving parts, and are inexpensive. However, they are inefficient because the restrictions cause pressure losses in the system, and their pressure versus flow rate is non-linear across the entire scale.

Positive Displacement Methods

Positive displacement (PD) devices are rotary instruments that mechanically make direct measurements to determine flow. They operate by separating the fluid into segments of known values, and passing them downstream through the pipe. Multiplying the count times the known volume of each segment provides a volumetric measure of flow.

Rotary-Vane Flowmeter The most common type of PD meter is the **rotary-vane flowmeter,** illustrated in Figure 12-10. This meter operates by fluid entering each chamber section through the inlet port. As fluid fills the chamber, it forces the rotor to turn clockwise, as shown in Figure 12-10(a). The chamber is separated by spring-loaded vanes located in channels of the rotor body, as shown in Figure 12-10(b). As the rotor turns, the vanes slide in and

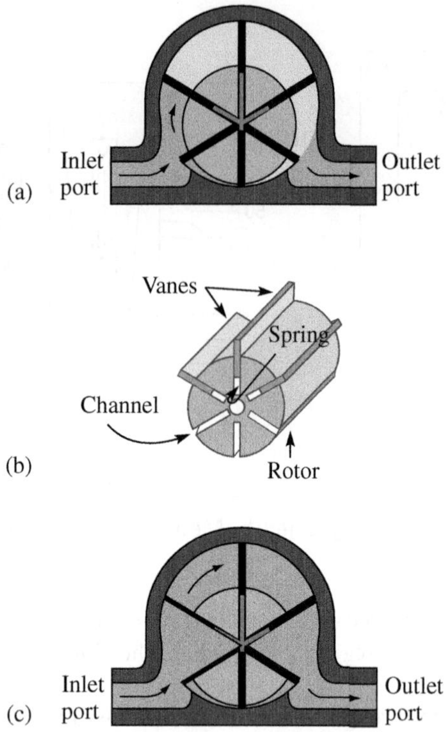

FIGURE 12-10 Rotary-vane flowmeter

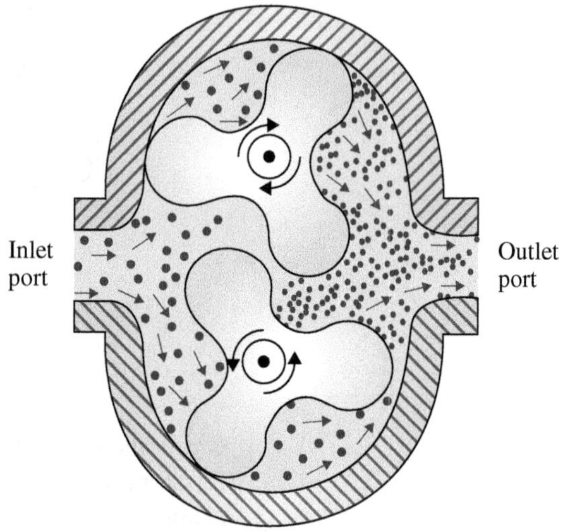

FIGURE 12-11 Lobed impeller flowmeter

out so that they make constant contact with the cylinder wall. The fluid is discharged when each chamber section reaches the outlet port, as shown in Figure 12-10(c). Since the volume of each revolution is known, the volumetric flow rate can be determined by multiplying the displacement times the revolutions per minute (RPM).

Lobed Impeller Flowmeter The **lobed impeller flowmeter,** illustrated in Figure 12-11, is a PD meter constructed with two carefully machined lobes that have very close clearances with each other and with the meter housing. The lobes are geared so that they rotate 90 degrees out of phase with each other. Since they are always in rolling contact with the housing and with each other, they form a seal. The force that turns the lobes is the flowing fluid. Since the volume of fluid required to turn the lobes one revolution is known, a volumetric measurement can be determined. By multiplying the displacement per revolution times the RPM, volumetric flow rate can be measured.

Limitations of PD Meters The measurements taken from PD meters are accurate. However, because they are self-powered, they extract some energy from the system. Also, since they consist of mechanical parts, they are prone to wear.

Velocity Meters

Velocity flowmeters measure the velocity of fluid flow directly. A volumetric flow measurement is determined by the formula $Q = VA$, where Q is volumetric flow rate, V is velocity, and A is area.

Turbine Flowmeter The most common type of velocity meter is the **turbine flowmeter,** illustrated in Figure 12-12. Fluid flow causes a rotation of the turbine that is proportional to the flow rate. The output of the turbine flowmeter is a pickup coil. Stationary flux lines extend from a permanent magnet placed inside the coil to the area in which the turbine blades turn. Each time one of the ferrous blades passes through the magnetic field, the flux lines

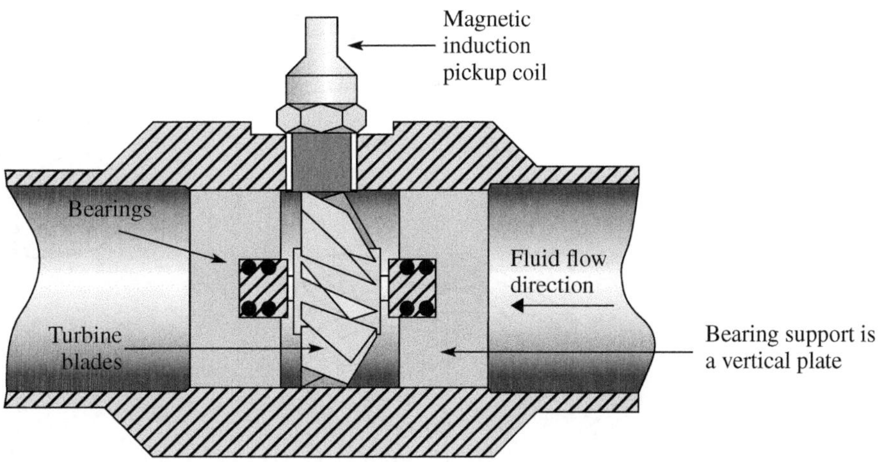

FIGURE 12-12 Turbine flowmeter that measures velocity

become distorted due to a change in reluctance. As the lines are being altered, they cut across the pickup coil, which generates a pulse voltage by induction. The frequency of the pulses is proportional to the rotational speed of the turbine.

The output of the turbine flowmeter indicates volumetric flow rate. These flowmeters can also be used to record *totalization* measurements, that is, the total volume of flow. Each pulse generated by the meter is equivalent to a measured volume of liquid. By feeding each pulse into a counter, the total volume that flows can be recorded and displayed. Totalization measurements are used in applications such as batch processes.

Turbine flowmeters provide high accuracy and repeatability, and their output is linear. However, they are limited to measuring low-viscosity fluids only.

Direct Reading Mass

Directly reading the mass, called **mass flow measurement,** provides the actual weight of the fluid during a given period of time. The readings are made directly, instead of using inferred data from other variables. Since the measurement data is independent of solids, temperature, pressure, viscosity, and other factors that affect fluids, the readings tend to be very accurate. The conveyor system illustrated in Figure 12-2 is an example of a mass flowmeter.

12-5 Electronic Sensors

Several electronic flowmeters have been developed: the Coriolis mass flowmeter, the thermal mass meter, the rotor flow detector, the electromagnetic flow detector, the thermal flowmeter, the vortex flowmeter, the ultrasonic flowmeter, and the time-of-flight meter.

Coriolis Meters

One type of device that measures mass flow of liquids is the **Coriolis mass flowmeter**. It features a U-shaped tube for fluids to flow through, as shown in Figure 12-13(a). Fluctuating currents are sent through coils mounted near the tube. The magnetic forces they generate cause the tube to vibrate, similar to a tuning fork, as shown in Figure 12-13(b). As fluids flow through the tube, kinetic energy is produced by its speed and mass. The energy from the liquid tends to resist the vibrating motion of the tube, causing it to twist sideways, as shown in Figure 12-13(c). The degree of deflection is directly and linearly proportional to the mass of liquid passing through the U-tube. Magnetic position sensors are mounted on both ends of the tube to measure the amount of twist. The outputs from each sensor are conditioned into standard signals before they are sent to display units or to control equipment, as shown in Figure 12-13(d).

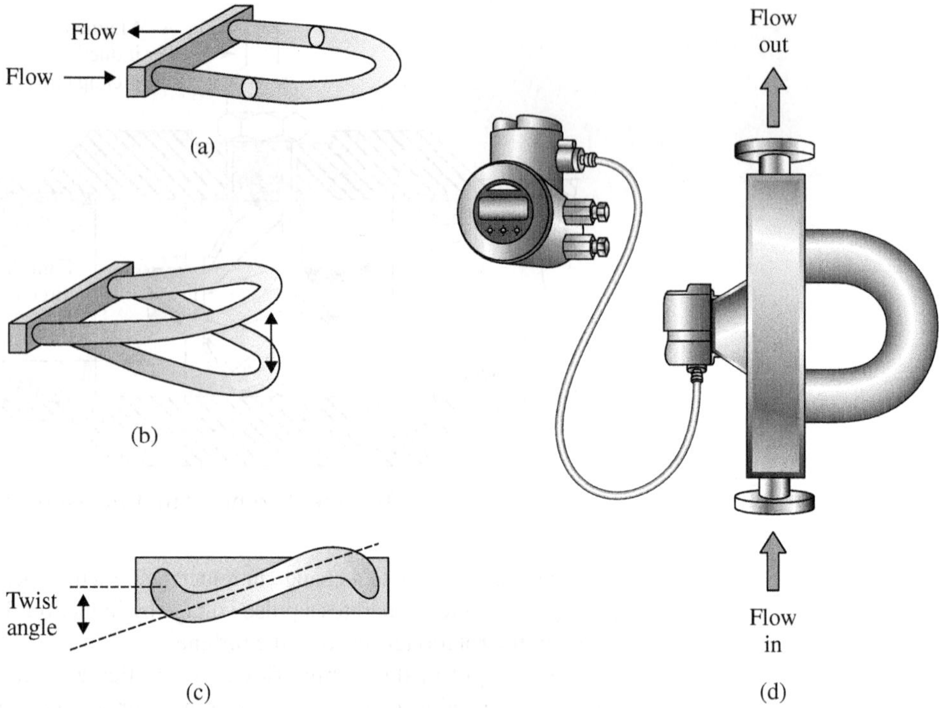

FIGURE 12-13 Coriolis mass flowmeter

Coriolis meters are capable of measuring the mass flow of all types of fluids. However, their accuracy can be diminished if exposed to mechanical noise vibration.

Mass Flowmeters

Mass flowmeters measure the actual quantity of mass of a flowing liquid. Unlike volumetric flowmeters, which are affected if the pressure and temperature of the medium they are measuring are altered, mass flowmeters are not affected as much when these changes occur. Processes that require mass flow measurements are combustion (which is based upon the mass flow rate of air and fuel), gas consumption, and the chemical reaction of ingredients. The two most common types of mass flow meters are the *Coriolis mass flowmeter* and the *thermal mass meter*.

Thermal Mass Meters

A **thermal mass meter** shown in Figure 12-14(a) uses thermal dispersion technology to monitor the flow of gases and liquids. It employs two probes, shown in Figure 12-14(b), that are in contact with the fluid as it flows through a pipe. One of the probes has an RTD sensor that measures the temperature of the fluid, which is the reference of the flow measurement. The second probe has both an active RTD sensor and a heating element (Figure 12-14 (c)). The function of the heater is to establish a temperature differential above the reference temperature. The active RTD measures the temperature of the heater.

As the stream passes over the probes, they are cooled by convection as the fluid absorbs heat. The electronic circuitry is designed to maintain a constant temperature difference between the probes by varying the power applied to the heating element. The power level is monitored because it is the measurement source of the thermal mass meter. The variation in power required to keep the probes at a constant differential temperature is proportional to the variations in mass flow rate. Under low mass flow conditions, there is minimal cooling of the heated element and minimal power required. As the mass flow increases, the cooling of the heated RTD increases, which decreases the differential temperature between the two RTDs.

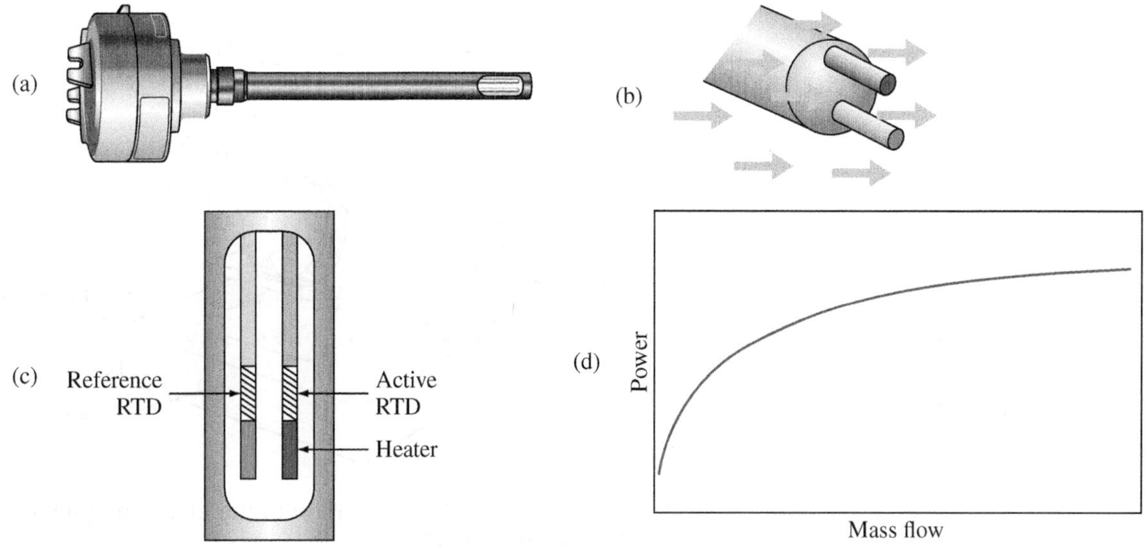

FIGURE 12-14 Thermal mass flowmeter

Therefore, more power to the element is required. The relationship of increasing power as the mass flow rate increases is shown in Figure 12-14(d).

An advantage of thermal mass meters over the mass flowmeter is that they do not require temperature or pressure compensation to make accurate measurements as conditions vary. For liquids, they are capable of making measurements for a wide variety of applications, including slurries, interface, high viscosity, turbulence, and the presence of foam. For gases, they can measure combustion airflow, compressed air/gas, and aeration airflow.

Rotor Flow Detectors

Rotor flow detectors are inserted inside pipes by using tee or saddle fittings. They use a simple paddle wheel design to provide flow indication. Figure 12-15 illustrates the rotor flow detector. A permanent magnet is embedded in each of the four rotor blades. Fluid flow causes a rotor rotation that is proportional to the flow rate. Each pass by a magnetized blade excites a Hall-effect device in the sensor body, producing a voltage pulse. The number of electrical pulses counted for a given period of time is directly proportional to flow volume.

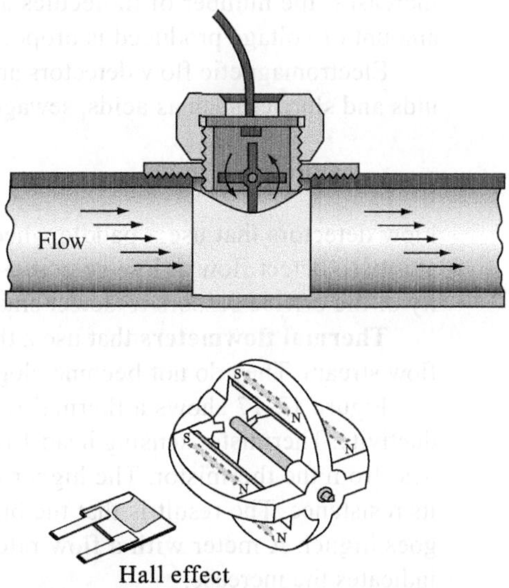

FIGURE 12-15 Rotor flow detectors

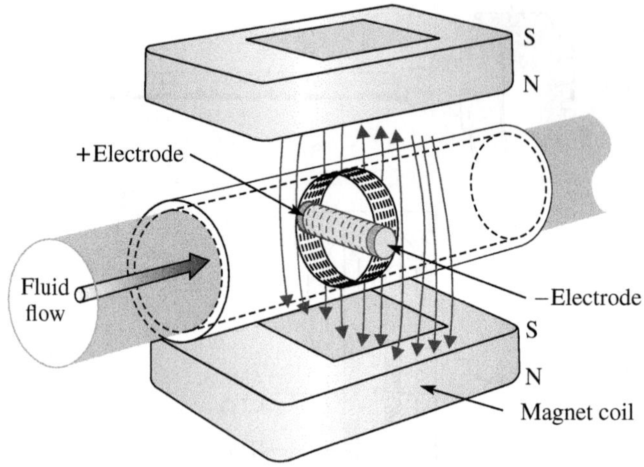

FIGURE 12-16 Simplification of the electromagnetic flowmeter principle

This sensor can measure the flow rate of a wide variety of liquids, including acids, solvents, and most corrosive fluids. They have a flow response of 0.3 to 10 feet per second in pipe sizes with diameters from 0.5 to 36 inches. Rotor flowmeters must be placed on the edge of the flow. If the entire rotor is placed in the flow, the paddle wheel may not turn at all. Turbine flowmeters can be totally immersed within the flow.

Electromagnetic Flow Detectors

The **electromagnetic flow detector** is a transducer that converts the volumetric flow rate of a conductive substance into voltage. Figure 12-16 shows the electromagnetic flow detector. Major components are a flow tube, two electromagnetic coils mounted across from each other outside the flow tube, and two electrodes inside the pipe wall.

The electromagnetic flow detector's principle of operation is based on Faraday's law of electromagnetic induction, which states that a voltage will be induced into a conductor when it moves through a magnetic field. The liquid serves as the moving conductor. The magnetic field is created by the energized coils, which produce flux lines perpendicular to the fluid flow. The induced voltage is measured by the two electrodes. This voltage is the summation of the voltage developed by each molecule in the flowing substance. As the fluid speed increases, the number of molecules a voltage is induced into also increases. Therefore, the amount of voltage produced is proportional to the flow rate.

Electromagnetic flow detectors are generally used to measure difficult and corrosive liquids and slurries such as acids, sewage, detergents, and liquid foods.

Thermal Flowmeters

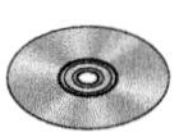

ELE1107 Thermal Flowmeters

Flow detectors that use a paddle wheel or an orifice are susceptible to clogging. Also, their ability to detect flow at low velocities is limited due to the inertia of the wheel or the inability of the orifice sensors to detect small differential pressures.

Thermal flowmeters that use a thermistor have only a sensor tip that is inserted into the flow stream. They do not become clogged and can detect very low flow rates.

Figure 12-17 shows a thermal flow detector. It works on the principle of thermal conductivity. Thermistor sensing head 1 is mounted inside a pipe. As fluid passes, it carries away heat from the thermistor. The higher the rate, the cooler the thermistor becomes, increasing its resistance. The result is that the bridge becomes more unbalanced and the output voltage goes higher. A meter with a flow rate scale is connected across the output terminals and it indicates the increase.

If the temperature of the fluid happens to change, so will the thermistor resistance. To prevent the flowmeter from giving a false reading, a second thermistor sensing head is used.

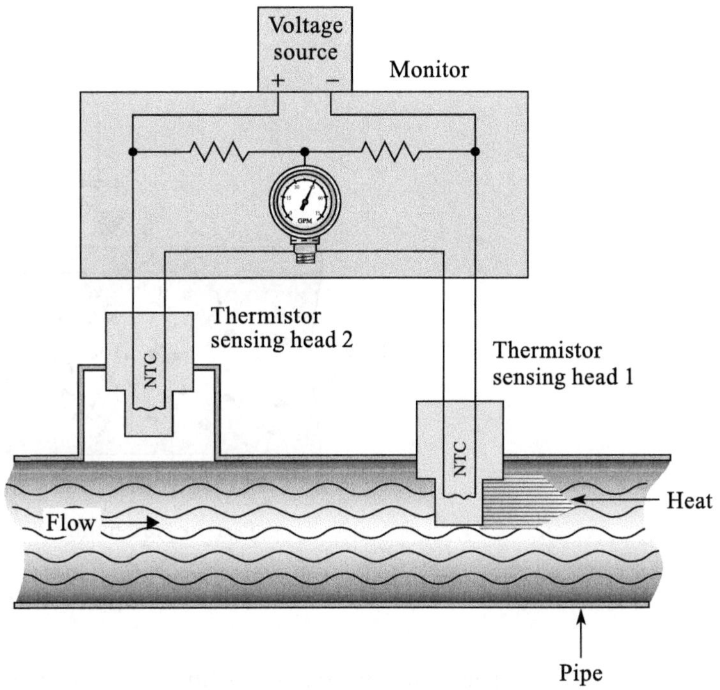

FIGURE 12-17 Thermal flowmeter

Since both thermistors are in the pipe, the fluid temperature affects their resistances equally. Their placement in the bridge causes the resulting voltage changes to cancel each other. Therefore, the only voltage at the output is the one caused by the flow rate. Thermistor 2 is shielded, so its resistance is not affected by the flow.

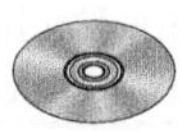

ELE1307
The Vortex
Flowmeter

Vortex Flowmeters

A liquid-measuring device called a **vortex flowmeter** is illustrated in Figure 12-18. A blunt unstreamlined object such as a bar or strut is placed in the flow path of the fluid. As the liquid is forced around the obstacle, viscosity-related effects cause a series of vortices to develop downstream. The swirls are shed from one side of the obstacle and then the other in a predictable pattern. Within a wide range of Reynolds numbers, the number of vortices that appear downstream in a given period is directly proportional to the volumetric flow rate. A pressure detector placed downstream from the blunt object detects the vortices. The sensing element converts the pressure fluctuations into electrical pulse signals. Within the sensor, the electronics convert the frequency into velocity, and velocity into a volumetric flow rate according to $Q = VA$.

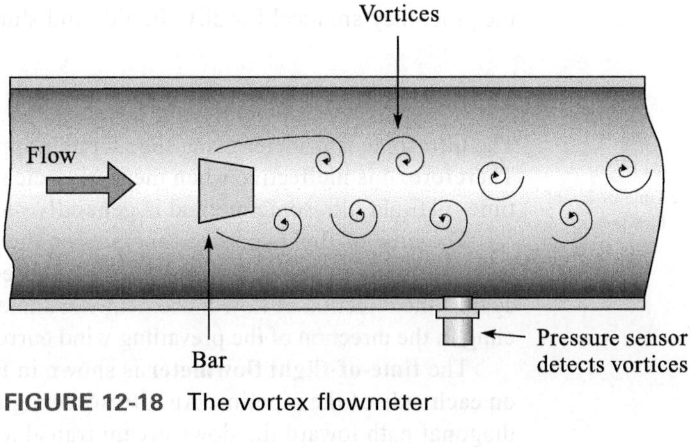

FIGURE 12-18 The vortex flowmeter

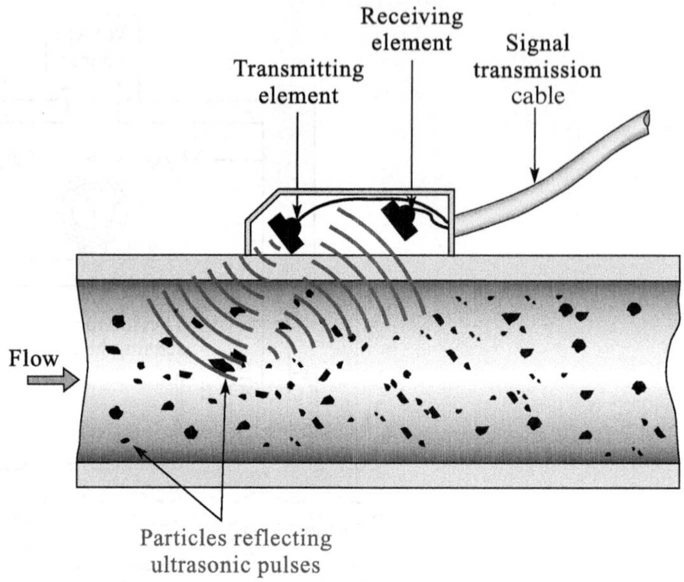

FIGURE 12-19 The Doppler ultrasonic flowmeter

The vortex meter requires little or no maintenance because it is rugged, simple, and has no moving parts. However, since it introduces an obstruction in the pipe, it is limited to measuring only clean liquids to avoid clogging the pipe.

Ultrasonic Flowmeters

ELE5208
Ultrasonic Flowmeters

A liquid-measuring device called an **ultrasonic flowmeter** is shown in Figure 12-19. It operates on a principle of sound propagation in a liquid called the *Doppler effect*. As current pulses are sent by an oscillator through a piezoelectric transducer, it vibrates and produces sound waves that are transmitted upstream into the flowing liquid. Each ultrasonic wave is reflected from particles or gas bubbles in the fluid back to a receiving element. The receiver is a piezoelectric device that detects pressure fluctuations created by the pulsating sound waves. This transducer converts the sound into electronic pulses that are processed by the measuring instrument circuitry.

Because the fluid is moving toward the receiver, the frequency of the reflected pulse received is higher than that of the transmitted pulse. The difference in frequency is proportional to the fluid velocity. As the velocity increases, the received frequency increases to create larger differences relative to the fixed frequency of the transmitter. Therefore, by measuring the frequency difference between the electrical pulses of the oscillator and the pulses at the receiver, flow rate can be recorded by the meter.

Ultrasonic flowmeters are not suited for clean fluids because they require particles from which the sonic pulses are reflected. For this reason, and because the sensor is placed outside the pipe, they are ideal for dirty liquids and slurries.

Time-of-Flight Flowmeter

ELE2207
The Time-of-Flight
Flowmeter

The ultrasonic flowmeter using the Doppler approach requires reflective objects in the fluid. Therefore, it is ineffective when measuring clean fluids. When very clean fluids are used, the time-of-flight ultrasonic method is generally recommended.

The time-of-flight approach operates on the principle that the speed of an ultrasonic sound wave will increase when transmitted in the direction of flow, and decrease when transmitted against the direction of flow movement. An analogy is that an airplane can fly faster when traveling in the direction of the prevailing wind current than it can when flying against the wind.

The **time-of-flight flowmeter** is shown in Figure 12-20. It contains two transducers, one on each side of the pipeline. An ultrasonic signal is sent from the upstream transducer on a diagonal path toward the downstream transducer. As it travels through the flowstream, the

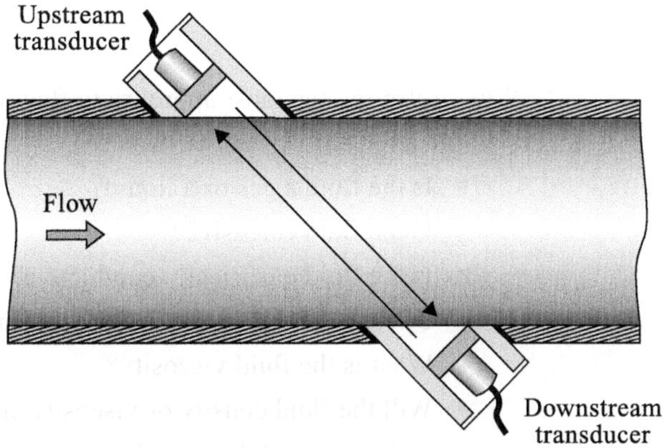

FIGURE 12-20 Time-of-flight flowmeter

natural velocity of the ultrasonic signal is increased by the speed at which the fluid flows. The time at which the signal travels from the upstream detector to the downstream detector is recorded. As soon as the downstream transducer receives the signal, it records the time and sends a signal back to the upstream transducer. By moving against the direction of flow, this signal travels at its natural speed minus the velocity of the fluid. When the upstream transducer receives the signal, the time is recorded. The difference in time it takes for the signals to move in both directions is a direct measure of fluid velocity. This information is electronically converted to volumetric flow rate.

One requirement for this type of meter is that the liquid being measured must be relatively clean. Any particles in the fluid may absorb or scatter the signal and make the reading inaccurate.

The advantages of ultrasonic flowmeters are that they are noninvasive and that the absence of obstructions does not create a pressure loss. Their limitations are that they are relatively expensive and are not as accurate as some types of flowmeters.

12-6 Flowmeter Placement

As fluid encounters obstacles, such as valves or other geometric obstructions in the piping, the flow profile may become distorted and swirl, as shown in Figure 12-21. One of the effects of swirl is that fluid flows in a direction that is not parallel to the pipe, but in a direction across the diameter to the pipe. Fluids flowing in such random paths can take more time to move past the point of measurement and cause the reading to become inaccurate.

To minimize these conditions, the measuring device should be placed 5 to 20 pipeline diameters downstream from an obstruction. One method used to eliminate destructive swirls is to use a flow straightener, as shown in Figure 12-22.

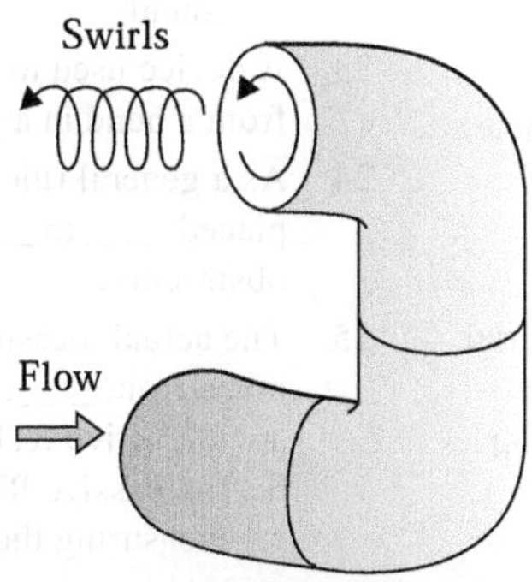

FIGURE 12-21 The swirling current produced by pipeline elbows

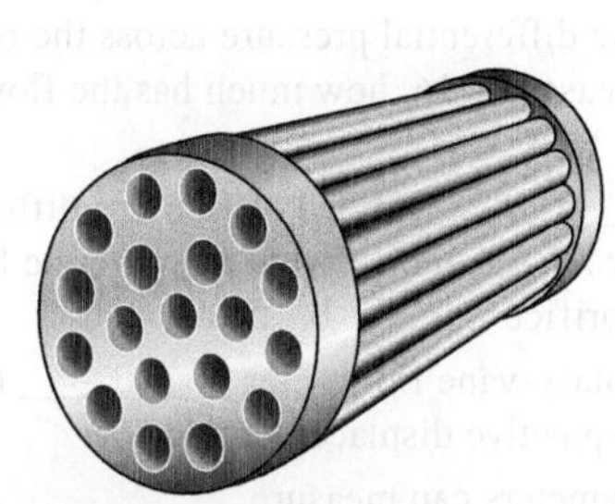

FIGURE 12-22 Flow straightener

12-7 Selecting a Flowmeter

When selecting the most appropriate flow-measuring device for a particular application, the following issues should be considered:

- Is the fluid a gas or a liquid?
- Is the fluid corrosive?
- Is the fluid electrically conductive?
- Does the fluid contain a slurry or large solids?
- What is the fluid viscosity?
- Will the fluid density or viscosity change?
- Is there a need for a noninvasive approach?
- What is the need for accuracy and repeatability?
- What is the cost?

Problems

1. List two units of measurement for volumetric flow rate.
2. List two units of measurement for mass flow rate.
3. A conveyor belt moves at 100 feet per minute and the weighing platform length is 5 feet. What is the mass flow rate if the cell measures 100 pounds?
4. The speed at which fluid moves through a pipe is called __________.
5. If water is forced through a pipe with an inside diameter of 4 inches and a flow velocity of 30 feet per second, what is the volumetric flow rate? What is the mass flow rate?
6. List two factors that influence the density of a fluid.
7. If liquid viscosity increases, the Reynolds number will __________ (increase, decrease) and its ability to flow will __________ (increase, decrease).
8. Explain the importance of using Reynolds numbers for fluid flow.
9. What is the Reynolds number under the following conditions?
   Volume = 3.528 m
   Inside pipe diameter = 0.0041 m
   Fluid density = 300 kg/m^3
   Viscosity = 0.0025
10. The Reynolds number that represents the transition range between laminar and turbulent flow is __________ to __________.
11. If the differential pressure across the orifice plate flowmeter increases by 16, how much has the flow rate increased if $K = 1.22$?
12. A __________-styled orifice in a differential pressure flowmeter can be used to measure the flow of slurry material.
    a. orifice   b. flow nozzle   c. Venturi tube
13. A rotary-vane flowmeter __________ (is, is not) classified as a positive displacement device.
14. Rotameters can measure __________.
    a. gases
    b. liquids
    c. both gases and liquids
15. The Coriolis mass flowmeter __________ (is, is not) capable of measuring slurry material.
16. A velocity meter __________ (is, is not) capable of measuring mass flow.
17. To avoid distorted liquid flow, the sensor should not be placed within __________ pipeline diameters of an obstacle.
18. An electromagnetic flow detector is ____ sensor.
    a. an invasive   b. a noninvasive
19. What kind of sensing element is used to detect the number of swirls in a vortex meter?
20. Thermal flowmeters use ____ to sense the temperature of the fluid flowing in the pipe.
    a. a thermocouple
    b. an RTD
    c. a thermistor
21. The time-of-flight flowmeter operates on the principle that the speed of ultrasonic sound waves ____.
    a. increases when transmitted in the direction of flow
    b. decreases when transmitted in the direction against the flow
    c. both a and b
22. The pulses transmitted upstream in an ultrasonic flowmeter are ____ waves.
    a. electromagnetic
    b. sound
    c. light
    d. infrared
23. A device used to eliminate destructive swirls downstream from a bend in a pipe is the __________ __________.
24. As a general rule, a flow measurement device should be placed ____ to ____ pipeline diameters downstream from an obstruction.
25. The actual measurement of a thermal mass flowmeter is taken from ____.
    a. the active RTD measurement
    b. the passive RTD measurement
    c. measuring the power level applied to the heater

CHAPTER

13

Level-Control Systems

OBJECTIVES

At the conclusion of this chapter, you should be able to:

- Define *level*.
- Describe the importance of measuring and controlling level in industrial processes.
- Define *interface* and list three types of interfaces that may be measured for level indication.
- List four level-measurement units.
- Define *direct level measurement,* and list types and applications of this method.
- Define *indirect level measurement,* and list types and applications of this method.
- Explain the difference between continuous and point level measurements.
- Describe the operation of the following level-indicator devices:

  Rod Gauge
  Sight Glass

- Describe the operation of the following mechanical measurement instruments used to determine level:

  Float
  Displacement
  Bubbler
  Paddle Wheel Detector
  Hydrostatic Pressure Detector
  Differential Pressure Detector
  Weight Detector

- Describe the operation of the following electronic sensors used to measure level:

  Conductive Probes
  Capacitive Probes
  Ultrasonic Sensors

- Select an appropriate level-measuring device for a particular application based on various considerations.

INTRODUCTION

In industrial process control, **level** refers to the height to which a material fills a container. The material can be either a liquid or a solid such as granules or powder. The container can be a tank, a bin, a hopper, a silo, or a vessel.

Two types of techniques are used to control the level of materials in a container: On-Off and proportional. The On-Off method activates a device used to fill a container when the level is too low. When the desired level is reached, the filling operation stops. The proportional method maintains a desired level by filling the container at the same rate at which the material it holds is removed.

Accurate measurements of level are essential to provide good control in the process industries. There are several reasons to monitor the level of materials in containers:

1. To ensure that enough material is available to complete a particular batch-production process.
2. To prevent an industrial accident caused by overfilling an open container. Spilling caustic, hot, or flammable materials could be catastrophic.
3. To prevent the overfilling of a closed container or an enclosed system. This situation could cause an overpressure condition that may result in a rupture or explosion.
4. To determine an inventory of the materials in stock.
5. To prevent a heating element from overheating and being destroyed by ensuring that a container holding heated liquid does not become empty.

13-1 A Level-Control System

An automated system is illustrated in Figure 13-1. It shows a liquid solvent–distribution system located in a factory. Its purpose is to release the solvent into the system after the product in a batch-process vessel has been emptied. As the solvent flows through connecting pipes, tubes, and vessels, it cleans the system in preparation for the next batch.

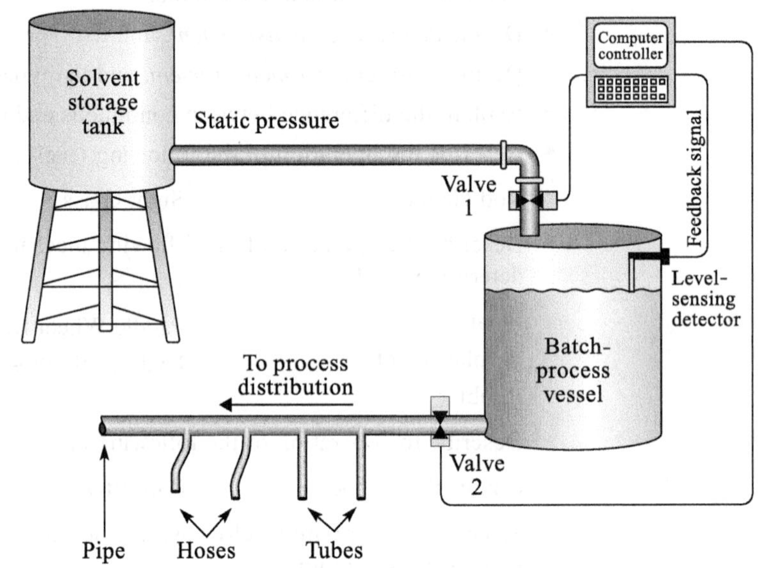

FIGURE 13-1 A level-determination system

Before some type of automatic control can be implemented, it is necessary to determine the level of the solvent with a mechanical or an electronic sensing detector. The control function is initiated by a signal from the sensing device, which indicates the actual level of the batch-process vessel. The solvent is transferred from a storage reservoir that functions as the source, through pipes that serve as the path, to the process vessel that is the load. Until the solvent level reaches the sensor, the controller keeps valve 1 open to allow the fluid to pass through. When the level reaches the sensor, valve 1 is shut off. The controller is programmed to keep the solvent in the vessel for a period of time while the chemical reaction of the liquid performs the necessary cleaning action. After the required time has elapsed, the controller opens valve 2 to enable the solvent to drain into and clean the distribution system. Work is performed when the solvent is in the vessel and again when it is released into the distribution system.

Power Sources

Sources that provide the force required to transfer materials include pumps, static-pressure tanks, and augers.

Pumps

A pump is primarily used to move liquids a required distance or to an elevated level. The pump is powered by an electric motor, a combustion engine, or a hydraulic system.

Static-Pressure Tanks

Static-pressure tanks are employed to transfer liquids and solids a required distance or to a lower level. The tank stores the material. Valves located near the output ports of the tank open and close the pathway, which controls the flow of material as needed. The pressure from the tank provides the force that moves the material. Gravity adds to the force if the flow is vertical.

Augers

The auger is used to move powders or granules in an upward, vertical direction. The auger is driven by an electric motor, a combustion engine, or a hydraulic system.

Transfer Systems

There are various types of distribution systems that serve as the pathway for materials to be transferred from one location to another.

Pipes

A pipe is the primary vehicle used for transporting solids and liquids that need to be moved upward, downward, or horizontally.

Conveyor Systems

The conveyor belt is often used to transfer solid materials. The belt is powered by a motor. The movement is horizontal or on an upward or downward slope.

13-2 Methods of Measurement

Level is measured by locating the boundary between two media, called the **interface.** The media can be liquid and gas, liquid and liquid, liquid and solid, or solid and gas. An example of a liquid and gas medium is water making contact with air in an open vessel. A liquid and liquid medium is two liquids that do not mix, such as oil floating on top of water. An example of a solid and gas medium is powder in contact with air in an open vessel.

Level can be measured *directly* or *indirectly.* The direct method includes measuring with a float or a dipstick. Direct measurement devices are also referred to as *invasive* devices because the sensor is in direct contact with the material. The indirect method, also known as *inferred measurement,* means that a variable other than level is measured and used to *infer,* or indicate, a level measurement through any conversion method. For example, the level of a vessel can be determined by weight, pressure, volume, buoyancy, or electrical properties. Indirect measurement devices are also referred to as *noninvasive* devices, because no part of the sensor comes in contact with the material. Noninvasive devices are preferred when the material is corrosive, hazardous, sterile, or at a high temperature or pressure.

Regardless of the method, level is measured either at a point value or continuously across a range.

Point Level Measurements

Point level measurements detect if the interface is at a predetermined point. Generally, this type of detection is used to signal either a low-level limit when a vessel needs to be refilled or a high-level limit beyond which a vessel will overflow. The output of point level measurement devices typically produces On-Off or 1- or 0-state digital signals.

Continuous Level Measurements

Continuous level measurement locates the interface point within a range of all possible levels at all times. The output of continuous level-measurement devices typically produces an analog signal between 4 and 20 mA, which is both proportional and linear to the level. The electrical signal can be converted into information that represents various quantitative values used to indicate levels. They include:

Height: In units of feet or meters

Percentages: Percent full or percent of measured span

Volume: In gallons, cubic feet, or liters

Weight: In pounds, tons, or kilograms

Continuous measurements are frequently used to maintain the level of a material at a given setpoint.

A factor that must be considered when taking level measurements is the shape of the container. If the vertical walls of a tank are parallel, its volume can easily be determined. If the sides are irregular and not parallel, volume is more difficult to calculate.

13-3 Level-Measurement Methods

Level measurements are made by a number of instruments that use different methods to make the readings. The instruments are classified as either visual observation systems or float and displacement systems, including buoyancy displacement, purge, hydrostatic head, differential pressure, weight, rotational suppression, electrical concepts, and ultrasonic radiation.

The selection of a specific method of measuring level is often based on the following considerations:

Material	Accessibility
Cost	Turbulence
Accuracy	Pressure
Level Range	

Visual Methods

Visual method simply means that a direct measurement is taken by observing the location of the material's top surface in the container.

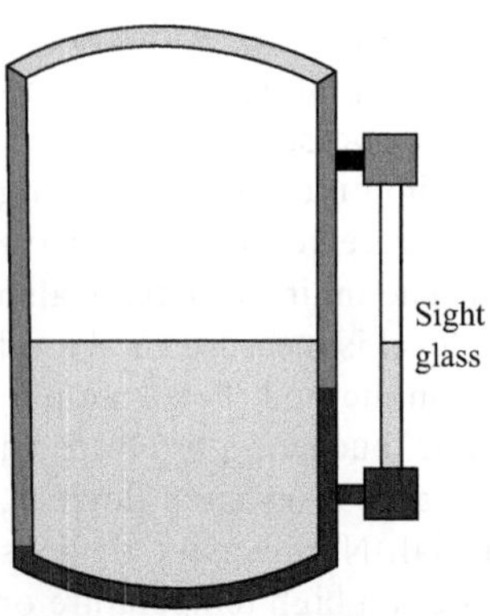

FIGURE 13-2 Level measurement by sight glass

Rod Gauge

A **rod gauge** is a dipstick that is inserted into the material being measured. It is the same type of device used to indicate oil level in a car. It has weighted line markings that indicate depth or volume. This device is very accurate and is often used during the calibration of other level-measurement devices. It is not used to measure hazardous materials, to produce remote indication, or in vessels that are pressurized.

Sight Glass

The **sight glass**, shown in Figure 13-2, is a transparent tube connected to the side of a vessel. As the tank level changes, so does the level in the sight glass. This device provides a direct, local, and continuous measurement. Sight glasses are used for high-pressure applications and for hot or corrosive fluids. They are not useful when foam or viscous liquids are used because visual quality is obscured.

Float and Displacement Methods

Rod gauges and sight glasses give visual indication of level. However, they do not produce a feedback signal in automated control applications. The remainder of the measurement devices described in this chapter can provide feedback information through mechanical linkages or electrical signals.

Buoyancy Method

Float-Type Level Indicator The float-type level indicator is a spherical or cylindrical element that rides on the surface of the liquid as a means of detecting the level. A **float** is a direct, invasive method that provides either point or continuous measurement. By using different types of linkages between the float and the indicating device, such as the gauges in Figure 13-3, a continuous measurement reading can be taken. Figure 13-3(a) shows a high level; Figure 13-3(b) shows a lower level.

A float providing the point method of measurement is often connected to an electrical switch. The switch activates a light or buzzer for high- and low-limit indication. One type of float switch is shown in Figure 13-4. It consists of a magnetic toroid as part of a buoyant device with a guide tube, or stem, through its center. One or more Reed switches are placed inside the guide at desired levels. As the liquid level changes, the float moves. When it comes close to the Reed switch, the magnetic field closes the switch's contacts. The switch closure indicates a point level measurement.

Float devices can also be used for automatic control. Connecting the float element to the stem of a flow-control valve with a linkage allows water to replenish a vessel if the level becomes too low. An example is the float mechanism used in a household toilet tank. To provide more-accurate control, a linkage can connect the float to the potentiometer of an electronic computing device. Small-diameter floats are used to measure higher-density fluids. Larger floats are used for liquid–liquid interface detection or for reading low-density materials.

Displacement Method

Displacement Sensor The **displacement** level sensor is somewhat different from a float sensor in that its probe is weighted so it actually sinks in the measured fluid. However, the

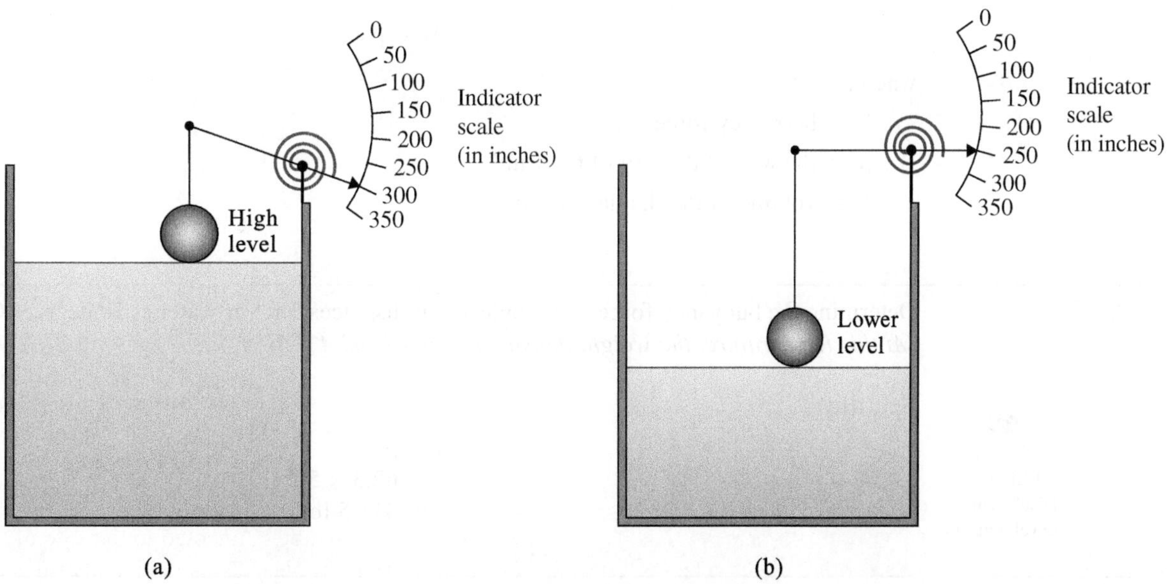

FIGURE 13-3 A float-type level instrument taking continuous measurements

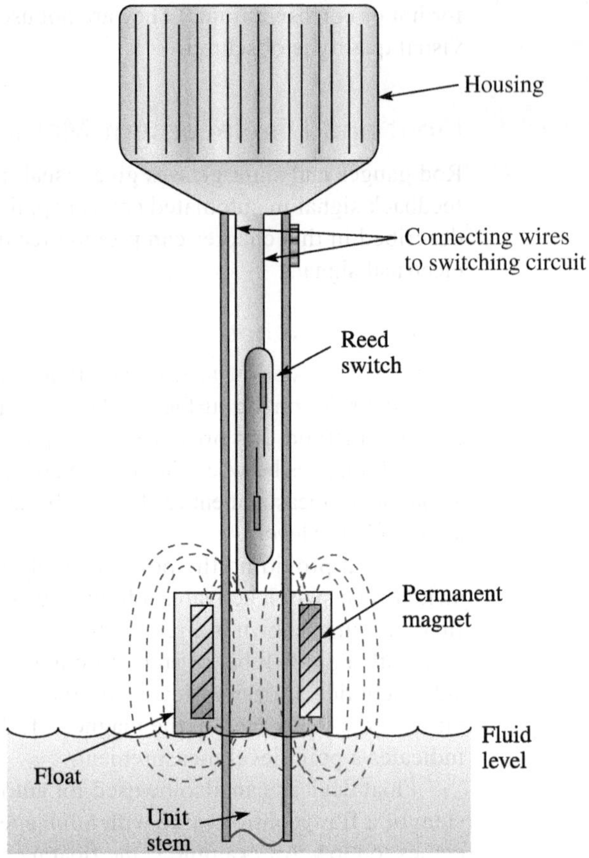

FIGURE 13-4 A magnetic float moves with the liquid level to actuate the magnetic Reed switch within the unit system

probe has bouyancy, meaning it tends to float as the liquid rises up over it. Displacement level sensors operate on Archimedes' principle: A body immersed in a fluid, either partially or fully, is buoyed up by a force equal to the weight of the fluid displaced. The mathematical equation for this principle is

$$B = pV$$

where,

B = buoyancy force

p = the weight density of the fluid

V = volume of the displaced fluid

EXAMPLE 13-1

Determine the buoyancy force on an object that displaces 5 ft^3 of water at 20 degrees Celsius. *(At this temperature, the weight density of water is 62.4 lb/ft³.)*

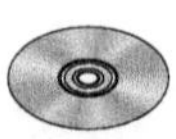

ELE2607
Displacement
Level Sensors

Solution

$$\begin{aligned} B &= pV \\ &= 62.3 \times 5 \text{ ft}^3 \\ &= 311.5 \text{ lbs} \end{aligned}$$

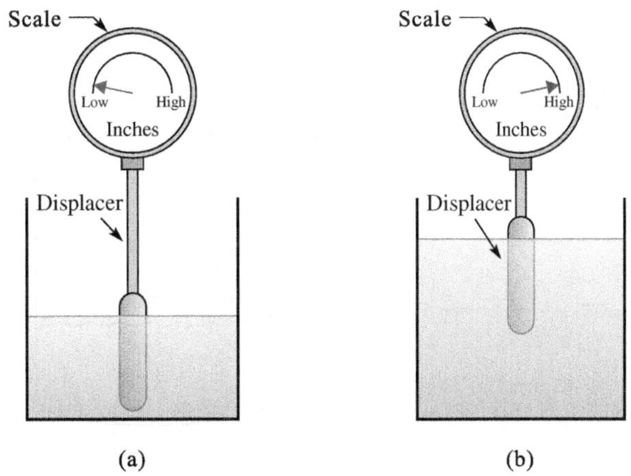

FIGURE 13-5 A displacement level sensor

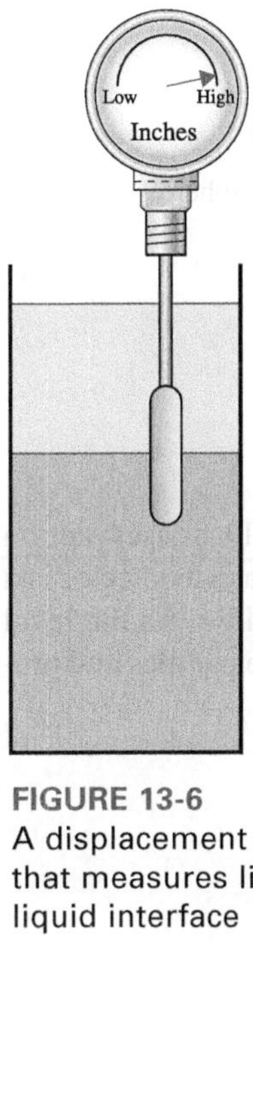

FIGURE 13-6
A displacement sensor that measures liquid–liquid interface

Figure 13-5 illustrates the operation of the displacement level sensor. The displacer displaces the same amount of liquid even when the depth of the liquid changes. This is similar to a boat that displaces the same amount of water regardless of the water's depth. As the level of liquid in the tank changes, the displacer moves by the same amount. The connecting rod slides into, or extends from, the enclosure on which the scale is mounted. As the rod goes up or down, it causes a mechanical linkage to move the needle across the scale on the meter to indicate the level. This device is limited to applications in open tanks and is capable of transmitting electronic signals for remote readings.

Displacement sensors are especially appropriate for measuring liquid–liquid interfaces, that is, two liquids that do not mix together, such as oil and water, different chemicals in the same container, and slurries. Two typical applications are monitoring water condensation in fuel storage tanks and separating chemical emulsions in process systems. Figure 13-6 shows the displacement method used to measure a liquid–liquid interface in a tank. The lighter liquid cannot support most of the weight of the displacer. Therefore, the heavier liquid, which is being measured, causes the displacer to float.

Displacement systems are very accurate, relatively simple, and easily understood and can be used at high temperatures and in pressurized vessels. Their disadvantages are that they are expensive and require the density of the liquid to be constant to make accurate readings.

Purge Method

Bubbler Another pressure-based system is commonly referred to as a *bubbler*. The **bubbler,** or purge, is one of the oldest methods of level determination. This type of system can measure such materials as water, oil, corrosive liquids, molten metal, pulp, and fine powders. A bubbler system at both low and high levels is shown in Figure 13-7. It consists of a dip-tube vertically immersed in the fluid of a tank with an open end placed close to the bottom. A tee connection joins the supply line, the bubbler pipe, and a pressure gauge.

The pressure at the bottom of the tank is hydrostatic head pressure. The more fluid in the tank, the higher the static head pressure. To make a level measurement, the air supply regulator must be adjusted so that its pressure is at least 10 psi higher than the highest hydrostatic pressure to be measured. The result is a flow of air that passes through the dip-tube, producing bubbles in the liquid. As the liquid rises, there is more static head pressure above the outlet of the tube. Therefore, back pressure at the bottom of the bubbler tube increases. The increase in backpressure allows less airflow through the bubbler tube and an increase in pressure at the gauge, causing a larger deflection. The increase in pressure is proportional to an increase in liquid level. Purge instruments are popular level detectors in the paper industry because they are self-cleaning and do not allow the pulp, measured in a vat, to clog the orifice.

ELE406
The Purge Level Measurement Method

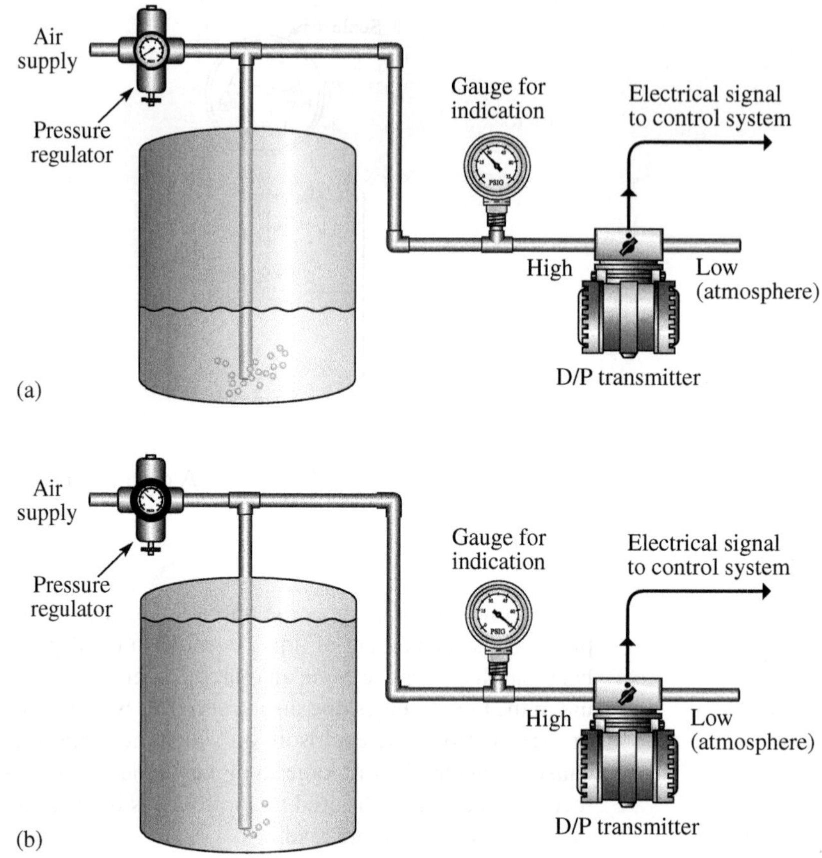

FIGURE 13-7 Bubble level detector

Rotational Suppression Method

Paddle Wheel Detector The **paddle wheel detector**, as shown in Figure 13-8, uses motion to measure the level of either granular or powdered solid material. This invasive detector is a point-measuring device that spins freely when the level is below the paddle. As the level rises, the presence of the material is detected when it makes contact with the paddles and prevents the wheel from turning.

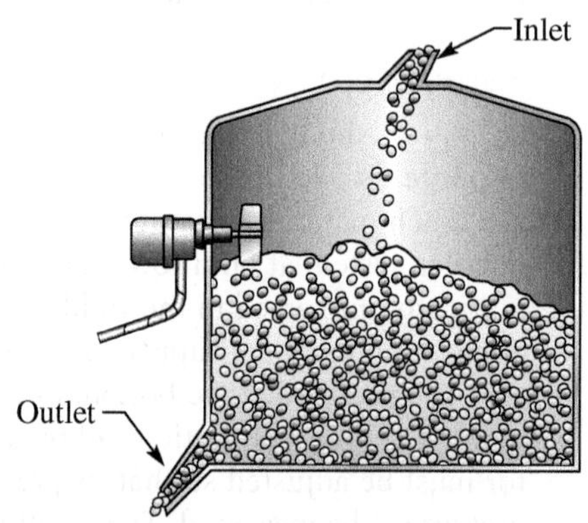

FIGURE 13-8 Paddle wheel detector

Optical Liquid-Level Sensor

An *optical liquid-level sensor* is an instrument that makes point-level measurements of liquids. Its operation is based on the principle that the direction of infrared or visible light is altered as it passes through the interface between two media. The change in direction is called *refraction*, which is what also causes a boat oar to appear bent when part of it is submerged in water.

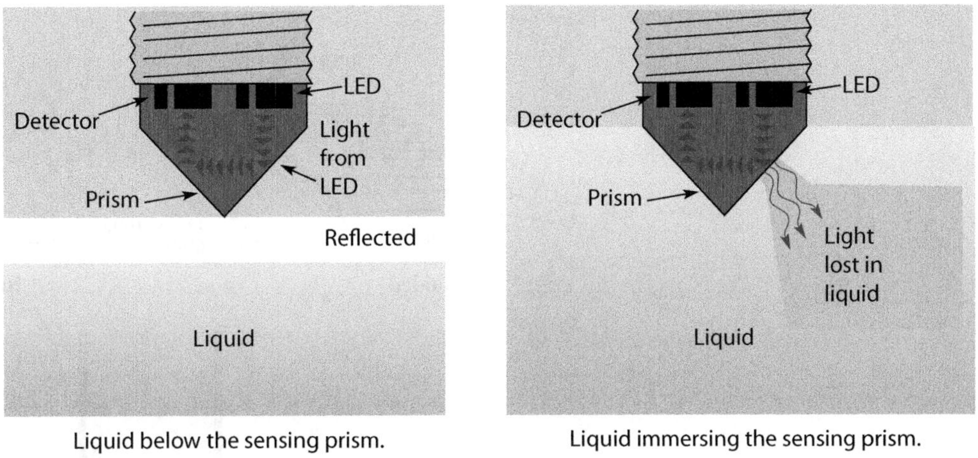

FIGURE 13-9 Optical liquid-level sensor

The sensor has an LED light source and a light detector mounted inside a glass or plastic cone-shaped prism, as shown in Figure 13-9. When the sensor is above the liquid, most of the light from the LED is reflected twice inside the prism and directed back to the detector. When the liquid level is high enough to submerge the prism, most of the light refracts into the liquid, and the amount of reflected light that reaches the detector drops substantially. Therefore, the reduction of the reflected light signal indicates that the liquid has made contact with the sensor.

Hydrostatic-Pressure Method

Hydrostatic-Head Level Detector

The **hydrostatic-head level detector** is a pressure detector that determines level in an open container using the indirect, or inferred, measurement technique. It operates on the principle that any column of material exerts a force at the bottom of the column due to its own weight. This force is called hydrostatic pressure or head pressure. Hydrostatic pressure is determined by the following formula:

$$\text{Pressure} = \text{Height} \times \text{Density}$$

As the height of the material changes, there is a proportional change in pressure. By placing a pressure gauge at the bottom of the vessel, the level of the material can be determined using the following formula:

$$\text{Height} = \frac{\text{Pressure}}{\text{Density}}$$

One common method of measuring the level of contents by weight in a vertical vessel is by using load cells. A load cell has a strain gauge bonded to a robust support capable of bearing the weight of the vessel, as shown in Figure 13-10. When weight is applied to the strain gauge, it deforms and causes its electrical resistance to change. By connecting the strain gauge to a bridge circuit, an output voltage is produced that is proportional to the force. This voltage can be converted by electronic circuitry to a level measurement when the dimension of the vessel and the weight of the particular contents are known. This conversion also requires subtracting the tare weight (weight of the empty vessel) from the total weight to make the calculation. The conversion process requires the following five steps:

1. Weigh the container tare weight.
2. Weigh the container with the contents.

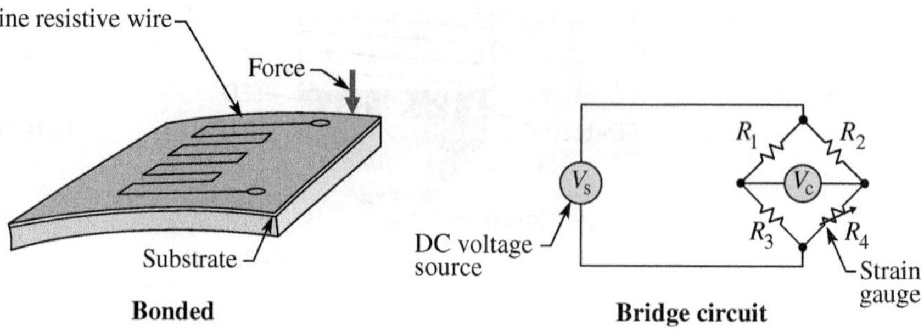

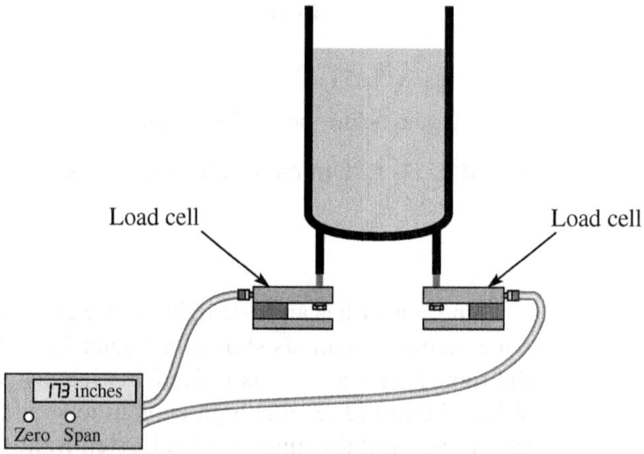

FIGURE 13-10 Load cell measurement

3. Determine the weight of the contents using the following formula:

$$\text{Contents Weight (lbs)} = \text{Measured Weight} - \text{Container Weight}$$

4. Determine the volume using the following formula:

$$\text{Volume (cubic feet)} = \frac{\text{Contents Weight (lbs)}}{\text{Density (lbs/ft}^3\text{)}}$$

5. Determine level using the following formula:

$$\text{Level (feet)} = \frac{\text{Volume (ft}^3\text{)}}{\text{Surface Area (ft}^2\text{)}}$$

EXAMPLE 13-2

Water at 60 degrees Fahrenheit is stored in an unpressurized tank. A pressure gauge that displays readings in pounds per square inch (psi) indicates a value of 100 psi. To determine the level of the water, divide the pressure by 0.43, which is the weight of a 1 inch × 1 inch column of water 1 foot high (at 60 degrees Fahrenheit).

Solution

$$\text{Level (Height)} = \frac{\text{100 psi (Pressure)}}{\text{0.43 psi per ft (Density)}}$$
$$= 232.56 \text{ ft}$$

This type of pressure detector is capable of measuring the level of solids and liquids.

Differential-Pressure Method

Differential-Pressure Level Measurement Hydrostatic pressure measurements determine the liquid level in an open container where the top is exposed to the atmosphere. When a liquid level inside a pressurized tank is determined using this method, the gauge will measure not only the fluid but the pressure above the liquid as well. If the tank is exposed to heat, the space above the liquid becomes a vapor. As the temperature to which the tank is exposed rises, the vapor pressure increases and pushes down on the top surface of the liquid, which causes the pressure at the bottom of the tank to increase also.

The vessel pressure in the vapor space above the liquid can be compensated for by using a **differential pressure transducer,** as shown in Figure 13-11. This measurement device has two inputs, one called the high-side pressure input and the other the low-side pressure input. The high-pressure input connects to the bottom of the tank to measure the head pressure. The low-pressure input of the D/P transmitter connects to the top of the tank to measure only vapor space pressure. The differential pressure transducer subtracts the vapor pressure from the high pressure to produce a reading that represents the hydrostatic head proportional to the liquid level.

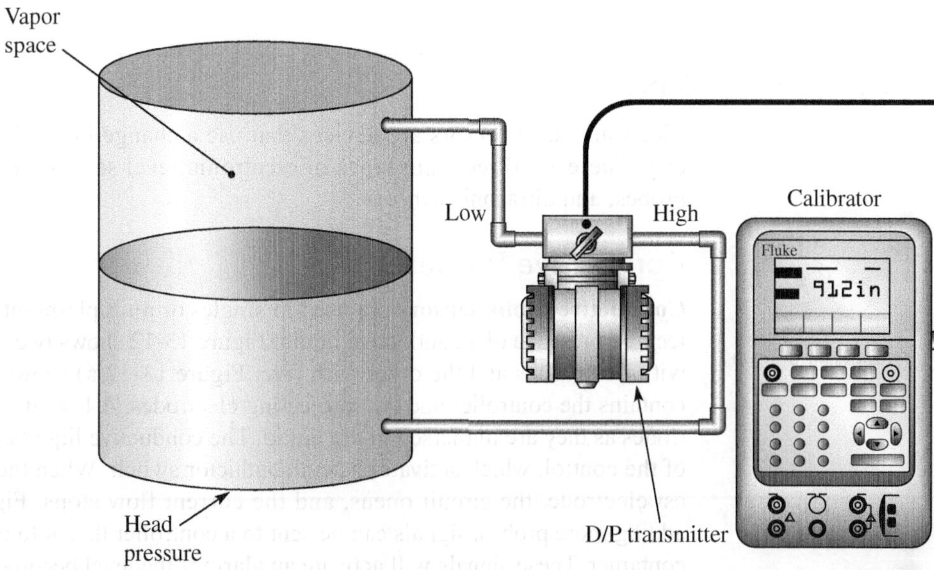

FIGURE 13-11 Differential-pressure level measurement

EXAMPLE 13-3

Determine the level of water inside a cylindrical container with an open top by measuring its weight.

Solution

Step 1: The container is 15 feet high and its inside diameter is 5 feet. It weighs 1,365 pounds.

Step 2: The weight of the container with its contents is 13,611 pounds.

Step 3: Determine the weight of the contents:

$$\text{Measurement Weight} - \text{Container Weight} = \text{Contents Weight}$$

$$13{,}611 - 1{,}365 = 12{,}246 \text{ lbs}$$

Step 4: Determine the volume of the water in cubic feet.

$$\text{Volume} = \frac{12{,}246 \text{ lbs}}{62.4} = 196.25 \text{ ft}^3$$

Step 5: Determine the level of the water.

$$\text{Level} = \frac{\text{Volume (ft}^3)}{\pi r^2}$$
$$= \frac{\text{Volume}}{3.14 \times 2.5^2}$$
$$= \frac{196.25 \text{ ft}^3}{19.625}$$
$$= 10 \text{ ft}$$

Weight is also detected by electronic sensors. The signals from these devices can easily be connected electronically to proportional weight, volume, and level readings for display. Weight level measurements can be used for both solids and liquids.

13-4 Electronic Sensors

Electronic level sensors are devices that use a change in level to change an electrical property. There are three main types of electronic level sensors: conductive probes, capacitive probes, and ultrasonic sensors.

Conductive Probes

ELE2507
Conductive Probe
Sensors

Conductive probe sensors are used in single- or multiple-point measurement systems to detect the presence of a conductive liquid. Figure 13-12 shows two conductive probe sensors, one with two probes and the other with five. Figure 13-12(a) shows the terminal housing, which contains the controller and two projecting electrodes. A low AC voltage is applied to the electrodes as they are immersed in the liquid. The conductive liquid completes the electrical circuit of the control, which activates a semiconductor switch. When the level drops below the shortest electrode, the circuit opens, and the current flow stops. Figure 13-12(b) shows that by adding more probes, signals can be sent to a controller that automatically supplies liquid to the container. These signals will activate an alarm if the level becomes too high or too low.

The advantage of conductive probes is their low cost and simple design. The disadvantage is that they are limited to point measurements and can only be used with conductive liquids. A material is considered conductive when it has a conductivity value greater than 10 micro-Siemens per centimeter.

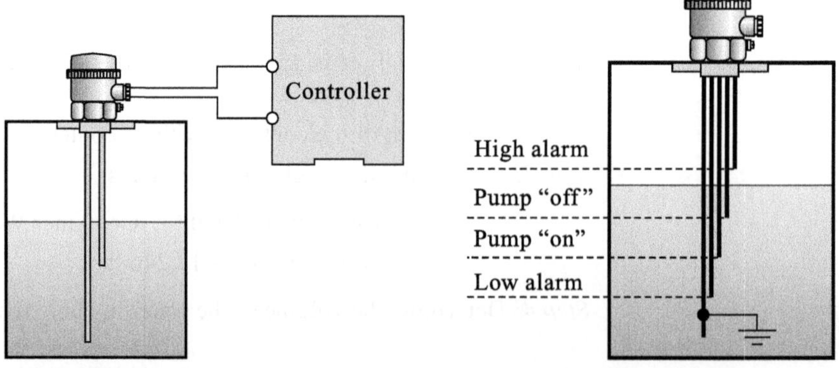

(a) (b)

FIGURE 13-12 Conductive level probes

ELE2407
Capacitive Probe Sensors

Capacitive Probes

Capacitive probe sensors are shown in Figure 13-13. They are used for continuous level measurement. The principle of operation is based on the theory of capacitance. According to this theory, the value of a capacitor can be changed by varying the size of one or more plates or by changing the dielectric. The probe and the metal wall of the tank form the two plates of a capacitor, and the contents in the tank is the dielectric. When a nonmetallic tank is used, a second electrode—referred to as a *counterelectrode*—is used.

When the tank is empty, the dielectric is the air. As the tank fills, the air is displaced by the liquid that has a different dielectric constant. The dielectric constant is a numerical value on a scale of 1 to 100. It refers to the ability of the dielectric material between the plates to store an electrostatic charge. The higher the number, the greater the charge. Air has a dielectric constant of 1, which means that most other materials will store a greater charge. As the dielectric constant value increases, the capacitance also increases. As the material in the tank rises, the higher the capacitance that develops. The capacitance is directly proportional to the level of the measured contents.

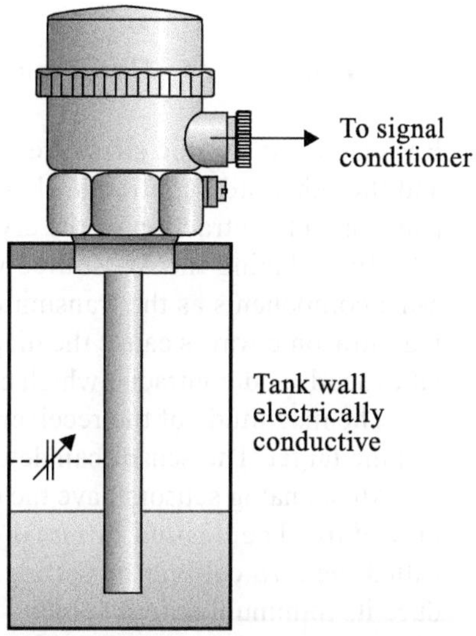

FIGURE 13-13 Capacitive probe

If the medium is conductive, the probe must be coated with an insulating material, usually Teflon, that becomes the dielectric. The wall is not coated, which causes the liquid to become the other plate of the capacitor. As the level of the liquid varies, the size of the capacitor plate becomes larger or smaller and causes the capacitance to change.

If the walls of the vessel are not parallel, the side of the container cannot be used as one of the plates. In this situation, a second probe must be used.

Capacitive level probes are simple to use and are relatively inexpensive. A limitation is that their accuracy is dependent on the condition of the liquid. The presence of solids in the liquid, or exposure to a large temperature change, will cause the dielectric to vary.

IAU106
Ultrasonic Sensors

Ultrasonic Sensors

The **ultrasonic sensor,** another type of continuous level detector, is shown in Figure 13-14(a). Ultrasonic sound waves (above the frequency heard by humans) are developed by an oscillator. They are emitted by a transmitter toward the top surface of the medium and are reflected back to the ultrasonic signal receiver. The time it takes the waves to travel from the transmitter to the target surface and back to the receiver is measured. The time lapse between transmission and detection is proportional to the distance. This data is calculated electronically and converted into a liquid-level measurement. The autofocus mechanism of a camera works on the same principle.

The ultrasonic sending unit consists of a piezoelectric crystal sandwiched between two metal plates. An AC voltage with a frequency of 20 kHz to 100 kHz is applied to the plates.

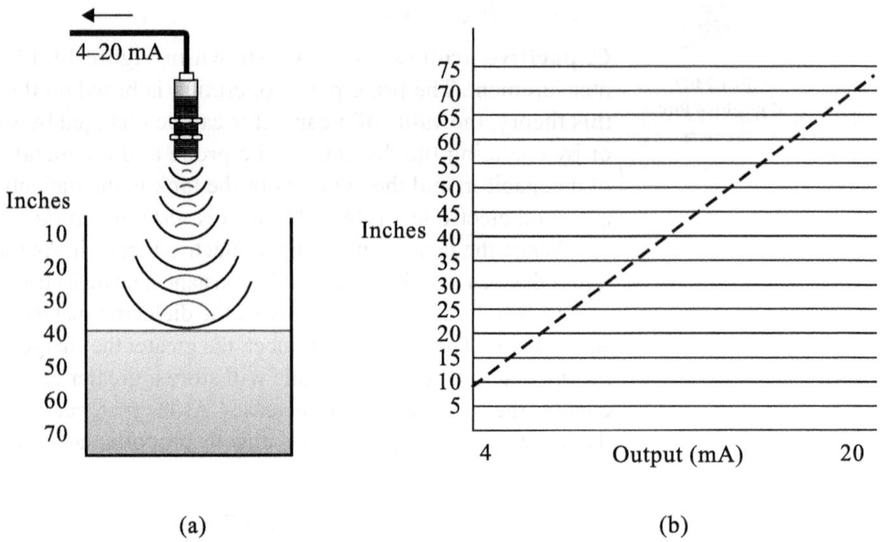

FIGURE 13-14 Ultrasonic sensors

Because of its atomic structure, the side of the crystal connected to one polarity expands, and the other side contracts when the opposite polarity is applied. The high-frequency expansion and contraction of the crystal causes the surrounding air to emit ultrasonic waves.

By replacing the AC source with a voltage amplifier, a second assembly, made of the same components as the transmitter, can operate as an ultrasonic receiving unit. The incoming ultrasonic waves cause the diaphragm to vibrate. The result is that the piezoelectric crystal expands and contracts, which creates a high-frequency AC voltage between the plates.

The magnitude of the receiver's output is proportional to the distance between the sensor and the target. The sensor can detect the object within a given range.

Most analog sensors have the capability of varying the range by performing a calibration procedure. The desired minimal distance at which the target is placed from the sensor is called the *zero* calibration setting. At this distance, the following types of sensors will produce its minimum output value:

Voltage Sensor: 0 V
Current Sensor: 4 mA

The desired maximum distance at which the target is placed from the sensor is called the *span* calibration setting. At this distance, the following types of sensors will produce its maximum output value:

Voltage Sensor: 10 V
Current Sensor: 20 mA

The specific distance of the zero and span settings can be programmed into a microcontroller incorporated into the sensor head.

After the zero and span settings are made, the sensor produces a linear output that is proportional to the distance the target is located from the sensor. For example, an analog voltage sensor will produce 5 volts when the target is halfway between the minimum and maximum sensing distances established during the calibration procedure. Figure 13-15 shows the relationship between the zero and span settings of an analog ultrasonic sensor, both pictorially and graphically.

The graph in Figure 13-14(b) shows that the output current produced by the ultrasonic sensor is proportional to the distance it measures. The controller to which the sensor is connected can be programmed to convert 20 mA to indicate that the tank is empty and 4 mA when it is at full capacity with 75 inches of material. Ultrasonic sensors should not be used in applications where a mist is present in the vapor space above the liquid. The reason is that sound travels through a mist at a different speed than it does through dry air. Inaccurate readings are also made if a layer of foam on the surface is detected.

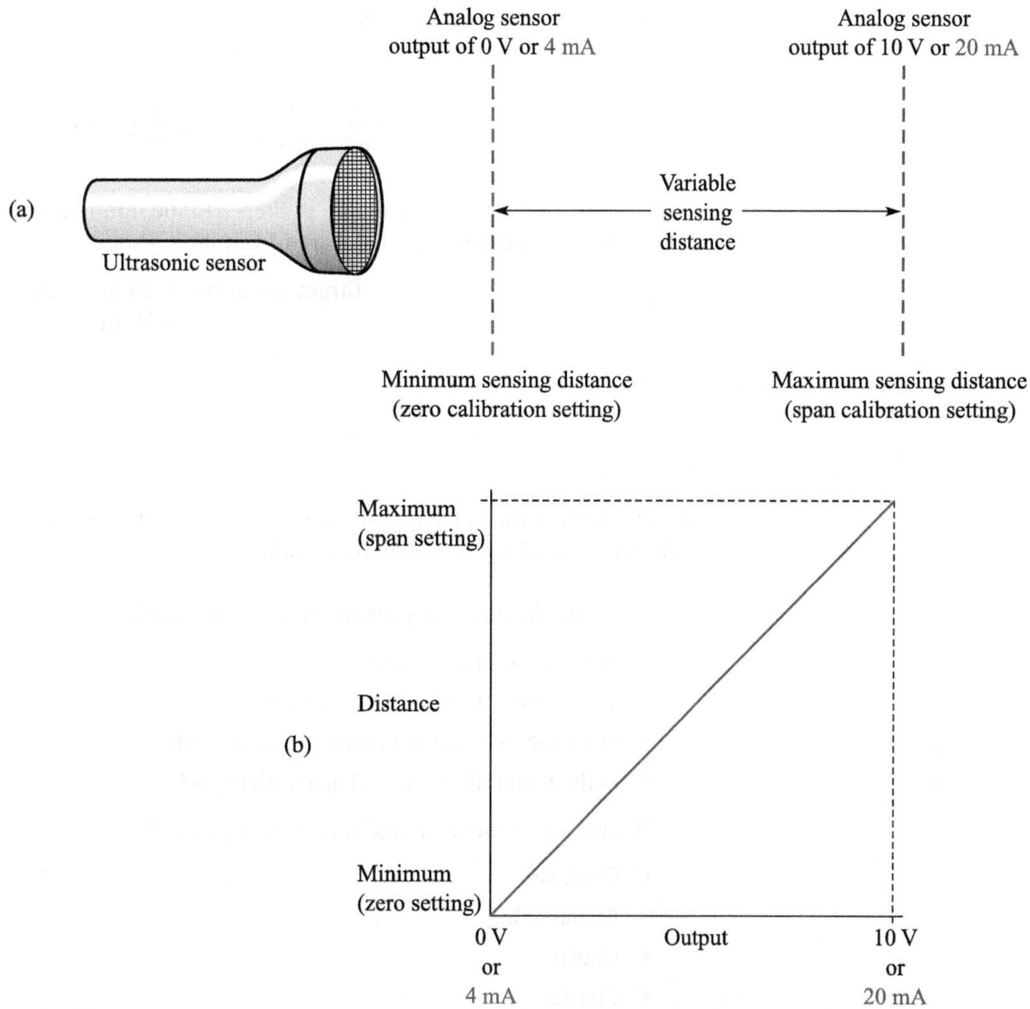

FIGURE 13-15 Zero and span settings of an analog ultrasonic sensor: (a) pictorial drawing; (b) graph

The advantages of ultrasonic sensors are that they are noninvasive, they have a long sensing range, and they have a long lifespan because there are no moving parts.

EXAMPLE 13-4

For an analog ultrasonic sensor that produces an output of 0 to 10 volts, at what distance is the target located when it produces 2 volts if the *zero* calibration setting is for 20 inches and the *span* setting is for 70 inches?

Solution

Step 1: Determine the unit volts per inch. This value represents how much the sensor's output voltage changes for each inch the target moves.

$$U_V \text{ (Unit Volts)} = \frac{\text{Sensor Voltage Range}}{\text{Target Range}}$$

$$= \frac{10\text{ V} - 0\text{ V}}{70\text{ in.} - 20\text{ in.}}$$

$$= \frac{10}{50}$$

$$= 0.2\text{ V}$$

Step 2: Divide the sensor's voltage by the unit volts value to determine how far the target is located from the minimum sensing range distance *(which is used as a reference).*

$$\frac{\text{Sensor Voltage}}{\text{Unit Volts }(U_V)} = \frac{2\text{ V}}{0.2\text{ V}} = 10\text{ in.}$$

Step 3: Add the distance calculated in Step 2 to the minimal sensing distance of 20 inches to determine how far the target is located from the sensor.

$$\begin{aligned}\text{Target Location} &= 10\text{ in.} + 20\text{ in.}\\ &= 30\text{ in.}\end{aligned}$$

13-5 Selecting a Level Sensor

When selecting the most appropriate level-measuring device for a particular application, the following considerations should be made:

What are the physical properties of the medium?

- Is it a solid or a liquid?
- Will foam, vapor, or mist be present?
- Does the material contain chunks or voids?
- Is the material prone to density changes?

What are the chemical and thermal properties?

- Corrosive
- Flammable
- Caustic
- Sterile

These factors are considered along with reliability, cost, and safety.

Problems

1. List two common units of measurement to describe height.
2. List two types of inferred values used to indicate level measurement.
3. List two reasons why it is essential to monitor the level of material in a container.
4. Explain the meaning of the term *interface*.
5. A dipstick is __________ (a direct, an indirect) measurement device.
6. Point measurements are used to determine ____.
   a. high-level limits
   b. low-level limits
   c. both high- and low-level limits
7. Using the weight of an object is ____ method of measuring level.
   a. an invasive
   b. a noninvasive
8. List an indicator method used for level measurements that does not provide a feedback signal.
9. A float is ____ method of determining level.
   a. an invasive
   b. a noninvasive
10. As fluid level rises, the weight of a displacement float indicator appears to ____.
    a. increase
    b. decrease
11. As the fluid level decreases, the back pressure developed in a purge level-measurement device __________ (increases, decreases).
12. Name a common mechanical instrument used to measure the level of solid materials.
13. Where is the pressure detector placed in a vessel to measure the level of a liquid?
14. A differential pressure to determine level is used in __________ (an open, a closed) vessel.
15. Weight measurements can be used to determine the level of ____ in a tank.
    a. solids
    b. liquids
    c. both solids and liquids

16. Capacitance measurements are made to determine the level by ____ method(s).
    a. point
    b. continuous
    c. both point and continuous
17. The ultrasonic instrument uses ____ method(s) to determine the level of foods and pharmaceuticals in a container.
    a. invasive
    b. noninvasive
    c. both invasive and noninvasive
18. Hydrostatic head pressure is measured from the ____ of a full tank.
    a. bottom
    b. middle
    c. top
19. In a purge level-measurement system, the rate at which the bubbles leave the tube ____ as the level of the contents rises.
    a. increases
    b. decreases
20. A conductive probe is capable of making ____ -type measurements.
    a. point
    b. continuous
    c. both a and b
21. Water is stored in a vessel at 60 degrees Fahrenheit. A pressure gauge at the bottom of the tank reads 40 psi. What is the height of the water?
22. At what distance is the target located from a 4–20-mA ultrasonic sensor that produces 13 mA if the zero calibration setting is 2 feet and the span setting is 10 feet?
23. Determine the buoyancy force on an object that displaces 3.2 ft^3 of water at 20 degrees Celsius.
24. When the optical liquid-level sensor is submerged, the intensity of the light source that reaches the detector is ____.
    a. minimal
    b. maximum

CHAPTER

14

Analytical Instrumentation

OBJECTIVES

At the conclusion of this chapter, you should be able to:

- Identify acidic and alkaline solutions and describe how to make them neutral.
- Explain how to treat solutions that are too conductive.
- Describe how proper combustion is achieved, and how to eliminate a volatile condition with a combustible gas.
- Explain how to increase or decrease the moisture content in air.
- Define the following terms:

  Aqueous Solution
  Absolute Humidity
  Combustion
  Conductivity
  Dew Point
  Dissociation
  Effluent
  Humidity
  Hydrocarbon Fuel
  Hygroscopic
  Influent
  Reagent
  Relative Humidity
  Process Stream

- Describe the operation of the following types of analytical measurement instruments:

  pH Electrodes
  Psychrometric Detector
  Conductivity Electrode Probe
  Hygrometric Detector
  Conductivity Inductive Probe
  Electronic Capacitance Detector
  Infrared Gas Analyzer
  Chilled-Mirror Detectors
  Optical Chilled-Mirror Hygrometer
  Aluminum-Oxide Sensor
  Sampling Measurement System
  Adiabatic Expansion Sensor

- List practical applications of the following types of analytical control systems:

  pH
  Conductivity
  Combustion
  Humidity

INTRODUCTION

Some types of industries use chemicals to manufacture their products. When the chemical composition of a product is a variable that must be controlled, a procedure called *analytical measurement and control* is performed. The process for which the chemical is used involves the flow of either a gas or a liquid, and is called a *process stream.* The instrument used to determine the condition of the chemical is called an *analyzer.* Process analyzers can be categorized as those that analyze gases and those that analyze liquids. In this chapter, information about the following types of analytical processes will be provided: pH, conductivity, combustion, and humidity.

14-1 pH Measurement and Control

The concentration of acids and alkaline (bases) in a chemical solution must be controlled in many industrial applications. The analytical process that performs this function is referred to as **pH control.**

The term *pH* represents a unit of measure that describes the degree to which a solution is acidic or alkaline. Most solutions become acidic or alkaline when a compound is mixed with water (anything mixed with water is referred to as an *aqueous solution*). A stable compound by itself is electrically neutral. When some compounds are combined with water, they *dissociate,* which means they break up into two or more charged particles. The charged particles formed are called *ions,* and are created by molecules that either gain or lose electrons. The degree to which a solution becomes acidic or alkaline is determined by the number of positive and negative ions that form, and the percentage of charged particles compared to the neutral, undissociated molecules with which they are combined. This relationship is called *dissociation (ionization) constant,* and is represented by the following formula:

$$K = \frac{(M+)(A-)}{(MA)}$$

where,

K = Dissociation Constant
M+ = Concentration of Positive Ions
A− = Concentration of Negative Ions
MA = Concentration of Undissociated Ions

The number of negative ions compared to positive ions determines whether the solution is acidic or alkaline. If the majority of ions are positive, it is an acidic solution. Conversely, a solution with predominately negative ions is alkaline. Pure water has an equal number of positive and negative ions, and therefore is considered neutral.

The water dissociates into equal concentrations of two types of ions, hydrogen (H+) and hydroxyl (OH−). However, the number of water molecules dissociated is very small in comparison to those undissociated. Its dissociation constant is only 1. When a compound such as hydrochloric acid is combined with water, it breaks up almost completely into many positive charged ions and few negatively charged ions. Therefore, its dissociation constant is practically infinity, which indicates it is a strong acid. When acetic acid is combined with water, fewer than 1 molecule ionizes for every 100 molecules that do not dissociate. Therefore, it has a low dissociation constant, which indicates it is a weak acid. When sodium hydroxide is combined with water, most of its molecules dissociate into negatively charged hydroxyl ions. Therefore, it becomes a strong base. A small concentration of hydroxyl ions form when ammonium hydroxide combines with water, forming a weak base solution.

The measured scale for pH is from 0 to 14. A pH value is directly related to the degree of acidity, or the number of hydrogen ions (H+) that form in a solution. One liter of pure water, for example, has 10^{-7} (0.0000001) gram equivalents of H+ ions. The power of 10 for this quantitative value is 7. Therefore, 7 is used as the pH number for water.

Table 14-1 shows the pH concentration, known as *activity,* through the pH range of 0 to 14. Any number lower than 7 indicates that a solution is acidic. Acid solutions increase in strength as the pH value becomes closer to 0. Any number greater than 7 indicates an alkaline solution. Alkaline solutions increase in strength as the pH value rises above 7 (up to 14).

The table also shows the number of H+ and OH− ions for each pH value. Recall that there are equal numbers of positive and negative ions in water. The table lists 10^{-7} hydrogen ions and 10^{-7} hydroxyl ions. The product of water concentration of H+ ions and OH− ions is 10^{-14}. Regardless of what other compounds are dissolved in water, the product of the concentration of H+ ions and OH− ions is always 10^{-14}. Note that there is an inverse relationship between the number of hydrogen and hydroxyl ions. If the concentration of one type of ion is known, the concentration of the other ion can be determined.

TABLE 14-1 pH Activity

	pH	*Hydrogen ion (H^+)*		*Hydroxyl ion (OH^-)*
Acid	0	(10^0=)	1	0.00000000000001
	1	(10^{-1}=)	0.1	0.0000000000001
	2	(10^{-2}=)	0.01	0.000000000001
	3	(10^{-3}=)	0.001	0.00000000001
	4	(10^{-4}=)	0.0001	0.0000000001
	5	(10^{-5}=)	0.00001	0.000000001
	6	(10^{-6}=)	0.000001	0.00000001
Neutral	7	(10^{-7}=)	0.0000001	0.0000001
Base	8	(10^{-8}=)	0.00000001	0.000001
	9	(10^{-9}=)	0.000000001	0.00001
	10	(10^{-10}=)	0.0000000001	0.0001
	11	(10^{-11}=)	0.00000000001	0.001
	12	(10^{-12}=)	0.000000000001	0.01
	13	(10^{-13}=)	0.0000000000001	0.1
	14	(10^{-14}=)	0.00000000000001	1

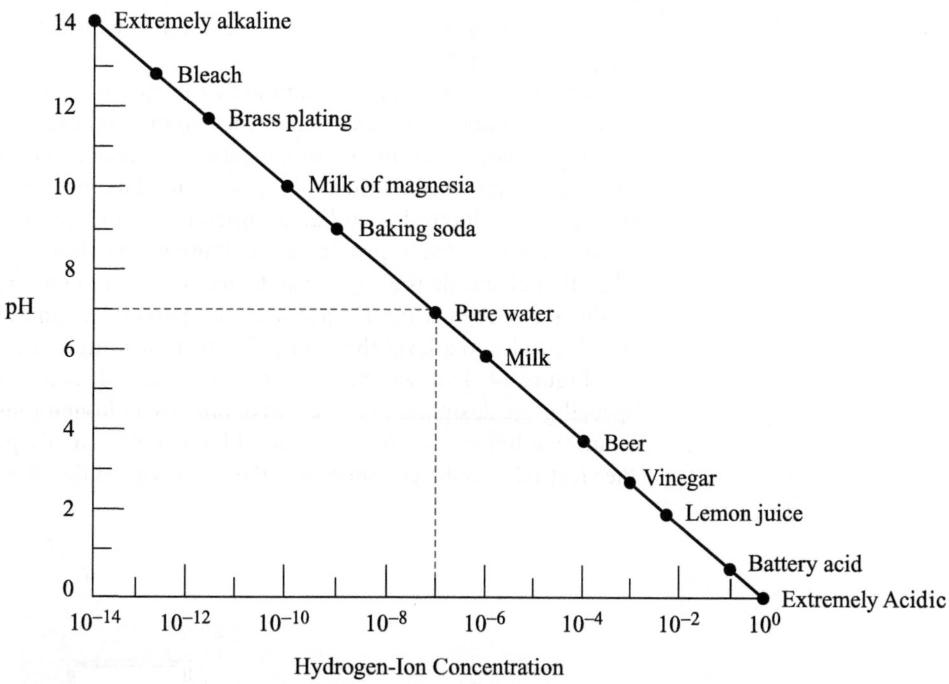

FIGURE 14-1 Relationship of pH and hydrogen-ion concentration

Table 14-1 also shows that a change of one pH unit represents a tenfold change in the hydrogen-ion concentration. Therefore, pH is a logarithmic function. This information helps explain how the term *pH* is derived. "p" refers to the mathematical symbol of the negative logarithm, and "H" is the symbol for hydrogen. Figure 14-1 graphically shows pH values of some common acidic and alkaline products manufactured in the process industry.

The values listed are based on the solution at a temperature of 25°C (77°F), which is referred to as *standard temperature.* As the temperature of the solution varies, the dissociation constant also changes. This situation causes the ratio of H+ to OH– ions to be altered, which changes the pH value. Table 14-2 shows the effect on pure water when it is exposed to temperatures ranging from freezing to boiling point. At the standard temperature of 25°C, the neutrality pH value is 7.0. However, the neutrality value goes above 7 when its temperature is lower than 25°C, and goes below 7 when its temperature is higher than 25°C.

TABLE 14-2 Effect of Temperature on the pH Value of Pure Water

Temperature		*Dissociation Constant*	*Neutral pH*
°C	°F		
0	32	14.94	7.47
25	77	14.00	7.00
50	122	13.26	6.63
75	167	12.69	6.35
100	212	12.26	6.13

To make accurate measurements, the effect of temperature must be considered. By measuring the temperature of the solution, a correction factor calculation can be used to compensate for any temperature effects. Many electronic pH analyzers have circuitry that is designed to automatically make the correction.

pH Measurements

ELE4208
pH Measurements

Devices used to measure pH values detect the concentration of hydrogen ions. Early techniques involved paper indicators. When submerged into the measurement sample, the indicator would produce a color change. The color to which it changed would indicate the value of the pH concentration.

One drawback of this method is that it cannot be used to measure solutions that are colored. Also, it does not provide an electrical measurement signal necessary for automatic control applications. For this reason, electronic sensors were developed to provide a continuous feedback signal used in closed-loop systems. The instrument used to measure pH consists of two separate electrodes and an amplifier (see Figure 14-2). One electrode is the *active* or sensing device, which produces a voltage proportional to the hydrogen-ion concentration. The other electrode is a *reference* device, which provides a potential against which the output of the active measuring electrode is compared. The amplifier boosts the small signal from the electrodes to a level that can either be transmitted or used for display.

Figure 14-3 shows the pH *active electrode.* It consists of a thin-walled tube made of a special glass designed to be sensitive only to hydrogen ions. The bottom of the tube contains a buffer solution, which is a liquid of known pH, usually potassium chloride. The cable from the electrode amplifier connects to the inner wire of the electrode, which is made of silver wire.

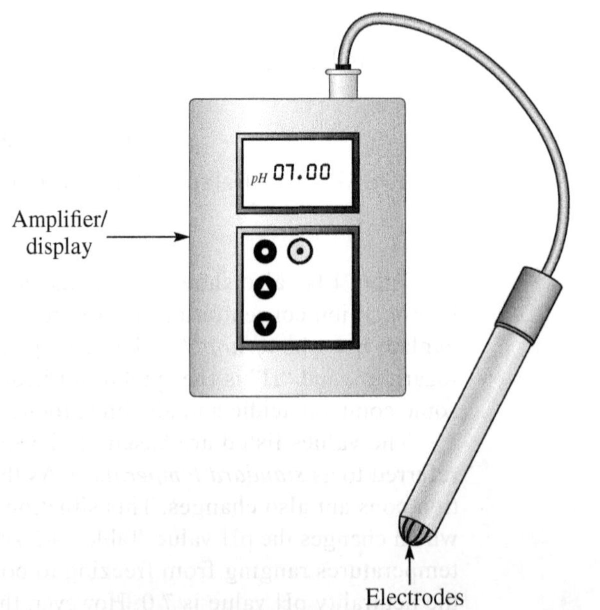

FIGURE 14-2 pH meter

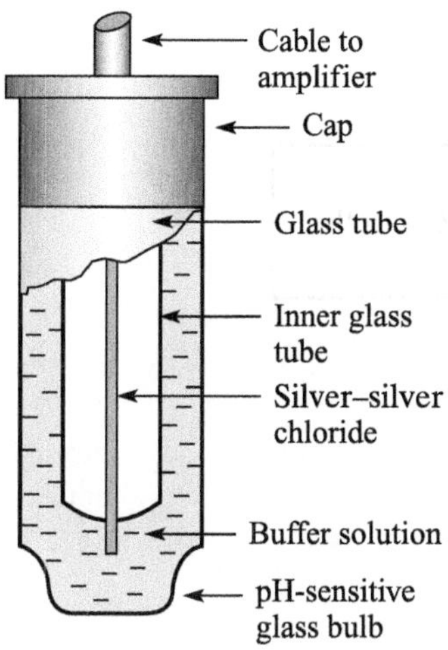

FIGURE 14-3 pH-sensing electrode

The wire is coated with a silver–silver chloride that makes the electrical connection between the inner wire and the buffer solution in which it is immersed.

Figure 14-4 shows the *reference electrode.* It also uses glass to form the outer shell of a tube. The cable between the electrode and amplifier connects to an inner tube wire made of silver that is coated with a paste made of silver chloride. The paste is in contact with potassium chloride (KCl), which is a solution that is an electrical conductor. The purpose of the KCl is to provide electrical contact between the pH process solution being measured and the amplifier. The electrode connection is made through a porous strand of ceramic material. This material acts as a wick and is fitted through the glass at the bottom of the electrode.

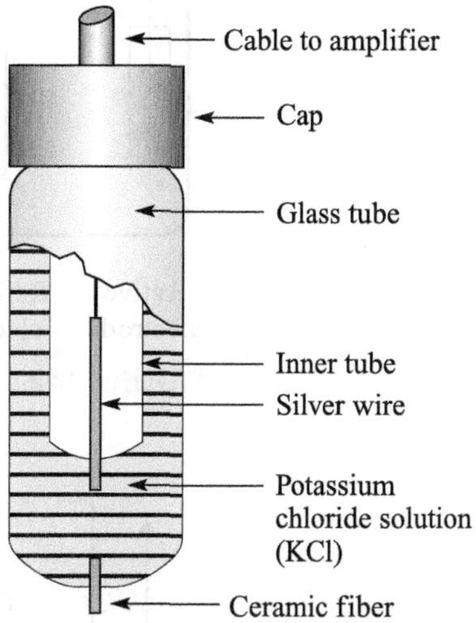

FIGURE 14-4 pH reference electrode

The pH active electrode operates on the principle that a potential (voltage) is formed between two solutions of different hydrogen-ion concentrations. The buffer solution within the electrode has a constant concentration of hydrogen ions at a pH of 7.0. Whenever the hydrogen-ion concentration of the process solution being measured is different from that of the buffer solution, there is a potential difference across the thin glass insulator of the electrode. The magnitude of the voltage is proportional to the pH of the process solution. For every pH unit of the process solution being detected, a potential of 59.2 mV is produced. If the process solution has a pH of 7.0, the potential difference is 0. When the pH is greater than 7.0, a voltage of one polarity is applied to the amplifier from the inner wire of the electrode. When the pH of the solution is less than 7.0, a voltage of the opposite polarity is applied to the amplifier.

The best way to understand the operation of the pH electronic sensor is to compare its operation to a battery measured by a voltmeter. The amplifier operates as the voltmeter. The wire inside the sensing electrode and the amplifier cable to which it is connected function as one lead of the voltmeter. A voltage measurement cannot be made with only one meter lead. The reading can be taken only if a second lead of a voltmeter is used. The reference electrode, which makes the electrical connection between the measured solution and the amplifier, functions as the second lead of the voltmeter. The buffer solution of the active electrode can be compared to one battery terminal, and the pH solution being measured is the other battery terminal.

Some pH amplifiers include a resistive element that is also immersed in the solution being measured. Its resistance changes with the temperature of the solution. Adding this element into the circuit enables the device to compensate for any changes in the pH measurement due to temperature variations. The sensing electrode, reference electrode, and the resistive element are shown in Figure 14-5. These three devices are often contained in the same housing instead of being separate.

Some manufacturers offer several different glass formulations from which probes are made. Each type is tailored to a particular process application. Figure 14-6 shows the classifications for which they are used.

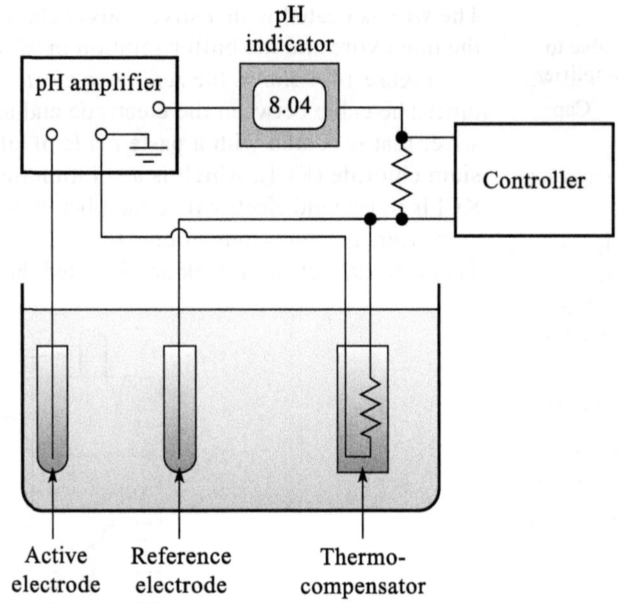

FIGURE 14-5 Electronic pH sensor

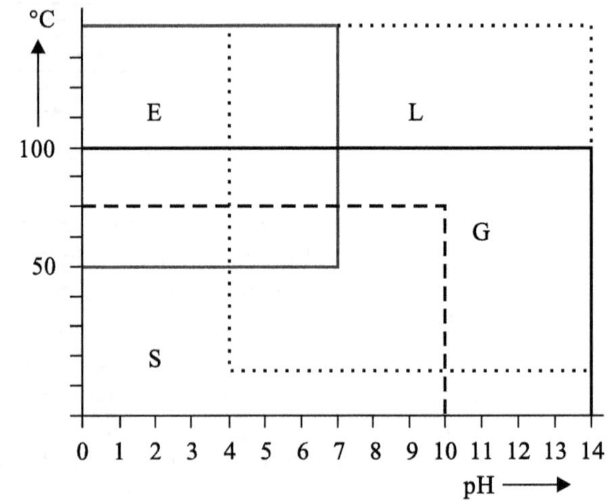

FIGURE 14-6 Glass characteristics

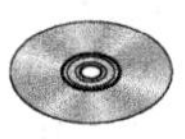

ELE5008 Controlling pH

Controlling pH

One of two conditions typically exists when controlling pH. Either a solution is too alkaline and an acid must be added to decrease the pH, or a solution is too acidic and a base must be added to increase the pH. In both situations, the corrective ingredient, called a **reagent,** must be added at a proper amount and rate to bring the pH to the desired level.

One objective of a pH-control system is to minimize the amount of reagent added to the solution. However, determining and feeding the exact amount is difficult because of the logarithmic characteristics of pH reaction in a solution. A problem that often arises is overshooting. By properly designing a control system, overshooting can be either eliminated or minimized. pH control may be implemented in batch or continuous systems.

Batch Systems

A batch system usually employs a tank to store the solution for treatment. Figure 14-7 shows a batch tank. The solution that enters the tank through an inlet is called an *influent*. After the tank is filled, the pH value of the solution is measured. A controller compares the feedback signal to the setpoint. If there is a difference, it sends a signal to an actuator, which causes a reagent to be applied. Overshoot is prevented and the control action is usually most accurate when the flow rate of the reagent is slow and a stirring action is provided by a mixer. To ensure accurate measurements, the electrode should be located away from the tank inlet and the feed point of the reagent. This separation is required to allow proper mixing before measurements are made. After the solution is treated, the tank is drained. This solution is called an *effluent,* and flows through an outlet port.

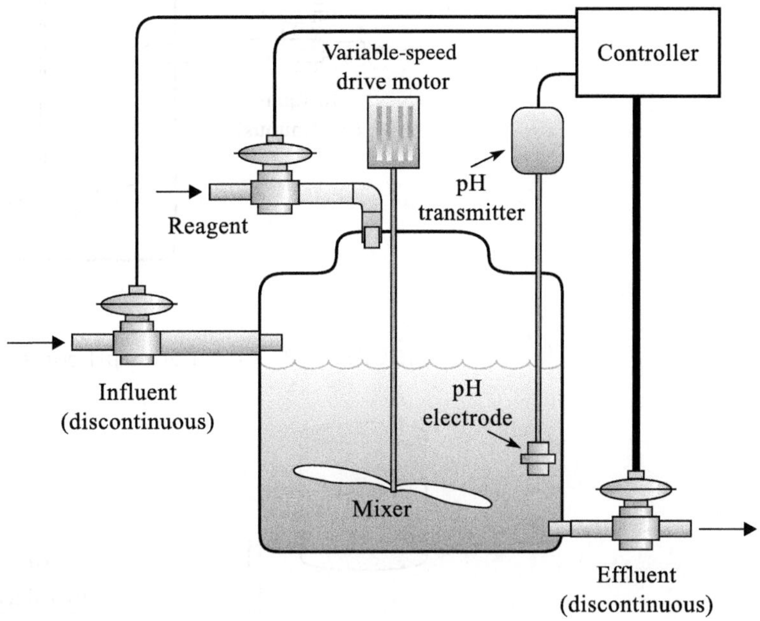

FIGURE 14-7 Batch pH-control system

Batch pH control is typically used when the volume of the solution to be treated is relatively small. A common application is the treatment of waste water. Liquids are collected efficiently in tanks. By adding a reagent and using a mixing action, the solution is neutralized to a pH level of 7.0.

Continuous Systems

Many continuous systems use a tank to treat a solution, as shown in Figure 14-8. A reagent is applied and mixed as a continuous flow of influent enters the tank and the treated effluent is discharged.

If the pH of the influent varies by a small amount in batch or continuous systems, an On-Off controller that drives a solenoid valve is used to apply the reagent. If the pH of the influent varies widely, a proportional controller that drives a control valve should be used.

One disadvantage of using a tank in a continuous-process system is that there is typically a long delay time. This characteristic can cause the process pH level to rise above and below the setpoint by a large amount. In applications that require a pH value to be within 4 and 10, a *static mixer* is used. Figure 14-9 shows a continuous system that uses this device. By injecting the reagent at the mixer input and placing the pH sensor at the mixer output, the reaction time becomes very short.

pH analyzers are used in practically every industry that uses water within its process. Applications can be found in water- and waste-treatment operations; power-generating plants; petroleum refineries; and in food processing, pulp and paper, metal production, pharmaceutical, and chemical industries.

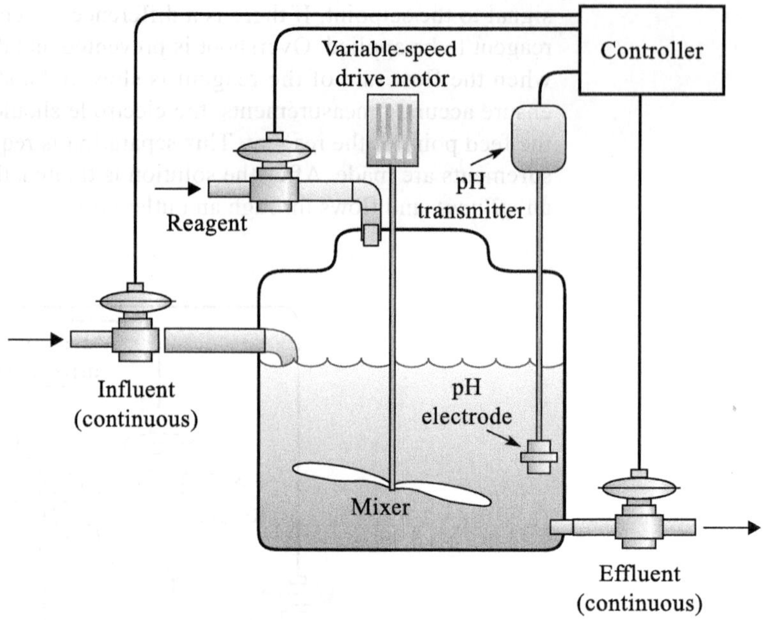

FIGURE 14-8 Continuous pH-control system

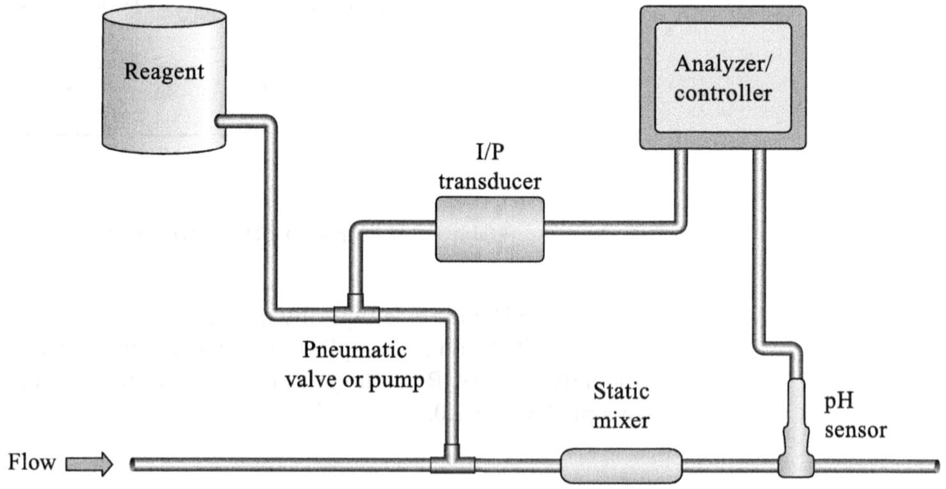

FIGURE 14-9 Continuous on-line control system using a static mixer

14-2 Conductivity

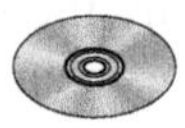

ELE3008
Conductivity

Any process that involves liquid requires flow. The flow may be continuous, or discontinuous, such as in a tank that is periodically filled or drained in a batch application. The liquid that is used in these processes is referred to as a *process stream.* Process streams contain either pure water, water, or solutions that are combined with other ingredients. In many manufacturing applications, it is necessary to determine either the purity of the water or the concentration level of the solutions dissolved in a liquid. This information may be obtained by using a measurement procedure called **conductivity.**

Conductivity refers to the ability of a material to pass electric current. The symbol for conductance is G, and its unit of measurement is siemens. The following formula shows that conductance is the reciprocal of resistance (R):

$$G = \frac{1}{R}$$

All liquids possess conductivity to some degree, ranging from low to very high. High conductivity indicates that electrons can flow easily through a liquid because it contains a large number of ions. These ions are formed by dissolving *electrolytes* such as acids, bases, and salts in water. Low conductivity indicates that less current flows because of the higher resistance caused by the presence of fewer ions in the liquid. An example is distilled or pure water. The only ions that are present are hydrogen and hydroxyl formed from the dissociation of water itself. Unlike pH measurements, which respond specifically to hydrogen ions, conductivity measurements respond to any and all ions in solutions. The degree of electrical conductivity of a liquid is determined by the following factors:

1. The concentration of an ingredient dissolved in water, ranging from zero to very high
2. The type of electrolyte contained in a dissolved ingredient
3. The temperature of the liquid

Several kinds of sensors, or probes, are used to measure the conductivity of a process stream. The two main types are the electrode probe and the inductive probe.

Conductivity Electrode Probe

Conductivity measurements can be obtained by immersing two plates in the process liquid. Shown in Figure 14-10(a), the plates, called *electrodes*, are placed parallel to each other. A known voltage is applied across the electrode, which causes a current to flow through the solution from one plate to the other. The current, which is linearly related to conductivity, is measured by the analyzer and converted by its circuitry to a conductivity reading for display.

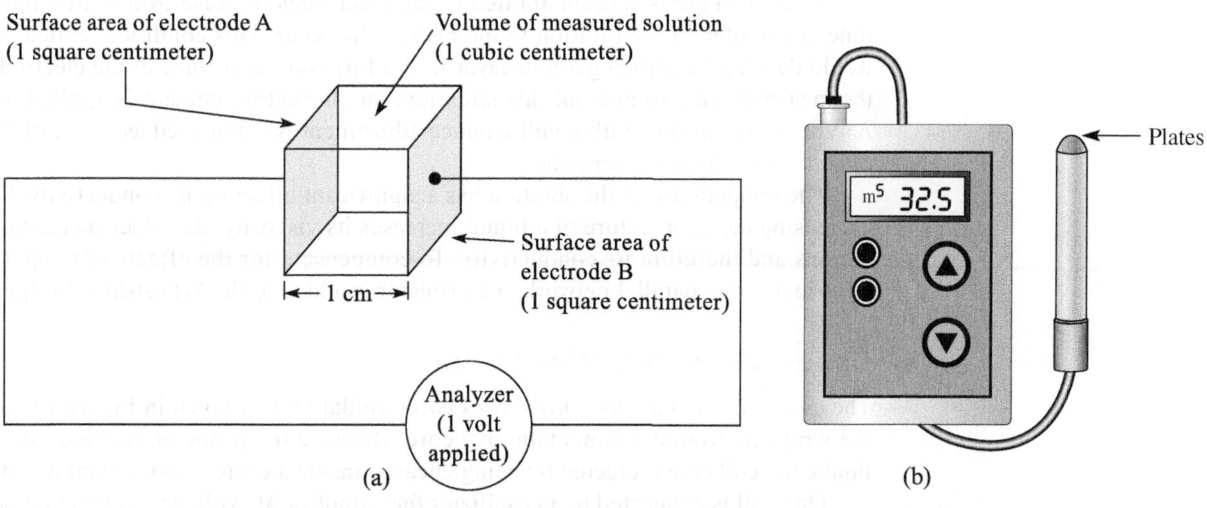

FIGURE 14-10 (a) Conductivity electrode probe; (b) Conductivity meter

The conductivity analyzer makes measurements at different ranges, similar to the way an ohmmeter reads resistance within several ranges. However, instead of having the same type of range switch used by an ohmmeter, conductive values are measured within ranges by changing the dimensions of the electrodes. By changing the area of the plates and the distance they are apart, the volume between them changes. Therefore, the current and indicated conductivity varies, although the applied voltage has not changed. The formula that shows the relationship of the dimensions to conductance is:

$$G = \frac{AK}{L}$$

where G is conductance (siemens), A is the area of each electrode (cm^2), L is the distance between the plates, and K is the specific conductivity of the material between the plates.

A standard to which conductivity measurements are compared uses the following criteria:

- The area of each plate is 1 cm^2.
- The distance between the plates is 1 cm.
- The solution consists of potassium chloride (KCl) at 25°C.

Under these constants, the volume of the solution is one cubic centimeter and the conductance is one siemen.

A theoretical reference value, which is shown in Figure 14-10, establishes a cell constant of 1. Cell constants range from 0.01 to 100 and vary by multiples of 10. Different cells must be used for different ranges so that the amount of current fed to the analyzer circuitry is within its operating requirements. Either the surface area of the plates or the distance between them is changed to each range. If the solution has low conductivity, sensors with plates that are either large or placed closely together are needed to increase the current between them. Conversely, a solution with high conductivity will use plates that are smaller or placed farther apart. Therefore, the cell constant used is determined by the range of conductivity of the liquid being measured. Solutions with an extremely high conductivity require a sensor with a cell constant greater than 1.0. Solutions with a low conductivity require a sensor with a cell constant less than 1.0.

The unit of measurement for conductivity is typically in siemens per centimeter. A sensor with a cell constant of 0.01 has a range of 1 to 10 microSiemens; a sensor with a cell constant of 0.1 has a range of 1 to 100 microSiemens; and a sensor with a cell constant of 1.0 has a range of 1 to 1000 microSiemens.

A DC voltage is seldom applied to the electrodes because over a prolonged period of time, a complete ion migration to the plates will occur. This condition, called *polarization,* would develop because a gaseous layer would form on the surface of the electrodes and alter the measurement. To prevent this situation, an alternating current is applied to the plates. A Wheatstone bridge with a null balance adjustment is often used as a part of the electrode circuit for calibration purposes.

The temperature of the solution has a significant effect on its conductivity. For example, decreasing the temperature of a liquid increases its viscosity, thus decreasing the mobility of its ions and therefore its conductivity. To compensate for the effects of temperature variations in a series-parallel network, a thermistor is added to the Wheatstone bridge.

Conductivity Inductive Probe

The *conductivity inductive probe* uses two toroidal coils. Shown in Figure 14-11, both coils are wrapped around a nonconductive core. Because the probe is inserted into the process liquid, the coils are protected by being encased inside a corrosion-resistant material.

One coil is connected to an oscillator that supplies AC voltage. Its function is to generate a fluctuating-current loop in the solution. A magnetic field formed around the current path cuts across the other (pickup) coil and induces a current into its circuit. The current induced in the pickup coil is directly proportional to the conductivity of the solution. The current from the pickup coil is converted by electronic circuitry into a conductivity reading for display.

Inductive probes are used in applications where the measured solutions are dirty or corrosive. These solutions are highly concentrated with impurities that contain many ions. Therefore, induction probes typically make high conductivity measurements. Conductivity values for some common process solutions are shown in Table 14-3.

Conductivity Applications and Control

Conductivity control is used in a variety of applications.

Closed Water System

To recycle the water used for equipment such as cooling towers and boilers, it is continually recirculated through the system as it flows. After a prolonged period of time, some of this

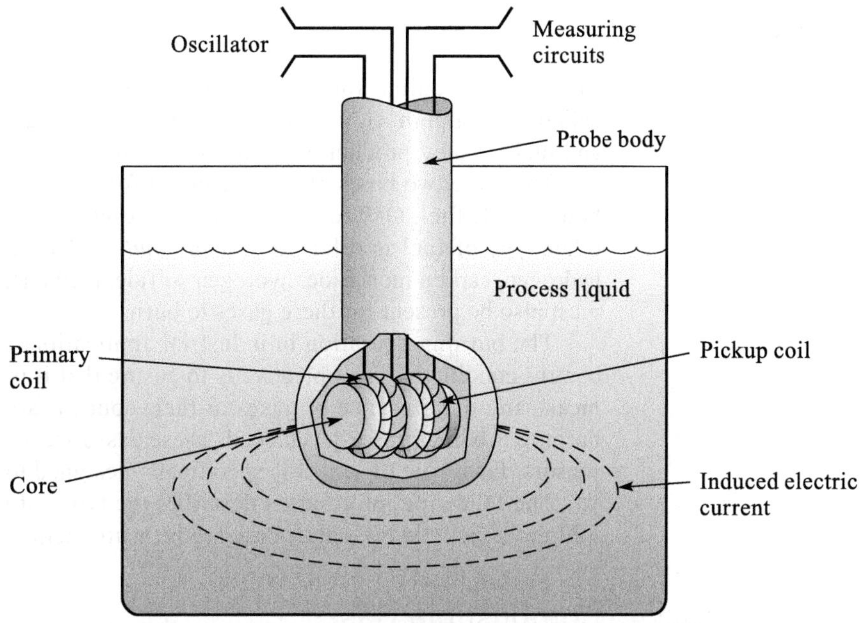

FIGURE 14-11 Conductivity inductive probe

TABLE 14-3 Conductivity Values of Common Solutions

Solutions	*Conductivity*
Water (theoretical)	0.055 μS/cm
Distilled Water	0.5 μS/cm
Power Plant Boiler Water	1.0 μS/cm
Pure Mountain Stream	10.0 μS/cm
City Tap Water	500–700 μS/cm
Ocean Water	54 mS/cm
10% Sodium Hydroxide	355 mS/cm
10% Sulfuric Acid	432 mS/cm
31% Nitric Acid	865 mS/cm

water evaporates, allowing the concentration of solids dissolved in the water to rise. If the concentration level becomes high enough, solids cause scaling to form, which can cause damage to the equipment. Conductivity control causes a system bleed valve to discharge the contaminants from a location where they collect.

Semiconductor Production

Water is used in the formation of semiconductor components. The water must be of very high purity and contain a negligible amount of ions. Conductivity sensors are used to measure the quality of the water. The information from these measurements is used to control the output of a *deionizer.* Water is neutralized as it passes through the device.

Conductivity control is also used in the following industries: textiles, brewing, mining, electroplating, food processing, paper, petroleum, and photographic development.

14-3 Combustion Analyzers and Control

The energy required for many industrial processes is obtained from a chemical reaction called **combustion.** Combustion, commonly known as *burning,* uses a combination of gases and fuel. As a by-product of the reaction, other gases are formed.

There are two types of fuels required for combustion to occur. One type is a hydrocarbon (fossil) fuel. Oxygen must also be present for this type of combustion to occur. The other type of fuel is referred to as a *combustible gas*. These combustible gases include hydrogen, carbon monoxide, hydrogen sulfide, methane, propane, butane, and ethane. Oxygen must also be present for these gases to burn.

The burning operation in industrial applications must be accomplished under precise control conditions. The objective is to ensure that it is performed efficiently and safely. By measuring the presence of gases or their concentration, it is possible to determine if the burning is being properly achieved. These gases are monitored by several types of analytical sensors. Each type of sensor is specifically designed to measure one kind of gas.

The following information describes the types of gases that are present in many industrial processes and the sensors used to measure them.

Combustible Gases

ELE3608 Combustible Gas Detectors

A combustible gas is used as a fuel to provide heat. These gases are usually confined under controlled conditions within a container, where they are stored, or as they burn. If they escape from their container, they can become very dangerous. For example, if they are exposed to a spark, they can create a fire or an explosion. These gases include hydrogen, carbon monoxide, hydrogen sulfide, methane, propane, butane, and ethane. Analytical sensors for combustible gases are designed to detect their presence to prevent damage or injury.

A common type of combustion gas analyzer is the *thermo-conductivity detector* (TCD). Its operation is based on the behavior of gases when they are exposed to heat. Each type of gas has the ability to conduct thermal energy. The type of gas and its concentration determines the rate at which the heat is conducted. The TCD uses a bridge network, as shown in Figure 14-12. One portion of the bridge is connected to two resistive heating elements (made of a tungsten–rhenium alloy) that are inside an enclosed chamber that contains a known reference gas. The other portion of the bridge has two resistive elements inside a chamber that is exposed to the measured gas. If this gas is the same type of gas as the one inside the

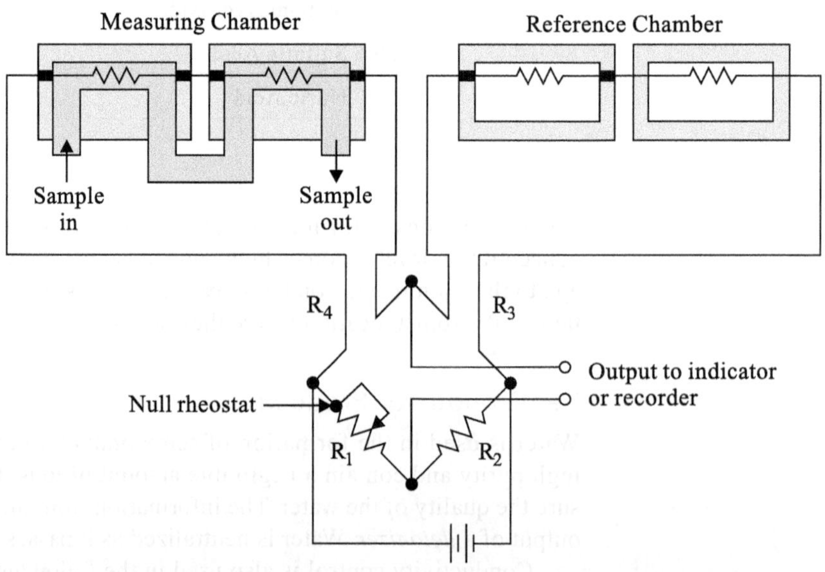

FIGURE 14-12 Thermo-conductivity detector

reference chamber, the bridge network will be balanced and its output will be zero volts. When a different gas flows through the measuring chamber, the rate of heat loss of the gas to which the heating element is exposed changes and its resistance becomes different. As a result, the bridge output changes. The reading is an inferred indication of which type of gas is present. Table 14-4 shows some common thermal conductivity readings of various combustible gases.

TABLE 14-4 Relative Thermal Conductivity (TC) of Gases at 100°C

Gas	*TC*
Air	1.0
Hydrogen	6.990
Helium	5.840
Carbon dioxide	5.3
Methane	1.450
Ethane	0.970
Nitrogen	0.996
Benzene	0.573

ELE3408 Hydrocarbon Gas Sensor

Hydrocarbon Gases

Combustion occurs when oxygen and *hydrocarbon fuels* are ignited. Together, they contain the elements of hydrogen, carbon, and oxygen. When they burn completely, their by-products are water and carbon dioxide. Complete combustion is the result of an efficient burning process. This action requires a proper fuel-to-air ratio. If the fuel does not burn completely, some of it becomes carbon monoxide. The carbon monoxide gas can be very dangerous in two ways. By itself, it is a combustible gas and can be reignited unintentionally. It is also very poisonous and can be harmful to humans if inhaled.

The presence and concentration of carbon dioxide and carbon monoxide can be detected by using an *infrared gas analyzer.* This type of sensor is often placed in the flue or stack of a combustion chamber to monitor the gases that are discharged.

An infrared gas analyzer is shown in Figure 14-13. It consists of six major elements:

1. An infrared light source, which supplies an electromagnetic signal outside the frequency range that can be seen by humans.
2. A sample chamber that holds the gas that is monitored. The chamber has glass windows that allow light to pass through.

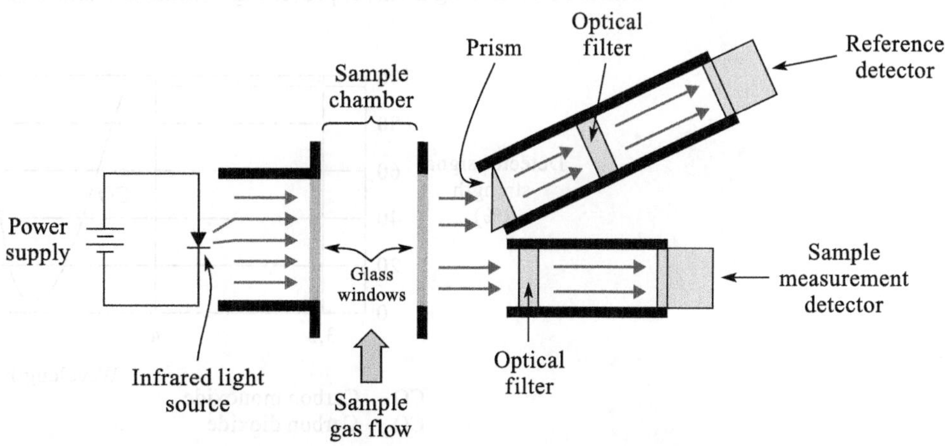

FIGURE 14-13 Split-beam optical gas analyzer

3. A prism, which splits a portion of the light from the sample into a second beam.
4. Two optical filters, each of which allows the light of only one particular wavelength to pass through.
5. Two detectors, each of which senses the amount of light that strikes its surface.
6. Electronic circuitry that converts the signal from the detector into a reading that indicates the consistency of the measured gas.

The analyzer operates as follows:

- Electromagnetic radiation containing many different wavelengths from a light source passes through the sample chamber. Each type of gas absorbs a certain wavelength of the light that passes through. Therefore, most of a certain wavelength is blocked by the gas inside the chamber, and the light of other wavelengths passes through.
- Light transmitted through the chamber is divided into a second beam by a prism.
- Both optical filters permit the radiation from the wavelengths that were not blocked to pass through. The wavelengths absorbed by a gas are referred to as an *absorption band.* Each type of gas has a different absorption band frequency. Figure 14-14 shows the absorption bands of carbon dioxide and carbon monoxide. Infrared light is used because the absorption bands of both gases fall within its frequency range.
- The upper optical filter permits light, which is at a frequency that will not be affected by the sample gas, to pass through. The frequency of this light is referred to as the *reference wavelength.* It strikes a light detector that produces an output voltage. The detector is designed to produce a steady amplitude that varies little or not at all with a change in the concentration of the gas.
- The light intensity of the measuring wavelength reaching its detector varies greatly with a change in the sample concentration. By comparing the signal strength of both detectors, the analyzer circuitry can compute the concentration of the gas. Figure 14-15 shows these wavelengths in relation to the absorption band of a gas.

Controlling Combustion

There are two ways in which combustion must be controlled in an industrial environment. Combustion must be controlled while it is occurring, or before it starts.

To achieve proper combustion while it is occurring, either measurements of gases used in the burning process are made or measurements of gases given off from the burn are taken. If the readings indicate an improper quantity, corrections can often be made by changing the fuel-to-air ratio.

Unwanted gases exposed to a spark can cause an explosion. These gases are usually present because of a leak. Sensors are used to detect their presence. Corrective action is often achieved by closing a valve, providing ventilation, and searching for the leak.

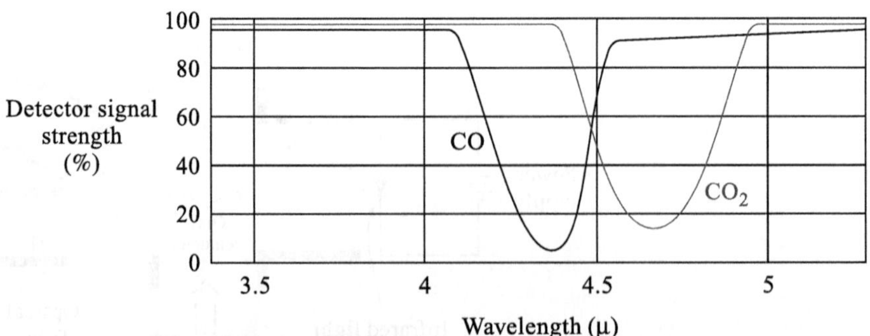

CO Carbon monoxide
CO_2 Carbon dioxide

FIGURE 14-14 Absorption bands of carbon monoxide and carbon dioxide

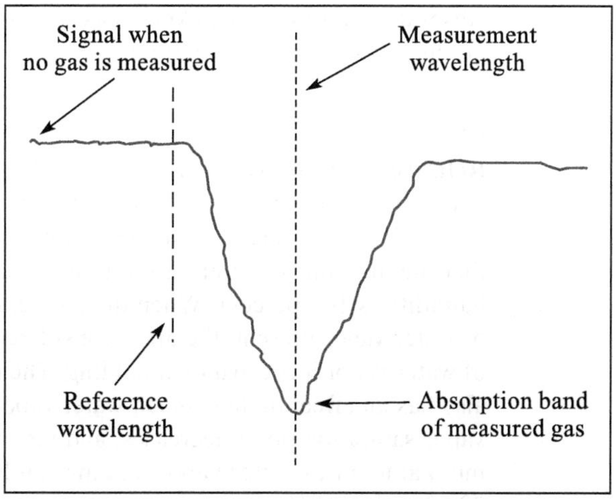

FIGURE 14-15 Absorption band signal in reference to associated wavelengths

14-4 Humidity

ELE4308 Humidity Measurements

Humidity is defined as the amount of moisture in the air. The air may be isolated or part of the atmosphere. In industry, where products are either being made or stored, the air to which they are exposed must contain humidity levels that are within certain parameters. The wrong amount of humidity can be potentially damaging to a process. Too much humidity can promote the growth of mold and mildew; dry air causes moisture to evaporate quickly from the surface of an object, creating a cooling effect that may be undesirable. Dry air also causes static electricity and with it the likelihood of an explosion if combustible materials are present.

Humidity affects *hygroscopic* materials. *Hygroscopic* refers to the tendency of a material to absorb and retain moisture. These materials include paper, wood, flour, textile fibers, and tobacco. As they take on or give off moisture, they undergo changes in dimensions, weight, quality, and usefulness.

Quantitative Measures of Humidity

The first step in controlling humidity within an industrial environment is to measure the quantity of moisture in the air. There are three different quantitative measures of humidity: *absolute, relative,* and *dew point.*

Absolute

Absolute humidity is defined as the mass of water vapor present in a particular volume of atmosphere. This definition is based on the number of water molecules that are present in a known amount of air or gas. The absolute humidity value, or vapor concentration, is expressed as the ratio of the mass of water vapor to the volume occupied by the air–water vapor mixture. The formula for the actual humidity ratio is written as:

$$W = \frac{P_W}{P_a}$$

where,

W = Absolute Humidity
P_W = Mass Density of Water
P_a = Mass Density of Air

Absolute humidity is also known as *specific humidity*. It is expressed in various units such as grains of water per pound of air or pounds of water per million standard cubic feet.

Relative

Relative humidity (RH) is defined as the actual amount of water vapor present as compared to the maximum amount of water vapor the air can hold at a given temperature. As the temperature of the air rises, the amount of water vapor it can hold increases. For example, suppose that the amount of water vapor at 50°F saturates the air early in the morning. The relative humidity is 100 percent. When the temperature rises to 100°F at noon with the same amount of water vapor present, the air is considered dry because it contains 19 percent of the amount of water vapor it is capable of holding. The relative humidity, therefore, is 19 percent. Pressure also has an effect on how much water vapor the air can hold. As pressure increases, the water vapor saturation point decreases, and vice versa. The graph in Figure 14-16 shows the maximum amount of water vapor that air can hold at various temperatures when the pressure is 29.92 inches of Hg.

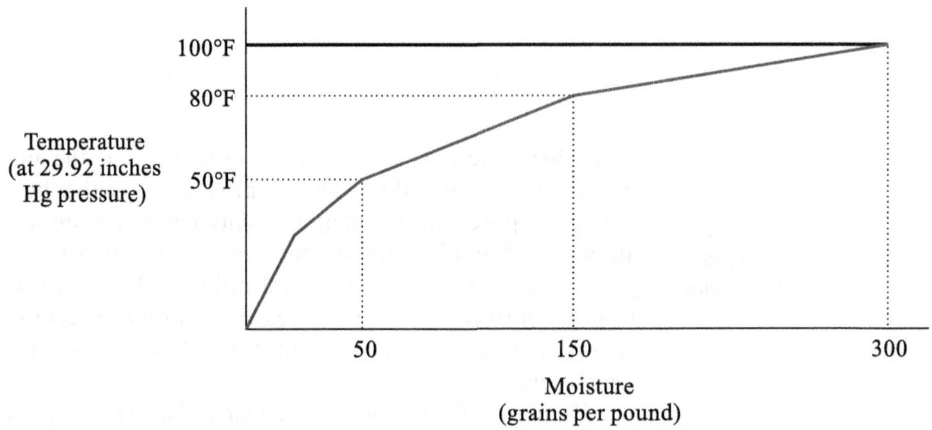

FIGURE 14-16 Maximum moisture content in air at various temperatures

Dew Point

Dew point is the temperature at which the air (or a gas) becomes saturated. When the air is cooled at a constant pressure, condensation of vapor will begin when the dew point is reached.

Absolute-Humidity Sensor

The most common type of device used to measure absolute humidity is the *aluminum-oxide sensor.*

Aluminum-Oxide Sensor

This detection device is the most widely used moisture sensor in gaseous and nongaseous liquid industrial processes. Shown in Figure 14-17, this sensor consists of an aluminum strip that is anodized to form a porous aluminum-oxide layer. A very thin permeable layer of gold is deposited over the oxide. Together, these three materials form a capacitor. The aluminum strip and gold layer each form an electrode, and the aluminum oxide is the dielectric. When the sensor is exposed to moisture, water vapor travels through the gold and equilibrates by penetrating into the pores of the oxide. The dielectric constant of the capacitor changes according to the amount of water vapor present. This condition affects the capacitor's value to represent the moisture reading.

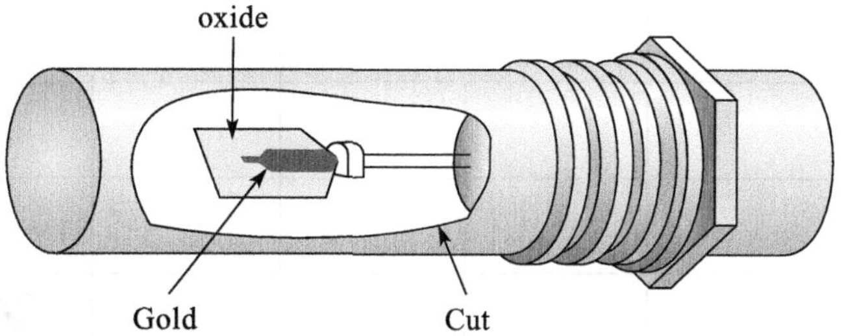

FIGURE 14-17 Aluminum-oxide absolute-humidity sensor

The aluminum-oxide sensor has a wide range of applications. For example, it will detect the presence of undesirably high moisture concentrations. These include:

- Compressed air stream that must be sufficiently dehumidified if the line runs outdoors. Too much moisture causes harmful condensation or freeze-up.
- Natural gas in pipelines contaminated with moisture can develop partial or even total restriction of flow due to the formation of hydrates, a combination of water and hydrocarbon molecules in a state similar to ice. In addition, moisture accelerates pipeline corrosion and reduces the energy value of the gas.

Relative-Humidity Detectors

The operation of relative-humidity detectors is based on the tendency of a material's physical or electrical properties to change in a predictable manner when exposed to humidity. Three common detection devices are the psychrometric, hygrometric, and electronic capacitance sensors.

Psychrometric Detector

This device uses two identical thermometers. One is called a *dry bulb* and the other a *wet bulb.* The dry bulb is directly exposed to the surrounding air. The wet bulb is covered by a wick that is kept moist. Air is passed over the wick by using one of two methods: either the wet bulb is whirled in a circular path, or it is stationary and the air is drawn past it by an air exchanger. When the air flows over the wick, evaporation cools the thermometer inside until saturated equilibrium is reached. The drier the air, the more cooling that occurs. Once wet-bulb and dry-bulb temperatures are known, relative humidity can be determined by reading tables, psychrometric charts such as the one shown in Figure 14-18, or data from a computer program.

Wet-dry-bulb psychrometers are very accurate, but they require a high degree of maintenance. Their accuracy diminishes when the RH is below 20 percent.

Hygrometric Detector

This type of detector determines humidity by measuring the change in dimension of hygroscopic materials, such as hair, cotton, paper, or a synthetic wick. Human hair is one of the most common hygroscopic materials used. The hair absorbs moisture from the surrounding air by an amount that is a function of the water vapor present. As the water content in the hair increases, its length increases. Expansion and contraction of the hair cause a pointer or a pen to move, displaying a continuous relative-humidity reading. Hygroscopic materials can also be used in conjunction with a humidistat switch to provide humidity controlled capabilities.

Humidity can also cause a change in the physical weight of hygroscopic materials. By measuring the weight as it absorbs or desorbs water, a relative-humidity reading can be taken.

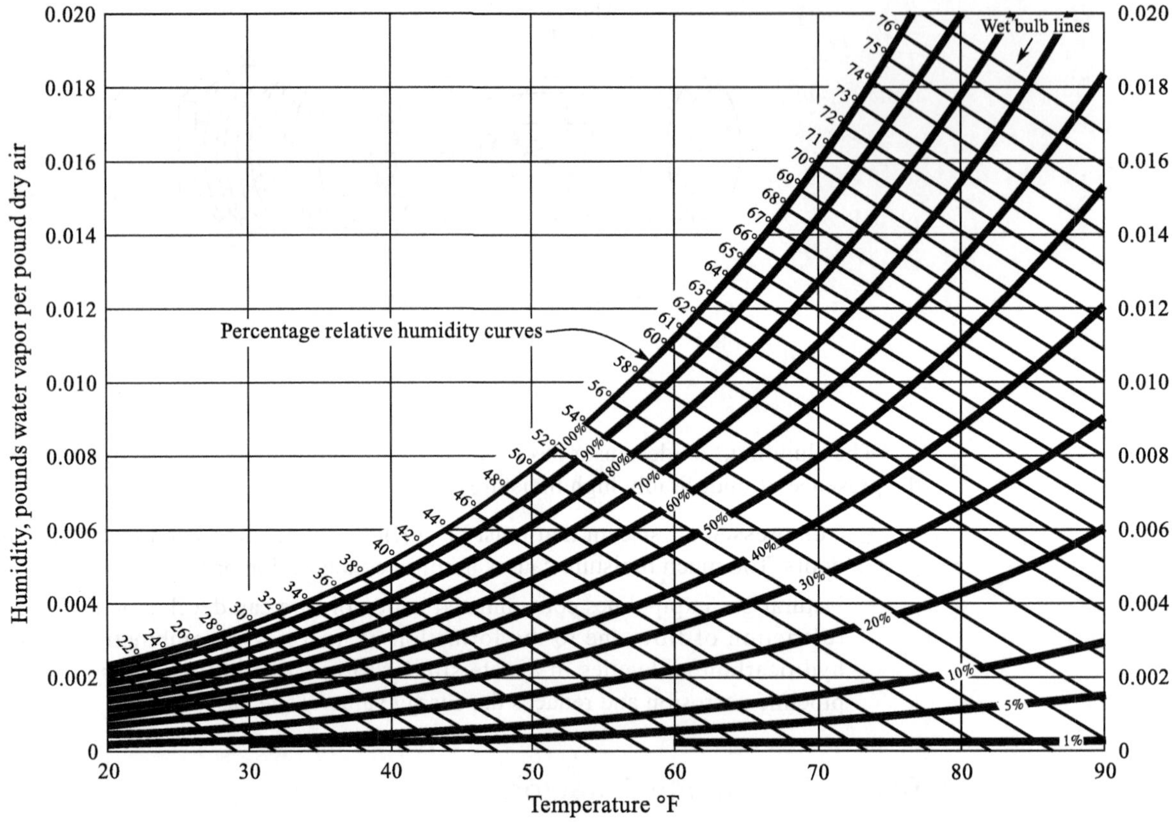

FIGURE 14-18 A psychrometric chart is constructed by using specific humidity, relative humidity, dry-bulb temperature, and wet-bulb temperature

Electronic Capacitance Detector

This type of sensor constitutes a large portion of the RH sensors sold. This device is a small capacitor consisting of a hygroscopic dielectric material placed between a pair of electrodes, as shown in Figure 14-19.

The plates are porous to allow moisture to pass through to the dielectric. The dielectric is typically made of plastic or polymer materials with a dielectric constant ranging from 2 to 15. When no moisture is present, the capacitor's value is determined by the plate geometry (plate size and distance between them) and the plate's dielectric constant.

When the sensor is exposed to humid conditions, moisture in the air is absorbed by the dielectric material. This action causes the dielectric constant to increase, which results in a change of the sensor's capacitance. Relative humidity is also a function of temperature. Therefore, it is necessary to combine a thermistor reading with the capacitance measurement. The sensor's electronics converts both of these measurements into a relative-humidity value.

Capacitance sensors are capable of measuring a range of relative humidity with an accuracy of 2 to 15 percent. Because these sensors require considerable time to change capacitance, they tend to be selected for applications where fast response time is not necessary.

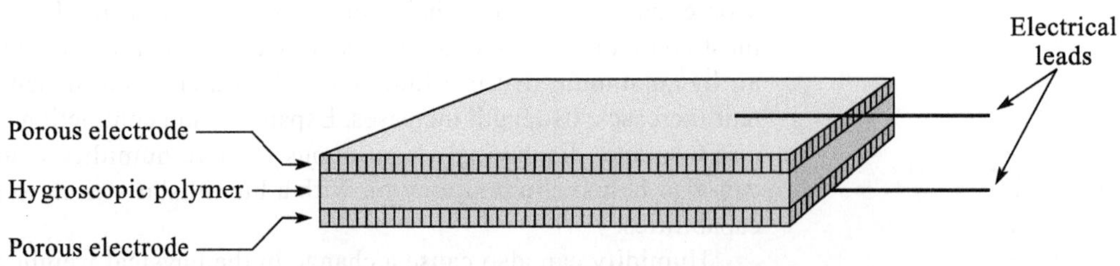

FIGURE 14-19 RH electronic capacitance detector

Dew Point Measurements

Three techniques used to measure dew point are the *manual chilled mirror* (dew cup), *adiabatic expansion sensing,* and the *optical chilled mirror.*

Manual Chilled Mirror

Also known as the *dew cup* technique, this method uses a polished cup made of chromium-plated copper. It is partially filled with acetone or methanol, and a thermometer is placed in the solution. Small cubes of dry ice are dropped into the solution until condensation forms on the outside of the cup. The dew point is measured by reading the thermometer when the condensation begins to appear. This method is a one-time measurement and its accuracy is dependent upon the skill of the operator.

Adiabatic Expansion

This method uses an instrument that draws an air sample into a chamber. The chamber is sealed and then pressurized to a predetermined level. Next, the chamber is unsealed and the air is released, causing a drop in temperature. By measuring the temperature and pressure, the instrument computes the dew point by determining their ratio. This method provides a one-time measurement.

Optical Chilled-Mirror Hygrometer

The optical chilled-mirror device is capable of providing a continuous on-line humidity measurement over a prolonged period of time. Shown in Figure 14-20, it contains the following elements:

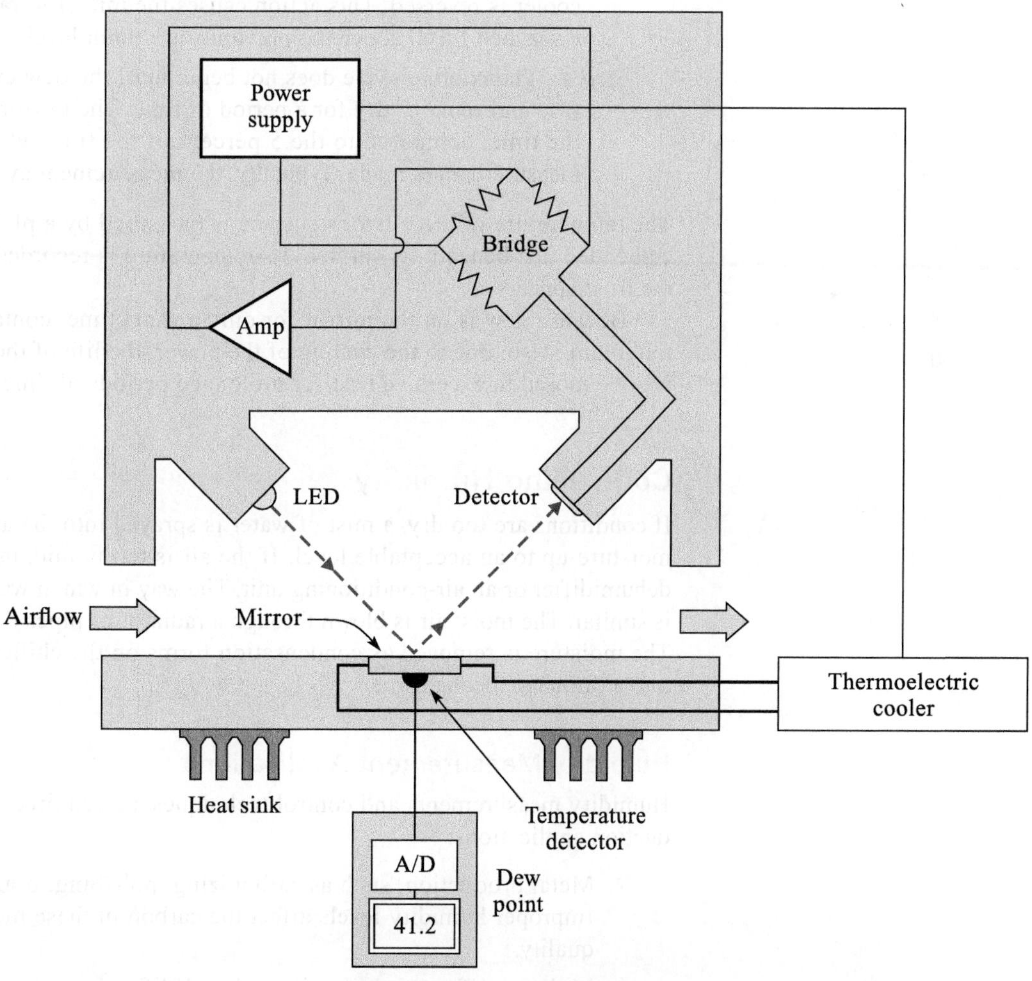

FIGURE 14-20 Optical chilled-mirror hygrometer

- Gold- or rhodium-plated copper mirror.
- A thermoelectric cooler used to control the temperature of the mirror.
- A high-intensity LED that shines its light on the mirror.
- A phototransistor or an optical detector used to measure the amount of light reflected from the LED off the surface of the mirror.
- The optical detector connected to one leg of an electronic bridge network that is coupled to an amplifier.

A flow of the sample air continuously passes over the surface of the mirror. When the mirror surface temperature is above the dew point, it is dry and highly reflective. Therefore, the maximum light is received by the optical detector. When the thermoelectric cooler reduces the mirror temperature, moisture will condense on its surface when the dew point is reached, causing the light to scatter due to refraction. Therefore, the light received by the detector is reduced. The change in the detector affects the bridge, which provides the feedback signal for closed-loop control. The purpose of the closed-loop system is to maintain the surface temperature on the mirror to within a few degrees of the dew point. This condition is performed by a four-step cycling process.

Step 1: The mirror is rapidly cooled from above ambient to 1.5°C above the last dew point.

Step 2: The cooling rate is decelerated to approach and cross the dew point as slowly as possible to allow dew to form in a uniform manner.

Step 3: When the dew detection is completed, the current through the thermoelectric cooler is reversed. This action causes the mirror to rapidly rise in temperature until it reaches 1.5°C above the previous dew point level.

Step 4: The cooling cycle does not begin until the dew evaporates from the mirror surface and remains dry for a period of time. The mirror is dry for about 95 percent of the time, compared to the 5 percent of the time when the dew is present and the measurement is made. Typically, the measurement cycle is once every 20 seconds.

The temperature of the mirror's surface is measured by a platinum resistance thermometer embedded just beneath its surface. Its temperature is recorded and displayed at the moment the frost appears.

Because dew is on the mirror for only a short time, contamination buildup is kept to a minimum. Also, due to the cycling of the power, the life of the sensor is extended because it is not exposed to excessive heat for prolonged periods of time.

Controlling Humidity

If conditions are too dry, a mist of water is sprayed into the air by a humidifier to bring the moisture up to an acceptable level. If the air is too humid, moisture is removed by using a dehumidifier or an air-conditioning unit. The way in which water is removed by both devices is similar. The moist air is blown through a radiator or pipes that are cooled to the dew point. The moisture is removed as condensation forms on the chilled surface areas and then drips into a drainage mechanism.

Humidity Measurement Applications

Humidity measurements and control techniques are required for the following types of production applications:

- Metal production, such as carbonizing, polishing, brazing, sintering, and annealing. Improper humidity levels affect the carbon in these metals and can diminish product quality.
- Moisture affects the integrity and shelf life of certain chemicals such as pharmaceutical products, especially pills or powder.

- Bakeries running large conveyor belts through an oven. Humidity levels outside certain limits cause improper baking.
- For industrial paper dryers, a constant humidity level throughout the drying process must be maintained to ensure batch-to-batch uniformity.

14-5 Sampling Measurement System

Whenever the temperature of the air being tested is above the level to which a sensor can be exposed, it is necessary to cool the air before a reading can be made. One method of achieving this requirement is to use a sampling system, as shown in Figure 14-21.

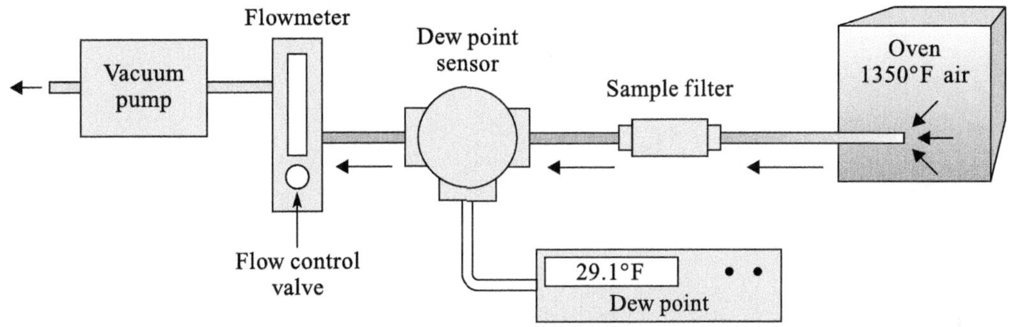

FIGURE 14-21 Sampling measurement systems

The interior of the oven is 1350°F, but the humidity sensor has a maximum rating of 250°F. The only way to measure the humidity of the air is to extract a sample from the furnace. This system consists of a small vacuum pump that draws out the air, a flowmeter that measures the gas flow rate, and a flow-control valve that regulates the flow rate. The air sample exits the oven through a stainless steel tube and flows into the humidity sensor. The tube functions as a very efficient heat exchanger. Even though the pipe is only a few feet in length, the sample temperature will be reduced from 1350°F to the ambient temperature that surrounds the tube outside the oven. Before the air enters the sensor, dust and dirt are removed by a filter.

Problems

1. The letter "p" in the term *pH* refers to __________, and the letter "H" refers to __________.
2. If the pH value of a solution is less than 7, it is a/an ____.
   a. base b. acid
3. The pH value of pure water is ____.
   a. 0 c. 7
   b. 1 d. 14
4. An acid solution is neutralized by adding ____.
   a. water
   b. a base
5. A ____ potential forms on the inside surface of the glass of a pH electrode when the process solution is alkaline.
   a. negative b. positive
6. Water that ____ has the greatest degree of conductivity.
   a. is pure
   b. contains impurities
7. T/F One factor that affects the conductivity of a liquid is its volume in a container.
8. ____ solutions contain ions.
   a. Acidic c. Conductive
   b. Alkaline d. All of the above
9. Electrode conductivity probes that measure conductivity make measurements at different ranges by ____.
   a. changing the applied voltage
   b. changing the dimensions of the plates
10. A ____ conductivity probe would be more likely to measure the conductivity of a solution with a large concentration of impurities.
    a. capacitive b. inductive
11. ____ gas is present if a hydrocarbon fuel is not completely burned.
    a. Carbon dioxide b. Carbon monoxide
12. The most likely reason why a hydrocarbon fuel does not completely burn is that the ____.
    a. flame is not hot enough
    b. fuel-to-air ratio is not correct

13. T/F A fossil fuel is considered a combustible gas.
14. T/F A specific type of gas can be identified by the light it absorbs at a specific frequency.
15. Static electricity is more likely present in ____ air.
    a. humid
    b. dry
16. What is the relative humidity when the temperature reaches the dew point?
17. The term __________ refers to the tendency of a material to absorb moisture.
18. T/F *Absolute humidity* refers to the amount of water vapor present at a specific temperature.
19. The wet-bulb and dry-bulb instruments measure ____.
    a. absolute humidity
    b. relative humidity
    c. dew point
20. The dew point is most likely to be reached when the air temperature ____.
    a. increases
    b. decreases
21. T/F The dew cup technique for measuring moisture is used in a closed-loop control system.

CHAPTER

15

Industrial Process Techniques and Instrumentation

OBJECTIVES

At the conclusion of this chapter, you should be able to:

- Describe the characteristics that pertain to the following types of manufacturing processes:
  - Batch
  - Chemical Reaction
  - Polymerization
  - Continuous
  - Separation
  - Product Composition
  - Mixing/Blending
- Define the following terms:
  - Endothermic
  - Sensitivity
  - Air-to-Open
  - Response Time
  - Static
  - Span
  - Precision
  - Exothermic
  - Hysteresis
  - Data Acquisition
  - Accuracy
  - Dynamic
  - Air-to-Close
  - Zero
  - Repeatability
  - Linearity
- Describe how varying heat and pressure levels affect a process.
- Explain how the following types of instruments and equipment operate:
  - Heat Exchanger
  - Agitator
  - Square Root Extractor
  - Indicator
  - Positioner
  - Final Control Element
  - Evaporator
  - Transducer
  - I/P, P/I, I/V, V/I Transducers
  - Reactor
  - Recorder
  - Transmitter
  - Alarm
- Provide the different types of standard electronic and pneumatic transmission signals and their numerical ranges.
- List the steps required for the calibration process of an instrument in their proper order.
- List the types of control valves, describe their characteristics, and explain applications for which they are used.

INTRODUCTION

The manufacturing industry provides a great diversity of products. The methods of production and types of equipment used to make a product vary greatly from one industry to another. Despite the differences, modern industries rely on the capabilities of automated systems to measure and control the manufacturing process.

In this chapter, instruments used in automated systems and process techniques commonly performed to achieve quality, efficiency, and safety standards will be discussed.

The industrial field is production-based. Its function is to transform raw materials into a final product. In the field of *process measurement and control,* raw materials are manipulated through various processes to manufacture goods and provide public services such as electrical energy or water treatment and purification. These products and services are provided either in *batch processes* or by *continuous processes.*

15-1 Batch Processes

A large variety of products, such as alcoholic beverages, explosives, pharmaceuticals, liquid detergents, foods, plastics, and metals, are produced by the batch process method of manufacturing. These products are made one batch at a time and usually in smaller quantities than the products produced by the continuous method of manufacturing. In batch processing, a sequence of steps is performed similar to the way a food recipe is followed. The product is made by putting ingredients into a vessel, called a *reactor,* and then causing them to react to form a product. The reactor, shown in Figure 15-1, can be described as a large kettle. The vessel is usually closed to keep contaminants from entering the atmosphere. When the *reaction* is completed, the finished product is discharged from the vessel. To ensure product quality, various control requirements must be achieved during the sequence of steps that takes place.

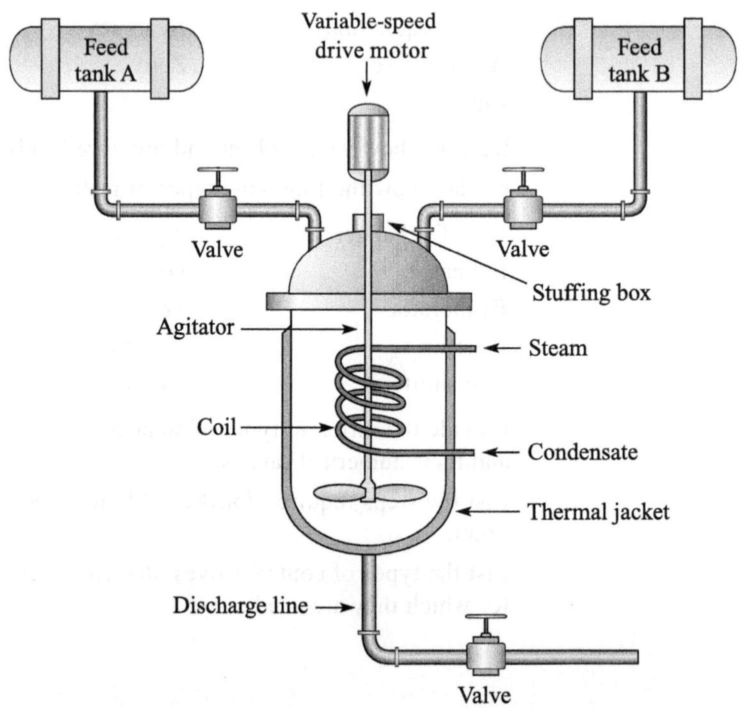

FIGURE 15-1 Batch reactor

Control Requirements

Controlling the Quantity of Raw Materials

In a batch process, exact quantities of raw materials are required for each batch. Measurements can be made by:

- Determining the weight of the ingredients inside the reactor with a pressure sensor.
- Using a level sensor to read the height of the ingredients in the vessel.
- Calculating the volumetric flow rate of a raw material being fed into the reactor by using a flowmeter.

Controlling the Process Variables During the Reaction Cycle

ELE806
Endothermic vs
Exothermic Reactions

During each step in the process, variables must be controlled. In batch production applications, temperature and pressure are the two most common variables that must be regulated to control the rate of reaction. Applying thermal energy and maintaining the temperature at a certain level is critical to most batch processes. The type of thermal energy applied, whether it be heating or cooling, is determined by the type of ingredients used in the recipe, resulting in an endothermic or an exothermic reaction. Processes that require a source of external heat while forming a product are called **endothermic.** The reaction will not take place unless heat is applied to the raw materials in the vessel. When heat is generated during the reaction phase, an **exothermic** process occurs. In this type of process, a source of cooling thermal energy must be applied to ensure product quality and to prevent overheating. A clogged residential plumbing fixture can be used as an example of why the reaction needs to be kept at lower temperatures. After the drain cleaner solution is applied, a chemical reaction takes place and produces enough heat to damage the pipe. By applying water, the temperature of the solution and the blockage is lowered to a safe level.

With some reactors, thermal energy is applied to the bottom of the vessel. However, this is not a very efficient way to apply the energy to the contents inside. Only the ingredients in contact with the bottom of the tank are directly exposed to the energy source, while the ingredients on top receive the energy from the thermal transfer that takes place along the sides of the vessel and through the medium itself. Most batch reactors use a thermal jacket, shown in Figure 15-1, that surrounds the sides and bottom of the vessel. By exposing thermal energy to a larger surface area, heating or cooling can be distributed more evenly and efficiently. Thermal energy in the form of heat is supplied by hot water or steam to the jacket. This is what happens in a double boiler used in the kitchen. When cold thermal energy is required, a cold water supply is circulated inside the jacket. In addition to using a jacket, circulation coils placed inside the reactor can be used to speed up the heating or cooling process.

Some endothermic processes require very high temperatures. However, the temperature of the material inside the reactor cannot be brought to a higher level than its boiling point. For example, pure water cannot be raised to a temperature above its boiling point of 212°F, regardless of how much heat is supplied. By increasing the pressure applied to a substance (above the level from the atmosphere), the boiling point is raised, thereby enabling it to become hotter. Pressurizing the reactor under controlled conditions provides a way to elevate the temperature of the material inside, causing the reaction time to speed up. This is the principle on which a pressure cooker operates.

Controlling Each Step in the Sequence

A batch process usually involves a series of sequential steps to form a product. These steps may include feeding, mixing, heating, cooling, reacting, discharging, and then cleaning the vessel after the product is removed. The duration of time, or the rate at which each of these operations occurs, is determined by the requirements of the particular recipe for the process.

Types of Batch Processes

The four basic categories of batch processing methods are:

- Mixing/Blending
- Chemical Reaction
- Separation
- Polymerization

Mixing/Blending

Mixing/blending is an operation that involves combining two or more ingredients together. This method may only require a one-step process of feeding the materials into a tank and then draining the container after a short period of time. To speed up the blending process, a mixing operation may be required by using an agitator to stir the ingredients, as shown in Figure 15-1. Some blending operations require precise control of the duration of the mixing cycle; this control is performed by a timer. Other types of operations require a precise speed at which the stirring action takes place to control the reaction rate. A variable-speed device is used to control the RPM at which the motor drives the agitator blades. Making paint is one example of a mixing/blending operation. In an operation that requires a pressurized condition, a mechanical device called a *stuffing box* is placed around the reactor's agitator shaft to provide a tight seal.

In some applications, the amount of agitation can be reduced, and the mixing time can be shortened by increasing heat. This concept can be illustrated by the simple example of making instant coffee. The granules are dissolved more quickly when they are stirred in hot water than when they are stirred in cold water.

Chemical Reaction

A **chemical reaction** is the process of combining two or more materials or reactants to form a product. The reaction usually occurs under the influence of temperature, pressure, and agitation and by introducing a catalyst. A large variety of products, such as fertilizer, antifreeze, and pesticides, are made by chemical reaction processes.

Separation

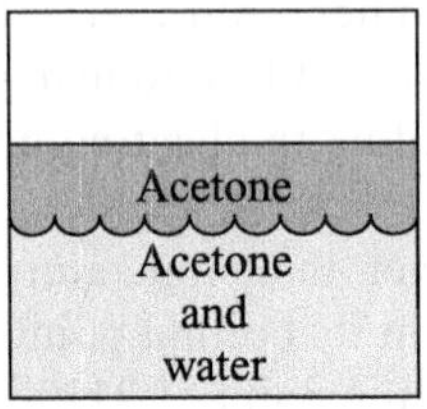

FIGURE 15-2 Separation of acetone and water

A **separation** operation is opposite to mixing/blending. During the separation process, an ingredient is removed from a mixture. One example of a separator process is shown in Figure 15-2, where acetone is removed from a mixture of acetone and water inside the reactor. Over a period of time, some of the acetone will separate and rise to the top. However, the time duration of this process can be shortened by heating (or cooling) the mixture.

Another type of separation process is crystallization. Crystallization is the formation of a solid material from a solution, vapor, melted material, or solid that is in a different phase of the reaction. Some types of pharmaceuticals are produced through this process.

Polymerization

Polymerization is a process in which a large number of molecules are combined to form a product. For polymerization to take place, temperature, pressure, and a catalyst are supplied under very precise, controlled conditions. Products such as plastics and synthetic materials are made using this type of process.

All four of these processes occur in a batch *reactor*. Some reactors must be made of specific materials. The type of material used to construct the reactor depends on such factors as:

- Its capability to withstand the corrosiveness of the ingredients inside.
- Its capability to not contaminate the contents inside. For example, stainless steel is often used to hold food products.

15-2 Continuous Processes

A large variety of products, such as petroleum, chemicals, paper, plastic garbage bags, and so forth, are produced by the continuous-process method of manufacturing. In a continuous process, raw materials are continuously passed through manufacturing equipment at a controlled rate, and the end product is continuously withdrawn. Unlike batch processing, where a relatively small amount of the product is made, continuous processing is designed to manufacture a large volume of a particular product. There are several types of continuous-process manufacturing equipment from which products are formed, such as screens and rollers on a paper machine, extruders that shape plastic bags, or an evaporator that processes liquids. To ensure product quality, several variables must be continuously controlled simultaneously by maintaining the process conditions at a constant *setpoint* for each one of them. The stability of the variable must be maintained despite changes in process conditions. There are several control requirements that must be achieved for continuous processing.

Control Requirements

Controlling the Quantity of Raw Materials

In a continuous process, exact quantities of raw materials are required as they are fed into the manufacturing equipment. These raw materials can be granules, powder, pulp, sewage, water, petroleum, and so on. The quantities of these materials are measured primarily by various types of flow sensors. Flow valves located in piping or at a discharge port of a gravity-fed storage tank are often used to vary the rate at which the raw materials are fed into the process.

Controlling Operating Parameters During the Process

As the raw materials are being fed through the manufacturing equipment, several variables must be monitored and kept at a constant value. For most continuous processing, these variables include temperature, pressure, level, flow, and product composition.

Temperature The temperature at which a continuous process takes place can be very critical. Some types of processes require heating, and others require cooling. Heat exchangers are often used to transfer the thermal energy (above and below ambient temperature) required for the process.

Figure 15-3 shows the construction of a heat exchanger. It consists of a shell, heads that are removable for maintenance, a tube section enclosed inside the shell, and inlet and outlet ports for both the shell and the tube. In this configuration, the product flows through the shell, and the thermal medium from which energy is obtained flows through the tube. Steam or hot water is usually supplied to the exchanger for heating, and cold water is supplied to the exchanger for cooling.

Pressure When the continuous process takes place within a covered vessel, the contents are often pressurized to raise the boiling temperature and shorten the reaction time. By maintaining pressure within a certain range, product quality standards and safety parameters can be achieved.

Level Some continuous manufacturing processes take place inside a vessel. One method used to determine the amount of material that is inside the vessel is to measure the level at which it fills the container. There are several reasons why the level of the contents must be monitored and regulated. If there is not enough liquid to cover a thermal exchanger heating element, it may become too hot and be damaged. Too much material can create a spillage problem by overflowing an uncovered vessel. If the contents become too high in a covered vessel, pressure within the air gap above the medium may also become too high and create an explosion. Various types of level sensors are used to detect high limits, low limits, or the entire depth range of the contents.

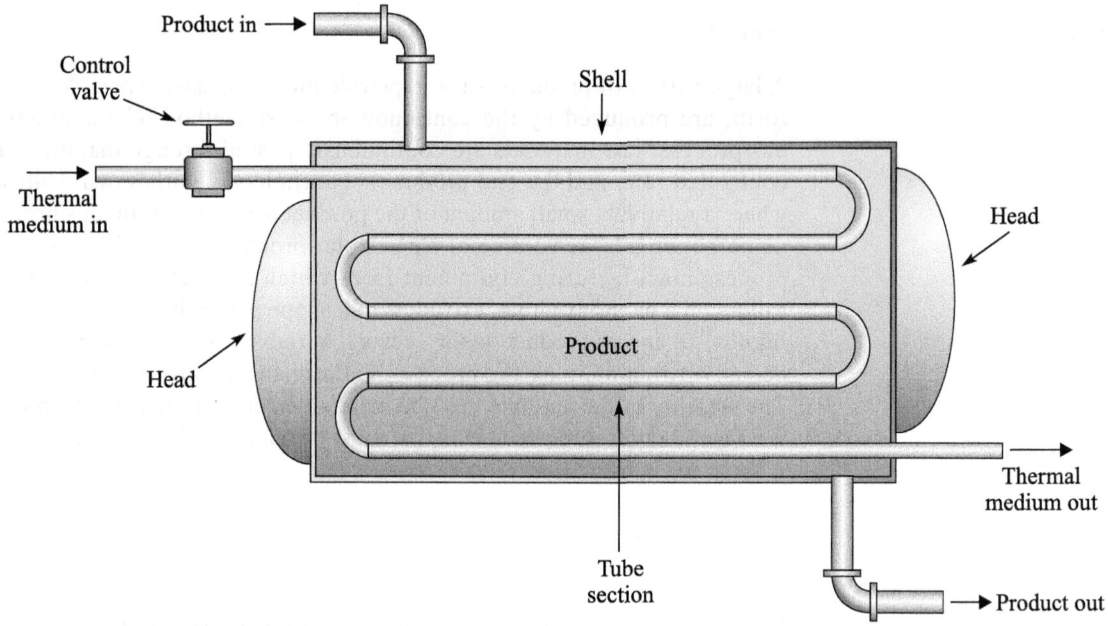

FIGURE 15-3 Heat exchanger

Flow The most common variable in a continuous process application that needs to be controlled is flow. Raw materials are continually fed into the machinery that produces the product. Conditions inside the machinery, such as temperature and pressure, must be at a certain value to control the reaction rate of the product. The temperature is maintained by controlling the flow of steam or fuel, and the pressure level can be regulated by the flow of air from a compressor or vacuum pump, which passes through a control valve. In summary, the rate of production and many variables that are essential elements of the process are controlled by flow.

Product Composition As the product stream of multiple ingredients is blended in a continuous process, the ingredients are often mixed, pressurized, heated, or cooled to cause the desired reaction. During this process, the composition of the material must be very precise. The *composition* refers to the conditions of the product solution, such as the concentration of solids in a liquid, completion of a chemical reaction, or consistency of a mixture. The status can be measured by reading boiling point temperature or by using a sensor called an *analyzer*. Controlling the composition of a product is also referred to as *analytical control*. The composition is often affected by changing the other variables: temperature, flow, level, and pressure.

The evaporator shown in Figure 15-4 can be used to illustrate how other variables can control the composition of a product. Liquid and paste are fed into a chamber, mixed, and boiled to produce a condensed solution. If the solution becomes too concentrated because the process is altered, the problem can be corrected by increasing the flow rate of the liquid, by decreasing the flow rate of the paste, or by reducing the flow rate of the heat supply that causes boiling to take place.

Some products can be manufactured by using either a batch process or a continuous process. If the volume of the product is small, a company will usually use a batch reactor similar to the one described in Figure 15-1. If the demand for the volume of the same product is large, the company can increase the production output by using continuous-process equipment, such as the machine shown in Figure 15-5.

Raw materials from the feed tanks A and B are mixed in an in-line mixer. Because the blended liquid must be heated to 100°F to cause a desired reaction, it passes through a heat exchanger. The end product is continuously withdrawn through a discharge port at the bottom of the vessel. If the temperature required for the process were greater than 212°F, the vessel could be pressurized to enable the liquid to be heated above the atmospheric boiling point.

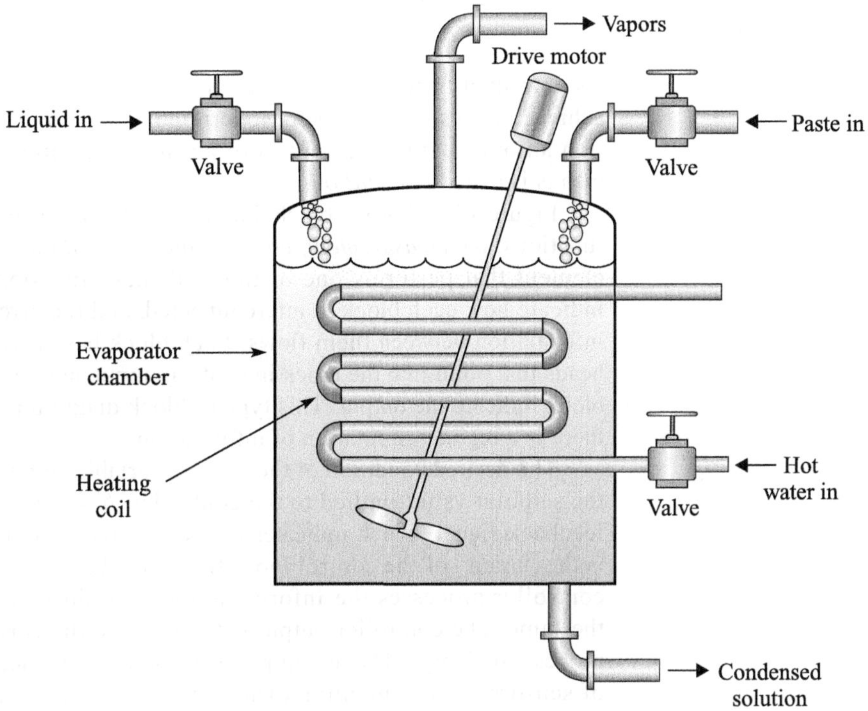

FIGURE 15-4 Heat blending process

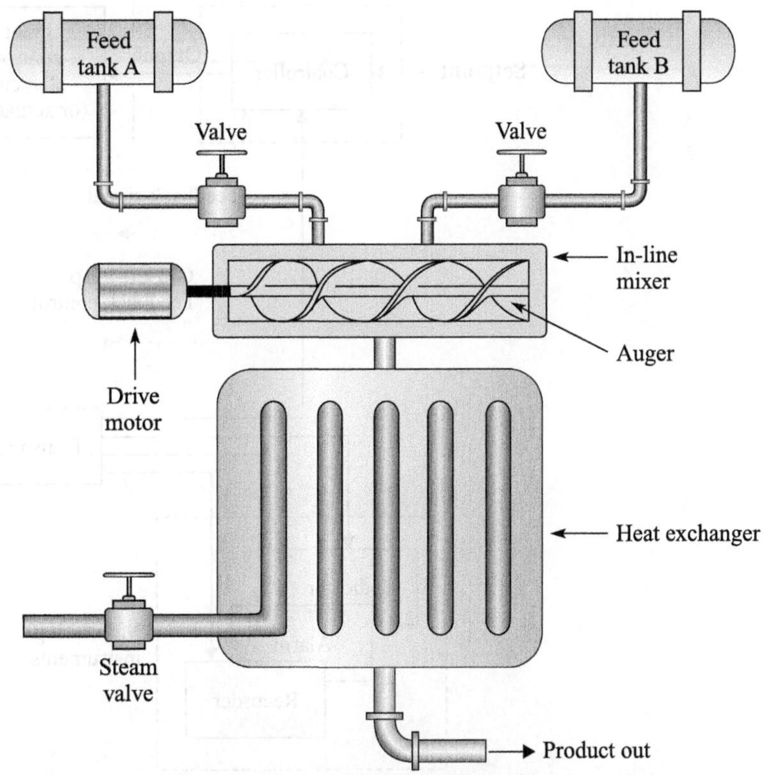

FIGURE 15-5 Continuous-process heat exchanger

15-3 Instrumentation

As a product is being manufactured, critical stages of the process must be manipulated to achieve the desired outcome. Modern industrial equipment performs the manipulation function automatically. The control of an industrial process by automatic rather than manual means is called *automation.*

Figure 15-6 shows a block diagram of an automated system. This system performs three functions: the *measurement, control,* and *manipulation* of the process. Each block is a basic element that performs one or more of these functions. The lines between the elements indicate how each block is interconnected, and the arrowheads show the direction in which information between them flows. Each block has at least one input and output. The arrowheads that point into the block indicate an input, and the arrowheads that point away from the block indicate the output. This type of block diagram is referred to as a *control loop* because there is a regular circulation of information.

The desired condition of the process variable being controlled is established by adjusting the setpoint value applied to the controller. A second input applied to the controller is the feedback signal, which indicates the actual condition of the process variable. The controller is the "brain" of the control loop. If the feedback signal is different from the setpoint, the controller processes the information and produces an appropriate output to make them the same. The controller output is the input to the final control element, which causes the process to change. The actual process condition is measured by a sensor. The large variety of sensors used to monitor various processes produce many types of output signals. An interface device is used to convert the sensor output into a standard signal that is compatible with the controller. The sensor interface device and its output line form the feedback portion of the loop. Monitoring instruments connected to the output of the interface device

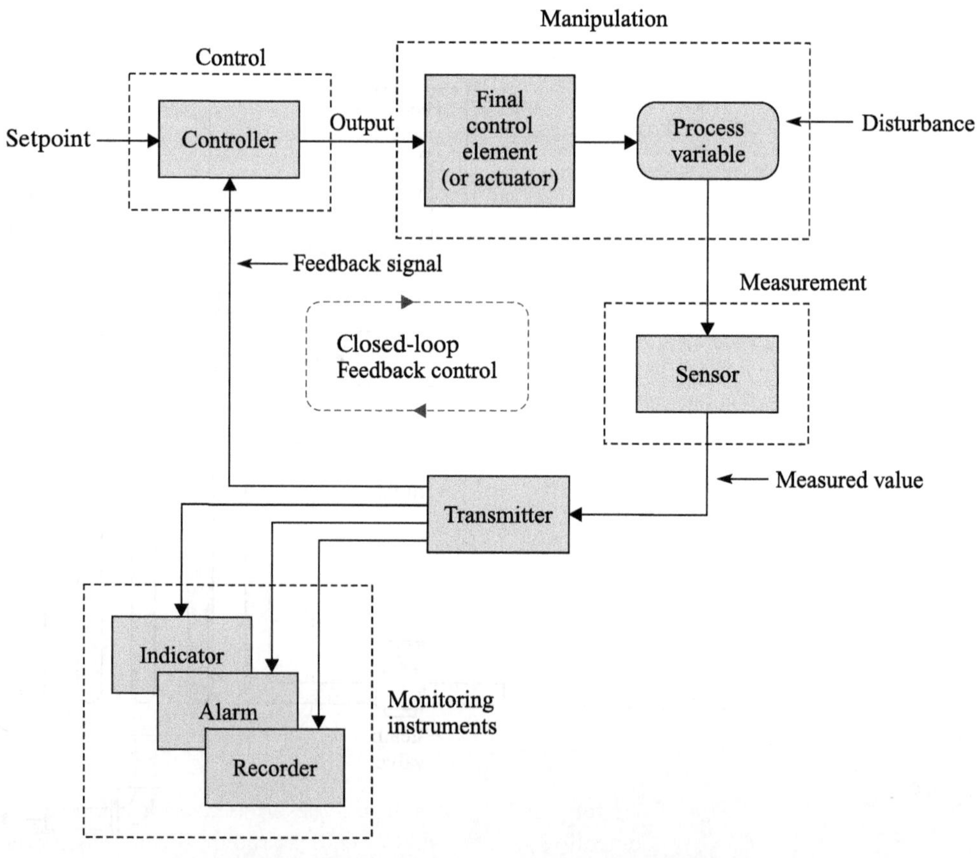

FIGURE 15-6 Block diagram of a closed-loop automated system

do not perform any of the feedback loop functions. Their purpose is to provide information for the human operator about how the closed-loop system is functioning.

By continuously measuring, controlling, and manipulating, the closed-loop system keeps a process variable, such as temperature, pressure, flow, or level, at a condition commanded by the setpoint. If a disturbance causes the process variable to change, the controller will detect this deviation of the feedback signal from the setpoint and cause the final control element to manipulate the variable back to the desired condition.

The functions of the elements in a closed-loop automated system are performed by various instruments, which are referred to as *instrumentation devices.* In most industrial process machines, several different variables must be controlled simultaneously to manufacture a product. A separate control loop is used for each one of them. The control functions of these loops are performed at intermediate stages of the production line to maintain quality-related standards of each variable.

15-4 Measurement Devices (Sensors)

An industrial process loop *begins* with measuring a variable. This function is performed by a *sensor,* which is located in the field near the process. A control loop is effective only if the sensor is reliable.

An ineffective loop can cause the quality of the variable to become unacceptable. The reliability of the sensor is affected by its characteristics, which are classified as either **dynamic** or **static.**

The *dynamic characteristic* refers to the transient response of the instrument, which is the time during which its output reaches a steady state after a new signal is applied to its input. The *static characteristic* refers only to the condition of the instrument when it is stable and not changing. The following explanations describe dynamic and static characteristics.

Dynamic

Response Time

Sensors do not respond to changes immediately. It takes a period of time for the sensor to produce the signal that represents the condition that it is detecting. The term **response time** is used to describe the amount of time the sensor takes to respond to a change in the measured variable. The response time is determined by several factors such as the design of the sensor and the type of variable being measured. For example, flow and level sensors respond almost immediately to any changes that occur. However, temperature sensors take longer because they must physically heat up or cool down, when a temperature change takes place, until they reach the same level as the measured variable. This effect, known as "temperature lag" or "thermal lag," can take from seconds to minutes when responding to changes. A sensor's proximity to the measured variable also affects the response time. For example, a temperature sensor inside a protection chamber called a *thermowell* will take longer to respond than a sensor directly exposed to the measured medium. Similarly, an air-pressure sensor inside a pipe will respond more quickly than an identical sensor connected to the pipe through a long length of tubing. The graph in Figure 15-7 shows a comparison of typical response times for different types of variables.

Since the controller receives an input from the sensor, it cannot react to changes in a time shorter than the response time of the sensing device. Therefore, the response time of the entire loop is affected by the sensor. Response time is a factor that must be considered if a control loop requires a very fast response to changes in the condition of a variable.

Static

Accuracy

The term **accuracy** is used to describe how closely a sensor measures the actual value of a controlled variable. Figure 15-8 is used to illustrate the accuracy of a temperature sensor as it takes five separate readings of a constant temperature. The graph shows several different

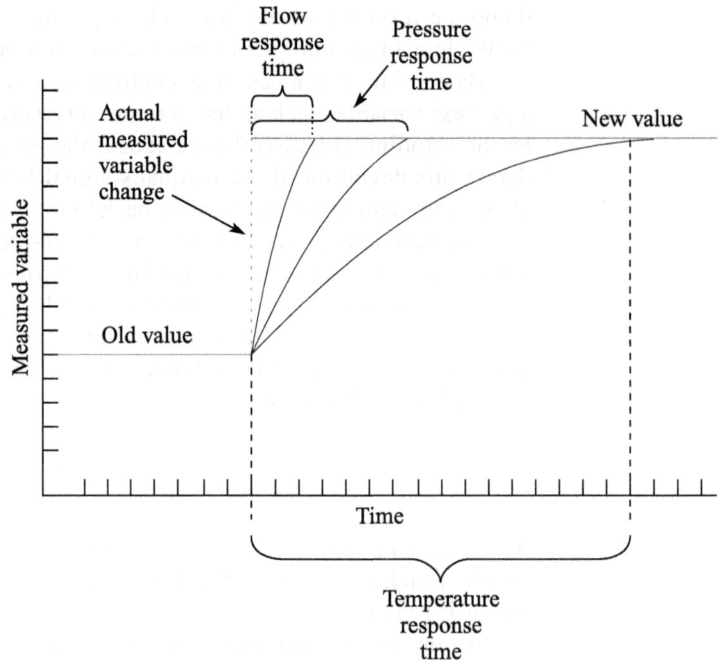

FIGURE 15-7 Response time

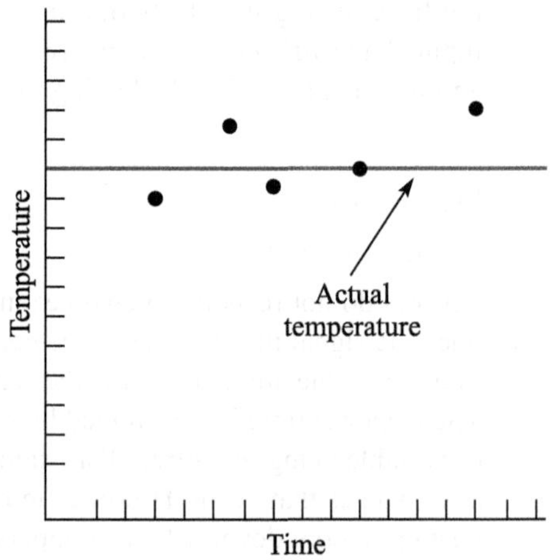

FIGURE 15-8 Accuracy

measurements. If each indication is within a tolerance range as specified by the manufacturer, the sensor is considered accurate.

Precision

The term **precision** is used to describe how consistently a sensor responds to the same input value. Figure 15-9 is used to illustrate the precision of a temperature sensor as it takes five separate readings of a constant temperature. The graph shows several different measurements. Even though the measurements are not exact, they are considered precise because they are consistent. Another term used to describe precision is "repeatability."

Linearity

The sensing device converts a physical quantity of the variable it measures into a signal such as pneumatic or electrical. With some types of sensing devices, the output produced is not

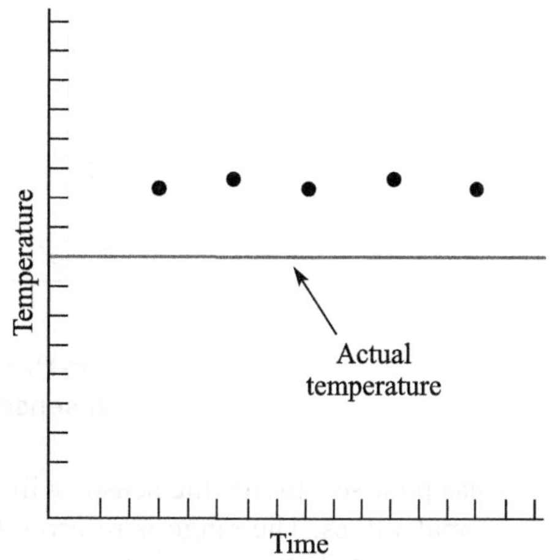

FIGURE 15-9 Precision

proportional to the actual condition measured. Instead, the signal produced is the square of the variable's physical quantity. If the input and output of the sensor are plotted graphically, as shown in Figure 15-10, a nonlinear line, or curve, will be produced.

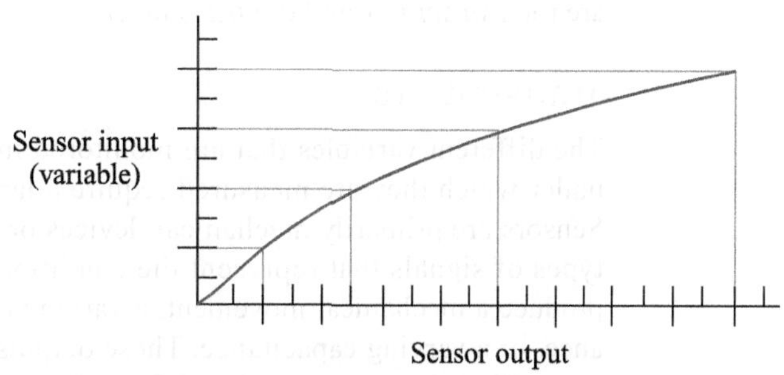

FIGURE 15-10 Nonlinear output of a sensor

Hysteresis

The graph in Figure 15-11 is used to show the **hysteresis** characteristics of some types of sensors. There are two curves identical in shape. The upward and downward arrows describe the way in which the output reading varies as the measured signal applied to its input increases and decreases, respectively. The illustration shows that the instruments produce different output values for equivalent low-to-high and high-to-low input changes. Hysteresis is the dissimilarity between these two curves.

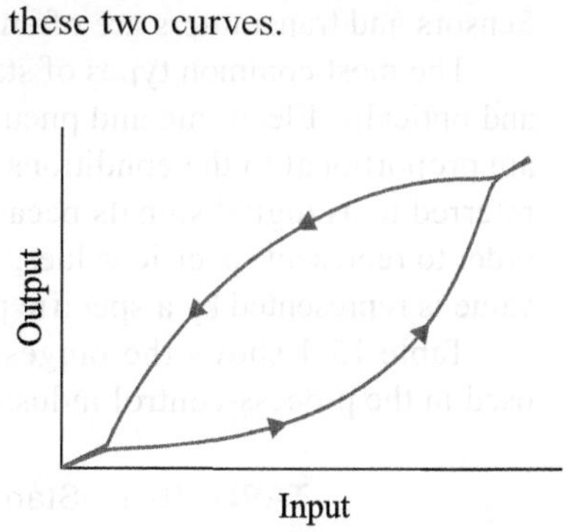

FIGURE 15-11 Hysteresis loop

Sensitivity

The **sensitivity** of a sensor is the ratio of its output change to a change in its input quantity, which represents the measurements. The graph in Figure 15-12 illustrates that, if a sensor

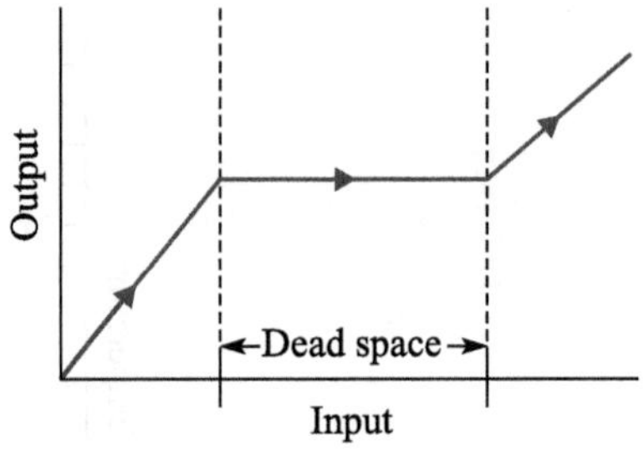

FIGURE 15-12 The dead space of a sensor that has poor sensitivity

has poor sensitivity, the sensor will not produce an output change in response to a range of input values. The range is referred to as *dead space.* The dead space will be smaller for a sensor with better sensitivity.

15-5 Feedback Loop Interface Instruments

The sensor is seldom connected directly to the controller. Instead, various interface instruments are usually added to the feedback loop that connects them together. These instruments are the *transmitter* and the *transducer.*

Transmitters

The different variables that are monitored in the process industry and the many conditions under which they are measured require a large variety of sensors to perform this function. Sensors are primarily mechanical devices or electronic instruments, and they produce many types of signals that represent the condition of the controlled variable. For example, they produce a mechanical movement, a varying current flow, a varying voltage, a varying resistance, or a varying capacitance. These outputs are sent to a **transmitter,** which has two functions. First, it converts a signal from the sensor into a standardized signal used in process systems. Second, since the sensor is often positioned at a remote location from the controller, it carries the signal through the distance between them. Some sensors produce very small voltages no greater than 1 mV. The transmitter contains an amplifier that boosts the signal high enough to overcome the resistance of long wires. Some transducers have a signal-filtering function, which removes a particular band of frequencies within a signal. For example, low-pass filtering may be required to remove the high-frequency noise component in a signal. This noise is any stray voltages that are induced into the sensor's conductor from magnetic lines that develop around motors and other high-current-carrying devices located close by. Sensors and transmitters are often combined into one unit.

The most common types of standard signals are transmitted electronically, pneumatically, and optically. Electronic and pneumatic signals are referred to as *analog* because their values are proportional to the conditions they represent within a standard range. Optical signals are referred to as digital signals because they are either in the "On" or "Off" state condition. In order to represent specific values, a series of On and Off pulses are optically transmitted. Each value is represented by a specific pattern of pulses.

Table 15-1 shows the ranges of standard electronic and pneumatic signals commonly used in the process-control industry.

IAU6407
Standard Transmission
Signals

TABLE 15-1 Standard Transmission Signals

Electronic	*Pneumatic*
4–20 mA DC	3–15 psi
0–20 mA DC	
0–10 VDC	

Electronic Signals

Analog electrical signals in a control system are direct current (DC) and can be divided into two categories: voltage and current.

Voltage Signals Voltage signaling is uncommon between transmitters and controllers within process industries. The most common application is to provide an input to display devices, recorders, and occasionally a controller. Voltage signaling is limited to short-distance transmission.

Current Signals The most commonly used electronic signals are current signals having current ranges from 4 to 20 mA and 0 to 20 mA. For signaling using 4 to 20 mA, the transmitter draws about 3 mA and therefore does not need a separate power supply. In a two-wire configuration, the same leads are used for signaling and to supply the power. The 4 mA makes the transmitter easy to calibrate because the lowest setting can be adjusted slightly lower than 4 mA if necessary. For signaling using 0 to 20 mA, the resolution is better than the 4 to 20 mA range. However, its lowest range for calibration is limited to 0 mA. Also, a separate power supply for the transmitter must be provided, which adds to the installation cost. Current signaling is commonly used for short-distance transmission but exclusively used for long-distance applications.

Pneumatic Signals

Pneumatic 3- to 15-psi (pounds per square inch) signals are often used for environmental conditions where a spark may cause an explosion.

Transducers

Most batch- and continuous-process-control machines have a large variety of instruments that do not respond to the same types of signals.

To enable these instruments to work together, some type of signal conversion is necessary. Transmitters perform this function and also provide long-distance transmission by using an amplifier. When long-distance transmission is not required, another instrument called a **transducer** is sometimes used to perform the signal-conversion function. The following examples show some of the most common types of transducers used in the process-control industry.

I/P Transducer

An I/P transducer converts an electrical current signal (I), produced by a sensor ranging from 4 mA to 20 mA, to a pneumatic signal (P) ranging from 3 to 15 psi, both of which are standardized values used in the process-control industry.

Pneumatic signals are often used to operate a pneumatic actuator that causes a valve to vary the fluid flow through a pipe. Figure 15-13 shows a control loop that uses an I/P transducer.

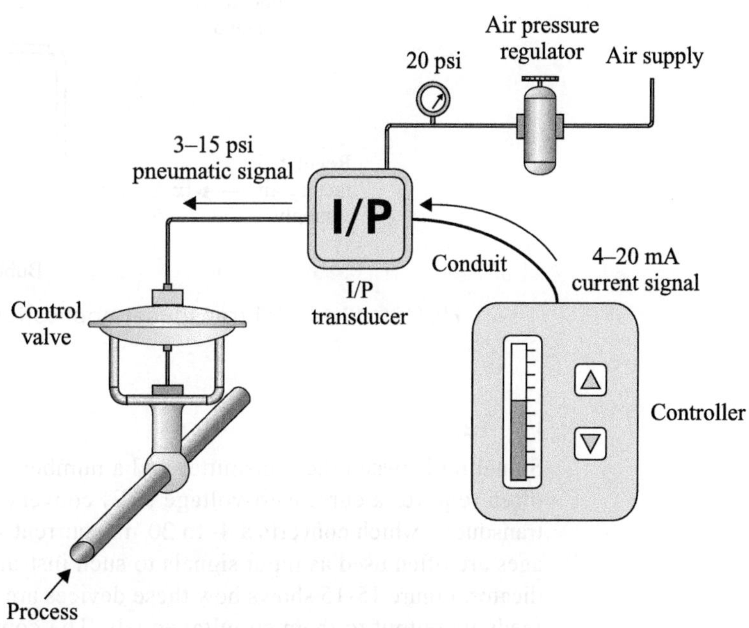

FIGURE 15-13 I/P transducer application

A pressure regulator supplies 20 psi of air to a transducer located near a pneumatic valve that controls the process flow. The transducer receives an electrical signal from a controller and produces a proportional analog pneumatic signal. For example, if a 4-mA input signal is received, 3 psi will be applied to the valve. Likewise, a 20-mA current signal will cause the transducer to apply 15 psi to the valve.

P/I Transducer

A P/I transducer converts a 3- to 15-psi pneumatic (P) signal to a 4- to 20-mA current (I) signal. Figure 15-14 provides an example of how the pneumatic signal is received from a liquid level sensor called a *bubbler*. Air is fed to an immersed dip-tube vertically inverted with an open end placed close to the bottom. The amount of air that is forced out of the tube is inversely proportional to the level of the fluid. For example, the higher the level, the more difficult it becomes to force air through the end of the tube. As a result, a larger amount of backpressure develops that represents the height of the liquid.

One application of a bubbler sensor is to measure the tank level of a flammable liquid because there are no electrical signals or connections that could create a spark to cause an explosion. The pressure signal is applied to the input of a P/I transducer that is capable of sending a current signal to the controller at a remote location. When the maximum level is detected, the bubbler creates a backpressure of 15 psi that is converted to 20 mA by the P/I transducer. At the minimum level in the tank, a backpressure of 3 psi would be converted to a 4-mA current signal for the controller.

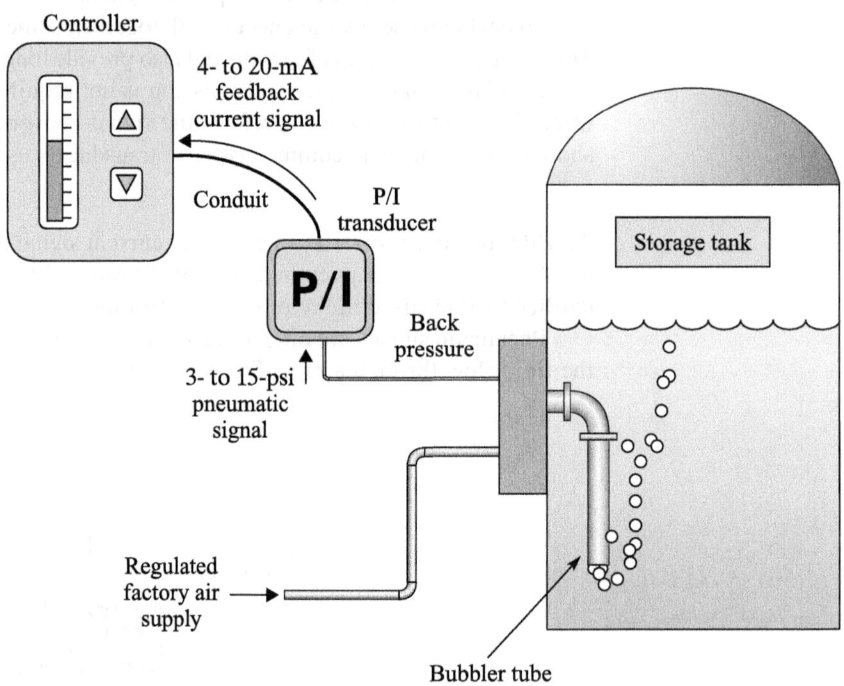

FIGURE 15-14 P/I transducer application

I/E Transducer

Signaling between the transmitter and a number of instruments located in the control room often requires a current-to-voltage (I/E) conversion. This function is performed by an I/E transducer, which converts a 4- to 20-mA current signal to a 1- to 10-V DC signal. DC voltages are often used as input signals to such instruments as a controller, a recorder, or an indicator. Figure 15-15 shows how these devices are connected in parallel when the transducer feeds its output to them simultaneously. The conversion process is achieved by using the

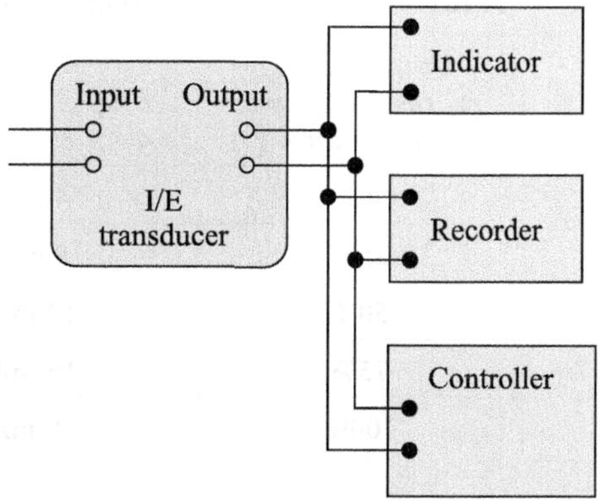

FIGURE 15-15 I/E transducer application

voltage drop across an internal resistor as the transducer output, which is proportional to the current that passes through from the transducer's input terminals.

Square Root Extractor

When a fluid passes through the orifice of a flowmeter, the differential pressure that develops across the restriction is nonlinear. Therefore, the magnitude of the signal produced by a differential pressure-type flowmeter is not proportional to the rate of flow. Instead, the flow rate is proportional to the square root of the pressure drop being measured. To compensate for this condition, it is necessary to linearize, or extract, the square root from the sensor's output. A transducer called a *square root extractor* is placed between the sensor and controller to perform this function. The symbol for this transducer is a circle with a square root symbol.

Analog-to-Digital Transducer

Most signals produced by measuring controlled variables are analog. However, most controllers are computer-based and process digital signals internally. An analog-to-digital (A/D) converter similar to the type described in Chapter 2 makes the conversion. A/D input modules are used by programmable controllers to perform this function, and computers use an I/O board.

Digital-to-Analog Transducer

As the controller processes information, such as comparing the digitized command "setpoint" signal with the measured "feedback" signal, it produces a digitized output that must be converted to an analog signal. This function is performed by a digital-to-analog (D/A) converter similar to the one described in Chapter 2. D/A output modules are used by programmable controllers to perform this function, and computers use an output card.

15-6 Controllers

The **controller** is the element in a closed-loop system that performs the decision-making function. By comparing a setpoint value that represents the desired condition in a process to a signal from a sensor that represents the actual condition, the controller determines if and how a correction needs to be made. The operations that most controllers perform are simple On-Off control and the more-sophisticated PID control. The gain settings are made during a tuning procedure, and a defined control algorithm programmed by the system designer determines how much and how rapidly the controller output will change in response to an error. The controller's output signal is sent to the final control element, which directly affects the process. There are two categories of controller outputs: *dimensional* and *nondimensional.* Dimensional outputs are absolute values such as 3 to 15 psi, 4 to 20 mA, or 0 to 10 volts DC. The output value of many controllers is also represented by nondimensional values, which are percentage figures between 0 and 100. For example, a percentage indicates that a valve is 70 percent open or a flow is 50 percent of the maximum rate. By using percentages, it is much easier to keep track of the huge diversity of absolute value signal ranges that

TABLE 15-2 0- to 100-Percent Output Signal Versus Common Dimensional Ranges

Output units represented by a percentage	*Output value for a 4- to 20-mA signal*	*Output value for a 0- to 10-VDC signal*	*Output value for a 3- to 15-psi signal*
0%	4 mA	0 V	3 psi
25%	8 mA	2.5 V	6 psi
50%	12 mA	5.0 V	9 psi
75%	16 mA	7.5 V	12 psi
100%	20 mA	10 V	15 psi

represent the condition of each variable. Table 15-2 shows how a 0- to 100-percentage output signal relates to common dimensional output values. The most common types of controllers are briefly described in the following text.

Pneumatic Controllers

Pneumatic controllers produce a pneumatic output signal that is applied to a flow-control valve. They are usually located in the field and are mounted near the point of measurement. These controllers are usually found in older systems and are capable of controlling only one loop.

Panel-Mounted Controllers

The panel-mounted controller is a microprocessor-based device which can measure, display, and control temperature, pressure, level, flow, and other process variables. This device is relatively inexpensive and is capable of performing simple On-Off as well as PID-type control operations. A diagram of this type of controller is shown in Figure 15-16.

Users can select the instrument functionality required from menus, using tactile membrane keys and the high-intensity LED-segment displays.

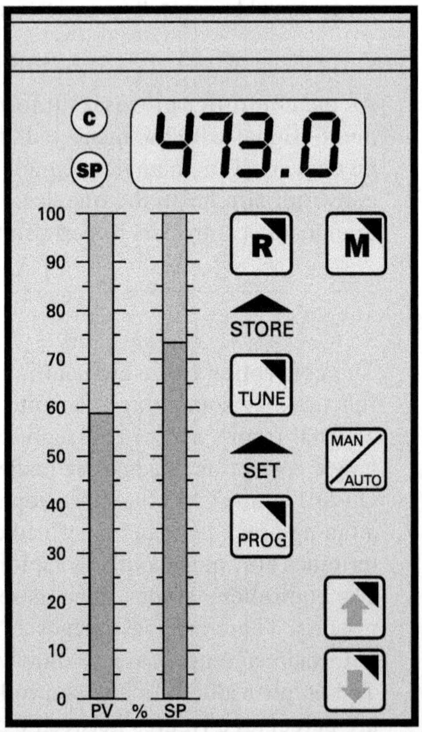

FIGURE 15-16 Panel-mounted controller

By pressing the **M** button, several modes of operation can be accessed, such as On-Off parameters, PID gain settings, and preset limits that activate an alarm when they are exceeded. The specific values for each mode are shown on the digital display as they are programmed by pressing the up/down arrows after the **PROG** key is pressed. Bar graph indicators on the left display the setpoint (**SP**) value selected and the run-time status of the process variable (**PV**) that is being controlled. The **TUNE** button is used for autotuning, which automatically selects the gain settings for each PID mode. The **R** button performs the reset function if the programmer needs to repeat the keypad entries.

An auto/manual button enables the operator to use the controller in either manual or automatic modes. When in manual mode, the control function is performed by the user. In automatic mode, the controller function is performed by the microprocessor-based circuitry.

Personal Computers

The personal computer (PC) can be used for small systems that are easy to control. Software programs provide the control of On-Off, PID, and multiple-loop operations. Some types of software programs show the real-time operations on a graph, as shown in Figure 15-17(a), or by a pictorial display on a computer screen that contains a drawing of the actual equipment, as shown in Figure 15-17(b).

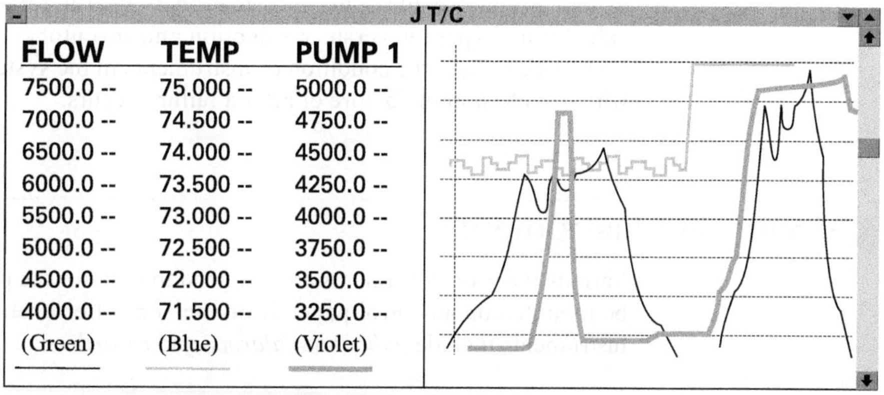

FIGURE 15-17(a) Paperless graph recorder

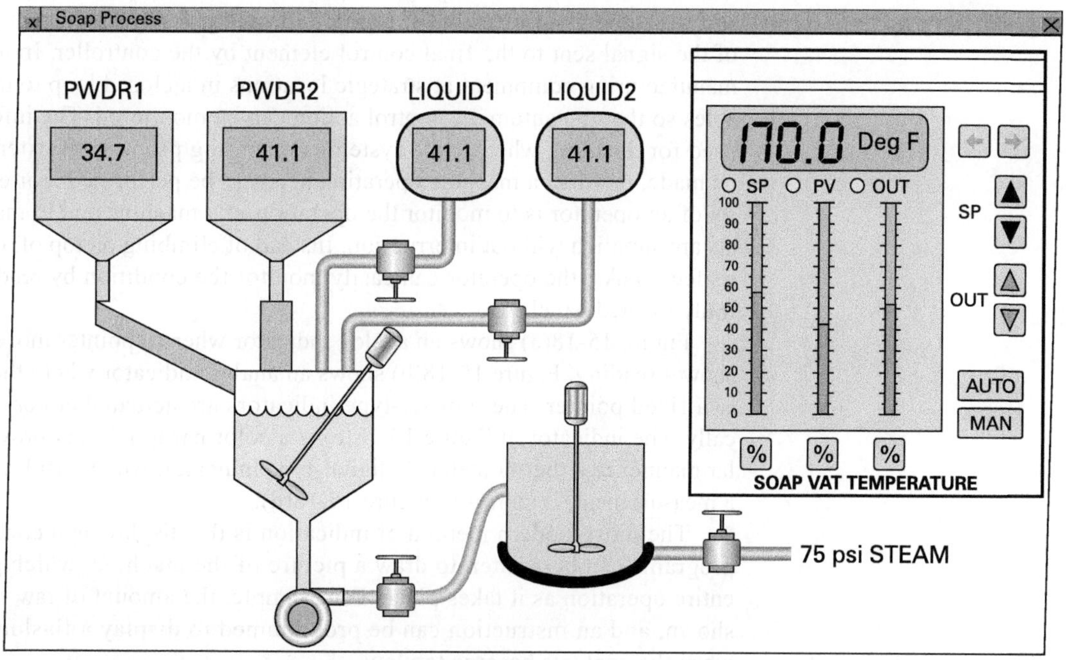

FIGURE 15-17(b) Pictorial computer display

Programmable Logic Controllers

The programmable logic controller (PLC) can perform most control operations for very complex systems. Discrete I/O modules perform On-Off batch operations. Input analog modules are capable of interfacing with sensors and transmitters, and output analog modules can send control signals to final control elements used in a continuous process. The monitor used by the PLC can display the same information as the PC.

Distributed Control Systems (DCS)

Factories that use large production machines have an entire room with a high-capacity computer dedicated to performing complex control operations. These computers can control hundreds of control loops, simultaneously enabling them to interact with one another to coordinate each part of the machine so that it performs the required manufacturing operation as one unit. The computer displays real-time information about each loop on indicators such as numerical displays, bar graphs, meters, and pictorial drawings on computer screens. Operators located at consoles inside the centralized control room observe the display screen to monitor the operation. These computers set off alarms if an undesirable condition arises and record historical data for analysis of each control loop at a later time. This data can also include the amount of raw material used, energy consumption required, and the amount of end product made. This information is useful for the accounting department when calculating expenses versus production and inventory. The DCS also provides diagnostic information about the condition of instruments in the system to assist maintenance personnel with troubleshooting before or after a failure occurs.

15-7 Monitoring Instruments

Various types of instruments are designed to monitor the process-control operation. They can be mounted on a control panel or on the controller itself or can be a dedicated unit. These instruments include *indicators, alarms,* and *recorders.*

Indicators

Indicators are used to display information for the operator or technician. They show data such as setpoint adjustments, the amplitude and polarity of an error signal, or the magnitude of the signal sent to the final control element by the controller. Indicators are placed on the manufacturing equipment at strategic locations in a closed loop to display the status of variables so that the automatic control action can be monitored. The information they provide is used for deciding whether the system is operating properly, if external adjustments need to be made, or when a machine operation needs to be performed. For example, the responsibility of an operator is to monitor the operation of a machine and to make sure that raw materials are supplied without interruption. Instead of climbing on top of, or around, a machine and its feed tanks, the operator can easily monitor the condition by reading indicators on a centrally located panel.

Figure 15-18(a) shows an analog indicator where a pointer moves across a fixed scale to show a reading. Figure 15-18(b) shows an analog indicator where the scale moves in relation to a fixed pointer. These meter-type indicators are actuated either electrically or pneumatically. The indicator in Figure 15-18(c) is a color bar graph that provides a reading in a similar manner to a thermometer. A digital-type indicator, which displays numbers that represent a measurement, is shown in Figure 15-18(d).

The most modern method of indication is the display on a computer monitor. Software programs can be written to draw a picture of the machine, which graphically displays the entire operation as it takes place. For example, the amount of raw material in a feed tank is shown, and an instruction can be programmed to display a flashing warning—"empty"—when the contents become too low.

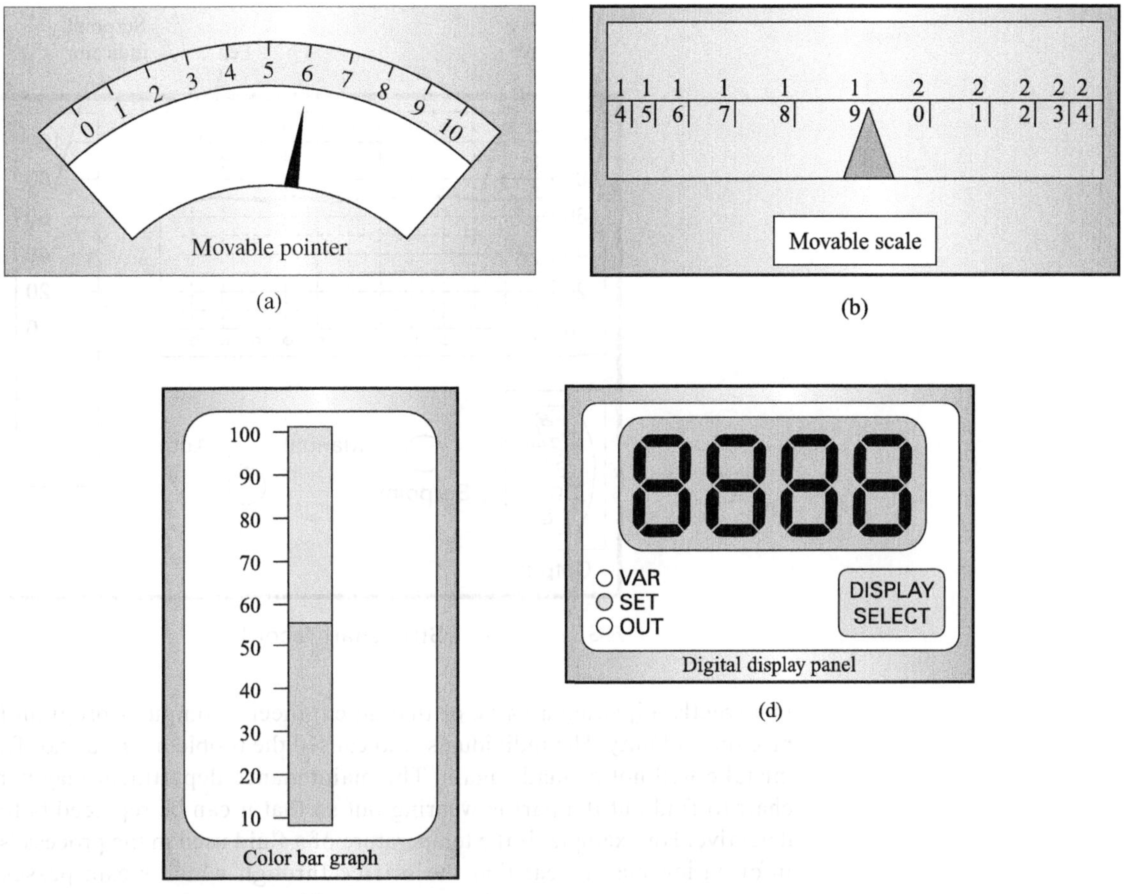

FIGURE 15-18 Indicators

Alarms

Control systems can be equipped with a variety of **alarm** features. Their purpose is to warn the operator or to initiate an action if an undesirable process condition develops. When a condition falls outside parameters that are programmed into a controller, it activates an alarm. The alarm action may be to turn on a light or electronic sound, to flash a message on the controller's display that describes the problem, or to shut the system down.

Examples of alarm operations are as follows:

1. A deviation alarm, which activates when the controlled variable differs from the setpoint by a certain value.
2. A rate-of-change alarm, which turns on when a controlled variable is increasing or decreasing more rapidly than desired.
3. A limit alarm, which is initiated if the controlled variable reaches or exceeds a predefined value. For example, the alarm system will turn off a feed pump if the level in a storage tank, to which it supplies a liquid, reaches a certain height.

Recorders

In some types of manufacturing equipment, such as a paper machine, a large number of variables must be continuously monitored and controlled. This information is often recorded to be read and analyzed at a later date. For example, suppose that some bad rolls of paper are discovered by the quality-control department. By reading data charts that were recorded while the paper was being made, it may reveal that an operator made a mistake by

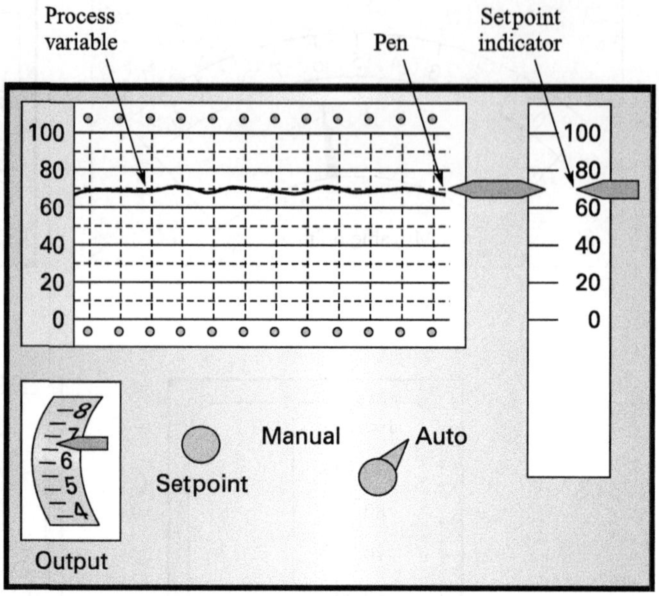

FIGURE 15-19 Strip chart recorder

incorrectly adjusting a valve or that an engineer's computer program for controlling the machine is faulty. The individuals who caused the problem will be notified so that the same mistake will not be made again. The maintenance department may periodically read the chart to find out if a part is wearing out so that it can be replaced before becoming fully defective. For example, if the temperature of a fluid used in the process is slowly rising, this information may reveal that the orifice through which steam passes inside a valve is becoming larger due to erosion.

Recording this type of information is called *data acquisition.* Some plants still use strip chart recorders, where a pneumatic element is connected by a linkage to a stylus. Figure 15-19 shows a chart **recorder** that displays the data on a controller/recorder. A pen that is connected to the stylus is in contact with a chart that is slowly rotated by a motor (at maybe only one revolution per hour). As the variable is sensed by the pneumatic element, the pen draws an ink line to graphically show its status. Modern data acquisition recorders use computers to display a chart recorder on its monitor. Instead of using a roll of paper to store historical data, the information is saved on disks or hard drives. This data can be retrieved at a later time and printed on a data sheet for examination.

15-8 Manipulation Devices (The Final Control Element)

In process-control systems, the **final control element** is the device that directly influences the process variable. There are several types of final control elements used in process applications, such as solenoids, electrical motors, and valves.

The Solenoid Valve

The solenoid valve, shown in Figure 15-20, is an electromagnetic device that converts On-Off electrical signals to On-Off flow control. The solenoid consists of a coil and a flow valve through which fluid passes. When electrical current flows through the coil, the electromagnetic field it produces moves an element inside the valve that allows fluid to pass through. When current is turned off, the element moves to another position that completely blocks the fluid through the valve. The electrical signal that controls the solenoid valve is primarily supplied by programmable logic controllers (PLCs) or microcontrollers.

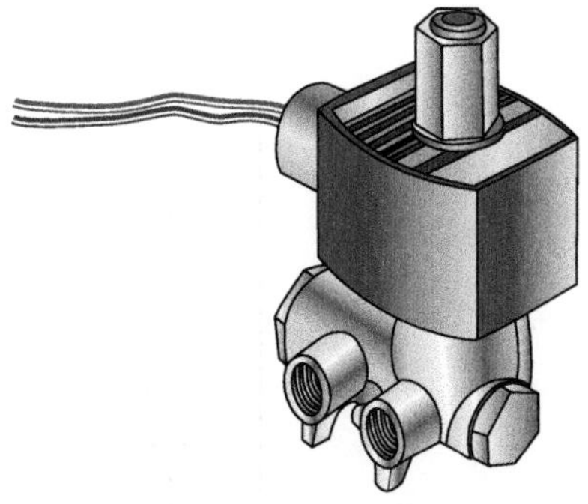

Figure 15-20 Solenoid valve

Solenoid valves are typically used in applications that provide On-Off control. The valve stops and starts fluid flow in pipelines for applications such as batch processes, automatic sequencing, or shutting down flow to prevent a safety hazard.

DC and AC Motors

In many process-control applications, an electric motor is the actuating device that powers a mechanical load. Motors are used in both On-Off control processes and variable-speed applications. Producing compressed air for a process is an example of an On-Off operation. When the pressure drops below a certain level, an electrical motor that drives a compressor will turn on. When the compressed air reaches a higher-pressure level, the motor driving the compressor will turn off. Another example of On-Off control is a food-processing application. Motor-driven stirrers are turned on for a predetermined amount of time to mix different ingredients that are blended together.

DC motors are often used in applications that require precise variable speeds with high-torque capabilities. For example, they are used to turn the auger in an extruding operation. AC motors are used to power the majority of machines in the process industry, such as driving a pump that forces liquids to flow through a pipe or to drive blowers and fans to move air. By themselves, AC motors turn at a constant speed, making it necessary to use control valves or dampers to vary the flow rate. At one point, the motor runs at its full speed continuously, and much energy is wasted when a control valve or a damper reduces the flow. To vary the speed of AC motors, they must be powered by AC variable-speed drives. An AC drive changes the speed of an AC motor by varying the frequency of the AC power it produces, as shown in Figure 15-21a. The amount of energy consumed by an AC motor is proportional to the speed it rotates. DC motors are capable of running at different speeds by varying the voltage of the applied DC power. For precise control, DC variable-speed drives are used to supply variable electrical DC power to DC motors, as shown in Figure 15-21b.

The Control Valve

The most widely used final control element for process systems is the *control valve*. The control valve is a mechanism that regulates the amount of fluid flow by varying the size of the passage through which fluid passes. Valves either pass flow at full capacity, throttle the flow (a condition in which the valve is partially open to restrict flow), or completely stop the flow of a fluid. A *fluid* is defined as a liquid, gas, or vapor.

The control valve, shown in Figure 15-22, is an assembly of components that is made up of three subassemblies.

1. The *valve body* is the housing that is a part of the main process line connected to the pipes through which the fluid in the system passes. The body consists of an inlet and outlet connection and internal trim elements that are in contact with the controlled

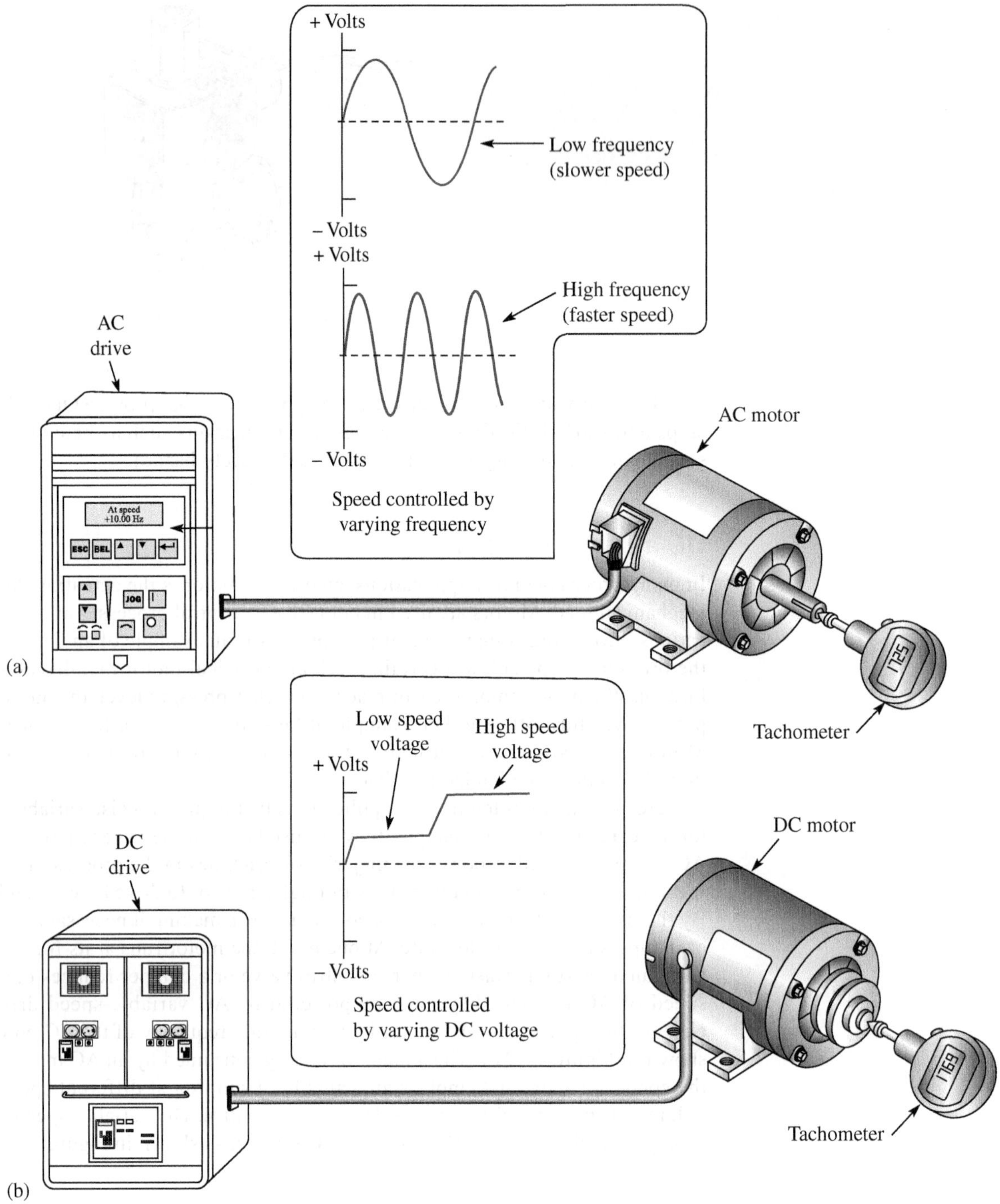

FIGURE 15-21 Variable-speed drives controlling motor speeds (a) AC drive and AC motor (b) DC drive and DC motor

fluid. These elements are the stationary valve-restrictor seat that is fixed to the valve body to form a port and a restrictor.

2. A movable restrictor called a *plug*. Its position controls the flow rate by changing the size of the passage at the seat. The plug may be a disk, a ball, or other element that opens, closes, or provides a variable restriction at the seat.
3. The *valve actuator*, which is connected to the plug. The actuator provides the force needed to overcome pressures that develop in the fluid and resist the movement of the plug.

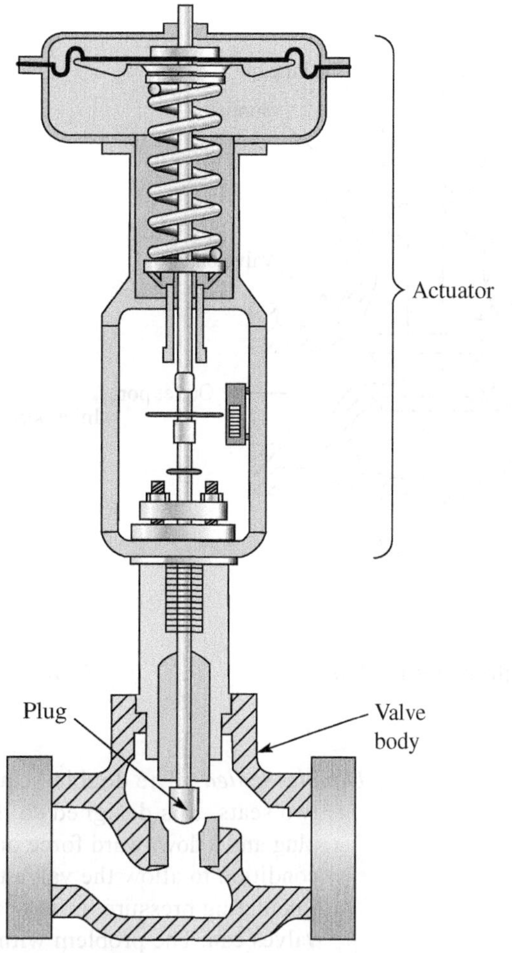

FIGURE 15-22 Control valve subassemblies

Control Valve Classifications

A control valve is usually classified on the basis of its *body style* or type of restrictor used to control the flow. *Sliding-stem globe valves* and *rotary valves* are the most common and versatile types of flow-control valves. Their popularity is the result of their rugged construction and multiple models to select from for satisfying different application needs.

Many styles of control valves have been developed through the years. The following summary describes some of the popular designs presently used in the process-control industry.

Sliding-Stem Globe Valves

The globe valve is the type most commonly used for controlling the flow of fluids. It gets its name from its globular-shaped cavity located around the port region. Three different types of globe valves are described as follows:

Single-Seated: The single-seated valve is shown in Figure 15-23. It consists of a single plug and seat. The fluid enters the port beneath the seat and creates an upward force against the plug. As long as the pressure from the process line does not exceed the force from the valve stem, tight shutoff will occur. For high-pressure applications, the valve plug is located below the valve seat, and the convoluted edge is around the top rim instead of the bottom rim. Also, the valve seat is located on the inlet side of the port, and the plug is pushed upward by the stem. This design causes the inlet pressure to push the valve plug upward and aids the force from the actuator.

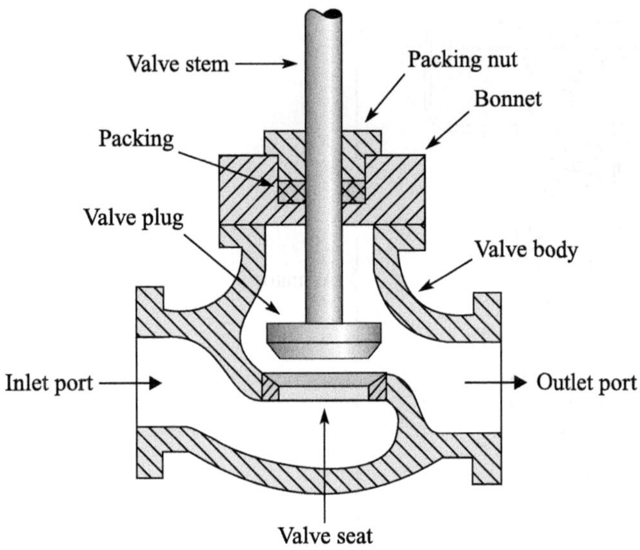

FIGURE 15-23 Single-seated valve

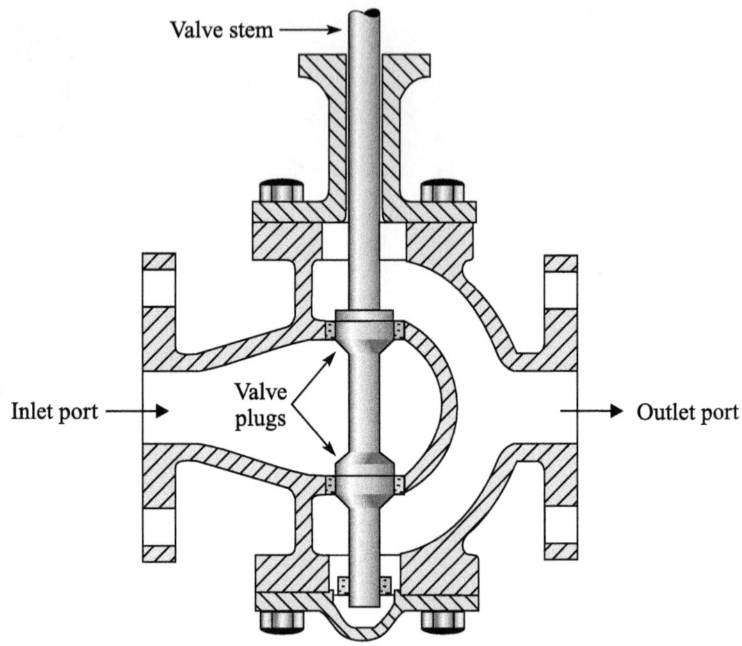

FIGURE 15-24 Double-seated valve

Double-Seated: The double-seated valve, shown in Figure 15-24, has two plugs and two seats. It is designed so that the line pressure creates an upward force on one plug and a downward force on the other plug. This configuration creates a balanced condition to allow the valve to be used for applications involving high pressure or fluctuating pressure, or to accommodate a higher flow capacity than single-seated valves can. The problem with double-seated valves is that over time, uneven wearing across the two ports occurs. This condition makes it impossible for the two sets of plugs and seats to shut at the same time.

Three-Way: The three-way valve has three external ports connected to three different pipes. It is used primarily for two types of applications: *mixing* (or *blending*) and *diverting.* Figure 15-25 shows its configuration for a mixing application. There are two inlet ports through which different fluids enter. Inside the valve body, the liquids converge and then exit the outlet port. When the valve plug is in the intermediate position, the fluids from ports A and B equally converge and mix before they exit port C. As the plug moves upward, the passage for port B becomes larger, and the passage for port A becomes smaller. The result is that a greater portion of the fluid is mixed from port B than from port A.

Figure 15-26 illustrates the configuration for a diverting application. The fluid enters an inlet port and splits inside the valve body before it exits two different outlet ports. Moving the plug up or down can change the portions of the diverted fluid.

Sliding-Stem Valves

Sliding-stem valves are designed without a globular cavity around the restrictor. Their internal valve port where the flow restriction action occurs aligns with the external pipes to which they are connected. The design allows the fluid to pass straight through, which minimizes turbulence that is common in a globe valve due to the shape of its internal cavity. Two different types of valves with this design are described as follows:

Knife Gate: The knife-gate valve, shown in Figure 15-27, uses a disk as its restrictor. Because fluid passes through the valve in a straight path and it has full port capabilities

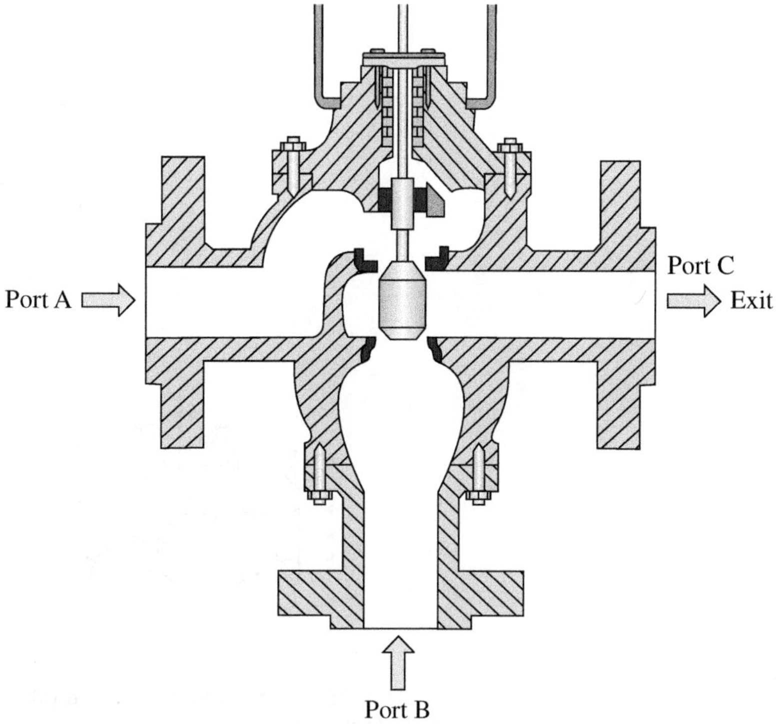

FIGURE 15-25 Mixing

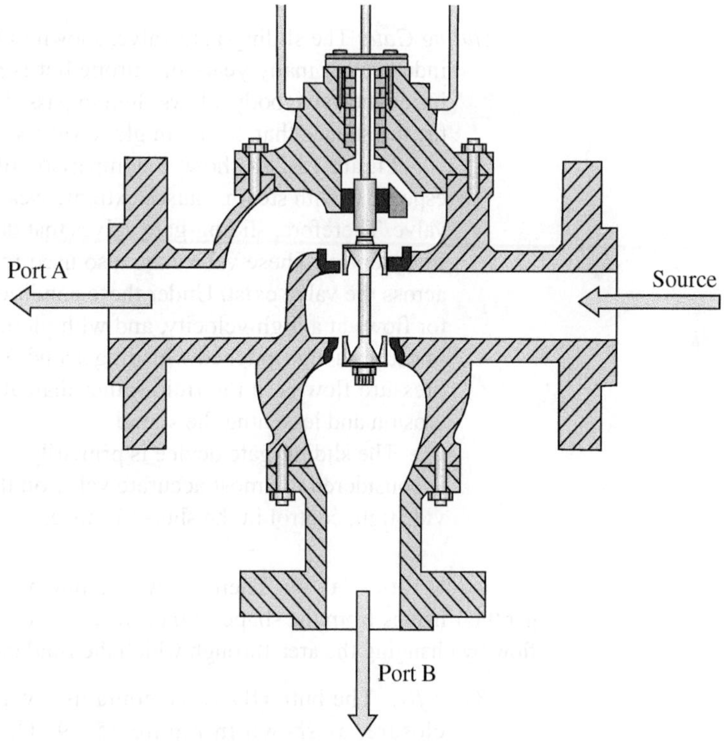

FIGURE 15-26 Diverting

FIGURE 15-27 Knife-gate valve

(meaning that it aligns with the external piping when fully open), particles do not get caught and clog the valve. Knife-gate valves are primarily used for viscous liquids (with a specific gravity greater than water) or fluids that are not consistent, such as slurries or oil with globs of grease. Because the range the disk travels is large and is equal to the diameter of the valve's internal port, a piston actuator must be used to move the disk to provide the large amount of travel that is required.

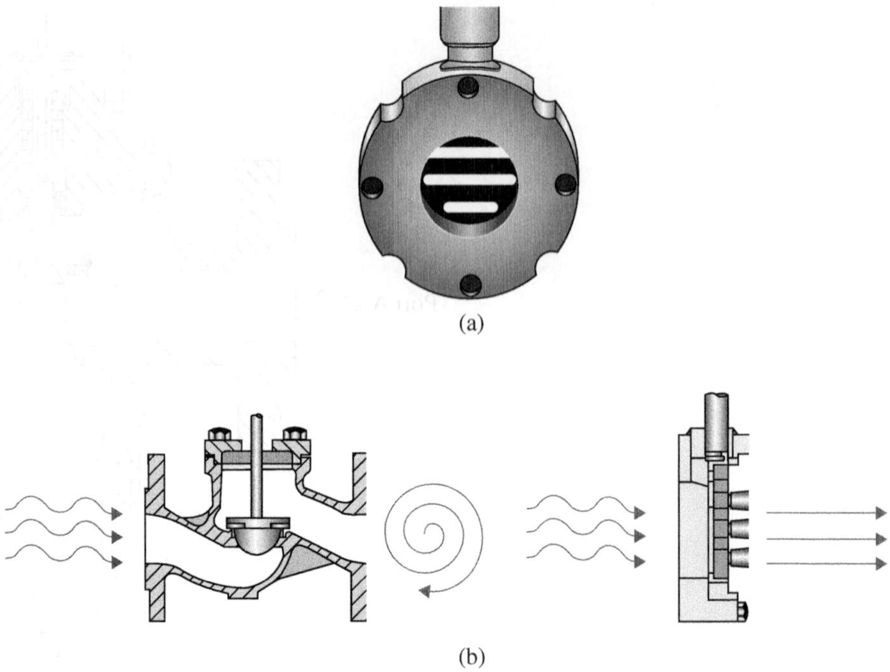

FIGURE 15-28 Sliding-gate valve (a) Valve restrictor (b) Comparing turbulent flow

Sliding Gate: The sliding-gate valve, shown in Figure 15-28a, has been used in the process industry for many years in Europe but is relatively new to the U.S. market. The port inside the valve body allows fluid to pass straight through. This characteristic minimizes the turbulence that occurs in globe valves.

Figure 15-28b shows a comparison of turbulence in both valves. Turbulent flow, especially with steam, causes extreme wear on the surface of the cavities inside a globe valve. Therefore, sliding-gate valves that do not generate turbulence are ideal for steam flow control. These valves are also used for applications in which high-pressure drops across the valve exist. Under these conditions, the fluid that passes through the restrictor flows at a high velocity, and with globe valves, cuts away at the surface in its path or causes a high-pitched whistling sound. With the sliding-gate valve, most of the high-pressure flow is in the fluid rather than at the inside surface of the pipe, minimizing erosion and lessening the sound.

The sliding-gate device is primarily used for proportional control applications and is considered the most accurate valve on the market (1/2 to 3 percent). They also provide tight control in the shut-off mode.

Rotary-Motion Valves There are two common types of rotary valves: *butterfly* and *ball.* They get their names from the shape of their restrictors. By rotating on a shaft, the restrictor alters the flow by changing the area through which the fluid passes. They are described as follows:

Butterfly: The butterfly valve contains a vane or a disk restrictor to provide the valve closure, as shown in Figure 15-29. The disk is rotated by a motor, spring-and-diaphragm, or spring piston. One advantage of the butterfly valve is that the valve body around the restrictor is small, allowing higher flow capacities for its size as compared to globe valves. Also, since the fluid passes straight through the valve, there is relatively little wear, and there is no accumulation of stock or sludge such as there is with globe valves that have pockets. Butterfly valves have flow characteristics ranging between linear and quick-opening, which means they exhibit a nonlinear relationship between the percent opening and the rate of flow. For example, when a butterfly valve is opened halfway, it allows a flow rate that is

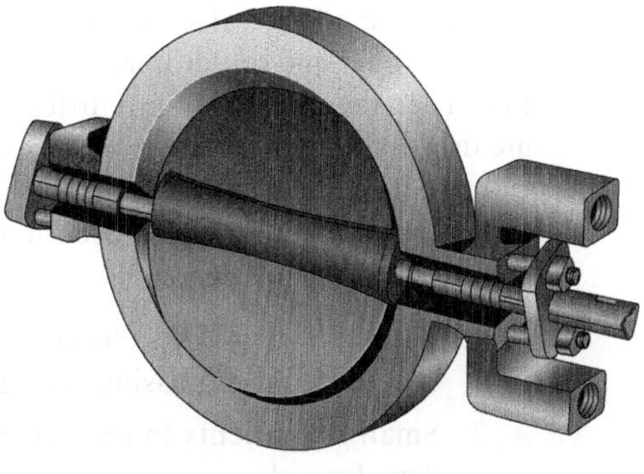

FIGURE 15-29 Typical wafer-type butterfly valve

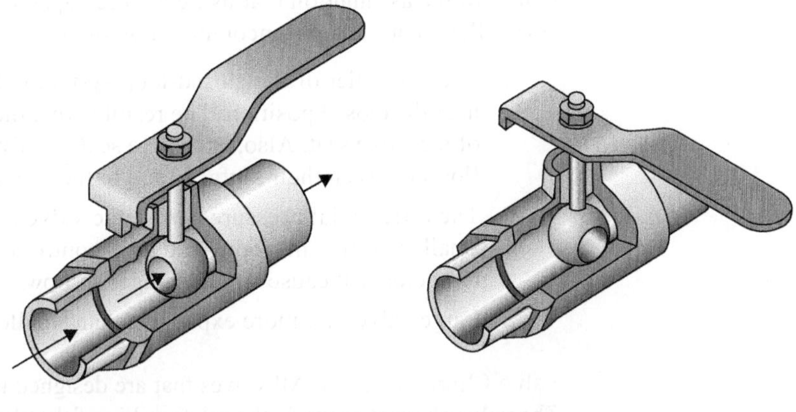

FIGURE 15-30 Flow diagram for a ball valve

near maximum flow capacity. The butterfly valve is capable of controlling all types of fluids and is used in applications where high static pressures with small pressure drops across the restrictor are desired, such as in a cooling water system.

Ball: The ball valve shown in Figure 15-30 contains a plug that is spherical in shape and has either a v-notch or a circular port. When it is activated, the plug rotates up to 90 degrees. At 0 degrees, the port lines up with the inner diameter of the pipe and is completely open, and at 90 degrees, the port aligns with the wall of the pipe and is completely closed. As the plug turns, it varies the flow from minimum to maximum capacity. Ball valves may be used for tight shutoff or for modulating flow applications. For their size, they have the greatest flow capacity of all control valves. Because there are no internal obstructions when the valve is fully open, ball valves have low maintenance requirements, and they withstand corrosive materials very well. These valves are often used to control fibrous flows such as pulp stock, paper stock, and slurries.

Selecting a Control Valve

The result of a survey conducted in recent years has revealed that improper control valves are used in roughly one-third of all installations. When the wrong valve is used, plant efficiency and product quality are degraded. When selecting the proper control valve for a particular application, two factors must be considered: *valve capacity* and *valve characteristics.*

Valve Capacity **Valve capacity** refers to the amount of material that a valve is capable of allowing to pass. Its capacity is influenced by the physical size of the flow passage within the body.

The size of the valve has an effect on the amount of system pressure that is dropped across its inlet and outlet ports. The smaller the valve, the larger the pressure that is developed. The amount of pressure differential needed for good control is a function of the pressure dropped across the valve with respect to the rest of the system. A rule accepted by many designers is that 50 percent of the system pressure should be dropped across the valve. Sizing valves incorrectly can cause them to operate at substandard levels. A valve that is too small creates two problems:

1. The controller of the closed-loop system will usually cause the valve to be fully open because it is not passing enough fluid.
2. Small movements in the valve restrictor may cause a change in flow that is greater than desired.

Historically, most valves are oversized because of inaccurate design procedures, inexperience, or the assumption that using a large-capacity valve than needed will meet all the necessary flow requirements encountered. A valve that is too large creates three problems:

1. The controller of the closed-loop system will cause the valve to always operate at or near the closed position. The result is that the valve plug may slam into or bounce out of the valve seat. Also, excessive seat or valve wear may result from the high-velocity flows between their surfaces due to the small amount of passage area.
2. The differential pressure across the valve (relative to the system pressure) becomes small, resulting in sloppy and slow control responses because large movements of the restrictor will cause small changes in flow.
3. Large valves are more expensive than smaller valves.

Valve Characteristics All valves that are designed to throttle flow have a valve characteristic. The **valve characteristic** is the relationship of the change in the valve opening to the change of flow through the valve. Figure 15-31 illustrates valve flow characteristics by using a graph. The valve-lift position is plotted as a percent of maximum lift along the horizontal axis. The graph shows three characteristic curves that represent the most common operation of control valves used in industry.

The top curve shows the flow characteristics of a *quick-open* valve. This valve is used predominately for On-Off control and not for throttling applications. A relatively small movement of the valve stem causes the maximum possible flow rate. For example, the curve shows that a 25-percent stem movement results in a flow rate of 70 percent of the maximum capacity. The middle curve shows the flow characteristics of a *linear* valve. This style of

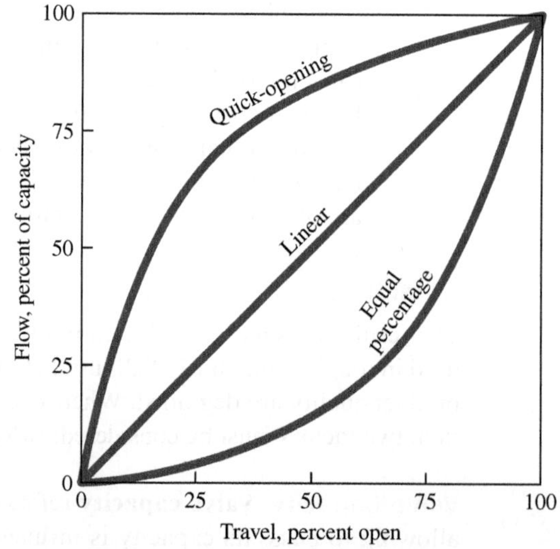

FIGURE 15-31 Characteristic curves of common valves

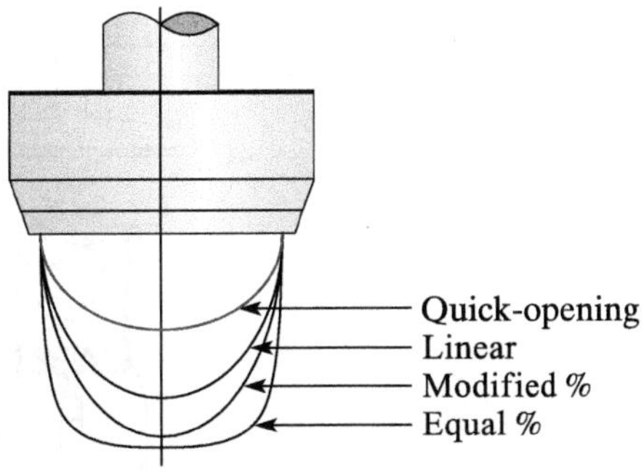

FIGURE 15-32 Contoured plug

valve has a flow rate that is directly proportional to the position of the valve stem. It is used in applications where most of the process system pressure drop is across its inlet and outlet ports. The bottom curve shows the flow characteristics of a nonlinear, *equal-percentage* valve. The name of this valve describes how it operates. A given percentage change in the stem position causes an equal percentage change in the flow. Observe on the graph that the valve flow coefficient increases slowly at first and then increases progressively more rapidly for equal changes in the stem position. This type of valve is used in applications where the valve pressure is high at low flows, or low at high flows. Its drawback is that, due to its limit of travel, it does not shut off flow completely.

These flow characteristics are based on the differential pressure developed across the valve, which is primarily determined by the shape and size of the housing, ports, and the restrictor. Figure 15-32 shows how the plug in a globe valve is contoured to cause the flow characteristics just described. In general, valves with equal percentage characteristics are used most often for throttling operations. In applications where there is a wide variation in setpoint, such as in flow-control systems, valves with linear characteristics should be used. For On-Off applications, quick-opening valves should be used. Whenever the load does not vary over a wide range, the flow characteristics become less critical.

There are two general guidelines that should be observed when selecting a throttle valve.

- Never use a valve that is less than half the pipe size.
- Avoid using the lower 10 percent and the upper 20 percent of the valve stroke because it is much easier to control in the 10- to 80-percent range. Below 10 percent, excessive seat wear and the bouncing off of the plug seat may occur. To allow for additional travel availability if flow conditions expand, the upper 20 percent should be avoided.

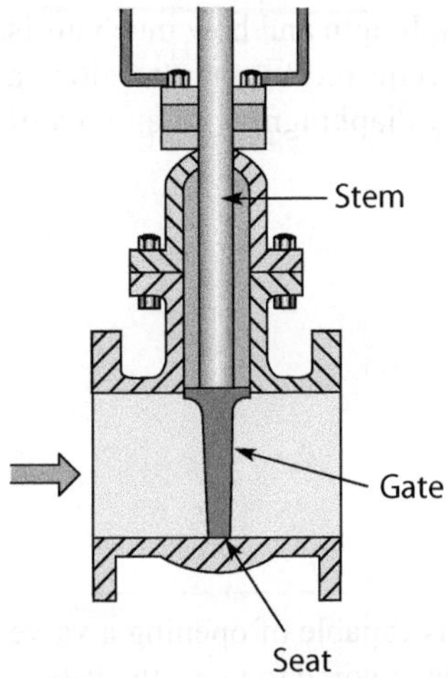

FIGURE 15-33 Gate valve

Computer programs are now available to help designers determine proper valve sizing and appropriate flow characteristic selection. Also, most vendors and manufacturers provide specialists who can assist designers with the selection process. For process applications that require On-Off service, a *gate valve* is often used. The gate valve, shown in Figure 15-33, is a single-seated valve that has a wedge-shaped plug that moves up or down. When it is closed, the plug wedges into a seat to create a tight shutoff. Because of its design, the gate valve should never be used for throttling operations; the plug would rapidly erode.

The Valve Actuator

ELE1907
The Spring-and-Diaphragm Actuator

The mechanism that converts an instrument signal into the linear or rotary motion of an element that restricts flow in a control valve is the *actuator*. There are two types of actuators: the *spring-and-diaphragm* and the *piston*. Because of its dependability and simplicity of design, the spring-and-diaphragm actuator shown in Figure 15-34 is used most frequently. The extent to which the actuator moves the restrictor is determined by the magnitude of a pneumatic or electric control signal. The pneumatic actuator is the most widely used.

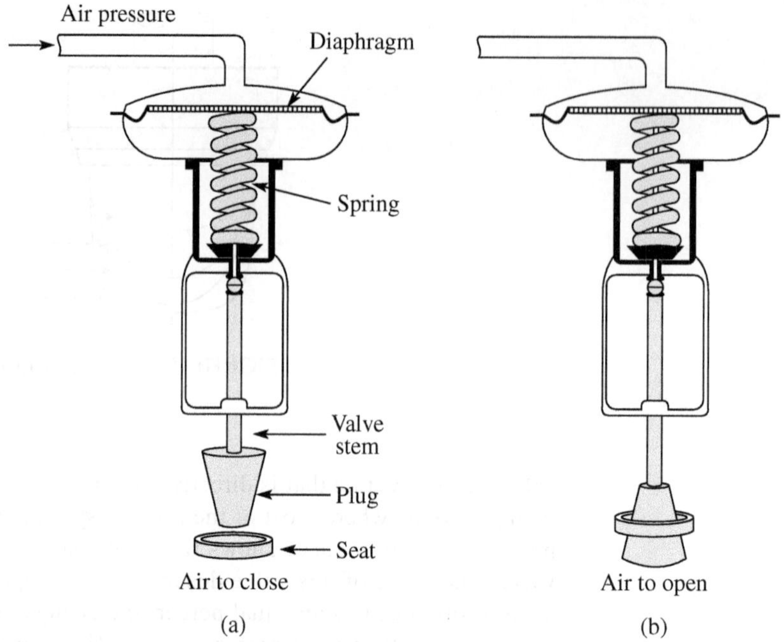

FIGURE 15-34 Valve actuator

In Figure 15-34(a), the air signal from the controller enters the actuator housing above the diaphragm. When the pressure of the control signal increases, the diaphragm (made of rubber or neoprene) is moved downward against a spring with a force equal to the air pressure multiplied by the area of the diaphragm. The diaphragm moves until the spring creates an equal and opposing upward force due to its compression. At this position, the motion stops, and the plug and valve stem to which it is connected are in a balanced state. For each different pressure of a controller signal, there is a corresponding plug position. When there is no air pressure, the valve stem is pushed upward by the spring, and when there is 15 psi pressure, the valve stem is forced downward. This type of valve is capable of exerting a large force. The amount of force depends on the size of the diaphragm and how much air is applied to it. For example, suppose a force of 100 pounds is required to open or close a valve. This requirement can be achieved by applying 10 psi to a diaphragm with an area of 10 square inches, according to the formula,

$$F = PA$$

where,

F = Force

P = Pressure

A = Area

EXAMPLE 15-1

Determine if an actuator with a diaphragm 3 inches in diameter is capable of opening a valve fully when 15 psi of air pressure is applied. A force of 50 pounds is required to move the valve.

Solution

$$\begin{aligned} F &= PA \\ &= 15\text{ psi} \times \pi^2 \times r^2 \\ &= 15 \times 3.14 \times 1.5^2 \\ &= 106\text{ lbs} \end{aligned}$$

The 3-inch actuator is large enough to provide the required force of 50 pounds.

ELE2007
Air-to-Open/Air-to-Close Valves

There are two different designs of the spring-and-diaphragm valve. Their action is generally defined as either "air-to-close" or "air-to-open." These terms indicate whether the plug will open or close the port at the valve seat when it is actuated by air. The illustrations in Figure 15-34 show how these valves operate. The valve style selected depends on the "fail-safe" position that is required if the control signal, which activates the valve, fails to occur.

To illustrate this guideline, assume that a material in a tank is heated by a heating jacket. The temperature is regulated by the amount of steam that flows through the control valve. If the material in the tank is damaged by overheating, an air-to-open valve will be used. By going to the closed position when the activating signal is absent, the steam is not allowed to pass through the valve to the heating jacket. On the other hand, if the material is damaged when the temperature drops below a certain level, an air-to-close valve will be used. With no activating signal applied to the valve, it will pass steam to the heating jacket and prevent the temperature from dropping. The air-to-open type, also referred to as a *fail-closed* valve, is used in most applications (about 80 percent).

There are three limitations associated with diaphragm actuators:

1. The length of the stroke is limited.
2. The material of the diaphragm will corrode when exposed to extreme temperatures and environments that are hostile to organic materials.
3. The amount of force the diaphragm develops cannot exceed the tensile strength of the spring.

To overcome the limitations of a diaphragm actuator, piston actuators are often used. Shown in Figure 15-35, the piston slides up or down within a cylinder. A specialized positioner device is used to convert a 3- to 15-psi control signal to an air pressure as high as 150 psig to produce a large force that moves the piston downward. To counter the downward pressure, a constant air pressure (called *air spring*) is maintained below the piston by an air regulator. When the variable air pressure applied to the top of the piston decreases, the air spring pushes the piston upward.

Actuation using the electrical method is performed in two different ways. When performing in the On-Off control mode, a solenoid is used to switch the valve, such as the one

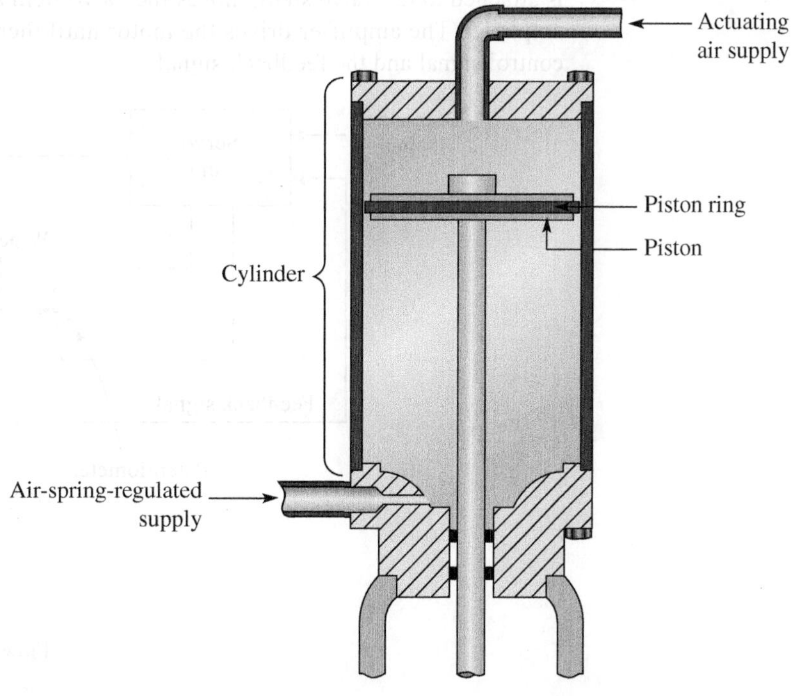

FIGURE 15-35 Piston actuator

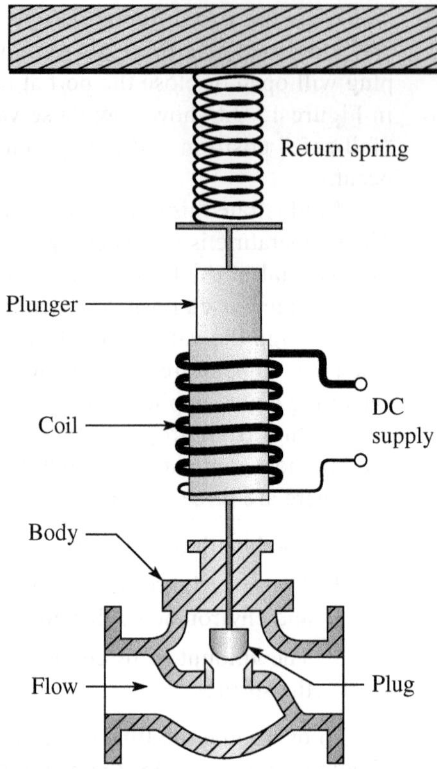

FIGURE 15-36 Using a solenoid

illustrated in Figure 15-36. When current flows through the coil, a magnetic field is generated, which moves the plunger downward against the spring. When the current stops, the spring pulls the plunger in the opposite direction.

When performing in the proportional control mode, a small motor is used to move the valve stem, as shown in Figure 15-37. A DC control signal is applied to a servo amplifier, which drives the gear motor to move the valve stem. The wiper arm of a potentiometer, which is attached to the valve stem, moves the valve stem and sends a feedback signal to the servo amplifier. The amplifier drives the motor until there is no longer a difference between the control signal and the feedback signal.

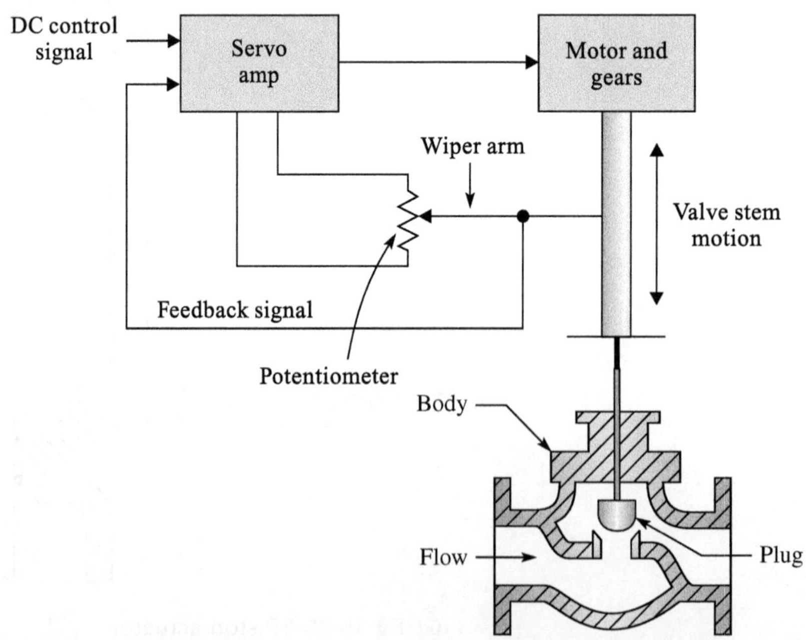

FIGURE 15-37 Using a motor or servomechanism

Valve Positioners

To ensure that the valve stem is at the precise position called for by the controller, an auxiliary instrument called a *positioner,* mounted on the top or side of the control-valve actuator, is used. Though they are not always required, positioners are very common because they can improve valve performance. Positioners are used to:

1. Overcome forces within the valve caused by friction or high pressure across the valve.
2. Gain speed and improve the frequency response in the closed loop.
3. Reduce actuator deadband and reduce the hysteresis effects of the spring-and-diaphragm.
4. Provide linear positioning of the actuator stem when dynamic imbalances are present in a valve.

One type of positioner that uses pneumatic power is the *force-balance* model, shown in Figure 15-38. The control signal is applied to a signal diaphragm that creates a force that is opposed by a feedback spring. The signal diaphragm is physically coupled to a spool valve. When the spool valve is centered, it blocks airflow between the ports of the chamber in which it is located.

Suppose the pressure of the input signal is increased. It pushes the signal diaphragm against the spring and causes the spool valve to move downward. The downward position of the spool valve creates a path between the ports of the air supply and the valve actuator. As the supply air pushes against the diaphragm of the valve actuator, it causes the control valve to open and moves the lever of the positioner cam downward. This action rotates the head of the cam in a counterclockwise direction, which causes the feedback spring to compress. When the pressure on the signal diaphragm from the control signal is equal to the pressure from the feedback spring, the spool becomes centered, the ports close, and the cam stops rotating.

If the pressure of the control signal is decreased, the pressure exerted by the feedback spring becomes greater. This condition causes the spool valve to move upward and create a path between the valve actuator and the exhaust ports. Since the force created by the spring at the bottom of the valve actuator diaphragm is greater than the force created by atmospheric pressure on the top, air is pushed out of the diaphragm chamber through the exhaust port. The result is that the control valve stem moves upward to close the control valve, causing the lever of the positioner cam to move upward. This action rotates the head of the cam in

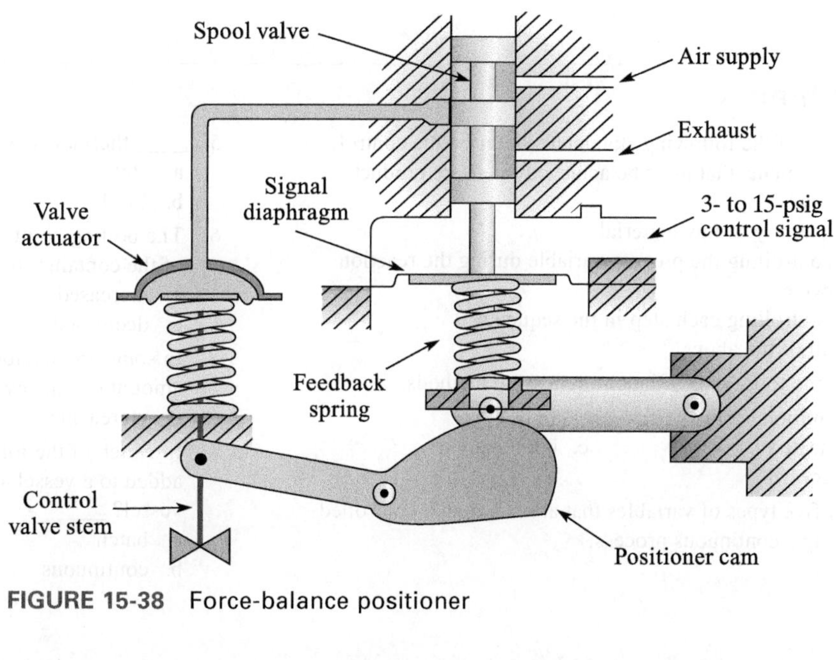

FIGURE 15-38 Force-balance positioner

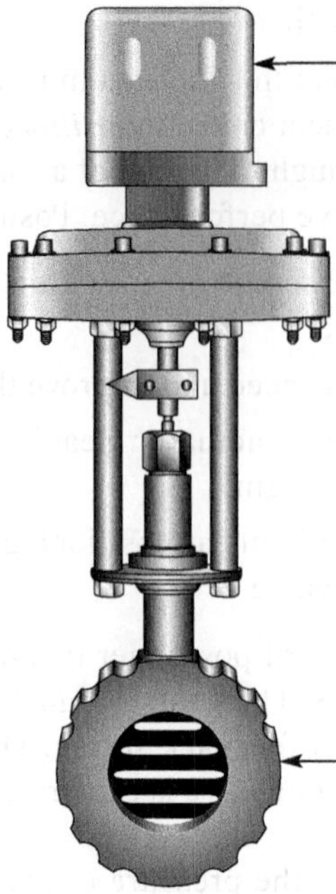

FIGURE 15-39 Intelligent positioner

the clockwise direction, which causes the feedback spring to decompress. When the pressure on the signal diaphragm from the control signal is equal to the pressure of the feedback spring, the spool valve becomes centered, the ports close, and the cam stops rotating.

In recent years, high-technology positioners called *intelligent positioners* have been developed. Shown in Figure 15-39, an intelligent positioner uses microprocessor technology to provide more-accurate control and shorter response times compared to conventional positioning devices. One function of the intelligent positioner is that it will determine if the position of the valve restrictor corresponds to the actuator signal it is monitoring. For example, if a 4- to 20-mA actuator signal is at 12 mA, the valve restrictor position should be 50 percent open. However, if the restrictor is at 60 percent, the positioner will cause the restrictor to move to its correct position. Intelligent positioners also reduce valve adjustments that typically take 5 to 20 seconds down to 1 or 2 seconds. Another feature of the intelligent positioner is that it has a calibration function programmed into its memory. By initiating a few keystroke entries, it automatically performs the zero-and-span procedure.

Problems

1. Which of the following statements describe the control requirements that must be achieved to ensure product quality? ____
   a. quantity of raw material
   b. controlling the process variable during the reaction cycle
   c. controlling each step in the sequence
   d. all of the above
2. List the four types of batch-processing methods.
3. Separation occurs when a mixture is ____.
   a. heated
   b. cooled
   c. both a and b
4. List five types of variables that are commonly controlled during a continuous process.
5. ____ thermal energy is supplied to heat exchangers.
   a. Hot
   b. Cold
   c. both a and b
6. The boiling point of a liquid can be increased if the pressure of the container in which it is held is ____.
   a. increased
   b. decreased
   c. both a and b
7. In some applications, ingredients can be mixed in a shorter amount of time by ____ heat.
   a. increasing
   b. reducing
8. In which of the following types of processes is material added to a vessel at the same time it is removed from the vessel? ____
   a. batch
   b. continuous
   c. discrete-parts manufacturing

9. In process-control applications, the setpoint is typically ____.
   a. changed frequently
   b. kept constant
10. The reliability of a sensor is determined by ____.
   a. response time
   b. accuracy
   c. precision
   d. all of the above
11. The term ____ is used to describe how consistently a sensor responds to the same input value.
   a. accuracy
   b. precision
   c. reliability
12. The ____ characteristic of a sensor refers to the time at which its output reaches a steady state after a signal is applied to its input.
   a. static
   b. dynamic
13. The two functions of a ____ are to convert the output of a sensor into a standardized signal and to send a signal to a distant location.
   a. transducer
   b. transmitter
14. The value for a standard transmission current signal is ____ when it is zeroed during calibration.
   a. 3 mA
   b. 4 mA
   c. 15 mA
   d. 20 mA
15. The value for a standard transmission pneumatic signal is ____ when it is spanned during calibration.
   a. 0 psi
   b. 3 psi
   c. 15 psi
   d. 20 psi
16. An I/P transducer converts ____ to ____.
   a. an input signal
   b. watts
   c. current
   d. pressure
17. The purpose of a square root extractor is to ____.
   a. multiply the magnitude of a signal
   b. linearize a signal
   c. remove noise from the signal
18. T/F When in the manual mode, the controller performs the decision-making process.
19. T/F Information obtained from recorders can be used to help troubleshoot a malfunction.
20. A term used to describe the process of storing historical data is referred to as ____.
   a. historical data
   b. data acquisition
   c. data history
21. The most widely used final control element for process control is the ____.
   a. AC drive
   b. electric motor
   c. control valve
22. The ____ control valve is recommended for high-pressure applications.
   a. single-seated
   b. double-seated
   c. butterfly
23. ____ valves have the greatest flow capacity of all control valves.
   a. Single-seated
   b. Three-way
   c. Butterfly
   d. Ball
24. If a control valve is sized too ____, small movements in the valve restrictor may cause a change in flow that is greater than desired.
   a. small
   b. large
25. If the valve should block flow when a control signal fails to occur, an ____ control valve should be used.
   a. air-to-close
   b. air-to-open
26. Which of the following functions is/are performed by valve positioners? ____
   a. overcome frictional forces
   b. increase the reaction speed of a valve
   c. reduce deadband and hysteresis
   d. provide linear positioning
   e. all of the above
27. How much force is exerted on the 4-inch-diameter diaphragm of a valve actuator if 12 psi of air pressure is applied?

CHAPTER

16

Instrumentation Symbology

OBJECTIVES

At the conclusion of this chapter, you should be able to:

- Define *P&ID*.
- Identify various instruments by the shapes of balloons that represent them.
- Identify and interpret functional identifiers in balloon symbols.
- Describe how tag numbers pertain to an instrumentation loop.
- Describe the function of line symbols.
- Identify the symbols for various actuators and valves.
- Read a simple loop on a P&ID.
- Describe the various types of information on a title block.

INTRODUCTION

In the field of electronics, schematic diagrams use symbols and lines to show which components are in a circuit and how they are connected. Likewise, in the field of process control, drawings called **piping and instrumentation diagrams,** or simply **P&ID**s, are used. P&IDs are a standard format used in all types of process-control fields, such as the petroleum, food, or utility industries. These drawings contain more information than just symbols and lines. Circles, letters, lines, numbers, and symbols are used to indicate which devices are included in a system, how these devices are arranged, where they are located, and the function each performs in the process. A thorough understanding of these diagrams will help the operator or technician monitor processes, do routine work more efficiently, and save time troubleshooting.

16-1 General Instrument Symbols

ELE4108
P&ID General
Instrument Symbols

Figure 16-1(a) shows *general instrument* (or functional) *symbols.*

Individual Instruments

To indicate an individual instrument in a process-control diagram, a circle called a *balloon* is used. The balloon contains letters, lines, and numbers that identify its location and its function in the process, and further specifies whether it is used to measure, indicate, record, or control the process variable. A circle by itself indicates a discrete stand-alone instrument, such as a transmitter, sensor, or alarm. If the symbol is a circle in a square, the instrument is described as a

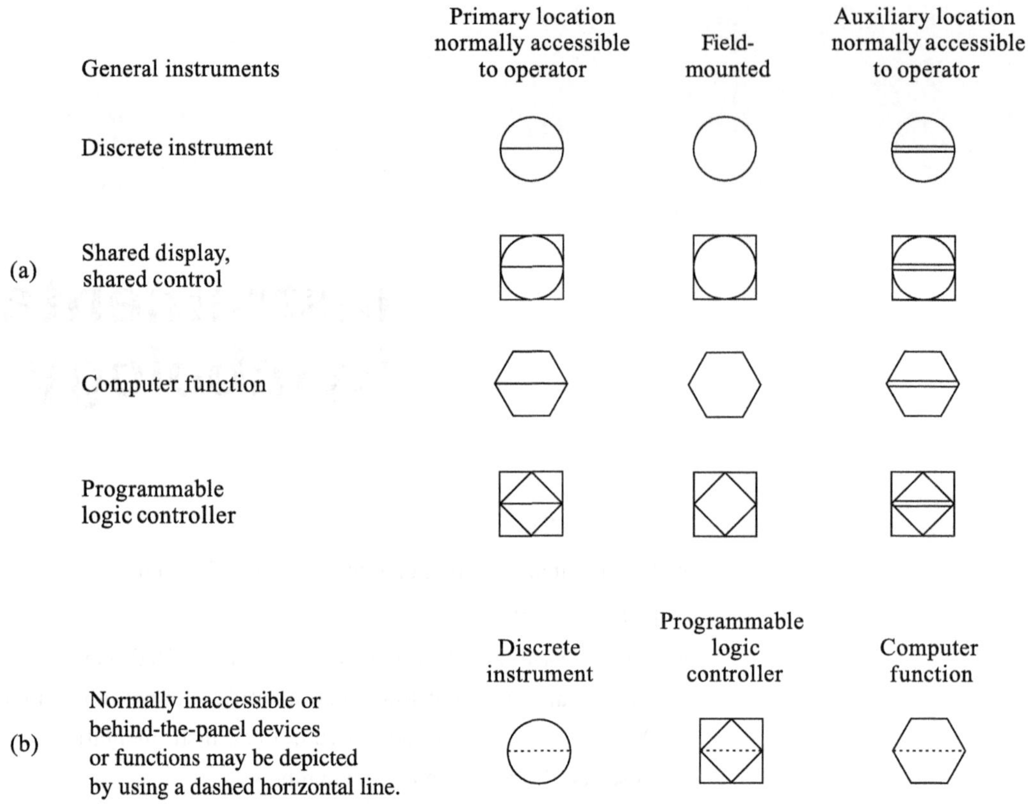

FIGURE 16-1 (a) and (b) General instrument or functional symbols

shared device, which means that, in addition to performing its specific function, it also displays or controls the process variable. If a hexagon is used instead of a balloon, it indicates a computer function. A programmable controller is identified by a diamond inside a square.

Figure 16-1(a) also shows that some symbols are divided in half by a single or double horizontal line, or are without any lines. These lines, or absence of lines, indicate how the instruments are mounted, or where they are located. A symbol without a horizontal line designates that it is installed in the field near the point of measurement or near the final control element. A single solid line indicates that the instrument is mounted on a panel board in a control room, usually among other instruments. Therefore, they are easily accessible to the operator or for routine maintenance. Double lines specify that the instrument is at an auxiliary location, away from the process.

Symbols with a single horizontal dashed line in Figure 16-1(b) denote an instrument located behind a panel, which may not be easily accessible.

16-2 Tag Numbers

ELE3905
P&ID Tag Numbers

Since there may be many different instruments used in a process, an alphanumeric code is placed inside each symbol to identify it. These instrument identifiers are called **tag numbers.** The code provides a variety of useful information about each instrument. Letters, called **functional identifiers,** are located in the top portion of the symbol. The sequence of letters designates the internal function.

First letter: The first letter denotes the measured or the initializing process variable. For example, **P** indicates *pressure*, **T** is *temperature*, **F** represents *flow*, and **L** means *level.*

Second letter: The second letter tells the function of the instrument. For example, **I** represents *indicate*, **R** is for *record*, **C** indicates *control*, and **T** means *transmit.*

When there are three- or four-letter identifiers, the second letter provides additional information about the first letter. For example, if **PDI** is used, **D** changes the measured variable, pressure (**P**), to differential pressure. The **I** represents an indicator, which means that the instrument displays differential pressure. An example of a four-letter identifier is **PDAH,** which indicates that the instrument is a differential pressure alarm that is activated when the pressure is too high. A list of standard functional identifiers is shown in Table 16-1.

EXAMPLE 16-1

Use Table 16-1 to determine the below.

Question: What does a balloon with the letters TR indicate?

Answer: A temperature recorder

The numbers located in the bottom portion of the symbol are the **loop identifiers.** A loop consists of one or more instruments arranged to measure and control a process variable. The loop ID identifies the loop where the instrument is located. All of the instruments in one loop are given the same loop number, regardless of the function or location of the instrument. When an instrument is used in two or more loops, it should be assigned the number of the most predominant loop. Figure 16-2 shows instruments used in a loop that measures and controls a flow process variable. In addition to providing specific information about each instrument contained in the loop, the number may also indicate a location within the plant or at another building. In some situations, tags and numbers are physically listed on the outer surface of the instrument encasement. This information makes it possible to verify which device in the field is the one designated on the diagram.

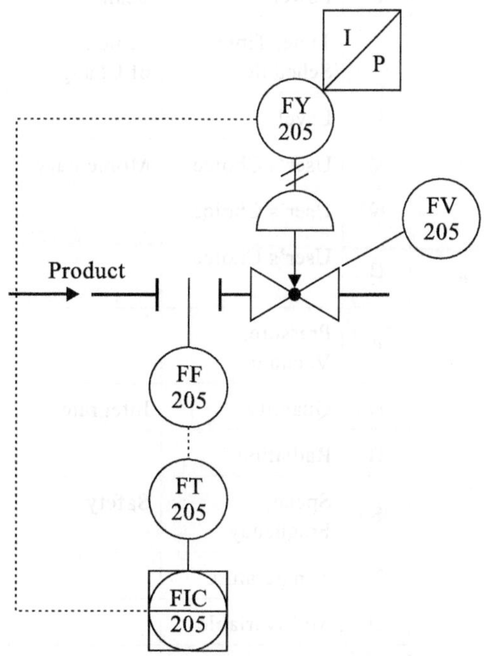

FIGURE 16-2 A P&ID of a closed-loop flow process system

In summary, the symbols, along with the letters and numbers inside them, provide the following information about the process system:

1. The measured variable controlled by the loop.
2. The function of the instrument.
3. How the instrument is mounted.
4. The loop number of the location where the operation is performed.

TABLE 16-1 Functional Identifiers

	First Letter		Succeeding Letters		
	Measured or Initiating Variable	***Modifier***	***Readout or Passive Function***	***Output Function***	***Modifier***
A	Analysis		Alarm		
B	Burner, Combustion		User's Choice	User's Choice	User's Choice
C	User's Choice			Control	
D	User's Choice	Differential			
E	Voltage		Sensor (Primary Element)		
F	Flow Rate	Ratio (Fraction)			
G	User's Choice		Glass, Viewing Device		
H	Hand				High
I	Current (Electrical)		Indicate		
J	Power	Scan			
K	Time, Time Schedule	Time Rate of Change		Control Station	
L	Level		Light		Low
M	User's Choice	Momentary			
N	User's Choice		User's Choice	User's Choice	User's Choice
O	User's Choice		Orifice, Restriction		
P	Pressure, Vacuum		Point Connection		
Q	Quantity	Integrate			
R	Radiation		Record		
S	Speed, Frequency	Safety		Switch	
T	Temperature			Transmit	
U	Multivariable		Multifunction	Multifunction	Multifunction
V	Vibration, Mechanical Analysis			Valve, Damper, Louver	
W	Weight, Force		Well		
X	Unclassified	X Axis	Unclassified	Unclassified	Unclassified
Y	Event, State	Y Axis		Relay, Compute	
Z	Position, Dimension	Z Axis		Driver, Actuator, Final Element	

16-3 Line Symbols

ELE4008
Line symbols

The symbols in a P&ID are interconnected by lines. Instead of using just solid lines, such as those used in an electronic schematic diagram, different types of lines are used. They differ in various ways to indicate if they are pipes, or lines that pass specific types of signals, and to show how the instruments are connected to the process and to one another. These lines may be solid or broken; their relative thickness may differ; and various markings may be added.

Figure 16-3 shows the different types of line symbols used in P&IDs.

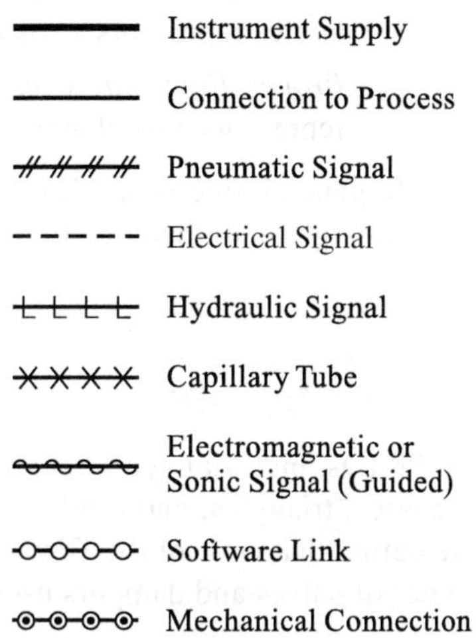

FIGURE 16-3 Line symbols

Solid Bold Line: The thicker solid line shows the pipes (referred to as *process piping*) that contain the process material such as steam, raw materials, or the final product.

Solid Fine Line: The thinner solid line indicates that an instrument is connected directly to the process. For example, Figure 16-4 shows how a field-mounted pressure orifice transmitter is connected directly to the process within the pipe.

Other types of line symbols on a P&ID represent lines that carry electronic, pneumatic, or optical information. Called signal lines, these carry information between two instruments electrically, through pneumatics, or by several other methods. They include:

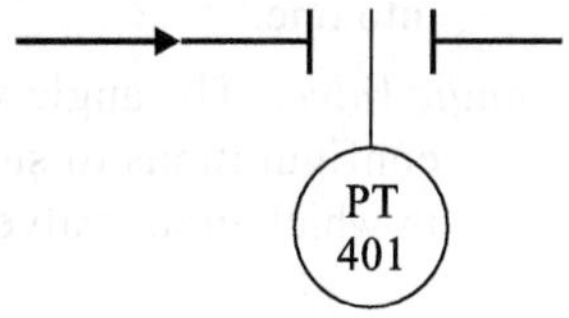

FIGURE 16-4 A thin, solid line indicates an instrument is connected directly to the process

Solid Fine Line with Double Diagonal Markings: This line carries pneumatic signals. An example of this type of connection is a transmitter that sends a pneumatic signal to a controller to indicate flow detected by a differential pressure sensor. The amount of pressure indicates how much the controller should vary the flow.

Dashed Line: A dashed (broken) line between two instruments indicates an electrical signal. These lines are usually preferred for carrying signals long distances.

Solid Fine Line with Ls: A solid fine line with Ls spaced apart represents a tube that passes a signal using hydraulic fluid.

Solid Fine Line with Xs: A solid fine line with Xs spaced apart represents a capillary tube. Capillary tubing is connected to a filled thermometer that measures the temperature of a process. The tube contains a fluid, which expands and contracts when subjected to temperature changes. The pressure the fluid exerts is often applied to a transmitter that converts the pressure to a proportional electrical signal.

Solid Fine Line with Sine Waves: A solid fine line with sine waves spaced apart represents electromagnetic transmitted signals.

Broken Fine Line Connected with Circles: A set of several fine lines connected with circles represents a software link.

Broken Fine Line Connected with Circles That Contain a Dot: This line symbol represents a mechanical connection.

In general, one signal line is used to indicate the connection between two instruments, even though more than one line may be used to make the physical connection.

16-4 Valve and Actuator Symbols

In P&IDs, pictorial symbols are used to show valves and the actuators that position them. Squares, triangles, and circles are used to show the different types of valves, dampers, and actuators that control the flow of fluids. Figures 16-5 and 16-6 show symbols for common types of valves and dampers used in process-control applications.

Control Valves

Control valves are positioned by the movement of the valve stem, either by a linear motion or a rotary motion.

Linear-Motion Valves

Globe Valve: The globe valve shown in Figure 16-5a is used in many industrial applications that require the tightest possible shutoff. In most applications, these valves are installed so that the flow tends to close the valve, and will shut off in the event of failure.

Three-way Valve: The three-way valve in Figure 16-5b is designed to divert fluid flow into two separate streams. Similar valves are designed to converge two streams into one.

Angle Valve: The angle valve in Figure 16-5c are single-seated valves with special body configurations to suit specific piping or flow requirements, such as turns or bends in which globe valves cannot be used.

Rotary-Motion Valves

Butterfly Valve: Shown in Figure 16-5d, the butterfly valve uses a rotating vane to provide closure to flow. Most often, they are used in applications with low or modest pressure.

Ball Valve: The ball valve shown in Figure 16-5e contains a plug that is spherical with a port through which the fluid flows. At zero degrees, the port is fully open and maximum fluid flows. The flow decreases as the ball is rotated. Ball valves require minimum maintenance and have the greatest flow capacity of all control valves.

Dampers: Dampers are rotary-action devices used to vary the volumetric flow rate of gas or air. Shown in Figure 16-6, they are typically used in heating, ventilation, and air-conditioning systems.

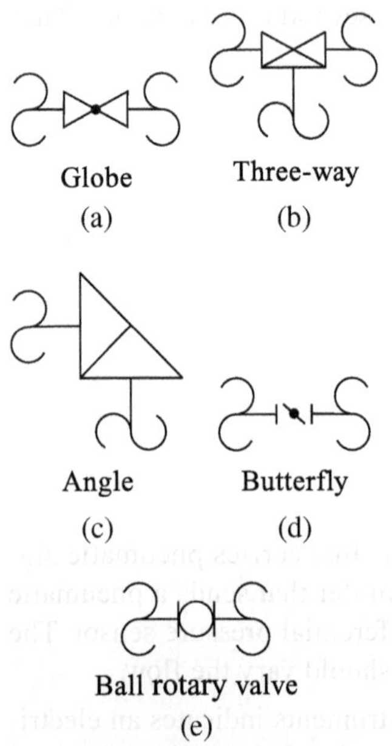

FIGURE 16-5 Valve symbols

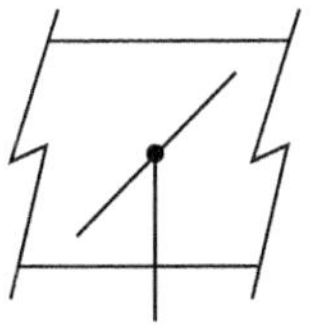

Damper or louver

FIGURE 16-6
Flow actuator

Actuators

An actuator is a component that is used to position another component, usually a final control element such as a valve or damper. There are many types and styles of actuators. Some are manual, but most are automatic. The type selected is usually based on the specific characteristics that best suit a particular application.

Most linear-motion valves are positioned automatically by a pneumatically activated diaphragm. The symbol for this type of actuator is shown in Figure 16-7. The diagram shows the actuator connected to a globe control valve. Notice that a fine line with two diagonal dash markings is used to indicate a pneumatic input signal to the diaphragm. A balloon included with the symbol has a tag number inside to identify the type of valve used and its function. Valves usually have a V in their functional identifier. The diagram indicates that a flow-control valve is in temperature-control loop 401.

Actuators can also be activated electrically. One example is a solenoid, which is identified by a square with an S inside the box, as shown in Figure 16-8(a). Another electrically activated actuator is a rotary motor, which is identified by an M inside a circle, as shown in Figure 16-8(b). The symbol for a manually activated valve is a T connected to the valve body (Figure 16-8(c)).

On the line between the actuator and valve body is an arrow that describes the valve's failure mode. Failure modes indicate the position of the valve when the signal to the actuator fails to be applied, usually because of a malfunction. An arrow that points upward, as shown in Figure 16-9(a), indicates that a failure of the control signal will cause the valve to be fully open. An arrow pointing downward, as shown in Figure 16-9(b), indicates a fail-closed valve.

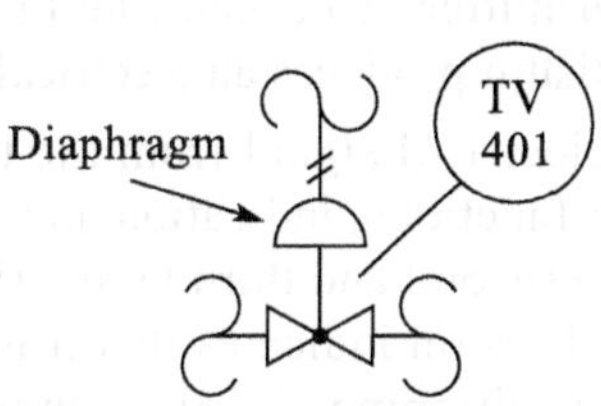

FIGURE 16-7 Pneumatically activated diaphragm symbol

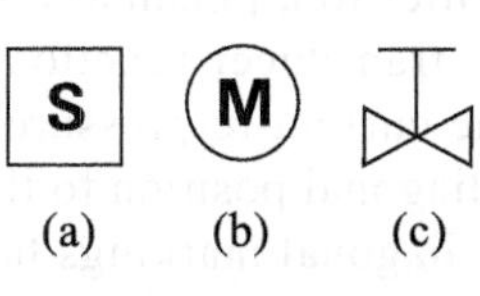

FIGURE 16-8 Identifiers for various actuators

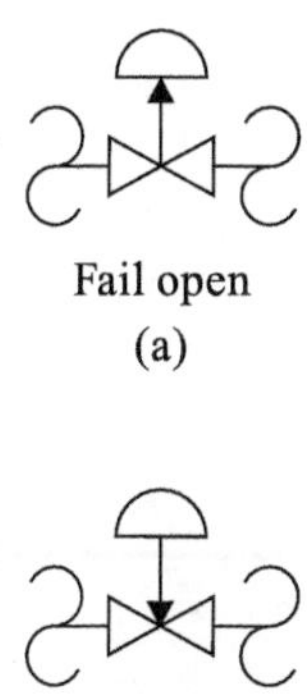

FIGURE 16-9 Symbols that show the failure mode of a valve

16-5 Reading a Single Loop

ELE3708
Reading P&ID Loop

The information provided on instrumentation symbology in this chapter is only a fraction of the amount of information on the subject. An entire manual is needed to provide a complete listing of the material. However, enough information has been given to read simple drawings, such as the one that contains a control loop for a heat exchanger application in Figure 16-10. The following steps illustrate how a P&ID provides information about the type of process variable used and how instruments work together to control the process:

Step 1: The tag numbers on the component symbols indicate that the portion of the P&ID is loop 401.

Step 2: The first letter of each functional identifier is a **T,** which means that it is a temperature loop.

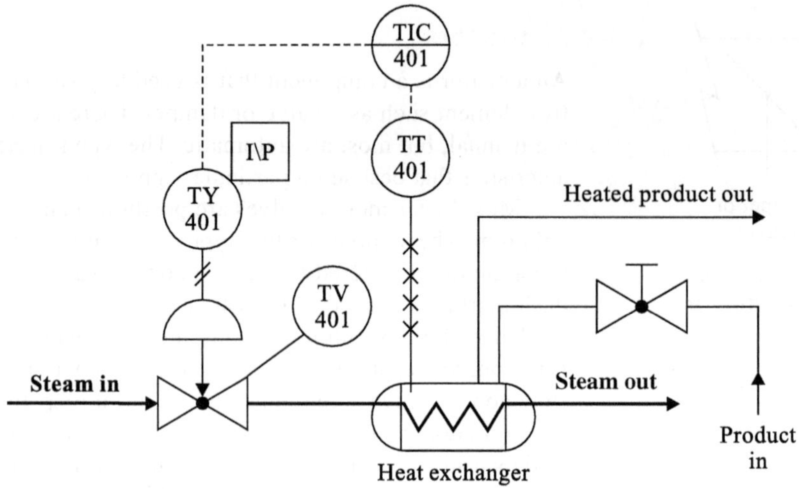

FIGURE 16-10 Control loop (401) for a heat exchanger application

Step 3: The instrument connected to the heat exchanger for measuring temperature is a transmitter, which is labeled with a **T** as the second letter of the functional identifier. The balloon has no line inside its symbol, which indicates that it is field-mounted. The line with Xs indicates that a capillary tube sends a signal from the exchanger to the transmitter. A dashed line that extends from the output of the transmitter indicates that it produces an electrical signal.

Step 4: The electrical signal from the transmitter is sent to an instrument that performs two functions, indication and control; each of these two functions is identified by the second and third letters (**IC**) of the functional identifier. The solid line inside the balloon indicates that it is board-mounted. The controller sends out an electrical positioning signal, as indicated by the dashed line extending from the left of the symbol.

Step 5: Since the actuator that varies the steam through the flow control valve is actuated pneumatically, a transducer is required to convert the electrical signal from the controller to a pneumatic signal. This function is performed by the I/P transducer. The transducer function is described by the second letter (**Y**) inside the balloon. The current-to-pressure conversion is identified by the small square, located in the diagonal position to the balloon, containing the letters I/P. The fine solid line with diagonal markings between the transducer and the actuator shows that it is a pneumatic line.

Step 6: The half-circle symbol of the actuator indicates that it consists of a pneumatically controlled diaphragm. The arrow that points downward between the actuator and globe valve symbol tells that it is a "fail-closed" type of valve.

16-6 Information Block

Besides the symbols and lines shown on the diagram, additional information about the system may be provided at the bottom of the drawing, as shown in Figure 16-11.

The Title Block: This section, shown in Figure 16-12, is located in the lower-right-hand corner of the drawing. It provides the title and the drawing identification number. The ID number should be checked to prevent problems resulting from work being done to the wrong system. On this sample, three signatures are required for validation. Additional information about the drawing may be included in other documentation.

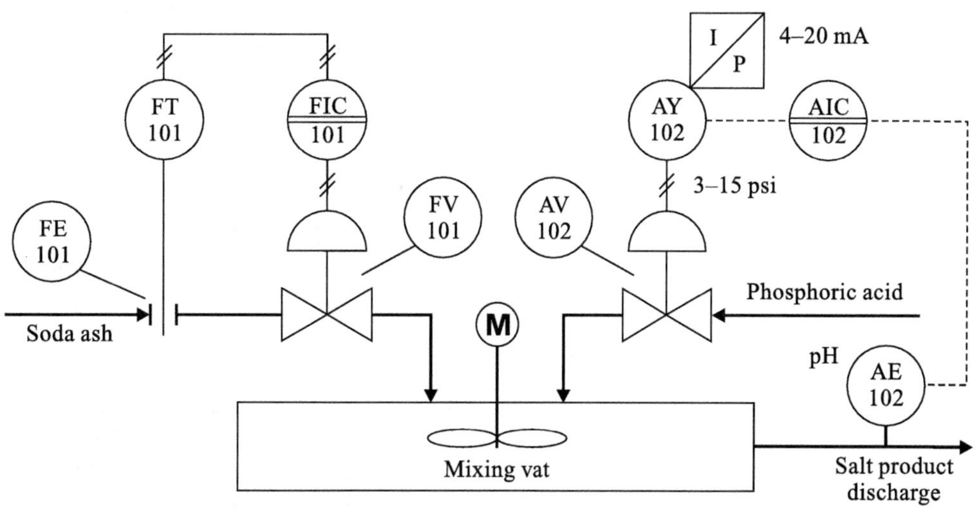

In loop 101, soda ash flows into the mixing vat. The flow is detected by an orifice flow sensor, which is shown connected directly to the process. It sends a pneumatic feedback signal to a flow/indicator/controller mounted behind the panel. FIC 101 sends a pneumatic signal to a diaphragm that controls the valve that regulates flow.

In loop 102, an analyzer sensor that monitors pH sends an electrical signal to an indicator/controller. An analyzer transducer converts the electrical signal from the controller to pressure, which causes a diaphragm to change the valve that controls the flow of phosphoric acid.

FIGURE 16-11 An information block example

ACME PROCESS CORP.	
MIXING SYSTEM #1	
DRAWN BY:	DATE
CHECKED BY:	DATE
APPROVED BY:	DATE
SHEET NO. 1 of 1	DRAWING NO. 3250-7A

FIGURE 16-12 The title block

Revisions: This supplement in Figure 16-13(a) provides information about any changes that have been made to the original drawing.

Material List: This supplement in Figure 16-13(b) contains a complete list of the components that are on the P&ID. It also includes a list of numbers for ordering parts when they need to be replaced.

Notes: Descriptive information about the diagram may be included in the form of notes. There are two types of notes: *general* and *local.* General notes pertain to the diagram, as shown at the bottom of Figure 16-11, or to the process system as a whole. Local notes contain information about an instrument or area of the diagram rather

REVISIONS					
REV #	DATE	DESCRIPTION	BY	CH'K	APRV

(a)

MATERIAL LIST			
TAG #	MANUFACTURER	MODEL	PART #
TV-302	FISHER	513RP	61121-41
PV-309	MASONEILAN	47-21134	54378-39
FIC-301	FOXBORO	130M	22447-12
PT-309	FISHER	4157	61247-33
TT-302	FOXBORO	45P-F2	22336-19
LV-305	FISHER	667F7	62458-20
PIC-308	MOORE	528M	14436-38
TIC-302	MOORE	528M	14436-38
LIC-305	FOXBORO	130M	22447-12
LT-305	TAYLOR	4807A	43741-80
LT-307	TAYLOR	4807A	43741-80
FV-308	MASONEILAN	48211-35	54267-37

(b)

FIGURE 16-13 (a) Revisions; (b) Material list

than the whole system. They are usually included on the diagram, and in some cases a line connects the note to the instrument or area to which it pertains.

If there is too much information to fit on the drawing, a local note will refer the reader to another document where it is located.

Problems

1. Which of the following types of information about instruments are not provided on a P&ID? ____
   a. which type of device b. location of devices
   c. function of the device d. how devices are arranged
   e. none of the above
2. The lines, letters, and numbers of a balloon identify the instrument's ____.
   a. location
   b. function in the process
   c. specific operation
   d. all of the above
3. Double lines inside the balloon indicate that the instrument is ____.
   a. mounted on a panel b. field-mounted
   c. mounted at an auxiliary location
4. A single line inside the balloon indicates that the instrument is ____ .
   a. mounted on a panel b. field-mounted
   c. mounted at an auxiliary location
5. The following symbol indicates a computer function that is ____ -mounted.

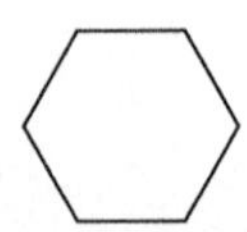

6. A broken line inside a balloon symbol indicates that the instrument is ____ .
   a. field-mounted b. inaccessible to the operator
   c. panel-mounted

7. The ____ letter of a functional identifier indicates the type of measured variable.
   a. first b. second
   c. third
8. Which symbol represents a stand-alone instrument?
   a. circle b. square
   c. hexagon d. diamond
   e. all of the above
9. A general instrument symbol without a horizontal line indicates that it is ____ .
   a. installed in the field b. board-mounted
   c. located behind a panel d. at an auxiliary location
10. Draw a shared instrument that is located at an auxiliary location.
11. A general instrument symbol that is a circle within a square indicates that it ____ .
    a. performs a measuring function
    b. displays or controls the process variable
    c. both a and b
12. T/F All of the instruments in one loop are given the same loop number.
13. A tag number consists of a ____.
    a. functional identifier b. loop number
    c. both a and b
14. Identify each letter of the following functional identifier: PDI.
15. T/F The term *process piping* refers to a pipe through which the process variable flows.
16. Which of these line identifiers represent an electronic signal? ____

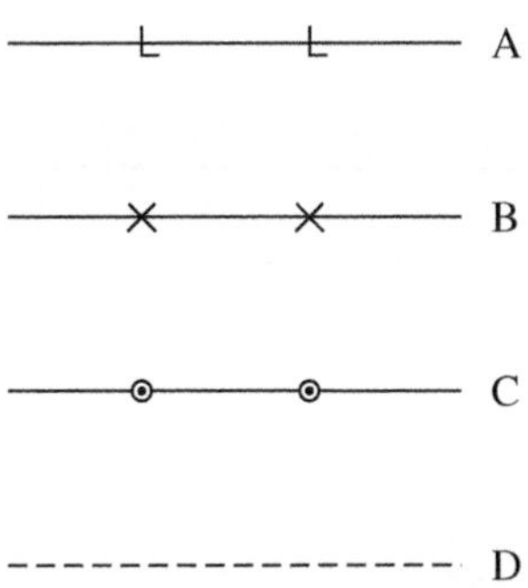

17. A pneumatic line is represented by which of the following identifiers? ____

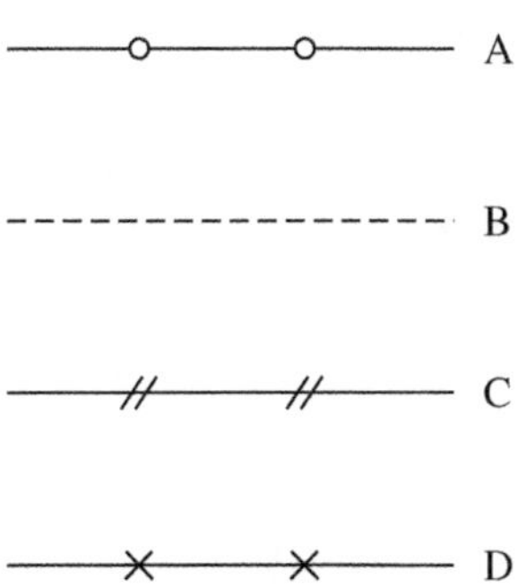

18. Which of the following devices are considered actuators?
    a. diaphragm b. piston
    c. solenoid d. all of the above
19. A valve symbol with an arrow pointing upward indicates that it is a fail-____ device.
    a. open b. closed
20. Which of the following symbols represents a globe valve?

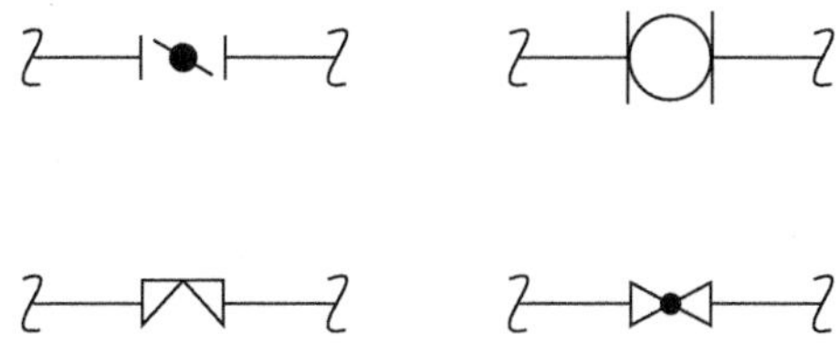

21. The device that positions the final control element is the ____ .
    a. damper b. valve
    c. actuator d. all of the above
22. A line symbol with individual circles spaced apart represents a/an ____ .
    a. pneumatic signal b. electronic signal
    c. capillary tube d. software link
23. A general instrument symbol with "TV" inside a circle indicates it is a ____ .
    a. variable transmitter b. valve in a temperature loop
    c. volume transducer
24. What does the phrase *panel mount* refer to?
25. Which of the following types of information are on the title block? ____
    a. diagram number
    b. the total number of P&ID sheets
    c. the name of the person who created the drawing
    d. all of the above

CHAPTER 17

Process-Control Methods

OBJECTIVES

At the conclusion of this chapter, you should be able to:

- List the operational characteristics of an open-loop system. Include the elements and the signals they produce.
- List the operational characteristics of a closed-loop system. Include the elements and the signals they produce.
- Define the following terms:

  Primary Element
  Heat Exchanger
  Feedback Loop
  Final Control Element
  Disturbance
  Manipulated Variable
  Controlled Variable
  Measured Variable
  Load Demand
  Time Proportioning Method
  Amplitude Proportional Method

- Identify which types of elements in an open- or closed-loop system are defined as instrumentation devices.
- List the factors that contribute to the dynamic response of a single-variable control loop.
- Define the following terms associated with the dynamic response of a control loop:

  Step Change
  Time Lag
  Dead Time
  First-Order Time Lag
  Pure Lag
  Static Inertia
  Time Constant
  Head Pressure

- Explain the operation and characteristics of an On-Off control system, and describe the function of the deadband.
- Describe the operational principles and characteristics of the following continuous-control modes:

  Proportional
  Integral
  Derivative

- Describe the operating principles and characteristics of the following advanced control techniques and give a practical application of each one:

  Cascade
  Feed-Forward
  Ratio
  Adaptive

- Define the following terms associated with various control techniques:

Proportional Gain	Remote Controller	Secondary Feedback Loop
Offset	Controlled Flow	Wild Flow
Reset Rate	Proportional Band	Hysteresis
Rate	Sensitivity	Deadband
Primary Feedback Loop	Integral Time	
Reset Time	Derivative Time	

INTRODUCTION

Many different operations are performed in an industrial machine to manufacture a product. For example, fluids flow through pipes at a certain rate, ingredients fill a vat to a required level, heat is applied to a vessel to cause a chemical reaction, or a vacuum pressure is applied to a confined tank to extract its contents. Each one of these operations is referred to as a *process*. Many of these individual processes are combined and run simultaneously to produce a finished product of a desired quality as rapidly and inexpensively as possible. To satisfy these requirements, each process must be precisely controlled, often by some type of automatic control device. The automatic operations performed by an industrial manufacturing machine are referred to as *process control.*

17-1 Open-Loop Control

Process control operations are performed automatically by either open-loop or closed-loop systems. If the process is controlled only by setpoint commands, without feedback measurement signals, the system is referred to as **open-loop.**

Open-loop control is used in applications where simple processes are performed. Timing functions are often the key factors used to control the operation. Examples of open-loop process machines are cafeteria dishwashers, commercial laundry machines, and printed circuit board burn-in chambers. These equipment run through a series of timed cycles, which are activated by controller devices such as relay ladder logic hardware, sequential drum controllers, programmable controllers, or computers.

The advantage of open-loop systems is that they are relatively inexpensive. Their main drawback is that without a feedback loop, there are no control capabilities to make corrections if the process deviates from its required state.

17-2 Closed-Loop Control

Closed-loop automatic control systems are more effective than open-loop systems. With the addition of a feedback loop, they become self-regulating. The diagram in Figure 17-1 is used to illustrate the operation of a temperature-type process-control closed-loop system. It shows a *heat exchanger,* which is used to heat a liquid to a temperature of 100°F. A source of steam supplies the thermal energy to the exchanger. The amount of steam that passes through a control valve determines the temperature at which the liquid is heated. The *primary element* (sensor) is used to detect the condition of the *controlled variable,* which is the temperature of the liquid leaving the exchanger. The sensor's output, which is called the *measured variable,* is conditioned by a transducer/transmitter into a standard signal before it is sent to the controller. The controller compares the feedback signal to the setpoint, and an error signal is developed if there is a difference. The controller uses the error signal to make computations to determine which type of *control signal* to produce at its output. The control signal is sent to the *final control element* (which is the control valve). This valve varies the steam flow into the exchanger. The steam is the *manipulated variable* that causes a change in the controlled variable. These actions of *measuring, comparing, computing,* and *correcting* go on continuously.

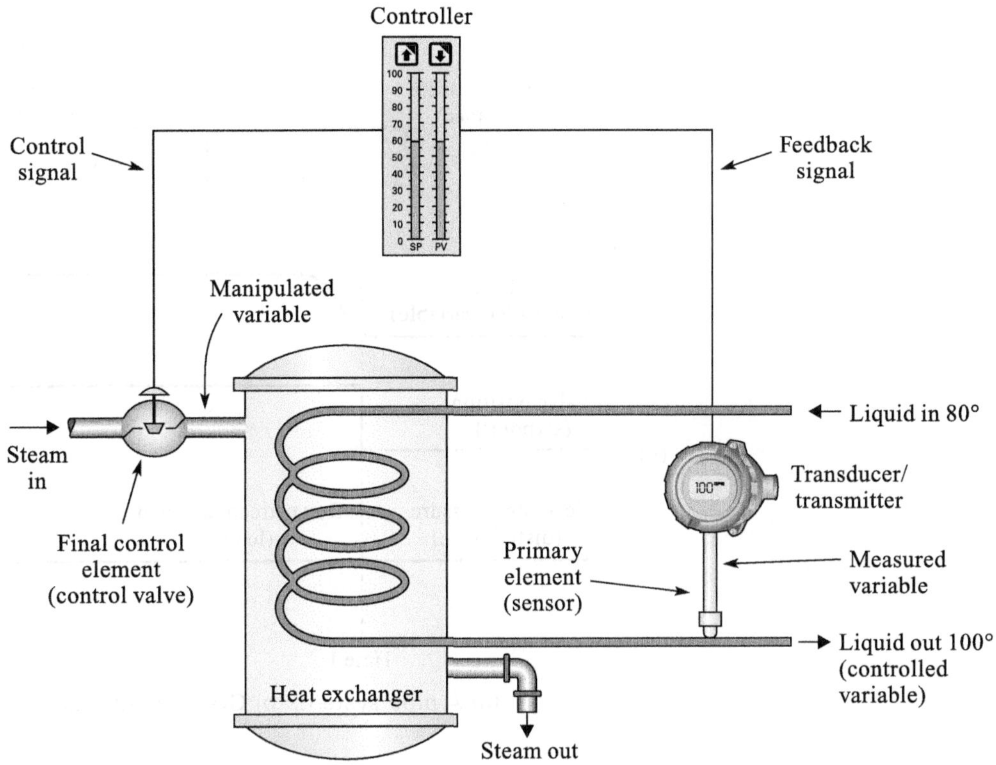

FIGURE 17-1 Closed-loop temperature process-control system

Process Behavior

The primary objective of process control is to cause a controlled variable to remain at a constant value at or near some desired setpoint. The term *variable* refers to the fact that an element varies when an influence to which it is exposed causes the variable to change. A change can happen when only one of the following conditions occurs:

- A disturbance appears
- Load demands vary
- Setpoints are adjusted

When a change does occur, the objective of the process-control system is to return the controlled variable to the setpoint as quickly as possible.

The behavior of a process system can be examined by observing the controlled variable's response after one of these influences abruptly changes. Referred to as a **step change,** it takes place over a small time interval and is plotted on a time graph (Figure 17-2(b)) as a vertical line.

Figure 17-2(a) shows how step changes develop in an actual application. The flow of fluid through a pipe system is the process. The fluid flow rate leaving the valve is the *controlled variable.* The position at which the valve is set is considered the *setpoint.* The flow rate governed by the position of the flow restrictor inside the valve is the *manipulated variable.* The demand for fluid downstream from the valve is the *load,* and the variance in upstream pressure is considered the *disturbance.*

Figure 17-2(b) is a time graph that shows how a setpoint change, disturbance, and a load variance affect the controlled variable. Flow, the controlled variable, is plotted on top by a measurement device such as a strip chart recorder. At time 1, the flow rate increases when the valve position is opened wider, while the upstream pressure remains constant. At time 2, with the valve position unchanged, the flow rate decreases because the upstream pressure drops. The flow rate would decrease in a similar fashion if the demand for fluid decreased downstream while the upstream pressure and the valve position remained constant.

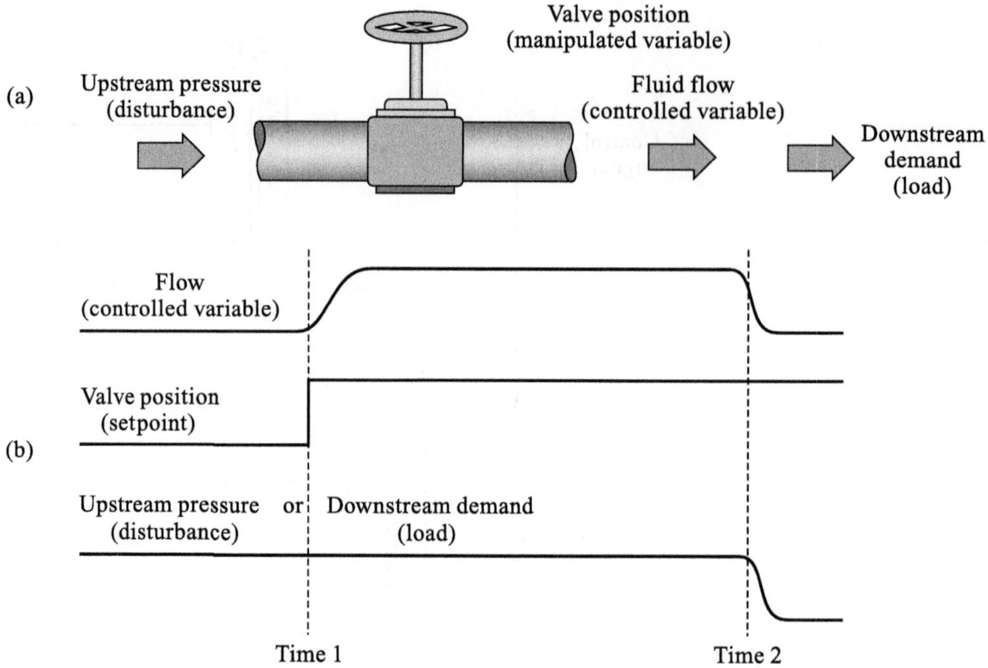

FIGURE 17-2 (a) A pipe system; (b) Graphic illustration of process behavior

17-3 Single-Variable Control Loop

Several process variables are typically controlled simultaneously in a machine that produces a product in the process industry. Usually, only one individual *feedback loop* is required to control each variable. Referred to as a *single-variable control loop,* it consists of the following elements:

- Measuring Device (Primary Element)
- Transducer/Transmitter
- Controller
- Final Control Element

These elements are also referred to as *instrumentation* devices.

When a step change takes place, there is not an immediate response by the control loop. The correcting action takes time. A measure of the loop's corrective action, as a function of time to the deviation, is referred to as its **dynamic response.** There are several factors that contribute to this delay.

Response Time of the Instruments

IAU12408
The Dynamic
Response of a
Closed-Loop System

All instruments have a **time lag.** This is the duration from the time a change is received at its input until the instrument produces an output response. Time lag also includes the duration of a signal passing from one instrument to the next. The following six factors contribute to the time delay caused by instruments:

- Response time of a sensor
- Time lag of the transducer
- The distance the feedback signal must travel from the transducer to the controller
- The time required for the controller to process information
- The distance the control signal travels from the controller to the final control element
- The time lag of the final control element

Pure Lag of the Controlled Variable

The controlled variable itself may contribute to the reaction time delay of the loop. For example, when a step change occurs, the loop reacts by causing the manipulated variable to be altered. The result is that the energy applied by the actuator to the controlled variable either increases or decreases. However, due to the static inertia of the material from which the controlled variable is made, it opposes being changed and creates a delay. Eventually, energy overcomes resistance and causes the process to reach its desired state. The delayed reaction is referred to as **pure lag.**

One factor that affects the pure lag time is the capacity (physical size) of the controlled variable. If the controlled variable has very little mass, the process will react instantly to a step change. A second factor that can affect the pure lag time is the physical properties of the controlled variable. For example, the temperature of a solvent will change more rapidly than an equal quantity of pure water when exposed to an equivalent thermal energy change. Another factor that can affect pure lag is the chemical properties of the controlled variable. For example, the hardness of water will affect the reaction rate when pH adjustments are made.

One common method used to analyze the pure lag of a controlled variable is to introduce a step change and observe the results. The storage tank in Figure 17-3(a) is used to describe such a response test, and the graph in Figure 17-3(b) illustrates the behavior.

The tank is being supplied with a liquid at a given rate, while liquid flows out through a drain line. The inflow is the manipulated variable, and the outflow is the controlled variable. The rate of outflow is determined by the amount of *head pressure* influenced by the depth of the water. Suppose the setpoint change causes the rate of inflow to increase abruptly, as shown at time T_1 on the graph. The volume of outflow does not immediately increase by the same amount as the new rate of inflow. Instead, it gradually increases as a result of the head

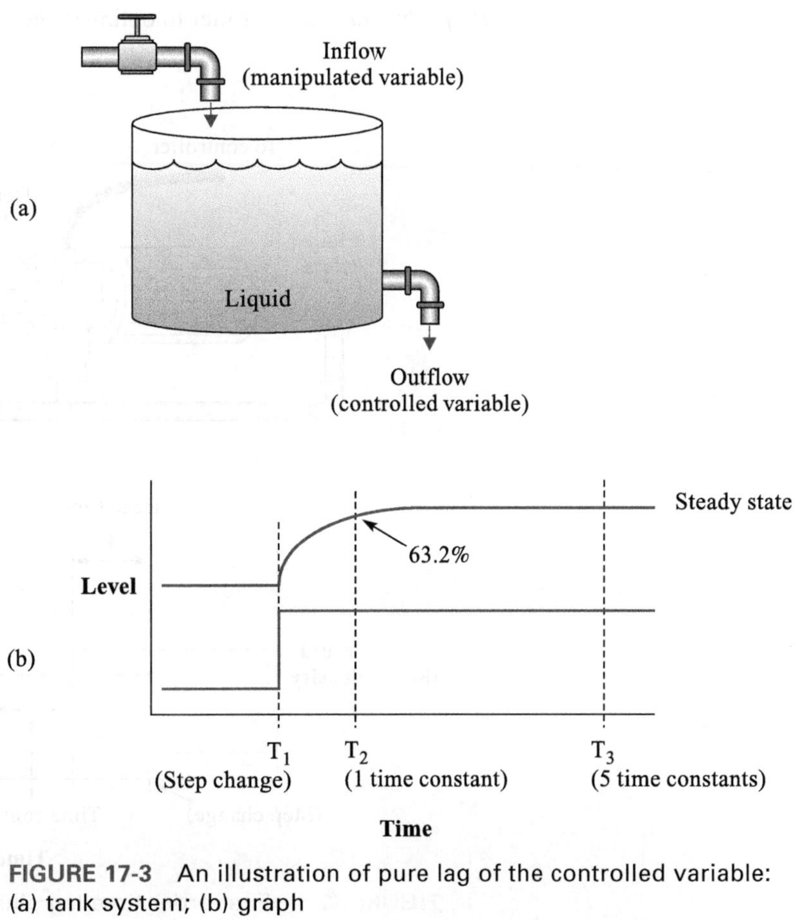

FIGURE 17-3 An illustration of pure lag of the controlled variable: (a) tank system; (b) graph

pressure that builds up from the accumulation of liquid within the tank. Eventually, equilibrium occurs when the liquid reaches a higher level, causing the head pressure to make the outflow equal the increased rate of inflow.

The graph shows that the rate at which the controlled variable changes is rapid at first and then tapers until it reaches its new steady state. The time it takes the controlled variable to change 63 percent of the new steady state is defined as one time constant. The 63-percent constant is based on a mathematical model that is commonly used to describe the dynamic behavior of physical objects exposed to an energy change. After five time constants, a new state is reached. This type of single-loop control delay is referred to as *first-order time lag.*

Dead Time

Another factor that contributes to time lag is **dead time.** This is the elapsed time between the instant a deviation of the controlled variable occurs and the instant corrective action begins. A process in which the density of a fluid is regulated can be used to illustrate dead time. The liquid enters the pipeline in Figure 17-4(a), and the fluid density is measured by a sensor/transmitter some distance downstream. If the composition of the fluid at the point of entry changes, there will be a time lapse before it reaches the point of measurement. The amount of time that elapses is based on the distance between the two points and the speed of flow. The diagram in Figure 17-4(b) graphically shows the effects of first-order time lag plus dead time.

Delays are inherent to each of these factors and cannot be avoided. However, the reaction time can be substantially reduced by maximizing the operation of an instrument in the control loop in two ways:

1. *Select a controller* with operational features that provide the kind of control action needed for a particular process.
2. *Properly tune* the controller to optimize the regulation of the process.

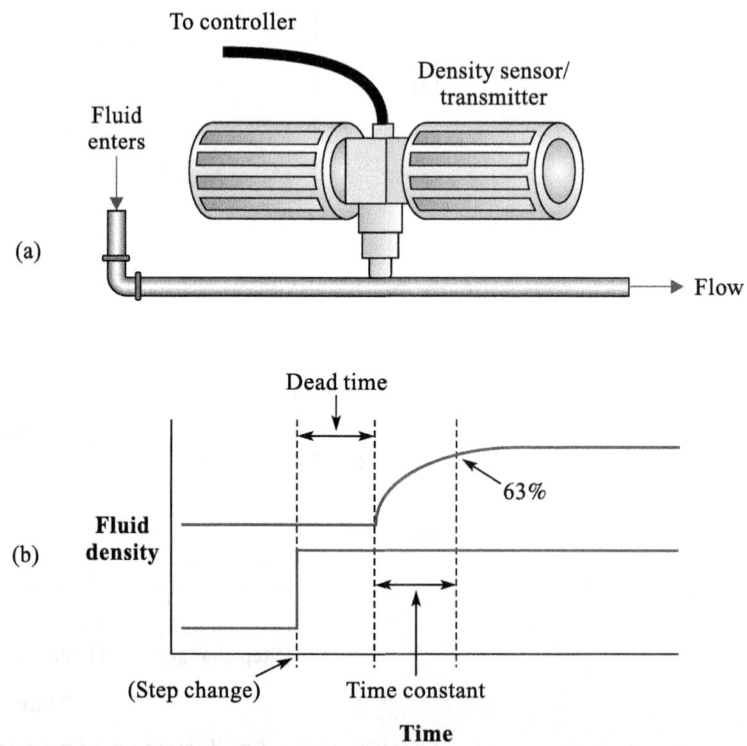

FIGURE 17-4 First-order time lag plus dead time

17-4 Selecting a Controller

Controllers are designed to operate by using different control modes. Each of these modes has specific characteristics that provide different types of control action. These control modes are:

- On-Off
- Proportional
- Integral
- Derivative

The mode or combination of modes that are selected by the design engineer are determined by the requirements of the process.

17-5 On-Off Control

IAU12308
On/Off Control of a Feedback System

In some types of process applications, the controlled variable changes very slowly. For example, the temperature of a large mass is difficult to raise or lower rapidly. Therefore, delays due to time lags in the control loop are unavoidable and, as a result, are usually tolerated.

The type of control mode often used for slow-acting operations is the one that provides *On-Off* action. This kind of action controls a final control element that has only two conditions, *fully on* or *fully off*. The controller cannot move the final control element to any intermediate position between the two extremes. One example of such a control system is a refrigeration unit. The controller compares the temperature (controlled variable) to the setpoint. When the temperature increases above the setpoint, the controller turns a compressor (final control element) fully on. As the temperature lowers below the setpoint, the compressor is turned off. The controlled variable cycles above and below the setpoint are shown in Figure 17-5(a).

One drawback of this system is that the rate at which the final control element switches on and off can be very high. This condition can result in excessive wear to equipment.

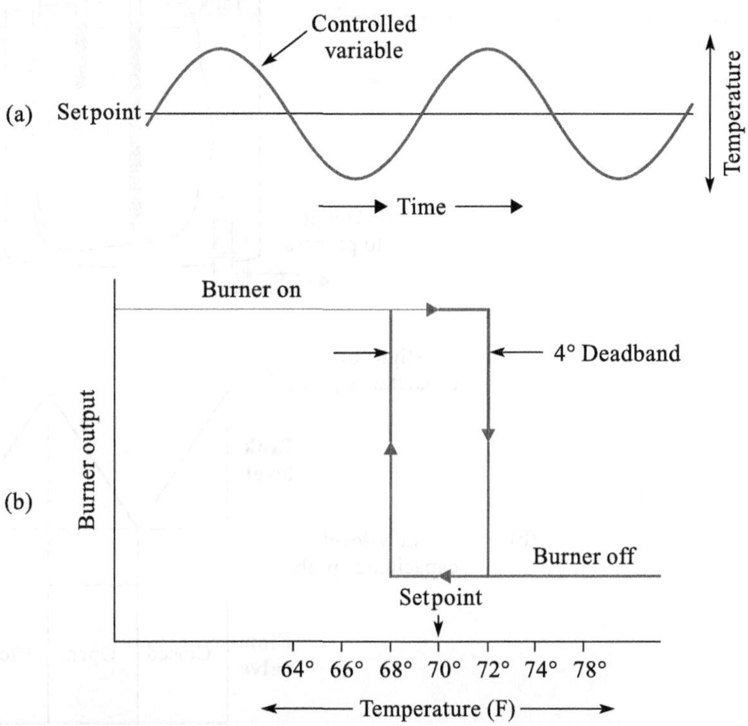

FIGURE 17-5 On-Off control: (a) cycling; (b) deadband

To reduce the switching rate, an On-Off differential, or *hysteresis,* is programmed into the controller. Also referred to as a *deadband,* it causes the controller to produce its on and off signals at different values around the setpoint. For example, a home heating thermostat may have a deadband of 4 degrees. If the temperature setting is 70 degrees, the furnace turns on at 68 degrees and turns off at 72 degrees, as shown graphically in Figure 17-5(b). The drawback of using a deadband is that the controlled variable will deviate from the setpoint by a larger amount than a system that does not use this method.

Another type of an On-Off system is the tank, shown in Figure 17-6(a), that stores a liquid. It contains two capacitance probe sensors (as primary elements), a flow valve (as the final control element), and a controller. The level of the liquid is the controlled variable. One sensor detects the high-level limit, and the other detects the low-level limit. As the fluid leaves the tank, the level lowers until it falls below the low-level sensor. When the controller detects a signal change from the sensor, it opens the valve. Since the inlet flow rate is greater than the outlet flow rate, the level in the tank rises. When it reaches the high-level sensor, the controller detects its signal change and causes the valve to close. Figure 17-6(b) graphically shows the action of the final control element and the controlled variable in relation to time.

On-Off controllers are also used in applications that limit the condition of a controlled variable. One example is a safety system that prevents a steam boiler from exploding by not allowing the temperature to rise above a certain level. The controller is programmed to close a fuel valve and activate an alarm if the high-limit temperature is reached. The alarm stays on, and the boiler remains shut down until the controller is manually reset.

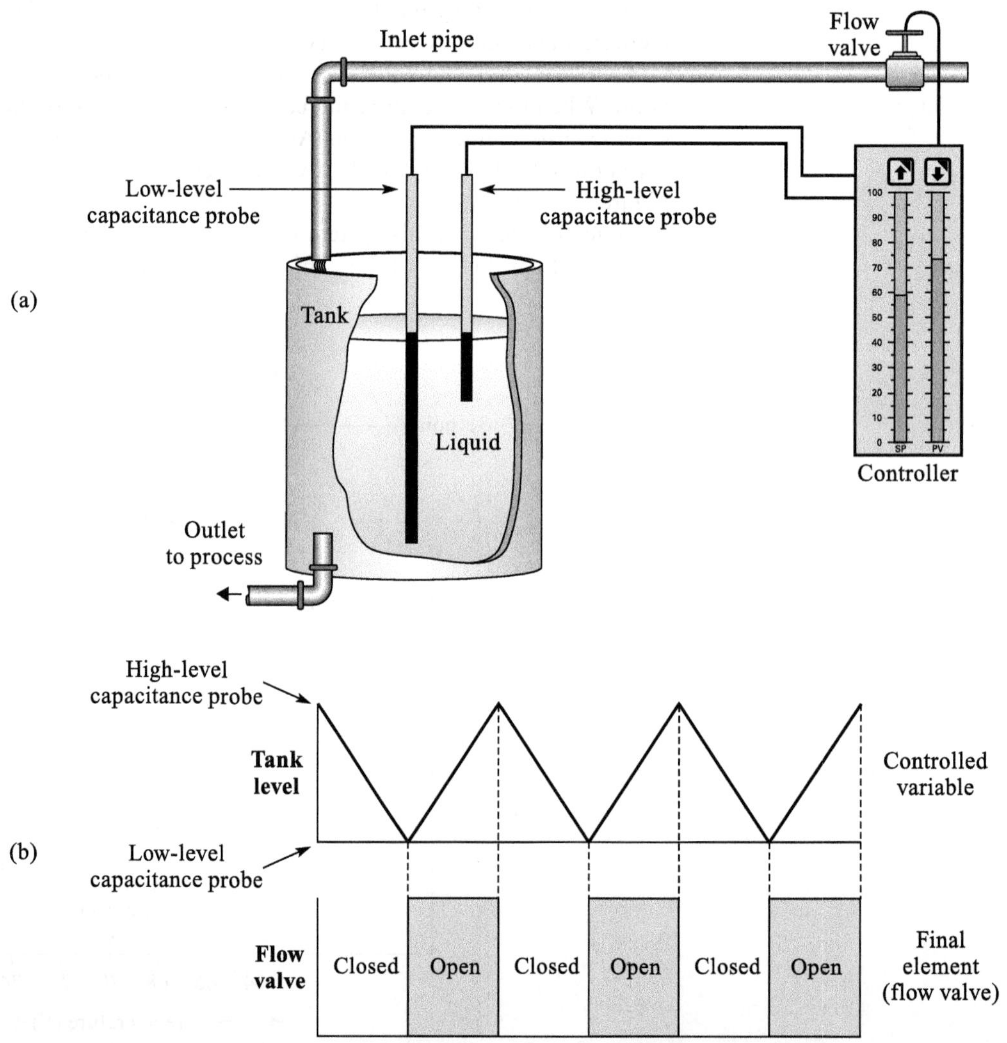

FIGURE 17-6 On-Off level control: (a) tank storage; (b) operation graph

The advantages of the **On-Off control** are that it is the least expensive closed-loop system and the easiest to design. Its limitation is that it cannot vary the controlled variable with precision. Its control action is to switch when extreme conditions exist, causing the final element to only turn fully on or fully off and the controlled variable to oscillate around the setpoint.

Case Studies—Using Different Control Modes

This chapter contains four case studies about a small chemical company that produces a specially blended product that is sold in bottles. Each study describes how the company uses different control modes to maintain the quality of its product as its production output capacity demand grows.

Case Study 1: On-Off Control

To produce the product, raw materials are made one batch at a time. Bulk materials and liquids are combined in a large tank, shown in Figure 17-7. A heat jacket that surrounds the tank is fed with steam because, for proper blending to occur, the temperature of the ingredients must be maintained at 80 degrees. If for some reason the temperature should rise above 85 degrees, one of the ingredients would become overactive and the product would be out of tolerance. On the other hand, if the batch is allowed to cool below 75 degrees, blending will be incomplete and the product will have to be reprocessed or discarded.

To regulate the flow of steam to the jacket, an On-Off controller, shown in Figure 17-8(a), is used to open or close the valve. The desired operating temperature, or setpoint adjustment, establishes the temperature at which the controller causes the steam valve to open or close. The setpoint of 80 degrees is adjusted by a knob, and the setting is displayed by the green

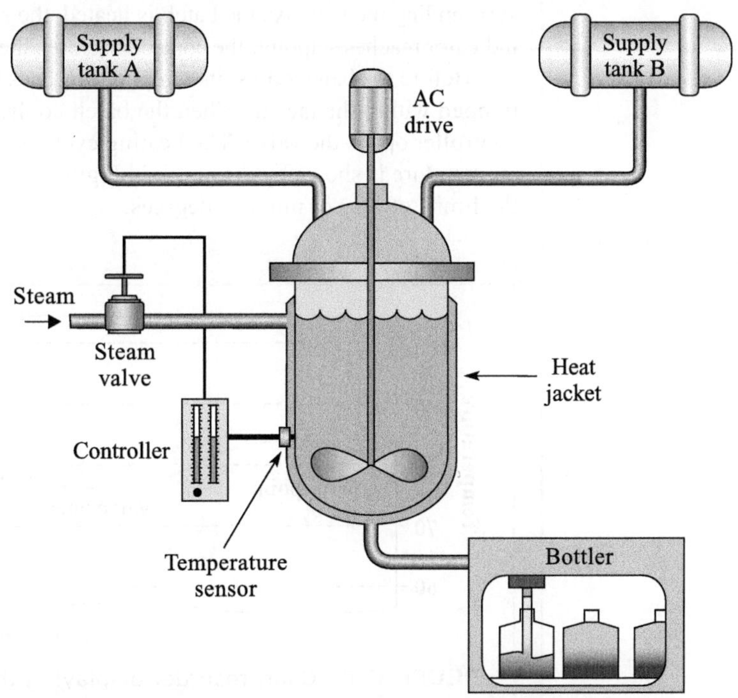

FIGURE 17-7 Batch mixer

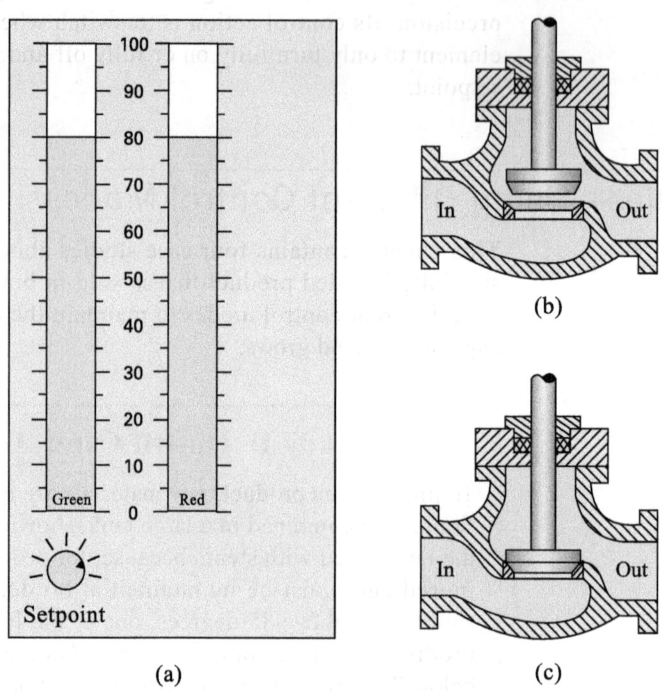

FIGURE 17-8 On-Off controller and steam valves

indicator on the left. The actual temperature is measured by a sensor and is displayed by the red indicator on the right.

At start-up, the ingredients in the batch are at a cool temperature. This condition causes the controller to open the steam valve, as shown in Figure 17-8(b). Steam enters the jacket surrounding the tank. As the batch is heated, the red indicator moves upward. When the red indicator reaches setpoint, the controller closes the steam valve, as shown in Figure 17-8(c).

Before the batch cools, its temperature rises to slightly above 80 degrees due to steam trapped within the jacket. When the batch cools, the indicator moves downward until the controller opens the valve. The heating cycle is then repeated. A continuous record of the temperature is shown by the graph in Figure 17-9. Note that the variations are well within the limits of plus or minus 5 degrees.

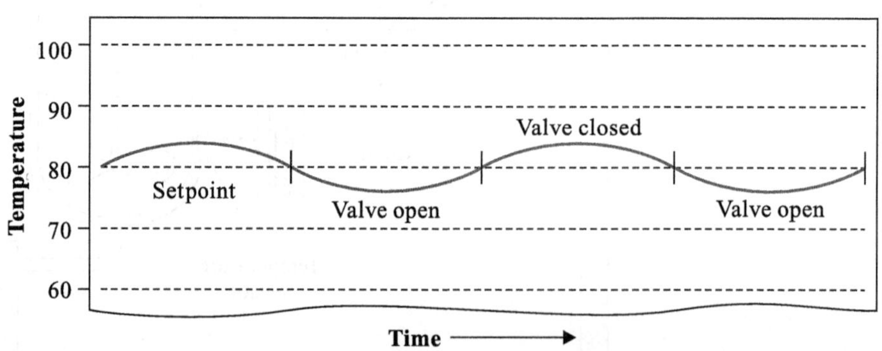

FIGURE 17-9 Chart recorder displaying the On-Off control mode operation

17-6 Continuous Control

On-Off control is suited to situations in which it is only necessary to keep a process variable between two limits. For continuous processes where the variable is required to be kept at a particular setpoint level, it becomes impractical. The controlled variable can be maintained only if the final control element is varied continuously over the entire range of its output. Systems that provide this function use any one, or a combination, of *proportional, integral,* and *derivative* control actions.

Proportional Mode

A proportional controller produces an output signal with a magnitude that is proportional to the size of the error signal (E) it is correcting. The error signal is the difference between the measured variable and the setpoint (desired value). A small error will cause the output to change by a small amount. Conversely, a large error will cause a larger output change. The output of the proportional controller moves the final control element to a definite position to attain a desired value of the controlled variable.

The proportional action can be accomplished in two different ways: by the *time proportioning* method and by the *amplitude proportional* method.

Time Proportioning

SSE4503
The Time Proportioning Operational Amplifier

SE4403
Time Proportioning Application

Time proportioning is a method in which the output at the controller is continually switched fully on and fully off. The average voltage produced is varied by changing the ratio of signal-on to signal-off. The ratio produced by the controller is determined by how much the measured variable differs from the setpoint. If the measured variable equals the setpoint, the output on-to-off ratio is 1:1, as shown in Figure 17-10(a). This ratio indicates that the on-time and off-time are the same. If the measured variable is below setpoint, the on-time will be longer than the off-time, as shown in Figure 17-10(b). The ratio increases as the error signal increases, producing a higher average output voltage. When the measured variable is above setpoint, the on-time will be shorter than the off-time, as shown in Figure 17-10(c), thus producing a smaller average voltage. For example, assume the output DC voltage of the controller to be 10 volts. When the controller is set for 100 percent, the output is on all of the time, and the average voltage produced is 10 volts. If the controller is set for 60 percent, the output is switched on for 60 percent of the time, and the average voltage is 6 volts (10 volts × 0.6 = 6 volts). When the controller is set for 0 percent, the output voltage is not on at all, and the average voltage is 0 volts.

Amplitude Proportional

IAU12008
Proportional Control Amplifier

The **amplitude proportional** method is the most common technique used to produce a proportional signal. The magnitude of the signal is proportional to the size of the error signal. The analog signal produced by a controller is either variable voltage or variable current.

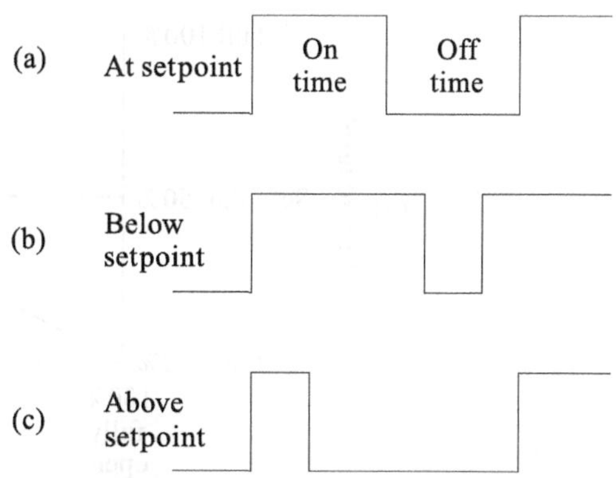

FIGURE 17-10 Time-proportioning signals produced by a controller

The controller also has the capability of amplifying the amount at which its output changes in proportion to the change applied to its input. There are two ways to refer to the amplification of a proportional controller: *proportional gain* and *proportional band.*

Proportional Gain **Gain** is the ratio of change in output to the change in input, as described mathematically by the following formula:

$$\text{Gain} = \frac{\text{Percentage Output Change}}{\text{Percentage Input Change}}$$

A float mechanism that regulates the level of fluid in a tank is used in Figure 17-11 to show the concept of gain. Referring to the gain formula, the liquid level represents the input, and the amount of fluid that flows through the inlet flow valve is the output. Suppose the load demand is increased by increasing the opening of the drain valve. The result is that the liquid level drops 5 percent. To prevent the level from dropping any further, the flow valve opens by 5 percent and increases the inlet flow 5 percent. According to the formula, the gain is 1.

$$\text{Gain} = \frac{5\%\ \text{Output Change}}{5\%\ \text{Input Change}} = 1$$

By repositioning the float along the rod toward the pivot point, as shown by Figure 17-12, the proportional gain is increased. Suppose that a load demand drops the level of the tank by 25 percent, from one-half to one-fourth of its capacity. With the new float alignment, the inlet

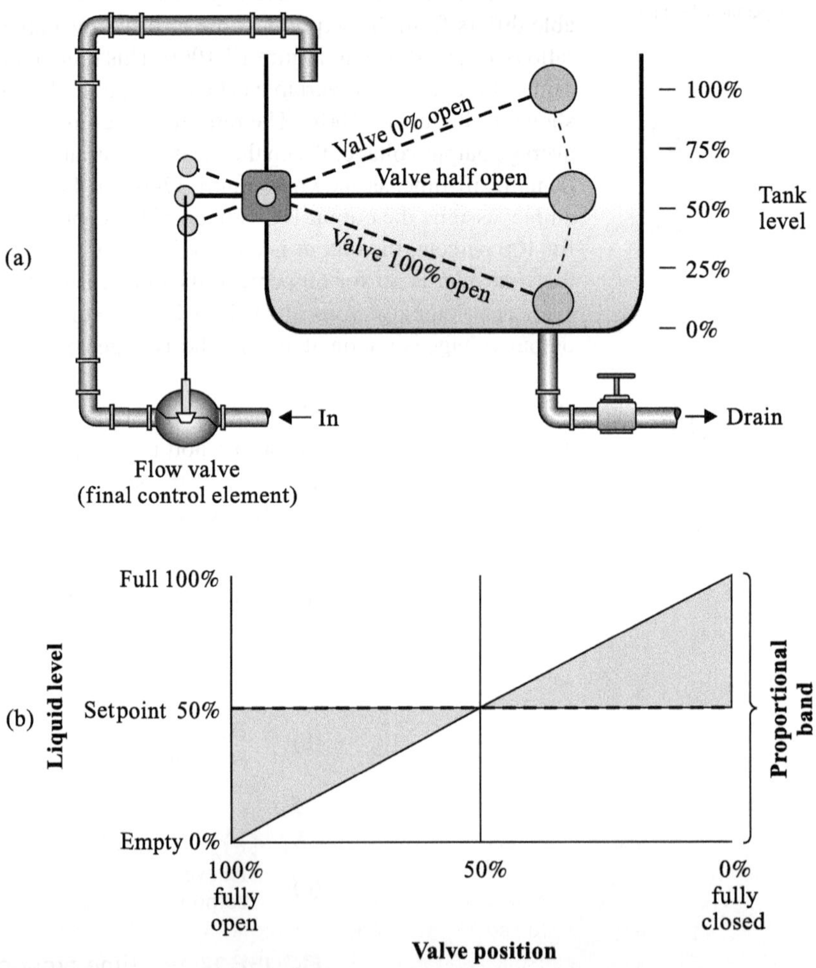

FIGURE 17-11 Level control at a gain of 1

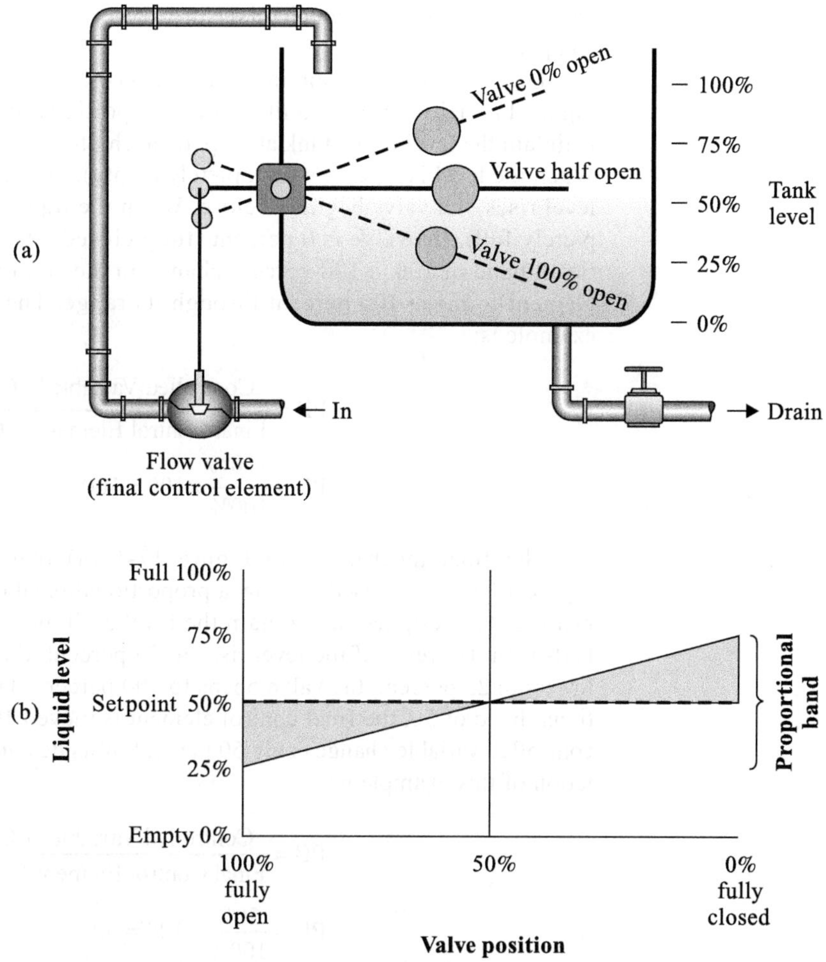

FIGURE 17-12 Level control at a gain of 2

valve changes 50 percent from the half-open to the fully open position, and the flow of inlet fluid increases. Therefore, the gain is 2.

$$\text{Gain} = \frac{50\%\ \text{Output Change}}{25\%\ \text{Input Change}} = 2$$

With a gain of 2, the controller responds to controlled variable changes with a greater output change than when the gain is 1. The result is that the process variable is restored to a desired value more quickly.

Proportional Band Amplification is also expressed as **proportional band (PB).** *Proportional band* is defined as the percentage change in the controlled variable that causes the final control element to go through 100 percent of its range. The proportional band can be determined mathematically by using the following formula:

$$\text{PB} = \frac{\text{Controlled Variable \% Change}}{\text{Final Control Element \% Change}} \times 100$$

The width of the proportional band determines how much of a controlled variable change is required to cause a given amount of movement by the final control element. For example, to cause a final control element to move 100 percent, a controller with a proportional band of 100 requires that the controlled variable change twice as much as one with a proportional band of 50. The float mechanism in Figures 17-11 and 17-12 can be used to make this comparison.

The float mechanism in Figure 17-11(a) that has a gain of 1 and the graph in Figure 17-11(b) are used to illustrate a proportional band of 100. The system is designed to maintain the level of the tank at 50 percent (half-full) by keeping the valve exactly half-open. When the level is 0 percent (completely empty), the valve is 100 percent (fully open). As the level rises, the valve begins to close. When the liquid reaches a level of 100 percent (completely full), the valve is 0 percent (fully closed). This example shows that with a proportional band of 100, a 100-percent change in the controlled variable causes the final control element to move 100 percent through its range. The mathematical representation of this example is:

$$PB = \frac{\text{Controlled Variable \% Change}}{\text{Final Control Element \% Change}} \times 100$$

$$PB = \frac{100\%}{100\%} \times 100 = 100$$

The float mechanism in Figure 17-12(a) that has a gain of 2 and the graph in Figure 17-12(b) are used to show a proportional band of 50. As in Figure 17-11, this mechanism is also designed to maintain the level at 50 percent of full capacity when the valve is half-open. However, if the level rises to 75 percent, the valve closes to 0 percent; if the level lowers to 25 percent, the valve opens to 100 percent. This example shows that with a proportional band of 50, the final control element is moved 100 percent through its range when the controlled variable changes only 50 percent of its designed range. The mathematical representation of this example is:

$$PB = \frac{\text{Controlled Variable \% Change}}{\text{Final Control Element \% Change}} \times 100$$

$$PB = \frac{50\%}{100\%} \times 100 = 50$$

Proportional action only occurs above and below the setpoint within the proportional band. The setpoint is located at the midpoint of the range of values in the proportional band. Outside the proportional band, the controller functions as if it were in the On-Off mode. When the controlled variable is below the proportional band, the final control element is fully on. When the controlled variable is above the proportional band, the final control element is fully off. Within the proportional band, however, the final control element is turned on at an amount that is proportional to the difference between the measured variable and setpoint.

Gain and proportional band are two different ways by which an adjustable amplification setting is made to the controller. Both values determine the magnitude at which the output changes in response to an input change. A larger gain or smaller PB causes a greater output response to an input change than a smaller gain or larger PB. The following formulas show how to convert between gain and PB values:

$$PB = \frac{1}{\text{Gain}} \times 100$$

$$\text{Gain} = \frac{1}{PB} \times 100$$

Note: Proportional band is in percent.

Sensitivity A common term used to describe a controller's ability to respond to input changes is **sensitivity.** The larger the gain or the narrower the PB, the more sensitive a controller is to input changes. The proportional controller can be too sensitive. If the float in Figure 17-12 is moved closer to the pivot point, the valve position will change too much in response to a small change in level. If the level drops below setpoint, the valve will open fully and allow an inrush of liquid, causing the level to rise too high. This situation makes the valve fully close and results in the liquid dropping below setpoint again. Since the system (and the controlled variable) constantly oscillates above and below setpoint, its operation is similar to On-Off control.

Proportional control is adequate when the inertia of the process is relatively large, the process reaction rate is relatively slow, and the process lag and dead time are relatively small. All of these characteristics permit the use of high gain (narrow PB) and provide fast corrective action for small to moderate load changes.

Case Study 2: Proportional Control

As the sales of the company product grow, blending tanks are added to increase production. Eventually, there is no space left in the plant for additional tanks and no surrounding property on which to increase the size of the building. The plant's maximum output peaks at 10,000 bottles a day. One day, a reliable customer offers a long-term contract that will generate an order for 20,000 items per day.

Instead of building a completely new plant at another location, the chemical company decides to remodel the existing building and installs a continuous-process machine to increase production. With the new machine, raw materials at a precise flow rate will be fed into an in-line mixer, as shown in Figure 17-13. An auger will provide rough

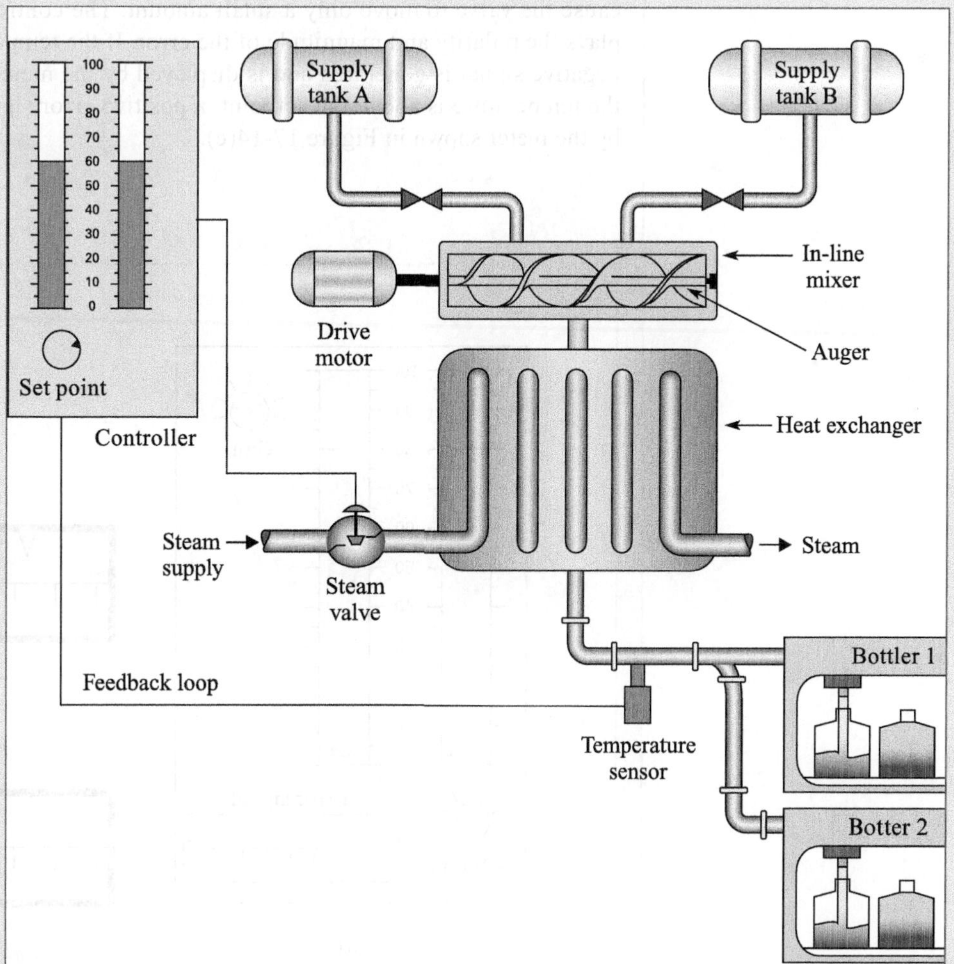

FIGURE 17-13 A continuous-process mixing/blending operation using a heat exchanger

blending, but the final blending will be performed by heating ingredients to 80 degrees as they flow through a heat exchanger. The blended end product will then flow through a pipeline to two automatic bottling machines.

To control the temperature, engineers install an On-Off controller. When the installation is complete and the process starts, they observe that the metering of the raw material and the machine's mixing operation is highly successful. However, as the controller turns on and off and fully opens or fully closes the steam valve, the temperature in the heat exchanger cycles first to 90 degrees and then to 70 degrees. The reason that the temperature in the heat exchanger fluctuates too much is that the rate of heating is too high. This condition develops because the mass of liquid in the heat exchanger is too small to absorb the energy change from the steam as it varies in temperature.

The engineers determine that if the valve is slowly adjusted until just the right amount of steam is admitted to the heat exchange, the temperature can be held within the desired range. This control action can be performed by using the proportional mode of the controller. By sending an output signal to open or close the valve to a position proportional to the temperature, the controller can maintain the heat at a more constant level.

Figure 17-14(a) shows the controller that provides proportional action. The magnitude of proportional action is adjusted by the dial called *gain.* With a high gain setting, a small temperature change causes a large valve adjustment. A low gain adjustment will cause the valve to move only a small amount. The controller also has a meter that displays the polarity and magnitude of the error. If the temperature is below the setpoint, a negative signal is generated and is displayed by the meter shown in Figure 17-14(b). If the temperature is above the setpoint, a positive error signal is generated and displayed by the meter shown in Figure 17-14(c).

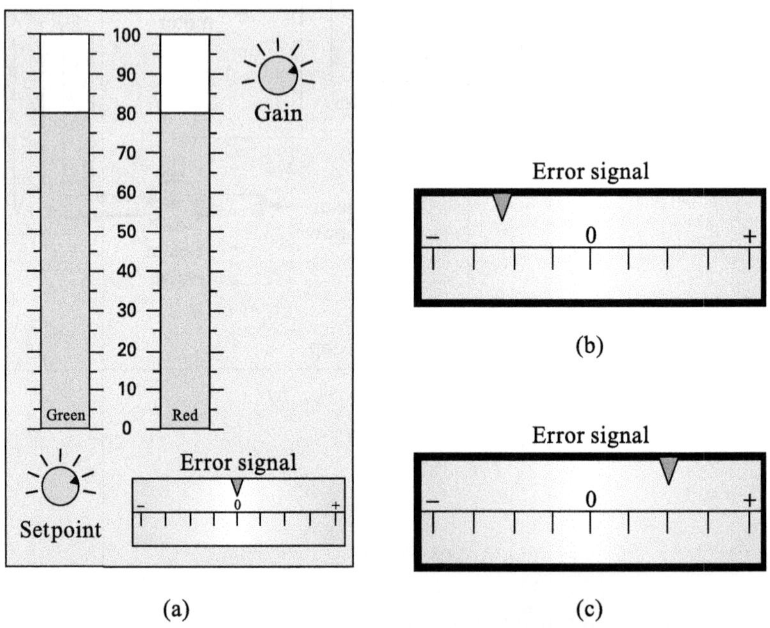

FIGURE 17-14 (a) Proportional controller; (b) Indicator showing a negative error signal; (c) Indicator showing a positive error signal

Integral Mode

IAU12108
Integral Control Mode Application

When the setpoint valve position in Figure 17-15 is exactly 50 percent of its full flow rate capacity, suppose a load change occurs from an increase of the outflow through the drain pipe. As the level drops, the float mechanism causes the opening of the inflow valve to increase. When the inflow equals the new outflow value, the system stabilizes. However, the level is below setpoint, as shown by the diagram. This error, a constant difference between setpoint (SP) and the controlled variable (CV), is called **offset.**

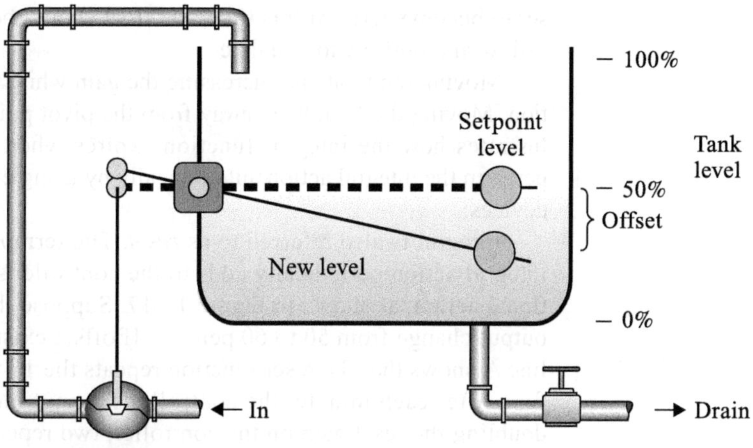

FIGURE 17-15 Offset example

Some systems cannot tolerate offset. They require that the controlled variable return to its original value. To eliminate offset, an **integral** function is added with the proportional mode to the controller. To perform the integral action, the controller senses that there is a difference between the SP and CV. As long as an error exists, the integral mode continuously causes the controller to adjust its output until the offset returns to zero. It performs this function by either adding to or subtracting from the controller output. The longer the error exists, the greater the integral becomes.

The graphs in Figure 17-16 show the functions of the proportional and integral actions in response to a gradual load change with a duration of 1 minute. During the first minute,

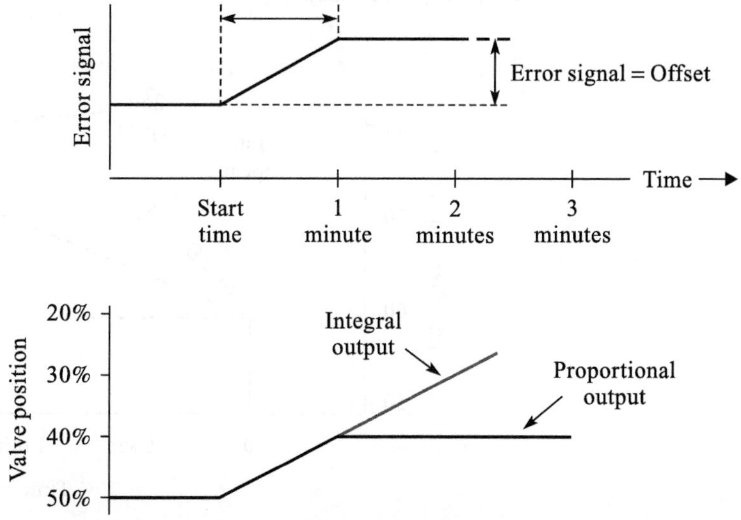

FIGURE 17-16 Proportional and integral action

the proportional output increases as the load changes. At 1 minute, the load stops changing and the proportional output stops increasing. However, there is an offset, and the integral mode continues to increase the amplitude of the controller's output. The result is that the final control element is adjusted until the controlled variable returns to the setpoint.

To better understand this concept, examine the tank system in Figure 17-15. When the system stabilizes after a load change, the new level is lower than the desired height of 50 percent. By manually moving the float toward the pivot point, an action that simulates an increase in controller gain, the valve opens further and causes the inflow to become greater than the outflow. Eventually, the liquid rises and returns to its desired level, causing the offset to become zero. At this time, the float should be moved back to a position that causes the inflow and outflow to equalize.

Moving the float and increasing the gain while there is an offset represents the integral action. Moving the float back away from the pivot point when the level returns to the setpoint illustrates how the integral function expires when there is no longer an offset. Controllers perform the integral action automatically by using electronic circuitry or microprocessor-based devices.

Integral is also referred to as *reset.* The term *reset* is derived from the way in which the integral action periodically adds to the controller's output by repeating the previous proportional action, as shown in Figure 17-17. Suppose the proportional action causes a 10-percent output change from 50 to 60 percent. If offset exists after the proportional action is finished, line A shows that the reset function repeats the 10 percent increase once each minute. Therefore, after each minute, the controller's output increases 10 percent. Line B shows that by doubling the reset gain on the controller, two repeats per minute occur, causing a 20-percent change each minute. The result is that the adjustment of the final element is doubled and causes the duration of the offset to be reduced by half.

Integral adjustments that affect the magnitude of the controller's output are labeled three ways:

Gain: Expressed as a whole number

Reset Rate: Expressed in repeats per minute

Reset or Integral Time: Expressed in minutes per repeat

Reset rate and reset time are reciprocals of each other, as shown by the following formula:

$$\text{Reset Rate} = \frac{1}{\text{Reset Time}} \qquad \text{Reset Time} = \frac{1}{\text{Reset Rate}}$$

For example, a reset rate of 10 repeats per minute is equal to an integral time of 0.1 minutes per repeat (6 seconds).

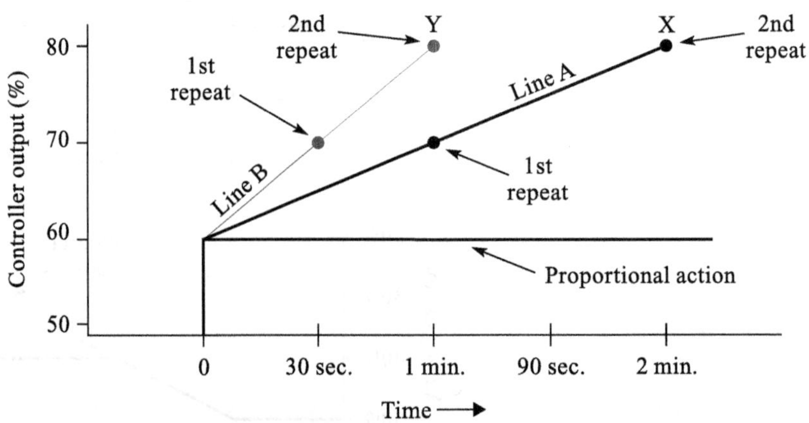

FIGURE 17-17 Reset action

The integral mode is used in process applications where the load varies slowly but by a large amount. The magnitude of integral should be large if the proportional gain is smaller (wider bandwidth) than when the proportional gain is larger (narrower bandwidth).

Case Study 3: Proportional-Integral Control

As the ingredients flow through the exchanger at a rate required to fill one bottling machine, the proportional controller is able to keep its temperature within the required range. When a second machine is started, the flow through the heat exchanger is doubled. However, the valve is positioned at that moment to emit only enough steam to heat one machine. Therefore, the product temperature drops when the second machine is added. The proportional controller senses the change and gradually increases the opening of the valve. When the temperature levels off, it is below the desired setpoint and an offset condition exists. To open the valve beyond that which is dictated by proportional action, it is necessary to add the integral mode function to the controller. As long as the error signal is present, the integral action causes the output of the controller to slowly increase. Eventually, the valve opening becomes large enough to pass the correct amount of steam, which allows the temperature to return to setpoint.

Once that happens, the error signal becomes zero and the integral action stops. The integral action is adjusted by the reset knob on the controller, shown in Figure 17-18. To obtain optimal performance, the engineer determines that the optimal setting for proportional is 5.5 and 1.5 for integral. Integral action is again required to eliminate an offset condition when the second bottling machine is shut off and the first machine remains in operation.

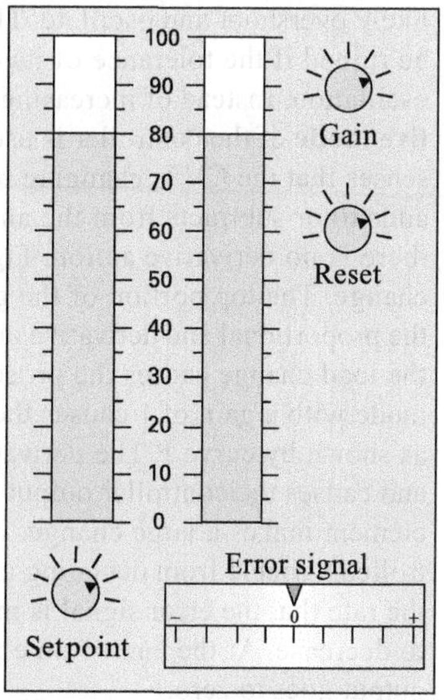

FIGURE 17-18 Proportional-integral controller

Derivative Mode

IAU9808
Derivative Control
Mode Analogy

If the setpoint or load changes suddenly, the controlled variable (CV) will deviate from the setpoint (SP) and create an error signal. The higher the initial rate of change, the further the CV is apt to move away from the setpoint. In some process applications, this situation is

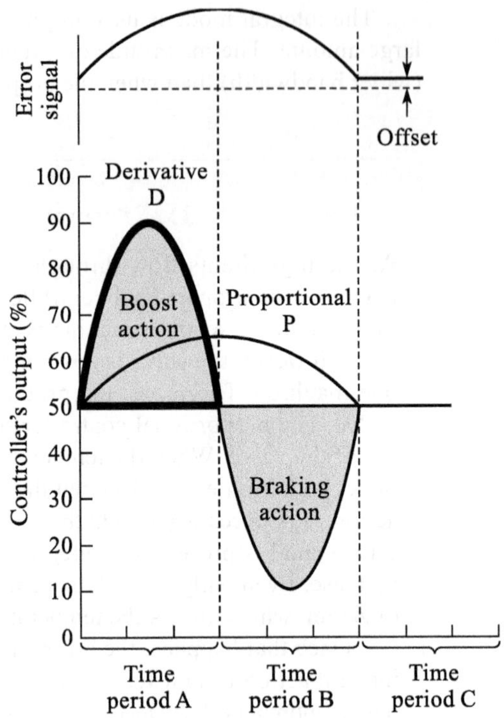

FIGURE 17-19 Proportional-plus-derivative action in response to an error signal

undesirable. If the proportional gain is increased to minimize this condition, the CV will likely overshoot and oscillate. This situation may not be acceptable because a product could be ruined if the tolerance of its condition is exceeded during the overshoot or peak of each oscillation. Instead of increasing the proportional gain to reduce the error signal, the **derivative mode** of the controller is used. The derivative function produces a response as soon as it senses that the CV is changing relative to SP. The response action is to produce a signal that adds to or subtracts from the amplitude of the proportional output. If the error is constant, there is no derivative action. Figure 17-19 shows how the derivative mode reacts to a load change. The top portion of the graph shows the error signal, and the bottom portion shows the proportional and derivative outputs of the controller. During the first half of time period A, the load change causes the process error to increase by almost 15 percent. The proportional mode with a gain of 1 causes the controller output to increase from 50 to almost 65 percent, as shown by curve P. The derivative mode shown by curve D adds to the proportional signal and causes the controller output to increase from 50 to 90 percent. The result is that the final element makes a large change, which causes a quick system response that prevents the controlled variable from deviating any further from the setpoint. At the middle of time period A, the rate that the error signal is increasing begins to drop off, and the derivative output begins to decrease. At the end of time period A, the error signal stops changing and the derivative output goes to zero.

At the beginning of time period B, the error signal starts returning to zero, and the proportional output reduces by the same amount. The derivative action responds to a reduction of the error signal by reversing its polarity, as shown by line D. During the first half of time period B, the amplitude of the derivative increases as it subtracts from the proportional output. The resulting decrease in controller output provides a braking action as the CV approaches SP. During the second half of time period B, the amplitude of the derivative output decreases, and the amount at which it subtracts from the proportional output diminishes. This action minimizes overshoot, a condition where CV goes past the SP value. At the beginning of time period C, the error stabilizes, the derivative signal goes to zero, and the proportional output is a small, constant value that is unable to overcome the offset.

The derivative mode is used when the controlled variable lags behind an alteration of the final control element and an error signal develops. This condition might occur in a slow-acting process-control application, such as regulating the temperature of a liquid in a large tank. If, for example, a new setpoint setting is made, the static inertia of the liquid does not allow its temperature to change immediately after the final control element is altered. The result is that a lagging condition develops. The derivative mode can be used by the controller to minimize the error signal that develops from this condition. An example of where derivative control would not be used is in a fast-acting application, such as an airflow-control system. Whenever a new setpoint setting is made, a flow-control valve changes and immediately alters the flow rate of the air. Since the response of the system is very fast, a lagging condition does not develop and the derivative action is not required. Instead, a two-mode controller (PI) is used.

The derivative-mode adjustment is called **derivative time.** Its setting determines the extent to which the derivative action changes the controller's output. A derivative time setting of 2 minutes will cause a response twice as fast as a setting at 1 minute.

There are several limitations to the derivative mode:

- The derivative mode is never used alone but in combination with proportional or proportional-plus-integral.
- Derivative action is unable to remove the offset present in proportional control. This offset is a steady-state error, which means that it is a "constant." Therefore, since there is no rate of change that occurs, the derivative action produced is zero.
- Derivative control is unsuitable for systems that are exposed to noisy environments. Noisy signals contain high-frequency components that are amplified by the derivative action. These amplified signals will appear at the controller output and may cause unwanted changes by the final control element.

Derivative control is beneficial in two types of process applications:

1. Those that have large and rapid load changes in a slow-response system. The derivative mode enables the controller to respond more rapidly and position the final control element more quickly than is possible with only proportional action.

ELE5308
PID Control

Case Study 4: Proportional-Integral-Derivative Control

Whenever the second bottling machine is started, the load change condition it creates is too sudden for proportional-integral control action to operate effectively. The controller does not increase the opening of the valve rapidly enough, which causes the temperature to rapidly drop out of the tolerance range. Before it eventually rises back to the setpoint value, 50 bottles are off-standard and have to be thrown out.

To open the valve more quickly, the derivative function mode is added to the controller. With this feature, the magnitude of the controller's output temporarily increases while it detects a fast error change. As the signal causes the opening of the valve to become larger, more steam is added to the heat jacket, and the temperature rises more quickly. The derivative action ends whenever the temperature levels off, and the valve returns to the position dictated by proportional and integral action. The amount of derivative action is adjusted by the rate knob on the controller, shown in Figure 17-20. A setting too low causes the system response to be sluggish, such as when only proportional-integral action is used. A setting too high causes the temperature to become too high by overshooting.

Derivative action also causes the system to respond more quickly when the second bottling machine is shut off.

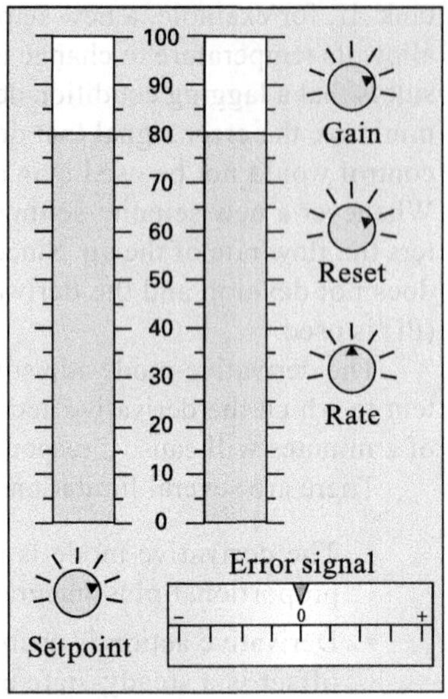

FIGURE 17-20 Proportional-integral-derivative controller

However, the rate time setting must be relatively low; otherwise, the controller may overreact and cause an unstable condition in the system.

2. Those in which systems are subject to frequent start-ups, such as batch processes.

Each of the three continuous-control modes has specific characteristics that can be advantageous in a control system. Table 17-1 provides a summary of how different mode combinations are used for various applications.

TABLE 17-1 Proportional, Integral, and Derivative Mode Summary

Mode Combinations	*Function*	*Applications*
Proportional (P)	To provide gain	For small setpoint or small load changes
Proportional-plus-Integral (PI)	To eliminate offset	For large and slow setpoint or load changes
Proportional-plus-Derivative (PD)	To speed up response and minimize overshoot	For sudden setpoint or quick load changes in a slow-response system
Proportional-Integral-Derivative (PID)	To speed up response, minimize overshoot, and eliminate offset	For large and sudden setpoint or load changes in a slow-response system

17-7 Advanced Control Techniques

Nearly all control systems are based on the principle of feedback control. The function of the control loop is to maintain the controlled variables as close to the setpoint as possible. By incorporating the characteristics of all three PID modes into a single loop, the degree of control may be adequate for most situations. However, some complex manufacturing applications often require more-advanced control to ensure a more-precise control of the process. There are four of these techniques frequently used: *cascade control, feed-forward control, ratio control,* and *adaptive control.* Cascade and feed-forward control techniques provide a faster and tighter response than PID control because they detect disturbances when they occur and begin correction early to overcome the effects of dead time and process-time constants. Ratio and adaptive control techniques are used to accommodate unique process and measurement situations.

Cascade Control

A single-loop feedback control system is designed to respond to changes in the controlled variable. In some types of applications that have long time constants, such as with temperature-related processes, the reaction time to disturbance is not fast enough. For example, Figure 17-21 shows a continuous process where two liquids are mixed and then slowly passed through a

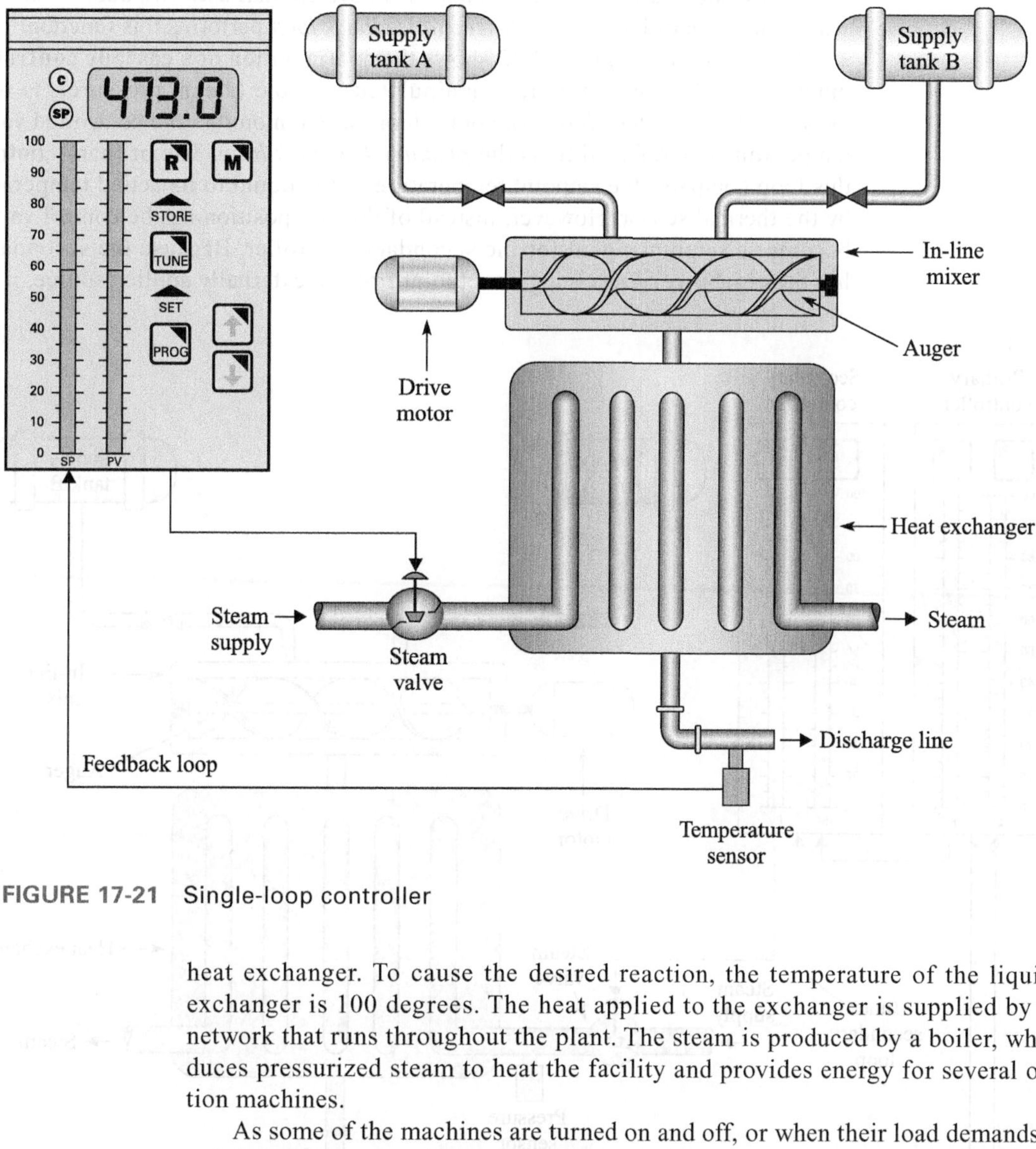

FIGURE 17-21 Single-loop controller

heat exchanger. To cause the desired reaction, the temperature of the liquid inside the exchanger is 100 degrees. The heat applied to the exchanger is supplied by a steam line network that runs throughout the plant. The steam is produced by a boiler, which also produces pressurized steam to heat the facility and provides energy for several other production machines.

As some of the machines are turned on and off, or when their load demands vary, the requirement for steam changes, which causes the pressure to fluctuate. If the pressure drops, the temperature of the steam will decrease. If the pressure increases, the temperature of the steam increases. The fluctuation of the steam pressure can cause the temperature of the liquid inside the exchanger to vary by as much as 5 degrees.

Suppose that a machine upstream in the boiler supply line is turned on and draws a large amount of steam. This situation causes the pressure in the line to decrease and the steam temperature inside the heat exchanger to drop. Due to the static inertia of the process liquid, it opposes a temperature change and creates a delay (pure lag) as the liquid retains thermal energy. Several minutes must pass before the liquid cools to the same temperature as the steam. As the controller detects a lower liquid temperature from the sensor located at the discharge line, it causes the steam valve to open by a larger amount. The increase in steam causes the temperature inside the exchanger to rise. Another pure lag develops as several minutes pass before the liquid rises to the same temperature as the steam.

The lagging effect in the transfer of thermal energy from the steam to the liquid can be undesirable, especially if the temperature drops below the level that provides an adequate chemical reaction. The lagging effect in the exchanger, shown in Figure 17-21, is due to the control method used in its design. In the system, the controller responds to changes in the controlled variable (the process liquid) instead of the upset that takes place in the manipulated variable (the steam). The process can be controlled more quickly by monitoring both the controlled and the manipulated variables. A *cascade control* system that uses two additional components, a sensor and a controller, to form a *second feedback loop* performs this function.

The diagram in Figure 17-22 shows the configuration of a **cascade control** system. The inner loop, which monitors the manipulated variable (steam pressure), is referred to as the *secondary feedback loop*. The outer loop, which monitors the controlled variable (liquid temperature), is referred to as the *primary feedback loop*. The primary controller used in this loop compares the setpoint temperature of the liquid to its actual temperature supplied by the thermal sensor. However, instead of directly positioning the control valve, its output becomes a setpoint signal for the secondary controller. Because the secondary controller has the capability of receiving its setpoint from an externally applied source, it is referred to

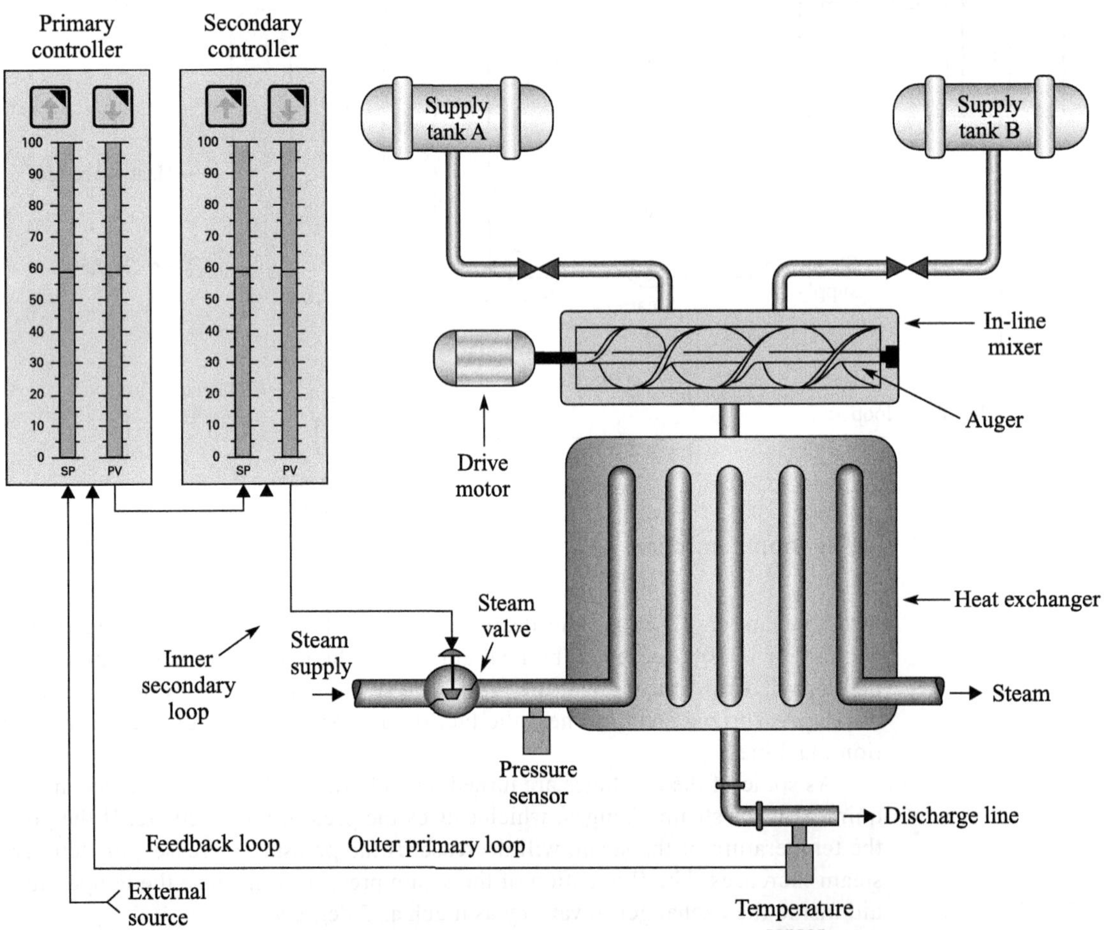

FIGURE 17-22 Cascade control system

as a *remote controller.* In addition to its setpoint input, it receives a feedback pressure signal from the sensor placed inside a steam chamber going into the heat exchanger. Based on the results when comparing the setpoint and feedback signal of the inner loop, the secondary controller positions the final control element (steam valve). There are two conditions that will cause the secondary controller to reposition the steam valve. First, if there are any fluctuations in steam pressure, an error signal develops due to the difference between the changed feedback signal and the stable setpoint signal. The result is that the valve position changes, which causes the steam pressure going into the heat exchanger to return to the desired level before the temperature of the liquid is substantially affected. The other condition that causes the secondary controller to adjust the valve is if the setpoint signal it receives from the primary controller varies. This situation occurs if there is a system setpoint change made to the primary controller or of the feedback signal from the outer loop varies due to a disturbance or load change of the controlled variable (liquid) temperature. Either situation creates a new error signal at the primary controller's output, which is fed to the secondary controller as a different setpoint.

In summary, the secondary control loop reacts quickly to changes in steam pressure. The secondary loop always causes a faster final control element reaction to changes in its variable than when variances occur in the primary loop. In addition, by only using the proportional mode in each control loop, cascade control often results in faster, more precise performance than with single-loop PID control systems. When tuning a cascade control system, always tune the secondary loop first. The proportional band of the secondary loop should always be narrower (higher gain) than the proportional band (lower gain) of the primary loop. To be effective, the time constant of the inner loop should always be three to ten times faster than the time constant of the outer loop.

Feed-Forward Control

IAU3406
The Feed-Forward
Control System

Feedback systems work on the principle that the process must deviate from setpoint before control action is applied. Figure 17-23 is used to illustrate this concept. A liquid passes through the tube section of a heat exchanger, which raises its temperature to a required level before it is discharged. The outflow, which is the controlled variable, is fed to a mixing tank for further processing. The heating medium is steam, which passes through the shell of the exchanger. The steam is the manipulated variable. A control valve is used as the final control element to vary the flow of steam and therefore the heat that is transferred to the product. The sensor located in the outlet line reads the temperature of the controlled variable. If a disturbance causes the outgoing fluid to deviate from setpoint, the controller causes the valve to vary the amount of steam that passes through until the situation is completely corrected. However, while the corrective action takes place, an erroneous product is being produced. Some processes cannot tolerate any deviation from setpoint. For example, if the temperature of the liquid that leaves the heat exchanger is too high when it enters the mixer, the resulting product will separate. If the temperature of the liquid is too low, the end product will be lumpy.

By measuring a variable that enters a process and by taking corrective action if it is affected by a disturbance, a deviation of the controlled variable from setpoint is reduced or eliminated. The operation of **feed-forward control** is based on this principle.

Figure 17-24 shows how the feedback heat exchanger system can be modified to perform feed-forward control. Under stable conditions, the system is tuned so that, as the fluid passes through the exchanger, its temperature is raised from 80 to 100 degrees. To maintain this operation, the temperature of the incoming fluid must be at 80 degrees, and the flow rate must be precisely controlled. One way the operation is disrupted is when the flow rate is constant but the temperature of the incoming fluid drops to 76 degrees. Since the exchanger is calibrated and tuned to raise the fluid's temperature 20 degrees, the outgoing fluid temperature will be only 96 degrees. Another way the operation is disrupted is when the incoming fluid is the required 80 degrees, but the flow rate increases. Since the fluid passes through the exchanger more quickly, the time during which the heat is transferred from the steam to the liquid is shortened, and its temperature is lower when the liquid is discharged. The feed-forward system reduces each type of disruption by using two sensors. One measures the temperature of the incoming liquid, and the other measures its flow rate. Based on inputs from the two sensors,

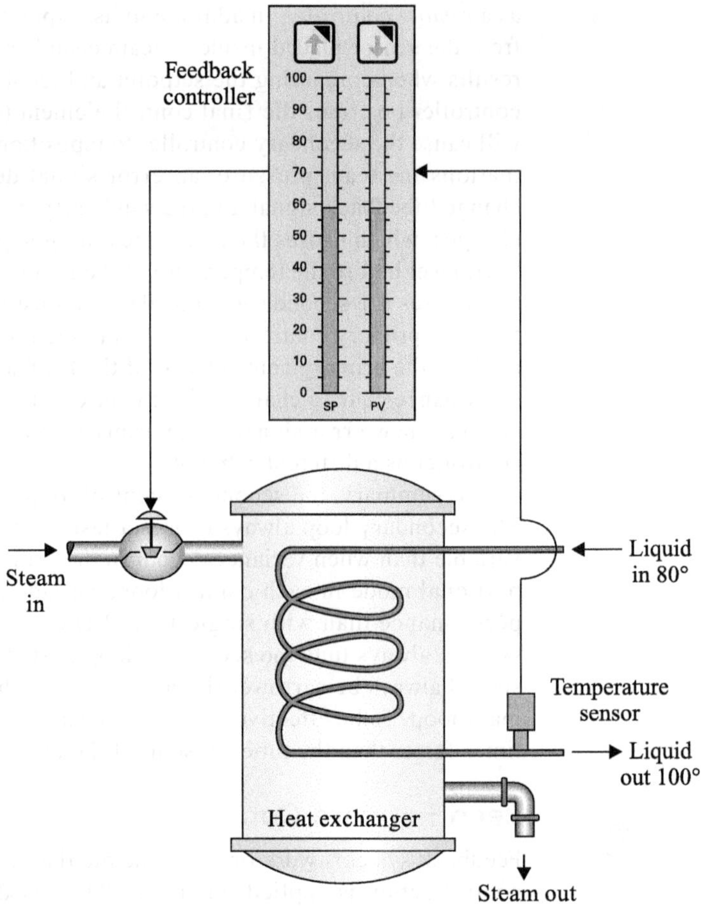

FIGURE 17-23 Feedback control system

the feed-forward controller calculates how much steam is required to maintain the controlled variable at setpoint.

The feed-forward control system does not operate perfectly. There are usually unmeasurable disturbances not compensated for by feed-forward control, such as a worn steam valve, sensors out of tolerance, or inexact mathematical calculations programmed into the controller. If any variances occur resulting from these problems, the controlled variable, such as the desired temperature of 100 degrees being discharged from the exchanger, will be affected.

To detect a deviation of the controlled variable from the setpoint due to an unmeasurable disturbance, feedback control is added to the system, as shown in Figure 17-25. The controller performs both the feed-forward and feedback operations. A temperature sensor is connected to the outflow line to form the feedback loop. The controller compares the feedback signal with the setpoint and causes the valve to alter the flow of steam if an error develops.

In summary, feedback systems determine a correction that needs to be made after the controlled variable deviates from setpoint; feed-forward systems detect a disturbance before the controlled variable can deviate from setpoint. Feed-forward control is used when no variation from setpoint can be tolerated in a process or when a system is very slow in responding to corrective action. Due to the inaccuracy of feed-forward control in some situations, it is seldom used without feedback control.

Ratio Control

IAU3906
Ratio Control

A wide range of industries use a mixing operation to blend two or more ingredients together to form an end product. To ensure that a quality product is produced, the quantity of each ingredient must be very precise. One method used for mixing applications is to proportionally control the flow of one ingredient based on the amount of flow of another ingredient. This procedure is called **ratio control.** With ratio control, the flow of one material is uncontrolled.

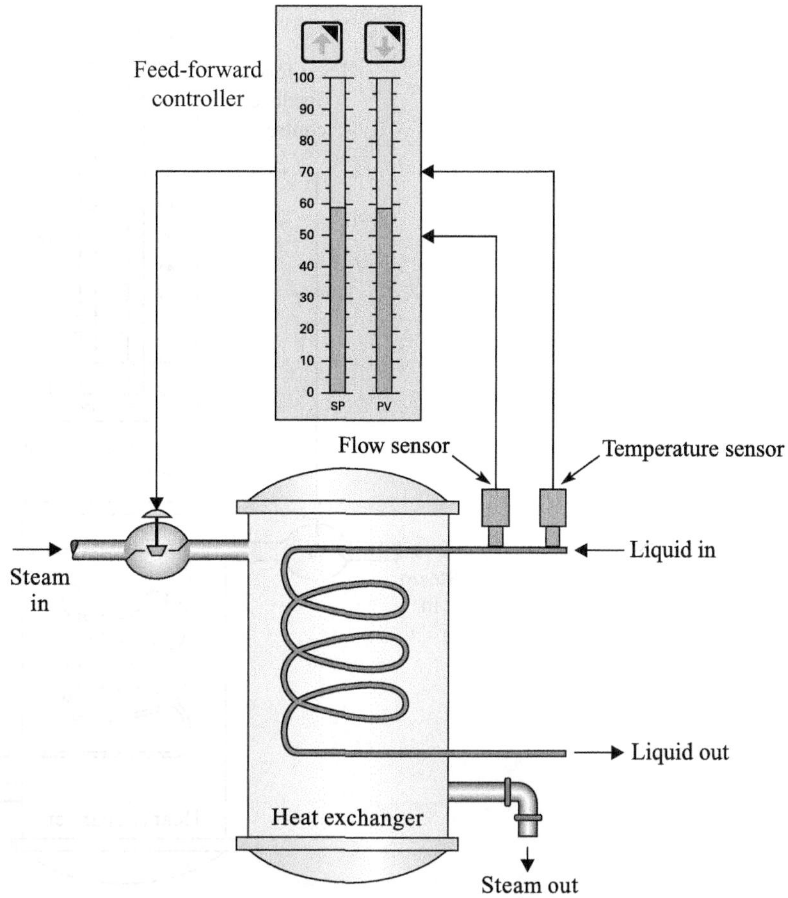

FIGURE 17-24 Feed-forward control system

This flow rate is commonly designated *wild flow*. The rate of the wild flow ingredients is measured and used as a reference to set the flow rate of the other material. This flow rate is referred to as *controlled flow*.

A ratio-control system is shown in Figure 17-26. One sensor is used to measure the flow rate of the wild flow, and another sensor is used to measure the flow rate of the controlled flow. The flow rate of each line is measured by reading the differential pressure across an orifice plate, which is nonlinear. A square root extractor transducer is connected between the each D/P flow sensor and the controller. Its function is to convert the nonlinear output of the pressure sensor to a linear signal that is proportional to the flow rate of each liquid. By using the flow signals from each line, the controller varies a valve position that adjusts the controlled flow and maintains it in proper ratio to the wild flow.

The ratio-control flow method is used for continuous blending applications and can be integrated to provide a specified volume for a batch process. An example of how ratio control is used in a continuous process, in which two materials are used, is a sewage treatment plant. When a sudden downpour of rain occurs, a rush of water flows through the facility. The flow rate of the untreated water is measured as the wild flow. A ratio controller is used to add a controlled percentage of chlorine to purify the water as it passes through the system. If two or more ingredients are combined in a process, such as when lead, dyes, and other additives are mixed with gasoline to form a specific blend, more than one ratio controller is used. An example of a batch process that uses ratio control is soft drink production, where syrup and water are mixed in a tank.

Adaptive Control

Most of the other control systems previously discussed in this chapter produce a control signal that is proportional to the size of the signal error. The graph in Figure 17-27(a) shows that the relationship between the process-measurement signal and the resulting signal

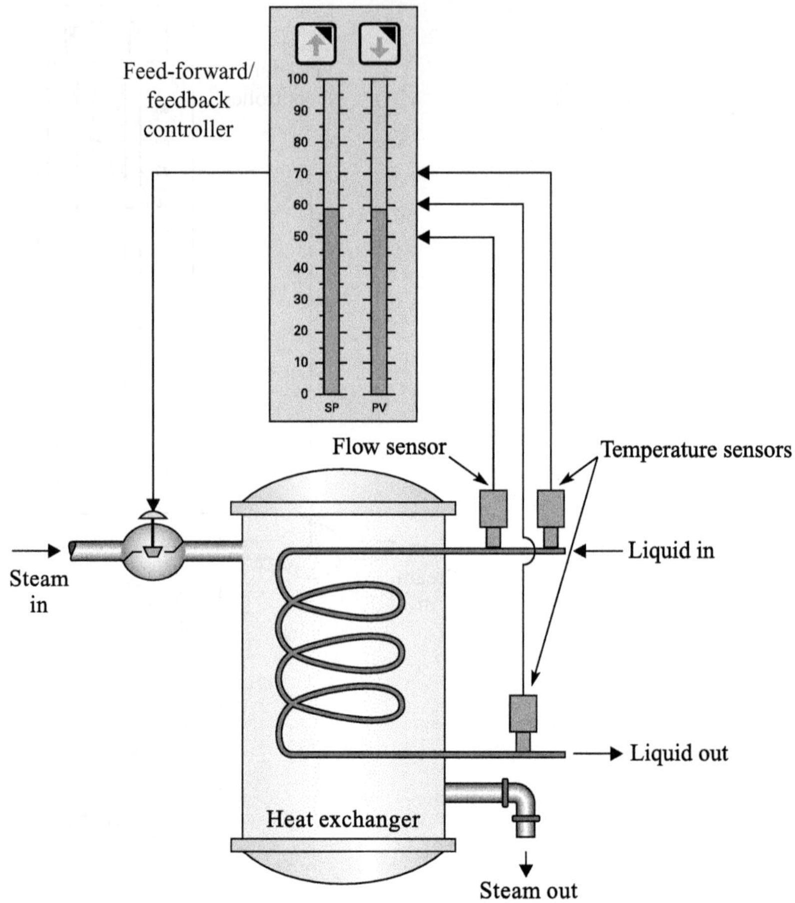

FIGURE 17-25 Feed-forward/feedback control system

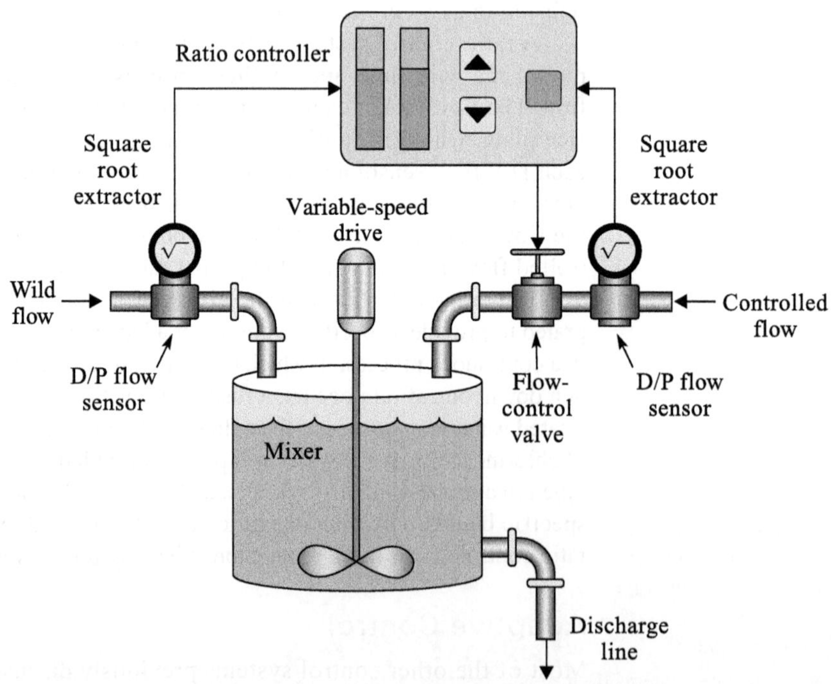

FIGURE 17-26 Ratio-control mixing process

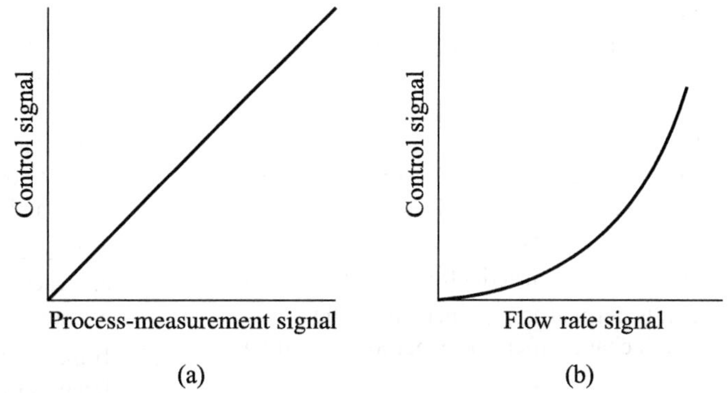

FIGURE 17-27 Relationship between measurement and control signals: (a) linear; (b) nonlinear

produced by the controller is linear. A relationship between the measurement and an output control signal that is not proportional is graphically represented by the diagram in Figure 17-27(b). The reason for the resulting nonlinear signal is the characteristic of the measurement device that reads the condition of a process variable. For example, the two pressures that develop on each side of the orifice plate of a differential flowmeter are not proportional to the flow rate, as shown in Table 17-2.

One way to compensate for the nonlinear characteristics of the differential pressure sensor is to use an instrument called a *square root extractor*. This specialized transducer converts the nonlinear output of the sensor into a linear signal used by the controller.

In addition to a differential pressure sensor that is nonlinear, some processes also have this characteristic. An example is an analytical control system that controls the *pH* of a solution. To accommodate a nonlinear process, a controller that has **adaptive control** capabilities is used. An adaptive controller uses a combination of software programming and microelectronics to compensate for nonlinear situations.

TABLE 17-2 Comparing an Actual Linear Flow Rate to the Nonlinear Output of a Flowmeter

Actual Flow Rate	*Differential Flowmeter Reading*
100%	100%
50%	70.7%
25%	50%
0%	0%

Problems

1. The automatic operation performed by an industrial manufacturing machine is referred to as ________.
2. Timing functions used to control an operation by industrial manufacturing machines are performed by ________.
   a. ladder logic hardware
   b. a sequential drum controller
   c. a programmable controller
   d. computers
   e. all of the above
3. T/F Open-loop systems do not provide automatic control of a machine operation.
4. Which of the following industrial operations is referred to as a process? ____
   a. fluid flow rate through a pipe
   b. ingredients filling a vat at a required level
   c. heat applied to a vessel to cause a chemical reaction
   d. a vacuum pressure applied to a confined tank
   e. all of the above
5. T/F The output signal of a sensor is referred to as the *controlled variable.*
6. List three influences that cause a controlled variable to change.

7. Dynamic response pertains to ____.
   a. the reaction time of instruments in a control loop
   b. the inertia of the controlled variable
   c. dead time
   d. all of the above
8. T/F Dead time is influenced by the mass of the material from which the controlled variable is made.
9. T/F The pure lag of a controlled variable is determined by the quantity and the type of material from which it is made.
10. One time constant is defined as ____ percent of the controlled variable's change after it is exposed to an energy change.
    a. 1 c. 63.2
    b. 36.2 d. 100
11. The dynamic response of a control loop can be improved by ____.
    a. selecting a controller with proper operation features that respond to the process
    b. properly tuning a controller
    c. both a and b
12. T/F The amount at which the final control element turns on in an On-Off system is proportional to the error signal.
13. What are two household applications that purposely have a deadband to turn a device on and off?
14. T/F The output of the proportional controller moves the final control element to a definite position for each value of the controlled variable.
15. The average voltage of a time proportioning controller ____ if the on-time decreases compared to the off-time.
    a. increases c. stays the same
    b. decreases
16. When the input to a proportional controller changes by 5 percent, the ____ will change by ____ percent if there is a gain of 5.
    a. 1 e. controller output signal
    b. 5 f. final control element
    c. 10 g. controlled variable
    d. 25 h. e, f, g and d
17. The proportional ____ setting affects how rapidly the controlled variable changes in response to a step change or load change.
    a. gain c. both a and b
    b. band
18. A controller with a proportional band of 50 will produce a proportional gain of ____.
19. When the controlled variable is above the proportional band, the proportional action will cause the final control element to be ____.
    a. fully off c. fully on
    b. partially on
20. A controller has more sensitivity if its proportional band is ____.
    a. narrower b. wider
21. What condition might occur if a controller is too sensitive? ____
    a. A sluggish response to a load change might occur.
    b. Excessive cycling will occur.
    c. There will be no signal change applied to the final control element.
22. A controller with what kind of control mode eliminates offset automatically? ____
    a. On-Off c. integral
    b. proportional d. derivative
23. T/F The magnitude of the integral output is proportional to the duration of a deviation between the setpoint and the controlled variable.
24. The ____ adjustment is made on a controller for integral.
    a. reset c. PB
    b. rate
25. If the reset rate adjustment on a controller is increased, the integral time will ____.
    a. increase c. stay the same
    b. decrease
26. A reset rate of 10 repeats per minute equals an integral time of ________ minutes per repeat.
27. What kind of controller action is related to the rate at which an error develops? ____
    a. On-Off c. integral
    b. proportional d. derivative
28. While the deviation between the setpoint and measured variable is increasing, the derivative action will ____ the control effort to compensate.
    a. increase b. decrease
29. While the deviation between the setpoint and measured variable is decreasing, the derivative action will exhibit a ____ action.
    a. braking b. boosting
30. The magnitude of the derivative output is directly proportional to the ____.
    a. duration of the deviation between setpoint and the measured variable
    b. rate at which the error signal changes
    c. both a and b
31. T/F The derivative mode is recommended in applications where the controlled variable lags behind the change of the final control element.
32. When tuning a cascade control system with two loops, which loop should be tuned first? ____
    a. the primary loop b. the secondary loop
33. Which of the following terms describes a control strategy in which the output of one controller is used to manipulate the setpoint of another controller? ____
    a. ratio c. feed-forward
    b. cascade d. adaptive
34. The ____ controller in a cascade system receives a feedback signal that represents the condition of the controlled variable.
    a. primary b. secondary
35. In a cascade system, the ____ loop is considered the primary loop.
    a. inner b. outer
36. Which of the following statements are true about feed-forward control? (More than one answer is possible.) ____
    a. It should be implemented with feedback control.
    b. It can compensate for unmeasurable disturbances.

c. Its output causes the manipulated variable to be changed.
d. It improves the speed at which corrections are made to a disturbance.

37. Mark each true statement about feed-forward control.

____ Feed-forward control is another name for feedback control.

____ Feed-forward control is based on observing changes in the controlled variable and then adjusting the manipulated variable to compensate.

____ Feed-forward control helps to prevent changes in the controlled variable before they occur.

38. T/F When two or more ingredients are blended in a mixing operation, the flow of each one is controlled by a ratio-control system.

39. A strategy in which the flow rate of one fluid stream is maintained at some proportion to the flow rate of another fluid stream is ____.

a. cascade control
b. feedback control
c. ratio control
d. adaptive control

40. T/F The wild flow ingredient in ratio control is measured.

41. In ratio control, the reference flow rate is referred to as ____ flow.

a. wild
b. controlled

42. An adaptive controller uses a combination of software programming and microelectronics to compensate for ____ measurements.

a. linear
b. nonlinear

43. Which of the following process measurements produce a nonlinear signal? (More than one answer is possible.)

____

a. differential pressure flowmeter
b. ultrasonic level meter
c. pH of a solution
d. none of the above

CHAPTER

18

Instrument Calibration and Controller Tuning

OBJECTIVES

At the conclusion of this chapter, you should be able to:

- List the circumstances that require calibration procedures.
- Properly assemble the instruments required to perform the calibration procedure.
- Explain three-step and five-step calibration procedures.
- Properly assemble the instruments required to perform the calibration procedure.
- Recognize zero shift and span errors so that they can be properly corrected.
- Describe process identification.
- Use the following methods to tune a controller:

Trial-and-Error	Autotuning
Ziegler-Nichols Reaction Curve	Ziegler-Nichols Continuous Cycling

- Define the following terms associated with controller tuning:

Bump Test	Proportional Gain (PG)	Effective Delay
Unstable	Proportional Band (PB)	Process Reaction Rate
Ultimate Period	Ultimate Proportional Gain Value	Unit Reaction Rate
Ultimate Proportional Value		

- List and describe three process models that result from a step change.

INTRODUCTION

To ensure the proper operation of a closed-loop feedback system, there are several requirements that must be met.

Proper Design. During the design and engineering phase, proper instruments throughout the system must be selected to perform the required operation. For example, an appropriate sensor must be selected to accurately monitor the variable that is being controlled; if a flow-control valve is used as the actuator, it must be properly sized.

Instrument Calibration. The instruments throughout the loop must accurately produce output signals that represent the value of the measured variable. Due to wearing or aging of components, the signals that are produced by various instruments may be altered and may create undesirable results. Many of these instruments are designed so that adjustments can be made to compensate for the deterioration of their internal components. The process of making these adjustments, called **calibration,** is typically performed by instrumentation technicians.

Controller Tuning. When a disturbance happens, a load demand varies, or a setpoint change is made, it is important that the system responds quickly and accurately. This action is primarily dependent on the controller, or the "brain," of the closed-loop. A proper response is dependent on how well the controller is tuned, which requires the proper settings of the proportional, integral, and derivative control modes. An engineer is typically responsible for tuning the controller, although properly trained technicians may also be given this responsibility.

After the closed-loop feedback system installation is refined and verified through a process called "start-up," its instruments are rarely changed unless they need to be replaced due to repair. However, it is necessary to constantly monitor the instruments by calibrating them to ensure their accuracy and to periodically re-tune the controller so that there is a quick and accurate response to any changes in the process variable or the setpoint.

This chapter presents information on instrument calibration and controller tuning, typically performed by instrumentation technicians and engineers.

18-1 Instrument Calibration

IAU6407
Standard Transmission Signals

The proper operation of any process depends on the accurate operation of each instrument in a closed-loop configuration. One of the instruments commonly used in a closed-loop system is a **transmitter.** Its function is to convert a signal from a sensor that monitors the condition of the process variable to a standard analog signal that is used by the controller as its feedback input. The two most common analog standard signals are electrical and air pressure. For electrical systems, the signals are transmitted by wires, and in an air-pressure system, the signals are carried by pipes or flexible tubing.

The most common electrical signal is current, which ranges from 4 to 20 mA. The most common air signal is a pneumatic signal that ranges from 3 to 15 psi. The lowest value of each signal range often represents the minimum condition of the process variable. For example, the lowest level of liquid in a tank is represented by a 4-mA current signal or a 3-psi pneumatic signal. The maximum level in a tank is represented by a 20-mA current signal or a 15-psi pneumatic signal.

Transmitters commonly have a feature that provides a way in which they can be adjusted to ensure that the output signals they produce accurately represent the input they receive. The procedure of making these adjustments is called *calibration.* The adjustments are made so that the transmitter's output will vary through the full range in proportion to the full range that the variable being measured changes. Calibration is performed to establish the *zero* and *span* settings for the transmitter. The zero-setting adjustment causes the transmitter to produce its lowest output signal when the variable is at its minimum condition. The span setting causes the transmitter to produce its largest output signal when the variable is at its maximum condition.

If the transmitter interfaces with electronic signals, it commonly has a small screw that can be turned to adjust a variable resistor located in the electronic circuitry of the instrument. If the transmitter is mechanical, a nut on a bolt may be turned and repositioned to change the leverage or range of movement, to control something like air pressure or flow. Smart transmitters, which are the newest version of this instrument, are programmable and use keypads to make the necessary adjustments.

18-2 Reasons for Performing Calibrations

There are various reasons why transmitters require calibrations. As previously explained, mechanical wear and aging of components will alter signals over a period of time. Environmental conditions such as temperature, humidity or pressure, and exposure to dirt or dust will also reduce the accuracy of the sensor/transmitter. To ensure that correct measurements continue over a period of time, transmitters require a scheduled calibration procedure on a routine basis to ensure they are operating properly and to make adjustments if they are not.

Calibration procedures are performed for other reasons, such as:

Before New Instrument Installation. Although they are factory-calibrated, the operation of new instruments may be altered due to rough handling in the shipping department or during transit.

After Extended Shutdown. Calibration settings can be adversely affected over a period of time due to a change in environmental conditions or prolonged exposure to process materials (especially if they are corrosive).

After Repair. Instruments must be calibrated to verify their accuracy.

The Product Fails to Meet Specifications. If the product being manufactured is out of tolerance, it may be the result of a transmitter not operating properly.

The calibration procedure should be documented so that a record of the instrument's calibration history is always available. It is also a common practice to attach calibration stickers to each instrument to provide a record of the last calibration.

18-3 Calibration Preparation

Before a calibration procedure is performed, several steps must be followed:

Step 1: Determine the full range of the variable being measured. For example, the full range of a level process being controlled is 10 inches to 90 inches.

Step 2: Establish what type of sensor is used to measure the process variable. This information is required because it is necessary to know what kind of signal is applied to the transmitter. For example, it may be pneumatic pressure, a variable voltage, or a variable resistance.

Step 3: Once the required signal is established, it is applied to the transmitter under test to simulate the process variable that is measured during manufacturing. This simulation signal may come from the actual sensor that measures the process or from a source that is independent of the actual sensor. When the signal is from the actual sensor of the system, the variable being controlled is changed manually to cause the sensor to produce a signal. Examples of a source that simulates the sensor's output is an adjustable air supply that represents a pneumatic pressure or a calibrator that provides a variable voltage to simulate a thermocouple output.

To apply the simulator signal to the transmitter, input connections are made, as shown in Figure 18-1. In this illustration, a temperature calibrator is used to simulate a variable resistance of a three-wire resistance temperature detector (RTD).

Step 4: Connect a power supply to the transmitter, as shown in Figure 18-1. All transmitters require a power source. As the transmitter produces an output signal, it uses power from the source to which it is connected. For example, if its output is pneumatic, the transmitter must be connected to an air supply.

Step 5: Connect a test instrument to the transmitter's output. As the signal that simulates the measured variable is applied to its input, the transmitter must produce a corresponding output. To ensure accuracy, a precision test instrument is used to measure the output signal.

Any precision test instrument used during calibration to monitor either the simulated signal at the transmitter's input or the corresponding signal produced at the output is called a *secondary standard* device. Ideally, the secondary standard should be at least ten times more accurate than the transmitter being tested.

The connection points for calibration of either electronic or pneumatic transmitters are similar in that both types of instruments require a simulated input, a power supply, and a method to measure the output value.

Calibration equipment connections must not introduce errors during the test procedure. To avoid this possibility, the connections, whether they are pneumatic or electronic, must be securely tightened.

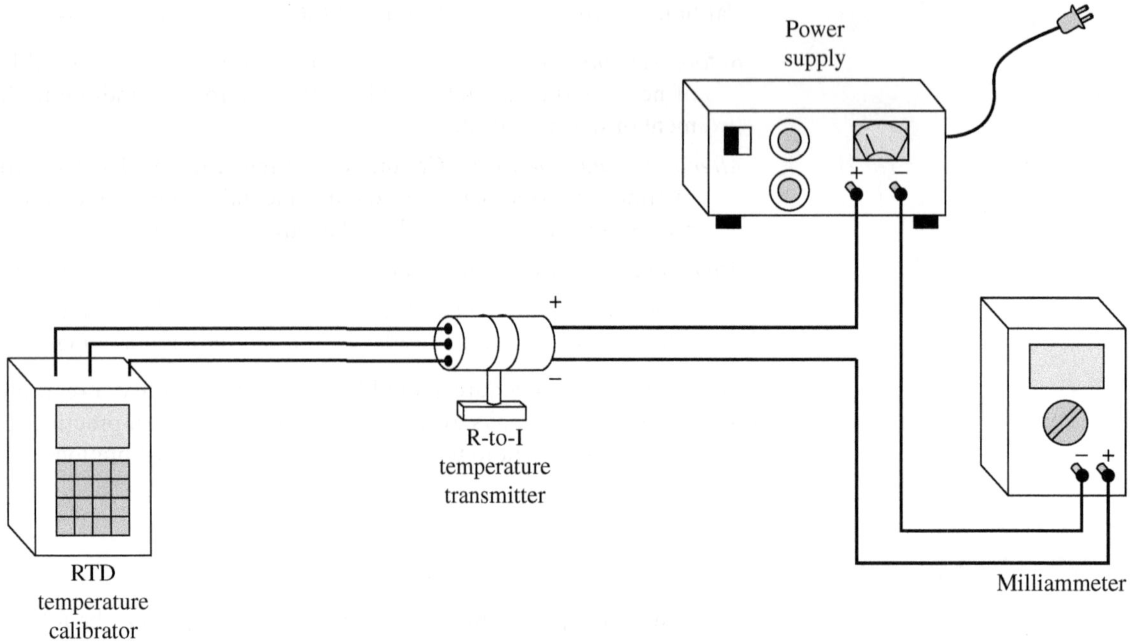

FIGURE 18-1 Calibration circuit

18-4 Standard Calibration Procedure

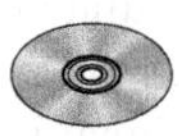

ELE1507
Standard Calibration
Procedure

Using the example of a level process that varies between 10 inches and 90 inches, the following steps are recommended to perform a standard calibration procedure. In the process shown in Figure 18-2, a float cable is attached to a rotary potentiometer.

The potentiometer is the level sensor that produces a variable voltage from 0 to 10 volts as the feedback signal. The float device is especially designed to measure water, so the amount of buoyancy is known. Therefore, to compensate for how much of it sinks, a mark exists on the side of the float to indicate the level as it rides on the surface.

The transmitter for this system converts a variable voltage at its input to a corresponding 4-mA to 20-mA output value.

Step 1: Empty the water from the Plexiglas tank.

Step 2: Manually move the float so that the indicator mark is at the minimum level of 10 inches. Assume that the potentiometer produces 1 volt at this position.

Step 3: Observe the output of the transmitter. If it is not 4 mA, make an adjustment to the transmitter's zero screw pot or keypad until it is.

Step 4: Move the float so that the indicator mark is at the maximum level of 90 inches. Assume that the potentiometer produces 9 volts at this position.

Step 5: Observe the output of the transmitter. If it is not 20 mA, make an adjustment to the transmitter's span screw pot or keypad until it is.

Step 6: Span adjustments may affect zero adjustments, and vice versa. Therefore, it is recommended that this calibration procedure be performed more than once.

Step 7: Verify the adjustments by measuring the 0-percent output at 4 mA when 10 inches are detected, a 50-percent output of 12 mA when 50 inches are detected, and a 100-percent output of 20 mA when 90 inches are detected.

This procedure, in which measurements at 0, 50, and 100 percent of the transmitter's input are made, is called a *three-point calibration check.*

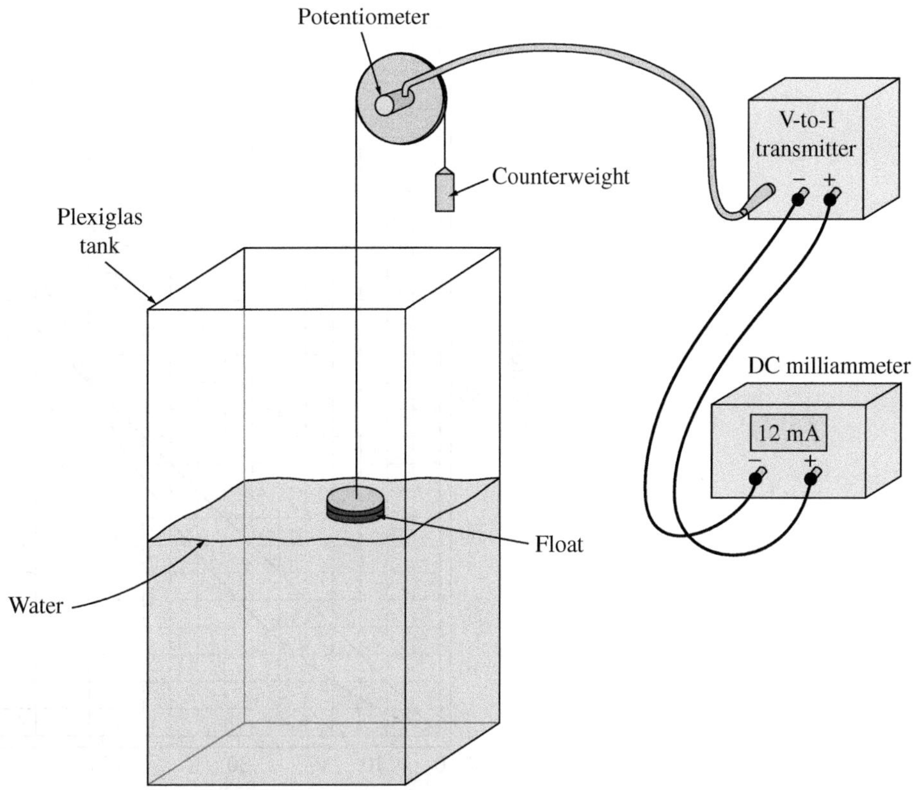

FIGURE 18-2 Calibrating a water-level process

18-5 Five-Point Calibration Procedure

The disadvantage of using the three-point calibration check is that it does not effectively show a linearity error condition. In a linearity error situation, the readings are in the form of an S-shaped calibration curve rather than being in a straight line throughout the range of the measurements, as shown in Figure 18-3. By adding two more steps to the procedure, a linearity problem can be detected.

Calibration Check

Although there are no firm rules on the procedure for a five-point calibration check, the following general guidelines are recommended. A P-to-I transmitter with an input that ranges from 0 to 25 psi, and an output of 4 to 20 mA, will be used as an example.

- The five input test points should be distributed over most of the instrument's entire range.
- Unless otherwise specified, the values of 10 percent and 90 percent are chosen to represent the low and high ends of an instrument's range. Zero percent should be avoided because some procedures require that each test point be approached from a lower value. Also, 100 percent should not be used because some procedures require the test points be approached from a higher level. To meet this requirement, the instrument's range would have to be exceeded.
- Test point values of 10, 30, 50, 70, and 90 percent of the instrument's input range are recommended. Once the test points have been established, it is necessary to calculate the input and output values that correspond to each percentage of the range.

Calculating the Input Values

The pressure applied to a P-to-I transmitter's input is 0 to 25 psi. Therefore, the range of the applied pressure is 25 psi. To determine the input pressure value that corresponds to

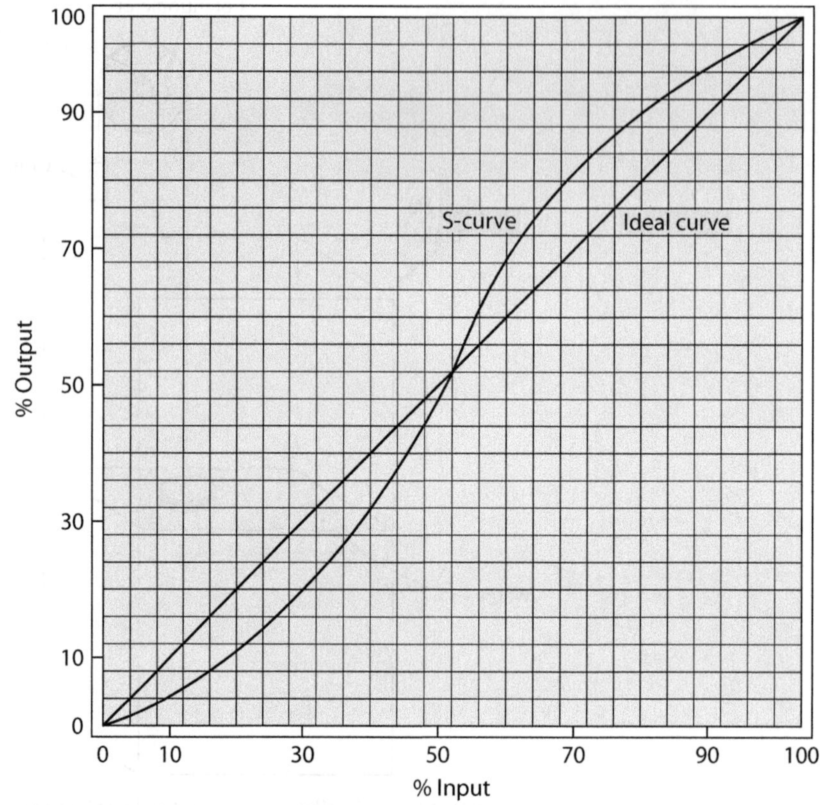

FIGURE 18-3 S-shaped calibration curve

10 percent, multiply 10 percent times the range (25) and add the product to the lower range limit of 0 psi. The 10 percent value is 2.5 psi. The remaining four test points are calculated using the same procedure for each test point percentage value.

Calculating the Output Values

The corresponding output test point values of the transmitter are calculated by using a similar procedure. For example, the range produced at the transmitter's output is 4 to 20 mA or 16 mA. To determine the output current value that corresponds to 30 percent, multiply 0.30 times 16 and add the product to the lower range limit of 4.0. The 30-percent value is $(0.30 \times 16) + 4.0 = 8.8$ mA. The remaining four test points are calculated the same way.

- A series of readings at each established test point should be taken on both upscale and downscale traverses. Upscale readings are approached from the low end, which means the check for each value is approached from a lower value. If the test point is exceeded, it is necessary to reduce the value below the desired check point to repeat the measurement. The same rules apply to downscale readings, which require that each check point be approached from the high end. Each series of readings is called a *calibration cycle.*
- A minimum of five calibration cycles should be made to ensure the measurements are valid.
- As each simulated value is applied to the input, record the corresponding test point measurements of the transmitter's output on an appropriate document, called a *data sheet.*

Analysis of Calibration Data

- After recording the data on a calibration check form, it is recommended that the measured output values recorded on the data sheet are plotted on a graph to visually analyze the results.
- The graph in Figure 18-4 shows that both the upscale and downscale values recorded at 10 percent were consistently lower than the ideal value. It also shows that the 90-percent measurements are very close to the ideal values.

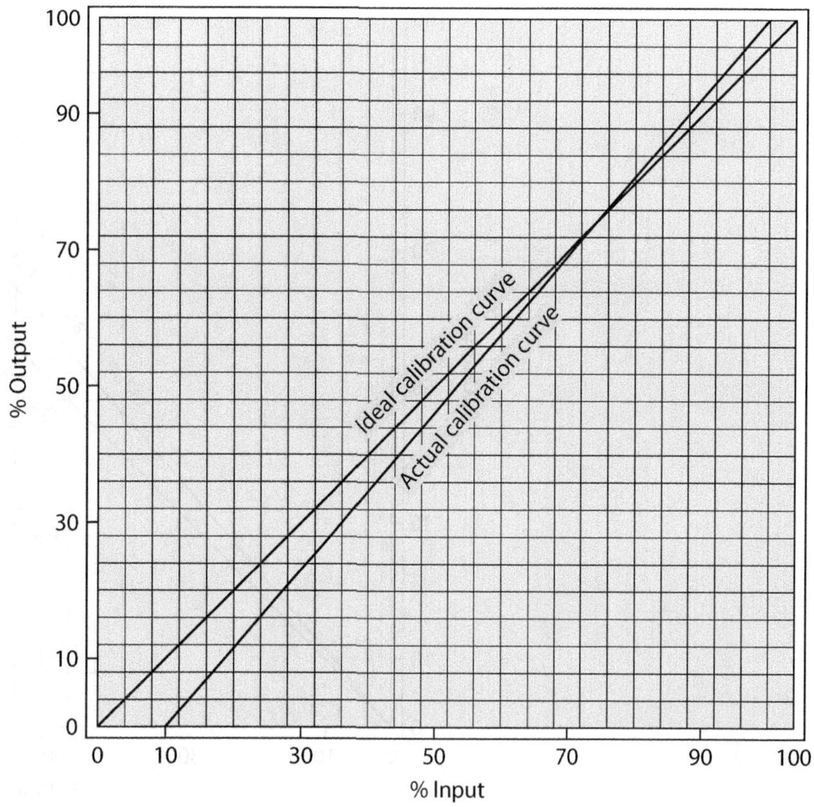

FIGURE 18-4 Combined zero shift and span error

It may appear that the only correction that is required is for the zero adjustment to be made higher. However, the 10-percent error should also cause the 90-percent reading to be low. Therefore, it can be concluded that the instrument's span setting will also need to be changed.

Transmitter Zero and Span Adjustments

- Instructions from the manufacturers who make transmitters are frequently available to explain how the calibration adjustments should be made. The zero adjustment is usually made first.
- After applying a simulated 10-percent value of 2.5 psi to the transmitter's input, make a zero adjustment until the transmitter's output reads 5.6 mA. This adjustment provides an accurate starting point for the span adjustment.
- Apply a simulated 90-percent value of 22.5 psi to the transmitter's input, and make a span adjustment until the transmitter's output reads 18.4 mA.

Verification of Adjustments

- Repeat the calibration check, because zero and span adjustments may affect each other. If necessary, readjust the zero setting first and then the span setting. Continue rechecking until no further adjustments are needed.
- Complete at least one calibration cycle to verify the adjustments eliminated the zero and span errors.

Another example of the need for a five-point calibration check is illustrated in Figure 18-5. Test points plotted on a graph are shown. It reveals that a zero shift does not occur, so the zero point is correct. However, the curve reveals that span error does exist because it is too high.

After the span adjustment is made, recheck the 10-percent value and change the zero adjustments if necessary. After the zero and span adjustments are properly made, run at least one calibration cycle to verify the results of the adjustments.

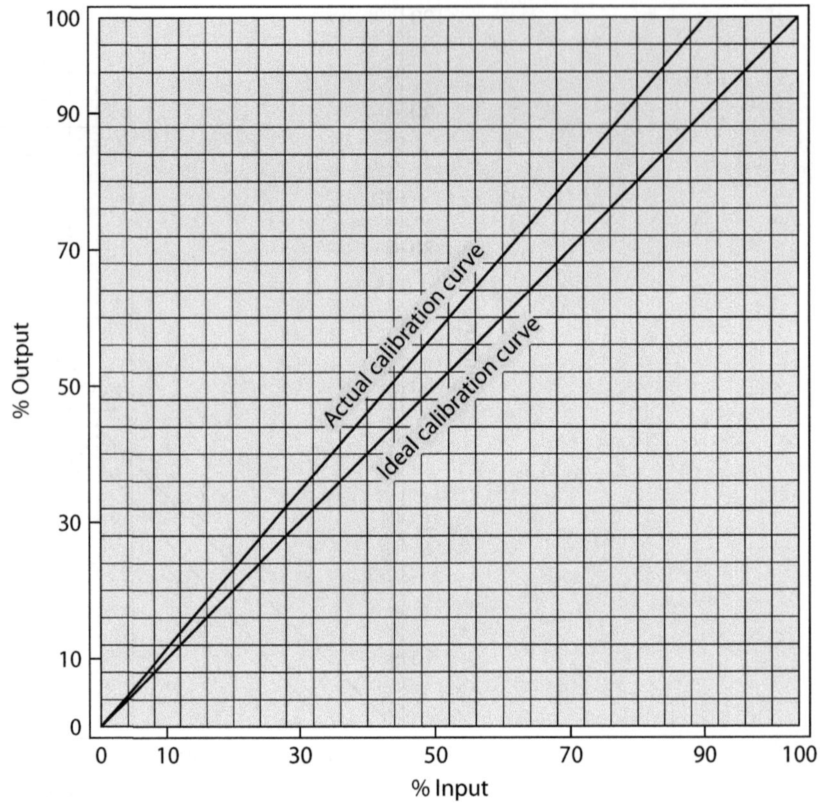

FIGURE 18-5 Span error

Documentation is usually provided by the manufacturer of transmitters on how to properly test and calibrate their products. The instructions usually include the following information.

- Diagrams that show the proper connection of the calibration equipment.
- Adjustment points for zero, span, and linearity.
- Requirements for input and output secondary standard devices.
- Recommended test conditions under which the calibration is made.

18-6 Process Calibrators

An instrument commonly used by technicians for calibration procedures is the **process calibrator**. It is used to calibrate both sensors and transmitters.

Sensor Calibration

The first requirement of a calibration procedure using the process calibrator is to verify the accuracy of the sensor. For example, to determine if an RTD is operating properly, measure the temperature of water at freezing and at the boiling point, or 0°C and at 100°C, as shown in Figure 18-6. The output of the RTD is connected to the calibrator, which then displays the temperature. To perform this measurement, the calibrator must be programmed using its keypad and its display to verify the accuracy of the RTD at the known temperatures. Process calibrators are capable of measuring other process sensors such as thermocouples, standard pressure sensors, and differential-pressure detectors.

Transmitter Calibration

After the sensor calibration steps are completed, a calibration procedure that verifies the accuracy of the transmitter is performed using the process calibrator. The function of the calibrator is to provide an input signal to the transmitter that simulates the output of the

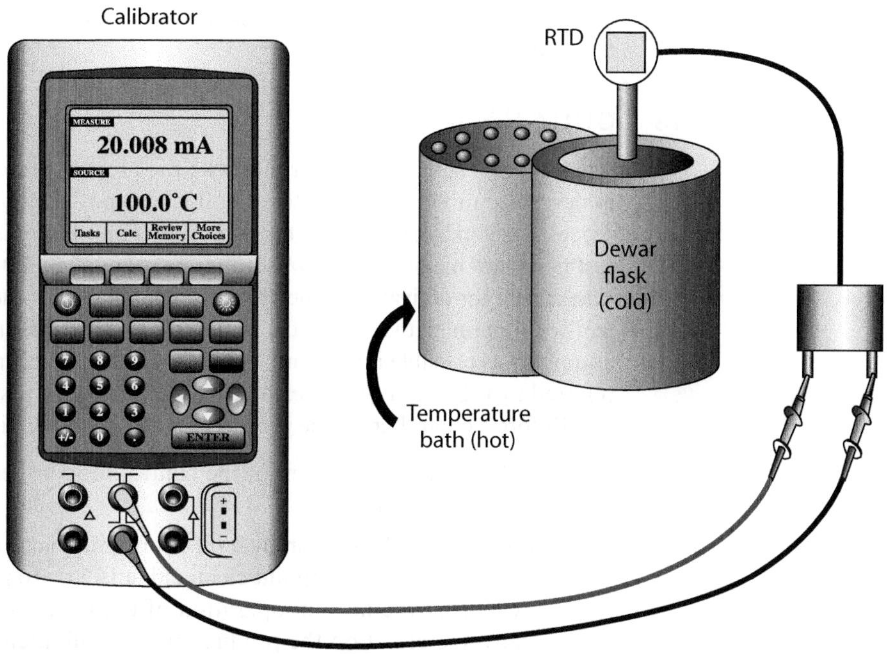

FIGURE 18-6 Sensor calibration with a process calibrator

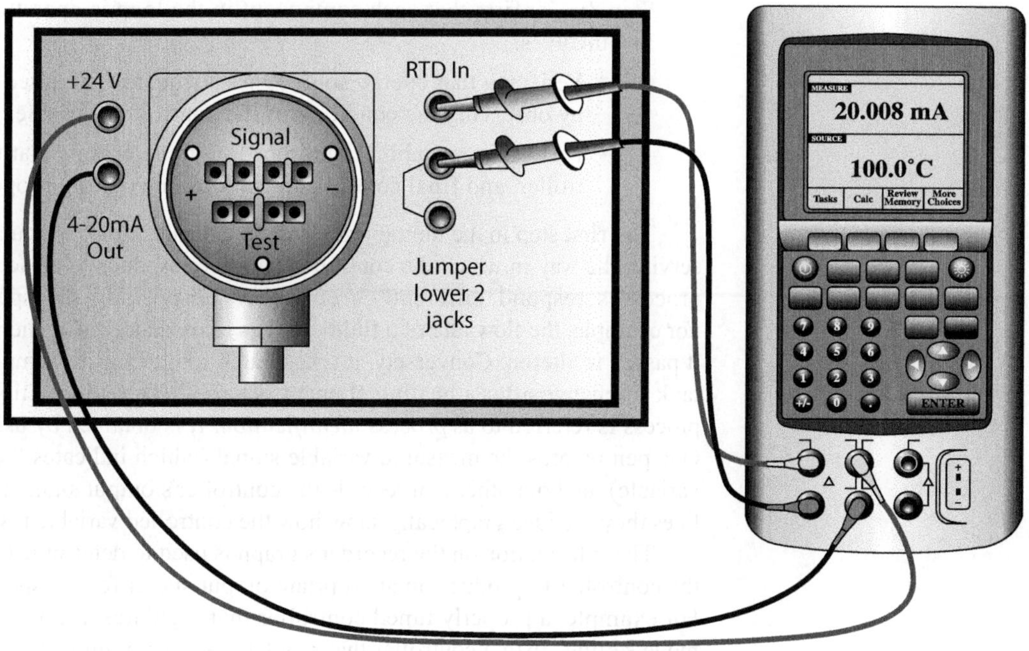

FIGURE 18-7 Transmitter calibration with a process calibrator

sensor at various measurement points. The connections from the calibrator to the transmitter are shown in Figure 18-7. When calibrating an RTD transmitter, for example, the simulated output from the calibrator simulates the output of an RTD's measurements ranging from 0°C to 100°C. A second set of leads is connected between the calibrator and the output of the transmitter. The signal from the transmitter provides a standard output signal, such as 4 to 20 mA, that represents the simulated RTD temperatures applied by the calibrator. The transmitter's standard output signal is displayed on the calibrator readout. During the calibration process, the displayed temperature and standard signal current values are used when making zero and span adjustments to the transmitter at the 0°C and 100°C levels.

The process calibrator is also capable of simulating pressure sensor outputs, thermocouple voltages, and so forth and measuring transmitter outputs such as 0 to 10 VDC, 0 to 15 psi, and so on. The keypad on the process calibrator is used to program the operation it is required to perform.

18-7 Tuning the Controller

In addition to selecting a controller with the proper control modes for a particular application, as described in Sections 17-4 to 17-6, it is also possible to minimize the dynamic response in a process system by fine-tuning the controller. *Fine-tuning* is a procedure in which the proper adjustments are made to the *gain* setting for proportional, the *reset* setting for integral, and the *rate* setting for derivative. When the optimal balance of mode adjustments is achieved, the system will minimize the size of the initial deviation and return the controlled variable to setpoint as quickly as possible if a disturbance or load change occurs. When a system reacts quickly, it is referred to as being *responsive*. When it reacts slowly, it is referred to as being *sluggish*.

Before a controller can be properly tuned, several preliminary steps should be performed:

Step 1: Study the diagram of the control loop to become familiar with its function and its components.

Step 2: Obtain the proper clearance for tuning activities. The tuning procedure often involves making a setpoint change that is 5 to 10 percent of its span. Since this change can alter the normal operation of the process, make sure that it will not have an adverse effect on the product or the equipment. The operator of the equipment is often the best person from whom to obtain information about the system and to make recommendations before tuning begins.

Step 3: Confirm that each component in the loop is operating correctly. This procedure involves:

- Verifying that energy sources are adequately supplied to the final control element by observing the condition of the controlled variable.
- Performing a calibration procedure to determine that the sensor, transmitter, controller, and final control element are functioning properly.

The first step in the tuning procedure is to analyze the particular system being tuned by observing the way in which the controlled variable responds to an actuator change. Some types of processes respond differently from others, especially the speed at which they change. For example, the flow rate of a fluid changes immediately after the valve position through which it passes is altered. Conversely, it takes much longer for the temperature of a liquid in a large tank to change after a heating element is varied. The information about the response of the process is referred to as *process identification*. It is obtained by using a two-pen chart recorder. One pen records the measured variable signal (which indicates the condition of the controlled variable), and the other pen records the controller's output signal (applied to the actuator). The lines they produce graphically show how the controlled variable responds to actuator changes.

The information on the recorder's graph is used to determine the proper settings that enable the controller to produce an appropriate output signal for the specific process it is regulating. For example, a properly tuned controller that regulates a flow process will require different mode settings than a controller that regulates a temperature process. The proper settings cause the controller to produce an output response signal that changes at an appropriate speed to match the dynamic response of the controlled variable. If the controller output is too slow, the system will be sluggish. If the controller output is too fast, overshoot or oscillations may occur.

There are two methods of process identification. One involves a closed-loop test where the process is forced into a sustained oscillating condition. The other method entails doing an open-loop step change of the actuator and then observing how the process responds. The data shown on the graph is either used in formulas to calculate proper controller settings or to provide the necessary response information to the person conducting a nonmathematical tuning procedure.

Modern computer programs are available that simulate the chart recorder operation on a computer monitor. These programs also enable the operator to make the controller mode settings by using a mouse and a computer screen.

There are many different methods used to tune a process-control loop. Some methods are mathematically based, some rely on the experience of the technician, and some can be performed automatically by the controller. Four common methods of tuning are covered in this section: *trial-and-error, Ziegler-Nichols continuous cycling, Ziegler-Nichols reaction curve,* and *autotuning.*

18-8 Trial-and-Error Tuning Method

The **trial-and-error method** of controller tuning does not use mathematical formulas. Instead, it involves using a chart recorder to observe the response of the controlled variable to a setpoint change or a load upset. Based on an interpretation of the observed response, adjustments are made to one or more of the controller mode settings. This method can give acceptable results in many situations. To perform this tuning procedure, the following steps are recommended:

- Place the controller in the manual mode.
- Turn the derivative mode off.
- Turn the integral mode off.
- Adjust the proportional mode to the minimum gain (or maximum PB) setting.
- Place the controller in the auto mode.
- Increase the proportional mode by increasing the gain from zero until a cycling action begins. Read the gain setting and readjust the setting to one-half that value. This should stop the cycling but also provide a reasonably fast response. If proportional band (PB) is used instead of gain (PG), begin with 100 percent and reduce the setting until oscillations begin. Read the PB setting and increase the adjustment to twice that value. If the value is 35 percent, for example, then double the setting to 70 percent.
- Produce a step change (setpoint change, also known as a *bump test*) of 5 to 10 percent of the span, and observe the system response on the chart recorder. If the system is too sluggish, as shown in Figure 18-8(a), increase the gain. If the system cycles continuously, as shown in Figure 18-8(b), reduce the gain. (Whenever a control loop produces excessive oscillations around the setpoint, it is referred to as being *unstable.*) Continue making proportional gain adjustments before repeating each bump test until the controlled variable cycles once or twice before stabilizing, as shown in Figure 18-8(c).
- If the process stabilizes with a constant difference between the setpoint and controlled variable, add integral until the offset is eliminated. It may be necessary to change the proportional setting each time the integral adjustment is made.
- When the proportional and integral adjustments are at their optimal settings, switch on and add derivative action to speed up the process response time.

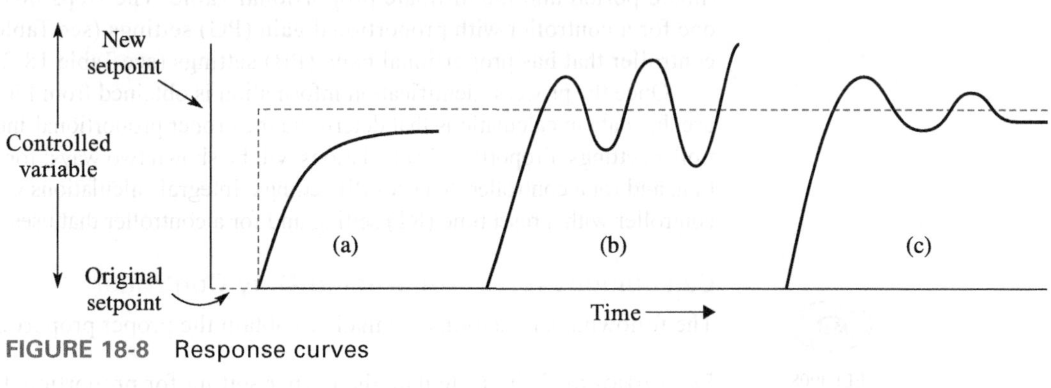

FIGURE 18-8 Response curves

The trial-and-error method can be very time-consuming because the settings of one mode affect the actions of the other modes. Therefore, after a change of one setting, it is usually necessary to readjust the settings of the other two modes. Time, patience, and the experience of the person performing the tuning procedure are often required before the optimal balance is found to achieve the desired output response of the controller.

Ziegler-Nichols Tuning Methods

In the early 1940s, two engineers who worked for Taylor Instrument Company, John Ziegler and Nathaniel Nichols, developed two formal procedures for tuning control loops. Their methods have proven to be effective in many applications and are still widely used throughout industry. The two tuning procedures are referred to as the **continuous-cycling method** and the **reaction-curve method.** Both methods provide a way to calculate controller settings mathematically. There are two major differences between them. The continuous-cycling method performs a closed-loop response test with the controller on automatic. The reaction-curve method performs an open-loop process-identification procedure with the controller in the manual setting.

18-9 Ziegler-Nichols Continuous-Cycling Method

ELE4708 Ziegler-Nichols Continuous Cycling Tuning Method

In the continuous-cycling method, the objective is to analyze the process response by forcing the controlled variable to oscillate in even, continuous cycles. This action is achieved by adjusting the proportional setting to a value that causes the cycling to take place. The time duration of one cycle, shown in Figure 18-9, is called an **ultimate period** (P_u). It is read from a chart recorder and recorded because it represents the response of the process to a change of the controller's output. The proportional setting that causes the sustained cycling is also recorded. This setting is called the **ultimate proportional gain value.** The ultimate period and the ultimate proportional value are used in mathematical formulas to calculate controller settings.

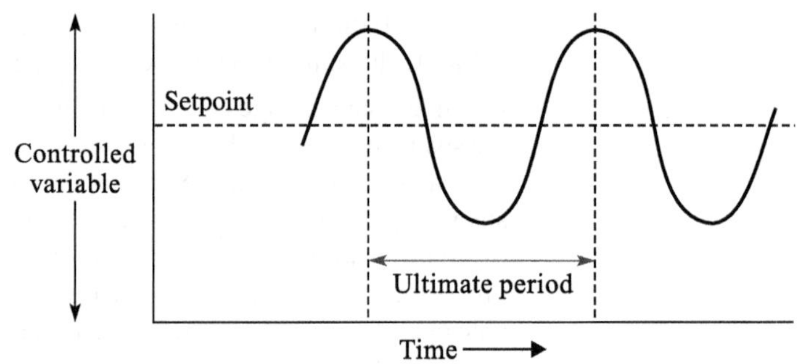

FIGURE 18-9 Determining the ultimate period

Process-Identification Procedure for the Continuous-Cycling Method

The process-identification procedure for the continuous-cycling method is performed as a closed-loop test with the controller in the automatic mode. The objective is to obtain the ultimate period and the ultimate proportional value. The steps involved are listed two ways, one for a controller with proportional gain (PG) settings (see Table 18-1) and the other for a controller that has proportional band (PB) settings (see Table 18-2).

Once the process-identification information is obtained from Figure 18-9, the next step is to use this data in calculations that determine the proper proportional-integral-derivative (PID) controller settings. Proportional calculations will be shown two ways: for a controller with a gain setting, and for a controller that uses PB settings. Integral calculations will be shown two ways: for a controller with a reset time (RT) setting and for a controller that uses a reset rate (RR) setting.

Calculations for a Proportional-Only Controller

ELE4908 Ziegler-Nichols Continuous Cycling Tuning Calculations

The following calculations are made to obtain the proper proportional setting.

Proportional Gain Calculate the proper setting for proportional gain by using the following Ziegler-Nichols formula:

$$K_c = G_u \times 0.5$$

where,

K_c = Proper Proportional Gain Setting and
G_u = Ultimate Proportional Gain Value

TABLE 18-1 Process-Identification Procedure for PG Settings

Step 1	Place the controller in the manual mode.
Step 2	Adjust the setpoint to the value most often used.
Step 3	Turn the integral and the derivative mode adjustments to a setting that will produce a minimal effect on the controller.
Step 4	Set the proportional gain to its lowest value, which is zero.
Step 5	Switch the controller from manual to automatic mode and introduce a small setpoint step change of 5 to 10 percent of the span. Return the setpoint to its original value as soon as the waveform is recorded.
Step 6	Watch the response of the process on the chart recorder. At the low setting of proportional gain, the process response curve will likely dampen out quickly as the oscillations stop and the process becomes stable, but with some offset, as shown in Figure 18-10. Return the setpoint to its original value.
Step 7	Slightly increase the proportional gain setting and make another step change. Continue repeating this procedure until the process cycles and produces the waveform on the chart recorder that is shown in Figure 18-9. Return the setpoint to its original value each time before repeating.
Step 8	Determine and record the time duration of one cycle that is displayed on the horizontal axis of the chart recorder shown in Figure 18-9. One cycle is referred to as an *ultimate period.* If the process is fast, such as flow, the ultimate period is typically read in seconds. If the process is slow, such as temperature, the ultimate period is typically read in minutes.
Step 9	Record the proportional gain setting that causes the sustained oscillations. This value is called the *ultimate proportional gain,* or G_u. The G_u value is used to mathematically find the appropriate setting for tuning the proportional mode.

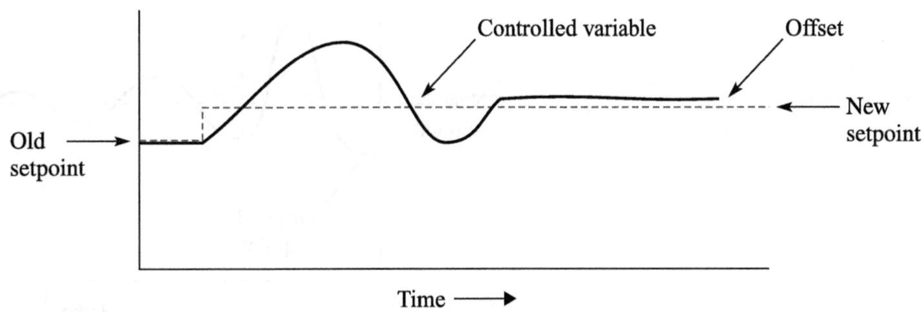

FIGURE 18-10 Response curve when the PG is too low or the PB is too high

Proportional Band Calculate the proper setting for proportional band using the following formula:

$$PB = PB_u \times 2$$

where

PB = Proper Proportional Band Setting and

PB_u = Ultimate Proportional Band Value

Verifying the Calculated Settings

The calculated value indicates the adjustment at which the proportional setting should be made. To verify that the correct value is used, introduce a step change that is 5 to 10 percent of the span, and observe the waveform on the chart recorder. The reaction curve should show a decay ratio of 1/4, as illustrated in Figure 18-11. The decay ratio indicates that the size

TABLE 18-2 Process-Identification Procedure for PB Settings

Step 1	Place the controller in the manual mode.
Step 2	Adjust the setpoint to the value most often used.
Step 3	Turn the integral and the derivative mode adjustments to a setting that will produce a minimal effect on the controller.
Step 4	Set the proportional band at its highest numerical value.
Step 5	Switch the controller from manual to automatic mode and introduce a small setpoint step change of 5 to 10 percent of the span. Return the setpoint to its original value as soon as the waveform is recorded.
Step 6	Watch the response of the process on the chart recorder. At a high proportional band setting gain, the process response curve will likely dampen out quickly as the oscillations stop and the process becomes stable, but with some offset, as shown in Figure 18-10 Return the setpoint to its original value.
Step 7	Slightly decrease the proportional band setting and make another step change. Continue repeating this procedure until the process cycles and produces the waveform on the chart recorder that is shown in Figure 18-9. Return the setpoint to its original value each time before repeating.
Step 8	Determine and record the time duration of one cycle that is displayed on the horizontal axis of the chart recorder shown in Figure 18-9. One cycle is referred to as an *ultimate period.* If the process is fast, such as flow, the ultimate period is typically read in seconds. If the process is slow, such as temperature, the ultimate period is typically read in minutes.
Step 9	Record the proportional band setting that causes the sustained oscillations. This value is called the *ultimate proportional band,* or PB_u. The PB_u value is used to mathematically find the appropriate setting for tuning the proportional mode.

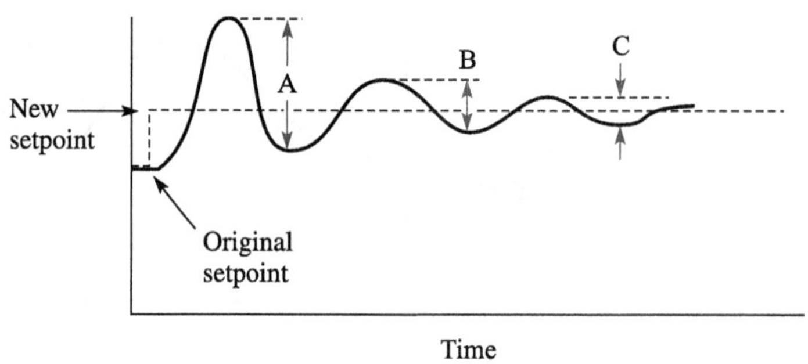

FIGURE 18-11 ¼ decay ratio

of any given peak is ¼ the size of the previous peak. The ¼ decay ratio is used by the Ziegler-Nichols tuning methods as a guideline for good control because it represents a compromise between the size of the initial deviation from the setpoint and a quick return to the setpoint.

If the curve dampens out too quickly, the proportional gain should be increased slightly, or the proportional band should be decreased slightly. If the curve has too much oscillation, the proportional gain should be decreased slightly or the proportional band increased slightly. When the oscillations stop, it is likely that a slight offset will be observed on the waveform.

Calculations for a Proportional-Integral Controller

The proportional-integral controller is also commonly referred to as a *two-mode controller*. To determine the proper settings from mathematical calculations for a two-mode controller, the *ultimate period* (P_u) and *ultimate proportional gain value* (G_u) or *ultimate proportional band value* (PB_u) are used in the formula.

Determining the Proportional Setting Since the proportional and integral actions of a two-mode controller interact, the settings for proportional gain (or proportional band) will be slightly different from the setting on a proportional-only controller. Therefore, the formulas to determine the proportional setting for the proportional-only and the two-mode controller are also different.

Proportional Gain

$$K_c = 0.45 \times G_u$$

where

K_c = Proper Proportional Gain Setting and
G_u = Ultimate Proportional Gain Value

Proportional Band

$$PB = 2.2 \times PB_u$$

where

PB = Proper Proportional Band Setting and
PB_u = Ultimate Proportional Band Value

Determining the Integral Setting The calculation to determine the proper integral setting is listed two ways: one for a controller with an RT adjustment and the other for a controller that has an RR adjustment.

Reset Time

$$T_i = \frac{P_u}{1.2}$$

where

T_i = Proper Reset Time and
P_u = Ultimate Period

Reset Rate

$$T_r = \frac{2}{P_u}$$

where

T_r = Proper Reset Rate and
P_u = Ultimate Period

When the process is very slow, *reset time* is used because the repeats are measured over a period of minutes. When the process is faster, *reset rate* is used as the controller's integral setting.

Verifying the Calculated Settings

The calculated values indicate the settings at which the proportional and integral adjustments should be made. To verify that the correct values are used, introduce a 5- to 10-percent step change, and observe the waveform on the chart recorder. The reaction curve should show a 1/4 decay ratio without offset. Slight readjustments of the proportional setting may be necessary to fine-tune the controller to produce the 1/4 decay ratio.

Calculations for a Proportional-Integral-Derivative Controller

The proportional-integral-derivative controller is also referred to as a *three-mode controller*. To determine the proper settings from mathematical calculations for a three-mode controller, the ultimate period and ultimate proportional gain (or ultimate proportional band) are used in the formula.

Determining the Proportional Setting Since the proportional, integral, and derivative actions of a three-mode controller interact, the proportional and integral settings will be slightly different from the settings on a two-mode controller. Therefore, the formulas used to determine the values of a three-mode controller are different from the formulas used for a two-mode controller.

Proportional Gain

$$K_c = 0.6 \times G_u$$

where

K_c = Proper Proportional Gain Setting and
G_u = Ultimate Proportional Gain Value

Proportional Band

$$PB = 1.7 \times PB_u$$

where

PB = Proper Proportional Band Setting and
PB_u = Ultimate Proportional Band Value

Determining the Integral Setting

Reset Time

$$T_i = \frac{P_u}{2}$$

where

T_i = Proper Reset Time and
P_u = Ultimate Period

Reset Rate

$$T_r = \frac{2}{P_u}$$

where

T_r = Proper Reset Rate and
P_u = Ultimate Period

Determining the Derivative Setting The following calculation is made to obtain the proper setting for derivative time (DT):

Derivative Time

$$T_d = \frac{P_u}{8}$$

where

T_d = Proper Derivative Time and
P_u = Ultimate Period

Verifying the Calculated Settings

The calculated values indicate the settings at which the proportional, integral, and derivative adjustments should be made. The proportional adjustment should be made first, the integral adjustment second, and the derivative adjustment last. To verify that the correct values are used, introduce a 5- to 10-percent step change, and observe the waveform on a chart recorder.

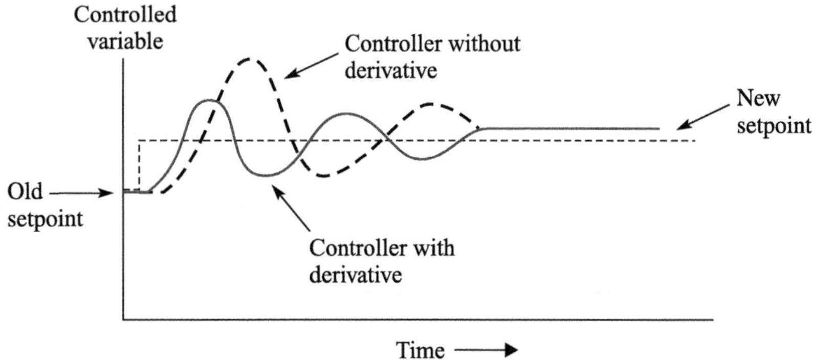

FIGURE 18-12 Reaction-curve observation to determine proper control mode settings

The reaction curve in Figure 18-12 shows no offset and a 1/4 decay ratio with a shorter duration and less amplitude during each peak of the oscillations than when a proportional-only test is performed. If the process overshoots too much when it reaches the new setpoint, the derivative time should be increased slightly. If the process is too slow in returning to the setpoint, the derivative time should be decreased slightly.

For convenience, the Ziegler-Nichols formulas are summarized in Table 18-3.

TABLE 18-3 Ziegler-Nichols Continuous-Cycling Formulas

Controller Mode	Proportional Gain K_c	Proportional Band PB	Reset Time T_i (Minutes per Repeat)	Reset Rate T_r (Repeats per Minute)	Derivative Time T_d
P	$0.5\ G_u$	$2\ PB_u$	N/A	N/A	N/A
PI	$0.45\ G_u$	$2.2\ PB_u$	$P_u/1.2$	$1.2/P_u$	N/A
PID	$0.6\ G_u$	$1.7\ PB_u$	$0.5\ P_u$	$2/P_u$	$P_u/8$

EXAMPLE 18-1

ELE4408
Continuous Cycling
Tuning Example

Continuous-Cycling Method

The example shows a slow-acting temperature process. The process-identification steps determine that the ultimate proportional gain value (G_u) is 2 when the process oscillates, as shown in Figure 18-13. The ultimate period (P_u) of the waveform is 10 minutes.

$$G_u = 2$$

Using the continuous-cycling method with a PID controller that has a proportional gain, reset time, and derivative time adjustments, determine the proper settings for each mode.

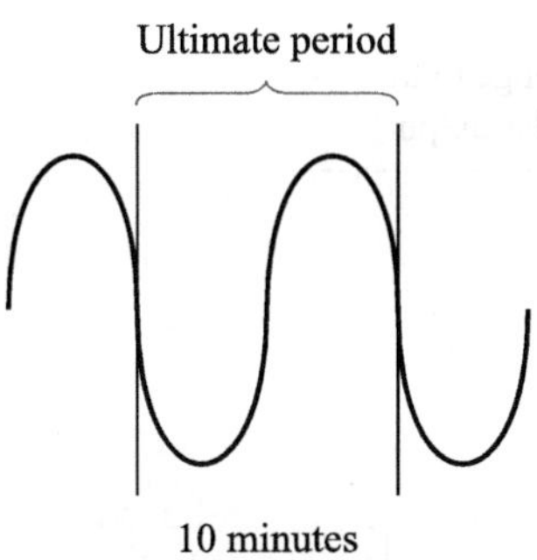

FIGURE 18-13 Determining the ultimate period

Solution

Proportional-Only Control

$$K_c = 0.5 \times G_u = 0.5 \times 2 = 1$$

Proportional-Integral

$$K_c = 0.45 \times G_u = 0.45 \times 2 = 0.9$$

$$T_i = \frac{P_u}{1.2} = \frac{10\ \text{min.}}{1.2} = 8.3 \text{ minutes per repeat}$$

Proportional-Integral-Derivative

$$K_c = 0.6 \times G_u = 0.6 \times 2 = 1.2$$

$$T_i = \frac{P_u}{2} = \frac{10 \text{ min.}}{2} = 5 \text{ minutes per repeat}$$

$$T_d = \frac{P_u}{8} = \frac{10 \text{ min.}}{8} = 1.25 \text{ minutes}$$

18-10 Ziegler-Nichols Reaction-Curve Tuning Method

The main drawback of the continuous-cycle tuning method is that the process is made to oscillate. This condition can be undesirable in some situations. Each time the cycle peaks, the process may go outside an acceptable tolerance range from the setpoint and cause, for example, a food product to be ruined. Cycling should also be avoided if it creates a safety hazard, such as too much pressure in a boiler or an excessive temperature in a nuclear power plant.

To avoid an oscillating condition, another version of Ziegler-Nichols tuning called the *reaction-curve tuning method* is used. This method involves doing a step change to the controller output and then observing the rate at which the process reacts on a chart recorder. The graph in Figure 18-14 shows the actuator output that represents the step change and the signal from a sensor that illustrates the reaction curve of the process.

The graph provides three different values that are used in mathematical calculations to determine the proper controller settings:

1. The **effective delay** (D), which is the time from when the step change is made until the process variable begins to react. This delay is caused by the process lag, dead time, or both.
2. The **process reaction rate,** which is defined as how much the process changes per unit of time. This value is obtained by calculating the slope of the process reaction curve. A curve with a steep slope indicates a faster reaction rate than a curve with a gradual slope.
3. The **unit reaction rate,** which is a measure of how much the process reacts for each percent of actuator change. To determine this value, the size of the step change must be taken from the graph. The size is read as a percentage of the actuator's span.

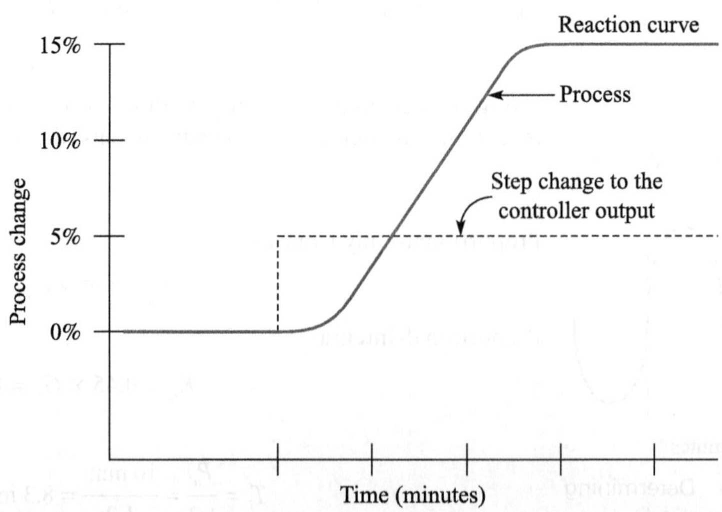

FIGURE 18-14 A process reaction curve produced by a step change

The reaction-curve tuning method performs an open-loop step change with the controller in the manual mode to obtain process-identification information.

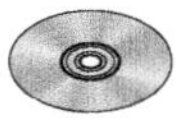

ELE4608
Ziegler-Nichols Reaction Curve Process Identification Procedure

Process-Identification Procedure for the Reaction-Curve Method

Before making calculations to determine proper controller settings, perform the following steps:

Step 1: Put the controller in the manual mode.

Step 2: Produce a step change by changing the controller output 5 to 10 percent, and observe the rate at which the process responds on the chart recorder.

Step 3: Find the maximum slope of the reaction curve, and draw a tangent line at this point, as shown in Figure 18-15(a).

Step 4: Calculate the slope of the tangent by drawing two lines on the graph. Line A is a horizontal line that begins at the starting point on the tangent, as shown in Figure 18-15(b). Line B is drawn vertically in an upward direction from line A to the end of the tangent, as shown in Figure 18-15(c).

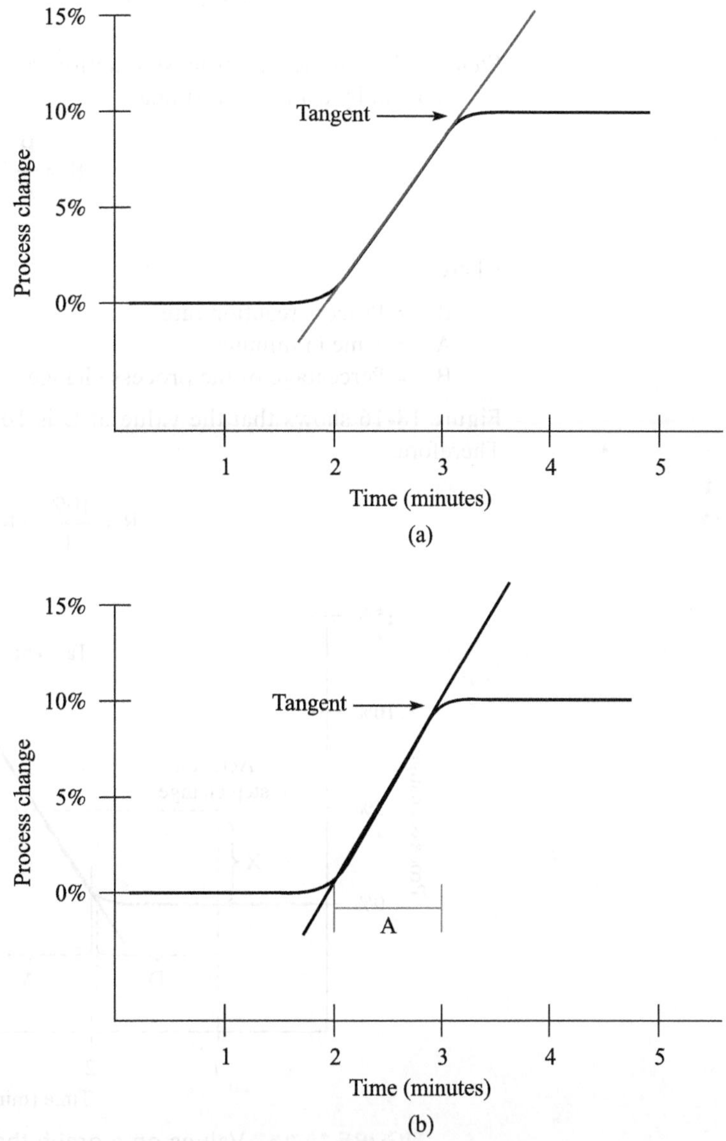

FIGURE 18-15 Reaction-curve tuning method

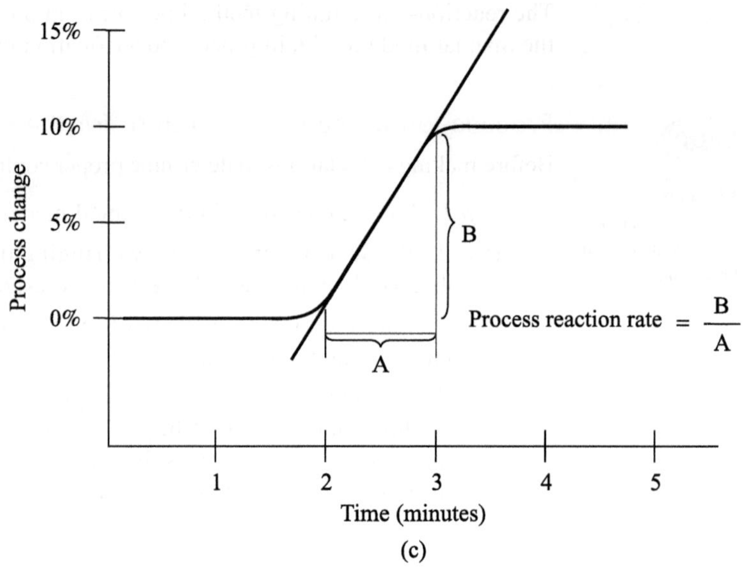

FIGURE 18-15 *(continued)*

Step 5: Determine the process reaction rate, which is indicated by the slope of the tangent, by using the formula,

$$R = \frac{B}{A}$$

where,

R = Process reaction rate
A = Time in minutes
B = Percentage of the process change

Figure 18-16 shows that the value of B is 10 percent and the value of A is 1 minute. Therefore,

$$R = \frac{10\%}{1} = 10\%$$

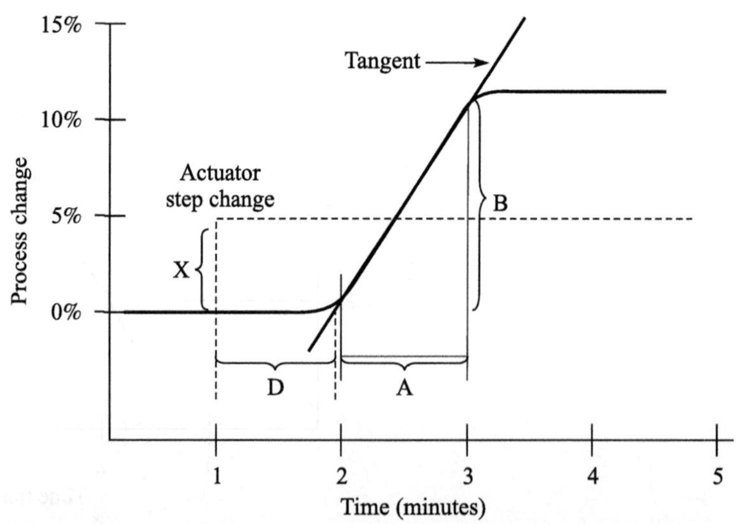

FIGURE 18-16 Values on a graph that are used to calculate controller settings

Step 6: Calculate the unit reaction rate (R_1). Divide the reaction rate (R) by the percentage of the actuator change (X).

$$R_1 = \frac{R}{X}$$

Figure 18-16 shows that the actuator changes 5 percent. The reaction rate is determined by Step 5. Therefore,

$$R_1 = \frac{10\%}{5\%} = 2$$

Step 7: Determine the effective delay (D). This delay is shown on the graph as the time from which the step change is made to where the tangent line crosses the line of initial controlled variable status.

Figure 18-16 shows that D = 0.9 minutes.

Calculating the Proper P, PI, and PID Controller Settings

ELE4808
Zeigler-Nichols
Reaction Curve Tuning
Calculations

Using the unit reaction rate and effective delay values from the formulas listed in Table 18-4, the proper settings for a P, PI, and PID controller can be determined. The proper settings are verified by doing a step change and observing a 1/4 decay ratio reaction curve on a chart recorder.

TABLE 18-4 Ziegler-Nichols Reaction-Curve Formulas

Controller Mode	Proportional Gain K_c	Proportional Band PB	Reset Time T_i (Minutes per Repeat)	Reset Rate T_r (Repeats per Minute)	Derivative Time T_d
P	$K_c = 1/R_1D$	$PB = 100R_1D$	N/A	N/A	N/A
PI	$K_c = 0.9/R_1D$	$PB = 110R_1D$	3.33D	0.3/D	N/A
PID	$K_c = 1.2/R_1D$	$PB = 83R_1D$	2D	0.5/D	0.5D

EXAMPLE 18-2

ELE4508
Reaction Curve Tuning
Example

Reaction-Curve Method

The values on the graph in Figure 18-17 are:

- Effective Delay (D) is 1.5 minutes
- Step Change is 10% of the actuator span
- Process Reaction is 27% each minute

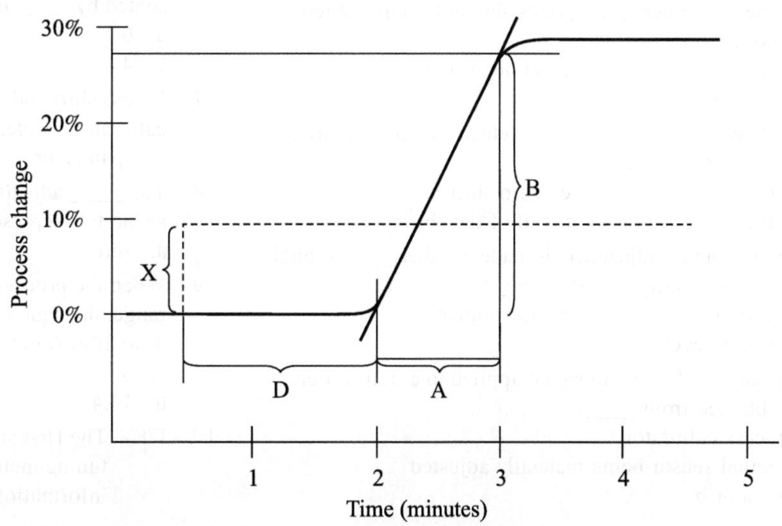

FIGURE 18-17 Reaction curve for Example 18-2

Use Table 18-4 to determine the proper settings for a PID controller with proportional band, reset time, and derivative time adjustments.

Solution

Step 1: Determine the slope of the tangent to find the process reaction rate.

$$R = \frac{B}{A} = \frac{27\%}{1\text{ min.}} = 27\%$$

Step 2: Determine the unit reaction rate (R_1).

$$R_1 = \frac{R}{X} = \frac{27\%}{10\%} = 2.7$$

Calculate the proportional band setting.

$$PB = 83R_1D = 83 \times 2.7 \times 1.5 = 336.15$$

Calculate the integral setting.

$$T_i = 2 \times D = 2 \times 1.5 = 3 \text{ minutes per repeat}$$

Calculate the derivative setting.

$$T_d = 0.5D = 0.5 \times 1.5 = 0.75 \text{ minutes}$$

18-11 Controller Autotuning

Modern microprocessor-based controllers are designed to calculate the appropriate mode settings automatically. When the autotuning function is activated, it causes the process to cycle. By measuring the duration of a cycle, it either adjusts the settings automatically for each mode or displays the recommended values for the operator to enter.

Problems

1. What is the purpose of instrument calibration?
2. During the ____ check, test points should be approached from below.
   a. upscale
   b. downscale
   c. either a or b
3. It is unlikely a ____ error can be detected using the three-point calibration check.
   a. span
   b. linearity
   c. zero shift
   d. all of the above
4. After an instrument adjustment is made, another ____ should be performed to verify the procedure.
   a. inspection
   b. calibration cycle
   c. adjustment
5. The simulation of an input signal applied to a transmitter can be obtained from ____.
   a. a process calibrator
   b. the actual sensor being manually adjusted
   c. either a or b
6. The lowest condition of a process variable is usually represented by ____ mA produced at the transmitter's output.
   a. 0
   b. 4
   c. 12
   d. 20
7. If zero shift and span errors are detected during a series of calibration cycles, the ____ is usually corrected first.
   a. span error
   b. zero shift
8. The ____ adjustment is made to the transmitter so that it accurately represents the highest value of a process.
   a. zero
   b. span
9. When the process being measured is at 90 percent of its range, the signal produced by a transmitter that has a 4- to 20-mA output is ____ mA.
   a. 4
   b. 14.4
   c. 18.4
   d. 20
10. T/F The first step in performing mathematically based tuning methods is to obtain *process-identification* information from a chart recorder graph.

11. The term *ultimate gain* (or *ultimate proportional band*) refers to the controller adjustment that ____.
    a. causes the process to continuously cycle
    b. is the proportional setting when the controller is tuned
12. Determine the proper settings for a two-mode controller using the Ziegler-Nichols continuous-cycling method and Table 18-3.
    Given: Ultimate Proportional Band = 3
    Ultimate Period = 2 minutes
    Proportional Setting __________
    Integral Setting (Reset Rate) __________
13. If a process reaction curve produced when the controller is tuned does not display a proper ¼ decay ratio because it dampens out too quickly, the proportional gain is set too ____.
    a. low b. high
14. The process-identification information for the Ziegler-Nichols reaction-curve method is observed on a chart recorder with the controller in the ____ mode, and the ¼ decay ratio is observed when the controller is in the ____ mode.
    a. manual b. automatic
15. T/F The *process reaction rate* value is obtained from the slope of the *process reaction curve*.
16. Using Table 18-4, determine the proper proportional, integral, and derivative controller settings by using the Ziegler-Nichols reaction-curve method, which provides the following process-identification information on a graph:
    Effective Delay (D): 0.5 minutes
    Step Change (X): 8%
    Slope of the Reaction Curve: 12%
    Process Reaction Rate = __________
    Unit Reaction Rate = __________
    Proportional Gain Setting = __________
    Integral Setting (Reset Time) = __________
    Derivative Time Setting = ________

SECTION

6

Detection Sensors

OUTLINE

Section 6 consists of two chapters on sensing devices that monitor the status or condition of controlled variables. In Chapter 19, various electronic sensors that produce both discrete and analog signals are described. Discrete sensors are devices that switch on or off as they detect an object and switch to the opposite state when the object is absent. Some manufacturers call these sensors *switches*, referring to the on or off signal they produce in response to the objects they detect. Examples of detection sensors in this section include inductive and capacitive switches, optical detectors, and Hall-effect sensors. These sensors are used for both motion-control and process-control applications. An analog sensor produces a variable current or voltage when it detects an object. The magnitude of its output is proportional to the distance the object is located from the sensor.

Chapter 19 also provides information on how to interface sensors. *Interfacing* refers to how the sensors are wired to other equipment, such as a programmable logic controller (PLC) or an industrial computer.

Chapter 20 addresses wireless devices that are used to sense and transmit data about the status of variables in the field. Information about architectural schemes, wireless technologies, network topologies, wireless standards, security, and power management are presented.

CHAPTER

19

Industrial Detection Sensors and Interfacing

OBJECTIVES

At the conclusion of this chapter, you should be able to:

- Define *industrial detection sensor* and provide several examples of the types of applications for which it is used.
- List the parts of a limit switch, provide examples of the types of functions it performs, and list precautions that should be followed when connecting it to machinery.
- Explain the operation of an inductive proximity detector, describe the function of the sensor circuitry, and provide examples of its applications.
- Explain the operation of a capacitive proximity detector, describe the function of the sensor circuitry, and provide examples of its applications.
- Explain the operation of a Hall-effect sensor and provide examples of its applications.
- Describe the operational theory of the three components that make up a photoelectric sensor.
- Describe the operational theory, characteristics, and application examples of the following photoelectric methods of detection:

Opposed Sensing	Diffuse Sensing	Specular Sensing
Retroreflective Scanning	Convergent Sensing	Color-Mark Sensing

- Properly interface electromechanical relays, solid-state relays, and analog sensor outputs to load devices.
- Define the following terms:

Excess Gain	Current Sinking	Target
Concentrator	Two-Wire System	Contrast
Negative Switching	Three-Wire System	Sensitivity
Positive Switching	Four-Wire System	Field-of-View
Current Sourcing	Hysteresis	Sensor Response Time
Load	Zero	Span

- List the factors that determine the sensing distance from which a target can be detected by the sensors that are described in this chapter.

INTRODUCTION

An industrial detection sensor is a specialized type of measurement device used in an automated system. Its function is to detect the absence, presence, or distance of an object from a reference point. The object to be detected is referred to as the **target.** When the target is detected, the function of the sensor is to send a signal to the load. A **load** is defined as a device to which the output of the sensor is connected. Examples of loads are programmable controller input modules, lamps, relays, buzzers, and other devices.

Applications of detection sensors are as follows:

1. Verifying when a machine part has reached a certain position.
2. Counting gear teeth or measuring the revolution of a shaft to determine rotational speed.
3. Verifying the proper placement of parts during an assembly line procedure.
4. Making edge guide measurements to detect the alignment of a manufacturing process.
5. Counting the number of products that are transferred on a conveyor belt.
6. Determining the size of a product passing an inspection point.

Industrial detection sensors are measurement devices that enable closed-loop equipment to perform process-control and motion-control operations. This chapter explains the operation of several types of industrial detection sensors, describes their characteristics, and provides practical examples of their use in industrial applications.

19-1 Limit Switches

The most fundamental detection sensor is the **limit switch.** By using some actuator type of lever, it converts mechanical motion into electrical signals. When the switch makes physical contact with a moving object, a set of electrical contacts is forced to either open or close. Limit switches are used in applications such as detecting the position of a machine shop carriage, counting parts on conveyor rollers, detecting size for sorting boxes on a conveyor belt, or limiting the travel of an elevator for safety reasons. In these applications, the limit switch causes a change in an operation such as starting, stopping, changing direction, recycling, slowing down, or speeding up.

Limit switches, shown as schematic symbols in Figure 19-1(a), have two main parts: the electrical contacts, usually sealed within an enclosed body, and an actuating mechanism. Three types of limit switches with different actuating mechanisms—a roller lever, a wobble stick, and a push-roller—are shown in Figure 19-1(b). When the mechanism is moved, it causes the electrical contacts to actuate.

The contacts are either normally open (N.O.), normally closed (N.C.), or a combination of both N.O. and N.C. The contacts are designed to open or close quickly so that rebounding and arcing is minimized.

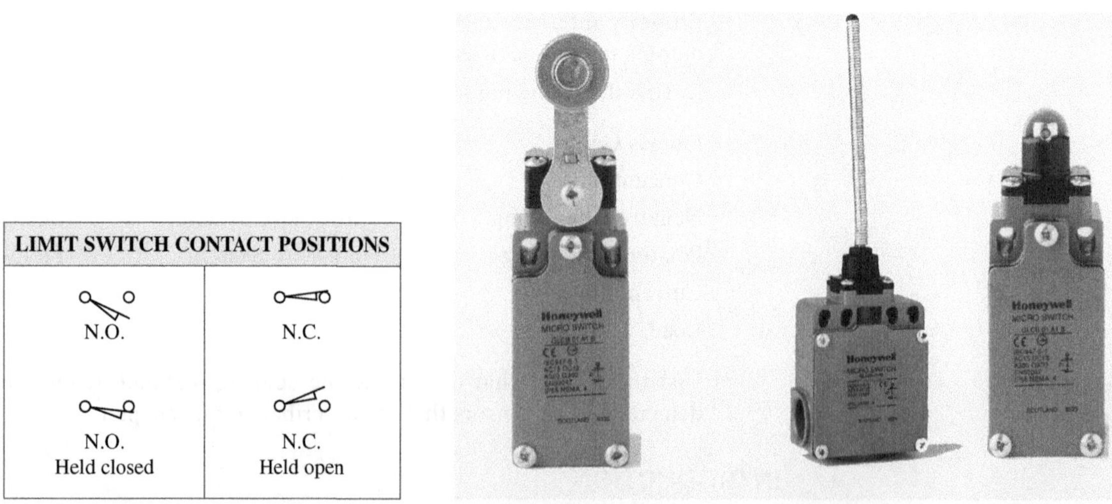

(a) (b)

FIGURE 19-1 Limit switches (Courtesy of Honeywell Micro Switch)

Certain rules should be observed when wiring the limit switch. It is important to make proper connections to multiterminal limit switches. Also, the current rating of the switch should be observed. For example, care should be taken not to exceed the current capacity of the switch. Failure to observe these considerations can damage the switch or alter the operation of equipment, which can cause breakage or injury. Most limit switches used today are connected to programmable logic controllers (PLCs). These switches require special contacts to make them compatible with PLCs that operate at low currents. Care should also be taken not to install the switch in a position where something may interfere with its operation. For example, it should not be placed where scrap material will move the lever or near liquid that flows between open contacts, because this can cause a false continuity condition.

19-2 Proximity Detectors

Proximity detectors are electronic sensors that indicate the presence of an object without making physical contact. The detector normally does not respond by producing a linear output signal proportional to the distance of the object to the sensor. Instead, the output turns on or off. That is why these devices are commonly called *proximity switches.*

The two major types of proximity detectors are inductive and capacitive.

19-3 Inductive Proximity Switches

IAU5707
Inductive Proximity Sensors

IAU5307
Internal Circuits of an Inductive Proximity Sensor

During World War II, limit switches were used in German tanks to detect the movements of parts. The switches worked adequately until the war expanded to the deserts of North Africa. Because sand granules became lodged between the contacts, the limit switches were unable to close. In their attempt to find a device impervious to sand, German engineers developed the inductive proximity detector. The **inductive proximity switch** detects the presence of ferrous or nonferrous metallic materials.

Parts of an Inductive Proximity Switch

Figure 19-2(a) shows a block diagram of the inductive proximity switch. It consists of an oscillator, a sensor head, a demodulator, a trigger, and an output stage.

Oscillator

The oscillator consists of a closed-loop op-amp configuration. It has a tank circuit connected to the noninverting input lead. One branch of the tank circuit contains a capacitor and the other branch an inductor. Once power is applied, resonance develops in the tank, and the oscillator will produce a frequency between 100 Hz and 1 MHz.

Sensor Head

The coil of the tank inductor is wound around a ferrite core. As the oscillating current flows through the inductor, an electromagnetic field continually expands and contracts around the coil. The construction of the core determines the pattern at which magnetic flux lines radiate from the sensor head. Figure 19-2(b) shows an *unshielded sensor,* which directs the flux lines to the front and side of the sensor head. This pattern enables the sensor to detect both lateral (from the side) and axial (directly in front) movements of the target. Because the flux lines are emitted from the side of the unshielded head, the sensor cannot be flush-mounted in metal. For correct operation, a minimum nonmetallic area is required around the sensor. Figure 19-2(c) shows a *shielded sensor* with a shielded metal band around the ferrite core. The shield directs the flux lines to the front of the sensor head. This flux pattern allows the sensor to be flush-mounted in metal without influencing the sensing range. This type of sensor head detects the *axial approach* of an object (an object directly in front of the sensor head).

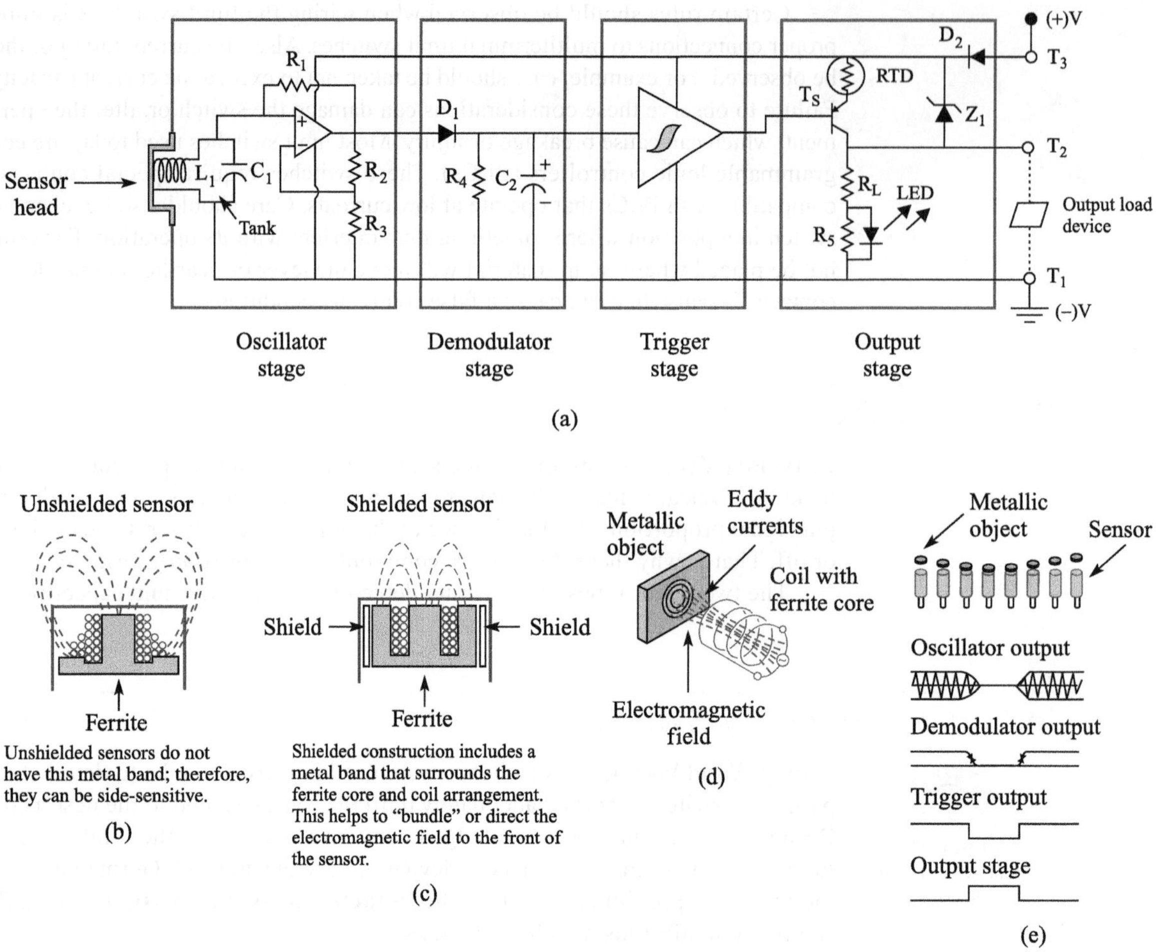

FIGURE 19-2 An inductive proximity switch

TABLE 19-1 Characteristics of Shielded and Unshielded Sensors

Unshielded
• A sensing range of 50% to 100% longer than a shielded sensor of the same size. • The best choice in most applications.
Shielded
• Tolerates metal in the mounting area. • Protected from physical damage by being flush-mounted. • Small targets are more easily detectable with better repeatability because of their narrow sensing zone.

The selection between shielded and unshielded inductive switches is primarily determined by the characteristics listed in Table 19-1.

Guidelines for mounting shielded and unshielded inductive sensors are shown in Figure 19-3.

Demodulator

The oscillating signal is fed into the demodulator section. The demodulator is a filtered rectifier which converts the AC signal into a DC voltage level.

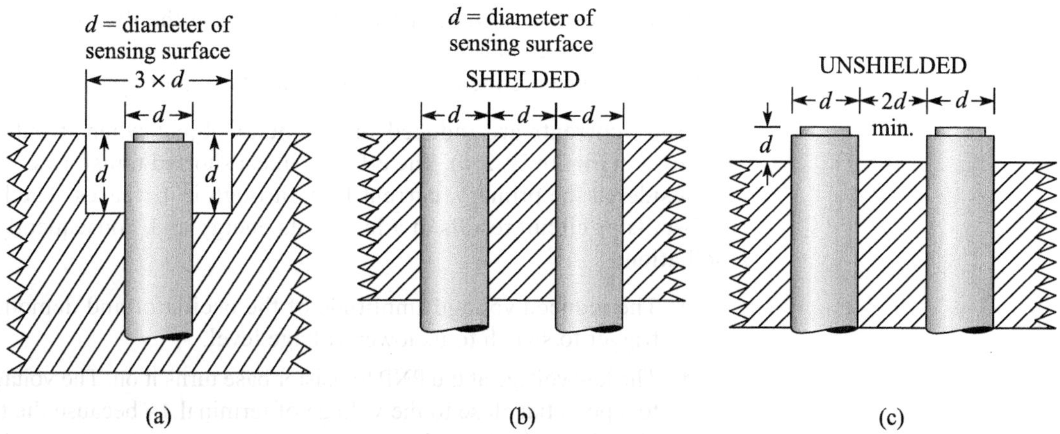

FIGURE 19-3 Guidelines for mounting inductive sensors

Trigger Stage

The output of the demodulator is connected to a Schmitt trigger input. The Schmitt trigger produces two voltages. If the demodulator decreases to a certain amplitude, it causes the trigger output to quickly change to a low-voltage level. When the demodulator increases again to a certain amplitude, the trigger output quickly changes to its higher-voltage level.

Output Stage

The function of the output stage is to drive a load *On* or *Off.* It consists of a PNP current-sourcing transistor circuit. The component functions depicted in Figure 19-2(a) are as follows:

Terminals

T_1	Ground terminal of sensor connected to the power supply Negative output terminal
T_2	Positive output terminal
T_3	Positive terminal of sensor connected to the power supply
T_S	Output-switching transistor
RTD	Overcurrent protection
R_L	Load resistor
Z_1	Zener diode to limit voltage peaks across transistor
D_2	Protection diode to block current flow if the power supply polarity is reversed
LED	Indicator (light-emitting diode), which turns on when the output terminals are energized

The circuitry in this stage is one of several possible configurations that produce a DC voltage to drive a load. The types of DC-operated loads to which they are connected include electromechanical relays, counters, solenoids, and modules that interface with logic systems or programmable controllers. Some inductive switches use other types of circuitry in this stage to produce an AC voltage. The types of AC-operated loads to which they are connected include relays, contactors, and solenoids.

Operation of an Inductive Proximity Switch

If metal is not detected, the following conditions will exist:

- The voltage amplitude of the oscillator and demodulator are at their highest level.
- The Schmitt trigger produces its high-voltage level.

- The positive voltage applied to the base of the PNP transistor turns it off. The open transistor condition causes the load to be de-energized because there is no potential difference between the output terminals T_1 and T_2.

Eddy currents are induced into a metal object when it enters the oscillating field, as shown in Figure 19-2(d). The effect is like a shorted transformer secondary. If the inductor is considered the primary, current flow through it increases. The high current drawn from the oscillator circuit results in a loss of its energy and, consequently, a smaller amplitude of oscillation.

- The reduced voltage amplitude of the oscillator and demodulator cause the Schmitt trigger to switch to its lower-voltage level.
- The low voltage at the PNP transistor base turns it on. The voltage of terminal T_2 is raised to a potential close to the voltage of terminal T_3 because the transistor functions like a closed switch. The voltage difference between output terminals T_1 and T_2 energizes the load. The term *load* refers to the device to which the sensor is connected.

IAU7707
Sensor Hysteresis

When the metal object leaves the sensing area, the oscillator regenerates, allowing the sensor to return to its normal state. The advantage of an inductive sensor is that it ignores nonferrous objects. Figure 19-2(e) pictorially shows the output signals of each stage when the object is at various distances from the sensor.

The proximity switch has a built-in characteristic called **hysteresis** or *differential travel.* This property means that the target must be closer to turn the sensor *on* than to turn it *off.* Figure 19-4 illustrates the advantage of hysteresis. Once an approaching object passes the On-point, the sensor remains on until it moves away to the Off-point. Without the differential gap, the output could potentially "chatter" on and off if a target was stationary at the fringe-sensing distance. The Schmitt trigger provides hysteresis. It switches high when the input increases to one threshold voltage level (Vth+), then switches low when the input reduces to a lower threshold (Vth–) than Vth+.

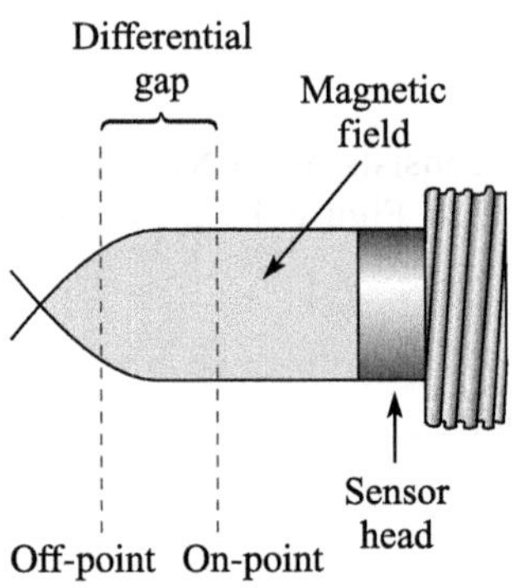

FIGURE 19-4 The hysteresis property of a proximity switch

Some inductive proximity sensors do not produce discrete on and off signals. Instead, they are designed to generate a 4- to 20-mA analog signal that is proportional to the distance of the target from the sensor. A 4-mA signal is produced when the target is at the minimum range, and a 20-mA signal when it is at the maximum range. These inductive sensors do not have a Schmitt trigger in their internal circuitry.

Guidelines for the Target

The ability of an inductive sensor to detect a target is determined by a number of factors:

1. The dimension of the sensor.
2. The dimension of the target and its shape.
3. The conductivity and permeability of the target, determined by the type of material of which it is made.
4. The distance and position of the target in front of the sensor head.

IAU5507
Inductive Proximity Sensor Target Considerations

Dimension of the Sensor

The range at which the sensor can detect a target is affected by the diameter of the sensor's coil. A sensor with a 30-mm diameter has a sensing distance that is more than that of a sensor with a 12-mm diameter. For example, the sensing distance of a shielded sensor with a 12-mm diameter is 2 mm, whereas the distance of a shielded sensor with a 30-mm diameter is 10 mm.

Dimension and Shape of the Target

When rating the distance at which a sensor detects an object, manufacturers base their data on a *standard target.* A standard target is a square piece of mild steel, 1 mm thick, with the length of each side equal to the diameter of the sensor head. Increasing the size of the target much beyond the sensor's diameter will not alter the sensing range, because the sensing field projection is limited to the area immediately in front of the sensor. However, decreasing the

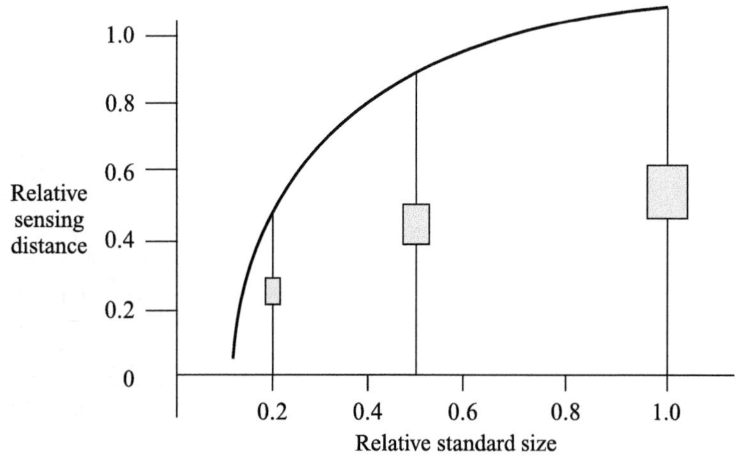

FIGURE 19-5 How target size affects sensor distance

size of the target, so that it is smaller than the sensor head, will alter the sensing range. This relationship is shown in Figure 19-5. The nonlinear curve shows that reducing the size by 80 percent decreases the sensing range by 50 percent.

The thickness of a standard calibration plate made of iron is 1 mm. Because the magnetic flux flows so easily through iron, due to its high permeability, significant eddy currents will not be established in the depths of the ferrous plate. Therefore, its sensing distance will be maintained regardless of thickness. However, other metals such as aluminum and brass become easier to detect if their thickness is below 1 mm. At the thickness of foil, they become as easy to detect as iron due to a phenomenon called *skin effect*. This term is derived from the way in which eddy currents are induced into the metal. These currents penetrate the surface by less than 1 mm. If the thickness of the metal is greater than 1 mm, additional layers of eddy currents form. Each eddy current cancels some of the current in the adjacent layers because the directions of current in alternate layers are opposite. The result is that a cumulative reduction of eddy currents flows in the target, thus drawing less power from the sensor's oscillator. This condition causes the sensing distance of a thick, nonferrous target to become smaller than the sensing distance of a foil target.

Types of Material

An inductive sensor can detect all materials that are good conductors. However, each type of metal has a different permeability rating, which affects how easily eddy currents can form. The ease at which these eddy currents flow determines how well the metal can be detected. Objects made from iron (iron is referred to as a *ferrous metal*) are the easiest targets to detect. To calculate the sensing distance of various nonferrous metals (those not containing iron) by a general-purpose inductive sensor, multiply the standard sensing distance of graphite, which is used as a reference, by a reduction factor. Typically, this value is 0.8 for stainless steel, 0.5 for brass, 0.4 for aluminum, and 0.3 for copper. The graph in Figure 19-6 shows the conductivity factor of various metals.

In addition to the general-purpose inductor, there are two special inductive sensors available, known as "nonferrous sensors" and "all-metal sensors." Nonferrous sensors will detect metals such as aluminum better than they sense iron. All-metal sensors will detect all metals, ferrous or nonferrous, at the same distance. The number of coils in the sensor head determines how the three types of inductive sensors detect the various types of materials. Nonferrous and all-metal types contain two or three separate coils, while the general-purpose sensor has only one. Consequently, nonferrous and all-metal sensors are larger and more expensive than their general-purpose counterparts.

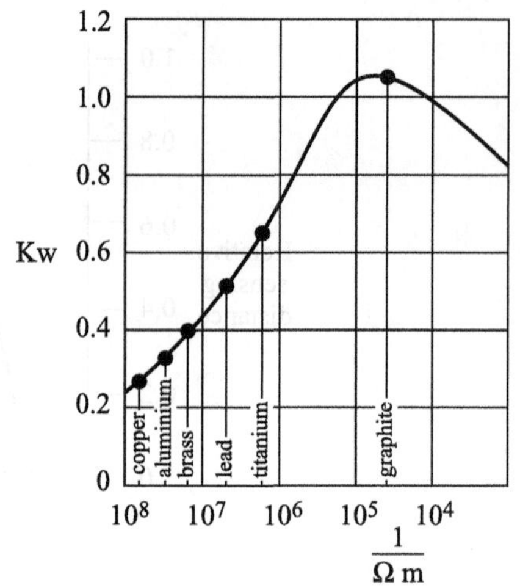

FIGURE 19-6 Conductivity factor of different materials

Distance and Position

The effective area in which an inductive proximity device senses a metal target is illustrated by two types of graphs, as shown in Figure 19-7. Figure 19-7(a) shows a sensing range graph for an axial approach (straight on) of a target, and Figure 19-7(b) shows a graph for the lateral approach (from the side) of a target. These diagrams are usually included in manufacturers' catalogs to show the relationship between switching action and range as a target enters the sensing zone in a lateral or axial direction. An understanding of these diagrams is very useful in predicting the distance at which proximity sensors switch in an actual application.

The switch-on function of a proximity device will take place only if it can induce eddy currents into the target. For this action to occur, the object must be within the range of its magnetic field. The rated maximum sensing range provided in most catalogs is listed as

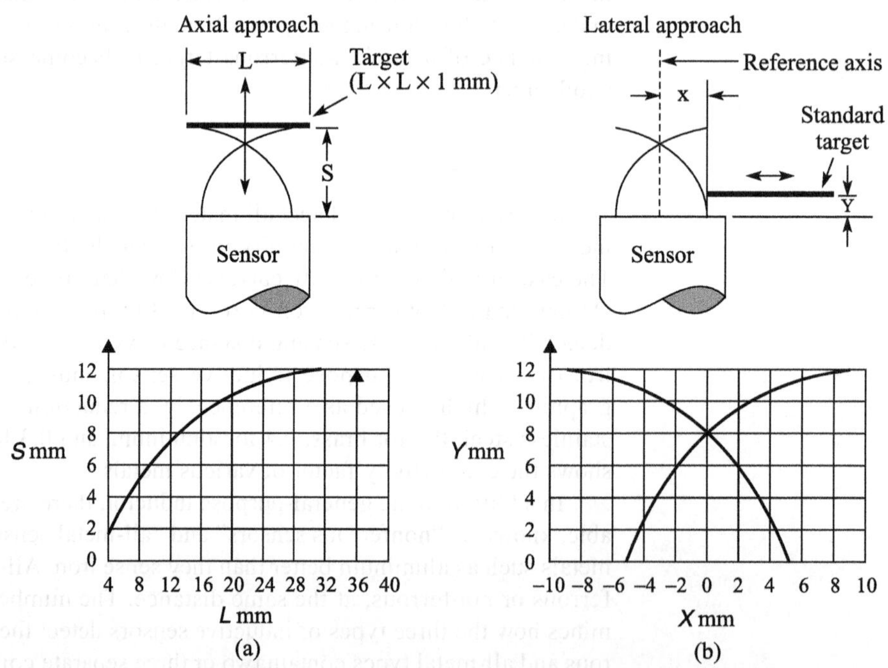

FIGURE 19-7 Switching action and range of an inductive proximity switch as a target enters the sensing zone

the *nominal sensing distance.* This information is identified on the sensing range diagram. The nominal sensing distance is at the flat, horizontal part of the curve. The recommended space between the target and the sensing head is half the nominal sensing distance.

When using an inductive sensor in repetitive applications, such as counting the teeth of a gear in Figure 19-8, a possibility exists that the switch will not have enough time to turn off because as the tooth leaves the sensing zone, the next tooth is about to enter the zone. To avoid this situation, the following guidelines should be followed:

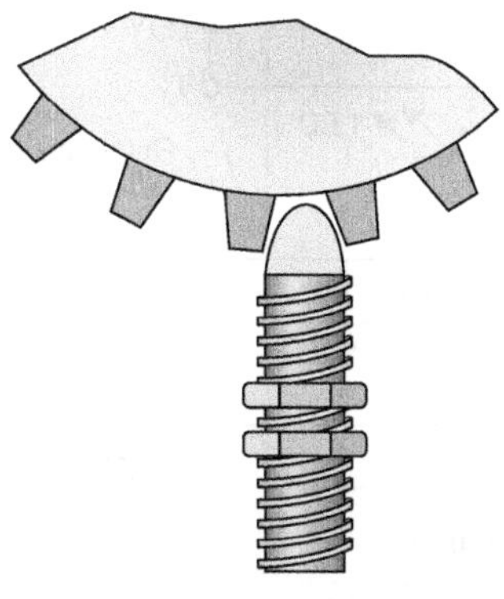

FIGURE 19-8 Repetitive sensing

1. Do not place the sensor head too close to the target. A recommended space between the sensor and the target is half the *nominal sensing distance.*
2. The diameter of the sensor head should not be larger than the width of the target. Sensors that are shielded should be considered because they have narrower magnetic fields.
3. Sensors that are smaller in size should be used for two reasons. Those that have smaller diameters have a narrower magnetic field. Also, the switching frequency of an inductive sensor is inversely proportional to size. The bigger it is, the longer it takes to lose energy when it turns off. Therefore, it is slower.
4. The width of the teeth should be at least half the diameter of the space between them.
5. Use DC switches because they operate more rapidly than AC switches. The maximum frequency of a DC switch is 5000 Hz.

IAU7307
Maximum Proximity
Sensor Switching
Speed

Applications

Inductive proximity limit switches are particularly well suited for operating conditions in which:

- Fast switching detection without contact bounce is required.
- Excess dirt, moisture, oil, shock, and vibration are present.
- Physical contact between the target and the sensor may cause damage (e.g., painted parts).
- Detection occurs at cold ambient temperatures.
- Metal objects are to be differentiated from nonmetal objects.
- Metal objects are to be detected through a nonmetal surface.

Typical applications include machine tools, production machinery, automated assembly operations, and material-handling equipment. Specific operations are counting, sorting, positioning, limiting, locating, inspection, starting, and stopping functions.

Analog Inductive Sensor

The operating principle of an analog inductive sensor is the same as that of the inductive switch. The only difference is that the analog version does not have the trigger circuit. Consequently, the output it produces is a variable analog signal. The amplitude of the signal is inversely proportional to the distance between the sensor and the target.

19-4 Capacitive Proximity Switches

IAU5607
Capacitive Proximity
Sensors

Capacitive proximity sensors exhibit most of the same operating characteristics as inductive sensors, with one major difference. Inductive devices are only capable of detecting metal objects. Capacitive switches can detect the presence of both metallic and nonmetallic targets. This sensor is activated when an object enters the electrostatic field of a capacitor.

Figure 19-9(a) shows a block diagram of its construction. Its design is very similar to the inductive counterpart just described. The major difference is the oscillator used. The sensing element C_1 is placed in the feedback loop of the oscillator. Its electrodes are concentrically positioned, as shown in Figure 19-9(b). The other significant difference is that an NPN current transistor is used in the output stage.

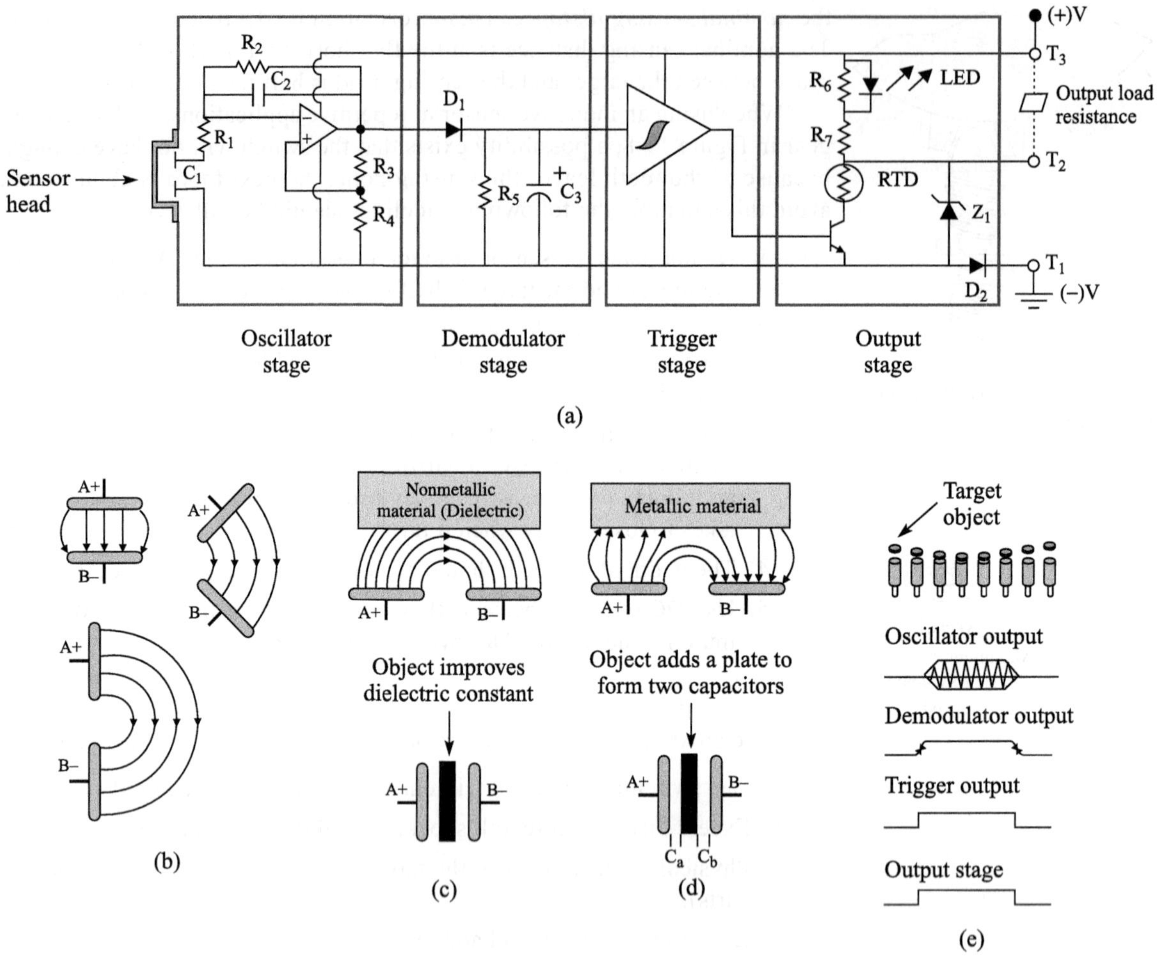

FIGURE 19-9 Capacitive proximity switch

Figure 19-9(c) shows a nonmetallic object in the field of the capacitor. The capacitance increases because the object's dielectric constant is greater than air. The value to which it increases depends on the dielectric constant of the material that is sensed. Many solids and liquids cause the capacitance to change.

Figure 19-9(d) shows a metallic material within the field. The object becomes another electrode and forms two capacitors. The capacitance increases because the effective distance between capacitor plates decreases.

Operation of a Capacitive Proximity Switch

IAU5207
Internal Circuits of a Capacitive Proximity Sensor

When an object is not present, the oscillator is inactive. The outputs of the demodulator and trigger are both 0 volts. When a 0-volt potential is present at the base of the transistor, there is no bias voltage across the emitter and base. The transistor is in the cutoff condition and causes the supply potential to drop across terminals T_1 and T_2. Since no voltage is developed across terminals T_2 and T_3, the detector's output is de-energized.

As an object approaches, the oscillator begins to oscillate. The closer the object gets to the electrode plates, the larger in magnitude the oscillations become. This action causes the Schmitt trigger to switch to its higher-voltage level. The positive potential forward-biases the base–emitter junction of the NPN transistor, causing it to turn on. The result is that the detector output energizes because most of the supply voltage drops across terminals T_2 and T_3. Figure 19-9(e) pictorially shows the output signals of each stage when the object is at various distances from the sensor.

IAU5407
Capacitive Proximity Sensor Target Considerations

The ability of a capacitive sensor to detect a target is determined by a number of factors:

1. The dimensions of the target and its shape.
2. The distance and position of the target in front of the sensor head.
3. The dielectric constant of the object.

Dimension and Shape

Ideally, the diameter of the sensor head and the object (target) should be the same. If the target is larger, the sensor will not produce a significant increase in the signal strength. A target that is smaller is more difficult to detect. When the target is made of a nonconductive material, the range at which it is detected is proportional to size. If the target is conductive, the sensing range is independent of the material's size and thickness.

The maximum amount of an object's material is present between the capacitor's plates if its sides are flat and it is rectangular or square. The sensing range is proportional to its size.

Distance and Position

The switching behavior of a capacitive sensor relative to distance is similar, but opposite to that of an inductive sensor. The difference is that the oscillator is on whenever the object is within the sensing range, and turns off when the distance is exceeded. The capacitive sensor also includes the hysteresis function. The On-point of the sensor is closer than the Off-point.

Dielectric Constant

The capacitive sensor detects not only metallic material, but also nonmetallic materials provided their dielectric constant is sufficiently high. Nonmetallic materials include plastic, glass, ceramics, oil, water, and all materials with a high moisture content, such as wood, paper, and foodstuffs. Figure 19-10 is a graph that shows relative dielectric constants of various materials. Materials with higher dielectric constant values are easier to sense than those with lower values. Air, which has the lowest dielectric constant of all materials, is used as a reference. Its rated dielectric constant value is 1. Water, which is very easy to detect, has a dielectric constant of 80. Other nonmetallic materials have dielectric constants that fall within this range. Metallic materials are not shown because they do not have a dielectric constant rating. They can affect the capacitance by decreasing the effective distance between the plates.

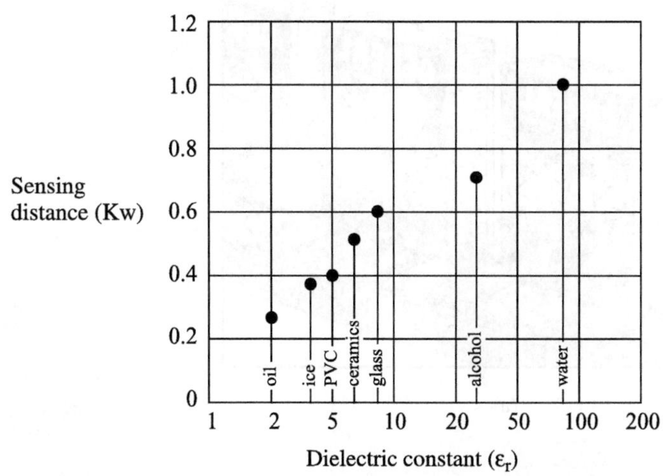

FIGURE 19-10 Relative dielectric constant of different materials

Capacitive sensors cannot tolerate a film of moisture (or condensate) on the sensor head surface. Because of the high dielectric constant of substances that contain water, the sensor operations will be adversely affected. For the same reason, a thick layer of dust should not be allowed to collect on the sensor body.

Applications

IAU5807
Sensitivity Adjustments of a Capacitive Sensor

In addition to sensing the presence of solids and liquids, capacitive switches can also detect powdered or granular materials. Most capacitor proximity switches also have a sensitivity control that is adjusted by a potentiometer. The sensitivity control provides two functions. This control can:

1. Change the range at which the target can be detected.
2. Enable targets with high dielectric constants to be sensed through the walls of containers made of materials that have lower dielectric constants. The difference in their dielectric constants makes it possible for a capacitive proximity sensor to detect the level of liquid inside a glass container. With the sensitivity properly set, the sensor turns on when it is placed below the water level, but not above it. If the sensor adjustment setting is too sensitive, it will detect the container that has a lower dielectric constant, and therefore it will not be able to distinguish between the glass and the water on the other side. The "see-through" capability enables a sensor to also detect cereal or other ingredients inside a box, or granular materials inside a hopper.

Capacitive sensors cannot detect materials through metal. Therefore, if the target is inside a metal container, the capacitive proximity switch can be used if a hole is made through the container.

Most sensing applications of a capacitive switch are to detect the high and low levels of materials in a tank. However, other applications include detecting the breakage of a textile web, sensing the presence of glass moving along a conveyor, counting boxes that pass by on a conveyor belt, or sensing metal at a much longer distance than would be possible with inductive switches.

Figure 19-11(a) shows a practical application of a capacitive proximity detector. Figure 19-11(b) shows a practical application of an inductive proximity detector. Table 19-2 compares capacitive and inductive proximity detectors.

Making sure the product is in the package

Detection of seal foil beneath cap

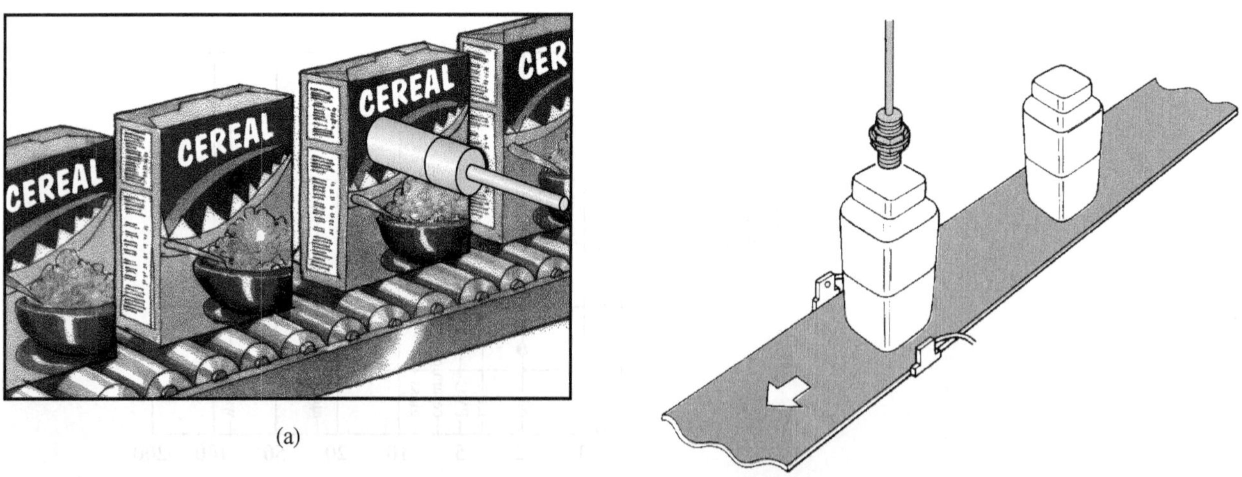

FIGURE 19-11 Proximity detector applications (Courtesy of efector, inc.)

TABLE 19-2 Proximity Detectors

Method of Detection	***Inductive Shielded***	***Inductive Unshielded***	***Capacitive***
Configuration	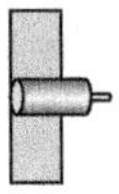 Designed for flush-mounting in metal. Available in cylindrical and limit switch sensor shapes.	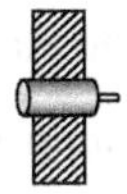 Requires clearance around sensing end to prevent false signals from surrounding metal. Available in cylindrical, limit switch, and small block, flat rectangular sensor shapes.	 Requires clearance around sensing end to prevent false signals from surrounding mounting materials. Available in cylindrical and flat rectangular sensor shapes.
Advantages	Detects ferrous (iron, mild steel, stainless steel) and nonferrous (brass, copper, aluminum) metals. Allows flush-mounting to prevent impact damage to the sensor. Color and surface conditions of the target do not affect sensing. Most cost-effective option where appropriate.	Detects ferrous (iron, mild steel, stainless steel) and nonferrous (brass, copper, aluminum) metals. Longer sensing distance than shielded sensors. Color and surface conditions of the target do not affect sensing.	Detects plastic, glass, liquids, leather, and wood, as well as metals. Can be used to detect materials inside nonmetallic containers.
Disadvantages	Reduced sensing distance. Usable only to 0.4 inch maximum or higher with rectangular switches.	Sensor is not protected from accidental impact damage. Usable only to 0.7 inch maximum or higher with rectangular switches.	Sensor is not protected from accidental impact damage. Usable to 0.9 inch maximum. Dirt, moisture may affect the operation.

19-5 Hall-Effect Sensor

IAU7407
Hall-Effect Sensors

A **Hall-effect** device is a sensor that detects the presence of a magnetic field. This sensor is a flat rectangular piece of P-type semiconductor (called a *Hall generator*) usually made of indium arsenide (InAs). Hall-effect sensors are four-terminal devices. The two end terminals are power supply connections. The terminals located on either side are the output leads.

When a power source supplies a constant current to the sensor, positively charged carriers flow uniformly through the material when a magnetic field is not present. In this condition, the voltage at the output is zero. Figure 19-12 shows what happens when the semiconductor is subjected to a perpendicular magnetic field from an electromagnet or permanent magnet. The charged carriers are deflected to one side of the flat piece by an effect called the *Lorentz force.*

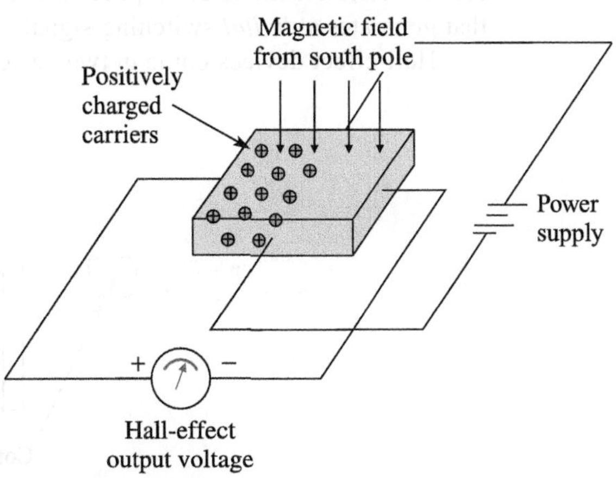

FIGURE 19-12 A Hall voltage produced by the Lorentz force

The result is that an EMF (called *Hall voltage*) develops across the output terminals. The side to which the charged carriers move becomes positive in respect to the other side. If the power supply or the magnetic field direction is reversed, the output of the Hall-effect device will be opposite in polarity.

Three variables determine the Hall voltage amplitude produced by the Hall generator:

1. The amount of current supplied by the power source.
2. The physical size of the semiconductor material. The voltage is directly proportional to the thickness and cross-sectional area to which the flux lines intersect at a perpendicular angle.
3. The magnetic field strength and orientation to which the Hall device is subjected.

Since the current is held constant and the physical size does not change, the magnetic field strength determines the voltage amplitude. As the magnet is brought near, the flux density rises, as shown in Figure 19-13(a), and the Hall voltage increases proportionally. This is known as the *Hall effect*, discovered by Edward H. Hall in 1879. Another factor that influences the voltage amplitude is the angle at which the flux lines pass through the Hall device. Figure 19-13(b) shows that the maximum output voltage is achieved when the lines are perpendicular to the sensor. At other angles, the output follows a cosine function.

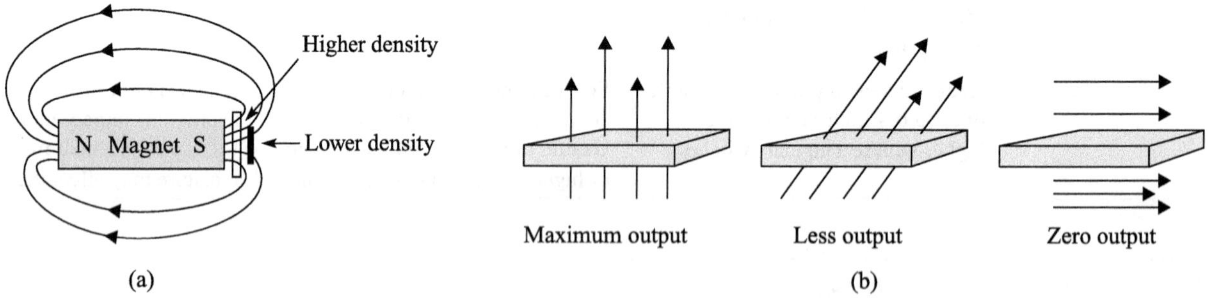

FIGURE 19-13 (a) As the distance from a pole increases, the flux density decreases; (b) Maximum output is achieved when the lines are perpendicular to the sensor

The output voltage produced by the Hall generator is not great enough to drive an output device to which it is connected. There are two methods commonly used to boost the signal. One method employs pieces of iron or other ferrous materials that are placed on either side of the Hall plate, as shown in Figure 19-14. These pieces, called *concentrators,* bend the local flux patterns so that a greater number of lines pass through the sensor. The greater concentration causes the Hall generator to produce a greater output signal in response to the flux lines. The other method of increasing the generator's output is to use an amplifier. There are two classifications of amplifiers used, one that produces a *linear* output, and the other that generates a *digital* switching signal.

Hall-effect devices come in two basic functional classifications: linear and digital.

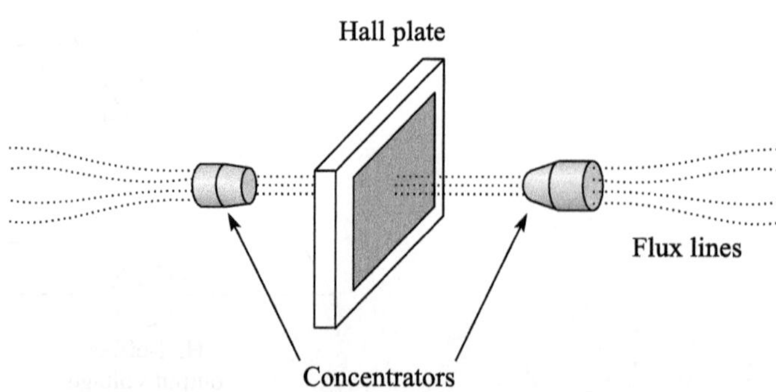

FIGURE 19-14 Concentrators used to boost the signal by directing a greater number of the flux lines through the device

Linear Hall-Effect Sensors

Figure 19-15(a) shows a diagram of a linear Hall-effect device. It consists of a Hall generator, linear amplifier, and emitter-follower output transistor. All three sections are fabricated onto an integrated circuit chip.

A power supply with good voltage regulation supplies a constant current flow. When flux lines intersect the semiconductor material, it produces an analog voltage that is directly proportional to the strength of the magnetic field. A differential op amp is used to boost the Hall generator voltage above the millivolt level. Its output controls the input bias to the output transistor. Note that emitter followers have no voltage gain. They are used to produce a current gain sufficient to drive an output indicator.

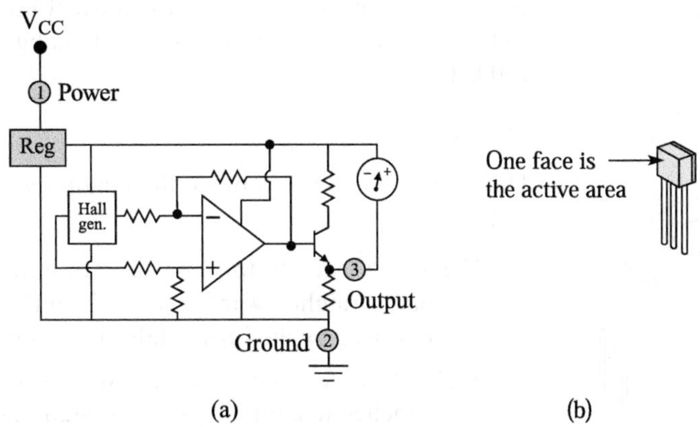

FIGURE 19-15 A linear Hall-effect device

Figure 19-15(b) shows the package style of a Hall-effect sensor. Typically, three leads are used because the negative leads of the power supply and output are connected. One face of the package is the Hall sensor's active area. To operate, the magnetic flux lines for a prescribed pole must be perpendicular to the face of the package. If the polarity of the magnet is reversed, the output voltage of the sensor will be opposite.

Hall-effect linear sensors are used to sense the motion, position, or change in field strength of a magnetized device. Applications include counting each tooth of a gear as it rotates, measuring the magnetic strength of a motor coil or a solenoid, or detecting level, pressure, and thickness.

Digital Hall-Effect Sensor

The linear Hall-effect sensor can be modified to make it compatible with digital circuitry. Figure 19-16 shows the output of the differential op amp. It is connected to a Schmitt trigger, which feeds a signal into an open-collector NPN output-switching transistor.

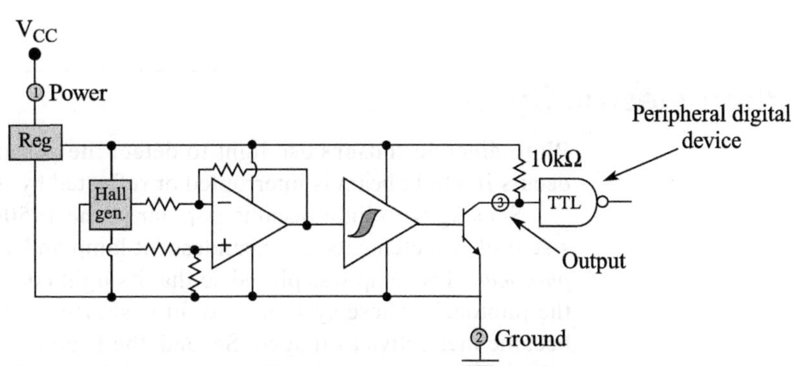

FIGURE 19-16 A schematic diagram of a digital Hall-effect device

As the magnetic south pole approaches, the Hall cell is exposed to increasing magnetic flux density. The Hall generator and the op amp outputs increase. At some point, the threshold input of the Schmitt trigger is reached. Its output switches from low- to high-level voltage, which turns the switching transistor on. The output (terminal 3) goes to zero. The level at which the flux density's strength causes the output to turn on is called the *operating point.*

If the magnet is moved away from the Hall cell, the flux density decreases, and the Hall generator and op amp voltages decrease. Because of the Schmitt trigger hysteresis, the Hall output will not return to a low level unless the magnetic flux density falls to a value far lower than the operating point. When it does, the switching transistor turns off and the output voltage equals the power supply. The value at which the flux density's strength causes the output to turn off is called the *release point.* The hysteresis property ensures that even if mechanical vibration or electrical noise is present, the switch output is fast, clean, and occurs only once per threshold crossing. The switch speed of the Hall-effect device is 100 kHz.

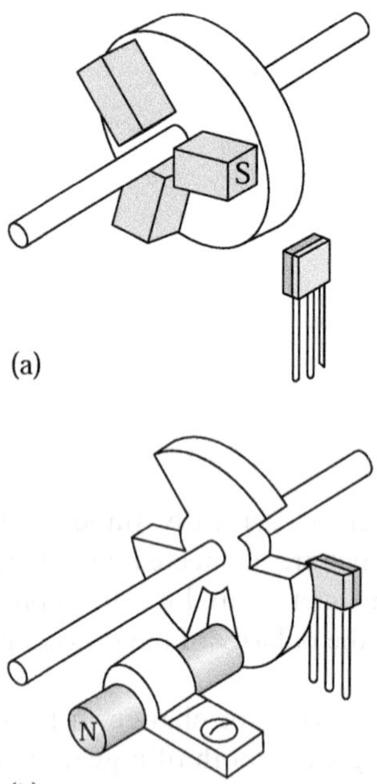

FIGURE 19-17 Magnetic rotor assemblies

IAU7607
Hall-Effect Sensor Application

Modes of Operation

There are three ways to move the magnet within the sensing range of the Hall-effect generator:

Head-on Mode. In this configuration, the magnetic pole moves along a perpendicular path straight toward or away from the face of the Hall device. This method is primarily used by the linear Hall-effect sensor.

Slide-by Mode. The configuration is shown in Figure 19-17(a). A disk with magnets is attached to a rotating shaft. A stationary Hall device is positioned so that it becomes activated by each lateral pass of a magnetic south pole. This method is primarily used by the digital Hall-effect sensor.

Stationary Mode. This configuration is shown in Figure 19-17(b). A magnet and Hall sensor are mounted at stationary positions with a small air gap between them. When the flux lines are uninterrupted, the Hall output is held *on* by the activating magnet. As the shaft spins, the vane will pass between the magnet and Hall-effect device. When it does, the Hall generator turns off because the vane will form a magnetic shunt that distorts the flux away from the sensor.

Applications

Hall-effect sensors are well suited for harsh environmental conditions. Because they are immune to dirt, they can be more effective than capacitive switches, optoelectronic sensors, and electromechanical limit switches. They are also capable of switching at higher frequencies than inductive or capacitive detectors because the response time of the Lorentz force-effect is shorter than the time it takes the oscillator to build or dampen the magnetic field. A frequent application of Hall-effect sensors is to generate a digital output indicating the velocity, displacement, or position of a rotating shaft.

19-6 Photoelectric Sensors

Photoelectric sensors use light to detect the absence or presence of an object. Detection occurs if a light beam is interrupted or reflected by the object being sensed.

Sensing with light became popular in the 1950s. The early photoelectric systems consisted of two elements: an incandescent lamp and a light-sensitive resistive device called a *photocell.* The lamp was placed so that its light could be projected across the sensing area to the photocell. These systems had three shortcomings. First, the bulb lost its intensity and became ineffective as it aged. Second, the filaments would break if exposed to temperature extremes or high vibration. Third, in order for the photocell to differentiate between the beam of the lamp and ambient light, both elements had to be carefully aligned and positioned within a limited distance. All of these problems were eliminated by the development of the semiconductor sensing devices presently used.

The advantages photosensors provide is that no physical contact with the target is required, and the object can be detected at varying distances ranging from one inch to several hundred feet. Photoelectronic sensors may be divided into three components: the light source, the light sensor, and the sensor circuitry.

Light Source

The **light source** supplies the light beam, which is transmitted to the light sensor. The light source is also referred to as an *emitter* or *transmitter.* Light-emitting diodes (LEDs) are used most frequently as light sources.

The LED, shown in Figure 19-18, is a PN junction semiconductor diode that emits light. When it is forward-biased, electrons in the N-type material enter the P-type material, where they combine with excess holes. As a hole and an electron combine, a packet of energy called a *photon* is generated, which escapes the surface as light radiation. The type of semiconductor material used determines the wavelength or color of the light. For example, gallium phosphide produces green; gallium arsenide phosphide produces red. Yellow and blue LEDs are also available. Semiconductors made of gallium arsenide produce infrared light, which is invisible to the human eye. Infrared LEDs produce approximately ten times as much light energy as visible red LEDs. Figure 19-19(a) shows a light spectrum chart that indicates the wavelength of each type of light generated by LEDs. The table in Figure 19-19(b) shows the uses for the various LEDs.

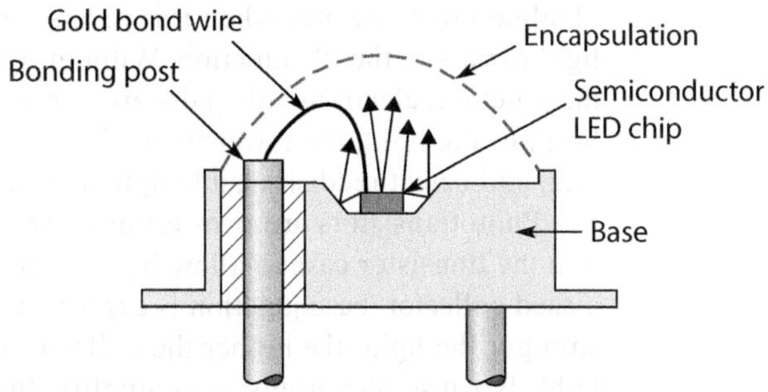

FIGURE 19-18 LED Light-emitting diode

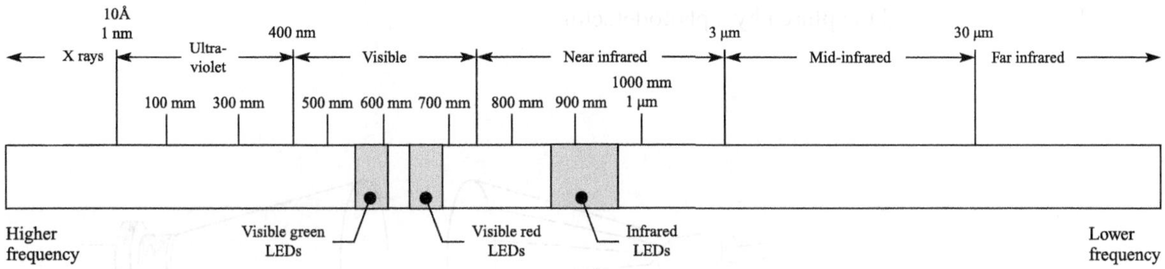

(a)

Color	Spectrum range	Use
Infrared	890–950 nm	Long distance, dusty or poor environment
Red	660–700 nm	General applications
Green	560–565 nm	Color mark detection

(b)

FIGURE 19-19 (a) The light spectrum; (b) Uses for various LEDs

Because LEDs are solid-state, they will last nearly forever. Their light intensity deteriorates very slowly, and they are immune to vibration and shock. LEDs produce only one wavelength, which is a distinct advantage in some applications, and turn on immediately without a warm-up time. The one disadvantage of an LED is that the light intensity it produces is only 1 percent of the light generated by an incandescent bulb of the same physical size. Its light, however, can be intensified by concentrating the waves into a fairly narrow angle through a lens.

Both visible and infrared LEDS are used as the light source for photoelectric transmitters. Visible LEDs can be monitored visually during their operation, are easy to align, and are used in applications where specific colors or contrasts must be detected. Infrared LEDs produce a stronger light than visible LEDs. This characteristic gives them the capability of penetrating certain objects, such as cardboard cereal boxes, in much the same way in which light from a flashlight shines through a hand. Also, photodiodes and phototransistors used in the sensing component respond only to the infrared wavelength and not to visible light waves. They are also popular for security applications because infrared light is invisible.

Light Sensor

The **light sensor** that detects the absence or presence of an object is called either the *detector* or the *receiver*. The most common types of light-sensitive components used for detection are photodiodes and phototransistors. A photodiode is a two-terminal junction device made of silicon that operates when it is reverse-biased. A lens is placed on the main body to allow light to pass to the PN junction. Without light, the resistance is very high. When light falls on the junction, electron–hole pairs are generated and the resistance of the diode decreases. The brighter the light, the lower the resistance. Photodiodes are linear, operate at high frequencies, and detect reddish visible light and infrared light waves.

Phototransistors are two-terminal, two-junction devices made of silicon. A lens is built into the transistor case to allow light to strike the collector–base junction. When the reverse-biased collector–base junction is exposed to light, the photons cause base current to flow. The stronger the light, the higher the collector current. Phototransistors respond well to infrared light. Because phototransistors amplify, they have greater light sensitivity than photodiodes but operate at much lower speeds, are nonlinear, and have large temperature coefficients. Focusing the waves through a lens, as shown in Figure 19-20, can maximize the amount of light captured by a photodetector.

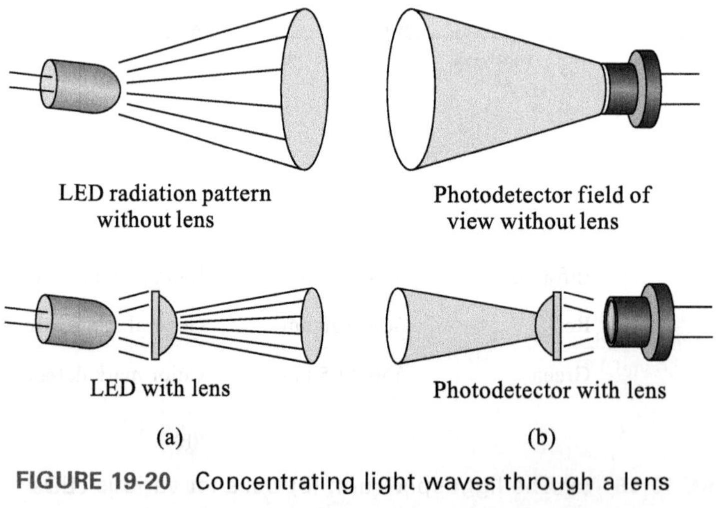

FIGURE 19-20 Concentrating light waves through a lens

The Sensor Circuitry

Figure 19-21(a) shows the block diagram of the emitter and detector for the photoelectric **sensor circuitry**.

Emitter Block

The amount of light generated by the LED of an emitter is determined by the amount of current it is conducting. To increase the distance the LED emits light, the amount of current must be increased. However, excessive current levels will cause the LED to become destroyed by generating too much heat. A technique called *modulation* provides a way to prevent the LED from overheating when conducting current at elevated levels. Modulating the LED simply means turning it on and off at a very high frequency. If the on-time of the power is 8 μsec compared to the off-time of 108 μsec, only 29 μA of a supplied 400 μA will be consumed. By tuning the amplifier in the receiver to the frequency of the modulation, only that frequency is amplified and others are rejected. (Receiving unwanted signals is called *cross talk*.) This principle of operation is similar to the way in which a radio receiver tunes to one station while rejecting all the other radio waves. In Figure 19-21(a), the oscillator circuit modulates the LED by turning it on and off at a high frequency. The frequency of the modulated wave ranges from 5 to 40 kHz, depending on the application. These frequencies are much higher than can be detected by the human eye. Using an operation called *synchronous detection* can maximize the modulation technique. In this method, the emitter

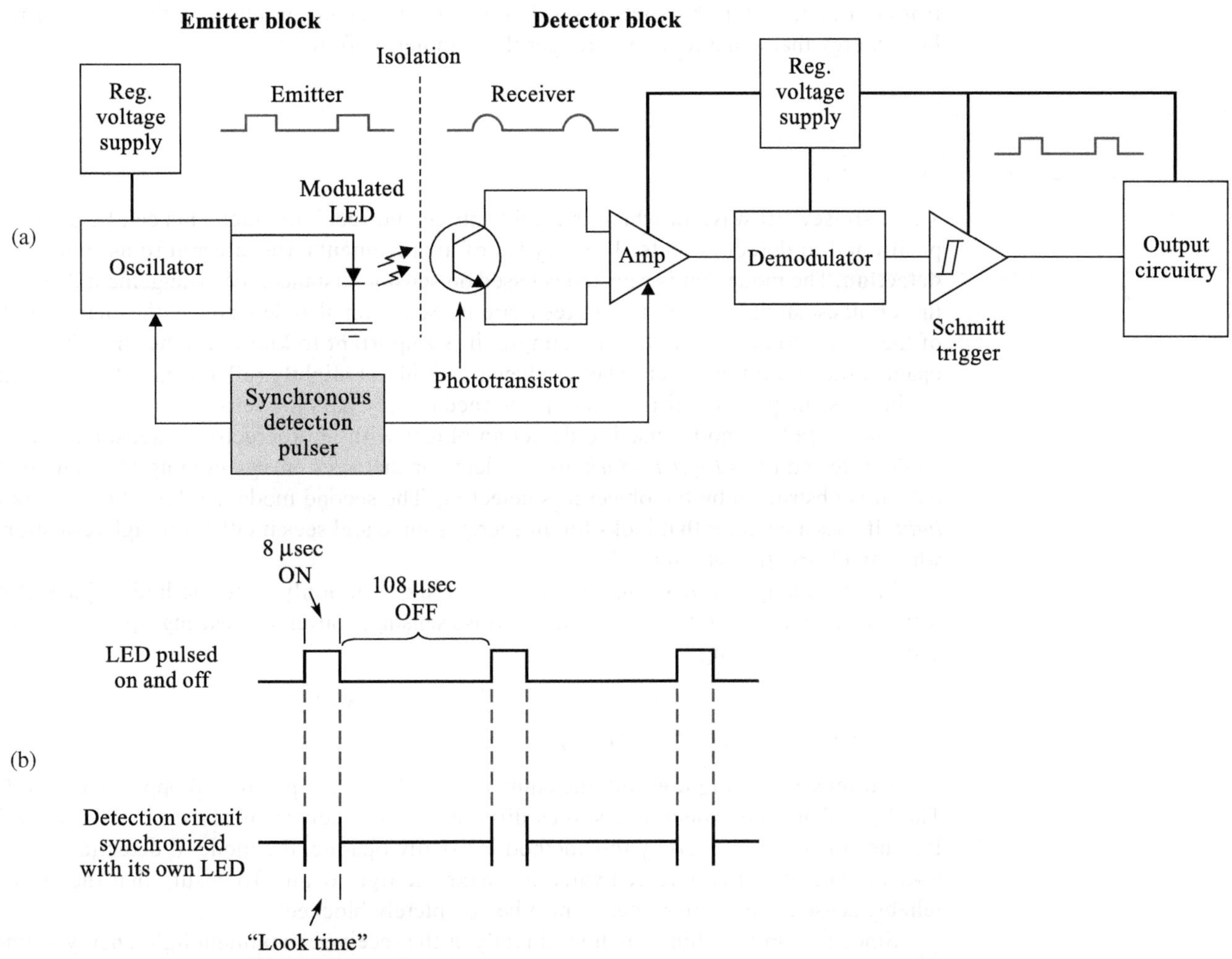

FIGURE 19-21 (a) A block diagram of a light sensor; (b) Synchronizing the emitter and detector

and detector are synchronized so that the receiver is activated to detect the signal from its own emitter only when the LED is turned on. The waveform in Figure 19-21(b) shows how the detector searches for its own LED when it is pulsed. Modulation allows the amount of current, and therefore the amount of emitted light, to far exceed what would be allowed if the LED were on continuously.

Detector Block

The detector portion of the sensor in Figure 19-21(a) shows that a phototransistor is used to detect the light. As the light wave strikes the transistor base, the signal produced is applied to the input of a high-gain amplifier tuned to the modulated frequency. The output of the amplifier is demodulated by a filtered rectifier. The demodulation function reduces the problem of critical alignment, and it allows the sensor to be used in areas where ambient light levels are relatively high or when dirt, oil, or smoke obscure the lens. A Schmitt trigger is used to increase the switching speed for very high-speed applications, for hysteresis, for noise immunity, and for logic level outputs.

The receiving sensor unit is available with or without the demodulator. A sensor without the demodulator has a gain that is limited to the point at which the receiver recognizes ambient light. Therefore, it requires critical alignment because of the long-focal-point lens it uses. In contrast, a demodulated receiver ignores ambient light and responds only to the modulated light source. As a result, the gain of the amplifier may be turned up to a very high level. The high-gain operation reduces the problem of critical alignment and enables the sensor to detect light effectively even when its lens is obscured by dirt, oil, or smoke. Some non-demodulated sensors are also referred to as *ambient receivers*. They are used to detect objects that emit their own light, such as red-hot metals, hot glass, or anything that emits infrared light energy that is many times stronger than ambient infrared light.

19-7 Methods of Detection

There are several ways in which the light source and receiving elements can be physically positioned to detect objects. These types of arrangements are referred to as **methods of detection.** The mode that is selected is based on sensing distance, the arrangement that yields the strongest signal, and mounting restrictions. Selection also depends on the characteristics of the object to be detected. For example, it is important to know whether the objects are opaque, translucent, or clear; whether they are highly or slightly reflective; and whether they are in the same position or randomly positioned as they pass the sensor.

There are two modes used to detect an object with photoelectronic sensors. The first mode, referred to as *light-to-dark,* uses a detector that sees energy coming from an emitter until it is obstructed by the object it is detecting. The second mode is referred to as *dark-to-light.* It uses a detector that looks for an energy source and sees it either through reflection or when an obstruction is removed.

Most sensing applications rely on one of six commonly used methods of detection: opposed sensing, retroreflective sensing, diffuse sensing, convergent sensing, specular sensing, and color-mark sensing.

Opposed Sensing Method

In the **opposed sensing method,** the emitter and detector are positioned opposite each other. The light from the transmitter shines directly at the receiver, as shown in Figure 19-22. Because the target sensed by this method is usually opaque, the mode of detection is light-to-dark. The object is detected when it breaks the light beam. To ensure that the target is reliably sensed, the emitter's beam must be completely blocked.

Since the emitter shines its light directly at the receiver, maximum light energy is transmitted to the detection device. This characteristic enables this sensor to perform some operations

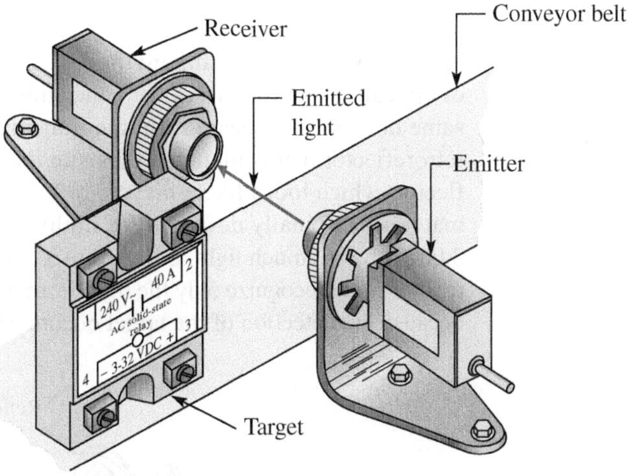

FIGURE 19-22 Opposed sensing method

better than other sensing techniques, such as penetrating air that has contaminants, overcoming lens contamination or misalignment, and providing longer sensing distances. Opposed sensors are used in applications that require the detection of small parts, accurate positioning, and parts counting. Whenever there is an application requiring the detection of small parts, the size and shape of the beam should be smaller than the smallest profile that will ever have to be sensed, while retaining as much lens area as possible. Often, the easiest way to detect small parts is to reduce the size of the beam by placing an aperture over the lens. The drawback of using an aperture is that the transmitted light energy is reduced. The opposed scanning method is also known as *direct scanning, beam break,* and *through beam.*

It is possible to measure the size of an object when using an array of opposed beams, as shown in Figure 19-23(a). This series of beams is referred to as a *light curtain.* Light curtains are used in other applications, for example counting objects such as towels, as shown in Figure 19-23(b), or operating as a machine guard to provide personal safety, as shown in Figure 19-23(c).

IAU4006 Calibrating an Opposed Optical Sensor

The opposed mode should be used whenever possible because it is the most reliable optical sensing system. This method should not be selected for translucent or transparent targets that cannot block light. It should also be avoided when the objects are too small to interrupt 100 percent of the beam.

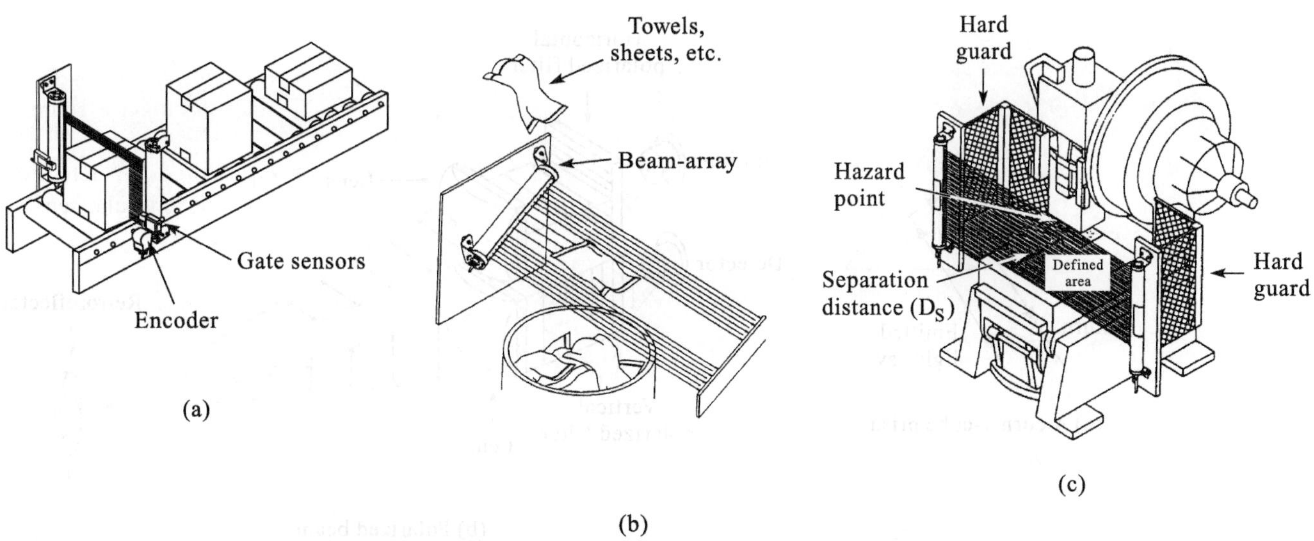

FIGURE 19-23 A series array of opposed sensors that operate as a light curtain (Courtesy of Banner)

IAU14808
Retroreflective
Optical Sensors

Retroreflective Sensing Method

The **retroreflective sensing method** (also called *retrosensing*) is by far the most popular method of optical sensing. In this system, the emitter and detector are mounted next to each other in the same housing. The beam from the transmitter is reflected back to the receiver by a prismatic retroreflector that is mounted opposite the sensor unit, as shown in Figure 19-24. The retroreflector, which looks like a bicycle reflector, is made of 3M's Scotchlight® and Reflectolite®, materials originally designed for highway signs and markers. A good reflector returns up to 3,000 times as much light as does a piece of white paper. This is why it is easy for a retroreflective sensor to recognize only the light from its retroreflector. The mode of detection is light-to-dark because the detection of the target occurs when the light beam is broken.

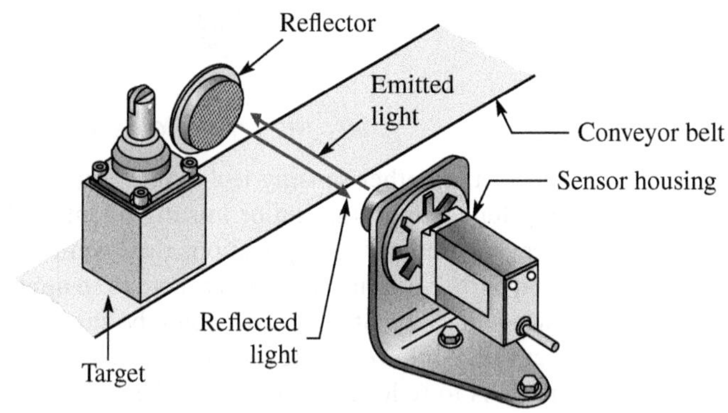

FIGURE 19-24 Retroreflective sensing method

Retrosensing is used in applications where the targets are large, the sensing environment is clean, and the scanning ranges are medium. The scanning range is typically from 2 to 10 feet. Specific examples include conveyors, automated storage and retrieval systems, and bar-code scanning.

Highly reflective objects may go undetected if they return a similar amount of light as the reflector. Three techniques are combined to prevent unwanted light from being reflected back to the receiver. The first technique affects the light emitted by the transmitter. A polarized filter inside the lens blocks the vertical light waves and passes only the horizontal waves. The second technique involves the retroreflector. The reflector is made up of many individual tiny specular inner surfaces called *corner cubes,* as shown in Figure 19-25(a). This configuration

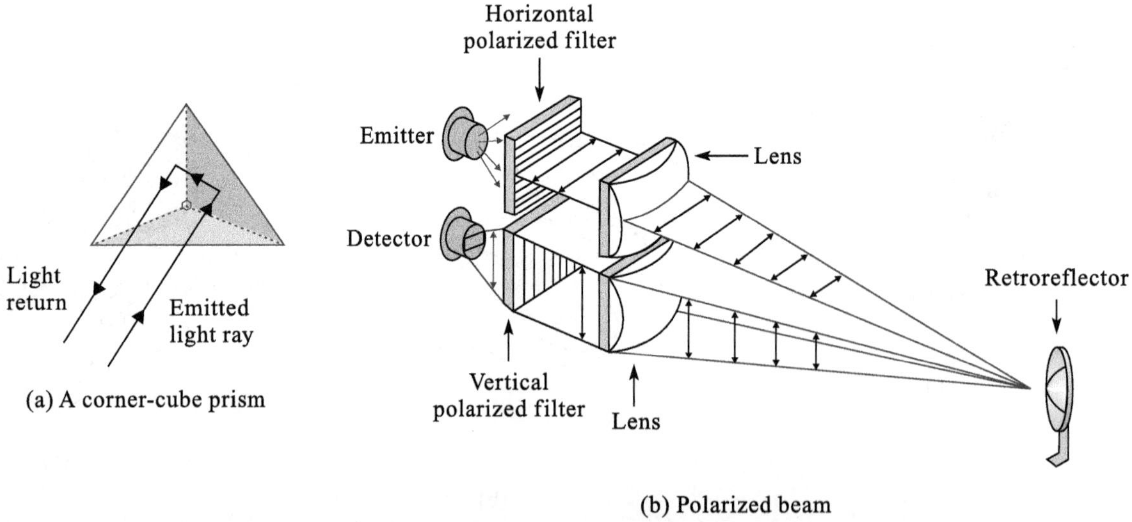

(a) A corner-cube prism

(b) Polarized beam

FIGURE 19-25 Retroreflective sensors with polarization filters

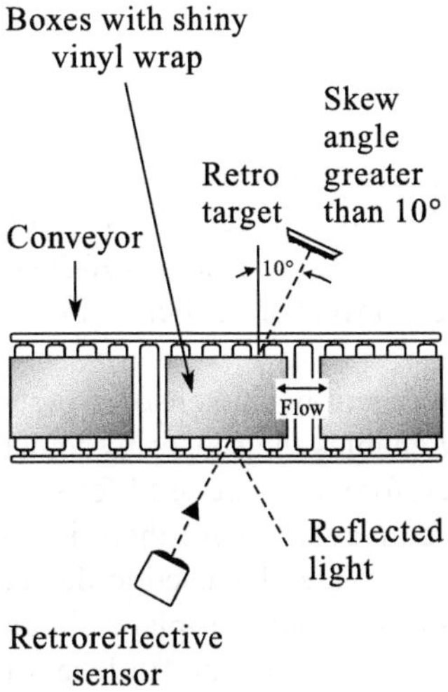

FIGURE 19-26 Use of skew angle to control reflection

is similar to the inside of a box that has been cut diagonally in half from one corner to another. The three sides of the cube are located 90 degrees from one another. When a light beam enters a cube, it reflects off one surface to another surface, then reflects back to its origin. The result is that the beam that enters is rotated by 90 degrees when it returns to the receiver. The third technique is that a polarized filter is positioned behind the receiver lens so that it blocks horizontal light waves and passes only vertical waves. When the three configurations are used together, as shown in Figure 19-25(b), they block any reflected light from the target and any ambient light waves. Another method of solving reflection problems is to position the sensor housing at a skew angle to the object's surface, as shown in Figure 19-26.

The retroreflective scanning method is also known as *retro* or the *reflex* method. This method should not be used for detecting translucent or transparent objects that do not block light or for sensing materials with shiny surfaces that may reflect as much light as the reflector.

Diffuse Sensing Method

In the **diffuse sensing method**, the transmitter and receiver are contained inside the sensor housing, as with the retro sensor. The target is sensed when its position is in line with the light beam. As the object is struck by the transmitter beam, the light is scattered (or diffused), with enough light being returned to the receiver to indicate its presence. This dark-to-light mode of detection is opposite to that of the opposed and retro sensing methods. Figure 19-27 shows a diffuse sensor. Because the light is scattered, only a small percentage of it reaches the receiver. Therefore the scanning distance is limited.

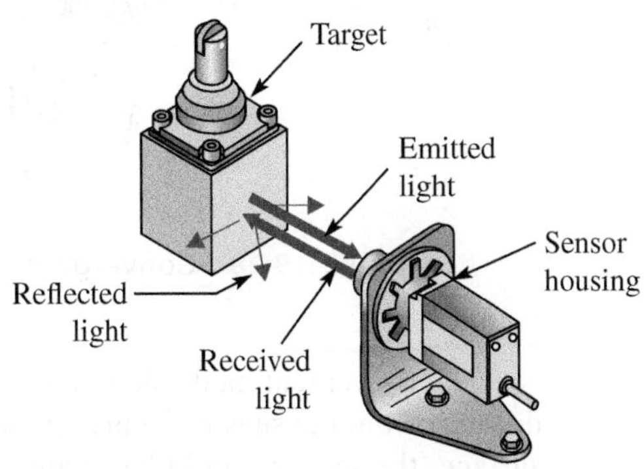

FIGURE 19-27 Diffuse reflection sensing method (Courtesy of Banner)

This scanning method is the first choice for sensing transparent or translucent (clear) materials when the distance from the scanner to the materials is not fixed. A common example of this type of application is clear web-break detection where web flutter is uncontrolled. This technique is predominantly used in the paper industry. There is not any good reason for this. It just seems to have evolved that way.

From an installation standpoint, diffuse sensing is the most convenient method, because detection is accomplished from one side of the process without the need of a reflector on the other side. However, a few requirements must be met to make this method effective:

1. The object must be reflective to diffuse enough light back to the receiver.
2. The surface area of the object must be large enough to reflect sufficient light back to the receiver.
3. The sensor must be placed so that the nearest background surface is three times the distance from the sensor to the target. This avoids the problem that may arise if the background reflectivity is greater than that of the object to be sensed. A matte-type light-absorbing background can also be used to avoid this problem.

Convergent Sensing Method

The **convergent sensing method** is similar to diffuse sensing. Both function on the principle of light being diffused from the target. Figure 19-28(a) shows the convergent sensor. The transmitter and receiver are set at the same angle from the vertical axis to capture the light. The sensor is designed to detect a target at one set distance. By using special lenses, the light source from the transmitter is focused to a narrow depth-of-field. (The **depth-of-field** is the distance on either side of the sensor's focus point.) Objects nearer or farther from the depth-of-field will not activate the sensor.

By focusing the light beam onto a very small area, a convergent beam sensor directs light much more intensely than a diffuse sensor. This property enables the sensor to detect objects that are very small or that have very low reflectivity. Therefore, they are used for specific applications such as counting bottles, jars, or cans on a conveyor, where there is no space between adjacent products, as shown in Figure 19-28(b). They are also used to detect height differential, as with moving parts on a conveyor; to inspect for "parts-in-place" in an assembly operation; to sense accurate position of clear materials; or to detect the fill level of materials in a open container.

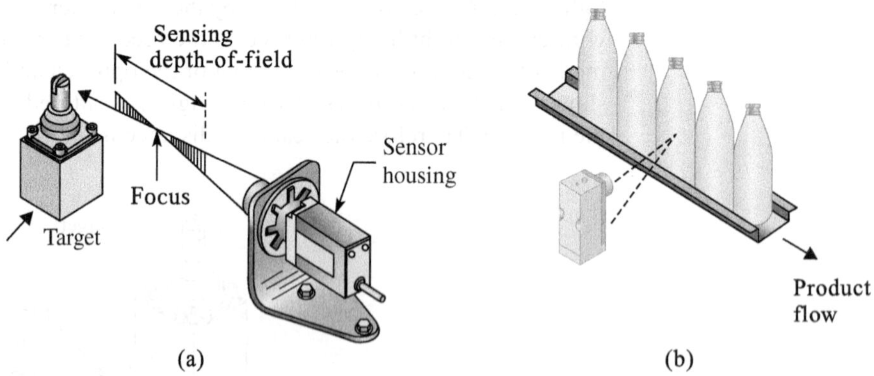

FIGURE 19-28 Convergent sensing method

The convergent mode should not be used to detect objects that pass at an unpredictable distance from the sensor. To prevent an unwanted light reflection from a shiny background surface, the sensor should be rotated at an angle away from a perpendicular position. The convergent optical method is also referred to as **fixed focus.**

Specular Sensing Method

The **specular sensing method** is used to detect objects with mirror-like surfaces. Figure 19-29 shows that the transmitter and receiver are placed at equal angles from the object. To operate effectively, the distance between the sensor and the target must remain constant. The angle between the light source and the receiver determines the depth of the sensing field. With a narrow angle, there is more depth and less accuracy than when a wide angle is used. In a fill level detection application, for example, the wider angle between the source and the receiver provides a more precise measurement.

The specular sensing method is used in applications in which it is necessary to differentiate between shiny and dull surfaces. A common example is detecting a break when printing a newspaper. As the paper goes from one reel to another, it passes over a stainless steel plate. By placing the emitter and receiver at equal angles over the newspaper, the receiver will detect the reflected light of the shiny plate from the emitter if a break in the paper occurs.

Color-Mark Sensing Method

Color-mark detection is different from other types of photoelectric sensor methods. Instead of scanning an object as it passes an inspection point, color-mark instruments detect the

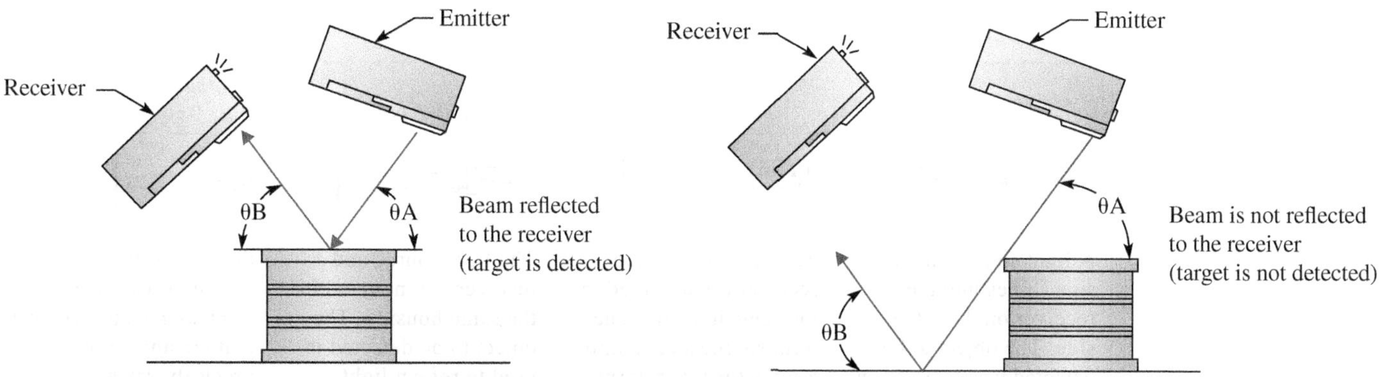

FIGURE 19-29 Specular sensing method, which detects light reflected off shiny surfaces

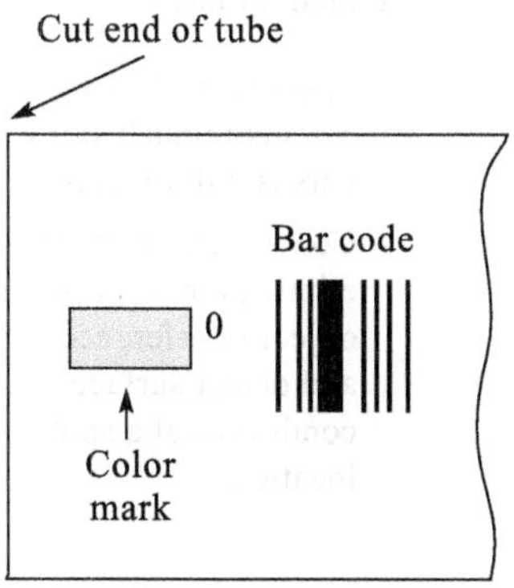

FIGURE 19-30 Color-mark detection example

IAU406
The Timing Functions of Optical Sensors

contrast between two colors. The two colors being sensed can be on the same surface or on separate objects.

Color marks are used extensively in packaging operations. For example, they are the reference point for indexing the cutoff location of packaging materials so that printed information always appears in the same location. Figure 19-30 shows a color mark printed on a toothpaste tube.

Convergent beam sensors with a visible light are commonly used to detect the color mark. The best source of light for color sensing is white, because it contains all colors. White is produced by clear LEDs or incandescent lamps. The greatest optical contrast is provided when a black color mark is printed on a white surface.

Visible LEDs with a red color are used most frequently to transmit the light in color-mark applications because they are spectrally matched to a phototransistor. However, for the combination of red, orange, or pink against a light color, a visible red LED should not be used, because reflected red light from a red mark will be almost as great as the amount reflected by the background surface. A green LED should be used instead, even though it is considered a low-powered color.

Table 19-3 provides a summary of advantages and disadvantages of some optical sensing methods.

19-8 Photoelectric Sensor Adjustable Controls

Some types of optical sensors are designed to activate an output when the monitored light source changes from dark to light, changes from light to dark, or produces a signal that exists for a different time deviation than when a target is detected. Figure 19-31(a) shows adjustable switches and potentiometers that can be changed on a sensor to customize its operation for various application requirements.

Light/Dark Operation

If the sensor turns on when the light level rises as the target enters the field-of-view, such as with retroreflective or diffuse optical detectors, it is referred to as a *light-operated device.* To produce an output when the sensor detects an increase in light intensity, the DK/LT DIP switch on the sensor is placed in the LT position. For applications that require the output to activate when the light intensity is reduced, the DIP switch is placed in the DK position.

Sensitivity

Sensitivity is a measure of the amount of change in light intensity that is required by the optical sensor to produce a switching action at its output. It is possible to vary the sensitivity

TABLE 19-3 Advantages and Disadvantages of Some Optical Sensing Methods

Method of Detection	***Opposed***	***Retroreflective***	***Diffuse Reflective***	***Specular Reflective and Convergent Beam***
Configuration	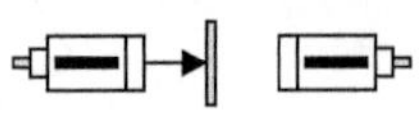 A light source on one side shines into the receiver on the other side. The object to be detected passes between the light source and receiver to break the light beam.	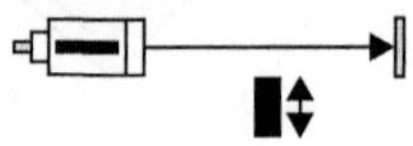 The light source and receiver are mounted in the same housing. The light source shines into a reflector that returns the beam to the receiver. The light source may be polarized for better detection of shiny objects.	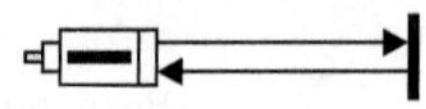 The light source and receiver are mounted in the same housing. The object to be detected is used to reflect light from the light source back to the receiver.	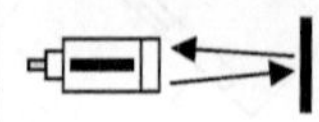 Specular reflective sensing uses a separate light source and receiver that are angled to precisely detect reflected light. Convergent beam sensing uses a diffuse reflective sensor and angled optics.
Advantages	• Long-distance detection • Small parts detection • Reliable performance in dusty and humid areas • Ideal for opaque objects • High detection accuracy	• Easy to install—only one piece requires wiring • More tolerant of vibration and less-than-exact reflector alignment • Polarized versions can detect shiny objects without false signals • Easy light beam alignment	• Easy to install—only one piece requires wiring • Alignment is not critical • Detects translucent and transparent objects	• Specular reflective can distinguish shiny versus dull objects • Convergent beam can eliminate background object interference and detect surface conditions at a spot location
Disadvantages	• Cannot detect transparent objects • Comparatively high wiring expenses	• More sensitive to dust and humidity contamination • Not good for small object detection	• Short sensing distance • Performance depends on the target's reflectivity	• Alignment is critical • Installation cost will be higher for specular reflective because of a separate light source and receiver

of the sensor by making a potentiometer adjustment to its *sensitivity potentiometer* (shown in Figure 19-31(a)) that changes the gain of its electronic amplifier circuitry. The sensitivity setting affects the ability of the sensor to detect a target.

On-Delay Operation

When a target is detected, a period of time passes before the output turns on if the on-delay function of the sensor is activated. Figure 19-31(b) graphically shows the on-delay operation. The duration of the delay is determined by placing the ON DIP switch to the S (short) or L (long) position, and adjusting the Delay On potentiometer. The span and the length of time are greater when the DIP switch is in the L position. If no time delay is desired, the potentiometer should be adjusted to the minimum setting.

Preventing an alarm from turning on unless a dangerous condition exists for a period of time is an example of an on-delay operation.

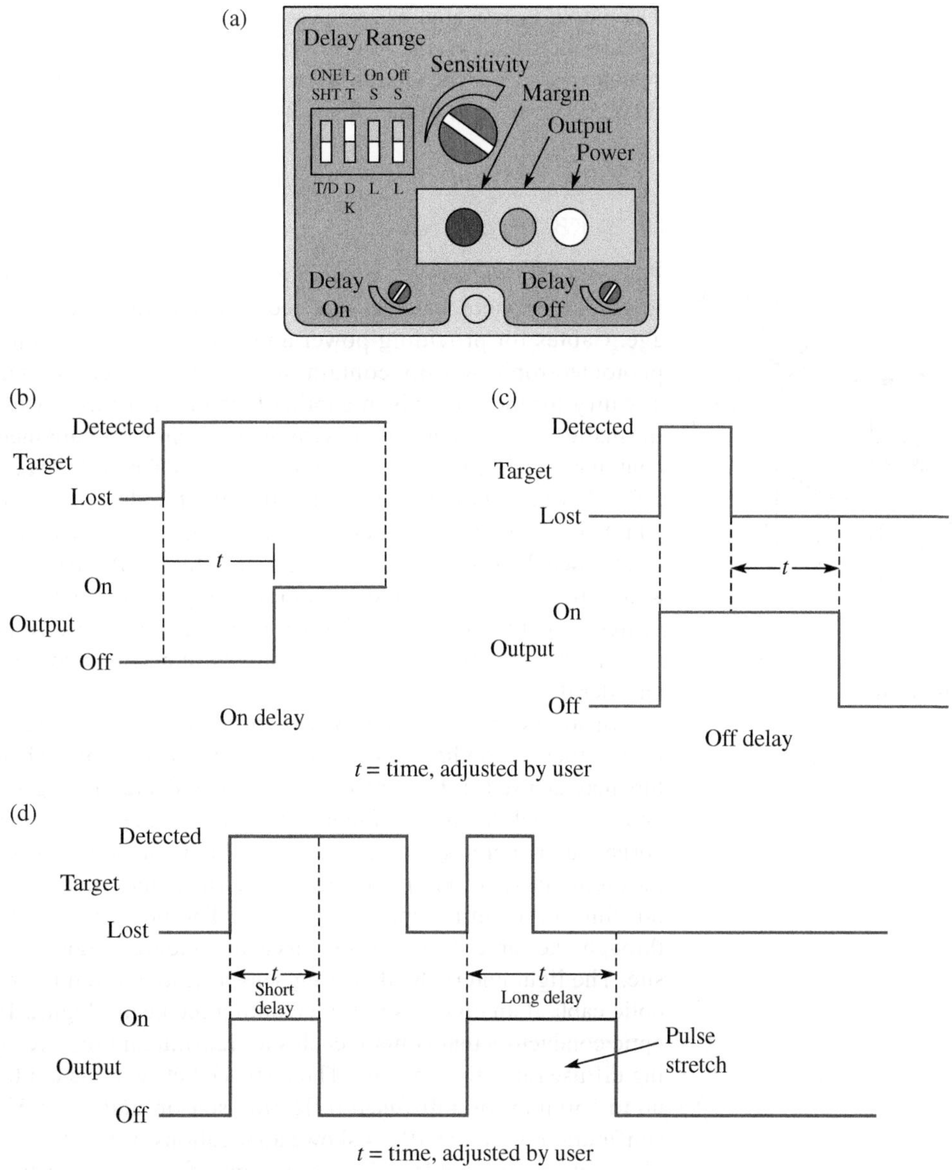

FIGURE 19-31 Photoelectric sensor adjustments: (a) Sensor; (b) On-delay function; (c) Off-delay function; (d) One-shot function

Off-Delay Operation

Figure 19-31(c) graphically shows the off-delay operation. When a target is detected, the output immediately turns on. When the target is lost, a period of time passes before the output of the sensor turns off. The duration of the delay is determined by placing the OFF DIP switch to the S (short) or L (long) position, and adjusting the Delay Off potentiometer. The span and length of time is greater when the DIP switch is placed in the L position. If no time delay is required, the potentiometer should be adjusted to the minimum setting. An application of an off-delay operation is to keep a blower fan running for a few minutes to exhaust fumes in a paint booth after the spraying operation has stopped.

One-Shot Operation

The function of a one-shot operation is to produce an output for a duration that is either longer or shorter than the time at which the target is detected, as graphically shown in Figure 19-31(d).

The output is activated when the sensor detects a target, and stays on for a specific period of time before it automatically turns off. To operate in the one-shot mode, the ONE SHT DIP switch is placed in the ONE SHT position. The time at which the output is on is determined by the adjustment made to the Delay Off potentiometer.

19-9 Photoelectric Package Styles

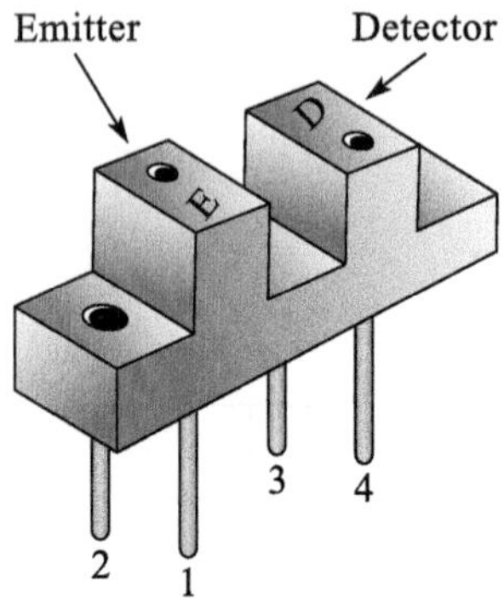

FIGURE 19-32 Opto interrupter

Photoelectric sensors are available in two package styles: self-contained and remote. The lenses, LED, photodetector, and electronic circuitry are all a part of the self-contained package. Cables for providing power and data transfer are connected to the package. Remote photoelectronic sensors contain only the optical components of the sensing system. The circuitry for the sensor is at another location. For this reason, remote sensors can be placed in smaller and more hostile environments than self-contained sensors. One specialized self-contained package is the *opto interrupter,* as shown in Figure 19-32. The housing provides a fixed alignment between the emitter and the detector with a sensing area, usually an air gap, between them. These devices are designed to sense the presence of an object and produce a switching signal each time one is detected. A variety of package styles are available to accommodate a range of size and mounting requirements. Applications for slotted optical switches include paper sensing for printing machines, motor speed tachometers, position sensing in computer disks, limit switching in machine control, and angular positioning using encoder disks.

In some situations, the space is too confined, the size of the target is too small, or the environment is too hostile even for remote sensors. For such applications, fiber-optic conductors may be used. **Fiber optics** are transparent strands of glass or plastic that transfer light to and from such locations. Fiber-optic "light pipes" are capable of conducting light around corners and operating when exposed to high temperatures or vibration. They are immune to magnetic noise. In Figure 19-33(a), the fiber-optic transmitter and receiver are in the same housing, performing opposed sensing. The beam is sent to the site from the transmitter through one cable. The light spreads at a 60-degree angle from the emitter sensing head at the site. The light enters the detecting sensor head and returns to the receiver through the other optic cable if an object is not interrupting the beam. Figure 19-33(b) shows bifurcated fiber-optic conductors that contain cables to transmit and receive signals simultaneously, performing diffuse reflective sensing. The distance between the emitter and detector sensing heads is up to 150 mm for a diffused reflection sensor, and up to 500 mm for the opposed sensing configuration. Figure 19-34 shows applications of fiber-optic photoelectric sensors for three different sensing modes. The limitations of fiber-optic sensors are the small scanning distance between the emitter and the receiver, and the limited length of fiber cables (no more than 30 feet in length).

Plastic fibers are used in applications where there is repeated flexing, such as on a reciprocating machine. However, their drawbacks are that they cannot be exposed to extreme temperatures, and they can be damaged if exposed to chemicals. Glass fibers can withstand high temperatures and exposure to chemicals, but they break if bent too often or too sharply.

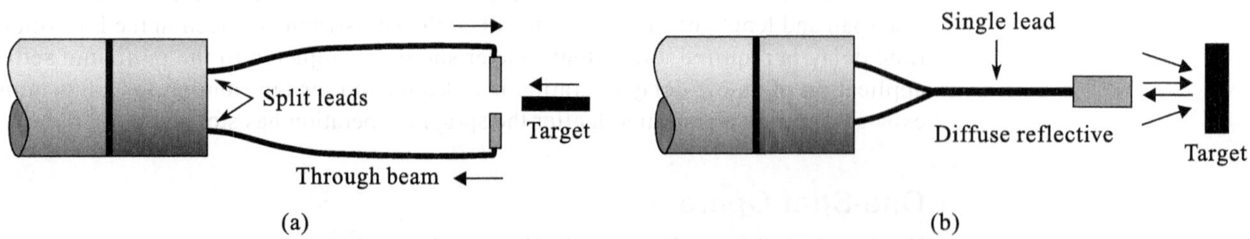

FIGURE 19-33 Fiber-optic cables used to perform photoelectric sensing methods

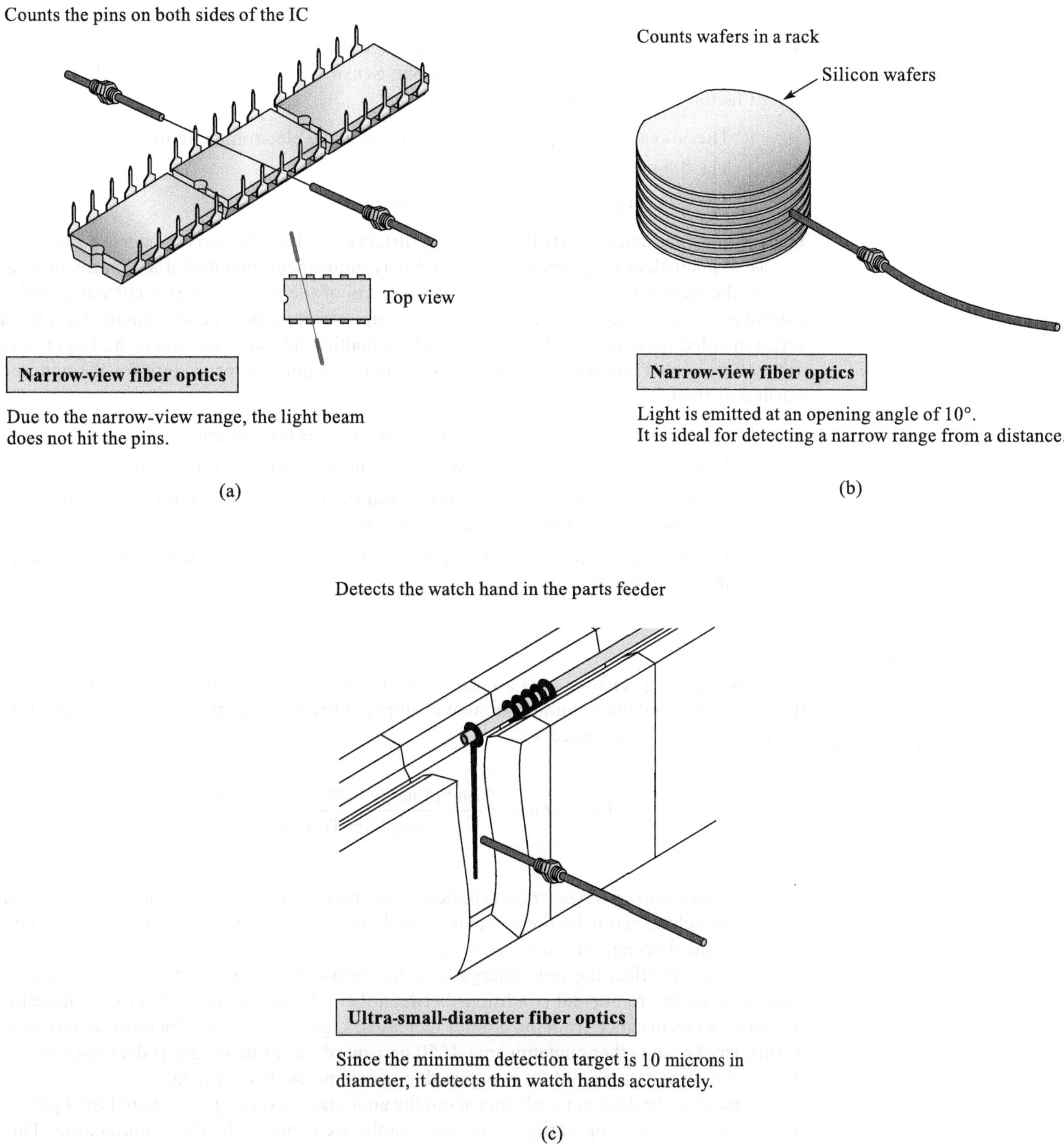

FIGURE 19-34 Applications of fiber-optic photoelectric sensors: (a) Opposed sensing (IC pin counting); (b) Retroreflective sensing (Wafer counting); (c) Diffuse sensing (Detection of watch hands)

19-10 Operating Specifications

Data specifications are available for each type of optical sensor. This information helps the user select the proper sensor by indicating how well it will operate under certain conditions. The data sheets provide information on *sensitivity, excess gain, field-of-view*, and *sensor response*.

Sensitivity

Sensitivity is a measure of the amount of change in light intensity that is required by the sensor to cause a switching action at its output. Sensitivity is the combined result of several design factors of the sensing device, such as:

1. The amount of amplification incorporated in the electronic circuitry.
2. The light power of the LED.
3. The size, shape, and quality of the lens.

Each of these variables affects the operating distance at which the sensor detects a target.

Most photoelectric systems have a sensitivity adjustment mounted at a convenient location on the enclosure. This adjustment sets the level of the phototransistor current at which control action takes place. To set the sensitivity adjustment properly, each manufacturer has a recommended procedure to follow. It involves making adjustments when the target is in view of the sensor and when it is not present. For example, the procedure for the opposed sensing method follows:

1. Start with the minimum setting when the target is not present.
2. Increase the sensitivity until the detector turns on. Record this setting.
3. Place the target between the emitter and the detector and increase the sensitivity until the detector turns on. Record this setting.
4. Readjust the setting to the optimum sensitivity, which is midway between the readings recorded in steps 2 and 3.

Excess Gain

IAU14308
Excess Gain of
Optical Sensors

The **excess gain** specification is a measurement of the amount of light energy that falls on the receiver beyond the minimum amount of light required to operate the sensor amplifier. In equation form, it is expressed as:

$$\text{Excess Gain} = \frac{\text{Light Energy Falling on Receiver}}{\text{Amplifier Threshold}}$$

An excess gain of 1× (*one times*) indicates the minimum amount of light energy required to cause switching action by the detector. If the light intensity is twice the minimum amount needed by the detector, the excess gain value is 2×.

Two factors affect the light energy from the emitter that reaches the detector: the distance and the environmental conditions between them. Light intensity seen by the detector decreases as its distance from the emitter increases. Light is attenuated by obstructions such as dirt, smoke, and other contaminants. If 50 percent of the emitted light is decreased due to unclear air, an excess gain of 2× is required to overcome the loss of light.

To provide the designer with data about the anticipated excess gain required for a particular sensor at various operating distances, graphs are supplied by the manufacturer. This information helps to determine if the sensor will operate effectively under a given condition. The graphs are plotted by taking measurements in a perfectly clean environment. Examples of three graphs are shown in Figure 19-35.

Figure 19-35(a) illustrates the characteristics of the opposed sensing detector. Its excess gain is related to scanning distance by the inverse square law. Increasing the distance by two will decrease the light energy received at the detector by a factor of four. Figure 19-35(b) shows the nonlinear characteristics of the retroreflective sensor. The top of the curve indicates that the excess gain is maximum at a given distance. To operate most effectively, the sensor should be positioned at this distance from the reflector. Figure 19-35(c) shows that the light energy received at the detector for the diffuse sensor drops to an excess gain of 1× at a much

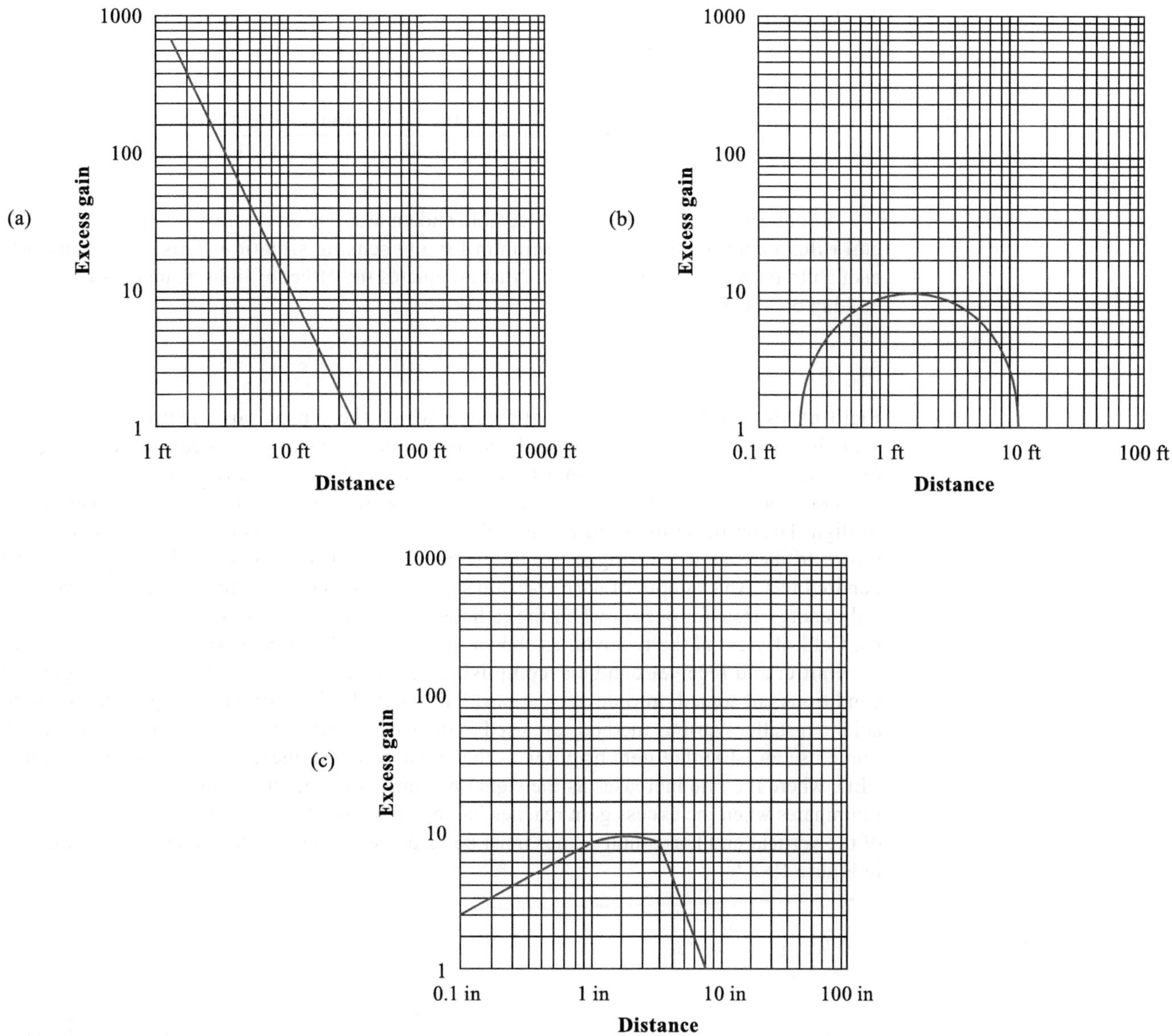

FIGURE 19-35 Excess gain graphs for three different detection modes: (a) Opposed; (b) Retroreflective; (c) Diffuse

shorter distance compared to the other modes. The low energy level is the result of the small amount of light bounced back from the target. The graph is plotted by using a white test card, arbitrarily rated at 90-percent reflectance, as a standard reference. Objects with reflectivity lower than 90 percent may not be detected.

For applications that require the beam to penetrate thin cardboard or similar materials, the opposed sensing method with an excess gain of at least 50× is required. An example is the level sensing of the contents of a cereal box.

Contrast

All photoelectric sensing applications require differentiating between two light levels received by the detector. These light levels pertain to the maximum light that shines on the detector (called the *light state*) and the minimum light that shines on the detector (called the *dark state*). The comparison between the two light levels is referred to as *contrast*. Also known as the *light-to-dark ratio*, it is represented by the following formula:

$$\text{Contrast} = \frac{\text{Light Level at the Receiver in the Light State}}{\text{Light Level at the Receiver in the Dark State}}$$

A contrast of 1.2 or less is undetectable and an alternative scheme is advised. As a general rule, a contrast of 3 is the minimum for any sensing situation. This value is usually enough to overcome a subtle variable that attenuates the effect of light changes, such as dirt buildup on a lens.

Field-of-View

Field-of-view is the dispersion angle within which the sensor can effectively sense light from the emitter. Some sensors are only capable of detecting a narrow beam within 2 degrees of where the light is aimed. Other types can sense light that is spread over a wide angle, exceeding 60 degrees. The broader the field-of-view of a photoelectric sensor, the easier it is to align. The narrower the field-of-view, the greater the light intensity becomes, enabling the sensor to increase its sensing range. Some narrow beam sensors detect light up to several hundred feet. An example of narrow-beam sensing is the opposed sensor method. Maximum light energy transfer occurs when the radiated energy from the light source is centered on the field-of-view of the receiver, as shown in Figure 19-36. In most cases, the alignment of the emitter and receiver can be accomplished satisfactorily by eye. For critical alignment conditions, advanced opto devices are available with built-in signal-strength indicators to aid the installer. Various methods are used to indicate when the proper alignment is achieved. Some use an LED that gets brighter as the signal strength increases. One uses a flashing LED, where the rate increases as the signal becomes greater. Others use a second LED that illuminates when the excess gain reaches 2×. When physically aligning sensors, adjustments of the vertical and horizontal planes must be made, as shown by the diffuse sensing method in Figure 19-37.

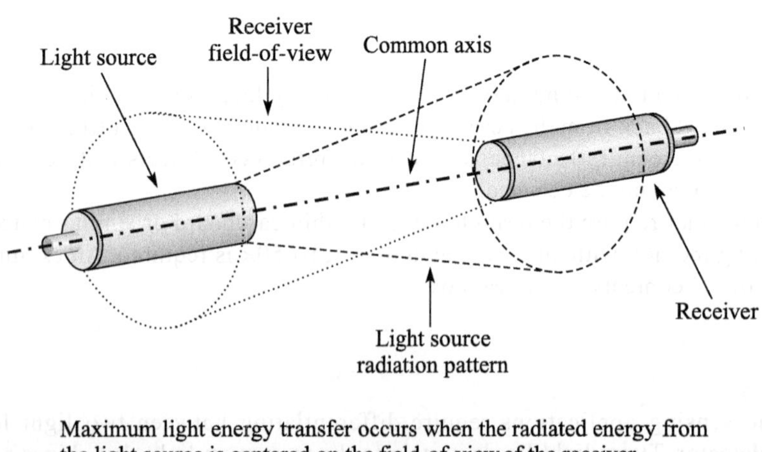

FIGURE 19-36 Alignment and field-of-view

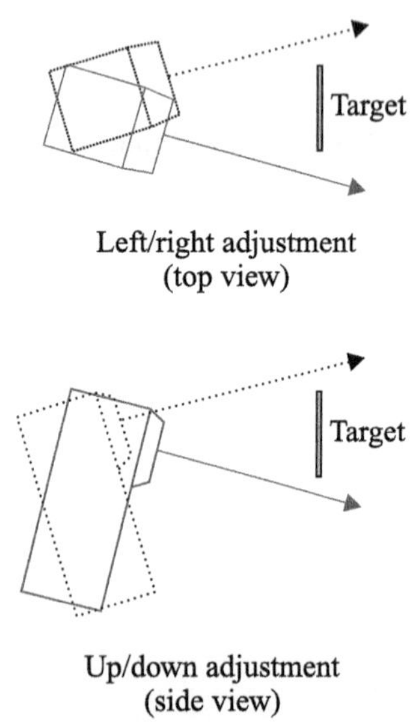

FIGURE 19-37 Horizontal and vertical alignment adjustments

Sensor Response

Whenever fast-moving objects are sensed, the response time of the sensor becomes important. **Sensor response time** is the maximum amount of time that elapses from an input transition mode (light-to-dark or dark-to-light) until the output switches. The required sensor response time is calculated by the following formula:

$$\text{Trequired} = \frac{W - D}{V}$$

where,

T = Required sensor response time
W = Size of the target
D = Beam diameter at the sensing location
V = Speed of the object that passes the sensor

EXAMPLE 19-1 Figure 19-38 shows an application of the sensor response calculation.

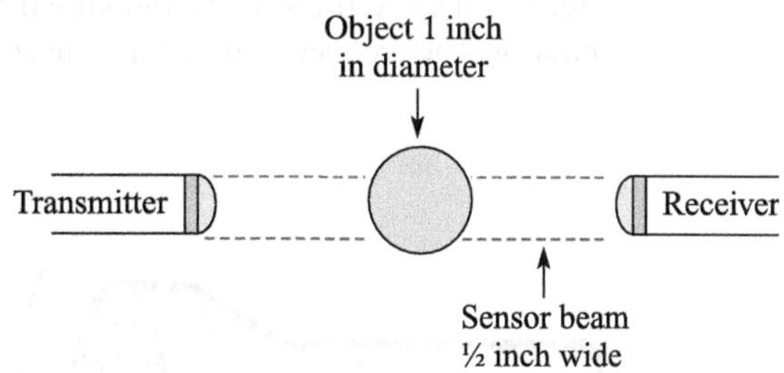

FIGURE 19-38 A sensor response example

- The diameter of the target (W) is 1 inch.
- The width of the light beam (D) is $^1/_2$ inch.
- The speed of the object (V) is 10 inches/second.

Solution

$$\text{Tresponse time} = \frac{W-D}{V} = \frac{1-0.5}{10} = 50 \text{ ms}$$

The sensor selected for this application must have a switching speed of at least 50 ms to operate effectively. There is a correlation between sensor response time and excess gain. As the gain is increased, its switching speed slows down.

Guidelines for Selecting an Optical Sensing Method

When selecting the most appropriate method for a particular application, the following operating conditions should be considered:

Size and shape of the target. The opposed method is often used to detect small objects because it has a narrow beam. The convergent sensing method is often used to detect the surface of an object that is not flat.

Distance between the emitter and the detector. In applications where the emitter and detector are far apart, the opposed sensing method has the longest range. When detecting an object at a specific distance, the convergent method is most effective.

Physical characteristics of the object. If the object is shiny, place a retroreflective opto sensor at a skewed angle instead of perpendicular. If the object surface is clear, use the diffuse sensing method.

Unwanted ambient light or background. Retroreflective sensors with polarized filters should be used when there is unwanted ambient light present.

Rate of speed at which the target passes the light beam. The opposed sensing method is often used for fast, repetitive operations, such as counting gear teeth, because it has a narrow beam.

19-11 Ultrasonic Sensors

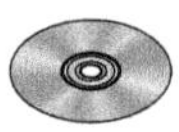

IAU106
Ultrasonic Sensors

An ultrasonic sensor is a device that uses high-frequency sound waves to detect objects. Sound waves produced by an oscillator are emitted toward the target from the sensor head, as shown in Figure 19-39. If there is an object from which the ultrasonic sound waves can be reflected, they will return to the signal receiver that is located in the same housing as the oscillator. The frequency of ultrasonic sound waves is above the range heard by humans.

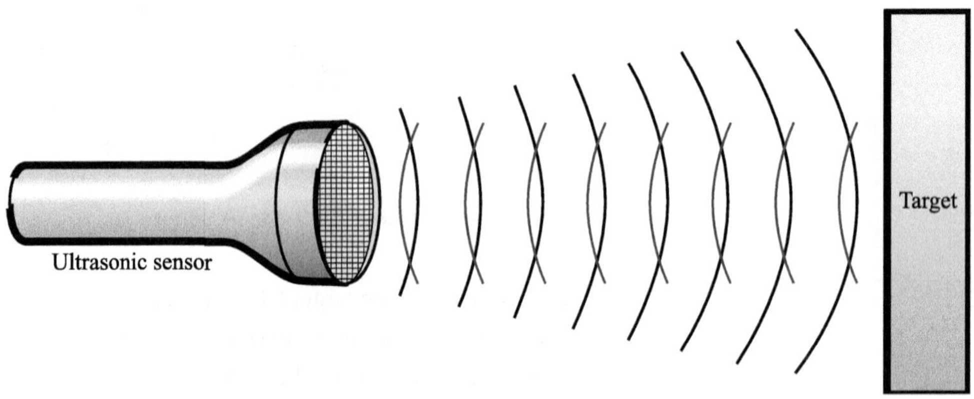

FIGURE 19-39 Ultrasonic sensor

There are two categories of ultrasonic sensors: *proximity switches* and *analog* sensors. The difference between them is the type of output signal they produce. The proximity switching ultrasonic sensor is used to detect the absence or presence of a target. When an object is present within its sensing range, the sensor output is switched on. The analog ultrasonic sensor produces a variable output voltage or current. The magnitude of the output is proportional to the distance between the sensor and the target. The sensor can detect the object within a given range.

Most analog sensors have the capability of varying the range by performing a calibration procedure. The desired minimal distance at which the target is placed from the sensor is called the *zero* calibration setting. At this distance, the following types of sensors will produce its minimum output value:

Voltage Sensor: 0 V
Current Sensor: 4 mA

The desired maximum distance at which the target is placed from the sensor is called the *span* calibration setting. At this distance, the following types of sensors will produce its maximum output value:

Voltage Sensor: 10 V
Current Sensor: 20 mA

The specific distance of the zero and span settings can be programmed into a microcontroller incorporated into the sensor head.

After the zero and span settings are made, the sensor produces a linear output that is proportional to the distance the target is located from the sensor. For example, an analog voltage sensor will produce 5 volts when the target is halfway between the minimum and the maximum sensing distance established during the calibration procedure. Figure 19-40 shows the relationship of the zero and span settings of an analog ultrasonic sensor both pictorially and graphically.

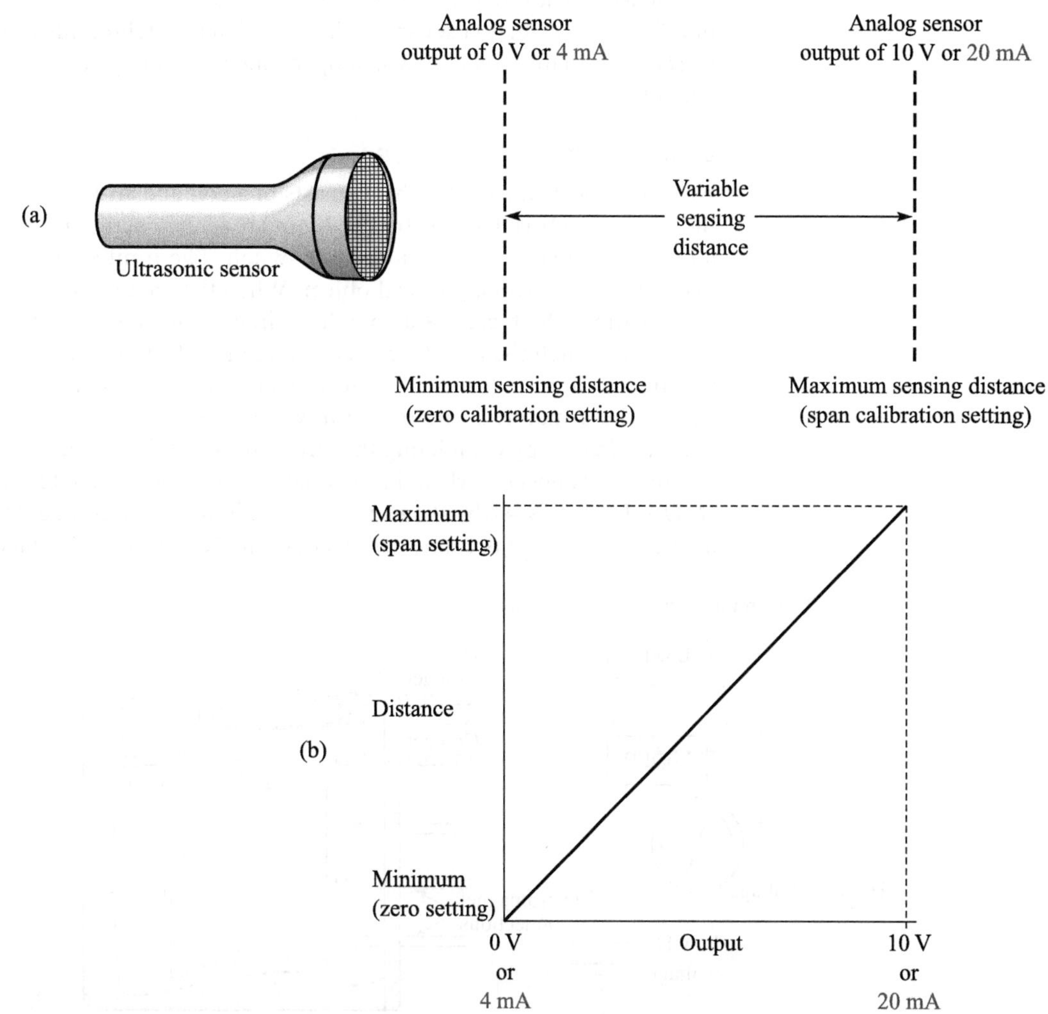

FIGURE 19-40 Zero and span settings of an analog ultrasonic sensor: (a) Pictorial drawing; (b) Graph

19-12 Sensor Interfacing

There are a large variety of sensors on the market and many types of load devices to which they can be connected. Unfortunately, not all of the sensors or load devices are compatible because they either do not use the same types of signals or operate within different voltage or current ranges. If the voltage and current requirements are not electrically matched, the sensor or load device either may be damaged or will not operate properly.

When selecting a sensor to use for a particular application, a key factor that must be considered is its electrical compatibility with the load to which it will be interfaced. The term *load* refers to the device or equipment to which the sensor signal is applied. The load may be an electromechanical device, such as a solenoid, clutch, brake, or contactor; a resistive load, such as a lamp or heater; or a solid-state input, such as a counter, logic circuit, electronic speed control, or PLC. The first interface consideration is to determine whether the load requires a switched signal or an analog signal from the sensor's output.

Switched Signal

When the load requires a switched signal, the sensor must provide an *on* and *off* digital output. These signals represent presence/absence, go/no-go, limit control, and counting applications.

Sensors that provide digital signals have an output switch of one type or another, which interfaces the detection sensor to the load. This switching function is performed by *electromechanical* (hard) *contacts* which open and close, or by *solid-state circuitry* which turns on and off.

Electromechanical Relays

An electromechanical relay that consists of a coil and armature and a set of contacts is shown in Figure 19-41(a). One portion of the relay is the energizing circuit that consists of a coil, a low voltage DC source, and a limit switch. The limit switch is open until it is closed by making contact with a physical object. When the energizing circuit is open, the armature is in a position which creates a complete circuit, causing current to flow through load A. The contact to which the armature is connected is referred to as normally closed (N.C.). When the limit switch is closed, the coil in the energizing circuit becomes an electromagnet and causes the armature to be pulled upward. The armature breaks contact with load A and activates load B by completing the circuit to which it is connected. The contact to which the armature is connected when the energizing circuit is activated is referred to as normally open (N.O.). Figure 19-41(b) shows the schematic of the circuit controlled by the relay. In most applications, a load is connected to either the N.C. or N.O. contact.

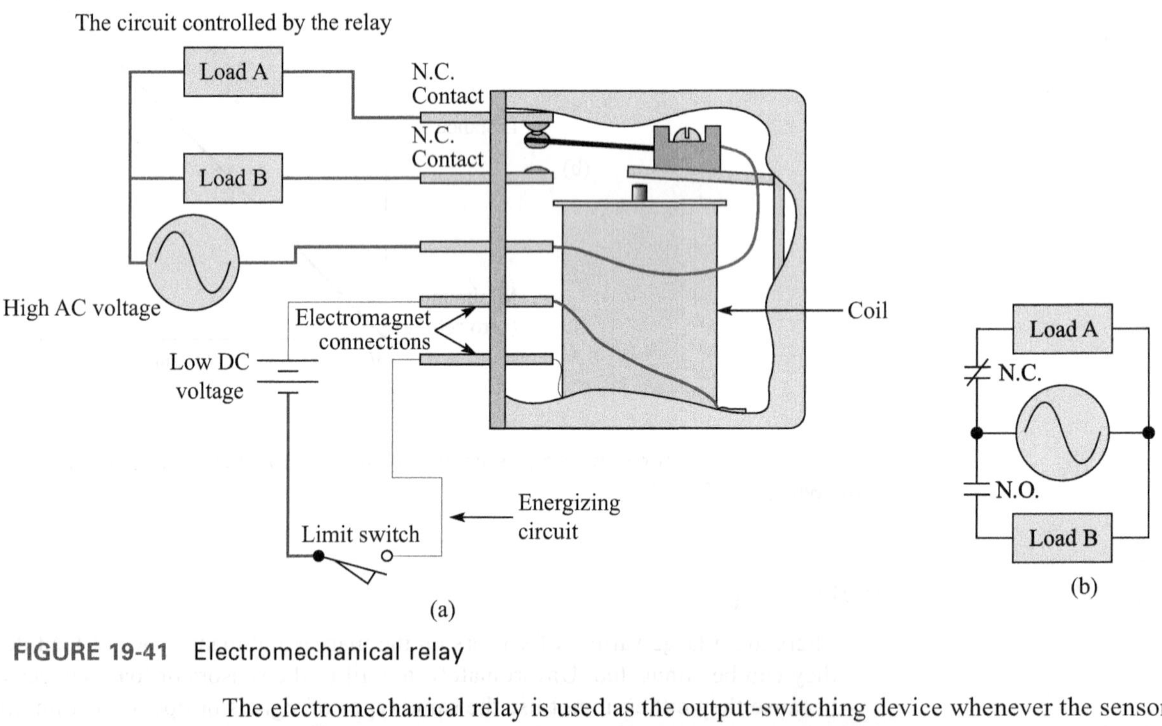

FIGURE 19-41 Electromechanical relay

The electromechanical relay is used as the output-switching device whenever the sensor provides direct control of a load that draws more contactor current than can be handled by solid-state switches. High-current situations are most common in control applications where the sensors directly start, stop, and divert mechanisms, such as clutches, brakes, and solenoids, at the point of sensing. Relay contacts are also used to interface the sensor to the load when their operating voltages are not the same. For example, if the relay coil is energized by a 12-volt DC source, its switching contacts can be connected to a load that operates at either

AC or another DC voltage level. Another advantage of electromechanical contacts over solid-state switches is that they can be wired in series to perform AND logic, or parallel to perform OR logic functions. Some solid-state sensors can also be used in these configurations, but with some limitations.

The major difference between various types of relay devices is the current levels at which they perform the switching operation. What determines their operating current levels is the type of material used on the contacts. For example, when small signals are being used, gold flashings on the contacts should be used to ensure reliable conductivity. When higher-current switching applications are required, contacts made of a silver alloy are often used. When switching large current or inductive loads, arcing can occur, which causes the contacts to be pitted.

The following subsections make recommendations to prevent arcing and thereby prolong the life of the contacts.

Resistive Loads When a large resistive load is controlled by an N.C. contact, a series R-C network is placed in parallel across the contacts, as shown in Figure 19-42. To prevent arcing when the contacts open, the capacitor shunts the voltage away from them. To prevent arcing when the contacts close, the resistor slows down the rate at which the capacitor discharges through them. The R-C network is called a *snubber.* The following formulas enable calculation of the component values:

Obtain the voltage value across the contacts when they are open: E

Obtain the holding current value of the load: I

To find the resistance value:

$$R = \frac{E}{10(Ix)}$$

$$\text{where } x = \left(1 + \frac{50}{E}\right)$$

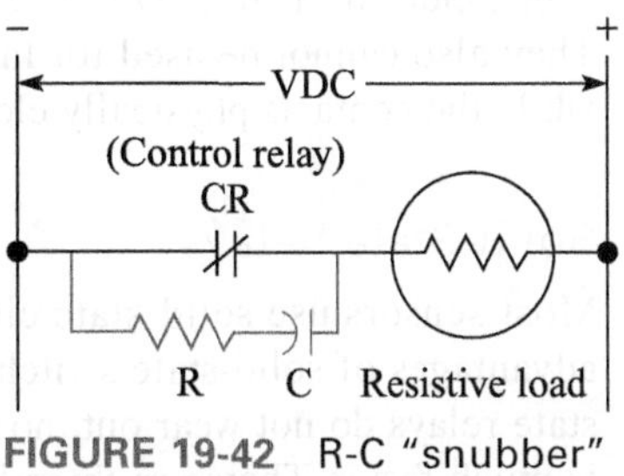

FIGURE 19-42 R-C "snubber"

The resistor should be 1/2 watt or larger.

To find the capacitor value:

$$C = \frac{I^2}{10} \text{ microfarads}$$

The capacitor's working voltage should exceed E.

Snubber networks are available in single packages.

Inductive Loads When an inductive load is controlled by a relay contact, a semiconductor clamping device is recommended to provide arc suppression.

When controlling a DC load, a diode is placed across the inductor with the anode connected to the "–" side, as shown in Figure 19-43(a). When the contact opens, the magnetic

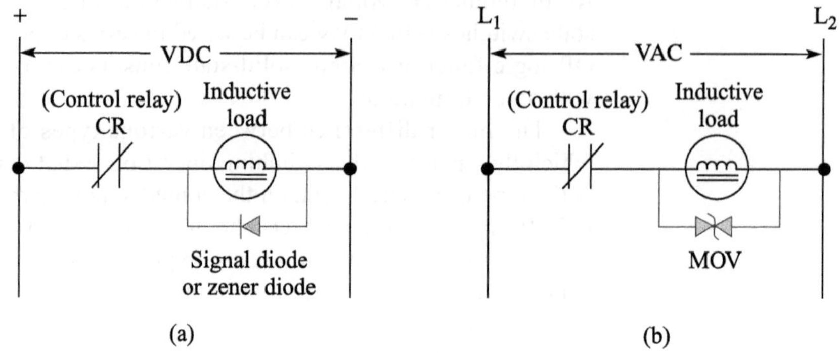

FIGURE 19-43 Arc-suppression devices for inductive loads

field around the inductor collapses as the coil attempts to keep the current flowing. The flux lines cut the coil at a very fast speed and create a momentary potential of up to several thousand volts, enough to cause arcing across the contacts as they open. Instead of arcing across the open contacts, current flows through the forward-biased diode until the magnetic field is gone. A clamping diode with a high surge current rating and peak reverse voltage (PRV) of twice the supply voltage should be used.

When controlling an AC load, a metal oxide varistor (MOV) is connected in parallel with the load inductor, as shown in Figure 19-43(b). The MOV has a very high impedance until a high-amplitude breakover voltage is reached, at which time it clamps on, its impedance reduces, and current passes through. When the contacts open, the magnetic field collapses, creating a voltage spike. The MOV clamps on and absorbs energy as induced current from the decaying magnetic field passes through, instead of arcing across the open contacts. An MOV with the proper clamping voltage and sufficient surge current capacity should be used.

There are several disadvantages to mechanical switches. Since they contain moving parts, they wear out. When using arc-suppression devices, the contacts still become pitted over a period of time. Due to contact bounce, they should not be used to input a counter. They also cannot be used for fast switching applications because a short time delay occurs while the contacts physically close.

Solid-State Relays

IAU806
Sensor Wiring
Configurations

Most sensors use solid-state circuits to perform their switching action. There are several advantages of solid-state switching over mechanical switches. Without moving parts solid-state relays do not wear out, no arcing or contact bounce occurs, and their switching action is much faster. There are three basic types of wiring configurations for solid-state sensors; *two-wire, three-wire,* and *four-wire.* These wires are used for making connections to the power supply and the load.

Most solid-state sensors are designated as *sourcing* or *sinking*. These two terms refer to the direction in which conventional current (+ to −) flows between the sensor's output terminal and the load. When conventional current flows from the sensor's output to the load, it is referred to as a **sourcing** sensor. When the conventional current flows from the load to the sensor output, it is referred to as a **sinking** sensor.

IAU606
Two-Wire Sensors

The Two-Wire System **Two-wire systems** are defined as sensors that have only two wires. They are connected in series with the load the same way that a mechanical switch is wired. Figure 19-44 is used to compare a mechanical to an electrical switch.

When the mechanical switch is open, the full supply voltage is dropped across it and zero current is allowed to pass through, as shown in the top portion of the diagram. However, an electronic switch requires at least a small amount of current to flow through it in the

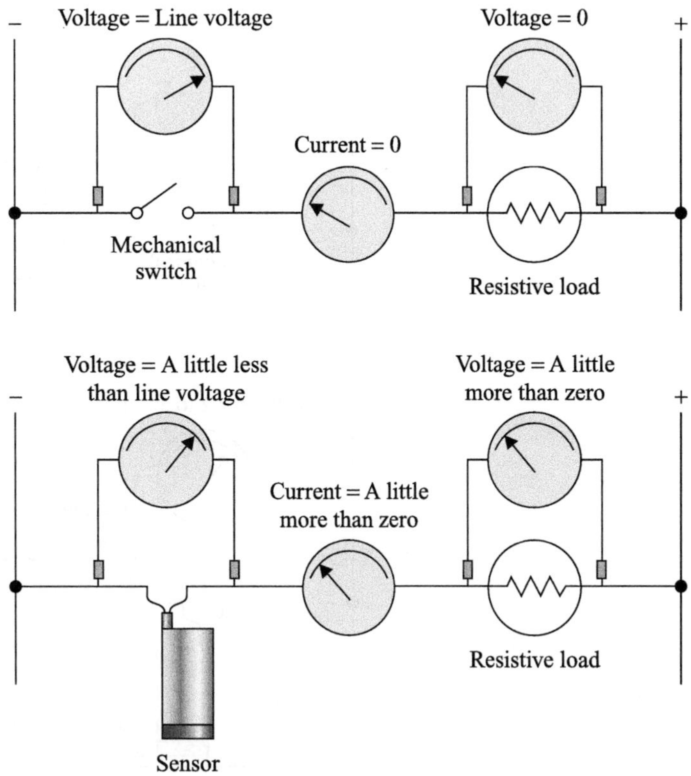

FIGURE 19-44 Comparison of mechanical and electrical open switches

off condition. For this reason, an electronic switch connected this way is referred to as a *load-powered device.* This residual current is called the sensor's *off-state leakage current* (I_{off}). The bottom portion of the diagram shows this current and the voltage drops developed when the electronic sensor is in the off condition. The leakage current must be low enough to not activate the load.

Figure 19-45 is used to compare a mechanical and electrical switch when they are in the closed condition. The top portion of the figure shows that a voltage does not drop across the switch when it is closed. As the electronic switch is activated (turned on) by a target, as shown in the bottom portion of the figure, a potential of a few volts drops across its terminals. An electronic switch requires some power to operate when it is closed. However, the voltage must be small enough so that the remaining potential from the power line is sufficient to activate the load. If the load is deprived of the voltage that it needs, it will either not actuate or not operate properly.

Two-wire sensors can be connected in series. However, there is a significant voltage drop (from 5 to 10 volts) developed across each sensor that is activated in the series configuration. Therefore, the number of series-connected sensors is limited so that the sum of their voltage drops cannot exceed the line voltage minus the voltage required to keep the load actuated.

Certain rules should be followed when connecting two-wire sensors in parallel. For example, when some types of loads are required to turn from the on to the off condition, the current flowing through them must reduce below a certain level, called a "holding current." This current will be exceeded and the load will remain on if there are too many branches, even if all of the sensors turn off. This condition exists because the sum of the leakage currents of each sensor that is turned off flows through the load.

Two-wire sensors are popular because they are simple to wire and the wiring configuration in which they are installed is a familiar design to engineers, technicians, and electricians.

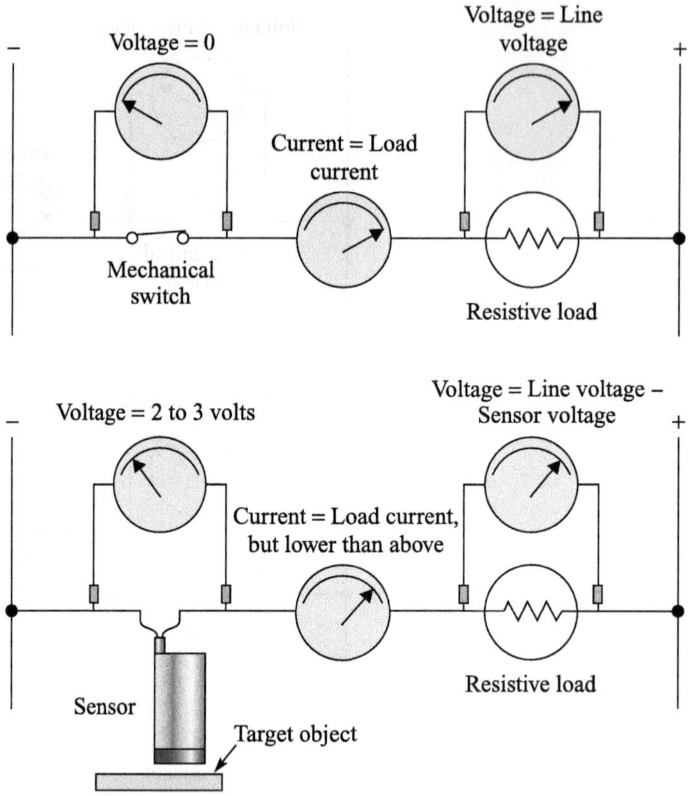

FIGURE 19-45 Comparison of mechanical and electrical closed switches

These electronic sensors also have an advantage over relay contacts. The mechanical switching properties of relays cause unreliability due to wear or eventual failure. The sensor performs its switching function electronically without mechanical deterioration.

Two-wire DC sensors are not *polarity-sensitive*. The term refers to the fact that the sensor will operate properly when the power source is connected at either polarity.

The wires of a two-wire sensor are color-coded based on IEC60947-5-2 standards. The blue wire is always connected to the negative of the power source, and the brown wire is connected to the positive of the power source. Depending on whether it is a sourcing or sinking sensor, its leads are connected to the power supply terminals either directly or through the load.

Two-Wire Sinking Sensor: The two-wire sinking sensor uses an NPN transistor. The brown terminal, shown in Figure 19-46(a), is its output and is connected to the load. The blue terminal is connected directly to the negative of the power supply. When a target is detected and the sensor turns on, conventional current flows from the positive of the supply, through the load, into the brown output wire through the sensor, and out of the blue wire to the negative of the supply.

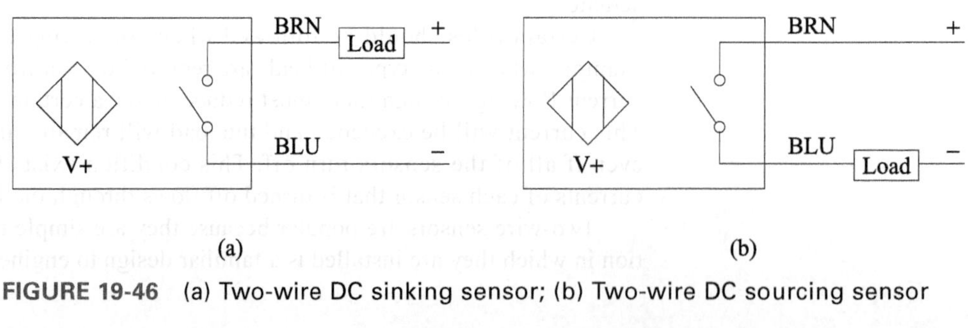

FIGURE 19-46 (a) Two-wire DC sinking sensor; (b) Two-wire DC sourcing sensor

Two-Wire Sourcing Sensor: The two-wire sourcing sensor uses a PNP transistor. The blue terminal, as shown in Figure 19-46(b), is its output and is connected to the load. The brown terminal is connected directly to the positive of the power supply. When a target is detected and the sensor turns on, conventional current flows through the brown wire from the positive of the supply, through the sensor, out of the blue output wire, and through the load to the negative of the supply.

IAU706
Three-Wire Sensors

The Three-Wire System **Three-wire systems** are defined as sensors that have three wires. One wire connects to a positive terminal of a DC power supply, the second wire is the output lead of the sensor, and the third wire is the DC common reference to which the other two are compared. Since the sensor gets the power it needs to operate from a power supply instead of using leakage current through the load (such as a two-wire sensor), it is called a *line-powered device.*

Three-wire sensors are divided into two categories by the way in which their outputs switch when they are activated:

1. *Negative Switching* If the voltage at the output lead decreases when the sensor is on, it uses the *negative switching* technique. Inside the sensor, shown in Figure 19-47, an NPN transistor is used as the switching element.

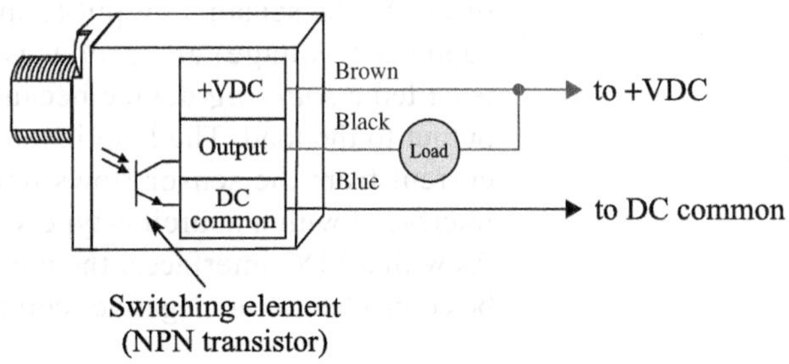

FIGURE 19-47 Negative-switching three-wire (sinking) sensor

When the transistor turns on, the sensor's output terminal switches to ground and produces a voltage potential across the load. The sensor is called a *sinking* device because conventional current flows into the sensor's output from the load in a manner similar to the way in which heat sink absorbs thermal energy. Since the conventional current is coming from the load, the load is referred to as a *sourcing* device. This designation is significant when sensors are connected to input modules of PLCs. In this situation, a sinking sensor is connected to a sourcing type input module of a PLC.

Some current sinking sensors have a resistor connected internally between the transistor collector and the sensor's power supply terminal. If the collector is not internally connected to anything else, as shown in Figure 19-48(a), it is called an *open collector*. The diagram in 19-48(b) gives an example of how an open-collector sensor is connected to a load, such as a TTL logic gate. It is necessary to connect a pull-up resistor between the collector output and the supply voltage of the load.

Most manufacturers have three-wire sinking sensors that are color-coded based on IEC60947-5-2 standards. A blue wire is connected to the DC common (negative terminal) of an external power supply, a brown wire is connected to the positive of the power supply, and a black wire is the output lead, as shown in Figure 19-47. The load is connected between the black output lead and the positive power supply terminal.

When the sensor is activated by a target, its internal NPN transistor turns on and conducts. Conventional current flows from the positive of the source, through the load into the black output lead, through the transistor inside the sensor, and out of the blue wire to the negative terminal of the power supply. When the target is not detected, the transistor is off and current does not flow through the load.

2. *Positive Switching* If the voltage at the output lead increases when the sensor turns on, it uses the *positive switching technique.* Inside the sensor, shown in

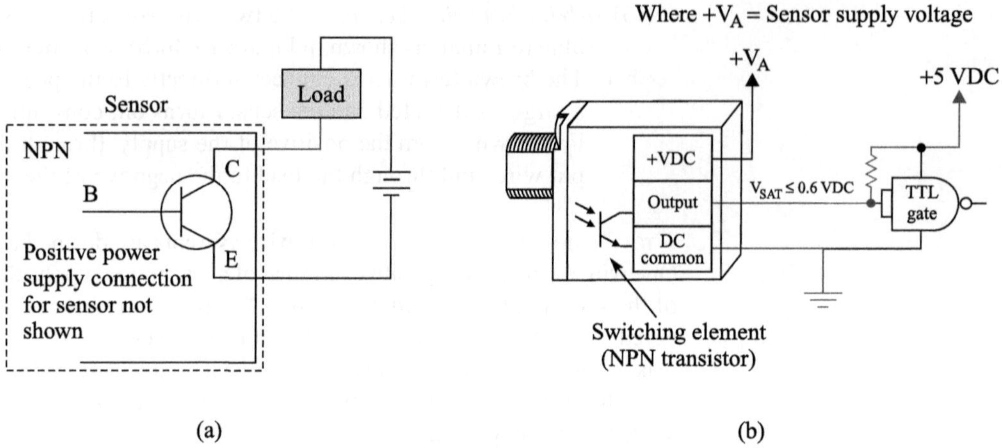

FIGURE 19-48 (a) Sinking sensor with an open collector; (b) Open collector connected to a TTL logic gate

Figure 19-49, a PNP transistor is used as the switching element. When the transistor turns on, the sensor's output terminal switches to a positive voltage (close to the supply's + voltage) and produces a voltage potential across the load. The sensor is called a *sourcing* device because conventional current flows from the sensor's output to the load. The load is referred to as a *sinking* device because conventional current from the sensor flows into it. Therefore, if a programmable controller is interfaced with a sourcing-type sensor, a PLC sinking input module must be used. As with all DC interfaces, the common terminals of the sensor and the load should be connected to the negative (common) terminal of the power supply.

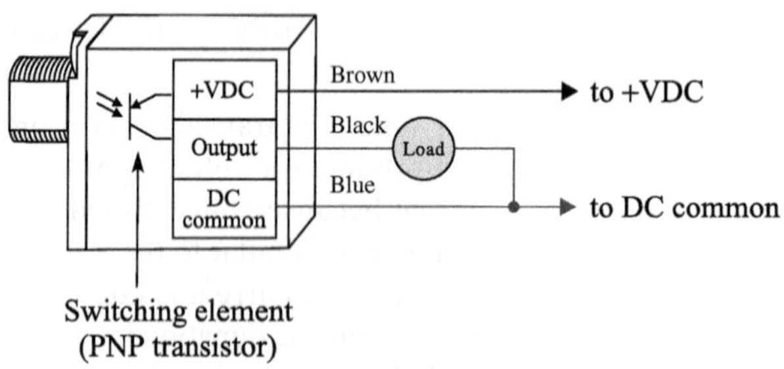

FIGURE 19-49 Positive-switching three-wire (sourcing) sensor

Most manufacturers have three-wire sourcing sensors that are color-coded based on IEC60947-5-2 standards. A blue wire is connected to the negative of an external power supply and a brown wire is connected to the positive of the power source, as shown in Figure 19-49. The load is connected between the black output lead and the negative (common) of the power supply.

When the sensor is activated by a target, its internal PNP transistor turns on and conducts. Conventional current flows from the positive of the power source, through the transistor inside the sensor, out of the black output wire through the load, and to the negative of the power supply.

Some current sourcing sensors have a resistor internally connected between the collector and ground. If the collector is not connected to anything else, it is an open collector, and a pull-down resistor should be placed between the sensor's output lead and ground.

In summary, the blue lead of a three-wire sensor is connected to the negative terminal of a power supply, the brown lead is connected to the positive terminal of the power supply, and the black output lead is connected to the load. This rule applies to both sensor types. The only difference is the polarity of the power supply that is connected to the load terminal opposite of the output lead. With the sinking sensor, the positive of the power supply is connected to the opposite terminal of the load; and with the sourcing sensor, the negative is connected to the opposite terminal.

IAU506
Four-Wire Sensors

The Four-Wire System Four-wire sensors are defined as sensors that have four wires. Two of the wires are connected to an external power supply. The other two wires are outputs, each of which is connected to a separate load.

Each of the four wires is a different color. A brown wire is connected to the positive of the power supply, and a blue wire is connected to the negative. One of the output wires is black. It is commonly referred to as *output 1*. The other output wire is white and is commonly referred to as *output 2*.

There are a variety of four-wire sensors. The difference between them is how the output wire connections are configured. A common four-wire sensor has outputs identified as N.O. (normally open) and N.C. (normally closed). The term *normally* refers to which type of electrical connection is made to the load when the target is not detected. The black wire is the N.O. output and the white wire is the N.C. output. When the target is detected, the output goes to the opposite switching state.

Sinking Four-Wire Sensors: The wires and the operation of a four-wire sinking sensor, shown in Figure 19-50(a), are identical to the three-wire sinking sensor. The brown and blue wires are connected in the same way to the power supply, and the connection of the black wire, is connected the same way to the load. When the target is not detected, the black output of the sensor does not pass sinking current from the load because its output is normally open. When the target is sensed, a transistor in the sensor turns on and conventional current flows from the positive of the power supply, through the load, and into the black output wire.

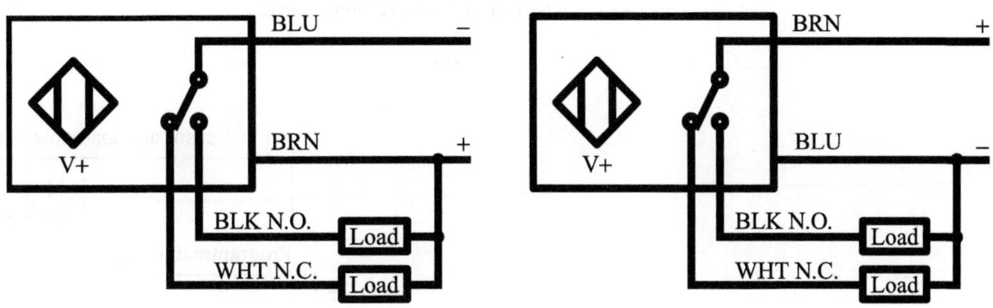

(a) (b)

FIGURE 19-50 (a) Four-wire DC sinking sensor; (b) Four-wire DC sourcing sensor

The connections made between the white output wire, load, and power supply are identical to the black wire output connections. The only difference is the switching action when the target is detected. When the target is not sensed, a transistor in the sensor is turned on. Conventional current flows from the positive of the power supply, through the load, and into the sinking white output wire. When the target is detected, the sensor transistor connected to the white wire turns off and current does not flow.

Sourcing Four-Wire Sensors: The wires and operation of a four-wire sourcing sensor, shown in Figure 19-50(b), are identical to the three-wire sourcing sensor. The brown, blue, and black wires are connected in the same way, and the black output operates the same way when the target is detected. The fourth wire is white, and is an output that operates the opposite way as the black output wire. When the target is not detected, a transistor in the sensor is turned on and conventional current flows from the white output, through the load, to the negative of the supply. When the target is detected, the sensor transistor connected to the white wire turns off and current does not flow.

Helpful Hints on How to Connect Three- and Four-Wire Sensors

- Always connect the brown wire to the positive of the power supply.
- Always connect the blue wire to the negative of the power supply.
- Always connect the black output wire to the load.
- The polarity of the power supply to which the load terminal opposite of the output lead is connected depends on whether it is a sourcing or a sinking sensor.
  - If it is a sinking sensor, the load is connected to the positive (+) terminal because conventional current flows from positive, through the load, and into the sensor's output sinking wire.
  - If it is a sourcing sensor, the load is connected to the negative (−) terminal because conventional current flows from the sourcing output wire, through the load, and into the negative of the supply.

Other types of four-wire sensors are shown in Figure 19-51.

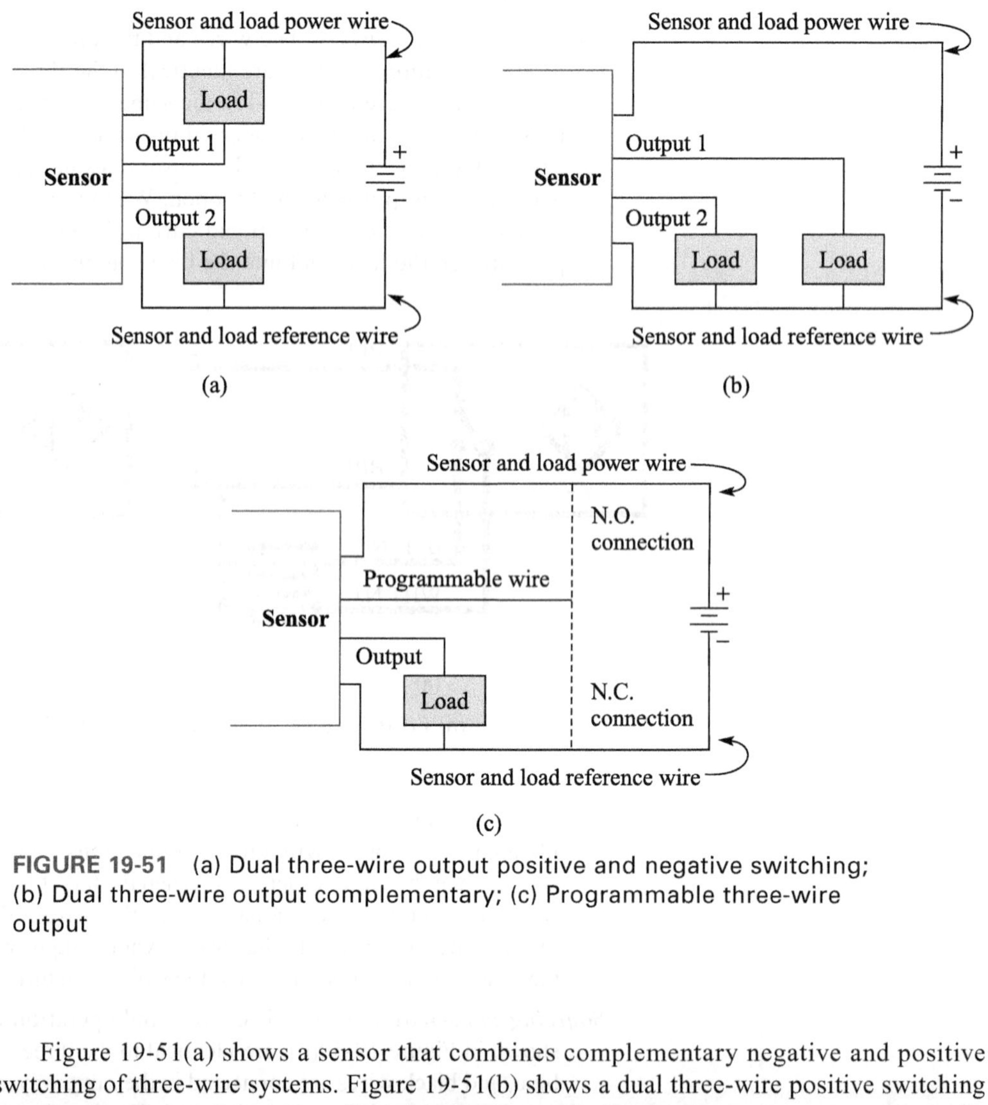

FIGURE 19-51 (a) Dual three-wire output positive and negative switching; (b) Dual three-wire output complementary; (c) Programmable three-wire output

Figure 19-51(a) shows a sensor that combines complementary negative and positive switching of three-wire systems. Figure 19-51(b) shows a dual three-wire positive switching sensor where two outputs are activated simultaneously. Dual output sensors are also available for negative switching. Figure 19-51(c) shows a fourth wire used as a programmable wire. If it is connected to the positive lead of the power supply, the switching transistor is off until it is turned on when the sensor is activated. When the programming wire is connected to the reference, the switching transistor is on until the sensor is activated.

Analog Signal

Sensors with analog signal outputs are usually used in process-control applications. For example, thermal sensors, or *thermocouples*, produce analog signals to indicate temperature. The signal a thermocouple generates is directly proportional to the temperature it is sensing: the higher the temperature, the greater the voltage the sensor produces. In contrast, one type of photoelectric sensor produces an analog signal that is inversely proportional to the translucency of an object it is measuring.

Electronic sensors that produce analog signals are described in Chapters 10 through 14 of Section 5, Process Control and Instrumentation.

Problems

1. The object to be detected by an industrial detection sensor is called a ____.
   a. retroreflector c. target
   b. detector d. emitter
2. Limit switches are used for which of the following applications? ____
   a. size detection c. counting parts
   b. limiting travel d. all of the above
3. A sensor that is designed to detect an object exclusively in front of the sensor head uses the ____ approach.
   a. radial b. axial
4. T/F The trigger stage performs the hysteresis function of the sensor.
5. As eddy currents are induced into a metal target by an inductive proximity detector, its oscillator circuit ____ energy.
   a. loses b. gains
6. Another term used for hysteresis is ________.
7. A standard target for an inductive proximity sensor is a square piece of mild steel, ________ mm thick, with a length of each side equal to the diameter of the ________.
8. The ____ of a target affects the distance at which it can be detected from the head of an inductive proximity sensor.
   a. size c. permeability
   b. position d. all of the above
9. The capacitance of the capacitive proximity detector ____ when a nonconductive material is within its range, and ____ when a conductive material is within its range.
   a. increases b. decreases
10. The ____ of a target affects the distance at which it can be detected from the head of a capacitive sensor.
    a. size d. dielectric constant
    b. shape e. all of the above
    c. position
11. A ____ proximity detector can reliably sense the exact position of a target that is coated with dirt.
    a. inductive b. capacitive
12. A voltage develops at the output terminals of a Hall-effect sensor when a magnetic field is ____ its sensing range.
    a. within b. outside
13. Which of the following variables affects the voltage amplitude of a Hall generator? ____
    a. output capacity of the power supply
    b. strength of the target's magnetic field
    c. size of the generator
    d. all of the above
14. The polarity of a Hall-effect sensor's output voltage will be reversed if the ____ reversed.
    a. power supply leads are c. both a and b
    b. polarity of the magnetic target is
15. Metal pieces on each side of a Hall-effect plate, called ________, increase the output voltage of a linear Hall-effect sensor.
16. Infrared LEDs produce ____ light energy than visible LEDs.
    a. weaker b. stronger
17. Phototransistors best respond to ____ light.
    a. visible b. infrared
18. The process of rejecting ambient light and selecting an On-Off frequency the same way a radio tuner operates is referred to as ________.
19. T/F Photodetectors with a high gain are capable of detecting the level of cereal inside a sealed box.
20. Match the following methods of light detection with the mode of detection that they use:
    a. light-to-dark b. dark-to-light
    ____ Opposed
    ____ Retroreflective
    ____ Diffuse
    ____ Convergent
    ____ Specular
21. The ________ optical sensing method is capable of detecting clear objects at a specific distance.
22. Which method of light detection provides the highest level of optical energy?
23. Which method of light detection uses a disk that looks like a bicycle reflector?
24. Which method of light detection is the best photoelectric sensor to use for monitoring very small parts?
25. Which method of light detection uses a series of sensors to form a light curtain?
26. What is the most common light detection method used to differentiate between shiny and dull surfaces?
27. Fiber-optic cables should be used to perform detection operations under which of the following conditions? ____
    a. The sensing space is confined.
    b. The environment is hostile.

c. The sensor is exposed to magnetic noise.
d. All of the above.

28. If emitted light is attenuated by 50 percent, an excess gain of ____× is required to overcome the loss of light.

29. Suppose the diameter of a target is 2 inches, the light beam width is ½ inch, and the speed of the object is 5 inches per second. What is the minimum response time required by the sensor?

30. The span of an analog ultrasonic sensor is the ____.
a. minimum distance of its sensing range
b. maximum distance of its sensing range
c. magnitude of its sensing range

31. The output of an analog ultrasonic sensor that produces 4 to 20 mA of current will be ____ mA when the target is located halfway between the minimum and the maximum calibrated sensing distance.

32. Calculate the values of the resistor and the capacitor in a snubber circuit when the voltage across the contacts is 200 volts and the holding current is 100 mA.

33. Conventional current flows ____ the output of a current sinking sensor and ____ the output of a current sourcing sensor.
a. into b. out of

34. Leakage current flows through a ____-wire sensor when it is off and it does not detect a target.
a. two c. four
b. three

35. If the voltage at the output lead ____ when the sensor turns on, it uses the negative switching technique.
a. increases b. decreases

36. A ____ switching sensor is also referred to as a sourcing sensor.
a. negative b. positive

37. Based on European standards, the ____ wire of a three-wire sensor is connected to the positive terminal of an external power supply, the ____ wire is connected to the negative terminal of the power supply, and the ____ wire is connected to the load.
a. blue c. black
b. brown

38. The lead connection to the load opposite to where the black wire of a three-wire ____ sensor is connected goes directly to the positive terminal of an external power supply.
a. sourcing b. sinking

CHAPTER 20

Industrial Wireless Technologies

OBJECTIVES

At the conclusion of this chapter, you should be able to:

- Explain what a radio wave is, and how it is produced.
- Identify and describe the characteristics of topologies used in industrial wireless systems.
- Define the following terms:

Duty Cycle	GSM	GPRS	CSMA
Energy Harvesting	TDMA	Attenuation	Radio Noise

- Describe the modulation and demodulation of digital data and analog signals.
- Identify and describe the applications of various types of wireless technologies.
- Describe power-management considerations for reducing the consumption of electrical power by wireless field devices.
- Explain how transmission power and distance affect the strength of a radio-frequency (RF) signal.
- List and explain the factors that affect the transmission of an RF signal.
- Describe the methods used to ensure the security of wireless transmissions.
- Identify industry standards for wireless technology and describe the application for each one of them.

INTRODUCTION

Many of the devices we commonly use are wireless. They include car radios, cell phones, satellite television, and accessing the Internet within Wi-Fi "hot spots."

Wireless devices are also used for industrial applications, and have existed for over 40 years to monitor oil and natural-gas pipelines. Over the past decade, wireless devices have been developed for various applications in manufacturing plant facilities. However, early attempts to adopt wireless technology for industrial applications met with limited success due to a variety of problems encountered. For example, electromagnetic noise from motors and welders, echoes from obstacles such as buildings and storage tanks, interfered with RF signals. Other factors such as limited battery life, crowded frequency space, undeveloped security standards, and lack of familiarity hampered the acceptance of this technology.

As technological advancements have been made to resolve these concerns, efforts to inform engineers, technicians, and maintenance personnel of the benefits of wireless devices have resulted in wireless devices becoming more widely accepted.

The major benefit of wireless systems is the amount of money that can be saved by using them. For example, the installation cost of wireless devices, such as sensors, is much less than comparable devices connected to cables. Running cables is very time-consuming and requires significant labor costs. Additional installation expenses include the cable, conduit, junction boxes, and digging trenches when some of the conductor connections are installed under roads or between buildings and other structures.

Table 20-1 shows a typical cost comparison of wired and wireless systems.

TABLE 20-1 Cost per Device

	Wired	*Wireless*
Hardwired cost	$1000	$1500
Installation cost	$2000 – 20,000	$1000
Maintenance cost	Negligible	$1000

Lower installation costs are only a part of the economic savings using wireless technology. Using a larger number of wireless field sensors than could previously be used because it was too cost-prohibitive, automated manufacturing can be more closely monitored and controlled. The result is higher profits due to higher quality and increased efficiency by eliminating the need for mobile workers to manually take hundreds of periodic readings of gauges, chart recorders, or other indicators at remote measuring points. In addition to the economic benefits they afford, sensors provide a variety of vital up-to-date information, such as:

- Determining how much inventory is being stored, such as in an oil tank farm to prevent outages.
- Provide diagnostic information that informs maintenance technicians that a part is wearing or that a malfunction has occurred.
- Improving environmental compliance capabilities. For example, a wireless sensor that monitors a pressure-relief valve can activate an alarm in the control room, and it can document when any harmful emissions occurred and how long they lasted.

20–1 Wireless Architecture

There are three basic components of an industrial communication system, whether it is wired or wireless.

- A transmitting **source** that sends a signal.
- The **medium** to carry the signal.
- The **receiver** to which the signal is sent and processed.

Figure 20-1(a) shows the typical wired communication architecture for an automated system. A sensor (field device) measures a process, such as temperature or flow. The signal from the sensor (source) is converted to a standard electrical signal, such as 4 to 20 mA, and is sent through a cable (medium) by a transmitter to a control system (receiver), such as a PLC, microcontroller, or computer.

A wireless system transmits data using radio waves as compared to hardwired devices that send electrical signals over cables, as shown in Figure 20-1(b). Electronic circuitry and antennas convert electronic signals from the sensor measuring the process into radio waves. At the receiver, another antenna and electronic circuitry convert the radio waves into information that indicates the measurement.

All industrial wireless systems use this basic model. However, there are a variety of architectures and wireless technologies used that are based on particular applications and

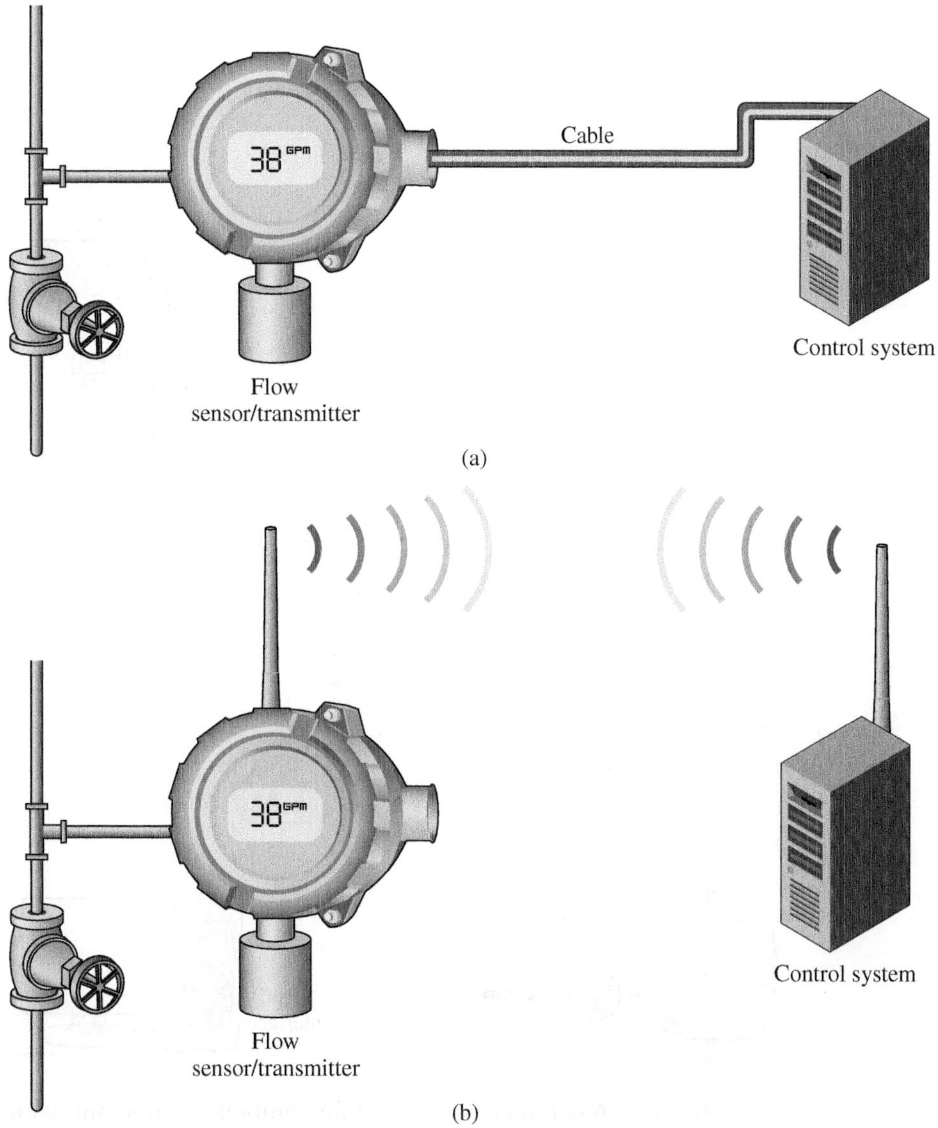

FIGURE 20-1 Typical wired communication architecture: (a) A sensor/transmitter connected to a control system using cables; (b) Wireless devices using radio waves to transmit data

specific needs. Figure 20-2 shows an example of a wireless system that uses one technology for a field network that carries data from sensors, and a different technology for the gateway connected to the controller (computer) that gives mobile workers easy access to control and maintenance information systems.

20-2 Wireless Signals

In a wireless system, signals travel from a transmitting source to a receiving device. The wireless signals are radio-frequency (RF) electromagnetic waves that can travel through air, a vacuum such as outer space, and even some solids and liquids.

Electromagnetic signals are emitted from a conductor. When electrical current flows through a conductor, such as a wire, an electromagnetic field builds up around it, as shown

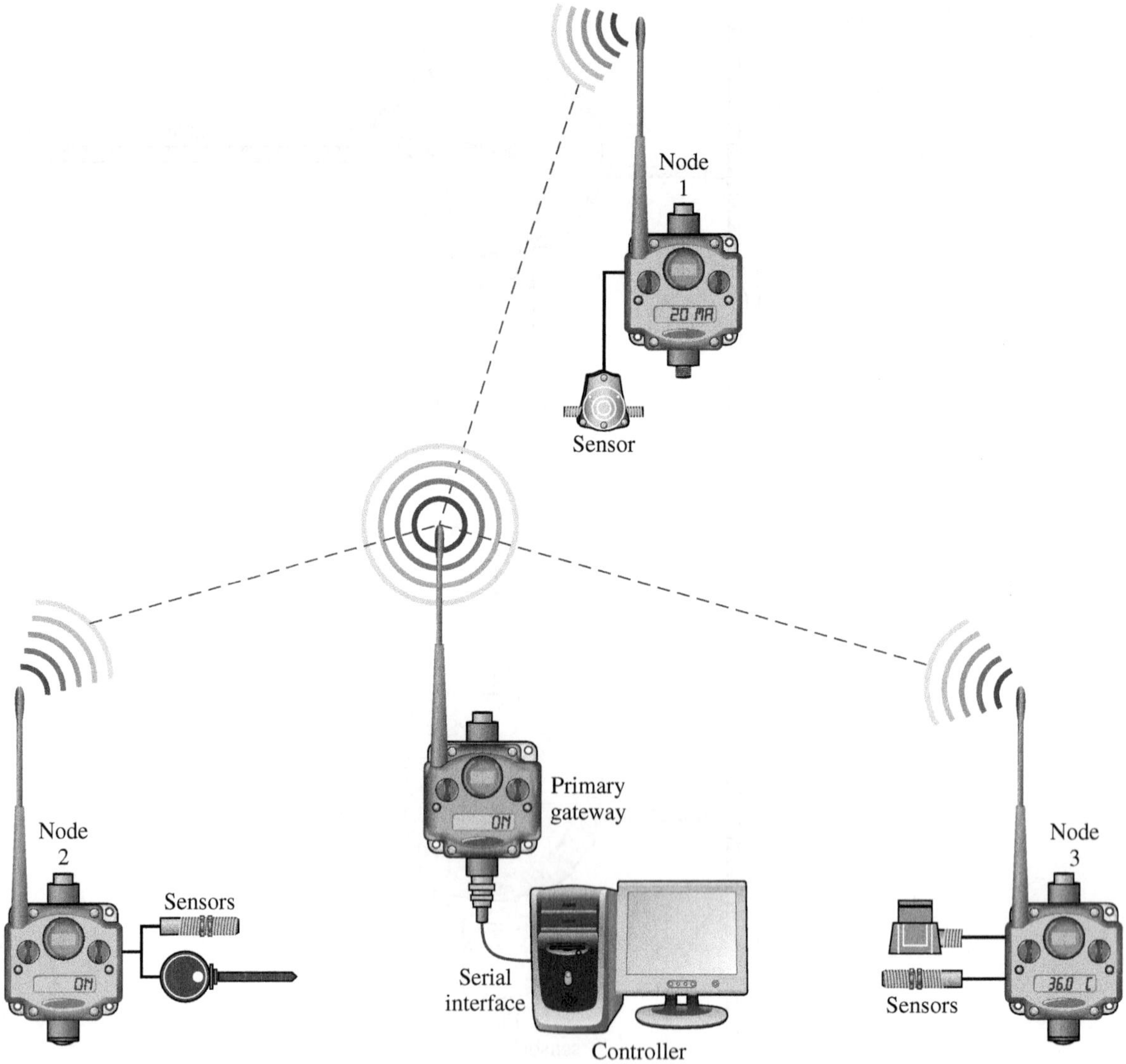

FIGURE 20-2 A communication system combining wired and wireless devices

in Figure 20-3(a). A transmitting antenna is a conductor. Each time a pulse of current flows through it, an electromagnetic wave is emitted, similar to the wave action that occurs when a rock is thrown into water, as shown in Figure 20-3(b). The rate at which the pulses occur is referred to as *frequency*. The number of pulses that are produced each second is called *hertz*. When an AM radio station operates at a frequency of 1450 kHz, 1450 thousand pulses of current are sent through its antenna each second, and the same number of electromagnetic waves are transmitted through the air.

When a radio wave reaches a receiving radio, a voltage in microvolts is induced into its antenna as the wave passes through it. The small voltage is then amplified by electronic circuitry before it is applied to the radio speaker. Electromagnetic waves from many transmitting station antennas are received simultaneously at a radio antenna. A tuner circuit in the radio electronically passes one desired frequency to its amplifier, and rejects all other frequencies.

The strength of the signal emitted from the transmitter is determined by the number of watts it is capable of producing. The higher the wattage, the greater the amount of current that is pulsed through its antenna and the stronger the magnetic field it develops. The stronger the magnetic field, the farther the radio wave will travel. Each radio wave emitted from the transmitter travels at approximately the speed of light, which is 186,000 miles per second.

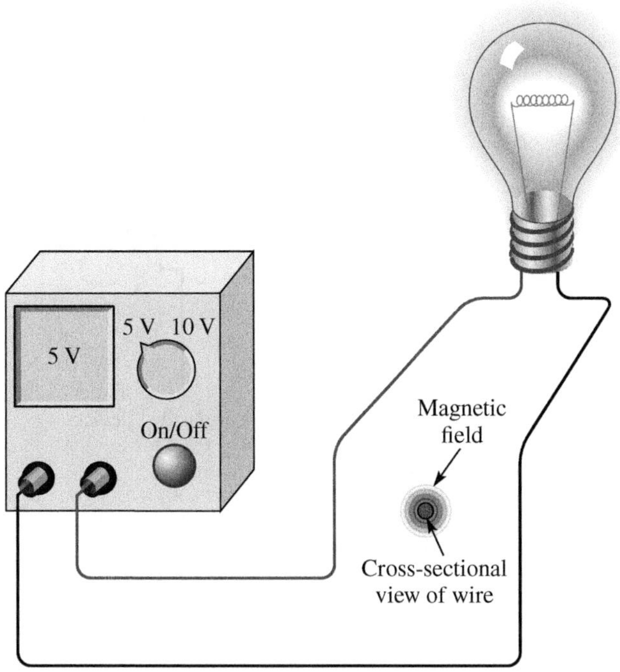

(a) When current flows through a wire, a magnetic field builds up around it.

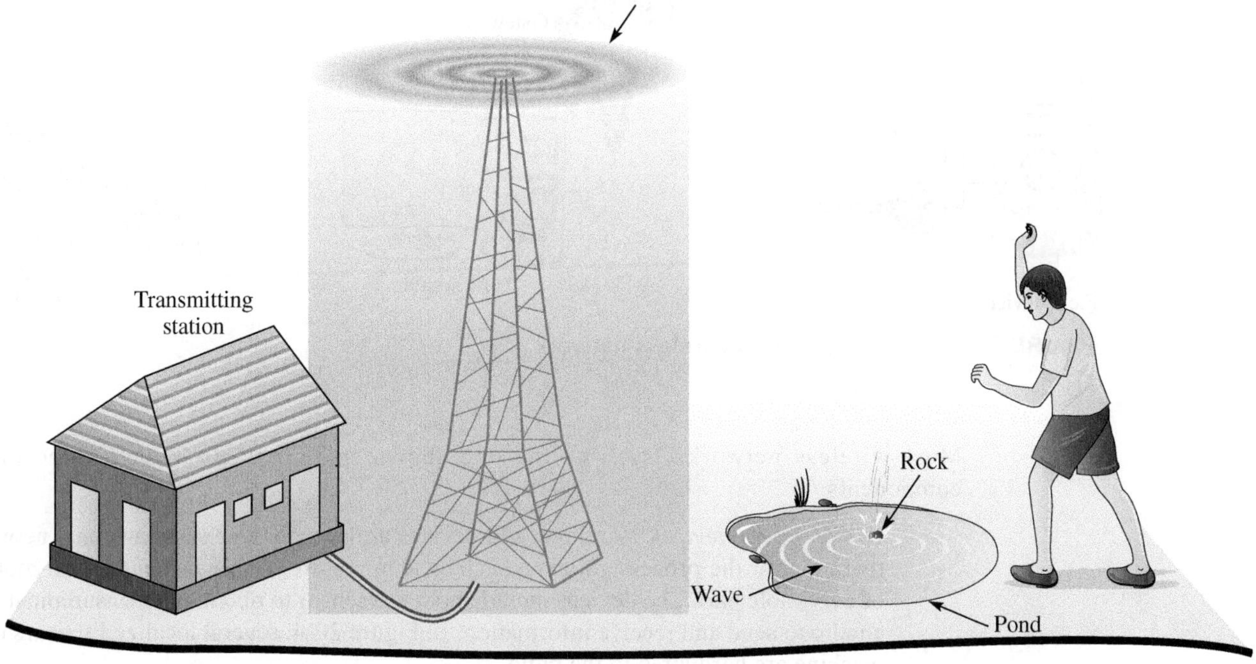

(b)

FIGURE 20-3 Electromagnetic radio waves

20-3 Wireless Topologies

The components in a wireless system are linked together into groups so that data can be exchanged between them. These groups of components are called a *network*, and the communication that takes place between them is called *networking*.

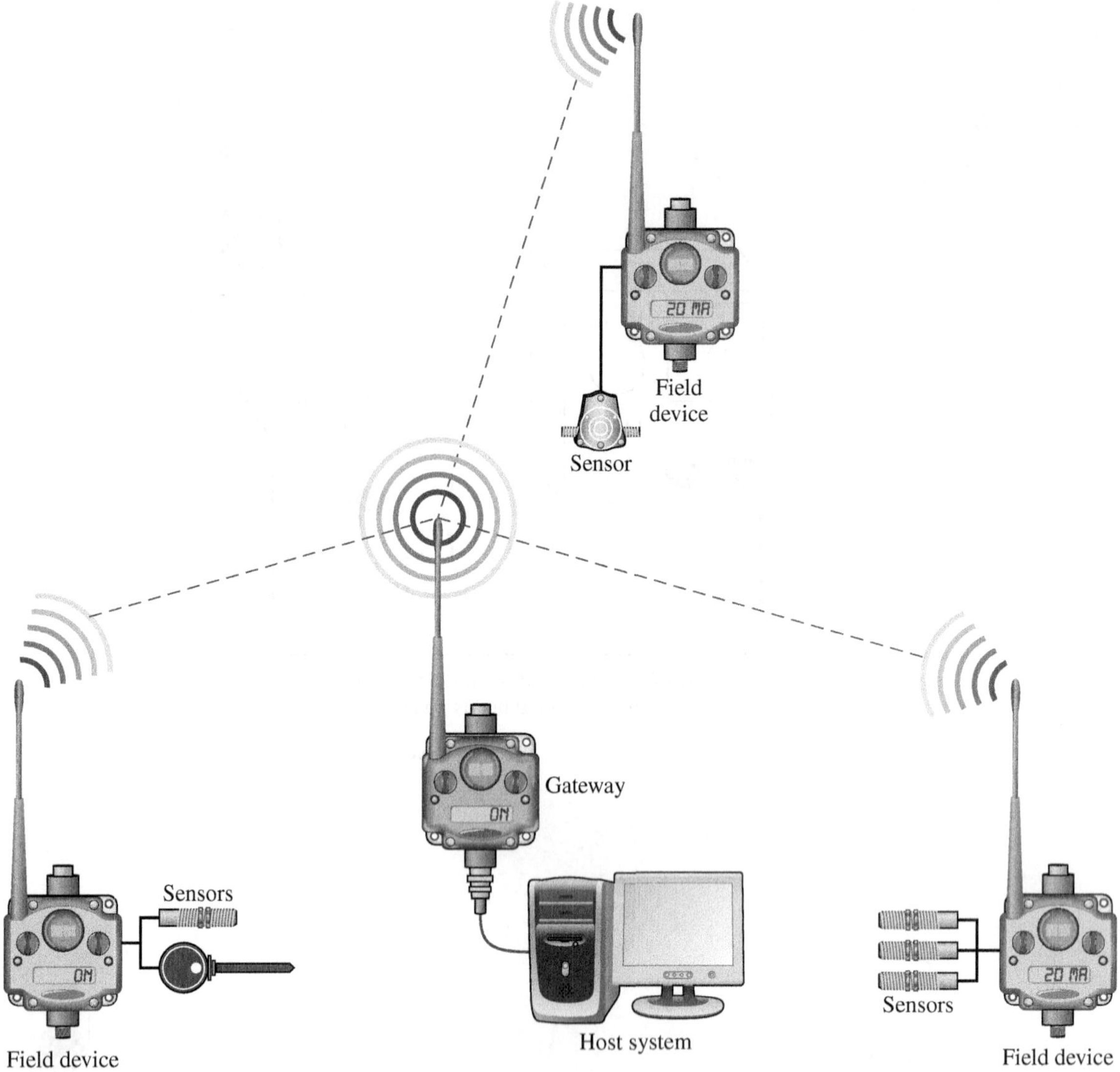

FIGURE 20-4 Components of a wireless network

Most wireless networks, such as the one shown in Figure 20-4, have three basic components.

- *Wireless field devices.* Wireless field devices are end-point components on a network that monitor the process variable, such as temperature, level, or the physical motion of a movable part. The devices include both the sensor to obtain the measurement and a radio to send and receive information. In Figure 20-4, several localized sensors on a machine are hardwired to the radio.
- *Host information systems.* The host information system is a computing device that performs various manufacturing functions, such as operating as the controller element of a closed-loop system, activating an alarm to prevent injury, compiling historical data that provides diagnostic information for troubleshooting, recording emissions to reveal environmental concerns, or providing information to help determine more efficient ways to operate a plant and manage its assets. Types of industrial computing devices include computers, PLCs, microcontrollers, or SCADA systems.

- *Gateway.* A gateway is a wireless device that is both a receiver and a transmitter. Its function is to interface between the field device and the host system, allowing them to communicate with each other.

The components in a network are arranged in various types of physical configuration, called *topologies*. The configuration that is selected is often based on such factors as:

- Reliability
- Flexibility
- Cost
- Power consumption
- Transmission distance
- Future growth
- Preventing the disruption of data flow

The term *topology* refers to the physical layout of the devices in a network and the paths that data travels between them. These devices include sensors, transmitters, routers, and gateways. There are three wireless topologies commonly used for in-plant wireless field network applications. They include *star, mesh,* and *cluster-free*. The characteristics of each layout, including strengths and weaknesses, must be considered to help determine which topology is best for a particular application.

Star Topology

A star topology, shown in Figure 20-5(a), resembles the hub of a wheel with spokes extending outward. All of the devices in the network communicate directly with the *hub,* which is a device more commonly referred to as a *gateway*. The signals from the sensor pass through the gateway to the host system, which is a collection point such as a central control room. In some star networks, signals are also sent from the gateway to individual sensors.

Because each device in a network communicates directly with the gateway, the star topology is sometimes referred to as a "point-to-point" or "line-of-sight" architecture. The drawback of the star topology system is that if something disrupts the direct path between a field device and its gateway, as shown in Figure 20-5(b), the radio wave will likely be blocked or hindered, and the data lost. Physical objects that impede the communication paths include storage tanks, piping, equipment, or even temporary obstructions such as a truck or construction equipment. Environmental factors such as rain, snow, or fog can also affect the

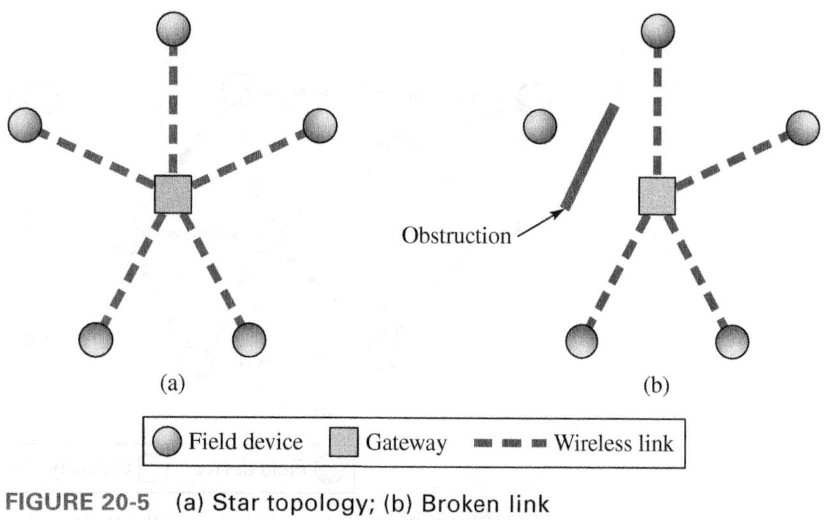

FIGURE 20-5 (a) Star topology; (b) Broken link

radio signals. Another factor that can cause an interruption to the signal is when there is a radio-frequency interference caused by a machine, such as a motor that emits electromagnetic waves.

Conducting a site survey during installation can minimize interference problems. The survey identifies where the components in a network can be placed to provide a greater likelihood for successful line-of-sight transmissions, and to ensure that the maximum range is not exceeded. If new construction or environmental changes occur that interrupt the line-of-sight transmission in the future, there are three options that may provide a solution.

1. Move the field device to provide a better path for the radio waves.
2. Move the gateway to provide a better path for the radio waves.
3. Increase the signal strength of the field device. However, this will shorten the sensor's battery life.

Star topology networks are limited to distances of 30 to 100 feet between the field device and the gateway. To conserve the amount of current drawn by the field devices, and to extend the life of the batteries that power them, sensors can be programmed to be in the sleep mode, where they draw minimal power, and then turn on only when data needs to be sent.

Mesh Topologies

Unlike the star topology configuration, in which each field device communicates only to the gateway, the devices in a mesh topology are equipped with radios that enable them to communicate with the other field devices in the network as well. A mesh topology configuration and its multiple transmission paths are shown in Figure 20-6(a).

There are two advantages of a mesh network over a star topology.

1. *Higher fault tolerance.* Each wireless field device has more than one transmission path to the gateway. If any path is blocked, as shown in Figure 20-6(b), the sensor's radio waves are sent to the other field devices in the network, which enables data to reach the gateway by alternative routes.
2. *Expansion capabilities.* If field devices are relocated or added, the only requirement is that they must be within the signal range of existing field devices in the network to communicate with the gateway.

There is one disadvantage of a mesh network. Unlike star topology field devices that operate in spurts to save battery life, mesh field devices need to always be on so that they can transfer

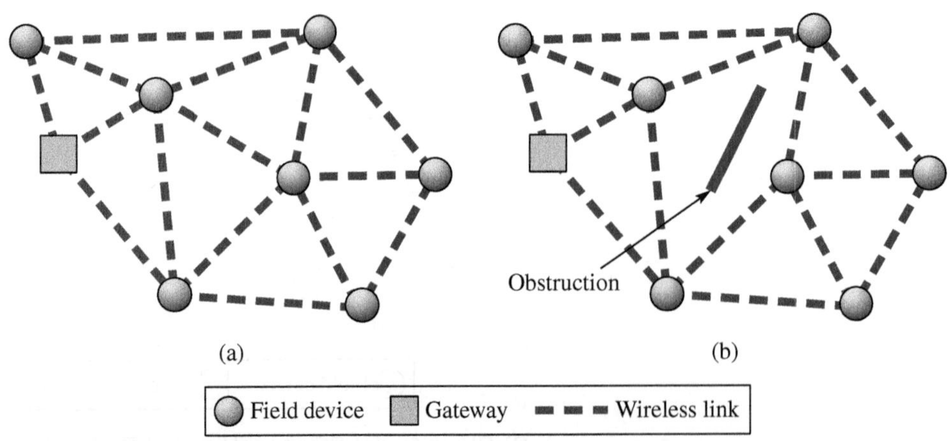

FIGURE 20-6 (a) Mesh topology; (b) Broken link

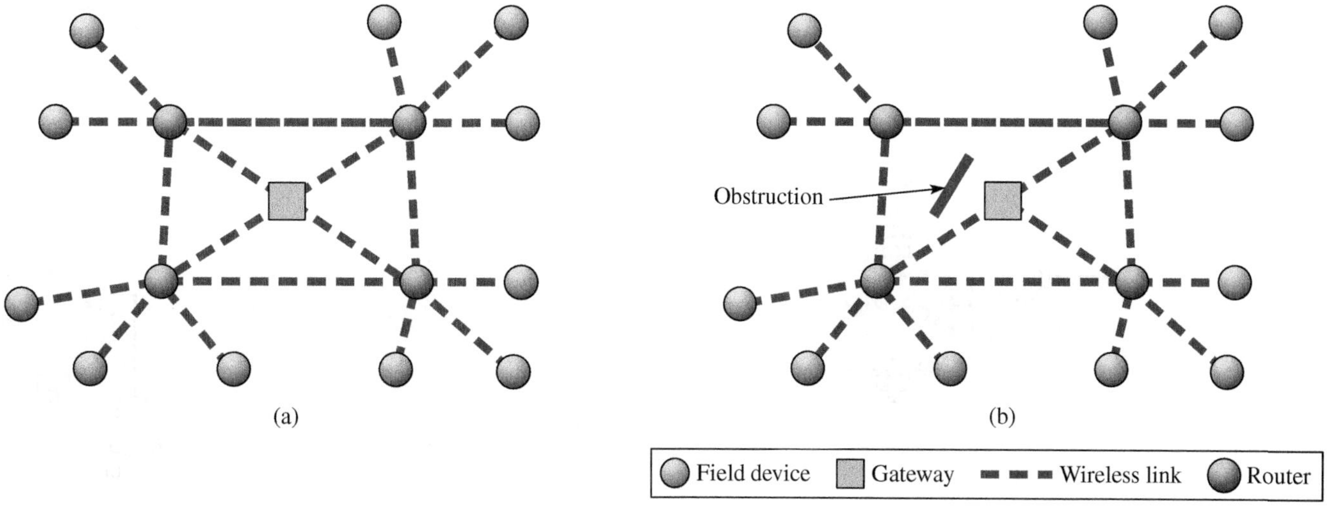

FIGURE 20-7 (a) Cluster-free topology; (b) Broken link

signals from other sensors. Since they never turn off, they continuously draw current, which reduces battery life.

Cluster-Free Topologies

A cluster-free topology is shown in Figure 20-7(a), and is a hybrid of star and mesh topologies. The field devices in star configurations are clustered around routers, which are repeaters that relay data. The routers and gateway, which are configured in a mesh topology, communicate with one another. Cluster-free configurations combine the advantages of both the star and mesh topologies: low power consumption, extended range, and fault tolerance when there is an obstruction, as shown in Figure 20-7(b).

20-4 Self-Organizing Networks

A self-organizing network consists of field devices that have microcomputers, which enables data to be moved more efficiently and reliably than other network options. Figure 20-8(a) shows the simplest communication paths possible in a self-organizing network, where the wireless device communicates directly with the gateway.

One function of the microcomputer is to figure out alternative transmission paths for field devices to the gateway. The microcomputer enables each device to operate as a router that relays data from other devices in the network. The routing function creates multiple communication paths, passing data along until it reaches the host information system. Suppose a truck enters and blocks the path between two field devices, as shown in Figure 20-8(b). A router will find alternative paths by using other routers to overcome the obstruction. The ability to automatically reconfigure the data paths without manual intervention gives the self-organizing network its name.

Because there are multiple paths for data to flow in self-organizing networks, collisions will occur when more than one message arrives at a field device router or gateway at the same time. When this situation occurs, data can be corrupted or lost. To avoid these problems, one of two different methods is used: CSMA (carrier sense multiple access) and TDMA (time division multiple access).

CSMA. With this method, the devices in the network transmit when they have data to send. If two or more transmissions occur simultaneously, messages often will collide. Each device tries to resend the data, either along a different path or at a

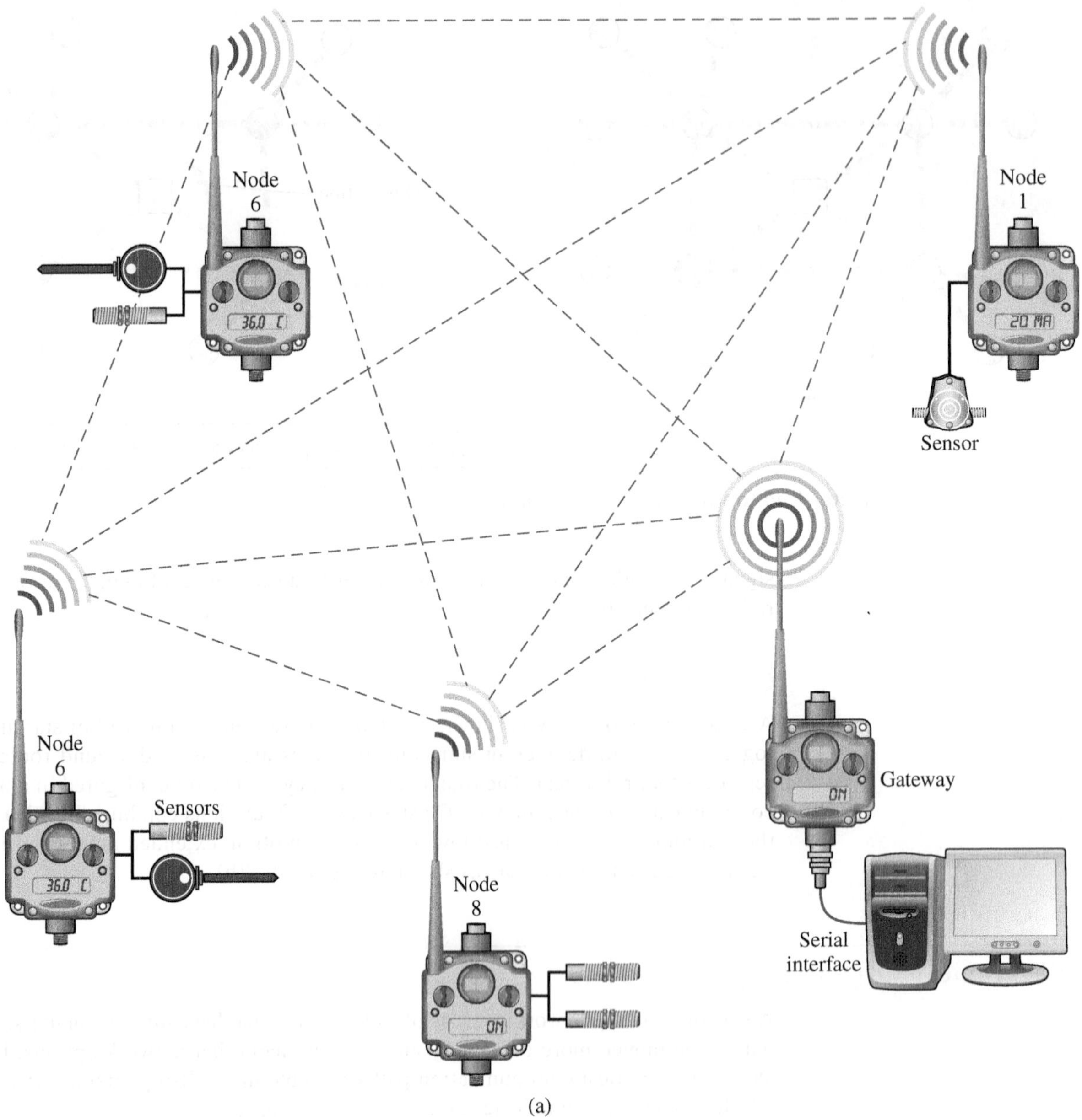

FIGURE 20-8 Self-organizing network: (a) Simplest communication path; (b) Blocked signal rerouted

different time. The drawback of CSMA is that the collisions and retries can bog a network down if they become excessive, which will occur if the network becomes large enough.

TDMA. With this method, each device in the network has its own specific time slot to transmit to another field device or gateway. Data is stored until its scheduled transmission. Because only one device transmits in any given time period, there are no collisions and resulting retries.

Because it is more reliable, can operate more rapidly by avoiding collisions, and has lower power consumption because it operates only during its scheduled time, the TDMA method is preferred over CSMA options.

Self-organizing networks are 99 percent reliable. *Reliability* refers to the percentage of transmitted messages that reach their final destination. This percentage increases as the network is expanded because adding more devices provide a potential for a larger number of communication paths.

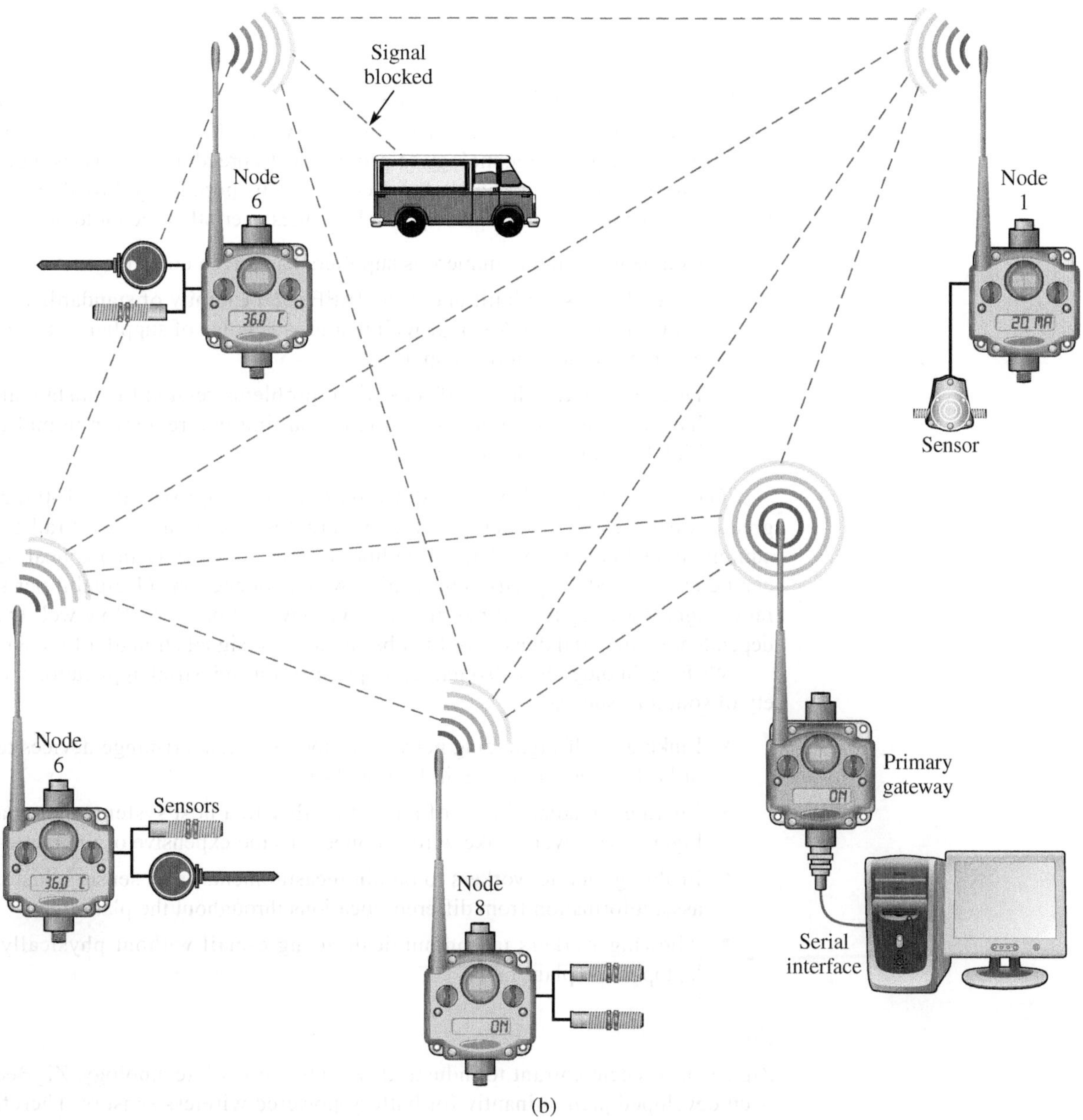

FIGURE 20-8 *(continued)*

20-5 Wireless Technologies

During the past several years, a broad array of wireless technologies have been developed and widely adopted, and continue to evolve at a rapid pace. Some of these wireless technologies are suitable for industrial applications. They include Wi-Fi, Zigbee, WiMAX, cellular, satellite, and RFID. Factors such as range, throughput, power consumption, and penetration are considered when determining which technology is best for specific applications.

Range. The distance that wireless devices can effectively communicate between the transmission source and the receiver.

Throughput. The rate at which wireless devices can transmit digital data. The more throughput, the more rapidly data is sent.

Power consumption. The amount of electrical power that is used by devices of a particular wireless technology.

Penetration. The capability of the radio signal to travel through obstructions.

Wireless Technologies

Wi-Fi

Wi-Fi (wireless fidelity) technology, commonly referred to as *wireless Ethernet*, is the most widely used wireless networking protocol in the world. Enormous progress has been made in the development of this technology to provide widespread use in offices, homes, and public Internet access points, or "hotspots." Because of its popularity, broad use has led to mass production that has created a variety of advantages over other technologies, such as:

- Competition among numerous suppliers has driven down the cost of Wi-Fi devices.
- Wi-Fi devices are built under the IEEE802.11 family of standards for wireless local area networks (WLANs), providing a broad choice of suppliers, easy integration, and assurance of long-term support.
- Due to countless hours of operation, problems related to substandard engineering have been identified and corrected, resulting in a reduction in malfunctions and a high level of reliability.

The range of typical Wi-Fi radio waves is up to approximately 160 meters, although greater distances up to several kilometers are possible with a 1-watt radio, depending on equipment and conditions. The throughput of these devices is also appealing because data can be transmitted at speeds approaching wired connections (1 gigabit per second). Wi-Fi radio signals are capable of passing through physical barriers. How well they penetrate is dependent on material density of the obstruction and signal strength of the RF signal.

Wi-Fi technology has also gained popularity for industrial applications, offering a variety of solutions such as:

- Linking a self-organizing network of low-power, short-range devices to a wired LAN or host system, as shown in Figure 20-9.
- Linking an isolated area of a plant facility to a host system where barriers such as highways or rivers make wired connections too expensive or impractical.
- Enabling mobile workers to obtain measurements from sensors, historical data, and asset information from different locations throughout the plant.
- Allowing workers to communicate using e-mail without physically connecting a laptop to the plant LAN.

ZigBee

ZigBee is a recent entrant to industrial wireless network technology. ZigBee devices have been developed predominantly for battery-powered wireless sensors. Therefore, they have been designed to draw very little power, enabling batteries to last for years. The drawback of ZigBee devices is that signals between two devices are limited to about 20 meters. To extend their operational range, up to 64,000 devices can be connected in a mesh topology network configuration to pass data from one unit to another. An additional drawback is that they have poor throughput because they are limited to 720 kilobits per second. ZigBee devices can use either 868/900 MHz when high penetration is required, or 2.4-GHz ISM bands.

WiMAX

WiMAX is an acronym for *worldwide interoperability for microwave access*. It is designed for metropolitan area networks (MAN), specifically to be used in cities to connect Wi-Fi hotspots with one another and with the Internet. Operating at a relatively low frequency range between 2 and 11 GHz, WiMAX radio signals are able to penetrate most physical obstructions, and also provide a high throughput up to 75 Mbps (megabits per second).

Cellular Technology

Cellular phones operate by connecting to the closest tower and automatically transferring phone connections from one tower network cell to another.

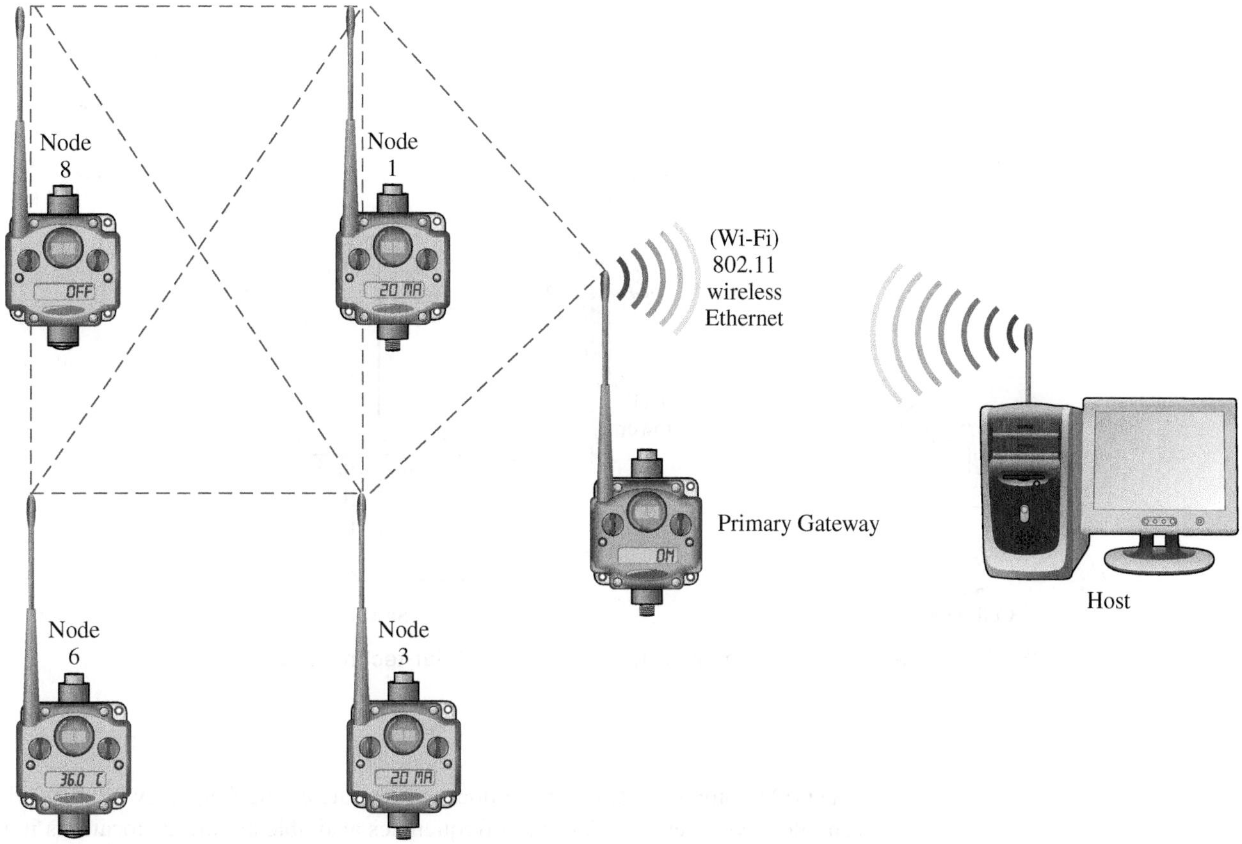

FIGURE 20-9 Wi-Fi field devices linked to a Wi-Fi host

The evolution of cellular technology has advanced rapidly due to the growth of its customer base. The result is that the technology is both highly dependable and affordable. Cellular service providers are constantly expanding their network of towers and other infrastructure so that coverage has reached many remote areas of the country.

To take advantage of this technology, manufacturers of industrial wireless systems are building devices that operate using existing public cellular networks. This technology provides a solution to the necessity of collecting data from remote and geographically dispersed locations, for example, monitoring inventory in remote storage tanks that were formerly prohibitive due to the high cost of long-distance cabling, or the high labor cost of going to remote sites and manually gathering the information. Figure 20-10 shows how a cellular radio transmits tank level measurements from a remote site to a nearby cell-phone tower. The data is transferred to a landline and travels over the phone system or the Internet to a secured server, where it is stored until accessed by an authorized user. The tank level data is then shown on a PC monitor or other type of display device.

Applications using cellular networks are currently limited to collecting periodic measurement data usage due to their high operating cost. For example, there may be a charge for each field device on a network. Also, cellular technology is not used for closed-loop control applications because of the high cost of continuous high-speed communication required to perform automation processes.

When cellular technology was first developed, there was not a universal communication standard established. The result was that various methods of transmitting data signals were created. As cellular technology has evolved, two formats have emerged that currently dominate the market: **GSM** and **GPRS**.

Global system for mobile communication, or GSM, is the most popular cellular technology for making phone calls. Its first widespread use was in Europe, and has now expanded to serve approximately 80 percent of the global market.

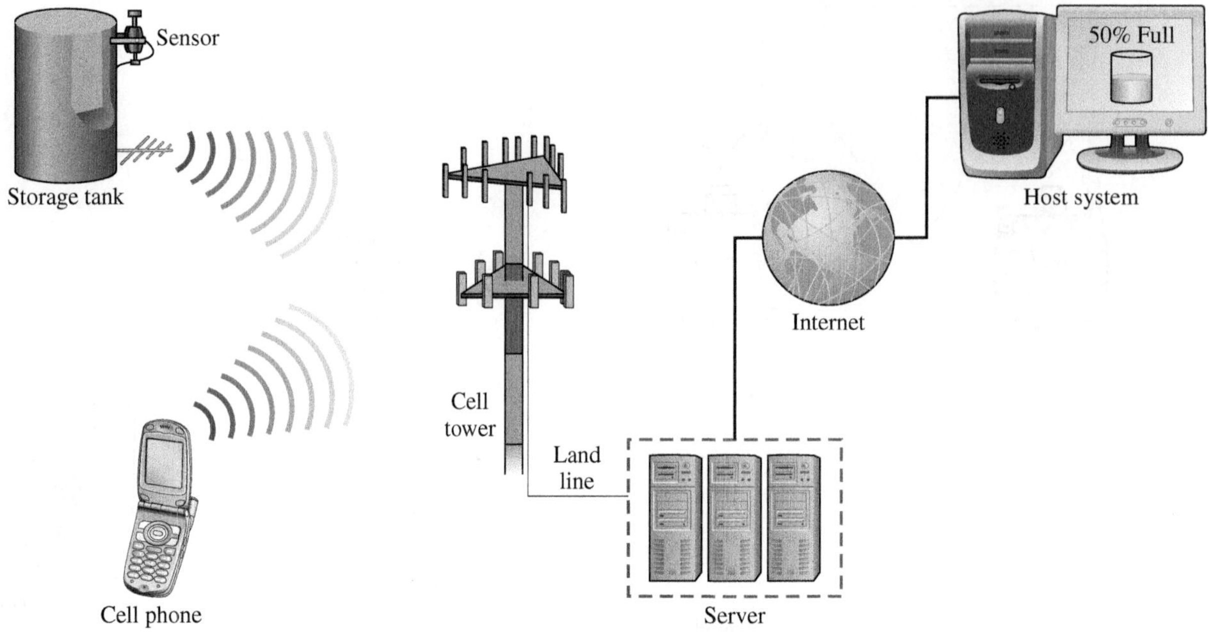

FIGURE 20-10 Remote sensor measurements using cellular technology

The GSM communication format does not operate at one frequency because when cellular technology was developed, certain frequencies available at various locations in the world were already used for different applications in other regions of the world. The result is that GSM operates at 850 and 1900 MHz in the US, and at 900 and 1800 MHz in Europe. For customers who travel internationally, cell phones with dual-band or quad-band capabilities are available for multiple world area usage.

General packet radio service, or GPRS, is the world's most popular wireless data service. Based on Internet protocols, GPRS is designed to layer "packets" of digital data on top of the digital voice signals used to make cell-phone calls. This is the same technology that is used to transmit e-mail messages to Blackberry phones.

Satellite Technology

For remote geographical areas that do not have cellular coverage, satellite communication is an option. Examples of these areas include offshore oil and natural-gas platforms, maritime operation, and mobile asset tracking in isolated areas. Satellite communication should only be used when cellular service is not available because it is so much more expensive. Satellite radios and antennas are also larger and require more power to operate than other wireless technologies because their signals travel such great distances, between the surface of the Earth and the satellite several hundred miles above.

RFID

RFID stands for *radio-frequency identification*. RFID devices are small radio-frequency transponders (called *RF tags*) that are electronically programmed with unique information used for tracking. Unlike the other wireless devices discussed in this chapter that are sensing devices, many manufacturers use RFID devices for inventory purposes, or to track products from the time they are made until they are removed from a warehouse.

RFID tags are activated when they receive an electromagnetic radio signal at a frequency of about 125 kHz. These radio signals come from a transceiver, which is both a radio transmitter and receiver (called a *reader*). When the radio signal arrives, it activates a circuit inside the tag that processes information, such as identification, and then returns a UHF signal at a frequency ranging from 840 to 960 MHz back to the reader.

FIGURE 20-11 Squiggle antenna on an RFID tag

There are two general categories of RFID tags: *active* and *passive*.

Active tags use internal batteries both to process information and to provide power for broadcasting information back to the reader.

Passive tags do not use a battery or external energy source to power their circuitry. Instead, they are able to use enough energy from the radio signal they receive to both process data and then return information back to the reader. Maximum energy is captured by using a conductor path on the miniature tag with a unique design, called a *squiggle antenna*, shown in Figure 20-11. As the electromagnetic wave passes through the antenna, a small voltage is induced that charges small capacitors, which gives the RFID its electrical power to transmit data. In addition to supplying transmission power to the tag, the signal activates identification and other relevant information stored in miniature memory cells. This data is applied to the squiggle antenna, and is transmitted back to the reader as electromagnetic waves.

Passive tags are commonly used for toll-road collection points that allow cars to pass without stopping to make the payment. The data that the tags return to the reader has identification information on the driver's account so that payment can be withdrawn from a checking or other type of account.

Since passive RFID tags rely entirely on the transceiver's signal for the energy to process and transmit data, there is only enough power to operate up to 20 feet. Active tags that have their own battery are capable of transmitting signals up to 300 feet away. Passive tags are less expensive than active tags, and are cheap enough to be disposable.

20-6 Radio Frequencies

The frequencies at which wireless RF signals are transmitted are called the *radio spectrum*. The radio spectrum is broken into groups of frequency ranges, such as HF (high frequency), VHF (very high frequency), and UHF (ultra-high frequency). Figure 20-12 graphically illustrates a portion of the radio spectrum with a logarithmic scale rather than linear.

The use of wireless devices, both commercial and industrial, is heavily regulated throughout the world. Most countries have a government agency responsible for establishing

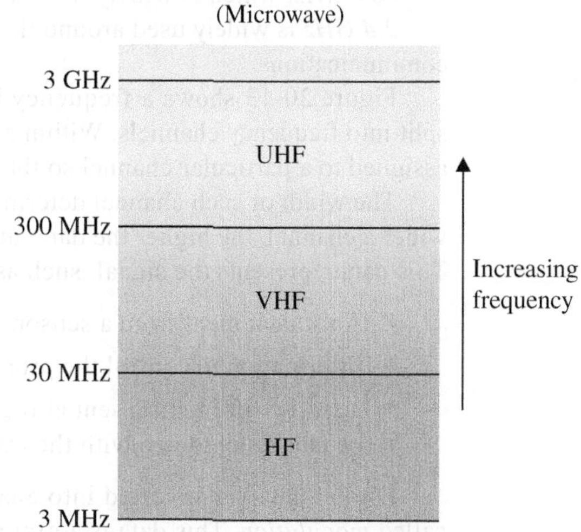

FIGURE 20-12 Groups of frequency ranges within the radio spectrum

and enforcing rules and regulations regarding the transmission of radio waves. In the United States, the government body with this responsibility is the Federal Communications Commission (FCC). This agency assigns frequencies for specific uses at certain power transmission levels within geographical regions to prevent broadcasts from being corrupted due to simultaneous transmitted signals interfering with one another. Most frequencies are already allocated for specific government applications, such as fire, military, police, or other specialized uses. Private companies who provide services to consumers pay substantial fees to license the frequencies they use. These service providers include cell-phone companies and satellite television. However, most countries have allocated parts of the radio frequency spectrum for open use, unlicensed and freely available.

Most industrial wireless products for short-range applications use the license-free areas of the spectrum, primarily to avoid the cost and delays related to obtaining a license. These license-free frequency ranges are known as *ISM bands* (industrial, scientific, and medical). Users of the ISM bands are required to comply with certain regulations, such as limiting the power of their transmissions. In many countries, including the United States, there are several ISM bands available. Common frequency bands for industrial wireless applications are shown in Table 20-2.

TABLE 20-2 Common frequency bands for international wireless applications

• 433 MHz band in Europe and some other countries—license-free
• 400–500 MHz—different parts are available in most parts of the world as licensed channels
• 869 MHz band in Europe—license-free
• 915 MHz band in North and South America and some other countries—license-free
• 2.4 GHz band allowed in most parts of the world—license-free

900 MHz, which is widely used on the American continent, is actually a frequency range of 902 to 928 MHz. This range of frequencies provides a high reliability of successful transmissions by overcoming various types of interferences. If a message sent at one frequency is corrupted, another frequency that provides a clear transmission will be used. Using alternative frequencies is called the *spread spectrum technique*. These 900-MHz frequencies are often selected for congested industrial environments because they are able to penetrate through obstructions better than radio signals at higher frequencies, such as 2.4 GHz. However, the 900 MHz has a smaller frequency band than the 2.4 GHz, which causes the data transmission rate to be slower. 900-MHz communication can reach up to approximately 40 miles.

869 MHz, which is widely used in Europe, is a single, fixed frequency.

2.4 GHz is widely used around the world for Wi-Fi technology and for wireless Ethernet communications.

Figure 20-13 shows a frequency band within the radio spectrum. The band is further split into frequency channels. Within a given geographical region, individual applications are assigned to a particular channel so that they do not interfere with one another.

The width of each channel determines how rapidly wireless data can be transmitted. The wider a channel, the higher the data rate. Data is comprised of digital binary bits (1s and 0s). This data represents the signal, such as:

- A measurement from a sensor.
- Information to control the proper handling of a message.
- Lead-in information sent at the beginning of a message to enable the receiver to lock on and synchronize with the transmitter.

Digital data is converted into analog frequencies within a channel by using a process called *modulation*. This data is a group of binary 1s and 0s that represent a data word. Each binary bit in the word is assigned a frequency. An analog generator produces the frequency

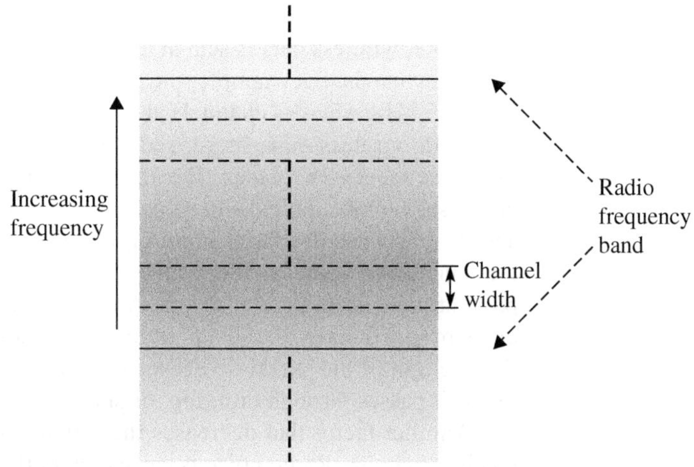

FIGURE 20-13 Radio-frequency band within the radio spectrum

when its corresponding bit is a logic 1, and the signal is not produced if the bit is a logic 0, as shown in Figure 20-14. All of the frequencies representing a 1 are transmitted simultaneously, and picked up by a receiver antenna. Each frequency is separated by a tuner, and through a process called *demodulation* is converted back to digital data. The demodulator produces a frequency representing a bit into a logic 1, and the absence of a frequency representing a bit into a logic 0. The wider a channel, the higher the data rate, because there are a greater number of frequencies that can be transmitted simultaneously.

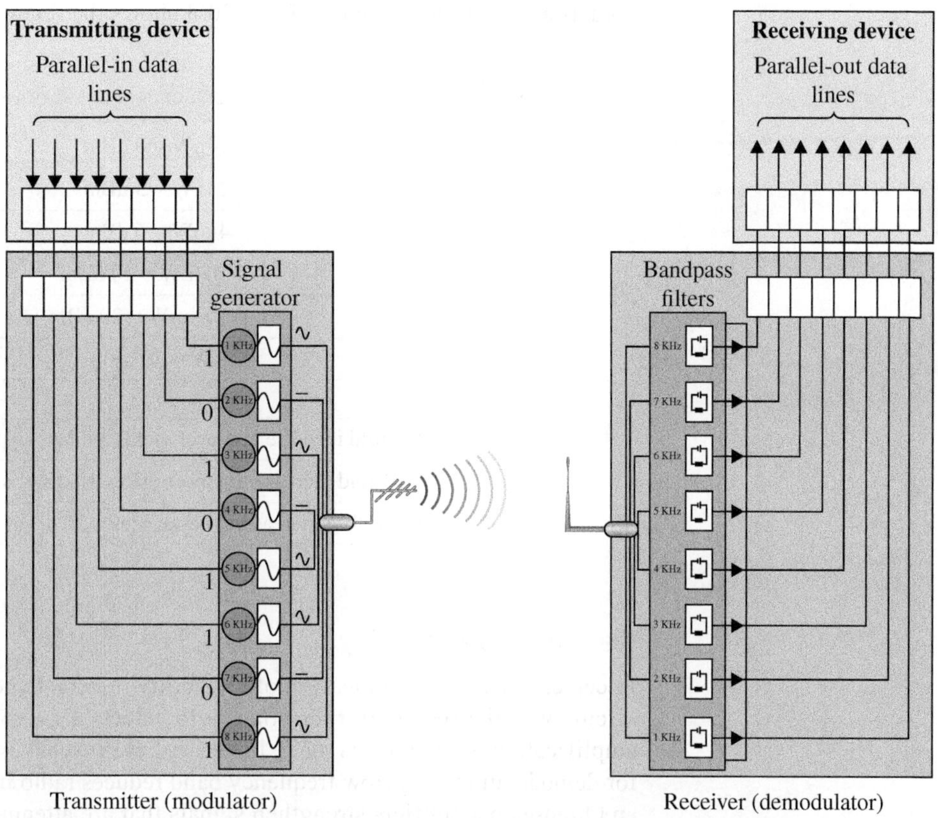

FIGURE 20-14 Radio transmitter sends modulated signals to a radio receiver

Higher frequencies have a larger channel width than is available at lower frequencies. Therefore, wireless data is sent at a greater rate over higher-frequency bands than over lower-frequency bands. For example, the channel widths of frequencies of 150 to 500 MHz are usually 12.5 kHz, whereas channels at 2.4 GHz can be hundreds of times larger.

With wireless data, there is always a trade-off between data rate and distance. As the radio frequency increases, the data rate increases, but the distance the signal travels decreases. For example, as the frequency is doubled, the distance a signal travels drops to ¼. The distance also decreases more rapidly than lower-frequency signals when higher frequencies encounter obstacles like walls or trees. Radio signals are attenuated (weakened) as they pass through air or another media. A radio signal that travels through a constant media attenuates proportionally to the square of the distance. For example, as the distance is doubled, the strength of the signal decreases to ¼. The strength of the signal is weakened at a greater rate if it passes through rain, fog, or snow.

Another factor that decreases the reliability of a wireless data signal from reaching the receiver is *noise*. Radio noise is caused by both natural and man-made factors. Lightning and radiation from the sun produced by solar flares are sources of natural noise. Man-made noise is caused by radio-frequency harmonics produced by devices that conduct high-level currents, such as electrical motors, transformers, and certain types of lighting.

There are several ways to overcome the factors that cause signal attenuation and noise.

RF Power

The strength of the RF signal emitted by an antenna is affected by the amount of power produced by its transmitter. The stronger the signal, the farther it will travel. Increasing the power level by a factor of four can double the distance. However, as the RF power is increased, so will the current demand on the system power supply.

RF power is measured in milliwatts (mW) or in a logarithmic scale of decibels (dB). Since RF power is a logarithmic function, the decibels per meter (dBm) scale is commonly used as a unit of measurement. Table 20-3 shows the relationship of RF power and decibels.

TABLE 20-3 The Relationship of RF Power and Decibels

1 mW = 0 dBm
2 mW = 3 dBm
4 mW = 6 dBm
10 mW = 10 dBm
100 mW = 20 dBm
1 W = 30 dBm

A 2-fold increase in power yields 3 dB of signal.
A 10-fold increase in power yields 10 dB of signal.
A 100-fold increase in power yields 20 dB of signal.

Receiver Sensitivity

Receiver sensitivity is a measure of the ability to detect and decode a radio signal. By using circuitry in the receiving radio that both selects a narrow frequency band and has high amplification, sensitivity is made better and even weak signals can be made strong enough for demodulation. A narrow frequency band reduces radio interference from nearby channels, and high-gain amplifiers strengthen signals that are attenuated by distance or environmental conditions, or partially corrupted by noise. Radio receivers that operate at lower frequencies have better sensitivity than radios that operate at higher frequencies.

Antenna Design and Selection

Antennas radiate and receive the radio signal. Choosing antennas with the right shape and size for a particular application can affect the transmission distance of a radio signal.

The shape of the antenna determines the direction a signal is transmitted. The signal is emitted equally in all directions by an isotropic antenna, shown in Figure 20-15(a). This type of antenna is used for multidirectional transmissions. A Yagi antenna in Figure 20-15(b) distorts the pattern of the signal so that it travels farther in the forward direction, and its distance is reduced in the opposite direction. This type of shape is used for transmitting signals farther than other antennas but in only one direction. The function of the elements is to reflect the signal, which causes the radiated energy to be emitted in a desired direction at the expense of other directions. In addition to transmitting RF signals farther, Yagi antennas also increase the distance at which it can receive a signal by a phenomenon known as *antenna reciprocity*.

Devices used in a mesh topology configuration use isotropic antennas because they communicate with other devices in all directions. Both isotropic and Yagi antennas are used in a star topology wireless configuration. The isotropic antenna is used by the gateway, and the Yagi antenna, which points to the gateway, is used by the field devices.

Some antennas are carefully built for specific frequencies. The antennas are most effective when their elements are the same size, or half the size of the wave for the frequency they transmit or receive. Low-frequency RF communication requires larger antennas than higher-frequency signals.

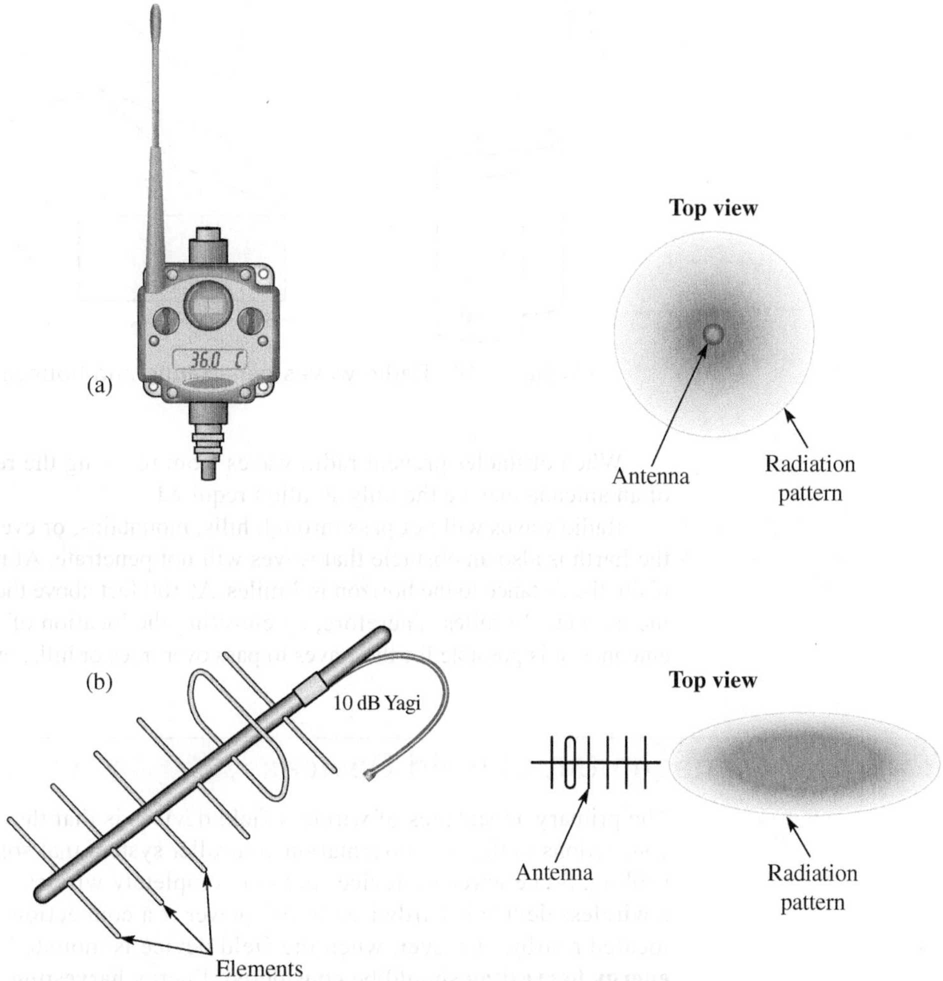

FIGURE 20-15 Wireless antennas: (a) Isotropic multidirectional antenna; (b) Yagi directional antenna

20-7 Characteristics of the Radio Path

The radio signals that travel between industrial wireless products take a "line-of-sight" path. "Line-of-sight" means that radio signals radiate in straight lines, unlike signals at lower frequencies that bend around the curvature of the Earth.

For reliable transmissions, a true line-of-sight path is not required. Radio signals will pass through some obstacles, such as buildings, trees, and other objects. However, the general rule is that obstacles decrease the reliable operating distance between the wireless transmitter and receiver.

Within a factory or at industrial sites that are heavily congested with objects that block line-of-sight radio paths, most waves are able to reach the receiver by either penetrating through or reflecting off obstacles, as shown in Figure 20-16.

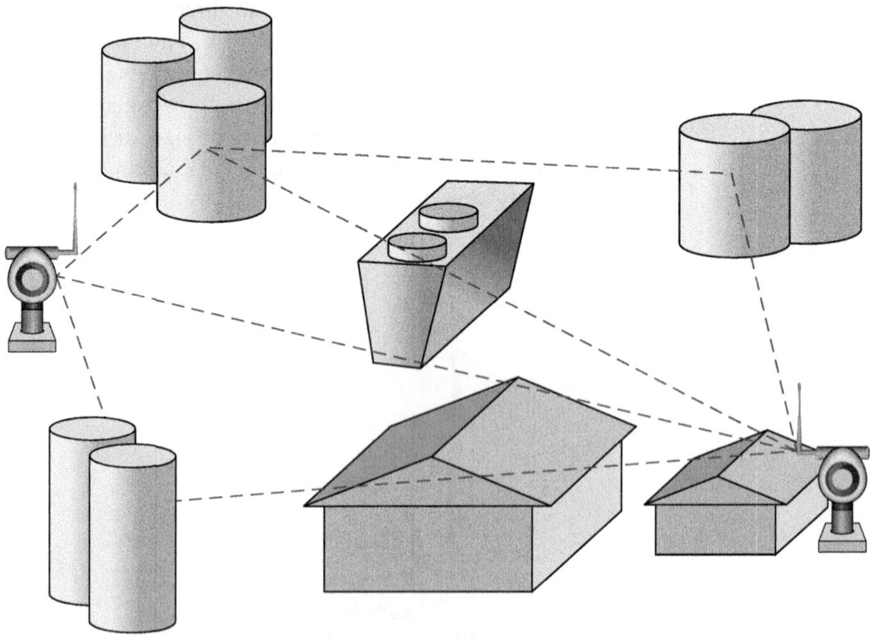

FIGURE 20-16 Radio waves penetrating and bouncing off obstacles

When obstacles prevent radio waves from reaching the receiver, changing the location of an antenna may be the only solution required.

Radio waves will not pass through hills, mountains, or even vegetation. The curvature of the Earth is also an obstacle that waves will not penetrate. At normal eye level across a level plain, the distance to the horizon is 3 miles. At 100 feet above the ground surface, the horizon is increased to 13 miles. Therefore, by elevating the location of the transmitting and receiving antennas, it is possible for the waves to pass over trees or hills, or travel farther over a flat plain.

20-8 Power Management of Field Devices

The primary advantages of wireless field devices is that they do not require the hardwired connections to the host information controller system that sometimes require long-distance cabling. Some wireless devices are not completely wireless, however. Whenever possible, a wireless device is hardwired to AC power if a connection point to the voltage source is located nearby. However, when the field device is mounted where there is no AC supply, **energy harvesting** should be considered. Energy harvesting, also known as *scavenging,* is the method of converting one energy source into electricity. For example, photovoltaic cells can be used to convert light energy from the sun into electrical energy. Converting thermal energy commonly found in factories into electricity is another potential energy source to supply power to the field device, its battery, or both. If energy harvesting is used for supplemental power for field devices, rechargeable batteries can be used.

Whenever energy harvesting is not conveniently available, batteries are used to power field devices. To extend the time between battery replacements, various power-management options must be considered, such as choosing the right kind of battery or reducing energy consumption. Lithium-ion batteries that are not rechargeable should be used because of their long life. There are two factors that impact battery life. The farther the radio signal is sent, the more power it consumes. Also, the more power a field device uses, the shorter its battery life will be. By using various technologies and methods to minimize power usage, battery life can be extended.

The wireless technology that is the most energy-efficient is the self-organizing network. For example, if a 1-watt field device radio is used in a point-to-point network to transmit a distance of 1000 meters, ten 1-mW (0.001-watt) field device radios can cover the same distance with 100-meter hops in a self-organizing network. Furthermore, power consumption can be impacted by the wireless technology that is selected. For example, Wi-Fi radios use 100 mW of power, compared to cellular devices that require 3 watts in order to transmit their signals over longer distances.

Duty cycling is a technique used to reduce power consumption by regulating how often and how long the power is on, as opposed to consuming more power by running the field device continuously. *Duty cycling* refers to how long a device is on over a given period of time. To conserve power, the field device is in the sleep mode and its power is reduced to almost zero. When it is time for its scheduled transmission, it turns on for a very short time, long enough for its sensor to take a measurement and for its radio to send the data. The unit returns to its sleep mode until it is time for the next scheduled measurement and transmission. Figure 20-17 graphically compares the duration of the measurement and radio transmission compared to the sleep mode, and how duty cycling reduces power consumption. To further reduce power consumption, the sleep modes can be extended as long as the measurements are frequent enough that the measurement data is not compromised.

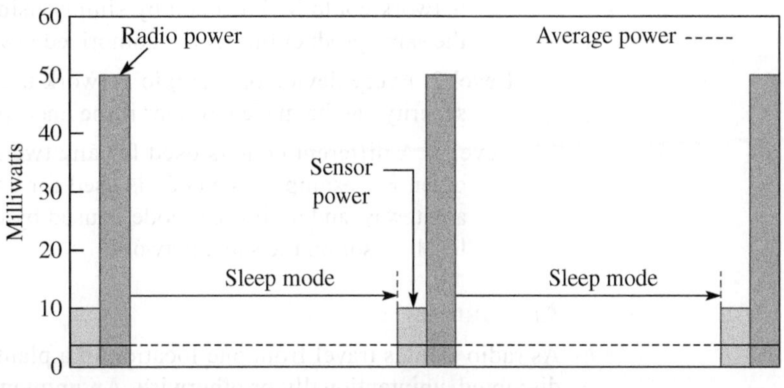

FIGURE 20-17 Duty cycling reduces average power consumption

20-9 Security

One of the primary reasons that wireless technology has not been more readily adopted for industrial usage is the concern about security. A common myth about wireless technology systems is that they are vulnerable to attacks because the transmission signals travel through airspace that is accessible by anyone. In response to the heightened awareness of security issues, manufacturers of wireless devices have designed their products to prevent potential problems.

To incorporate security into their products, the following functions have been implemented.

Authentication

This function requires that the sending and receiving devices identify themselves by exchanging "keys." Similar to a password, keys are programmed into each device to validate that they are authorized to operate on the network. The keys are entered into the device during installation. Whenever data is transmitted, both the transmitter and receiver exchange their ID information to prevent data exchange with an unauthorized device.

Encryption

The best protection against wireless espionage is the encryption of data. Encryption is a technique in which the transmitting device alters the data before it is sent so that it is unreadable if intercepted during transmission. When the scrambled data arrives at the intended receiver, it is unscrambled back to its original state. The process of scrambling the data is called *encrypting,* and the unscrambling process is called *deciphering*. To perform both processes, each device on the network must be programmed with the required code that makes them compatible.

Two common types of data encryption methods used by wireless devices are:

AES (advanced encryption standard) This method is used by organizations with high security requirements, such as Federal Reserve banks that transfer money among themselves. Considered "unbreakable," this standard requires heavy computing resources that can significantly slow down the operation of wireless devices.

XTEA (extended timing encryption algorithm) This method is less powerful than AES, but is still secure enough to prevent data from being breached.

Key Management

This function is the process of creating, distributing, authenticating, and storing encryption codes so that data can be properly encrypted and deciphered. There are three basic security levels of key management.

Level 1: Every wireless device from a manufacturer has the same encryption code. Even though it provides some measure of security, it is vulnerable to an attack because the network could be breached by simply using an unauthorized wireless device from the same product line as the authorized device.

Level 2: Every device on a single network uses its own encryption code. This level of security can be made stronger if the encryption code is changed periodically.

Level 3: A different code is used for any two devices that transmit data between each other. For example, one code is used for a temperature sensor to communicate with a gateway, and a different code is used by the same gateway to communicate with a level sensor on the same network.

Anti-Jamming

As radio signals travel from one location in a plant to another, data communication can be disrupted, unintentionally or otherwise. An unintentional disruption could occur from electromagnetic waves emitted from an electrical motor. An intentional disruption could occur if a hacker intends to cause harm by using a radio transmitter.

To overcome the disruption of radio waves, an anti-jamming technique called *spread spectrum* has been developed. Spread spectrum radios use multiple frequencies within a continuous band during a transmission. Each message of a wireless device is in the form of digital data made up of binary data bits. These bits, each either a logic 1 or a 0, are converted to RF analog signals using the spectrum of frequencies within an assigned radio band.

While there are various types of spread spectrum methods, the two most common anti-jamming techniques are FHSS (frequency-hopping spread spectrum), and DSSS (direct-sequence spread spectrum).

FHSS

During a transmission using the FHSS technique, the frequency of the transmitter is automatically changed from one to another within the frequency band it uses. The transmitter hops frequencies according to a preset sequence that is based on a code. The radio receiver, which has the same code as the transmitter, is able to select the same sequence of frequencies as the transmitter to assemble the original message. This scheme requires low bandwidth.

DSSS

During a transmission using the DSSS method, each binary bit of a message is sent over a separate frequency. Multiple frequencies spread across a wide channel are sent simultaneously. The frequencies used during the transmissions are randomly generated by a code called a "chipping key." When the radio waves arrive at the receiver, it decodes the frequencies so that the data is properly arranged to reconstruct the original message. This scheme requires high bandwidth.

20-10 Wireless Standards

In the United States, a standard exists for electrical power. The suppliers that produce and provide power, and different manufacturers who make devices and equipment that run on electricity, must comply to the standard of 120 volts AC at a frequency of 60 hertz to ensure everything is compatible.

Standards have also been established to enable wireless products from multiple vendors to communicate and work with one another. The result is that field devices and host systems from different manufacturers on the same network are compatible. Two types of groups primarily sponsor these standards.

- Professional and official organizations that publish documents on a broad spectrum of existing standards. These organizations include ISA, IEEE, and IEC.

ISA Formerly the Instrumentation Society of America, the organization is now called the Instrumentation, Systems, and Automation Society. Comprised of approximately 30,000 members, this professional organization has published more than 150 standards related to the process-automation industry. ISA has formed a committee made up of suppliers and end-users to create standards on wireless technologies, called *SP100*. Categories of these standards include:

SP100.14—pertains to establishing standards for industrial applications such as monitoring, data-logging, and alerting.

SP100.11—pertains to establishing standards for industrial control applications, ranging from open-loop operations that require manual control to closed-loop regulatory control required for automation.

SP100.21 and 22—pertains to RFID devices for warehouse inventory management.

IEEE Institute of Electrical and Electronics Engineers, Inc., is a nonprofit organization of professionals that has developed standards for both public and industrial applications, including wireless networks. Categories of these standards include:

IEEE802.11—pertains to specific requirements for wireless Ethernet devices using DSSS within the 2.4-GHz and 5-GHz ISM frequency bands. Currently, most wireless networks operate using this standard. There are three versions: b, a, and g.

802.11b operates in the unlicensed 2.4-GHz radio spectrum. This standard is used for Wi-Fi devices. Its primary limitation is its relatively slow operating speed of 11 Mbps, and its signal may be interfered with mobile phones and Bluetooth devices.

802.11a operates in the unlicensed 5-GHz frequency range (5.15 to 5.35 GHz), which is a less-populated band. Its advantages over 802.11b are that at 54 Mbps, its data rate is better, and it is less prone to interference.

802.11g is the latest 802.11 standard. Its characteristics are that it operates at high speeds and has good penetration.

Table 20-4 compares the three different 802.11 standards.

TABLE 20-4 IEEE 802.11 Wireless Standards

Standard	Data Rate	Frequency	Throughput	Range	Power Drain
802.11a	54 Mbps	5 GHz	30 Mbps	36 ft	High
802.11b	11 Mbps	2.4 GHz	6.5 Mbps	209 ft	Moderate
802.11g	54 Mbps	2.4 GHz	30 Mbps	55 ft	Moderate

IEEE802.15.4—pertains to specific requirements for wireless sensor networks using 868/900-Hz or 2.4-GHz ISM bands. A high-level layer of this standard, commonly known as *ZigBee*, is used for wireless mesh networks in which field products pass information from one device to another.

IEC The International Electrotechnical Commission is an international organization based in Switzerland. Its function is to establish standards for electrical, electronics, and related technologies. Categories of these standards include:

IEC61511—pertains to safety applications.

IEC61158—pertains to standards for fieldbus.

There is a Consortium of companies that have formed alliances to focus on establishing standards on specific technologies. These organizations include Fieldbus Foundation, Hart Communication Foundation, and Wi-Fi Alliance.

Fieldbus Foundation. This foundation is a nonprofit organization made up of over 350 companies. Its purpose is to help members follow fieldbus standards so that their products are compatible and work together on the same network.

Hart Communication Foundation. This foundation is a nonprofit organization made up of over 150 companies. Its function is to help its members follow standards for the Hart protocol so that the low-power field devices one company makes can communicate with the devices built by other companies in a self-organizing mesh-structure wireless network.

Wi-Fi Alliance. This is a nonprofit organization made up of over 250 companies. Its function is to help its members adopt the IEEE802.11 standard so that the Wi-Fi wireless networking products they make are compatible and can communicate with one another.

In addition to developing standards used in process and automation industries, the consortium-type organizations test and verify the proper operation of equipment, and provide technical support. The professional organizations do not offer these additional services.

Problems

1. List the three basic components of a wired or wireless industrial communication system.
2. What is a radio signal, and how is it produced?
3. The number of _____ transmitting antenna each second are measured in hertz.
   a. electrical pulses that flow through a
   b. electromagnetic waves emitted by a
   c. both a and b
4. The distance a radio wave travels is determined by the _____.
   a. amount of wattage produced by the transmitter
   b. type of transmitting antenna
   c. height of the transmitting antenna
   d. obstructions in its path
   e. all of the above

5. The components in a wireless network are arranged in various physical configurations, called _____.
6. The drawback of a _____ topology network is that its data is lost when something disrupts the direct path between a field device and its gateway.
7. The most efficient way to link field devices and the host system of a wireless system is to use _____ technology.
   a. Wi-Fi b. cellular c. satellite
8. What is the difference between the GSM and GPRS communication formats?
9. RFIDs use energy from _____ to both process and transmit data.
   a. a battery
   b. the RF signal it receives
10. The _____ is the most efficient and reliable network.
    a. star topology
    b. mesh topology
    c. cluster-free topology
    d. self-organizing
11. Describe the operation of the CSMA and TDMA data transmission methods.
12. Which of the following users of wireless technology do not require a license to use the airwaves?
    a. government agencies
    b. cell-phone companies
    c. ISM-band users
    d. all of the above
13. Channel bands at lower frequencies transmit data at a _____ rate than channels that broadcast at higher frequencies.
    a. lower b. higher
14. Radio signals at _____ frequencies are better at penetrating physical objects.
    a. lower b. higher
15. If the distance that an RF wave travels doubles, its signal strength _____.
    a. doubles b. decreases to ½ c. decreases to ¼
16. _____ refers to how long a device is turned on over a given time period.
    a. Energy harvesting
    b. Duty cycle
    c. Spectrum averaging
17. As the frequency of a radio signal is doubled, the distance it travels _____.
    a. doubles b. decreases to ½ c. decreases to ¼
18. _____ signals are primarily used for Wi-Fi technology.
    a. 900-MHz b. 867-MHz c. 2.4-GHz
19. Radio noise is caused by _____.
    a. RF waves produced by motors
    b. solar flares from the sun
    c. both a and b
20. _____ is the process of unscrambling transmitted data at the receiver.
    a. Authentication b. Encrypting c. Deciphering
21. Describe the difference between FHSS and DSSS anti-jamming techniques.
22. The type of organization that develops wireless standards, tests the operation of wireless equipment, and provides technical support is the _____.
    a. professional b. consortium-type
23. The ISA standard that pertains to closed-loop operations is _____.
    a. SP100.14 b. SP100.11 c. SP11.21 and 22
24. The IEC standard that pertains to safety applications is _____.
    a. IEC61511 b. IEC61158
25. The IEEE standard that pertains to Wi-Fi communication is _____.
    a. IEEE802.11 b. IEEE802.15-4

SECTION

7

Programmable Controllers

OUTLINE

Much of the automated equipment used in industry is controlled by programmable controllers. The programmable controller is a specialized computer designed for many types of manufacturing applications. These machines use a microprocessor that provides artificial intelligence enabling them to perform various functions formerly performed by people. The primary advantage of programmable controllers over hardwired circuits is their programmability. For example, an equipment function or manufacturing process can be altered by making a few entries on a keyboard instead of making costly rewiring changes within the controller section.

Because of the important role they play in industrial control, this three-chapter section describes the operation of programmable controllers. Chapter 21 describes the principles of relay ladder diagrams and hardware circuitry of the equipment. Chapter 22 provides a generic approach to ladder diagram programming. Chapter 23 examines advanced programming instructions, applications, interfacing, and trouble-shooting of the programmable controller.

CHAPTER 21

Introduction to Programmable Controllers

OBJECTIVES

At the conclusion of this chapter, you should be able to:

- Identify the symbols used in relay ladder circuits.
- Draw a ladder diagram and label the parts of the configuration.
- Design a ladder diagram that will perform a specified operation.
- Draw a block diagram of the PLC and describe the function of the following sections:

Power Supply	Input/Output Modules
Processor Unit	Programming Unit

- Describe the operation of the internal interface circuit inside the following input/output (I/O) modules:

AC Input Module	DC Output Module

- Draw the proper wiring connections of field devices to the terminals of input and output modules.
- Describe the relationship between a terminal on an I/O module and a corresponding address location in memory.

INTRODUCTION TO PLC FUNCTIONS

Programmable controllers are computer-based devices capable of controlling many types of industrial equipment and entire automated systems. The programmable controller is also commonly referred to as a **programmable logic controller (PLC)** because its operation often requires many logic functions to be performed.

Prior to the late 1960s, automated equipment was primarily controlled by discrete, inflexible circuits consisting of electromechanical relays and coils hardwired on panels. (*Hardwiring* means that all of the components were manually connected by wires.) These devices are still used for many industrial control applications.

21-1 Industrial Motor Control Circuits

Industrial companies use many different types of machinery to produce the product that they manufacture. The machinery may be used for machine tooling, injection molding, textile production, die casting, woodworking, or packaging. Some of the less sophisticated machines are operated manually, using levers, switches, or other input devices. The operation of complex machines that perform more advanced functions is controlled by electrical circuits. Many of these "electrical control" circuits perform operations that are required in applications that demand greater speed, higher efficiency, and wider versatility than the manually controlled operation. The objective of this part is to introduce hardwired electrical control circuits used for operating industrial machinery. Even though electrical control circuits operate actuator devices that are hydraulic, pneumatic, and mechanical, they are often referred to as *industrial motor control circuitry,* because the primary actuator used is the electric motor.

Wiring diagrams are used to provide information about electrical control circuits. The wiring diagram serves two main functions:

1. It is the source of written communication about the operation that is conveyed from the designer to the technician.
2. It serves as a troubleshooting guide.

A good wiring diagram for electrical control circuits should contain the following features:

1. A drawing that shows the physical layout of all electrical devices in the circuit.
2. A method that shows how the devices are physically connected to one another.
3. Information that shows the manner in which all of the control devices function electrically with respect to one another.

Wiring diagrams are made up of lines connecting certain symbols. Symbols are used as a sort of shorthand that allows many electrical functions to be identified and sketched out in a small area. To effectively work with electrical motor control wiring diagrams, the technician must become familiar with the symbols used. The symbols presented in this chapter have been standardized for electrical control diagrams through the joint efforts of the National Machine Tool Builders Association (NMTBA) and automotive industry representatives.

The types of wiring circuits used in electrical control circuits are referred to as **relay ladder logic diagrams.** The term *ladder* is derived from the appearance of the diagrams, which resemble the rungs of a ladder. The term *logic* is derived from the decision-making function that is performed by relays. A ladder diagram is shown in Figure 21-1. Two vertical lines labeled L1 and L2 represent the potential differences between two voltages that supply power to the circuit. These vertical lines are referred to as *rails* because they resemble the vertical sides, or rails, of a ladder. The various components used in the circuit are located on the horizontal lines between L1 and L2. The input field components are located on the left portion of the rung, and the output devices are located on the right portion of the rung.

The ladder circuit consists of four elements.

1. A power source.
2. An input control device, such as a switch.
3. A load device at the output, such as a light or motor.
4. Interconnecting wires.

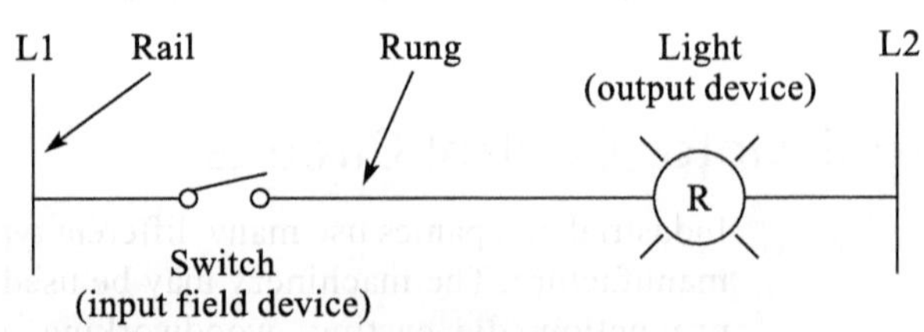

FIGURE 21-1 A ladder diagram

When a continuous, unbroken path exists between L_1 and L_2, the circuit is *energized.* When the input field device opens and breaks the path, the circuit is *de-energized.*

The following information lists the most common schematic symbols used in ladder wiring diagrams and describes their operation.

IAU3006
Ladder Logic
Schematic Symbol
Flashcards

Wiring

- Vertical lines L1 and L2 are thicker than the horizontal lines.
- Any lines that cross are not connected unless they show a dot where they intersect.

Mechanical Connections

- To show a physical connection between two or more points, use a dashed line.

\- - - - - - - - -

Components Used as Input Devices

Manual Switches

Manual switches are used in the input section of a ladder diagram to cause an action to occur at a desired time. These switches are activated by a machine operator or technician.

Momentary Switches A momentary switch is usually a push button that actuates a contact when it is pressed and reverts the contact back to its original position when the pressure on the button is released. Types of momentary switches include:

Normally Open Push Button This type of push button is spring-loaded and closes its contacts for whatever period of time the button is held down. Start buttons are commonly made of a normally open (N.O.) push button.

N.O.

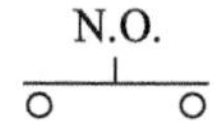

Normally Closed Push Button This type of push button consists of contacts which open for whatever period of time the button is held down. Stop buttons are commonly made of a normally closed (N.C.) push button.

N.C.

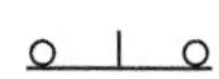

Multiple-Pole Push Button These push buttons are available with more than one set of contacts. For example, the double-pole push button shown below has one set of contacts that are normally closed and one set that are normally open.

N.O. & N.C.

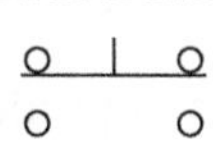

Palm-Operated This push button is internally designed the same as N.C. push-button devices. Also called *mushroom-head switches*, they are often used for emergency stop switches when the operator needs to physically hit this switch hard and not be injured.

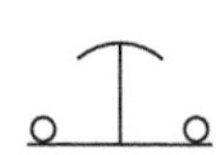

Foot Switches Many machines or processes require frequent actuation of a pilot device to energize a circuit. If the operator's hands are engaged in performing some function of the process, a foot switch is often used to activate the pilot device in the same way as a push button. Foot switches are heavy-duty devices usually found in N.O. and N.C. configurations.

Maintained Switches These switches make contact when pressed and remain in position until released by pressure from another button through a mechanical linkage.

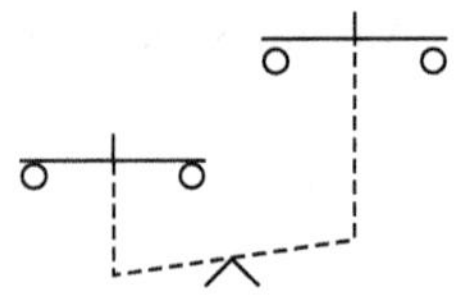

Selector Switches Selector switches are multipositional devices that connect more than one set of contacts. The two- and the three-position selector switches are most common. The symbols for the switches label the two sets of contacts as A1 and A2. They also have an arrow that points to positions J and K for a two-position selector, or J, K, and L for a three-position switch. A truth table indicates which contacts are made by placing an X in a column that shows the corresponding position.

For example, the A1 contacts of the two-position switch are made when the switch is in position J, and the A2 contacts are made when the switch is in position K. The K position for the three-position switch allows for a shut-off condition.

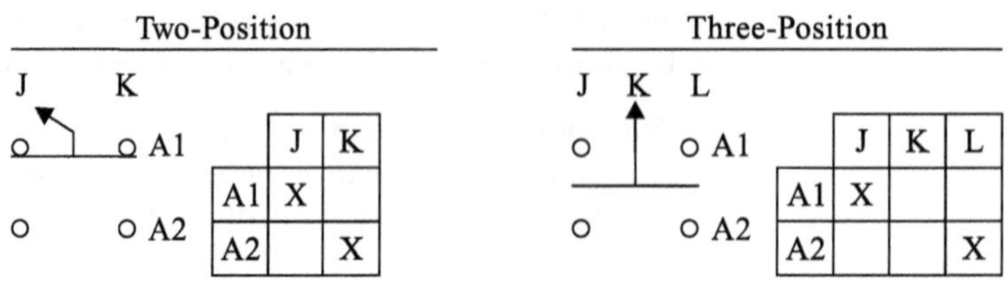

	J	K
A1	X	
A2		X

	J	K	L
A1	X		
A2			X

X — Contact closed

Automatic Switches

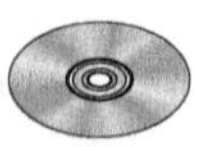

IAU11408
Selector Switch
Labeling

Automatic switches activate N.O. or N.C. contacts when they sense the presence or absence of some physical material or condition. Such devices include:

Mechanical Limit Switches These switches detect the presence of an object when physical contact is made. The switch symbol is shown on the diagram in its *unoperated* condition when physical contact is not made, as illustrated in Figure 21-2(a). Figure 21-2(b) pictorially shows how the limit switches are affected in their *operated* condition when physical contact is made with an object. Figure 21-2(c) shows a limit switch that combines N.O. and N.C. contacts. The broken line indicates that they are mechanically joined together.

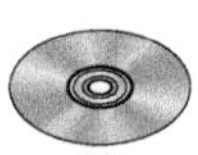

IAU11908
Limit Switches

Proximity Limit Switches These switches detect the absence or presence of solid or liquid material without making physical contact. There are two types of proximity switches, inductive and capacitive. Inductive proximity switches can only detect ferrous (magnetic) and nonferrous metals.

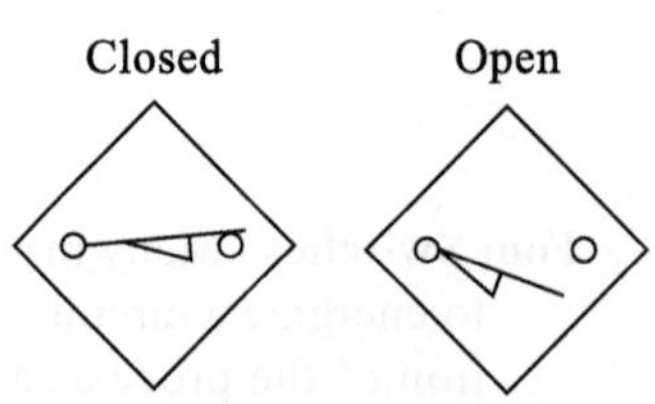

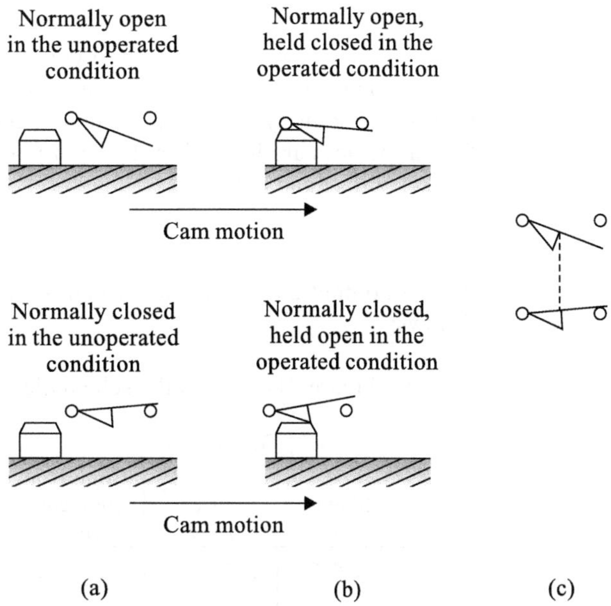

FIGURE 21-2 Mechanical limit switches

Capacitive proximity switches can detect any type of solid or liquid material. These detectors use solid-state components that are inside the sensor body. A diamond is placed around the limit switch symbol to indicate that it is a proximity device.

Pressure Switches Pressure switches are commonly used in pumps, air compressors, and lubrication systems. They are used to detect the pressure of a medium, such as air, water, or oil.

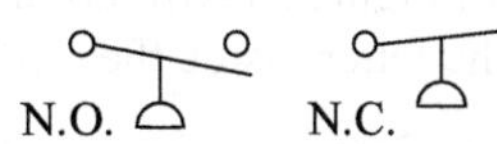

Flow Switches These switches are used to detect the flow of liquids (e.g., water, oil) and gases in a pipe or duct. One application example of a flow switch is to detect if a clogged air filter is not allowing enough air flow to properly cool a heating coil. Another example of a flow switch application is to detect if enough lubricating oil is being supplied to a drill on a drilling machine.

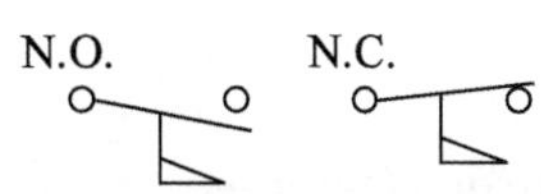

Level Switches These switches are used to detect the level of liquids, slurries, or solids such as powder or paste. One application example of a level switch is to turn off an electric motor when it has pumped a required amount of water into a storage tank. Level switches are sometimes referred to as *float switches*.

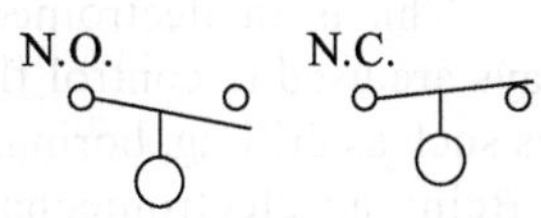

Temperature Switches These switches are used to detect the temperature of a solid, liquid, or gas. One application of a temperature switch is to open and cause a heater to turn off when the heat has reached a certain level, or to close and cause the heater to turn on when the heat is below a certain temperature.

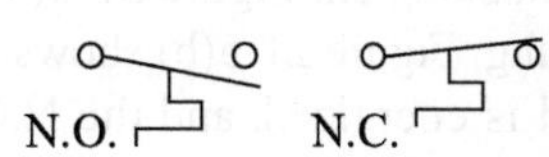

Components Used as Output Devices

The input device provides a path of continuity from a power line source to an output device when a rung is energized. Output devices are also referred to as *actuators*. Their function is to produce some type of action, such as turning on an indicator lamp or running a motor. Output devices are located in the right portion of each rung. There are two types of output devices, direct and indirect.

Direct Devices

A direct output device actuates when its rung is energized. The most common types are as follows:

Motors Figure 21-3 shows the schematic symbols of DC and AC motors.

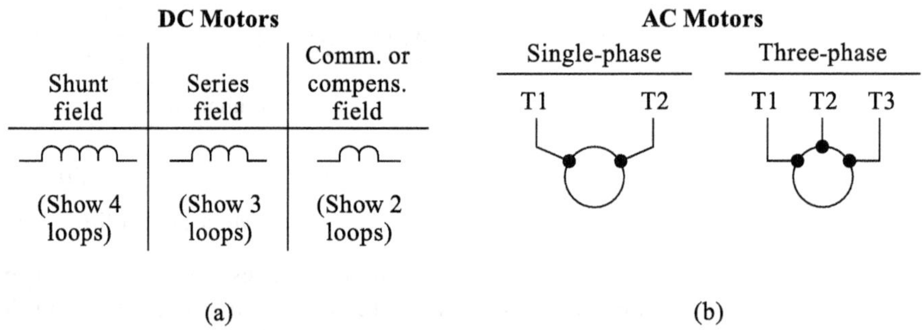

FIGURE 21-3 Motor symbols

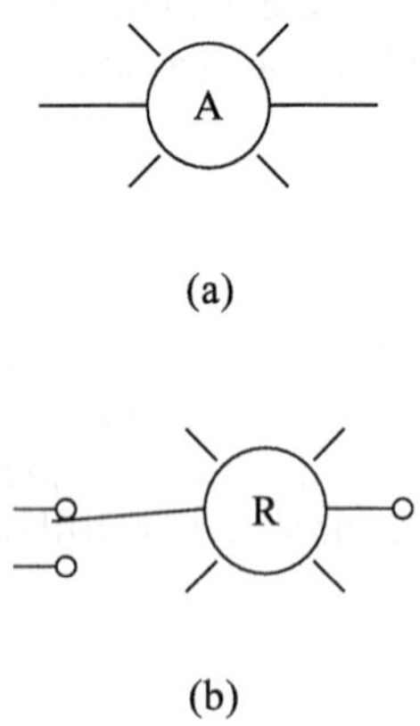

FIGURE 21-4 Pilot lights

Pilot Lights A light can be used as an indicator to show the status of a ladder diagram rung. A stand-alone lamp in Figure 21-4(a) indicates that the rung is energized. Another type of lamp in Figure 21-4(b) combines with a switch and illuminates when it is closed.

The letter inside the circle indicates the color of the lamp. For example:

G = Green A = Amber
R = Red Y = Yellow

Heating Elements These actuators are used to heat things such as containers of liquid or to liquefy solids into liquids (such as plastics).

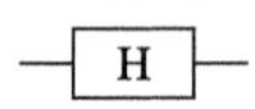

Indirect Devices

These output devices actuate N.O. or N.C. contacts when they are energized. Also referred to as *auxiliary contacts,* they are electrically isolated from the input and usually control output devices at a different location in the ladder diagram. Indirect components include:

Relay This is an electromechanical device that acts basically as a communication carrier. Relays are used to control fluid power valves and as machine sequence controls for operations such as drilling, boring, milling, and grinding.

Being an electromechanical device, the relay has a coil and a plunger that moves through an opening in the coil. There are contacts, which are connected to the plunger. These contacts move from a closed to an open position, or from an open to a closed position. Contacts are referred to as normally closed or normally open in relation to the condition of the de-energized coil. When the coil is energized, N.O. contacts close and N.C. contacts open. Figure 21-5(a) shows the symbol of the energized coil located at the output of a rung. Figure 21-5(b) shows the symbol of the N.O. contact, which closes if the corresponding coil is energized, and the N.C. contact, which opens if the coil is energized. Figure 21-5(c)

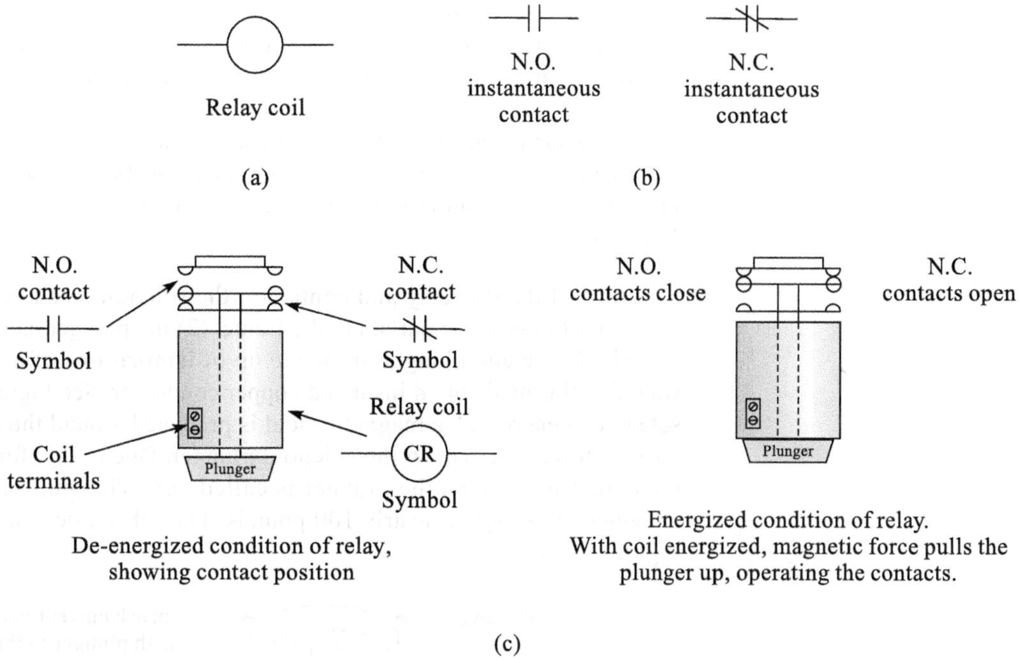

FIGURE 21-5 Electromechanical relay

provides a drawing that shows how the wiring diagram symbols correspond to the physical parts of a relay.

Time Delay Relay Timing devices are used in industry to regulate many types of machine operations. This function is often performed by a timing relay that causes contacts to open or close a certain period of time after its coil is energized or de-energized. The symbols for the timing relays are shown in Figure 21-6.

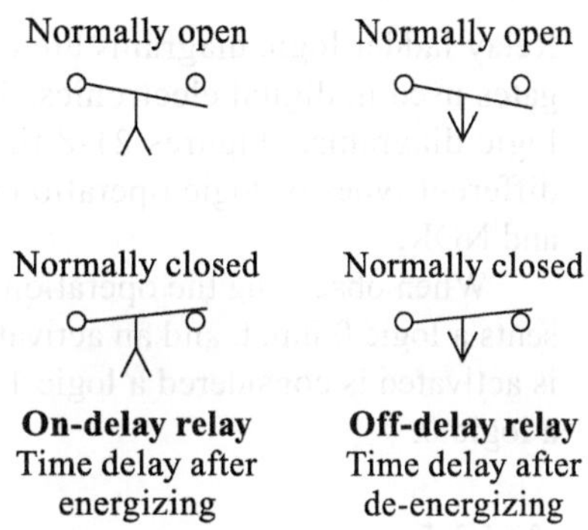

FIGURE 21-6 Time-delay relays

When power is applied to the coil of an on-delay relay, a period of time passes before its contact changes states. For an N.O. type, a period of time passes before the contact is closed. For an N.C. type, a period of time passes before the contact opens.

The off-delay relay also controls N.O. and N.C. contacts. When a voltage is applied to the coil, the contacts change states immediately, as they do in a normal control relay. An N.O. contact closes and remains closed as long as voltage is applied to the coil. After the voltage is removed, the contact remains closed for the delay period of the timer and then opens. An N.C. contact opens when the coil is energized. When the coil is de-energized, the contact remains open until the relay times out.

Contactor In general, a contactor operates in a similar fashion to the relay. Like the relay, it is an electromechanical device. The major difference between them is that the contactor is capable of carrying larger amounts of current than a control relay.

The most common function of a contactor is to switch power that is applied to resistance heating elements, magnetic brakes, or heavy industrial solenoids. Contactors are also used to switch motors, but require separate overload protection. They are then called **motor starters.** Published ratings on these devices are generally given in continuous amperes (usually 9 or 10 amps).

Solenoid Like the relay and contactor, the solenoid is an electromechanical device. It is made up of three basic parts: (1) the frame, (2) the plunger, and (3) the coil.

The frame and plunger are made up of laminations of a high-grade silicon steel. The wound coil is made of an insulated copper conductor. See Figure 21-7(a). When the coil of a solenoid is energized, a magnetic field is produced around the coil. This magnetic field produces a force that acts on the solenoid plunger. Due to this force, the plunger moves inside the coil. The force on the plunger is called *pull.* This pull may be as low as a fraction of an ounce or as high as nearly 100 pounds. The schematic symbol of a solenoid is shown in Figure 21-7(b).

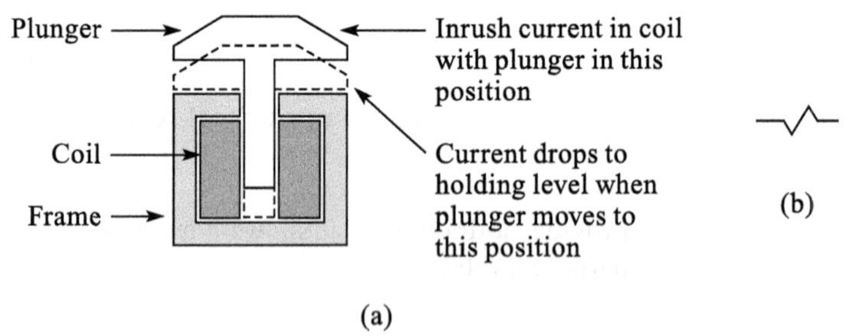

FIGURE 21-7 Solenoid

21-2 Relay Ladder Logic Circuits

Relay ladder logic diagrams are capable of performing logic functions similar to the logic gates used in digital electronics. Therefore, ladder diagrams are also referred to as "ladder logic diagrams." Figures 21-8 through 21-12 show five ladder circuits that perform five different types of logic operations. These logic functions include AND, OR, NOT, NAND, and NOR.

When observing the operation of these circuits, assume that a nonactivated switch represents a logic 0 input, and an activated switch represents a logic 1 input. An output device that is activated is considered a logic 1 state, and an output device that is not activated represents a logic 0.

IAU3206
Equivalent Gates/
Ladder Logic Circuits

AND Function

In Figure 21-8, switches 1 and 2 are considered input devices and lamp 1 is the output device. With S1 *and* S2 connected in series, they must remain activated, or closed, for the lamp to remain lit. If either switch is deactivated, or opened, the lamp will turn off. This circuit performs the AND function.

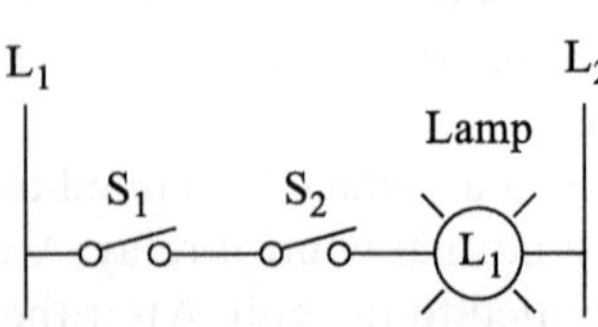

FIGURE 21-8 A relay ladder circuit that performs the logic AND function

OR Function

In Figure 21-9, N.O. pressure switch 1 and N.O. flow switch 2 are considered input devices and a solenoid is the output device. With S1 and S2 connected in parallel, the solenoid will become energized if S1 or S2 is closed (or both S1 *and* S2 are closed). If both switches are opened, the solenoid will not be energized. This circuit performs the OR function.

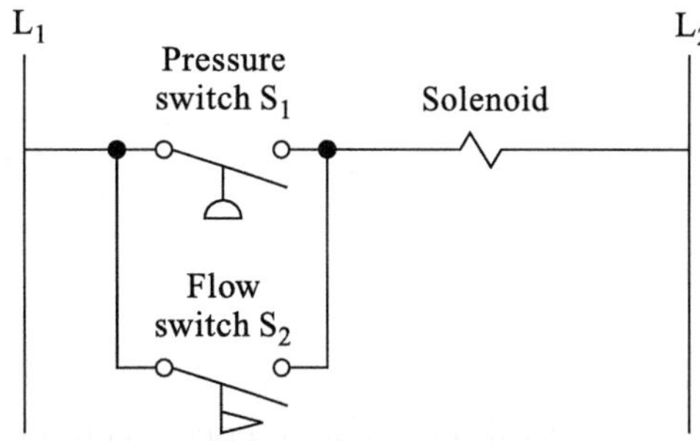

FIGURE 21-9 A relay ladder circuit that performs the logic OR function

NOT Function

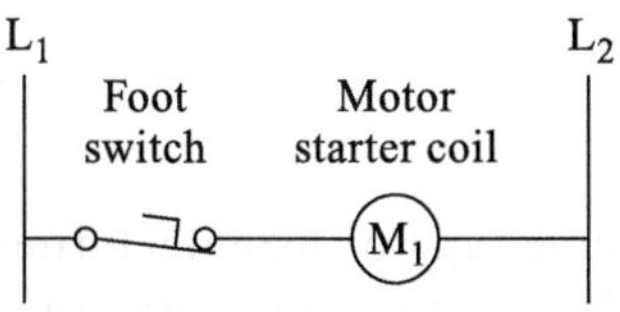

FIGURE 21-10 A relay ladder circuit that performs the logic NOT function

In Figure 21-10, an N.C. foot switch is the input connected in series with a motor starter coil that operates as the output. When the input switch is *not* activated, the coil is energized. When the foot switch is pressed, the output is *not* energized. This circuit performs the NOT, or *inverter,* function because the output is always opposite the input.

NAND Function

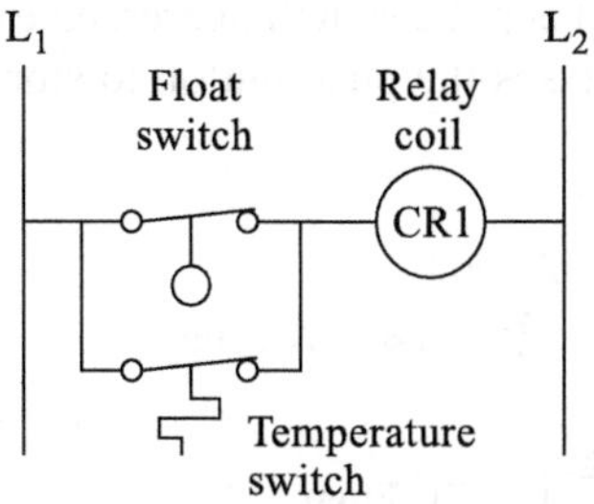

FIGURE 21-11 A relay ladder circuit that performs the logic NAND function

In Figure 21-11, an N.C. float switch and an N.C. temperature switch in parallel are the inputs, and a magnetic relay coil is the output. When one or both switches are not activated, the relay coil will be energized. If both switches are opened simultaneously, the output will be de-energized. The circuit performs the NAND function.

NOR Function

In Figure 21-12, three N.C. push buttons in series serve the inputs, and a timer is the output. When all of the switches are not activated, the timer will be energized. By pressing one or more of the stop buttons, the circuit will de-energize the load. The circuit performs the NOR function.

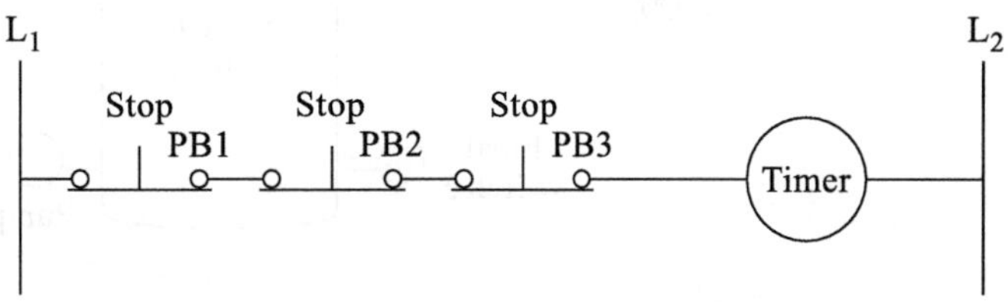

FIGURE 21-12 A relay ladder circuit that performs the logic NOR function

START/STOP Interlocking Circuit

IAU8707
Interlocking
Ladder Diagrams

The circuit in Figure 21-13 illustrates two very important functions of a ladder circuit: *stopping* and *interlocking.* It consists of an N.C. stop push button, an N.O. start button, a control relay coil, and an N.O. contact activated by the relay. The contact connected in parallel with the start button is called a **branch** connection.

When power is applied to the circuit, the relay coil cannot energize because there is an incomplete path for current to flow due to the N.O. start button and N.O. contact CR1. When the start button is pressed, a path for current develops from L_1 through the N.C. stop button, through the start button, and through the relay coil to L_2. Because control relay 1 is energized, it closes the N.O. contact also labeled CR1. If the start button is released, the output coil remains energized, because there is an alternate path through contact CR1 for current to flow. By maintaining an energized circuit after the momentarily closed start button is

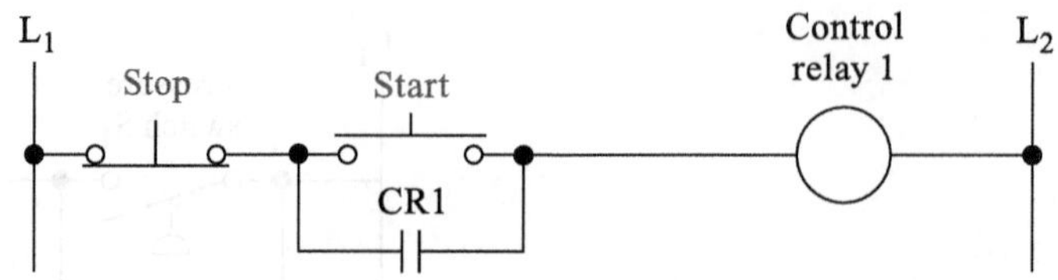

FIGURE 21-13 Ladder diagram for a basic STOP/START circuit

released, a *latching* or *interlocking* function is performed. The circuit becomes de-energized when the series-connected N.C. stop button is pressed.

Whenever possible, components in a ladder diagram are shown in order of importance. For safety reasons, the stop button has a higher importance than the start button. That is why it is placed in front of the start button.

21-3 Building a Ladder Diagram

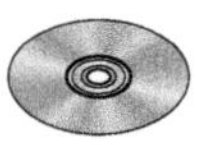

IAU3106
Designing a
Ladder Diagram

Ladder circuits are primarily used to control a machine or some type of manufacturing process. To help the reader understand how to develop a ladder diagram, a drawing of a pumping station and the sample circuit that controls its operation are shown in Figure 21-14.

The function of the pumping station shown in Figure 21-14(a) is to pump water from a storage tank into a pressure tank. The ladder diagram that controls the pump system is shown in Figure 21-14(b). An N.O. push-button switch is in series with the pump motor. Whenever the water level in the pressure tank is too low, the operator must keep the switch depressed to make the pump run until the tank is full. The operator then releases the push button to stop the flow of water into the pressure tank.

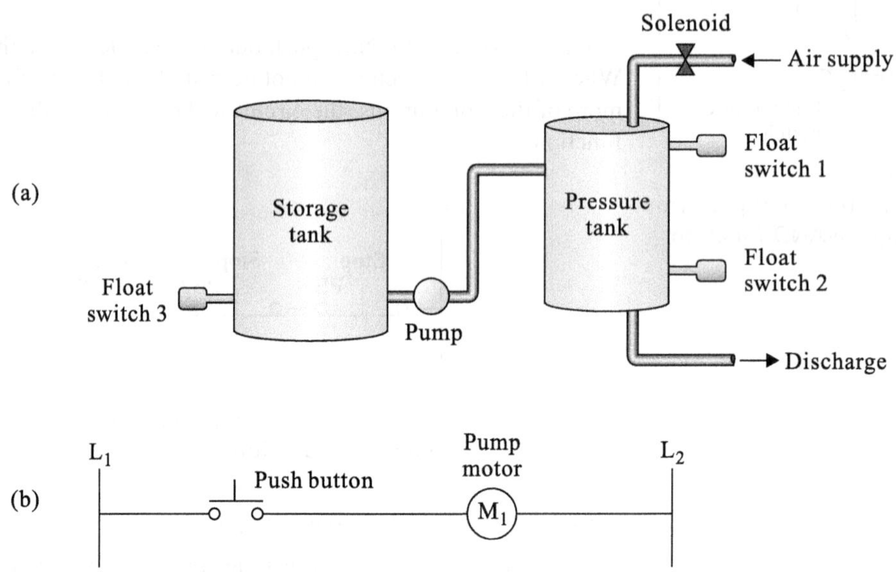

FIGURE 21-14 A water-pumping system

Modification 1

Figure 21-15 shows a modification to the original circuit: the operator is not required to manually keep the push button pressed while the pressure tank is filling up with water. Instead, the operator momentarily depresses the start button, which completes the current path from L_1 through the pump motor to L_2. The pump motor M_1 performs two functions. First, it pumps the water into the pressure tank. Second, it also has a coil that, when energized, closes contact M_1, which is in parallel with the start button, thus maintaining the current to coil M_1.

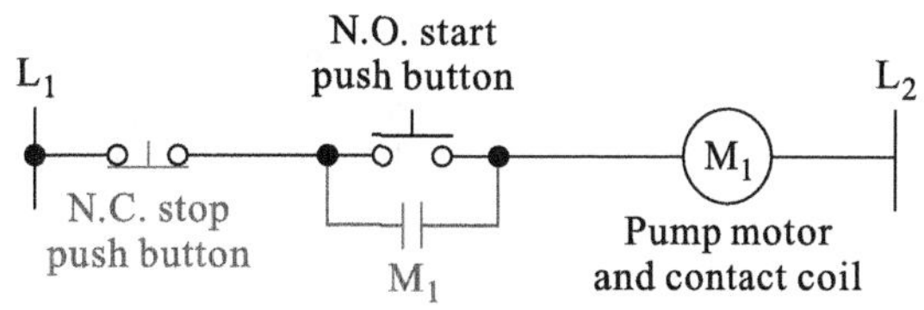

FIGURE 21-15 Circuit modification 1

When the pressure tank is full, the operator pushes the N.C. stop button, which opens the circuit, stopping the pump.

Modification 2

By installing a float switch near the top of the pressure tank, the operator is not required to push the stop button. Instead, the operator needs only to push the start button, thus energizing the pump and starting water flowing into the tank. When the level of the water has reached N.C. float switch one (FS_1), its contacts will open, thus stopping the pump and the flow of water. The function performed by the float switch is that of an automatic stop. The ladder diagram in Figure 21-16 shows an N.C. contact added in series to fulfill the new operational requirements. The manual stop button remains connected in the circuit so that the operator can stop the pump at any time.

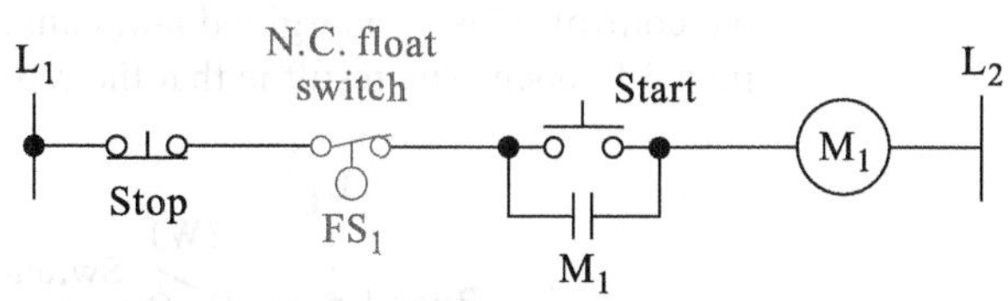

FIGURE 21-16 Circuit modification 2

Modification 3

By installing an N.C. float switch at the lower section of the tank, the pump system will also turn on automatically. This modification causes the pump to start whenever the water reaches a predetermined low level, thereby causing the N.C. contacts to close. Figure 21-17 shows the changes that are required on the control circuit to perform the operation. The contact (FS_2) is connected in parallel with the original start button, which enables it to start the motor. The parallel configuration provides a way for the manual start button to operate independently.

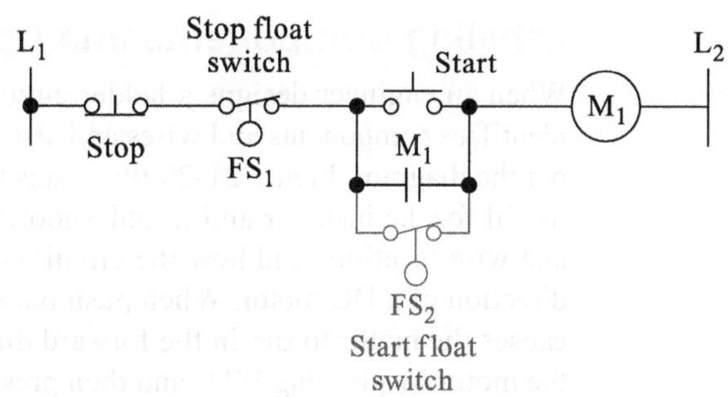

FIGURE 21-17 Circuit modification 3

Modification 4

Modification 3 enabled the pump to start automatically whenever the water level in the pressure tank reached a predetermined low level. However, suppose the water in the storage tank has dropped to a level too low to enter the pump orifice. A pump will be damaged if it is run without water flowing through it. As a fail-safe measure to protect the pump, an N.O. float

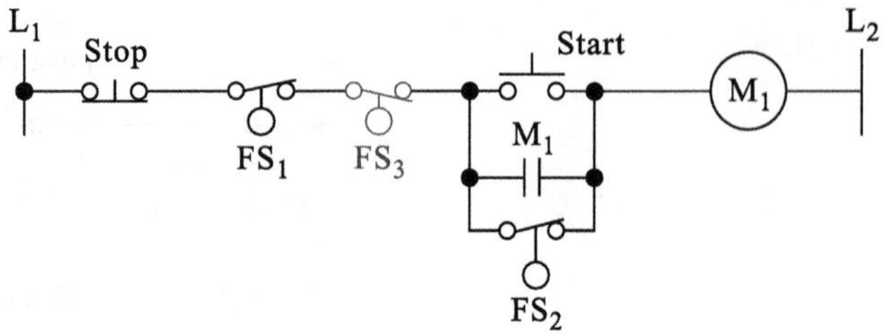

FIGURE 21-18 Circuit modification 4

switch (FS_3) is placed at an extreme low level of the storage tank. Its contacts will open whenever the water level drops to the set level of the float switch. Figure 21-18 shows that it is wired in series with the other start switches in the control circuit.

Multi-Rung Control

IAU8607
Multi-Rung Control in a Ladder Circuit

In many ladder circuits, an output relay coil in one rung will control the operation of circuits in one or more other rungs. An example is shown in Figure 21-19. A control relay is connected in rung 1, and the contacts that it controls are located in rungs 2 and 3. When SW1 is open, there is no current path between L_1 and L_2 to flow in rung 1. Since the control relay is de-energized, the N.O. contact in rung 2 remains open, and the N.C. contact in rung 3 stays closed. Therefore, the red light is on and the green light is off. When switch SW1 is closed, the control relay is energized and causes the contact in rung 2 to close and the contact in rung 3 to open. The result is that the red light turns off and the green light turns on.

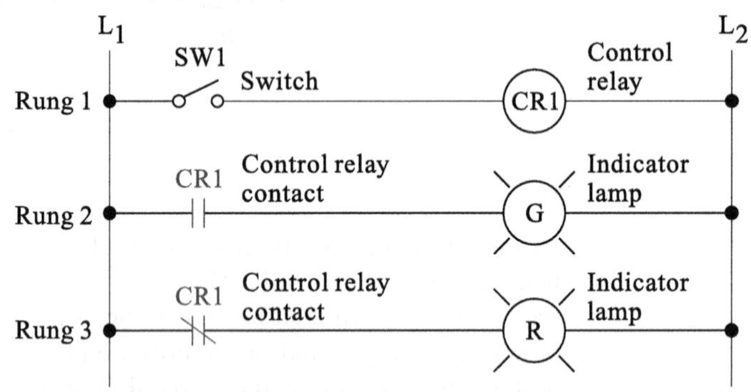

FIGURE 21-19 A multi-rung controlled circuit

Labeling Components and Connections in a Ladder Circuit

When an engineer designs a ladder circuit, it is useful to develop a labeling scheme, which identifies components and wires and also indicates how outputs control components throughout the diagram. Figure 21-20 illustrates how a forward/reverse circuit is labeled. Labeling is useful for the installer and troubleshooter because it provides information about component and wire locations and how the circuit operates. The function of the circuit is to control the direction of a DC motor. When push button PB2 is pushed, output coil CR1 is energized and causes the motor to run in the forward direction. To reverse the motor, it is necessary to stop the motor by pressing PB1, and then press PB3 to energize output coil CR2.

The boxed numbers to the left of rail L1 identify the horizontal rungs of the ladder diagram. Each rung has numbers that identify the wires that connect components to each other. For example, rung 1 shows numbers 1 through 5 to indicate five interconnecting wires. The leftmost wire begins with a 1 and the remaining wire segments to the right are labeled by increasing numbers. Common wires have the same numbers, even if they are interconnected with other rungs. Any components that are alike in the ladder diagram are labeled in numerical order from left to right and from top to bottom. For example, the push button located at the upper-left corner is labeled PB1, the push button to its right is labeled PB2, and the next push

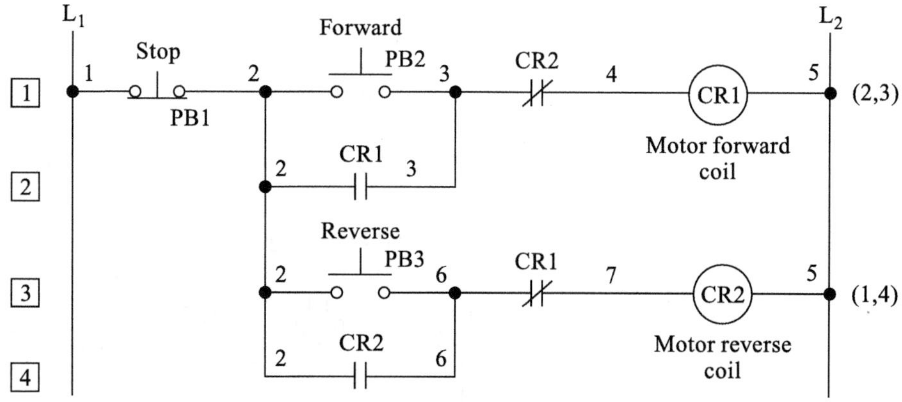

FIGURE 21-20 Labeling in a ladder circuit

button in the diagram placed on rung 3 is labeled PB3. Numbers in parentheses are also placed at each rung just to the right of rail L2. This number gives the location (rung number) of components that are controlled by its output device. For example, the number (2,3) at rung 1 indicates that its output CR1 controls N.O. contact CR1 in rung 2, and N.C. contact CR1 in rung 3.

IAU8507 Forward/Reverse Ladder Circuits

Summary of Rules That Apply to Ladder Diagrams

The following statements provide a list of rules that apply to understanding how to properly read ladder diagrams. Refer to Figure 21-20 as you read them.

1. The vertical lines or rails represent the power lines. The voltage potential may be either AC or DC. The left rail is normally labeled L_1. L_1 is the hot lead for AC and the positive line for DC. The right rail is normally labeled L_2. L_2 is the neutral lead for AC and the negative lead for DC voltages.
2. The horizontal lines or rungs are labeled in numerical order, from top to bottom.
3. The ladder diagram is read like a book, from left to right, and from top to bottom.
4. Whenever possible, the components are labeled in numerical order, from left to right, and from top to bottom.
5. The components are shown in their normal condition, which means they are de-energized.
6. Contacts will always have the same letter and number designation as the device that controls them. These control devices include relay coils, timers, or motor starters.
7. An N.O. contact closes when the device that controls it is energized. An N.C. contact opens when the device that controls it is energized.

21-4 Motor Starter Control Circuits

IAU9007 Motor Starter Control Circuits

Electrical control systems are used with many types of loads. The most common electrical loads in industry are three-phase AC motors. A motor control circuit for a three-phase motor is shown in Figure 21-21. Its primary functions include (1) on and off control, and (2) overload protection.

On the top of the diagram are three-phase power lines labeled L1, L2, and L3. The three-phase power lines are connected to a *line disconnect* which contains the following three devices (within the modular body) that perform three separate functions:

1. *Three-Phase Contact.* This device operates the same way as a triple-pole single-throw switch. Each pole switch turns a three-phase line on or off simultaneously with the other two lines. When the three-way switch is closed, it provides power for the control section of the motor starter circuit, and to the three-phase motor, if it is actuated by the control section.

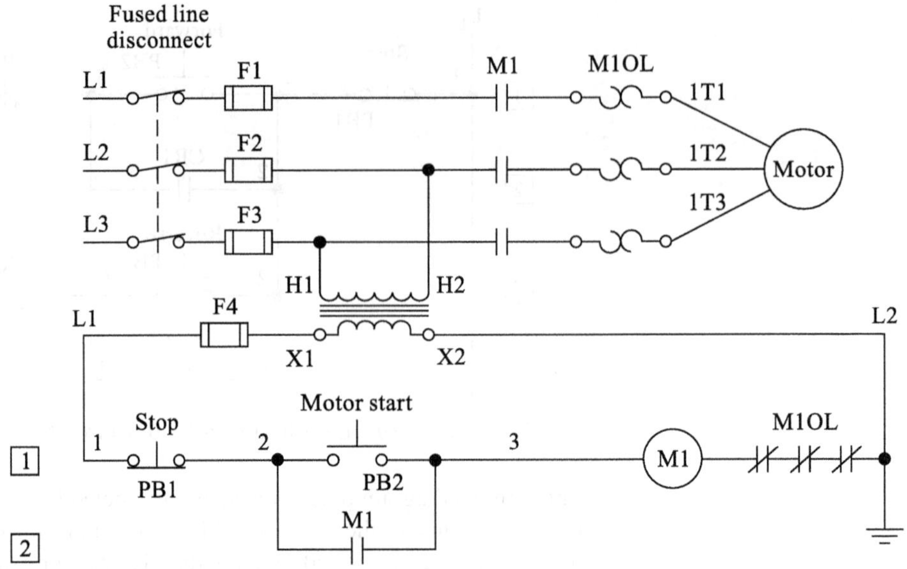

FIGURE 21-21 Motor starter circuit

2. *Arc Chutes.* When a set of contacts are opened under load, there is a short period of time during which the contacts are not fully in touch with each other, yet not completely separated. As they continue to separate, as shown in Figure 21-22, the contact surface area decreases, increasing the electrical resistance. With full-load current passing through the increasing resistance, a substantial temperature rise is created on the surface of the contacts. This temperature rise is often high enough to cause the contact surfaces to become molten and emit ions of vaporized metal into the gap between the contacts. Since this hot, ionized vapor is electrically conductive, it will permit the current to continue to flow in the form of an arc, even though the contacts are completely separated. These arcs in turn produce additional heat which, if continued, can damage the surfaces. To suppress these arcs, a device called an *arc chute* is used. An arc chute is a trap that confines the arc inside a separate chamber with a cover. The trap also vents any ionized gases inside the chamber, which would extend the flammability of the arc. Furthermore, the metal contacts are made of compounds that retard arcing. To prevent arcing between the three contacts, vertical barriers are placed between them.

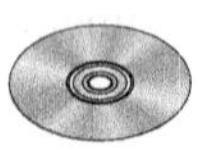

IAU10508
Motor Protection

3. *Short-Circuit and Ground-Fault Protection:* This overcurrent protection device is a fuse (labeled F_1, F_2, and F_3) or mechanical circuit breaker, which instantaneously opens the supply lines if excessive current flows. When the motor starts, a surge of inrush current develops because there is little or no counterelectromotive force (CEMF). Therefore, the circuit breaker must be rated at six to eight times the normal running current of the motor during the start-up interval. However,

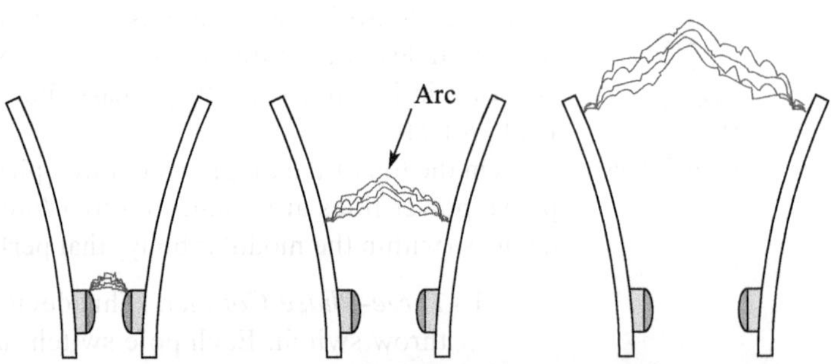

Operation: When a set of contacts are opened, an electrical arc will be created between the sets of contact tips as they separate.

FIGURE 21-22 Arc chute

the breaker must open instantly if the current exceeds its rated value due to a short that develops somewhere down the line. Excessive current conditions that occur are the result of:

- A short that develops somewhere down the line.
- Single-phasing (one of the three power lines opens due to a loose connection or a motor coil becomes defective) during start-up.
- A ground fault condition.

To the right of the line disconnect are three N.O. contacts labeled M1. When the contacts are closed, they enable the three-phase motor to be supplied with line power, causing it to run. When they are open, the motor shuts down. These contacts are controlled by the control section of the motor starter circuit.

Located to the right of the three M1 contacts are three thermal sensing elements called *overload heaters* that are labeled M1OL. Their function is to protect the motor if it becomes overloaded due to being overworked by the load (the load is too heavy for its horsepower capabilities) or because of a condition that does not allow the motor to turn, such as a broken part that prevents the motor load from moving. Instead of breaking the circuit immediately after excessive current is sensed, such as an inrush of current flow during motor start-up, this circuit breaker opens only after a *sustained* amount of excessive current (115 percent of rated full-load current) is drawn by the motor. Therefore, it must have a time delay to allow temporary overloads without disrupting the circuit. The overload device does not provide short-circuit protection. This is the function of the circuit breakers or fuses located in the disconnecting switch enclosure.

The overload circuit breaker is usually a relay that provides some means of resetting the circuit once the overload condition has ended. The metal alloy relay is a popular type of overload protection element. The melting alloy assembly is shown in Figure 21-23. It consists of wire screw connections and N.C. contacts. As a solder pot conducts the motor current, it also holds a ratchet wheel in place. If the current becomes excessive, the solder melts. The upward pressure from the spring causes the N.C. contacts to open as it forces the ratchet to turn in the molten pool. A cooling-off period is required to allow the solder pot to become solid again before the overload relay can be reset.

Control Circuitry Section

The starting and stopping of the three-phase motor are performed by an auxiliary circuit, often located some distance from the motor. Auxiliary starting and stopping are activated by pilot

IAU8807
Motor Starter
Protection

IAU10308
Sizing Motor
Overloads

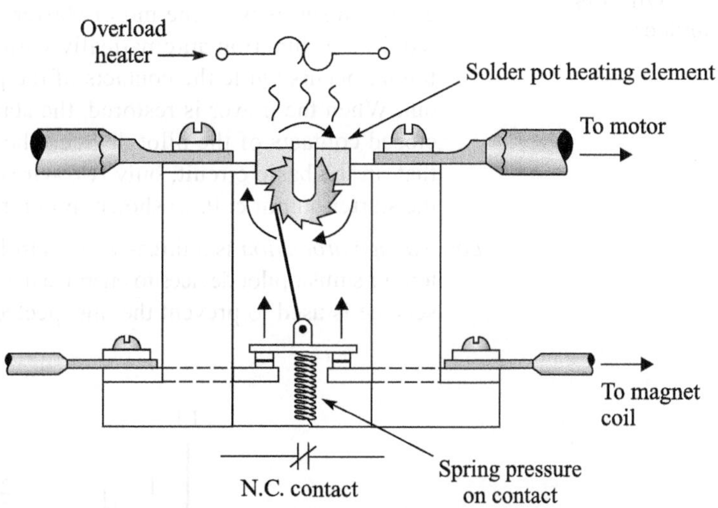

Operation: Melting alloy thermal overload relay. Spring pushes contact open as heat melts alloy allowing ratchet wheel to turn freely. Note electrical symbols for overload heater and normally closed control contact.

FIGURE 21-23 Thermal overload

switches such as push buttons, float switches, timing relays, and so forth. The auxiliary function is performed by supplying the primary of a transformer (labeled H1 and H2 at the terminals) with power tapped from two of the three-phase lines. The primary voltage is induced into a secondary coil (labeled X1 and X2 at the terminals), which supplies the voltage for the control circuitry section of the ladder diagram. The control section is de-energized by fuse 4 and a separate circuit protector if a malfunction develops that causes excessive current to flow.

When the motor start button PB2 is depressed, it completes the path of current from L1 to L2. With current flowing through the M1 coil, N.O. contact M1 in rung 2 closes and causes the circuit to latch. Also, the three N.O. contacts labeled M1 in the three-phase circuit (known as the *motor starter contacts*) also close and supply power to the three-phase motor. By pressing button PB1, coil M1 de-energizes and opens the latching contact and the contacts in the three-phase circuit, causing the motor to stop.

Three N.C. contacts labeled M1OL are located on rung 1 to the right of coil M1. These contacts are controlled by the three thermal-overload circuit breakers in the three-phase motor circuit. These breakers provide running protection for the motor. If the motor becomes overloaded, it may draw three times the full-load current (because the CEMF decreases), which is not enough to trip the main circuit breaker. The thermal overload is designed so that it will not trip during the momentary heavy surge that occurs during starting. Instead, the heat-up after 115 percent of sustained full-load current occurs and eventually trips, causing one or more of these contacts to open, thus shutting everything down. The same protection occurs if a single-phased condition of the three-phase motor develops. Single phasing takes place if a wire to one of the stator windings opens due to a break in a conductor connection or a coil becoming defective. In this situation, the motor continues to run, but usually at a lower speed if it is under load. The result is that a lower CEMF is developed and the current through the remaining two lines increases. To protect the motor from excessive current, the thermal overload trips and shuts the motor off.

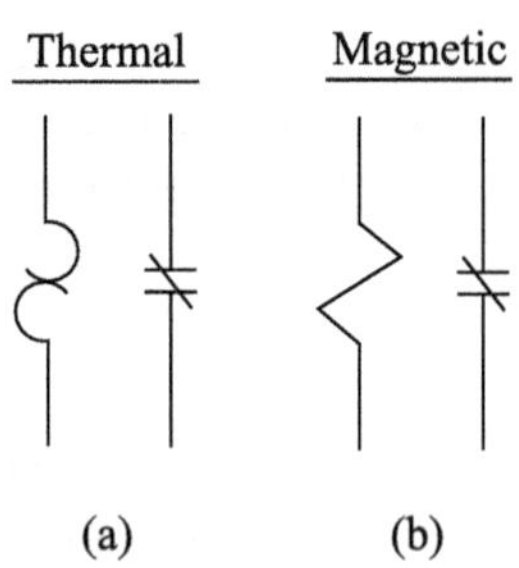

FIGURE 21-24 Symbols of overload devices

Figure 21-24(a) shows the symbol of the thermal overload device. Figure 21-24(b) shows the symbol of a magnetic overload device. When excessive current flows, a magnetic field develops around the energizing coil to cause its auxiliary contacts to pull open. The three-phase circuit and the control circuit also shut down if either the line disconnect switches are opened or the circuit breakers in the line disconnect assembly are tripped open.

There are two basic types of auxiliary control circuits, **low-voltage release** and **low-voltage protection.**

Low-voltage release is a two-wire control scheme using a maintained contact pilot device in series with the motor starter coil. This scheme is used when a starter is required to function automatically without the attention of an operator. If a power failure occurs while the contacts of the pilot device are closed, the starter will drop out. When the power is restored, the starter will pick up automatically through the closed contacts of the pilot device. The term *two-wire* control arises from the fact that, in the basic circuit, only two wires are required to connect the pilot device to the starter output coil, as shown in Figure 21-25.

Low-voltage protection is a three-wire control scheme using momentary-contact push buttons or similar pilot devices to energize the motor starter coil. Shown in Figure 21-26, this scheme is used to prevent the unexpected automatic starting of motors, which could

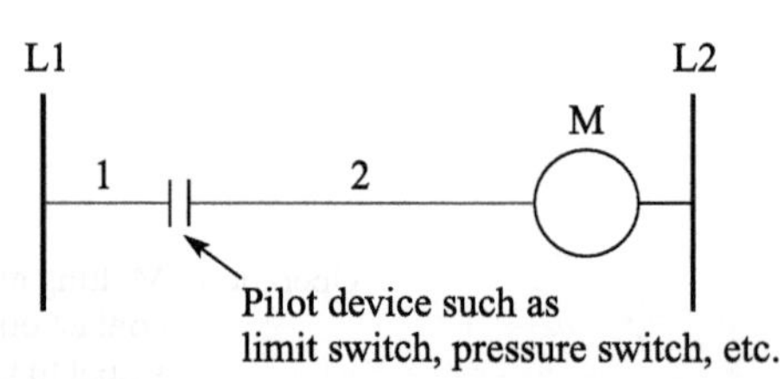

FIGURE 21-25 Two-wire low-voltage release

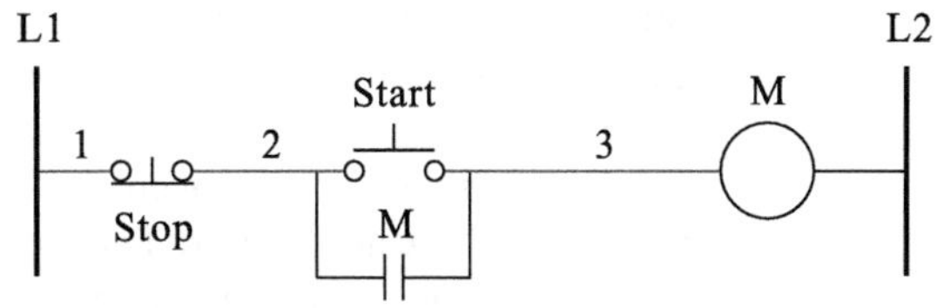

FIGURE 21-26 Three-wire low-voltage protection

result in possible injury to machine operators or damage to machinery. The starter is energized by pressing the start button. An auxiliary *holding circuit* interlock in the circuit forms a parallel circuit around the start button. These contacts keep the coil energized after the start button is released. If a power failure occurs, the magnetic field around the starter coil will collapse and open the interlocking circuit. Upon resumption of power, the start button must be pressed again before the motor will restart. The term *three-wire* refers to the fact that, in the basic circuit, three wires are required to connect the input pilot device to the output starter coil.

All control circuits, regardless of how complex they may be, are merely variations and extensions of these two basic types of circuits. The term *low-voltage,* used for both circuits, refers to the stepped-down voltage of the control circuit that is obtained from the high-voltage three-phase power lines.

INTRODUCTION TO PLC COMPONENTS

Even though relays and other electromechanical hardwire devices have served well for many years, there are disadvantages to using them. For example, they have moving parts that eventually wear out. They also have to be hardwired to perform specific functions. Whenever an operational requirement has to be changed or modified, the circuits must be rewired. It was this situation that caused the automobile industry to require the replacement of control circuits that were made exclusively from hardwired configurations.

As annual model changes were made, modifications of the relay circuitry were necessary. This effort was costly because it required shutting down an assembly system while the panels were rewired or replaced by new ones. The need for reduced downtime by a more rapid changeover scheme became a priority.

In the mid-1960s, Hydramatic, a division of General Motors Corporation, envisioned that a computer could be used to perform the logic functions then performed by relays. The company reasoned that circuit changes could be made using a keyboard, without performing extensive rewiring. The engineering team wrote a list of features of the proposed computing device. GM initiated the development of the computing device by specifying certain design criteria, including:

- The device must be durable so that it can operate in the harsh environments (dirty air, humidity, vibration, electrical noise, etc.) encountered in a factory.
- It must provide flexibility by implementing circuit modifications quickly and easily through software changes.
- It must be designed to use a programming language in ladder diagram form already familiar to technicians and electricians.
- It must allow field wiring to be terminated on input/output terminals of the controller.

GM used this list of specifications when it solicited interested companies to develop a device that met its design requirements. In 1968, the ruggedly constructed computer-based control, which became the programmable controller, was delivered to GM by Modicon.

The first PLCs were large and expensive. They were capable of On-Off control only, which limited their application to operations that required repetitive movements. Yet their initial design exceeded Hydramatic's requirements. For example, they could be installed easily, consumed less space and power than the wiring panels they replaced, and provided diagnostic indicators useful in troubleshooting.

Innovations and improvements in microprocessor technology and software programming techniques have added more features and capabilities to the PLC. These enhancements enable the PLC to perform more complex motion- and process-control applications, and with greater speed. Special-purpose modules have expanded their capacity to include operations associated with PID control, bar-coding, vision systems, radio frequency communication, voice recognition, and voice synthesizers.

Presently, more than a dozen manufacturers produce PLCs. Most of these companies make several models that vary in size, cost, and sophistication to meet the needs of specific applications. Regardless of size, cost, or complexity, all PLCs share the same basic components and functional characteristics.

All PLC systems are comprised of the same basic building blocks that detect incoming data, process it, and control various outputs. The following sections discuss these blocks:

- Rack Assembly
- Power Supply
- Programming Unit
- Input/Output (I/O) Section
- Processor Unit

With the exception of the rack assembly, Figure 21-27 shows all of these sections and how they are connected.

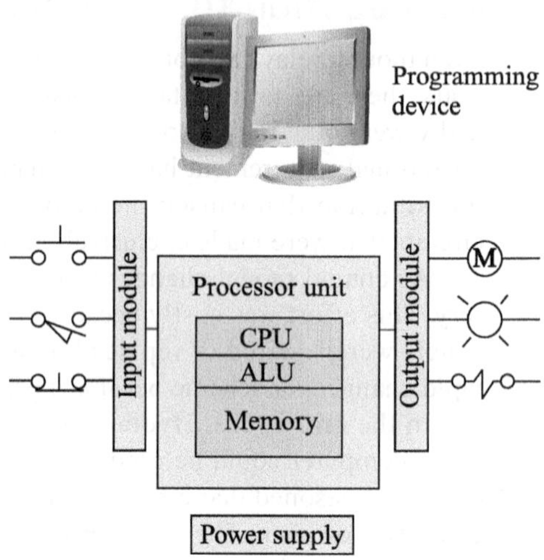

FIGURE 21-27 Major sections of a programmable controller

21-5 Rack Assembly

Most programmable controllers that have a large number of input and output terminals are constructed by using a variety of modules. These modules include the power supply, processor unit, and input/output modules. The modules are installed in a rack, such as the one shown in Figure 21-28. The PLC rack serves several functions. It physically holds the modules in place, and it also provides electrical connections between the modules by using a printed circuit board, called a **back plane,** at the back of the rack assembly.

The modules are easily inserted into channels on the rack. They fit into sockets mounted on the motherboard to make electrical contact with the other circuitry. The ability to plug modules into the rack allows maintenance personnel to replace defective units quickly.

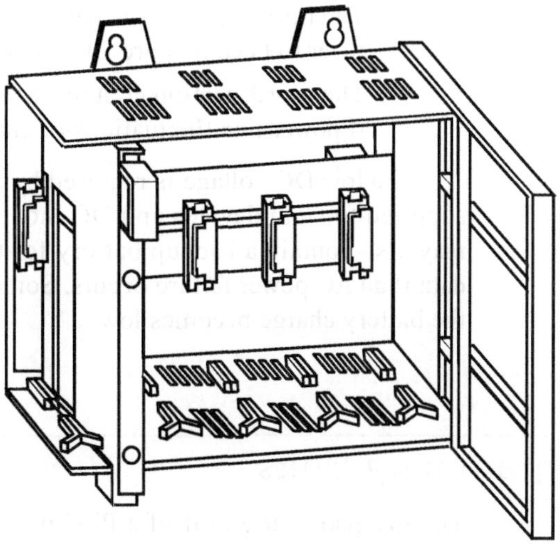

FIGURE 21-28 The rack assembly of a programmable controller

21-6 Power Supply

The **power supply** provides voltages that are necessary to operate the circuitry throughout the programmable controller. Some sections of the PLC require an AC voltage, such as AC input and output modules or the field devices that are connected to them. Other sections require a low-level DC voltage source. Figure 21-29 shows the section of the power supply that converts AC power to the DC voltage level required by the internal circuitry of the programmer.

1. The first section, on the far left, uses a step-down transformer that reduces the incoming line power of 120 or 240 volts to a lower level.
2. The second section uses two full-wave bridge rectifiers to convert the transformer secondary AC voltage to pulsating DC voltages. The top bridge develops a positive voltage, and the bottom bridge develops a negative voltage.

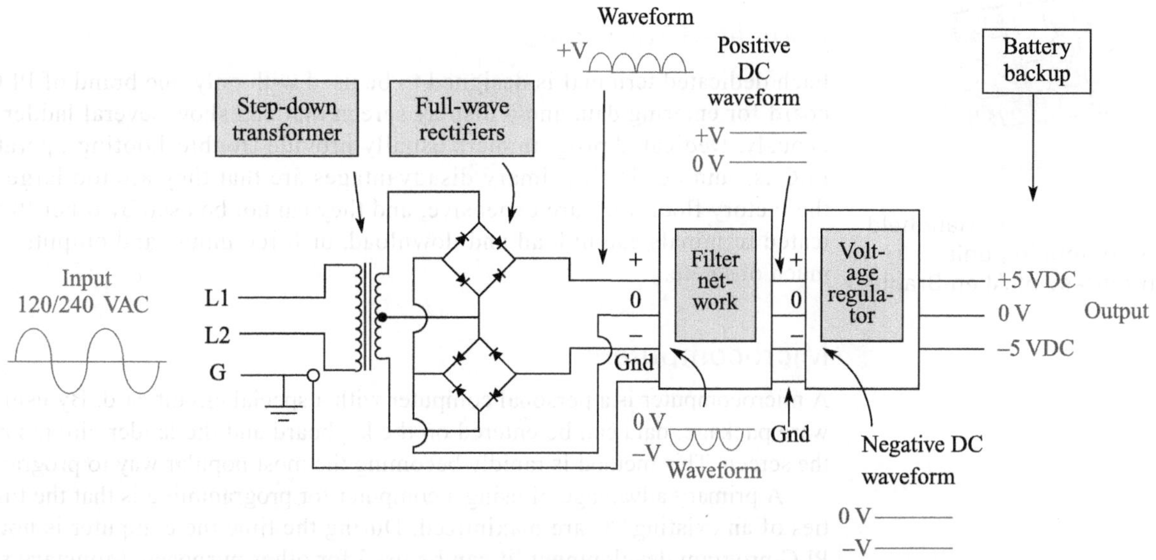

FIGURE 21-29 Power supply

3. The third section uses filter circuits to condition the pulsating DC output of the rectifier to become pure DC voltages.
4. The fourth section contains a voltage regulator that maintains a constant DC voltage if power line fluctuations or changing load demands occur.

The low DC voltage is required by the internal circuitry of the processing unit. It is also supplied to field devices and DC modules to which they are connected. The power supply may also contain a backup battery for the memory devices in the processing unit to retain data if an AC power failure occurs. Some PLCs include a battery indicator, which indicates if the battery charge becomes low.

21-7 PLC Programming Units

The **programming unit** of a PLC provides a way for the user to enter data and to edit and monitor programs stored in the processor unit. The programming unit communicates with the processor unit by using a data communication link that transfers data in a serial or parallel fashion. Most programmers also perform troubleshooting procedures by simulating signals from input devices or by forcing output devices to energize through keyboard entries. This method is called *forcing* inputs and outputs.

Some programming units also have the capability of storing programs in their memory chips once they are written, and then loading them into the PLC at a later time. This process is called *uploading* and *downloading*.

Types of programming units used to perform these operations include handheld programmers, dedicated terminals, and microcomputers.

Handheld Programmers

FIGURE 21-30 Handheld programming unit (Courtesy of Allen-Bradley)

Handheld programmers, such as the one shown in Figure 21-30, are very small, inexpensive devices. They typically have membrane keys for entering data and LCD displays that show one line of a ladder diagram. They are useful for programming simple ladder diagrams or for troubleshooting on the factory floor because they are easy to carry. Their primary disadvantage is that the ladder diagrams are difficult to follow on the screen because only a small segment is shown at any given time. Many handheld programmers are capable of forcing inputs and outputs, and uploading and downloading.

Dedicated Terminals

Each dedicated terminal is designed to be used with only one brand of PLC. It has a keyboard for entering data and a display screen that can show several ladder rungs simultaneously. Dedicated programmers usually provide troubleshooting operations while the PLC is running. Their primary disadvantages are that they are too large to carry out to the factory floor, they are expensive, and they cannot be used by other PLC brands. Dedicated terminals can upload and download, or force inputs and outputs. They are pretty much obsolete.

Microcomputers

A microcomputer is a personal computer with a special circuit card. By using a special software package, data can be entered on the keyboard and the ladder circuits are displayed on the screen. This method is rapidly becoming the most popular way to program PLCs.

A primary advantage of using a computer for programming is that the time-use capabilities of an existing PC are maximized. During the time the computer is not being used for PLC program development, it can be used for other purposes. Monetary savings are also possible because it is not necessary to purchase a handheld or dedicated programming unit.

21-8 Input/Output Sections

There are many types of external field devices and circuitry connected to the programmable controller. The purpose of the I/O section of the PLC is to interface its internal circuitry to outside equipment.

The I/O section contains **input/output modules** that serve four basic functions:

1. *Termination.* Each I/O module provides terminal connections to which field devices can be connected. Each terminal is assigned an identification number.
2. *Signal Conditioning.* Most of the voltages used by field devices are not compatible with the low DC voltage data signals processed inside the programmable controller. The module converts the external signal to a voltage suitable for the PLC.
3. *Isolation.* The factory floor is a very noisy electrical environment. Electrical noise is created by stray magnetic fields produced by devices such as large motors, welding equipment, and contactors used to switch high currents on and off. If the flux lines created by these devices cut across any PLC conductors, unwanted voltage transients can be induced that may be falsely recognized as data pulses by the processor unit. A noise signal may be interpreted by the PLC as an instruction, which could make it operate erratically or cause an output device to move when it should not. Noise picked up by output lines could be fed back to the processor unit and cause damage. By using optical coupler circuitry, there is no physical connection between the input or output field devices and the circuitry inside the PLC. The module isolates the processor unit by blocking the noise and passing only those signals that are valid.
4. *Indication.* Each terminal has an associated indicator. Its function is to illuminate when a voltage is applied to that terminal. I/O modules typically use LEDs as indicators. Older units use neon light bulbs as indicators.

The most common type of signal interfaced to a PLC input, or from its output, is the discrete voltage. The discrete voltage is either present or absent. The signals applied to the input terminal are produced by On/Off, Open/Closed, or energized/de-energized conditions of field devices. Input field devices—which include push buttons, photoelectric switches, proximity detectors, and level sensors—are external instruments connected to the PLC. The controller interprets discrete field signals as logic states. For example, when a field device is on, it is interpreted as a logic 1 state, while an off condition is interpreted as a 0 state.

A logic 1 state produced by the controller turns an output field device on. A logic 0 turns it off. These signals from the controller are interfaced to discrete output devices external to the PLC. These field devices include relays, solenoids, valves, alarms, motors, and indicator lamps.

Input Modules

The PLC receives discrete incoming signals from switching devices. When a switching device is closed, the input interface module senses a voltage and converts it to a logic 1 for the internal circuitry of the processor unit. When the device is open, a voltage is not sensed at the input terminal. The module converts this input as a logic 0 for the processor unit.

There are three common forms of discrete input signals—120 VAC, high DC voltage, and low DC voltage–that are compatible with transistor–transistor logic (TTL) integrated circuits. An input interface module that converts 120 VAC to logic signals used by the PLC is discussed below.

AC Input Module

The AC input module converts the presence of 120 volts applied to its input terminals to a logic 1 state. The absence of an AC voltage produces a logic 0. A typical circuit used to perform this function is shown in Figure 21-31(a) and discussed in the following list:

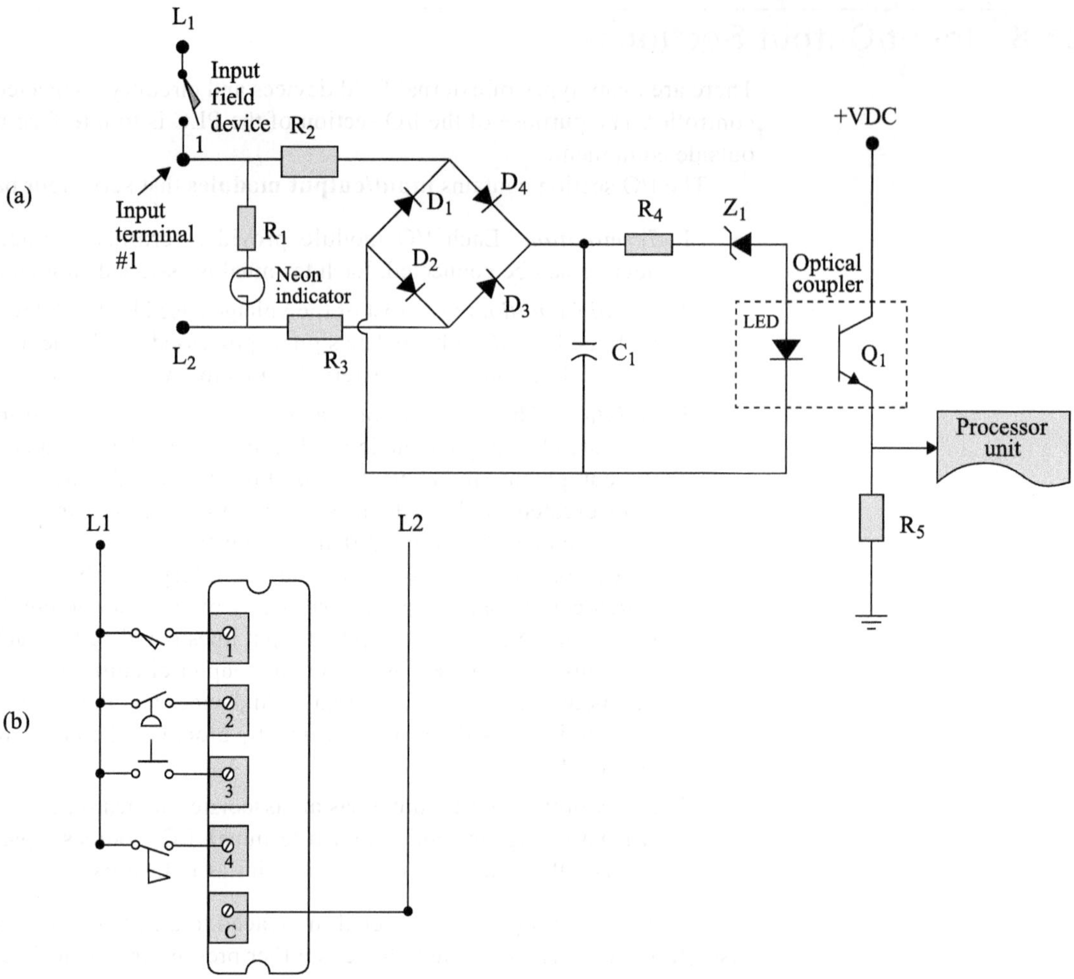

FIGURE 21-31 AC input module

1. An N.O. limit switch is connected between L_1 and the input terminal. An AC potential difference between L_1 and L_2 is supplied to the circuit when the switch is closed.
2. A neon lamp and a series resistor are connected across the input terminal and L_2. The lamp is illuminated only when a voltage is present, thus performing the indication function. Since the lamp lights at approximately 50 volts, the remaining potential drops across R_1, which also is a current-limiting resistor.
3. Resistors R_2 and R_3 are used to drop most of the incoming voltage. A remaining voltage of 5 to 12 volts is applied to a full-wave bridge rectifier that converts the AC voltage to a pulsating DC voltage.
4. The capacitor and resistor, R_4, is used to debounce the input signal, filter out electrical noise from the input line, and filter the DC pulsations to a pure DC signal.
5. The zener diode Z_1 and resistor R_4 form a threshold circuit that breaks over and produces a sudden regulated stable DC voltage when the voltage of each pulsating alternation has reached a certain amplitude.
6. When a valid signal is produced by the threshold components, it forward-biases the LED in the optical coupler. As the LED illuminates, the light it transmits turns on the phototransistor Q_1. The optical coupler performs the isolation function in the module.
7. As the LED turns on, the phototransistor turns on. The switching action of Q_1 provides continuity and allows current to flow through R_5. As a positive potential near 5 volts develops across R_5, a logic 1 state is supplied to the processor unit.

8. An AC potential is no longer present between L_2 and the numbered input terminal if the limit switch opens. Therefore, the LED in the coupler does not light and the phototransistor is off. Because Q_1 is in an open condition, there is no voltage dropped across R_5. A voltage that is near ground potential is supplied to the processor unit and is recognized as a logic 0 state.

One input module can have 4, 6, 8, 10, 16, or 32 interface circuits, providing the same number of individual connections to receive input signals. Figure 21-31(b) shows how external connections are made to the AC input module. The connector (labeled "C") indicates that AC line L_2 is secured to a common terminal. Inside the module, the bottom input lead of each interface circuit shown in Figure 21-31(a) is connected to this terminal. The top lead of each input circuit is connected to one of the numbered input terminals. Several types of field-switching devices are connected between AC line L_1 and the numbered terminals. A switch closure provides the continuity required to energize the input circuit.

Output Modules

The outgoing discrete signals from the processor unit are transmitted to output field devices by the output module assembly. An example of an output interface module follows.

DC Output Module

The DC output module interfaces the logic signal from the processor with a DC output field device that operates at a potential greater than +5 volts. A typical DC module circuit is shown in Figure 21-32(a).

1. When the processor generates a logic 1, the LED in the optical coupler is forward-biased. The light from the LED turns on the phototransistor.
2. As the phototransistor switches on, the potential at the base of the PNP transistor Q_1 becomes more negative, which causes it to turn on. As the transistor switches on, current flows through R_3 and a voltage near 5 volts develops at output terminal 4. The voltage drop that forms across R_3 activates the solenoid and also causes the LED indicator to light. The optical coupler performs the isolation function in the module.
3. If the processor's signal is a logic 0, the optical coupler's LED does not light and the phototransistor is off. Since the base current will not flow at Q_1, its off condition

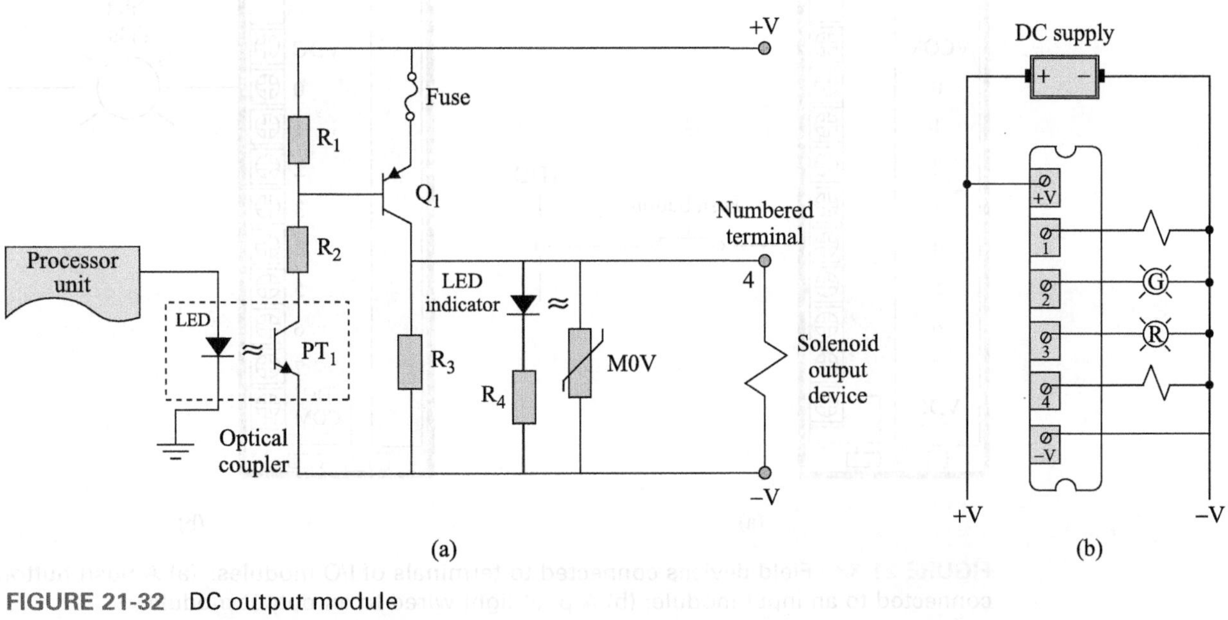

FIGURE 21-32 DC output module

creates a voltage of 5 volts from its emitter to its collector. The load potential from the –V terminal to terminal 4 reduces to a 0-volt level, causing the solenoid to de-energize and the LED indicator to turn off.

Figure 21-32(b) shows how external field devices are connected to the DC output terminal. When one of the output circuits is energized by the processor unit, it enables current to flow from –V through the output field device, through Q_1, to +V.

Module Termination

I/O modules are inserted into the PLC rack assembly. A single input module is shown in Figure 21-33(a). The bottom portion of the module contains eight terminals to which field devices are connected. Each terminal has an identification number. For example, the push button is connected to terminal 4 of the module. The identification number is also used to identify the instruction on the display screen that is controlled by the device through the terminal connection.

On the top portion of each module are indicator lamps that are associated with the terminals below them. A lamp will turn on when a voltage is present at the terminal to which it is connected. Each lamp is labeled with the same number as the terminal that causes its illumination. Figure 21-33(b) shows an output module with an output field device connected to terminal 0.

Some PLC modules have as many as 16 terminals, while other types have fewer than 8. AC or DC power is supplied to each module. The type of power supplied is determined by the type of power used by the field device. A common ground is connected to most types of modules.

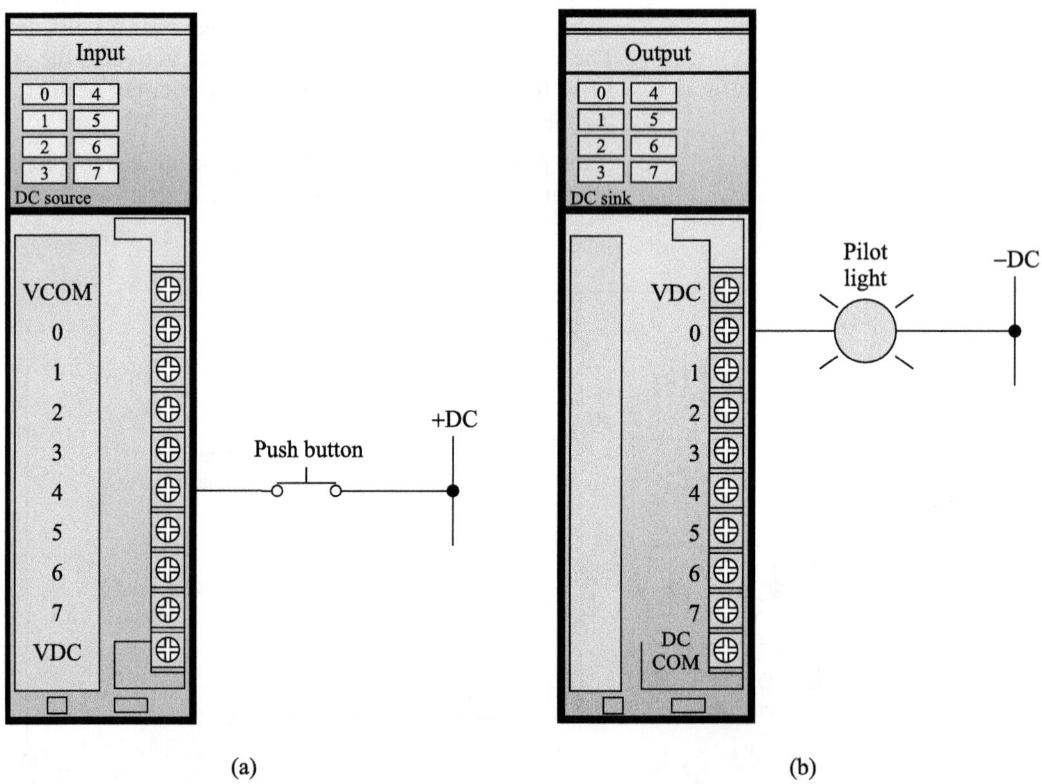

FIGURE 21-33 Field devices connected to terminals of I/O modules: (a) A push button connected to an input module; (b) A pilot light wired to an output module

21-9 Processor Unit

The **processor unit** coordinates and controls the operation of the entire programmable controller system. A processor module is usually located at one side of the rack assembly. It contains integrated circuit chips that include one or more microprocessors, memory chips, and circuits that enable data to be stored in and retrieved from memory. Some processor units have communication circuitry that provides interfacing between the processor and peripheral devices such as programmers, printers, and personal computers.

The processor is composed of three main sections: the central processing unit (CPU), the arithmetic logic unit (ALU), and the memory, as shown by the block diagram in Figure 21-34.

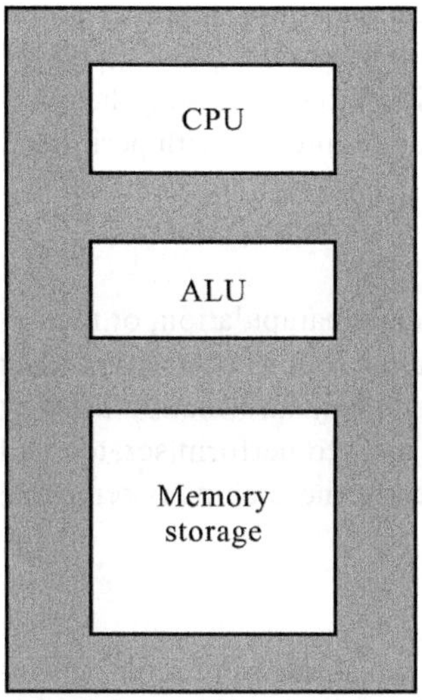

FIGURE 21-34 Block diagram of the processor unit

Central Processing Unit

The **central processing unit** is the brain of the PLC. The intelligence it provides is performed by one or more microprocessors, which are integrated circuits with tremendous computing and control capabilities.

The principal function of the CPU is to interpret and execute computer-based programs that are permanently stored in the processor's memory. These programs are written by the PLC manufacturer to enable the PLC to perform ladder logic instead of other programming languages. The CPU also coordinates the operation of the ALU and the memory. For example, based on the software program, the CPU determines what should be done in the ALU and the memory, and when it should be done. The CPU also performs other functions, such as self-diagnostic routines to determine whether the PLC is operating properly and communication with peripheral devices and other processors.

Arithmetic Logic Unit

The function of the **arithmetic logic unit** is to perform mathematical calculations and make logic decisions.

Memory

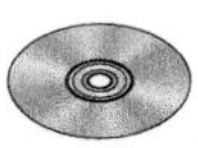

IAU4607
AB SLC-500
Memory Structure

The **memory** function of the processor stores programs and data that the CPU needs to perform various operations. The memory is organized into several sections according to the functions they perform. Figure 21-35 shows a block diagram of these sections.

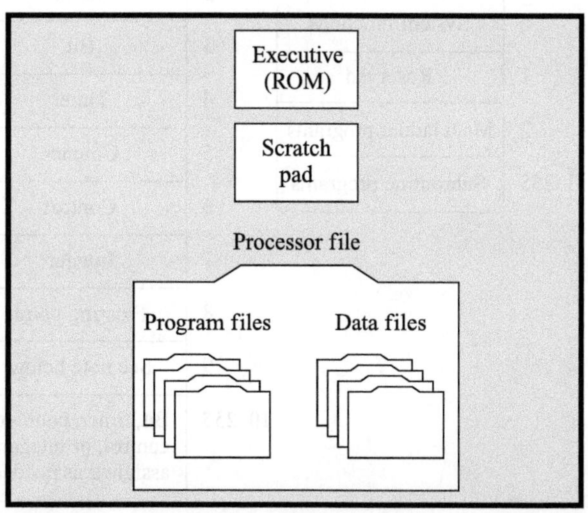

FIGURE 21-35 Block diagram of the PLC memory

Executive

The **executive** is a collection of system programs permanently stored in ROM devices. *ROM* refers to *read-only memory,* which contains information that is permanently stored. The system programs enable the CPU to understand the commands it receives from the program instructions written by the operator. For those readers who are familiar with PCs, the ROM contains the same type of information as a readable disk. In addition to supervising the operation of the PLC by executing control programs, the executive communicates with peripheral devices and performs system housekeeping functions.

Scratch Pad

As the CPU performs various operations such as logic analysis, data manipulation, or mathematical functions, it is necessary to temporarily hold data as calculations are performed or decisions are made. The work area used to temporarily store the binary information used by the processor is the **scratch pad.** RAM-type memory chips are used to perform scratch pad operations. RAM is volatile, which means that if power supplied to these chips is removed, the contents will be lost.

Processor File

The control functions performed by the CPU are determined by the software program that is written. The memory block in which the programmer stores and manipulates the software is called the **processor file.** The CPU can hold one processor file at a time. The following specifications refer to the Allen-Bradley SLC 500 PLC. The processor file is made up of *program files* (up to 256 per controller), and *data files* (up to 256 per controller). There are 4 classifications of program files and 11 classifications of data files, as shown in Figure 21-36.

Program Files Program files are categorized into four types: *system functions, reserved, main ladder programs,* and *subroutine programs.* An explanation of the four types follows:

1. File 0—*System Functions.* This file is used for a personal password, program identification, and user-programmed information such as identifying which type of processor module is used and which I/O modules are inserted into the slots of the rack.

	Program files
0	System functions
1	Reserved
2	Main ladder programs
3–255	Subroutine programs

	Data files
0	Output image
1	Input image
2	Status
3	Bit
4	Timer
5	Counter
6	Control
7	Integer
8	Floating point
9	See note below
10–255	Bit, timer, counter, control, or integer; assigned as needed

Note: Data file 9 can be used as an ordinary data file if the processor is not on a network or for network transfer if devices exist on the RS-485 link.

FIGURE 21-36 Classifications of program files and data files

2. File 1—*Reserved.* If a block of memory is needed at some time in the future, this file is kept open for that purpose.
3. File 2—*Main Ladder Logic.* The programming instructions, which are written by the PLC user, are stored in this type of memory. The information contained in these programs is used by the CPU to perform the control functions of the PLC. The user program instructions create the main ladder diagrams used to perform logic functions, mathematical calculations, data manipulations, and other operations.
4. File 3—*Subroutine Programs.* Subroutine programs are created by the user. They are accessed during the run mode from the main ladder program when subroutine instructions are used. Subroutine programs help conserve memory or reduce scan time.

Most of your work with program files will be in file 2, the main program files. This file contains your ladder logic program, which you create to control your application.

IAU5907
SLC-500 Data File Elements

Data Files Data files contain information used by the program files. This information includes the status of input or output field devices, the accumulated value of a counter, the time expired in a timer, the result in an arithmetic operation, and so on. These files can also be used to store recipes, standard recurring instructions, and tables (if necessary). There are 256 data files; they are organized by the type of data they contain. Each one has a separate address to distinguish it from the others. Usually, a *number* and a *letter* are both used to indicate the address.

Each file has registers that store 16-bit words. The registers contain data that is organized into segments called *elements*. These elements are assigned numerical addresses. Depending on the type of file, the elements have different sizes. For example, file 3 in Figure 21-37(a), which is the bit file, uses elements that are one word in length. File 3 holds 256 words that are addressed 0 to 255. File 4 in Figure 21-37(b), which is the timer file, contains

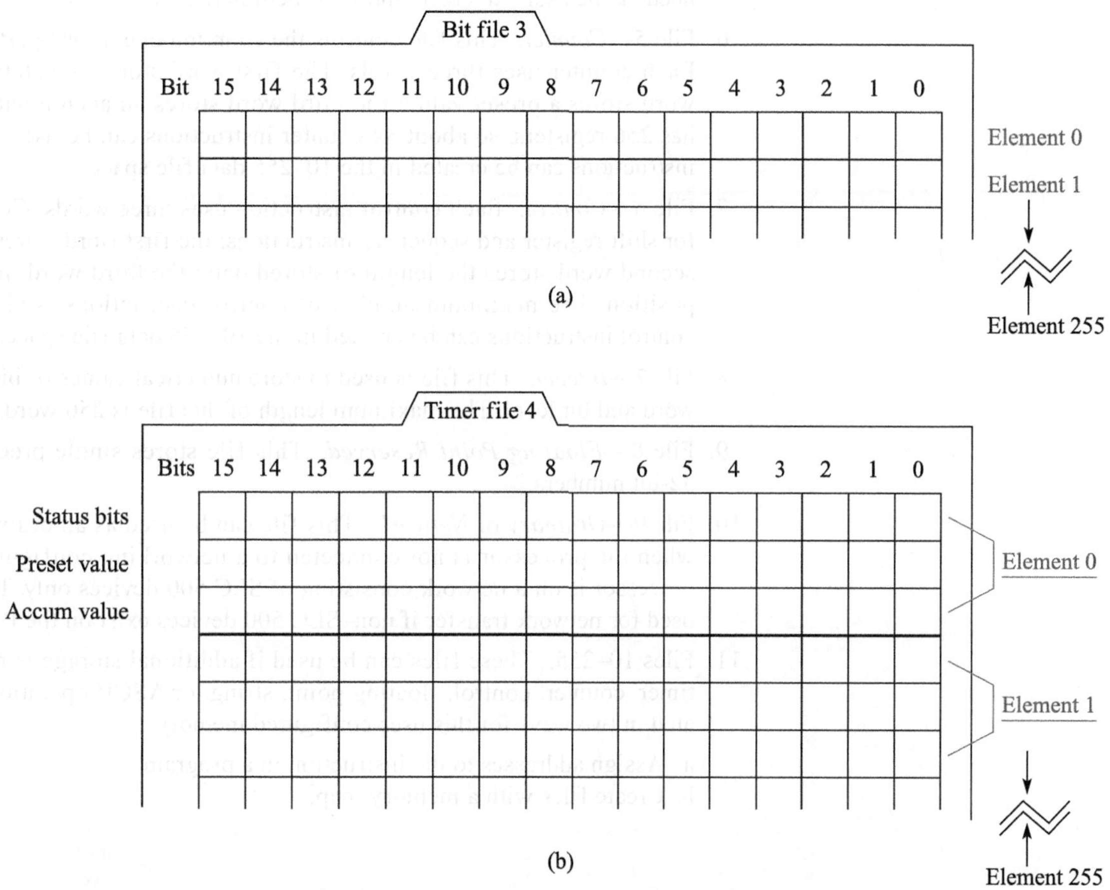

FIGURE 21-37 Examples of elements in different data files

768 registers. Each timer instruction uses elements that contain three 16-bit memory word registers. Each element is assigned a numerical address to distinguish it from another element and to indicate where it is located in data file 4. File 4 holds 256 elements that are addressed 0 to 255.

The following list shows how the 256 data files are organized:

1. File 0—*Output.* This file contains all of the output data bits generated by the ladder circuit. These bits are transferred to the terminals of the output module and are applied to field devices as logic 0 or 1 states. This file uses 256 16-bit elements.
2. File 1—*Input.* This file stores the logic signals, produced by input field devices that are connected to the terminals of the input module, for the controller to process. It uses 256 16-bit word elements.
3. File 2—*Status.* This file identifies faults and stores information concerning the processor operation. Arithmetic status bits and the math register are contained in this file. It has 256 16-bit word registers.
4. File 3—*Bit.* This file is used primarily for internal relay logic instructions. Internal relay instructions pertain to input and output contacts that are created on a ladder diagram but are not directly associated with external field devices. Instead, they are primarily used by a rung to perform logic functions and to affect input contacts on other rungs. Bit files are also used for shift register and sequencer instructions.
5. File 4—*Timer.* This file controls timing operations performed by the PLC. Each timer instruction consists of three words stored in three consecutive registers. The first word stores the status bits; the second word stores a preset value; the third word stores an accumulated value. This file has 768 registers. Since each timer instruction uses three words, 256 timer instructions can be used. If additional timers are needed, the user can create more of them in the 10–255 data file space.
6. File 5—*Counter.* This file controls the counting operations performed by the PLC. Each counter uses three words. The first word stores the status bits; the second word stores a preset value; the third word stores an accumulated value. This file has 256 registers, so about 84 counter instructions can be used. Additional counter instructions can be created in the 10–255 data file space.
7. File 6—*Control.* Each control instruction uses three words. Control files are used for shift register and sequencer instructions: the first word stores the status bits; the second word stores the length of stored data; the third word indicates the pointer position. The maximum number of control instructions is about 84. Additional control instructions can be created in the 10–255 data file space.
8. File 7—*Integer.* This file is used to store numerical values or bit information at the word and bit level. The maximum length of this file is 256 words.
9. File 8—*Floating Point Reserved.* This file stores single precision nonextended 32-bit numbers.
10. File 9—*Ordinary* or *Network.* This file can be used as an ordinary data file either when the processor is not connected to a networking configuration, or when the processor is on a network consisting of SLC 500 devices only. This file can also be used for network transfer if non–SLC 500 devices exist on the RS-485 link.
11. Files 10–255. These files can be used if additional storage is required for the bit, timer, counter, control, floating point, string, or ASCII operation. Data can be created in two ways for this user-configured memory.

    a. Assign addresses to the instruction in a program.
    b. Create files with a memory map.

21-10 Addressing

When programming a ladder diagram, the memory locations to be used by each instruction must be assigned an address. The address identifies various types of information, such as the type of file being addressed and the specific word, element, and/or bit number it is assigned to in the file. To address data files, alphanumeric characters separated by delimiters are used. Alphanumeric means that some of the characters are letters of the alphabet, and the other characters are numbers. Delimiters include characters such as a colon, slash, or period. A number or prefix will follow the delimiter. The type of delimiter used indicates what type of information follows it. For example:

- A number that follows a colon represents which element is used in a file.
- A number that follows a slash represents which bit is used in a file.
- A number that follows a period represents which word is used in an element (consisting of three words). For instructions that use three-word elements, such as counters, a three-letter prefix follows the period to indicate which word is being addressed. For example, the prefix *PRE* indicates that the word being addressed in a counter is its preset value.

When an input for an instruction symbol is placed on a ladder rung during programming, it is followed by writing the assigned address. There are four types of formats used for addressing symbols: *word, element, bit,* and *input/output.* The address format that is used depends on the type of file instruction. These formats pertain to the characters that must be entered on the keyboard for each symbol placed on the rung. The alphanumeric instruction displayed at the symbol after the data is entered will not always be exactly the same as what is entered on the keyboard.

Most addresses begin with a letter, called an *identifier,* followed by a corresponding file number. Together, they indicate the type of data file represented. Figure 21-38(a) shows the identifier and corresponding file number for each type of data file. Figure 21-38(b) shows the identifiers allowed to be stored in files 10–255 for situations when additional storage space is required.

Each data file is capable of storing only 256 elements into memory. Usually, this storage capacity is large enough for most programs. When programming an instruction, it is necessary to enter the address in one of the four types of formats.

(a)

File Type	***Identifier***	***File Number***
Output	O	0
Input	I	1
Status	S	2
Bit	B	3
Timer	T	4
Counter	C	5
Control	R	6
Integer	N	7
Float	F	8
Ordinary/Network		9

(b)

USER-DEFINED FILES		
File Type	***Identifier***	***File Number***
Bit	B	10–255
Timer	T	
Counter	C	
Control	R	
Integer	N	

FIGURE 21-38 Addressing by using identifiers and file numbers

Word Address

This format is used for PLC instructions that contain three words. Counter, timer, and control file instructions use this format of addressing. An example format of a timer instruction address is:

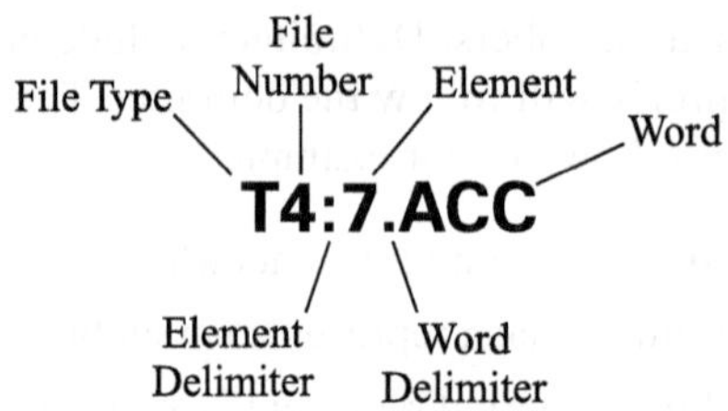

This address reads:

Identifier T (Timer File)
File Number 4
Element 7 (The 8th timer instruction in the file.)
The word accessed for this instruction is the accumulator.

In this example, the timer is specified by the identifier T and its corresponding data file number 4. A colon, which is an element delimiter, is followed by a 7, which specifies that element 7 is used. This indicates that timer 7 in the data file is used in this instruction. A period, which is a word delimiter symbol, is followed by *ACC,* which is an abbreviation for an accumulator. This indicates that the word being addressed is the accumulated value of a timer instruction.

To access another word for the same timer instruction, the preset value for example, the same identifier, file number, and element number are used. The only difference is to use the abbreviation *PRE* after the delimiter indicated by the period.

Examples of word address identifiers, corresponding file numbers, and the range of addresses where the element of various instructions can be stored are as follows:

Timer T4:0 through T4:255
Counter C5:0 through C5:255
Control R6:0 through R6:255

Element Address

This format is used when the programmer needs to read a word in an integer or a control file. The following format illustrates an element address:

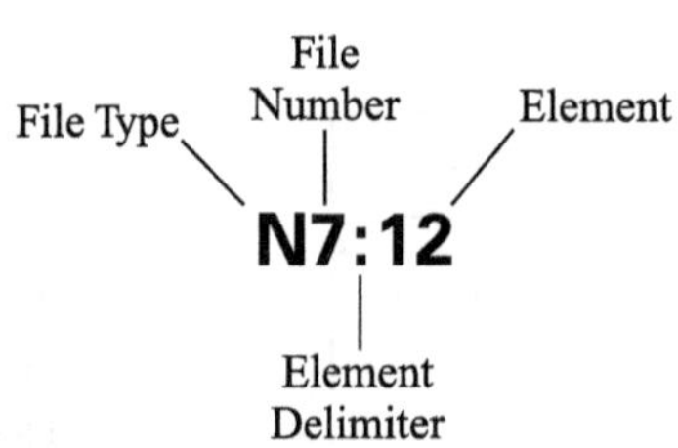

This example reads:

Identifier N (Integer File)
File Number 7
Element 12 (Word 12 in the file.)

Bit Address

This format is used to address a single bit in file 3. There are two addressing versions that can be used, the *sequential bit number technique* or the *long form technique.*

When using the sequential bit number technique, a single bit can be accessed by identifying a sequential number from 0 to 4095 in the file. By entering B3/28, for example, bit 28 is accessed from the file, as shown in Figure 21-39. Notice in the diagram that the first bit in the first register is addressed 0. There are 256 16-bit registers in this file. By incrementing the bits in sequential order through the last 16-bit register in the file (255), the identification address of the last bit is 4095.

TOP OF BIT FILE #3																
15	*14*	*13*	*12*	*11*	*10*	*9*	*8*	*7*	*6*	*5*	*4*	*3*	*2*	*1*	*0*	*Bit number*
15	14	13	12	11	10	9	8	7	6	5	4	3	2	1	0	Word 0
31	30	29	28	27	26	25	24	23	22	21	20	19	18	17	16	Word 1
47	46	45	44	43	42	41	40	39	38	37	36	35	34	33	32	Word 2
63															48	Word 3
79															64	Word 4
95															80	Word 5
111															96	Word 6
127															112	Word 7

FIGURE 21-39 One type of bit address format

The second addressing method is the long form technique. To access the same bit in the addressing example as was given for the sequential bit number technique, the following format is entered on the keyboard:

B3:1/12

This address reads:

Identifier B (Bit File)
File Number 3
Element 01 (Word 1)
Bit 12 (of Word 1)

The format displayed on the screen will be different. B3:1 will be above the symbol, and 12 will be located below.

Either addressing technique works, but the sequential number technique is shorter and therefore is the preferred method.

Input/Output Address

The status of an external input field device is represented by the status bit in the *input data file* (file 1) of memory. Likewise, the condition of an external output field device is controlled by a status bit in the *output file* (file 0) of memory. Each status bit in these files has an address. The programmer specifies the desired address of the file and bit when entering an instruction in the ladder diagram that uses the corresponding field device.

An example of how an input address is formatted is as follows:

I:e/b

I = The alphabetical *identifier* that indicates the Input Data File
: = Slot Delimiter
e = Slot Number of the Input Module
/ = Bit or Terminal Delimiter
b = Terminal Number of the module to which the Input Device is connected

Notice that the file number does not follow the identifier in this type of addressing format.

The slot refers to the position where the I/O module is placed in the rack assembly. Typically, the power supply is placed on the far left position of the rack. The first slot position to its immediate right is assigned number 0, in which the processor module is usually inserted. The slot number where the next module is inserted is assigned number 1, and so on. A variety of racks exist with differing numbers of slots, depending on how many devices are controlled by the PLC and how many functions it must perform. The terminal number in the address also corresponds to the register bit where data is stored.

An example of how an output address is formatted is as follows:

O:e/b

O = The identifier for the Output Data File
: = Slot Delimiter
e = Slot Number of the Output Module
/ = Bit or Terminal Delimiter
b = Terminal Number to which the Output Device is connected

Example 1:

Entered on the keyboard: I: 1/0, *which represents,*

Input, Slot 1, Terminal 0

The characters displayed are I:1 above the symbol, and 0 located below.

Example 2:

Entered on the keyboard: O: 2/4, *which represents,*

Output, Slot 2, Terminal 4

The characters displayed are O:2 above the symbol, and 4 located below.

Each I/O module is assigned to a register in memory. The register address is the same number as the slot in which the module is inserted. Each terminal of the module is assigned a bit address in the register. Whenever an input field device, for example, supplies power to a terminal, the corresponding bit in the register becomes a 1. The status of an activated input field device connected to terminal 4 inserts a 1 into bit 4 of the register. The absence of power to the terminal causes the bit to be a 0. Output modules are also connected to registers. If the status of a register bit is a 1, it typically energizes the corresponding output device. Likewise, a zero state de-energizes the field device.

21-11 Relationship of Data File Addresses to I/O Modules

IAU1206
The Relationship of Data File Addresses to I/O Modules

Figure 21-40 illustrates the relationship between the number assigned to the data files in memory and the number used by the I/O terminals. The figure contains an input module, a partial memory map of the data file, an output module, and field devices. The ladder diagram at the bottom shows two input switching contacts connected in series with an output symbol. The ladder diagram is programmed into the user program memory (program file 2). A push button is wired to terminal 1, and a limit switch is wired to terminal 5 of the input module located in slot 2. Pressing the push button will cause the corresponding status bit 1 in the input data file to go from 0 to 1, and the limit switch closure causes bit 5 to be set to a 1. This input status creates a True logic condition in the rung and causes bit 2 of the output data file to become a 1. As a result, corresponding output terminal 2 of the module located in slot 4 becomes energized and turns on the output field device to which it is connected. If the push button or the limit switch is open, the corresponding bit in the data file will be a 0, and the logic condition of the rung becomes False and makes bit 2 of the output data file a logic 0. Note that the number following the colon in an I/O address is the same number as the slot position where the module to which the corresponding field device is connected is located. None of the addresses in a data file where the elements are stored can be used more than once.

PLCs are not limited to interfacing with discrete field devices. There are input modules, for example, that convert analog voltages to proportional digital binary values, which are stored in the PLC memory. The digits in the memory register may be pure binary numbers, or BCD numbers that represent decimal digits.

Unlike discrete inputs that require a single memory bit, an analog input requires an entire 16-bit word to represent its value. A pure binary number can represent 65,536 (2^{16}) different voltage levels, and a four-digit BCD number can represent 10,000 (0000 to 9999) different analog voltage levels. Data that represents discrete signals and analog voltages are both stored in the input data file, as shown in Figure 21-41.

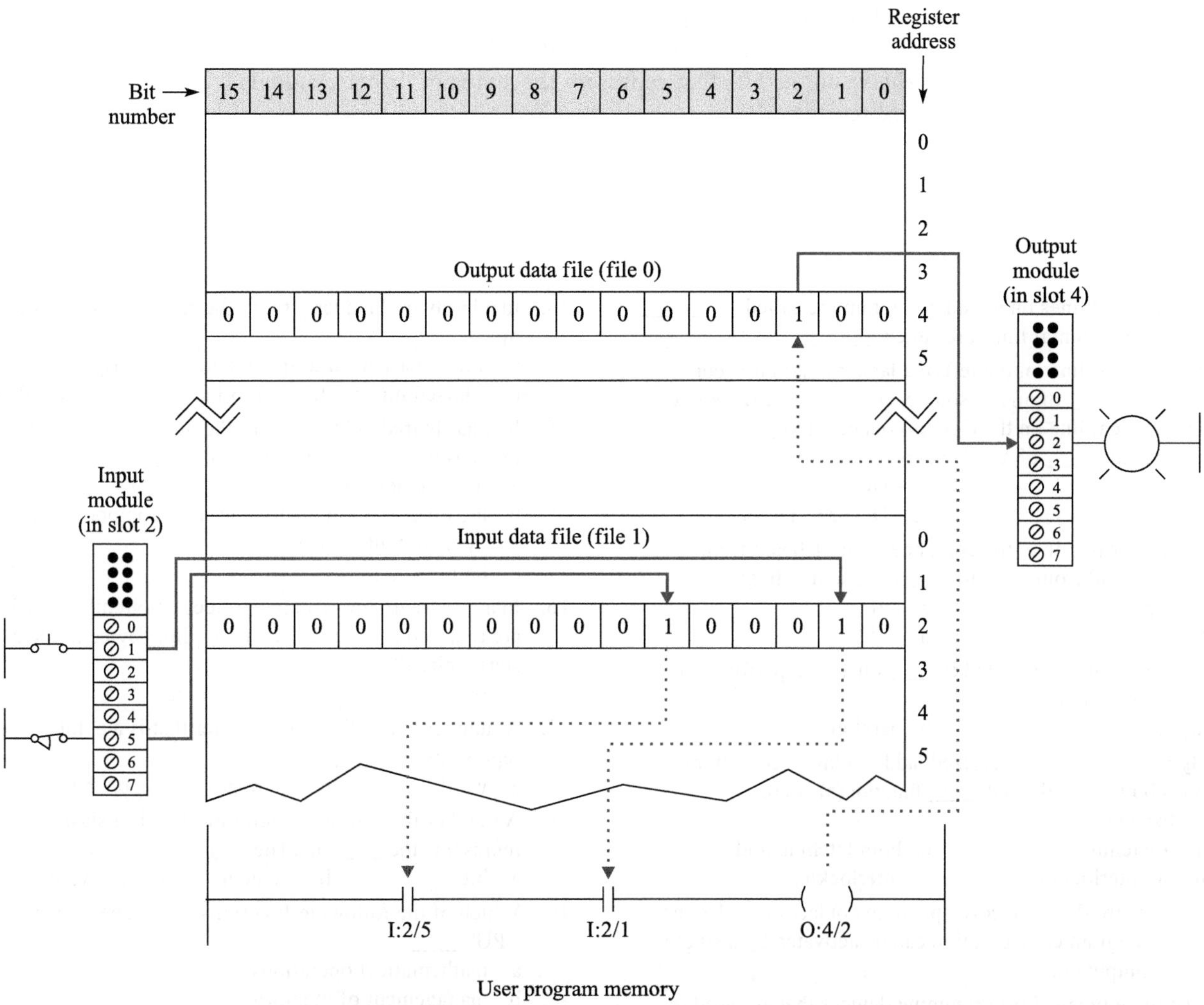

FIGURE 21-40 Relationship of data file addresses to input and output devices

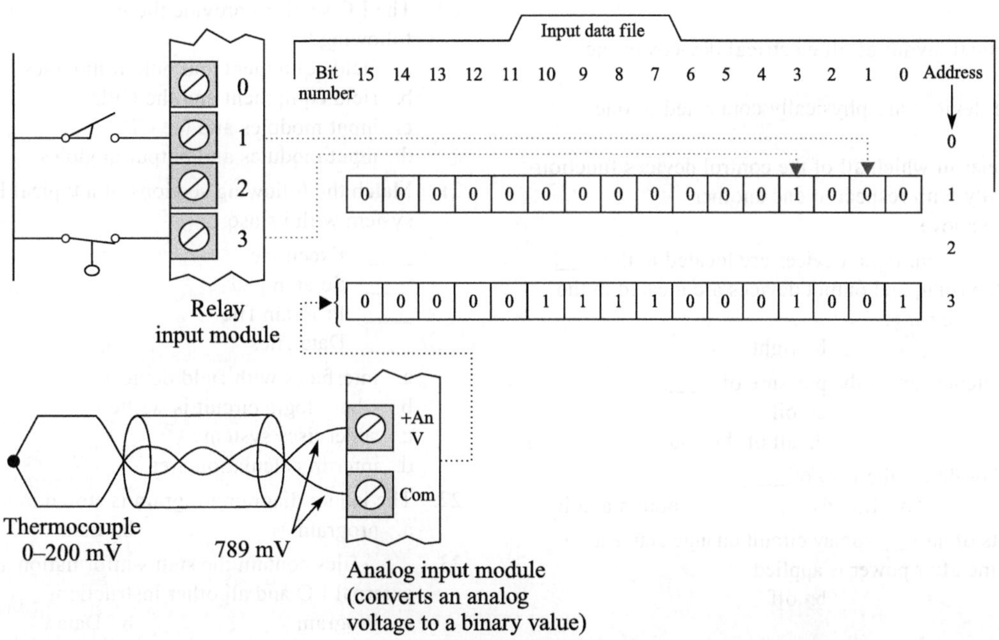

FIGURE 21-41 Analog and discrete relay signals stored in the input data file

There are also output modules that convert binary numbers from the PLC into proportional analog voltages used by field devices. The binary numbers that represent the analog values and the bits that represent the status of discrete signals can never be stored in the same register.

Problems

1. The vertical lines on a ladder diagram are called ________ and the horizontal lines are called ________.
2. The vertical line on the left of a ladder diagram is connected to ____ of an AC source, and ____ of a DC source. The vertical line on the right is connected to ____ of an AC source, and ____ of a DC source.
   a. L_1 c. +DC
   b. L_2 d. −DC or Common
3. Whenever possible, the components are labeled from ____ to ____, and from ____ to ____ on a ladder diagram.
   a. top c. left
   b. bottom d. right
4. Input switches connected in ____ on a rung perform an AND operation.
   a. series b. parallel
5. By maintaining an energized ladder relay circuit after a push button is released, ____ function(s) is/are performed.
   a. a latching c. both latching and interlocking
   b. an interlocking
6. T/F More than one corresponding contact in the ladder diagram configuration can be activated by a single output coil.
7. List three types of programming devices that are used to enter a program into the PLC.
8. List the four functions of the I/O module.
9. A good wiring diagram for electrical control circuits shows ____.
   a. the physical layout of all electrical devices in the circuit
   b. how the devices are physically connected to one another
   c. the manner in which all of the control devices function electrically with respect to one another
   d. all of the above
10. In a ladder diagram, input devices are located in the ____ portion of the rung, and output devices are located in the __ portion of the rung.
    a. left b. right
11. Pressure switches detect the pressure of ____.
    a. air c. oil
    b. water d. all of the above
12. Flow switches detect the flow of ____.
    a. air b. liquids c. both a and b
13. The contacts of an ____-delay circuit change states a period of time after power is applied.
    a. on b. off
14. In a hardwired ladder circuit, the rung number is identified by ____.
    a. a number placed to the outside of the right rail
    b. a boxed number located to the outside of the left rail
15. In a hardwired ladder circuit, the number in parentheses to the right of each rung indicates ____.
    a. its rung address
    b. the other rung numbers that have components activated by its output device
    c. both a and b
16. In a ____-wire low-voltage protection circuit, two wires are required to connect the input pilot device to the output starter circuit.
    a. two b. three
17. A number that follows a delimiter that is a colon represents a/n ____.
    a. word b. element c. bit
18. A number that follows a delimiter that is a slash represents the ____ in a file.
    a. bit b. element c. word
19. Which of the following functions are not performed by the CPU? ____
    a. mathematical operations
    b. management of memory
    c. self-diagnostic routines
    d. communication with other processor units
    e. none of the above
20. The I/O sections provide the interface between which of the following?
    a. field equipment and output modules
    b. field equipment and the CPU
    c. input modules and the CPU
    d. input modules and output modules
21. Match the following sections of a typical PLC memory system with its use:
    ____ Executive
    ____ Scratch pad
    ____ Program file
    ____ Data files
    a. interfaces with field devices
    b. where logic circuit is written
    c. supervises system
    d. interim calculation area
22. The ladder diagram program is stored in a ____ file.
    a. program b. data
23. ____ files contain the status information associated with external I/O and all other instructions.
    a. Program b. Data

24. The second instruction in a timer file has ____ as its element address.
    a. 1
    b. 2
    c. 4
    d. 4 through 6
25. A *counter* instruction has ____ element(s) and a bit instruction has ____ element(s).
    a. 1
    b. 3
26. The status file uses ____-word elements and a control file uses ____-word elements.
    a. one
    b. three
27. The processor file is made up of ____ program files and ____ data files.
    a. 4
    b. 11
    c. 256
28. The highest element address number of an instruction in a control file is ____, and the highest element address number of an instruction in a bit file is ____.
    a. 83
    b. 84
    c. 255
    d. 256
29. The following data file address for an I/O instruction indicates that the input module is located in slot ____ and the field device is connected to terminal ____. **I:2/4**
    a. 0
    b. 1
    c. 2
    d. 4
30. A personal password is stored in the ____ file.
    a. system function
    b. reserved
    c. subroutine
31. Each data file is capable of storing ____ 16-bit words in its memory.
    a. 4
    b. 11
    c. 256
32. Word addresses include ____.
    a. counters
    b. timers
    c. control files
    d. all of the above
33. T/F The accumulator and the preset value of the same timer instruction use the same identifier and data file number.
34. The bit address using the *long form technique* is: **B3:01.11** What is the equivalent bit address using the *sequential number technique*?
35. Write the word address of the preset register for the twelfth instruction in the counter file.

CHAPTER 22

Fundamental PLC Programming

OBJECTIVES

At the conclusion of this chapter, you should be able to:

- Describe a typical processor scan cycle.
- Describe the purpose of software used by PLCs.
- Identify a True and False logic condition at the rung of a ladder diagram.
- Understand the Examine-On instruction.
- Understand the Examine-Off instruction.
- Program a PLC using the following types of instructions:

Relay Logic Inputs and Outputs	Timers	Data Manipulation
	Counters	Arithmetic

INTRODUCTION

The ability of a programmable controller to achieve machine control is the result of the program written into its memory and the power of its CPU. The **program** is simply a list of instructions that guide the CPU. The execution of the program is performed by the CPU through a three-step process called a **processor scan cycle.**

This chapter provides a more detailed explanation of the processor scan. The remainder of the chapter examines the different types of ladder logic instructions required to write a program.

22-1 PLC Program Execution

IAU6807
Slow PLC Scan Time

IAU1306
The Processor Scan Cycle

Unlike the relay ladder circuit, which monitors every input and output device throughout the circuit simultaneously, the PLC performs the same function in a sequence of steps called a *processor scan cycle*. Figure 22-1 shows one rung of a relay ladder diagram to help explain the following three steps that take place during the scanning process.

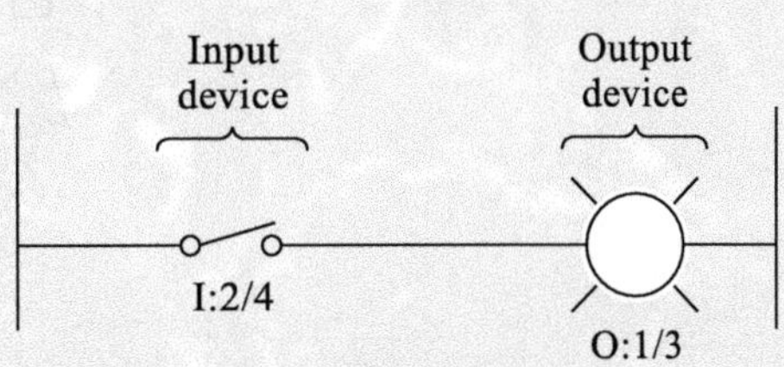

FIGURE 22-1 A standard one-rung ladder diagram

When the PLC is turned on, a program is fed into the CPU from the executive memory to run an internal self-diagnostic check on the system. If any part of the processor is not functioning properly, its fault-indicator light illuminates. The diagnostic self-check also determines if there is a faulty memory or an improper connection with the I/O module.

If the self-diagnostic check determines that the system is operating properly, it starts its scanning operation.

Step 1: The first step of the scanning process is to *Update the Input Image Table* by sensing the voltage levels of the input terminals. The status of the input field devices is recorded in memory. A 0 or 1 is stored into the bit of the memory location designated for a particular input terminal. In Figure 22-1, a 0 is stored into memory bit I:2/4 when no voltage is sensed, and a 1 is loaded when a voltage is present.

Step 2: The second step of the scanning process is to *Scan Program Instructions* located in file 2 (user program) of the program file in the user memory. The process for each instruction involves three operations for each rung in a sequential order:

1. First, the CPU makes a reference to the input image table to find out the status of the inputs. In Figure 22-1, it finds out the condition of input I:2/4.
2. Second, the CPU reads the instruction. The CPU makes decisions based on whether the input conditions are met and the type of logic function specified by the program. The program for Figure 22-1 would read:

   If a 1 is present at memory bit 4 of the module in slot 2, then write a 1 at bit location 3 of the output module in slot 1.
3. Third, according to the decisions made, the CPU updates the output image table by recording a 0 or 1 at the relevant bit location in memory.

   The processor executes the entire list of instructions, rung by rung, in ascending rung order.

Step 3: The third step of the scanning process is to *Update the Output Terminals.* The CPU takes data from the output data file and sends it out to the real world through the terminals of the output modules.

The input/output updates and the program instruction step are separate, independent functions. Any status changes occurring in an external input device during the program instruction step are not accounted for until the next I/O scan. Similarly, data changes associated with external outputs are not transferred to the outputs until the next scan. Figure 22-2 graphically summarizes the three steps of the scan.

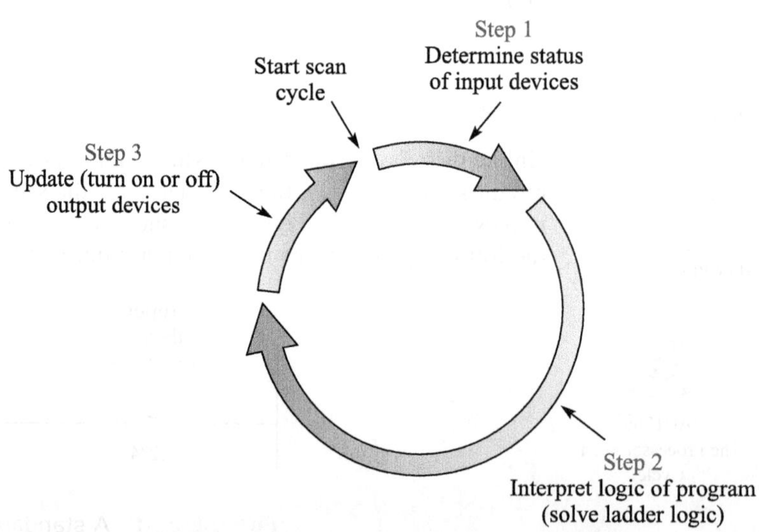

FIGURE 22-2 Three steps of a processor scan cycle

The three-step scanning process is continuous and is repeated many times each second. The time it takes to complete one scan depends on the size of the program and the clock speed of the microprocessor used by the PLC. If, for example, a program is written that uses 1K words of memory, the time to scan the instructions once may take 5 ms. Therefore, there will be 200 scans each second.

In some applications, the response time of 5 ms by the output device to a change of an input condition may be too slow. To solve this high-speed requirement, many PLCs have special programs available that permit the scan to be interpreted so that the I/O devices can be updated immediately.

22-2 Ladder Diagram Programming Language

IAU1406
PLC Software Circuit Capabilities

A program written into the user memory provides a way for the user to communicate with the PLC. The program is a control plan that tells the processor what to do when certain conditions exist. Several different forms of programming languages have been developed by PLC manufacturers to enable instructions to be written. The most common type of language is the ladder diagram.

22-3 Ladder Diagram Programming

The **ladder logic language** closely resembles hardwired relay circuits. It is composed of symbols that are inserted onto ladder rungs. The symbols represent an instruction set that performs different types of On-Off operations. In general, the input conditions are represented by contact symbols, and the output instructions are represented by coil symbols. Figure 22-3 compares a relay diagram to a ladder diagram rung. Figure 22-3(a) shows a limit switch LS1 and relay contact CR2 that must be closed to energize relay coil CR3. Figure 22-3(b) shows an equivalent PLC ladder diagram with the input devices and the output device identified with their respective data table bit addresses. The address numbers correspond to the location of the I/O modules and the terminal to which each field device is wired.

The format of the rung dictates which type of logic control is performed. For example, if the input contacts are placed in series, an AND function is performed. A parallel configuration performs OR function. When the conditions of the input devices allow a true logic path in the rung, the rung condition is True. When there is no true logic path, a False condition exists. A True rung condition energizes an output coil, which activates an output device that performs a desired operation. Table 22-1 summarizes five major instruction categories.

TABLE 22-1 Five Categories of Instructions

Relay Logic
Timers
Counters
Data Manipulation
Arithmetic

The remainder of the chapter describes the common instructions and their symbols used to perform relay logic operations. Example diagrams will be provided periodically to show how symbols are used in an actual ladder circuit. Each symbol in the diagram will be assigned a reference number, which indicates the address of a corresponding bit located in memory.

The ladder logic instructions discussed in the following pages are generic examples, using the Allen-Bradley SLC 500 (and the MicroLogix 1000) as samples. For detailed information on the instruction set of a specific controller, consult its manufacturer's manual.

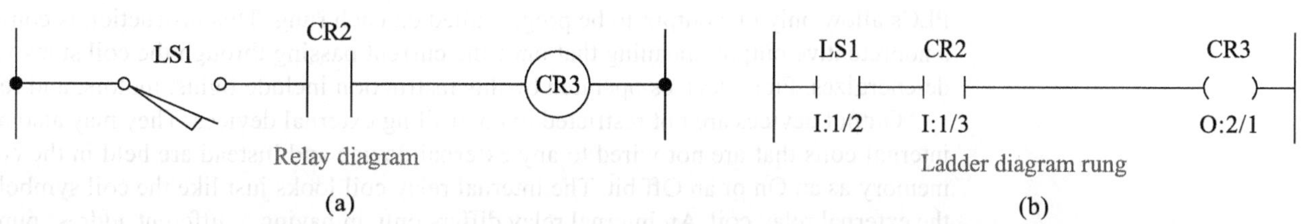

FIGURE 22-3 Comparing a relay diagram to a ladder diagram

The manual will provide information on how to use each input and output instruction that is available in the PLC.

22-4 Relay Logic Instructions

The most frequently used instructions in PLC programs are the relay ladder logic operations. The function of these instructions and the symbols used to represent them on the rungs of a ladder diagram are described in this part. The relay-type instructions are Examine-On Contact, Energize Output Coil, Examine-Off Contact, Latch Output, and Unlatch Output.

IAU5107
The PLC Examine-On Instruction

Symbol: –| |–

Name: Examine-On Contact

This symbol represents an input condition instruction. It tells the CPU to examine a bit at the memory location specified by the address number listed with the symbol. The referenced address may represent the status of an external input affected by the condition of a field device, or the logic states of an internal memory bit.

- The Examine-On instruction is True when the addressed memory bit is a 1, meaning that a corresponding external input field device is closed and supplies a voltage at the terminal of an input module.
- The Examine-On instruction is False when the addressed memory bit is a 0, meaning that a corresponding external field device is open and does not supply a voltage at the input terminal.

This symbol is also referred to as an *Examine-If-Closed* (*XIC*) instruction because its corresponding memory bit is a logic 1 (True condition) when the field device that controls it is closed.

An Examine-On instruction is often a single input on a rung. Several of them can also be programmed in series to perform an AND operation, or in parallel to perform an OR operation. These instructions are located in the left portion of the rung. The field devices that provide input signals include switches, push buttons, and sensors.

Symbol: –( )–

Name: Energize Output Coil

This symbol represents an output action instruction. This instruction will be performed only if the condition (input) instructions preceding it provide a path of logic continuity. The Energize Output instruction tells the CPU to *set a bit to 1* or *reset a bit to 0* at the memory location specified by the address listed with the symbol. The referenced bit provides the action command signal for the output.

- The output instruction sets the memory bit to a 1 when the rung condition is True. The result is that the corresponding field device energizes.
- The output instruction resets the memory bit to a 0 when the rung condition is False. The result is that the corresponding field device de-energizes.

Output instructions are located on the right portion of the ladder diagram rungs. Most PLCs allow only one output to be programmed on each rung. This instruction is considered as a nonretentive output, meaning that once the current passing through the coil stops flowing, it de-energizes. Field devices operated by this instruction include lights, motors, and relay coils.

Output devices are not restricted to controlling external devices. They may also operate as internal coils that are not wired to any external device and instead are held in the computer's memory as an On or an Off bit. The internal relay coil looks just like the coil symbol used for the external relay coil. An internal relay differs only in having a different address number than the real output. Internal relays are used when it is only necessary to control contacts in other

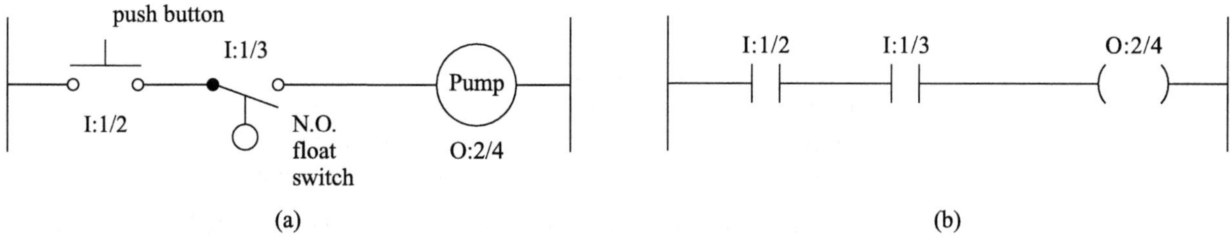

FIGURE 22-4 A rung with two Examine-On inputs and one output

rungs and not real external devices. The advantage of internal relay coils is that, in large ladder programs, they reduce the number of output I/O modules required for an application.

- When the output in the form of a coil is On, an Examine-On contact with the same address will be True, and an N.C. (Examine-Off) contact will be False.
- When the output in the form of a coil is Off, an Examine-On contact with the same address will be False, and an N.C. (Examine-Off) contact will be True.

There can be more than one corresponding contact in the ladder diagram configuration that is activated by only one output coil.

Figure 22-4 shows a one-rung relay diagram (Figure 22-4(a)) and its equivalent ladder diagram (Figure 22-4(b)) with two Examine-On inputs and one Energize Coil output. An N.O. push button is connected to terminal I:1/2 of the input module, and an N.O. float switch is connected to terminal I:1/3. A pump is connected to terminal O:2/4 of the output module. The float switch closes when the level of liquid in a storage tank lowers to a given height, and the pump switch is manually closed by an operator. These actions energize the pump to add more water to the tank.

Branching

If one or more sets of parallel input conditions are required to energize an output device, **branching** instructions are used to develop the program. A branch configuration is shown in Figure 22-5. It performs a logic OR function. If the push button *or* the N.O. float switch is closed, the output will be energized.

To write this diagram into the user memory, a programming unit is used to enter the required data. Before writing the program, the processor must be in the program mode. There are limitations to the number of branches that can be entered for a single ladder rung. The exact limitations are dependent on the particular type of PLC being used.

After the circuit entries on the programming device have been completed, the program must be downloaded, and the processor must be placed into the RUN mode before the program can actually perform its operation.

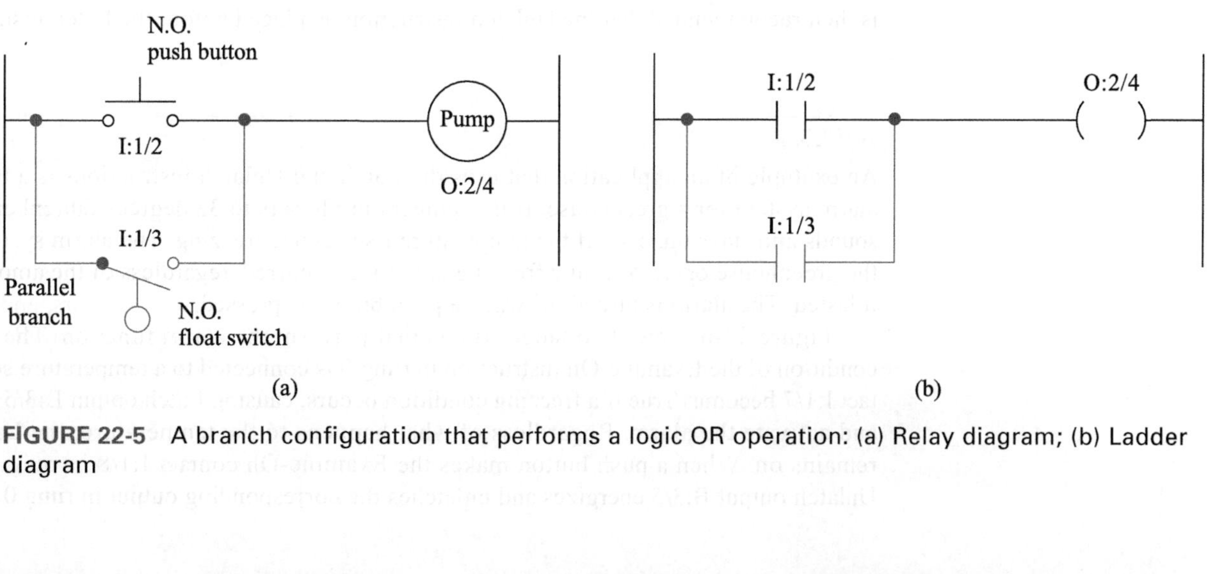

FIGURE 22-5 A branch configuration that performs a logic OR operation: (a) Relay diagram; (b) Ladder diagram

IAU5007
The PLC Examine-Off Instruction

Symbol: –| / |–

Name: Examine-Off Contact

This symbol represents an input condition instruction. It tells the CPU to examine a bit at the memory location specified by the address listed with the symbol. The referenced address may represent the status of an external input or an internal program bit. This instruction is used when the absence of the signal is needed to turn an output on.

- The Examine-Off instruction is True when the addressed memory bit is a 0, meaning that a corresponding external input field device does not allow a voltage to be present at the input terminal. The result is that the Examine-Off input will be True and allow logical continuity in its rung.
- The Examine-Off instruction is False when the addressed memory bit is a 1, meaning that a corresponding external input device allows voltage to be present at the input terminal. The result is that the Examine-Off input will be False. Therefore, logical continuity in its rung does not exist.

This symbol is also referred to as an *Examine-If-Open* (*XIO*) instruction because its corresponding memory bit is a logic 0 (True condition) when the field device that controls it is open.

Symbol: –(L)–

Name: Latch Output

This symbol represents a retentive output action instruction. When its rung becomes True, it tells the CPU to set the bit to a 1 at the memory location specified by the address listed with the symbol. It is called a *retentive output* because, once the rung condition goes False, the latch bit remains set. The bit becomes reset to a 0 by an Unlatch instruction programmed with the same reference address, or if system power is lost. Although some PLCs use Latch commands for internal and external outputs, other PLCs are restricted to latching internal outputs only.

Symbol: –(U)–

Name: Unlatch Output

This symbol is used to reset a latched output with the same address. It is the only automatic means of resetting a latched output. Therefore, the Latch and Unlatch instructions should be used in pairs with each other. When its rung becomes True, the Unlatch instruction tells the CPU to reset the corresponding latch bit. In some situations, the programmer may want the program to run through an entire scan cycle before the Latch instruction is reset. It is then recommended that the Unlatch instruction be placed before the Latch instruction.

APPLICATION 1

An example of an application that uses the Latch and Unlatch instructions is a temperature alarm system for a greenhouse. If the temperature lowers to 32 degrees fahrenheit, an alarm sounds and stays on, even if the temperature rises above freezing. The alarm system notifies the greenhouse operator if any freezing condition occurred, regardless of the amount of time it lasted. The alarm is turned off when a push button is pressed.

Figure 22-6 shows two ladder rungs that perform the alarm function. The True/False condition of the Examine-On instruction in rung 0 is connected to a temperature sensor. Contact I:1/7 becomes True if a freezing condition occurs, causing Latch output B:3/5 to energize and activate the alarm. Regardless of what happens to the temperature, the Latch output remains on. When a push button makes the Examine-On contact I:1/8 in rung 1 True, the Unlatch output B:3/5 energizes and unlatches the corresponding output in rung 0.

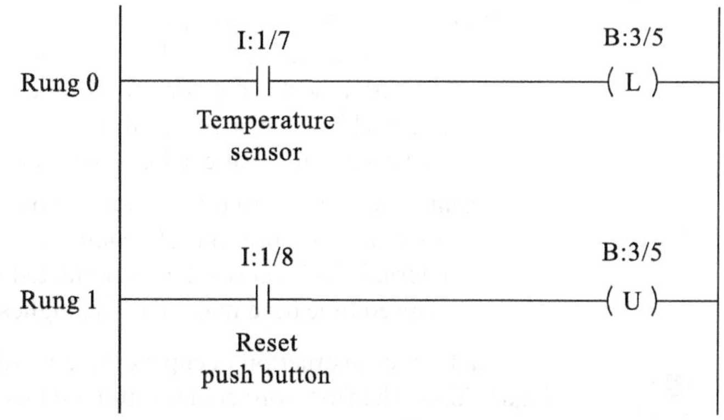

FIGURE 22-6 Using Latch and Unlatch instructions for a temperature alarm system

22-5 Timer Instructions

IAU6007 PLC Timer Instructions

Timers are output instructions that are internal to the programmable controller. They are capable of providing timed control of devices that they activate or deactivate. Timing operations are used in many industrial applications. PLC timers perform various functions such as delaying an action, causing an operation to run a predetermined period of time, or recording the total accumulated time of continuous or intermittent events. They can also operate as astable or one-shot multivibrators.

IAU6907 Programming a PLC Timer

A timer is activated by a change in the logic continuity of its rung. The rung condition is most often controlled by an Examine instruction. After the timed interval has expired, the timer output is energized, causing an N.O. or N.C. contact it controls to change its logic state. These contacts can be used throughout the program as many times as necessary. The contacts associated with a specific timer are identified by using the same address number.

As each timing instruction is programmed into the PLC, as shown in Figure 22-7, the following information must be entered:

1. Symbol and Address
2. Time Base
3. Preset Value
4. Accumulated Value

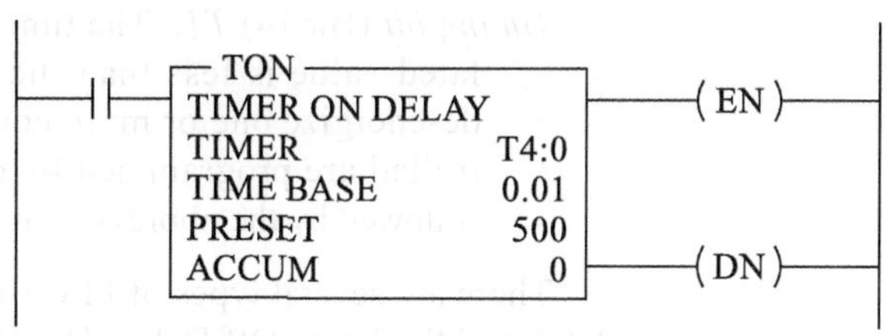

FIGURE 22-7 A timer instruction for the PLC

Symbol and Address. As a timer instruction on the programmer is entered, a rectangle with an abbreviation indicating which type of timer has been selected appears on the screen. A file address ranging from 0 to 255 must be entered after the *Timer* appears to identify the location where the instruction is stored. Figure 22-7 shows that an address of T4:0 is entered. An address file is reserved in memory for timers. It is recommended that the first timer programmed be loaded into the lowest address of the file. Each additional timer should then be entered into succeeding memory locations.

Time Base. After the address is entered, program the time base value. A time base is a fixed time interval, usually 1.0 seconds, 0.1 seconds, or 0.01 seconds in length. One of the numbers is selected by the user.

Preset Value. After the time base data is entered, the user is prompted to enter a three-digit number to the right of the term *PRESET,* which represents *Preset Value.* The value entered indicates the amount of time that needs to expire before action is taken by the timer. This value specifies the number of time base intervals to be counted. The highest preset value that can be programmed into the timer instruction is 32,767.

Accumulated Value. Another display shows the abbreviation *ACCUM* followed by a 0. This information represents *Accumulated Value equals 0.* When the timer is activated, an internal clock causes the accumulated value to begin incrementing from 0 as it counts elapsed time base intervals. The highest value that can be accumulated is 32,767.

IAU6607
The Timer 3-Word Element

Each timer instruction occupies three words of an element in the timing file, as shown in Figure 22-8. The first word contains three status bits that are used to control various functions in a program. The second and third words store the preset and accumulator values. They include:

	15	14	13	
Word 1	EN	TT	DN	INTERNAL
Word 2	PRESET VALUE			
Word 3	ACCUMULATED VALUE			

TIMER INSTRUCTION (3 WORD ELEMENT)

FIGURE 22-8 A timer instruction in a file

Enable bit (Bit 15) *EN.* The enable bit is set when the rung condition is True and resets when the rung condition is False. The enable bit is used to energize or de-energize one or more contacts in other rungs. The contacts to be controlled are programmed by entering the address of the timer that controls them, followed by the abbreviation *EN.*

Done bit (Bit 13) *DN.* This bit is considered the primary output of the timer. When the accumulated value equals the preset value, the timer has timed out. The output bit then sets on or resets off. The on or off condition depends on the type of timer instruction selected. The output bit is used to energize or de-energize one or more Examine-On or Examine-Off contacts located at other rungs. The contacts to be controlled are programmed by entering the address of the timer that controls them, followed by the abbreviation *DN.*

Timing bit (Bit 14) *TT.* The timing bit is on whenever the timer is on, but the accumulated value is less than the preset value. The timing bit is used to energize or de-energize one or more contacts located at other rungs. The contacts to be controlled are programmed by entering the address of the timer that controls them, followed by the abbreviation *TT.*

There are several types of PLC timer instructions available. These include the Timer-On Delay and the Timer-Off Delay. The choice of which to use is dependent on the type of operation to be performed.

Symbol: –(TON)–

Name: Timer-On Delay

When the rung condition becomes True, the timer begins causing the accumulator value to increment and bits 14 and 15 to set. When AC=PR, the timer stops timing and the output energizes as bit 13 sets. At the same time, bit 14 resets. If logic continuity of the rung is lost before or after the timer has timed out, the accumulator value goes to zero, and any bit that is set, resets. The timing diagram in Figure 22-9 graphically shows the operation of the TON timer.

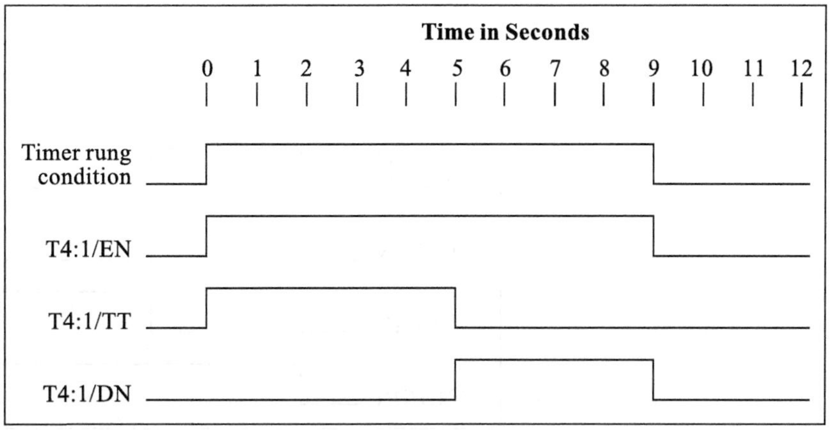

FIGURE 22-9 A timing diagram for a TON timer instruction

The Timer-On Delay output is used to provide time-delayed action or to measure the duration for which some event is occurring.

IAU6507
A Timer-On Application

APPLICATION 2

An example of an application that uses a Timer-On instruction is a high-speed transfer line of canned foods. An inductive proximity switch is used to detect the movement of the cans. If the cans become jammed upline, the movement stops. Since the cans will not be passing the detection point, the timer will time out. The output bit of the timer is programmed either to set off an alarm or to stop the machine until the problem is corrected.

Figure 22-10 shows the ladder diagram that controls the process. The inductive proximity switch causes the Examine-Off contact I:1/3 to become True each time there is a gap between the cans. The True condition of the rung starts timer T4:0, and the accumulated value increments until the next can is picked up by the sensor. Each time a can is present, the Examine-Off contact goes False and causes the accumulated value to reset. If a new can does not pass within 40 ms, the preset value 4 is reached by the accumulated value in the timer. The moment AC=PR, its DN bit sets, creating a False condition at contact T4:0(DN) in rung 2. A False condition of rung 2 also develops, de-energizes output O:2/8, and stops the transfer line until the problem is corrected.

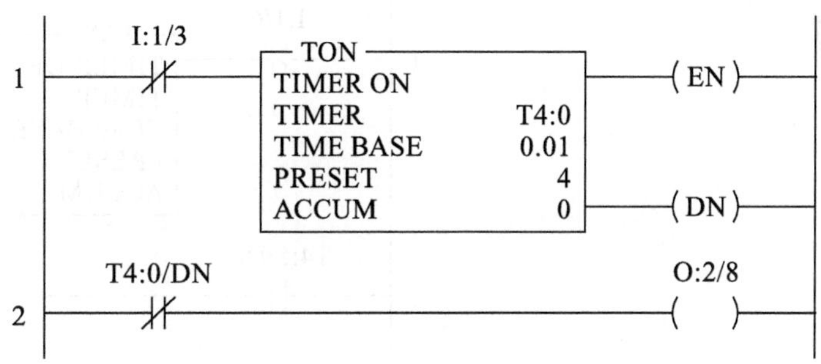

FIGURE 22-10 Control circuit for a transfer line

Symbol: –(TOF)–

Name: Timer-Off Delay

While the condition of a rung is True, the accumulator value of a Timer-Off Delay is reset to 0 and bits 15 and 13 set. Once logic continuity of the rung is lost, the timer resets the EN bit to 0, the timing bit 14 (TT) sets, and this causes the accumulator value to begin incrementing. When AC=PR, the timer stops timing and the output energizes as bits 13 and 14 reset. If logic continuity of the rung occurs before or after the timer has timed out, it will not count and the accumulator value resets to 0.

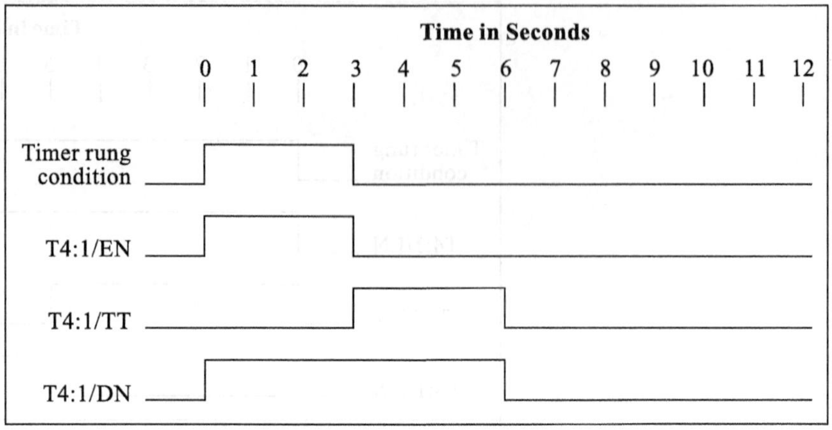

FIGURE 22-11 Timing diagram for the TOF timing instruction

The Timer-Off Delay timer output is used to provide time-delayed action. A timing diagram in Figure 22-11 graphically helps to show the operation of the TOF timer.

IAU4807
A Timer-Off Application

APPLICATION 3

A security light for a garage-door opener is an example of an application for a Timer-Off instruction. Whenever the garage door is open, the light is on. After the garage door is closed, the light remains on for 20 seconds while the driver walks to the house from the car.

Figure 22-12 shows an Examine-Off instruction addressed I:1/6 that is controlled by a N.O. limit switch and a Timer-Off output addressed T4:5. When the garage door opens, the limit switch also opens and makes the logic condition of rung 1 True, causing the DN output bit of the timer to set. The Examine-On DN contact becomes True and provides continuity to output O:2/3, which turns the light on. As soon as the garage door shuts, the limit switch closes, the condition of rung 1 becomes False, and the Timer-Off instruction begins to time. While the timer is timing, the DN bit remains on, causing the light to stay lit. When the timer times out, output bit 13 is reset, causing rung 2 to go False and the light to turn off.

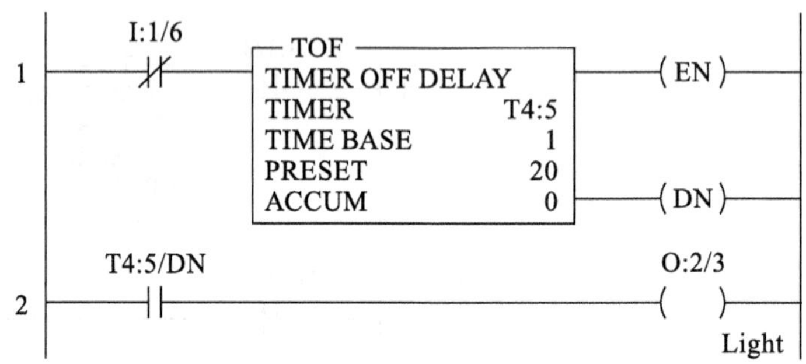

FIGURE 22-12 Security light control

Symbol:

Name: Retentive Timer-On Delay

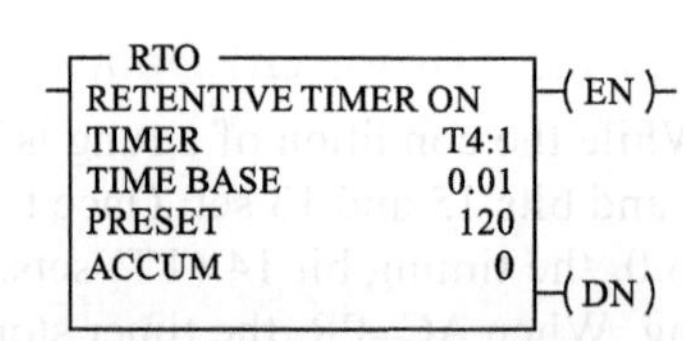

The operation of a *Retentive Timer-On Delay* (RTO) instruction is similar to the TON instruction with one exception. Instead of losing its accumulated value when its rung condition goes False, it is retained. The accumulated value will continue timing where it left off after the

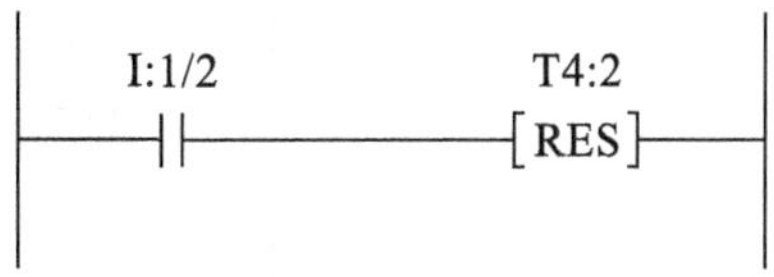

FIGURE 22-13 A symbol for the reset [RES] instruction

rung returns to a True condition. All of the status bits work the same way as in a TON instruction.

Because the timer does not reset when its rung goes False, a corresponding reset instruction assigned to it is required to clear the accumulated value. This reset instruction is located in another rung as an internal output device. The symbol for a reset instruction is shown in Figure 22-13. The particular timer it is required to reset is identified by the address located above the symbol. The reset operation takes place whenever the rung on which the reset instruction is located becomes True.

The timing diagram in Figure 22-14 graphically shows the operation of this device as it responds to the circuit conditions of the ladder diagram above it.

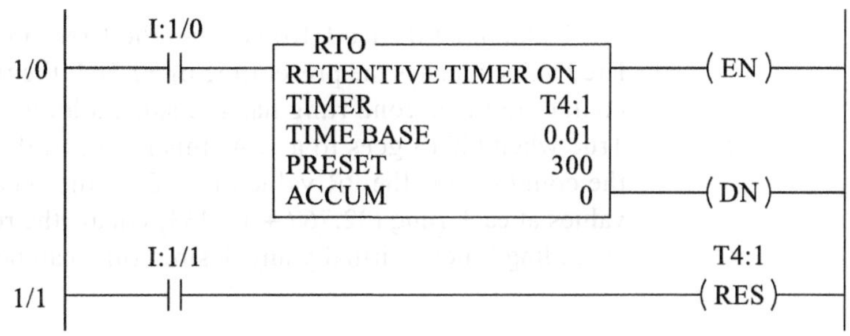

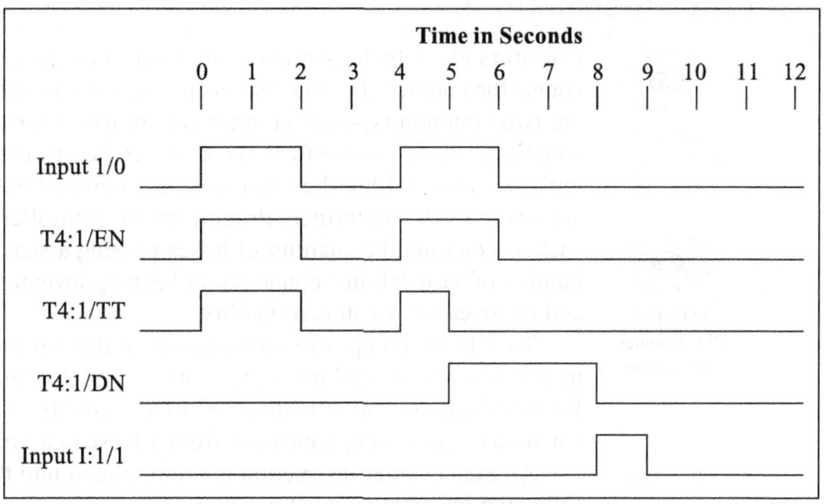

FIGURE 22-14 A timing diagram for an RTO timer instruction

Cascading Timers

IAU6707
Cascading Timers

If an application requires more time than the maximum seconds available from a single timer, two or more timers can be used to increase the maximum time value. The timer instruction for an SLC 500 with a 5/01 processor has a maximum count of 32,767. The method of programming timers together to extend the time range is called **cascading.**

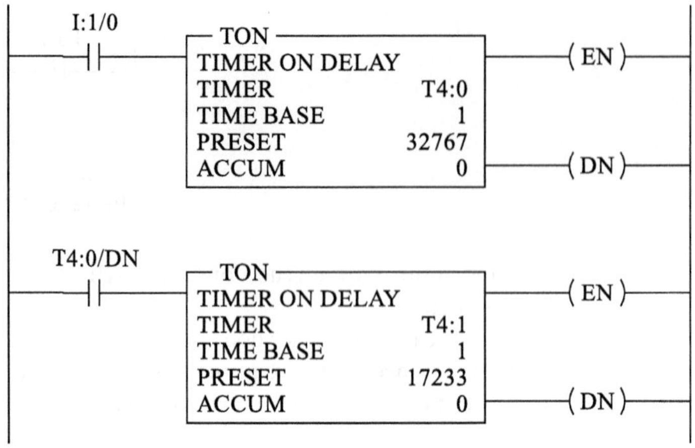

FIGURE 22-15 Cascading timers

Figure 22-15 shows two timers that are cascaded to achieve a time of 50,000 seconds. In this circuit, input device I:1/0 controls the timer in the top rung. The timer in the bottom rung is controlled by an internal Examine-On input that is activated by the timer on the top rung.

If the input device I:1/0 is True, the first timer starts incrementing. When the accumulated value reaches the preset time of 32,767, the timer times out and sets bit 13. Because the contact in the second rung has the same address as the timer in the first rung, it becomes True when bit 13 goes to a 1. At this moment, the second timer begins to increment. When the count equals the PR value of 17,233, the second timer stops. The sum of accumulated values at each rung (32,767 + 17,233) equals the required time period of 50,000 seconds. By cascading timers, virtually any desired time can be achieved.

22-6 Counter Instructions

IAU15008
PLC Counter
Operation

IAU6107
PLC Counter
Instructions

Counters are output instructions internal to the programmable controller. A counter simply counts the number of events that occur, then stores and displays the accumulated value. There are two common types of counter functions performed by PLCs: up-counting and down-counting. (PLCs also perform the Counter-Reset instruction, which simply clears any accumulated values within the counter.) The choice of which one to use depends on the type of application to be performed. Programmable controller counters perform various applications such as counting the quantity of boxes passing a sensor on a conveyor belt, determining the number of parts left in a container, or keeping inventory of items in stock that are loaded into and removed from a storage facility.

To activate an up- or down-counter, either an external device or a software command must be used to control the logic continuity of the counter's rung. Each count occurs when a False-to-True transition is detected by the counter. The transition is sensed by the counter when its rung continuity changes from a False to a True logic condition.

As each counter instruction is programmed into the PLC, as shown in Figure 22-16, the following information must be entered:

1. Symbol and Address
2. Preset Value
3. Accumulated Value

COUNT UP (CTU)

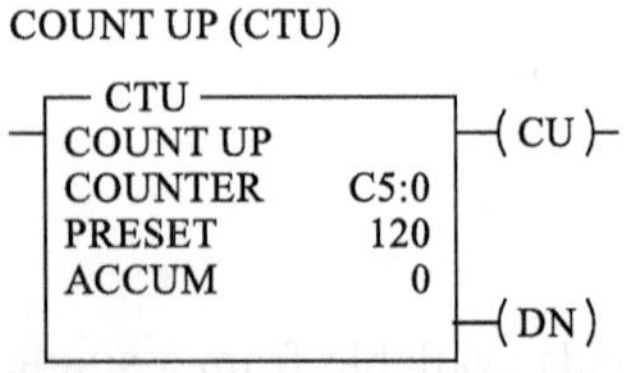

FIGURE 22-16 A symbol for an up-counter instruction

Symbol and Address. As a counter instruction is entered, a rectangle appears on the screen along with an abbreviation indicating which type of counter has been selected. A file address ranging from 0 to 255 must be entered after the word *Counter* to identify the address location where the instruction is stored. Figure 22-16 shows that address C5:0 is entered. A file is reserved in memory for counters. It is recommended that the first counter programmed be loaded into the lowest address of the file. Each additional counter should then be entered into succeeding memory locations. Each counter instruction occupies three words of an element in the counter file, as shown in Figure 22-17. The contents of words two and three are the same as for timer instructions. They hold the preset and the accumulated value.

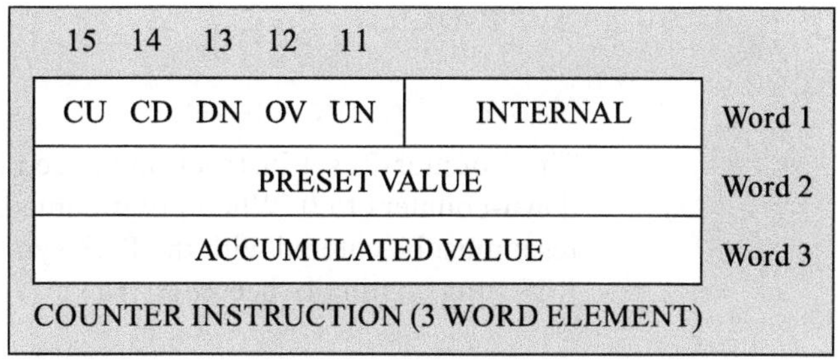

FIGURE 22-17 A counter instruction in a file

Preset Value. After the address has been programmed, a number ranging from 0 to 32,767 at the right of the word *PRESET* can be selected. The value entered indicates the number of counts that occur before action is taken by the counter.

Accumulated Value. After the preset value has been entered, the cursor prompts the user to enter a multidigit number from 0 to 32,767 at the right of the abbreviation *ACCUM.* Every time the counter is activated by a False-to-True state in its rung condition, it changes by one count. An up-counter increments and a down-counter decrements.

The first word of the element in a memory file contains five status bits. Each bit is assigned a number. These bits are often used to change the logic states of an input contact symbol in another rung. The address of the counter controlling it, along with the abbreviation of the bit, is placed above the symbol.

IAU3706 Overflow and Underflow for Signed Binary Numbers

IAU3606 Signed Binary Numbers

Up-Counter Enable bit (CU). This enable bit (15) is set when the condition of a rung where an up-counter is located becomes True and resets when the rung condition is False.

Down-Counter Enable bit (CD). This enable bit (14) is affected by a down-counter and sets when the rung where its symbol is located is in a True condition.

Count Complete bit (DN). The Done bit (13) logic status is determined by comparing the accumulated value to the preset value. It sets when these two values are equal. This bit 0 is usually considered the primary output bit of the counter. Its status can cause a device that the counter controls to turn on or off after reaching a certain count.

Overflow (OV)/Underflow (UN) bit. Unlike timers, which stop counting when the accumulated value equals the preset value, a counter will continue to count up (or down). The Overflow bit sets when the accumulated value is greater than the maximum count of 32,767, and the Underflow bit sets when it falls below the count of –32,768.

Symbol:

Name: Up-Counter

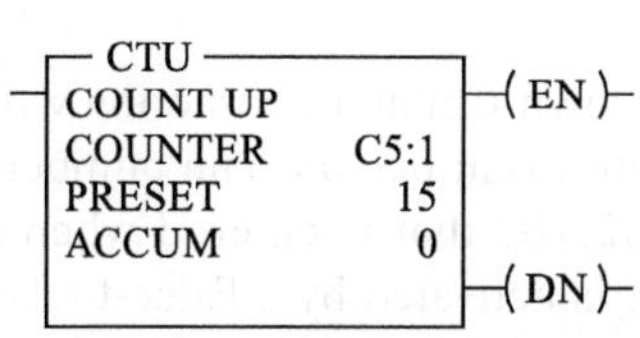

The Up-Counter instruction will increment by 1 each time a counted event occurs. The number of events is recorded in the accumulator. The counts are activated by a False-to-True change of the counter's rung condition. When the accumulated value reaches the preset value, the count complete bit (13) will set. Bit 14 will remain set if the counting continues beyond a value greater than the preset number. If the count goes beyond 32,767, the Overflow bit (12) will set. A Reset instruction is required to clear the accumulated value.

An application of an up-counter is counting the desired cereal boxes loaded into a case to indicate when it is filled to capacity.

Symbol: C5:0 -(RES)-

Name: Counter-Reset

The Counter-Reset instruction is used to clear the accumulated value of up-counters and down-counters to 0. When programmed, the reference address number of the counter to be reset must be entered with the RES symbol. The referenced counter is cleared to 0 when the RES rung continuity becomes True.

APPLICATION 4

IAU9307
The Counter-Reset Function

IAU9407
How to Reset a Counter

Count-up instructions are often used when repeating a process. In Figure 22-18, for example, suppose that input device I:1/0 in rung 0 drills five holes into a part. The number 5 is programmed as the preset value. Each time a hole is drilled, the count is incremented. When the accumulated value reaches 5, the DN bit will set. The corresponding Examine contact C5:0 (DN) in rung 1 becomes True and turns on output O:2/0. When energized, this output activates a robotic arm, which removes the finished part and places a new part beneath the drill. It also makes contact O:2/0 a logic True so that continuity is provided for the Counter-Reset output of rung 2. When activated, the RES instruction resets the accumulated value of the counter in rung 0 before the first hole is drilled into the new part.

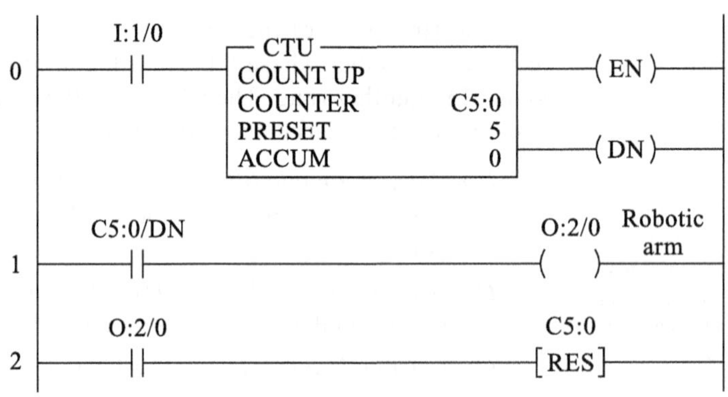

FIGURE 22-18 A drilling operation that uses an up-counter

Symbol:

Name: Down-Counter

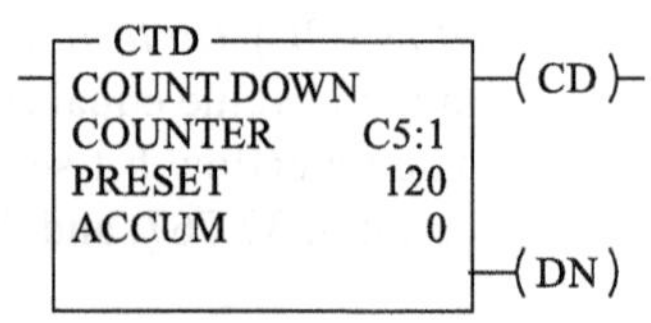

The Down-Counter instruction will decrement the count in the accumulator by one each time a count event occurs. The number in the accumulator begins at any value between –32,768 and 32,767 that is entered when the CTD instruction is programmed into the PLC. Each count is activated by a False-to-True transition of the rung. Any preset value can be entered between –32,768 and 32,767 during the programming of the CTD instruction. Bit 13 is set until the accumulated value goes below the PR number, at which time it resets. If the count goes below –32,768, the Underflow bit (11) will set. A reset instruction is required to clear the accumulator value.

APPLICATION 5

The Down-Counter instruction is often used to end a cycle. For example, suppose that only 50 parts are to be produced by an automation process. The ladder diagram in Figure 22-19 can be used to stop the process when the desired number of parts is completed. The count-down preset value is set to 1 and the accumulated value to 50. When the PLC is put into the run mode, the DN bit for timer C5:0 sets, causes the Examine-On contact in rung 1 to become True, and activates the conveyor motor. A sensor that detects a finished part as it passes an inspection point momentarily makes contact I:1/0 True. Each time a finished part passes the sensor, the accumulated value decreases by 1. When the accumulated value equals 0, the count-down-done bit (13) resets. As the Examine-On condition of contact C5:0 on rung 1 becomes False, the conveyor motor is de-energized.

IAU7107
Cascading Counters

IAU4907
Up/Down-Counter
Application

IAU 15308
Automatic PLC
Counting

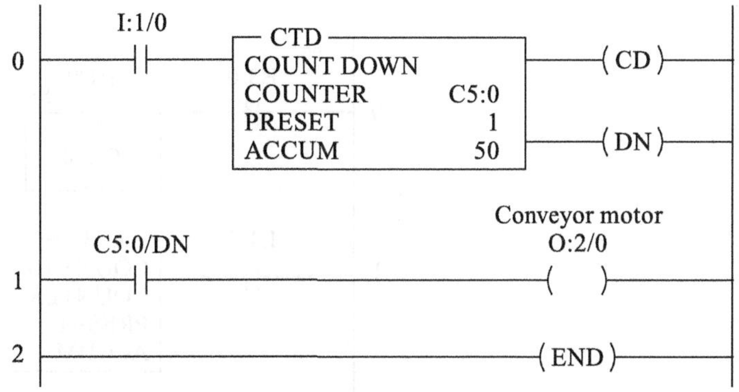

FIGURE 22-19 A conveyor operation that uses a down-counter

22-7 Data-Manipulation Instructions

Data manipulation is a category of instructions that enables words to be moved within the memory. These instructions enable the PLC to perform more complex operations than relay-type instructions, because they use multi-bit data to control outputs rather than only one bit. Data manipulation is divided into three categories: **data transfer, data conversion,** and **data compare.**

Data Transfer

Data Transfer instructions result in moving contents stored in one memory register to another memory location. Only the memory addresses accessible to the user can be used by these types of instructions.

To carry out the data-transfer function, a *MOVE* (MOV) instruction is used to read the data from one register and store the contents in another register. Figure 22-20 illustrates a data-transfer operation. The symbol for a MOV instruction contains a source and destination. Source is the address from which the instruction moves the data. Destination (DEST) is the address to which the instruction moves the data. When Examine-On contact I:3/0 is closed,

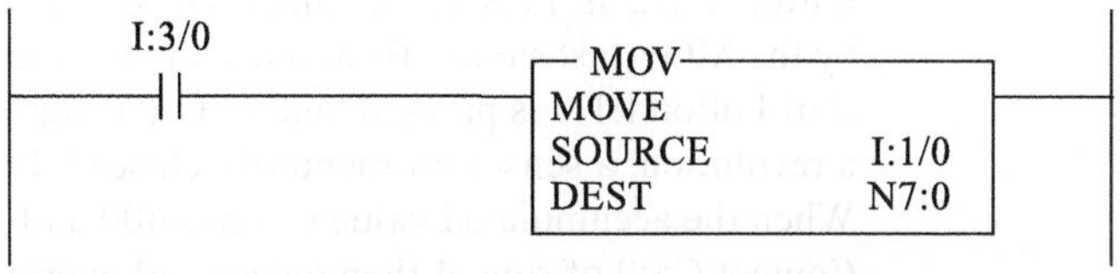

FIGURE 22-20 A MOV instruction

the rung on which the MOV instruction is located goes True, and the data is transferred from the source to the destination.

The types of applications for which data-transfer instructions can be used are presetting numbers into the registers of timers or counters, comparing data, displaying a number selected from a thumbwheel switch, or performing arithmetic operations.

APPLICATION 6

An example of data transfer is to preset a counter. An up-counter is used in a factory that manufactures inductor coils. The counter controls the number of windings of wire wound around a core. When a 5-millihenry (mH) inductor is produced, a lathe makes 400 revolutions to wind the coil. If the machine makes a 10-mH coil, the lathe must make 800 turns before stopping. Figure 22-21 shows how to change the preset value for each different inductor by using push buttons.

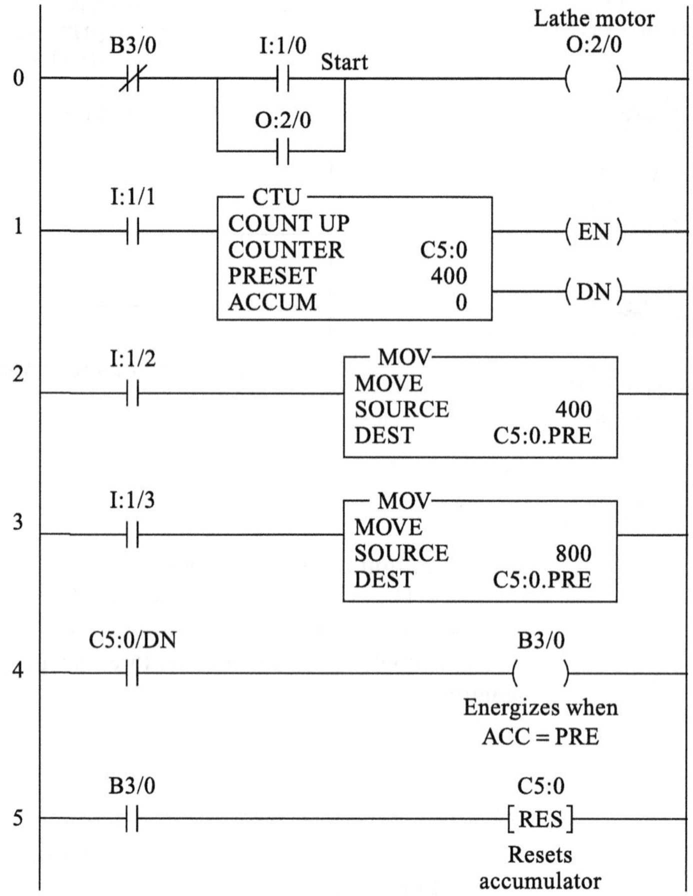

FIGURE 22-21 Ladder diagram control circuit for winding inductors

Up-counter C5:0 is initially programmed with a preset value of 000. The MOV instructions in rungs 2 and 3 are addressed to transfer the number from their source registers to the preset register of up-counter C5:0. To prepare the machine to make the 5-mH coil, push button I:1/2 is pressed to make rung 2 a logic True condition. The 400 is transferred by the MOV instruction from the source to the counter preset C5:0.PRE (Destination). When start button I:1/0 is pressed, output O:2/0 starts the lathe turning. Every time the lathe makes a revolution, a sensor momentarily closes I:1/1, causing the counter to increment one count. When the accumulated value reaches 400 and AC=PR, the DN bit of counter C5:0 goes high. Contact C5:0 of rung 4 then closes and energizes output coil B3/0. The result is that contact B3/0 of rung 0 becomes False, de-energizes output O:2/0, and stops the lathe. It also causes a

True condition at contact B3/0 in rung 5, which activates the RES output so that the accumulator value of the counter clears to 000.

To make a 10-mH coil, the same procedure is repeated after the push button I:1/3 is pressed to load 800 into the counter, and then start button I:1/0 is pressed.

IAU3506
The Binary-Coded Decimal

Data Conversion

Decimal is the most familiar number system. Therefore, digital-based equipment is designed with decimal input devices, such as keyboards, and decimal output devices, such as numerical displays. However, the internal circuitry of digital equipment operates by using binary bits. Therefore, a method of converting binary to and from decimal is needed. This conversion is made possible by converting between pure binary bits and groups of four bits called *BCD numbers*. The 4-bit groups range from 0000 to 1001, and each one represents one of the numerical decimal digits, 0 to 9. Two different PLC instructions exist that enable the conversion between binary and equivalent values.

Symbol:

Name: Convert to BCD (TOD)

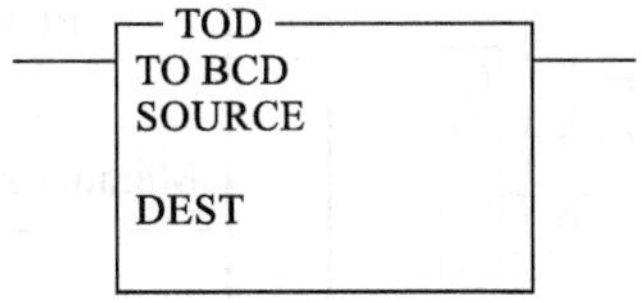

Convert to BCD is an output instruction. It changes a pure 16-bit binary number into an equivalent numerical BCD number in the form of four separate groups of four bits. Each group represents a separate decimal digit. The instruction contains a source and a destination address. *Source* is where a pure binary integer resides until converted into a BCD value and moved. *DEST* is the address to which the instruction moves the BCD value. This instruction is performed when the rung on which it is located becomes True.

Symbol:

Name: Convert from BCD (FRD)

Convert from BCD is an output instruction. It converts four separate groups of BCD values into an equivalent binary value consisting of 16 bits. The instruction contains a destination and a math register address identified as Source. *DEST* is the address to which the instruction moves the data. Source is where the BCD value resides until converted to the binary integer and moved. The instruction is performed when the rung on which it is located becomes True.

Thumbwheel Switch and BCD Display

Four thumbwheel switches are shown in Figure 22-22(a). A thumbwheel switch is a rotary knob with ten different positions representing decimal digits 0 through 9. Each switch has four terminals that produce a four-digit BCD output that corresponds to the equivalent decimal setting of each knob. The switches shown in the diagram are connected to a PLC input module.

The BCD display has four 7-segment digit modules that are capable of producing a pattern to show any decimal number from 0 through 9. Each digit that is displayed is the result of a corresponding BCD value applied to the four inputs of each module. The display modules shown in the diagram are connected to a PLC output module.

The thumbwheel switch is a common device used to input decimal numerical values into a PLC. Likewise, the BCD display modules are commonly used to display decimal numerical values stored in PLC memory. The four-digit number set by the thumbwheel switch can be read on the four-digit display by using the MOV and BCD conversion instructions in Figure 22-22(b).

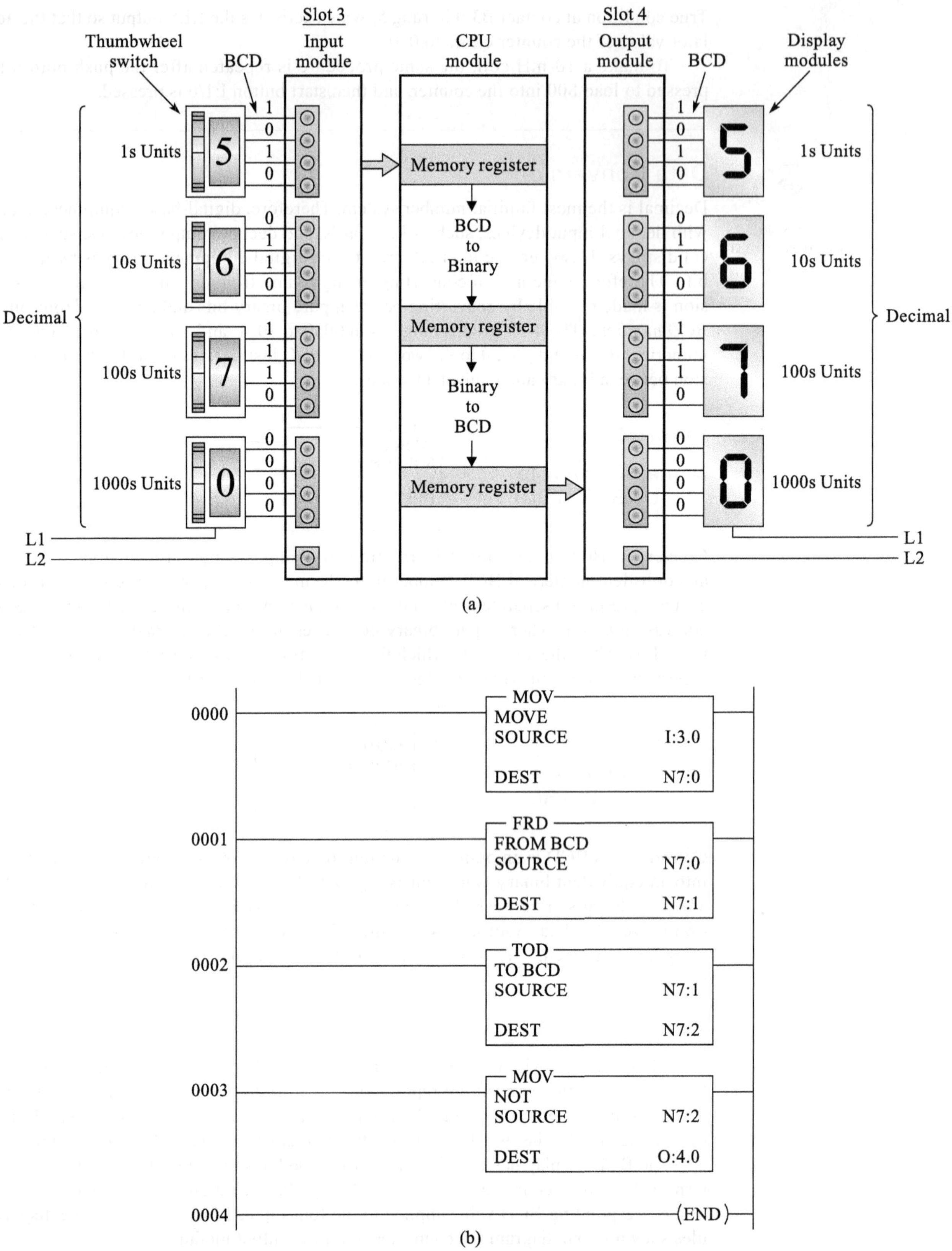

FIGURE 22-22 Converting BCD and binary values within memory to display decimal numbers loaded with thumbwheel switches

IAU4106
Data Compare Instructions

Data Compare

The third category of data-manipulation operations is data-compare instructions. These commands instruct the processor to compare the numerical contents of two registers and to make decisions based on their values and the type of instructions used. The compare instructions are used as inputs on a rung. Based on the result of the comparison, the rung's output device can be activated or deactivated, or some other operation can be performed. The values compared are the contents of one address, such as a counter, timer, or memory location, with the contents of another address, or with a constant value.

The descriptions of six comparison instructions are provided with their respective symbols. Each symbol has a three-letter abbreviation to indicate the type of comparison, a short description to specify which type of comparison it makes, and two identifiers labeled *source A* and *source B*.

When one of these instructions is programmed onto a rung, the contents that are compared are listed inside the symbol after the source A and source B identifiers. Source A must be a word address. Source B can be either an address or a program constant.

Symbol:

Name: Compare Equal (EQU)

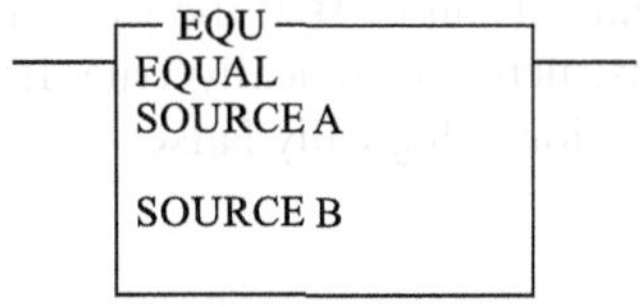

The EQU instruction compares two numerical values that follow the source A and source B indicators. Source A must be a word address. Source B can be either a word address or a program constant. If source A and source B are equal, the instruction is logically True. If these values are not equal, the instruction is logically False.

Symbol:

Name: Compare Not Equal (NEQ)

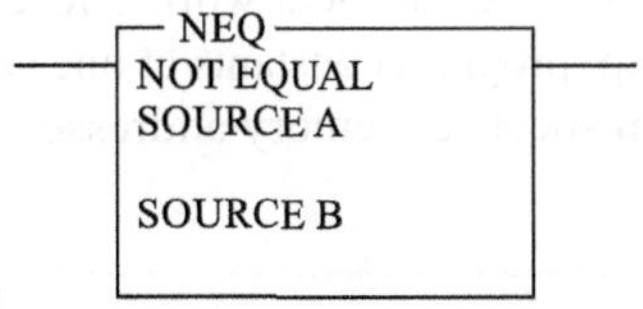

The NEQ instruction is the opposite of the EQU instruction. If source A and source B are not equal, the instruction is logically True. If these values are equal, the instruction is logically False.

Symbol:

Name: Compare Less Than (LES)

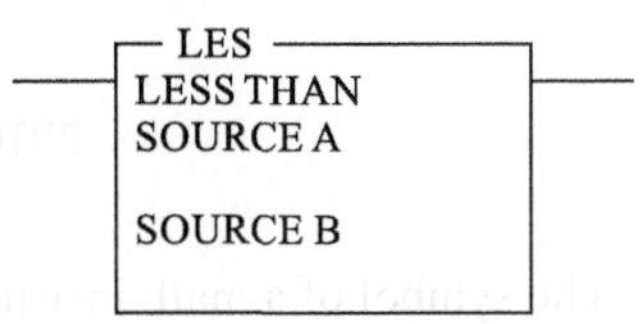

The LES instruction determines if the contents of source A are less than the contents of source B. If the value at source A is less than the value at source B, the instruction is logically True. If the value at source A is greater than or equal to the value at source B, the instruction is logically False.

Symbol:

Name: Compare Less Than or Equal (LEQ)

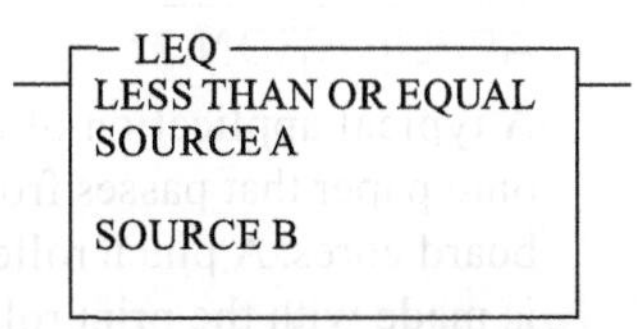

The LEQ instruction determines if the contents of source A are less than or equal to the contents of source B. If the contents of source A are less than or equal to the source B value, the instruction is logically True. If they are not, the instruction is False.

Symbol:

Name: Compare Greater Than (GRT)

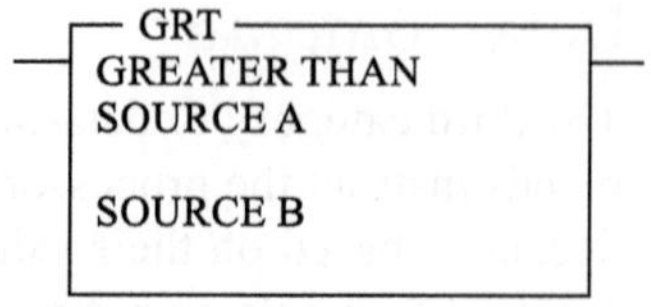

The GRT instruction examines if the contents of source A are greater than the contents of source B. If the value at source A is greater than the value at source B, the instruction is logically True. If the value at source A is less than or equal to the value at source B, the instruction is logically False.

Symbol:

Name: Compare Greater Than or Equal (GEQ)

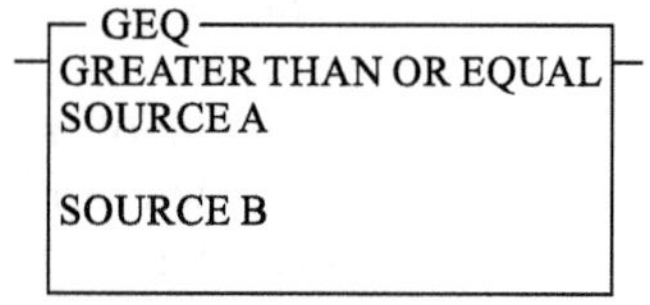

The GEQ instruction tests whether the contents of source A are greater than or equal to the contents of source B. If the value at source A is greater than or equal to the value at source B, the instruction is logically True. If the value at source A is less than the value at source B, the instruction is logically False.

22-8 Arithmetic Operations

Most PLCs have the ability to carry out **arithmetic operations.** Instructions are used to tell the processor to perform the four basic mathematical functions: addition, subtraction, multiplication, and division. To make the calculations, the numbers from two sources are used. The source can be a word address or a program constant. All of the math operations will accept program constants at one source, but not at both of them. The contents are taken directly from specific memory addresses, counters, timers, or any other accessible word location.

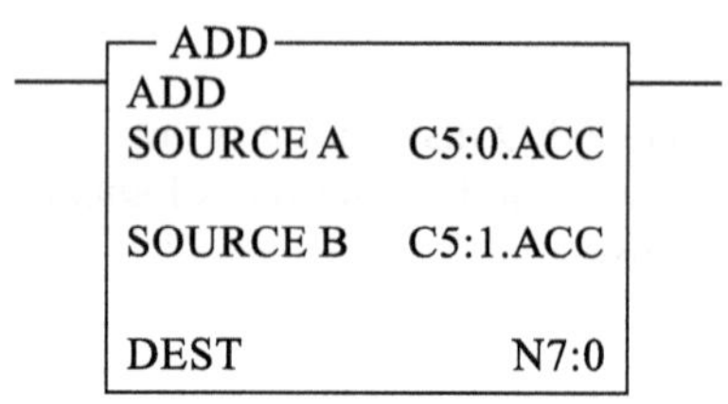

FIGURE 22-23 Arithmetic symbol

The symbol of a math instruction is shown in Figure 22-23. It has an abbreviation at the top of the symbol to indicate the type of math operation it performs, the title of the math operation, source A and B indicators, and a destination identifier. The addresses of the contents that are calculated upon are listed behind each source identifier. The result of the mathematical operation is placed in the address listed after the destination identifier.

APPLICATION 7

A typical application of an arithmetic function is shown in Figure 22-24. A machine prints onto paper that passes from a large supply roll to smaller rolls, which are wound around cardboard cores. A pinch roller places a small pressure onto the paper to ensure that good contact is made with the print roller. All four rollers rotate at variable speeds. For example, when the winding is started, the RPM increases until full speed is achieved. When the receiving roll is almost full, the rollers decelerate until they come to a complete stop. When an empty core replaces the full roll on the receiver shaft, the printing machine starts up again.

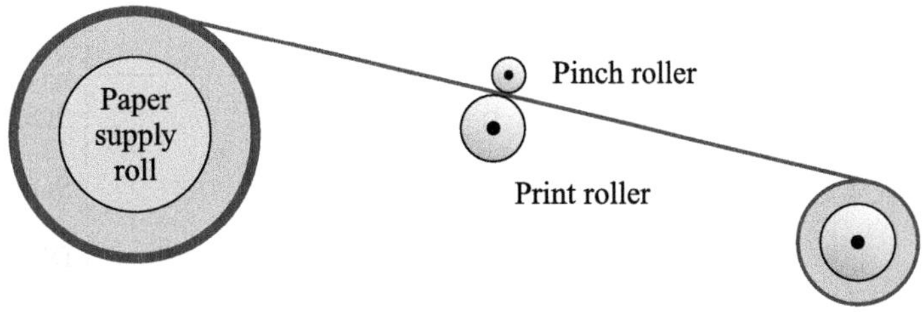

FIGURE 22-24 A printing machine that uses an arithmetic operation to control the speed

As the printing roller turns, a sensor detects when it completes each revolution. The sensor signal is transformed into an RPM reading that is fed to a memory address in the PLC. Since the pinch roller is exactly half the circumference of the print roller, it must rotate at exactly twice the speed. To ensure that the proper RPM of the pinch roller is maintained, the PLC takes the RPM of the print roller and multiplies it times two during every scan cycle. The product of the calculation is fed to a converter that gives the motor of the pinch roller the proper current to rotate at twice the RPM of the print roller.

The following instructions describe a method to perform the arithmetic calculations. All arithmetic operations are output instructions. If there is an Examine instruction in the rung, it must precede the arithmetic symbol, and it must be True before a calculation can be performed.

After each math instruction is performed, the arithmetic status bits in the status file of the data table are updated as follows:

C (carry) S:0/0. Bit 0 is set if a carry is generated; otherwise, it is reset.

V (overflow) S:0/1. Bit 1 is set if the result does not fit in the destination register; otherwise, it is reset. The maximum numbers that can be loaded into the registers are +32,767 or −32,768.

Z (zero) S:0/2. Bit 2 is set if the result of the math calculation is 0; otherwise, it is reset.

S (sign) S:0/3. Bit 3 is set if the result is negative; otherwise, it is reset.

S:13 and S:14. These are multiple bits containing 16-bit registers. They store word values generated by multiplication and division instructions. Together, they will store a full 32-bit integer result of the multiplication operation. The least significant word is stored in S:13, and the most significant word is stored in S:14.

Some examples of the use of math instructions include combining parts counts, subtracting detected defects, calculating run rates, and logging or counting product.

Symbol:

Name: Addition Instruction (ADD)

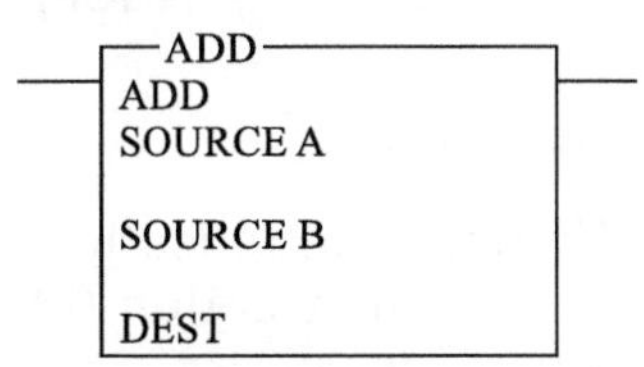

The addition operation adds two values specified after the source A and source B indicators. The sum is stored at the destination address. The following status bits are affected:

Bit 0. The C bit sets if a carry is generated.
Bit 1. The V bit sets if the sum does not fit into the destination address.
Bit 2. The Z bit sets if the result is 0.
Bit 3. The S bit sets if the result is negative.

An example of an addition operation is shown in Figure 22-25. When the input instruction I:1/0 is True, the PLC will enable the ADD instruction. The data in source A (a program

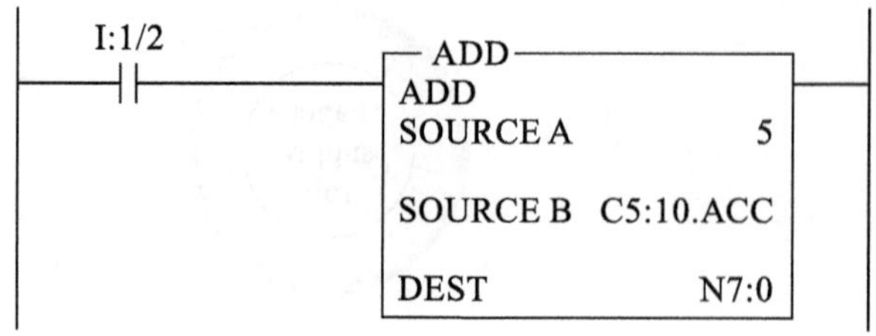

FIGURE 22-25 Addition operation

constant of 5) will be added with the data in source B (the accumulated value of counter C5:10), with the result being placed in the DEST (destination), N7:0.

Symbol:

Name: Subtraction Instruction (SUB)

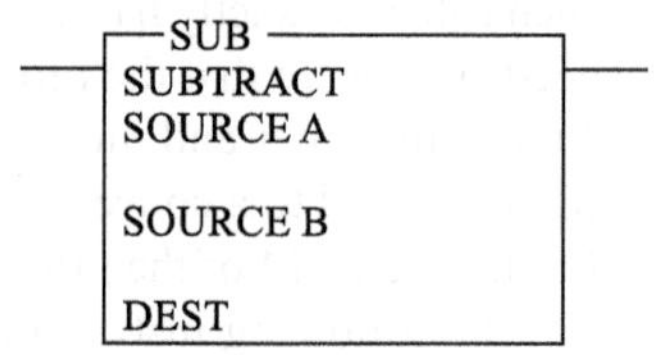

The subtraction operation calculates the difference between two values specified after the source A and source B indicators. The value at source B is subtracted from the value at source A and stored at the destination. The following bits are affected:

Bit 0. The C bit is set if a borrow is generated.
Bit 1. The V bit is set if the result underflows.
Bit 2. The Z bit is set if the result is 0.
Bit 3. The S bit sets if the result is negative.

An example of a subtraction operation is shown in Figure 22-26. When the input instruction I:1/3 is True, the PLC will enable the SUB instruction. The data in source B (the program constant) will be subtracted from the data in source A (the accumulated value of counter C5:10), with the result being placed in the DEST (destination), N7:1.

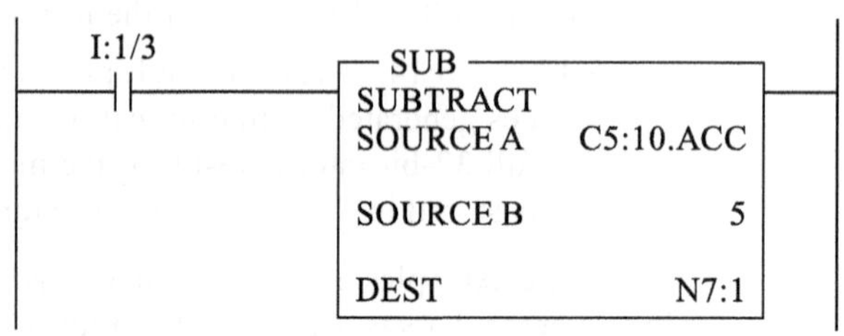

FIGURE 22-26 Subtraction operation

Symbol:

Name: Multiplication Instruction (MUL)

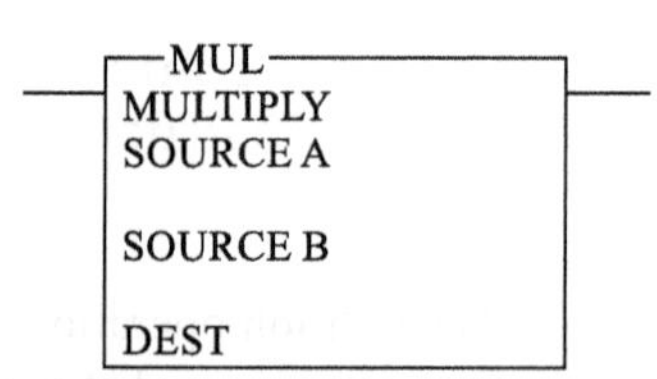

The multiplication operation performs multiplication on two values specified after the source A and source B indicators. The value at source A is multiplied by the value at source B and the answer is stored at the destination. The following bits are affected:

Bit 0. The C bit is always set.
Bit 1. The V bit is set if the result overflows at the destination address.
Bit 2. The Z bit is set if the result is 0.
Bit 3. The S bit sets if the result is negative.

S:14, S:13. These bits contain the full 32-bit integer result of the multiplication operation. This value is used in the event of an overflow.

An example of a multiplication operation is shown in Figure 22-27. When the input instruction I:1/4 is True, the PLC will enable the multiplication (MUL) instruction. The data in source A (the program constant 20) will be multiplied by the data in source B (the accumulated value of counter C5:10), with the result being placed in the DEST (destination), N7:2.

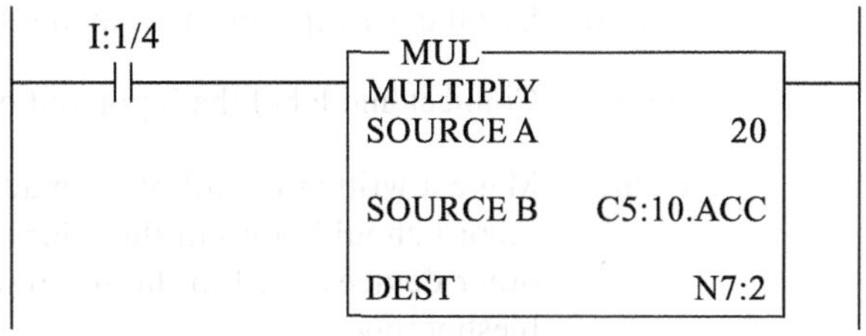

FIGURE 22-27 Multiplication operation

Symbol:

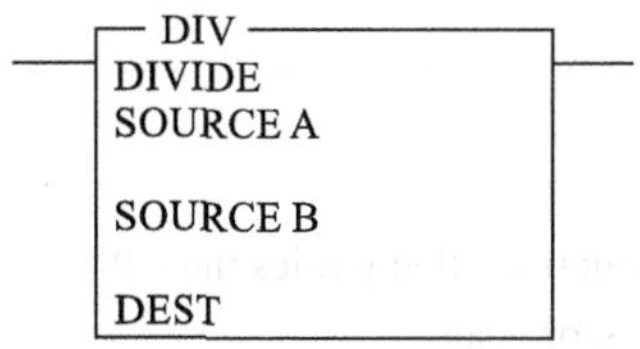

Name: Division Instruction (DIV)

The division operation divides the value of source A by the value in source B. If the quotient is rounded, it is stored in the destination. The unrounded quotient is stored in bit register S:14, and the remainder is stored in bit register S:13. The following bits are affected:

Bit 0. The C bit is always cleared.
Bit 1. The V bit is set if the result overflows at the destination address.
Bit 2. The Z bit is set if the result is 0.
Bit 3. The S bit is set if the result is negative.

An example of a division operation is shown in Figure 22-28. When the input instruction I:1/5 is True, the PLC will enable the DIV instruction. The data in source A (the accumulated value of counter C5:10) will be divided by the data in source B (the program constant 2) with the result being placed in the DEST (destination), N7:3.

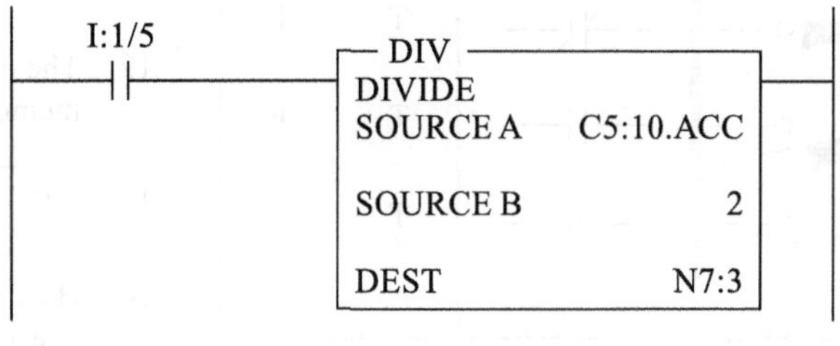

FIGURE 22-28 Division operation

22-9 Writing a Program

When you write a program for a particular application, there are many different ways to achieve the same results. If other PLC users need to read the program you have written, it is important to document your work to help them understand the sequence of instructions. Use the following steps as a guideline to use your time efficiently and to ensure that the information you provide helps others follow your program.

Step 1: Choose the sequence in which you want the input and output devices to operate. Decide what the devices must do and what conditions must be True before these devices can begin operating.

Step 2: Write a description and make a drawing that shows the sequence and conditions for energizing each output device. The diagram of the mechanical and electrical components will help visualize the operation, making it easier to write the program.

Step 3: Use the description and drawing to write the ladder diagram program.

Step 4: Connect and label the input and output devices.

Step 5: Make a written record of each address used and what each address represents. The record should contain the addresses of all inputs, outputs, timers, counters, and other devices used in the program. This information is especially useful in troubleshooting.

Step 6: Enter the program into the programmable controller.

Problems

1. A __________ is a list of instructions that guides the CPU.
2. List the three steps of a processor scan.
3. If 10K words of memory are scanned in 10 ms, how many scans will occur every second? __________
4. An AND function is performed by the PLC when two or more inputs are connected in ____.
   a. series   b. parallel
5. Using Figure 22-29, indicate the logic condition of the program instruction when the corresponding input switching device in the left column is in the position shown.

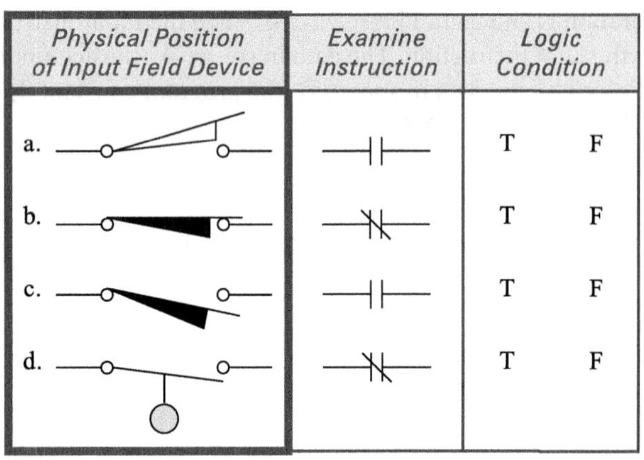

FIGURE 22-29

6. When an Examine-Off symbol is the only input on a ladder rung, the output will be energized when a corresponding N.O. push button is ____.
   a. open   b. closed
7. Branching instructions are used to ____.
   a. create a parallel circuit
   b. develop a logic AND circuit
   c. create a series circuit
   d. enable all of the above
8. The corresponding memory bit of an Examine-If-Open (XIO) instruction is a logic 0 when the field device that controls it is ____.
   a. open   b. closed
9. Choose all of the following statements that apply to a single coil: ____
   a. It could be used as an internal output.
   b. Its contacts could be used as an Examine-Off input in a rung.
   c. It could be used to drive several field output devices in more than one rung.
   d. All of the above.
10. The address of internal output devices begins with the letter ____.
    a. B   b. O   c. I
11. T/F An address for a given input or output can only be used once in a ladder program unless additional wiring is done.
12. The output instruction __________ (sets, resets) the memory bit to a 1 when the rung condition is ________ (True, False).
13. T/F A Latch output instruction can be used to replace an interlocking branch circuit.
14. How many timer instructions can be programmed into timer file T4? __________
15. Indicate which of the following functions is performed by a timer circuit:
    a. Delay-On action.
    b. Cause an operation to run a specific period of time.
    c. Record accumulated time of a continuous event.
    d. Record accumulated time of intermittent events.
    e. Operate as an astable multivibrator.
    f. Operate as a one-shot multivibrator.
    g. All of the above.
16. List four types of information that must be programmed along with the timer symbol.

17. The ____ bit of a TON timer is set when the rung on which it is located is True only when the accumulated value is less than the preset number, and resets when the rung is False.
    a. EN b. TT c. DN
18. Bit ____ of a TON timer sets when the accumulated value equals the preset number.
    a. 13 b. 14 c. 15
19. When more time is needed than the maximum amount provided by one timer, two or more timers can be programmed together. This programming technique is called ____.
    a. stacking
    b. cascading
    c. multiplying
    d. piggyback
20. The TOF instruction begins to time after the rung on which it is loaded becomes ____.
    a. True b. False
21. The EN bit of the ____ timer is set when the condition of the rung on which it is located is True, and resets when it is False.
    a. TON b. TOF c. both a and b
22. T/F An RTO timer needs a separate instruction to clear the accumulated value.
23. List three types of information that must be programmed along with the counter symbol.
24. Each counter instruction occupies ____-word elements in the counter file.
    a. 1 b. 3 c. 85
25. The Counter-Reset instruction is used to clear the accumulated value of the ____ to 0.
    a. up-counter
    b. down-counter
    c. both a and b
26. T/F The bits of timers and counters are used to change the logic states of input contacts in other rungs.
27. An up-counter ____ and a down-counter ____ each time its rung changes from a ____ condition.
    a. increments
    b. decrements
    c. True-to-False
    d. False-to-True
28. List three categories of data-manipulation instructions.
29. T/F One application of an MOV instruction is to load a number into the preset value of an accumulator.
30. A thumbwheel switch produces a ____ output that corresponds to its decimal digit position.
    a. binary b. BCD c. hexadecimal
31. A/An ____ instruction converts a pure binary number into a BCD value.
    a. FRD b. TOD
32. A BCD display produces a pattern that will show which single-digit numbers?
33. Numbers to be compared by compare instructions can be from ____.
    a. timers
    b. counters
    c. specific memory locations
    d. constant values
    e. all of the above
34. When the two values compared by an EQU instruction are the same, the rung on which it is located becomes ____.
    a. True b. False
35. List four types of arithmetic functions performed by the PLC.
36. Numbers operated on by arithmetic instructions are taken from which of the following sources?
    a. specific memory addresses
    b. counters
    c. timers
    d. thumbwheel switches
    e. all of the above

CHAPTER 23

Advanced Programming, PLC Interfacing, and Troubleshooting

OBJECTIVES

At the conclusion of this chapter, you should be able to:

- List reasons for using jump instruction commands.
- Describe how data bits are moved through a shift register to perform automated control functions.
- Develop a function table that can be used to program a sequencer to perform data-manipulation operations.
- Draw wiring diagrams to show how sourcing and sinking I/O modules are wired.
- Describe the operation of analog input and output modules.
- List the different types of special-purpose I/O modules and describe what types of functions they perform.

INTRODUCTION

Programmable controllers were initially designed to replace hardwired relay logic circuits. However, their operation is no longer restricted to relay equivalent On-Off functions. As technological advancements have been made through software, hardware, and microprocessor developments, the capabilities of the PLC have expanded. Advanced programming languages make PLCs more powerful and easier to use. Many of the operations they must now perform require some of the same capabilities as a computer system. Advanced PLCs may contain specialized modules enabling them to perform very specific operations for a variety of motion- and process-control applications.

In this chapter, various PLC modules and the operations they perform, interfacing with field devices, and troubleshooting procedures will be described.

23-1 Jump Commands

Most programmable controllers have instructions that allow the normal sequential program execution to be altered if certain conditions exist. Output instructions that perform this function are referred to as *override* commands. An example of an override command is a **jump** instruction. When this command is performed, a portion of the program is skipped over,

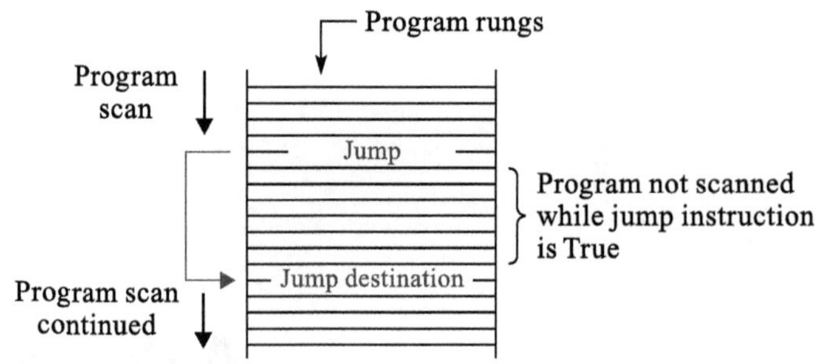

FIGURE 23-1 Jump operation

as shown in Figure 23-1. The advantage of the jump instruction is that it allows the PLC to hold more than one program and scan only the portion of the program needed to perform the desired action. This reduces scan time, allowing more scans to take place within a given period of time, so that program information is updated more frequently.

A set of rung conditions always precede the output symbol with a JMP label. To activate a jump command, the JMP output is energized if the rung condition is True. A decimal number ranging from 0 to 255 preceded by a Q2:, located above the JMP symbol, instructs the PLC where it should resume the program execution after the jump has taken place. The label input instruction (LBL) with the same decimal reference number as the JMP symbol identifies the destination of the jump operation, as shown in Figure 23-2. It is placed as the first condition instruction in the rung and is always in the True condition. A total of 256 jump commands are allowed per program. The number assigned to the first jump and label (JMP and LBL) programmed into the ladder diagram should be 1. Any additional numbers assigned are based on the number of preceding jump and label instructions. It is possible to jump forward or backward. However, backward jumps may very well generate a watchdog timer fault. The watchdog timer is a special internal timer that monitors the amount of time to complete a program scan and is reset at the end of each scan. It must be reset every few hundred milliseconds. If the program scan is caught jumping backward in the ladder too long, it will fault the processor and stop the program execution of the PLC. There can be more than one jump instruction pointing toward the same label instruction. Since the controller will not scan program rungs between a True jump instruction and its label instruction, the operation of timers may be thrown off if they are located on rungs that are jumped over.

An application example of a jump and label instruction is a multistep operation where some steps are not always needed. Suppose several types of parts are moving down a conveyor and each must be assembled in a different way. An optical sensor is used to determine which kind of part passes a detection point. Each type of part inspected makes a different rung True. The rung

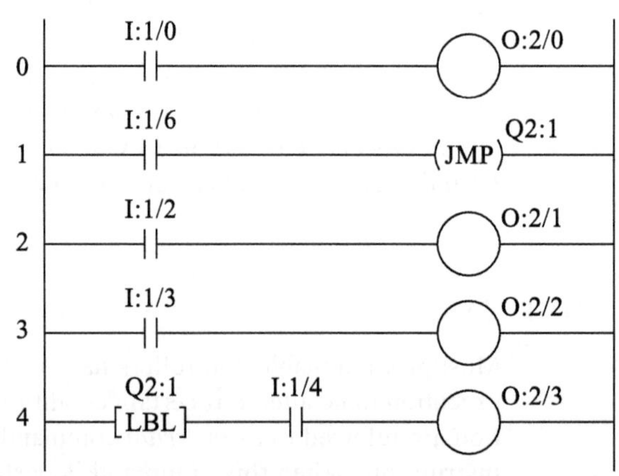

FIGURE 23-2 Jump and label operation

that is activated causes the program to jump to the subroutine that contains the assembly operation for the part that is present.

Another type of jump operation is performed by the subroutine (JSR) instruction. This command is used to direct the controller to a miniprogram file within the program if certain conditions exist. A unique file number ranging from 3 to 255 above the JSR label tells the controller where to jump. The rung condition must be True to activate this command. An SBR instruction with the same reference number is located at the first rung of the subroutine.

An instruction labeled RET located in the last rung of the subroutine file tells the processor that the subprogram has been scanned and executed. When encountered, the controller returns to the main program and begins the ladder rung immediately following the JSR instruction that initiated the subroutine. Figure 23-3 shows a sketch of how the controller scans through memory when a subroutine instruction is encountered.

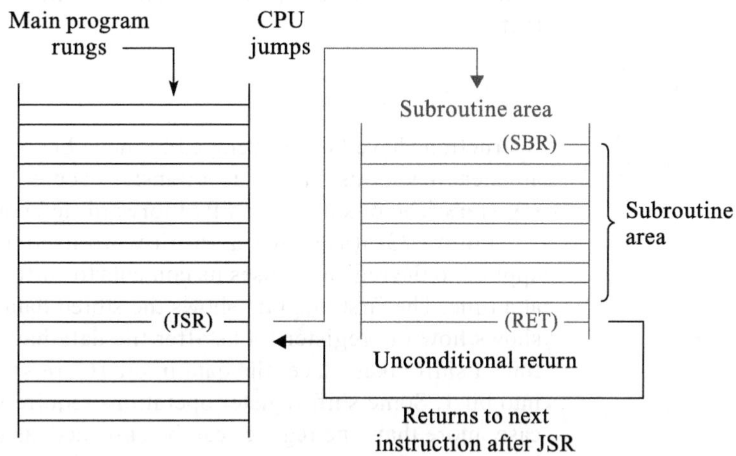

FIGURE 23-3 Jump to subroutine

A subroutine instruction may be used in a program that controls stoplights at an intersection with a crosswalk button. The main program controls the light sequence to allow cars to pass first in one direction and then the other. This cycle continues until the main program is interrupted by the closing of the crosswalk button. The button is an Examine-On contact, which makes the rung with the JSR instruction True. As the program jumps to a subroutine, it turns on the crosswalk light and the stoplights turn red for 20 seconds. When the time expires, the processor exits the miniprogram and returns to the main program to resume normal control of the lights at the intersection.

Up to four levels of subroutines can be nested. The program can jump to any one of four subroutines, one after another, and then return in reverse order, retracing the jumps. Figure 23-4 shows three nested subroutine levels.

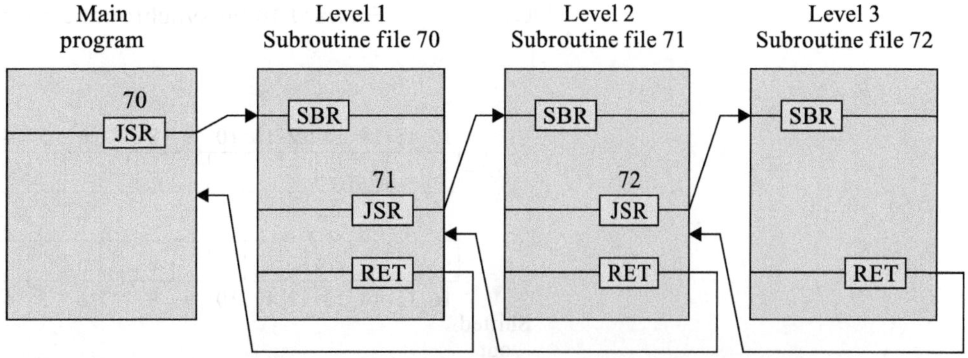

FIGURE 23-4 Three nested subroutine levels

23-2 Data Manipulation

To energize the terminal of an output module using ladder instructions, a single bit in a memory register must be a 1. The binary state is usually determined by the conditions of input examine instructions and the circuit configuration to perform the desired logic function. Instead of using individual input contacts and logic circuits to control field devices, a set of instructions that maneuver bits in a chosen register can perform the same function. These commands insert bits one at a time and set or clear individual bits. They can also move all the bits in one or more registers simultaneously. For example, suppose a sign that uses 256 light bulbs to display different patterns is controlled by a PLC. Sixteen registers that contain 16-bit words control the lights instead of 256 rungs.

Instructions that perform these multiple control operations are used extensively in all types of automated systems. Shift registers and drum/controller/sequencer instructions perform these types of functions. These are discussed in detail in the following text.

Shift Registers

IAU15108
PLC Shift Registers

Instructions have been developed to move bits within a memory word or from one word to another. When this type of data transfer occurs, the memory devices are referred to as **shift registers.** The bits may be shifted forward (left) or reverse (right).

Figure 23-5(a) shows a 16-bit word used as a forward shift register. A clock pulse applied to the register causes its contents to shift. Data can be entered from the right, one bit at a time. The first diagram shows the stored data prior to a clock pulse, and Figure 23-5(b) shows how the register looks after the data have been shifted one place to the left. Each time a shift takes place, the data in bit 16 are shifted out and lost and a data bit is entered into bit 1. Some shift register operations require words that are greater than 16 bits. In this case, more than one register can be cascaded, as shown in Figure 23-6. Data is shifted into bit 1 of word 1. Instead of the data being lost out of bit 16, it is transferred to bit 1 of the second shift register.

Shift registers are used primarily to control a process on a conveyor system. As an object is inspected, it causes data to be entered into the register. The inspection results

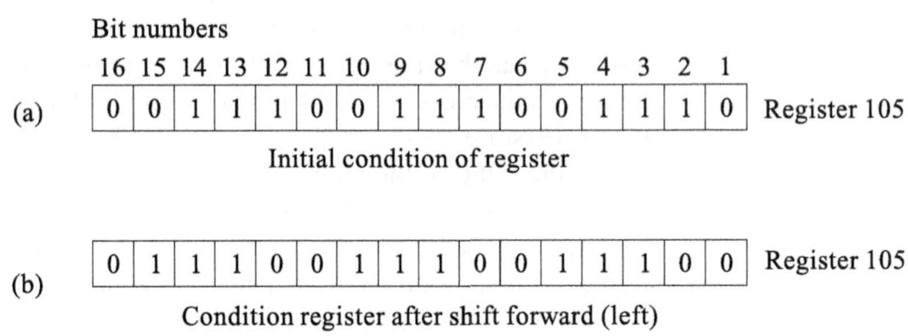

FIGURE 23-5 Forward 16-bit synchronous shift register

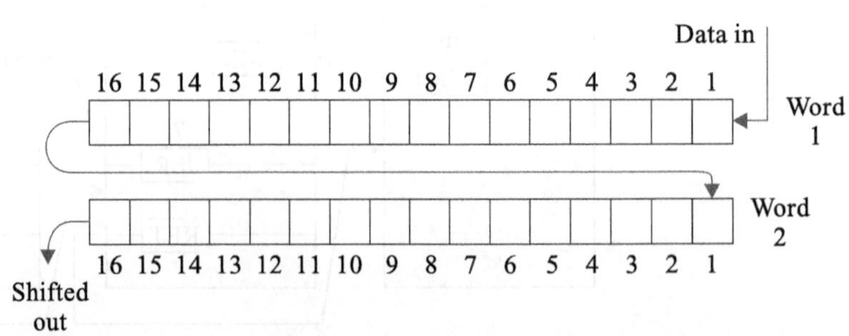

FIGURE 23-6 Cascaded two-word forward shift register

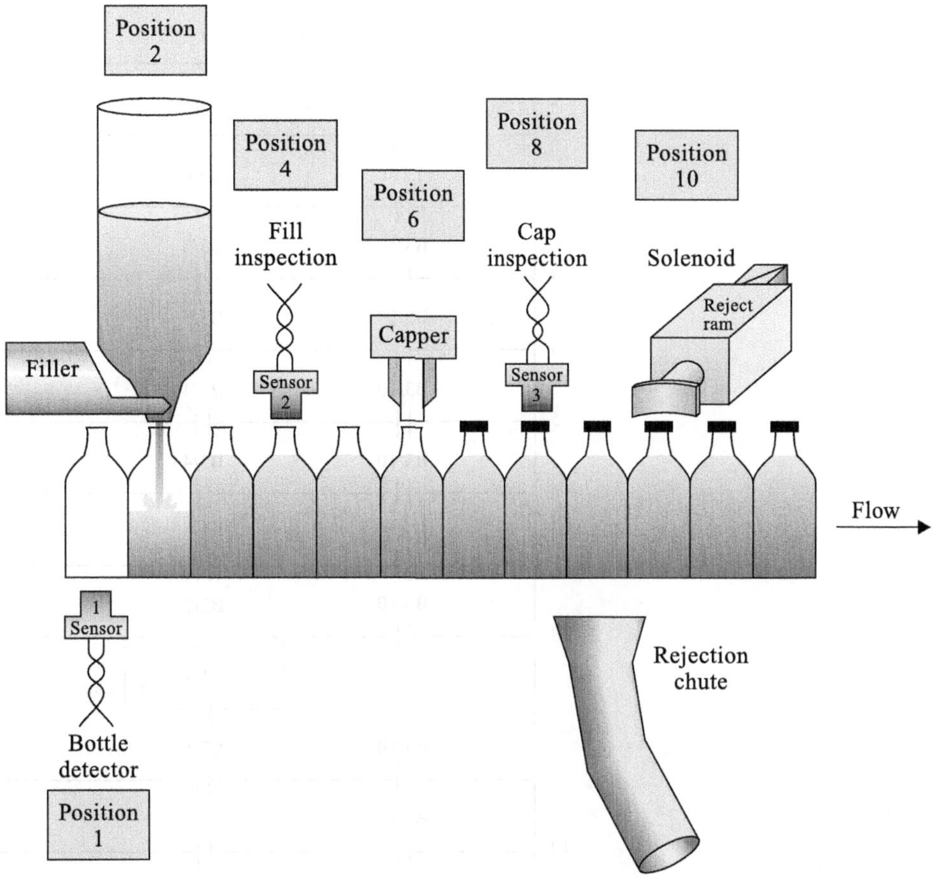

FIGURE 23-7 Conveyor line in a brewery

determine which logic state is entered. For example, the presence of an object produces a logic 1, and the absence generates a 0. An example of a practical application for a serial shift register is the filling and crowning of bottles on a conveyor line in a brewery, as shown in Figure 23-7.

The bottles physically touch each other as they travel down the line. They are all stopped simultaneously for 2 seconds when a bottle is being filled. The conveyor runs only long enough to move the distance of one bottle diameter before it stops to fill the next bottle. Light sensor 2, placed two bottle positions from the filler, detects if the bottle is full. Each bottle is capped two bottle positions from sensor 2. Light sensor 3, placed two bottle positions from the capper, determines if the cap is placed securely on the bottle. A pneumatic ram, two bottle positions from sensor 3, pushes a bottle off the conveyor if it is rejected because the bottle is not full enough or the cap is not secure.

Figure 23-8 shows a ladder diagram of the logic circuit that is used to inspect and reject the filling and capping processes.

Step 1: Sensor 1 detects when a new bottle passes a reference point on the conveyor line. When the neck of a bottle is sensed, contact I:1/0 of rung 0 turns on and latches output B3/0. When the bottle is past the sensor, the condition of rung 2 becomes True and energizes output coil B3/10. The energized coil causes input contacts B3/10 from rungs 3 to 17 to turn on during the remainder of the scan cycle. This action is similar to providing a clock pulse to all flip-flops in a synchronous shift register used in digital electronics. The duration of the pulse ends when the Unlatch output is energized in rung 17. Another pulse will not occur until the next bottle is detected by sensor 1.

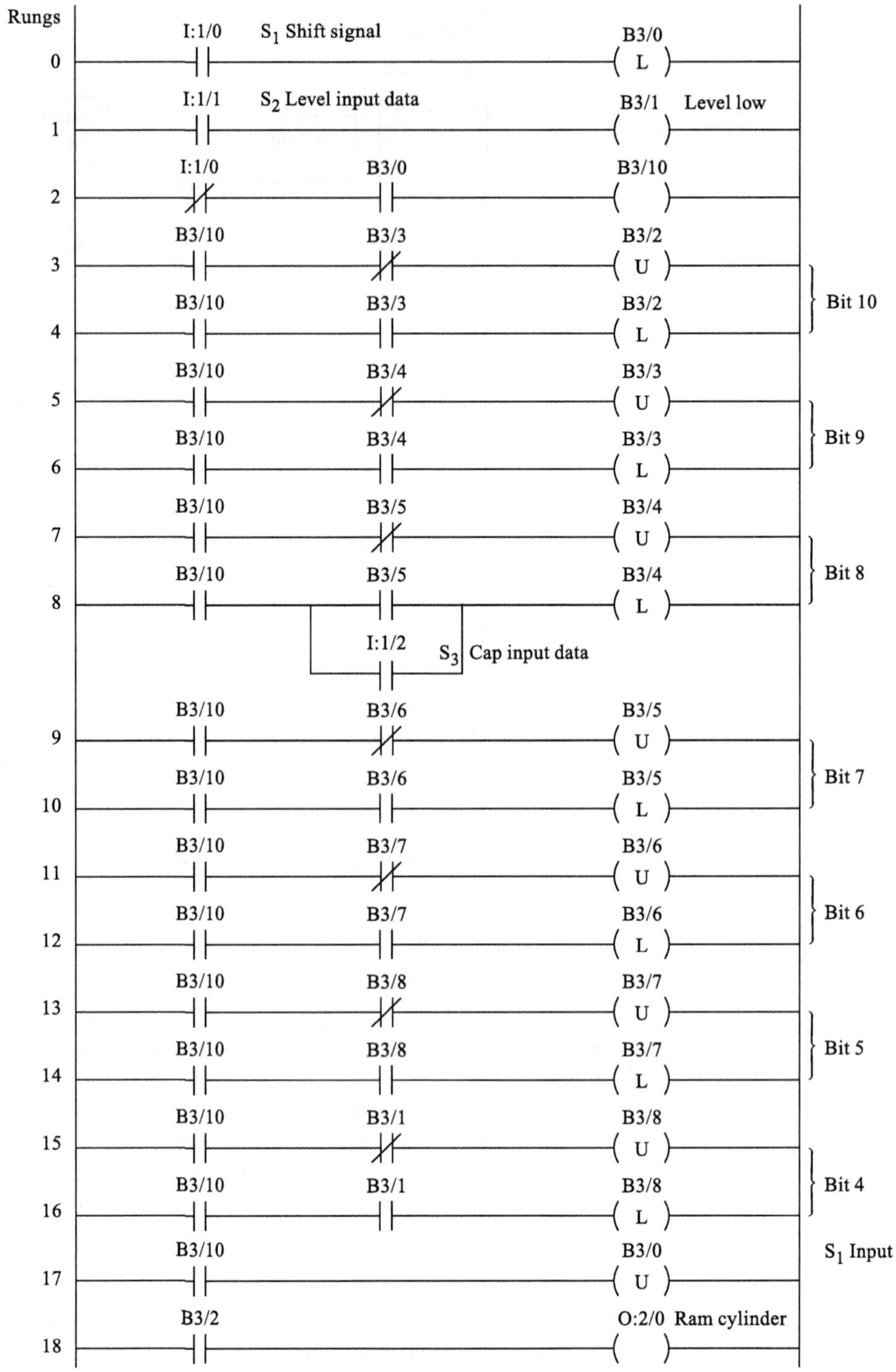

FIGURE 23-8 Ladder diagram of a control circuit for a brewery conveyor line

Step 2: The operation of a serial shift register is shown in Figure 23-9(a). Whatever logic level is present at a data input of each flip-flop, it is entered after the clock pulse arrives. Since the outputs of the flip-flop are fed into the data input of the adjacent flip-flops, all data are shifted one position to the left every time a clock pulse arrives.

Each pair of Latch and Unlatch output coils with the same number in rungs 3 through 16 perform the same function as a flip-flop. When the latch rung is activated, the flip-flop sets to a 1. When the unlatch rung is activated, the flip-flop resets to a 0.

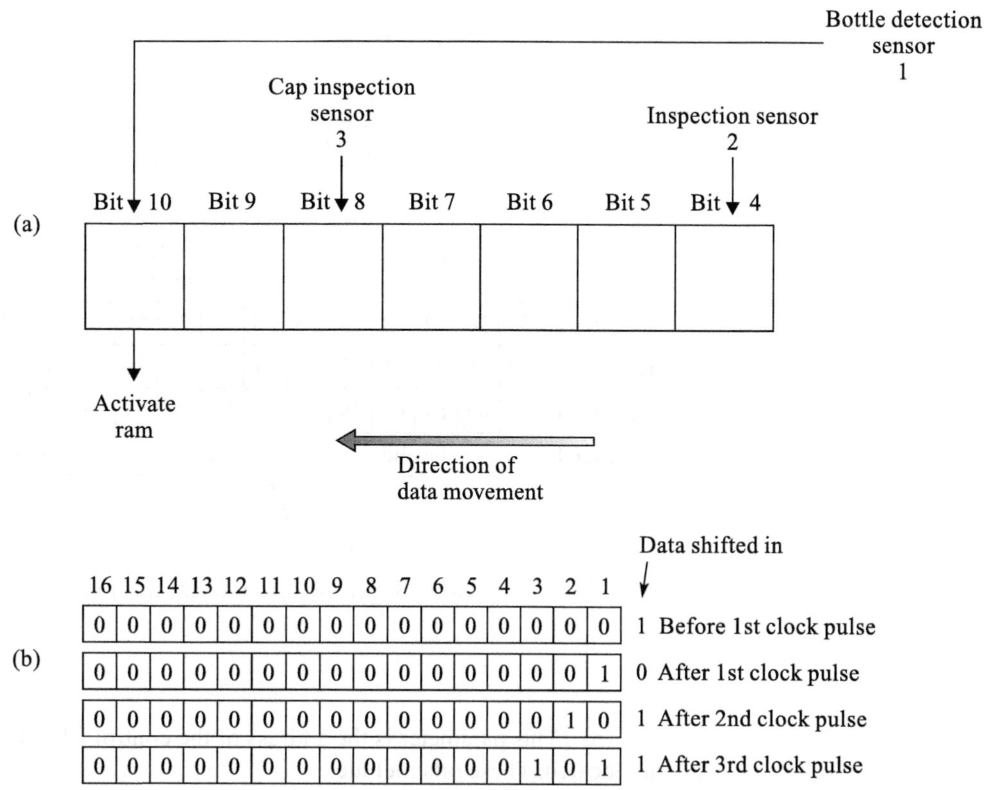

FIGURE 23-9 The operation of a shift register

Step 3: Whenever a bottle that is not full passes sensor 2, contact I:1/1 of rung 1 turns on and energizes output coil B3/1. Contact B3/1 closes and causes latch B3/8 to activate when the clock pulse reaches rung 16 during the scan cycle. This condition loads a 1 into bit 4 of the register shown in Figure 23-9(a). If the bottle is full, a 0 is loaded into bit 4 when a clock pulse arrives. Any 1 is shifted every time a clock pulse arrives, as shown in Figure 23-9(b).

Step 4: Rung 8 of the ladder is an OR function. Whenever a cap is not properly placed on the bottle, Sensor 3 will set bit 8 by energizing latch B3/4. If a 1 is already shifted into the bit from Sensor 2, the bit will remain latched.

Step 5: Whenever any 1 state bits are shifted into bit 10, output B3/2 is latched on rung 4. This latched condition closes contact B3/2 of rung 18 to activate a pulse valve that causes a spring-return ram cylinder suddenly to extend and retract. The extension movement rejects the bottle by pushing it off the conveyor into a chute.

Another way of shifting data is by using two instructions: *BSL,* which stands for *bit shift left,* and *BSR,* which stands for *bit shift right.* Both are output instructions, and the data is shifted one bit position each time the rung on which it is located goes from False to True.

Name: Bit Shift Left (BSL)

This instruction shifts data to the left from a specified bit location in an array of registers to a higher bit number (in ascending order). Various parameters must be programmed into the symbol. Using Figure 23-10 as a reference, these parameters are described as follows:

File This parameter is the address in the bit file where data is shifted. The address number listed follows #, and indicates the data word in which data is entered. The reference shows that File #B3:1 is used, and that data is loaded into register 1.

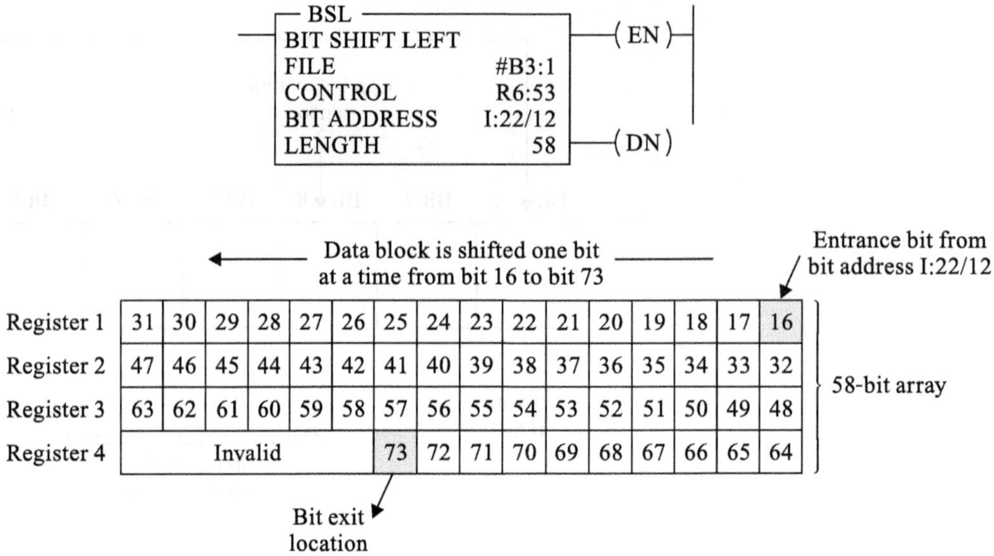

FIGURE 23-10 Bit shift left (BSL) instruction operation

Control This parameter is the address of the control file where status bits of the instructions are located. These bits include:

EN (Bit 15). The enable bit, which is set when the rung condition is True.

DN (Bit 13). The done bit, which sets the shift operation and is completed on all the bits in the file.

UL (Bit 10). The unloaded bit, where data exits the shift register. This bit sets when a 1 is shifted into it and resets when a 0 is shifted to its location.

Bit Address This parameter is the bit address of the input file where data is inserted into the register. The reference shows that data is entered through I:22/12.

Length This parameter specifies the desired size of the array. The maximum number of register file bits that can be used is 2047. A length of 0 causes the input data to be transferred directly to the UL bit. The reference shows that 58 bits are used in the array.

Name: Bit Shift Right (BSR)

This instruction shifts data to the right from a specified location in an array of registers to a lower bit number (in descending order). Figure 23-11 is used to illustrate how the parameter loaded into the BSR instruction affects the register in a bit file.

File—#B3:2 This information indicates that register 2 of the bit file 3 is used. The first bit in register 2 is bit 32.

Control—R6:54 This is the location in the control file where the status bits of the instruction are located.

Bit Address—I:23/06 This is the bit address of the input file where data is inserted into the register.

Length—38 This is the number of bits used in the array. Therefore, data enters bit 32 and exits 38 bit locations later through bit 69.

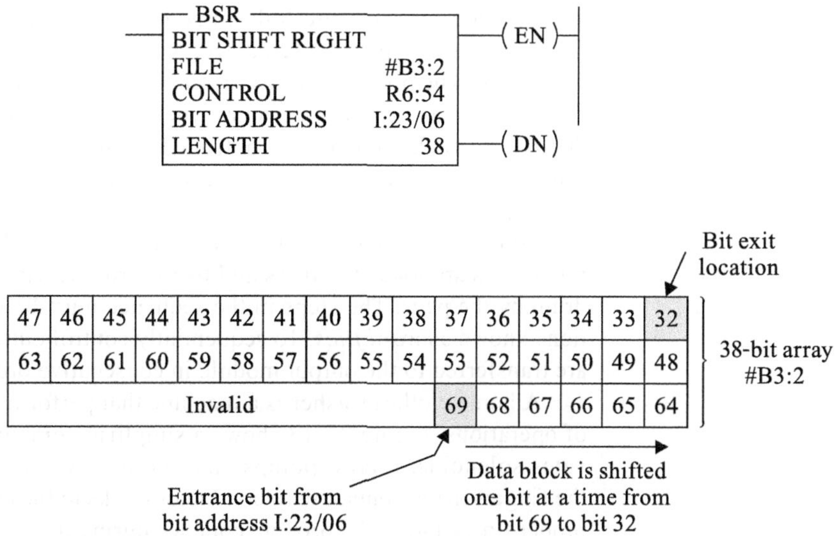

FIGURE 23-11 Bit shift right (BSR) instruction operation

DIG6604
Sequencer Output Instruction for the SLC-500 PLC

Sequencers

Before PLCs were developed, the control function of an automatic assembly machine was often performed by an electric drum **sequencer,** shown in Figure 23-12. The type of control function that took place required specific on or off patterns of outputs to be continuously repeated. The drum sequencer is a cylinder with pegs placed in holes at varying horizontal positions around the outside. As a motor turns the drum, switches aligned with the pegs close when they make contact, and at positions where pegs do not exist, the corresponding switches remain open.

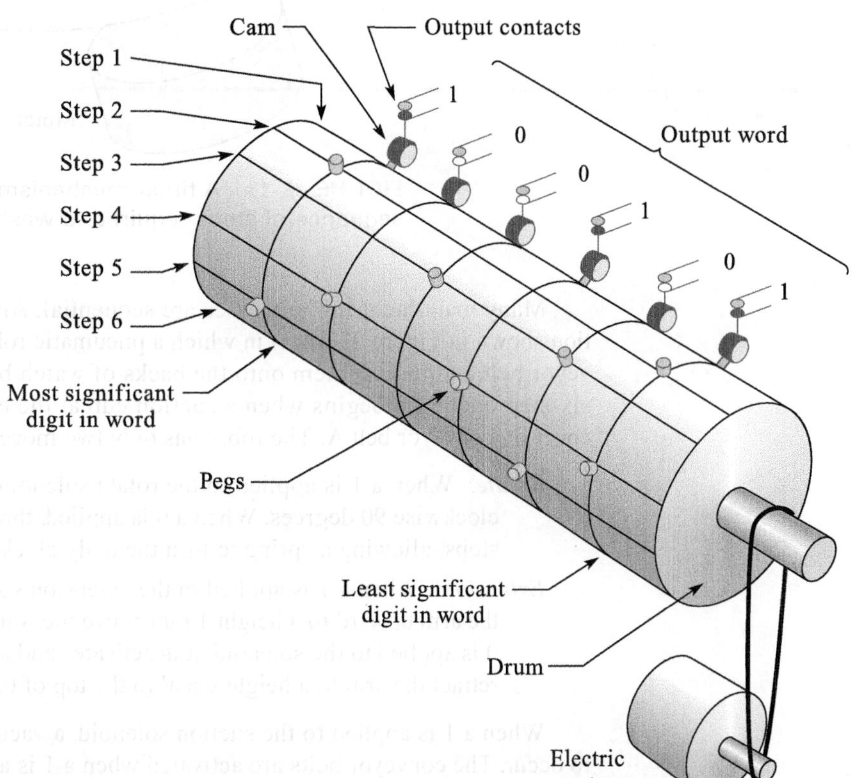

FIGURE 23-12 Sequencer drum

Pegs are inserted into holes, one row at a time, to program the drum sequencer. Each row provides a step in the sequence. A closed switch represents a 1 and an open switch represents a 0. The operation of the sequencer provides a direct relationship between action and time. Action at the output occurs when the pegs cause the contacts to open or close. The duration of each sequence is controlled by the speed of the motor. Since each step takes the same amount of time, consecutive pegs are installed if the output must be on longer than one step.

Many sequential operations can also be controlled by words in a PLC memory. The binary bits are loaded with 1s and 0s to form the same kind of pattern programmed into the drum by the pegs. The steps in the sequence must be loaded into consecutive memory locations known as *word files.* To sequentially control output devices, words from each word file are transferred to the output module in consecutive order.

A home clothes washer is a machine that performs an automatically controlled sequence of operations. Figure 23-13 shows a simplified drawing of the timing mechanism that activates solenoids, relays, pumps, and motors. As the electric motor slowly turns the disk, the fingers make contact with conductor tracks to turn the controlled devices on and off at the proper times and in the proper sequence during the wash cycle.

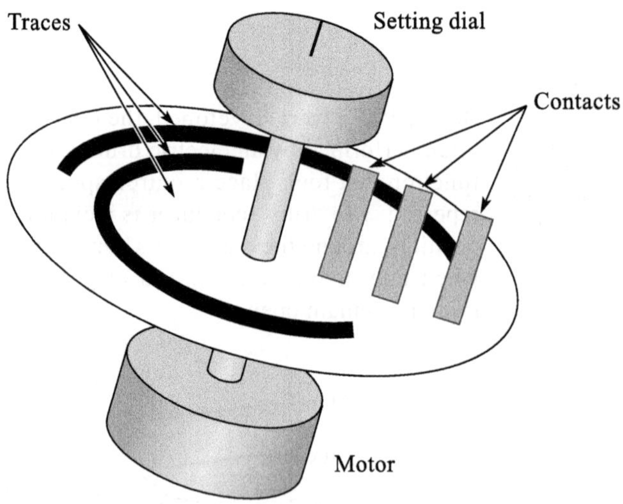

FIGURE 23-13 A timer mechanism that controls the sequence of steps required to wash a load of laundry

Many manufacturing processes are sequential. An example is a pick-and-place operation shown in Figure 23-14(a) in which a pneumatic robot takes watch covers from one conveyor belt and inserts them onto the backs of watch bodies on another conveyor belt. The six-step operation begins when a suction cup at the end of the arm makes contact with a cover on conveyor belt A. The robot has only two movements: *rotate* and *extension.*

Rotate: When a 1 is applied to the rotate solenoid, air forces the body to turn counterclockwise 90 degrees. When a 0 is applied, the solenoid deactivates and air pressure stops, allowing a spring to turn the body clockwise 90 degrees.

Extension: When a 1 is applied to the extension solenoid, air forces a cylinder to extend the arm upward to a height 1 mm above the watch covers on conveyor belt A. When a 0 is applied to the solenoid, it deactivates and air pressure stops, allowing a spring to retract the arm to a height equal to the top of the watch bodies on conveyor belt B.

When a 1 is applied to the suction solenoid, a vacuum port is activated, causing suction to occur. The conveyor belts are activated when a 1 is applied to the driver motor. The speed of the motor is set so that the belts travel the correct distance during the sequence steps. Figure 23-14(b) shows the sequencer steps that control the assembly procedure.

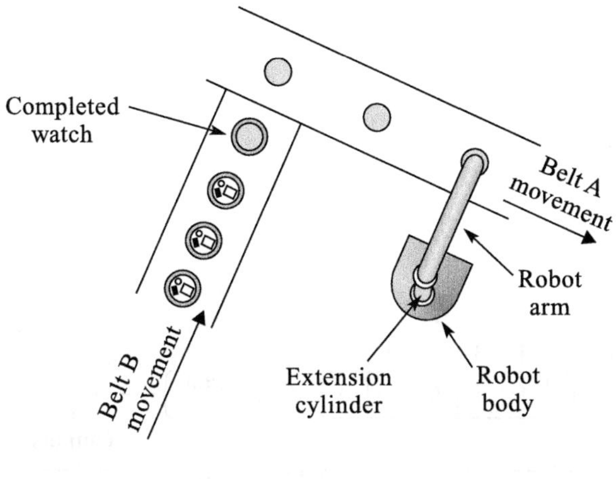

(a)

Step	Suction pickup	Robot body activated	Extension cylinder	Belt A	Belt B
1	1	0	1	0	0
2	1	1	1	0	0
3	1	1	0	0	0
4	0	1	1	0	0
5	0	0	1	0	0
6	0	0	1	1	1
7	1	0	1	0	0

(b)

FIGURE 23-14 A robotic watch assembler

Step 1: The cylinder extends over belt A and the suction at the end of the arm picks up the watch cover.

Step 2: While the cylinder is extended and the cover is held in place, the solenoid for the robot body is actuated, causing the arm to swing counterclockwise by 90 degrees.

Step 3: While the cover is held in place and the robot arm is positioned over conveyor belt B, the extension solenoid is deactivated, causing the cylinder to lower until the cover snaps securely onto the watch body.

Step 4: While the robot body is in position at conveyor belt B, the suction pressure is removed and the extension solenoid is activated, causing the arm to rise.

Step 5: When the cylinder fully extends the arm, the body solenoid is deactivated, causing the arm to rotate clockwise by 90 degrees.

Step 6: While the arm is extended and positioned above conveyor belt A, both conveyor belt motors are activated and run until the end of the sequence step.

Step 7: A cover on conveyor belt A and a body on conveyor belt B are in the required positions to repeat the entire sequence of steps.

Figure 23-15(a) shows an example of a PLC sequencer instruction symbol. Inside the box is a list of parameters that must be programmed before the instruction can perform its desired operation.

File This parameter is the designated memory location within the PLC that forms the 16-bit pattern of the outputs during the sequence. One area that can store this information is the bit data file. Figure 23-15(b) shows the data for each step of the sequence loaded at addresses B3:0 through B3:6.

Mask When a sequencer operates on an entire word, there may be outputs associated with the word that do not need to be controlled. To prevent the sequencer from controlling these bits of the output word, a *mask* word is used to selectively screen out data as it is transferred from a word file to the output word. For each bit of the output that the sequencer is to control, the corresponding bit of the mask word must be set to a 1. All other bits of the mask are 0.

Dest This parameter is the output destination to which the data in the sequencer is transferred.

(a)

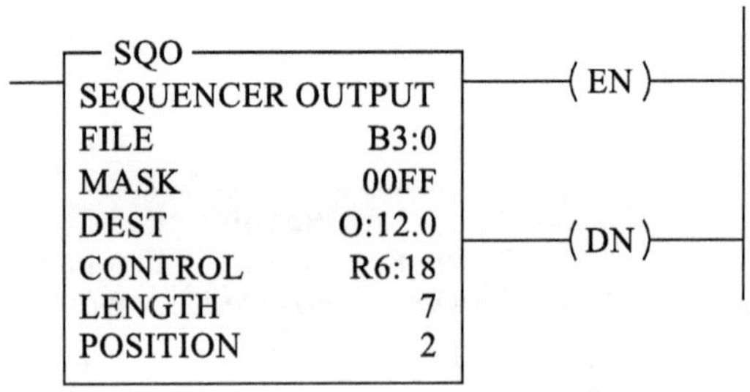

(b)

	15	14	13	12	11	10	9	8	7	6	5	4	3	2	1	0	
	0	0	0	0	0	0	0	0	0	1	0	1	1	0	1	0	Destination O:12
	0	0	0	0	0	0	0	0	1	1	1	1	1	1	1	1	Mask value 00FF

BIT DATA FILE

Address																	(Position)
B3:0	0	0	0	0	0	0	0	0	0	0	0	0	0	0	0	0	Step 0
B3:1	0	0	0	1	0	1	0	0	0	1	0	0	1	1	0	1	1
B3:2	1	0	0	1	1	1	1	0	1	0	1	1	0	1	0	1	2
B3:3	0	1	1	0	1	0	1	1	0	1	0	1	1	0	1	0	3
B3:4	1	0	0	0	1	1	1	1	0	0	0	0	1	1	0	1	4
B3:5	1	1	0	0	1	1	0	0	1	1	0	0	1	1	0	0	5
B3:6	0	1	1	1	0	0	0	1	1	1	0	0	0	1	0	0	6

Outputs	
00	
01	← ON
02	
03	← ON
04	← ON
05	
06	← ON
07	
08	
09	
10	
11	
12	
13	
14	
15	

FIGURE 23-15 Architecture of sequencer output instruction

Figure 23-15 shows data being transferred from word file B3:3, through the mask value of 00FF, to the destination designated as output file O:12. This diagram also illustrates how the masking operation screens data as it transfers through. Notice that the mask value of 00FF creates a pattern where the eight most significant bits are 0s, and the eight least significant bits are 1s. The only data from the bit data file that is transferred to the destination are those that pass through the bits of the mask that are designated as 1s.

When entering a hexadecimal number greater than 9FFF, it is necessary to precede this number with a 0. For example, to set the mask of FFFF, 0FFFF must be entered.

Control This parameter indicates the address of where the control bits of the instruction are located.

Length This is the number of steps of the sequencer file starting at position 1. All SQO instructions start at position 0 when the PLC is switched from the program to the run mode. Due to this extra state, the file length is always +1 of words. The instruction resets to position 1 at the end of the sequence, not position 0.

Position This parameter shows the actual step in the sequence that is being transferred from the bit to the output file.

The sequencer instruction can be either *event-driven* or *timer-driven*. An event-driven instruction is preceded by an input instruction on the rung that turns on and off. The sequencer advances one step each time the rung goes False to True. A timer-driven instruction is preceded by a Timer DN input on the rung, and increments after a preset timer period.

PROGRAMMABLE CONTROLLER INTERFACING

PLC input/output (I/O) modules provide the electrical connection between external devices (called *field devices*) and the internal processor unit. These modules are needed because most field devices cannot be directly connected to the processor unit's circuits. Through various interface circuitry inside the module, they become compatible.

Unfortunately, not all field devices use the same type of signal. Therefore, PLC manufacturers make several types of modules to work with each of these signals. To properly interface these I/O modules to field devices, it is important to have a general understanding of how they operate and an awareness of certain operating specifications.

Two groups of I/O modules are used in most applications, *discrete* and *analog*.

23-3 Discrete Input/Output Modules

The word *discrete* means "separate" or "different." Connected to discrete *input* modules are field devices that provide different switching signals based on whether they are in the on or off condition. Discrete *output* modules turn field devices to which they are connected on or off. Table 23-1 lists several discrete input/output field devices.

TABLE 23-1 Types of Field Devices to which I/O Modules are Connected

Input Devices	*Output Devices*
Push Buttons	Lights
Limit Switches	Alarms
Photoelectric Sensors	Valves
Circuit Breakers	Motor Starters
Proximity Switches	Solenoids
Motor Starter Contacts	Fans
Relay Contacts	Pumps
Level Switches	Mixers

Discrete Input Modules

Two common types of field devices that operate in either the on or off switching condition are **relay contacts** and **solid-state relays.**

Relay Contacts: Relays operate in the On-Off condition as their mechanical contacts open or close. Because they can accommodate a large variety of signals, such as AC, DC, or a range of voltages, they are popular and useful in control environments that have a broad diversity of electrical I/O requirements.

Solid-State Relays: It is common for discrete input devices to use solid-state circuitry that operate in either the on or off condition. Specifically, the circuitry consists of transistors, which are driven into saturation or cutoff to perform the required switching action.

PLC manufacturers produce several types of discrete input modules to which these field devices are wired. To ensure electrical compatibility and that particular application requirements are met, the following specifications must be considered when selecting the proper module:

1. *Type of Current (AC or DC).* The first step in selecting an input module is to determine which voltage is required for an application, alternating current (AC) or direct

current (DC). Most input modules are designed to operate with either AC or DC, but not both.

2. *Voltage Level.* After the type of voltage required is determined, select an input module that matches the voltage level of the field device to which it is connected. Table 23-2 shows the voltage levels commonly used by most PLC input modules on the market.

TABLE 23-2 Voltage Levels Commonly Used by PLC Input Modules

Type of Current	*Voltage Level*
DC	5
DC	12–24
AC	120
AC	220

3. *Number of Inputs.* When selecting an input module, determine how many input field devices are to be used. PLC modules with 4, 8, 16, or 32 inputs are available. In manufacturers' catalogs, each input is referred to as a *point*. For example, a 4-point module has 4 inputs.

   For future expansion capabilities, it is a common practice to select a module with several more inputs than are initially needed. This alternative is less expensive than purchasing another input module when additional inputs are required.

4. *Active-High or Active-Low.* Before a field device is wired to a PLC system, it must be determined whether the module requires a logic-high or logic-low signal to activate (turn on) its inputs. These *active-high* and *active-low* specifications are dictated by how the PLC interprets logic states. Some PLCs recognize a "1" state logic signal as a high-level voltage (near 5 volts, or +24 volts). Others recognize a "1" state logic signal as a low-level voltage (near ground).

An example of how relay contact switches provide these signals is shown in Figure 23-16. In Figure 23-16(a), the relay's normally open contacts are connected to input 02 of an active-high module and the positive terminal of a +24-volt DC supply. If the relay's coil is not energized, the N.O. contacts remain open and 0 volts are applied to the module's input terminal. The module considers the input off, and the PLC's processor unit interprets the input terminal deactivated. When the relay's coil becomes energized, the N.O. contacts close. As +24 volts are applied to terminal 02, the module considers the input on and the PLC's processor unit interprets the input to be activated.

Figure 23-16(b) shows the N.O. contacts of the relay connected to input 01 of an active-low input module and the negative lead of a 24-volt DC supply. If the relay's coil is not energized, the contacts remain open, +24 volts are present at the input terminal (through the internal circuitry of the PLC module), and the PLC's processor unit considers the input off. When the relay coil becomes energized, the N.O. contacts close, 0 volts are present at the input terminal, and the PLC's processor unit considers the input activated.

Many types of solid-state sensors are interfaced to the PLC's input module. These include inductive, capacitive, photoelectric, and others. Some electronic sensors use only two wires. These detection devices are wired to the input module the same way as the relay contact shown in Figure 23-16. Depending on the manufacturer, some two-wire sensors are polarity-sensitive, and others are not. Most electronic sensors use three wires. Inside these sensors, a transistor switch is used to send a signal to the PLC. There are two types of transistors used in three-wire sensors, PNP and NPN. When the PNP sensor is not activated, its output is *open*, as shown in Figure 23-17(a). The PNP transistor provides a logic-high signal when the sensor is activated, as shown in Figure 23-17(b). Therefore, these sensors must be used with PLC input modules that need active-high signals.

Sensors with NPN transistors operate just the opposite of sensors with PNP transistors. When the sensor is off, as shown in Figure 23-18(a), its output terminal is an open circuit. When

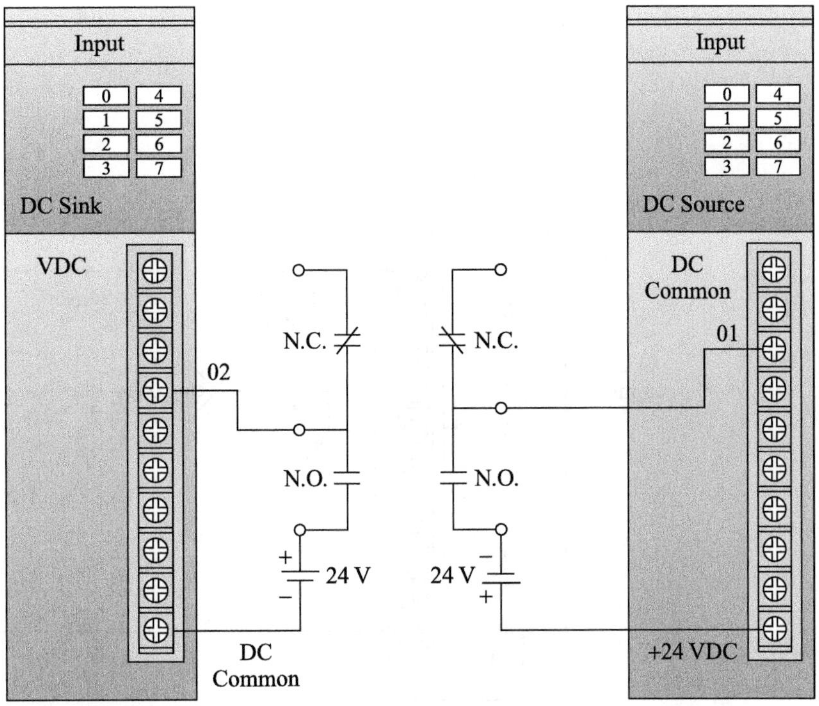

(a) (b)

FIGURE 23-16 Relay contact input modules: (a) Active-high module; (b) Active-low module

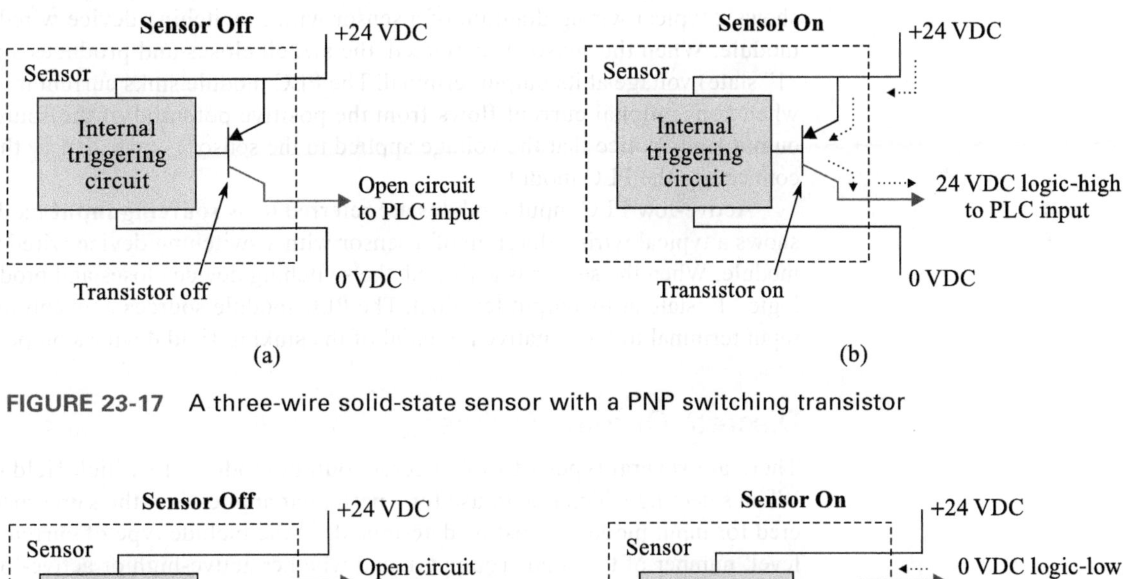

(a) (b)

FIGURE 23-17 A three-wire solid-state sensor with a PNP switching transistor

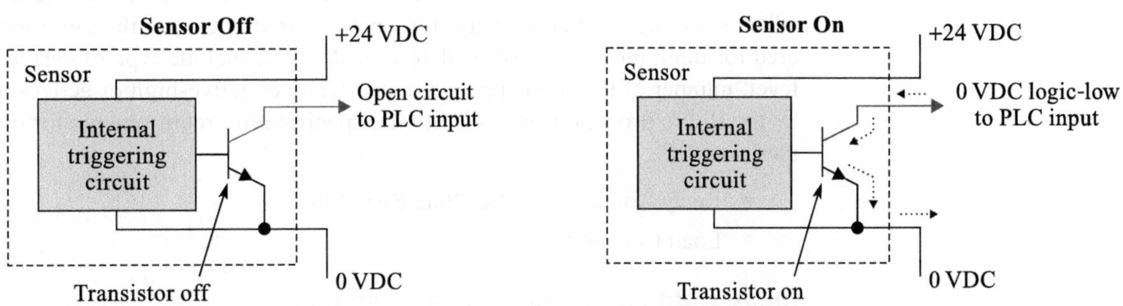

(a) (b)

FIGURE 23-18 A three-wire solid-state sensor with an NPN switching transistor

IAU1006
PLC Sinking Input
Modules

IAU1106
PLC Sourcing Input
Modules

IAU6307
Sinking/Sourcing
Input Modules

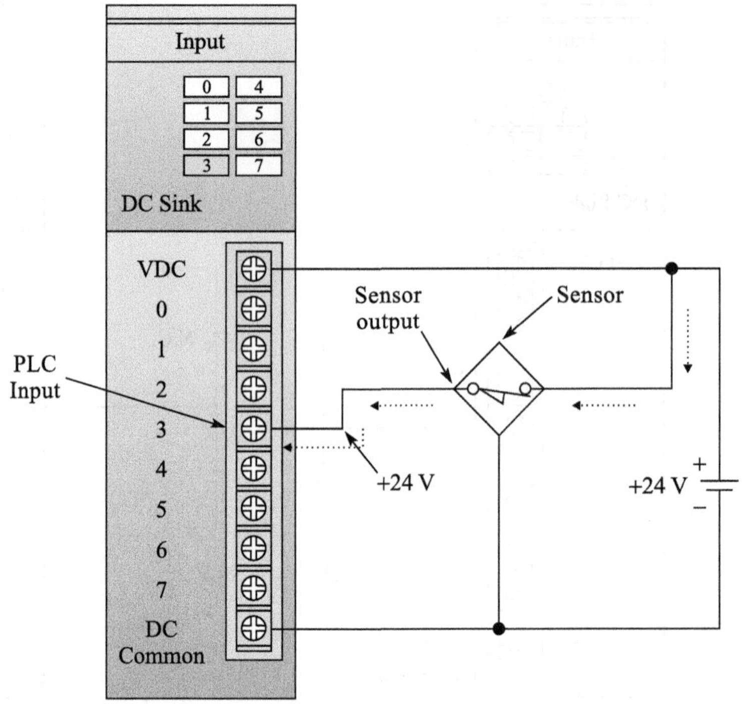

FIGURE 23-19 Sourcing electronic sensor with a PNP transistor wired to a PLC sinking input module

the sensor is activated, as shown in Figure 23-18(b), its output terminal goes low. Therefore, these sensors must be used with PLC input modules that need active-low signals.

Active-high PLC input modules are referred to as **sinking input** modules. Figure 23-19 shows a typical wiring diagram of a sensor with a switching device wired to a sinking input module. When the sensor is activated, the switch closes and produces a positive DC (logic "1" state) voltage at its output terminal. The PLC module sinks current into its input terminal when conventional current flows from the positive potential of the sourcing field device's output lead. Notice that the voltage applied to the sensor is provided by the same supply that connects to the PLC module.

Active-low PLC input modules are referred to as **sourcing input** modules. Figure 23-20 shows a typical wiring diagram of a sensor with a switching device wired to a sourcing input module. When the sensor is activated, the switching device closes and produces a low-voltage logic "1" state at its output terminal. The PLC module sources conventional current from its input terminal to the negative potential of the sinking field device's output lead.

Discrete Output Modules

There are several types of PLC discrete output modules to which field devices terminate. When selecting which one to use for a particular application, the same specifications considered for input modules must be determined. These include type of current (AC/DC), voltage level, number of terminals (outputs), and whether active-high or active-low signals are used by the PLC's processor unit. Additional specification requirements for discrete output modules are:

- Relay Contact or Solid-State Switching
- Load Capability

Relay Contact or Solid-State Switching

Output modules energize field devices by closing an internal switch to complete a circuit. These internal switches are either relay contacts or solid-state devices.

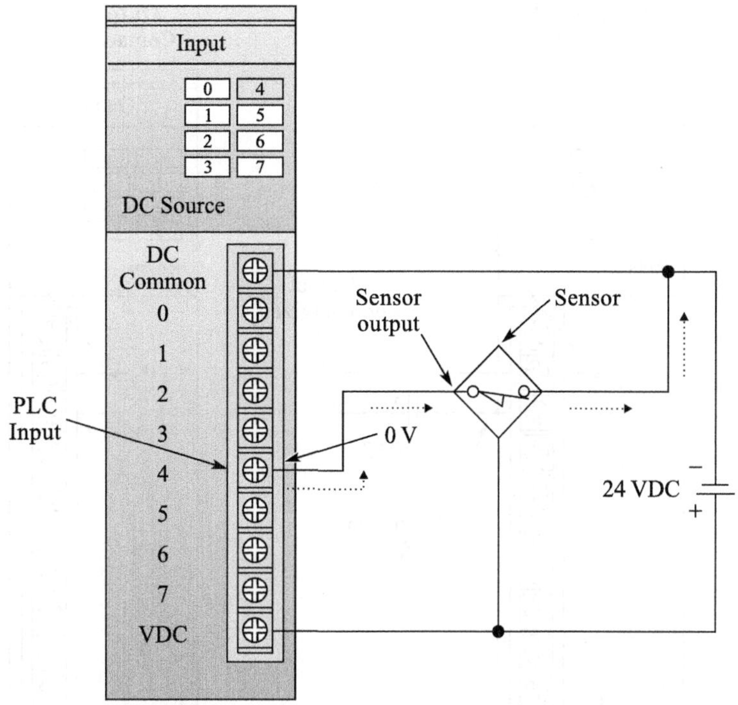

FIGURE 23-20 Sinking electronic sensor with an NPN transistor wired to a PLC sourcing input module

Relay Contact Switching The relay contact output module has physical relay contacts inside that open or close to make or break the circuits to which its terminals are connected. Figure 23-21 shows that a field device and power source are connected to a set of terminal pairs. Figure 23-21(a) shows that when the output is off, the contact is open and no current flows through the field device. Figure 23-21(b) shows the device energized and the corresponding indicator lamp on when the output is on.

Relay modules have several pairs of terminals, each set connecting to relay contacts. Relay contacts are very easy to interface because their terminals are not polarity-sensitive, they can conduct DC or AC current, and they can operate at a variety of voltage levels.

Solid-State Switching Solid-state output modules are available to energize field devices that use DC or AC current. The difference between them is the type of solid-state device they use to perform the switching action. DC solid-state output modules use transistors for switches. When the transistor turns on, it closes the output circuit path to energize the field device. When the transistor is turned off, the output circuit is open. There are two types of DC solid-state output modules. The difference between them is the type of transistor they use to switch a load: one requiring an active-high output, or one that is energized by an active-low output.

Figure 23-22 shows a module that uses a PNP transistor. It is referred to as a *current sourcing* module. When the output is activated, the transistor turns on and completes the output circuit. As the module's output terminal produces a logic level "1" state (24 volts), it supplies (sources) conventional current through the field device to the negative terminal of the power supply. When the output is not activated, the transistor is off and the field device does not energize, as shown in Figure 23-22(b). Both drawings in Figure 23-22 show that the negative lead of the power supply is connected to the common terminal of the module.

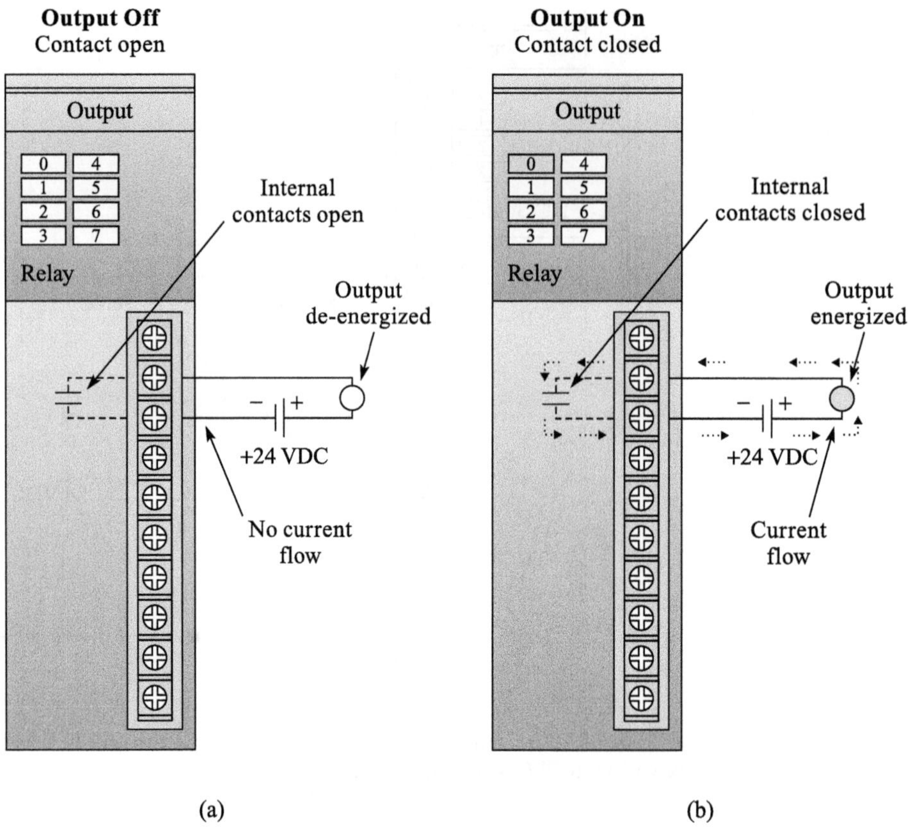

FIGURE 23-21 Wiring of a relay contact type of output module

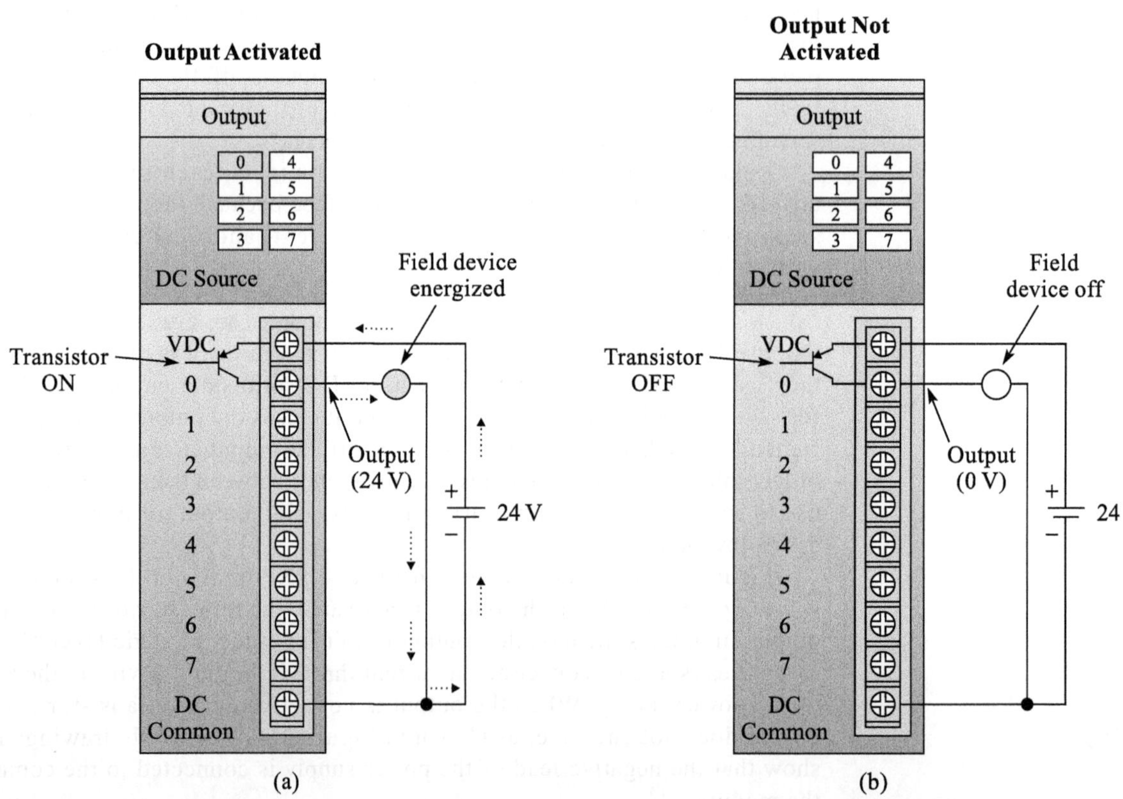

FIGURE 23-22 Wiring of a sourcing solid-state output module with an active-high operation

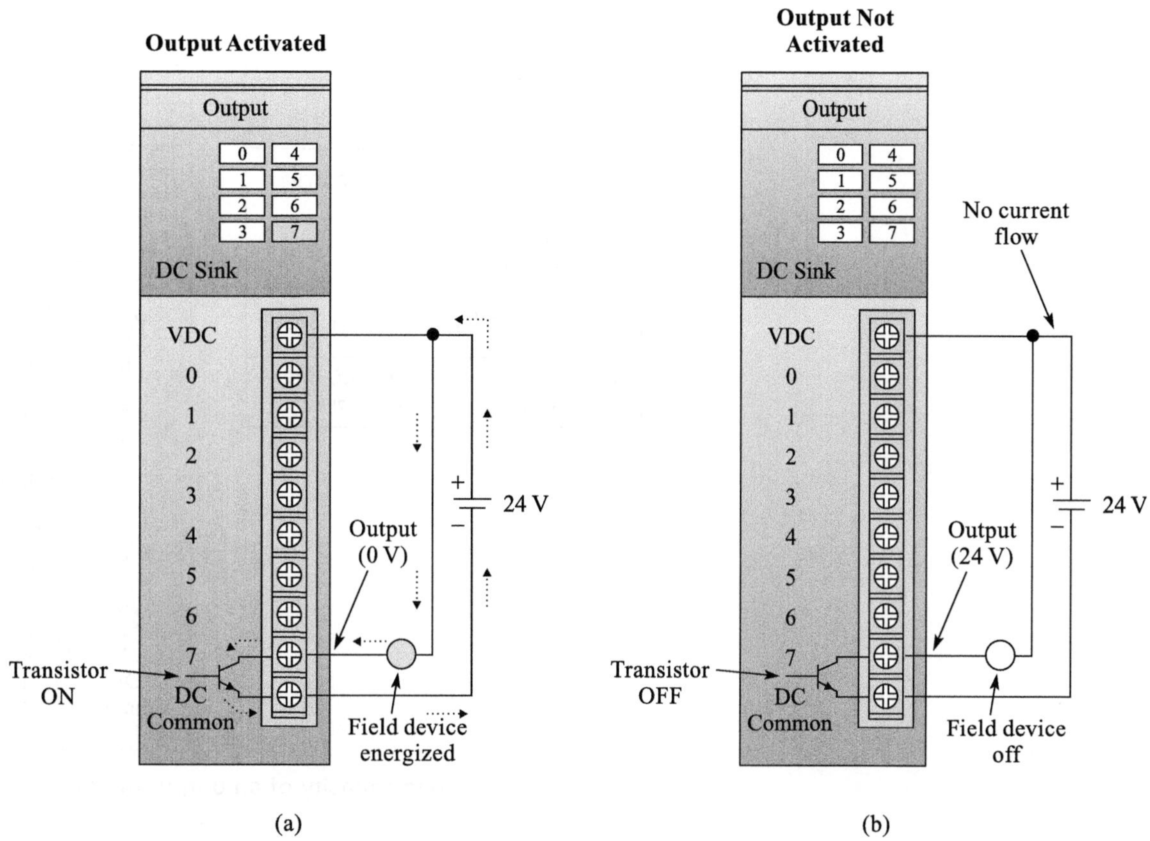

FIGURE 23-23 Wiring of a sinking output module with an active-low operation

Figure 23-23 shows a *current sinking* module that uses an NPN transistor to produce an active-low logic "1" state. When the output is activated, the transistor turns on and completes the output circuit. A low-voltage (near ground) logic "1" state is produced at the output terminal. Conventional current flows from the positive terminal of the supply through the field device, into the (sinking) output terminal, and through the module to the negative lead of the supply. When the PLC output is not activated, the output terminal is high, so current does not flow through the field device.

Solid-state AC output modules use a special type of solid-state device called a *triac*. When the output is activated, a corresponding triac turns on and passes current in both directions through the field device.

IAU906
PLC Output Modules

Load Capability

IAU6207
Sinking/Sourcing
Output Modules

The PLC's output module is capable of operating at limited current levels. Manufacturers supply a specification data sheet, similar to the one shown in Figure 23-24, that shows its current-handling capacity. It specifies the maximum amount of current that can flow through each one of its output terminals. Damage to the module will occur if these values are exceeded.

23-4 Troubleshooting I/O Interfaces

When the wiring connections between the field devices and the PLC I/O module have been completed, the next step is to determine that everything functions properly.

Discrete Interface Testing

Three steps are used to test the proper operation of discrete input interfacing.

OUTPUT MODULE SPECIFICATIONS	
Type of Current	***DC***
Level of Voltage	+10–50 VDC
Number of Outputs (Points)	8
Transistor or Relay	PNP Transistor
Load Capability per Output	1.0 A @ 30°C 0.5 A @ 60°C

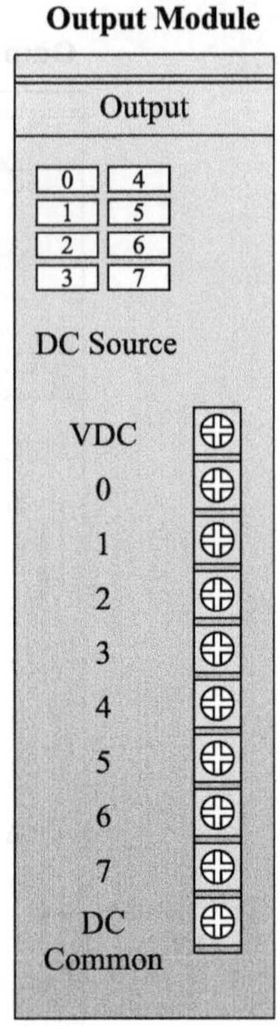

FIGURE 23-24 Load capacity of an output module

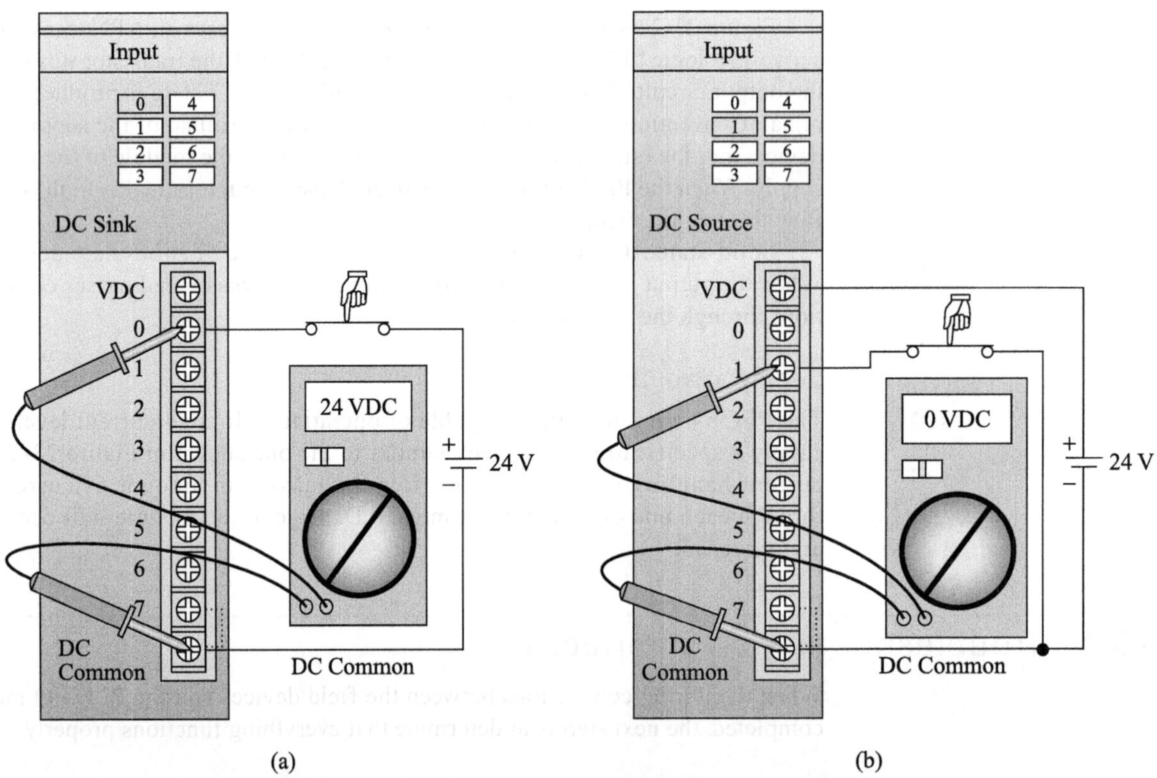

FIGURE 23-25 Meter connection for testing: (a) An active-high input module; (b) An active-low input module

Input Testing

In Figure 23-25, a simple normally open push button is connected to the input module of a PLC that uses 24 VDC.

Step 1: The first step is to observe the operation of the indicator light on the PLC input module. If the light turns on and off when the push button is pressed and released, respectively, the interface operation is correct.

Step 2: If the light does not turn on and off, a voltmeter should be used to test the signal at the input terminal of the module to which the push button is connected. When the push button is pushed, the signal at the terminal should change. Figure 23-25(a) shows the type of reading that should be observed at the terminal of an active-high input module when the push button is pressed. Figure 23-25(b) shows that 0 volts should be displayed by a meter when testing the input of an active-low module as the push button is pressed.

Step 3: If an improper reading is observed when the push button is pressed, the next step is to disconnect it from the terminal of the module and the power source. By using an ohmmeter, a continuity test can determine the condition of the N.O. push button. If the push button is operating properly, 0 ohms are displayed when the push button is pressed and infinity when it is released.

If the field device is a solid-state sensor instead of a switch, specific testing procedures are often recommended by the manufacturer. Swapping a potentially defective sensor with a replacement sensor is often the easiest way to determine its condition.

Output Testing

Step 1: The first step in testing the PLC output interface configuration is to activate the output terminal of the module. This function can be performed by running a program that controls the corresponding output. It can also be accomplished by an operation mode called *forcing,* which allows the technician to activate a specific output by addressing its location. When the output is activated, the indicator light should turn on.

Step 2: If the indicator lamp does light and the field device does not energize, connect a voltmeter to the corresponding terminal of the output module. Figure 23-26(a) should indicate a logic-high voltage when the active-high output module is turned on, and a logic-low voltage when it is off. Figure 23-26(b) should indicate a logic-low voltage level when the active-low output module is turned on, and a logic-high voltage level when it is off. If the voltmeter does not indicate the proper voltage when the output module is turned on, disconnect the output device from the module. Recheck the voltage on the module. If the proper voltage is still not present, the output module is defective. If the voltage is present, either the field device is loading the circuit and is defective, or it is wired improperly.

If the output module has N.O. relay contacts, the supply voltage should be read across its pair of terminals when the output is off, and 0 volts as the contacts close when the output is on.

Step 3: If the proper meter readings are displayed when the indicator remains off at both voltage levels, the lamp is defective. If the terminal remains in its inactive state and the lamp remains off, the program is written incorrectly, the wrong address is written into the *force* instruction, the output module is defective, or the processor unit is faulty.

23-5 Analog Input and Output Signals

Programmable controller circuits that are programmed with relay logic instructions operate by using discrete signals. The True or False output signals they produce are based on the on or off conditions applied to the input and the design of the logic circuit.

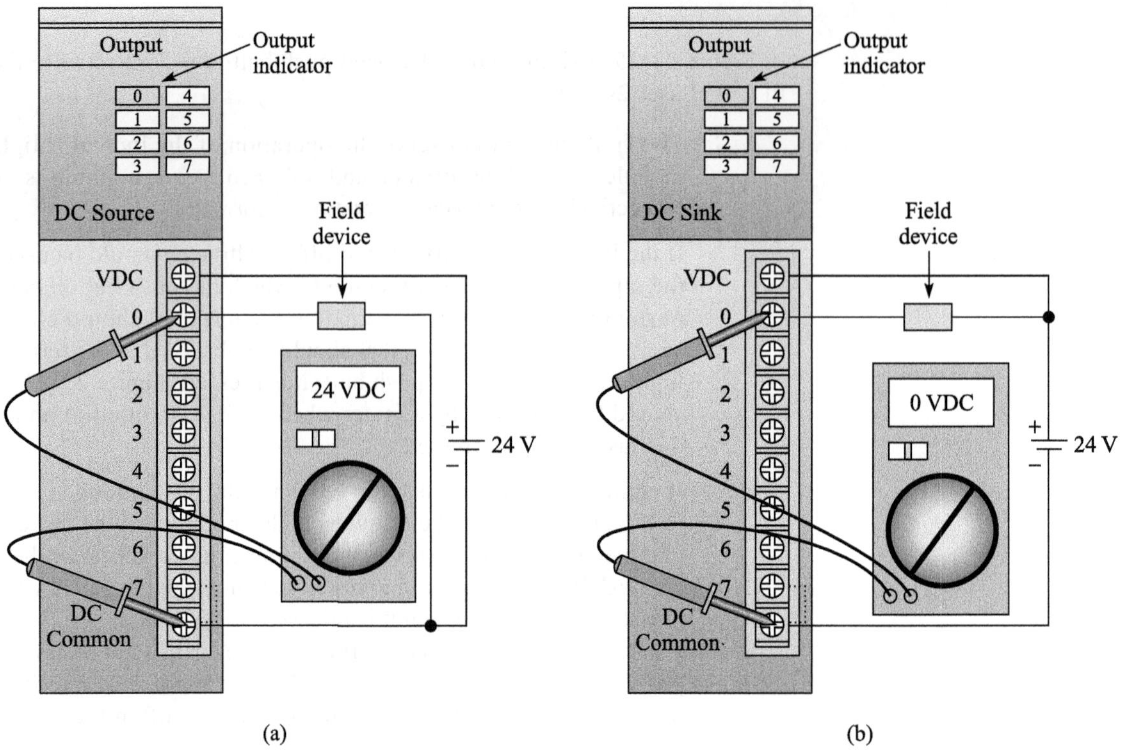

FIGURE 23-26 Meter connections for testing: (a) An active-high DC output module; (b) An active-low DC output module

In many industrial applications, especially process control, the PLC receives analog electrical signals. These signals represent a varying condition anywhere between two limits. An example is a variable voltage that indicates the liquid level of a tank. In contrast, discrete circuits usually indicate the extreme limits, such as the low level or high level of the tank. Field devices that produce analog signals include thermal indicators, pressure transducers, flow rate detectors, and other sensors. When analog output signals are produced by PLCs, they are able to control a variable process. An example is a motor speed application: The motor may be driven at 0 RPM when 0 volts are produced, at maximum speed when the largest voltage is generated, or at any speed in between as the PLC analog output voltage varies. In contrast, a discrete output would drive a motor at two extremes, 0 RPM or maximum speed.

Analog Input Module

The analog signal being monitored is fed to an **analog input module.** Its function is to convert the amplitude of the analog signal into proportional digital data in the form of a numerical value. An analog-to-digital converter similar to the type described in Chapter 2 performs this operation. Each analog module is inserted into a slot of a PLC rack.

The input module must have an electrical range compatible with the sensor or signaling device that feeds it. Standard analog input signals are:

0 to 20 mA
4 to 20 mA
0 to 5 VDC
1 to 5 VDC
0 to 10 VDC
−10 to +10 VDC

Most modules have the capability of inputting any one of these analog signals. Some modules have DIP switches for selecting the desired current or voltage range (Figure 23-27(a)). Figure 23-27(b) indicates the required settings of the DIP switches for each selection.

(a)

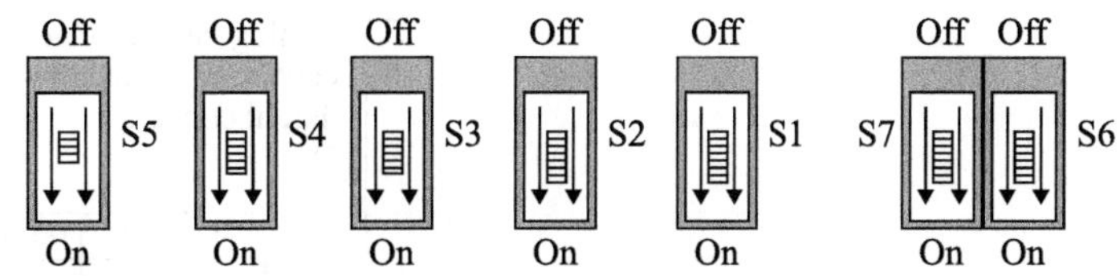

X=On
O=Off

(b)

User range			S1 to S4	S5	S6	S7
			X	X	O	O
1	to	5 V	O	X	X	O
0	to	10 V	O	O	X	X
−10	to	+10 V	O	O	O	X
4	to	20 mA	X	X	X	O

FIGURE 23-27 DIP switch settings to select desired voltage range

Most modules have the capability of receiving more than one analog signal. Figure 23-28 shows a module with four groups of three input terminals. The plus and minus terminals for the respective channels indicate the polarity of the analog input signal. The SLD terminal is provided for the shield wire of each pair of input cables. All four minus terminals share a common analog ground.

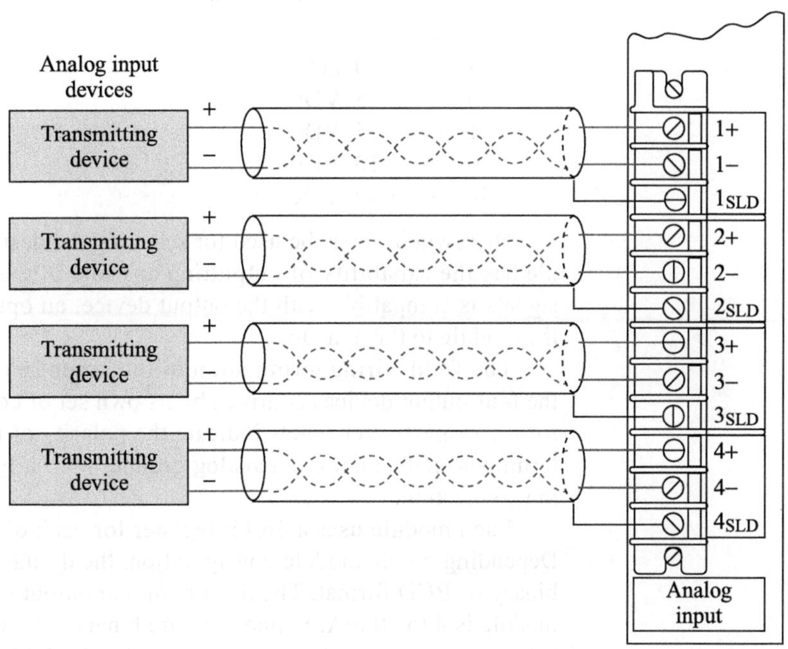

FIGURE 23-28 Field wiring terminations for an analog input module

Each module uses a 16-bit register for each of the channels, as shown in Figure 23-29. The analog current or voltage being measured is converted into pure binary format before it is loaded into the register.

The most common input signal supplied to the PLC analog module by sensors is 4 to 20 mA. Suppose that a module is programmed to convert the analog current value into a proportional 16-bit pure binary word. It would be capable of producing 2^{16} (65,536) different numbers to represent the range of 4 to 20 mA. A digital number of 1111 1111 1111 1111 would be loaded into the register by the digital-to-analog converter if 20 mA were fed to the input. If 4 mA were applied to the input, the module converts the analog signal to a number of 0000 0000 0000 0000. The number 0000 would be displayed.

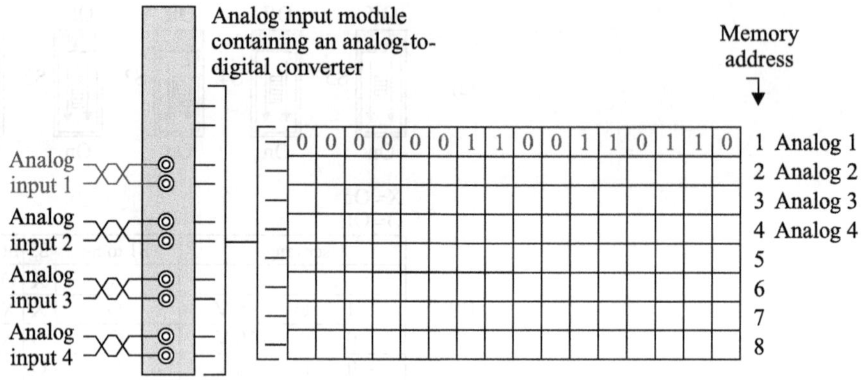

FIGURE 23-29 A binary number that represents an analog input voltage is sorted in a memory register

Analog Output Module

The digital data in a register can be converted into an equivalent analog signal by the **analog output module.** A digital-to-analog (D/A) converter similar to the type described in Chapter 2 is used by the module to transform the digital number to the analog signal. Each analog output module is inserted into a slot of the PLC rack.

The output module must have an electrical range compatible with the output device that it activates. Standard analog output signals are:

0	to	20	mA
4	to	20	mA
0	to	5	VDC
1	to	5	VDC
0	to	10	VDC
−10	to	+10	VDC

DIP switches may be used for selecting the desired current or voltage ranges, if the module has the capability of outputting any one of these analog signals. If none of these four signals is compatible with the output device, an optional amplifier may be used to interface the module to the actuator.

The field wiring termination of the output module is shown in Figure 23-30. Each of the four output devices is driven by its own set of connections. The plus and minus terminals for the respective channels indicate the polarity of the analog output signal. The four minus terminals share a common analog ground internally. The SLD terminal is used for a shielding connection.

Each module uses a 16-bit register for each of the channels, as shown in Figure 23-31. Depending on the module configuration, the digital number in the register can be in the pure binary or BCD format. The most common output signal supplied by the PLC analog output module is 4 to 20 mA. Suppose a pure binary format is used. The module would produce 2^{16} (65,536) different analog output current levels. A binary value of 1111 1111 1111 1111 would produce 20 mA. A maximum BCD value of 1001 1001 1001 1001 may also produce 20 mA, and would display 9999 on a numerical readout if groups of four binary bits were used to represent each decimal digit. A binary or BCD value of 0000 0000 0000 0000 would produce 4 mA.

Scale with Parameters (SCP) Instruction

The Allen-Bradley SLC 500 has an instruction that programs the output of analog-to-digital converters (ADC) and digital-to-analog converters (DAC) for use in a variety of applications. The instruction, shown in Figure 23-32, is referred to as a **scale-with-parameter (SCP)** output. Its function is to produce a scaled value.

The circuit-and-ladder diagram in Figure 23-32 is used to illustrate its operation when producing a scaled value for a circuit that has an analog-to-digital converter. A sensor measures pressure in a tank ranging from 0 to 1000 psi. Its analog output is connected to

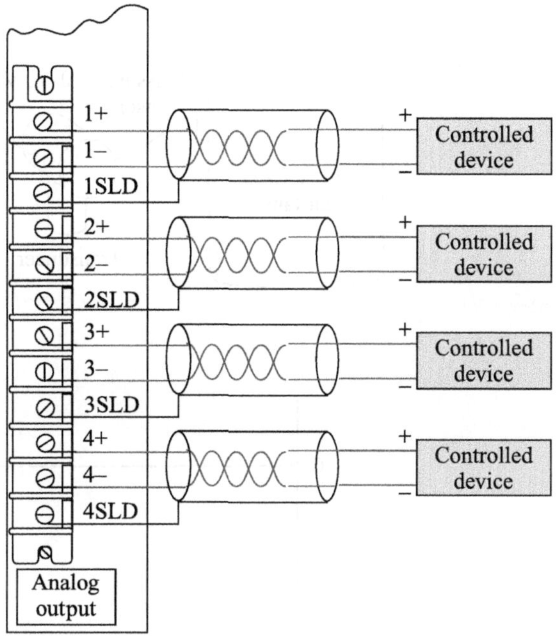

FIGURE 23-30 Field wiring terminations for an analog output module

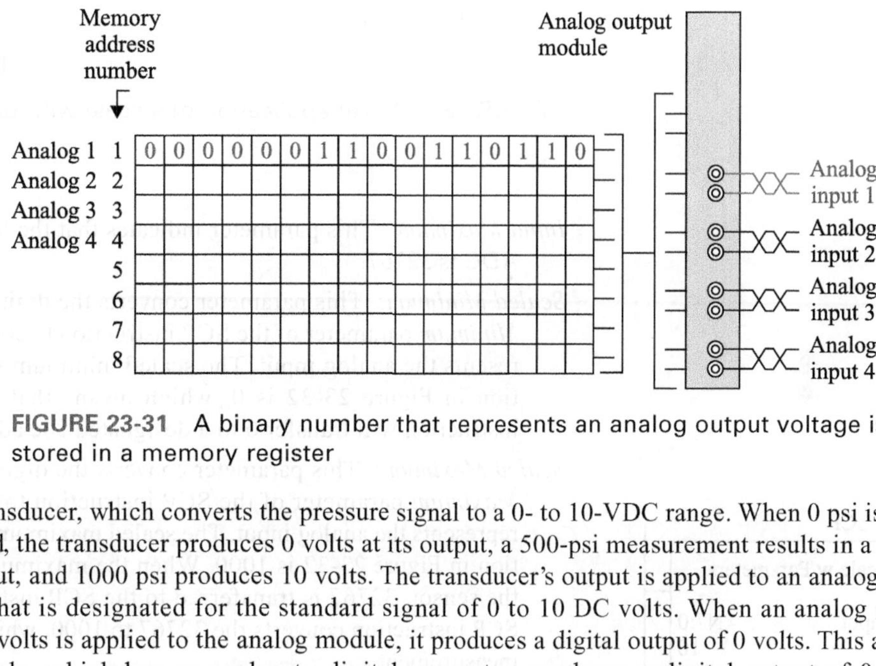

FIGURE 23-31 A binary number that represents an analog output voltage is stored in a memory register

a transducer, which converts the pressure signal to a 0- to 10-VDC range. When 0 psi is measured, the transducer produces 0 volts at its output, a 500-psi measurement results in a 5-volt output, and 1000 psi produces 10 volts. The transducer's output is applied to an analog module that is designated for the standard signal of 0 to 10 DC volts. When an analog signal of 0 volts is applied to the analog module, it produces a digital output of 0 volts. This analog module, which has an analog-to-digital converter, produces a digital output of 0 when 0 volts are applied to its analog input. When 10 volts are applied, a digital signal of 32767 is produced. The function of the SCP instruction is to scale the digital values ranging from 0 to 32767 into a range of 0 to 1000 (representing the psi measurements) to be viewed on a display module.

There are several parameters inside the SCP instruction symbol that must be programmed.

Input: This parameter indicates the memory address to which the ADC sends its digital output. In this example, the output of the analog module supplies its signal to I:6.0, which is the entire 16-bit word of register 6 in the input file.

Input Minimum: This parameter indicates that the smallest digital value produced by the ADC module is zero.

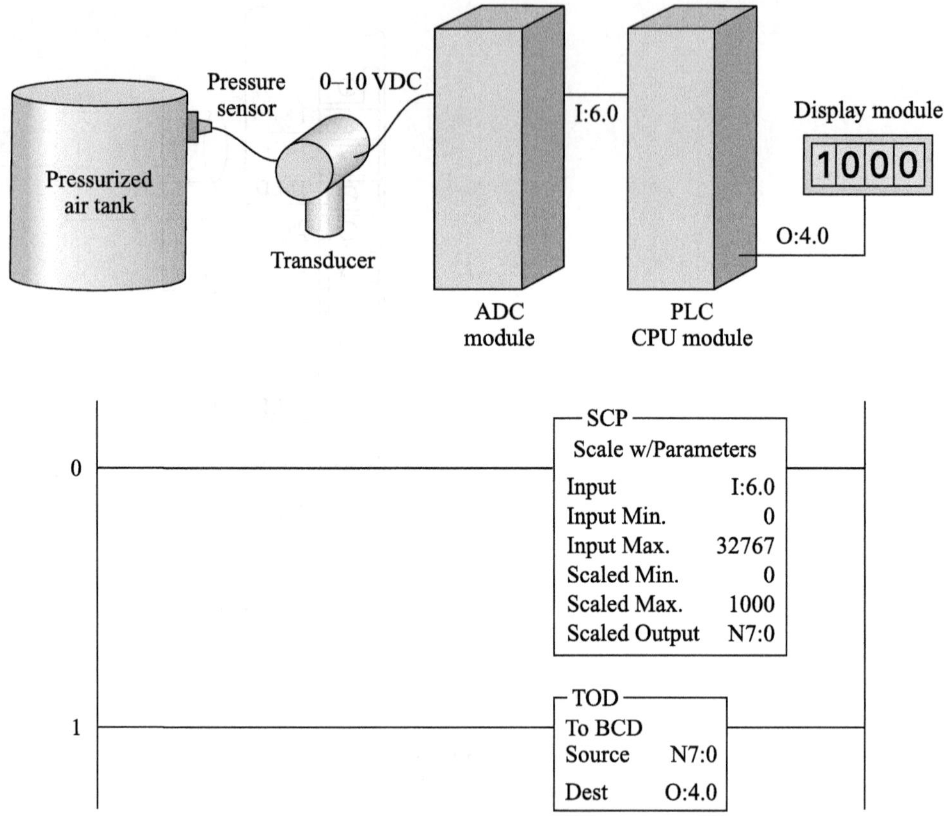

FIGURE 23-32 An application of a scale-with-parameters (SCP) instruction

Input Maximum: This parameter indicates that the largest digital value produced by the ADC is 32767.

Scaled Minimum: This parameter converts the digital value programmed into the *Input Minimum* parameter of the SCP instruction to some desired minimum value that represents the analog input. The scaled minimum value programmed into the instruction in Figure 23-32 is 0, which means that when 0 psi is measured, the SCP instruction will transfer 0 to a designated file address.

Scaled Maximum: This parameter converts the digital value programmed into the *Input Maximum* parameter of the SCP instruction to some desired maximum value that represents the analog input. The scaled maximum value programmed into the instruction in Figure 23-32 is 1000. When the maximum value of 1000 psi is measured by the sensor, 32767 is transferred to the SCP instruction from the ADC module. The SCP instruction converts the 32767 to 1000, which corresponds to the maximum psi measurement.

Scaled Output: This parameter identifies a file address to which the scaled output is sent. In this example, a 16-bit word is transferred to register 0 of integer file 7.

When the SCP instruction is used with a DAC, the parameters must be programmed, as shown by the example in Figure 23-33.

Input: This parameter is the memory address to which the analog signal of the DAC is sent. The example shows the input is N7:0.

Input Minimum: This parameter indicates the lowest analog value produced by the DAC. The example shows this analog value is zero.

Input Maximum: This parameter indicates the highest analog value produced. The example shows this analog value is 9800.

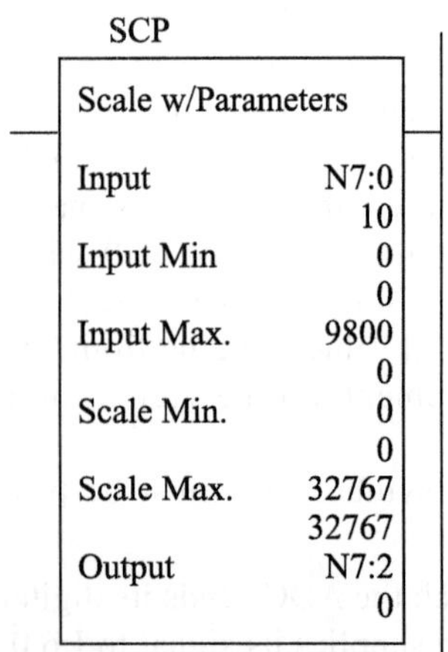

FIGURE 23-33 Scale digital input 0–9800 to 0–32767 counts

Scaled Minimum: This parameter converts the analog value programmed at the *Input Minimum* parameter of the SCP instruction to some desired minimum digital value. The example shows this value as 0.

Scaled Maximum: This parameter converts the analog value programmed at the *Input Maximum* parameter of the SCP instruction to some desired maximum digital value. The example indicates that when an analog signal of 9800 is produced by the DAC, the maximum digital value it is converted to by the SCP instruction is 32767.

Scaled Output: This parameter identifies a file address to which the scaled digital output is sent. The example shows a digital value ranging from 0 to 32767 is sent to register 2 of integer file 7.

23-6 Special-Purpose Modules

Through advancements in integrated circuit technology and software programming, PLCs have evolved into sophisticated equipment capable of controlling most industrial operations. However, some operations cannot be performed by using standard I/O modules. Therefore, special I/O modules have been developed to interface the processor unit with field devices in certain applications.

These special I/O interfacers, sometimes called **preprocessing modules,** may incorporate an on-board microprocessor that performs complete processing tasks independent of the CPU or processor scan. By performing control operations by itself, the preprocessing module enables the processor unit to operate more efficiently. Other special-purpose modules simply condition input signals, such as low-amplitude voltages or high-frequency data.

Special-purpose I/O modules are used by medium to large PLCs. The following section provides information about some of the common special-purpose I/O modules used.

Bar-Code Modules

Bar-code modules are primarily used to gather information about different types of products or parts. Bar codes were originally developed for inventory control by libraries in England to keep track of books. Soon afterward, they appeared on products in retail stores and supermarkets. Eventually, they were adopted by industry for product identification, production, and inventory control.

Additional bar-code applications include:

Item sorting on high-speed conveyor lines.
Item tracking to identify the location of a product on an assembly line.
Tool crib checking to keep track of the assignment of tools.
Product verification to determine which products are put into containers.

A bar code is a series of rectangular bars and spaces representing letters, numbers, or symbols. It is usually printed on the outside surface of the object being identified. Figure 23-34(a) shows a block diagram of a bar-code reader. To read the bar code, a light source from an optical scanner illuminates the symbol. The series of light bursts reflected back to the scanner indicates the pattern of the symbol. A photodetector converts the light bursts into an equivalent electronic signal, as shown in Figure 23-34(b). After the electronic pulses are amplified and conditioned into a digital signal, this signal is sent to a decoder. The decoder interprets the pulses as letters, numbers, or symbols, and changes them into an ASCII code. This code is processed by a microprocessor after it is received from the scanner.

There are three types of bar-code systems: *unattended, attended,* and *scanning.*

Unattended

Unattended systems are automatic scanners that do not require a human operator. As items move on an assembly line or conveyor, a light beam is reflected off an oscillating mirror.

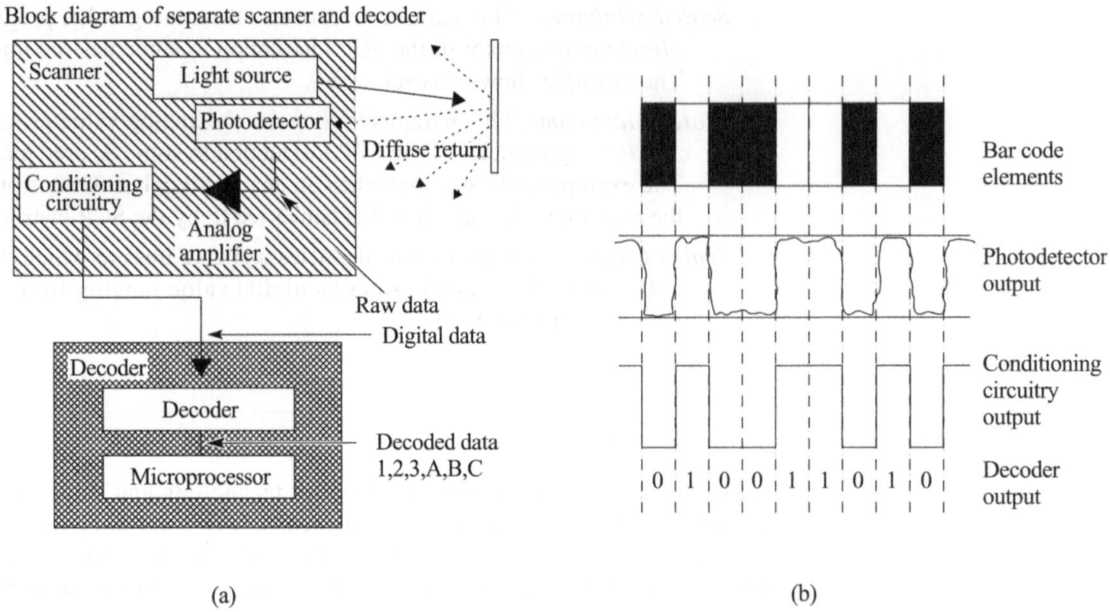

FIGURE 23-34 A bar-code system (Courtesy of Allen-Bradley)

This action causes the beam to sweep back and forth, making the beam appear as a line of light. As the symbol moves into range, the beam sweeps the entire symbol to provide the required information.

Attended

Attended systems require a human operator to trigger the scanner, to move the scanner across the symbol, or to move the symbol into the scanner depth-of-field. Attended scanners fall into two categories: stationary and portable. A stationary scanner is mounted in place and the operator moves an object marked with the symbol past the beam. A portable scanner can be moved by the operator to cause the beam to pass the label on an object.

Scanning

The scanning system uses a scanner in the form of a wand- or gun-type laser designed to be handheld and pointed at a bar code. A flexible cable connected to the input module provides the electrical interface.

Radio Frequency Module

Some programmable controllers use radio-frequency (RF) technology to perform manufacturing operations. Radio frequencies are used to pass information back and forth between the processor unit and the product. The PLC uses such information as product identification, product tracking, data collection, and product control.

To perform the required communication, small tags are attached to a product. These tags are either *active* or *passive.* An active tag contains a battery, which enables it to operate as a transmitter. The signal it sends is received by an RF module connected to the PLC. The active tag can also receive a radio signal from the RF module. The passive tag does not have a battery and is only capable of receiving radio signals. Each tag contains 100 to 2000 bytes of memory. Information about the product is stored in the tag. For example, a tag might be used on a factory assembly line that makes sofa furniture. Information is read from the tag at each station to ensure that the proper options are installed for a special order. The tag provides instructions on such items as the type of wood, the stain color, the fabric, and the hardware.

Vision System Modules

Some programmable controllers can control manufacturing operations by vision. A vision system typically consists of a camera, lighting, a vision input module, a processor unit, and a video monitor (Figure 23-35).

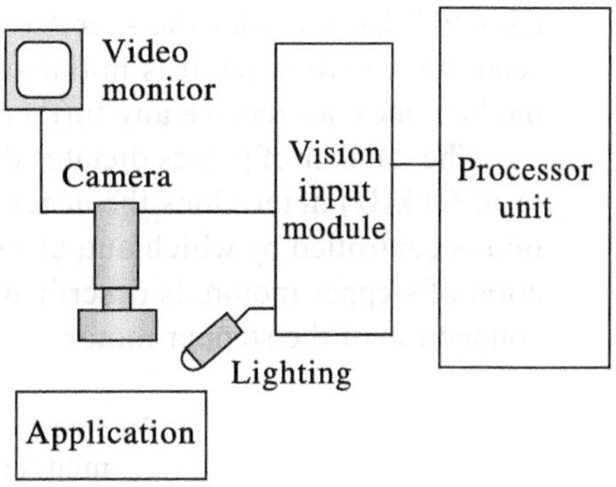

FIGURE 23-35 Block diagram of a typical vision system

Vision systems are used primarily for part-presence sensing, measurement, alignment, and other inspection functions. An example of a vision application is the placement of surface-mount integrated circuits onto a printed circuit board by a robotic arm. During machine setup, two steps must be "taught" to the vision system by the programmer. First, a good part is placed in front of the camera. The camera provides a picture of the image, which is then converted into binary data that is stored in memory. Next, the robot arm is physically moved to a precise location where the IC must be placed. This data is also entered into the vision system memory.

During the assembly operation, the vision system controller determines the part orientation by comparing it with data previously taught to the controller. If, for example, the vision controller recognizes that a lead is broken, it will reject the part. A good part is placed on the board precisely at the location that the vision system was taught.

Vision systems can inspect parts at the rate of several hundred a minute very accurately. Because bad parts are rejected before being inserted, not as many completed boards become scrap, resulting in significant cost savings.

PID Control Modules

Some programmable controllers are capable of performing proportional, integral, and derivative (PID) control. PID systems are used to perform automated closed-loop operations in both motion or process manufacturing operations. The module has an on-board microprocessor. As setpoint and feedback signals are fed into a PID module, the microprocessor performs computer software PID calculations. The appropriate output signal is sent from the PID module to the actuator that controls the desired operation. PID control is explained in Chapter 4.

Fuzzy Logic Modules

Some industrial applications require more accurate control than PID closed-loop systems can provide. Fuzzy logic is capable of providing very complex control operations that were previously impossible to perform. Some programmable controllers can perform fuzzy logic control. Setpoint, feedback, and output control signals are processed by an on-board microprocessor in the fuzzy logic module.

Stepper Motor Module

Stepper motors are used in motion-control applications where position locations need to be very precise, such as in the movement of each axis on a robotic arm.

The purpose of a stepper motor output module is to produce a pulse train that is compatible with stepper motor translators. The pulses sent to the translator control the rotation, distance, speed, and direction of the stepper motor. The module receives command data from the CPU that provides the specific values for these three parameters. The module begins to send out signals when it is initialized by a start command. Once the motor is in motion, the module does not receive any further input from the CPU.

The number of pulses dictates the distance the motor will turn. The step rate of the pulses (1 to 60 kHz) determines the acceleration, deceleration, and speed of the motor. The direction is controlled by which output line (forward or reverse) transmitted the signals. The operation of stepper motors is described in Chapter 7. Figure 23-36 illustrates a typical interface connection of the stepper motor.

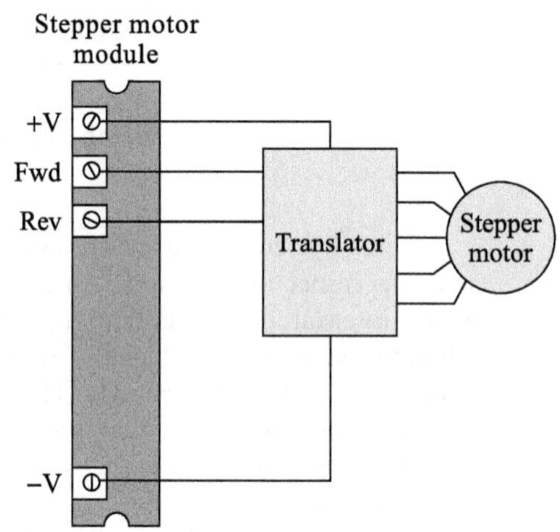

FIGURE 23-36 Typical stepper motor connection diagram

Thermocouple Modules

Thermocouples are used as measuring devices for temperature-control applications. Since they produce a very small voltage (approximately 45 mV at maximum temperature), their signals must be conditioned by an input interface module. The thermocouple input module filters, amplifies, and digitizes the signal through an A/D converter so that it can be used by the processor unit. Thermocouple devices are described in Chapter 11.

23-7 Troubleshooting Programmable Controllers

When a fault develops in a PLC installation, two methods are used to troubleshoot the system: by trial-and-error, or by a logical procedure. Trial-and-error takes too much time because it often involves arbitrarily checking connections and field devices. The best way to proceed is to observe the system in operation and determine the symptoms. Then, using these observations, logically identify the source of trouble.

There are six areas in a PLC system where a fault is likely to occur:

1. I/O field device failure
2. Incorrect wiring of the I/O field devices
3. Input module
4. CPU or power supply

5. Programming error
6. Output module

The most likely cause of a PLC problem is a programming error. These mistakes are usually the result of either loading the wrong address or an improper programming design scheme that causes the operation to perform incorrectly. This problem is more likely to occur in extremely long programs. To verify the accuracy of a program, the documentation that provides the designed ladder diagram should be compared to the diagram on the display screen.

If the circuit on the monitor is correct, the following procedure provides a logical sequence of steps that will help determine where a fault exists in one of the five remaining problem areas:

Step 1: Observe the LED indicator on the input module when someone activates a field device, such as closing a push button. If the LED does not light, the problem lies in the push button, the power supply, or incorrect wiring. The fault can be determined by taking voltmeter readings.

Step 2: If a voltage is present at the input terminal when the field device is activated, the corresponding indicator lamp should illuminate and the contact on the display screen should intensify. If the indicator lamp turns on and the display contact does not intensify, the input module probably has failed. It is a good practice to have replacement modules available.

Step 3: If the monitor shows that the input contact symbol becomes brighter, the input field device and input module are functioning properly. By energizing the input, the output symbol should also intensify if the rung is in a True condition. If the symbol does not become brighter, the CPU may be the problem. Some PLCs have a fault register that indicates an internal malfunction. The register holds a binary-based number. Each bit represents a specific type of fault, such as a memory, timer, software, or bus error. Most internal problems require repair by the manufacturer.

Step 4: If the output on the monitor intensifies and the LED in the output module does not turn on, the problem is probably in the output interface module. To verify the problem, the module should be replaced.

Step 5: If the LED indicator lamp on the output module illuminates and the output field device does not energize, the module is probably functioning properly. To determine if the field device is defective, use a voltmeter to see if a signal is present at the output module terminal. If it is, check the connections to the device to determine whether it is properly wired. If there are no incorrect wiring connections, repair or replace the field device.

The proper operation of output modules and output field devices can also be determined by using a software troubleshooting function that is referred to as *forcing.* This feature enables the technician to energize or de-energize the output and then verify the signal status with a meter. Inputs can also be turned on or off by the forcing function. Input forcing is helpful when many outputs that are controlled by one input need to be checked.

Problems

1. T/F If the scan bypasses a rung with a timer instruction due to a jump command, it is necessary to reprogram the timer.
2. T/F A subroutine is bypassed if the rung condition of a JSR instruction is False.
3. Jumping from one subroutine to another is called ________.
4. Using a jump command, it is possible to go in a ____ direction.
   a. forward
   b. backward
   c. both a and b
5. T/F A jump instruction can reduce scan time.

6. T/F The LBL input instruction with the same reference number as a JMP symbol identifies the destination of a jump operation.
7. When a forward shift takes place in a shift register, the data moves to the ________ (left, right).
8. T/F Data can be entered into a shift register only at bit 1.
9. T/F The method of cascading is limited to only two registers.
10. Data is shifted one bit position in a bit shift ____ instruction each time the rung on which it is located goes False to True.
    a. left b. right c. both a and b
11. The parameter labeled *length* in a bit shift right instruction refers to the number of ____.
    a. word files programmed into the operation
    b. bits used in an array
    c. outputs that are controlled
12. When several steps are programmed into a sequencer, the data is loaded into consecutive memory locations known as ____.
    a. module groups c. files
    b. storage blocks d. all of the above
13. T/F One advantage of sequencers is that several outputs can be activated simultaneously.
14. Which of the following is the most common method used to extend the time that an output is activated by a sequencer?
    a. Connect a gear reducer to the output.
    b. Slow the clock pulse speed of the PLC.
    c. Place a logic state that activates the output in the bit of consecutive memory locations.
    d. Program a timer delay subroutine.
15. A sequencer steps ____ when the rung on which it is loaded goes False to True.
    a. to the next register
    b. through all of the registers programmed with a sequence of data
16. A/n ____ -driven sequencer instruction increments after a preset time period.
    a. event b. time
17. If 00FF is programmed into the Mask parameter of a sequencer instruction, the name of the eighth ____ significant bit from the file parameter will be transferred to the destination address.
    a. least b. most
18. List four specifications that must be considered when selecting the proper input interface module.
19. T/F Some PLCs recognize a "1" state logic signal as a low-level voltage (near ground).
20. Referring to I/O PLC modules, the direction of sourcing and sinking current is based on ____ current flow.
    a. electron b. conventional
21. A three-wire solid-state sensor with a PNP transistor is wired to a ____ input module.
    a. sourcing b. sinking
22. A ________ (low, high)-level signal applied to a sourcing DC input module is recognized as a logic "1" state by the PLC.
23. A sinking solid-state PLC output module produces an active-____ signal to energize a field device.
    a. low b. high
24. A PLC that operates by observing high-level voltages as logic "1" states uses a ________ (sourcing, sinking) output module.
25. A forcing operation is used to test an ____ module.
    a. input c. both a and b
    b. output
26. List four standard signals that are commonly used by analog I/O modules.
27. How many different pure binary number combinations can the 16-bit register in an analog I/O module use to represent analog values?
28. The output of an analog-to-digital converter module is directly loaded to an ____ file.
    a. integer b. input
29. T/F The function of the *Scaled Max* parameter of an SCP instruction is to convert the number programmed in the *Input Maximum* parameter to some desired maximum value that represents the analog input.
30. The instruction labeled ________, located on the last rung of a subroutine file, tells the processor to return to the rung following the JSR instruction that initiated the subroutine.
31. Which of the following function statements describes the reason for using specialty modules?
    a. Signals are conditioned to a usable form.
    b. They enable the processor unit to operate more efficiently.
    c. They enable small PLCs to become more powerful.
    d. They can perform control functions independent of the processor unit.
32. Place the appropriate letter next to the following specialty modules to indicate which type of interface function each performs.
    a. Input c. Input and Output
    b. Output

    ____ Thermocouple
    ____ Vision
    ____ Bar Code
    ____ Radio Frequency
    ____ Stepper Motor
    ____ PID
    ____ Fuzzy Logic

SECTION

8

Motion Control

OUTLINE

As industries compete in the world market, they must strive to increase production with lower costs. To be competitive, many manufacturing companies are using motion-control automated equipment to make their products. In a motion system, the physical movement or position of an object is controlled. One example is an industrial robot that performs welding operations and assembly procedures. As technology advances, this equipment produces products more quickly and with higher quality. To enable these systems to operate effectively, the technicians who install, repair, and calibrate them must understand their operation. Motion-control systems are also known as *servo systems*. The term servo, as in servomechanisms, comes from the Latin word *servus* meaning slave. The mechanism is the slave in the motion system because it is driven by a controller as it reacts to an error signal produced by the feedback signal from the sensor.

There are three characteristics that are common to motion-control systems. First, the position, velocity (speed), and acceleration or deceleration of an object are controlled. Second, the motion or position of the object being controlled is monitored by some measurement device. Third, the response time of motion devices to command-signal changes or disturbances is typically in fractions of a second, rather than in seconds or minutes, as in process control. Also, many of the terms used and the characteristics of a process-control system are also used in servo systems. As an example, the terminology and the operation of open-loop and closed-loop control are used by both.

This section of *Industrial Automated Systems* provides information about motion control equipment. Chapter 24 describes the operation of each element in a closed loop and the parameters controlled in a motion system. Chapter 25 explains the operation of feedback devices that measure the motion of an output actuator or the load it is moving. Chapter 26 discusses different types of motion-control systems and control loops used for various application. Chapter 28, which is on the disk located on the inside cover of this textbook, provides examples of how concepts covered in chapters throughout the text are used in practical industrial applications.

CHAPTER

24

Elements of Motion Control

OBJECTIVES

At the conclusion of this chapter, you should be able to:

- Explain the terms *servo* and *servomechanisms.*
- Describe the difference between open-loop and closed-loop motion-control systems.
- Describe the four motion-control parameters.
- Identify the elements of a closed-loop motion-control system.
- List and describe the operation of devices and equipment used to perform the function of each motion-control element.
- Define the following terms associated with motion control:
  - Following Error
  - Traverse Rate
  - Loop Gain
  - Indexing
  - Damping
  - Position
  - Tracking
  - Backlash
  - Velocity
  - Home Position
  - Bandwidth
  - Acceleration/Deceleration
  - End Point
  - Holding Torque
  - Torque
- Identify six different mechanical systems that transmit force from a rotary actuator to a load, and describe their operation.
- List and explain different types of position movements used in motion-control applications.

INTRODUCTION

Motion control is the process whereby a system converts input commands into controlled mechanical movements. The movements can be linear, rotary, or a combination of the two. Both open-loop and closed-loop systems produce these movements. The commands can be provided by mechanical cams or gears, a simple potentiometer, or a complex computer.

The purpose of this chapter is to describe the mechanical movements, the power transmission, and the parameters controlled in a motion system. It also identifies each element of a closed-loop motion-control system and explains its function. Many types of equipment and devices covered earlier in the book are used by the elements. Brief explanations of their operation and how they are used by each element are included.

24-1 Open-Loop and Closed-Loop Servo Systems

The term *servo,* or *servomechanism,* does not apply to a particular motion device, such as a specialized motor. Instead, it refers to a motion-control operation. There are two categories of motion-control systems, open-loop and closed-loop. The closed-loop system performs the operation of a servomechanism.

Open-Loop Systems

The block diagram of an open-loop system is shown in Figure 24-1. It consists of three basic elements common to all such systems: a *controller,* an *amplifier,* and an *actuator.* The system also may have an add-on operator interface device, such as a computer, that provides a way to enter information that indicates the desired movement of the load.

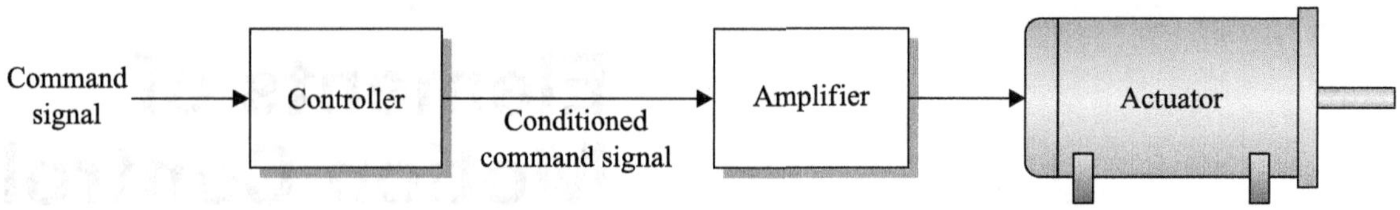

FIGURE 24-1 Block diagram of an open-loop system

In an open-loop system, there is only one signal used, which is the command that goes from the controller to the actuator. There is no signal from the actuator or the load to inform the controller that motion has occurred. Therefore, the open-loop system is unable to correct for deviations of the desired motion.

A stepper motor is often used in an open-loop system. Each pulse from the controller moves the stepper motor one increment, representing a certain number of degrees. Figure 24-2 shows a pick-and-place application using a stepper motor drive. The function of the system is to automatically place parts into bins A, B, and C. When the controller triggers 10 pulses, the rotation of the motor shaft will place a part into bin A; 20 pulses cause a shaft rotation and the placement of a part in bin B, and 30 pulses for a part placement in bin C. If for some reason the stepper does not move, for example due to jamming, the controller is unaware of the problem.

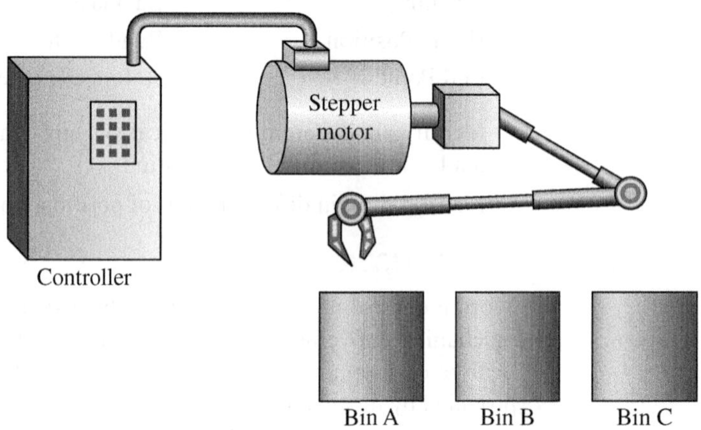

FIGURE 24-2 An open-loop pick-and-place application

The primary advantage of open-loop systems is that they are simple to wire and are less expensive than more complex closed-loop systems. The weakness of an open-loop system is that it has no way of monitoring the load.

Open-loop systems are ideal for applications requiring high torque; low-speed, continuous operations; or short, rapid, repetitive movements produced by stepper motors.

Closed-Loop Systems

For many applications, a more precise level of control is required than that of an open-loop system. By adding a device that measures the condition of the load, the system can adjust itself when a disturbance alters the operation. The correction is made by sending an output

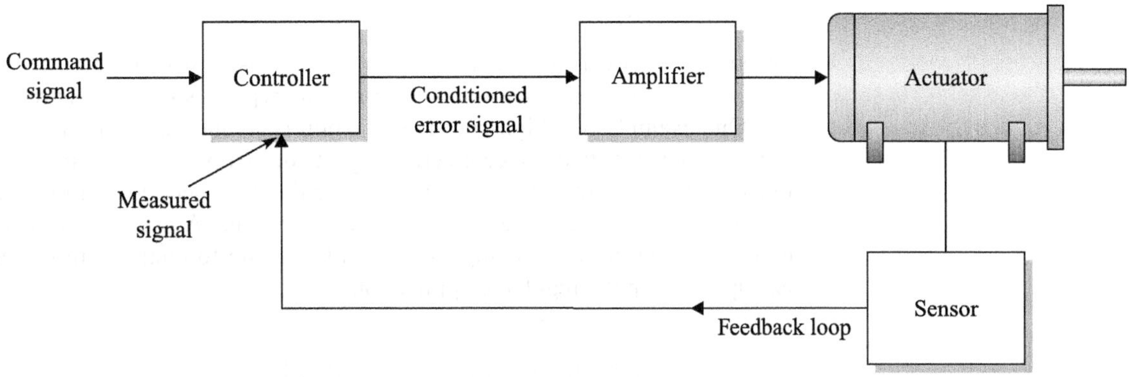

FIGURE 24-3 Block diagram of a closed-loop system

signal from the measuring device to the controller, as shown is Figure 24-3. The connection between the measurement device and the controller is called the *feedback loop*. By comparing the measured signal and the command signal, the controller produces a low-power error signal that is applied to the amplifier. These low-power signals must be amplified because high voltages and current are required to drive the actuator at appropriate speeds and torque when moving heavier loads. If the load condition is altered due to a new command signal or when a disturbance occurs, the feedback device will produce a signal that indicates the change. The result is that the controller will produce a different error signal, which causes the actuator to automatically make the proper adjustment to reach the desired condition. The feedback correction makes the system a closed-loop configuration that performs the operation of a servomechanism.

The primary difference between an open-loop and a closed-loop system is the signal produced by the controller. In the open-loop system, the command signal, which is processed and conditioned by the controller, is used. This comparison occurs continuously. The error signal either increases or decreases, depending on the difference between the feedback and command signals. As it does, the actuator causes the system to reduce the error until an equilibrium point is reached.

24-2 Motion-Control Parameters

The purpose of a motion-control system is to control any one, or any combination, of the following parameters:

Position
Velocity
Acceleration/Deceleration
Torque

The parameters that are controlled depend on the performance requirements of the system.

Position Control

This parameter controls the motion displacement of an object to a specific location. This control method is achieved by both open- and closed-loop systems.

In a closed-loop system, the load is moved from one known, fixed position to another known, fixed position. A key element to this method of control is feedback. In an open-loop system, position is controlled by an indexer, which provides a specific number of pulses to a stepper motor. The stepper motor moves in precise, small, incremental steps equal to the number of pulses it receives.

Velocity Control

Speed, or velocity, is the quantity of motion that takes place with respect to time. The particular manufacturing application determines the type of speed that is required to perform the job. One example of a high-speed application is an automated insertion machine that makes traverse movements to place electronic parts onto a printed circuit board. In some applications, it is important to have good speed regulation. Speed regulation is the ability to maintain a constant speed under varying load conditions. A machine tool spindle that requires constant speed under varying load conditions due to changes in the cutting action is an example of a speed-regulated application.

Acceleration/Deceleration Control

Acceleration and deceleration are changes in speed with respect to time. Their rates are affected by inertia, friction, and gravity.

Inertia. Inertia is a measure of an object's opposition to a change in velocity. It is also a function of an object's mass. The larger the inertial load, the longer it takes to accelerate and decelerate.

Friction. Friction is the resistance to motion caused by surfaces rubbing together. Mechanical parts of a machine and the loads that they drive exhibit some frictional forces. Friction is caused by factors such as bearing drag, sliding friction, and system wear in the vertical direction.

Gravity. The gravitational force on an object affects its acceleration and deceleration. In horizontal movement applications, gravitational force has no effect. It only applies to vertical movements.

Torque Control

Force is a quantity that causes motion. Force can exist whether it is moving an object or not. When measuring motion produced by a linear actuator, force as a unit of measurement is referred to as ounces or pounds.

Torque is a twisting type of force that causes an object to rotate. It is used as a unit of measurement for rotary actuation, such as with electric or hydraulic motors. The English unit of measurement is inch-ounces or foot-pounds. In design applications, torque calculations must be made to determine if the size of a motor actuator provides enough acceleration torque, constant velocity torque, or deceleration torque to drive the load. Torque is also the force that can prevent an object from moving by holding it in a fixed position.

24-3 Motion-Control Elements

Figure 24-4 shows the block diagram of a closed-loop motion-control system. Each block represents an element that performs a specific function. There are many different types of devices available to perform the functions of each element. Most of these devices have been described in previous chapters. This chapter explains the operation of a motion-control system, its elements, and the role the devices play in the overall scheme.

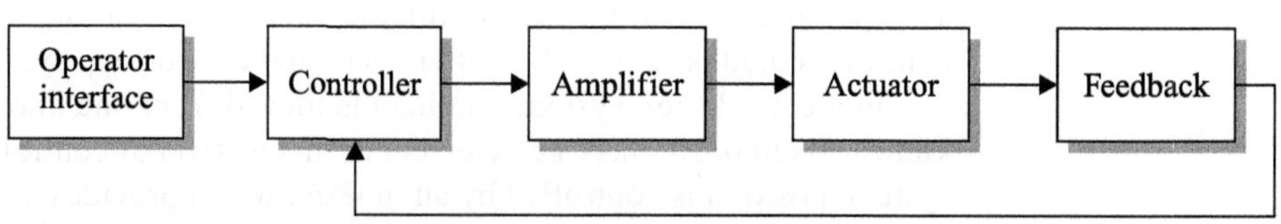

FIGURE 24-4 Elements of a motion-control system

Operator Interface

The primary function of the operator interface element is to provide the command signal that indicates the desired condition of the parameters being controlled. It is used to communicate with the other elements of a motion system in two ways.

The operator interface is used to communicate with the other elements of a motion-control system. Interfaces such as keyboard terminals, handheld keypads, thumbwheels, and switch panels are used for data input. Display terminals such as CRT screens, LCD displays, and indicator lamps provide a way to visually read the programs, messages, or other information useful for the programmer.

Controller

The controller is the "brain" of the system. It consists of memory where data instructions from the user are stored and microprocessor-based circuitry that processes information. When activated, the controller translates the instructions in the program memory into a series of analog or digital signals that are fed to the amplifier. In a closed-loop system, it compares the measured feedback signal with the command signal. By processing this data, either electronically or through mathematical computations using software, it produces an output that causes the actuator to produce the desired motion.

Amplifier

The amplifier receives signals from the controller and converts them into power sufficient for the actuator to drive the load.

Actuator

The actuator receives power from the amplifier. It provides the "muscle" required to drive the load so that the desired physical motion is achieved.

Feedback

The feedback element closes the loop in a control system. It provides information about the condition of the mechanical load under control. The information is in the form of an analog signal or digital data.

This feedback information is compared by the controller with the command signal to determine if adjustments should be made to maintain the load within preset parameters.

24-4 Terminology

Before describing each block of the motion-control system in more detail, an understanding of the following terms is necessary.

Following error is the positional error during motion resulting from the load movement's lagging behind the desired movement specified by the command signal.

Tracking is the movement of the load by the actuator as it attempts to follow a changing command signal in a motion-control position system.

Home position is a reference position from which movements are measured in a motion-control system.

End point is the desired location to which a load is moved by a motion-control position system.

Traverse rate is a fast rate of speed at which the load is moved from one position to another in a point-to-point position servo system.

Damping is the prevention of overshoot of the load past the end point in a motion-control system. Damping is achieved by reducing the overall gain of the system to slow the movement. The greater the damping, the slower the movement.

Backlash is the movement within the space between mating parts or gears. Normally, gears need some space to accommodate lubrication. However, it is desirable to have minimal backlash; otherwise, positioning errors may occur.

Bandwidth is the measure of how quickly the controlled quantity tracks and responds to the command signal. The greater the bandwidth, the more rapidly the system responds.

Holding torque is the peak force that can be applied to the shaft of an energized motor at standstill without causing it to move. Also known as *static torque,* it is the result of the interaction between magnetic fields of the stator and rotor.

Loop gain is a measurement value based on the ratio of output speed to the following error and is expressed in inches per minute per mil of error. A mil equals 0.001 inches. The loop gain is improved by increasing the gain of the proportional loop of the system.

24-5 Operator Interface Block

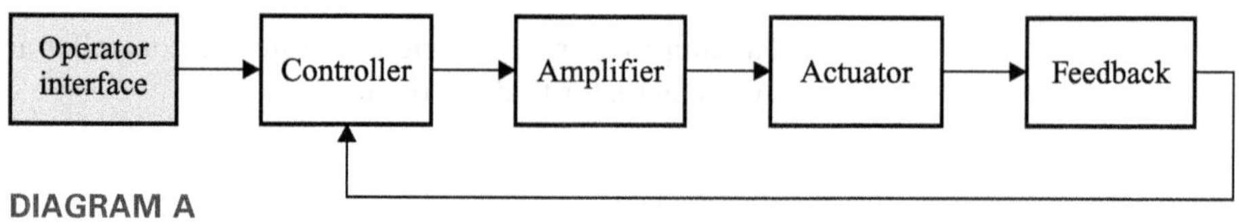

DIAGRAM A

The operator interface block, highlighted in Diagram A, is the equipment used by the operator to communicate with the motion-control system. An input device provides a means for the user to enter information. Data can be entered from a keyboard or a thumbwheel switch, or downloaded from a host computer. The information entered forms the control parameters for the command signals that cause the system to make the desired motion. The terminal also includes a monitor screen unit. It enables the user to examine and program various types of information. The main parameter menu in Figure 24-5(a) provides an easy way to program the parameter values. The parameters include:

- The direction, distance, and any other information that specifies the end-point position.
- The velocity data that instructs the system how rapidly to get to the desired location.

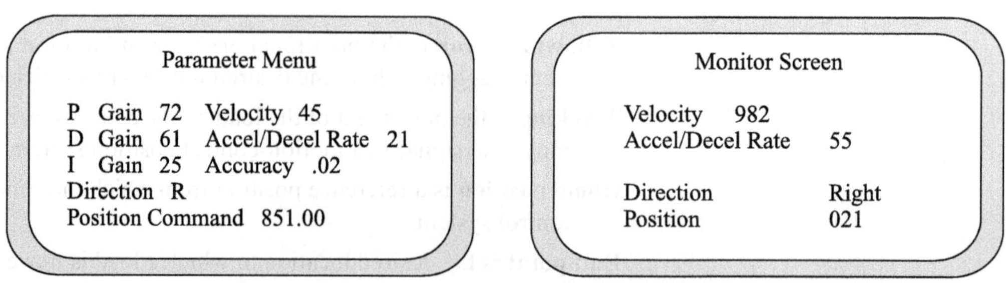

(a) (b)

FIGURE 24-5 Operator interface screen

- The gain setting for the derivative function that controls the acceleration and deceleration rates when starting, stopping, or deviating from a desired speed.
- The gain setting for the integral function that controls the tolerance for positioning and speed.

The status display in Figure 24-5(b) enables the operator to examine various drive parameters at any time while the system is operating. The screen is constantly updated.

24-6 Controller Block

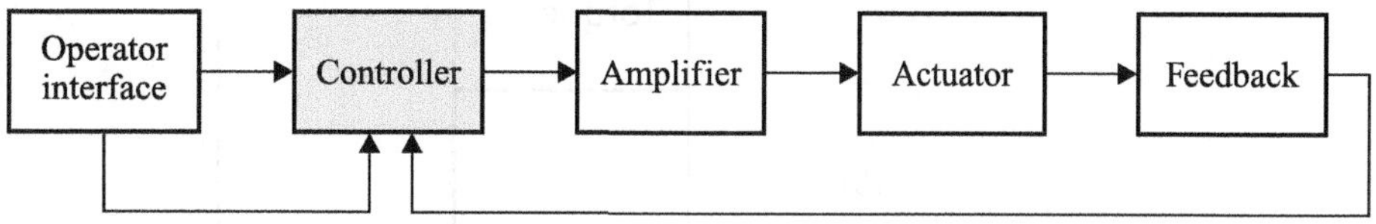

DIAGRAM B

The controller, highlighted in Diagram B, is the brain of the system. Its primary function is to produce electrical signals, either as digital pulses or as analog voltages, that cause the actuator to perform its desired operation. The desired motion profile the controller produces is the result of collecting information like command signals, feedback signals, parameter adjustments (such as gain settings), and other data. This information is processed by the controller before it sends the proper output voltages to the amplifier. The processing operation can be performed by analog electronic circuitry or by digital equipment.

A typical analog motion controller requires the adjustment tuning of many potentiometers by trained technicians during installation. Because of component aging, these circuits must be retuned periodically. A digital motion controller requires tuning by inputting calibration data from a keyboard or by downloading from a computer. The computer software processes command signals, feedback signals, and parameter settings to produce the desired output. A typical motion profile generated by the output is shown in Figure 24-6. The velocity profile in Figure 24-6(a) resembles a trapezoid. It is characterized by constant acceleration, constant velocity, and constant deceleration. The torque profile in Figure 24-6(b) shows that a positive force causes acceleration, a small force causes constant speed, and a negative force causes deceleration. A small torque exists to hold the load in a fixed position. Figure 24-6(c) shows the resultant positioning of the object being driven. The parameters established at setup and the size of the load determine the profile.

The profile parameters are normally programmed during setup. However, they can be established during normal operation or changed under emergency conditions. The profile can be maintained by derivative and integral control, or adjusted by fuzzy logic control as load conditions change.

There are various types of motion controllers available. One type is a dedicated controller that is very specialized and designed to perform a specific task. These controllers perform operations such as computerized numerical control, welding, and laser cutting. Another type of controller system, which provides signals to cause positioning movements, is called an **indexer.** The indexer produces a series of pulses to a stepper drive and motor. Each pulse causes the shaft of the stepper motor to turn. Depending on the stepper motor, the rotation ranges from a fraction of a degree to several degrees. These controlled movements are referred to as **indexing.** Depending on the design and size of the stepper motor used, it can deliver from 200 to 100,000 steps per revolution. One specialized controller is a unit that is in chip form or constructed on a single printed circuit board. These controllers are often placed inside the machine enclosure of the equipment they operate. General-purpose controllers also exist. These higher-level units are very versatile. They are capable of coordinating

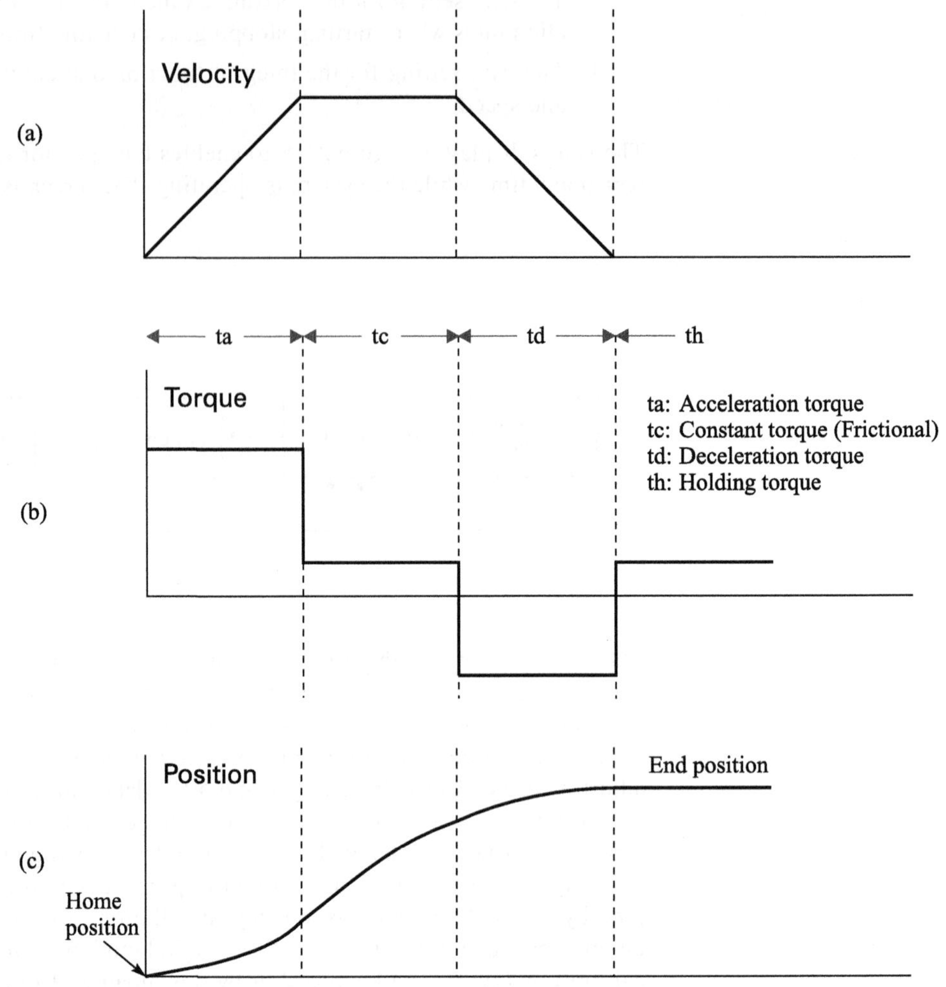

FIGURE 24-6 Typical motion profile

several control operations simultaneously, performing high-speed mathematical calculations, and communicating with other controllers. The operations they are capable of performing include X-Y positioning, palletizing, and computer-integrated manufacturing (CIM). These units are also able to manage faults in the system by shutting down power or activating an alarm if a problem develops. Programmable controllers and computers are considered general-purpose controllers.

24-7 Amplifier Block

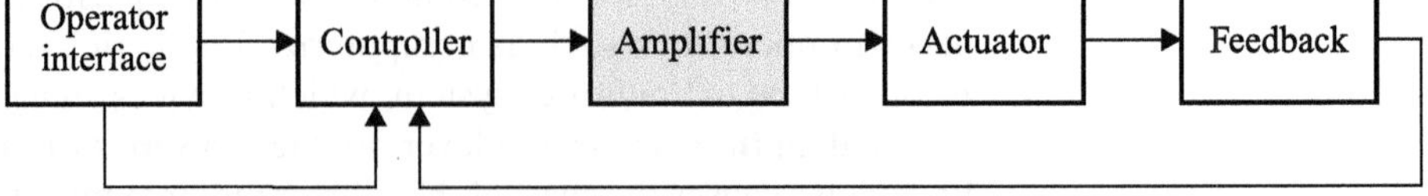

DIAGRAM C

The amplifier, highlighted in Diagram C, converts the low-level output signal from the controller to a level of electrical power sufficient to drive the actuator. In most motion applications, the actuator is a motor. The selection of an amplifier is based on how closely the characteristics of its outputs match the voltage and current requirements of the actuator.

The most common types of amplifiers include:

DC Servo Amplifier. A DC servo amplifier is a linear amplifier that is capable of producing a bidirectional DC voltage for powering a brush-type DC motor. When it outputs a positive voltage, the motor rotates in one direction. A negative voltage causes the motor to rotate in the opposite direction. DC amplifiers may consist of a simple circuit with a power transistor or power operational amplifier, or a DC drive that consists of complex circuitry.

Brushless DC Amplifier. A brushless DC amplifier consists of electromagnetic stator coils that are arranged in a cylindrical shape. Its rotor is made of a permanent magnet that is free to turn inside the stator. By switching the DC power applied to the coils on and off sequentially, a rotating magnetic field is created. The rotor turns because of its magnetic interaction with the rotating stator field.

AC Servo Amplifier. The AC servo amplifier is used to supply power to a single-phase squirrel-cage induction motor called an *AC servo motor*. It provides AC power to two sets of stator coils, the main winding and the auxiliary winding. The power from an AC line is fed directly to the main winding. Power to the auxiliary winding is 90 degrees out of phase from the main power and is produced by an amplifier circuit. Its amplitude is determined by the error signal that results from the difference between a command signal and the feedback signal. By varying the amplitude to the auxiliary winding, the speed and position of the rotor can be controlled.

AC Inverters. AC inverters, also known as *AC drives*, are used to drive single-phase and three-phase induction motors. They control the speed of the motor by changing the frequency and amplitude of the simulated AC waveforms they supply. The two most popular AC drives are PWM and vector drives. Both drives use a pulse-width modulation amplifier that produces a simulated AC wave to drive AC induction motors. Microprocessor-based circuits produce small signals that switch the amplifier's output power on and off to produce squarewave signals. As the duty cycle of the resulting square waves is changed by the switching signals, the output frequency and voltages can be varied to turn the motor at the desired speed and torque and to the desired position.

Stepper Motor Amplifiers. Stepper motors are position motors that are energized by electrical pulses. As the pulses are applied to a series of coils that are positioned in a cylindrical pattern, a permanent magnet rotor is forced to turn. Each incoming pulse causes the rotor to turn a specified angular distance. The rate at which the pulses are applied controls the rotor velocity. The pulses are commonly supplied by programmable controller modules or other peripheral devices that interface digital equipment with the motor. The pulses originate from a microprocessor-based device. Since the pulses are low-voltage, they are used to switch amplifier transistors on and off. The output of each transistor is connected to a stator coil in the stepper motor. Stepper motor amplifiers are also referred to as *indexers*.

24-8 Actuator Block

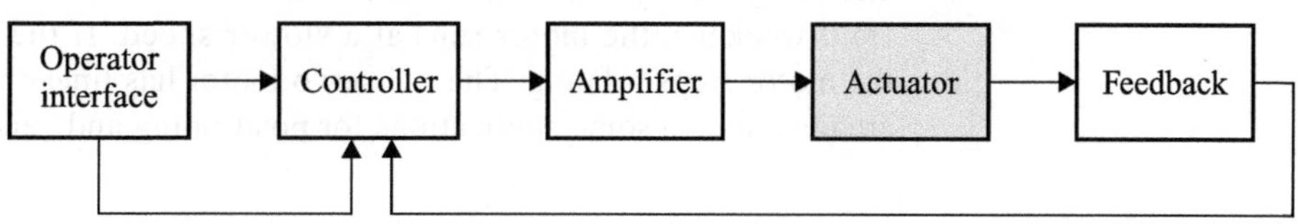

DIAGRAM D

The actuator, highlighted in Diagram D, provides the actual physical linear or rotary motion. Motion-control actuators are usually electric-powered motors, hydraulic motors, or hydraulic cylinders. The motor's rotary force is transmitted by the actuator transmission system.

Electric Motors

There are many types of electric motors used to drive the load in motion-control applications. The most common types are brush-type DC motors, brushless DC motors, AC servo motors, induction motors, stepper motors, and hydraulic actuators.

Brush-Type DC Motors

The most widely used actuator in servo applications is the DC motor that uses brushes. Among its advantages are proven performance, availability at a variety of specification ranges, and favorable cost. Its primary disadvantage is its higher maintenance requirement due to brush wear.

Its application characteristics are:

- Low-to-medium speed
- Light-to-heavy loads
- Brute-force On-Off control
- Low-to-medium acceleration/deceleration

Brushless DC Motors

The rotor of the brush-type DC motor consists of the commutator, an iron core, and wire coils coupled to the same shaft. Because these elements are made of metal and are therefore heavy, there is a high rotor inertia. High starting and stopping inertia is an undesirable characteristic in positioning operations. The brushless DC motor reduces rotor inertia by replacing the mechanical commutation with an electronic commutator and by using permanent magnets instead of heavy electromagnet windings. The stators, which are made of polyphase windings, create a rotating field as they are energized by semiconductor switching circuits. The interaction of the stator field and the permanent magnets causes the rotor to turn. The brushless DC motor is used in applications that require low inertia, high torque, and a wide variable-speed range desirable in positioning operations.

Its application characteristics are:

- High speed—short index moves
- Heavy loads—high torque control
- Short duty-cycle moves
- High power-to-size ratio
- High acceleration/deceleration capability

AC Servo Motors

The operation of an AC servo motor is similar to a split-phase induction motor. Two phases displaced by 90 degrees applied to two sets of stator coils cause a resultant rotating field. One phase is supplied to the main winding from the AC power source. The second phase is supplied to the auxiliary windings by a servo drive amplifier. The movement of the flux lines causes the rotor to turn. The strength of the magnetic field at the auxiliary winding can be altered by the servo amplifier. As it changes, the speed of the motor can be varied. When the field weakens, the motor runs at a slower speed. If the field strength is reduced to zero, the motor stops rotating. The AC servo motor has linear torque-speed characteristics, which are desirable in some applications for positioning and velocity.

Induction Motors

Until the vector drive was developed, induction motors were not used in servo systems because inductance lag in the rotor circuit caused slow response characteristics. The vector drive controls the speed and position of the AC motor by varying the synchronous frequency and voltage to the stator. For the rotor to turn, current must be induced into its coils from the rotating stator field. The magnetic induction takes place when the rotor coils turn at a rotation speed slower than the synchronous speed. At low synchronous frequencies, the induction action does not work effectively. Therefore, the induction motor is usually limited to high velocities and high-speed positioning applications.

Stepper Motors

A number of motion applications use open-loop systems to perform their operation. One of the most reliable open-loop actuators is the stepper motor. It is an extremely accurate device for positioning and does not require a feedback device to operate.

Power ratings of stepper motors are typically restricted to 1 hp and below, and their maximum RPM is 2000, which is considered a relatively low speed. If the torque requirements of the load exceed the capability of the stepper motor, a position error resulting from losing steps is introduced, which cannot be corrected. Also, its speed performance limits its usefulness in velocity-control applications. The clear advantage of an open-loop system is the low cost of the equipment.

For critical applications, the accuracy of the step position can be verified by adding a feedback device to form a closed loop. Stepper motors that use feedback devices, such as encoders, only perform end-point verification. They do not verify the position on the fly. For example, suppose the command signal indicates an end-point value of 200 and the encoder indicates a position count of 198 when the motor stops. An error signal causes the stepper to move the extra two steps.

Its application characteristics are:

- Light load moves
- Speed torque dependency
- No encoder required
- High power-to-size ratio
- Low-to-medium acceleration/deceleration capabilities

Hydraulic Actuators

Hydraulic actuators are playing an increasingly important part in high-performance machine tools, steel mill machinery, metal fabrication machinery, and other types of motion-control applications. Hydraulic actuation may be linear or rotary depending on the load requirements.

Linear actuation is performed by a cylinder with a piston and a rod. The heart of a hydraulic actuator system is the electrohydraulic servo valve. Figure 24-7 shows the cross-sectional view of a servo valve. The bottom portion of the armature is attached to a push rod that moves a metal spool. When the valve is de-energized, the armature is vertical and the spool is centered. In this position, the spool blocks the flow to the cylinder. When the servo valve is energized, a DC electrical signal is applied across points A and B of the coil wrapped around the armature. Because the armature is made of iron, it becomes an electromagnet. If point A is more positive than point B, the top of the armature becomes a north pole and the bottom becomes a south pole. The poles of the armature interact with the poles of two permanent magnets mounted inside the body of the servo valve. Because like poles repel and unlike poles attract, the armature moves a few degrees in the clockwise direction and pushes the spool to the left. Fluid from a pump flows into P_4 (inlet port 4) and through P_2 to the cap end of the piston. As fluid is forced out of the cylinder from P_1 through P_3 and to the tank, the cylinder rod extends. If the opposite voltage polarity is applied to points A and B, the change of current direction reverses the poles of the electromagnet. This causes the armature to rotate a few degrees in the counterclockwise direction, thus causing the spool to move to the right. With the spool in this position, fluid flows into P_4 and out of P_1, and produces a force against the rod end of the piston head, pushing it to the right. The fluid on the cap end is forced out of the cylinder from P_2 through P_5 and to the tank. This causes the rod to retract. By varying the amount of current to the coil, the servo valve is capable of controlling infinitely variable positioning in both directions and controlling the flow rate to regulate the extension or retraction speed. By replacing the cylinder with a hydraulic motor, the servo valve is capable of controlling the speed and direction of rotary actuation.

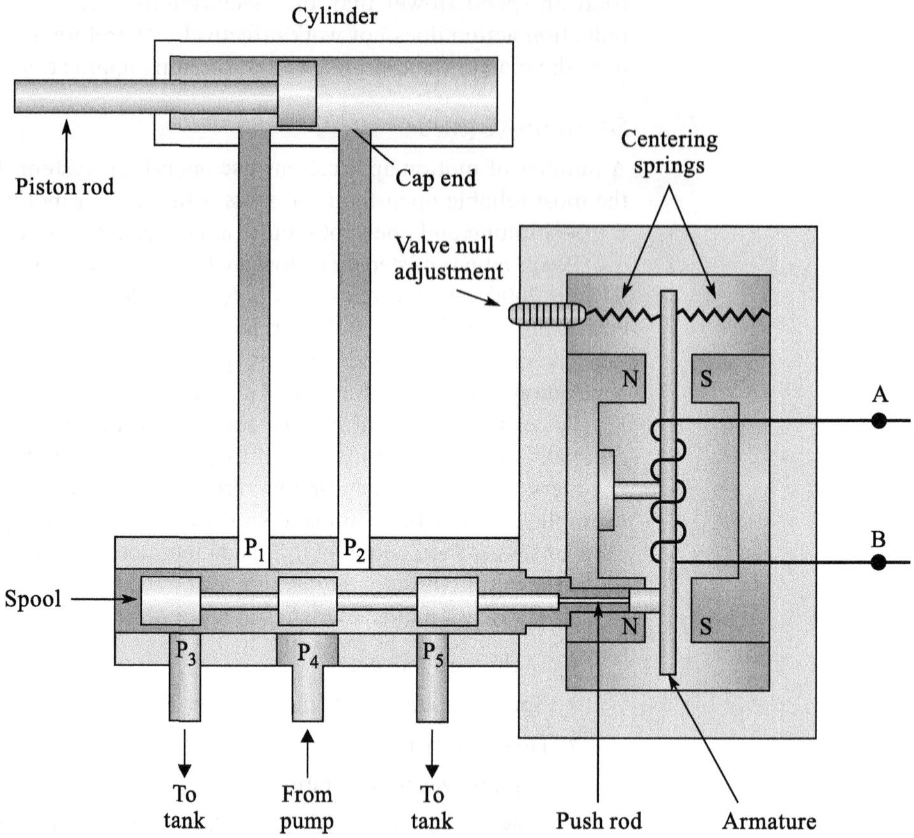

FIGURE 24-7 A cross-sectional view of a servo valve

Figure 24-8 shows the block diagram of a basic electrohydraulic closed-loop servo system. As the command signal is compared to the feedback signal, an error signal is generated. The error signal is amplified and becomes a proportional electrical current that operates the electrohydraulic servo valve. The output of the servo valve is a flow of hydraulic fluid that is proportional to the applied electrical current. A change in polarity will cause the direction of fluid flow to reverse. This flow drives the hydraulic actuator, which in turn drives the load. A transducer measures the load output and develops a proportional electrical signal, which is fed back and compared to the input signal. If the system is a linear position servo, an LVDT is used as the feedback device. The core of the LVDT is attached to the piston rod. If the system is a rotary velocity servo, a tachometer is used. The tachometer is attached to the

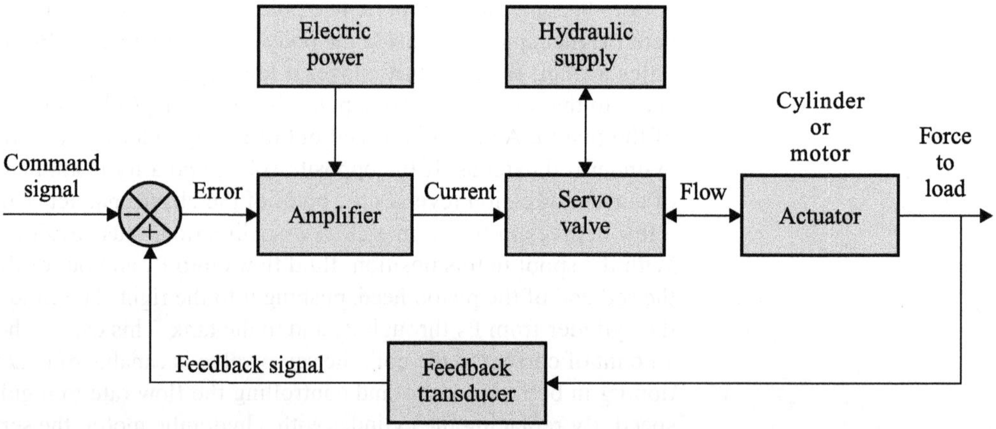

FIGURE 24-8 Block diagram of basic electrohydraulic closed-loop servo system

shaft of the rotary actuator, such as a hydraulic motor. The error signal will cause corrective action to be taken until the output meets the requirements of the command signal.

There are three basic types of electrohydraulic servo systems: position control, velocity control, and force control. An example of a position control application is a numerically controlled machine tool that uses a piston to position the cutting head. Here, the cutting head will move to a position where the feedback signal equals the command signal established by a numerical value from a computer. An example of velocity control is where the feed rate of a machine tool application, such as grinding, boring, or drilling, is performed. Hydraulic cylinders are capable of force amplification. An example of force control is the maintenance of a constant tension in a rolling or winding operation.

Actuator Transmission Systems

There are several types of mechanical systems used for transmitting the rotary force of the motor to cause the load to move. The following six systems—direct drive, gearbox, leadscrew, rack-and-pinion, worm-and-wheel, and belt drive—are used in the vast majority of applications.

Direct Drive

The simplest method of transmitting motion from the rotary actuator to the load is by using the direct drive. The configuration in Figure 24-9 shows how the motor shaft connected to a load pulley transmits the force directly.

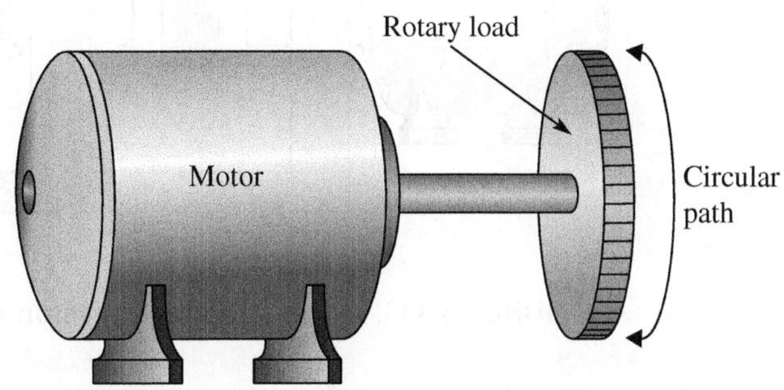

FIGURE 24-9 Direct drive transmission system

Gearbox

The gearbox consists of two or more gears. Basically, a gear is a wheel with teeth cut into its circumference. One gear, called the *driver gear,* is directly connected to the shaft of the motor. Another gear, called the *follower gear,* is connected directly to the shaft that drives the load. The gears mesh together because their teeth are of the same size, as shown in Figure 24-10. When the motor turns under controlled conditions, force is transmitted from its shaft, through the gears, and to the follower shaft, which causes the load to move at the desired speed and direction.

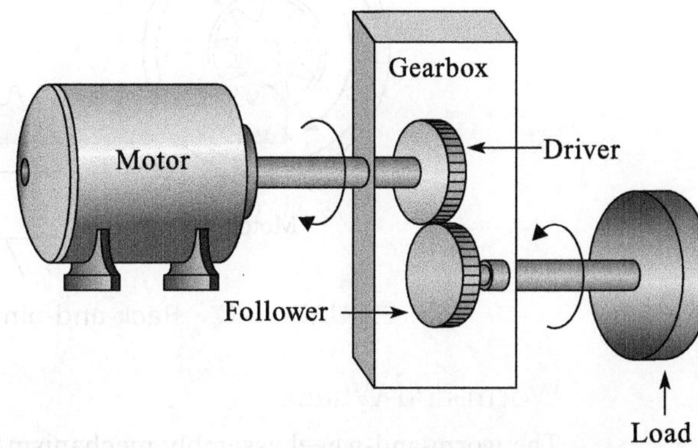

FIGURE 24-10 Gearbox transmission system

Gears mount directly to motor shafts, thereby eliminating the need for belts, pulleys, or gear reducers in an application. The benefits include increased resolution, increased torque, or reduced lag inertia to help meet the requirements of many applications.

Leadscrew

The leadscrew is widely used for converting rotary motion into linear motion. The mechanism is shown in Figure 24-11. The screw element is directly coupled to a motor shaft. A block element, which is often referred to as a *table,* has internal threads that mesh with the screw threads. As the motor runs, the screw turns and causes the table to move laterally. The screw is normally mounted at each end for support, and a guide shaft prevents the table from turning. Each degree the screw turns causes the table to move a precise distance. To eliminate backlash and reduce friction, ball bearings can be placed between the threads of the block and leadscrew. This system is referred to as a *ball screw.*

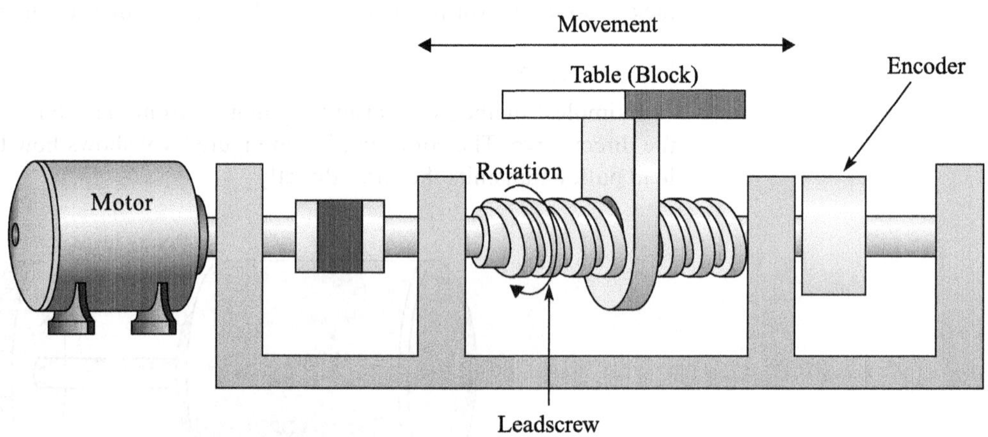

FIGURE 24-11 Leadscrew transmission system

Rack-and-Pinion

Another transmission system that converts rotary motion to linear motion is the rack-and-pinion shown in Figure 24-12. A *rack* is a straight bar with involuted gear teeth cut into one surface. A pinion gear is attached to the shaft of a motor. As the motor runs, it turns the pinion and forces the rack to slide. The direction of motor rotation determines the direction of linear movement by the rack. The RPM of the motor determines how rapidly the rack moves.

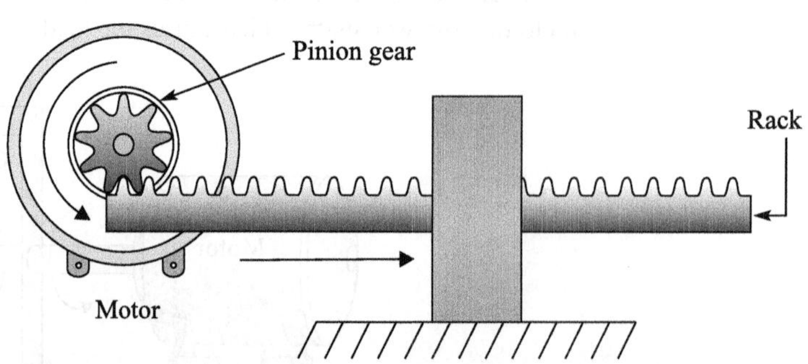

FIGURE 24-12 Rack-and-pinion transmission system

Worm-and-Wheel

The worm-and-wheel assembly mechanism is shown in Figure 24-13. It consists of a worm-pinion cylinder with a helical thread cut on its outer surface. The worm is meshed with a specially designed gear wheel. The wheel is manufactured with curved tooth tips so that it

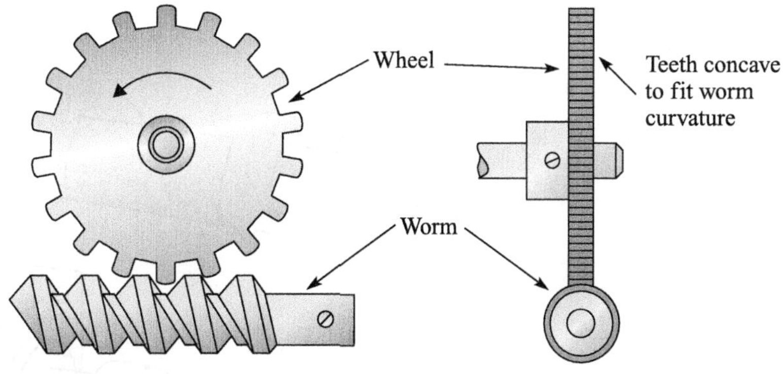

FIGURE 24-13 Worm-and-wheel transmission system

will fit well into the worm. Worm-and-wheel assemblies are able to transmit relatively large loads because the faces of the teeth on each gear are in full contact. They are used in applications where precision movements of heavy loads are required. This mechanism is also self-braking and can be used to hold the load in place. They are not used in high-speed operations because of the high friction that is developed at the contacting surfaces of the teeth.

Belt Drive

Belts and pulleys are used in a wide variety of mechanical drive systems. One type of belt drive example, shown in Figure 24-14, is the tangential system that moves a load perpendicular to the axial direction of the motor shaft. The motor shaft is coupled to a drive pulley. Another pulley, called the *follower*, is connected to an idler shaft for support. The rotary torque from one pulley is transferred to the other pulley by a belt made of durable fabric. A common application of a belt drive is a conveyor system.

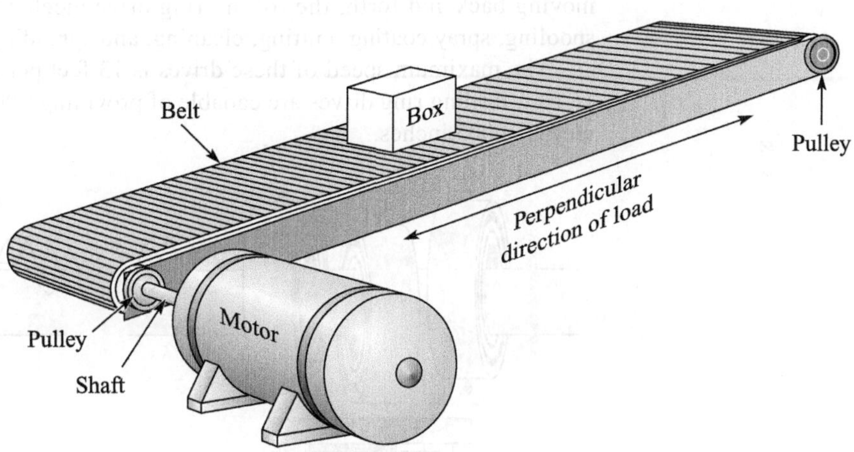

FIGURE 24-14 Belt drive transmission system

Rolling Ring Drive

Rolling ring drives are mechanical devices that convert the constant-speed rotation of a plain round shaft into linear movement. Shown in Figure 24-15, a rolling ring drive consists of a cylindrical shaft that has a smooth surface and a cube-shaped housing. Payloads attached to the housing move along the entire shaft in either direction or remain stationary. The shaft turns at a constant speed in one direction. There are three rolling rings within the housing. They have spring-loaded ball bearings inside the ring that are precisely contoured to be in continuous contact with the shaft. A pitch lever mounted to the housing is used to adjust the pitch angles of the three rings.

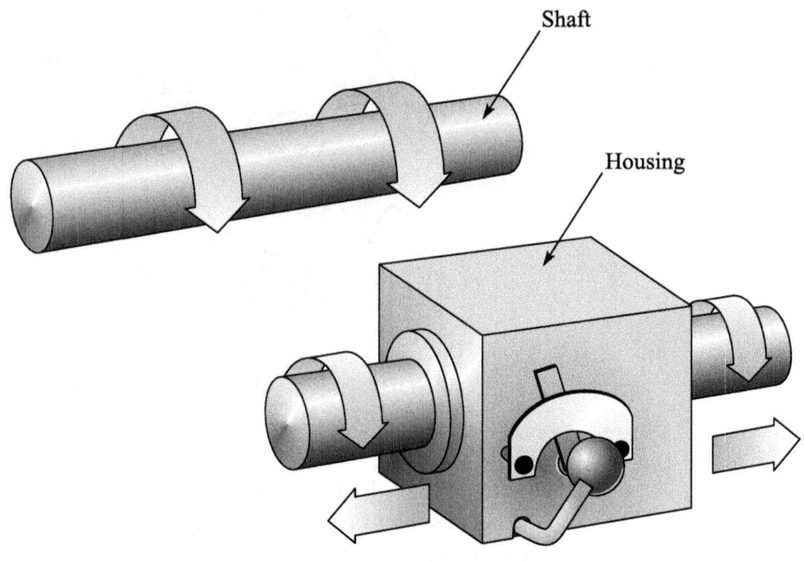

FIGURE 24-15 Rolling ring drive

When the rings are perpendicular to the shaft, the housing will be stationary. But when turned at an angle to the shaft, as shown in Figure 24-16(a), the rings are compressed and the friction created causes them to act as a nut on a threaded leadscrew, which causes the housing to move to the right. Figure 24-16(b) shows how the rings can be swiveled in alternating left-hand and right-hand pitches. The direction they travel is dependent on how the rings are swiveled. By varying the pitch angle, the speed of travel can be precisely controlled. By using a reverse mechanism attached to the housing, the swivel angle of each ring is changed to the opposite angle when the end of the travel is reached. This action causes the housing to reverse direction and provides an automatic reciprocating-action capability. With the housing moving back and forth, the rolling ring drive mechanism is ideal for applications such as spooling, spray coating, cutting, cleaning, and spreading.

The maximum speed of these drives is 13 feet per second with traverse lengths of up to 16 feet. Rolling ring drives are capable of providing 800 pounds of axial thrust with an accuracy of 0.005 inches.

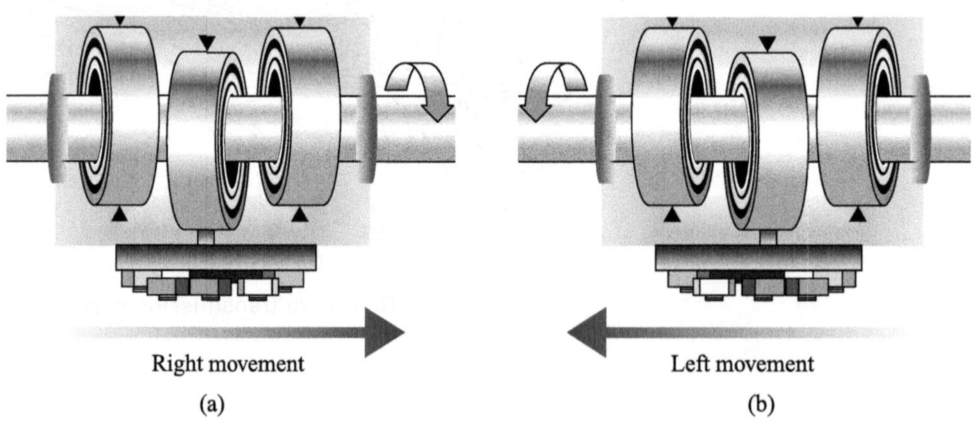

FIGURE 24-16 Rolling rings inside the housing

Positioning Movements

Actuators are used to move loads to specified locations. The movement of an object from one position to another is accomplished in one of three ways: single-axis, multiple-axis, and contouring. The three axes are identified as X, Y, and Z. The X and Y axes are at right angles to each other on the same plane. The Z axis is perpendicular to the plane of the X–Y axis.

Single-Axis Positioning

The simplest and perhaps the most common method of moving a load is by using single-axis motion control. The single-axis system causes objects to make linear movements in both directions. An example is moving a spindle as it rotates into or out of a stock during boring or drilling operations, as shown in Figure 24-17.

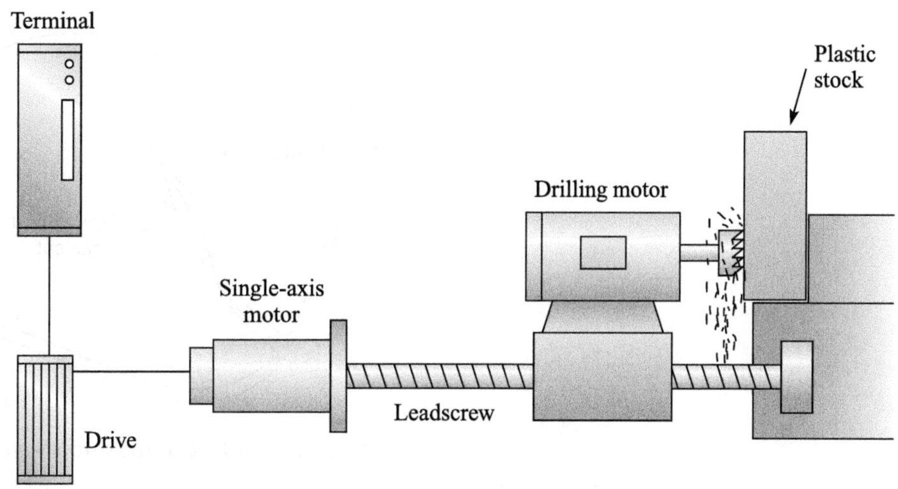

FIGURE 24-17 Single-axis positioning

Multiple-Axis Positioning

If the single-axis system is expanded to more than one axis, more complex operations can be accomplished. An example of multiple-axis positioning is the drilling operation shown in Figure 24-18, where holes are bored into a metal disk at precise locations. The X- and Y-axis movements control horizontal positioning, and the Z axis causes the drill to move vertically into the disk.

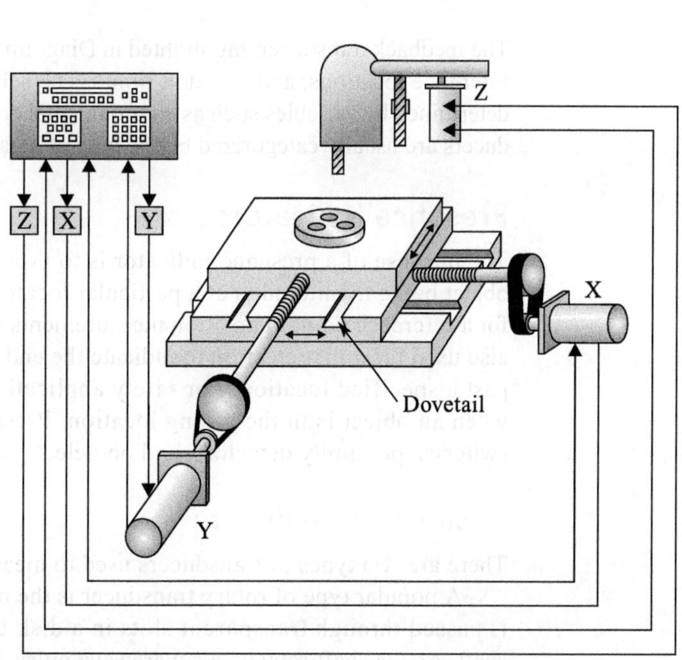

FIGURE 24-18 Multiple-axis positioning

Contouring

Multiple-axis applications that require continuous path control instead of end positioning use a positioning system known as *contouring*. To perform contouring, a complex control process is necessary. An example of contouring is the cutting of fabric in a circular arc shown in Figure 24-19. This action requires mathematical computations at intermediate positioning points to ensure that one axis corresponds to the position of the other axis.

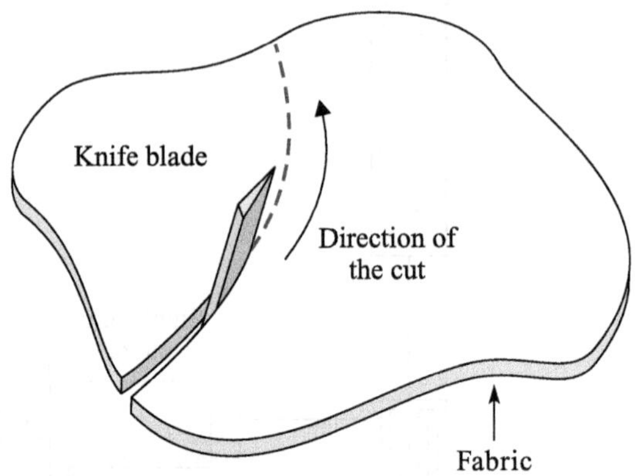

FIGURE 24-19 Contouring movement application

24-9 Feedback Transducer Block

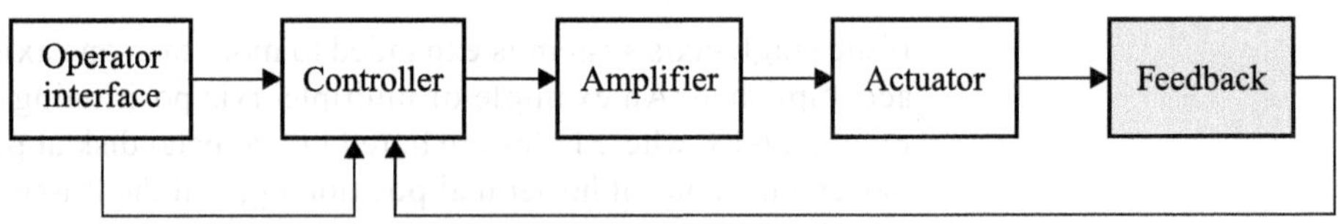

DIAGRAM E

The feedback transducer, highlighted in Diagram E, is used to indicate position, direction, speed, reference locations, and position limits. The selection of which type of device to use is often determined by variables such as environmental considerations, accuracy, and cost. Motion transducers are usually categorized by the type of measurements they make: presence or position.

Presence Indicators

The purpose of a presence indicator is to provide a signal that tells the controller when an object being monitored is at a particular location. For example, a home position that is used for a reference point in position measurements must be detected by a sensor. Transducers are also used for limit detection to indicate the end of travel or to prevent an object from moving past a specified location. For safety applications, equipment is disabled to prevent injury when an object is in the wrong location. Presence indicators include limit switches, reed switches, proximity detectors, and optoelectronic sensors.

Position Transducers

There are two types of transducers used to measure position: rotary and linear.

A popular type of rotary transducer is the optical encoder. By using a light emitter, light is passed through transparent slots in a disk to a light detector. The disk is mounted to a shaft. As the shaft rotates, a squarewave pulse is produced by the detector every time a light shines through a passing slot. Optical encoders are either absolute or incremental. Each type produces a different squarewave signal because the patterns on their disks are not the same. The signals indicate position, direction, and velocity.

The other widely used rotary transducer is the resolver. The resolver is an absolute position-sensing device that uses phase- and amplitude-modulated sine waves to indicate position. Resolver-to-digital (R/D) converters are required to convert the analog signals they produce into position and velocity information.

Linear measurements can also be made by rotary detectors. For example, Figure 24-11 shows a machine table that travels horizontally. As the leadscrew turns, the table travels in one direction. If the rotation is reversed, the table travels in the opposite direction. The encoder rotates when the gear that drives the shaft turns. The measurement system is calibrated to detect the linear distance for every degree of rotary movement.

There are several types of detectors that are designed specifically to make linear positioning measurements. These devices include inductosyns, LVDTs, and linear displacement transducers. An inductosyn has a fixed primary coil energized by an AC signal. The secondary is attached to the object being measured. As the object moves, a resolver-type output is produced and is translated into linear information by an R/D converter. An LVDT has fixed primary and secondary coils. The iron core of the transformer is attached to the object being measured. As the object moves, the amount of voltage induced into the secondary from the primary changes. The movement signal is calibrated to indicate position based on the varying secondary output. A linear displacement transducer (making linear measurements) determines the travel time of a strain pulse between the object being measured and an element in the sensor head. Displacement can be determined because it is proportional to time.

Problems

1. The actuator of a/an ______ -loop system is activated when the command signal or a disturbance occurs.
   a. open
   b. closed
   c. both a and b
2. An open-loop configuration (is/is not) considered a servo system.
3. Which of the following parameters primarily maintain speed regulation?
   a. position control
   b. velocity control
   c. acceleration/deceleration control
   d. torque control
4. List three factors that impede acceleration or deceleration.
5. A twisting force is referred to as ______.
6. List the five major elements of a closed-loop motion-control system.
7. Match the following terms with the definitions.
   a. Holding torque
   b. Home position
   c. Bandwidth
   d. End point
   e. Damping
   f. Traverse rate

   ____ The prevention of overshooting.

   ____ A measure of how quickly the actuator responds to the command signal.

   ____ The desired location to which a load is moved.

   ____ A reference point from which a load is measured.

   ____ The force that causes the load to maintain a standstill position.

   ____ The speed at which a load is moved from one location to another.
8. List three devices that are used to enter data into an operator interface device.
9. Indexers perform the operation of which element in a motion-control closed-loop system?
   a. controller
   b. amplifier
   c. feedback
   d. actuator
10. AC drives perform the operation of which element in a motion-control closed-loop system?
    a. controller
    b. amplifier
    c. feedback
    d. actuator
11. An inductosyn measures ______ (linear, rotary) motion.
12. Stepper motors are primarily ______ -loop devices.
    a. open
    b. closed
13. A hydraulic actuator is a ______ device.
    a. linear
    b. rotary
    c. linear or rotary
14. A servo valve controls ______ action.
    a. linear
    b. rotary
    c. linear or rotary
15. A leadscrew mechanism converts ______ motion into ______ motion.
    a. linear
    b. rotary
16. Which of the actuator transmission systems are capable of multiplying rotary speed from the input shaft to the output mechanism?
    a. direct drive
    b. gearbox
    c. leadscrew
    d. rack-and-pinion
    e. belt drive
    f. belt actuator
17. In an XYZ system, if two axes movements control horizontal positioning, which axis controls the vertical position? ______

18. ______ is the movement of the load by the actuator as it attempts to follow a changing command signal in a motion-control system.
    a. Following error
    b. Tracking
19. The load movement lagging behind the desired movement specified by the command signal is called ______.
    a. following error
    b. tracking
    c. indexing
20. A rack-and-pinion transmission system converts ______ motion into ______ motion.
    a. rotary
    b. linear
21. A common application of a ______ is to run a conveyor belt.
    a. rack-and-pinion
    b. worm-and-wheel
    c. belt drive
    d. leadscrew
22. The ______ is self-braking, which enables it to hold a load in a stationary position.
    a. rack-and-pinion
    b. worm-and-wheel
    c. gearbox
    d. direct drive
23. T/F Linear measurements can also be made by rotary detectors.
24. The speed at which the housing of a rolling ring drive moves is determined by the ______.
    a. pitch of the rings
    b. RPM of the rotating shaft
25. The direction that the housing of a rolling ring drive moves is determined by the ______.
    a. direction in which the rings are pitched
    b. direction in which the shaft rotates

CHAPTER

25

Motion-Control Feedback Devices

OBJECTIVES

At the conclusion of this chapter, you should be able to:

- Describe the differences among angular velocity, angular displacement, linear velocity, and linear displacement feedback devices.
- Explain the operation of the following motion-control feedback devices:

Tachometers	Inductosyns	Linear Velocity Transducer
Potentiometers	LVDTs	Angular Displacement Transducer
Optical Encoders	Linear Displacement Transducers	
Resolvers		

- List the factors that determine the voltage produced by a tachometer.
- Describe the difference between incremental and absolute optical encoders.
- Convert Gray code to binary and binary to Gray code.
- Describe the difference between ratiometric-tracking and phase-digitizing resolvers.

INTRODUCTION

Automation in motion-control applications is only possible if the controller section receives information about conditions in the manufacturing process. These conditions include displacement, position, speed, acceleration, and deceleration. The devices capable of monitoring these conditions are called **transducers.** The transducer performs the measurement of the condition and produces a feedback that provides information on the results.

Some transducers produce an analog output signal. The amplitude of the signal is proportional to the measured quantity. Other transducers produce a digital output consisting of a group of 1s and 0s called *words.* The words are numerical values that represent the measurement.

Motion-control detectors are used to measure conditions in a wide variety of applications. For example, they measure the precise depth of a hole bored by a drill bit; they sense the precise position of the work piece in an automatic milling operation; they measure the acceleration and speed of a robotic arm performing a welding operation, and they detect if there is a stall condition.

This chapter explains the operating principles of several types of motion feedback transducers. Each device falls under one of the following feedback transducer categories: angular velocity, angular displacement, or linear displacement.

To enable the reader to more clearly understand the characteristics of motion-control feedback devices, it is important to know various terms used to describe them. The following short glossary provides a list of these words and their definitions.

Glossary

Accuracy For positioning purposes, *accuracy* is defined as the maximum deviation between the desired position and the actual position.

Life The minimum rated lifetime of the measurement device while maintaining positioning specifications.

Range The maximum distance the measurement device can travel.

Resolution The smallest theoretical movement of the device when it indicates a change of position.

Linearity A given mechanical movement produces a given range of a variable, such as a resistance or a voltage. Linearity occurs because the variable changes evenly along the range of the measurement device.

Displacement The difference between the initial position of an object and any later position.

25-1 Angular Velocity Feedback Devices

Tachometers

A **tachometer** is a device that measures the angular velocity of a rotating shaft. There are two types of tachometers: the *DC generator,* and the *toothed rotor*. The most common unit of measurement for indicating angular velocity is revolutions per minute (RPM).

The speed of shaft rotation is indicated in two different ways.

- The angular velocity is indicated by the *magnitude* of a generated voltage, which is produced by the DC generator tachometer.
- The angular velocity is indicated by the *frequency* of pulses produced by the toothed rotor tachometer.

The most common type of tachometer is the DC generator, which resembles a miniature motor. Its shaft is connected to the rotary device being measured. The tachometer has a permanent magnet stator that produces magnetic flux lines, a conventional DC armature consisting of coils wound around the shaft, brushes, and a commutator. It has a permanent magnet stator that produces magnetic flux lines. A conventional DC armature using a commutator and brushes is connected to the rotary object being measured. When the object spins, the armature cuts the flux lines and produces an analog voltage. The equation for generated voltage is:

$$V_g = KB(RPM)$$

where,

V_g = Generated voltage
K = A constant, which represents the physical construction (armature length, armature diameter, etc.)
B = Strength of the magnetic field
RPM = Angular velocity measured in revolutions per minute

Since the physical size of the armature and the field strength are constant, the generated voltage is proportional to the angular velocity of the shaft. If connected to the armature, a voltmeter with an RPM faceplate can be calibrated to indicate the speed of a shaft's rotation. As the tachometer rotates in one direction, the polarity of the voltage is positive. The polarity reverses if the tachometer turns in the opposite direction. The result is that the tachometer output indicates both direction and speed.

DC generator tachometers have voltage ratings for increments of every 1000 RPM. For example, a tachometer may produce 3 volts per 1000 RPM. If its speed is rated from 0 to 5000 RPM,

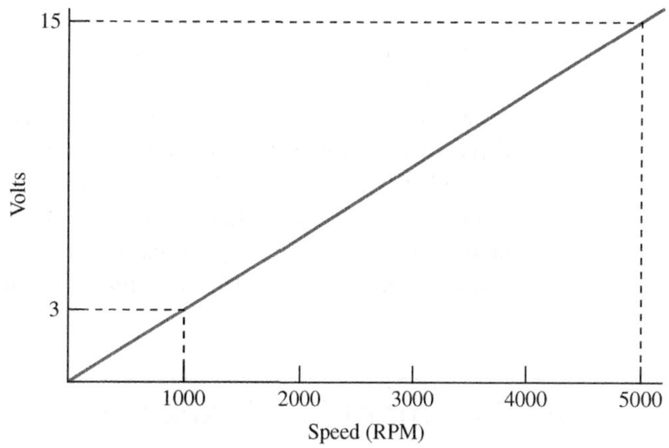

FIGURE 25-1 Tachometer output proportional to speed

the tachometer will produce 15 volts at its fastest velocity, as shown in Figure 25-1. These tachometers are very linear, with a deviation of only 0.2 percent. This means that the tachometer's reading will only be 0.2 percent higher or lower than its actual speed. Suppose the tachometer reads 4000 RPM. The RPM measurement of the tachometer can range from 3992 to 4008 RPM. DC tachometers are commonly used for closed-loop applications because their DC output can be directly connected to the controller of a servo system without the need for signal conditioning.

The toothed rotor is another common type of tachometer. It consists of a rotor with ferromagnetic teeth that rotate past a permanent magnet, which has a coil wrapped around it, as shown in Figure 25-2(a). The rotor is connected to the shaft of the motor being monitored. As the shaft turns, the rotor teeth pass by the magnet. When a tooth is close to the magnet, the strength of the magnetic field increases because the magnet reluctance is reduced. When a gap between two teeth is close to the magnet, the reluctance increases and causes the magnetic field strength to decrease. The variation of field strength as each tooth passes induces a voltage into the coil and creates a pulse. There is one pulse produced by each tooth, or six pulses per revolution, as shown in Figure 25-2(b). The number of RPM can be determined by multiplying the revolutions per second by 60 and then dividing the answer by 6.

The disadvantage of toothed-rotor tachometers is that they require a signal processing circuit that converts the pulses into analog signals.

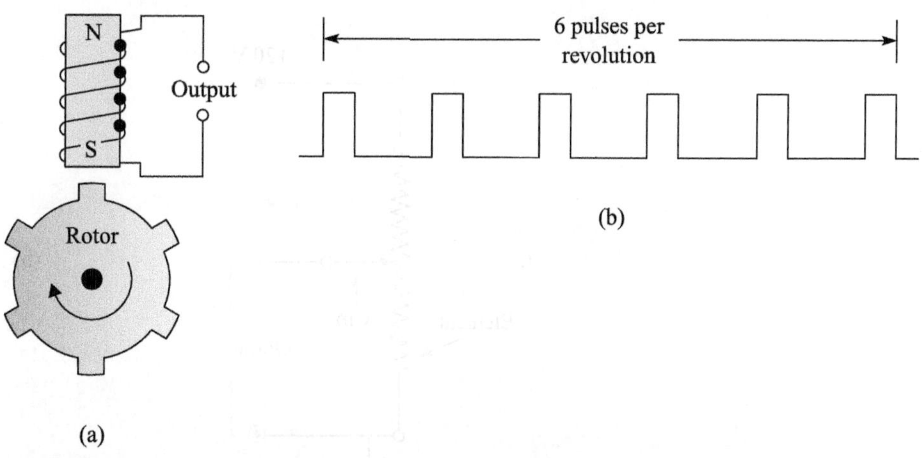

FIGURE 25-2 The toothed-rotor tachometer: (a) Pictorial diagram; (b) Output pulses

Tachometers are often used to provide speed regulation for velocity servo systems. Another term used to indicate speed regulation is *stability*. In this type of system, the tachometer provides a feedback signal that subtracts from the command voltage. The difference between the command signal and the feedback signal is the error signal, which is amplified to drive a DC motor. The tachometer is coupled to the motor shaft to form the feedback loop.

Whenever an increased load condition occurs, the motor slows to below its desired speed and reduces the tachometer output. This condition creates a larger difference between the command signal and the feedback signal, causing the error signal to become greater. A larger error signal causes the motor current to rise so that it returns to the original speed before the load is changed.

25-2 Angular Displacement Feedback Devices

Potentiometers

Perhaps the simplest device used as a feedback position indicator is the **potentiometer.** It is capable of converting mechanical motion into an electrical voltage variation.

The potentiometer consists of a fixed resistive element with an AC or DC voltage connected across its ends. Ideally, the resistance of a feedback resistor is uniform across its entire length. A conductive metal slide called an *arm* moves across the element. A voltage divider output is developed from one end terminal of the element to a terminal connected to the arm.

The arm is mechanically coupled to the object that is being measured. As the object moves, the resistance between the terminals varies. A voltage proportional to the resistance forms, indicating position.

Figure 25-3 shows two types of potentiometers. The type in Figure 25-3(a) has a straight resistive element. It is used to measure linear displacement. Suppose the object being measured has a linear motion range of 2 inches. If the resistive element is also 2 inches long and has 20 volts applied across it, then an output of 10 volts will indicate that the object has moved 1 inch, or halfway through its travel. When the arm is positioned at the top end of the element, an output of 20 volts will indicate that the object has moved the entire 2 inches of travel. As the object moves toward the other end, the voltage will decrease linearly until it moves 2 inches (where a reading of 0.0 volts will occur). The potentiometer illustrated in Figure 25-3(b) has a circular resistive element. It produces an output voltage that is proportional to the angular displacement of the shaft. If 20 volts is applied across the element terminals, an output of 10 volts will indicate that the object has turned halfway through its rotation. Since the resistive element does not make a complete circle, rotary measurement is limited to about 350 degrees.

Multiple-turn potentiometers are available to provide better resolution. They can have anywhere from 3 to 40 revolutions before hitting internal stops. Potentiometers are classified

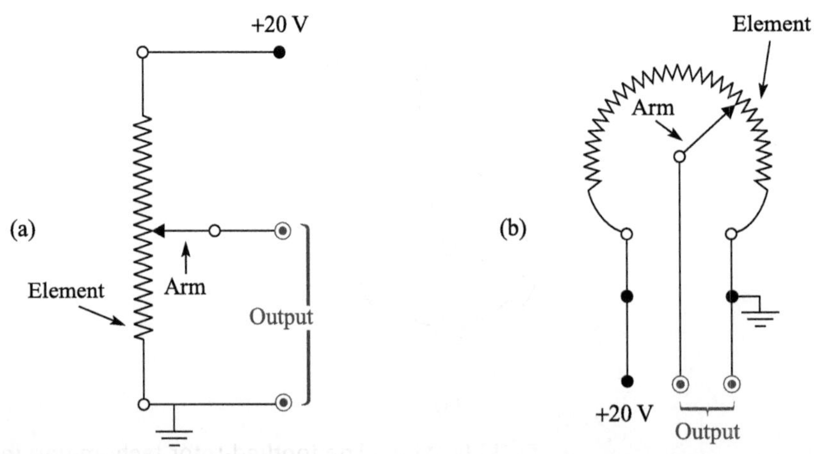

FIGURE 25-3 Potentiometer symbols

as "passive" transducers, which means an excitation voltage must be applied across the terminals of the fixed resistive element.

The potentiometer has a reputation for lacking accuracy because, as the arm moves across the element, both the arm contact and the element wear. They also become dirty as a result of the residue buildup from friction and oxidation. The resistance of the potentiometer therefore changes, reducing its accuracy.

Potentiometers are making a comeback because development of hard, conductive plastic is improving their performance. This plastic material provides a long life because it does not wear or oxidize and can withstand exposure to an extensive temperature range, harsh chemicals, or oils. High-performance potentiometers for servo applications have a guaranteed life rating of 250×10^6 revolutions. Since the plastic element is a continuous material, it yields infinite resolution.

One advantageous characteristic of a potentiometer is that it provides an absolute positioning measurement on power-up. This means that the actual position is indicated by a specific reading when the voltage across the element terminals is turned on. One of the limitations of a potentiometer is that it cannot be used for measuring a continuous rotating application, such as an electric motor.

The resistance value of a potentiometer is often chosen to match the output impedance of the power supply connected across the element terminals. Matching these resistances reduces power consumption. Typical values for servo pots are 1 to 10 kΩ.

Angular Displacement Transducer

An angular displacement transducer (ADT) uses capacitance technology to make rotary position measurements. Figure 25-4 shows the construction of an ADT. It has a differential capacitor with a movable rotor connected to a shaft, and electronic circuitry that conditions voltage and process signals. The capacitive sensor within the transducer consists of a transmitting board, a receiving board, and the rotor that is placed between them with a clearance by only a few thousandths of an inch. One end of the transducer's stainless steel shaft is connected to the rotor hub and the other end is connected to the rotary object being measured. This enables the shaft to rotate with the object being measured to determine its position.

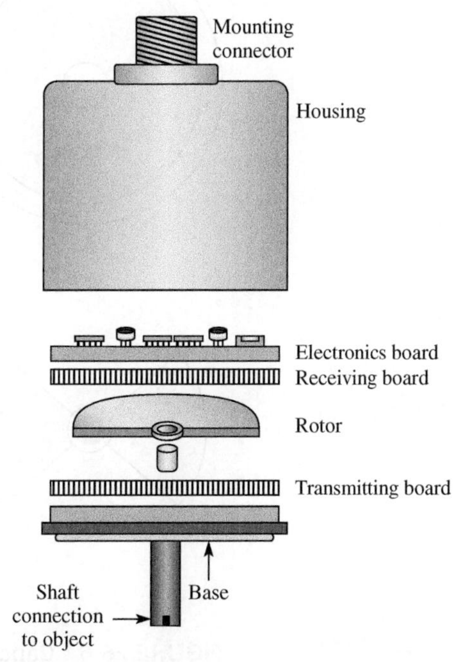

FIGURE 25-4 Exploded view of the angular displacement transducer

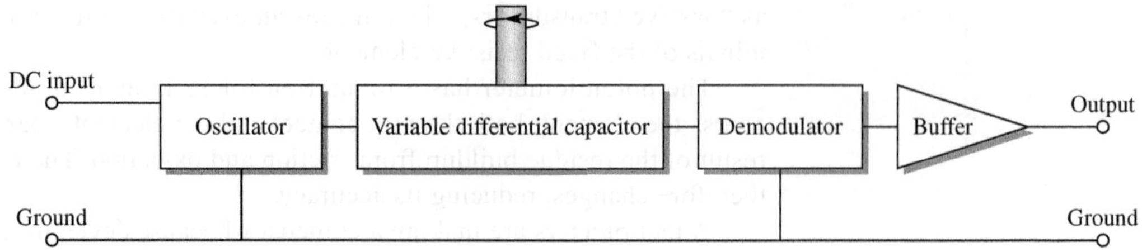

FIGURE 25-5 Block diagram of the angular displacement transducer's electronic circuitry

Figure 25-5 shows a block diagram of the ADT. A DC input voltage supplies the power to a solid-state oscillator in the capacitance portion of the transducer, which generates a high-frequency (400–500 kHz) AC voltage. The AC signal is used to excite one plate of the capacitance element, called the transmitting board, which is shown in Figure 25-6(a). The receiving board functions as the other plate of the capacitor. The receiving board is divided into sections, as shown in Figure 25-6(b). Sections A and D are electrically connected, as are sections B and C. The rotor, shown in Figure 25-6(c), is electrically grounded through a mechanical contact. Changing the position of the rotor causes the capacitance level between the pair of sections on the receiving board to change. By differencing the capacitance between the pair of sections, an output proportional to the angular movement of the shaft is generated. For example, when the rotor is positioned equally over the top and bottom halves of the receiving board, as shown in Figure 25-6(d), the capacitance of each pair of sections will be equal, and when differenced it will be zero. As the rotor moves CW or CCW, the capacitance of one pair of sections becomes greater than the other pair. The result is that the AC voltage at one pair of sections becomes greater than the AC voltage at the other pair of sections. Rectifier and filter circuits in the demodulator convert each AC voltage to DC. The differential output formed by the two DC voltages is proportional to the angular distance moved from the zero-point position.

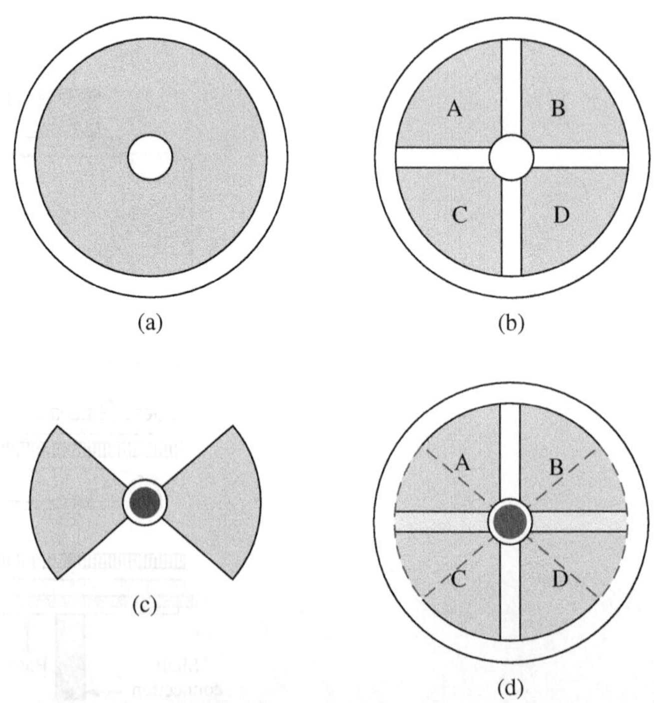

FIGURE 25-6 Capacitive elements of an angular displacement transducer: (a) Transmitting board; (b) Receiving board; (c) Rotor; (d) Rotor at the 0-degree position

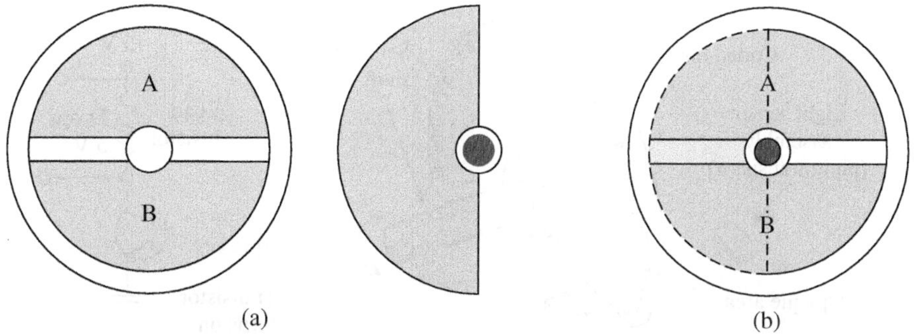

FIGURE 25-7 (a) ADT with a 60-degree stroke; (b) Rotor at 0-degree position

The ADT with four sections is capable of measuring maximum strokes of 30 degrees. To increase the stroke to 60 degrees, a receiving board with two sections and a rotor that is a half circle is used, as shown in Figure 25-7.

When the rotor is in the position that covers each section equally, as shown in Figure 25-7(b), the capacitance and voltage at each segment are equal. The result is that the differential output voltage is equal, which indicates a zero-point angular position. As the rotor moves CW or CCW, the capacitance of one section becomes greater than that of the other section, and the differential output voltage becomes proportional to the angular distance moved from the zero-point position.

An ADT is an absolute measuring device. This means that if power to the transducer is turned off, the actual position will be measured as soon as power is restored without going back to the 0-degree reference position.

Optical Encoders

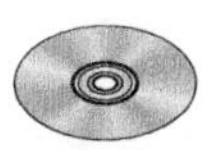

IAU8207
Optical Encoders

An **optical encoder** is an electromechanical device that is used to monitor the direction of rotation, position, or velocity of a rotary or linear operating mechanism. This device typically has four major elements: a light source, a light sensor, an optical disk, and signal conditioning circuitry. They are shown in Figure 25-8.

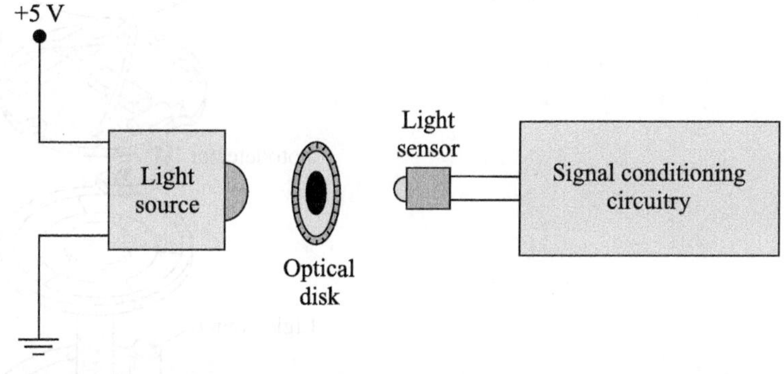

FIGURE 25-8 The four major elements of an optical encoder

The light source is usually a light-emitting diode (LED) that transmits infrared light. The light sensor is a phototransistor that is more sensitive to infrared energy than to visible light. The optical disk is connected to the shaft being measured so that they rotate together. The disk is made of plastic, Mylar®, glass, or metal. It has opaque and translucent regions. The disk is placed between the light source and the light sensor as shown in Figure 25-9(a). As the disk is rotated, light passes through the translucent segments and is blocked by the opaque areas. Figure 25-9(b) shows that when light passes through these segments, it strikes the base of the

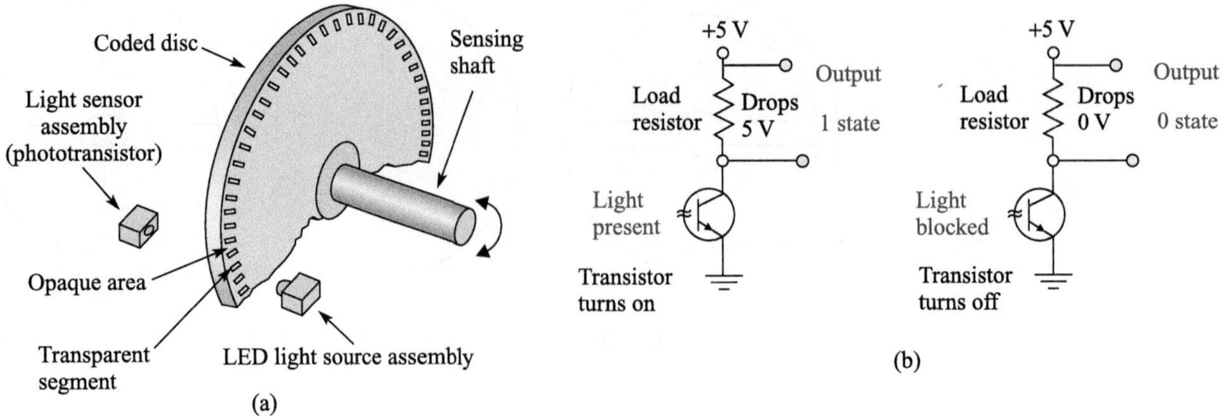

FIGURE 25-9 Operation of an optical encoder

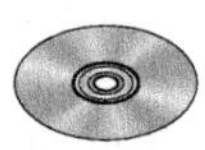

IAU8107
Optical Encoder Classifications

transistor which conducts. The voltage at the detector output goes high. When light is blocked, the transistor is off and a low is formed at the detector output. The conditioning circuitry uses Schmitt triggers to convert pulses from the phototransistor into square waves. An exploded view of an optical encoder assembly is shown in Figure 25-10.

Optical encoders are classified as either *incremental* or *absolute.*

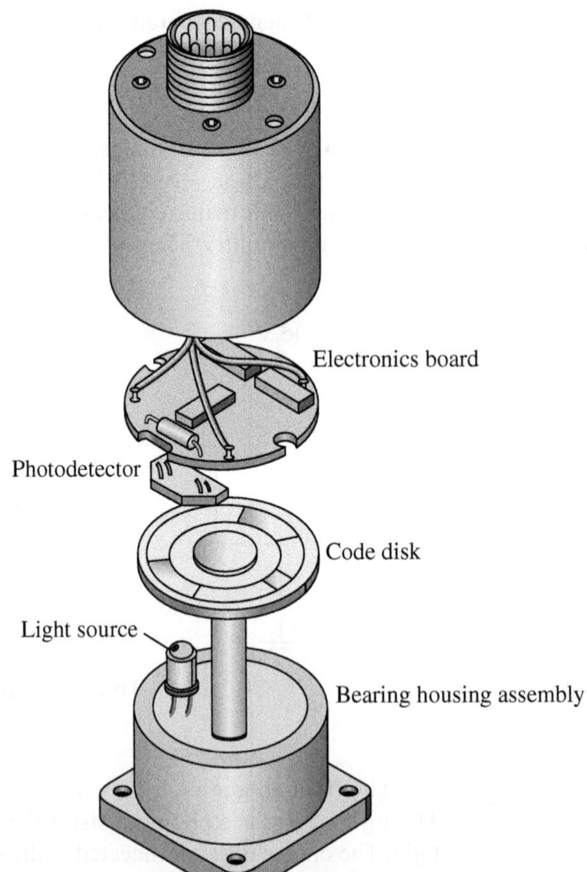

FIGURE 25-10 Exploded view of an optical encoder assembly

IAU8407
Incremental Optical encoders

Incremental Encoders

Figure 25-11 shows an **incremental optical encoder** disk in its simplest form. It has one track of imprinted marks precisely positioned around the outside of the disk. The quantity of marks is equivalent to the number of outputs for every 360-degree revolution of the encoder. As the disk rotates, light passes through the translucent slots, generating a series of electrical pulses. The exact number of pulses produced corresponds to how much the disk has rotated. For example, if there are 360 slots on a disk, the encoder will supply 360 pulses per revolution. The number of degrees the disk is rotated can be determined by using a digital counter to record the number of pulses. For example, 36 counts indicate the disk has turned 36 degrees, 180 counts indicate 180 degrees, and so on.

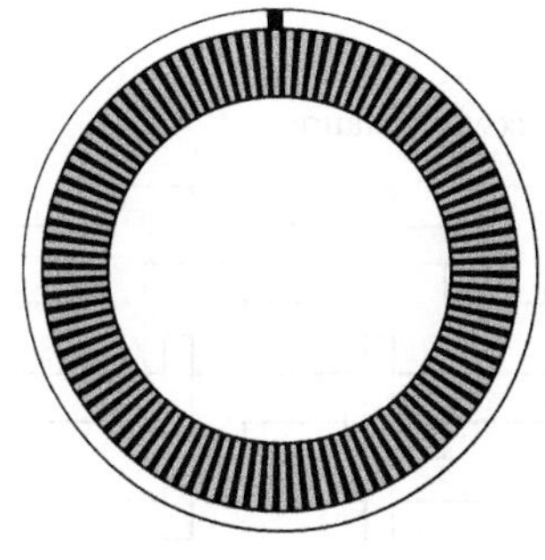

FIGURE 25-11 Single-track incremental disk

Single-track encoders are typically used as tachometers to measure the RPM of a rotating device. The velocity information is determined in one of two ways: by measuring the time interval between pulses or by counting the number of pulses within a time period. The disadvantage of using one track is that the direction cannot be determined.

Figure 25-12 shows a partial diagram of a two-track encoder. The two tracks (labeled A and B) provide the pulse train that indicates displacement. The number of pulses is recorded by a digital counter circuit. A separate light source and detector are used for each. The tracks are in quadrature, meaning that they are 90 degrees out of phase with each other. A continuous square wave is produced by each track as the encoder rotates. Track A lags track B by 90 degrees when the shaft rotates in a CW direction. Track A leads track B by 90 degrees when the shaft rotates in the CCW direction. Therefore, the direction is indicated by the relationship of which track leads or lags.

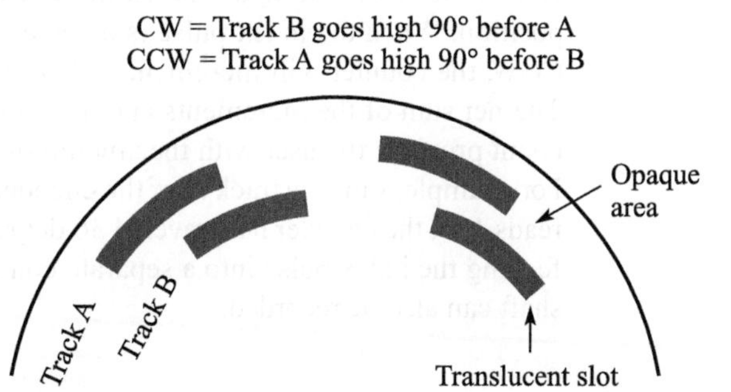

FIGURE 25-12 A two-track incremental encoder

Figure 25-13(a) shows a circuit that counts the number of degrees the incremental encoder rotates. It also indicates which direction the encoder is turning. When the encoder rotates CCW, the count increments. When the encoder turns CW, the count decrements. The circuit consists of the encoder that produces the A and B outputs that are phase-shifted by 90 degrees from each other. The A output of the encoder is connected to the input of the one-shot. Each time the A signal goes low high, the one-shot is activated and produces a temporary high-level pulse.

The waveforms produced by each component in the circuit as the encoder turns are shown in Figure 25-13(b). Each time the output from track A goes high, a pulse is applied to the NAND gates by the one-shot. The B output of the encoder is connected directly to the CD (count-down) NAND and the inverter. The output of the inverter is connected to the CU (count-up) NAND gate. Each NAND gate requires that both inputs be high to produce an active-low output. The direction of rotation determines which of the two NAND gates is active. Before operating the encoder, it is necessary to turn it to the 0-degree indexing position and reset the counter to 0000. The waveforms show that when the encoder rotates CCW, the CU NAND gate is active. It supplies pulses to counter input CU, causing it to increment. When the encoder rotates CW, the CD NAND gate feeds pulses to counter input CD, causing it to decrement.

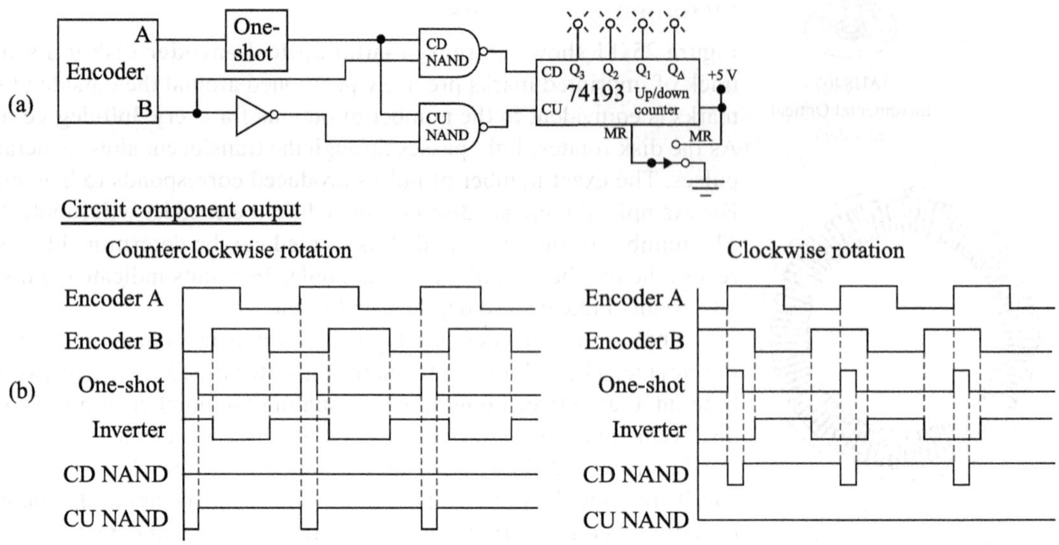

FIGURE 25-13 Incremental encoder conditioning circuitry

To indicate the amount of distance traveled from a reference point, a third track is added to the incremental encoder. Track C contains only one translucent segment. The signal produced is called the *index pulse*, or *zero reference pulse*. This pulse is generated once per revolution in the CW or CCW direction, as shown in Figure 25-14. It is used to establish a reference or home position. When the index pulse is detected, the counter is reset to zero. As the encoder rotates CCW, the counter will increment. When the encoder rotates CW, the counter will decrement. The net sum of the increments and decrements is indicated by the number in the counter. The count provides the user with the amount of travel from the reference point at any point in time. For example, suppose track A of the encoder produces 1000 pulses per revolution. If the counter reads 100, the encoder has traveled 36 degrees; 150 would indicate 54 degrees; and so forth. By feeding the index pulse into a separate counter, the number of revolutions made by the encoder shaft can also be recorded.

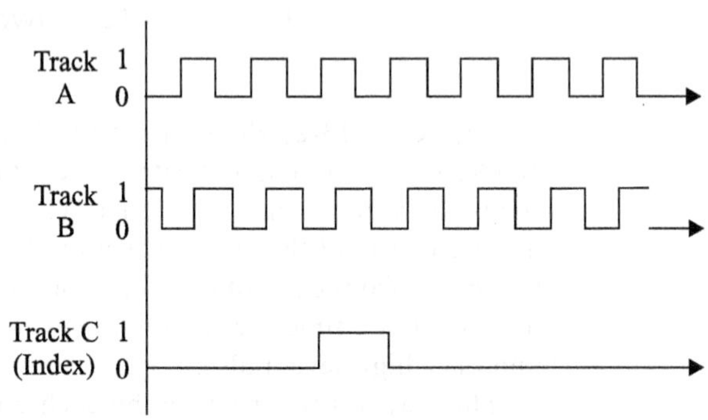

FIGURE 25-14 Quadrature encoder output

The number of degrees per count is called **resolution.** The larger the number of translucent segments on an encoder wheel, the fewer degrees per count, thus the better the resolution. The number of pulses per revolution is limited by the physical line spacing of the segments and by the quality of light transmission. Higher resolutions are available through various interpolation (multiplication) techniques.

One method uses track A and track B and counts in three modes called X1, X2, and X4. The X represents multiplication. Figure 25-15 summarizes the three modes of operation.

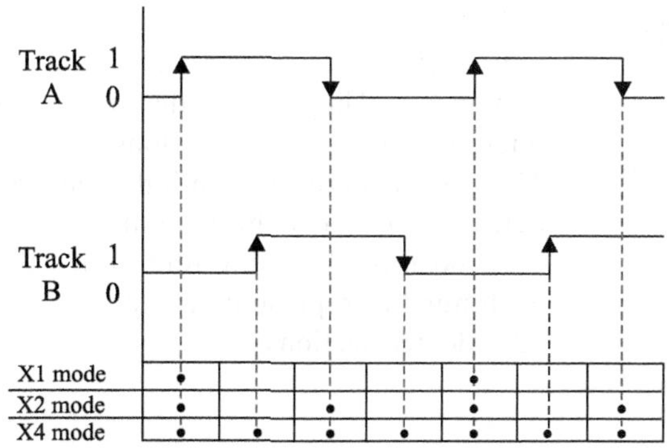

FIGURE 25-15 Quadrature modes

IAU10008
Incremental Encoder
Quadrature Modes

In the X1 mode, the number of pulses generated is equal to the segments of track A on the encoder wheel. An encoder with 500 segments will generate 500 counts per revolution. Only the off-to-on transitions in track A are counted. In the X2 mode, both the off-to-on and the on-to-off transitions in track A are recognized. Therefore, an encoder with 500 segments will generate 1000 counts per revolution. In the X4 mode, the off-to-on and on-to-off transitions of both tracks are used by the external counter. An encoder with 500 segments on each track will generate 2000 counts per revolution.

Rotary incremental encoders are also capable of making linear measurements. The linear motion is converted to rotary motion by using rack-and-pinion or nut-ball-screw mechanisms. Linear measurements are also made by linear encoders that have optical tracks in a straight line. Linear incremental encoders operate on the same principle as rotary types, except that linear grating scales are used instead of round disks. Either the scale or the light source is moved, while the other is stationary. The output produced is the same as that of the rotary encoder.

Incremental encoders are used for high-speed applications from small printers to large, multiaxis machine tools. The disadvantage of the incremental encoder is that the contents will be lost if power to the counter is interrupted. Once power is restored, the reference point must be re-established.

There is a maximum speed at which an incremental encoder can operate. The limitation is typically determined by the speed at which the light sensor can switch on or off, or how fast the signal conditioning circuitry can react to the frequency of the pulses. If the speed is excessive, the signal from the encoder disk will be lost.

The following formula is used for solving the frequency produced by a two-track incremental encoder disk with 1000 translucent slots per circular track, when the maximum speed of the rotary device is measuring 3000 RPM, and it has an interpolation factor of 4.

$$f = \frac{S \times I_F \times RPM}{60}$$

where,

f = Counts per second
S = Number of slots on a track
I_F = Interpolation factor
RPM = Revolution per minute of the device being monitored
60 = Numerical constant

$$f = \frac{1000 \times 4 \times 3000}{60}$$

$$= 200{,}000 \text{ counts per second}$$

It is common for signal conditioning circuitry to operate as high as 4 MHz.

Absolute Encoders

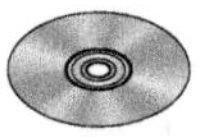

IAU8007
Absolute Optical
Encoders

Figure 25-16 shows an **absolute optical encoder** disk. It has four concentric tracks that vary in size. They are smaller at the outside edge and become larger toward the center. The tracks on the disk form sections with unique code patterns to represent particular positions. This code is derived from light sensors located at each track. A parallel output from these detectors produces highs and lows to form a 4-bit pure binary *whole word* that indicates position. By using four digits, the binary count of 0000 through 1111 is possible. If each number represents a sector of equal size, the encoder disk is divided into sixteen 22.5-degree sections.

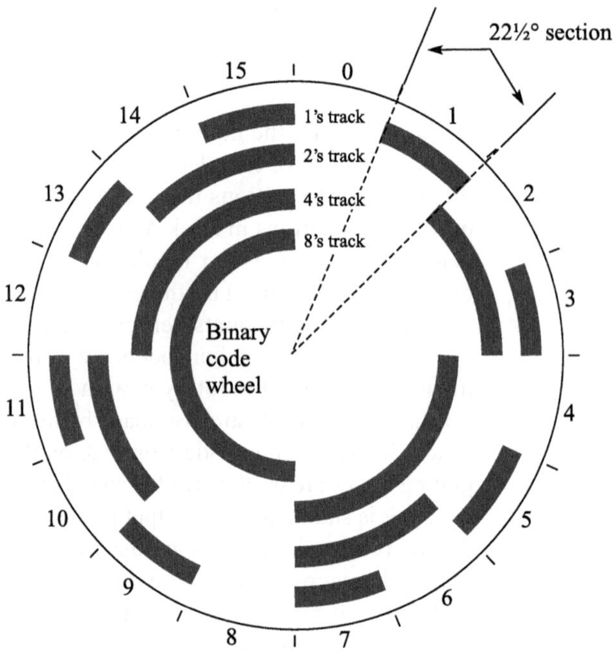

FIGURE 25-16 A 4-bit absolute binary encoding disk

Using a pure binary encoder sometimes presents a problem. If all the bits do not change at exactly the same instant as the encoder crosses over from one sector to the next, a false number may temporarily be detected. For example, when going from the seventh section to the eighth, all four bits on the encoder must change at exactly the same time and as the count increments from 0111_2 to 1000_2. However, if the most significant bit changes slightly earlier than the other three bits, the encoder output would temporarily read 1111_2. This represents a 180-degree error of the encoder position. Since it is extremely difficult to construct a disk with enough precision to prevent such slight differences in the bit-switching times, another numbering system, called the **Gray code,** is frequently used.

The Gray code is a system that also uses 1s and 0s. Table 25-1 shows a comparison between the standard binary system and the Gray code. The table shows that when counting is done in the standard binary system, it is not uncommon to have more than one bit change at a time. The Gray code, however, is designed so that only one bit changes at a time. This characteristic makes the Gray code system an ideal choice for encoder disks. Even with a disk that lacks precision, if the bit changes too soon or too late, the temporary error is insignificant. The 4-bit Gray code encoder disk is divided into 16 sectors, as shown in Figure 25-17. To operate as a 4-bit optical encoding device, four separate light sources with four separate phototransistors are used for each of the four tracks.

TABLE 25-1 Gray code and binary code comparison

Decimal	*Gray Columns ABCD*	*Binary* $2^3 2^2 2^1 2^0$
0	0000	0000
1	0001	0001
2	0011	0010
3	0010	0011
4	0110	0100
5	0111	0101
6	0101	0110
7	0100	0111
8	1100	1000
9	1101	1001
10	1111	1010
11	1110	1011
12	1010	1100
13	1011	1101
14	1001	1110
15	1000	1111

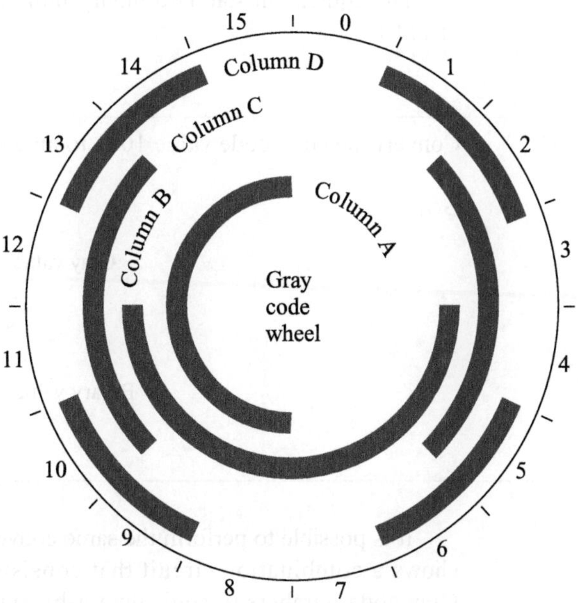

FIGURE 25-17 A 4-bit Gray code encoding disk

IAU8307
Converting Between Gray and Binary Codes

Gray-Code-to-Binary-Code Conversion One method used to convert a Gray code number to its equivalent binary value is shown in Figure 25-18. This table is used to convert 4-bit Gray code numbers to 4-bit standard binary numbers. Similar tables can be used to convert numbers that contain greater or fewer than four bits.

The conversion from Gray code to binary code can be accomplished by following these five steps:

1. Enter the Gray code number to be converted into the Gray section of the table. The number 1110 is entered. From Table 25-1, which compares the standard binary system and the Gray code, note that the most significant bit (MSB) is the same for both.

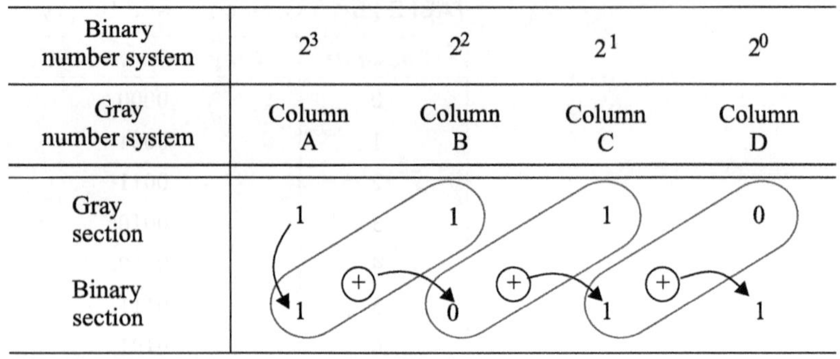

FIGURE 25-18 Gray-code-to-binary-code conversion table

2. Because the MSB of the Gray code number is 1, place another 1 in column A of the binary section of the table.
3. Exclusive-OR the bit in column A of the binary section with the bit in column B of the Gray section. Place the result in column B of the binary section of the table.
4. Exclusive-OR the bit in column B of the binary section with the bit in column C of the Gray section. Place the result in column C of the binary section of the table.
5. Exclusive-OR the bit in column C of the binary section with the bit in column D of the Gray section. Place the result in column D of the binary section of the table.

The equivalent standard binary number 1011_2 is generated from the Gray code number of 1110.

EXAMPLE 25-1

Convert the Gray code value 1001 to an equivalent binary number.

Solution

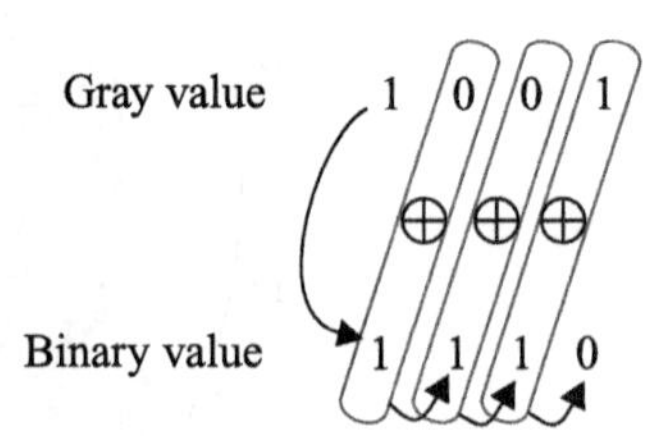

It is possible to perform the same conversion process by using logic circuitry. Figure 25-19 shows a combination circuit that consists of three exclusive-OR gates that convert 4-bit Gray code numbers to equivalent 4-bit standard binary numbers. As the Gray code number 1110 is placed into the register on the left, the figure illustrates how the 1s and 0s are manipulated by each gate to generate binary number 1011_2 at the output.

Binary-Code-to-Gray-Code Conversion Some applications require that binary numbers be converted to equivalent Gray code values. Figure 25-20 shows how this process is performed.

Follow these steps to convert binary code to Gray code:

1. Enter binary 1011_2 number into the binary section of the table.
2. Since the MSB of the standard binary number is 1, the same value is placed in column A of the Gray section.
3. Exclusive-OR the binary bits of columns A and B, and place the result in column B of the Gray code section.

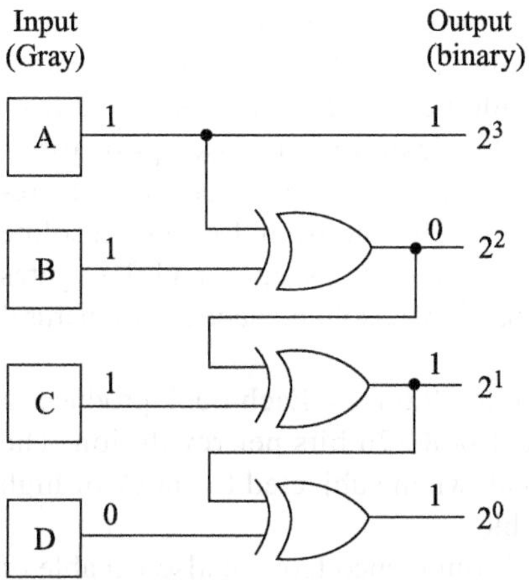

FIGURE 25-19 Gray-code-to-binary-code conversion circuit

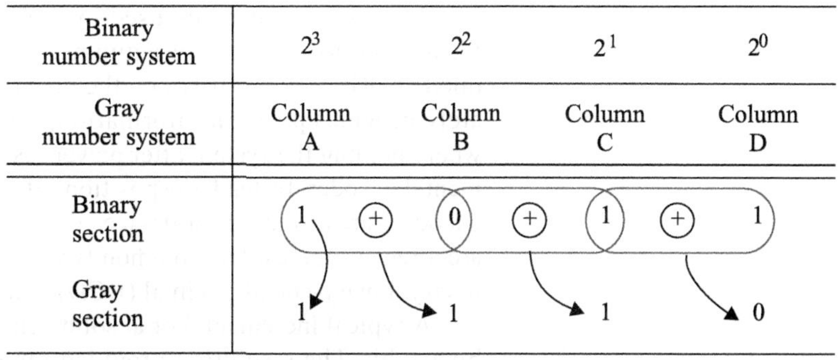

FIGURE 25-20 Binary-code-to-Gray-code conversion table

4. Exclusive-OR the binary bits of columns B and C, and place the result in column C of the Gray code section.
5. Exclusive-OR the binary bits of columns C and D, and place the result in column D of the Gray code section.

The Gray code number 1110 is generated from the standard binary value of 1011_2.

EXAMPLE 25-2 Convert the binary number 1010 to its Gray code equivalent.

Solution

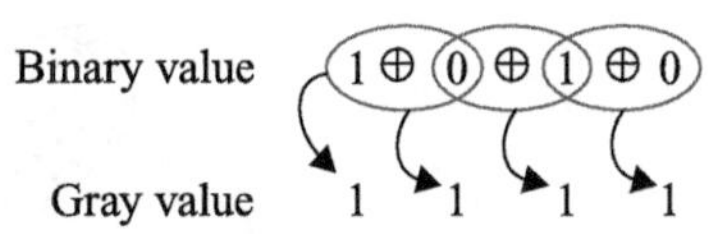

The binary-to-Gray conversion function can be performed by a combination circuit consisting of three exclusive-OR gates, as shown in Figure 25-21. The placement of 1s and 0s at the inputs and outputs of the gates shows how the binary number 1011_2 is converted to its equivalent Gray code number (1110) by the circuit.

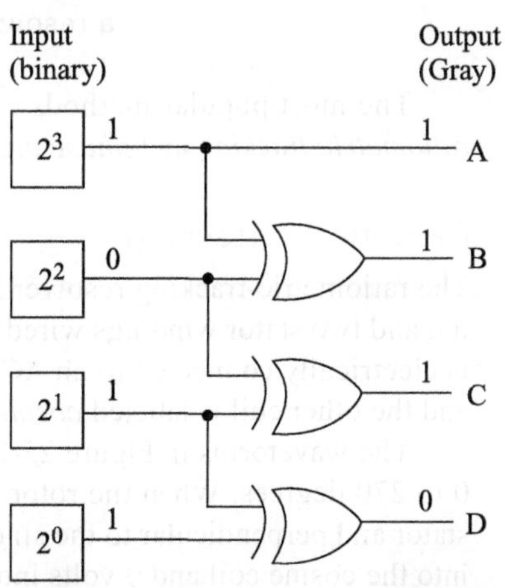

FIGURE 25-21 Binary-code-to-Gray-code conversion circuit

Absolute encoders are used in applications where devices being measured are inactive for long periods of time or equipment power is frequently turned off. When power is reapplied, the Gray code pattern on the disk that is aligned with the emitter and detector immediately provides position information. This characteristic is desirable in those applications where homing the system after power loss is not feasible, such as when required by the incremental encoder. Instead, the position information is immediately read from the disk. Absolute encoders are also used for devices that are relatively slow. Cranes, telescopes, and floodgates are a few examples. The common types of binary codes used by absolute encoders are natural binary, binary-coded-decimal (BCD), and Gray code.

A typical incremental or absolute encoder disk has 6 to 20 tracks. Each track produces a binary bit. The resolution provided ranges from 64 to 1,048,576 bits per revolution. The disadvantages of optical encoders are that they can break when subjected to shock or high temperatures and their light sources are not always reliable.

In addition to detecting an angular position, some absolute encoders are also capable of measuring velocity.

Resolvers

The **resolver** is a rotary transformer capable of making precise position measurements and, in some applications, velocity measurements. The resolver consists of a rotor winding that is rigidly attached to the rotating shaft being measured, and of stator windings that are stationary. Depending on the manufacturer, the rotor or the stator coils are used as the primary. The primary winding is typically driven by a reference voltage at a frequency ranging from 400 Hz to several kHz. As the shaft rotates, the relative position between the rotor and stator windings changes, which causes the amplitude or the phase shift of the induced signal at the secondary to vary. The amplitude of the voltage, or the phase shift, corresponds to the shaft position. The physical configuration of a resolver is shown in Figure 25-22.

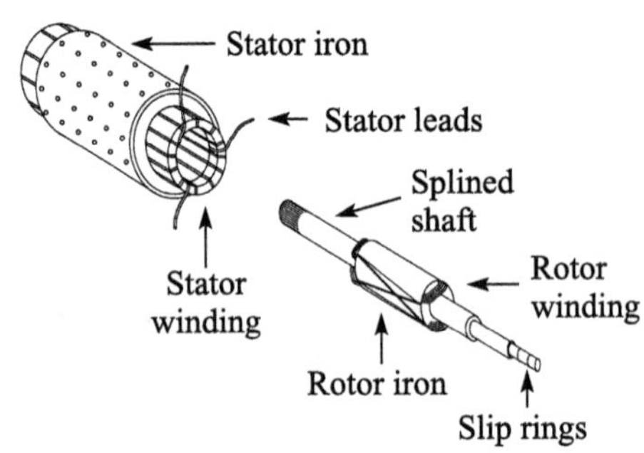

FIGURE 25-22 Physical configuration of a resolver

The most popular methods of determining rotational measurements by a resolver are *ratiometric tracking* and *phase digitizing.*

Ratiometric Tracking

The ratiometric-tracking resolver is shown in Figure 25-23(a). It consists of one rotor winding and two stator windings wired 90 degrees apart. The rotor winding connected to the shaft is electrically connected to an AC supply through slip rings. One stator coil is labeled *sine,* and the other coil is labeled *cosine.*

The waveforms in Figure 25-23(b) depict the operation of the resolver as it rotates from 0 to 270 degrees. When the rotor coil is in the 0-degree position, it is parallel to the cosine stator and perpendicular to the sine stator coil. The result is that maximum voltage is induced into the cosine coil and 0 volts into the sine coil. As the rotor moves from the 0-degree position toward the 90-degree position, it induces progressively less AC into the cosine winding and progressively more into the sine winding. At 90 degrees, there is no voltage induced into

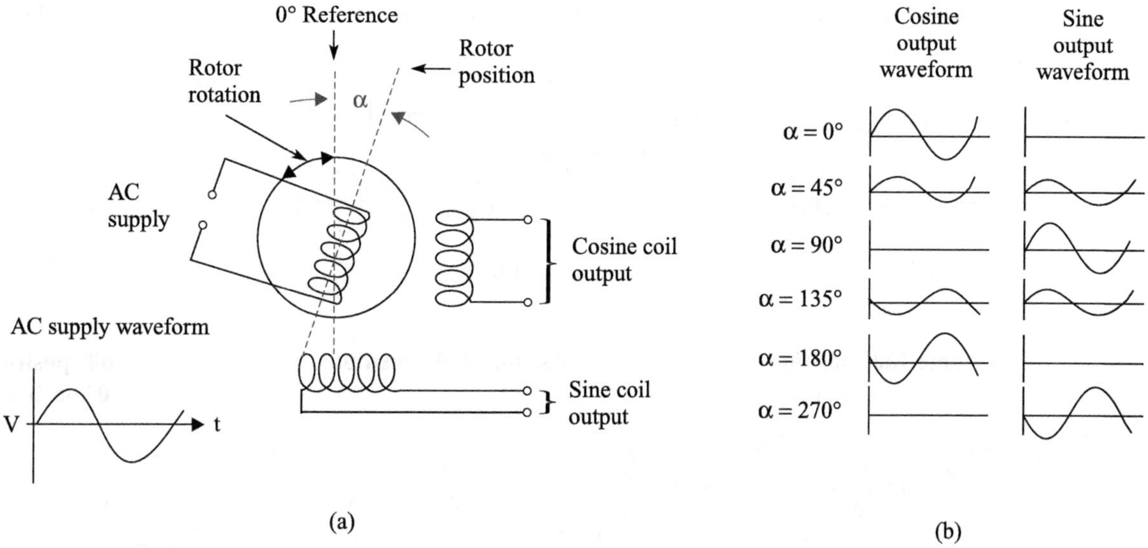

FIGURE 25-23 A ratiometric-tracking resolver

the perpendicular cosine winding, and the AC voltage at the sine coil is maximum. As the rotor turns from 90 to 180 degrees, the sine voltage reduces to 0 volts. The cosine voltage increases until it is at the maximum level, but it is now 180 degrees out of phase with the rotor signal. Therefore, at any 360-degree position, the angle of the shaft is determined by the unique waveforms produced at the sine and cosine coils.

By applying the sine and cosine output stator-winding signals to an analog-to-digital converter, referred to as a ratiometric tracking resolver-to-digital converter (R/D), absolute position can be determined. The R/D calculates the ratio of the sine and cosine voltages by the following formula:

$$\frac{V_{S_1}}{V_{S_2}} = \frac{\text{VR Sin } \theta}{\text{VR Cos } \theta} = \text{Tan}$$

The division function finds the resultant tangent value that is used to determine the shaft angle. A 10-bit R/D will convert one resolver revolution into 1024 digital parts.

Phase-Digitizing Method

The phase-digitizing resolver operates on the principle that a phase shift between a reference signal and the voltage induced into the rotor is a direct measure of the shaft position. A simplified diagram of this resolver is shown in Figure 25-24(a). Two stator windings are mechanically fixed at a 90-degree angle to one another. The rotor winding is free to rotate, and the shaft around which it is wound is connected to the physical object being measured. Both stator windings operate as the primary of a transformer. A 2 kHz 10-V P-P excitation signal is applied to the stator winding on the left. This signal is referred to as the RPO, which means *reference phase output* signal. Another excitation signal, which lags the RPO by 90 degrees, is applied to the winding on the bottom. This signal is referred to as the QPO, or *quadrature phase output,* which means 90 degrees out of phase. The rotor winding operates as a transformer secondary. It moves 360 degrees in either direction as the shaft rotates. The voltage induced into it by both stator windings produces a signal that represents position.

Figure 25-24(a) shows the resolver at 0 position. The voltage induced by the RPO winding into the rotor coil is maximum because the coils are parallel to each other. No voltage is induced by the QPO winding into the rotor winding because they are perpendicular to each other. Therefore, the rotor output signal will be in phase with the reference (RPO) signal.

Figure 25-24(b) shows the resolver at the 45-degree position. The rotor coil is at a 45-degree angle to both stator windings. An equal amount of voltage is induced by the RPO and

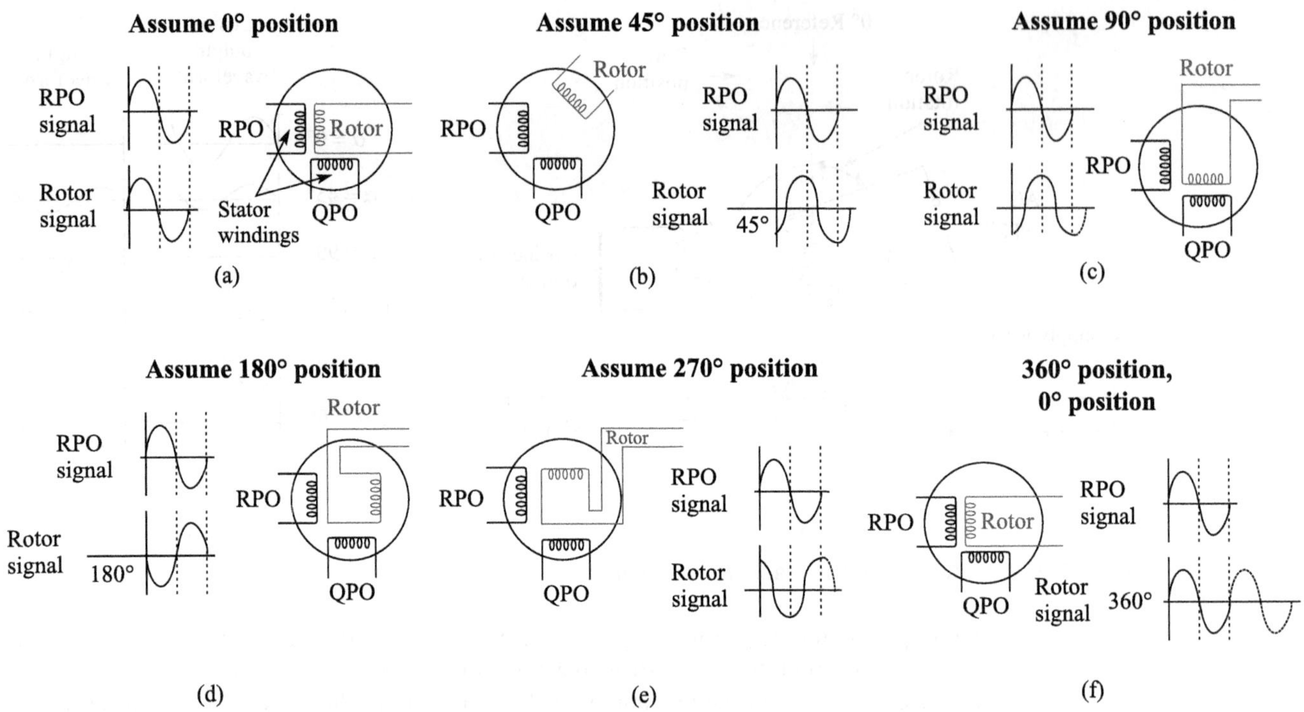

FIGURE 25-24 The waveforms of a resolver at various positions

QPO windings into the rotor coil. The rotor output shifts to the right and lags the reference (RPO) signal by 45 degrees.

Figure 25-24(c) shows the resolver at the 90-degree position. The voltage induced by the QPO winding into the rotor coil is maximum because they are in parallel. No voltage is induced into the rotor by the RPO winding because they are perpendicular to each other. Therefore, the rotor output signal will be in phase with QPO or lag the reference (RPO) by 90 degrees.

Figure 25-24(d) shows the resolver at the 180-degree position. The rotor is in parallel with RPO, but it has been rotated 180 degrees. Therefore, a phase shift of 180 degrees is developed. Figure 25-24(e) shows that at 270 degrees, the rotor coil is in parallel with QPO, but it is 180 degrees out of phase, resulting in a 270-degree phase shift with the reference RPO. At 360 degrees, or 0 position, RPO and the rotor will be in phase again, as shown in Figure 25-24(f).

Figure 25-25 shows the resolver connected to two operation modules. The first, the digitizing reference module, generates and synchronizes the excitation signals necessary for phase digitizing. An 8-MHz clock is the heart of the system. A frequency divider reduces this high-speed oscillator signal into a lower frequency of 2 kHz. This signal becomes the reference phase output, which is connected to the RPO winding. It is also fed into a phase-shift circuit, which produces a signal that lags RPO by 90 degrees. This becomes the 2-kHz signal that is applied to the QPO winding.

To determine the phase shift between the reference (RPO) signal and the rotor output signal, a second device, called the *position detection module,* is used. It consists of a high-speed counter, a zero-crossover circuit that monitors the RPO signal, and another zero-crossover circuit that monitors the rotor output signal. When RPO crosses the zero reference line in the positive direction, it is detected by the zero-crossover circuit, which starts the counter. The clock pulses fed to the counter are supplied from the digitizing reference module at a frequency of 8 MHz. The counter increments until the rotor output crosses the 0-volt reference line in the positive direction. When it does, it is detected by the second zero-crossover circuit, which stops the counter. The number will be trapped, which represents position information called *feedback units*. Feedback units are simply measurements of how far the shaft has turned.

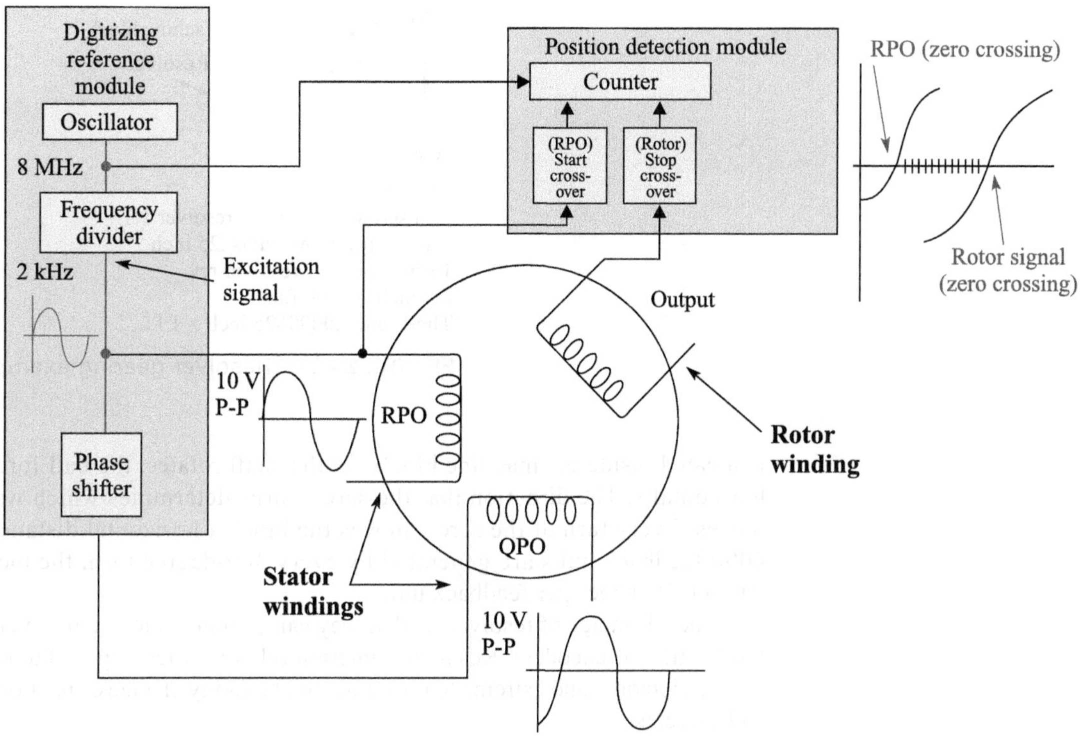

FIGURE 25-25 A resolver connected to the operation modules

The counter is capable of incrementing from 0 to 3999, or 4000 counts. If the rotor is at the 0 position, there will be no phase difference between the reference signal and resolver output. Thus, the number in the counter will be 0. If the rotor turns somewhere between 359 and 360 degrees, the phase shift will be nearly one period of a cycle. Eight MHz is 4000 times greater than 2 kHz. Therefore, nearly 4000 counts, called *feedback units,* will increment the counter during one cycle. Thus, the number 3999 represents a shaft revolution of just under 360 degrees.

Figure 25-26 shows that, since 4000 feedback units equal one 360-degree revolution, 1000 represents a 0- to 90-degree CW rotation; 2000 indicates a 180-degree position; and a 3000 count is a 270-degree rotation. The counter is continuously updated to indicate any change in position. As the resolver is turned CCW, the count becomes smaller.

Figure 25-27 shows how a resolver monitors linear position. A ball-screw mechanism is physically coupled to the resolver shaft. The ball, which rides along a groove in the screw,

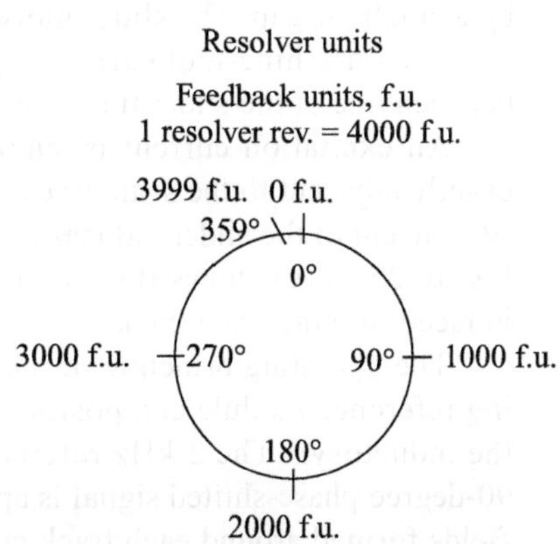

FIGURE 25-26 Resolver units

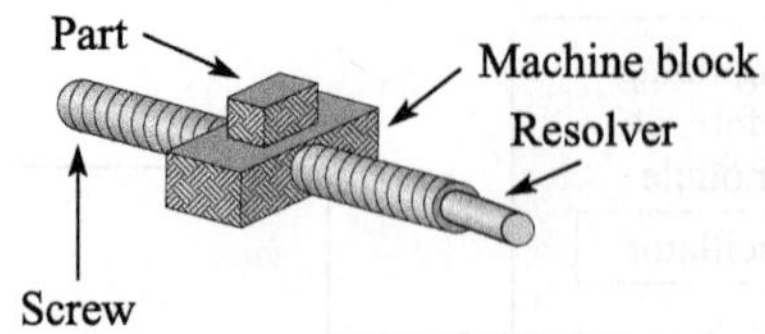

Assume:
Ball screw is geared to resolver 1:1
1 screw rev. moves axis .25 inch
1 screw rev. = 1 resolver rev.
.25 inch = 4000 f.u.
Therefore: .0000625 inch = 1 f.u.

FIGURE 25-27 Resolver gearing example

is located inside the machine block. As the shaft rotates, the ball forces the block to move horizontally. The direction that the screw turns determines which way the machine block moves. Every turn of the screw moves the block a horizontal distance of 0.25 inch. Since 4000 feedback units are generated by every 360-degree turn, the measuring resolution is 0.0000625 inches per feedback unit.

The advantage of resolvers is that they can operate under environmental conditions that are too hostile for encoders. Resolvers function reliably when exposed to continuous mechanical shock, vibration, and extreme temperature and humidity changes, and around oil mists, coolants, and solvents.

25-3 Linear Displacement Feedback Devices

Inductosyn

Although encoders and resolvers are ideal for measuring rotary position, they can make linear position measurements by employing lead screws (ball screws). However, ball-screw mechanisms lack accuracy due to backlash and are expensive. A device that overcomes these drawbacks by making linear position measurements directly is called an **inductosyn.**

The inductosyn is shown in Figure 25-28. It consists of a 10-inch continuous rectangular pattern called a *scale.* The scale is a conductive printed circuit track covered by an insulating layer. The function of the track can be considered the same as the windings of a coil. The scale remains fixed to a solid axis, such as a machine-tool bed.

Another part of the inductosyn is the *slider.* It has two separate but identical printed circuit tracks (bonded to the surface of the Plexiglas board) that move parallel to the scale. These two tracks have a rectangular pattern identical to the pattern on the scale. The slider tracks are insulated from each other. The patterns are also offset one quarter of a cycle, or 90 degrees, from each other. The slider is about 4 inches long and is separated from the scale by a small air gap. The slider moves along the scale with the device that is being measured, such as a machine-tool carrier. Figures 25-28(b) and (c) illustrate the magnetic coupling between one of the slider tracks and the scale track.

An excitation current is supplied to the slider track. When a slider track is located exactly adjacent to the scale track, the inductive coupling is maximum. The result is that the AC current of the slider induces a maximum AC signal into the scale track (Figure 25-28(b)). Figure 25-28(c) shows that when the slider and scale patterns are opposite, no signal is induced into the scale track.

The operating principle of the inductosyn is similar to that of the resolver. The digitizing reference module and position detection module used for the resolver are also used by the inductosyn. The 2-kHz reference signal is applied to one slider track, labeled RPO. A 90-degree phase-shifted signal is applied to the other slider track, labeled QPO. The magnetic fields formed around each track cut into the scale track. Like the rotor of a resolver, an AC voltage is induced into the scale track. The relative position of the slider tracks to the scale

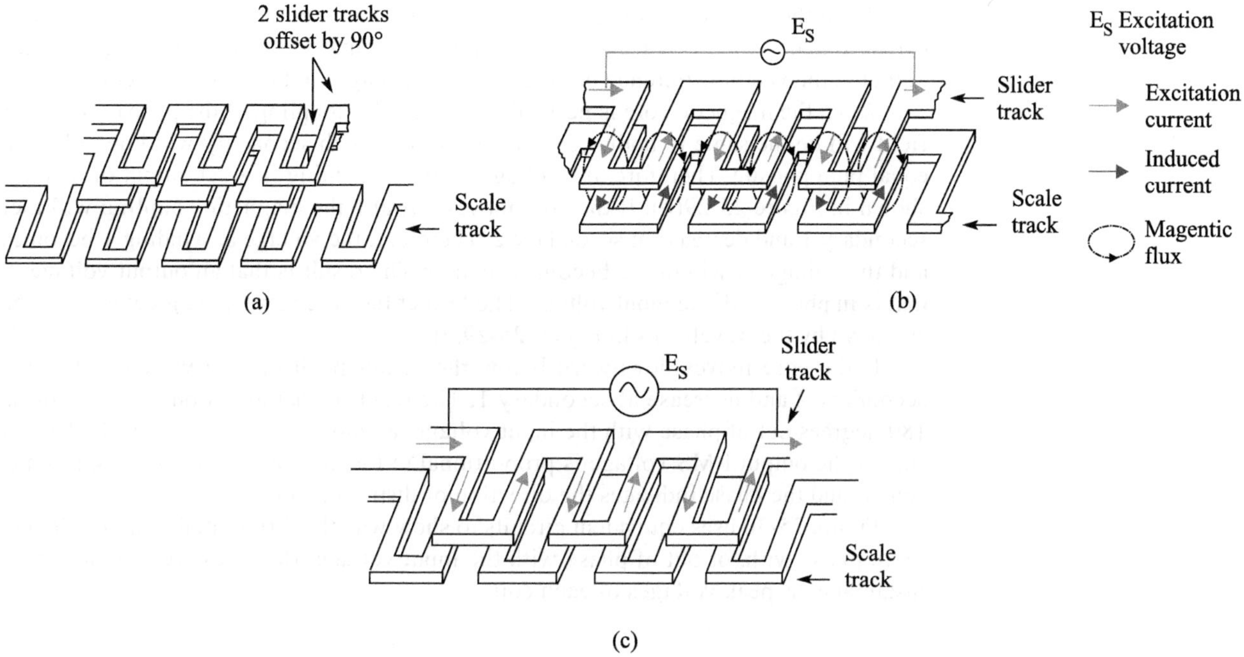

FIGURE 25-28 Linear inductosyn

track determines the phase difference between the RPO signal and the scale output. When the RPO signal crosses the zero reference point in the positive direction, the counter begins to count. The counter is stopped when the scale output crosses the zero reference point in the positive direction.

Suppose the length of each inductosyn section (the same pattern as on square wave) is 2 mm. If the slider moves 1 mm from the 0-degree reference position, the counter will increment to 2000.

Linear Variable Differential Transformer (LVDT)

IAU7807
The LVDT: A Linear Voltage Differential Transformer in Action

A **linear variable differential transformer** (**LVDT**) is an electromechanical transducer. It produces an AC voltage proportional to the linear physical displacement of an object. In its simplest form, an LVDT consists of one primary and two secondary coils wound around a hollow tube. A ferrous metal cylinder called a *core* is placed inside the tube. It provides a path for the magnetic flux linking the coils. Figure 25-29(a) shows the basic winding configuration of the device.

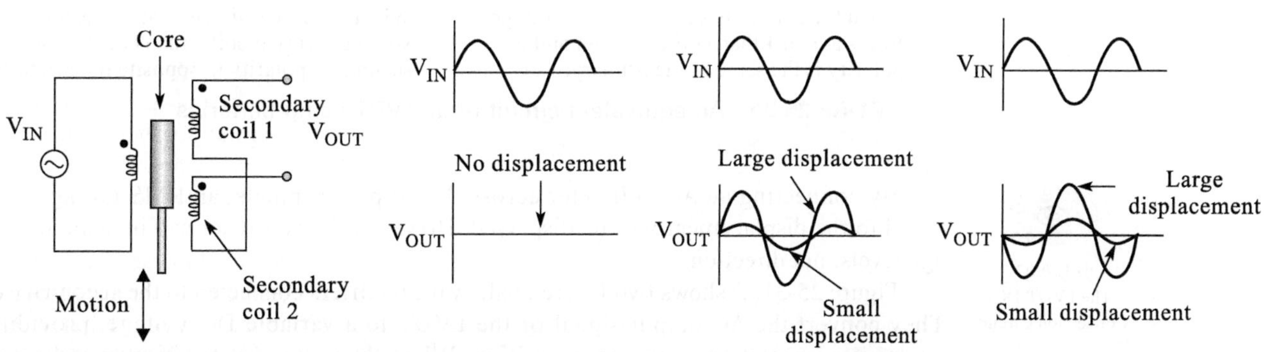

FIGURE 25-29 LVDT

When the primary coil is energized with alternating current, voltages are induced in the two secondary coils. The secondary windings are connected series opposing. The transducer output is the vector difference between the two voltages induced in the secondaries.

When the magnetic core is perfectly centered, the coupling of the primary coil's magnetic field will be the same to both secondary windings. Both secondary winding voltages will be equal but opposite. Therefore, the voltages cancel, resulting in a net output of 0 volts. This is shown in Figure 25-29(b). If the core moves upward, the magnetic coupling will increase at secondary 1 and decrease at secondary 2. Therefore, the voltage at winding 1 becomes larger, and the voltage at winding 2 becomes smaller. The result is that an output voltage develops that is in phase with the input voltage. The farther the core moves, the greater V_{OUT} becomes, as shown by the waveforms in Figure 25-29(c).

If the core moves downward below the center position, the voltage will increase at secondary 2 and decrease at secondary 1. The result is that an output voltage forms that is 180 degrees out of phase with the input voltage, as shown in Figure 25-29(d). Therefore, the size of the output RMS voltage is proportional to the amount of the core displacement from center, and the phase indicates the direction of displacement.

Figure 25-30 uses equivalent circuits to show why the differential output voltage is equal to, in phase with, or out of phase with the input voltage. Batteries are used to represent the instantaneous peak voltages of each coil.

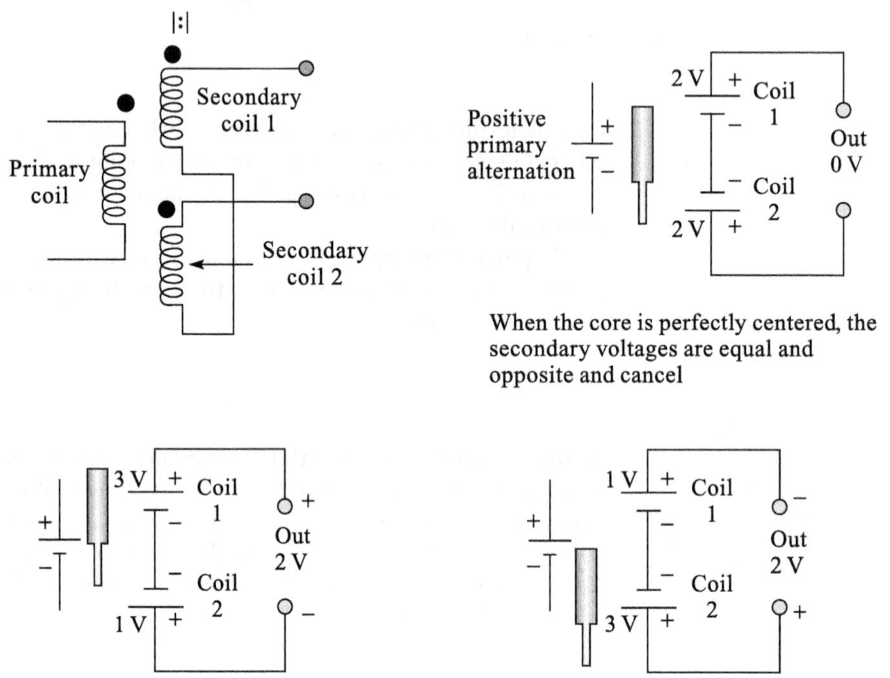

When the core is perfectly centered, the secondary voltages are equal and opposite and cancel

When the core is upward, a greater voltage forms at coil 1 than coil 2. The secondary polarity is the same as the primary.

When the core is downward, a greater voltage forms at coil 2 than coil 1. The secondary polarity is opposite the primary.

FIGURE 25-30 An equivalent circuit of an LVDT using batteries

IAU7907
The LVDT DC Conversion Circuit

By connecting an AC voltmeter across the output terminals, an RMS voltage proportional to the displacement will be displayed. However, the voltmeter can indicate only voltage levels, not direction.

Figure 25-31(a) shows two filtered half-wave rectifiers connected to the secondary coils. They convert the AC output signal of the LVDT to a variable DC voltage, providing an analog representation of the core position. When the voltage is positive, it indicates the positive displacement of the core. A 0-volt output is produced when the core is centered. A negative voltage indicates the negative displacement of the core. The graph in Figure 25-31(b)

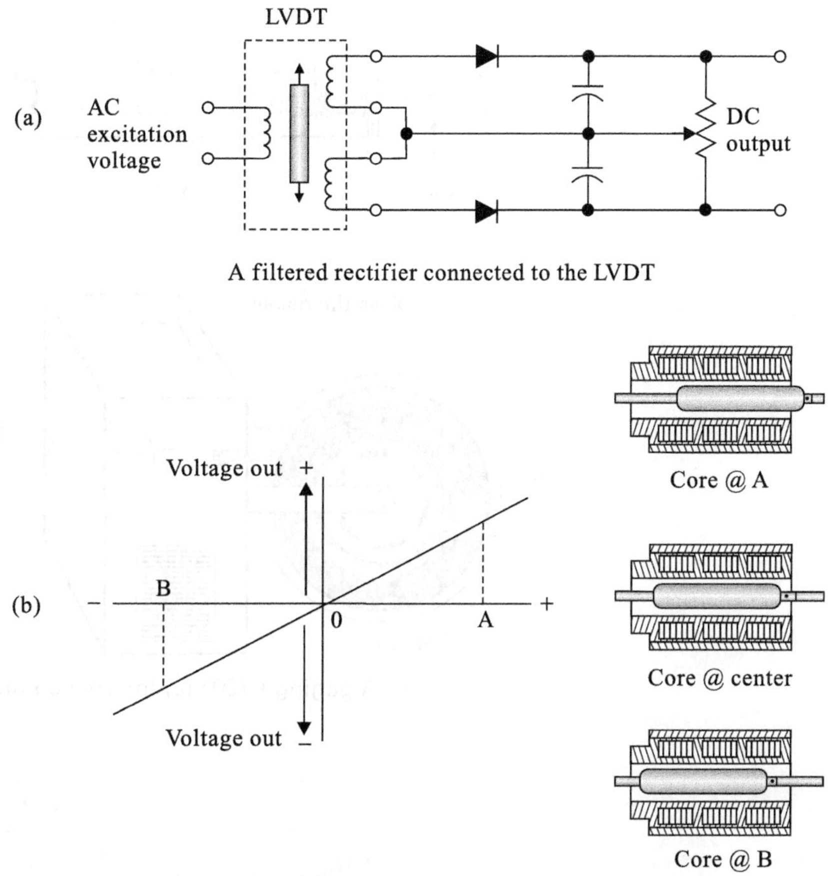

FIGURE 25-31 A filtered rectifier connected to the LVDT

shows how the DC output voltages form at three different core positions—core at A, core at center, and core at B.

The frequency of the excitation voltage ranges from 60 to 20,000 Hz at less than 10 volts. The unfiltered output voltages can range from 0.5 to 10 V AC. Most LVDTs have a displacement range of about 1 inch from center in each direction. If a displacement greater than 1 inch is required, a ratioing mechanical gear apparatus can be used.

In applications where the shaft of the LVDT core cannot be connected to the object being measured, a *gaging transducer* is used. A gaging LVDT, shown in Figure 25-32(a), consists of a high-quality spring, which pushes the end of the shaft that protrudes from the housing against the object being measured. Figure 25-32(b) shows how it is used to inspect the surface of a tire by searching for defects or nonconformances.

Linear Displacement Transducer

IAU4206
Linear Displacement
Transducer

A device that applies magnetorestrictive technology for measuring linear position is the **linear displacement transducer**. Figure 25-33 shows a drawing of this device. It has a housing that contains its electronics. A hollow tube called a *waveguide* is mounted to the housing. A conducting wire inside the waveguide runs the entire length of the transducer tube. The end of the waveguide inside the housing is spring mounted. The other end is hard mounted to a stationary object. Also inside the housing are two strain tapes made of a special nickel alloy through which magnetic lines can easily pass. They are welded to the waveguide. The flux lines from the bias magnets also pass through the tapes. A coil wrapped around each tape is placed within the field of two stationary bias magnets. A ring-shaped permanent magnet passes along the outside of a protective tube that surrounds the waveguide, generating a magnetic field.

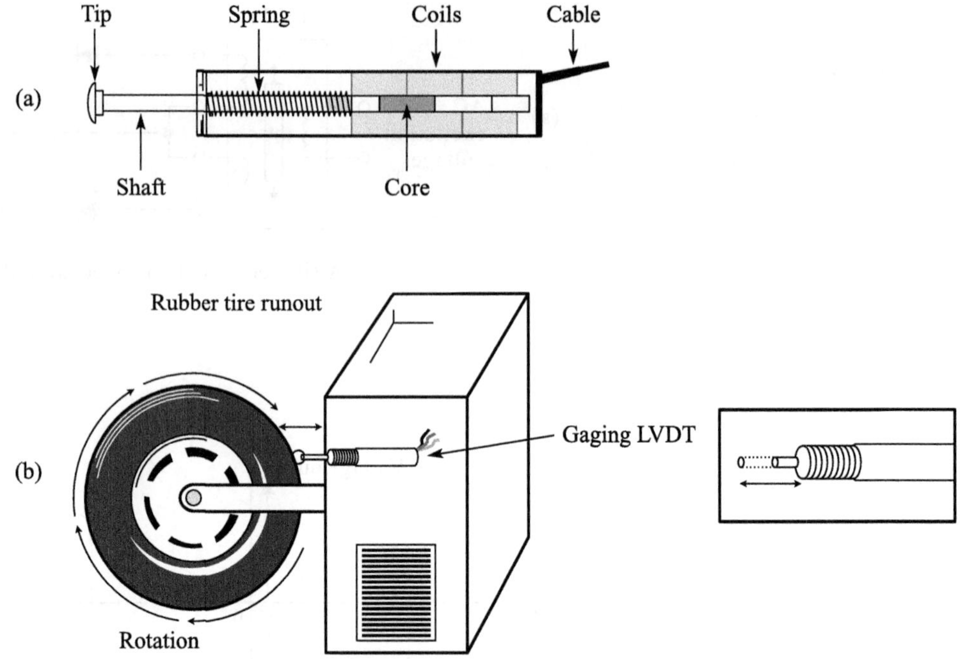

FIGURE 25-32 A gaging LVDT: (a) Internal components; (b) Application example

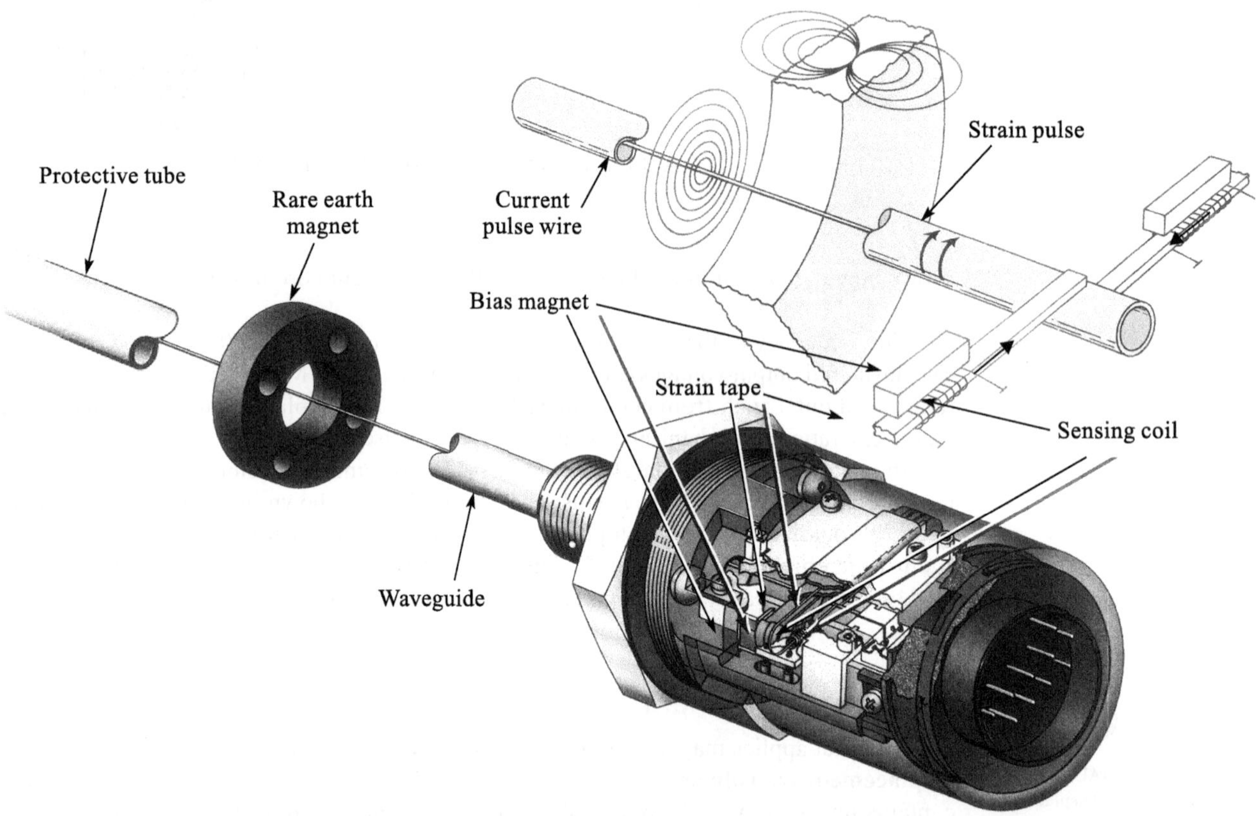

FIGURE 25-33 Linear displacement transducer (Courtesy of Temposonics™)

The ring (magnet) is attached to the object being measured. A conducting wire inside the waveguide runs the entire length of the transducer tube. The measurements are taken along the tube from an indexing point near the housing to the location of the ring.

To measure linear position, a current pulse is launched from the housing along the wire inside the waveguide at a known speed. This is the *interrogation pulse*. As the current travels down the wire from the sensor head, a magnetic field that envelopes the current moves along with it. When the field from the pulse enters the field created by the ring magnet, a momentary interaction between them takes place that produces a helical field. The interaction causes a twisting motion, or torsional strain pulse, in the waveguide. This twist action is known as the *Wiedemann effect*. The strain pulse propagates along the waveguide as an ultrasonic wave at the speed of sound, or over 9000 feet per second. When the pulse arrives at the housing, it is detected by the two tapes. The pulse puts a strain on the tapes, which causes their reluctance to momentarily change. This change makes it more difficult for the flux lines of the magnet to pass through the tape. The result is that the flux lines emitted from the magnet weaken and collapse. As they cut across the sensing coils, a momentary electrical pulse is induced and sent to the detector. This is known as the *Villari effect*. After the strain on the tapes expire, the reluctance of the tapes returns to normal, and the flux lines expand back to their original pattern.

The position of the ring magnet attached to the target is determined by measuring the lapsed time between the launching of the current pulse and the arrival of the strain pulse sensed by the coils. If the target attached to the ring magnet moves farther away from the sensor body, the elapsed time from when the interrogation pulse is launched and the strain pulse is returned to the sensor head will be longer. The longer time duration is converted by the sensor's electronics into a reading that indicates the target moved a greater distance from the reference point.

This information is converted by the electronics inside the housing into various options of electrical output signals. These include:

Analog

Voltage: 0 to +10 VDC
Current: 0 to 20 mA or 4 to 20 mA

Digital

Binary
BCD
Gray code
Pulse width modulation (PWM)

These outputs can be directly connected to a variety of controllers, meters, or other devices to make positioning easy and straightforward in both electronic and hydraulic systems. Zero- and span-positioning adjustments can also be made.

The linear displacement transducer is very durable because there are no moving parts to wear out. Also, it is very accurate because its resolution is 80 millionths of an inch (0.00008) at a speed of 10,000 measurements per second.

In addition to making linear position measurements, this transducer can also measure velocity. Suppose a transducer that produces an analog 0- to 10-DC voltage is used to measure velocity. The amplitude of the voltage it produces is proportional to the speed at which the object being measured moves. The polarity of the signal indicates the direction of travel. For example, the output velocity voltage will be +10 VDC at full (specified maximum) speed when the magnet is moving away from the sensor head. At full velocity toward the sensor head, a −10 VDC will be produced. The velocity output will be 0 VDC when the target magnet is stationary.

The linear displacement transducer is used for a variety of applications such as:

- To control the amount of plastic or glass being fed by an injection molding machine.
- Hydraulic cylinder positioning.
- To control the bending angle of metal forms.
- To determine where to perform a cutting action by a shearing press.
- When there are high levels of dust and contamination.

Linear positioning and velocity measurements can be made over a length of 1 inch up to 25 feet.

Wire Rope Transducer

The **wire rope actuated transducer** is a device used to measure linear position and linear velocity. Shown in Figure 25-34, it consists of a wire rope, tension spring, and sensing element.

Operationally, the base of the device is mounted at a fixed location. The movable object that is being monitored for measurement is attached to the end of the wire rope. As the object moves away from the transducer, the rope extends as it unwinds. An internal spring coupled to the capstan creates a constant tension on the wire rope and also causes it to retract.

As movement occurs, the capstan that turns also rotates a sensing element to which it is coupled. The sensing element produces an electrical output that represents the position or velocity of the device to which the wire rope is attached. For absolute positioning, a potentiometer is used as the sensing element. An incremental encoder is the sensing element that provides digital signals, which represent position. A tachometer is used for velocity measurements.

The attributes of a wire rope actuator include compact size, little or no critical alignment requirements, durability, and low cost. It provides a multitude of applications in many different industries, such as crash-testing automobiles, bone densometers in the medical field, and pump jacks used in crude oil extraction for valve positioning and injection molding machines.

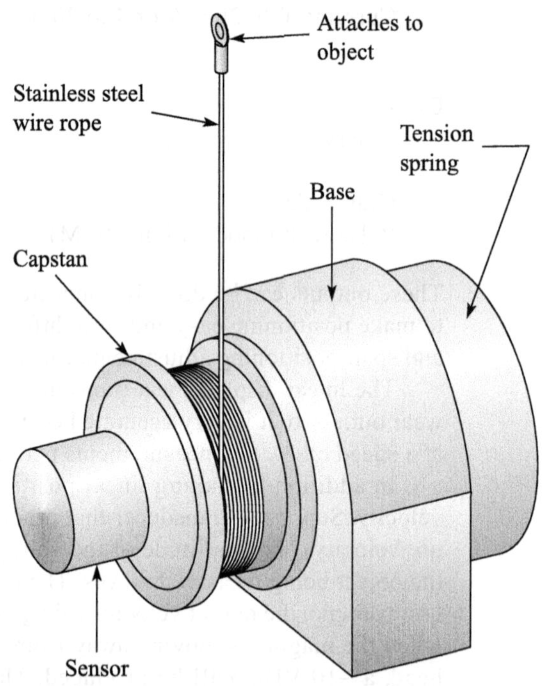

FIGURE 25-34 Wire rope transducer

Linear Velocity Transducer

A linear velocity transducer (LVT) uses inductive technology to measure the speed at which a physical object moves in a straight line. Its operation is based on the principle of Faraday's and Lenz's laws, where moving a magnet through a coil of wire will induce a voltage in the coil. The voltage produced is proportional to the velocity of the magnet and its field strength.

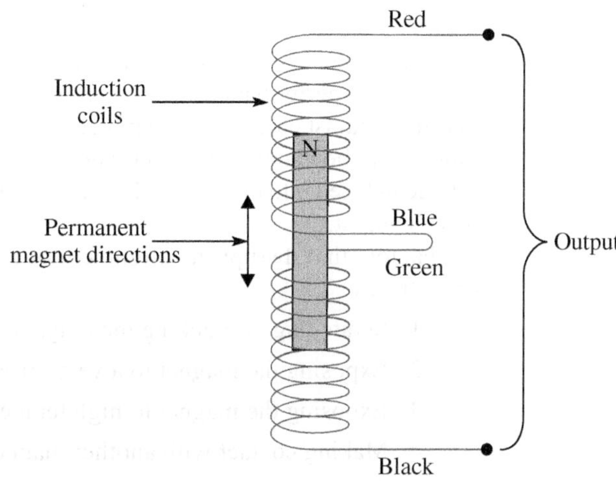

FIGURE 25-35 Linear velocity transducer

The LVT consists of a permanent magnet and a fixed geometric coil, as shown in Figure 25-35. The magnet, which is free to move concentrically inside the coil, is connected to the object it is measuring. During the operation within the working range of the transducer, both ends of the magnet are inside the coil. The coil is divided into sections. With a single coil, a 0-volt output would be produced because the voltage generated by one pole of the magnet would cancel the voltage generated by the other pole. With a two-section coil configuration, the N (north) pole of the magnet will induce a voltage in one coil and the S (south) pole will induce a voltage in the other coil. The two coils are connected in a series-aiding configuration to obtain a DC voltage output proportional to the magnet's velocity.

Figure 25-36 shows that the two coils are in series by connecting the blue and green wires. Figure 25-36(a) shows the polarity of the voltages that are induced in each coil as the magnet is moved in the upward direction. The voltages are summed and the output of the transducer is produced at the terminals of the black and red wires. Figure 25-36(b) shows the voltage polarities at each coil and the output terminals when the magnet moves in the downward direction.

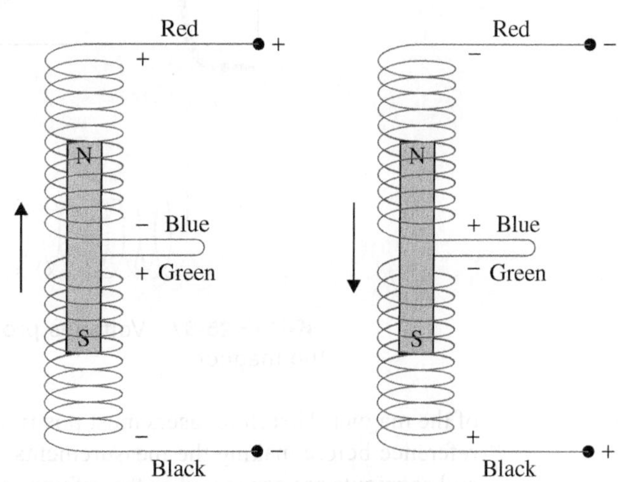

FIGURE 25-36 Polarities produced by the direction of the magnet

Magnets

The performance of the LVT is directly associated with the condition of the magnet. To ensure the optimum performance of the transducer, the integrity of the magnet must be preserved. By guarding against changes in the molecular alignment and the magnetic field, the strength of the magnet can be maintained. However, if the magnetic field strength is altered, the amount of voltage induced into the coils will be changed and alter the output voltage produced at the transducer's output.

Factors that disrupt the alignment of molecules, resulting in the loss of magnetic field strength, are:

1. Scratching or cracking the magnet.
2. Exposing the magnet to a very strong electromagnetic field.
3. Exposing the magnet to high temperatures.
4. Making contact with another magnet
5. Subjecting the magnet to a physical shock.

The output of the LVT is a DC voltage. Figure 25-37 shows the difference of the output voltages it produces when the magnet moves at different speeds. Since a voltage is generated only when the magnet is moving, it is impossible to electrically identify the correct location

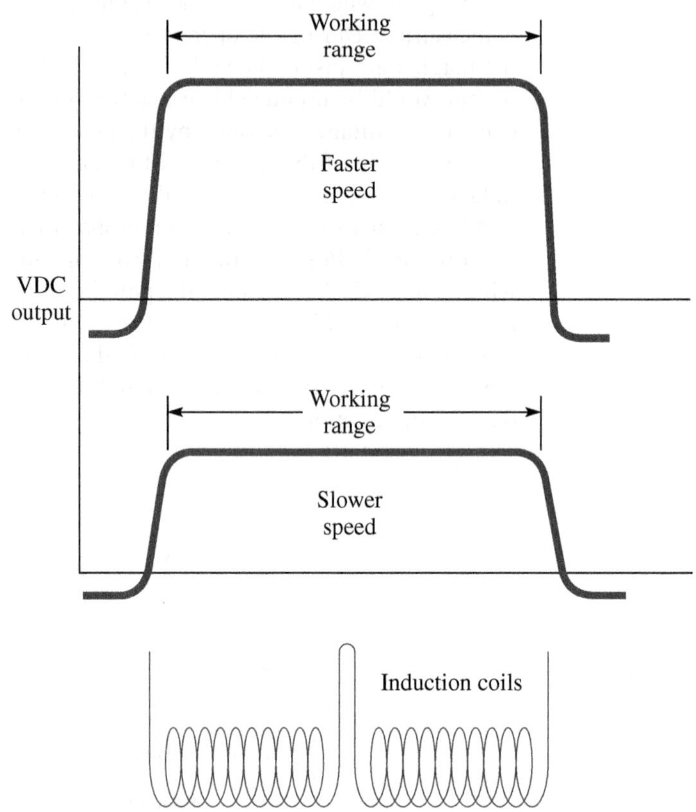

FIGURE 25-37 Voltages produced by different speeds of the magnet

of the magnet. Therefore, users must position the device connected to the magnet at a known reference before making the measurements. It is common to use the junction where the two coil segments are connected as the reference point. When the center of the magnet is aligned with the junction, the magnet can be moved by half of the linear range in either direction.

The LVT is an ideal measuring device for monitoring hydraulic ram speed, drilling rates, and any other applications that require an instantaneous velocity measurement.

Problems

1. Which of the following parameters is measured by a tachometer?
   a. position c. direction
   b. RPM d. distance
2. The faster a toothed-rotor tachometer turns, the greater the ____ it produces to indicate speed.
   a. DC voltage
   b. frequency of pulses
3. A term used to indicate speed regulation is ____.
   a. resolution c. linearity
   b. stability d. accuracy
4. T/F A potentiometer is capable of measuring velocity.
5. T/F A feedback potentiometer is a linear feedback device.
6. T/F A potentiometer is considered a precision measuring device.
7. The function of the demodulator in the angular displacement transducer is to perform a ____ operation.
   a. rectification c. both a and b
   b. filtering
8. Where is the rotor positioned over the sections of the receiving board when the angular displacement transducer produces 0 volts?
9. Which of the following parameters is measured by the optical encoder?
   a. direction c. velocity
   b. position d. all of the above
10. What is the function of the Schmitt trigger in an optical encoder?
11. List two ways an incremental optical encoder can measure velocity.
12. T/F A one-track incremental encoder is capable of measuring direction.
13. The number of degrees per count in an incremental optical encoder is called ________.
14. T/F The disadvantage of an incremental encoder is that if power to the counter is removed, the count is lost.
15. T/F An advantage of an absolute optical encoder is that if power is lost and then restored, it is not necessary to re-establish a reference point.
16. The advantage of the ________ code over pure binary numbers is that only one bit changes at a time when its numbers increment or decrement.
17. Convert Gray code number 1000 to an equivalent binary number.
18. A 5-bit Gray code wheel can be divided into ________ sectors of equal size that are each ________ degrees.
19. Convert the binary number 0101_2 to an equivalent Gray code number.
20. Logic circuits that convert binary numbers to Gray code, or vice versa, use ________ gates.
21. Which of the following parameters is measured by a resolver? ____
    a. position c. velocity
    b. direction d. all of the above
22. When a rotor operates as a transformer primary, it will induce ________ (minimum, maximum) voltage into a parallel stator winding, and ________ (minimum, maximum) voltage into a perpendicular stator.
23. The best type of position device to use in conditions where there is extreme heat, humidity, and dirt is the ____.
    a. potentiometer c. resolver
    b. encoder
24. T/F The operation of an inductosyn is similar to that of a resolver.
25. A resolver is typically used to measure ____ movement, and an inductosyn measures ____ movement.
    a. rotary b. linear
26. T/F The LVDT uses transformer action to measure position.
27. The Wiedemann effect is a ____ action caused by the interaction between two magnetic fields.
    a. pushing b. pulling c. twisting
28. The direction at which the object being measured moves in relation to the housing of a linear displacement transducer is indicated by the ____.
    a. direction that the waveguide twists
    b. polarity of the voltage produced
29. A wire rope transducer uses a(n) ____ to indicate velocity.
    a. incremental encoder
    b. potentiometer
    c. tachometer
30. The index pulse of an incremental optical encoder is used to ____.
    a. establish a reference position
    b. count the number of revolutions made
    c. both a and b
31. When the core is perfectly centered between the primary coil and the two secondary coils of an LVDT, the output produced is ____.
    a. a positive voltage
    b. a negative voltage
    c. either a positive or a negative voltage
    d. 0 V
32. The ____ of an LVDT is physically attached to the object being measured for position.
    a. coil b. center tap c. iron core
33. The linear velocity transducer produces a voltage that is ____ proportional to the speed of the magnet.
    a. inversely b. directly

CHAPTER

26

Fundamentals of Servomechanisms

OBJECTIVES

At the conclusion of this chapter, you should be able to:

- Explain the operation of a closed-loop velocity servo and describe its speed-regulating properties.
- Summarize the characteristics of bang-bang, proportional, and digital position servo controls and explain their operation.
- List and describe the static and dynamic characteristics of a servomechanism.
- Summarize the characteristics of a point-to-point servo and a contouring position servo.
- List and explain the function of the multi loop feedback amplifiers used in position servo systems.
- Describe the characteristics of PID modes and feed-forward.
- Explain the procedure for tuning a servomechanism.
- Describe the operation of a master–slave servo system.

INTRODUCTION

Numerous types of motion-control systems help manufacture products. Each is uniquely designed to perform the required operation. When developing a system, the designer's objective is to build a machine with the particular characteristics needed for the application.

The terms *systems* and *control* describe servomechanisms. A system is defined as the organization of parts that are connected together to operate as a complete unit. To function properly, the system must be controlled to perform its desired operation. Control is achieved by both analog and digital devices that perform both open-loop and closed-loop operations.

In this chapter, examples of analog and digital open-loop and closed-loop systems are examined. Digital control is being used more extensively in modern control applications. Motion-control characteristics are also identified. Different types of circuitry and control modes that are capable of providing each characteristic are described. This information is useful when developing velocity and position servomechanisms.

26-1 Closed-Loop Velocity Servo

IAU4506
The Closed-Loop Velocity Servo System

The control of velocity is a common operation in motion-control applications. The purpose of velocity control is to establish the desired rotary or linear speed of a moving object and to provide speed regulation under varying load conditions.

A velocity-control system is shown in Figure 26-1. Its angular velocity is proportional to the applied reference input voltage on the left of the diagram. Suppose that a 5-volt command

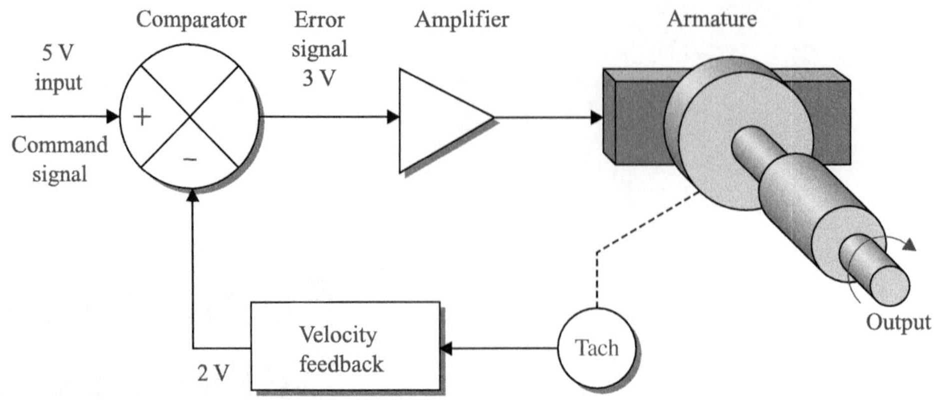

FIGURE 26-1 A closed-loop velocity-control system operating under normal load conditions

signal is applied to the input. The system responds by supplying armature current that causes the DC motor to rotate. If the command voltage is increased, the motor runs more rapidly. As the motor runs, a tachometer mechanically coupled to its shaft also turns.

The DC tachometer generates a DC voltage that is proportional to its shaft velocity. For this tachometer, the output is 1 volt per 1250 RPM. This voltage changes polarity when the direction of shaft rotation is reversed. The tachometer provides the feedback signal of the closed-loop system and applies it to the summing junction of the comparator. The 2-volt feedback signal applied to the inverting input of the comparator is subtracted from the 5-volt command signal applied to the noninverting input of the comparator. The result is that a 3-volt error signal is produced at the output of the comparator. After it is amplified, the error signal is used to drive the motor.

The primary function of the tachometer is to provide speed regulation (which is referred to as *stability*) for the system. Suppose that the motor in Figure 26-1 turns a drive pulley on a conveyor. The motor RPM is 2500. If a disturbance is introduced, such as a heavy box being placed on the belt, the current supplied to the armature is not sufficient to keep the motor turning at the same speed. As the motor slows down to 1250 RPM, the tachometer voltage decreases to 1 volt, as shown in Figure 26-2. The comparator responds by increasing the error voltage to 4 volts. With an increased error signal, a larger current flows through the armature, which produces the higher torque required to meet the increased demand of the load.

Now suppose the disturbance is removed. The elevated current through the armature causes the motor to speed up to 3750 RPM. As the tachometer turns more rapidly, it generates a feedback signal of 3 volts, as shown in Figure 26-3. The comparator responds by dropping

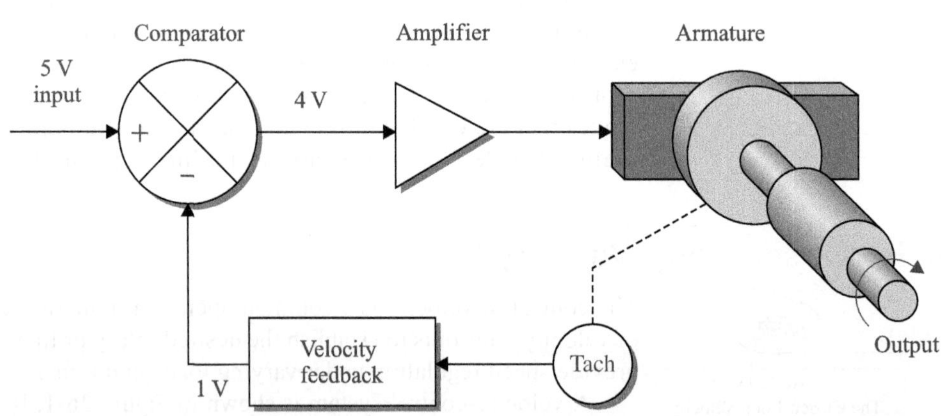

FIGURE 26-2 The velocity-control system's reaction to an excessive load condition

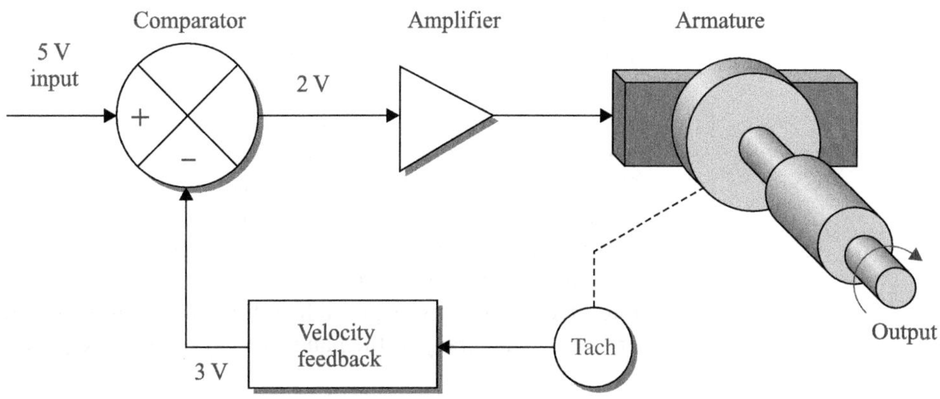

FIGURE 26-3 The velocity system's reaction when the heavy load is removed

the error signal to 2 volts. The reduced current supplied to the armature causes the motor to slow down. When the system eventually stabilizes, the error signal returns to 3 volts and the motor speed goes back to 2500 RPM.

26-2 Bang-Bang Position Servo

A **bang-bang position servo** is an inexpensive system that is rather easy to implement. An example of a bang-bang servo is the system in Figure 26-4(a), which uses a cart to transfer grain from one location to another.

The operation begins by moving the cart into the load position so that it can be filled with the incoming material. Once a strain gauge under the track senses that the cart is full, the loading process stops. The cart then immediately moves down the rail as quickly as possible after power is applied to its motor. As the cart hits the "dump" limit switch, power is removed from the motor and it coasts to a stop over a dump chute. A bottom door on the cart opens and dispenses the load as outgoing material. Once the strain gauge on the track senses the cart has emptied, the door closes, power is restored to its motor, and the cart moves toward the load position as rapidly as possible. As the cart hits the "load" limit switch, power is removed from the motor and it coasts to a position for reloading.

Figure 26-4(b) shows a graphical representation of the operation when the cart moves to the dump position. The "dump" limit switch represents the upper limit. When it is hit by the

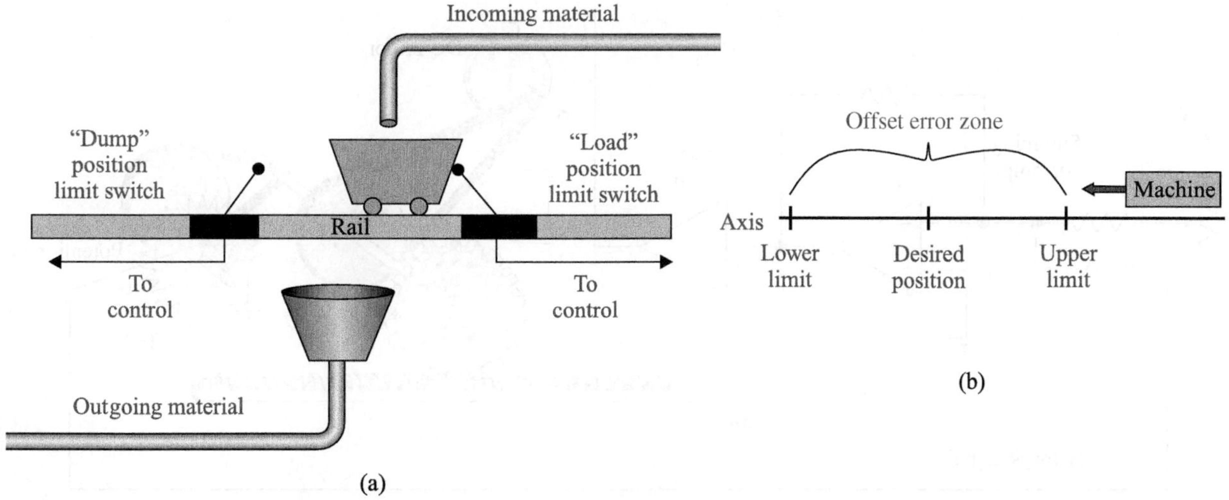

FIGURE 26-4 Bang-bang servo control

cart moving to the left, motor power shuts down. The cart will coast into the desired position somewhere within the offset error zone. If the inertia of the cart is too large due to an excessive load or speed, it will coast past the lower limit and out of range. To correct this problem, it is necessary to lighten the load, reduce the gain of the motor amplifier, or widen the area in which the cart stops. The most convenient option is to lower the amplifier gain. The term "bang-bang" is derived from the manner in which the load moves (at full speed) until it reaches one limit of travel or the other.

The bang-bang servo is used in high-productivity applications that do not require close tolerances. For example, it can be used in a concrete rebar operation where the stock is fed into place as quickly as possible before being cut to size. If the tolerances are within an inch or two, the accuracy of the bang-bang system is adequate.

26-3 Proportional Position Servomechanisms

IAU4406 Analog Position Control

With the proportional closed-loop system, the offset error zone can be made much narrower. Figure 26-5 shows that only the desired position exists. There are many applications where proportional position control is used, such as an automated insertion machine that places parts onto a printed circuit board, a robotic arm that makes a weld on an automobile frame, or a numerical-control machine tool that cuts patterns into a metal stock.

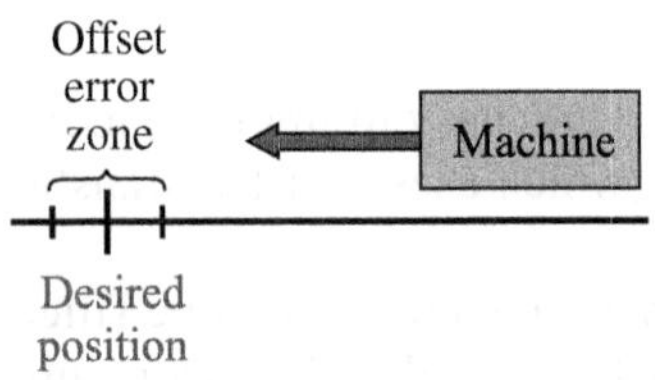

FIGURE 26-5 Offset error of a proportional position servo

Analog Position Servo

An analog position servo is shown in Figure 26-6. The actuator is a permanent-magnet DC motor. Through a series of pulleys, belts, and gears, the motor drives a rack horizontally in both directions. The rotation of the mechanism moves the wiper on a potentiometer through a 300-degree arc. The potentiometer is the feedback transducer of the system. The voltage on the wiper is an indication of the position. The feedback signal is fed back to a summing amplifier where it cancels the setpoint signal that indicates the desired command position. The summing amplifier operates as the controller and produces the error signal of the system.

When the voltage from the position feedback potentiometer equals the setpoint voltage, there is no error signal. The motor does not move because there is no voltage applied to the power amplifier. To move the rack mechanism to the left, the setpoint voltage is increased. Because the positive setpoint amplitude becomes greater than the negative feedback voltage, a negative error signal is produced by the summing amplifier. The result is that a positive voltage is produced by the inverting power amplifier. This potential causes the motor to

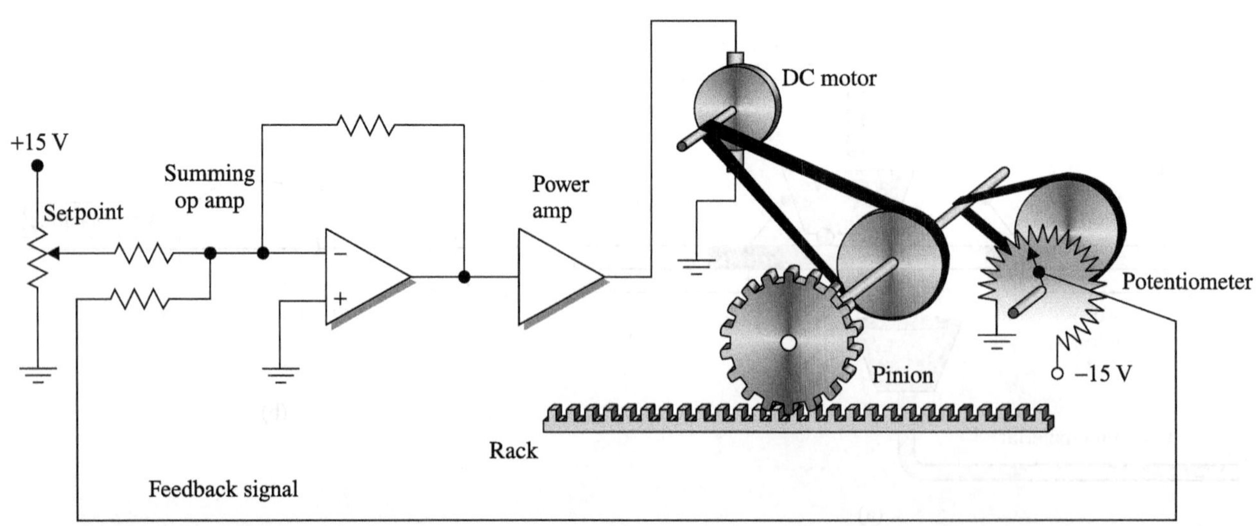

FIGURE 26-6 An analog position servo

rotate CW and drive the pinion gear CW, which moves the rack to the left. As the pinion moves CW, so does the potentiometer. As the feedback voltage becomes more negative, the error signal becomes smaller and the motor slows down. Eventually, the negative feedback voltage from the potentiometer cancels the positive setpoint voltage and the error signal drops to zero. The rack stops moving.

By reducing the setpoint voltage, the same sequence of events occurs. However, because the polarities and mechanical movements are opposite, the rack moves to the right.

26-4 Digital Position Control

IAU4306 Digital Position Control

Figure 26-7 illustrates how the positioning of a linear-motion device can be controlled by a digital command that is entered by a keypad input with numbers 0 through 9. A bar mechanism positioned by a rack-and-pinion assembly has 10 position regions with mechanical stops at both ends of travel. The bar is shown at position 0 and its location is detected by a Gray code wheel mounted on the same motor shaft as the gear. When a number greater than 0 is entered into the keypad, the gear turns in a CW direction, causing the bar to travel to the left. When the bar reaches the region that is the same as the number entered into the keypad, the gear stops and a light turns on. If a second number is entered greater than the first, the gear will again rotate CW until the bar reaches the new location. However, if the second number entered is less than the first entry, the gear will turn CCW, pushing the bar to the right until its position equals the number entered. At that moment, the gear will stop and the light will go on.

To better explain the operation of positioning, functional descriptions of the various sections are provided.

Input Section (Keypad). To move the bar mechanism to any one of 10 locations, a command is entered by a keypad that has 10 different keys numbered 0 through 9. As the desired value is inserted by pressing one of the keys, a 10-ms negative reset pulse ($\overline{R}$) clears the flip-flops. Then, the active-low numerical outputs are activated when an equivalent binary number is parallel-loaded into a 4-bit storage register.

Decision Section (Magnitude Comparator). The control portion of the positioner is a digital circuit called a *magnitude comparator*. As described in Chapter 2, it is made up of two separate 4-bit inputs and three single-lead outputs. The function of the circuit is to compare 4-bit input A to 4-bit input B. If the binary-coded-decimal numerical value applied to input A is larger than B, the $A > B$ output line will go high while the other two outputs go low. If the numerical value of input B is larger than A, the $A < B$ line will go high while lows are generated at the other two outputs. If the binary numerical values of inputs A and B are the same, the $A = B$ line will go high while the other outputs go low.

Amplifier Section. The amplifier section consists of two separate inputs and one output line. One of the inputs is connected to the $A > B$ output of the magnitude comparator. The other input is connected to the $A < B$ output. The output line of the amplifier section provides an amplified potential that is sufficient to drive the DC positioning motor of the circuit.

When the $A < B$ output of the comparator goes high, it is applied to the input of the op amp. Since the op amp is set at a gain of 1, its only function is to invert the +5-volt signal to −5 volts. As the negative 5 volts are applied to the summing power amplifier input, it is amplified and inverted to a positive potential.

When output $A > B$ of the comparator goes to a +5-volt high, it is applied directly to the power amplifier, which amplifies and inverts the signal to a negative potential.

Actuator Section. The device that causes the actual physical positioning in the system is the DC motor. The potential used to activate the motor is provided by the summing power amplifier. When the power amplifier output is positive, electron current flows

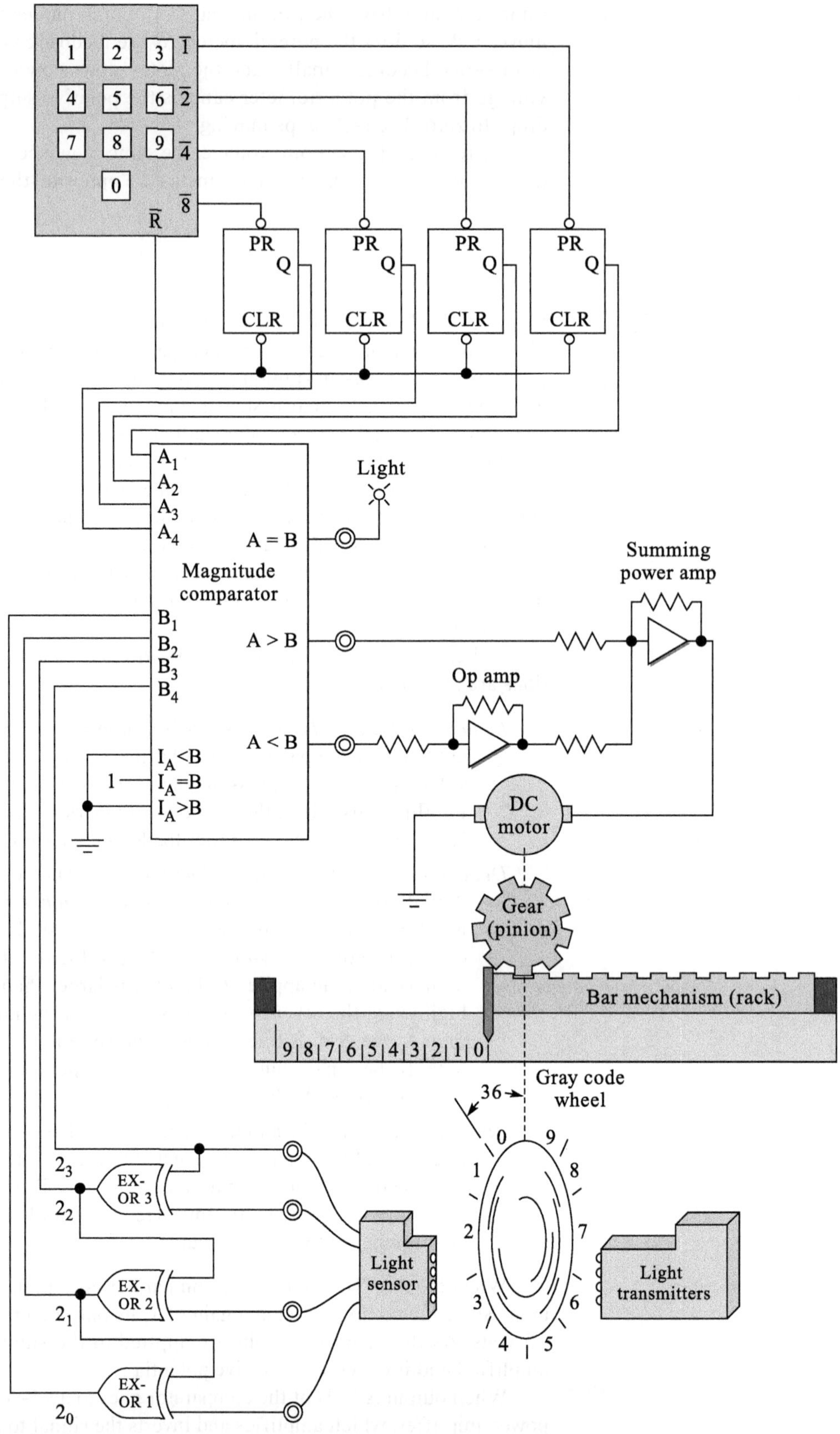

FIGURE 26-7 Digital control of physical position

from ground through the motor to the output terminal. When the power section output goes negative, electron current flows from its output through the motor and then to ground. Since current can flow through the motor in two different directions, it can rotate either CW or CCW.

Sensing Section. Attached to the shaft of the motor is a sensing device that detects the positioning of the system. Figure 26-7 shows that the sensor consists of a disk, four lights, and four optical light sensors. Slots on the disk are arranged in a pattern of 1s and 0s in a Gray code format. (A diagram of a gray code disk is shown in Figure 25-17.) It reveals that the Gray code is divided into 10 pie-shaped sectors of equal size, each representing a 36-degree region.

Rack-and-Pinion. Attached to the end of the motor shaft is a pinion gear that meshes with the teeth on the bar mechanism. When the gear rotates CW, the bar moves to the left; when the gear rotates CCW, the bar moves to the right.

Conversion Circuit. The Gray code numbers generated by the four sensors represent the 10 different physical locations of the digital positioner. However, before the 4-digit Gray code numbers are applied to input B of the magnitude comparator, they are converted into their equivalent pure binary values. Conversion is necessary because a binary number is applied to input A of the magnitude comparator and the comparator must compare values from the same number system to operate properly. The conversion process is accomplished by the three exclusive-OR gates (1, 2, and 3) that act as a Gray-code-to-binary encoder.

Digital Position Control Operation

Refer to Figure 26-7 during the following explanation.

1. Suppose that the bar mechanism is located at position 0. The Gray code wheel mounted on the motor shaft generates a 4-bit output of 0000 that is applied to the encoder, which also generates an output of 0000. Also assume that the number stored in the register is 0000.
2. Since the register output applied to input A of the comparator is the same as the encoder output applied to input B, the A = B output is high and turns on the light.
3. If key 6 is pressed on the keypad, number 0110 is loaded into the register. With the bar mechanism still at location 0, the A input is greater than the B input. Therefore, the A > B output of the magnitude comparator goes high, and the A = B output goes low.
4. The +5-volt signal from the A > B output is applied to the power amplifier, which becomes inverted, and is amplified to produce an adequate potential to drive the motor.
5. As electron current flows through the motor from the negative output of the power amplifier to ground, the motor rotates in a CW direction. Because the disk and gear are mounted on the same motor shaft, they rotate together. The length of each region at the outer portion of the gear and disk is the same size as the corresponding region on the bar mechanism. Therefore, as the gear and disk turn 36 degrees, the bar mechanism moves the distance of one region.
6. When the bar mechanism reaches region 6, the Gray code wheel generates a 0101, which is converted into a binary 0110 by the encoder.
7. As the register supplies a 0110 to input A of the magnitude comparator, and the encoder output supplies a 0110 to input B, the comparator's A > B output goes low and stops the motor. Because the gear-and-bar mechanism also stops, the A = B output goes high and turns on the light.

Electrohydraulic Linear Position Servo

The electrohydraulic linear position servo, shown in Figure 26-8, accepts a digital command signal that represents a numerical value. After receiving the command input, the circuit causes a hydraulic piston to move to a linear position that corresponds to the number entered on the keypad. The number of different positions is limited to a range from 0 to 9 inches. Each count causes the piston to move 1 inch.

Each decimal number entered into the keypad is also loaded into a 4-bit register, consisting of four flip-flops, as a binary-coded-decimal value. The BCD number is then changed to a proportional analog voltage by the D/A converter. The analog voltage, which is either 0 or a positive voltage, is applied to one input of a summing op amp that operates as a comparator. Each time the BCD value increments, the D/A output increases by 1 volt.

The piston (linear actuator) is mechanically coupled to an LVDT. The LVDT produces an analog voltage, which is either 0 or a negative voltage. This voltage is applied to a second

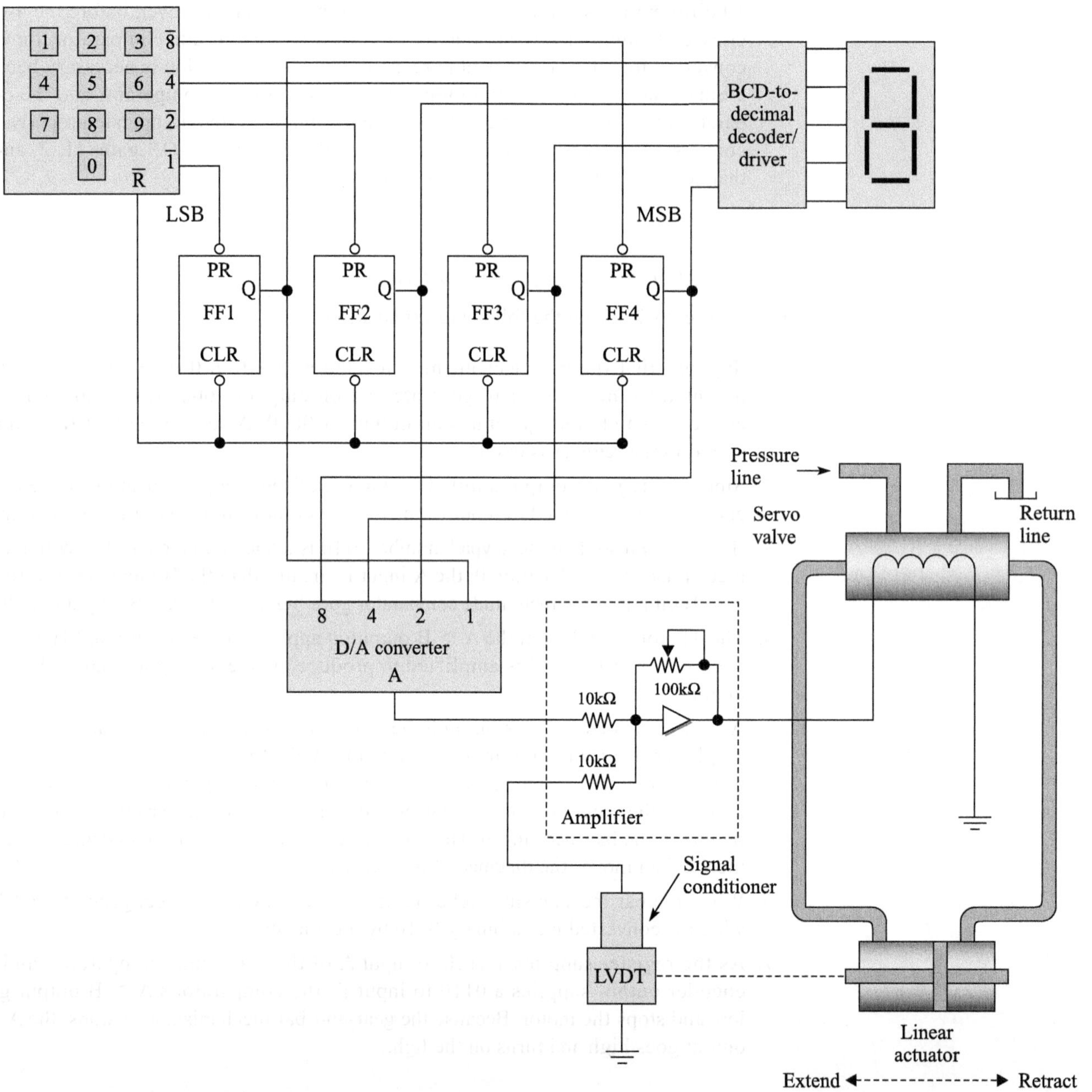

FIGURE 26-8 Electrohydraulic linear position servo

input of the summing op amp comparator. When the piston is retracted at the home position, the LVDT produces a 0-volt output. Each inch it extends from the home position increases the LVDT output by –1 volt.

Operation of the Electrohydraulic Linear Position Servo

Refer to Figure 26-8. Assume that the piston is at the home position and the register contents is 0. Because the output of the D/A converter is 0 volts and the LVDT output is 0 volts, the servo-valve spool is in the center position.

Suppose that the machine operator wants the piston to extend 5 inches. A description of how the circuit performs this function follows:

1. The operator presses key 5.
2. The push-button closure activates a one-shot multivibrator inside the keypad circuitry that generates a momentary 10-ms 0-state pulse at the $\overline{R}$ output. This negative pulse is sent to the clear inputs of flip-flops FF1–FF4, which causes them to reset simultaneously.
3. The completion of the one-shot pulse during the closure of the number 5 key causes outputs $\overline{1}$ and $\overline{4}$ to go low and outputs $\overline{2}$ and $\overline{8}$ to remain high. Therefore, the parallel-loaded storage register is preset to a BCD 5.
4. The LED readout displays a 5 as the Q outputs of each flip-flop are applied to the BCD-to-decimal decoder driver.
5. The count in the register causes the D/A converter to produce a +5-volt output. This voltage is applied to the top input of the summing op amp, and 0 volts from the LVDT are applied to the bottom input lead. The result is that the op amp produces a –5 volts and moves the servo-valve spool to the left, which causes the piston to extend. When the piston has moved 5 inches, the –5 volts from the LVDT cancels the +5 volts of the D/A converter, causing the spool to move to the center position.
6. The operator presses key 2.
7. After the one-shot pulse clears the count, a BCD 2 is loaded into the register. This value causes +2 volts to be applied to the top summing op amp input.
8. +3 volts are produced by the summing op amp because the +2 volts cancel out some of the –5 volts from the LVDT. The result is that the solenoid action in the servo valve moves the spool to the right and causes the fluid to retract the piston.
9. When the piston is at the 2-inch position, the +2 D/A voltage and the –2 LVDT voltage cancel. As the servo spool centers, the fluid is blocked and the piston rod stops moving.

The primary difference between a closed-loop position servo system and a closed-loop velocity servo system is what happens to the error signal when equilibrium is reached. Equilibrium in a position system is attained when the error signal becomes zero and the actuator stops. Equilibrium in a velocity system does not occur when the error signal becomes zero. Instead, it happens when the magnitude of the error signal is adjusted to cause the actuator to return to the desired speed after a disturbance occurs, or when a command signal change is made. If the error signal were to become zero, as in a position system, the actuator causing the velocity would stop.

26-5 Characteristics of a Servomechanism

The designer attempts to ensure that a motion-control servomechanism possesses the characteristics necessary to function effectively. These characteristics fall under two general categories: static and dynamic.

The **static** characteristics relate to the steady-state behavior of the servo system. This condition exists when the load is stationary, such as when it is in the home position or has reached the end point. The **dynamic** characteristics relate to the behavior of the system when a dramatic change in the input signal is applied. This condition exists when the load is moving.

The static characteristics of a typical servomechanism are accuracy, resolution, repeatability, and stiffness. They usually relate to a position-type servomechanism.

Accuracy is the degree to which an output will attempt to match the input command.

Resolution is the smallest discrete output change of position that the servo can make.

Repeatability is the range in which the output position of the servosystem will come to rest whenever a given input command signal is repeated.

Stiffness is the reluctance of an actuator to deviate from the desired position specified by the command signal. A system exhibits good stiffness characteristics if it responds quickly to a command signal change or springs back to its stationary position when a load disturbance is introduced.

The dynamic characteristics of a typical servosystem are stability and transient response.

Stability is the maximum amount of allowable overshoot past the desired position. Instability occurs when the load continuously oscillates instead of standing still.

Transient response is the response time of the actuator to sudden changes of speed or position command signals. This value is defined as the response time required to go from 10 percent to 90 percent of the final value. The term *transient* is also known as *bandwidth.* The wider the bandwidth, the faster the response time.

26-6 Designing a Position Servo

Position-control applications typically fall into two basic categories: *point-to-point* and *contouring.*

Point-to-Point

The command signal for many point-to-point control applications is a step voltage applied to the input of the system. A step voltage is characterized by its sudden discrete change in amplitude from one level to another.

Dynamic characteristics such as stability and transient response are of major importance in the design of a servomechanism. To be useful, the system must be stable. Therefore, the amplifier gain must be limited to a value that will give the required degree of stability. To determine stability, a transient response test is performed by using the circuit in Figure 26-9. To perform the test, the transient switch is opened while a new command signal is selected by adjusting the control pot 1 to a new level. The test is performed by observing the motion of the output when the switch is closed. At the moment the step input voltage occurs, the error signal at the feedback potentiometer becomes maximum. The stability criterion is defined as the maximum allowable overshoot. The speed of the response is defined as the time required to move from 10 percent to 90 percent of the final value. Most point-to-point applications require that the traverse movement is swift without causing excessive overshoot.

By connecting a chart recorder across the motor terminals, the transient response test can be displayed graphically. Figure 26-10 shows three examples of the tests. In Figure 26-10(a), the ideal condition is attained by providing the shortest possible rise time with no overshoot. The system is *critically damped* when this condition is achieved. In Figure 26-10(b), an *underdamped* condition occurs as overshoot develops when the rise time is shortened. This is an unstable or oscillatory condition. In Figure 26-10(c), an *overdamped* case occurs where the rise time is lengthened and no overshoot develops. Overdamping is undesirable because the response time is too long.

To achieve a critically damped condition, the amplifier gain must be precisely set to produce the required torque to move the load. If the gain is too high, an underdamped condition exists. When it is too low, overdamping occurs. If inertia, friction, or gravitational factors on the load are changed in a critically damped system, amplifier gain adjustments alone might not be sufficient to overcome these conditions. For example, suppose the weight of the load is increased. The increased weight creates a higher static inertia. Consequently, the system

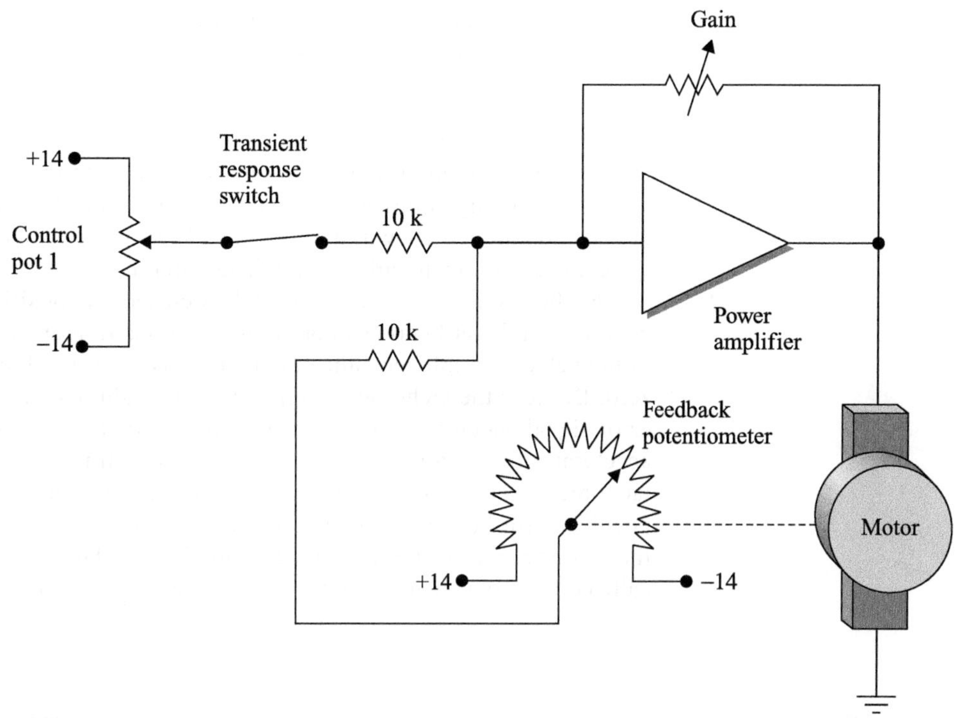

FIGURE 26-9 Circuit to demonstrate the transient response test

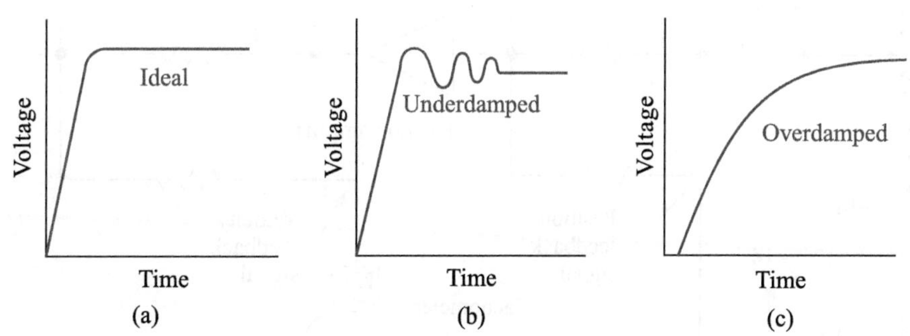

FIGURE 26-10 Examples of transient response tests

will take longer to reach the desired velocity when an input change is introduced because the acceleration is decreased, as verified by the following formula:

$$a = \frac{T}{J_s}$$

where,

a = Rotational acceleration
T = Applied torque
J_s = System rotation inertia

The system will also require a longer time to travel a given distance:

$$\text{Displacement} = \text{Velocity} \times \text{Time}$$

To achieve a transient response time similar to that of the original load, the gain is increased to overcome the static inertia. However, when the endpoint position is reached, the dynamic inertia will cause the load to move past the desired position. To overcome various load conditions, system modifications that provide necessary characteristics are made, such as the tachometer feedback described below.

Tachometer Velocity

One method of improving the transient response time for various load conditions is to add a velocity summing amplifier with tachometer feedback. This velocity loop is placed inside the position loop as shown in Figure 26-11. By using two amplifiers, the overall gain of the system is greater than with one amplifier circuit.

At the instant the transient switch is closed, the position error signal produced by the position amplifier becomes maximum. The position error signal is multiplied by the gain of the velocity amplifier. Since the motor has not started to move, the tachometer output is zero. Because the tachometer signal does not subtract from the position error, the velocity error signal becomes maximum. The large current that results creates a torque that quickly accelerates the motor, providing a fast response time. As soon as the motor speeds up, the tachometer sends a feedback signal to the velocity amplifier. The combined effect of the position error becoming small and the velocity feedback signal subtracting from it causes the velocity error signal to reduce dramatically. The result is a deceleration braking action called *damping,* which prevents the load from overshooting.

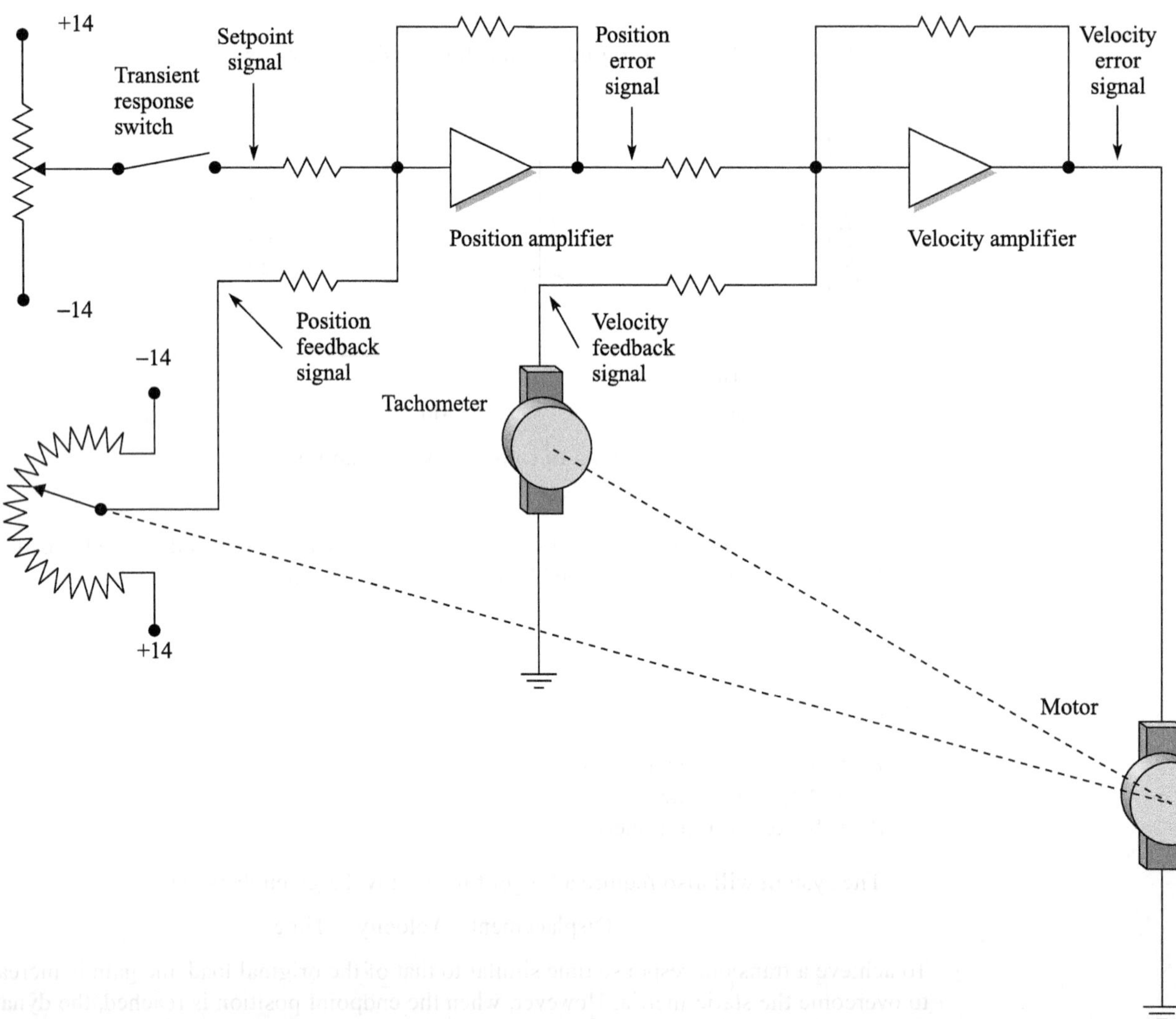

FIGURE 26-11 Tachometer velocity feedback

Tachometer feedback provides several additional benefits to a position servo. For example, in a position error system without the tachometer, as the load approaches the desired position, the position feedback voltage changes until it almost equals the command signal voltage but at the opposite polarity. The resulting position error signal is reduced to a level too small to overcome the friction of the load. The result is that the motor slows down until it stops just short of its desired position. Therefore, the error signal does not reach zero, leading to a steady-state error. When the tachometer loop is added, the error signal from the position amplifier is multiplied by the amplifier gain of the velocity amplifier to overcome the load friction more quickly. The motor movement rotation continues until the end position is reached. A simple test can be made to demonstrate the impact of a velocity loop. If the shaft of a motor is forced out of the command position without a velocity loop, there is very little resistance. If the velocity loop is added, the shaft of the motor will turn much harder. With increased overall amplification, the system attempts to correct any deviation from the desired position. When the shaft is released, it springs back to the command position. Therefore, the higher system gain provides good stiffness. It also improves accuracy and repeatability characteristics.

FIGURE 26-12 A multistep trapezoidal command signal

In some point-to-point applications, the one-step voltage command signal is not used. Instead, the command signal consists of a series of incremental voltage steps, as shown in Figure 26-12. Its shape resembles a trapezoid profile and is referred to as a *position profile.* The inclined slope portion of the profile indicates acceleration. The acceleration stops when the desired speed is reached. The speed is decelerated as the desired position is approached. The acceleration and deceleration rates are typically the same. The voltage reduces to zero when the end point has been reached. As the command steps change, the actuator tracks behind. The advantage of the trapezoid profile signal over a step voltage signal is that less stress is put on the actuator because it can accelerate gradually instead of abruptly.

Contouring

The command signal for contouring control is a series of smaller steps than the trapezoidal command supplies to point-to-point servos. The contouring application requires that the actual position precisely tracks the command position steps with a minimal error signal throughout the movement.

26-7 Digital Controller

Digital controllers that use complex computer programs are required to provide multistep command signals. These computer systems can provide the required control characteristics for various load applications.

A common structure of a high-performance digital motion controller is illustrated in Figure 26-13. This cascaded controller consists of:

1. An innermost current loop that provides torque to the motor
2. A velocity loop around the current loop to control speed
3. A position loop around the velocity loop that monitors the location of the load

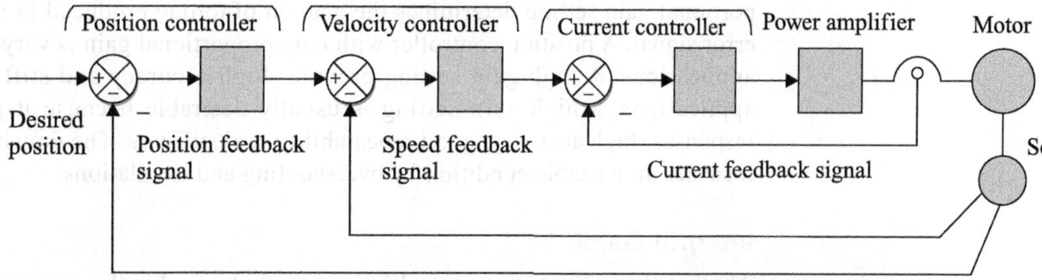

FIGURE 26-13 Digital multi loop control for high-performance motion control

The speed with which the control functions are performed matches the structure of the circuit. For example, the fastest response time is performed by the current loop function, then the velocity loop event, and then the position loop. The current loop has the fastest bandwidth, followed by the velocity loop; the position loop has the lowest bandwidth. Therefore, when tuning the system, it is necessary to start with the innermost loop and work outward.

Current Loop

The torque power to the armature of a DC motor is supplied by a power amplifier. The signal that regulates the power is provided by the current controller. In many applications, the motor speed must be held constant under varying load conditions. When an operation requires an extremely fast response to a load change, a current inner loop is placed in the system to keep the speed stable. It consists of an extra summing junction, an amplifier, and a sensor that produces a feedback signal by detecting the motor's armature current. This current inner-loop circuitry is usually contained within a DC drive unit when a DC motor is the actuator, and in an AC drive when an AC motor is used. For example, if a DC motor is loaded down, the CEMF reduces and armature current increases. The current sensor may be a fractional-ohm resistor placed in series with the armature. As the current changes, a proportional voltage change occurs, which is used to produce the current feedback signal. Another current sensor may be a Hall-effect device placed next to the armature conductor. A magnetic field proportional to the armature current is present around the conductor. As the field changes, the sensor produces a proportional feedback voltage.

The current feedback signal is compared to the velocity error signal at the summing junction of the current comparator. The output of the current junction, called the *current error,* is applied to the power amplifier that drives the motor.

The advantage of the current loop is its quick response time. The response time of the two outer loops is a little longer because they are related to the load. Since the response time of the current loop is so quick, it can replace tachometers, which are used for fast reduction applications.

Velocity Loop

For position applications that require good stability characteristics, tachometers usually provide adequate speed feedback. To create a stable position servo loop without using an analog tachometer, software programming can digitally simulate a tachometer's damping effect. The velocity feedback signal is provided by a feedback device such as an incremental encoder. The speed is determined by calculating the rate of change of the position. The actual velocity is subtracted from the position error signal at a comparator to generate the velocity error. The velocity error signal is applied to the current amplifier as a current command input. The comparing takes place at the junction of a summing op amp that functions as a proportional amplifier.

Proportional Gain

The proportional mode provides the primary control function in the velocity loop. The proportional gain setting determines the amount of torque produced in response to the position error signal. A position controller with only proportional gain is very common in contouring applications. A high gain setting provides high accuracy and stiffness. In point-to-point applications, a high gain setting is usually desirable because it provides fast transient responses, high accuracy, good repeatability, and stiffness. The drawback of high gain is that it causes an unstable condition by overshooting and oscillations.

Integral Gain

Another type of velocity control loop commonly used is the proportional-plus-integral (PI) regulator, as shown in Figure 26-14.

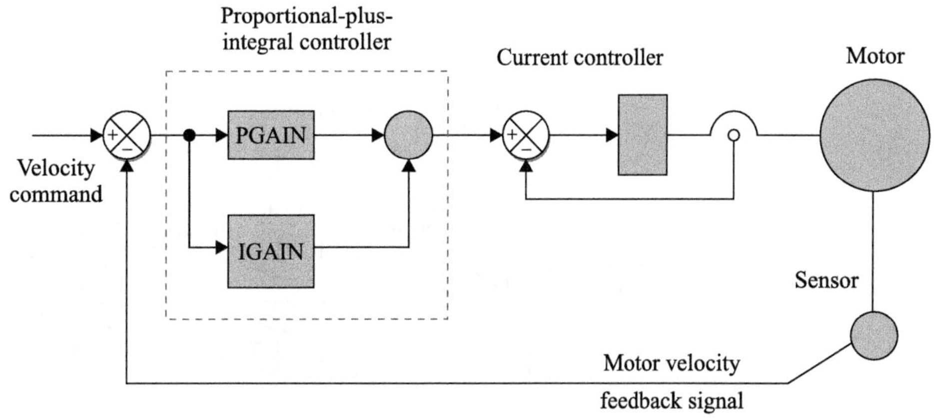

FIGURE 26-14 A velocity-control proportional-plus-integral closed-loop regulator

If an integral-mode control network is added to the system, the output of its amplifier accumulates as long as there is a steady-state error signal applied to its input. Eventually, its output amplitude becomes great enough for the motor to produce the torque required to move the load the remaining distance, eliminating the steady-state error.

When the gain setting of the integral parameters is adjusted properly, the overall positioning accuracy is improved. It also provides stiffness to load torque disturbances. If the gain setting is too high, a low-frequency oscillation may occur around the command position. Also, the integral function results in a longer settling time due to a phase shift that it introduces into the system. To reduce the phase shifting in point-to-point applications where the movements are very rapid, the integral gain is adjusted to a low level. In contouring applications where the movements are slower, the phase shifting is not a factor. Therefore the gain can be adjusted to a significant value to provide high stiffness during the travel.

Position Loop

In contouring applications, it is common to use only the proportional mode in the position loop. As the command signal increments through a series of step changes, it is constantly compared to the position feedback signal by the summing junction of the proportional op amp. The resulting output that the op amp produces is the position error, which becomes the command signal for the velocity loop.

When a position loop is used for point-to-point applications, it is not designed to follow paths. Instead, it must provide fast transient response time, minimal overshoot, and a good velocity profile. In general, a point-to-point regulator is more complicated than the simple proportional gain-only configuration used for contouring movements.

The multi loop point-to-point position loop structure shown in Figure 26-15 uses proportional-integral-derivative (PID) mode and feed-forward. Each mode and feed-forward has unique operational characteristics. Together, they provide maximum performance.

PID is a relatively sophisticated compensation technique used to improve the performance of a servo. A servo with a good internal velocity loop will provide many of the benefits of PID. If no velocity (tachometer) loop exists, PID should be considered. It is always best to keep designs simple and introduce complexity only when necessary.

A brief explanation of the purpose for each control mode in a PID with feed-forward position loop follows.

Proportional Mode

The proportional mode is the primary control function. It is the simplest of the four functions. The proportional amplifier generates a velocity command proportional to position error. For example, high velocity results when the following error (the amount that the

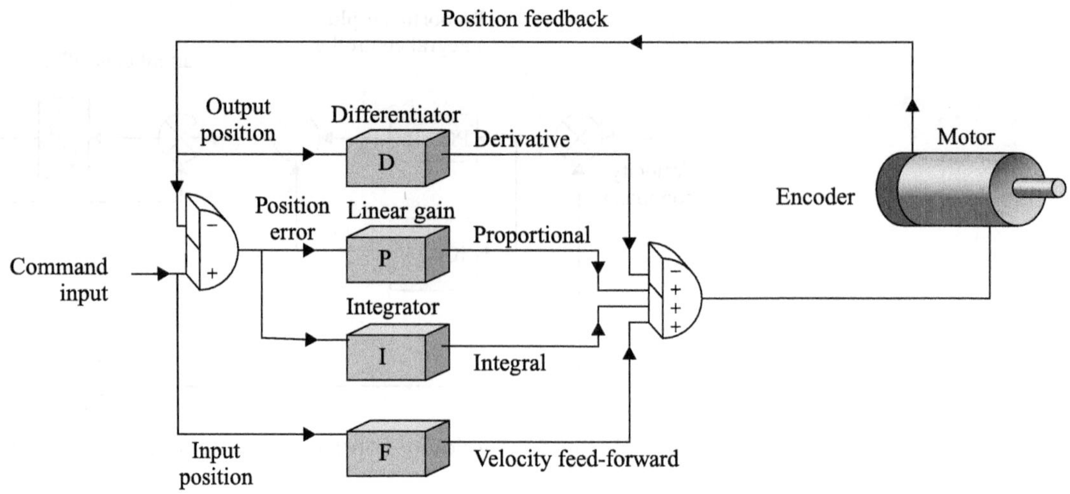

FIGURE 26-15 A PID and feed-forward point-to-point position loop

load lags behind the command signal) is large. By increasing the proportional gain, the system can allow the following error to be reduced while still achieving a fast transient response. However, the higher gain may cause the actual position to overshoot the commanded position.

Feed-Forward

When the command signal is introduced in a position servo, an error signal develops. The error signal results in a proportional velocity command that causes the load to move to the endpoint position. In some applications, the system operation is more responsive if the following error is reduced to near zero during the entire time the load is being moved. By adding a feed-forward mode into the system, this requirement is achieved.

Velocity feed-forward is produced by altering the command signal with the addition of a boost signal. The larger command signal causes the motor to speed up and therefore minimize the following error. Figure 26-16(a) illustrates how a feed-forward command signal is developed.

C = Command signal
FF = Feed-forward signal
F = Feedback signal
E = Error signal

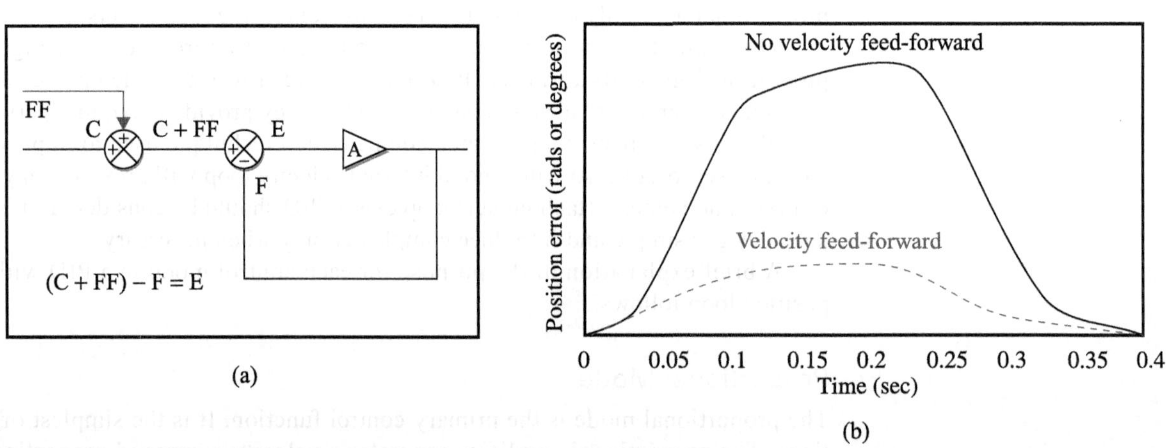

FIGURE 26-16 Feed-forward control

To keep the error signal at zero, C must equal F, and FF must equal E. Because the command velocity and the gain are known, the error can be computed and added to the command to generate a modified command signal, which will cause C and F to be the same. The diagram shows an ideal system where the feed-forward provides the exact velocity command without the need for any position error. In practice, a conservative approach is taken to setting a feed-forward value so that it is not quite high enough to completely eliminate an error signal. (If the feed-forward gain setting is too high, position overshooting will result.) Therefore, a minimal following error exists. To provide stability, a proper gain setting of the proportional control loop is required.

An example of the effect a feed-forward velocity command has on a position following error when making a trapezoidal velocity profile move is shown in Figure 26-16(b).

Derivative Mode

The derivative mode performs an operation similar to that of the velocity feedback tachometer. Recall that as a one-step input command signal is applied, the position error is maximum. Since the position error signal is maximum and the tachometer is not moving, the multiplication factor of the position and velocity amplifier is at the highest level. The result is that the acceleration rate is very high, which improves the transient response time of the system. Once the motor is turning at full speed, the tachometer feedback cancels the velocity command signal developed from position error. This action dampens the overall gain of the system and causes the load to decelerate. As the load approaches the end point, the error reduces and the motor slows down, stopping when it reaches the command position. The disadvantage of tachometer velocity feedback is that the deceleration action begins early during the travel time of the movement.

The derivative amplifier refines the system by replacing the tachometer. The input of the derivative amplifier is the actual position feedback signal. A system that uses the derivative mode requires a trapezoidal command signal. As the command signal is ramped upward, the error signal rapidly increases. During the ramping, the output of the derivative amplifier is maximum and it adds to the proportional output to accelerate the motor. When the command signal is constant and at the top of the trapezoid, the derivative is zero and only the proportional output is applied to the motor. When the command signal is ramped downward, the derivative amplifier produces a maximum output but at the opposite polarity. As the derivative voltage cancels the proportional voltage, the motor decelerates quickly. The overall effect of the derivative action is that it quickly boosts the motor speed at the beginning of travel and provides a braking action at the end of travel.

The derivative mode is used for damping in two different situations. The first occurs when the velocity loop is replaced by a current servo. The second situation occurs when it is necessary to reduce overshooting because the proportional gain must be set high.

Integral Mode

The purpose of the integral mode is to provide a velocity command signal that reduces static position error to zero and to provide stiffness for the system. Since integral control is provided by the velocity loop, it is normally not used in the position loop.

26-8 Tuning a Servomechanism

When developing a servomechanism, it is necessary to tune the system so that it will function effectively. The tuning involves making gain adjustments of various control mode parameters. By making proper gain settings, the system matches the unique conditions of the load and overcomes load disturbances and friction. The choice of which control mode to use in a particular case is made based on the application's static and dynamic characteristic requirements.

In practice, system tuning is rarely determined by calculations. Instead, a trial-and-error technique is used with the motor connected to the actual or simulated load. The practical approach is to make gain adjustments, excite the system, and then view the response on an oscilloscope or chart recorder. Each parameter is tuned by starting with a low value and increasing the gain until the desired response is achieved. The gain may be adjusted by a potentiometer or by selecting a parameter value on a setup menu of computer software. Achieving accuracy for desired movements often becomes an exercise in matching several gains.

26-9 Master–Slave Servosystem

Master–slave is a form of coordinated motion. In some applications, a number of axes need to be synchronized to enable the entire system to operate properly. The most precise way of achieving this type of motion control is to monitor the primary axis, or master axis, and to slave all the other axes' motions to it. Whenever the operator changes speed or position of the master axis, each slave axis will maintain its relationship to the master by making the necessary adjustments. Historically, gears and line shafts were mechanically coupled to start, move, and stop in precise synchronism to perform the required movements. These control functions are now coordinated digitally by the controller element of a closed-loop system.

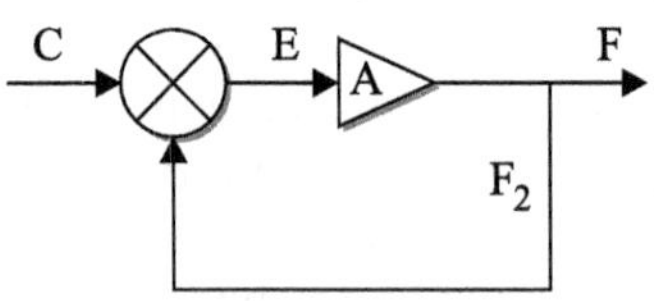

FIGURE 26-17 A basic closed-loop motion control

The concept of master–slave operation is based on a closed-loop motion control system shown in Figure 26-17. The loop's function is to make the output (feedback) follow the command. When the feedback F coincides with the command C, the error E is zero. Figure 26-18 shows a closed-loop slave network that receives its command signal (C_2) from the output (F_1) of a master closed-loop network. Assume that A_2 is a typical drive and motor controller. The error (E_2) that develops causes the motor to adjust its speed proportional to the error amplitude. In this system, output (F_2) lags behind the command signal (C_2) by the amount of the following error (E_2) that develops. The greater the following error, the higher the speed at which the motor will run. A master–slave system does not operate in this fashion. Instead, F_2 must precisely follow the master F_1.

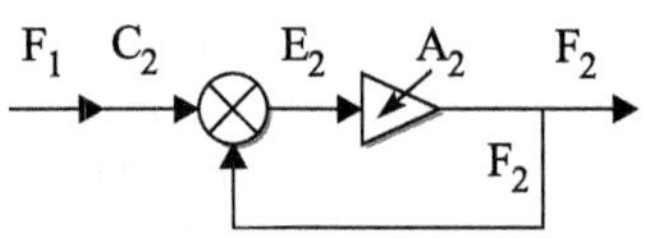

FIGURE 26-18 Slave fed from master feedback

Figure 26-19 shows how the closed-loop system in Figure 26-18 can be modified to operate as a master–slave network. A software module (S) is inserted between F_1 and C_2.

A_2 = A typical drive-and-motor combination
F_1 = Output of a master network fed into the slave network
C_2 = The modified command signal produced by the software module
E_2 = Error signal
F_2 = Output of the slave network

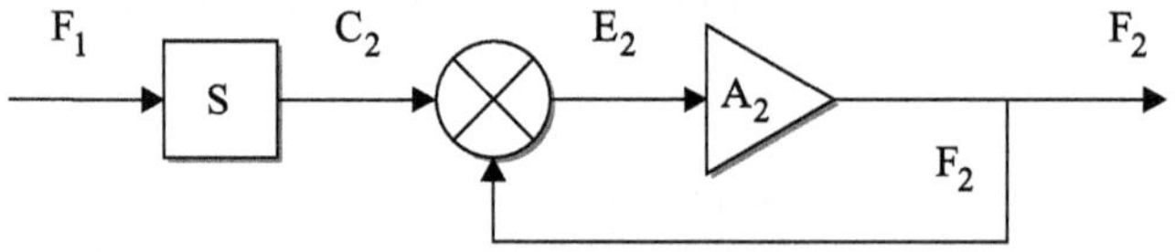

If S = (+E_2) and $F_1 = F_2$, then $C_2 = F_1 + E_2$

FIGURE 26-19 Master–slave network

By supplying the module with the known gain of A_2, its software calculations can predict the E_2 value at any velocity value supplied by F_1. Data entered for computation into the software are the gain of A_2 and the output signal F_1 from the master. The program predicts the error signal that would develop based on these two input values. By adding the predicted error signal E_2 to F_1, C_2 is boosted so that F_2 coincides with F_1.

A rewinder machine that does not use master–slave control is shown in Figure 26-20. The purpose of the machine is to transfer paper from a large supply roll to make smaller rolls. The supply roll is identified as the feed reel and the smaller roll is labeled the take-up reel. The motor that drives the feed reel runs at a constant RPM and the take-up motor runs at a variable speed. A tension roller rides on the paper and turns as the paper passes under it.

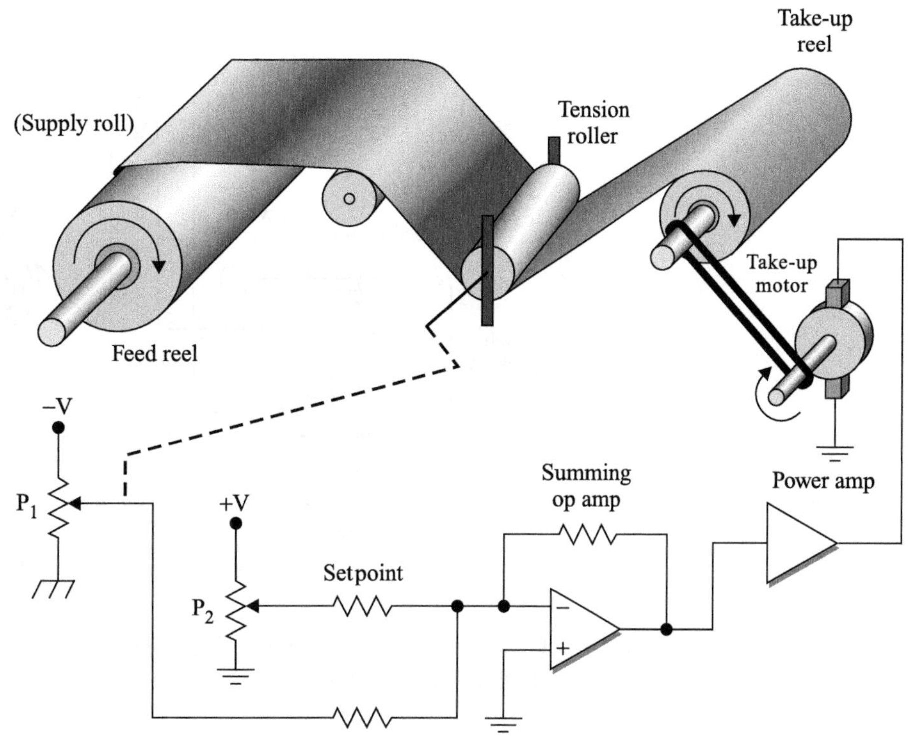

FIGURE 26-20 A rewinder paper machine

As the tension on the paper changes, the roller responds by moving upward if it tightens, or downward if the tension slackens. Mechanically coupled to the roller is the wiper of a potentiometer. Together, the tension roller and the potentiometer form a tension transducer that produces a DC feedback voltage proportional to the tension on the paper.

Suppose that at the beginning of a winding operation, a full supply of paper is mounted on the shaft of the feed reel, and a cardboard tube is placed on the take-up reel. As the rewinding process starts, the wiper arm is positioned at the bottom of potentiometer body (P_1). Since the wiper arm does not feed any cancelling voltage into the summing op amp, the take-up motor will run at a high speed. As paper builds up on the take-up reel, the tension will begin to increase. The roller begins to move upward, causing the potentiometer's wiper to slide in the same direction. As the wiper moves, the negative voltage it supplies to the summing op amp becomes greater. By canceling some of the setpoint voltage, it causes the take-up motor's speed to reduce and slow down the rate at which the tension changes. The reason the motor reduces speed is to ensure that the take-up reel winds the same amount of paper that is unwound by the feed reel. When the take-up roll is full, it is replaced with an empty tube and the process is repeated.

As the supply roll unwinds, its diameter reduces. The drawback of this action is that the rate at which it feeds the paper slows down because the RPM of the feed reel remains constant. A more desirable condition exists when the feed rate remains constant as the diameter of the supply roll changes. A velocity control system called a master–slave network (also known as a *master–follower*) is capable of performing this function. In a master–slave network, several motors and controllers are linked together to perform coordinated operations.

In the master–slave rewinding machine shown in Figure 26-21, one motor makes the required speed adjustments by tracking the speed of the other. By convention, the motor that is tracking is called the *slave* and the motor being tracked is called the *master*. In the rewinding operation, for example, the operator programs the master controller to run the wind motor at a speed that causes paper to be wound at a constant feed rate. An RPM signal is sent to the master controller by a tachometer. The controller uses the data to make complex mathematical ratio calculations. The result of the calculations determines the speed rate at which the slave motor

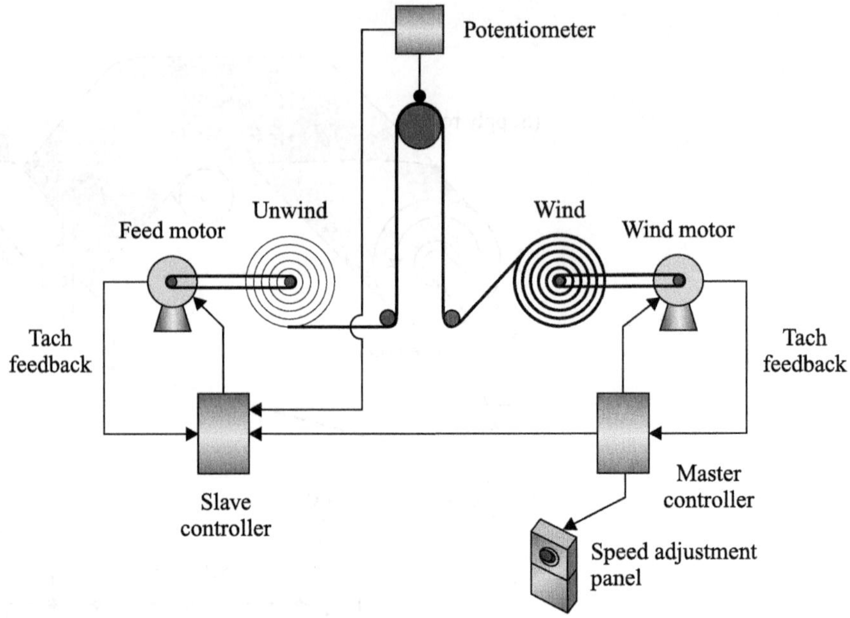

FIGURE 26-21 A master–slave network used to control two motors in a rewinding machine

must turn to unwind the paper. This information is sent to the slave controller, which converts the data to the appropriate electrical signals supplied to the feeder motor. To provide better speed and torque accuracy, a roller that rides on the paper is connected to a potentiometer. The voltage from the potentiometer is used to produce a feedback signal that indicates paper tension.

Problems

1. Another term for speed regulation of a closed-loop velocity control system is _____.
2. In a closed-loop velocity servosystem with a tachometer, if the load slows the motor down, the error signal _________ (increases, decreases).
3. A bang-bang position servo is used in _________ (high-speed, precision) applications.
4. A (bang-bang, proportional) _________ position servo has a zero bandwidth.
5. The feedback tachometer in a closed-loop servomechanism performs which of the following functions?
   a. provides speed regulation
   b. provides stiffness
   c. provides stability
   d. all of the above
6. At the moment a transient command signal is applied to a closed-loop position servo, the error signal is _________ (minimum, maximum).
7. In Figure 26-6, if the setpoint potentiometer applies +7 V to the summing op amp when the feedback potentiometer applies –12 V to the other summing amplifier input, the rack will move ____.
   a. left   b. right

Refer to Figure 26-7 to answer problems 8 and 9.

8. What is the purpose of the three exclusive-OR gates?
9. When input A of the magnitude comparator is less than input B, the motor turns ________ (CW, CCW, nowhere).
10. The linear motion-control feedback transducer used by the electrohydraulic system in Figure 26-8 is a(n) _________.
11. In Figure 26-8, the LVDT produces a voltage that is the ____ polarity as the output of the D/A converter.
    a. same
    b. opposite
12. In Figure 26-8, if a 3 replaces a count of 9 in the register, the linear actuator will ____.
    a. retract
    b. extend
13. When equilibrium is reached in a ____ servo system, the error signal becomes zero.
    a. velocity   b. position
14. The ________ (static, dynamic) characteristics of a servomechanism pertain to the condition of the load when it is moving.
15. The static condition of a servosystem occurs when it is ____.
    a. stationary   b. moving
16. List the four types of static characteristics.

17. The smallest movement that a position servo can make is referred to as ____.
    a. accuracy
    b. resolution
    c. repeatability
    d. stiffness
18. In a transient response test, the system is ____ when it provides the shortest possible rise time without overshoot.
    a. underdamped
    b. overdamped
    c. critically damped
19. It is possible to correct an underdamped condition of a position servo by __________ (increasing, decreasing) the system gain.
20. Assuming that the system gain of a position servo remains unchanged, the transient response time __________ (increases, decreases) if the load inertia is decreased.
21. A decelerating braking action called __________ prevents the load in a position servo from overshooting its end point.
22. The following error in a contouring position control application is __________ (minimal, large).
23. A current loop in a position-control application has a __________ (low, high) bandwidth.
24. The ____ mode control eliminates steady-state error.
    a. proportional
    b. integral
    c. derivative
25. In a slow-moving contouring application, the ____ mode provides high stiffness during travel.
    a. proportional
    b. integral
    c. derivative
26. A ____ provides a better braking action in a point-to-point position system.
    a. tachometer
    b. derivative amplifier
27. The derivative mode provides a ____ action.
    a. boost
    b. braking
    c. both a and b
28. When a setpoint command signal is held constant and the motor that the system is driving is rotating at a steady speed, the ____ amplifier output is zero.
    a. proportional
    b. integral
    c. derivative
29. Feed-forward __________ (increases, decreases) the following error signal in a position servosystem.
30. List the two ways in which gain adjustments can be made when tuning a servosystem.
31. If a master axis slows down on a rewinder paper machine in a master–slave velocity servo, it is likely that the slave axis will ____.
    a. slow down
    b. speed up

SECTION

9

Industrial Networking

OUTLINE

Section 9 consists of only one chapter. The ability to retrieve, exchange, and analyze data has increased tremendously in our homes and at work. The same digital communication technology is now being incorporated into the industrial manufacturing sector in a variety of ways. Field devices, such as smart transmitters, control valves, variable speed drives, PLCs, and a variety of other instruments, can all be connected to a network inside the plant. By using software specifically designed for industrial networking, automation functions have become more sophisticated, devices can be programmed and calibrated remotely, and diagnostic data is provided to assist with troubleshooting.

Chapter 27 provides a variety of information about industrial networking, which includes the hierarchy of networks, topologies, data communication schemes, backbone devices that direct traffic, and hardware. It also gives an overview of common network protocol products that are used for industrial applications, such as AS-I, HART, Foundation Fieldbus, Modbus, Profibus, and NetLinx (Rockwell Automation).

CHAPTER

27

Industrial Networking

OBJECTIVES

At the conclusion of this chapter, you should be able to:

- Define the following terms:

Configured	Patch	Transmission Media
Proprietary Network	Crossover	Protocol
Foreign Devices	Subnets	Fieldbus
ASCII	RTU	Interoperability
Bus	Legacy	I/O Interface Device
Segments	Intrinsically Safe	

- List the hierarchy of industrial networks and describe the typical functions each one performs.
- List the network topologies typically used for industrial communications, describe how each one operates, and explain the advantages and disadvantages of each one.
- Describe the following terms associated with data flow management:

Master–Slave	Token-Passing	Publisher–Subscriber
Source–Destination	CSMA	Unicasts
Full Duplexing		

- Identify the types of hardware used to transmit communication data signals and power.
- List the following backbone devices used for industrial networks and describe how they operate:

Hubs	Switches
Bridges	Gateways

- Describe the following industrial communication protocols and explain their primary networking functions:

Modbus	Foundation Fieldbus
HART	DeviceNet
Profibus-DP	ControlNet
Profibus-PA	Ethernet/IP
Profibus-FMS	AS-Interface
ProfiNet	

27-1 Introduction

A stand-alone computer is capable of performing many tasks. For example, by loading various software programs, manuscripts can be written or complex computer graphics can be created. However, the functions the computer is capable of performing are limited. For example, there can be no e-mails exchanged, and the inaccessibility of the Internet makes it necessary to physically transfer the files for the written document and graphics to another location using a thumb drive or a disk.

Connecting the computer to a network vastly increases the functionality of the computer. A network is a group of computers and other devices that exchange information over some type of transmission media to which they are connected. Transmission media are copper wires, fiber optic cables, or the atmosphere used by wireless devices. A common example of a network is an office where computers are connected together by an Ethernet cable. Multiple users can exchange data such as graphics, spreadsheet programs, or files, send e-mails to one another, or print documents to the same printer. The functionality of the office computers can be further increased by connecting the LAN (local area network) to which they are connected to other networks, especially the Internet. E-mail communications are then expanded to anywhere in the world, files can be sent electronically to customers in another state, and research data needed by a company's legal department is accessible.

The exchange of information through networking has also proven to be beneficial for the industrial manufacturing sector, such as:

- Diagnostic information used for troubleshooting can be obtained from a remote location.
- The accumulated amount of time that a lathe runs is monitored so that its blade is sharpened when a predetermined number of hours is reached.
- Devices can be configured (programmed or calibrated) from a remote location.
- Data can be shared among controllers, enabling one programmable logic controller (PLC) that detects a catastrophic event to instruct PLCs controlling other sections of a machine to shut down.

The networks to which instruments on a factory machine are connected are also capable of merging with higher-level networks that have the capacity to perform a variety of sophisticated functions, enabling the plant to run more efficiently. For example, if a sensor detects that bulk material in a storage tank drops to a predetermined level, data is sent to the network that supports the business office, informing the purchasing department to have the tank refilled.

27-2 Hierarchy of Industrial Networks

Networks used for industrial automation are categorized in several ways, using various terms. A generic term commonly used to identify industrial networks is *fieldbus*. Note that there is a specific network known as Foundation Fieldbus that is often referred to as just "fieldbus." Industrial networks are also categorized by using different levels. Each level represents different communication functions that a network is capable of performing. Figure 27-1 uses a pyramid to show the hierarchy of industrial networks.

Sensor Bus Networks

The lowest level of the pyramid is the category representing sensor bus networks, which are the most primitive and the first ones that were developed for industrial applications. On these networks, field devices that produce discrete input signals to a controller, such as limit switches, or level optical sensors, are connected to one network cable. Sensor bus networks are also capable of transmitting output signals from a controller over one cable to indicator lamps, alarms, or actuator devices such as solenoid valves. These networks replace conventional I/O systems that have individual wires connected from each field device to a set of terminals on a module inserted into a PLC rack, which is a more costly option due to expenses resulting from labor-intensive installations and extra wiring.

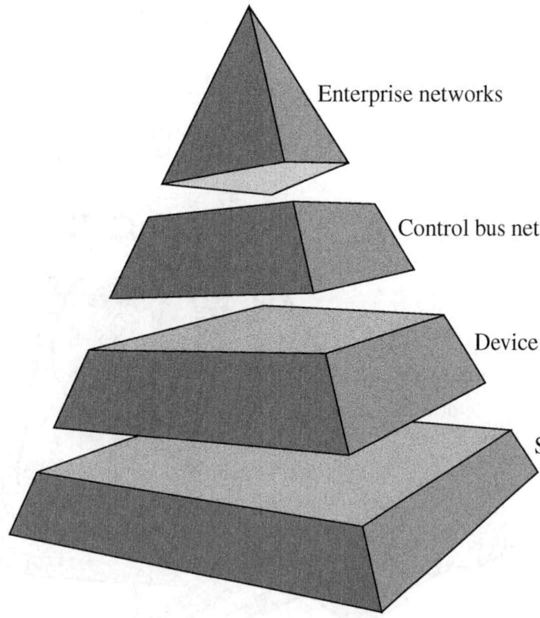

FIGURE 27-1 Hierarchy of industrial networks

Instead of being connected directly to the bus cable, most field devices on a sensor bus network are connected to an I/O interface device. The interface device, which is electrically connected to the bus, can support as many as four field devices. There are three versions of the I/O interface device. One version is designed only to accept input signals from sensors. A second version can only support output actuator devices, such as solenoid valves, alarms, and indicator lamps. A third version supports both input and output field devices.

One end of the bus network cable is terminated at a scanner module. The scanner can distinguish one field instrument on a network cable from another by using an addressing scheme. Each field device is assigned a unique address. The addresses are programmed into I/O interface devices by using a portable handheld programming device shown in Figure 27-2a, which is temporarily connected to the interface device during installation, by setting DIP switches, as shown in Figure 27-2b, or by an automatic addressing software operation performed by some scanners.

Newer-generation sensor bus networks are also capable of supporting analog data representing 4 to 20 mA. AS-I (actuator–sensor interface) is the most common sensor bus network worldwide.

Device Bus Networks

Moving up in complexity, device buses both provide control of complex discrete devices and supply equipment power over a single cable. Device bus networks are typically used in areas with a large population of sensors and actuators, and connect to variable-speed drives and motor-control centers. The most commonly used device bus networks include DeviceNet and Profibus-DP.

Control Bus Networks

Control bus networks are the most advanced networks on the factory floor. On these networks, data communication occurs at a very high level, enabling PLCs to communicate with one another. For example, as an incremental encoder monitors the RPM of a roller on a large printing press, this information is fed to one PLC that connects to a second PLC that controls the speed of another roller so that a proper tension on the paper is maintained. HMIs (human interface panels) also connect to these networks, making it possible to configure (program and calibrate) all the instruments on the same network. There are also "smart" instruments that can be

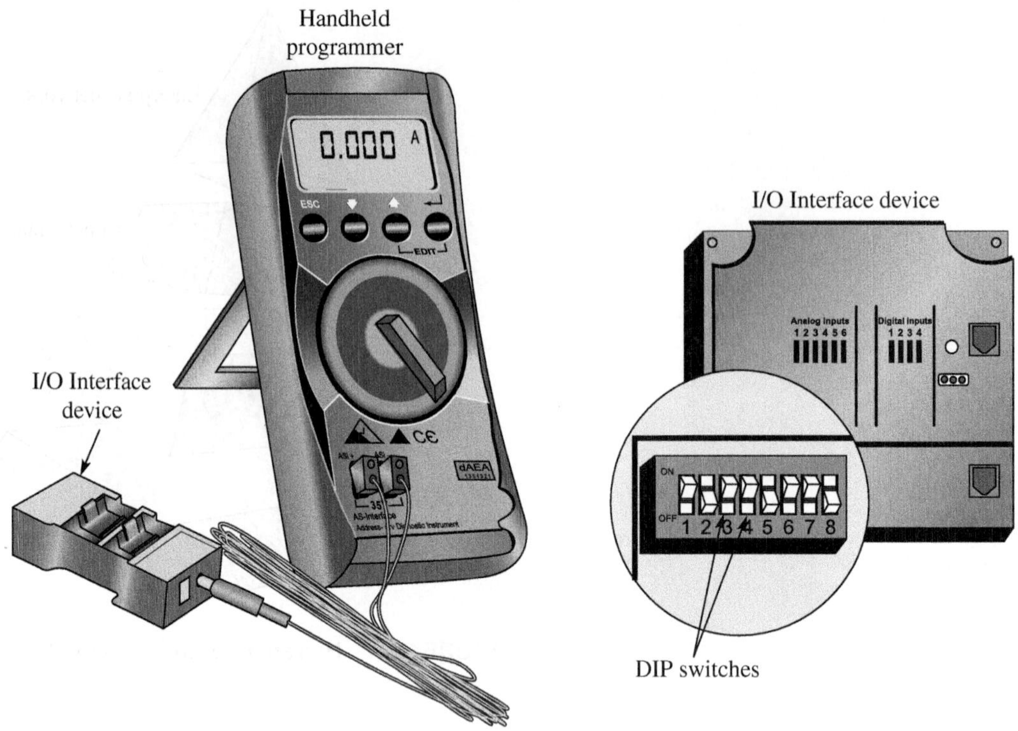

(a) (b)

FIGURE 27-2 Programming I/O interface devices: (a) With a handheld programming device; (b) Using DIP switches

connected to these networks, and are capable of performing complex operations. For example, they can send a warning signal to the controller about excessive wear to a valve.

The high-level communication capabilities of these networks enable them to exchange data representing various types of information, such as analog measurements, or the condition of actuators in the process industry. Communication requirements for closed-loops with On-Off and PID functions, or to operate industrial robots in a work cell are also provided for at this control bus network level. The most commonly used control bus networks include Profibus-PA and ControlNet.

Enterprise Networks

The enterprise-level network, also known as the *information level,* shares data among all departments in a company, such as accounting, sales, purchasing, and production. These networks are computer-driven, and have the ability to collect and deliver vast amounts of information. Common uses are data collection, data monitoring, file transfers, and e-mail exchange.

An example of how an enterprise network functions is when an order for ten cylinders for paper rewinders is placed at a machine tool company. The order is entered into a computer by the sales department, and then sent to the file server. Information is then sent from the file server to other departments on the company network to initiate the following actions:

- Ten bar stocks are needed. If there is not enough material in inventory, more steel needs to be ordered from a supplier.
- Production specifications for the particular cylinders that were ordered are sent to a PLC to set up the manufacturing equipment.
- Production data is sent to shipping to schedule transport on the date the cylinders will be completed.
- Data goes to the billing department, which will send an invoice when the products are shipped.

27-3 Network Topologies

CIS5408 Network Topologies

Networks are assembled in a number of different physical configurations called **topologies**. They consist of cables that carry data and provide power, hardware connections, and a variety of components called *nodes*. Nodes include computers, switches, hubs, transmitters, I/O interface devices, sensors, valves, and so on. The physical configuration used is often selected based on the capabilities each one has to offer, such as:

- Reliability
- Flexibility
- Cost of adding a node
- Future growth
- Potential of data flow disruption

Network topologies common in industrial applications are *bus, star,* or *ring*. Networks can also be configured using a combination of these three topologies.

Bus Topology

A bus network, also known as a *trunk/drop,* has each node attached to a cable called a bus (also referred to as a *trunk*). Shown in Figure 27-3, nodes are connected to the trunk using a tap and a drop cable. A tap is a device that physically attaches and electrically connects to the bus. The drop cables provide the electrical connection between the bus and the nodes. There is a maximum length allowed for both the bus and the cables. At each end of the trunk, it is necessary to use a terminator, which has a resistor that connects electrically to the conducting wires. The terminator resistor prevents electrical signals from bouncing back (called *reflections*) off the ends of the trunk by absorbing energy.

As communication data is sent down the trunk, all of the nodes receive information. However, only the node or nodes that are programmed with addresses contained in the data message receive the data. The primary advantage of a bus topology network is that it is easy to add nodes by simply plugging them into the trunk line. The primary disadvantage is that if an open or short occurs, a portion of or the entire network could become inoperative or damaged. Also, the more users on a bus topology, the more slowly it operates.

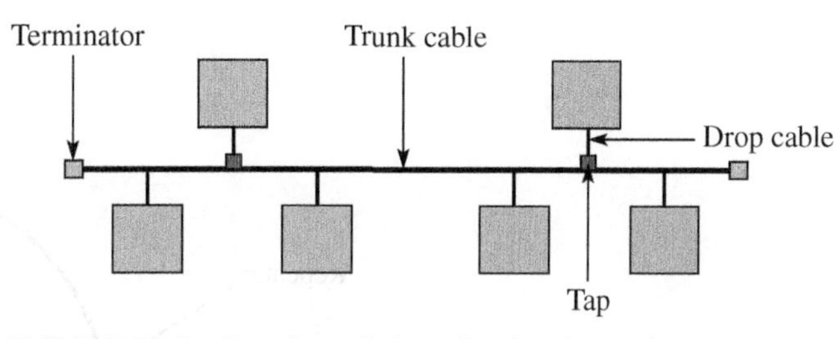

FIGURE 27-3 Trunk-and-drop (or bus) topology

Ring Topology

The nodes of a ring topology network are connected in series, as shown in Figure 27-4, and make up a continuous loop. Data in the form of packets is passed from one node to the next one. Each packet contains the address of the destination node on the ring to which the message is being sent. When data arrives at a node, it will be passed on if the message does not have the assigned node address. However, if the address in the data packet matches the node address, the data will be extracted, stored, and not passed on to the next node.

One advantage of a ring network is that its configuration provides redundancy, which means the message has two different paths. If a network wire is broken, the message can reach its destination by reversing direction.

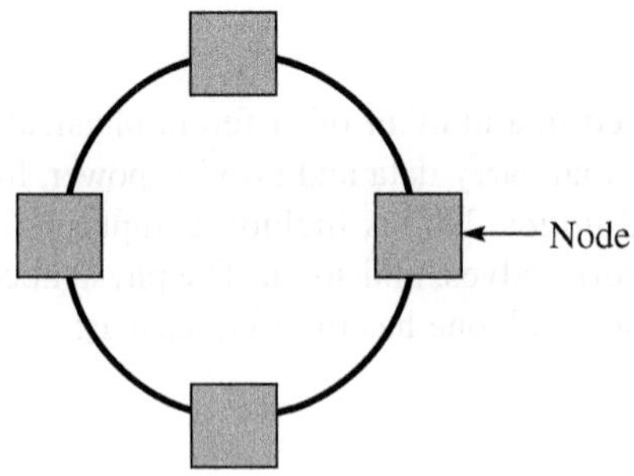

FIGURE 27-4 Ring topology

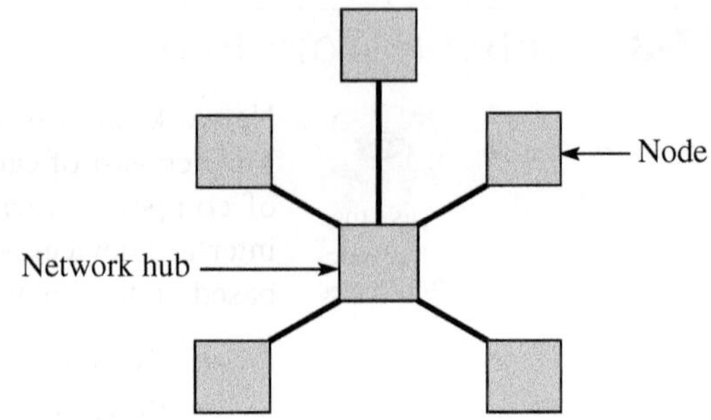

FIGURE 27-5 Star topology

Star Topology

The star topology, shown in Figure 27-5, resembles the hub of a wheel with spokes extending outward. The hub is an electrical connection to all of the nodes on the network. All of the data transmitted in the network passes through the hub on its way to the intended destination. Before data is sent, the destination address must be listed in the data packet by the transmitting node. When the hub receives a data message, it reads the destination address so that it knows where to send the information.

Advantages of the star topology configuration are that it is one of the least expensive in terms of cost per node, and adding nodes to the hub is easy when future growth is required. Star networks are also easy to troubleshoot. Each node has its own cable, and if one goes down, the others are not affected. The disadvantages of star configurations are that they require the most cable. Even when two nodes are next to each other, they both must be cabled back to the central unit. Also, if the hub device fails or loses power, the whole network goes down.

Combination Topologies

To utilize the advantages of the different topologies, many larger networks incorporate a combination of one or more of them, as shown in Figure 27-6.

Each network within a larger multiple network is called a *segment*.

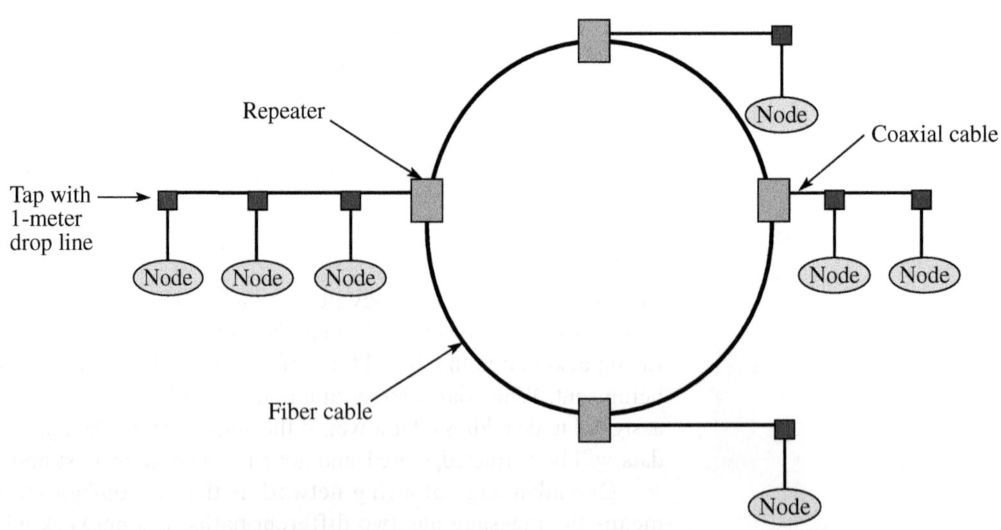

FIGURE 27-6 Combination topologies

27-4 Data Flow Management

There are various methods used to transmit data in an industrial network. Most of the signals that flow through network buses are in a digital format, and are organized into packets that contain addresses and information relevant to the production process. Whenever all the bus users are connected in parallel, as in the case for bus topology networks, a multicast transmission is performed. This means that all the other nodes on the network receive the message transmitted by the sending device. With this type of transmission, two requirements must be fulfilled:

1. In the message, the address of the intended users of the data must be included. Each node is assigned a unique address to distinguish itself from other nodes.
2. The nodes must recognize if a collision occurs and how to respond. A collision is identified as when there are two or more messages sent simultaneously over a network, as shown in Figure 27-7. A data collision results in the corruption of a message, which then becomes a lost transmission. When a collision occurs and is recognized, the senders are given a random time delay, each one different, before they are allowed to re-broadcast.

CIS6408
How the Ethernet Works

A common method of transmitting data in a star network is to use the *master–slave* model. Also referred to as "*source–destination*," the master (source) device is the node from which the message is sent. When the master sends messages to one of the slaves, a reply is returned to the master to verify that it was received. These transmissions that contain both the source and destination addresses are called *unicasts* because they are sent to one destination instead of all the nodes on the network. Also, the transmissions are cyclic, which means that when the master is finished communicating with one device, it then goes to the next node. Each node gets its turn during a scheduled short time period during each cycle.

Both the bus and ring topologies support a multicast management structure. There are two multicast versions, *token-passing* and *CSMA*.

Token-Passing

With a multimedia network, all users are equivalent. One node after another gets the master function for a short amount of time. After a message is sent following a specific time period, the user passes the right to send (token) on to the next user. This form of management is called "token-passing" and is cyclic, which means after the last user on the network gets its turn, the sequence begins all over again. This process is called *deterministic,* which means that the transmissions are scheduled and predictable.

CSMA

In the CSMA (carrier sense multiple access) network, all of the users are equivalent. The users are permitted to send a message as soon as the bus is not busy. CSMA is event-driven, which means that a user on the network transmits data when a message needs to be sent (such

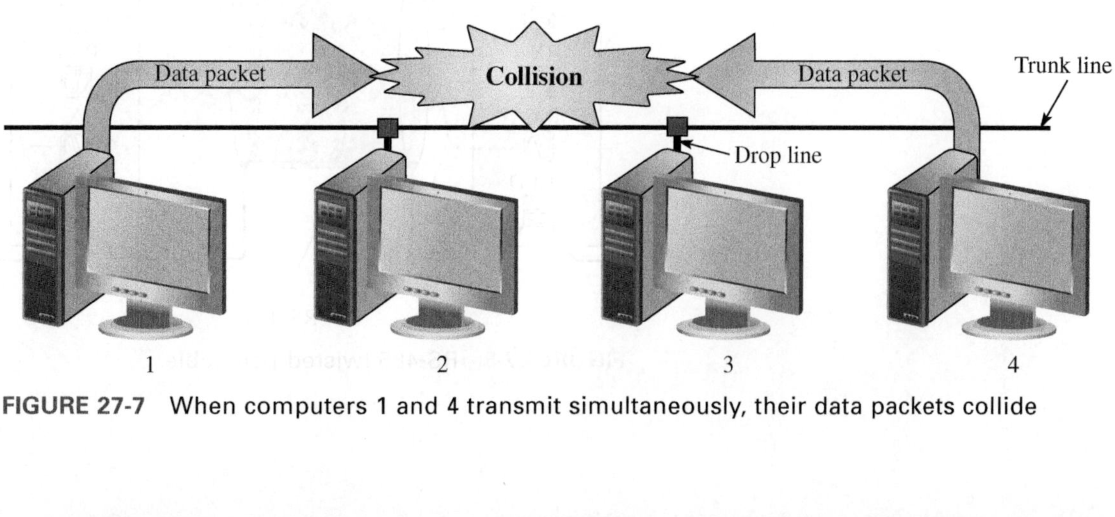

FIGURE 27-7 When computers 1 and 4 transmit simultaneously, their data packets collide

as when an emergency button is pressed) instead of having to wait for a scheduled transmission time. Since the time at which a message is transmitted cannot be predicted, the communication method is referred to as being *nondeterministic*.

The CSMA method uses the "producer–consumer" model. As a message is being sent by a transmitting node (producer), the data packet it sends includes the transmitter's address, called an *identifier*. As the data packet arrives at each of the other nodes on the network, the identifier is read to determine if the message is intended for them. Any receiving node that uses the message is referred to as a "consumer."

27-5 Transmission Hardware

The cables used for industrial network applications are designed to operate in various environmental conditions in a factory. For example, they must withstand extreme temperatures, moisture, and liquids that are corrosive, and they must block out electromagnetic waves (called *noise*), emitted from high-current conductors, which distort and corrupt data transmission signals. The cables used in networks basically fall into two categories: copper conductors and fiber optics.

Copper conductor cables are relatively inexpensive and are easy to work with. Examples are twisted-pair, coaxial, and quick-connect cables.

Twisted-Pair

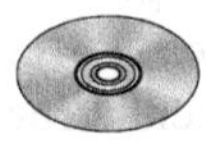

CIS6308 Unshielded Twisted-Pair Cables

A common type of twisted-pair conductor is the RS-485 cable. It consists of two twisted pairs of wires, as shown in Figure 27-8. The data is carried by the voltage difference between each twisted pair. The pairs labeled TX+ and TX− are the wires over which data is transmitted. The other pairs, labeled RX+ and RX−, are the wires over which data is received. An overall shield is electrically connected to terminals labeled SHLD. These cables are commonly referred to as *blue hose* because the insulation casing is blue.

Cables that provide communication for high-level industrial networks using Ethernet typically consist of four twisted pairs of wires, as shown in Figure 27-9. Each end of the cable is usually terminated with RJ-45 connectors, which look like oversized telephone jacks. To accommodate the high speeds of digital data transfer, these cables, identified as 100baseT,

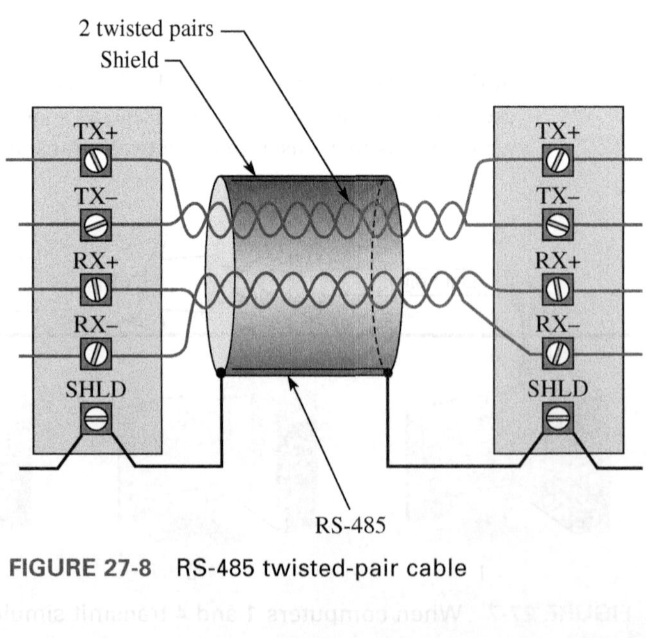

FIGURE 27-8 RS-485 twisted-pair cable

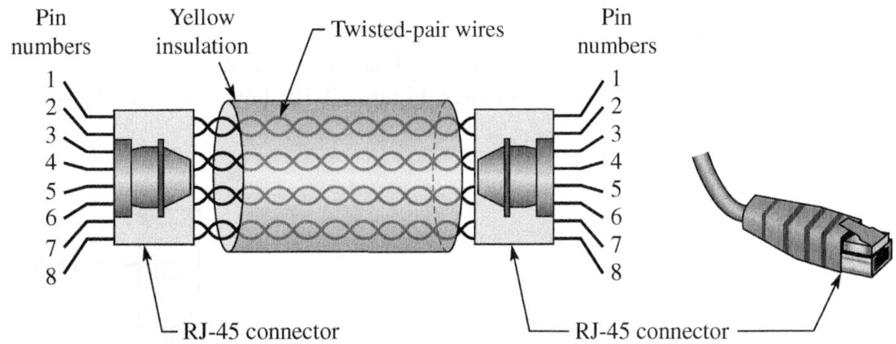

FIGURE 27-9 Ethernet cable

are capable of transmitting at 100 Mbps. The speed of Ethernet transmissions can be increased by a technique called *full duplex.* Full duplexing involves sending data in both directions simultaneously over the TX and RX pairs of cables.

There are two categories of 100baseT cables, *patch* and *crossover*. With a patch cable, the wire connected to pin 1 of one connector goes to pin 1 of the connector at the other end. In crossover cables, each pair of wires terminated at the send connections are terminated at the receive connections on the other end of the conductors, as shown in Figure 27-10. Patch and crossover cables are labeled for identification because they are not interchangeable. The insulators of Ethernet cables are yellow.

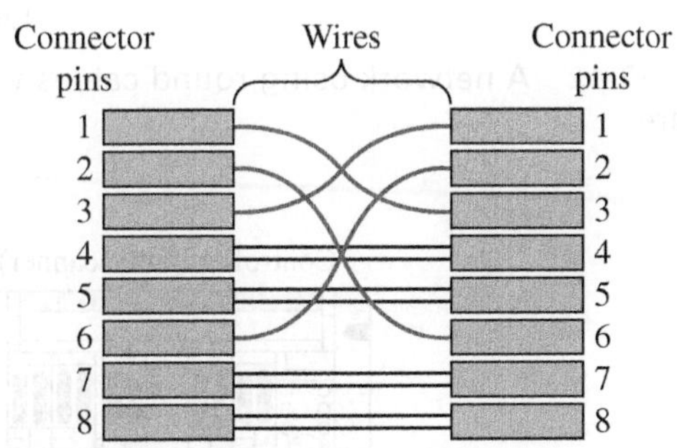

FIGURE 27-10 A crossover cable reverses the sending and receiving pairs

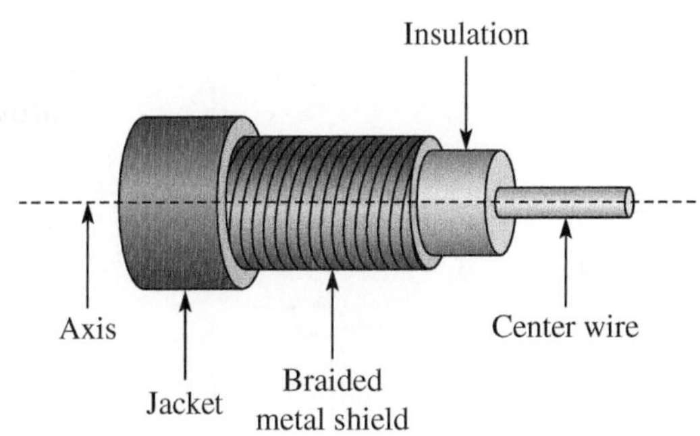

FIGURE 27-11 Coaxial cable

Coaxial Cable

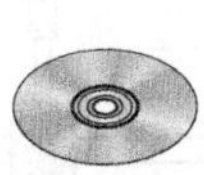

CIS5208
Coaxial Cables

Coaxial cable contains a wire that is surrounded by an insulating material, which in turn is surrounded by a braided metal shield. A protection jacket surrounds the metal shield. See Figure 27-11.

In this type of cable, more commonly referred to as *coax,* the center conductor and metal shield share the same axis (center line). The wire in the center is the electrical conductor through which network data is transmitted. The function of the braided metal shield is to prevent electromagnetic waves from corrupting the data by shorting the noise to ground. One type of coaxial cable, labeled RG-8, is used for Ethernet communication. The maximum speed at which data can be transmitted over RG-8 cable is 10 Mbps. Having a yellow insulating jacket identifies RG-8.

CIS5308
Fiber Optic Cables

Quick-Connect Cables

Cables with quick-connect hardware are available for making easy connections, especially for drop lines. Figure 27-12 shows an example of a quick-connect plant trunk line with drop

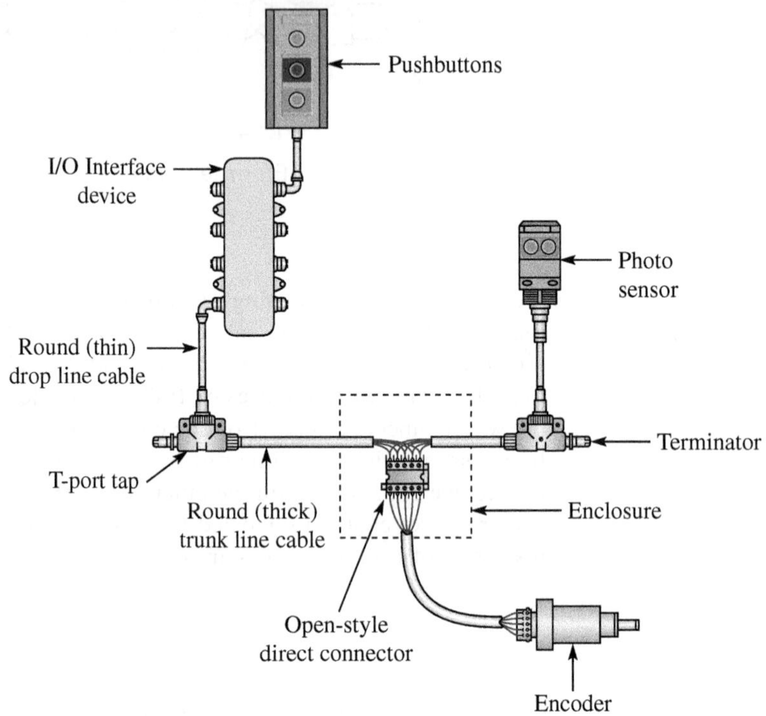

FIGURE 27-12 A network using round cables with quick-connect hardware

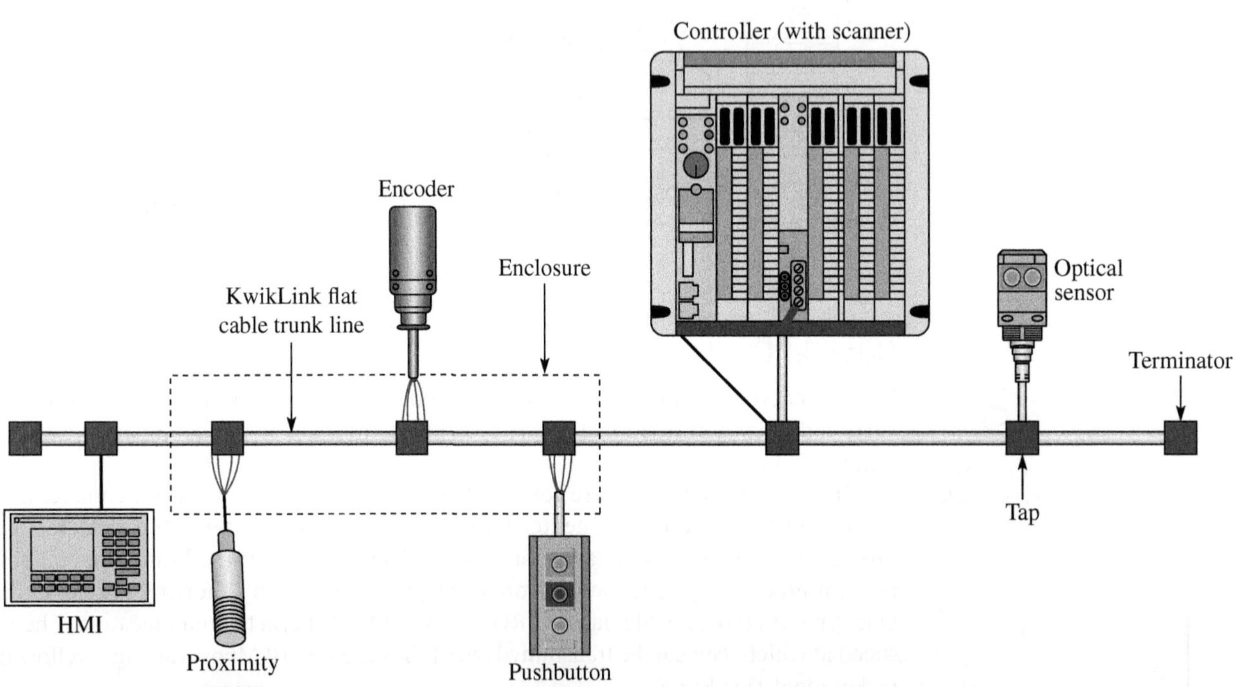

FIGURE 27-13 A network using flat cables with quick-connect hardware

lines using round cables. Flat cables with quick-connect hardware are also available for networking, as shown in Figure 27-13.

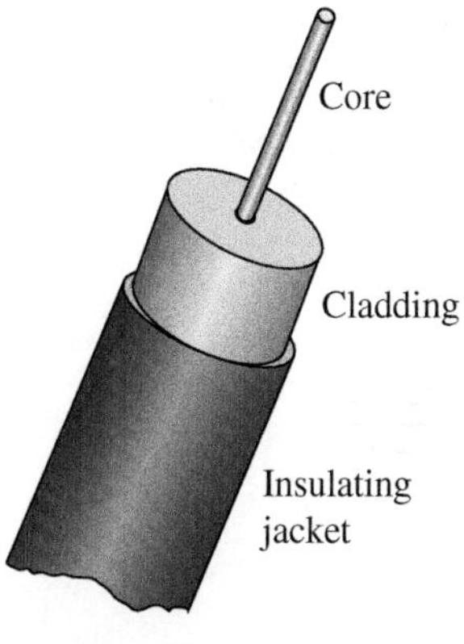

FIGURE 27-14 Fiber optic cable

Fiber Optic

Unlike copper cables that transmit data in the form of electrical signals, data in the form of light waves are sent through fiber optic cables that consist of strands of glass or plastic. The fiber optic cable has three components: the fiber core through which the light signals are sent, the cladding that surrounds the core, and the insulating jacket, as shown in Figure 27-14.

Fiber cables are more expensive than copper conductors and are harder to work with. However, compared to copper, a larger volume of data can be sent over them within a given time period, the transmission distances are greater, and they are immune to electromagnetic noise.

27-6 Network Backbones

When designing an industrial network, one objective is to determine the best way to move data. For example, if the distance of a bus line becomes large enough, the data signal weakens to a point where it is not strong enough at the receiving end. A solution to this situation is to use a repeater, which is a signal amplifier. A repeater, shown in Figure 27-15, is inserted into the bus line to boost the incoming signal to acceptable amplitude for devices further downline.

Based on the characteristics of the different types of data-flow methods, various devices called *backbones* are available to pass signals between networks and smaller networks (subnets). These backbones are called *hubs, bridges,* and *gateways*. Each one transfers data, often in the form of data packets, in a different way.

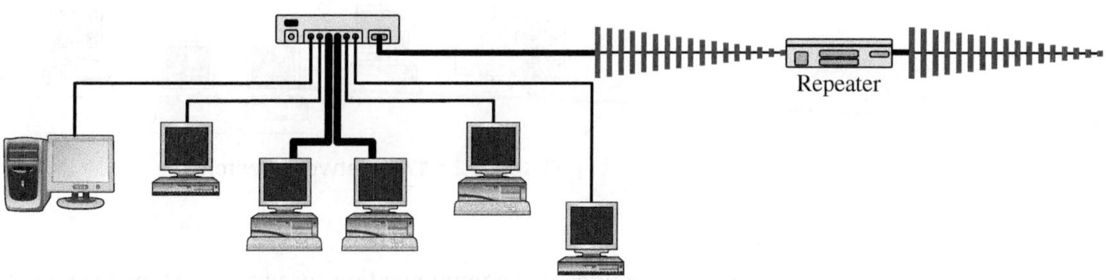

FIGURE 27-15 A repeater boosts the transmission signal

Hubs

A hub, shown in Figure 27-16, is a central connecting device. As it receives data packets from devices on a network, it re-broadcasts the signal to all the other nodes connected to it on the network. A hub is commonly the center of a star topology. Since all of the devices on a hub compete for transmission usage on the network media, collisions can occur when two or more send transmissions at the same time. Hubs should not be used when immediate messaging is required.

Bridges

A bridge is a device that connects two network segments together. Bridges are smart devices that record and process information about traffic in the network. By building and maintaining address tables of the nodes, they are able to find the optimal data path between a transmitting device and the receiving device, instead of sending the message to all the other devices on the network.

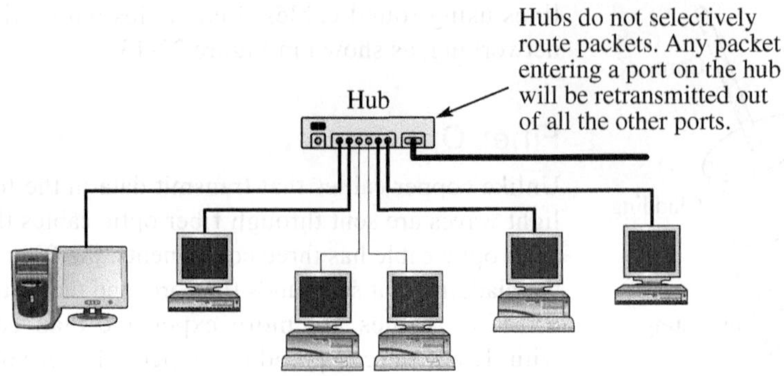

FIGURE 27-16 A network hub

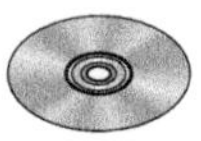

CIS5508
Switches and Hubs

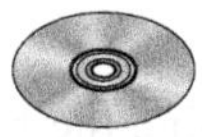

CIS5608
How Switches
Increase Speed

Switches

A switch, shown in Figure 27-17, is a multiport bridge that selectively forwards data packets straight through to their proper destinations. Incoming messages are transmitted only to the desired destination nodes. Operating speeds are very fast because switches allow more than one transmission to occur at any given moment. Switches are also full-duplex devices that allow data signals to flow in both directions simultaneously. By being able to pass multiple signals at the same time, the collisions that commonly occur with Ethernet communications are avoided.

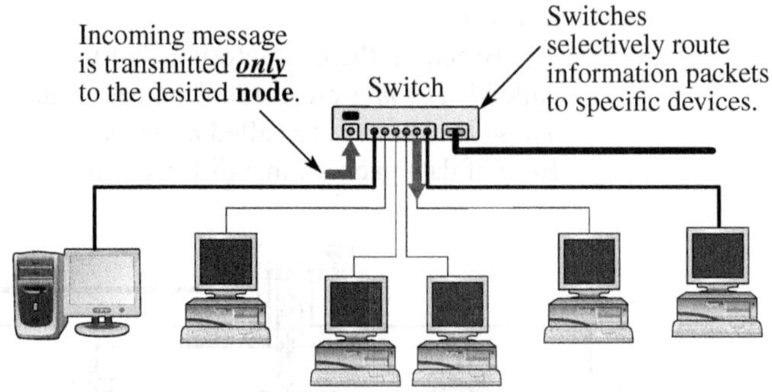

FIGURE 27-17 Network switch

Gateway

In some situations, the communication format used by one segment of a network may be different than the format used by another segment. For example, a computer on an Ethernet network may use the TCP/IP protocol, and a PLC on a subnet may be using the ControlNet protocol. Unless there is a way in which to interpret messages between both protocols, they cannot communicate. A gateway, shown in Figure 27-18, is a device that performs the conversion function rather than supporting one protocol format within another.

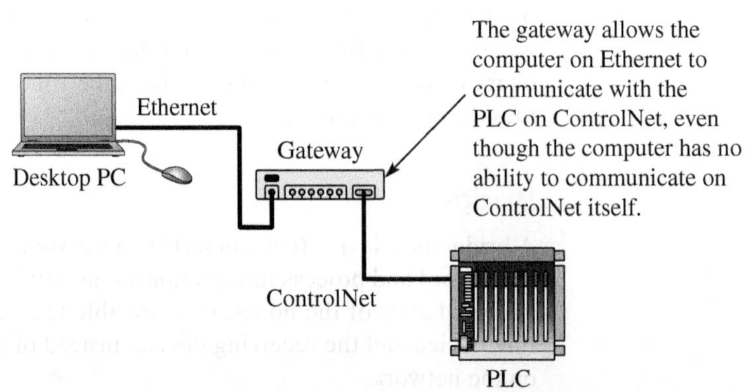

FIGURE 27-18 Network gateway

27-7 Network Communication Standards

Many different industrial networks currently exist, for a variety of reasons. When the development of industrial networking began in the 1980s and 1990s, many of the competing suppliers who build instruments for factory machines developed their own versions of networking products and communication software protocols that enabled everything in their product line to work together. Called *proprietary networks,* they were not compatible with instrumentation built by other companies, which are called *foreign devices*. This meant that customers were forced to purchase networking products from one vendor.

To address this situation, nonprofit independent organizations such as Foundation Fieldbus and HART were formed to establish standard communication protocols that enabled devices from multiple vendors to work together. A **protocol** is a standard, or set of rules that determine the format of digital data and the method used for data transmissions over networks. The standards determine how much information is sent, such as how many bits of data are in a message, how often it is sent, how to recognize transmission errors, and so forth. Eventually, the vendors with proprietary products recognized the value of being interoperable with foreign devices. The result is that they made their networking software open-source, which enables foreign devices from other vendors to be compatible.

27-8 Fieldbus Networks

In this part, an overview of the more prevalent open-communication-protocol products used in industrial networking will be provided. They include Modbus, AS-Interface, HART, Foundation Fieldbus, Profibus, and NetLinx.

Modbus

In 1978, Modicon, a small company in California that built PLCs, had its engineers develop a communication protocol that enabled equipment in a factory to exchange information. This protocol, called *Modbus,* became the industrial standard for exchanging data and communication information in a PLC system. Today, it is reported that there are well over seven million Modbus devices in North America and Europe.

Modbus devices communicate by transferring data over RS-232 and RS-432/485 hardware that physically and electrically connects PLCs, I/O, and other devices. Over the years, Modbus has used three different modes to package and transmit information. These three transmission modes are:

1. ASCII (American Standard Code for Information Interchange). This mode was the original transmission format used when Modbus was first created. This code uses 7 bits that represent 128 unique numbers and letters. ASCII has been used by computer peripheral equipment for years. When a letter on a keyboard is pressed, the closure action is converted to a corresponding ASCII code and sent to the computer for processing. The ASCII code representing the key is sent from the computer to the printer.

ASCII codes are actually 8 bits in length. However, only 7 bits are used because the MSB (most significant bit) is always a 0. Table 27-1 provides a listing of the ASCII codes and alphanumeric characters they represent.

When the ASCII mode is used for Modbus communications, they are sent as two 8-bit bytes. The primary advantage of ASCII is that an interval between characterless transmissions can last up to 1 second without causing communication errors. Its primary drawback is that it uses a time-consuming method called *cyclic redundancy check* (CRC) that verifies if information was transmitted without error.

2. RTU (remote terminal unit). As computer technology evolved, better methods to send binary data in 1s and 0s were developed. Modbus adopted a transmission mode that incorporates the hexadecimal number system, which uses all of the 8 bits in 2-byte data transmissions. By using 8 bits instead of 7 bits, RTU provides a higher throughput than ASCII for the same baud rate. The RTU mode uses the *longitudinal redundancy check* (LRC) method, which is a faster way than CRC to verify data is sent without errors. Because of these two advantages, RTU is more widely used than ASCII.

TABLE 27-1 The 7-bit ASCII Code

		MSBs (Columns)							
LSBs (Rows)		000 (0)	001 (1)	010 (2)	011 (3)	100 (4)	101 (5)	110 (6)	111 (7)
0000	(0)	NUL	DLE	SP	0	@	P	‘	P
0001	(1)	SOH	DC1	!	1	A	Q	a	q
0010	(2)	STX	DC2	“	2	B	R	b	r
0011	(3)	ETX	DC3	#	3	C	S	c	s
0100	(4)	EOT	DC4	$	4	D	T	d	t
0101	(5)	ENQ	NAK	%	5	E	U	e	u
0110	(6)	ACK	SYN	&	6	F	V	f	v
0111	(7)	BEL	ETB	‘	7	G	W	g	w
1000	(8)	BS	CAN	(	8	H	X	h	x
1001	(9)	HT	EM	)	9	I	Y	i	y
1010	(A)	LF	SUB	*	:	J	Z	j	z
1011	(B)	VT	ESC	+	;	K	[	k	{
1100	(C)	FF	FS	,	<	L	\	l	|
1101	(D)	CR	GS	–	=	M	]	m	}
1110	(E)	SO	RS	.	>	N	^	n	~
1111	(F)	SI	US	/	?	O	—	o	DEL
Bit numbers	Hex code	Machine commands		Characters most often used				Characters least often used	

} Bit number
} Hex code

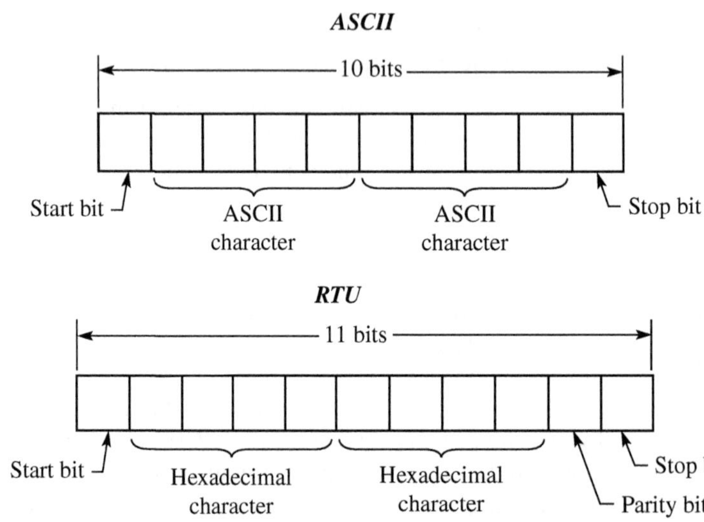

FIGURE 27-19 Formats of the Modbus ASCII and RTU transmission modes

Figure 27-19 shows the data format of both the Modbus ASCII and the RTU transmission modes.

Modbus devices communicate by serial transmissions using the master–slave (request/response) technology. This means that one device on the network operates as the master, while the others act as slaves. When Modbus data is transmitted, each message contains the address of the slave, the command that specifies a particular function (e.g. "read register" or "write register"), the specific data that will be read or written into a register, and a checksum operation to verify that the data does not contain errors. Figure 27-20 shows the protocol (sequence of steps) that occurs during a Modbus transmission using the ASCII or RTU modes.

3. TCP/IP (transmission control protocol/Internet protocol). The most recent and advanced Modbus communication mode is called Modbus/TCP. This mode allows data generated following the ASCII or RTU Modbus protocols to be packaged within the same TCP/IP protocol used for the Internet, Ethernet networks, and intranet structures. The result is that TCP/IP is usually compatible with the communication networks already present in the plant, even if legacy Modbus is still being used.
In addition to operating using the master–slave communication method, Modbus/TCP also supports peer-to-peer communications.

The Modbus ASCII and RTU communication formats are not compatible. Therefore, all components on a Modbus network must be set for one or the other transmission format.

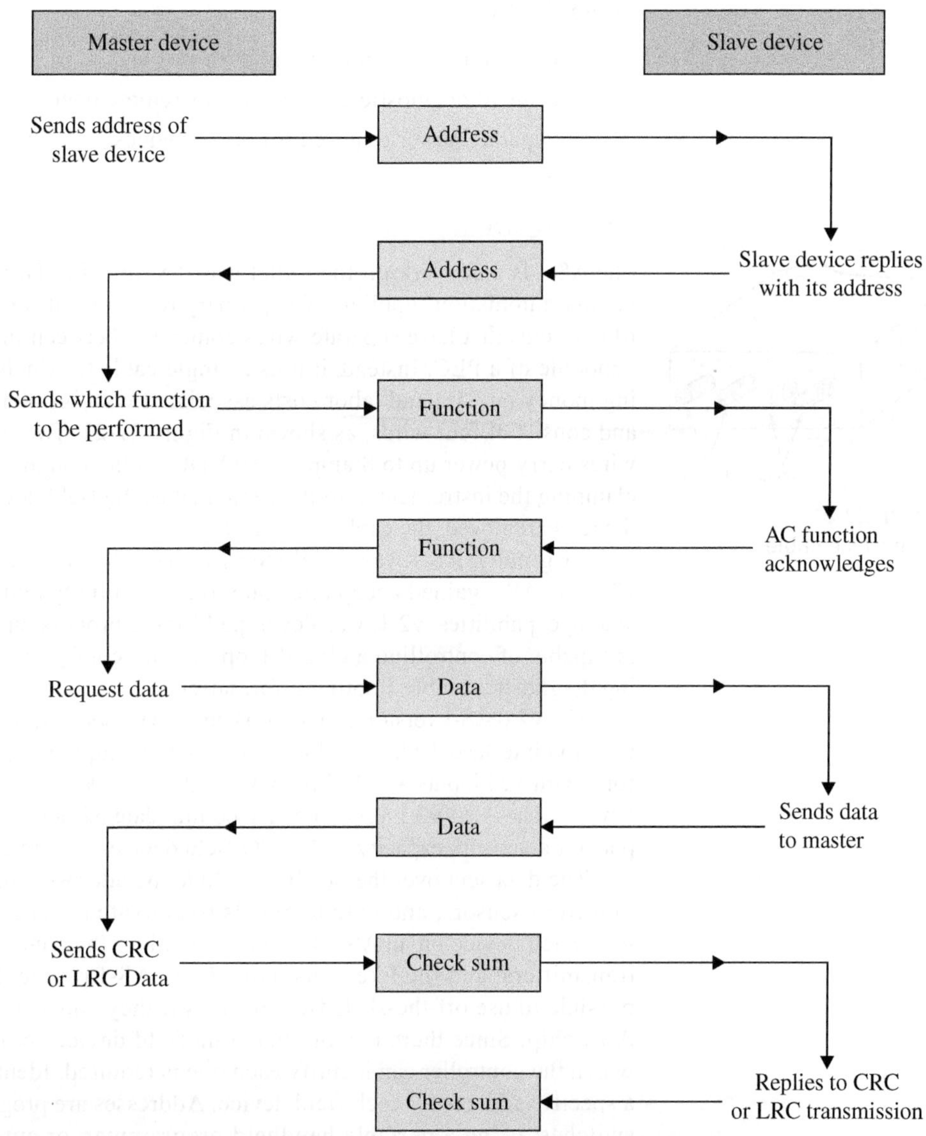

FIGURE 27-20 Modbus transmission protocol using ASCII or RTU modes

However, Modbus/TCP is able to communicate with either ASCII or RTU. Another situation where incompatibility will arise is when two devices on the network have different physical layers, for example, one device using RS-232 and another using RS-485. To resolve this situation, converters are available that allow a seamless flow from an RS-232 connection to an RS-485, and back again.

For master–slave networks, Modbus is limited to 254 connections per master. However, the number of connections is unlimited when using Modbus/TCP. There are several advantages to using Modbus communications:

- *It is free*. Early on, Modbus opened the protocol to all users, eliminating licensing fees or upgrade fees.
- *Interoperability*. The way in which Modbus packages information enables effective communications with most types of industrial equipment or operating systems on the market today.
- *Ease of usage*. Modbus software is relatively simple to work with, and the installation of devices on a network is very easy.
- *Closes the technology gap*. Modbus TCP/IP enables legacy (older technology) equipment to operate with newer equipment.

Its drawbacks are:

- Limited use as a device bus
- Limited diagnostic capabilities for remote devices on the network
- Separate power required for field devices

AS-Interface

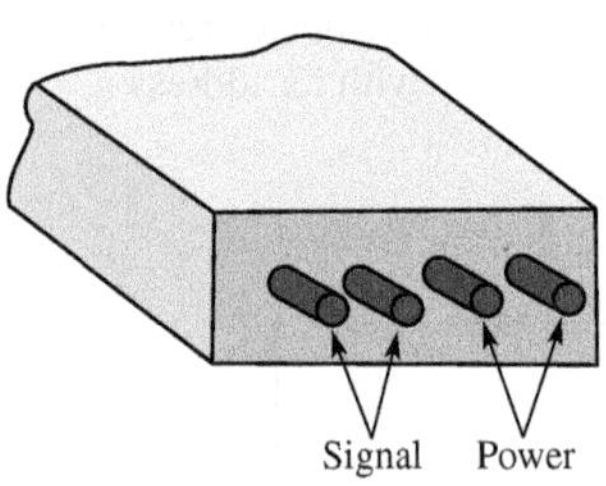

FIGURE 27-21
AS-Interface cable

The AS-I is a network protocol that was developed in Germany in 1994 by a consortium of factory automation suppliers. The primary reason for its creation was to replace conventional I/O systems that have separate wires connected between individual sensors and terminals on a module of a PLC. Instead, it uses a single cable to which field devices are connected, saving money on wire and labor costs associated with installations. AS-I cables are rectangular and consist of four wires, as shown in Figure 27-21. Two wires carry data, and the two other wires carry power up to 8 amps at 30 VDC, high enough to activate solenoid valves. Simply clamping the instrument onto the cable makes the field device connections. Whenever a field device is removed, the cable seals itself.

Originally, AS-I was developed for discrete field components. This version is called v2.0. As AS-I gained acceptance due to its simplicity and cost savings, a new version with analog capabilities, v2.1, was developed for the process-automation industry. Neither version is capable of controlling a closed-loop system, configuring instruments remotely, or providing diagnostic troubleshooting information.

The v2.0 AS-I version can network up to 31 nodes, which are slave field devices. Each node is an I/O interface device capable of accommodating four inputs and four output devices. Therefore, with 124 inputs and 124 outputs, each network has a maximum capacity of 248 I/O field devices. The AS-I v2.1 version can accommodate 62 nodes, providing 248 inputs and 186 outputs for a network capacity of 434 I/O field devices. Figure 27-22 shows an AS-I network.

The data sent over the bus line include the addresses of the field devices, input information from sensors, and output signals to activate actuators. Data is updated every 5 ms or less. Each device on an AS-I network, including the controller, I/O interface modules, smart transmitters, and the like, must contain an AS-I protocol chip to communicate. It is also possible to use off-the-shelf field devices if they connect to an intelligent node with a slave AS-I chip. Since there is more than one field device connected to a single cable, a way in which the controller can identify each one is required. Identification is achieved by assigning a specific address for each field device. Addresses are programmed into a node by using DIP switches, using a portable handheld programmer, or automatic addressing performed by some master device on the network.

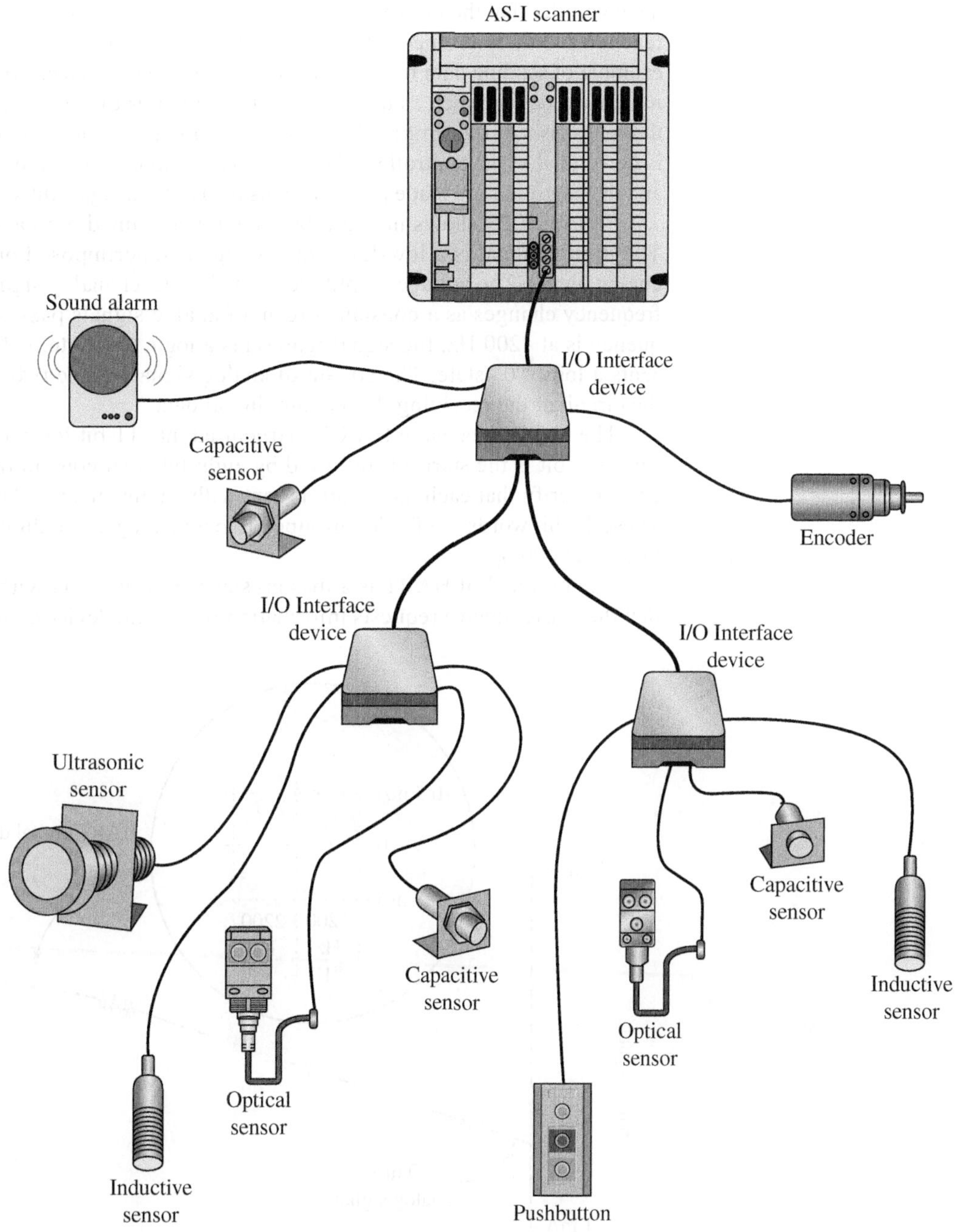

FIGURE 27-22 AS-Interface networks

AS-I networks can be configured in a variety of topologies, such as star, ring, or line-and-drop, and is easily expandable. The maximum length of cable runs is limited to 100 meters per node, and can be expanded to 300 meters when two repeaters are added.

HART

HART (highway addressable remote transducers) protocol is a hybrid communication system that combines both analog and digital signals on the same transmission wire. The analog signal is typically 4- to 20-mA measurements from a sensor output that indicates the status of a process variable, such as temperature or pressure. The digital signal can be data sent in both directions, between the field device and the controller's processor. Data sent from a field device to the processor may be diagnostic information. For example, after extended use, a valve will inform the controller that the predetermined travel distance of the stem connected to the plug has been reached and therefore needs to be lubricated or recalibrated. Configuring a field device is an example of sending data from the processor to the field device. Configuration involves calibrating a field device, such as using known temperatures to make the zero and span settings for a thermal sensor or transmitter. The sensor calibration operation is done remotely, as the procedure is performed at the controller. The data transmissions that communicate between field devices and the controller are made possible by using asset-management software.

Figure 27-23 shows how digital data are combined with a 4- to 20-mA current signal. The digital signal is a low-level analog signal superimposed on the standard 4- to 20-mA analog signal. The voltage amplitude of the low-level analog signal does not vary. Instead, its frequency changes as a constant stream of analog signals pass over the wire. When the frequency is at 1200 Hz, the signal represents a logic “1” state. A frequency of 2200 Hz represents a logic “0” state. The stream of analog signals at the two frequencies is converted by the circuit at the receiving device into digital data.

The digital data using HART is structured into 11-bit formats, as shown in Figure 27-24. The first bit is the start bit, followed by eight bits that contain the data. An odd parity bit is used to verify that each data word is sent without any errors. The eleventh bit is the stop bit. These 11-bit words are further grouped into a message. The digital signals are sent at a baud rate of 24,000 bps.

The protocol of HART is a master–slave configuration with token-passing. This means that the master device requests information from field devices, one at a time. The field device

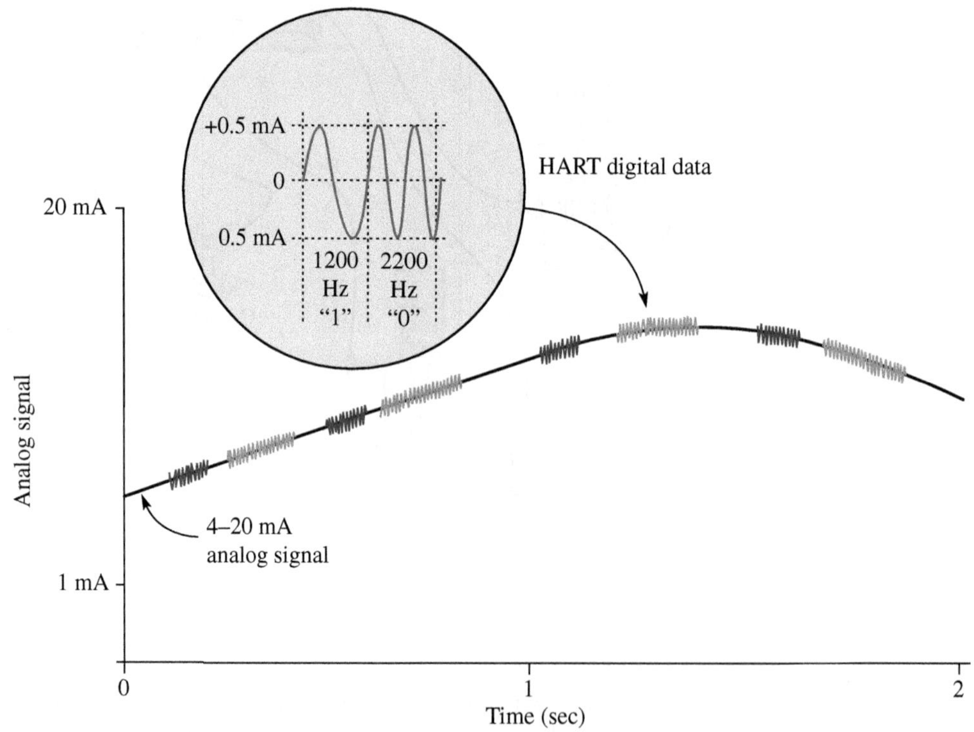

FIGURE 27-23 HART digital data superimposed on a 4- to 20-mA HART analog signal

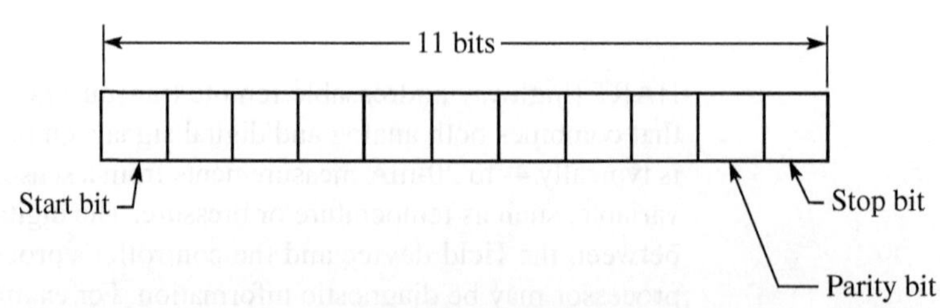

FIGURE 27-24 Format of HART digital data

cannot transmit data until its turn, and when it does, there is a time limitation. A drawback of this communication procedure is that a sensor cannot immediately provide information about a fault condition at the moment it arises.

HART is most prevalent in process manufacturing because 4 to 20 mA is the most common standard signal used by instruments in that industry. Approximately 80 percent of all instruments used in process manufacturing have HART connectivity. Even though many customers do not use the HART protocol, most device manufacturers incorporate HART into their products rather than making HART and non-HART instruments separately. The primary advantage of HART is that it can be readily adopted with existing equipment that operates at 4 to 20 mA.

Foundation Fieldbus

Foundation Fieldbus was created in 1994 as an open protocol by a group of suppliers of process-automation equipment. It is a nonprofit foundation consisting of over 150 member companies, users, and suppliers around the globe. The mission of Foundation Fieldbus is to establish common standards that enable hardware and software products from different manufacturers to work together. Foundation Fieldbus instruments must undergo rigorous tests to prove interoperability and earn "registered" status, and the right to carry a Foundation Fieldbus logo.

Foundation Fieldbus signals are carried bidirectional in a serial format over a single cable between a controller and multidrop buses to various types of devices, such as sensors, transmitters, control valves, and so on. To identify which information is associated with a particular device, each device has a unique address that is assigned automatically online by a configuration tool or host system as each device is connected to the segment. There are two versions of Foundation Fieldbus: H1 and HSE.

Foundation Fieldbus H1

H1 works at 31.25 Kbps and generally connects to field devices. It provides both communication and power over its twisted-pair wiring. H1 segments operate at 9 to 32 VDC at 15 to 20 mA. Up to 32 devices can be connected per segment, although 4 to 16 devices are more typical based on demands such as power draw and execution speed. The segments can handle multiple signals from one field device. For example, a smart temperature transmitter is capable of communicating the measurements from eight temperature sensors. An example of the two-way communication capability of H1 is sending a signal from a controller to a flow control valve, and then a signal returning to the controller from the valve to indicate the actual position, providing for greater accuracy in a closed-loop system. Another function of Foundation Fieldbus is that field devices are capable of informing the controller if it is operating properly, and if the process information it is sending is valid. Foundation Fieldbus is also capable of executing On-Off and PID control functions for a closed-loop system. Foundation Fieldbus also performs configuration operations, alarm, event, and trend data, diagnostics, and information for operator displays. Foundation Fieldbus is especially appealing to industries with remote monitor requirements, such as oil and gas pipelines and platforms.

Foundation Fieldbus HSE

HSE (high-speed Ethernet) operates at 100 Mbps, sending data between I/O subsystems, host systems, gateways, and field devices using standard Ethernet cabling. However, power is not supplied over the cable.

All of the communication over Foundation Fieldbus networks is executed on a regular, repeating cycle. The timing is determined by a link-active master scheduler that resides in a host system, or one of the devices on the same segment. These scheduled communications use a publisher–subscriber method at specific intervals, which means data is sent (or published) once on a bus, and all the devices that need the information listen (subscribe) to the transmission. These communications are *determinant,* which means they occur on a predetermined schedule.

The primary difference between Foundation Fieldbus and other fieldbus systems is that its software contains function blocks that allow its operation to occur without using a PLC or DCS. Also, since their devices operate at very low power levels, Foundation Fieldbus is intrinsically safe, which means they can work in hazardous areas prone to explosions that are common in process industries.

Profibus

Profibus is a network protocol that was created in 1989 in Germany by a consortium of factory automation suppliers. One well-known company that has adopted Profibus is Siemens, a high-profile German-based company. Today, there are 1100 member Profibus companies worldwide, mostly in Europe. Originally, it was developed for automated factory operations. Since then, its use has expanded into process manufacturing and for enterprise-wide applications.

Profibus is based on network communication over bus lines between a master controller (host) and an intelligent field device (slave), and several slave devices can communicate with one another. Slave devices are typically remote I/O devices, master control centers, control valves, variable-speed drives, and transmitters. Whenever a Profibus network performs closed-loop On-Off and PID operations, a separate controller or computer is required. Profibus encompasses several industrial products, including Profibus-DP, Profibus-PA, Profibus-FMS, and ProfiNet.

Profibus-DP

This product is a device-level bus that supports both analog and discrete signals. Originally, it was developed to replace conventional I/O systems that required separate wires between a controller and field devices that carry 4- to 20-mA, 24-VDC signals. It is designed to handle time-critical communications, in which each device is assigned a predetermined time period to communicate.

Profibus-DP communicates over RS-485 media at speeds of 9.6 Kbps to 12 Mbps at distances up to 100 to 1200 meters, and power is supplied to field devices separately from the communication bus.

Profibus-PA

This product was introduced in 1997 as a version of Profibus-DP to support process-automation applications. Profibus-PA has device description and functional block capabilities, and can manage smart process instrumentation. The bus that carries the communication data also supplies power for the device to which it is connected. Data is transmitted at speeds of 31.25 Kbps and has a maximum length of 1900 meters per segment.

Profibus-FMS

This product is used for communicating among the upper level, cell level, and field-device level of the Profibus hierarchy. Upper-level equipment includes DCS and PLC systems. At the cell level, the production operations are controlled in each individual cell. Profibus-FMS uses the fieldbus message specifications to execute communications between its hierarchical levels over RS-485 media.

ProfiNet

ProfiNet is the newest-generation Profibus product, and allows communications across Ethernet networks that have TCP/IP protocol and IP services. Not only does it perform the highest level of operations for the Profibus family, it is also able to operate in large networks by using the other versions at lower-level subnetworks.

There are two versions of ProfibusNet: *IO* and *CBA*. ProfibusNet IO implements communication between controllers and IO devices with RT and IRT requirements. RT is real-time communication restricted to subnetworks. IRT allows scheduled communications between networks. ProfibusNet CBA (component-based automation) is suitable for communicating among controllers.

The hierarchy of Profibus networks is shown in Figure 27-25.

NetLinx Networks

NetLinx is an architecture used by Rockwell Automation that provides open networking technology for seamless integration of all the components in an automated system. The term

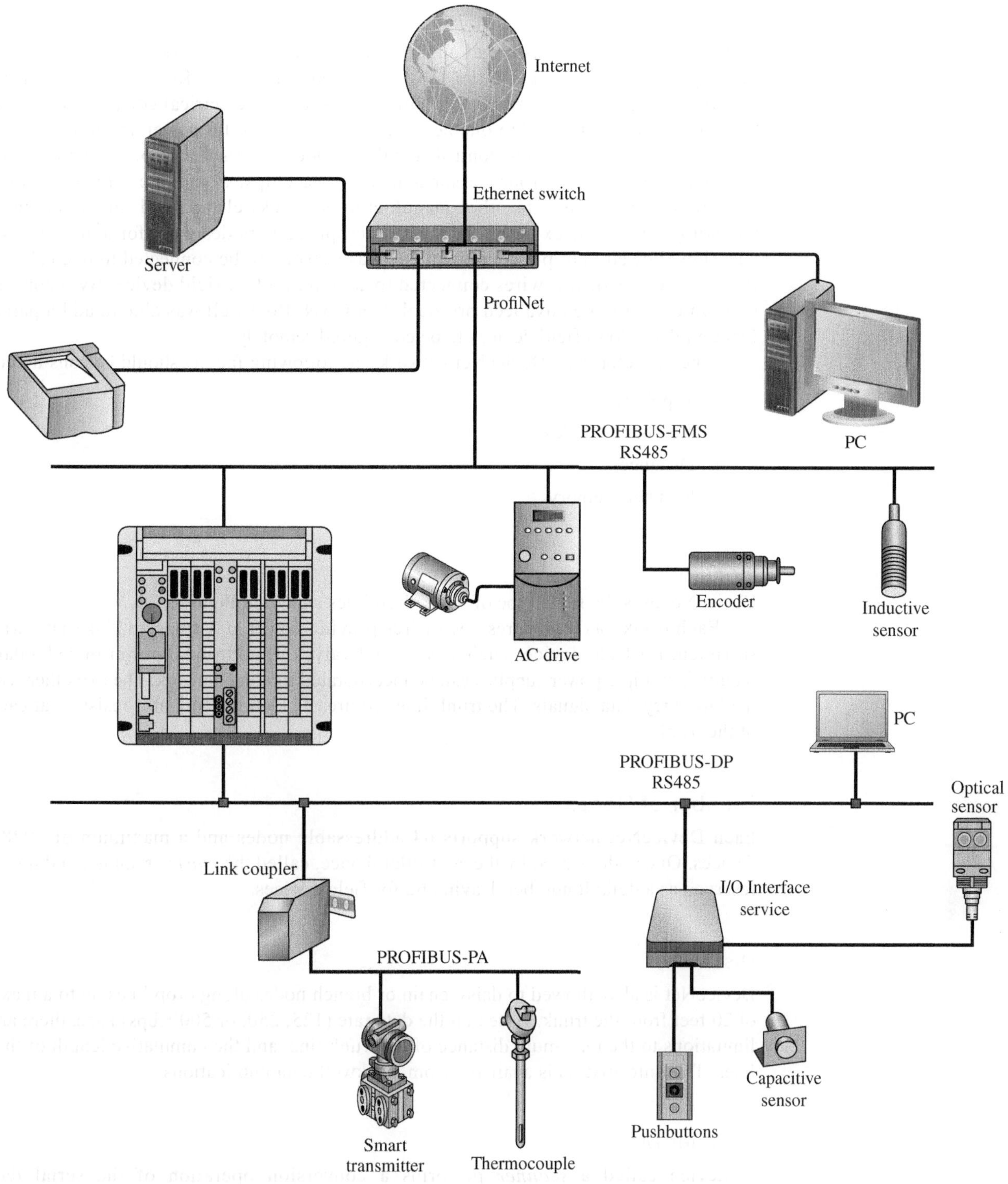

FIGURE 27-25 Profibus hierarchy

seamless means that it is not necessary to do programming of devices between networks in order to communicate through them. NetLinx uses three Rockwell networks to make up the NetLinx architecture: DeviceNet, ControlNet, and Ethernet/IP. Each shares the same protocol, called CIP (common industrial protocol), which enables any of the NetLinx networks to control, configure, and collect information and data without the need to translate information as it passes between them.

DeviceNet

The DeviceNet network is based on CAN (controller area network) technology. CAN was developed and refined during the 1980s and 1990s in Europe for automobile wiring, not manufacturing. Instead of running separate wires to each electrical component in a car, the components are connected to one network cable that provides power and transmits control signals. The functions of the control signals include actuating components and performing diagnostics using a microprocessor and interface chips. Sometime after CAN protocol became an open source to other manufacturers, Rockwell adopted this communication technology using the existing chips and microprocessors designed for automobiles. This adoption allowed multiple equipment, such as sensors, to be connected to one cable rather than having sets of two wires connected to each individual field device. By incorporating CIP, which is a more advanced protocol than CAN, Rockwell was able to add a parameter function that allows field devices to be configured remotely.

When developing a DeviceNet network, the following factors should be considered:

- Topology
- Number of nodes
- Distance
- Scanner memory

Topology

DeviceNet uses the same type of trunk/drop lines shown in Figure 27-3.

Each trunk has four wires. Two wires provide power at 8 amps and 24 VDC, which is sufficient for field devices such as solenoid valves. If additional power or redundancy is required, multiple power supplies can be incorporated into the network. The two other wires on the bus carry data signals. The trunk line requires 121-ohm terminator resistors at each end of the trunk.

Number of Nodes

Each DeviceNet network supports 64 addressable nodes and a maximum of 2048 field devices. One node is used by the controller device, called the *master scanner,* and node 63 is reserved as a default number, leaving 62 for field devices.

Distance

DeviceNet is also allowed to daisy-chain or branch nodes along drop lines up to a maximum of 20 feet from the trunk. Based on the data rate (125, 250, or 500 Kbps) used, there are also limitations to the maximum distance of the trunk line, and the cumulative length of the drop lines. This information is available from Rockwell data publications.

Scanner Memory

A device called a *scanner* performs a conversion operation of the serial data on the network to a format usable by the controller's processor. The DeviceNet scanner module has a limited amount of I/O data that it can handle. Therefore, the cumulative I/O size of the field device cannot exceed the capability of the scanner, or it will not work properly.

DeviceNet supports master/slave, peer-to-peer, and multicasting network models. The protocol used by DeviceNet uses a CSMA/BA (carrier sense multiple access/bitwise arbitration) method that ensures that the highest-priority I/O data message always gets transmitted. DeviceNet is typically used to communicate between I/O interface devices, and variable-speed drives. It is an open network maintained by ODVA (Open DeviceNet Vendor Association) that is actively supported by over 250 vendors.

ControlNet

The ControlNet is a more powerful network model than DeviceNet, which is designed to primarily communicate between field devices and the controller. Instead, ControlNet is a control system network that supports multiple controllers, remote I/O racks, HMIs, and DeviceNet subnetworks. Data is transmitted at high speed (5 Mbps), and all devices on the network are capable of communicating with one another. All communication transmissions on the network are scheduled at predetermined intervals. However, an unscheduled transmission can occur when needed if a MSG instruction is initiated. Also, if bandwidth is available, other data functions can be performed. The types of data that ControlNet handles are:

- Time-critical (scheduled) I/O data
- Non-time-critical programming
- Uploading and downloading of programs
- Messaging

The media over which ControlNet data is transmitted are coaxial R6/U cables or fiber optic cables. Both have high noise immunity, which is desirable for the types of data they pass. These cables do not carry power, so electrical power must be provided to field devices externally.

When developing a ControlNet network, the following factors should be considered:

- Topology
- Number of nodes
- Distance
- Connections

Topology

ControlNet uses trunk-and-drop, star, and ring topologies when utilizing repeaters.

Number of Nodes

Each ControlNet network supports up to 99 nodes, but can be expanded by using controllers on which multiple networks can be connected.

Distance

The maximum distance of a ControlNet network depends on the number of nodes that are connected. However, distance can be increased by using repeaters.

Connections

The communication cards on a controller have a limited number of connections that can be made with the devices on a system. This number cannot be exceeded when terminating I/O modules, remote modules, or remote racks.

Ethernet/IP

Ethernet/IP is Rockwell's enterprise network that uses standard Ethernet and TCP/IP protocols. The use of Ethernet for networking in an industrial environment was a natural progression as plant floor information increased, such as:

- The capacity to transfer large data files
- Devices and network prognosis and diagnostics
- The ability to perform controller operations, such as operating a control loop
- Device configurations and programming
- To accommodate products that have grown in complexity, such as processors, HMIs, and software
- Speed of data flow

Ethernet incorporates open-application CIP inside the standard protocols used by the Internet (TCP/IP and UDP). The TCP/IP is used for general messaging information-exchange services, and UDP for I/O messaging services to support control applications. CIP is also the application layer used by the Rockwell DeviceNet and ControlNet products, enabling all three networks to work with one another. The protocol used by Ethernet/IP means that data is broken up into packets that contain addresses of the sender and receiver, data used in an operation, and information about the placement of packets in one message for when they are reassembled at the receiving end. An advantage of using TCP/IP is that the packets are routed the best way to their destination and avoid bottlenecks.

To minimize collisions, switches are used at the hub of the network topology, and the industrial networks can be isolated from the office Ethernet network. The high-speed baud rate at which data is sent over Ethernet/IP media has two options: 10 Mbps and 100 Mbps.

When designing an Ethernet network, the following factors should be considered:

- Topology
- Distance
- Connections

Topology

Because Ethernet complies with IEEE802.3/TCP/UDP/IP standards, off-the-shelf, readily available hardware such as cables and connectors can be used. The simplest topology option is the star configuration shown in Figure 27-26. The diagram shows that point-to-point connections are made from each field device to the hub using a switch. Connecting the switches from each star network together can expand a star network. Category 5 (CAT 5) unshielded twisted-pair cables with RJ-45 connectors connect the field devices to the switch. Fiber optic cables are also used, especially for the connections between the switches.

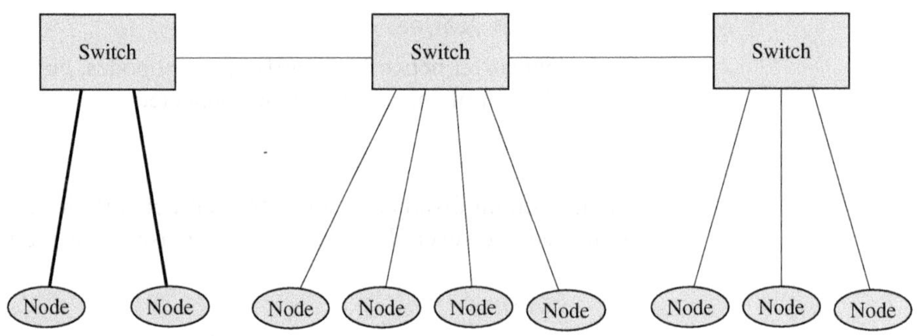

FIGURE 27-26 Sample of Ethernet/IP system star topology

Distance

Distances vary, depending on whether CAT 5 cables or fiber optic conductors are used. The maximum distance between the switch and node is 100 meters when CAT 5 is used.

Connections

The number of terminations on a communication card used by equipment, such as switches or controllers, determines how many connections can be made. Messages sent over Ethernet/IP networks to devices such as controllers or HMIs are unscheduled. An unscheduled communication is triggered by a MSG instruction, which allows data to be sent or received when needed. Each message uses one connection, regardless of how many devices are in the message path. However, the option of configuring one message to read from or write to multiple devices is available.

The hierarchy of NetLinx networks is shown in Figure 27-27.

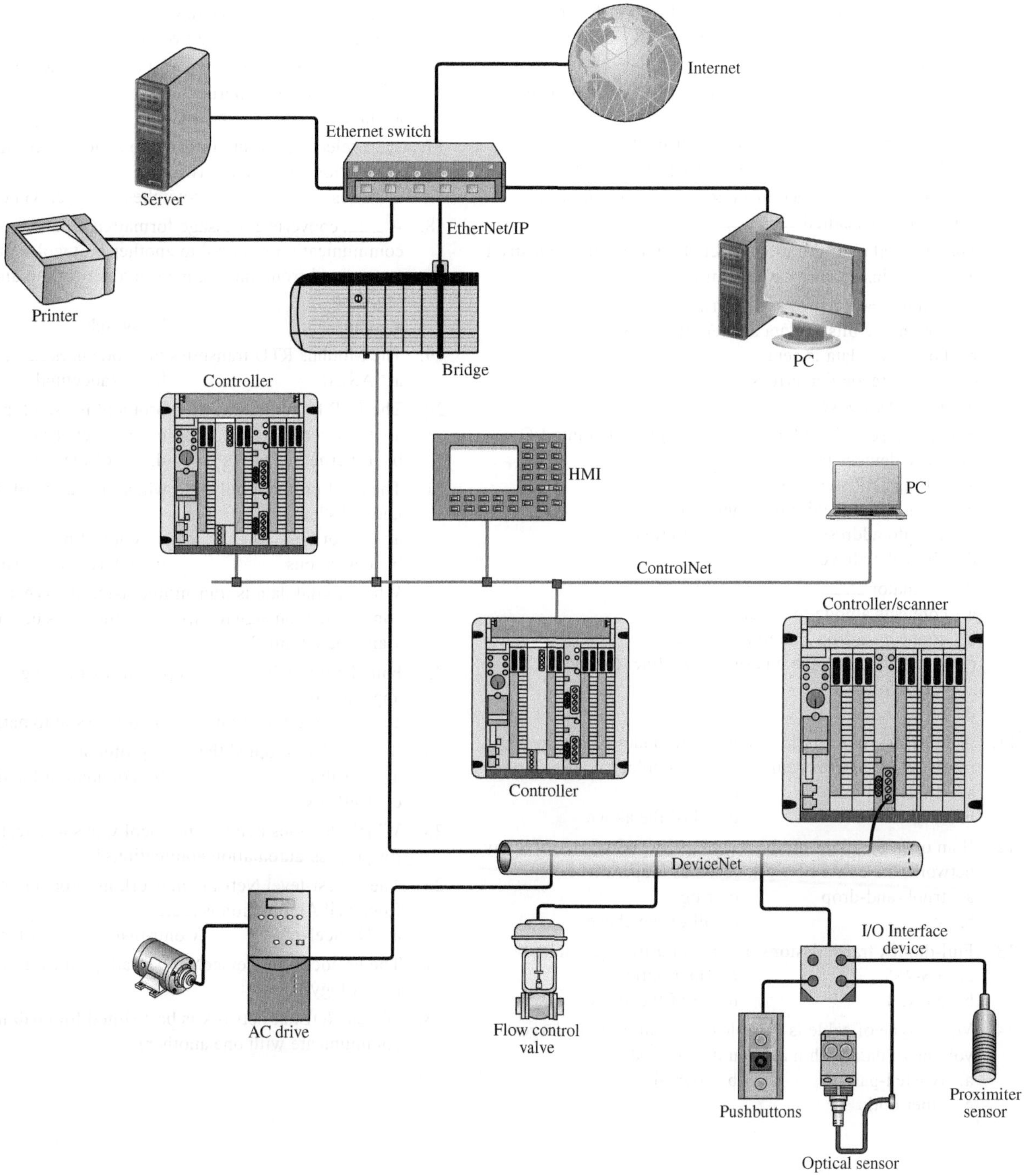

FIGURE 27-27 NetLinx hierarchy

Problems

1. Which of the following are transmission media through which data communication signals pass?
   a. copper conductors
   b. fiber optic cables
   c. atmosphere
   d. all of the above
2. List two nonprofit independent organizations that were formed to establish standard communication protocols.
3. Match the following networks with the level at which they operate.

a. device bus	____ Level 1 (highest)
b. sensor bus	____ Level 2
c. enterprise bus	____ Level 3
d. control bus	____ Level 4 (lowest)

4. How does a controller on a network distinguish between one field device and another when they are all connected to only one cable?
5. The company server is connected to which hierarchical level?
   a. sensor bus c. control bus
   b. device bus d. enterprise network
6. Networks are assembled in a number of different physical configurations called ________.
7. The physical configuration selected for a specific industrial network is based on what capabilities?
8. A communication protocol determines ____.
   a. how many bits of data are sent in a message
   b. how often data is sent
   c. how to recognize errors
   d. all of the above
9. The addresses of field devices are programmed into I/O interface devices by ____.
   a. setting DIP switches
   b. using a handheld programming device
   c. an auto addressing software operation
   d. all of the above
10. A terminator ____.
   a. connects a node to a trunk line
   b. connects a drop line to a trunk line
   c. is connected to the end of a trunk line to prevent reflections
   d. all of the above
11. A multicast transmission in which all nodes on a network receive a broadcast occurs on which topology configuration?
   a. trunk-and-drop c. star
   b. ring d. all of the above
12. If an open develops on the main bus line of a ____ topology network, the entire network becomes inoperative.
   a. trunk-and-drop c. ring
   b. star d. all of the above
13. Full-duplex transmissions occur over a(n) ____ cable.
   a. RS-485 c. fiber optic
   b. coaxial d. all of the above
14. Which type of cable is capable of transmitting the largest volume of data within a given time period?
   a. twisted-pair b. coaxial
   c. fiber optic
15. What is the name of the device that extends the distance a signal can be transmitted over a bus line?
16. Any data packet entering a port on a ____ will be transmitted out of all the other ports.
   a. hub b. switch c. gateway
17. ____ selectively route information packets to specific destinations on a star network.
   a. Hubs b. Switches c. Gateways
18. A ____ converts a message formatted on one type of communication protocol to another protocol as it is transmitted from one segment of a network to another segment.
   a. hub b. switch
19. The Modbus RTU transmission mode uses ____ characters.
   a. ASCII b. hexadecimal
20. The TCP/IP communication protocol is used for ____.
   a. the Internet c. intranet structures
   b. Ethernet networks d. all of the above
21. The AS-I protocol falls into which hierarchy of industrial networks?
   a. sensor bus c. control bus
   b. device bus d. enterprise networks
22. When digital data is transmitted using the HART communication signals, what distinguishes between logic 1 and logic 0 bits?
23. Foundation Fieldbus is an open protocol designed for ____ applications.
   a. motion-automation b. process-automation
24. Siemens has adopted the ____ protocol.
   a. Profibus b. Foundation Fieldbus
   c. NetLinx
25. Which Profibus network protocol was specifically designed for process-automation applications?
26. The lowest-level NetLinx networking product developed by Rockwell Automation is ____.
   a. DeviceNet b. ControlNet c. Ethernet/IP
27. The protocol for DeviceNet was adopted from which technology?
28. Which NetLinx network is best suited for enabling PLCs to communicate with one another?

Answers to Odd-Numbered Problems

Section 1

CHAPTER 1

1. motion, process or open-loop, closed-loop
3. negative
5. *Motion Control* — Hall-effect speed sensor; *Process Control* — Float
7. c. feedback loop
9. feedback signal
11. error, error
13. b. manipulated;
    a. controlled
15. d. control signal
17. d. all of the above
19. True
21. True
23. b. feed-forward

CHAPTER 2

1. $R_F/R_{IN} = 5\text{ K}/1\text{ K} = 5$
3. +2.4 V
5. squarewave
7. a. < b. = c. >
9. It determines if one binary number is greater than, less than, or equal to the other binary number.
11. reverse
13. $2^5 - 1 = 32 - 1 = 31 \quad 15\text{ V}/31 = 0.4838\text{ V}$
15. $2^5 - 1 = 32 - 1 = 31 \quad 10\text{ V}/31 = 0.3225\text{ V}$
17. 2.56
19. high
21. $f = \dfrac{1.44}{(R_A + 2R_B)C} = \dfrac{1.44}{120\text{k}\Omega \times 10\ \mu\text{fd}} = 1.2\text{ Hz}$

CHAPTER 3

1. c. switching
3. b. off
5. c. 0° to 180°
7. a. on
9. c. holding current falls below the minimum point
11.

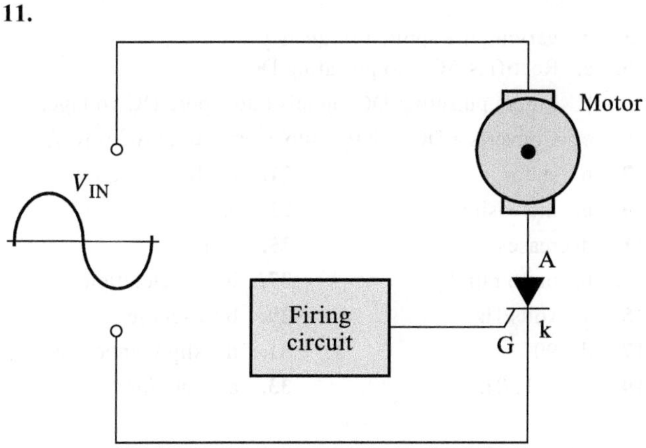

360° A–k Voltage waveform

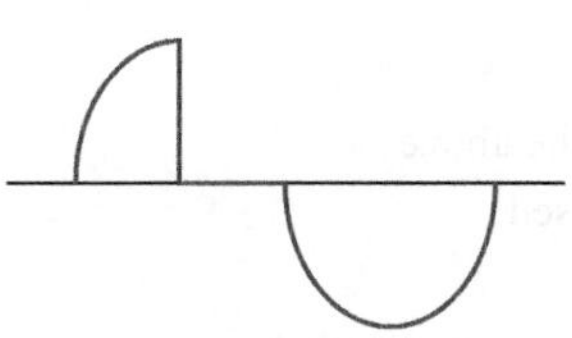

13. V_E: decreases
    I_E: increases
    B_1: increases
    B_2: decreases
15. 100
17. a. diac
19.

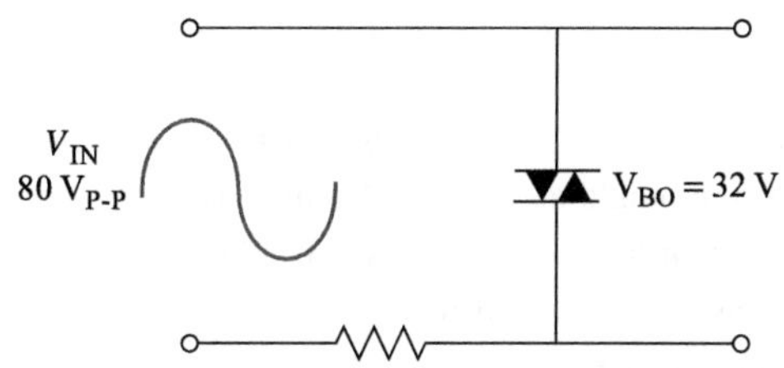

B. 

21. e. all of the above
23. c. a triac

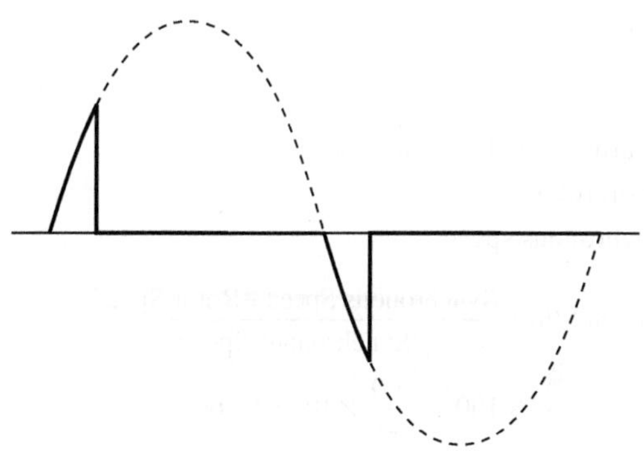

25. a. gate

Section 2

CHAPTER 4

1. control section
3. process disturbances
   The controller cannot adjust the output to match the process demand.
5. $\%\text{ Differential Gap} = \dfrac{\text{Differential Gap}}{\text{Total Control Range}} = \dfrac{8}{80} = 0.1 \times 100 = 10\%$
7. $\text{PB} = \dfrac{25}{100} \times 100 = 0.25 \times 100 = 25$

9. True
11. offset
13. True
15. 25
17. b. reset
19. c. both a and b
21. False
23. subtracts from
25. c. both a and b
27. Proportional: Inverting opting
    Integral: Integrator
    Derivative: Differentiator
29. a. on
31. c. 15V

Section 3

CHAPTER 5

1. torque
3. b. at a right angle to
5. True
7. increases
9. c. either a or b
11. b. 559.5
13. a. increases armature current
15. b. a constant speed rating
17. d. decrease, increase
19. True
21. c. high starting torque
23. c. a higher starting torque than a shunt motor
25. c. more constant speed
27. True

CHAPTER 6

1. b. brushes and commutator
3. stator, rotor
5. synchronous speed
7. $$\text{Percent Slip} = \frac{\text{Synchronous Speed} - \text{Rotor Speed}}{\text{Synchronous Speed}} \times 100 = \frac{200}{3600} \times 100 = 5.56\%$$
9. a. rotating field
11. b. 75–80, d. opens, g. disconnect, i. auxiliary
13. b. series/auxiliary; c. parallel/main
15. two
17. b. shaded pole
19. universal
21. b. armature
23. blown fuse
    broken connection
25. False
27. True
29. synchronous
31. overexcited
33. False
35. b. each phase

CHAPTER 7

1. True
3. forklift
   wheelchair
   motorscooter
5. tape transport systems
   computer peripheral devices
7. b. collector
9. stepper
11. b. counterclockwise
13. $$n = \frac{Y \times S}{6} = \frac{12 \times 360}{6} = 720$$
15. stepping rate
17. a. increasing the stepper rate applied to the stator
19. c. both a and b
21. b. the auxiliary winding

Section 4

CHAPTER 8

1. high starting torque
   higher horsepower
3. energy savings
   stress reduction
5. conveyor system
   extruder
7. a. motor's armature
9. e. all of the above
11. True
13. b. to operate as a variable-resistance device
15. increases, sooner
17. b. doubles
19. a. armature voltage; b. field strength
21. Constant-torque load:
    c. hoist; e. extruder
    Constant-horsepower load:
    a. machine tool lathe
    Variable-torque load:
    b. pump; d. centrifugal fan

CHAPTER 9

1. False
3. c. varying the applied frequency
5. a. Rectifies AC into pulsating DC.
   b. Filters pulsating DC signals into a pure DC voltage.
   c. Converts a DC voltage into a simulating AC waveform.
7. b. stator
9. b. increasing
11. decreases
13. b. more rapidly
15. a. volts/Hz
17. d. 90
19. a. volts/Hz
21. b. Inverter duty
23. a. 2
25. True
27. b. acceleration
29. b. S-curve
31. d. slip-compensation
33. a. variable

Section 5

CHAPTER 10

1. e. all of the above
3. b. decreased
5. less
7. b. liquid(s), a. gas(es)

9. b. decrease
11. 6.28 (S.G.) × 0.433 = 2.72 for 1 ft. (pressure)
51.3 psi ÷ 2.72 = 18.86 ft
13. b. decreases
15. a. compression
17. Atmospheric pressure: Gage
Absolute zero (vacuum): Absolute
19. 29.92
21. 3 × 0.491 = 1.473 psig
23. higher
25. pins 2 and 4
27. compressed air
29. gravity

CHAPTER 11

1. thermal energy
3. a. conductance
5. Blast Furnace: Glass, steel, cement
Fossil Fuel Furnace: Heat-treating metals
Arc Furnace: Melting materials in a foundry
Resistance Furnace: Burn-in chamber for ICs
Induction Furnace: Melting iron in a foundry
7. b. absorbing heat from the contents
9. 5 × 10 = 50
11. C = 5/9 (74 − 32)
= 0.555 (42)
= 23.33 degrees C
13. ceramic kiln
15. decreased
17. cold
19. directly
21. $a = \frac{R_{100} - R_0}{100(R_0)} = \frac{69.25 - 50}{100(50)} = \frac{19.25}{5000} = 0.00385$
23. over-current protection; motor starters
25. negative
27. surge suppression
29. by a drilled hole because only emitted energy is radiated from the hole
31. several thermocouples connected in series
33. ratio pyrometer

CHAPTER 12

1. cubic feet, gallons, liters
3. $F = \frac{WS}{L} = \frac{100\text{ lb} \times 100\text{ ft/min}}{5\text{ feet}} = 2{,}000\text{ lb/min}$
5. Volume = 156.6 ft^3/min
Mass = 9,772 lb/min
7. decrease, decrease
9. $R = 1{,}693$
11. $Q = K\sqrt{16} = 1.22\sqrt{16} = 1.22 \times 4 = 4.9$
13. is
15. is
17. 5 to 20
19. pressure
21. c. both a and b
23. flow straightener
25. c. measuring the power level applied to the heater

CHAPTER 13

1. feet, meters
3. to determine if there is enough material to complete a job.
to determine inventory.
to prevent a container from underfilling.
5. a. direct
7. b. a noninvasive
9. a. an invasive
11. decreases
13. at the bottom
15. c. both solids and liquids
17. b. noninvasive
19. b. decreases
21. $\text{Level} = \frac{40\text{ psi}}{0.43} = 93\text{ feet}$
23. 62.3 × 3.2 ft^3 = 199.36 lbs

CHAPTER 14

1. negative logarithm;
hydrogen
3. c. 7
5. b. positive
7. False
9. b. changing the dimensions of the plates
11. b. Carbon monoxide
13. False
15. b. dry
17. hygroscopic
19. b. relative humidity
21. False

CHAPTER 15

1. d. all of the above
3. c. both a and b
5. c. both a and b
7. a. adding
9. b. kept constant
11. b. precision
13. b. transmitter
15. c. 15 psi
17. b. linearize a signal
19. True
21. c. control valve
23. d. Ball
25. b. air-to-open
27. $F = PA$
$= 12\text{ psi} \times \pi \times r^2$
$= 12 \times 3.14 \times 4$
$= 150.7$

CHAPTER 16

1. e. none of the above
3. c. mounted at an auxiliary location
5. field
7. a. first
9. a. installed in the field
11. c. both a and b
13. c. both a and b
15. True
17. c. loop number
19. a. open
21. c. actuator
23. b. valve in a temperature loop
25. d. all of the above

CHAPTER 17

1. process control
3. False
5. False
7. d. all of the above

9. True
11. c. both a and b
13. furnace, refrigerator
15. b. decreases
17. c. both a and b
19. a. fully off
21. b. Excessive cycling will occur.
23. True
25. b. decrease
27. d. derivative
29. a. braking
31. True
33. b. cascade
35. b. outer
37. X Feed-forward control helps to prevent changes in the controlled variable before they occur.
39. c. ratio control
41. a. wild
43. a. differential pressure flowmeter
c. pH of a solution

CHAPTER 18

1. to ensure that the output accurately represents the signal applied to the input
3. b. linearity
5. c. either a or b
7. b. zero shift
9. c. 18.4
11. a. causes the process to continuously cycle
13. a. low
15. true

Section 6

CHAPTER 19

1. c. target
3. b. axial
5. a. loses
7. 1; sensor head
9. a. increases; a. increases
11. a. inductive
13. d. all of the above
15. concentrators
17. b. infrared
19. True
21. convergent
23. retroreflective
25. opposed sensing method
27. all of the above
29. $T = \frac{W-D}{V} = \frac{2-0.5}{5}$
$= 300 \text{ ms}$
31. 12
33. a. into; b. out of
35. b. decreases
37. b. brown; a. blue; c. black

CHAPTER 20

1. Source—a sensor
Medium—air or cable
Receiver—computer
3. c. both a and b
5. topologies
7. a. Wi-Fi
9. b. the RF signal it receives
11. CSMA—When two or more messages are sent simultaneously and collisions occur, each transmitter resends the data, either along a different path or at a different time.
TDMA—To avoid collisions, each transmitting device in the network has its own time slot to send its data.
13. a. lower
15. c. decreases to 1/4
17. c. decreases to 1/4
19. c. both a and b
21. FHSS—The frequency of the transmitter is automatically changed from one to another as it sends a message.
DSSS—Each bit of a message is sent over a separate frequency. All the bits in the message are transmitted simultaneously.
23. b. SP100.11
25. a. IEEE802.11

Section 7

CHAPTER 21

1. rails; rungs
3. c. left; d. right; a. top; b. bottom
5. c. both latching and interlocking
7. handheld programmer
dedicated terminal
microcomputer
9. d. all of the above
11. d. all of the above
13. a. on
15. b. the other rung numbers that have components activated by its output device
17. b. element
19. e. none of the above
21. c. Executive
d. Scratch pad
b. Program file
a. Data files
23. b. data
25. a. 1; a. 1
27. c. 256; c. 256
29. c. 2; d. 4
31. c. 256
33. True
35. C5:11.PRE

CHAPTER 22

1. program
3. 100 scans
5. a. True c. False
b. False d. True
7. a. create a parallel circuit
9. d. all of the above
11. False
13. True
15. g. all of the above
17. b. TT
19. b. cascading
21. c. both a and b
23. address
preset value
accumulated value
25. c. both a and b
27. a. increments; b. decrements; d. False-to-True
29. True
31. b. TOD
33. e. all of the above
35. addition subtraction
multiplication division

CHAPTER 23

1. True
3. nesting
5. True
7. left
9. False
11. b. bits used in an array
13. True
15. a. to the next register
17. b. most
19. True
21. b. sinking
23. a. low
25. b. output
27. 65,536
29. True
31. a. Signals are conditioned to a usable form.
    b. They enable the processor unit to operate more efficiently.
    d. They can perform control functions independent of the processor unit.

Section 8

CHAPTER 24

1. b. closed
3. b. velocity control
5. torque
7. e. damping
   c. bandwidth
   d. end point
   b. home position
   a. holding torque
   f. traverse rate
9. a. controller
11. linear
13. c. linear or rotary
15. b. rotary; a. linear
17. *Z*-axis
19. a. following error
21. c. belt drive
23. True
25. a. direction in which the rings are pitched

CHAPTER 25

1. b. RPM; c. direction
3. b. stability
5. True
7. c. both a and b
9. d. all of the above
11. by measuring the time interval between pulses
    by counting the number of pulses within a timer period
13. resolution
15. True
17. 1111
19. 0111
21. d. all of the above
23. c. resolver
25. a. rotary; b. linear
27. c. twisting
29. c. tachometer
31. d. 0 V
33. b. directly

CHAPTER 26

1. stability
3. high-speed
5. d. all of the above
7. b. right
9. CCW
11. b. opposite
13. b. position
15. a. stationary
17. b. resolution
19. decreasing
21. damping
23. high
25. b. integral
27. c. both a and b
29. decreases
31. b. speed up

Section 9

CHAPTER 27

1. d. all of the above
3. c. Level 1 (highest)
   d. Level 2
   a. Level 3
   b. Level 4 (lowest)
5. d. enterprise network
7. reliability
   flexibility
   cost of adding a node
   future growth
   potential of data flow disruption
9. d. all of the above
11. a. trunk-and-drop; b. ring
13. a. RS-485
15. repeater
17. b. Switches
19. b. hexadecimal
21. a. sensor bus
23. b. process-automation
25. Profibus-PA
27. the CAN (Controller Area Network) technology developed for the automobile industry

A

absolute humidity the mass of water vapor present in a particular volume of atmosphere.

absolute optical encoder a device used primarily to measure the angular position of a rotary object.

absolute pressure 1. a pressure scale that uses absolute zero, which is the complete absence of pressure, as a reference. 2. the gas pressure above a perfect vacuum.

AC motor a device that converts adjustable alternating current frequency and voltage to rotating mechanical energy.

AC servo motor an electric motor that uses AC command signals, which cause its rotating member to rotate to a desired position.

accuracy 1. how closely a sensor measures the actual value of a controlled variable. 2. the degree to which an output will attempt to match the input command signal.

actuator an element of a control system that converts electrical, hydraulic, or pneumatic energy into work.

adaptive control a control technique that uses a combination of software programming and microelectronics to compensate for nonlinear situations.

alarm an instrument that produces a warning signal when an undesirable process condition develops.

amplitude proportional a method in which the output of the controller is proportional to the size of the error signal.

analog input module an interface module of a programmable controller that converts the amplitude of an analog signal into proportional digital data.

analog output module an interface module of a programmable controller that converts a digital value into an equivalent analog signal.

analog-to-digital converter (ADC, A/D) a device that converts an analog voltage applied to its input into a proportional digital output.

analytical measurement and control a procedure that monitors and regulates the composition of a chemical.

aqueous solution water that is mixed with another ingredient.

arithmetic logic unit (ALU) the section of the programmable controller that performs the arithmetic and logic operations.

arithmetic operations instructions of the programmable controller that performs the four basic mathematical functions: addition, subtraction, multiplication, and division.

armature a laminated iron core with wire wrapped around it in which an EMF is induced as it rotates inside the stator field.

armature reaction the distortion of the main field flux lines due to interaction with the magnetic field around the armature.

astable multivibrator a circuit that generates a continuous squarewave output.

auto boost a circuit in the electronic AC motor drive that causes current and torque to increase when the motor is running at a low RPM.

B

backlash the movement within the space between mating parts or gears.

back plane a printed circuit board located at the back of a PLC rack assembly that provides electrical connections between modules.

bandwidth the measure of how quickly the controlled quantity tracks and responds to the command signal.

bang-bang position servo a position servo that moves the load from one end position to another end position very rapidly and without the need for extreme accuracy.

base speed the rated speed at which the motor operates.

batch processing a sequence of timed operations executed on a product being manufactured.

branch a circuit path connected in parallel.

branching input instruction symbols placed in parallel on a ladder diagram.

brushless DC motor (BDCM) a DC electric motor with a permanent magnet rotor and several fixed stator windings to which control voltages are applied to create a rotating magnetic field.

bubbler a tube through which air is forced to determine level by counting the number of bubbles emitted within a specified period of time.

bus line the output terminal of the intermediate section in an AC electronic motor drive.

C

calibration a procedure of making adjustments to a transmitter so that its output varies through its full range in proportion to the full range that the variable being measured changes.

capacitive probe sensor a sensor that uses a capacitive probe to detect the level of contents inside a tank.

capacitive proximity sensor an electronic sensor that detects the presence of both conductive and nonconductive targets.

capacitor-start motor an AC single-phase electric motor that has a main winding and a start winding connected to a capacitor.

cascade control a method that controls a process by monitoring both the controlled and the manipulated variable.

cascading connecting the output of one timer to activate another timer to extend its maximum range.

central processing unit (CPU) the brain of the programmable controller.

chemical reaction the process of combining two or more materials or reactants to form a product.

closed-loop a method of control in which feedback is used by a system to produce a controlled process dictated by a command signal.

closed-loop automatic control a self-regulating operation commonly used in automated control systems that incorporates a feedback loop.

color-mark detection *See color mark sensor.*

color-mark sensor a photoelectric sensor that detects the contrast between two colors.

combustion also known as *burning,* a chemical reaction that occurs when heat, gases, and fuel are combined.

commutation the switching action of the brushes and commutator that causes the armature to rotate.

commutator 1. part of a DC motor that is constructed as a split ring, each segment of which is connected to an end of a corresponding armature coil. 2. a device that employs an opening and closing switch action that occurs when the brushes and commutator segments make and break contact with each other. This process always causes current to flow through the armature in the proper direction.

comparator a device that produces various output signals by comparing the signals applied to its inputs.

compensating windings small windings wired in series with the armature windings that cancel the magnetic field of the armature windings and eliminate armature reaction.

compound motor a DC electric motor that has two field windings, one connected in series and the other connected in parallel with the armature windings.

compression 1. the process of storing additional gas into a confined container. 2. reducing the size of the confined container that holds a fixed quantity of gas.

concentrators ferrous metals that bend magnetic flux lines to boost the output of a Hall-effect sensor.

conduction the process by which heat is transferred by a solid.

conductive probe sensor a sensor that uses a conductive probe to detect the level of a conductive liquid.

conductivity a process used to determine the purity of a liquid by measuring the amount of current that is able to flow through it.

continuous-cycling method a controller tuning method in which a closed-loop response test is made with the controller on automatic.

continuous level measurement a method of locating the interface point within a range of all possible levels at all times.

continuous process one or more operations performed simultaneously as a product is produced during a manufacturing process.

contouring a motion-control method that causes the load to make continuous movements along curved or straight lines.

control circuit 1. a circuit in an AC drive that controls when the inverter's switching devices will turn on or off. 2. the section of an electronic AC or DC motor drive that provides a way for the operator to preset operational parameters and control speeds.

controlled variable the actual process that is being controlled by an open- or closed-loop system, such as temperature or pressure.

controller an element that is considered the "brain" of a closed-loop system. An instrument in a closed-loop system that performs the decision-making function.

convection the transfer of heat through fluids such as liquids and gases.

convergent sensing method a photoelectric sensor in which the light source and receiver are mounted next to each other at the same angle from the vertical axis. The object to be detected is used to reflect light at one set distance.

converter the section of an electronic AC drive that converts AC line voltage to a pulsating DC voltage.

coriolis mass flowmeter a mass flowmeter consisting of specially formed tubing that is oscillated at a 90 degree angle to the flowing mass fluid.

Coriolis meter a U-shaped tube instrument that measures flow by determining at how much of an angle it is twisted as fluid passes through.

counterelectromotive force (CEMF) a voltage generated inside an electric motor that is always of opposite polarity to the applied voltage.

counters output instructions of a programmable controller that counts, stores, and displays the number of events that occur in an operation.

critically damped a system that is tuned so accurately that it causes the controlled variable to reach its end position or desired state very quickly without overshoot.

current limiting a circuit in an electronic DC drive that prevents excessive current from flowing throughout the components when the motor is loaded down.

custody transfer measuring flow to determine how much of a product is passed from the supplier to the customer.

D

damping the prevention of overshoot of the load past the end point in a motion-control system or the desired state in a process-control system.

data compare the process of comparing the numerical contents of two PLC registers and making decisions based on their value and the type of instructions used.

data conversion instructions to the programmable controller that make conversions between binary and decimal numbers.

data manipulation a category of instructions that enable words to be moved within the memory of a PLC.

data transfer the process of moving the contents stored in one memory register to another memory location in a PLC.

DC bus line the output terminal of the intermediate section of a DC electronic motor drive.

deadband *See differential gap.*

dead time the elapsed time between the instant a deviation of the controlled variable occurs and the corrective action begins.

density 1. the weight of a certain volume of liquid. 2. the weight per unit volume of a fluid.

depth-of-field the distance on either side of a photoelectric sensor's focus point.

derivative control a control scheme whereby the controller produces an output that is proportional to the rate that the error signal changes. This function is also called *rate control.*

derivative mode *See derivative control.*

derivative time a setting on a controller used in a closed-loop operation to determine the extent to which the derivative action will occur.

dew point the temperature at which the air (or gas) becomes saturated.

diac a bidirectional, two-terminal, solid-state device that is used to trigger a triac.

difference operational amplifier an amplifier that produces an output signal that is the algebraic difference between two input voltages.

differential gap the range above and below the setpoint reached by the controlled variable before the controller element turns an actuator on or off.

differential pressure 1. a difference in pressure between two measured points. 2. the difference in gas pressure between any two points in a system.

differential pressure flowmeter an instrument that measures flow rate by comparing two different pressures developed across a restrictor.

differential pressure level detector a device that detects the level of a material inside a confined container by measuring the difference between hydrostatic pressure and the pressure above the material.

differentiator an amplifier circuit that produces an output proportional to the rate of change of the input signal. It performs the derivative function.

diffuse sensing method a photoelectric sensor in which the light source and receiver are mounted next to each other. The object to be detected is used to reflect light from the emitter back to the receiver.

digital-to-analog converter (DAC, D/A) a device that translates digital data into an analog voltage.

displacement the amount of material replaced by a sensor probe as it measures level.

dissociation compounds that break up into charged particles, called *ions,* when they are combined with water.

disturbance a factor that upsets the manufacturing process, causing a change in the controlled variable.

drive an electronic device that controls the speed, position, acceleration, deceleration, and torque of electric motors.

drive controller a device that converts the fixed source voltage and frequency of a power line to an adjustable voltage and frequency.

duty cycle the ratio of time a squarewave signal is high to the total time period of one cycle.

dynamic 1. the state in which the controlled variable is moving or changing. 2. the characteristic of an instrument while it is changing.

dynamic braking 1. a method in which a resistor is used to absorb kinetic energy to stop a motor quickly.

2. a technique of stopping a DC motor quickly by using a resistor to absorb kinetic energy from the rotating armature and load to which it is connected.

dynamic response the time a closed-loop system takes to perform a corrective action.

E

effective delay the time that expires from when a change is made until the process begins to react.

efficiency the ratio of the power produced by a motor's shaft to the power supplied by the electrical source.

effluent a treated solution that flows out of the tank in a pH batch-process system.

electromagnetic flow detector a transducer that converts volumetric flow rate of a conductance substance into voltage.

element a programmable controller memory location that consists of one or three registers where data is organized and addressed.

end point the desired location to which a load is moved by a motion-control position system.

endothermic processes that require a source of heat while forming a product.

energy harvesting a method of converting one energy source into electricity.

error detector (comparator) 1. the element of a closed-loop control system that produces an error signal by comparing the setpoint to the feedback signal. 2. a device that produces various output signals by comparing the signals applied to its inputs.

error signal the difference between the desired response and the actual response.

excess gain a measure of the amount of light energy that falls on the receiver beyond the minimal amount of light required to operate a photoelectric amplifier.

executive a collection of system programs permanently stored in ROM devices.

exothermic a process where heat is generated during the reaction phase.

F

feedback signal the signal or data fed to the comparator of a closed-loop system from an actuator or processor to indicate the response to the command signal.

feed-forward the process of providing information to the controller element device, which indicates that a change is going to occur.

feed-forward control *See feed-forward.*

fiber optics transparent strands of glass or plastic that transfer light for photoelectric sensing.

field current speed control a method used to control the speed of an electrical shunt motor by varying the current applied to its field coil.

field-of-view the dispersion angle at which a photoelectric sensor can effectively detect light from the emitter.

field poles the electromagnets in a motor that are stationary.

final control element also referred to as an *actuator,* a device that directly influences the process variable.

fixed focus an optical sensor capable of detecting an object at a specific distance.

float a spherical element that rides on the surface of the material it is measuring to determine its level.

flow the transfer of material from one location to another.

flow rate the measurement of flow that is determined by how rapidly a material is moving past a given point.

flow velocity a measurement of the distance a material travels within a given unit of time.

fluid a liquid or gas commonly used in a process-control system.

flux vector control a method in which a microcontroller in an AC variable-speed drive precisely controls motor speed, torque, acceleration, deceleration, and position.

following error the proportional error during motion resulting from the load movement lagging behind the desired movement specified by the command signal.

4-quadrant control a method in which a microcontroller in an AC variable-speed drive provides full speed control of a motor in both directions.

four-wire system a sensor that has four wires.

full load the maximum power a motor can provide to drive its rated mechanical load.

functional identifier letters located in the top portion of a P&ID diagram symbol to identify the types of functions it performs.

G

gage pressure 1. a pressure scale that uses atmospheric pressure as the reference point. 2. the gas pressure above or below atmospheric pressure.

gain the ratio of change in output to the change in input.

Gray code a number system that uses multibits of 0s and 1s, with only one of the bits changing when incrementing or decrementing the count.

H

Hall-effect sensor a sensor that detects the presence of a magnetic field.

head a term commonly used to describe the height of a liquid above the measurement point.

high/low speed adjustment the adjustment made to prevent a motor from attaining full maximum speed or absolute minimum speed, either of which may cause the motor to short out.

holding torque the amount of force required to keep the rotor of a motor from falling out of a stationary position.

home position the reference position from which movements are measured in a motion-control system.

humidity the amount of moisture present in air.

hydrocarbon fuel a material that will burn when combined with fuel and heat.

hydrostatic pressure exerted equally in all directions at points within a confined fluid.

hydrostatic head level detector a sensor that determines level by measuring the weight of the contents in the container.

hydrostatic pressure the resultant pressure obtained from multiplying the height times the density of a liquid.

hygroscopic the ability of a material to absorb moisture.

hysteresis the characteristic evidenced when a target causes a sensor to turn on at one distance and off at another distance. *Also see process lag time.*

I

incremental optical encoder an optical rotary encoder that has tracks of equally spaced slots that indicate position or speed. Position is determined by counting the number of slots that pass by a photo sensor. Speed is determined by counting the number of slots that pass by a photo sensor within a period of time.

indexing controlled positioning movements caused by signals sent from a controller.

indexer a controller used in motion control applications that provides signals to cause positioning movements of a stepper motor.

indicator an instrument used to display information about a process.

induction motor an electric motor that operates on three-phase AC power.

inductive proximity switch an electronic sensor that detects the presence of metallic materials.

inductosyn a linear feedback device that produces a specific code or pulse for each position.

industrial controls the automated equipment that monitors and controls the operation of a manufacturing process.

inferred a procedure whereby some other variable is measured and then translated into the reading that is required.

inferred measurement a condition wherein one type of measurement is taken to find the value of another type of measurement.

influent an untreated solution that enters the tank through an inlet port in a pH batch-process system.

input/output module a type of module that interfaces the internal circuitry of a PLC to outside equipment.

instability an action that occurs in a motion-control operation when the device being positioned oscillates because of overshoot.

instrumentation a term commonly used to describe process control; it refers to the instruments that control and monitor the condition of the process.

insulated gate bipolar junction transistor (IGBT) a type of transistor commonly used in AC variable-speed drives. It is capable of turning on and off at an extremely high rate of speed.

integral (also reset) a mode of control in which the magnitude of the output by a controller is proportional to the length of time the controlled variable has been away from the setpoint.

integral control a control scheme whereby the controller produces an output that is proportional to the length of time an input signal has been applied. This function is also called *reset control*.

integrator an amplifier circuit that continuously increases gain over a period of time. It performs the integral function.

interface 1. the boundary between two media, such as water and air. 2. the connecting together of two different circuits.

intermediate circuit the section of an electronic AC drive that filters the pulsating DC voltage from the converter and varies the DC voltage supplied to the bus line.

interpoles small windings wired in series with the armature, which counteract armature reaction by producing a local field that restores the main flux lines to the original neutral plane.

inverter a term commonly used to identify an electronic AC motor drive. It is also the section of the drive that provides high currents to power the motor.

IR compensation a method in which an electronic DC drive detects armature current to use as a feedback control signal.

J

jump a change in the normal sequence of program execution in a PLC.

L

ladder logic language a software programming format used by programmable controllers that resembles hardwired relay circuits.

level the height at which a material fills a container.

life the minimum rated lifetime of the measurement device while maintaining positioning specifications.

light sensor the element of a photoelectric sensor that detects the absence or presence of an object. It is also referred to as a *detector* or a *receiver*.

light source the element of a photoelectric sensor that supplies the light beam to a light sensor. It is also referred to as an *emitter* or *transmitter*.

limit switch a mechanical switch activated by physical contact with a moving object.

linear displacement transducer a type of linear-motion position sensor that detects the distance from a reference point to a measured object by counting the time interval between a launching pulse and a return pulse.

linear variable differential transformer (LVDT) a type of linear-motion position sensor that uses transformer action to produce a signal that is proportional to distance.

linearity the output signal produced by a sensor that is not proportional to the actual condition of a variable being measured.

load 1. the type of device or equipment to which the sensor output signal is applied. 2. the demand on an actuator by the device to which power is delivered.

lobed impeller flowmeter a rotating device with lobed impellers that measures volumetric flow rate of fluid that passes through it by multiplying displacement times the RPM.

loop gain the ratio of output speed to the following error.

loop identifier numbers located in the bottom portion of a symbol in a P&ID diagram to identify the control loop where it is located.

low-voltage protection a three-wire control scheme using momentary contact push buttons to energize a starter coil.

low-voltage release a two-wire control scheme using a maintained contact pilot device in series with the motor starter.

M

magnitude comparator a logic circuit that compares two binary numbers and produces an output signal that indicates either which one is greater or if they are equal.

main field the magnetic field that forms between two poles of a magnet, which interacts with the magnetic field of the armature.

manipulated variable the fuel or energy that is physically altered by the actuator to change the condition of the controlled variable.

manufacturing process the operation performed by an actuator to control a physical variable, such as motion or a process.

mass flow measurement a method used to measure flow by reading the actual weight of a fluid during a given period of time.

mass flow rate the measurement of flow that is determined by the weight of materials that move during a specific time period.

master–slave a servo system in which coordinated motion is achieved between two control systems by one (slave) system monitoring the condition of the other (master) system.

measured variable the condition of the controlled variable, usually detected by a sensor.

measurement device an element in a closed-loop control system that detects a controlled variable and produces an output signal, which represents its status. Other terms used are *detector, transducer,* and *sensor*.

memory a location where data is stored by a PLC, which is used by the CPU to perform various functions.

methods of detection the various physical arrangements of a light source and its receiving elements, which allow an object to be detected by a sensor.

mixing/blending an operation that involves combining two or more ingredients together.

monostable multivibrator a circuit that produces a temporary logic level voltage after an activating signal is applied to its input. Also known as a *one-shot*.

motion control an industrial control system that controls the physical motion or position of an object.

motor action the conversion of electrical energy to mechanical energy resulting from the interaction of two or more magnetic fields.

motor starters an electrical control circuit used to start, stop, and protect motors from excessive current.

moving coil motor (MCM) a DC electric motor that uses several permanent magnets as its stator and a thin disk with copper conductive tracks as its rotor.

multiple axis positioning a motion-control method that causes a load to make many movements horizontally and vertically.

N

negative temperature coefficient the characteristic of a sensor in which its resistance decreases when the ambient temperature to which it is exposed increases.

neutral plane 1. the axis that is at a right angle to the main field flux lines. 2. the plane that is perpendicular to the flux lines of the motor's main field.

no load a condition during which the motor operates when the physical load is disconnected from the motor shaft.

O

offset the error that remains between the setpoint and the desired output condition after the controller element has caused a transient response. *See steady-state error.*

On-Off control the most basic type of control system in which the actuator is either fully on or fully off.

open-loop a method of control where there is no feedback to initiate self-correcting action for the error of the desired operational conditions.

operational amplifier an integrated circuit that performs several types of linear circuit operations.

operator control an electronic device that provides the operator a means of controlling the start, stop, directional, and speed functions of a motor.

opposed sensing method a photoelectric sensor in which the light emitter and detector are positioned opposite each other. The target is detected when it blocks the light beam.

optical encoder a rotary feedback device that produces a specific code or series of pulses for each position.

optoelectronic device an interface device that uses light energy to pass signals from one circuit to another circuit.

overdamped a system that is sluggish and responds to a changed command signal by causing the controlled variable to reach its end position or desired state very slowly and without overshoot.

overload the condition in which an electric motor stalls because it does not have enough power to move the load it is driving.

P

paddle wheel detector a paddle that detects when the material it is measuring reaches a specified level by stopping it from turning as contact is made.

partial-load a condition in which the physical load a motor is driving is reduced from the full-load condition.

pH control the analytical process in which acid and alkaline levels are controlled.

photodiode a semiconductor device that conducts current when it is exposed to light.

photoelectric sensor an electronic sensor that uses light to detect the absence or presence of an object.

photo SCR a semiconductor SCR that switches on when it is exposed to light.

photo triac a semiconductor triac that switches on when it is exposed to light.

phototransistor a semiconductor device that switches on when it is exposed to light.

PID a controller that uses proportional, integral, and derivative control in one unit.

pipe size the diameter of a pipe that carries fluid.

piping and instrumentation diagram (P&ID) a standard drawing format used in all types of process-control fields.

point level measurement a measuring method that detects if the interface is at a predetermined level.

polymerization process in which a large number of molecules are combined to form a product.

positive temperature coefficient (PTC) the characteristic of a sensor in which its resistance increases when the ambient temperature to which it is exposed increases.

potentiometer a variable resistor capable of converting mechanical motion into an electrical voltage variation.

power supply a section of the programmable controller that provides the voltages necessary to operate the circuitry throughout the entire PLC.

precision the degree of consistency with which a sensor responds to the same repeated input value.

pressure the force exerted over a surface area.

process calibration an instrument used to calibrate devices such as sensors and transmitters that operate in process systems.

process control an industrial control system that regulates one or more variables during the manufacturing of a product.

process lag time the response lag time of a control system to a setpoint change or a disturbance.

process reaction rate a measure of how much the process changes per unit of time after a step change is made.

process stream the flow of either a gas or liquid in analytical process-control applications.

processing modules special purpose modules used to interface the processor unit with field devices.

processor file a memory block in the central processing unit where the programmer stores and manipulates software.

processor scan cycle a sequence of steps performed by the microprocessor in a PLC during which the input conditions are examined, logic decisions are performed, and appropriate signals are applied to the output.

processor unit the "brain" of the PLC, which coordinates and controls the operation of the entire system.

program a list of instructions that guide the PLC through a desired operation.

programmable logic controller (PLC) a solid-state control system that has a user programmable memory for the storage of instructions to implement specific functions such as logic operations, timing, counting, arithmetic, and data manipulation.

programming unit a module in a PLC that provides a way for the user to enter data, edit, and monitor the program stored in the processor unit.

proportional band the range at which a controller produces an output signal proportional to the error signal applied to its input.

proportional control a control scheme whereby the controller produces a signal that is proportional to the error signal.

proportional gain the ratio of change in output to the change in input.

protocol a set of rules that determines the format and transmission methods of data flow in an industrial computer network.

proximity detector an electronic sensor that indicates the presence of an object without making physical contact.

proximity switch a sensor that detects the absence or presence of an object without making physical contact.

pull-out torque a condition in which the physical load connected to a synchronous motor causes the rotor to fall out of synchronism with the stator field.

pulse width modulation (PWM) a method in which the output of the inverter section of a PWM or AC vector drive is pulsated to produce a simulated AC waveform.

pulse width modulation AC drive an AC variable-speed drive that varies the RMS voltage applied to the motor by changing the width of pulses produced by its output.

pure lag the opposition of a controlled variable being changed due to the material from which it is made.

pyrometer an instrument that measures temperature by detecting the amount of thermal energy radiated from the surface of the measured body.

R

radiation the transfer of thermal energy through a vacuum.

radiation thermometry a process where temperature measurements are made from thermal energy radiated from the surface of the measured body.

range the maximum amount of distance a measurement device can travel.

rate a control scheme whereby the controller produces an output that is proportional to the rate that the error signal changes. This function is also called *derivative control*.

rate time an adjustment made to a controller for the derivative mode.

rated-load torque the torque developed by a motor when driving its rated mechanical load.

ratio control a method of controlling one variable based on the measured condition of another variable.

reaction-curve method a controller tuning method in which a closed-loop response test is made with the controller in the manual setting.

reagent the corrective ingredient of an acid or a base that is added to bring the pH to the desired level.

recorder a data acquisition instrument that stores information about a process.

regenerative braking a motor braking action that returns power from kinetic energy to the power supply.

regenerative feedback a method that regulates speed or stops a motor quickly by generating a voltage that opposes the supply voltage.

relative humidity the actual amount of water vapor present as compared to the maximum amount of water vapor the air can hold at a given temperature.

relay contacts an electromechanical device that operates in the On-Off condition as its mechanical contacts open or close.

relay ladder logic a programming language used by programmable controllers.

relay ladder logic diagrams the type of wiring circuit used in electrical control circuits.

repeatability the range in which the output position of the servosystem will come to rest whenever a given input command signal is repeated.

reset a control scheme whereby the controller produces an output that is proportional to the length of time an input signal has been applied. This function is also known as *integral control*.

resistance-start induction-run motor an AC single-phase electric motor that has a main winding and an auxiliary winding for starting.

resistance temperature detector (RTD) a thermal sensing device made of two metals that increases in resistance when the temperature to which it is exposed increases.

resolution 1. the number of degrees per count produced by an optical encoder. 2. the number of equal divisions into which a digital-to-analog converter divides the reference voltage.

resolver a transducer that converts rotary or linear position into an electrical signal by the interaction of electromagnetic fields between movable and stationary coils.

response time the amount of time a sensor takes to respond to a change in the measured variable.

retroreflective sensing mode a photoelectric sensor in which the light source and receiver are mounted next to each other. The light source shines into a reflector that returns the light beam to the receiver. The target is detected when it blocks the light beam.

Reynolds number (R number) a numerical scheme that assigns values to express fluidity of a moving fluid. It represents the ratio of the liquid's inertial force to its drag (viscous) forces.

rod gauge a dipstick inserted into the material being measured to determine level.

ROM Read-only memory, where information is permanently stored for the central processing unit.

rotameter a flow-measurement device that is based on the proportionality of the rise of a float in a tapered tube, arranged vertically, placed in the flow system.

rotary-vane flowmeter a rotary device with spring-loaded vanes that measures volumetric flow rates of the fluid that passes through it by multiplying displacement times the RPM.

rotor the armature assembly of the motor that physically rotates to drive the shaft.

rotor flow detector an instrument that uses the rate at which a paddle turns to determine flow rate.

runaway a condition in which the physical load driven by a DC motor is disconnected, causing the motor to accelerate until it breaks apart.

S

scale-with-parameter (SCP) a programmable controller output instruction that produces a scaled value.

Schmitt trigger a device that converts sine waves or arbitrary waveforms into crisp square-shaped signals.

scratch pad temporary memory storage to hold data while calculations are performed by the central processing unit.

sensitivity the ratio of a sensor's output change to a change of its input quantity that represents a measurement.

sensor circuitry the circuitry of a photoelectric sensor.

sensor response time the maximum amount of time that elapses from an input detection signal until the output switches.

separation an operation that involves removing an ingredient from a mixture.

sequencer a controller in a PLC that operates an application through a sequence of events.

series motor a DC electric motor with its field windings connected in series with its armature windings.

servo motor a specialty motor that uses a closed-loop feedback signal to control its position and speed.

servo valve a transducer that converts a low-energy electrical signal into a proportional, high-energy hydraulic output.

servomechanism a closed-loop motion device that automatically controls velocity and position.

setpoint the input applied to the elements of a control system that represents the desired value of the controlled variable.

shaded-pole motor an AC single-phase motor that uses a metal ring to delay the shifting of the field flux lines.

shift register a memory device in a PLC that is used to store or move binary words within itself.

short cycling a condition in which a device is frequently turned on and off.

shunt motor a DC electric motor with its field winding connected in parallel with its armature winding.

sight glass a transparent tube connected to the side of a vessel to measure level.

signal processors special devices that change or modify signals applied to their inputs.

silicon-controlled rectifier (SCR) a three-terminal unidirectional solid-state device that passes current in one direction when it is triggered.

single-axis positioning a motion-control method that causes a load to make linear movements in both directions.

single-phased a three-phase motor condition in which one of the power lines is open.

sinking the type of sensor connection that draws conventional current into its output terminal through the load device to which it is connected.

sinking input active-low PLC input modules into which conventional current flows from field devices.

slip the difference between rotor speed and synchronous speed in an AC motor.

soft start when a DC drive accelerates the motor slowly instead of abruptly.

solid-state relay discrete solid-state circuits that operate in either the on or off condition.

sourcing the type of sensor connection that supplies conventional current from its output to the load device to which it is connected.

sourcing input active-high PLC input modules out of which conventional current flows to field devices.

specific gravity the relative weight of any liquid when compared to water at 60 degrees Fahrenheit.

specular sensing method a photoelectric sensor in which the transmitter and receiver are placed at equal angles from an object. The object to be detected is used to reflect light at one set distance. This method is used to differentiate between shiny and dull surfaces.

speed regulation the ability to maintain a constant speed under varying load conditions.

stability the characteristic of good speed regulation in a closed-loop velocity-control system.

stable a closed-loop system that is under control and does not oscillate around setpoint.

static 1. a state in which the controlled variable does not move or change appreciably within an arbitrary time interval. 2. the condition of an instrument, which has stabilized after changing.

static head the force developed at the bottom of a tank that results from the weight of fluid placed above it.

static inertia the opposition to change of movement by a stationary body.

stator a term that refers to the main field pole assembly.

steady-state error the error that remains between the setpoint and the desired output condition after a transient response by the controller element is completed. Also known as *offset*.

step angle the number of degrees per arc that the rotor of a stepper motor moves per step.

step change an abrupt setpoint change.

stepper motor an electrical motor that converts electronic digital signals into mechanical motion in fixed increments.

stepping rate the maximum number of step movements the stepper motor can make in one second.

stiffness the reluctance of an actuator to deviate from the desired position specified by the command signal.

synchronous motor a three-phase electric motor that operates on the principle that its rotor turns at the same speed as its rotating field.

synchronous speed the rate at which the magnetic field in the stator rotates.

T

tachometer a device that measures the angular velocity of a rotary shaft.

tag numbers an alphanumeric code used for identification that is placed inside a symbol of a P&ID diagram.

target the object detected by a sensor.

Thermaldynamics a condition in which thermal energy migrates from a hot area of an object to a colder area.

thermal energy molecular movement that creates heat.

thermal flowmeter a detector that measures the flow rate of liquids by using the principle of thermal conductivity.

thermal mass meter a mass flowmeter consisting of two RTD temperature probes and a heating element that measures the heat loss of the fluid mass.

thermistor a temperature sensor that exhibits a large change in resistance when subjected to a small change in temperature.

thermocouple a temperature-sensing device made of dissimilar metals that converts heat into a voltage.

thermowell a protective device that encloses a temperature sensor.

three-phase power AC power consisting of three alternating currents of equal frequency and amplitude, each differing in phase from the others by one-third of a period.

three-wire system a sensor that has three wires.

thyristor a four-layer semiconductor device.

time duration the amount of time at which a signal passes from one instrument to the next instrument in a closed-loop system.

time lag the duration from when a change is received at the input of an instrument to when it produces an output response.

time-of-flight flowmeter a liquid-measuring device that uses ultrasonic waves to determine flow rate.

time proportioning a method in which the output of the controller is continually switched fully on and fully off.

timers output instructions of a programmable controller that perform functions such as delaying an action, causing an operation to run a predetermined amount of time, or recording the accumulated time of continuous or intermediate events.

topologies the physical configuration of devices that are connected on an industrial computer network.

torque the measure of a motor's rotary force.

tracking the movement of the load by the actuator as it attempts to follow a changing command signal in a motion-control position system.

transducer an instrument that converts one type of signal into another. Also, a device that monitors various conditions in the manufacturing process, such as displacement, speed, position, and acceleration.

transient response the response time of the actuator to sudden changes of speed or position signals.

transmitter an instrument that converts a signal from a sensor into a standardized signal in process systems.

traverse rate the fast rate of speed at which a load is moved from one position to another in a point-to-point servo system.

triac a three-terminal bidirectional solid-state device that passes AC current when it is triggered.

trial-and-error method a method of tuning a controller by interpreting a step response on a chart recorder to determine proper control mode settings.

triggering a process in which the application of a voltage causes a solid-state device to turn on.

turbine flowmeter a rotating device with turbine blades that measure the velocity or total volume of fluid flow.

two-wire system a sensor that has two wires.

U

ultimate period the duration of one cycle.

ultimate proportional gain value the proportional controller setting that causes a sustained cycling to occur.

ultrasonic flowmeter a liquid-measuring device that operates on the principle of sound propagation to measure flow rate.

ultrasonic level sensor a sensor that uses sound waves to detect the height of the contents in a container.

underdamped a system that is tuned with too much gain and responds to a changed command signal by causing the controlled variable to react too quickly and overshoot its end position or desired state.

unijunction transistor (UJT) a three-terminal solid-state device that is used primarily to trigger an SCR.

unit reaction rate a measure of how much the process reacts for each percent of actuator change.

universal motor an electric motor that can operate on either AC or DC voltages.

unstable a closed-loop system that is out of control and oscillates around setpoint.

V

vacuum the absence of a gas inside a container. Any reduction of pressure compared to atmospheric pressure is called a *partial vacuum*.

valve capacity the amount of material that a flow control valve is capable of allowing to pass.

valve characteristics the relationship of the change in the valve opening to a change of flow through a flow control valve.

variable capacitor pressure detector a sensor that has a bellow that changes the position of a capacitor plate when air pressure causes it to expand or contract.

variable reluctance (VR) stepper motor a motor that uses electrical reluctance to rotate the rotor instead of using the attraction or repulsion of magnetic fields.

variable-voltage drive a DC variable-speed drive that controls motors by regulating the amount of voltage applied to the armature windings, shunt field windings, or both.

velocity the speed at which an object or a material moves.

velocity flowmeter an instrument that directly measures fluid flow to determine volumetric flowrate.

viscosity 1. the ability of a liquid to flow and take the shape of a container. 2. the ease with which a liquid flows.

volts/Hertz a function of a variable DC drive, which causes the bus voltage to be varied in direct proportion to the inverter output frequency so that a constant current and torque of the motor are maintained.

voltage-to-frequency (V/Hz) ratio a method in which the bus line voltage of an AC drive is varied in direct proportion to the inverter output frequency to maintain constant current and motor torque.

volumetric flow rate the measurement of flow, which is determined by the volume of material that flows during a given time period.

vortex flowmeter an instrument that measures the pressure of vortices that form downstream from a blunt object to determine flow.

W

weight detector a device that uses an inferred method of determining level by measuring the weight of a material within a container.

wire rope actuated transducer a device used to measure the linear position and linear velocity of an object.

wound armature PM motor a DC motor that has armature windings and a stator made of permanent magnets.

wound rotor a rotor that has coils wound around an iron core in an AC motor.

wound-rotor induction motor a three-phase induction electric motor that uses three coils in place of conductor bars in the construction of its rotor.

Index

NOTE: Page numbers followed by *f* refer to figures and *t* refer to tables.

D

N

O